Foundations in
Microbiology
BASIC PRINCIPLES

Foundations in
Microbiology
BASIC PRINCIPLES

EIGHTH EDITION

Kathleen Park Talaro
Pasadena City College

Barry Chess
Pasadena City College

Connect
Learn
Succeed

The McGraw-Hill Companies

McGraw Hill

Connect
Learn
Succeed™

FOUNDATIONS IN MICROBIOLOGY: BASIC PRINCIPLES, EIGHTH EDITION

Published by McGraw-Hill, a business unit of The McGraw-Hill Companies, Inc., 1221 Avenue of the Americas, New York, NY 10020. Copyright © 2012 by The McGraw-Hill Companies, Inc. All rights reserved. Previous editions © 2008, 2005, and 2002. No part of this publication may be reproduced or distributed in any form or by any means, or stored in a database or retrieval system, without the prior written consent of The McGraw-Hill Companies, Inc., including, but not limited to, in any network or other electronic storage or transmission, or broadcast for distance learning.

Some ancillaries, including electronic and print components, may not be available to customers outside the United States.

This book is printed on acid-free paper.

1 2 3 4 5 6 7 8 9 0 DOW/DOW 1 0 9 8 7 6 5 4 3 2 1

ISBN 978–0–07–734280–7
MHID 0–07–734280–1

Vice President, Editor-in-Chief: *Marty Lange*
Vice President, EDP: *Kimberly Meriwether David*
Senior Director of Development: *Kristine Tibbetts*
Sponsoring Editor: *Lynn M. Breithaupt*
Senior Developmental Editor: *Kathleen R. Loewenberg*
Marketing Manager: *Amy L. Reed*
Senior Project Manager: *Jayne L. Klein*
Senior Buyer: *Laura Fuller*
Senior Media Project Manager: *Jodi K. Banowetz*
Designer: *Tara McDermott*
Cover Designer: *Elise Lansdon*
Cover Image: © *Luke Jerram*
Lead Photo Research Coordinator: *Carrie K. Burger*
Photo Research: *Danny Meldung/Photo Affairs, Inc*
Compositor: *Aptara, Inc.*
Typeface: *10/12 Times New Roman*
Printer: *R. R. Donnelley*

All credits appearing on page or at the end of the book are considered to be an extension of the copyright page.

Library of Congress Cataloging-in-Publication Data

Talaro, Kathleen P.
 Foundations in microbiology : basic principles / Kathleen Park, Barry Chess. — 8th ed.
 p. cm.
 Includes index.
 ISBN 978–0–07–734280–7 — ISBN 0–07–734280–1 (hard copy : alk. paper) 1. Microbiology. 2. Medical microbiology. I. Chess, Barry. II. Title.
 QR41.2.T354 2012
 579–dc22

2010036381

www.mhhe.com

Brief Contents

About the Authors

Kathleen Park Talaro is a microbiologist, educator, author, and artist. She has been nurturing her love of microbiology since her youth growing up on an Idaho farm where she was first fascinated by tiny creatures she could just barely see swimming in a pond. This interest in the microbial world led to a biology major at Idaho State University, where she worked as a teaching assistant and scientific illustrator for one of her professors. This was the beginning of an avocation which she continues today—that of lending her artistic hand to interpretation of scientific concepts. She continued her education at Arizona State University, Occidental College, California Institute of Technology, and California State University.

She has taught microbiology and major's biology courses at Pasadena City College for 30 years, during which time she developed new curricula and refined laboratory experiments. She has been an author of, and contributor to, several publications of the William C. Brown Company and McGraw-Hill Publishers since the early 1980s, first illustrating and writing for laboratory manuals and later developing this textbook. She has also served as a coauthor with Kelly Cowan on the first two editions of *Microbiology: A Systems Approach*.

Kathy continues to make microbiology a significant focus of her life and is passionate about conveying the significance and practical knowledge of the subject to everyone, regardless of their profession or position. In addition to her writing, she keeps current attending conferences and participating in the American Society for Microbiology and its undergraduate educational programs. She is gratified by the many supportive notes and letters she has received over the years from book adopters and students.

She lives in Altadena, California with husband Dave Bedrosian and son David. Whenever she can, she spends time with her daughter Nicole, who lives in Wyoming. In her spare time she enjoys photography, reading true crime books, music, crossword puzzles, and playing with her seven rescued kitties.

Kathy Talaro (right) and her daughter, Nicole.

Dedication

We wish to dedicate this book to microbes, those ingenious beings that beckon us into another realm that exists beyond our naked eyes. We marvel at their fantastic variety and wild, exotic ways of life. And even after many lifetimes of study, we still have much to learn from the tiny "animalcules" that Leeuwenhoek first saw over 300 years ago in "such enormous numbers that all the water seemed to be alive."

About the Authors

The addition of two proven educators makes a great learning system even better

Barry Chess has been teaching microbiology at Pasadena City College for 14 years. He received his Bachelor's and Master's degrees from California State University, Los Angeles, and did several years of post-graduate work at the University of California, Irvine, where his research focused on the expression of eukaryotic genes involved in the development of muscle and bone.

At Pasadena City College, Barry developed a new course in human genetics and helped to institute a biotechnology program. He regularly teaches courses in microbiology, general biology, and genetics, and works with students completing independent research projects in biology and microbiology. Over the past several years, Barry's interests have begun to focus on innovative methods of teaching that lead to greater student understanding. He has written cases for the National Center for Case Study Teaching in Science and presented talks at national meetings on the use of case studies in the classroom. In 2009, his laboratory manual, *Laboratory Applications in Microbiology: A Case Study Approach*, was published. He is thrilled and feels very fortunate to be collaborating with Kathy Talaro, with whom he has worked in the classroom for more than a decade, on this eighth edition. Barry is a member of the American Society for Microbiology and regularly attends meetings in his fields of interest, both to keep current of changes in the discipline and to exchange teaching and learning strategies with others in the field.

Writing a textbook takes an enormous amount of time and effort. No textbook author has the time to write a great textbook and also write an entire book's worth of accompanying digital learning tools—at least not with any amount of success or accuracy. In the past, this material has often been built after the text publishes, but hopefully in time for classes to start. With the new digital era upon us, it is time to begin thinking of digital tools differently. In classrooms across the country, thousands of students who are visual learners and

have been using computers, video games, smart phones, music players, and a variety of other gadgets since they could talk are begging for an interactive way to learn their course material. Enter the digital author. With this eighth edition, we are excited to add professor **Heidi Smith** from Front Range Community College to the Talaro/ Chess team. Heidi teaches microbiology and anatomy & physiology and has worked hand-in-hand with the textbook authors, creating online tools that truly complement and enhance the book's content. She ensured that all key topics in the book have interactive, engaging activities spanning levels of Bloom's taxonomy, and tied to Learning Outcomes in the book. Instructors can now assign material based on what they cover in class, assess their students on the Learning Outcomes, and run reports indicating individual and/or class performance on a variety of data. Because of Heidi, we can now offer you a robust digital learning program, tied to Learning Outcomes, to enhance your lecture and lab, whether you run a traditional, hybrid, or fully online course.

*"I am gratified to introduce **Barry Chess**, a professor at Pasadena City College, as my coauthor on this new edition. He promises to bring a fresh eye to this project along with his own expertise in genetics and molecular biology, and a commitment to crafting a high quality product. Barry has an easy, very reader-friendly writing style that complements my own. He is astute and knowledgeable, with a rare ability to get to the heart of complex principles yet keep the reader involved and interested along the way. He often incorporates anecdotes, mnemonic devices, case studies, and analogies for helping students to learn and understand more difficult and abstract concepts."* —Kathleen Park Talaro

Integrated Learning System Customized to your Course Outcomes

McGraw-Hill Higher Education and Blackboard® have teamed up! What does this mean for you?

1. **Your life, simplified.** Now you and your students can access McGraw-Hill's Connect™ and Create™ right from within your Blackboard course – all with one single sign-on. Say goodbye to the days of logging in to multiple applications.

2. **Deep integration of content and tools.** Not only do you get single sign-on with Connect™ and Create™, you also get deep integration of McGraw-Hill content and content engines right in Blackboard. Whether you're choosing a book for your course or building Connect™ assignments, all the tools you need are right where you want them—inside of Blackboard.

3. **Seamless Gradebooks.** Are you tired of keeping multiple gradebooks and manually synchronizing grades into Blackboard? We thought so. When a student completes an integrated Connect™ assignment, the grade for that assignment automatically (and instantly) feeds your Blackboard grade center.

4. **A solution for everyone.** Whether your institution is already using Blackboard or you just want to try Blackboard on your own, we have a solution for you. McGraw-Hill and Blackboard can now offer you easy access to industry leading technology and content, whether your campus hosts it, or we do. Be sure to ask your local McGraw-Hill representative for details.

Introducing ConnectPlus Microbiology

McGraw-Hill ConnectPlus™ Microbiology integrated learning platform provides auto-graded assessments, a customizable, assignable eBook, an adaptive diagnostic tool, and powerful reporting against Learning Outcomes and level of difficulty—all in an easy-to-use interface.

". . . I and my adjuncts have reduced the time we spend on grading by 90 percent and student test scores have risen, on average, 10 points since we began using Connect!"

—*William Hoover,* Bunker Hill Community College

Save time with auto-graded assessments and tutorials

Fully editable, customizable, auto-graded interactive assignments using high quality art from the textbook, and animations and videos from a variety of sources take you way beyond multiple choice. Assignable content is available for every Learning Outcome in the book. Extremely high quality content, created by digital author Heidi Smith, includes case study modules, concept mapping activities, animated learning modules, and more!

Generate powerful data related to student performance against Learning Outcomes, specific topics, level of difficulty, and more.

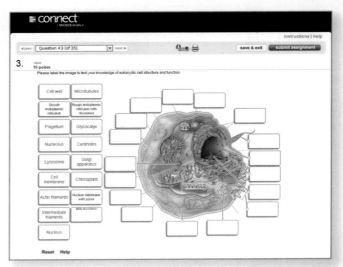

Presentation Tools Allow Instructors to Customize Lectures

Everything you need, in one location

Enhanced Lecture Presentations contain lecture outlines, FlexArt, art, photos, tables, and animations embedded where appropriate. Fully customizable, but complete and ready to use, these presentations will enable you to spend less time preparing for lecture!

FlexArt—fully editable (labels and leaders) line art from the text, with key figures that can be manipulated even further. Take the images apart and put them back together again during lecture so students can understand one step at a time.

Animations—over 100 animations bringing key concepts to life, available for instructors and students.

Animation PPTs—animations are truly embedded in PowerPoint® for ultimate ease of use! Just copy and paste into your custom slide show and you're done!

Chapter	Enhanced Lecture PPTs Complete with Lecture Outlines, Flex Art (Fully editable images from the text and key figures that can be manipulated), art, photos, tables and animations embedded where appropriate. Also available without animations embedded.	Animation PPTs Animations embedded into PowerPoint. Allows for easy presentation display and simple copying/pasting to your custom slide show.	Image PPTs All art, tables and photos from the text, optimized for presentation.	Art, Photos & Tables Art files, tables and photos as they appear in the text. Also provided as base art files without labels or leader lines.	Animations Animation files that can be easily downloaded and played.
All Chapters	Enhanced PPT Enhanced Keynote Enhanced PPT without animations	Enhanced PPT Enhanced Keynote	Image PPT	Labeled Images Unlabeled Images	Animations

Take your course online—*easily*—with one-click Digital Lecture Capture

McGraw-Hill Tegrity Campus™ records and distributes your lecture with just a click of a button. Students can view them anytime/anywhere via computer, iPod, or mobile device. Tegrity Campus indexes as it records your slideshow presentations and anything shown on your computer so **students can use keywords to find exactly what they want to study**.

Student Study Tools to Fit Individual Needs

24/7 access with a customizable, interactive eBook

McGraw-Hill ConnectPlus™ eBook takes digital texts beyond a simple PDF

With the same content as the printed book, but optimized for the screen, ConnectPlus has embedded media, including animations and videos, which bring concepts to life and provide "just in time" learning for students. Additionally, fully integrated, self-study questions and in-line assessments allow students to interact with the questions in the text and determine if they're gaining mastery of the content. These questions can also be assigned by the instructor.

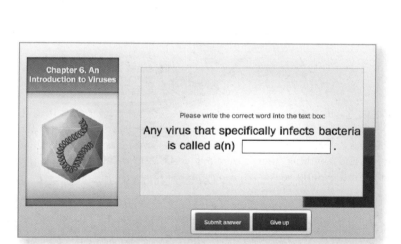

"Use of technology, especially LearnSmart, assisted greatly in keeping on track and keeping up with the material."

—*student,* Triton College

LearnSmart—A Diagnostic, Adaptive Learning System to help you learn—smarter

McGraw-Hill LearnSmart™ is an adaptive diagnostic tool, powered by Connect Microbiology, which is based on artificial intelligence and constantly assesses a student's knowledge of the course material.

Sophisticated diagnostics adapt to each student's individual knowledge base in order to match and improve what they know. Students actively learn the required concepts more easily and efficiently.

"I love LearnSmart. Without it, I would not be doing well."

—*student,* Triton College

The **Profile** of an Expertly Crafted Learning Tool

Art and organization of content make this book unique

Carefully crafting a textbook to be a truly useful learning tool for students takes time and dedication. Every line of text and every piece of art in this book is scrutinized for instructional usefulness, placement, and pedagogy, and then reexamined with each revision. In this eighth edition, the authors have gone through the book page by page, with more depth than ever before, to make sure it maintains its instructional quality; fantastic art program; relevant and current material; and engaging, user-friendly writing style. Since the first edition, the goals of this book have been to explain complex topics clearly and vividly, and to present the material in a straightforward way that students can understand. The eighth edition continues to meet these goals with the most digitally integrated, up-to-date, and pedagogically important revision yet.

Like a great masterpiece hanging in a museum, *Foundations in Microbiology, Basic Principles,* is not only beautiful, but also tells a story, composed of many pieces. A great textbook must be carefully constructed to place art where it makes the most sense in the flow of the narrative; create process figures that break down complex processes into their simplest parts; provide explanations at the correct level for the student audience, and offer pedagogical tools that help all types of learners. Many textbook authors write the narrative of their book and call it a day. It is the rare author team indeed, who examines each page and makes changes based on what will help the students the most, so that when the pieces come together, the result is an expertly crafted learning tool—a story of the microbial world.

A unique feature of this text's format is the early survey of microbial groups and their taxonomy (chapters 4, 5, 6). By using general and specific names for microbes from the very beginning students

> "This textbook is thorough and informative with exceptional illustrations. The various illustrations and summary tables help organize the large amount of material, which helps students study."
>
> —*Danita Bradshaw-Ward,* Eastfield College

Process Figure 9.6 The Assembly line of DNA replication in a circular bacterial chromosome. (a) A bacterial chromosome showing the overall pattern of replication. There are two replication forks where new DNA is being synthesized. (b) An enlarged view of the left replication fork to show the details of replication.

Kathy Talaro introduces new art to a revision by carefully sketching out what she envisions in precise detail, with accompanying instructions to the illustrator. The result is accurate, beautifully rendered art that helps difficult concepts come to life.

develop a working background that eases them into the later chapters. Students have a far greater appreciation for later topics of nutrition, metabolism, genetics, and microbial control if they recognize the main characters—bacteria, viruses, and eukaryotic microorganisms—and already know significant facts about them.

Another different feature of this text is chapter 17, "Procedures for Identifying Pathogens and Diagnosing Infections." It brings together in one place the current methods used to diagnose infectious diseases. The chapter starts with collecting samples from the patient and details the biochemical, serological, and molecular methods used to identify causative microbes.

The Structure of an Expertly Crafted Learning Tool

Chapter opening case files

Each chapter opens with a Case File, which helps the students appreciate and understand how microbiology impacts their lives. Line art, micrographs, and quotes have been added to the chapter-opening page, where appropriate, to help the students pull together the big picture and grasp the relevance of the material they're about to learn. The questions that directly follow the Case File challenge students to begin to think critically about what they are about to read, expecting that they'll be able to answer them once they've worked through the chapter. A **new Continuing the Case** feature now appears within the chapter where relevant, to help students follow the real-world application of the case. The Case File Perspective wraps up the case at the end of the chapter, pulling together the applicable content and the chapter's topics. Nearly all Case Files are new in the eighth edition, including hot microbiological topics that are making news headlines today.

CHAPTER 3

Tools of the Laboratory
Methods of Studying Microorganisms

"A matter of life or death"

The meningococcus: A million of these tiny culprits could fit on the head of a pin, yet they can knock out a healthy adult in a few hours.

CASE FILE 3 Battling a Brain Infection

One Saturday evening in 2007, a 50-year-old woman began to suffer flulike symptoms, with fever, aching joints, sore throat, and a headache. Feeling miserable but not terribly concerned, she took some ibuprofen and went to bed. By the following morning revealed no sign of pneumonia, and a blood test indicated an elevated white blood cell count. To rule out a possible brain infection, a puncture of the spinal canal was performed. As it turned out, the cerebrospinal fluid (CSF) the technician extracted appeared normal, microscopi... saying, "Her medical situation was so critical that our intervention was truly a matter of life or death."

Because her symptoms pointed to a possible infection of the central nervous system, a second spinal puncture was performed. This time, the spinal fluid looked cloudy. A Gram stain was performed right away, and cultures were started.

▶ What are signs and symptoms of disease? Give examples from the case that appear to be the most diagnostically significant.
▶ Why is so much importance placed on the CSF and its appearance?

To continue the case, go to page 74.

NEW!

CONTINUING CASE FILE 6

Beginning with the first diagnoses in March 2009, the influenza outbreak exploded into a pandemic in only six weeks. Cases rapidly appeared in Canada, Central and South America, then Europe and Asia, and eventually more than 200 countries. By CDC estimates, from April to November 2009 in the United States alone, there were 50 million cases and close to 10,000 deaths. Deaths were particularly high among young children and pregnant women whose treatment had been delayed. Fortunately, the disease experienced by most people was milder than the usual seasonal flu, and it cleared up ...cations. The common symptoms are fever, m...problems with breathing and coughing that s...two weeks. The most serious complication is ...group that seemed to be less susceptible to H...virus were people 60 years or older.

■ What is a pandemic?
■ Why would some people be more resistan...

For a wrap-up, see the Case File Perspective on page ...

CASE FILE 9 PERSPECTIVE

The source of the infection in the first case was most likely the ventilator that controlled the woman's breathing. Medical devices are readily contaminated by patients and healthcare workers. Any case acquired during a stay in a hospital is defined as a nosocomial infection. These infections are most problematic in very compromised patients, but stringent disinfection procedures can greatly reduce their incidence. The second case probably came from the soiled piece of shrapnel rather than the field hospital.

To say that a microbe has resistance to a drug means that it naturally possesses or has acquired a genetic mechanism to avoid the effects of the drug. Usually, its genome carries one or more genes that can eliminate the drug or prevent it from acting on the cells of the microbe. This leaves the microbe free to grow and infect even in the presence of that drug.

> "I love the case study intro to the chapter. I feel the students (no matter what discipline/major they are pursuing) benefit from information presented in a real-life scenario. The information is more engaging and relevant than straight lecture and often leads to great group discussions."
>
> —*Tracey M. Steeno,* Northeast Wisconsin Technical College

> "I think the case study at the beginning of each chapter is wonderful because it introduces the students to the real life scenarios they will be involved in when they go into the allied health profession. So in a sense, these are 'practice' studies."
>
> —*Carroll W. Bottoms,* Collin County Community College

The **Art** of an Expertly Crafted Learning Tool

Author's experience and talent transforms difficult concepts

Truly instructional artwork has always been a hallmark feature of *Foundations in Microbiology, Basic Principles*. Kathy Talaro's experiences as a teacher, microbiologist, and illustrator have given her a unique perspective and the ability to transform abstract concepts into scientifically accurate and educational illustrations. Powerful artwork that paints a conceptual picture for students is more important than ever for today's visual learners. *Foundations in Microbiology, Basic Principles* art program combines vivid colors, multi-dimensionality, and self-contained narrative to help students study the challenging concepts of microbiology.

Process Figure 8.11 One type of genetic control of enzyme synthesis: enzyme repression. (1), (2), (3), (4), (5) Genetic controls are active and the enzyme is synthesized continuously until enough product has been made. (6), (7) Excess product reacts with a site on DNA that regulates the enzyme's synthesis, thereby inhibiting further enzyme production.

"The illustrations are excellent! They are clear and concise, well drawn, and visually quite attractive. They are designed to aid students' comprehension. Without exception, I consider them among the best illustrations I have seen."

—*James Doyle,* Paradise Valley Community College

Process Figures

Many difficult microbiological concepts are best portrayed by breaking them down into stages that students will find easy to follow. These process figures show each step clearly marked with a yellow, numbered circle and correlated to accompanying narrative to benefit all types of learners. Process Figures are clearly marked next to the figure number. The accompanying legend provides additional explanation.

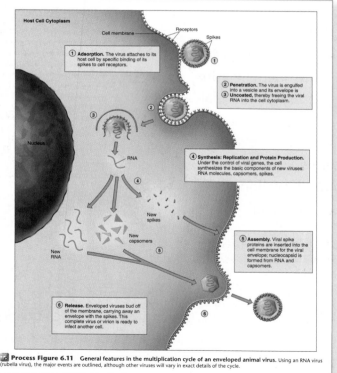

Process Figure 6.11 General features in the multiplication cycle of an enveloped animal virus. Using an RNA virus (rubella virus), the major events are outlined, although other viruses will vary in exact details of the cycle.

The **Purpose** of an
Expertly Crafted Learning Tool

NEW! **Learning Outcomes and Assess Questions**

Every section in the book now opens with Expected Learning Outcomes and closes with a summary and assessment questions (Check & Assess). The Learning Outcomes are tightly correlated to digital material. Instructors can easily measure student learning in relation to the specific learning outcomes used in their course. You can also assign Assess questions to students through the eBook with McGraw-Hill ConnectPlus™ Microbiology.

Check & Assess **Section 9.6**

✔ The genetic material of viruses can be either DNA or RNA occurring in single, double, positive-, or negative-message strands. Viruses contain genes primarily for taking over the cell machinery and synthesizing new virus components.

28. Explain why some viruses must enter the nucleus to replicate whereas others do not.
29. Explain the difference between positive strand and negative strand viral genomes.

Expected Learning Outcomes

7. Present an overview of the main aspects of the flow of genetic information in cells.
8. Explain the relationship between the structure of DNA and the structure of proteins.
9. Describe the different types of RNA and their basic functions in genetic expression.
10. Explain what happens during transcription.
11. Describe the genetic code, codons, and anticodons, and how they relate to one another.
12. Relate the participants and steps in translation (protein synthesis).
13. Distinguish major points of difference between prokaryotic and eukaryotic transcription and translation.

NEW! **Animated Learning Modules**

Certain topics in microbiology need help to come to life off the page. With animations, video, audio, and text combine to help students understand complex processes. Many figures in the text have a corresponding Animation Learning Module available for students and instructors online through Connect. Key topics now have an animated Learning Module assignable through Connect. A new icon in the text indicates when these learning modules are applicable.

Food Pathogens and Temperatures feedback key terms

Growth Factors

Temperature

2. After six generations, how many bacteria would have formed from the reproduction of one bacterium?

Food Pathogens next: submit ►

The **Framework** of an Expertly Crafted Learning Tool

Pedagogy created to promote active learning

INSIGHT 9.1

The Packaging of DNA: Winding, Twisting, and Coiling

Packing the mass of DNA into the cell involves compacting the DNA molecule by means of supercoils or superhelices. In the simpler system of prokaryotes, the circular chromosome is packaged by the action of a special enzyme called a topoisomerase (specifically, DNA gyrase). This enzyme coils the chromosome into a tight bundle by introducing a reversible series of twists into the DNA molecule.

The system in eukaryotes is more complex, with three or more levels of coiling. First, the DNA molecule of a chromosome, which is linear, is wound twice around the histone proteins, creating a chain of nucleosomes. The nucleosomes fold in a spiral formation upon one another. An even greater supercoiling occurs when this spiral arrangement further twists on its radius into a giant spiral with loops radiating from the outside. This extreme degree of compactness is what makes the eukaryotic chromosome visible during mitosis.

The condensation of DNA and protein into chromatin was once thought to function mainly as a way for such a long molecule to fit into a cell. However, new research has shown that the coiling of DNA also serves to make certain segments of the genetic program either more or less available to the cell. This is a major control on the expression of many traits, especially in eukaryotes. Having an additional regulatory function creates an entirely new set of information to guide phenotypic expression. The realization that the phenotype of a cell is due to more than just the order of nucleotides in the DNA has led to a rethinking of some of the most basic aspects of genetics (see Insight 9.3).

Because genetic codes cannot be properly accessed when DNA is supercoiled, what other system would a cell need to manage DNA? Answer available at http://www.mhhe.com/talaro5

The packaging of DNA. Eukaryotic DNA is intimately associated with a variety of proteins and small molecules. This relationship modulates the expression of many genes while at the same time allowing for the ordered condensation of DNA into a compact molecule.

Terminology

Learning the terminology of microbiology can be a daunting task. To make this task easier, key terms are noted with an asterisk and the pronunciation and definition are provided at the nearest section break.

* *vesicle* (ves'-ik-l) L. *vesios,* bladder. A small sac containing fluid.
* *lysosome* (ly'-soh-sohm) Gr. *Lysis,* dissolution, and *soma,* body.
* *vacuole* (vak'-yoo-ohl) L. *vacuus,* empty. Any membranous space in the cytoplasm.

Notes

"Take Note" call-outs appear, where appropriate, throughout the text. They give students helpful information about various terminologies, exceptions to the rule, or provide clarification and further explanation of the prior subject.

TAKE NOTE: A CARBON CLARIFICATION

It seems worthwhile to emphasize a point about the *extracellular source* of carbon as opposed to the *intracellular function* of carbon compounds. Although a distinction is made between the type of carbon compound cells absorb as nutrients (inorganic or organic), the majority of carbon compounds involved in the normal structure and metabolism of all cells are organic.

Insight Readings

Found throughout each chapter, current, real-world readings allow students to see an interesting application of the concepts they're studying.

Footnotes

Footnotes provide the reader with additional information about the text content.

5. The terms *ciliate* and *flagellate* are common names of protozoan groups that move by means of cilia and flagella.

Tables

This edition contains numerous illustrated tables. Horizontal contrasting lines set off each entry, making them easy to read.

TABLE 4.2	Comparison of the Two Spiral-Shaped Bacteria				
	Overall Appearance	**Mode of Locomotion**	**Number of Helical Turns**	**Gram Reaction (Cell Wall Type)**	**Examples of Important Types**
Spirilla	Rigid helix	Polar flagella; cells swim by rotating around like corkscrews; do not flex; have one to several flagella; can be in tufts	Varies from 1 to 20	Gram-negative	Most are harmless; one species, *Spirillum minor,* causes rat bite fever.
Spirochetes	Flexible helix	Periplasmic flagella within sheath; cells flex; can swim by rotation or by creeping on surfaces; have 2 to 100 periplasmic flagella	Varies from 3 to 70	Gram-negative	*Treponema pallidum,* cause of syphilis; *Borrelia* and *Leptospira,* important pathogens

"This is one of those rare textbooks that I would carry around in my car and read during lunches, dinners, while on the road or when I have a block of time to read. It is well written, easy to follow, attractive and well-illustrated."

—*Ronald A. Weiss,* Marian University, Indianapolis, Indiana

The **Planning** of an Expertly Crafted Learning Tool

Pedagogy designed for varied learning styles

The End of Chapter material for the eighth edition has been carefully planned to promote active learning and provide review for different learning styles and levels of Bloom's Taxonomy. The consistent layout of each chapter allows students to develop a learning strategy and gain confidence in their ability to master the concepts, leading to success in the class!

Chapter Summary with Key Terms

4.1 Basic Characteristics of Cells and Life Forms
 A. All living things are composed of **cells,** which are complex collections of macromolecules that carry out living processes. All cells must have the minimum structure of an outer **cell membrane, cytoplasm,** a **chromosome,** and **ribosomes.**
 B.

Multiple-Choice Questions

Select the correct answer from the answers provided. For questions with blanks, choose the combination of answers that most accurately completes the statement.

1. Which structure is not a component of all cells?
 a. cell wall c. genetic material
 b. cell membrane d. ribosomes

2. Viruses are not considered living things because
 a. they are not cells
 b. they car
 c. they lac
 d. All of th

Case File Questions

1. What is true of the condition endocarditis?
 a. It occurs in the heart muscle.
 b. It is caused by microbes growing in the internal organs.
 c. It is an infection of the heart valves and lining.
 d. It can be transmitted to others.

 t made the biofilm originate?
 c. from the surgery
 d. from the patient's wife

Writing to Learn

These questions are suggested as a *writing-to-learn* experience. For each question, compose a one- or two-paragraph answer that includes the factual information needed to completely address the question. Any question listed in a section's Check and Assess may be considered as a writing-to-learn exercise.

1. Label the parts on the bacterial cell featured here and write a brief description of its function.

Chapter Summary with Key Terms
A brief outline of the main chapter concepts is provided for students with important terms highlighted. Key terms are also included in the glossary at the end of the book.

Multiple-Choice Questions
Students can assess their knowledge of basic concepts by answering these questions. Other types of questions and activities that follow build on this foundational knowledge. The ConnectPlus eBook allows students to quiz themselves interactively using these questions!

NEW! Case File Questions
These questions deepen the real-life experience students embarked upon at the start of the chapter and allow instructors to assess students on the case file material.

Writing to Learn Questions
These questions are suggested as a writing experience. Students are asked to compose a one- or two-paragraph response using the factual information learned in the chapter.

> "This text is highly readable, sustaining the reader's interest with plenty of real-life examples and current information. At the same time, it is complete enough to serve as a valuable reference for students going into a variety of healthcare fields."
>
> —*Randall K. Harris, Ph.D.,* William Carey University

The **Innovation** of an Expertly Crafted Learning Tool

Concept Mapping Exercises

Three different types of concept mapping activities are used throughout the text in the end-of-chapter material to help students learn and retain what they've read. Concept Mapping exercises are also now made interactive on ConnectPlus Microbiology!

Concept Mapping

Appendix E provides guidance for working with concept maps.

1. Construct your own concept map using the following words as the *concepts*. Supply the linking words between each pair of concepts.

genus species
serotype domain
Borrelia *burgdorferi*
spirochaetes phylum

> "I LOVE concept maps! And, McGraw-Hill has great website resources to support and expand upon these great learning tools."
>
> —*Jackie Reynolds,* Richland College

Critical Thinking Questions

Using the facts and concepts they just studied, students must reason and problem-solve to answer these specially developed questions. Questions do not have a single correct answer and thus open doors to discussion and application.

Critical Thinking Questions

Critical thinking is the ability to reason and solve problems using facts and concepts. These questions can be approached from a number of angles, and in most cases, they do not have a single correct answer.

1. What is required to kill endospores? How do you suppose archaeologists were able to date some spores as being thousands (or millions) of years old?

2. Using slew, demonstrate how cocci can divide in several planes and ...ion. Show how the arrangements of ...es.

Visual Challenge

Visual Challenge questions take images and concepts learned in previous chapters and ask students to apply that knowledge to concepts newly learned in the current chapter.

Visual Challenge

1. From **chapter 3, figure 3.27b.** Which bacteria has a well-develo[ped] capsule: *"Klebsiella"* or *"S. aureus"*? Defend your answer.

> "I think that the Visual Challenge questions that have been incorporated into the textbook are an excellent idea."
>
> —*Mark Pilgrim,* College of Coastal Georgia

The **Revision** of an Expertly Crafted Learning Tool

What's New in the Eighth Edition?

Changes in the Eighth Edition of *Foundations in Microbiology, Basic Principles*

First and foremost in every revision of *Foundations in Microbiology, Basic Principles* is the careful reading of reviews and correction of any errors followed by the updating of content to ensure that the textbook is at the top of its class in being up-to-date. For example, in the eighth edition, areas of technology and antimicrobic drugs were revised and expanded upon, and all disease statistics have been updated.

Case Files

- All of the chapter Case Files are new except 7, 10, and 19, which have been revamped or expanded.
- The Case Files are now more integrated into the chapter with "Continuing the Case" boxes, a final "Case Perspective", and end-of-chapter Case File questions.

Expected Learning Outcomes and Checkpoints

- The chapter overviews have been replaced with expected learning outcomes that begin every major section of a chapter. These direct the student's learning towards the most important topics in that section.
- Each section of a chapter ends with assess questions that focus on the Expected Learning Outcomes.

Additional Areas of Change

- For the chapters focusing on the agents of diseases, we have added new "Pathogen Profiles." These are abbreviated snapshots of the major pathogens in the chapter and include a micrograph, a description of the microscopic morphology, identification descriptions, habitat information, virulence factors, primary infections/disease, and control and treatment.
- Technology, antimicrobic drugs, and disease statistics have been updated.
- The eighth edition has over 200 new photographs, over 20 new or greatly revised figures, and a number of new illustrated tables.

- Many chapters have had changes in organization and modifications in their sections and headings.
- Numerous new assessment, writing-to-learn, and critical thinking questions have been added.
- A number of figures now contain insets of micrographs superimposed over a macroscopic photograph.
- Most chapters have new visual challenge questions.
- Figures have been evaluated to improve labeling and proportion and to clarify legends.
- Newly designed chapter opening pages tie in microbes more tightly with the Case Files.

Chapter 1

- New Case File about the exploration for new microbes in the oceans
- Figure on evolution has more illustrations added to the timeline
- Table on the work of microbiologists is revised
- The Introduction to Cells section is reorganized so that all coverage appears in the same area
- New photographs and examples for importance of microbes in natural environments
- New figures of microbes used in bioremediation and biotechnology
- New insight on emerging infections with update and figure on influenza
- Figure on scientific methods has been simplified
- Eight new photographs have been added to this chapter

Chapter 2

- New Case File about the search for life on Mars

Chapter 3

- New Case File introduces basic aspects of diagnosing meningitis
- Figure for tools of the laboratory and methods of studying microorganisms has been revised and an additional step called information gathering has been added to the 5 "I"s
- A new table to accompany this figure summarizes the steps in lab techniques
- Chapter now starts with microscopy, then isolation, identification, culturing and media
- A new introduction to identification techniques and keys has been added
- New information on unculturables has been added to the Insight reading

- Figure on dyes and staining has been revised
- Eleven new photographs have been added to this chapter

Chapter 4

- New Case File about a biofilm infection of a heart valve
- The section on bacterial taxonomy now contains photographs of examples
- Section on characteristics of life has been rewritten and shortened
- A description of type IV pili and motility has been added
- Thirteen new photographs have been added to this chapter

Chapter 5

- New Case File covers the neglected eukaryotic parasites
- The figure of eukaryotic taxonomy has been revised
- Ten new photographs have been added to this chapter

Chapter 6

- New Case File deals with the 2009 epidemic of H1N1 Influenza
- Several virus illustrations have been revised
- The figure on viral penetration has been updated
- Converted the table of virus families to an illustrated table
- Updated the Insight reading on creation of new viruses
- Seven new photographs have been added to this chapter

Chapter 7

- Chapter title changed to: Microbial Nutrition, Ecology, and Growth
- Updated Case File covers the Berkeley Pit with expanded information
- New anchoring figure 7.1 provides overview of the relationship of microbes to the environment
- We consolidated tables and moved text into a single table summarizing the functions of bioelements in microbial physiology
- Discussion on active transport was revised
- New information on *Deinococcus* was added to Insight on life in the extremes
- Six new photographs have been added to this chapter

Chapter 8

- New Case File covers the importance of microbes to ruminants

The **Effort** of an Expertly Crafted Learning Tool

- Revised the section on estimated amounts of ATP production
- Most of the chapter has been edited to improve flow and accuracy

Chapter 9
- New Case File details the developing drug resistance in *Acinetobacter*
- Extensive rewriting of sections on DNA replication, epigenetics, and regulation of RNA
- New figure 9.6 on DNA replication
- Added a clarification of detecting mutations
- New material on recombination in microbes
- Converted the boxed reading on genetics of animal viruses to regular text

Chapter 10
- Updated Case File on identifying the victims of the World Trade Center disaster
- New introduction to genetic engineering
- Revised figures on actions of endonucleases
- Added FISH figure
- Revised the text and a figure on DNA sequencing
- Added a short section on new "omics"
- Revised tables on genetically modified plants and animals
- Updated discussion on gene therapy
- Revised the descriptions of DNA fingerprinting
- Five new photographs have been added to this chapter

Chapter 11
- New Case File on the outbreak of hepatitis C linked to a Las Vegas clinic
- More details of the levels of resistance among microorganisms
- New discussion of how to select antimicrobial agents
- New figure to show the overall effects of temperature on a microbe's growth and survival
- New figure on the electromagnetic spectrum
- Revised figure on glutaraldehyde
- Replaced tables on thermal effects on microbes with tables of applications of physical agents
- Updated Insight reading on use of antibacterial substances
- Revised discussion of use of germicides
- Nine new photographs have been added to this chapter

Chapter 12
- New Case File about the aftermath of a needle-stick from an AIDS patient

- Revision of figure on effects of penicillin
- Added a Note on special strategies in drug therapy
- Revised the section on anti-HIV drugs
- Revised the table on actions of antiviral drugs
- Added a new figure on transfer of drug resistance
- Revised a figure on natural selection for drug resistance
- Added bacteriophage therapy to Insight reading on alternative therapies
- Updated information on drug resistance
- Four new photographs have been added to this chapter

Chapter 13
- Changed the chapter title to: Microbe-Human Interactions: Infections, Disease, and Epidemiology
- New Case File on outbreak of *Salmonella* food infection
- Added a new Insight reading on the role of the appendix
- New figure on the stages in infection
- New figure to show mechanism of invasion into host cells
- Revision of section on virulence factors
- Added mode of transmission to table on zoonoses
- Revised discussion on incidence and prevalence
- New figure on the percentage of nosocomial infections and the major infectious agents involved in them
- New figures to compare epidemiologic data
- Three new photographs have been added to this chapter

Chapter 14
- Changed the chapter title to: An introduction to Host Defenses and Innate Immunities
- New Case File on chronic granulomatous disease
- Added new material on defensins
- Revised introduction to recognition and surveillance
- Added a Note on chronic edema and filariasis
- Added clarifying information on MALT
- Moved toll-like receptor figure and discussion to phagocytosis section
- Rewrote the section on edema
- Added a new Note on neutrophil NETS that trap microbes
- More coverage on reactive oxygen intermediates
- Simplified coverage of the classical complement pathway

- Added photomicrographs of real WBCs to the figure on blood development
- New figure on lymphatic system
- New figure comparing blood and lymphatic circulations
- Added a dendritic cell to the macrophage maturation figure
- Six new photographs have been added to this chapter

Chapter 15
- New Case File on rabies and rabies immunization
- Moved section on natural, artificial, active, passive immunity to the section following T cell functions
- Revised figure on primary and secondary immune responses
- Replaced cancer cell photograph with two new ones
- Updated vaccination tables
- Clarified antigen and immunogen
- Added explanation of what accounts for the speed of the secondary immune response

Chapter 16
- New Case File about a transfusion reaction and its aftermath
- Six new photographs have been added to this chapter

Chapter 17
- Changed chapter title to: Procedures for Identifying Pathogens and Diagnosing Infections
- New Case File outlines a nosocomial *Vibrio* infection and includes tables used to narrow the identification of the pathogen involved
- Expanded flowchart for genera in infections
- New figure on pulse-field gel electrophoresis
- Expanded section on selection of media for isolation
- Six new photographs have been added to this chapter

Find the Right Fit for You

Create what you've only imagined

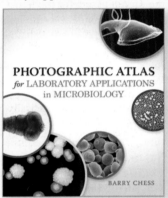 Introducing **McGraw-Hill Create™**—a new, self-service website that allows you to create custom course materials—print and eBooks—by drawing upon McGraw-Hill's comprehensive, cross-disciplinary content. Add your own content quickly and easily. Tap into other rights-secured third party sources as well. Then, arrange the content in a way that makes the most sense for your course. Even personalize your book with your course name and information! Choose the best format for your course: color print, black and white print, or eBook. The eBook is now even viewable on an iPad! And, when you are done you will receive a free PDF review copy in just minutes!

Visit McGraw-Hill Create – **www.mcgrawhillcreate.com** – today and begin building your perfect book.

Finally, a way to quickly and easily create the course materials you've always wanted.

Imagine that.

Need a lab manual for your microbiology course?

McGraw-Hill offers several lab manuals for the microbiology course. Contact your McGraw-Hill representative for packaging options with any of our lab manuals:

Brown: Benson's Microbiological Applications: Laboratory Manual in General Microbiology, Twelfth Edition

 Short Version (978-0-07-337527-4)

 Complete Version (978-0-07-730213-9)

Chess: Laboratory Applications in Microbiology: A Case Study Approach, Second Edition

 (978-0-07-340237-6)

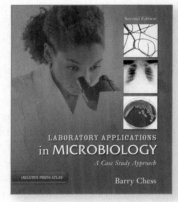

Chess: Photographic Atlas for Laboratory Applications in Microbiology, First Edition

 (978-0-07-737159-3)

Harley: Laboratory Exercises in Microbiology, Eighth Edition

 (978-0-07-729281-2)

Kleyn: Microbiology Experiments, Sixth Edition

 (978-0-07-299549-7)

Morello: Lab Manual and Workbook in Microbiology: Applications to Patient Care, Tenth Edition

 (978-0-07-352253-1)

Acknowledgments

We find ourselves excited and very enthusiastic about the transformations we have made in this edition, along with the addition of a dedicated CONNECT website that fully integrates with the features of the book. As with prior revisions, it has been a pleasure and a comfort to work with an energized and talented publishing team, most of whom are familiar friends. The finished product would not have been possible without the able support and input of senior developmental editor, Kathleen Loewenberg, sponsoring editor Lynn Breithaupt, and marketing manager Amy Reed. We are fortunate once again to have the highly competent Jayne Klein as senior project manager. Other dedicated and hard-working personnel are the photo research coordinator, Carrie Burger; photo researcher, Danny Meldung at Photo Affairs; C.J. Patterson, the copy editor; and the book designer, Tara McDermott, who has so artfully showcased the microbes throughout this edition. Special appreciation goes to Heidi Smith for her capable and diligent efforts to develop the digital assets for the new CONNECT website.

We owe a debt to the hundreds of reviewers who, through the years, have provided valuable insights into chapter organization, content, accuracy, and "teachability," and who have made a lasting imprint on many facets of this book. This revision is no exception. We have been fortunate in having a wide spectrum of microbiology specialists with helpful and insightful critiques and valuable feedback. Several of these reviewers deserve particular mention for providing substantive reviews above and beyond the usual expectations. Many thanks to Benjie Blair, *Jackonville State University;* Susan Bornstein-Forst, *Marian University;* Deborah V. Harbour, *College of Southern Nevada;* Luis Materon, *University of Texas, Pan American;* Mark Pilgrim, *College of Coastal Georgia;* Luis Rodriguez, *San Antonio College;* David J. Schwartz, *Houston Community College;* Kristine Snow, *Fox Valley Technical College;* James Doyle, *Paradise Valley Community College;* and Louise Thai, *University of Missouri.*

For the users of this book, we hope that you enjoy your explorations in the world of microbiology and that this fascinating science will leave a lasting impression on you. Although the book has been carefully inspected to weed out errors, no work in progress is ever perfect, and there will always be a few that slip through. If you detect any missing or misspelled words, missing labels, mistakes in content, or other errata, do not hesitate to contact the publisher, sales representative, or authors (ktalaro@aol.com or bxchess@Pasadena.edu).

—*Kathy Talaro* and *Barry Chess*

Reviewers

Joel Adams-Stryker, *Evergreen Valley College*
Michelle Alexander, *Baptist College of Health Sciences*
Lois Anderson, *Minnesota State University*
Sandra Barnes, *Housatonic Community College*
Melody Bell, *Vernon College*
Benjie Blair, *Jacksonville State University*
Ramaraj Boopathy, *Nicholls State University*
Susan Bornstein-Forst, *Marian University*
Carroll Bottoms, *Collin County Community College*
Danita Bradshaw-Ward, *Eastfield College*
Ana L. Dowey, *Palomar College*
James Doyle, *Paradise Valley Community College*
P. K. Duggal, *Maple Woods Community College*
Frances Duncan, *Pensacola Junior College*
Susan Finazzo, *Georgia Perimeter College*
Christina Gan, *Highline Community College*
Constance Hallberg, *University of Kansas*
Deborah Harbour, *College of Southern Nevada*
Julie Harless, *Lone Star College – Montgomery*
Randall Harris, *William Carey University*
Amy Helms, *Collin County Community College*
Jennifer A. Herzog, *Herkimer County Community College*
Phyllis Higley, *College of Saint Mary*
Kendricks Hooker, *Baptist College of Health Sciences*
Sheela Huddle, *Harrisburg Area Community College*
Dena Johnson, *Tarrant County College, Northwest*
Dennis Kitz, *Southern Illinois University*
Marcie Lehman, *Shippensburg University*
Terri J. Lindsey, *Tarrant County College South*
Danny Loosemore, *Northcentral Technical College*
Luis Materon, *The University of Texas Pan American*
Ethel Matthews, *Midland College*
Elizabeth McPherson, *The University of Tennessee*
Steven Obenauf, *Broward College*
Jean Petri, *Western Technical College*
Marcia Pierce, *Eastern Kentucky University*
Mark Pilgrim, *College of Coastal Georgia*
Teri Reiger, *University of Wisconsin–Oshkosh*
Jackie Reynolds, *Richland College*
Luis Rodriguez, *San Antonio College*
Benjamin Rowley, *University of Central Arkansas*
Mark A. Schneegurt, *Wichita State University*
David Schwartz, *Houston Community College, Southwest*
Timothy Secott, *Minnesota State University*
Heidi R. Smith, *Front Range Community College*
Kristine Snow, *Fox Valley Technical College*
Tracey Steeno, *Northeast Wisconsin Technical College*
Louise Thai, *University of Missouri*
Sanjay Tiwary, *Hinds Community College*
Diane Vorbroker, *Cincinnati State Technical and Community College*
Delon Washo-Krupps, *Arizona State University*
Ronald Weiss, *Marian University*

A Note to the Student

Tips on learning to gain understanding

Most of you are probably taking this course as a prerequisite to nursing, dental hygiene, medicine, pharmacy, optometry, physician assistant, or other health science programs. Because you are preparing for professions that involve interactions with patients, you will be concerned with infection control and precautions, which in turn requires you to think about microbes and how to manage them. This means you must not only be knowledgeable about the characteristics of bacteria, viruses, and other microbes, their physiology, and primary niches in the world, but you must also have a grasp of disease transmission, the infectious process, disinfection procedures, and drug treatments. You will need to understand how the immune system interacts with microorganisms and the effects of immunization. All of these areas bring their own vocabulary and language—much of it new to you—and mastering it will require time, motivation, and preparation. A valid question students often ask is: "How can I learn this information to increase my success in the course as well as retain it for the future?"

Right from the first, you need to be guided by how your instructor has organized your course. Since there is more information than could be covered in one semester or quarter, your instructor will select what he/she wants to emphasize and construct a reading and problem assignment that corresponds to lectures and discussion sessions. Many instructors have a detailed syllabus or study guide that directs the class to specific content areas and vocabulary words. Others may have their own website to distribute assignments and even sample exams. Whatever materials are provided, this should be your primary guide in preparing to study.

The next consideration involves your own learning style and what works best for you. To be successful, you must commit essential concepts and terminology to memory. A list of how we retain information called the "pyramid of learning" has been proposed by Edgar Dale: We remember about 10% of what we read; 20% of what we hear; 50% of what we see and hear; 70% of what we discuss with others; 80% of what we experience personally; and 95% of what we teach to someone else.

> "The Talaro textbook not only gives a clearly-written, accurate verbal account of the workings of the microbial world, but also incorporates detailed and colorful figures and tables to give the students a visual picture of the life of microbes and the illnesses of microbiology. I love this textbook."
>
> —*Diane K. Vorbroker,* Cincinnati State Technical and Community College

There are clearly many ways to go about assimilating information—but mainly, you need to become involved in reading, writing, drawing simple diagrams, and discussion or study with others. This means reading alone will not gather the most important points from a chapter. You must attend lecture and laboratory sessions to listen to your instructors or teaching assistants explain the material. Notes taken during lecture can be rewritten or outlined to organize the main points. This begins the process of laying down memory. You should go over concepts with others—perhaps a tutor or study group—and even take on the role of the teacher-presenter part of the time. It is with these kind of interactions that you will not just rote memorize words but *understand* the ideas and be able to apply them later.

A way to assess your understanding and level of learning is to test yourself. You may use the exam questions in the text, on the CONNECT website, or make up your own. LearnSmart, available within the CONNECT site, is an excellent way to map your own, individualized learning program. It tracks what you know and what you don't know and creates questions just for you based on your progress.

Another big factor in learning is the frequency of studying. It is far more effective to spend an hour or so each day for two weeks than a marathon cramming session on one weekend. If you approach the subject in small bites and remain connected with the terminology and topics, over time it will become yours and you will find that the pieces begin to fit together. In the final analysis, the process of learning comes down to self-motivation and attitude. There is a big difference between forcing yourself to memorize something to get by and really wanting to know and understand it. Therein is the key to most success and achievement, no matter what your final goals. And though it is true that mastering the subject matter in this textbook requires time and effort, millions of students will affirm how worthwhile it has been in their professions and everyday life.

Contents

CHAPTER 8

An Introduction to Microbial Metabolism: The Chemical Crossroads of Life 217

CHAPTER 9

Microbial Genetics 254

CHAPTER 10

Genetic Engineering: A Revolution in Molecular Biology 291

CHAPTER 15

Adaptive, Specific Immunity and Immunization 452

CHAPTER 16

Disorders in Immunity 486

CHAPTER 17

Procedures for Identifying Pathogens and Diagnosing Infections 517

Beginning the Journey

The beauty of microorganisms

If you were taking a survey of people on the street about their impression of microorganisms, you would not have to look very far to find someone with a rather negative vision, probably gleaned from commercials, of unsightly, disease-causing germs swarming around everywhere, lying in waiting for unsuspecting victims. But we'd like to think that, after looking through a microscope focused on a drop of pond water, most people would have their eyes opened, both literally and figuratively, by the astonishing images they see. A glimpse of the microbial world often surprises us with unusual and bizarre forms of great beauty and complexity, from the crystalline perfection of a virus to fantastic colors and shapes of algae. It is with this realization that we are featuring magnified views of interesting and striking microorganisms set in the context of a case study at the start of most chapters. Even our cover, depicting a glass sculpture of the H1N1 influenza virus, reflects a trend of embracing microorganisms as an inspiration for artistic works. Although a major intent of this textbook is to promote your understanding of the effects of microbes on humans and their involvement in diseases, we also aspire to communicate our appreciation and awe for these tiniest creatures and to provide insights into the tremendous impact they have on every facet of the earth. It is our hope that, in time, you too will become an advocate for microorganisms and help educate others about their importance, and perhaps even their beauty. Happy reading. . .

—*Kathleen Park Talaro* and *Barry Chess*

"I would have to say that this text is sophisticated, logically written, illustrated effectively, and very comprehensive. The chapters I reviewed were well written and very comprehensive. The chapter on metabolism was superb."

—*Luis A. Rodriguez*, San Antonio College

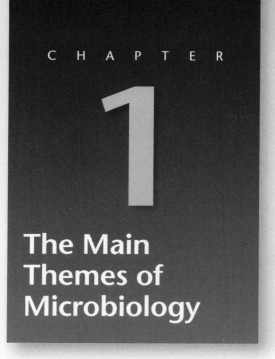

The Main Themes of Microbiology

This is a colorized view of bacterial plankton taken from the Sargasso Sea

"Peering through the microscope into a drop of seawater is like looking at stars with a telescope on a clear night." *–Dr. Victor Gallardo, ocean researcher*

CASE FILE **1** ## Microbes Rule the Earth

In 2003, a 100-foot sailboat called the *Sorcerer II* embarked on a highly unusual fishing expedition in the Sargasso Sea. What was most striking about this voyage was that it did not involve actually catching fish with hooks or nets. Instead, the targets were tiny floating microbes "hooked" by an exceedingly sophisticated and specific technology.

This project was the brainchild of Dr. Craig Venter, a prominent genetics researcher,* and its primary goal was to survey in detail the microbial population of ocean water. Scientists aboard the vessel randomly collected surface water about every 200 miles, extracted the tiniest forms of microscopic plankton, primarily bacteria, and sent samples back to Venter's laboratory. It was here that his scientific crew engaged in a new and powerful way of examining the world. Instead of painstakingly locating and identifying the individual microbes in the sample, as might have been done in the past, they extracted the genetic material (DNA) from the samples and analyzed the

DNA using state-of-the-art molecular techniques and computers.** Their stunning and somewhat unexpected discovery was that the variety and numbers of microbes living in the ocean exceeded by far the levels found in any previous ocean studies.

This ambitious undertaking was just the beginning. It was followed by several additional voyages by Dr. Venter's ship along with marine microbiologists from the Marine Biological Institute in Woods Hole, Massachusetts, and is continuing today all over the globe. Even though microbiologists had previously described around 5,700 different types of bacteria, the evidence from these studies showed that this number represented only the tiniest "drop in the ocean." Some of the data uncovered evidence of more than 20,000 different kinds of microorganisms in just a single liter of seawater, most of them unknown. Realizing that the ocean is a vast space with endless nooks and crannies for organisms to hide in, by one estimate, it could easily contain 5 million to 10 million

different microscopic creatures, each of them having unique characteristics and roles in the ocean environment. According to Dr. David Thomassen, Chief Scientist, U.S. Department of Energy, "Microbes rule the earth. Scientists estimate that there are more microbes on earth than there are stars in the universe—an estimated nonillion (one followed by 30 zeros). Microbes and their communities make up the foundation of the biosphere and sustain all life on earth."

▶ *Which groups of microorganisms would likely be found in the plankton?*

▶ *What fields of microbiology could be involved in the further study of these microbes and in uncovering their basic characteristics?*

To continue the case, go to page 9.

*Dr. Venter was one of the main individuals behind the mapping of the human genome in 2001.

**This technique, called metagenomic analysis, will be discussed in chapter 10.

1.1 The Scope of Microbiology

Expected Learning Outcomes

1. Define *microbiology* and *microorganisms,* and identify the major organisms included in the science of microbiology.
2. Name and define the primary areas included in microbiological studies.

As we observe the natural world, teeming with life, we cannot help but be struck by its beauty and complexity. But for every feature that is visible to the naked eye, there are millions of other features that are concealed beyond our sight because of their small size. This alternate microscopic universe is populated by a vast microbial menagerie that is equally beautiful and complex. To sum up the presence of microbes in one word, they are **ubiquitous.*** They are found in all natural habitats and most of those that have been created by humans. As scientists continue to explore remote and unusual environments, the one entity they always find is microbes. They exist deep beneath the polar icecaps, in the ocean to a depth of 7 miles, in hot springs and thermal vents, in toxic waste dumps, and even in the clouds.

Microbiology is a specialized area of biology that deals with tiny life forms that are not readily observed without magnification. Such **microscopic*** organisms are collectively referred to as **microorganisms, microbes,*** or several other terms, depending upon the purpose. Some people call them "germs" or "bugs" in reference to their role in infection and disease, but those terms have other biological meanings and perhaps place undue emphasis on the disagreeable reputation of microorganisms. There are several major groups of microorganisms that we will be studying. They are **bacteria, viruses, fungi, protozoa, algae,** and **helminths** (parasitic worms). As we will see in subsequent chapters, each group exhibits a distinct collection of biological characteristics. The nature of microorganisms makes them both easy and difficult to study. Easy, because they reproduce so rapidly and can usually be grown in large numbers in the laboratory. Difficult, because we can't observe or analyze them without special techniques, especially the use of microscopes (see chapter 3).

Microbiology is one of the largest and most complex of the biological sciences because it integrates subject matter from many diverse disciplines. Microbiologists study every aspect of microbes—their genetics, their physiology, characteristics that may be harmful or beneficial, the ways they interact with the environment, the ways they interact with other organisms, and their uses in industry and agriculture.

See **table 1.1** for an overview of several areas of basic and applied microbiology.

Each major discipline in microbiology contains numerous subdivisions or specialties that deal with a specific subject area or field (table 1.1). In fact, many areas of this science have become so specialized that it is not uncommon for a microbiologist to spend an entire career concentrating on a single group or type of microbe, biochemical process, or disease.

Among the specialty professions of microbiology are:

- geomicrobiologists, who focus on the roles of microbes in the development of earth's crust;
- marine microbiologists, who study the oceans and its smallest inhabitants;
- medical technologists, who do the tests that help diagnose pathogenic microbes and their diseases;
- nurse epidemiologists, who analyze the occurrence of infectious diseases in hospitals; and
- astrobiologists, who study the possibilities of organisms in space **(Case file 2).**

Studies in microbiology have led to greater understanding of many general biological principles. For example, the study of microorganisms established universal concepts concerning the chemistry of life (see chapters 2 and 8), systems of inheritance (see chapter 9), and the global cycles of nutrients, minerals, and gases (see chapter 26).

1.2 General Characteristics of Microorganisms and Their Roles in the Earth's Environments

Expected Learning Outcomes

3. Describe the basic characteristics of prokaryotic cells and eukaryotic cells and their evolutionary origins.
4. State several ways that microbes are involved in the earth's ecosystems.
5. Describe the cellular makeup of microorganisms and their size range, and indicate how viruses differ from cellular microbes.

The Origins of Microorganisms

For billions of years, microbes have extensively shaped the development of the earth's habitats and the evolution of other life forms. It is understandable that scientists searching for life on other planets first look for signs of microorganisms.

The fossil record dating from ancient rocks and sediments points to bacterialike cells that existed at least 3.5 billion years ago **(figure 1.1).** These simple cells were the dominant cells on earth for about 2 billion years. They were very small and lacked complex internal structures. One of these structures was a nucleus. The term that is used to define these types of cells is **prokaryotic,*** referring to the lack of a nucleus (karyon). About 1.8 billion years ago, there appeared in the fossil record a more complex cell, which

* *ubiquitous* (yoo-bik′-wih-tis) L. *ubique,* everywhere and *ous,* having. Being, or seeming to be, everywhere at the same time.
* *microscopic* (my″-kroh-skaw′-pik) Gr. *mikros,* small, and *scopein,* to see.
* *microbe* (my′-krohb) Gr. *mikros,* small, and *bios,* life.

* *prokaryotic* (proh″-kar-ee-ah′-tik) Gr. *pro,* before, and *karyon,* nucleus. Sometimes spelled procaryotic and eucaryotic.

TABLE 1.1 A Sampling of Fields and Occupations in Microbiology

A. Immunology

This branch studies the complex web of protective substances and reactions caused by invading microbes and other harmful entities. It includes such diverse areas as blood testing, vaccination, and allergy (see chapters 15, 16, and 17).

Figure A A specialist in the CDC special pathogens unit reads a microscopic test to screen for infection that is based on an immune reaction.

B. Public Health Microbiology and Epidemiology

These branches monitor and control the spread of diseases in communities. Some of the institutions charged with this task are the U.S. Public Health Service (USPHS) and the Centers for Disease Control and Prevention (CDC). The CDC collects information and statistics on diseases from around the United States and publishes it in a newsletter, *The Morbidity and Mortality Weekly Report* (see chapter 13).

Figure B Public health microbiologists examine mice and take samples to determine if they carry the hantavirus, one of the emerging pathogens that concerns the CDC.

C. Biotechnology

This branch is defined by any process that harnesses the actions of living things to arrive at a desired product, ranging from beer to stem cells. It includes industrial microbiology, which uses microbes to produce and harvest large quantities of such substances as vaccines, vitamins, drugs, and enzymes (see chapters 10 and 27).

Figure C A biotechnology technician prepares a bioreactor for vaccine production.

D. Genetic Engineering and Recombinant DNA Technology

These interrelated fields involve deliberate alterations of the genetic makeup of organisms to create novel microbes, plants, and animals with unique behavior and physiology. This is a rapidly expanding field that often complements biotechnology (see chapter 10).

Figure D A geneticist at the US Department of Agriculture examines a wheat plant that has been genetically engineered to resist a fungal pathogen.

TABLE 1.1 (*continued*)

E. Food Microbiology, Dairy Microbiology, and Aquatic Microbiology

These branches examine the ecological and practical roles of microbes in food and water.

- Food microbiologists are concerned with the effects of microbes, including such areas as food spoilage, food-borne diseases, and production.
- Aquatic microbiologists explore the ecology of natural waters as well as the impact of microbes on water purity and treatment.

F. Agricultural Microbiology

This branch is concerned with the relationships between microbes and domesticated plants and animals.

- Plant specialists focus on plant diseases, soil fertility, and nutritional interactions.
- Animal specialists work with infectious diseases and other associations animals have with microorganisms.

G. Branches of Microbiology

Branch	Chapter	Involved in the Study of:
Bacteriology	4	The bacteria—small single-celled prokaryotic organisms
Mycology	5, 22	The fungi, a group of eukaryotes that includes both microscopic eukaryotes (molds and yeasts) and larger organisms (mushrooms, puffballs)
Protozoology	5, 23	The protozoa—animal-like and mostly single-celled eukaryotes
Virology	6, 24, 25	Viruses—minute, noncellular particles that parasitize cells
Parasitology	5, 23	Parasitism and parasitic organisms—traditionally including pathogenic protozoa, helminth worms, and certain insects
Phycology or Algology	5	Simple photosynthetic eukaryotes, the algae, ranging from single-celled forms to large seaweeds
Morphology	4, 5, 6	The detailed structure of microorganisms
Physiology	7, 8	Microbial function (metabolism) at the cellular and molecular levels
Taxonomy	1, 4, 5	Classification, naming, and identification of microorganisms
Microbial Genetics, Molecular Biology	9, 10	The function of genetic material and biochemical reactions that make up a cell's metabolism
Microbial Ecology	7, 26	Interrelationships between microbes and the environment; the roles of microorganisms in the nutrient cycles and natural ecosystems

contained a nucleus and other complex internal structures (**figure 1.2***a*). These types of cells and organisms are defined as **eukaryotic*** in reference to their "true" nucleus. The early eukaryotes, probably similar to algae and protozoa, started lines of evolution that eventually gave rise to fungi, plants, and multicellular animals such as worms and insects. You can see from figure 1.1 how long that took! The bacteria preceded even the earliest animals by about 3 billion years. This is a good indication that humans are not likely to, nor should we try to, eliminate microorganisms from our environment. They are the ultimate survivors.

The Cellular Organization of Microorganisms

As a general rule, prokaryotic cells are smaller than eukaryotic cells, and in addition to lacking a nucleus, they lack other complex internal compartments called **organelles.** Organelles are structures in cells that are bound by one or more membranes. Examples such as mitochondria and Golgi complex perform specific functions in eukaryotic cells. Prokaryotes also perform specific functions, but they lack the dedicated organelles to carry them out.

The body plan of most microorganisms consists of a single cell or clusters of cells (**figure 1.3**). All prokaryotes are microorganisms and include the bacteria and archaeons (see figure 1.14). Only some of the eukaryotes are microorganisms: primarily algae, protozoa, molds and yeasts (types of fungi), and certain animals such as arthropods and worms. These last two groups may not be microscopic, but they are still included in the study because worms can be involved in infections and may require a microscope to identify them. Some arthropods such as fleas and ticks may also be carriers of infectious diseases. Additional coverage on cell types and microorganisms appears in chapters 4 and 5.

* *eukaryotic* (yoo″-kar-ee-ah′-tik) Gr. *eu,* true or good, and *karyon,* nucleus.

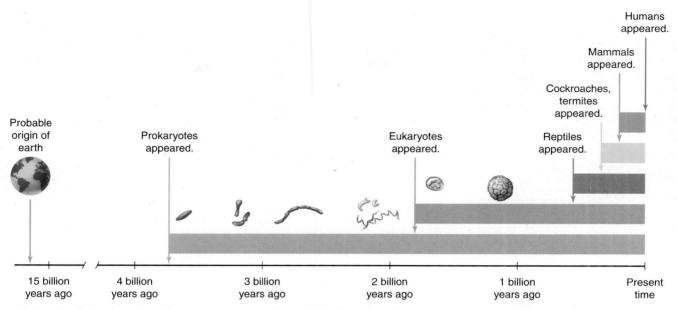

Figure 1.1 **Evolutionary time line.** The first simple prokaryotes appeared approximately 3.5 billion years ago. They were the only form of life for half of the earth's history.

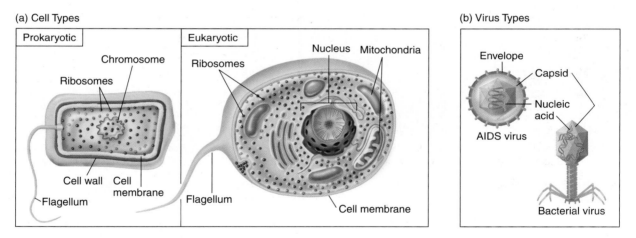

Figure 1.2 **Basic structure of cells and viruses.** **(a)** Comparison of a prokaryotic cell and a eukaryotic cell. **(b)** Two examples of viruses. These cell types and viruses are discussed in more detail in chapters 4, 5, and 6.

TAKE NOTE: VIRUSES

Viruses are considered one type of microbe because they are microscopic and can cause infections and disease, but they are not cells. They are small particles that exist at a level of complexity somewhere between large molecules and cells (**figure 1.4**). Viruses are much simpler than cells; they are composed essentially of a small amount of hereditary material wrapped up in a protein covering. Some biologists refer to viruses as parasitic particles; others consider them to be very primitive organisms. One thing is certain: They are highly dependent on a host cell's machinery for their activities.

Microbial Dimensions: How Small Is Small?

When we say that microbes are too small to be seen with the un-aided eye, what sorts of dimensions are we talking about? This concept is best visualized by comparing microbial groups with some organisms of the macroscopic world and also with the molecules and atoms of the molecular world (**figure 1.4**). The dimensions of macroscopic organisms are usually given in centimeters (cm) and meters (m), whereas those of most microorganisms fall within the range of micrometers (µm) and, sometimes, nanometers (nm) and millimeters (mm). The size range of most microbes extends from the smallest viruses, measuring around 10 nm and actually not much bigger than a large molecule, to protozoans measuring 3 to 4 mm and visible with the naked eye.

Bacteria: *Mycobacterium tuberculosis,* a rod-shaped cell (15,500x).

Reproductive spores

Fungi: *Thamnidium,* a filamentous fungus (400x)

Algae: desmids, *Spirogyra* filament, and diatoms (golden cells) (500x).

A single virus particle

Virus: *Herpes simplex,* cause of cold sores (100,000x).

Protozoa: A pair of *Vorticella* (500x), stalked cells that feed by means of a whirling row of cilia.

Helminths: Cysts of the parasitic roundworm, *Trichinella spiralis* (250x) embedded in muscle.

Figure 1.3 **The six basic types of microorganisms.** Organisms are not shown at the same magnifications so approximate magnification is provided.

Microbial Involvement in Energy and Nutrient Flow

The microbes in all natural environments have lived and evolved there for billions of years. We do not yet know everything they do, but it is likely they are vital components of the structure and function of these ecosystems and critical to the operations of the earth.

Microbes are deeply involved in the flow of energy and food through the earth's ecosystems.[1] Most people are aware that plants carry out **photosynthesis,** which is the light-fueled conversion of carbon dioxide to organic material, accompanied by the formation of oxygen. But microorganisms were photosynthesizing long before the first plants appeared. In fact, they were responsible for changing the atmosphere of the earth from one without oxygen to one with

oxygen. Today photosynthetic microorganisms (including algae) account for more than 50% of the earth's photosynthesis, contributing the majority of the oxygen to the atmosphere **(figure 1.5a).**

Another process that helps keep the earth in balance is the process of biological **decomposition** and nutrient recycling. Decomposition involves the breakdown of dead matter and wastes into simple compounds that can be directed back into the natural cycles of living things **(figure 1.5b).** If it were not for multitudes of bacteria and fungi, many chemical elements would become locked up and unavailable to organisms. In the long-term scheme of things, microorganisms are the main forces that drive the structure and content of the soil, water, and atmosphere. For example:

• Earth's temperature is regulated by "greenhouse gases," such as carbon dioxide and methane, that create an insulation layer in the atmosphere and help retain heat. A significant proportion of these gases is produced by microbes living in the environment and the digestive tracts of animals.

1. Ecosystems are communities of living organisms and their surrounding environment.

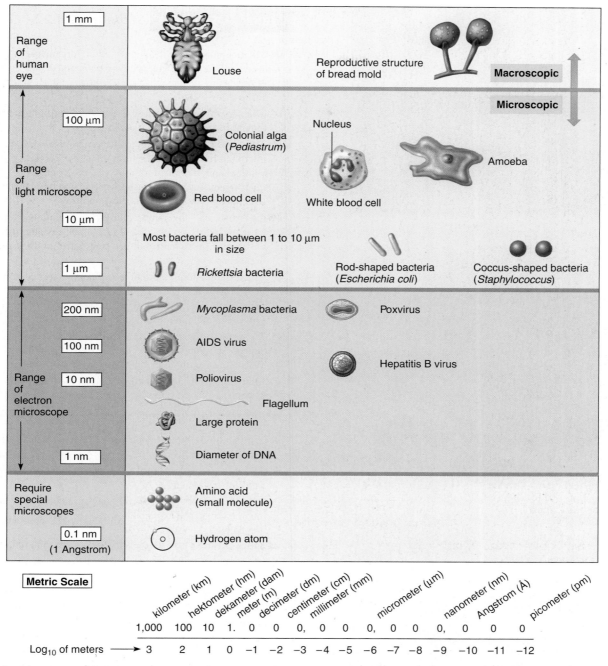

Figure 1.4 The size of things. Common measurements encountered in microbiology and a scale of comparison from the macroscopic to the microscopic, molecular, and atomic. Most microbes encountered in our studies will fall between 100 µm and 10 nm in overall dimensions. The microbes shown are more or less to scale within size zone but not between size zones.

- Recent estimates propose that, based on weight and numbers, up to 50% of all organisms exist within and beneath the earth's crust in soil, rocks, and even the frozen Antarctic **(figure 1.5c)**. It is increasingly evident that this enormous underground community of microbes is a major force in weathering, mineral extraction, and soil formation.

- Bacteria and fungi live in complex associations with plants. They assist the plants in obtaining nutrients and water and may protect them against disease. Microbes form similar interrelationships with animals, notably as residents of numerous bodily sites.

(a)

(b)

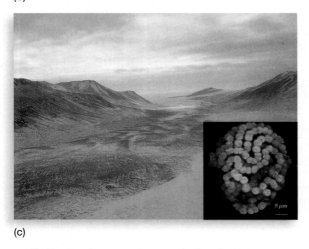

(c)

Figure 1.5 A microscopic wonderland. (a) A summer pond is heavily laden with surface scum that reveals golden-colored diatoms and blue green bacteria in strands (600× magnification). **(b)** A rotting tomato being decomposed by a furry forest of mold including *Rhizopus*, bearing tiny sacs of spores on a stalk (250×). **(c)** Even a dry lake in Antarctica, one of the coldest places on earth (−35°C), can harbor microbes under its icy sheet. Here we see a red cyanobacterium, *Nostoc* (3,000×), that has probably been frozen in suspended animation there for 3,000 years. This is one kind of habitat on earth that may well be a model for conditions on Mars.

Check & Assess Sections 1.1–1.2

✔ Microorganisms are defined as "living organisms too small to be seen with the naked eye." Among the members of this huge group of organisms are bacteria, fungi, protozoa, algae, viruses, and parasitic worms.

✔ Microorganisms live nearly everywhere and influence many biological and physical activities on earth.

✔ The scope of microbiology is incredibly diverse. It includes basic microbial research, research on infectious diseases, study of prevention and treatment of disease, environmental functions of microorganisms, and industrial use of microorganisms for commercial, agricultural, and medical purposes.

✔ Two basic cell lines appeared during evolutionary history: prokaryotes, which are small and lack a nucleus and organelles, and eukaryotes, which are larger and have both a nucleus and organelles.

✔ Viruses are not cellular and are sometimes called particles rather than organisms. They are included in microbiology because of their small size, their close relationship with cells, and their involvement in numerous infectious diseases.

✔ Most microorganisms are measured in micrometers, with two exceptions. The helminths are measured in millimeters, and the viruses are measured in nanometers.

✔ Microorganisms are essential to the operation of the earth's ecosystems, as photosynthesizers, decomposers, and recyclers.

1. Explain the important contributions microorganisms make in the earth's ecosystems.

2. Describe five different ways in which humans exploit microorganisms for our benefit.

3. Identify the groups of microorganisms included in the scope of microbiology, and explain the criteria for including these groups in the field.

4. Observe figure 1.4 and place the microbes pictured in a size ranking, going from smallest to largest. Use the magnification as your gauge.

5. Construct a table that displays all microbial groups based on what kind of cells they have or do not have.

6. Where do you suppose viruses came from? Why must they exist inside host cells?

7. Explain this statement: Microorganisms—we need to live with them because we can't live without them.

1.3 Human Use of Microorganisms

Expected Learning Outcome

6. Discuss the ways microorganisms may be applied to solve human problems.

The incredible diversity and versatility seen in microbes make them excellent candidates for solving human problems. By accident or choice, humans have been using microorganisms for thousands of

years to improve life and even to further human progress. Yeasts, a type of microscopic fungi, cause bread to rise and ferment sugar to make alcoholic beverages. Historical records show that households in ancient Egypt kept moldy loaves of bread to apply directly to wounds and lesions, which was probably the first use of penicillin! When humans manipulate microorganisms to make products in an industrial setting, it is called biotechnology. One newer application farms algae to extract a form of oil (biodiesel) to be used in place of petroleum products **(figure 1.6a).**

Genetic engineering is a newer area of biotechnology that manipulates the genetics of microbes, plants, and animals for the purpose of creating new products and genetically modified organisms.

(a)

(b)

Figure 1.6 **Microbes at work.** (a) Algae specialists peer from their biodiesel bioreactor, a platform of hanging bags containing cultures of single-celled algae (see inset 750×). Their research will test the capacity of algae to mass produce oil that could be used as an alternative to fossil fuels. So far, this alternate way of growing renewable fuels looks very promising and may truly provide a "green" source of energy. (b) Workers remove waste from the Hanford Nuclear Facility in Washington state that has been dangerously contaminated with spent radioactive substances. The cleanup of this site is expected to take several years. A newly discovered bacterium, *Shewanella* (inset 5,000×), is being tested as a bioremediation measure. It is capable of reducing and detoxifying even dangerous elements such as uranium.

One powerful technique for designing new organisms is termed **recombinant DNA.** This technology makes it possible to deliberately alter DNA[2] and to switch genetic material from one organism to another. Bacteria and fungi were some of the first organisms to be genetically engineered, because their relatively simple genetic material is readily manipulated in the laboratory. Recombinant DNA technology has unlimited potential in terms of medical, industrial, and agricultural uses. Microbes can be engineered to synthesize desirable proteins such as drugs, hormones, and enzymes (see table 1.1c).

Among the genetically unique organisms that have been designed by bioengineers are bacteria that contain a natural pesticide, yeasts that produce human hormones, pigs that produce hemoglobin, and plants that are resistant to disease (see table 1.1d). The techniques have also paved the way for characterizing human genetic material and diseases.

Another way of tapping into the unlimited potential of microorganisms is the relatively new science of **bioremediation.*** This process introduces microbes into the environment to restore stability or to clean up toxic pollutants. Bioremediation is required to control the massive levels of pollution that result from human activities. Microbes have a surprising capacity to break down chemicals that would be harmful to other organisms. Agencies and companies have developed microbes to handle oil spills and detoxify sites contaminated with heavy metals, pesticides, and even radioactive wastes **(figure 1.6b).** The solid waste disposal industry is focusing on methods for degrading the tons of garbage in landfills, especially plastics and paper products. One form of bioremediation that has been in use for some time is the treatment of water and sewage. With dwindling clean freshwater supplies worldwide, it will become even more important to find ways to reclaim polluted water.

CONTINUING CASE FILE **1**

The genetic technology featured in this case is currently focused on mapping the microbial diversity of habitats such as soil, icebergs, hot springs, deep-ocean sediments, and the human body. This will allow us to maximize our understanding of these complex communities and to comprehend how they are involved in the functions of the earth and all of its life forms.

- What will likely be the result of probing all of the habitats on earth with this type of technology?

- Other than discovering and attempting to identify hidden communities of microbes, what could be some other possible rationales for modern-day microbe hunting?

For a wrap-up, see the Case File Perspective on page 23.

2. DNA, or deoxyribonucleic acid, the chemical substance that comprises the genetic material of organisms.

* *bioremediation* (by′-oh-ree-mee-dee-ay″-shun) *bios,* life; *re,* again; *mederi,* to heal. The use of biological agents to remedy environmental problems.

1.4 Microbial Roles in Infectious Diseases

Expected Learning Outcomes

7. Review the roles of microorganisms as parasites and pathogens that cause infection and disease.
8. Define what is meant by emerging and reemerging diseases.

It is important to remind you that the large majority of microorganisms are relatively harmless and highly beneficial and essential. They live a free existence in the array of habitats on earth and are able to derive their requirements for life from the nonliving environment. Much of the time, they form cohesive communities with other organisms, sharing habitat and nutrients. Examples include the natural partnerships that are found in symbiosis and biofilms[3].

Some microbes have adapted to a non-free-living lifestyle called parasitism. A **parasite** lives in or on the body of a larger organism called the **host** and derives most of its requirements from that host. A parasite's actions can damage the host through infection and disease. Another term that can be used to specify this type of microbe is **pathogen.***

Humanity is plagued by nearly 2,000 different microbes that can cause various types of disease. Infectious diseases still devastate human populations worldwide, despite significant strides in understanding and treating them.

The most recent estimates from the World Health Organization (WHO) point to around 10 billion infections of all types across the world every year. There are more infections than people because many people acquire more than one infection. Infectious diseases are also among the most common causes of death in

3. A biofilm is a complex network of microbes and their secretions that form in most natural environments, discussed further in chapter 4.

* *pathogen* (path′-oh-jen) Gr. *pathos,* disease, and *gennan,* to produce. Disease-causing agents.

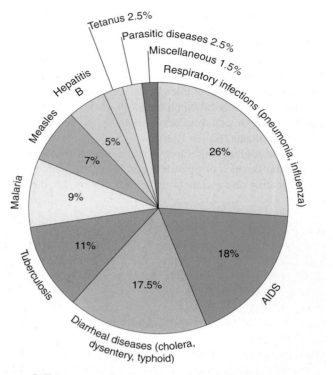

Figure 1.7 Worldwide infectious disease statistics. This figure depicts the 10 most common infectious causes of death.

much of humanity, and they still kill a significant percentage of the US population. **Table 1.2** depicts the 10 top causes of death per year (by all causes, infectious and noninfectious) in the United States and worldwide. The worldwide death toll from infections is about 12 million people per year. In **figure 1.7,** you can see the top infectious causes of death displayed in a different way. Note that many of these infections are treatable with drugs or preventable with vaccines.

Those hardest hit are residents in countries where access to adequate medical care is lacking. One-third of the earth's inhabitants

TABLE 1.2 Top Causes of Death—All Diseases			
United States	**No. of Deaths**	**Worldwide**	**No. of Deaths**
1. Heart disease	696,950	1. Heart disease	8.12 million
2. Cancer	557,270	2. Stroke	5.51 million
3. Stroke	162,670	3. Respiratory infection	3.88 million
4. Chronic lower respiratory disease	124,800	4. Cancer	3.33 million
5. Unintentional injury (accidents)	106,740	5. HIV/AIDS	2.78 million
6. Diabetes	73,250	6. Chronic lower respiratory disease	2.75 million
7. Influenza and pneumonia*	65,680	7. Diarrheal disease	1.80 million
8. Alzheimer disease	58,870	8. Tuberculosis	1.57 million
9. Kidney problems	40,970	9. Malaria	1.27 million
10. Septicemia (bloodstream infection)	33,865	10. Accidents	1.19 million

*Diseases in red are those most clearly caused by microorganisms, although cancer and other diseases may be associated with infections.

live on less than $1 per day, are malnourished, and are not fully immunized. Take the case of malaria, which kills more than a million people every year worldwide. It is caused by a microorganism transmitted by mosquitoes, and currently the most effective way for citizens of developing countries to avoid infection is to sleep under a bed net, because the mosquitoes are most active in the evening. Yet even this inexpensive solution is beyond the reach of people in many developing countries who cannot afford the $3 to $5 for nets to protect their family.

One of the most eye-opening discoveries in recent years is that many diseases once considered noninfectious probably do involve microbial infection. The most famous of these is gastric ulcers, now known to be caused by a bacterium called *Helicobacter* (see Chapter 21). But there are more. A connection has been established, between diabetes and the coxsackievirus, and between chronic fatigue syndrome and a new retrovirus. Diseases as disparate as multiple sclerosis, obsessive compulsive disorder, and coronary artery disease have been linked to chronic infections with microorganisms. It seems that the golden age of microbiological discovery, during which all of the "obvious" diseases were characterized and cures or preventions were devised for them, should more accurately be referred to as the *first* golden age. We're now discovering the roles of microorganisms in quiet but slowly destructive diseases. These include female infertility caused by *Chlamydia* infection and malignancies such as liver cancer (hepatitis viruses) and cervical cancer (human papillomavirus). Most scientists expect that, in time, many chronic conditions will be found to have some association with microbial agents.

Another important development in infectious disease trends is the increasing number of patients with weakened defenses who are kept alive for extended periods. We are becoming more susceptible to infectious disease precisely because of advances in medicine. People are living longer. Sicker people are staying alive much longer than in the past. Older and sicker people have heightened susceptibility to what we might call "garden-variety" microbes.

Among the more significant factors in the overall picture of infectious diseases are emerging and reemerging diseases **(Insight 1.1). Emerging diseases** are newly identified conditions that are being reported in increasing numbers. Since 1969, at least 26 novel infectious agents have arisen within the human population. Some of them have been associated with a specific location (Ebola fever virus), whereas others have become pandemics, meaning they spread across continents (human immunodeficiency virus—HIV). A number of them cause zoonoses, which are diseases spread to humans from other animals (an example is West Nile fever, spread by mosquitoes). Recently a new strain of influenza virus (H1N1) that is a combination of swine, bird, and human viruses caused a significant pandemic that first erupted in 2009.

Reemerging diseases have existed since some time in the past but are on the rise once again. Among the most common resurgent infectious diseases are tuberculosis (TB), influenza, malaria, cholera, and hepatitis B. Tuberculosis, which has been known since ancient times, still causes 8 million new infections and kills 1 million to 2 million people every year. As you will see, numerous factors play a part in the tenaciousness of infectious diseases, but fundamental to all of them is the formidable capacity of microbes to adapt to alterations in the individual, community, and environment.

Check & Assess Sections 1.3–1.4

✔ Humans use the versatility of microbes to make improvements in industrial production, agriculture, medicine, and environmental protection.

✔ Most microorganisms are harmless, free-living species that perform vital functions in both the environment and larger organisms. Comparatively few species are agents of disease.

✔ Microbiologists have identified the causative agents for over 2,000 infectious diseases.

✔ Some infectious diseases are currently emerging and reemerging. These are on the rise because of rapid travel, the opening up of undeveloped geographic areas, questionable agricultural practices and food handling, drug resistance, and increases in people with chronic medical conditions.

8. Describe several ways that the beneficial qualities of microbes greatly outweigh their roles as infectious agents.

9. Look up each disease shown on figure 1.7 in the index and see which ones could be prevented by vaccines or cured with drugs. Are there other ways (besides vaccines) to prevent any of these?

10. Name some factors that could cause older diseases to show an increase in the number of cases.

1.5 The Historical Foundations of Microbiology

Expected Learning Outcomes

9. Outline the major events in the history of microbiology, including the major contributors to the early development of microscopy, medical advances, aseptic techniques, and the germ theory of disease.

10. Explain the main features of the scientific method, and differentiate between inductive and deductive reasoning and between hypothesis and theory.

If not for the extensive interest, curiosity, and devotion of thousands of microbiologists over the last 300 years, we would know little about the microscopic realm that surrounds us. Many of the discoveries in this science have resulted from the prior work of men and women who toiled long hours in dimly lit laboratories with the crudest of tools. Each additional insight, whether large or small, has added to our current knowledge of life forms and processes. This section summarizes the prominent discoveries made in the past 300 years: microscopy, the rise of the scientific method, and the development of medical microbiology, including the germ theory and the origins of modern microbiological techniques. See table A.3 in appendix A, which summarizes some of the pivotal events in microbiology from its earliest beginnings to the present.

INSIGHT 1.1

The Changing Spectacle of Infectious Diseases

The middle of the last century was a time of great confidence in science and medicine. The introduction of antibiotics in the 1940s and a lengthening list of vaccines for preventing numerous diseases caused many medical experts to declare a victory. For a short time, there was a sense that infectious diseases were going to be completely manageable. But this optimistic viewpoint was on a collision course with a vast army of very tiny invaders that are everywhere—namely, microorganisms. Because humans are constantly interacting with microbes, we serve as a handy incubator for infectious diseases, both those newly recognized and older ones previously identified.

All together, government agencies are keeping track of a total of 75 emerging and reemerging infectious diseases.* The newer or emerging diseases usually erupt suddenly with no warning (SARS respiratory syndrome). Older or reemerging diseases demonstrate just how difficult it is to eradicate microbes and the diseases they cause, even though we are very aware of them and often have drugs and vaccines to combat them. Only smallpox has been completely eliminated, although we are very close to eradicating polio.

What are some of the reasons that account for this dilemma with our microbial coinhabitants? A major contributing factor is increased mobility and travel, especially by air. An infected person can travel around the world before showing any symptoms of infection. He can carry the infectious agent to many far-flung locations, exposing populations along the way, who in turn can infect their contacts. The recent outbreak of H1N1 influenza (swine flu) provides a dramatic example. After being first diagnosed in Mexico in March 2009, it spread around the world in 6 short weeks! Within 6 months, it had caused millions of cases and thousands of deaths and had been reported in at least 200 countries. See the figure to view this rapid emergence of a new disease.

Other significant effects involve our expanding population and global food-growing practices. As we continue to encroach into new territory and wild habitats, there is potential for contact with exotic pathogens, as we saw with Ebola fever, Lyme disease, and hantavirus pulmonary syndrome. Our agricultural practices can unearth microbes that were lying dormant or hidden. A bacterium carried in the intestine of domestic cattle, *Escherichia coli* O157:H7, the agent of a serious kidney disease, has been associated with hundreds of thousands of infections from food and water contaminated with cattle feces. Much mass-produced fresh food can also travel around the world, infecting people along the way. Several large outbreaks of salmonellosis, shigellosis, and listeriosis have been traced to contaminated dairy and poultry products, and vegetables. In some areas of the world, farmers knowingly or unknowingly

*http://www.3.niaid.nih.gov/topics/emerging/list.htm

Influenza Positive Tests Reported to CDC by U.S. WHO/NREVSS Collaborating Laboratories, National Summary, 2008-09

The emergence of a new disease: Graph shows the progression of influenza cases from 2008–2009. Note that about April 2009, there is a sudden appearance of a the H1N1 strain orange, which soon becomes an epidemic and later, a pandemic (spreading to several continents).

use fecally contaminated water and fertilizer on fresh produce that will be on your table the next week.

And then there is the matter of the incredible resistance of microbes. You know about their capacity to live where no other living things could—volcanoes, salt lakes, radioactive waste pits—so it has not been a surprise to discover how readily they can adapt to drugs we use to treat them. The emergence of drug-resistant "superbugs" has become a massive problem in medicine. Some forms of *Staphylococcus aureus* (MRSA) and *Mycobacterium tuberculosis* are resistant to so many drugs that there are few, and sometimes no, choices left. In a recent report from a small African village, 52 out of 53 persons infected with extreme drug-resistant TB died within a short time because none of the drugs was effective.

As hard as we may try to manage microbes, we keep coming up against a potent reality, a sentiment summed up by the renowned microbiologist Louis Pasteur 130 years ago when he declared: "Microbes will have the last word."

Why do you think it is unlikely that infectious diseases can ever be completely eradicated? *Answer available at http://www. mhhe.com/talaro8*

The Development of the Microscope: "Seeing Is Believing"

It is likely that from the very earliest history, humans noticed that when certain foods spoiled, they became inedible or caused illness and yet other "spoiled" foods did no harm and even had enhanced flavor. Even several centuries ago, there was already a

sense that diseases such as the black plague and smallpox were caused by some sort of transmissible matter. But the causes of such phenomena were vague and obscure because the technology to study them was lacking. Consequently, they remained cloaked in mystery and regarded with superstition—a trend that led even well-educated scientists to believe in spontaneous generation **(Insight 1.2).**

Figure 1.8 An oil painting of Antonie van Leeuwenhoek (1632–1723) sitting in his laboratory. J. R. Porter and C. Dobell have commented on the unique qualities Leeuwenhoek brought to his craft: "He was one of the most original and curious men who ever lived. It is difficult to compare him with anybody because he belonged to a genus of which he was the type and only species, and when he died his line became extinct."

(a)

(b)

Figure 1.9 Leeuwenhoek's microscope. (a) A brass replica of a Leeuwenhoek microscope and how it is held. (b) Examples of bacteria drawn by Leeuwenhoek.

True awareness of the widespread distribution of microorganisms and some of their characteristics was finally made possible by the development of the first microscopes. These devices revealed microbes as discrete entities sharing many of the characteristics of larger, visible plants and animals. Several early scientists fashioned magnifying lenses, but their microscopes lacked the optical clarity needed for examining bacteria and other small, single-celled organisms. The most careful and exacting observations awaited the simple single-lens microscope hand-fashioned by **Antonie van Leeuwenhoek,** a Dutch linen merchant and self-made microbiologist **(figure 1.8).**

Paintings of historical figures such as the one of Leeuwenhoek in figure 1.8 don't always convey a meaningful feeling for the event or person depicted. Imagine his dusty shop in Holland in the late 1600s filled with bolts of linens for draperies and upholstery. Between customers, Leeuwenhoek retired to the workbench in the back of his shop, grinding glass lenses to ever-finer specifications. He could see with increasing clarity the threads in his fabrics. Eventually, he became interested in things other than thread counts. He took rainwater from a clay pot, smeared it on his specimen holder, and peered at it through his finest lens. He found "animals appearing to me ten thousand times less than those which may be perceived in the water with the naked eye."

He didn't stop there. He scraped plaque from his teeth, and from the teeth of some volunteers who had never cleaned their teeth in their lives, and took a close look at that. He recorded: "In the said matter there were many very little living animalcules, very prettily a-moving. . . . Moreover, the other animalcules were in such enormous numbers, that all the water . . . seemed to be alive." Leeuwenhoek started sending his observations to the Royal Society of London, and eventually he was recognized as a scientist of great merit.

Leeuwenhoek constructed more than 250 small, powerful microscopes that could magnify up to 300 times **(figure 1.9).** Considering that he had no formal training in science and that he was the first person ever to faithfully record this strange new world, his descriptions of bacteria and protozoa (which he called "animalcules") were astute and precise. Because of Leeuwenhoek's extraordinary contributions to microbiology, he is sometimes considered the father of bacteriology and protozoology.

From the time of Leeuwenhoek, microscopes became more complex and improved, with the addition of refined lenses, a condenser, finer focusing devices, and built-in light sources. The

INSIGHT 1.2

The Fall of Superstition and the Rise of Microbiology

For thousands of years, people believed that certain living things arose from vital forces present in nonliving or decomposing matter. This ancient idea, known as **spontaneous generation,** was continually reinforced as people observed that meat left out in the open soon "produced" maggots, that mushrooms appeared on rotting wood, that rats and mice emerged from piles of litter, and that other magical phenomena occurred. Though some of these early ideas seem quaint and ridiculous in light of modern knowledge, we must remember that, at the time, mysteries in life were accepted and the scientific method was not widely practiced.

Even after single-celled organisms were discovered during the mid-1600s, the idea of spontaneous generation continued to exist. Some scientists assumed that microscopic beings were an early stage in the development of more complex ones.

Over the subsequent 200 years, scientists waged an experimental battle over the two hypotheses that could explain the origin of simple life forms. Some tenaciously clung to the idea of **abiogenesis** (ah-bee″-oh-jen-uh-sis), which embraced spontaneous generation. On the other side were advocates of **biogenesis,** saying that living things arise only from others of their same kind. There were serious proponents on both sides, and each side put forth what appeared on the surface to be plausible explanations of why their evidence was more correct. Gradually, the abiogenesis hypothesis was abandoned, as convincing evidence for biogenesis continued to mount. The following series of experiments were among the most important in finally tipping the balance.

Among the important variables to be considered in challenging the hypotheses were the effects of nutrients, air, and heat, and the presence of preexisting life forms in the environment. One of the first people to test the spontaneous generation theory was **Francesco Redi** of Italy. He conducted a simple experiment in which he placed meat in a jar and covered it with fine gauze. Flies gathering at the jar were blocked from entering and thus laid their eggs on the outside of the gauze. The maggots subsequently developed without access to the meat, indicating that maggots were the offspring of flies and did not arise from some "vital force" in the meat. This and related experiments laid to rest the idea that more complex animals such as insects and mice developed through abiogenesis, but it

did not convince many scientists of the day that simpler organisms could not arise in that way.

Redi's Experiment

Closed — Meat with no maggots Open — Maggots hatching into flies

The Frenchman **Louis Jablot** reasoned that even microscopic organisms must have parents, and his experiments with infusions (dried hay steeped in water) supported that hypothesis. He divided an infusion that had been boiled to destroy any living things into two containers: a heated container that was closed to the air and a heated container that was freely open to the air. Only the open vessel developed microorganisms, which he presumed had entered in air laden with dust. Regrettably, the validation of biogenesis was temporarily set back by **John Needham,** an Englishman

Jablot's Experiment

Infusions

Covered — Dust Uncovered — Dust

Remains clear; no growth Heavy microbial growth

prototype of the modern compound microscope, in use from about the mid-1800s, was capable of magnifications of 1,000 times or more, largely because they had two sets of lenses for magnification. Even our modern laboratory microscopes are not greatly different in basic structure and function from those early microscopes. The technical characteristics of microscopes and microscopy are a major focus of chapter 3.

The Establishment of the Scientific Method

A serious impediment to the development of true scientific reasoning and testing was the tendency of early scientists to explain natural phenomena by a mixture of belief, superstition, and argument. The development of an experimental system that answered questions objectively and was not based on prejudice marked the beginning of

true scientific thinking. These ideas gradually crept into the consciousness of the scientific community during the 1600s. The general approach taken by scientists to explain a certain natural phenomenon is called the **scientific method.** A primary aim of this method is to formulate a **hypothesis,** a tentative explanation to account for what has been observed or measured. A good hypothesis should be in the form of a statement. It must be capable of being either supported or discredited by careful, systematic observation or experimentation. For example, the statement that "microorganisms cause diseases" can be experimentally determined by the tools of science, but the statement that "diseases are caused by evil spirits" cannot.

There are various ways to apply the scientific method, but probably the most common is called the **deductive approach.** Using this approach, a scientist constructs a hypothesis, tests its validity by outlining particular events that are predicted by the

who did similar experiments using mutton gravy. His results were in conflict with Jablot's because both his heated and unheated test containers teemed with microbes. Unfortunately, his experiments were done before the realization that heat-resistant microbes are not usually killed by mere boiling. Apparently Jablot had been lucky; his infusions were sterile.

Additional experiments further defended biogenesis. **Franz Shultze** and **Theodor Schwann** of Germany felt sure that air was the source of microbes and sought to prove this by passing air through strong chemicals or hot glass tubes into heat-treated infusions in flasks. When the infusions again remained devoid of living things, the supporters of abiogenesis claimed that the treatment of the air had made it harmful to the spontaneous development of life.

Shultze and Schwann's Test

- Air inlet
- Flame heats air.
- Previously sterilized infusions remain sterile.

Then, in the mid-1800s, the acclaimed microbiologist **Louis Pasteur** entered the arena. He had recently been studying the roles of microorganisms in the fermentation of beer and wine, and it was clear to him that these processes were brought about by the activities of microbes introduced into the beverage from air, fruits, and grains. The methods he used to discount abiogenesis were simple yet brilliant.

To further clarify that air and dust were the source of microbes, Pasteur filled flasks with broth and fashioned their openings into elongate, swan-neck-shaped tubes. The flasks' openings were freely open to the air but were curved so that gravity would cause any airborne dust particles to deposit in the lower part of the necks. He heated the flasks to sterilize the broth and then incubated them. As long as the flask

remained intact, the broth remained sterile, but if the neck was broken off so that dust fell directly down into the container, microbial growth immediately commenced.

Pasteur summed up his findings, "For I have kept from them, and am still keeping from them, that one thing which is above the power of man to make; I have kept from them the germs that float in the air, I have kept from them life."

Pasteur's Experiment

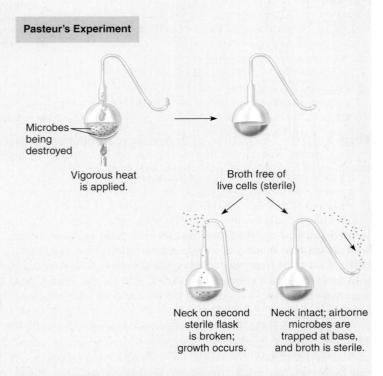

Microbes being destroyed

Vigorous heat is applied.

Broth free of live cells (sterile)

Neck on second sterile flask is broken; growth occurs.

Neck intact; airborne microbes are trapped at base, and broth is sterile.

Why do you think Jablot and Pasteur's closed infusions turned out to be sterile, unlike those of several other experimenters?
Answer available at http://www.mhhe.com/talaro8

hypothesis, and then performs experiments to test for those events **(figure 1.10).** The deductive process states: "*If* the hypothesis is valid, *then* certain specific events can be expected to occur."

A lengthy process of experimentation, analysis, and testing eventually leads to conclusions that either support or refute the hypothesis. If experiments do not uphold the hypothesis—that is, if it is found to be flawed—the hypothesis or some part of it is reconsidered. This does not mean the results are invalid; it means the hypothesis may require reworking or additional tests. Eventually, it is either discarded or modified to fit the results of the experiment. If the hypothesis is supported by the results from the experiment, it is not (or should not be) immediately accepted as fact. It then must be tested and retested. Indeed, this is an important guideline in the acceptance of a hypothesis. The results of the experiment must be published and then repeated by other investigators.

In time, as each hypothesis is supported by a growing body of data and survives rigorous scrutiny, it moves to the next level of acceptance—the **theory.** A theory is a collection of statements, propositions, or concepts that explains or accounts for a natural event. A theory is not the result of a single experiment repeated over and over again but is an entire body of ideas that expresses or explains many aspects of a phenomenon. It is not a fuzzy or weak speculation, as is sometimes the popular notion, but a viable declaration that has stood the test of time and has yet to be disproved by serious scientific endeavors. Often, theories develop and progress through decades of research and are added to and modified by new findings. At some point, evidence of the accuracy and predictability of a theory is so compelling that the next level of confidence is reached, and the theory becomes a **law,** or principle. For example, the *germ theory of disease* has been so thoroughly tested that it has clearly passed into the realm of law.

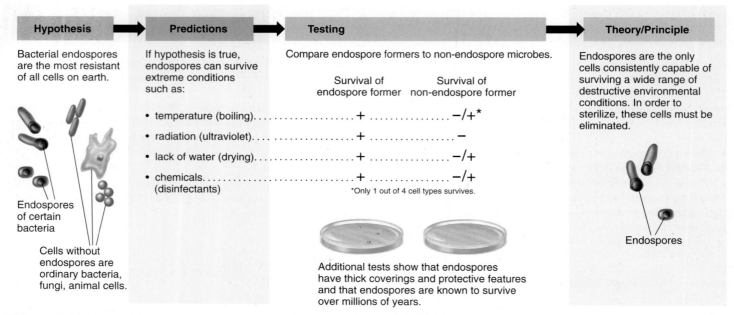

Hypothesis	Predictions	Testing	Theory/Principle

Figure 1.10 **The pattern of deductive reasoning using two examples.** The deductive process starts with a general hypothesis that predicts specific expectations that may or may not be borne out by testing. This example shows the reasoning behind a well-established principle that has been thoroughly tested over the past 150 years.

Science and its hypotheses and theories must progress along with technology. As advances in instrumentation allow new, more detailed views of living phenomena, old theories may be reexamined and altered and new ones proposed. It is for this reason that scientists do not take a stance that theories or even laws are absolutely proved.

The characteristics that make scientists most effective in their work are curiosity, open-mindedness, skepticism, creativity, cooperation, and readiness to revise their views of natural processes as new discoveries are made. The events described in Insight 1.2 provide important examples.

The Development of Medical Microbiology

Early experiments on the sources of microorganisms led to the profound realization that microbes are everywhere: Not only are air and dust full of them, but the entire surface of the earth, its waters, and all objects are inhabited by them. This discovery led to immediate applications in medicine. Thus the seeds of medical microbiology were sown in the middle to latter half of the nineteenth century with the introduction of the germ theory of disease and the resulting use of sterile, aseptic, and pure culture techniques.

The Discovery of Spores and Sterilization

Following Pasteur's inventive work with infusions (Insight 1.2), it was not long before English physicist **John Tyndall** provided the initial evidence that some of the microbes in dust and air have very high heat resistance and that particularly vigorous treatment is required to destroy them. Later, the discovery and detailed description of heat-resistant bacterial endospores by **Ferdinand Cohn,** a German botanist, clarified the reason that heat would sometimes fail to completely eliminate all microorganisms. The modern sense of the word **sterile,** meaning completely free of all life forms including spores and viruses, had its beginnings here (see chapter 11). The capacity to sterilize objects and materials is an absolutely essential part of microbiology, medicine, dentistry, and some industries.

The Development of Aseptic Techniques

From earliest history, humans experienced a vague sense that "unseen forces" or "poisonous vapors" emanating from decomposing matter could cause disease. As the study of microbiology became more scientific and the invisible was made visible, the fear of such mysterious vapors was replaced by the knowledge and sometimes even the fear of "germs." About 125 years ago, the first studies by **Robert Koch** clearly linked a microscopic organism with a specific disease. Since that time, microbiologists have conducted a continuous search for disease-causing agents.

At the same time that abiogenesis was being hotly debated, a few budding microbiologists began to suspect that microorganisms could cause not only spoilage and decay but also infectious diseases. It occurred to these rugged individualists that even the human body itself was a source of infection. **Dr. Oliver Wendell Holmes,** an American physician, observed that mothers who gave birth at home experienced fewer infections than did mothers who gave birth in the hospital, and the Hungarian **Dr. Ignaz Semmelweis** showed quite clearly that women became infected in the maternity ward after examinations by physicians coming directly from the autopsy room.

The English surgeon **Joseph Lister** took notice of these observations and was the first to introduce **aseptic* techniques** aimed at reducing microbes in a medical setting and preventing wound infections. Lister's concept of asepsis was much more limited than our modern precautions. It mainly involved disinfecting the hands and the air with strong antiseptic chemicals, such as phenol, prior to surgery. It is hard for us to believe, but as recently as the late 1800s, surgeons wore street clothes in the operating room and had little idea that hand washing was important. Lister's techniques and the application of heat for sterilization became the bases for microbial control by physical and chemical methods, which are still in use today.

The Discovery of Pathogens and the Germ Theory of Disease

Two ingenious founders of microbiology, **Louis Pasteur** of France (**figure 1.11**) and **Robert Koch** of Germany (**figure 1.12**), introduced techniques that are still used today. Pasteur made enormous contributions to our understanding of the microbial role in wine and beer formation. He invented pasteurization and completed some of the first studies showing that human diseases could arise from infection. These studies, supported by the work of other scientists, became known as the **germ theory of disease.** Pasteur's contemporary, Koch, established *Koch's postulates,* a series of proofs that verified the germ theory and could establish whether an organism was pathogenic and which disease it caused (see chapter 13). About 1875, Koch used this experimental system to show that anthrax was caused by a bacterium called *Bacillus anthracis.* So useful were his postulates that the causative agents of

20 other diseases were discovered between 1875 and 1900, and even today, they serve as a basic premise for establishing a pathogen-disease link.

Numerous exciting technologies emerged from Koch's prolific and probing laboratory work. During this golden age of microbiology, he realized that study of the microbial world would require separating microbes from each other and growing them in culture. It is not an overstatement to say that he and his colleagues invented many of the techniques that are described in chapter 3: inoculation, isolation, media, maintenance of pure cultures, and preparation of specimens for microscopic examination. Other highlights in this era of discovery are presented in later chapters on microbial control (see chapter 11) and vaccination (see chapter 15).

* *aseptic* (ay-sep′-tik) Gr. *a,* no, and *sepsis,* decay or infection. These techniques are aimed at reducing pathogens and do not necessarily sterilize.

Figure 1.11 **Photograph of Louis Pasteur (1822–1895), the father of microbiology.** Few microbiologists can match the scope and impact of his contributions to the science of microbiology.

Check & Assess Section 1.5

✔ Our current understanding of microbiology comes from the cumulative work of thousands of microbiologists whose contributions spanned over 300 years and did much to advance science and medicine.

✔ The microscope made it possible to see microorganisms and thus to identify their widespread presence, particularly as agents of disease.

✔ Antonie van Leeuwenhoek is considered the father of bacteriology and protozoology because he was the first person to produce precise, correct descriptions of bacteria and protozoa.

✔ The theory of spontaneous generation of living organisms from "vital forces" in the air was disproved once and for all by Louis Pasteur.

✔ The scientific method is a process by which scientists seek to explain natural phenomena. It is characterized by specific procedures that either support or discredit an initial hypothesis.

✔ Knowledge acquired through the scientific method is rigorously tested by repeated experiments by many scientists to verify its validity. A collection of valid hypotheses is called a theory. A theory supported by much data collected over time is called a law.

✔ Medical microbiologists developed the germ theory of disease and introduced the critically important concept of aseptic technique to control the spread of disease agents; Koch's postulates are still used today to pinpoint the causative agent of a specific disease.

✔ Louis Pasteur and Robert Koch were the leading microbiologists during the golden age of microbiology (from the mid-1800s to early 1900s).

11. Outline the most significant discoveries and events in microscopy, culture techniques, and other methods of handling or controlling microbes.

12. Differentiate between a hypothesis and a theory.

13. Is the germ theory of disease really a law, and why?

14. Why was the abandonment of the spontaneous generation theory so significant? Using the scientific method, describe the steps you would take to test the theory of spontaneous generation.

Figure 1.12 **Photograph of Robert Koch looking through a microscope with colleague Richard Pfeiffer looking on.** Robert Koch won the Nobel Prize for Physiology or Medicine in 1905 for his work on *M. tuberculosis*. Richard Pfeiffer discovered *Haemophilus influenzae* and was a pioneer in typhoid vaccination.

1.6 Taxonomy: Organizing, Classifying, and Naming Microorganisms

Expected Learning Outcomes

11. Define taxonomy and its supporting terms classification, nomenclature, and identification.

12. Explain how the levels of a taxonomic scheme relate to each other. Give the names of the levels, and place them in a hierarchy.

13. Describe the goals of nomenclature and how the binomial system is structured. Know how to correctly write a scientific name.

Students just beginning their microbiology studies are often dismayed by the seemingly endless array of new, unusual, and sometimes confusing names for microorganisms. Learning microbial **nomenclature*** is very much like learning a new language, and occasionally its demands may be a bit overwhelming. But paying attention to proper microbial names is just like following a baseball game or a theater production: You cannot tell the players apart without a program! Your understanding and appreciation of microorganisms will be greatly improved by learning a few general rules about how they are named.

The formal system for organizing, classifying, and naming living things is **taxonomy.*** This science originated more than 250 years

ago when **Carl von Linné** (also known as Linnaeus; 1701–1778), a Swedish botanist, laid down the basic rules for taxonomic categories, or **taxa.** Von Linné realized early on that a system for recognizing and defining the properties of living things would prevent chaos in scientific studies by providing each organism with a unique name and an exact "slot" in which to catalog it. This classification would then serve as a means for future identification of that same organism and permit workers in many biological fields to know if they were indeed discussing the same organism. The von Linné system has served well in categorizing the 2 million or more different types of organisms that have been discovered since that time.

The primary concerns of taxonomy are classification, nomenclature, and identification. These three areas are interrelated and play a vital role in keeping a dynamic inventory of the extensive array of living things. **Classification** is an orderly arrangement of organisms into groups that indicate evolutionary relationships and history. **Nomenclature** is the process of assigning names to the various taxonomic rankings of each microbial species. **Identification** is the process of determining and recording the traits of organisms to enable their placement in an overall taxonomic scheme. A survey of some general methods of identification appears in chapter 3.

The Levels of Classification

The main taxa, or groups, in a classification scheme are organized into several descending ranks called a **hierarchy.** It begins with **domain,** which is a giant, all-inclusive category based on a unique cell type, and ends with **species,*** the smallest and most specific taxon. All the members of a domain share only one or few general characteristics, whereas members of a species are essentially the same kind of organism—that is, they share the majority of their characteristics. The order of taxa between the top and bottom levels is, in descending order: **domain, kingdom, phylum*** or **division,**[4] **class, order, family, genus,*** and **species.** Thus, each domain can be subdivided into a series of kingdoms, each kingdom is made up of several phyla, each phylum contains several classes, and so on. Because taxonomic schemes are to some extent artificial, certain groups of organisms do not exactly fit into the eight taxa. In that case, additional levels can be imposed immediately above (super) or below (sub) a taxon, giving us such categories as superphylum and subclass.

To illustrate the fine points of this system, we compare the taxonomic breakdowns of a human and a protozoan (**figure 1.13**). Humans and protozoa belong to the same domain (Eukarya) but are placed in different kingdoms. To emphasize just how broad the category kingdom is, ponder the fact that humans belong to the same kingdom as jellyfish. Of the several phyla within this kingdom, humans are in the Phylum Chordata, but even a phylum is rather all-inclusive, considering that humans share it with other vertebrates as well as with creatures called sea squirts. The next level, Class Mammalia, narrows the field considerably by grouping only

* *nomenclature* (noh′-men-klay″-chur) L. *nomen,* name, and *clare,* to call. A system of naming.

* *taxonomy* (tacks-on″-uh-mee) Gr. *taxis,* arrangement, and *nomos,* name.

* *species* (spee′-sheez) L. *specere,* kind. In biology, this term is always in the plural form.

* *phylum* (fy′-lum) pl. phyla (fye′-luh) Gr. *phylon,* race.

4. The term *phylum* is used for protozoa, animals, bacteria, and fungi. Division is for algae and plants.

* *genus* (jee′-nus) pl. genera (jen′-er-uh) L. birth, kind.

Domain: Eukarya (All eukaryotic organisms)

Kingdom: Animalia

Lemur

Sea squirt

Sea star

Phylum: Chordata

Class: Mammalia

Order: Primates

Family: Hominoidea

Genus: *Homo*

Species: *sapiens*

(a)

Domain: Eukarya (All eukaryotic organisms)

Kingdom: Protista
Includes
protozoa
and algae

Phylum: Ciliophora
Only protozoa
with cilia

Class: Hymenostomea
Single cells with
regular rows of
cilia; rapid
swimmers

Order: Hymenostomatida
Elongate oval cells with
cilia in the oral cavity

Family: Parameciidae
Cells rotate while swimming
and have oral grooves.

Genus: *Paramecium*
Pointed, cigar-shaped cells with
macronuclei and micronuclei

Species: *caudatum*
Cells cylindrical, long, and pointed
at one end

(b)

Figure 1.13 **Sample taxonomy.** Two organisms belonging to the Eukarya domain, traced through their taxonomic series. **(a)** Modern humans, *Homo sapiens.* **(b)** A common protozoan, *Paramecium caudatum.*

those vertebrates that have hair and suckle their young. Humans belong to the Order Primates, a group that also includes apes, monkeys, and lemurs. Next comes the Family Hominoidea, containing only humans and apes. The final levels are our genus, *Homo* (all races of modern and ancient humans), and our species, *sapiens* (meaning wise). Notice that for both the human and the protozoan, the categories become less inclusive and the individual members more closely related and similar in overall appearance. Other examples of classification schemes are provided in sections of chapters 4 and 5 and in several later chapters.

We need to remember that all taxonomic hierarchies are based on the judgment of scientists with certain expertise in a particular group of organisms and that not all other experts may agree with the system being used. Consequently, no taxa are permanent to any degree; they are constantly being revised and refined as new information becomes available or new viewpoints become prevalent. Because this text does not aim to emphasize details of taxonomy, we will usually be concerned with only the most general (kingdom, phylum) and specific (genus, species) levels.

Assigning Specific Names

Many larger organisms are known by a common name suggested by certain dominant features. For example, a bird species may be called a red-headed blackbird or a flowering species a sweet pea. Some species of microorganisms (especially pathogens) are also

called by informal names, such as the gonococcus (*Neisseria gonorrhoeae*) or the TB bacillus (*Mycobacterium tuberculosis*), but this is not the usual practice. If we were to adopt common names such as the "little yellow coccus" (for *Micrococcus luteus**) or the "club-shaped diphtheria bacterium" (for *Corynebacterium diphtheriae**), the terminology would become even more cumbersome and challenging than scientific names. Even worse, common names are notorious for varying from region to region, even within the same country. A decided advantage of standardized nomenclature is that it provides a universal language, thereby enabling scientists from all countries on the earth to freely exchange information.

The method of assigning the **scientific** or **specific name** is called the **binomial (two-name) system of nomenclature.** The scientific name is always a combination of the generic (genus) name followed by the species name. The generic part of the scientific name is capitalized, and the species part begins with a lower-case letter. Both should be italicized (or underlined if italics are not available), as follows:

Staphylococcus aureus

Because other taxonomic levels are not italicized and consist of only one word, one can always recognize a scientific name. An organism's scientific name is sometimes abbreviated to save space, as in *S. aureus,* but only if the genus name has already been stated. The source for nomenclature is usually Latin or Greek. If other languages such as English or French are used, the endings of these words are revised to have Latin endings. In general, the name first applied to a species will be the one that takes precedence over all others.

An international group oversees the naming of every new organism discovered, making sure that standard procedures have been followed and that there is not already an earlier name for the organism or another organism with that same name. The inspiration for names is extremely varied and often rather imaginative. Some species have been named in honor of a microbiologist who originally discovered the microbe or who has made outstanding contributions to the field. Other names may designate a characteristic of the microbe (shape, color), a location where it was found, or a disease it causes. Some examples of specific names, pronunciations, and origins are:

- *Staphylococcus aureus* (staf′-i-lo-kok′-us ah′-ree-us) Gr. *staphule,* bunch of grapes, *kokkus,* berry, and Gr. *aureus,* golden. A bacterial pathogen of humans with a yellow pigment.
- *Campylobacter jejuni* (cam′-peh-loh-bak-ter jee-joo′-neye) Gr. *kampylos,* curved, *bakterion,* little rod, and *jejunum,* a section of intestine. One of the most important causes of intestinal infection worldwide.
- *Lactobacillus sanfrancisco* (lak″-toh-bass-ill′-us san-fran-siss′-koh) L. *lacto,* milk, and *bacillus,* little rod. A bacterial species used to make sourdough bread.

* *Micrococcus luteus* (my″-kroh-kok′-us loo′-tee-us) Gr. *micros,* small, and *kokkus,* berry; L. *luteus,* yellow.
* *Corynebacterium diphtheriae* (kor-eye″-nee-bak-ter′-ee-yum dif′-theer-ee-eye) Gr. *coryne,* club, *bacterion,* little rod, and *diphtheriae,* the causative agent of the disease diphtheria.

- *Giardia lamblia* (jee-ar′-dee-uh lam′-blee-uh) for Alfred Giard, a French microbiologist, and Vilem Lambl, a Bohemian physician, both of whom worked on the organism, a protozoan that causes a severe intestinal infection.

When you encounter the name of a microorganism in the chapters ahead, it is helpful to take the time to sound it out one syllable at a time and repeat until it seems familiar. You are much more likely to remember the names that way—and they will become part of the new language you will be learning.

Check & Assess Section 1.6

✔ Taxonomy is the formal filing system scientists use to organize living organisms. It puts every organism in its place and makes a place for every living organism.

✔ The taxonomic system has three primary functions: classification, nomenclature, and identification of species.

✔ The eight major taxa, or groups, in the taxonomic system are (in descending order): domain, kingdom, phylum or division, class, order, family, genus, and species.

✔ The binomial system of nomenclature describes each living organism by two names: genus and species.

✔ Taxonomy groups organisms by phylogenetic similarity, which in turn is based on evolutionary similarities in morphology, physiology, and genetics.

15. Differentiate between taxonomy, classification, and nomenclature.

16. What is the basis for a phylogenetic system of classification?

17. Explain the binomial system of nomenclature and give the correct order of taxa, going from most general to most specific. Create a mnemonic (memory) device for recalling this order.

18. Give some reasons to explain the benefits of using scientific names for organisms.

1.7 The Origin and Evolution of Microorganisms

Expected Learning Outcomes

14. Discuss the fundamentals of evolution and how they are used in studying organisms.

15. Outline some of the primary evidence used to verify evolutionary trends.

16. Explain how trees of life are constructed, and tell what characteristics are used in organizing the organisms on these trees.

17. Indicate where the major groups of microorganisms fall on these trees.

18. Explain how the domains are classified and how they differ; cite several representatives of each domain.

As we indicated earlier, *taxonomy,* the classification of biological species, is a system used to organize all of the forms of life. In biology today, there are different methods for deciding on taxonomic categories, but they all rely on the degree of relatedness among organisms. The natural relatedness between groups of living things is called their **phylogeny.** So biologists use phylogenetic relationships to create a system of taxonomy.

To understand the relatedness among organisms, we must understand some fundamentals of evolution. **Evolution** is an important theme that underlies all of biology, including microbiology. Put simply, evolution states that living things change gradually through hundreds of millions of years and that these evolvements result in various types of structural and functional changes through many generations. The process of evolution is selective: Those changes that most favor the survival of a particular organism or group of organisms tend to be retained, and those that are less beneficial to survival tend to be lost. Space does not permit a detailed analysis of evolutionary theories, but the occurrence of evolution is supported by a tremendous amount of evidence from the fossil record and from the study of **morphology** (structure), **physiology** (function), and **genetics** (inheritance). Evolution accounts for the millions of different species on the earth and their adaptation to its many and diverse habitats.

Evolution is founded on two preconceptions: (1) that all new species originate from preexisting species and (2) that closely related organisms have similar features because they evolved from common ancestral forms. Usually, evolution progresses toward greater complexity, and evolutionary stages range from simple, primitive forms that are close to an ancestral organism to more complex, advanced forms. Although we use the terms *primitive* and *advanced* to denote the degree of change from the original set of ancestral traits, it is very important to realize that all species presently residing on the earth are modern, but some have arisen more recently in evolutionary history than others.

The phylogeny, or evolutionary relatedness, of organisms is often represented by a diagram with a branching, treelike format. The trunk of the tree represents the main ancestral lines, and the branches show offshoots into specialized groups of organisms. This sort of arrangement places the more ancient groups at the bottom and the more recent ones at the top. The branches may also indicate origins, how closely related various organisms are, and an approximate timescale for evolutionary history (**figures 1.14 and 1.15).**

Systems for Presenting a Universal Tree of Life

The first phylogenetic trees of life were constructed on the basis of just two kingdoms (plants and animals). In time, it became clear that certain organisms did not truly fit either of those categories, so a third kingdom for simpler organisms that lacked tissue differentiation (protists) was recognized. Eventually, when significant differences became evident even among the protists, a fourth kingdom was proposed for the bacteria. Robert Whittaker

built on this work and during the period of 1959 through 1969 added a fifth kingdom for fungi. Whittaker's five-kingdom system has been the standard for a simple working model.

One example of a phylogenetic tree of life based on the five kingdoms is shown in figure 1.14. The relationships considered in constructing the tree were based on structural similarities and differences, such as cellular organization and the way the organisms get their nutrition. These methods produced a system with five major kingdoms: the monera, fungi, protists, plants, and animals. Because these kingdoms easily accomodate the prokaryotic and eukaryotic cell types, this plan of classification has proved useful.

Applying, newer methods for determining phylogeny has led to the development of a differently shaped tree—with important implications for our understanding of evolutionary relatedness. The new techniques come from *molecular biology* and include a study of genes—both their structure and function—at the molecular level. Molecular biological methods have demonstrated that certain types of molecules in cells, called small ribosomal ribonucleic acid (rRNA), provide a "living record" of the evolutionary history of an organism. Analysis of this molecule in prokaryotic and eukaryotic cells indicates that certain unusual cells called **archaeons** (originally archaebacteria) are so different from the other two groups that they should be included in a separate superkingdom. Many archaeons are characterized by their ability to live in extreme environments, such as hot springs or highly salty environments. Under the microscope they resemble bacteria, but molecular biology has revealed that the cells of archaeons, though prokaryotic in nature, are actually more closely related to eukaryotic cells than to bacterial cells (see table 4.5).

To reflect these relationships, Carl Woese and George Fox have proposed a system that assigns all organisms to one of three domains, each described by a different type of cell (see figure 1.15). The prokaryotic cell types are placed in the Domains **Archaea** and **Bacteria.** Eukaryotes are all placed in the Domain **Eukarya.** It is believed that these three superkingdoms arose from an ancestor most similar to the archaeons. This new system is still undergoing analysis, and it somewhat complicates the presentation of organisms in that it disposes of some traditional groups, although many of the traditional kingdoms (animals, plants, fungi) still work within this framework. The original Kingdom Protista may also be represented as a collection of protozoa and algae in several separate kingdoms (discussed in chapter 5).

This new scheme does not greatly affect our presentation of most microbes, because we discuss them at the genus or species level. But be aware that biological taxonomy—and more important, our view of how organisms evolved on earth—is in a period of transition. Keep in mind that our methods of classification reflect our current understanding and are constantly changing as new information is uncovered.

Please note that viruses are *not* included in any of the classification or evolutionary schemes, because they are not cells and their position cannot be given with any confidence. Their special taxonomy is discussed in chapter 6.

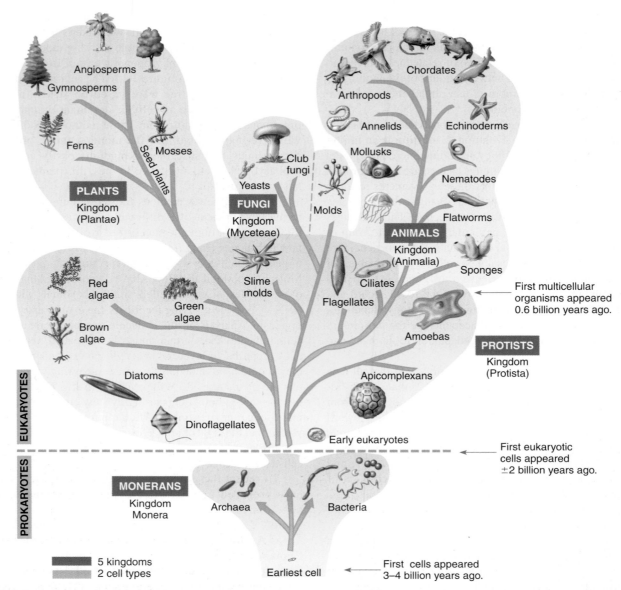

Figure 1.14 Traditional Whittaker system of classification. In this system, kingdoms are based on cell structure and type, the nature of body organization, and nutritional type. Bacteria and Archaea (monerans) have prokaryotic cells and are unicellular. Protists have eukaryotic cells and are mostly unicellular. They can be photosynthetic (algae), or they can feed on other organisms (protozoa). Fungi are eukaryotic cells and are unicellular or multicellular; they have cell walls and are not photosynthetic. Plants have eukaryotic cells, are multicellular, have cell walls, and are photosynthetic. Animals have eukaryotic cells, are multicellular, do not have cell walls, and derive nutrients from other organisms.

Check & Assess Section 1.7

✔ Through a well-documented process termed evolution, living things have undergone gradual changes over billions of years in response to alterations in their environment.

✔ Evolution has given rise to hundreds of millions of different species whose history demonstrates their ancestry and relatedness to other organisms.

✔ Evolution states that new organisms can arise only from preexisting organisms. Species are subject to conditions that select for individuals with more successful adaptations and a greater ability to survive and reproduce.

✔ Evolutionary patterns show a treelike branching from simple, primitive life forms to complex, advanced life forms.

✔ The Whittaker five-kingdom classification system places all prokaryotes in the Kingdom Monera and subdivides the eukaryotes into Kingdoms Protista, Myceteae, Animalia, and Plantae.

✔ The Woese-Fox classification system places all eukaryotes in the Domain Eukarya and subdivides the prokaryotes into the two Domains Archaea and Bacteria.

Figure 1.15 **Woese-Fox system.** A system for representing the origins of cell lines and major taxonomic groups as proposed by Carl Woese and colleagues. They propose three distinct cell lines placed in superkingdoms called domains. The first primitive cells, called progenotes, were ancestors of both lines of prokaryotes (Domains Bacteria and Archaea), and the Archaea emerged from the same cell line as eukaryotes (Domain Eukarya). Selected representatives are included. Some of the traditional kingdoms are still present with this system (see figure 1.14). Further details of classification systems are in chapters 4, 5, and the Appendix.

19. Evolution accounts for the millions of different species on the earth and their adaptation to its many and diverse habitats. Explain this and cite examples in your answer.

20. Looking at the tree of life (figure 1.14), determine which kingdom or kingdoms humans are most closely related to.

21. Archaea are often found in hot, sulfuric, acidic, salty habitats, such as the early earth's conditions. Speculate on the origin of life, especially as it relates to the archaea.

22. Compare the new domain system with the five-kingdom system. Does the newer system change the basic idea of prokaryotes and eukaryotes? What is the third cell type?

CASE FILE **1** PERSPECTIVE

Plankton is the assemblage of living organisms floating with the currents in upper regions of aquatic environments such as lakes and oceans. In addition to the bacteria (called bacterioplankton) targeted by the study, it also includes microscopic algae (the phytoplankton) and protozoa and tiny invertebrates (the zooplankton). One would also expect to find some archaeons and viruses. Molds and yeasts are not prominent members of the planktonic community.

Some kinds of microbiology specialists who could be involved in further studies are marine ecologists, bacterial physiologists, molecular biologists, biochemists, geomicrobiologists, taxonomists, and geneticists.

As information begins pouring in from many other habitats being sampled, it is clear that after 300 years of observing the microbial world, we have really just scratched the surface. For quite some time, microbiologists have known that the human body harbors nine bacterial cells for every one of our own cells. But now we can have much greater insight into the identity and function of these resident microbes. So far, genomic studies of various habitats of the human body have revealed thousands of new microbes that were unknown because of our inability to grow them in the laboratory.

Much of the information gathered will help us understand the ecology of these communities, how the members interact, what kinds of nutrients they use or waste products they give off. Discoveries may lead to applications for tapping energy sources, uses in bioremediation, climate change, and soil fertility, and even new medicines and industrial products.

Chapter Summary with Key Terms

1.1 The Scope of Microbiology

A. **Microbiology** is the study of **bacteria, viruses, fungi, protozoa,** and **algae,** which are collectively called **microorganisms,** or microbes. In general, microorganisms are **microscopic** and, unlike **macroscopic** organisms, which are readily visible, they require magnification to be adequately observed or studied.

B. The simplicity, growth rate, and adaptability of microbes are some of the reasons that microbiology is so diverse and has branched out into many subsciences and applications. Important subsciences include **immunology, epidemiology,** public health, food, dairy, aquatic, and industrial microbiology.

1.2 General Characteristics of Microorganisms and Their Roles in the Earth's Environments

Microbes live in most of the world's habitats and are indispensable for normal, balanced life on earth. They play many roles in the functioning of the earth's ecosystems.

A. Microbes are **ubiquitous.**

B. There are many kinds of relationships between microorganisms and humans; most are beneficial, but a few are harmful.

C. Microbes are involved in nutrient production and energy flow. Algae and certain bacteria trap the sun's energy to produce food through **photosynthesis.**

D. Other microbes are responsible for the breakdown and recycling of nutrients through **decomposition.** Microbes are essential to the maintenance of the air, soil, and water.

E. Microbial cells are either the small, relatively simple, nonnucleated **prokaryotic** variety or the larger, more complex **eukaryotic** type that contain a nucleus and **organelles.**

F. **Viruses** are microbes but not cells. They are smaller in size and infect their prokaryotic or eukaryotic hosts in order to reproduce themselves.

G. **Parasites** and **pathogens** are microorganisms that invade the bodies of **hosts,** often causing damage through infection and disease.

1.3 Human Use of Microorganisms

Microbes have been called upon to solve environmental, agricultural, and medical problems.

A. **Biotechnology** applies the power of microbes toward the manufacture of industrial products, foods, and drugs.

B. Microbes form the basis of **genetic engineering** and **recombinant DNA** technology, which alter genetic material to produce new products and modified life forms.

C. In **bioremediation,** microbes are used to clean up pollutants and wastes in natural environments.

1.4 Microbial Roles in Infectious Diseases

A. Nearly 2,000 microbes are **pathogens** that cause infectious diseases. Infectious diseases result in high levels of mortality and morbidity (illness). Many infections are **emerging,** meaning that they are newly identified pathogens gaining greater prominence. Many older diseases are also increasing.

B. Some diseases previously thought to be noninfectious may involve microbial infections (e.g., *Helicobacter,* causing gastric ulcers, and coxsackieviruses, causing diabetes).

C. An increasing number of individuals have weak immune systems, which makes them more susceptible to infectious diseases.

1.5 The Historical Foundations of Microbiology

A. Microbiology as a science is about 300 years old. Hundreds of contributors have provided discoveries and knowledge to enrich our understanding.

B. With his simple microscope, Leeuwenhoek discovered organisms he called animalcules. As a consequence of his findings and the rise of the **scientific method,** the notion of **spontaneous generation,** or **abiogenesis,** was eventually abandoned for **biogenesis.** The scientific method develops rational **hypotheses** and **theories** that can be tested. Theories that withstand repeated scrutiny become **laws** in time.

C. Early microbiology blossomed with the conceptual developments of **sterilization, aseptic techniques,** and the **germ theory of disease.** Prominent scientists from this period include Robert Koch, Louis Pasteur, and Joseph Lister.

1.6 Taxonomy: Organizing, Classifying, and Naming Microorganisms

A. **Taxonomy** is a hierarchical scheme for the **classification, identification,** and **nomenclature** of organisms, which are grouped in categories called **taxa,** based on features ranging from general to specific.

B. Starting with the broadest category, the taxa are **domain, kingdom, phylum** (or **division**), **class, order, family, genus,** and **species.** Organisms are assigned **binomial scientific names** consisting of their genus and species names.

1.7 The Origin and Evolution of Microorganisms

A. All life on earth **evolved** from simple cells appearing in ancient oceans about 3.5 billion years ago.

B. Evolutionary change occurs when the environment places pressure on organisms that selects for survival of those with more adaptive inheritable traits.

C. All new species are the products of preexisting species, and their ancestry may be traced by examining fossils, morphology, physiology, genetics, and other scientific forms of investigation.

D. The records of **phylogeny** are displayed in the form of a tree of life that shows organisms' relatedness.

E. An alternative classification scheme uses a five-kingdom organization developed by Whittaker:
 1. Kingdom Prokaryotae (Monera), containing eubacteria and archaeons;
 2. Kingdom Protista, containing primitive unicellular microbes such as algae and protozoa;
 3. Kingdom Myceteae, containing the fungi;
 4. Kingdom Animalia, containing animals; and
 5. Kingdom Plantae, containing plants.

F. The latest classification scheme for living things is based on the genetic structure of their ribosomes. The Woese-Fox system often recognizes three **Domains: Archaea,** simple prokaryotes that often live in extreme environments; **Bacteria,** typical prokaryotes; and **Eukarya,** all types of eukaryotic organisms.

Multiple-Choice Questions

Select the correct answer from the answers provided. For questions with blanks, choose the combination of answers that most accurately completes the statement. Answers are in Appendix F.

1. Which of the following is not considered a microorganism?
 - a. alga
 - b. bacterium
 - c. protozoan
 - d. mushroom

2. An area of microbiology that is concerned with the occurrence of disease in human populations is
 - a. immunology
 - b. parasitology
 - c. epidemiology
 - d. bioremediation

3. Which process involves the deliberate alteration of an organism's genetic material?
 - a. bioremediation
 - b. biotechnology
 - c. decomposition
 - d. recombinant DNA

4. A prominent difference between prokaryotic and eukaryotic cells is the
 - a. larger size of prokaryotes
 - b. lack of pigmentation in eukaryotes
 - c. presence of a nucleus in eukaryotes
 - d. presence of a cell wall in prokaryotes

5. Which of the following parts was absent from Leeuwenhoek's microscopes?
 - a. focusing screw
 - b. lens
 - c. specimen holder
 - d. condenser

6. Abiogenesis refers to the
 - a. spontaneous generation of organisms from nonliving matter
 - b. development of life forms from preexisting life forms
 - c. development of aseptic technique
 - d. germ theory of disease

7. A hypothesis can be defined as
 - a. a belief based on knowledge
 - b. knowledge based on belief
 - c. a scientific explanation that is subject to testing
 - d. a theory that has been thoroughly tested

8. Which early microbiologist was most responsible for developing standard microbiology laboratory techniques?
 - a. Louis Pasteur
 - b. Robert Koch
 - c. Carl von Linné
 - d. John Tyndall

9. Which scientist is most responsible for finally laying the theory of spontaneous generation to rest?
 - a. Joseph Lister
 - b. Robert Koch
 - c. Francesco Redi
 - d. Louis Pasteur

10. When a hypothesis has been thoroughly supported by long-term study and data, it is considered
 - a. a law
 - b. a speculation
 - c. a theory
 - d. proved

11. Which is the correct order of the taxonomic categories, going from most specific to most general?
 - a. domain, kingdom, phylum, class, order, family, genus, species
 - b. division, domain, kingdom, class, family, genus, species
 - c. species, genus, family, order, class, phylum, kingdom, domain
 - d. species, family, class, order, phylum, kingdom

12. By definition, organisms in the same _____ are more closely related than are those in the same _____.
 - a. order, family
 - b. class, phylum
 - c. family, genus
 - d. phylum, division

13. Which of the following are prokaryotic?
 - a. bacteria
 - b. archaea
 - c. protists
 - d. both a and b

14. Order the following items by size, using numbers: 1 = smallest and 8 = largest.
 - _3_ AIDS virus
 - _1_ amoeba
 - _4_ rickettsia
 - _2_ protein
 - _8_ worm
 - _5_ coccus-shaped bacterium
 - _6_ white blood cell
 - _1_ atom

15. Which of the following is not an emerging infectious disease?
 - a. avian influenza
 - b. SARS
 - c. common cold
 - d. AIDS

16. How would you categorize a virus?
 - a. as prokaryotic
 - b. as eukaryotic
 - c. as an archeon
 - d. none of these choices

 Explain your choice for question 16.

Case File Questions

1. What is plankton?
 - a. nutrients dissolved in seawater
 - b. organic matter in the ocean
 - c. organisms living in the ocean sediments
 - d. organisms free-floating in the aquatic environment

2. What aspect of the plankton was actually analyzed by the marine researchers to arrive at their conclusions?
 - a. the total number of cells
 - b. the types of eukaryotes
 - c. the genetic material of cells
 - d. the types of viruses

Writing to Learn

These questions are suggested as a *writing-to-learn* experience. For each question, compose a one- or two-paragraph answer that includes the factual information needed to completely address the question. Note: The assess questions can also serve as a writing-to-learn assignment.

1. What does it mean to say microbes are ubiquitous?

2. What is meant by diversity with respect to organisms?

3. What events, discoveries, or inventions were probably the most significant in the development of microbiology and why?

4. Explain how microbiologists use the scientific method to develop theories and explanations for microbial phenomena.

5. Explain how microbes are classified into groups according to evolutionary relationships, provided with standard scientific names, and identified by specific characteristics.

6. a. What are some of the sources for "new" infectious diseases?
 b. Comment on the sensational ways that some tabloid media portray infectious diseases to the public.

Concept Mapping

Appendix E provides guidance for working with concept maps.

1. Supply your own linking words or phrases in this concept map, and provide the missing concepts in the empty boxes.

Critical Thinking Questions

Critical thinking is the ability to reason and solve problems using facts and concepts. These questions can be approached from a number of angles, and in most cases, they do not have a single correct answer.

1. What do you suppose the world would be like if there were cures for all infectious diseases and a means to destroy all microbes? What characteristics of microbes would prevent this from happening?

2. How would you describe the types of scientific reasoning in the various experiments for supporting and denying spontaneous generation.

3. Give the technical name of a microbiologist who researches or works with: protozoa in a termite's gut, bacteria that live in volcanoes, the tapeworms of dogs, molds that cause food poisoning, emerging viral diseases, the metabolism of bacteria that live in acid swamps, the classification of *Paramecium*. pg 4

4. Why do you think genetic screening of samples as described in the case file was more effective in mining the ocean's microbes than traditional methods? pg 1

5. What does it mean to say that the human body is 90% prokaryotic?

6. What is the ultimate way that microbes will, as Pasteur said, have the "last word"?

7. Can you develop a scientific hypothesis and means of testing the cause of stomach ulcers? (Are they caused by an infection? By too much acid? By a genetic disorder?)

8. Construct the scientific name of a newly discovered species of bacterium using your name, a pet's name, a place, or a unique characteristic. Be sure to use proper notation and endings.

Visual Challenge

Refer to figure 1.10 and use the following pattern to construct an outline of the scientific reasoning that was involved in developing the germ theory of disease.

Observations

↓

Hypothesis

↓

Testing the Hypothesis

↓

Germ Theory of Disease

↓

Accepted Principle or Law

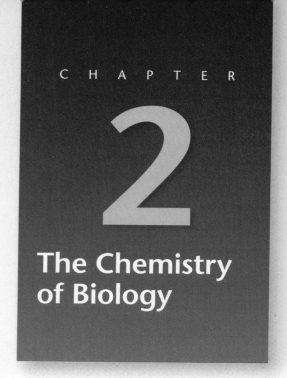

2

The Chemistry of Biology

The mysteries of Mars

An ancient Martian meteorite (center) viewed with an electron microscope yields up tantalizing little cylinders that look like bacterial cells (right).

"Some locations on the planet may have "hot spots" deeper in the crust where liquid water exists part of the time."

CASE FILE 2 Tasting and Sniffing Martian Soil and Ice

Throughout history, humans have gazed at the night sky and wondered about the nature of the universe. What has most intrigued us revolves around one profound question: Does other life exist out there? The development of telescopes allowed closer glimpses of celestial bodies, but even early telescopes could not pick up the details one would need to verify life. So beginning about 46 years ago, the U.S. National Aeronautics and Space Administration (NASA) began to carry out some actual voyages to directly explore our solar system. Astronomers knew from the outset that most of the planets in our solar system had far too extreme temperatures and toxic conditions to support life. So attention was focused on Mars, a close neighbor at a mere 422 million miles away. It was known to be very arid and cold (the range is 20°F to −120°F), but there was a reasonable possibility that at least simple organisms, similar to earthly microorganisms, could have developed there. A series of space explorations was launched to collect evidence for potential signs of life on Mars. From these studies sprang a new science called astrobiology, which applies principles from biology, chemistry, and geology to investigate the possibilities of extraterrestrial life.

Among the significant questions tackled by astrobiologists have been: what are the basic requirements of life, and what would signs of life look like? As with so many ideas of a biological nature, they turned to chemistry for answers. One of their assumptions was that since some of the same chemical substances on earth would also exist on Mars, then Martian life forms would be expected to have a similar profile. Attention was turned to essential substances that accompany all life on earth, mainly water and carbon-based chemicals.

The main goals of one of the first Mars projects, *Viking Explorer*, were to sample for water, carbon dioxide, and organic compounds, and to culture microbes from Martian soil. The results were mostly negative or inconclusive, except for one—that the atmosphere contained carbon dioxide. A later study turned up tiny "rods" buried in a Martian meteorite from the Antarctic that is several billion years old. This tantalizing image bears a resemblance to cells, but there is still a controversy about whether these are ancient microbes or geologic artifacts.

▶ *What chemical features of water make it essential for living processes?*

▶ *What characteristics of carbon make it the central building block of life?*

To continue the case, go to page 40.

2.1 Atoms: Fundamental Building Blocks of All Matter in the Universe

Expected Learning Outcomes

1. Describe the properties of atoms and identify the relationships of the particles that they contain.

2. Characterize elements and their isotopes.

3. Explain the differences between atomic number, mass number, and atomic weight.

4. List the major elements that are associated with life.

5. Describe electron orbitals and energy shells, and how they are filled.

The universe is composed of an infinite variety of substances existing in gaseous, liquid, and solid states. All such tangible materials that occupy space and have mass are called **matter.** The organization of matter—whether air, rocks, or bacteria—begins with individual building blocks called atoms. An **atom** is defined as a tiny particle that cannot be subdivided into smaller substances without losing its properties. Even in a science dealing with very small things, an atom's minute size is striking; for example, an oxygen atom is only 0.0000000013 mm (0.0013 nm) in diameter, and 1 million of them in a cluster would barely be visible to the naked eye.

Although scientists have not directly observed the detailed structure of an atom, the exact composition of atoms has been well established by extensive physical analysis using sophisticated instruments. In general, an atom derives its properties from a combination of subatomic particles called **protons** (p^+), which are positively charged; **neutrons** (n^0), which have no charge (are neutral); and **electrons** (e^-), which are negatively charged. The relatively larger protons and neutrons make up a central core, or **nucleus,**[1] that is surrounded by 1 or more electrons (**figure 2.1**). The nucleus makes up the larger mass (weight) of the atom, whereas the electron region, sometimes called the "electron cloud," accounts for the greater volume. To get a perspective on proportions, consider this: If an atom were the size of a football stadium, the nucleus would be about the size of a marble! The stability of atomic structure is largely maintained by

- the mutual attraction of the protons and electrons (opposite charges attract each other), and
- the exact balance of proton number and electron number, which causes the opposing charges to cancel each other out.

At least in theory, then, isolated intact atoms do not carry a charge.

Different Types of Atoms: Elements and Their Properties

All atoms share the same fundamental structure. All protons are identical, all neutrons are identical, and all electrons are identical.

But when these subatomic particles come together in specific, varied combinations, unique types of atoms called **elements** result. Each element is a pure substance that has a characteristic atomic structure and predictable chemical behavior. To date, 118 elements

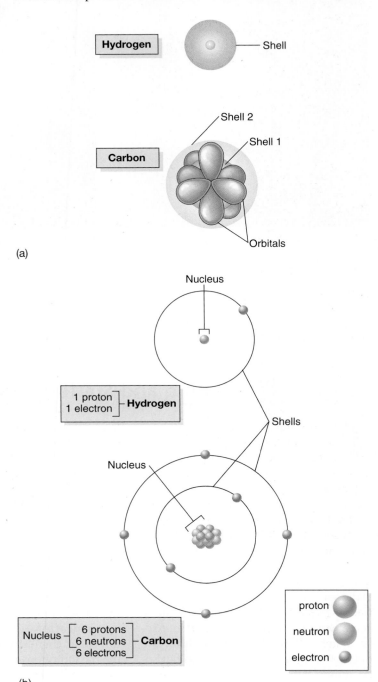

(a)

(b)

Figure 2.1 Models of atomic structure. (a) Three-dimensional models of hydrogen and carbon that approximate their actual structure. The nucleus is surrounded by electrons in orbitals that occur in levels called shells. Hydrogen has just one shell and one orbital. Carbon has two shells and four orbitals; the shape of the outermost orbitals is paired lobes rather than circles or spheres. **(b)** Simple models of the same atoms show the numbers and arrangements of shells and electrons, and the numbers of protons and neutrons in the nucleus. (Not to accurate scale.)

1. Be careful not to confuse the nucleus of an atom with the nucleus of a cell.

TABLE 2.1	The Major Elements of Life and Their Primary Characteristics				
Element	Atomic Symbol*	Atomic Number	Atomic Mass**	Examples of Ionized Forms	Significance in Microbiology
Calcium	Ca	20	40.1	Ca^{2+}	Part of outer covering of certain shelled amoebas; stored within bacterial spores
Carbon	C	6	12.0	CO_3^{2-}	Principal structural component of biological molecules
Carbon·	C-14	6	14.0		Radioactive isotope used in dating fossils
Chlorine	Cl	17	35.5	Cl^-	Component of disinfectants; used in water purification
Cobalt	Co	27	58.9	Co^{2+}, Co^{3+}	Trace element needed by some bacteria to synthesize vitamins
Cobalt·	Co-60	27	60		An emitter of gamma rays; used in food sterilization; used to treat cancer
Copper	Cu	29	63.5	Cu^+, Cu^{2+}	Necessary to the function of some enzymes; Cu salts are used to treat fungal and worm infections
Hydrogen	H	1	1	H^+	Necessary component of water and many organic molecules; H_2 gas released by bacterial metabolism
Hydrogen·	H-3	1	3		Tritium has 2 neutrons; radioactive; used in clinical laboratory procedures
Iodine	I	53	126.9	I^-	A component of antiseptics and disinfectants; contained in a reagent of the Gram stain
Iodine·	I-131, I-125	53	131, 125		Radioactive isotopes for diagnosis and treatment of cancers
Iron	Fe	26	55.8	Fe^{2+}, Fe^{3+}	Necessary component of respiratory enzymes; some microbes require it to produce toxin
Magnesium	Mg	12	24.3	Mg^{2+}	A trace element needed for some enzymes; component of chlorophyll pigment
Manganese	Mn	25	54.9	Mn^{2+}, Mn^{3+}	Trace element for certain respiratory enzymes
Nitrogen	N	7	14.0	NO_3^- (nitrate)	Component of all proteins and nucleic acids; the major atmospheric gas
Oxygen	O	8	16.0		An essential component of many organic molecules; molecule used in metabolism by many organisms
Phosphorus	P	15	31	PO_4^{3-} (phosphate)	A component of ATP, nucleic acids, cell membranes; stored in granules in cells
Phosphorus·	P-32	15	32		Radioactive isotope used as a diagnostic and therapeutic agent
Potassium	K	19	39.1	K^+	Required for normal ribosome function and protein synthesis; essential for cell membrane permeability
Sodium	Na	11	23.0	Na^+	Necessary for transport; maintains osmotic pressure; used to prevent food spoilage by microbes
Sulfur	S	16	32.1	SO_4^{-2} (sulfate)	Important component of proteins; makes disulfide bonds; storage element in many bacteria
Zinc	Zn	30	65.4	Zn^{2+}	An enzyme cofactor; required for protein synthesis and cell division; important in regulating DNA

*Based on the Latin name of the element. The first letter is always capitalized; if there is a second letter, it is always lowercased.

**The atomic mass or weight is equal to the average mass number for the isotopes of that element.

·An isotope of the element.

have been described. Ninety-four of them are naturally occurring, and the rest were artificially produced by manipulating the particles in the nucleus. By convention, an element is assigned a distinctive name with an abbreviated shorthand symbol. The elements are often depicted in a periodic table. **Table 2.1** lists some of the elements common to biological systems, their atomic characteristics, and some of the natural and applied roles they play in the field of microbiology.

The Major Elements of Life and Their Primary Characteristics

The unique properties of each element result from the numbers of protons, neutrons, and electrons it contains, and each element can be identified by certain physical measurements.

Each element is assigned an **atomic number (AN)** based on the number of protons it has. The atomic number is a valuable measurement because an element's proton number does not vary, and knowing it automatically tells you the usual number of electrons (recall that a neutral atom has an equal number of protons and electrons). Another useful measurement is the **mass number (MN)**, equal to the number of protons and neutrons. If one knows the mass number and the atomic number, it is possible to determine the numbers of neutrons by subtraction. Hydrogen is a unique element because its common form has only one proton, one electron, and no neutron, making it the only element with the same atomic and mass number.

Isotopes are variant forms of the same element that differ in the number of neutrons and thus have different mass numbers. These multiple forms occur naturally in certain proportions. Carbon, for example, exists primarily as carbon 12 with 6 neutrons (MN = 12); but a small amount (about 1%) consists of carbon 13 with 7 neutrons and carbon 14 with 8 neutrons. Although isotopes have virtually the same chemical properties, some of them have unstable nuclei that spontaneously release energy in the form of radiation. Such *radioactive isotopes* play a role in a number of research and medical applications. Because they emit detectable energy, they can be used to trace the position of key atoms or molecules in chemical reactions, they are tools in diagnosis and treatment, and they are even applied in sterilization procedures (chapter 11).

TAKE NOTE: MASS, WEIGHT, AND RELATED TERMS

Mass refers to the quantity of matter that an atomic particle contains. The proton and neutron have almost exactly the same mass, which is about 1.66×10^{-24} grams, a unit of measurement known as a Dalton (Da) or unifed atomic mass unit (U). All elements can be measured in these units. The terms *mass* and *weight* are often used interchangeably in biology, even though they apply to two different but related aspects of matter. Weight is a measurement of the gravitational pull on the mass of a particle, atom, or object. Consequently, it is possible for something with the same mass to have different weights. For example, an astronaut on the earth (normal gravity) would weigh more than the same astronaut on the moon (weak gravity). Atomic weight has been the traditional usage for biologists, because most chemical reactions and biological activities occur within the normal gravitational conditions on earth. This permits use of atomic weight as a standard of comparison. You will also see the terms *formula weight* and *molecular weight* used interchangeably, and they are indeed synonyms. They both mean the sum of atomic weights of all atoms in a molecule.

Another important measurement of an element is its **atomic mass** or **weight.** This is given as the average mass numbers of all isotopic forms (table 2.1). You will notice that this number may not come out even, because most elements have several isotopes and differing proportions of them.

Electron Orbitals and Shells

The structure of an atom can be envisioned as a central nucleus surrounded by a cloud of electrons that constantly rotate about the nucleus in pathways (see figure 2.1). The pathways, called **orbitals,** are not actual objects or exact locations but represent volumes of three-dimensional space in which an electron is likely to be found. Electrons occupy energy shells, proceeding from the lower-level energy electrons nearest the nucleus to the higher-energy electrons in the farthest orbitals.

Electrons fill the orbitals and shells in *pairs,* starting with the shell nearest the nucleus:

- The first shell contains one orbital and a maximum of 2 electrons.
- The second shell has four orbitals and up to 8 electrons.
- The third shell with nine orbitals can hold up to 18 electrons.
- The fourth shell with 16 orbitals contains up to 32 electrons.

The number of orbitals and shells and how completely they are filled depend on the numbers of electrons, so that each element will have a unique pattern. For example:

- Helium (AN = 2) has only a filled first shell of 2 electrons.
- Oxygen (AN = 8) has a filled first shell and a partially filled second shell of 6 electrons.
- Magnesium (AN = 12) has a filled first shell, a filled second one, and a third shell that has only one orbital, so is nearly empty.

As we will see, the chemical properties of an element are controlled mainly by the distribution of electrons in the outermost shell. **Figure 2.2** presents various simplified models of atomic structure and electron maps, superimposed over a partial display of the periodic table of elements.

Check & Assess Section 2.1

✔ Protons (p^+) and neutrons (n^0) make up the nucleus of an atom. Electrons (e^-) orbit the nucleus. The number of protons (+) exactly balances the number of electrons (−).

✔ All elements are composed of atoms but differ in the numbers of protons, neutrons, and electrons they possess.

✔ Each element has unique characteristics and is identified by an *atomic number, mass* or *weight,* and *mass number.*

✔ Isotopes are varieties of one element that contain the same number of protons but different numbers of neutrons.

✔ The number of electrons in an element's outermost orbital (compared with the total number possible) determines its chemical properties and reactivity.

Figure 2.2 Examples of biologically important atoms. Featured element boxes contain information on symbol, atomic number, atomic mass, and electron shell patterns. Simple models show how the shells are filled by electrons as the atomic numbers increase. Chemists depict elements in shorthand form (red Lewis structures) that indicate only the valence electrons, because these are the electrons involved in chemical bonds. In the background is a partial display of the periodic table of elements showing the position of these elements.

1. How are the concepts of an atom and an element related? What causes elements to differ?

2. What are subatomic particles and how do they contribute to the structure and character of atoms?

3. How are mass number and atomic number derived? What is the atomic mass or weight?

4. Using data in table 2.1, give the electron number of nitrogen, sulfur, calcium, phosphorus, and iron.

5. What is distinctive about isotopes of elements, and why are they important?

6. How are the concepts of molecules and compounds related?

7. Compute the molecular weight of oxygen and methane.

8. Why is an isolated atom neutral?

9. Describe the concept of the atomic nucleus, electron orbitals, and shells.

2.2 Bonds and Molecules

Expected Learning Outcomes

6. Explain how elements make chemical bonds to form molecules and compounds.

7. State the relationship among an atom, molecule, and compound.

8. Identify the differences between covalent, ionic, and hydrogen bonds.

9. Summarize the concepts of valence, polarity, and diatomic elements.

10. Describe ionization and distinguish between anions and cations.

11. Compare oxidation and reduction and their effects.

Most elements do not exist naturally in pure, uncombined form but are bound together as molecules and compounds. A **molecule*** is a distinct chemical substance that results from the combination of two or more atoms. Molecules come in a great variety. A few of them such as oxygen consist of the same element (see the note about diatomic elements). But most molecules, for example carbon dioxide (CO_2) and water (H_2O), contain two or more different elements and are more appropriately termed **compounds.** So compounds are one major type of molecule. Other examples of compounds are biological molecules such as proteins, sugars, and fats.

When atoms bind together in molecules, they lose the properties of the atom and take on the properties of the combined substance. In the same way that an atom has an atomic weight, a molecule has a **formula mass** or **molecular weight**[2] (MW), which is calculated from the sum of all of the atomic masses of the atoms it contains.

Chemical bonds of molecules are created when two or more atoms share, donate (lose), or accept (gain) electrons (**figure 2.3**). This capacity for making bonds, termed **valence,*** is determined by the number of electrons that an atom has to lose or share with other atoms during bond formation. The valence electrons determine the degree of reactivity and the types of bonds an element can make. Elements with a filled outer orbital are relatively nonreactive because they have no extra electrons to share with or donate to other atoms. For example, helium has one filled shell, with no tendency either to give up electrons or to take them from other elements, making it a stable, inert (nonreactive) gas. Elements with partially filled outer orbitals are less stable and are more apt to form some sort of bond.

Many chemical reactions are based on the tendency of atoms with unfilled outer shells to gain greater stability by achieving, or at least approximating, a filled outer shell. For example, an atom such as oxygen that can accept two additional electrons will bond readily with atoms (such as hydrogen) that can share or donate electrons. We explore some additional examples of the basic types of bonding in the following section.

In addition to reactivity, the number of electrons in the outer shell also dictates the number of chemical bonds an atom can make. For instance, hydrogen can bind with one other atom, oxygen can bind with up to two other atoms, and carbon can bind with four.

Covalent Bonds: Molecules with Shared Electrons

Covalent (cooperative valence) **bonds** form between atoms with valences that suit them to sharing electrons rather than to donating or receiving them. A simple example is hydrogen gas (H_2), which

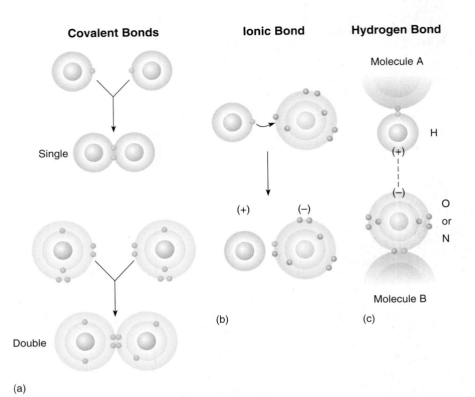

Covalent Bonds

Single

Double

(a)

Ionic Bond

(+) (−)

(b)

Hydrogen Bond

Molecule A

H

(+)

(−)

O
or
N

Molecule B

(c)

Figure 2.3 **General representation of three types of bonding. (a)** Covalent bonds, both single and double. **(b)** Ionic bond. **(c)** Hydrogen bond. Note that hydrogen bonds are represented in models and formulas by dotted lines, as shown in *c*.

consists of two hydrogen atoms. A hydrogen atom has only a single electron, but when two of them combine, each will bring its electron to orbit about both nuclei, thereby approaching a filled orbital (2 electrons) for both atoms and thus creating a single covalent bond (**figure 2.4a**). Covalent bonding also occurs in oxygen gas (O_2), but with a difference. Because each atom has 2 electrons to share in this molecule, the combination creates two pairs of shared electrons, also known as a double covalent bond (**figure 2.4b**). The majority of the molecules associated with living things are composed of single and double covalent bonds between the most common biological elements (carbon, hydrogen, oxygen, nitrogen, sulfur, and phosphorus), which are discussed in more depth in chapter 7. A slightly more complex pattern of covalent bonding is shown for methane gas (CH_4) in **figure 2.4c.**

TAKE NOTE: DIATOMIC ELEMENTS

You will notice that hydrogen, oxygen, nitrogen, chlorine, and iodine are often shown in notation with a 2 subscript—H_2 or O_2. These elements are *diatomic* (two atoms), meaning that in their pure elemental state, they exist in pairs, rather than as a single atom. In every case, these elements are more stable as a pair of atoms joined by a covalent bond (a molecule) than as a single atom. The reason for this phenomenon can be explained by their valences. The electrons in the outer shell are configured so as to complete a full outer shell for both atoms when they bind. You can see this for yourself by observing figures 2.2 and 2.4. It is interesting that most of the diatomic elements are gases.

* *molecule* (mol´-ih-kyool) L. *molecula,* little mass.

2. Biologists prefer to use this term.

* *valence* (vay´-lents) L. *valentia,* strength. The binding qualities of an atom dictated by the number of electrons in its outermost shell.

(a) A hydrogen molecule is formed when two hydrogen atoms share their electrons and form a single bond.

Molecular oxygen (O_2)

Double bond

(b) In a double bond, the outer orbitals of two oxygen atoms overlap and permit the sharing of 4 electrons (one pair from each) and the saturation of the outer orbital for both.

Methane (CH_4)

(c) Simple, three-dimensional, and working models of methane. Note that carbon has 4 electrons to share and hydrogens each have one thereby completing the shells for all atoms in the compound and creating 4 single bonds.

Figure 2.4 **Examples of molecules with covalent bonding.** Note that a and b also show the formation of diatomic molecules.

Polarity in Molecules

When atoms of different electronegativity[3] form covalent bonds, the electrons will not be shared equally and are pulled more toward one atom than another. This force causes one end of a molecule to assume an overall negative charge and the other end to assume an overall partial positive charge. A molecule with such an unequal distribution of charges is termed **polar** and shows polarity, meaning it has positive and negative poles.

Observe the water molecule shown in **figure 2.5** and note that, because the oxygen atom is larger and has more protons than the hydrogen atoms, it will have a stronger attraction for the shared electrons than the hydrogen atoms. Because the electrons will spend more time near the oxygen, it will express a negative charge. The electrons are less attracted to the hydrogen orbits, causing the positive charge of their single proton to dominate. The polar nature of water plays an extensive role in a number of biological reactions, which are discussed later. Polarity is a significant property of many large molecules in living systems and greatly influences both their reactivity and their structure.

When covalent bonds are formed between atoms that have the same or similar electro-negativity, the electrons are shared equally between the two atoms. Because of this balanced distribution, no part of the molecule has a greater attraction for the electrons. This

3. Electronegativity—the ability of an atom or molecule to attract electrons.

(a) Sodium atom (Na) Chlorine atom (Cl)

(b) Na :Cl: ⟶ [Na]⁺ [Cl]⁻

(c)

(d)

Figure 2.6 **Ionic bonding between sodium and chlorine.**
(a) During the reaction, sodium loses its single outer orbital electron to chlorine, thereby filling chlorine's outer shell. **(b)** Simple model of ionic bonding. **(c)** Sodium and chloride ions form large molecules, or crystals, in which the two atoms alternate in a definite, regular, geometric pattern. **(d)** Note the cubic nature of NaCl crystals at the macroscopic level.

details do not negate the fundamental reaction as described here. The outcome of this reaction is a solid crystal complex that interlinks millions of sodium and chloride atoms **(figure 2.6c, d).** The binding of Na^+ and Cl^- exists in three dimensions. Each Na is surrounded by 6 Cls and vice versa. Their charges balance out, and no single molecule of NaCl is present.

Ionization: Formation of Charged Particles

Compounds with intact ionic bonds are electrically neutral, but they can produce charged particles when dissolved in a liquid called a solvent. This phenomenon, called **ionization,** occurs when the ionic bond is broken and the atoms dissociate (separate) into

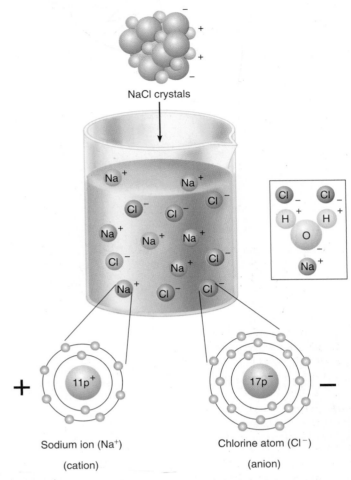

Figure 2.7 **Ionization.** When NaCl in the crystalline form is added to water, the ions are released from the crystal as separate charged particles (cations and anions) into solution. (See also figure 2.12.) In this solution, Cl^- ions are attracted to the hydrogen component of water, and Na^+ ions are attracted to the oxygen (box).

unattached, charged particles called **ions (figure 2.7).** To illustrate what imparts a charge to ions, let us look again at the reaction between sodium and chlorine. When a sodium atom reacts with chlorine and loses 1 electron, the sodium is left with 1 more proton than electrons. This imbalance produces a positively charged sodium ion (Na^+). Chlorine, on the other hand, has gained 1 electron and now has 1 more electron than protons, producing a negatively charged ion (Cl^-). Positively charged ions are termed **cations,*** and negatively charged ions are termed **anions.*** (A good mnemonic device is to think of the *t* in cation as a plus (+) sign and the first *n* in anion as a negative (−) sign.) Substances such as salts, acids, and bases that release ions when dissolved in water are termed **electrolytes** because their charges enable them to conduct an electrical current.

Owing to the general rule that particles of like charge repel each other and those of opposite charge attract each other, we can

* *cation* (kat′-eye-on) A positively charged ion that migrates toward the negative pole, or cathode, of an electrical field.
* *anion* (an′-eye-on) A negatively charged ion that migrates toward the positive pole, or anode.

Figure 2.8 **Hydrogen bonding in water.** Because of the polarity of water molecules, the negatively charged oxygen end of one water molecule is weakly attracted to the positively charged hydrogen end of an adjacent water molecule.

expect ions to interact electrostatically with other ions and polar molecules. Such interactions are important in many cellular chemical reactions, in the formation of solutions, and in the reactions microorganisms have with dyes.

Hydrogen Bonding

Some types of bonding do not involve sharing, losing, or gaining electrons but instead are due to attractive forces between nearby molecules or atoms. One such bond is a **hydrogen bond,** a weak electrostatic force that forms between a hydrogen covalently bonded to one molecule and an oxygen or nitrogen atom on the same molecule or on a different molecule. This phenomenon occurs because a hydrogen atom in a covalent bond tends to be positively charged. Thus, it can attract a nearby negatively charged atom and form an easily disrupted bridge with it. This type of bonding is usually represented in molecular models with a dotted line. A simple example of hydrogen bonding occurs between water molecules **(figure 2.8).** More extensive hydrogen bonding is partly responsible for the structure and stability of proteins and nucleic acids, as you will see later on.

Weak molecular interactions similar to hydrogen bonds that play major roles in the shape and function of biological molecules are **van der Waals forces.** The basis for these interactions is also an attraction of two regions on atoms or molecules that are opposite in charge, but van der Waals forces can occur between nearly any types of molecules and not just those containing hydrogen, oxygen, and nitrogen. These forces come into play whenever the electrons in molecules move about their orbits and become unevenly distributed. This unevenness leads to short-term "sticky spots" in the molecule—some positively charged and others negatively charged. When such regions are located close together, their opposite charges pull them together. These forces can hold even large molecules together because of the cumulative effects of numerous sites of interaction. They not only function between molecules but may occur within different regions of the same

large molecule. Van der Waals forces are a significant factor in protein folding and stability (see figure 2.22, steps 3 and 4).

Electron Transfer and Oxidation–Reduction Reactions

The metabolic work of cells, such as synthesis, movement, and digestion, revolves around energy exchanges and transfers. The management of energy in cells is almost exclusively dependent on chemical rather than physical reactions because most cells are far too delicate to operate with heat, radiation, and other more potent forms of energy. The outer-shell electrons are readily portable and easily manipulated sources of energy. It is in fact the movement of electrons from molecule to molecule that accounts for most energy exchanges in cells. Fundamentally, then, cells must have a supply of atoms that can gain or lose electrons if they are to carry out life processes.

The phenomenon in which electrons are transferred from one atom or molecule to another is termed an **oxidation reduction** (shortened to **redox**) **reaction.** The term *oxidation* was originally adopted for reactions involving the addition of oxygen, but this is no longer the case. In current usage, *oxidation* includes any reaction that causes an atom to lose electrons. Because all redox reactions occur in pairs, it follows that *reduction* is the result of a different atom gaining these same electrons. Keep in mind that because electrons are being added during reduction, the atom that receives them will become more negative; and that is the meaning of *reduction* in this context. It does not imply that an atom is getting smaller. In fact, reduction often results in a greater complexity of the molecule.

To analyze the phenomenon, let us again review the production of NaCl but from a different standpoint. When these two atoms, called the redox pair, react to form sodium chloride, a sodium atom gives up an electron to a chlorine atom. During this reaction, sodium is oxidized because it loses an electron, and chlorine is reduced because it gains an electron **(figure 2.9).** With this system, an atom such as sodium that can donate electrons and thereby reduce another atom is a **reducing agent.** An atom that can receive extra electrons and thereby oxidize another molecule is an **oxidizing agent.** You may find this concept easier to keep straight if you think of redox agents as partners: The reducing partner gives its electrons away and is oxidized; the oxidizing partner receives the electrons and is reduced.[5]

Redox reactions are essential to many of the biochemical processes discussed in chapter 8. In cellular metabolism, electrons are

5. A mnemonic device to keep track of this is *LEO* says *GER: L*ose *E*lectrons *O*xidized; *G*ain *E*lectrons *R*educed.

Reducing agent gives up electrons. Oxidizing agent accepts electrons. Oxidized cation Reduced anion

Figure 2.9 **Simplified diagram of the exchange of electrons during an oxidation–reduction reaction.** Numbers indicate the total electrons in that shell.

frequently transferred from one molecule to another as described here. In other reactions, oxidation and reduction occur with the transfer of a hydrogen atom (a proton and an electron) from one compound to another.

2.3 Chemical Reactions, Solutions, and pH

Expected Learning Outcomes

12. Classify different forms of chemical shorthand and types of reactions.
13. Explain solutes, solvents, and hydration.
14. Differentiate between hydrophilic and hydrophobic.
15. Describe the pH scale and how it was derived; define acid, base, and neutral levels.

Formulas, Models, and Equations

The atomic content of molecules can be represented by a few convenient formulas. We have already been using the molecular formula, which concisely gives the atomic symbols and the number of the atoms involved in subscripts (CO_2, H_2O). More complex molecules such as glucose ($C_6H_{12}O_6$) can also be symbolized this way, but this formula is not unique, because fructose and galactose also share it. Molecular formulas are useful, but they only summarize the atoms in a compound; they do not show the position of bonds between atoms. For this purpose, chemists use structural formulas illustrating the relationships of the atoms and the number and types of bonds (**figure 2.10**). Other structural models present the three-dimensional appearance of a molecule, illustrating the orientation of atoms (differentiated by color codes) and the molecule's overall shape (**figure 2.11**). Many complex molecules such as proteins are now represented by computer-generated images (see figure 2.22, step 4).

Molecules, including those in cells, are constantly involved in **chemical reactions,** leading to changes in the composition of the matter they contain. These changes generally involve the breaking and making of bonds and the rearrangement of atoms. The chemical substances that start a reaction and that are changed by the reaction are called the **reactants.** The substances that result from the reaction are called the **products.** Keep in mind that all of the matter in any reaction is retained in some form, and the same types and numbers of atoms going into the reaction will be present in the products. Chemists and biologists use shorthand to summarize the content of a reaction by means of a **chemical equation.** In an equation, the reactant(s) are on the left of an arrow and the product(s) on the right. The number of atoms of each element must be balanced on either side of the arrow. Note that the numbers of reactants and products are indicated by a coefficient in front of the formula (no coefficient means one). We have already reviewed the reaction with sodium and chloride, which would be shown with this equation:

$$2Na + Cl_2 \rightarrow 2NaCl$$

Most equations do not give the details or even exact order of the reaction but are meant to keep the expression a simple overview of

(a)

(b)

(c)

(d)

Figure 2.10 Comparison of molecular and structural formulas. (a) Molecular formulas provide a brief summary of the elements in a compound. **(b)** Structural formulas clarify the exact relationships of the atoms in the molecule, depicting single bonds by a single line and double bonds by two lines. **(c)** In structural formulas of organic compounds, cyclic or ringed compounds may be completely labeled, or **(d)** they may be presented in a shorthand form in which carbons are assumed to be at the angles and attached to hydrogens. See figure 2.16 for structural formulas of three sugars with the same molecular formula, $C_6H_{12}O_6$.

the process being shown. Some of the common reactions in organisms are syntheses, decompositions, and exchanges.

In a **synthesis* reaction,** the reactants bond together in a manner that produces an entirely new molecule (reactant A plus reactant B yields product AB). An example is the production of sulfur dioxide, a by-product of burning sulfur fuels and an important component of smog:

$$S + O_2 \rightarrow SO_2$$

Some synthesis reactions are not such simple combinations. When water is synthesized, for example, the reaction does not really involve one oxygen atom combining with two hydrogen atoms, because elemental oxygen exists as O_2 and elemental hydrogen exists as H_2. A more accurate equation for this reaction is:

$$2H_2 + O_2 \rightarrow 2H_2O$$

In *decomposition reactions,* the bonds on a single reactant molecule are permanently broken to release two or more product molecules. A simple example can be shown for the common chemical hydrogen peroxide:

$$2H_2O_2 \rightarrow 2H_2O + O_2$$

* *synthesis* (sin'-thuh-sis) Gr. *synthesis,* putting together.

(a) (b)

(c)

Figure 2.11 Three-dimensional, or space-filling, models of (a) water, (b) carbon dioxide, and (c) glucose. The red atoms are oxygen, the white ones hydrogen, and the black ones carbon.

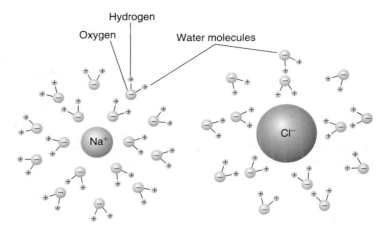

Figure 2.12 Hydration spheres formed around ions in solution. In this example, a sodium cation attracts the negatively charged region of water molecules, and a chloride anion attracts the positively charged region of water molecules. In both cases, the ions become surrounded by spherical layers of specific numbers and arrangements of water molecules.

During **exchange reactions,** the reactants trade portions between each other and release products that are combinations of the two.

$$AB + XY \rightleftharpoons AY + XB$$

This type of reaction occurs between an acid and a base when they react to form water and a salt:

$$HCl + NaOH \rightarrow NaCl + H_2O$$

The reactions in biological systems can be reversible, meaning that reactants and products can be converted back and forth. These reversible reactions are symbolized with a double arrow, each pointing in opposite directions, as in the exchange reaction shown earlier. Whether a reaction is reversible depends on the proportions of these compounds, the difference in energy state of the reactants and products, and the presence of **catalysts** (substances that increase the rate of a reaction). Additional reactants coming from another reaction can also be indicated by arrows that enter or leave at the main arrow:

$$X + Y \xrightarrow{\quad CD \quad C \quad} XYD$$

Solutions: Homogeneous Mixtures of Molecules

A **solution** is a mixture of one or more substances called **solutes** uniformly dispersed in a dissolving medium called a **solvent.** An important characteristic of a solution is that the solute cannot be separated by filtration or ordinary settling. The solute can be gaseous, liquid, or solid, and the solvent is usually a liquid. Examples of solutions are salt or sugar dissolved in water and iodine dissolved in alcohol. In general, a solvent will dissolve a solute only if it has similar electrical characteristics as indicated by the rule of solubility, expressed simply as "like dissolves like." For example, water is a polar molecule and will readily dissolve an ionic solute such as NaCl, yet a nonpolar solvent such as benzene will not dissolve NaCl.

Water is the most common solvent in natural systems, having several characteristics that suit it to this role. The polarity of the water molecule causes it to form hydrogen bonds with other water molecules, but it can also interact readily with charged or polar molecules.

When an ionic solute such as NaCl crystals is added to water, the Na^+ and Cl^- are released into solution. Dissolution occurs because Na^+ is attracted to the negative pole of the water molecule and Cl^- is attracted to the positive poles. In this way, they are drawn away from the crystal separately into solution. As it leaves, each ion becomes **hydrated,** which means that it is surrounded by a sphere of water molecules (**figure 2.12**). Molecules such as salt or sugar that attract water to their surface are termed **hydrophilic.*** Nonpolar molecules, such as benzene, that repel water are considered **hydrophobic.*** A third class of molecules, such as the phospholipids in cell membranes, are considered **amphipathic*** because they have both hydrophilic and hydrophobic properties.

Because most biological activities take place in aqueous (water-based) solutions, the concentration of these solutions can be very important (see chapter 7). The **concentration** of a solution expresses the amount of solute dissolved in a certain amount of solvent. It can be figured by percentage or molarity. Percentage is

* *hydrophilic* (hy-droh-fil´-ik) Gr. *hydros,* water, and *philos,* to love.
* *hydrophobic* (hy-droh-fob´-ik) Gr. *phobos,* fear.
* *amphipathic* (am´-fy-path´-ik) Gr. *amphi,* both.

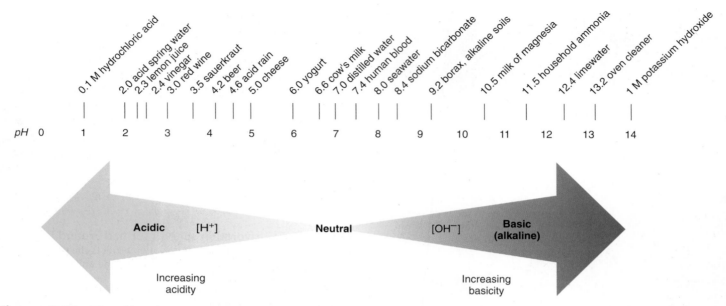

Figure 2.13 The pH scale. Shown are the relative degrees of acidity and basicity and the approximate pH readings for various substances.

the ratio of solute in solution expressed as some combination of weight or volume.

A common way to calculate concentration by percentage is to use the weight of the solute, measured in grams (g), dissolved in a specified volume of solvent, measured in milliliters (ml). For example, dissolving 3 g of NaCl in the amount of water to produce 100 ml of solution is a 3% solution; dissolving 30 g in water up to 100 ml of solution produces a 30% solution; and dissolving 3 g up to 1,000 ml (1 liter) produces a 0.3% solution.

A frequent way to express concentration of biological solutions is by its molar concentration, or *molarity* (M). A standard molar solution is obtained by dissolving 1 *mole,* defined as the molecular weight of the compound in grams, in 1 L (1,000 ml) of solution. To make a 1 M solution of sodium chloride, we would dissolve 58 g of NaCl to give 1 L of solution; a 0.1 M solution would require 5.8 g of NaCl in 1 L of solution.

Acidity, Alkalinity, and the pH Scale

Another factor with far-reaching impact on living things is the concentration of acidic or basic solutions in their environment. To understand how solutions become acidic or basic, we must look again at the behavior of water molecules. Hydrogens and oxygen tend to remain bonded by covalent bonds, but under certain conditions, a single hydrogen can break away as an ionic H^+, or **hydrogen ion,** leaving the remainder of the molecule in the form of an OH^-, or **hydroxide ion.** The H^+ ion is positively charged because it is essentially a hydrogen that has lost its electron; the OH^- is negatively charged because it remains in possession of that electron. Ionization of water is constantly occurring, but in pure water containing no other ions, H^+ and OH^- are produced in equal amounts, and the solution remains neutral. By one definition, a solution is considered **acidic** when one of its components (an acid) releases

excess hydrogen ions.[6] A solution is **basic** when a component (a base) releases excess hydroxide ions, so that there is no longer a balance between the two ions. Another term used interchangeably with basic is **alkaline.**

To measure the acid and base concentrations of solutions, scientists use the **pH** scale, a graduated numerical scale that ranges from 0 (the most acidic) to 14 (the most basic). This scale is a useful standard for rating the relative acid or base content of a substance. Use **figure 2.13** to familiarize yourself with the pH readings of some common substances. It is not an arbitrary scale but actually a mathematical derivation based on the negative logarithm (reviewed in appendix B) of the concentration of H^+ ions in moles per liter (symbolized as $[H^+]$) in a solution, represented as:

$$pH = -\log[H^+]$$

Acidic solutions have a greater concentration of H^+ than OH^-, starting with: pH 0, which contains 1.0 moles H^+/liter (L). Each of the subsequent whole-number readings in the scale reduces the $[H^+]$ by tenfold:

- pH 1 contains [0.1 moles H^+/L].
- pH 2 contains [0.01 moles H^+/L].
- Continuing in the same manner up to pH 14, which contains [0.00000000000001 moles H^+/L].

These same concentrations can be represented more manageably by exponents:

- pH 2 has an $[H^+]$ of 10^{-2} moles.
- pH 14 has an $[H^+]$ of 10^{-14} moles **(table 2.2).**

6. Actually, it forms a hydronium ion (H_3O^+), but for simplicity's sake, we will use the notation of H^+.

TABLE 2.2	Hydrogen Ion and Hydroxide Ion Concentrations at a Given pH		
Moles/L of Hydrogen Ions	**Logarithm**	**pH**	**Moles/L of OH⁻**
1.0	10^{-0}	0	10^{-14}
0.1	10^{-1}	1	10^{-13}
0.01	10^{-2}	2	10^{-12}
0.001	10^{-3}	3	10^{-11}
0.0001	10^{-4}	4	10^{-10}
0.00001	10^{-5}	5	10^{-9}
0.000001	10^{-6}	6	10^{-8}
0.0000001	10^{-7}	7	10^{-7}
0.00000001	10^{-8}	8	10^{-6}
0.000000001	10^{-9}	9	10^{-5}
0.0000000001	10^{-10}	10	10^{-4}
0.00000000001	10^{-11}	11	10^{-3}
0.000000000001	10^{-12}	12	10^{-2}
0.0000000000001	10^{-13}	13	10^{-1}
0.00000000000001	10^{-14}	14	10^{-0}

It is evident that the pH units are derived from the exponent itself. Even though the basis for the pH scale is $[H^+]$, it is important to note that, as the $[H^+]$ in a solution decreases, the $[OH^-]$ increases in direct proportion. At midpoint—pH 7 or **neutrality**—the concentrations are exactly equal and neither predominates, this being the pH of pure water previously mentioned.

In summary, the pH scale can be used to rate or determine the degree of acidity or alkalinity of a solution. On this scale, a pH below 7 is acidic, and the lower the pH, the greater the acidity; a pH above 7 is basic, and the higher the pH, the greater the alkalinity. Incidentally, although pHs are given here in even whole numbers, more often, a pH reading exists in decimal form; for example, pH 4.5 or 6.8 (acidic) and pH 7.4 or 10.2 (basic). Because of the damaging effects of very concentrated acids or bases, most cells operate best under neutral, weakly acidic, or weakly basic conditions (see chapter 7).

Aqueous solutions containing both acids and bases may be involved in **neutralization** reactions, which give rise to water and other neutral by-products. For example, when equal molar solutions of hydrochloric acid (HCl) and sodium hydroxide (NaOH, a base) are mixed, the reaction proceeds as follows:

$$HCl + NaOH \rightarrow H_2O + NaCl$$

Here the acid and base ionize to H^+ and OH^- ions, which form water, and other ions, Na^+ and Cl^-, which form sodium chloride. Any product other than water that arises when acids and bases react is called a salt. Many of the organic acids (such as lactic and succinic acids) that function in **metabolism*** are available as the acid and the salt form (such as lactate, succinate), depending on the conditions in the cell (see chapter 8).

* *metabolism* (muh-tab′-oh-lizm) A general term referring to the totality of chemical and physical processes occurring in the cell.

Check & Assess Sections 2.2 and 2.3

✔ Atoms combine by forming several types of bonds.

✔ General terms for bound atoms are molecules and compounds.

✔ Covalent bonds are chemical bonds in which electrons are shared between atoms. Equally distributed electrons form nonpolar covalent bonds, whereas unequally distributed electrons form polar covalent bonds.

✔ Ionic bonds are chemical bonds resulting from opposite charges. The outer electron shell either donates or receives electrons from another atom so that the outer shell of each atom is completely filled.

✔ Hydrogen bonds are weak chemical attractions that form between covalently bonded hydrogens and either oxygens or nitrogens on different molecules.

✔ Chemical equations express the chemical exchanges between atoms or molecules that occur during chemical reactions such as synthesis or decomposition.

✔ Solutions are mixtures of solutes and solvents that cannot be separated by filtration or settling.

✔ The pH, ranging from a highly *acidic* solution to highly *basic* solution, refers to the concentration of hydrogen ions. It is expressed as a number from 0 to 14.

10. Distinguish between the general reactions in covalent, ionic, and hydrogen bonds.
11. Which kinds of elements tend to make covalent bonds?
12. Distinguish between a single and a double bond.
13. What is polarity?
14. Which kinds of elements tend to make ionic bonds?
15. Differentiate between an anion and a cation, using examples.
16. Differentiate between oxidation and reduction, and between an oxidizing agent and a reducing agent, using examples.
17. Review the types of chemical reactions and the general ways they can be expressed in equations.
18. Define solution, solvent, and solute.
19. What properties of water make it an effective biological solvent, and how does a molecule like NaCl become dissolved in it?
20. How is the concentration of a solution determined?
21. What is molarity? Tell how to make a 1 M solution of $Mg_3(PO_4)_2$ and a 0.1 M solution of $CaSO_4$.
22. What determines whether a substance is an acid or a base? Briefly outline the pH scale.

2.4 The Chemistry of Carbon and Organic Compounds

Expected Learning Outcomes

16. Describe the chemistry of carbon and the difference between inorganic and organic compounds.

17. Identify functional groups and know some examples.

18. Relate what macromolecules, polymers, and monomers are.

Linear

Branched

Ringed

(a) (b)

Figure 2.14 The versatility of bonding in carbon. In most compounds, each carbon makes a total of four bonds. **(a)** Single, double, and triple bonds can be made with other carbons, oxygen, and nitrogen; single bonds are made with hydrogen. Simple electron models show how the electrons are shared in these bonds. **(b)** Multiple bonding of carbons can give rise to long chains, branched compounds, and ringed compounds, many of which are extraordinarily large and complex.

So far, our main focus has been on the characteristics of atoms, ions, and small, simple substances that play diverse roles in the structure and function of living things. These substances are often lumped together in a category called **inorganic chemicals.** A chemical is usually inorganic if it does not contain both carbon and hydrogen. Examples of inorganic chemicals include H_2O, O_2, NaCl (sodium chloride), $Mg_3(PO_4)_2$ (magnesium phosphate), $CaCO_3$ (calcium carbonate), and CO_2 (carbon dioxide). In reality, however, most of the chemical reactions and structures of living things occur at the level of more complex molecules, termed **organic chemicals.** The minimum requirement for a compound to be considered organic is that it contains a basic framework of carbon bonded to hydrogens. Organic molecules vary in complexity from the simplest, methane (CH_4; see figure 2.4c), which has a molecular weight of 16, to certain antibody molecules (produced by an immune reaction) that have a molecular weight of nearly 1 million and are among the most complex molecules on earth. Most organic chemicals in cells contain other elements such as oxygen, nitrogen, and phosphorus in addition to the carbon and hydrogen.

The role of carbon as the fundamental element of life can best be understood if we look at its chemistry and bonding patterns. The valence of carbon makes it an ideal atomic building block to form the backbone of organic molecules; it has 4 electrons in its outer orbital to be shared with other atoms (including other carbons) through covalent bonding. As a result, it can form stable chains containing thousands of carbon atoms and still has bonding sites available for forming covalent bonds with numerous other atoms. The bonds that carbon forms are linear, branched, or ringed, and it can form four single bonds, two double bonds, or one triple bond **(figure 2.14).**

CONTINUING CASE FILE **2**

Wanting to move from speculation to scientific facts, NASA set up a series of missions that could perform more detailed investigations. One of the most recent projects, a probe called *Phoenix*, arrived on the northern pole of the red planet in the spring of 2008. Its program was to analyze the Martian surface geology and soil, and to use a small robotic explorer to grab samples of rock and other deposits. An important component of its design was to explore more thoroughly the idea that conditions conducive to life exist on Mars now or did in the past. Rather than trying to isolate microorganisms, the lander's instruments were designed to "taste and sniff" the soil and ice to detect pH, salts, and other chemicals. The lander's robotic arm dug half a meter into the soil to sample just beneath the surface. Early tests show that the soil is salty and slightly acidic. Perhaps the most definitive evidence has been that deposits of ice were shown to be frozen water (and not carbon dioxide) and that the soil may have harbored films of liquid water fairly recently.

A significant amount of methane also turned up in the samples, but more complex organic compounds were not clearly evident. Some scientists feel that the severe atmospheric conditions on the surface of the planet would break them down. It is thought that sampling deeper in a future mission to Mars would increase the chance of yielding organic chemicals.

- Is inorganic carbon by itself a sign of life? In what ways are organic carbon compounds considered evidence of life?

- What other elements present in earth microorganisms would one expect to find if life exists on Mars?

For a wrap-up, see the Case File Perspective, page 54.

Functional Groups of Organic Compounds

One important advantage of carbon's serving as the molecular skeleton for living things is that it is free to bind with a variety of special molecular groups or accessory molecules called **functional groups.** Functional groups help define the chemical class of certain groups of organic compounds and confer unique reactive properties on the whole molecule **(table 2.3).** Because each type of functional group behaves in a distinctive manner, reactions of an organic compound can be predicted by knowing the kind of functional group or groups it carries. Many synthesis, decomposition, and transfer reactions rely upon functional groups such as R—OH or R—NH_2. The R designation on a molecule is shorthand for remainder, and its

TABLE 2.3 Representative Functional Groups and Organic Compounds That Contain Them

Formula of Functional Group	Name	Can Be Found in
R* — O — H	Hydroxyl	Alcohols, carbohydrates
R — C (=O) OH	Carboxyl	Fatty acids, proteins, organic acids
R — C (H)(H) — NH_2	Amino	Proteins, nucleic acids
R — C (=O) O — R	Ester	Lipids
R — C (H)(H) — SH	Sulfhydryl	Cysteine (amino acid), proteins
R — C (=O) H	Carbonyl, terminal end	Aldehydes, polysaccharides
R — C (=O) — C —	Carbonyl, internal	Ketones, polysaccharides
R — O — P (=O)(OH) — OH	Phosphate	DNA, RNA, ATP

*The R designation on a molecule is shorthand for remainder, and its placement in a formula indicates that what is attached at that site varies from one compound to another.

placement in a formula indicates that the group attached at that site varies from one compound to another.

Organic Macromolecules: Superstructures of Life

The compounds of life fall into the realm of **biochemistry.** Biochemicals are organic compounds produced by (or components of) living things, and they include four main families: carbohydrates, lipids, proteins, and nucleic acids **(table 2.4).** The compounds in these groups are assembled from smaller molecular subunits, or building blocks, and because they are often very large compounds, they are termed **macromolecules.** All macromolecules except lipids are formed by polymerization, a process in which repeating subunits termed **monomers*** are bound into chains of various lengths termed **polymers.*** For example, amino acids (monomers), when arranged in a chain, form proteins (polymers). The large size and complex, three-dimensional shape of macromolecules enable them to function as structural components, molecular messengers, energy sources, enzymes (biochemical catalysts), nutrient stores, and sources of genetic information. In the following section and in later chapters, we consider numerous concepts relating to the roles of macromolecules in cells. Table 2.4 will also be a useful reference when you study metabolism in chapter 8.

2.5 Molecules of Life: Carbohydrates

Expected Learning Outcomes

19. Define *carbohydrate* and know the functional groups that characterize carbohydrates.
20. Distinguish among mono-, di-, and polysaccharides, and describe how their bonds are made.
21. Discuss the functions of carbohydrates in cells.

The term **carbohydrate** originates from the way that most members of this chemical class resemble combinations of carbon and water. Carbohydrates can be generally represented by the formula $(CH_2O)_n$, in which *n* indicates the number of units of this combination of atoms. The basic structure of a simple carbohydrate is a backbone of carbon bound to two or more hydroxyl groups. Because they also have either an aldehyde or a ketone group, they are often designated as *polyhydroxy* **aldehydes** or **ketones (figure 2.15).** In simple terms, a sugar such as glucose is an aldehyde with a terminal carbonyl group bonded to a hydrogen and another carbon. Fructose sugar is a ketone with a carbonyl group bonded between two carbons.

Carbohydrates exist in a great variety of configurations. The common term sugar **(saccharide*)** refers to a simple carbohydrate such as a monosaccharide or a disaccharide that has a sweet taste. A **monosaccharide** is a simple polyhydroxy aldehyde or ketone

* *monomer* (mahn′-oh-mur) Gr. *mono,* one, and *meros,* part.
* *polymer* (pahl′-ee-mur) Gr. *poly,* many; also the root for polysaccharide and polypeptide.
* *saccharide* (sak′-uh-ryd) Gr. *sakcharon,* sweet.

TABLE 2.4	Macromolecules and Their Functions	
Macromolecule	**Description/Basic Structure**	**Examples/Notes**
Carbohydrates		
Monosaccharides	3- to 7-carbon sugars	Glucose, fructose / Sugars involved in metabolic reactions; building block of disaccharides and polysaccharides
Disaccharides	Two monosaccharides	Maltose (malt sugar) / Composed of two glucoses; an important breakdown product of starch
		Lactose (milk sugar) / Composed of glucose and galactose
		Sucrose (table sugar) / Composed of glucose and fructose
Polysaccharides	Chains of monosaccharides	Starch, cellulose, glycogen / Cell wall, food storage
Lipids		
Triglycerides	Fatty acids + glycerol	Fats, oils / Major component of cell membranes; storage
Phospholipids	Fatty acids + glycerol + phosphate	Membranes
Waxes	Fatty acids, alcohols	Mycolic acid / Cell wall of mycobacteria
Steroids	Ringed structure	Cholesterol, ergosterol / Membranes of eukaryotes and some bacteria
Proteins		
Polypeptides	Amino acids bound by peptide bonds	Enzymes; part of cell membrane, cell wall, ribosomes, antibodies / Metabolic reactions; structural components
Nucleic acids	Nucleotides, composed of pentose sugar + phosphate + nitrogenous base	Purines: adenine, guanine; Pyrimidines: cytosine, thymine, uracil
Deoxyribonucleic acid (DNA)	Contains deoxyribose sugar and thymine, not uracil	Chromosomes; genetic material of viruses / Inheritance
Ribonucleic acid (RNA)	Contains ribose sugar and uracil, not thymine	Ribosomes; mRNA, tRNA / Expression of genetic traits
Adenosine triphosphate (ATP)	Contains adenine, ribose sugar, and 3 phosphate groups	A high-energy compound that gives off energy to power reactions in cells

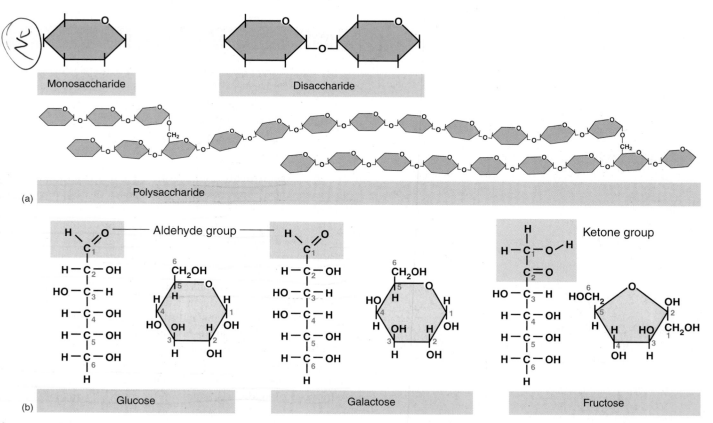

Figure 2.15 **Common classes of carbohydrates.** **(a)** Major saccharide groups, named for the number of sugar units each contains. **(b)** Three hexoses with the same molecular formula ($C_6 H_{12} O_6$) and different structural formulas. Both linear and ring models are given. The linear form emphasizes aldehyde and ketone groups, although in solution the sugars exist in the ring form. Note that the carbons are numbered in red so as to keep track of reactions within and between monosaccharides.

molecule containing from 3 to 7 carbons; a **disaccharide** is a combination of two monosaccharides; and a **polysaccharide** is a polymer of five or more monosaccharides bound in linear or branched chain patterns (figure 2.15). Monosaccharides and disaccharides are specified by combining a prefix that describes some characteristic of the sugar with the suffix **-ose.** For example, **hexoses** are composed of 6 carbons, and **pentoses** contain 5 carbons. **Glucose** (Gr. sweet) is the most common and universally important hexose; **fructose** is named for fruit (one of its sources); and xylose, a pentose, derives its name from the Greek word for wood. Disaccharides are named similarly: **lactose** (L. milk) is an important

component of milk; **maltose** means malt sugar; and **sucrose** (Fr. sugar) is common table sugar or cane sugar.

The Nature of Carbohydrate Bonds

The subunits of disaccharides and polysaccharides are linked by means of **glycosidic bonds,** in which carbons (each is assigned a number) on adjacent sugar units are bonded to the same oxygen atom like links in a chain **(figure 2.16).** For example:

- Maltose is formed when the number 1 carbon on a glucose bonds to the oxygen on the number 4 carbon on a second glucose.

Figure 2.16 Glycosidic bond in three common disaccharides. (a) General scheme in the formation of a glycosidic bond by dehydration synthesis. **(b)** Formation of the 1,4 bond between two α glucoses to produce maltose and water. **(c)** Formation of the 1,2 bond between glucose and fructose to produce sucrose and water. **(d)** A 1,4 bond between a galactose and glucose produces lactose.

INSIGHT 2.1

Better Living through Bacteria?

Probably the most prolific chemical factories are not massive facilities with high-tech processing machinery, but minute bacterial cells that are just doing what comes naturally. We learned in chapter 1 that microbes have the power to decompose everything from garbage to toxic pollutants, but it turns out that they can also be used to synthesize a variety of useful organic chemicals. A recent discovery is a form of cellulose synthesized by a common soil bacterium called *Acetobacter xylinum*. The composition of bacterial cellulose is similar to that from plants, being a polymer of glucose, except that bacterial cellulose is composed of even finer fibers. This makes it an effective choice for a number of medical applications that require dense, strong materials. For example, it can double as a skin replacement for severe burn patients, and it could be adapted to repair small blood vessels. It would also work well as a

dressing impregnated with antibiotics and other drugs to deliver the medicines directly into an incision or wound. Scientists at the University of Texas are currently developing a method for mass production of this material. Actually, harnessing the chemical versatility of bacteria is nothing new. For decades, the biotechnology industry has been using bacteria for synthesizing hundreds of common, everyday chemicals such as antibiotics, steroids, enzymes, vitamins, alcohols, and amino acids. Chapter 27 covers this aspect of applied microbiology.

Most bacteria exist as single cells that can make only a tiny amount of any one chemical. How does industry get these tiny organisms to make millions of pounds of these same chemicals? *Answer available at http://www.mhhe.com/talaro8*

- Sucrose is formed when glucose and fructose bind oxygen between their number 1 and number 2 carbons.
- Lactose is formed when glucose and galactose connect by their number 1 and number 4 carbons.

In order to form this bond, one carbon gives up its OH group and the other (the one contributing the oxygen to the bond) loses the H from its OH group. Because a water molecule is produced, this reaction is known as **dehydration synthesis**, a process common to most polymerization reactions (see proteins, page 47). Three polysaccharides (starch, cellulose, and glycogen) are structurally and biochemically distinct, even though all are polymers of the same monosaccharide—glucose. The basis for their differences lies primarily in the exact way the glucoses are bound together, which

greatly affects the characteristics of the end product (**figure 2.17**). The synthesis and breakage of each type of bond require a specialized catalyst called an enzyme (see chapter 8).

The Functions of Carbohydrates in Cells

Carbohydrates are the most abundant biological molecules in nature. They play numerous roles in cell structure, adhesion, and metabolism.

Polysaccharides typically contribute to structural support and protection and serve as nutrient and energy stores. The cell walls in plants and many microscopic algae derive their strength and rigidity from **cellulose,** a long, fibrous polymer (**figure 2.17a** and **Insight 2.1**). Because of this role, cellulose is probably one of the most common organic substances on the earth, yet it is digestible

(a) **Cellulose**

(b) **Starch**

Figure 2.17 **Polysaccharides.** **(a)** Cellulose is composed of β glucose bonded in 1,4 bonds that produce linear, lengthy chains of polysaccharides that are H-bonded along their length. This accounts for fibrous support structures of plants. **(b)** Starch is also composed of glucose polymers, in this case α glucose. The main structure is amylose bonded in a 1,4 pattern, with side branches of amylopectin bonded by 1,6 bonds. The entire molecule is compact and granular.

only by certain bacteria, fungi, and protozoa that produce the enzyme cellulase. These microbes, called decomposers, play an essential role in breaking down and recycling plant materials (see figure 7.2). Some bacteria secrete slime layers of a glucose polymer called *dextran*. This substance causes a sticky layer to develop on teeth that leads to plaque, described later in chapter 21.

Other structural polysaccharides can be conjugated (chemically bonded) to amino acids, nitrogen bases, lipids, or proteins. **Agar,** an indispensable polysaccharide in preparing solid culture media, is a natural component of certain seaweeds. It is a complex polymer of galactose and sulfur-containing carbohydrates. **Chitin** (ky-tun), a polymer of glucosamine (a sugar with an amino functional group), is a major compound in the cell walls of fungi and the exoskeletons of insects. **Peptidoglycan*** is one special class of compounds in which polysaccharides (glycans) are linked to peptide fragments (a short chain of amino acids). This molecule provides the main source of structural support to the bacterial cell wall. The cell wall of gram-negative bacteria also contains **lipopolysaccharide,** a complex of lipid and polysaccharide responsible for symptoms such as fever and shock (see chapters 4 and 13).

The outer surface of many cells has a delicate "sugar coating" composed of polysaccharides bound in various ways to proteins (the combination is called mucoprotein or glycoprotein). This structure, called the **glycocalyx,*** functions in attachment to other cells or as a site for *receptors*—surface molecules that receive and respond to external stimuli. Small sugar molecules account for the differences in human blood types, and carbohydrates are a component of large protein molecules called antibodies. Some viruses have glycoproteins on their surface for binding to and invading their host cells.

Polysaccharides are usually stored by cells in the form of glucose polymers such as **starch (figure 2.17b)** or **glycogen** that are readily tapped as a source of energy and other metabolic needs. Because a water molecule is required for breaking the bond between two glucose molecules, digestion is also termed **hydrolysis.*** Starch is the primary storage food of green plants, microscopic algae, and some fungi; glycogen (animal starch) is a stored carbohydrate for animals and certain groups of bacteria and protozoa.

2.6 Molecules of Life: Lipids

Expected Learning Outcomes

22. Define *lipid, triglyceride, phospholipid, fatty acid,* and *cholesterol.*
23. Describe how an ester bond is formed.
24. Discuss major functions of lipids in cells.

The term **lipid,** derived from the Greek word *lipos,* meaning fat, is not a chemical designation but an operational term for a variety of substances that are not soluble in polar solvents such as water (recall that oil and water do not mix) but will dissolve in nonpolar solvents such as benzene and chloroform. This property occurs because the substances we call lipids contain relatively long or complex C—H (hydrocarbon) chains that are nonpolar and thus hydrophobic. The main groups of compounds classified as lipids are triglycerides, phospholipids, steroids, and waxes.

Important storage lipids are the **triglycerides,** a category that includes fats and oils. Triglycerides are composed of a single molecule of glycerol bound to three **fatty acids (figure 2.18). Glycerol** is a 3-carbon alcohol[7] with three OH groups that serve as binding sites. Fatty acids are long-chain unbranched hydrocarbon molecules with a carboxyl group (COOH) at one end that is free to bind to the glycerol. The bond that forms between the —OH group and the —COOH is defined as an **ester bond (figure 2.18a).** The hydrocarbon portion of a fatty acid can vary in length from 4 to 24 carbons and, depending on the fat, it may be *saturated* or *unsaturated.* A saturated fatty acid has all of the carbons in the chain bonded to hydrogens with single bonds. Fatty acids having at least one carbon—carbon double bond are considered unsaturated **(figure 2.18b).** Fats that contain such fatty acids are described with these terms as well. The structure of fatty acids is what gives fats and oils (liquid fats) their greasy, insoluble nature. In general, solid fats (such as beef tallow) are more saturated, and oils (or liquid fats) are more unsaturated.

In most cells, triglycerides are stored in long-term concentrated form as droplets or globules. When the ester linkage is acted on by digestive enzymes called lipases, the fatty acids and glycerol are freed to be used in metabolism. Fatty acids are a superior source of energy, yielding twice as much per gram as other storage molecules (starch). Soaps are K^+ or Na^+ salts of fatty acids whose qualities make them excellent grease removers and cleaners (see chapter 11).

Membrane Lipids

The **phospholipids** serve as a major structural component of cell membranes. Although phospholipids also contain glycerol and fatty acids, they differ significantly from triglycerides. Phospholipids contain only two fatty acids attached to the glycerol, and the third glycerol binding site holds a phosphate group. The phosphate is in turn bonded to an alcohol, which varies from one phospholipid to another **(figure 2.19a).** This class of lipids has a hydrophilic region from the charge on the phosphoric acid–alcohol "head" of the molecule and a hydrophobic region that corresponds to the long, uncharged "tail" (formed by the fatty acids). When exposed to an aqueous solution, the charged heads are attracted to the water phase, and the nonpolar tails are repelled from the water phase **(figure 2.19b).** This property causes lipids to naturally assume single and double layers (bilayers), which contribute to their biological significance in membranes. When two single layers of polar lipids come together to form a double layer, the outer hydrophilic face of each single layer will orient itself toward the solution, and the hydrophobic portions will become immersed in the core of the bilayer. The structure of lipid bilayers confers characteristics on membranes such as selective permeability and fluid nature.

Miscellaneous Lipids

Steroids are complex ringed compounds commonly found in cell membranes and animal hormones. The best known of these is the sterol (meaning a steroid with an OH group) called **cholesterol.**

* *peptidoglycan* (pep-tih-doh-gly′-kan).
* *glycocalyx* (gly″-koh-kay′-lix) Gr. *glycos,* sweet and *calyx,* covering.
* *hydrolysis* (hy-drol′-eye-sis) Gt. *hydro,* water, and *hydrein,* to dissolve.

7. Alcohols are hydrocarbons containing an —OH functional group.

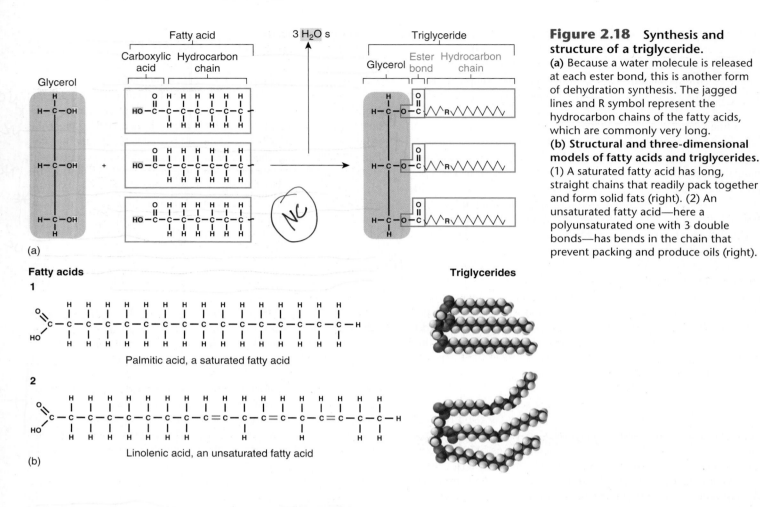

Figure 2.18 Synthesis and structure of a triglyceride.
(a) Because a water molecule is released at each ester bond, this is another form of dehydration synthesis. The jagged lines and R symbol represent the hydrocarbon chains of the fatty acids, which are commonly very long.
(b) Structural and three-dimensional models of fatty acids and triglycerides. (1) A saturated fatty acid has long, straight chains that readily pack together and form solid fats (right). (2) An unsaturated fatty acid—here a polyunsaturated one with 3 double bonds—has bends in the chain that prevent packing and produce oils (right).

(figure 2.20). Cholesterol reinforces the structure of the cell membrane in animal cells and in an unusual group of cell-wall-deficient bacteria called the mycoplasmas (see chapter 4).

The cell membranes of fungi also contain a unique sterol, called ergosterol. *Prostaglandins* are fatty acid derivatives found in trace amounts that function in inflammatory and allergic reactions, blood clotting, and smooth muscle contraction. Chemically, a *wax* is an ester formed between a long-chain alcohol and a saturated fatty acid. The resulting material is typically pliable and soft when warmed but hard and water-resistant when cold (paraffin, for example). Among living things, fur, feathers, fruits, leaves, human skin, and insect exoskeletons are naturally waterproofed with a coating of wax. Bacteria that cause tuberculosis and leprosy produce a wax (wax D) that contributes to their pathogenicity.

Figure 2.19 Phospholipids—membrane molecules.
(a) A model of a single molecule of a phospholipid. The phosphate-alcohol head lends a charge to one end of the molecule; its long, trailing hydrocarbon chain is uncharged. **(b)** Phospholipids in water-based solutions become arranged (1) in single layers called micelles, with the charged head oriented toward the water phase and the hydrophobic nonpolar tail buried away from the water phase, or (2) in double-layered systems with the hydrophobic tails sandwiched between two hydrophilic layers.

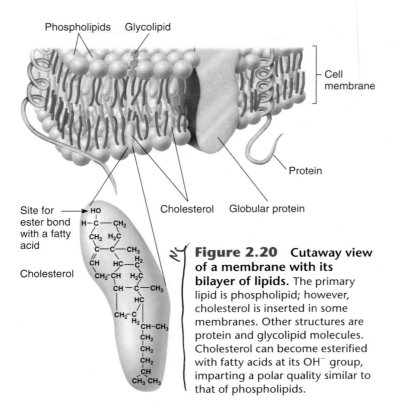

Phospholipids Glycolipid

Cell membrane

Protein

Cholesterol Globular protein

Site for ester bond with a fatty acid

HO
H–C—CH₂
CH₂ H₂C
C—C—CH₃
CH HC
CH₂—CH H₂C
CH—C—CH₃
HC
CH₂—C
H₂ CH–CH₃
CH₂
CH₂
CH₂
CH
CH₃ CH₃

Cholesterol

Figure 2.20 Cutaway view of a membrane with its bilayer of lipids. The primary lipid is phospholipid; however, cholesterol is inserted in some membranes. Other structures are protein and glycolipid molecules. Cholesterol can become esterified with fatty acids at its OH⁻ group, imparting a polar quality similar to that of phospholipids.

TAKE NOTE: THE STRUCTURE OF MEMBRANES

The word **membrane** appears frequently in descriptions of cells in chapters 4 and 5. The word itself can be used to indicate a lining or covering including such multicellular structures as the mucous membranes of the body. At the cellular level, however, a membrane is a thin sheet of molecules composed of phospholipids and sterols (averaging about 40% of membrane content) and proteins (averaging about 60%). All cells have a membrane that completely encases the cytoplasm. Membranes are also components of eukaryotic organelles such as nuclei, mitochondria, and chloroplasts, and they appear in internal pockets of certain prokaryotic cells. Even some viruses, which are not cells, can have a membranous envelope.

A standard model of membrane structure has the lipids forming a continuous bilayer. The polar lipid heads face toward the aqueous phases and the nonpolar tails orient toward the center of the membrane. Embedded at numerous sites in this bilayer are various-sized globular proteins (figure 2.20). Some proteins are situated only at the surface; others extend fully through the entire membrane.

Membranes are dynamic and constantly changing because the lipid phase is in motion and many proteins can migrate freely about. This fluidity is essential to such activities as engulfment of food and discharge or secretion by cells. The structure of the lipid phase provides an impenetrable barrier that accounts for the selective permeability and transport of molecules. Membrane proteins function in receiving molecular signals (receptors), in binding and transporting nutrients, and as enzymes, topics to be discussed in chapters 7 and 8.

Check & Assess Sections 2.4–2.6

✔ Biologists define organic molecules as those containing both carbon and hydrogen. Inorganic chemicals are all other substances that lack these elements.

✔ Carbon is the backbone of biological compounds because of its ability to form single, double, or triple covalent bonds with itself and many different elements.

✔ Functional (R) groups are specific arrangements of organic molecules that confer distinct properties, including chemical reactivity, to organic compounds.

✔ Macromolecules are very large organic molecules (polymers) built up by polymerization of smaller molecular subunits (monomers).

✔ Carbohydrates are biological molecules whose monomers are simple sugars (monosaccharides) linked together by glycosidic bonds to form polysaccharides. Their main functions are protection and support (in organisms with cell walls), and nutrient and energy stores.

✔ Lipids are biological molecules such as fats that are insoluble in water and contain special ester linkages. Their main functions are cell components, cell secretions, and nutrient and energy stores.

23. What atoms must be present in a molecule for it to be considered organic?
24. Name several inorganic compounds.
25. What characteristics of carbon make it ideal for the formation of organic compounds?
26. What are functional groups?
27. Differentiate between a monomer and a polymer. How are polymers formed?
28. What is the structure of carbohydrates and glycosidic bonds?
29. Differentiate between mono-, di-, and polysaccharides, and give examples of each.
30. What are some of the functions of polysaccharides in cells?
31. Draw simple structural molecules of triglycerides and phospholipids to compare their differences and similarities. What is an ester bond?

2.7 Molecules of Life: Proteins

25. Describe the structures of peptides and polypeptides and how their bonds form.

26. Characterize the four levels of protein structure and describe the pattern of folding.

27. Summarize some of the essential functions of proteins.

The predominant organic molecules in cells are **proteins,** a fitting term adopted from the Greek word *proteios,* meaning first or prime. To a large extent, the structure, behavior, and unique qualities of each living thing are a consequence of the proteins they contain. To

best explain the origin of the special properties and versatility of proteins, we must examine their general structure.

The building blocks of proteins are **amino acids,** which exist in 20 different naturally occurring forms **(table 2.5).** Various com-binations of these amino acids account for the nearly infinite variety of proteins. Amino acids have a basic skeleton consisting of a carbon (called the α carbon) linked to an amino group (NH_2), a carboxyl group (COOH), a hydrogen atom (H), and a variable

TABLE 2.5 Twenty Amino Acids and Their Abbrevations*

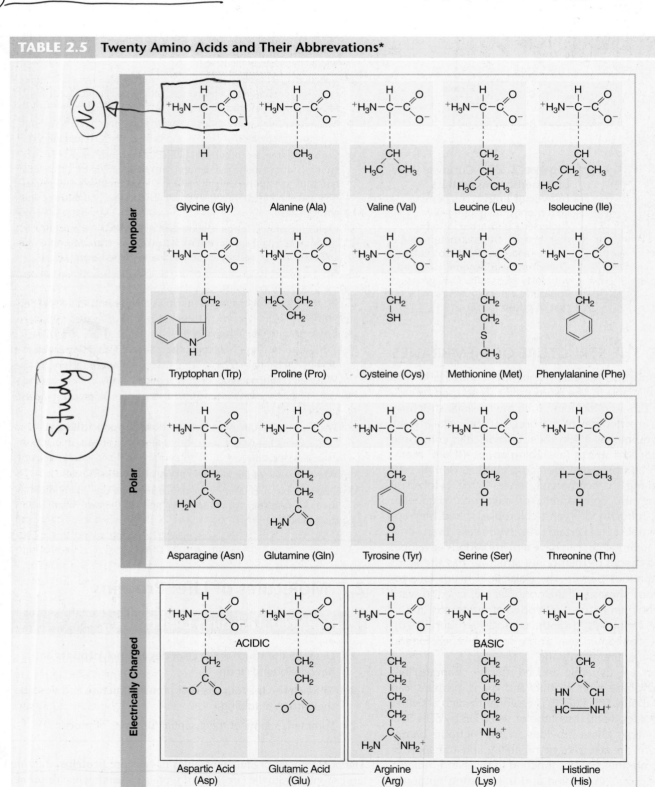

*The basic skeleton is in yellow; R groups are in purple, blue, or green, depending on the nature of their composition.

Bond forming

Figure 2.21 The formation of peptide bonds in a tetrapeptide.

R group. The variations among the amino acids occur at the R group, which is different in each amino acid and imparts the unique characteristics to the molecule and to the proteins that contain it. A covalent bond called a **peptide bond** forms between the amino group on one amino acid and the carboxyl group on another amino acid. As a result of peptide bond formation, it is possible to produce molecules varying in length from two amino acids to chains containing thousands of them.

Various terms are used to denote the nature of compounds containing peptide bonds. **Peptide*** usually refers to a molecule composed of short chains of amino acids, such as a dipeptide (two amino acids), a tripeptide (three), and a tetrapeptide (four) (**figure 2.21**). A **polypeptide** contains an unspecified number of amino acids but usually has more than 20 and is often a smaller sub-unit of a protein. A protein is the largest of this class of compounds and usually contains a minimum of 50 amino acids. It is common for the terms *polypeptide* and *protein* to be used interchangeably, though not all polypeptides are large enough to be considered proteins. In chapter 9, we see that protein synthesis is not just a random connection of amino acids; it is directed by information provided in DNA.

Protein Structure and Diversity

The reason that proteins are so varied and specific is that they do not function in the form of a simple straight chain of amino acids (called the primary structure). A protein has a natural tendency to assume more complex levels of organization, called the secondary, tertiary, and quaternary structures (**figure 2.22**).

The **primary (1°) structure** of a protein is the fundamental chain of amino acids just described, but proteins vary extensively in the exact order, type, and number of amino acids. It is this quality that gives rise to the unlimited diversity in protein form and function.

A polypeptide does not remain in its primary state, but instead, it spontaneously arranges itself into a higher level of complexity

called its **secondary (2°) structure.** The secondary state arises from numerous hydrogen bonds occurring between the C=O and N—H groups of peptide bonds. This bonding causes the whole chain to coil or fold into regular patterns. The coiled spiral form is called the α helix, and the folded, accordion form is called the β-pleated sheet. Polypeptides ordinarily will contain both types of configurations.

Once a chain has assumed the secondary structure, it goes on to form yet another level of folding and compacting—the **tertiary (3°) structure.** This structure arises through additional intrachain[8] forces and bonds between various parts of the α helix and β-pleated sheets. The chief actions in creating the tertiary structure are additional hydrogen bonds between charged functional groups, van der Waals forces between various parts of the polypeptide, and covalent disulfide bonds. The disulfide bonds occur between sulfur atoms on the amino acid cysteine,* and these bonds confer a high degree of stability to the overall protein structure. The result is a complex, three-dimensional protein that is now the completed functional state in many cases.

The most complex proteins assume a **quaternary (4°) structure,** in which two or more polypeptides interact to form a large, multiunit protein. The polypeptide units form loose associations based on weak van der Waals and other forces. The polypeptides in proteins with quaternary structure can be the same or different. The arrangement of these individual polypeptides tends to be symmetrical and will dictate the exact form of the finished protein (**figure 2.22, step 4**).

The most important outcome of bonding and folding is that each different type of protein develops a unique shape, and its surface displays a distinctive pattern of pockets and bumps. As a result, a protein can react only with molecules that complement or fit its particular surface features. Such a degree of specificity can provide the functional diversity required for many thousands of different cellular activities. **Enzymes** serve as the catalysts for all chemical reactions in cells, and nearly every reaction requires a different enzyme (see chapter 8). **Antibodies** are complex glycoproteins with specific regions of attachment for bacteria, viruses, and other microorganisms. Certain bacterial toxins (poisonous proteins) react with only one specific organ or tissue. Proteins embedded in the cell membrane have reactive sites restricted to a certain nutrient. Some proteins function as *receptors* to receive stimuli from the environment.

The functional, three-dimensional form of a protein is termed the *native state,* and if it is disrupted by some means, the protein is said to be *denatured.* Such agents as heat, acid, alcohol, and some disinfectants disrupt (and thus denature) the stabilizing intrachain bonds and cause the molecule to become nonfunctional, as described in chapter 11.

8. **Intra**chain means within the chain; **inter**chain would be between two chains.

* *cysteine* (sis´-tuh-yeen) Gr. *Kystis,* sac. An amino acid first found in urine stones.

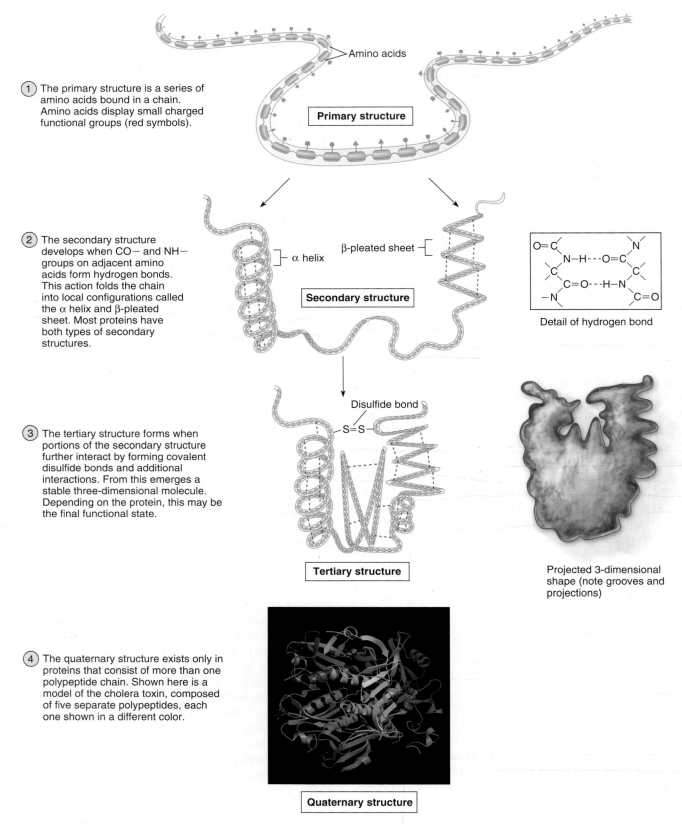

1. The primary structure is a series of amino acids bound in a chain. Amino acids display small charged functional groups (red symbols).

Amino acids

Primary structure

2. The secondary structure develops when CO— and NH— groups on adjacent amino acids form hydrogen bonds. This action folds the chain into local configurations called the α helix and β-pleated sheet. Most proteins have both types of secondary structures.

α helix

β-pleated sheet

Secondary structure

Detail of hydrogen bond

3. The tertiary structure forms when portions of the secondary structure further interact by forming covalent disulfide bonds and additional interactions. From this emerges a stable three-dimensional molecule. Depending on the protein, this may be the final functional state.

Disulfide bond

S=S

Tertiary structure

Projected 3-dimensional shape (note grooves and projections)

4. The quaternary structure exists only in proteins that consist of more than one polypeptide chain. Shown here is a model of the cholera toxin, composed of five separate polypeptides, each one shown in a different color.

Quaternary structure

Process Figure 2.22 Formation of structural levels in a protein.

2.8 The Nucleic Acids: A Cell Computer and Its Programs

28. Identify a nucleic acid and differentiate between DNA and RNA.

29. Describe the structures of nucleotides and list the nitrogen bases.

30. Explain how the DNA code may be copied, and describe the basic functions of RNA.

The nucleic acids, **deoxyribonucleic*** **acid (DNA)** and **ribonucleic*** **acid (RNA),** were originally isolated from the cell nucleus. Shortly thereafter, they were also found in other parts of nucleated cells, in cells with no nuclei (bacteria), and in viruses. The universal occurrence of nucleic acids in all known cells and viruses emphasizes their important roles as informational molecules. DNA, the master computer of cells, contains a special coded genetic program with detailed and specific instructions for each organism's heredity. It transfers the details of its program to RNA, "helper" molecules responsible for carrying out DNA's instructions and translating the DNA program into proteins that can perform life functions. For now, let us briefly consider the structure and some functions of DNA, RNA, and a close relative, adenosine triphosphate (ATP).

Both nucleic acids are polymers of repeating units called **nucleotides,*** each of which is composed of three smaller units: a **nitrogen base,** a pentose (5-carbon) sugar, and a *phosphate* (**figure 2.23a**). The nitrogen base is a cyclic compound that comes in two forms: *purines* (two rings) and *pyrimidines* (one ring). There are two types of purines—**adenine (A)** and **guanine (G)**—and three types of pyrimidines—**thymine (T), cytosine (C),** and **uracil (U) (figure 2.24).** A characteristic that differentiates DNA from RNA is that DNA contains all of the nitrogen bases except uracil, and RNA contains all of the nitrogen bases except thymine. The nitrogen base is covalently bonded to the sugar *ribose* in RNA and *deoxyribose* (because it has one fewer oxygen than ribose) in DNA. Phosphate (PO_4^{3-}), a derivative of phosphoric acid (H_3PO_4), provides the final covalent bridge that connects sugars in series. Thus, the backbone of a nucleic acid strand is a chain of alternating phosphate-sugar-phosphate-sugar molecules, and the nitrogen bases branch off the side of this backbone (**figure 2.23b, c**).

The Double Helix of DNA

DNA is a huge molecule formed by two very long polynucleotide strands linked along their length by hydrogen bonds between complementary pairs of nitrogen bases. The pairing of the nitrogen bases occurs according to a predictable pattern: Adenine ordinarily pairs with thymine, and cytosine with guanine. The bases are

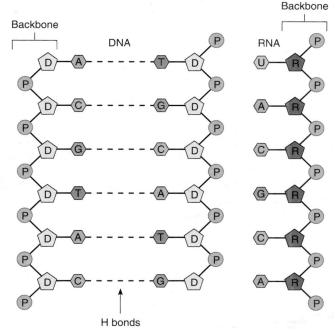

(a) A nucleotide, composed of a phosphate, a pentose sugar, and a nitrogen base (either A,T,C,G, or U) is the monomer of both DNA and RNA.

(b) In DNA, the polymer is composed of alternating deoxyribose (D) and phosphate (P) with nitrogen bases (A,T,C,G) attached to the deoxyribose. DNA almost always exists in pairs of strands, oriented so that the bases are paired across the central axis of the molecule.

(c) In RNA, the polymer is composed of alternating ribose (R) and phosphate (P) attached to nitrogen bases (A,U,C,G), but it is usually a single strand.

Figure 2.23 **The general structure of nucleic acids.**

attracted in this way because each pair shares oxygen, nitrogen, and hydrogen atoms exactly positioned to align perfectly for hydrogen bonds (**figure 2.25**).

For ease in understanding the structure of DNA, it is sometimes compared to a ladder, with the sugar-phosphate backbone representing the rails and the paired nitrogen bases representing the steps. The flat ladder is useful for understanding basic components and orientation, but in reality, DNA exists in a three-dimensional arrangement called a *double helix*. A better analogy may be a spiral staircase. In this model, the two strands (helixes) coil together, with the sugar-phosphate forming outer ribbons, and the paired bases sandwiched between them (figure 2.25). As is true of protein, the structure of DNA is intimately related to its function. DNA molecules are usually extremely long, a feature that satisfies a requirement for storing genetic information in the sequence of base pairs the molecule contains. The hydrogen bonds between pairs can be disrupted when DNA is being copied, and the fixed complementary base pairing is essential to maintain the genetic code.

* *deoxyribonucleic* (dee-ox″-ee-ry″-boh-noo-klay′-ik).

* *ribonucleic* (ry″-boh-noo-klay′-ik) It is easy to see why the abbreviations are used!

* *nucleotide* (noo′-klee-oh-tyd) From nucleus and acid.

Figure 2.24 The sugars and nitrogen bases that make up DNA and RNA. (a) DNA contains deoxyribose, and RNA contains ribose. (b) A and G purine bases are found in both DNA and RNA. (c) Pyrimidine bases are found in both DNA and RNA, but T is found only in DNA, and U is found only in RNA.

Figure 2.25 A structural representation of the double helix of DNA. At the bottom are the details of hydrogen bonds between the nitrogen bases of the two strands.

Making New DNA: Passing on the Genetic Message

The biological properties of cells and viruses are ultimately programmed by a master code composed of nucleic acids. This code is in the form of DNA in all cells and many viruses; a number of viruses are based on RNA alone. Regardless of the exact genetic makeup, both cells and viruses can continue to exist only if they can duplicate their genetic material and pass it on to subsequent generations. **Figure 2.26** summarizes the main steps in this process in cells.

During its division cycle, the cell has a mechanism for making a copy of its DNA by **replication,*** using the original strand as a pattern (figure 2.26). Note that replication is guided by the double-stranded nature of DNA and the precise pairing of bases that create the master code. Replication requires the separation of the double strand into two single strands by an enzyme that helps to split the hydrogen bonds

along the length of the molecule. This event exposes the base code and makes it available for copying. Free nucleotides are used to synthesize matching strands that complement the bases in the code by adhering to the pairing requirements of A—T and C—G. The end result is two separate double strands with the same order of bases as the original molecule. With this type of replication, each new double strand contains one of the original single strands from the starting DNA.

RNA: Organizers of Protein Synthesis

Like DNA, RNA consists of a long chain of nucleotides. However, RNA is a single strand containing ribose sugar instead of deoxyribose and uracil instead of thymine (see figure 2.23). Several functional types of RNA are formed using the DNA template through a replicationlike process. Three major types of RNA are important

* *replication* (reh″-plih-kay′-shun) A process that makes an exact copy.

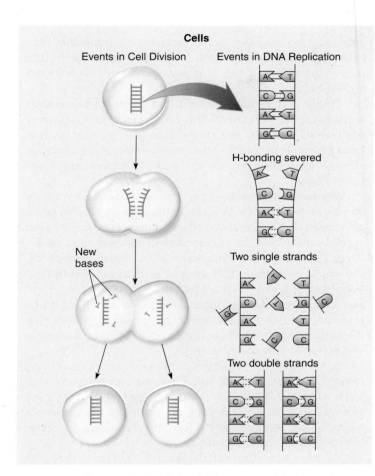

Figure 2.26 **Simplified view of DNA replication in cells.** The DNA in the cell's chromosome must be duplicated as the cell is dividing. This duplication is accomplished through the separation of the double DNA strand into two single strands. New strands are then synthesized using the original strands as guides to assemble the correct new complementary bases.

Figure 2.27 **Model of an ATP molecule, the chemical form of energy transfer in cells. (a)** Structural formula: The wavy lines connecting the phosphates represent bonds that release large amounts of energy when broken. **(b)** Ball-and-stick model shows the arrangement of atoms in three dimensions.

for protein synthesis. Messenger RNA (mRNA) is a copy of a gene from DNA that provides instructions for the order of amino acids; transfer RNA (tRNA) is a carrier that delivers the correct amino acids during protein synthesis; and ribosomal RNA (rRNA) is a major component of ribosomes (described in chapter 4). More information on these important processes is presented in chapter 9.

ATP: The Energy Molecule of Cells

A relative of RNA involved in an entirely different cell activity is **adenosine triphosphate (ATP).** ATP is a nucleotide containing adenine, ribose, and three phosphates rather than just one **(figure 2.27).** It belongs to a category of high-energy compounds (also including guanosine triphosphate, GTP) that give off energy when the bond is broken between the second and third (outermost) phosphate. The presence of these high-energy bonds makes it possible for ATP to release and store energy for cellular chemical reactions. Breakage of the bond of the terminal phosphate releases energy to do cellular work and also generates adenosine diphosphate (ADP). ADP can be converted back to ATP when the third phosphate is restored, thereby serving as an energy depot. Carriers for oxidation-

reduction activities (nicotinamide adenine dinucleotide [NAD], for instance) are also derivatives of nucleotides (see chapter 8).

Check & Assess Sections 2.7 and 2.8

✔ Proteins are biological molecules whose polymers are chains of amino acid monomers linked together by peptide bonds.

✔ Proteins are called the "shapers of life" because of the many biological roles they play in cell structure and cell metabolism.

✔ Protein structure determines protein function. The primary structure is dictated by amino acid composition. Proteins undergo increased levels of folding and complexity, due to internal bonds, called the secondary, tertiary, and quaternary structures. The final level retains a particular shape that dictates its exact function.

✔ Nucleic acids are biological molecules whose polymers are chains of nucleotide monomers linked together by phosphate–pentose sugar covalent bonds.

✔ Double-stranded nucleic acids such as DNA are linked together by hydrogen bonds.

✔ Nucleic acids are information molecules that direct cell metabolism and reproduction.

✔ Nucleotides such as ATP also serve as energy transfer molecules in cells.

32. Describe the basic structure of an amino acid and the formation of a peptide bond.

33. Differentiate between a peptide, a polypeptide, and a protein.

34. Explain what causes the various levels of structure of a protein molecule.

35. What functions do proteins perform in a cell?

36. a. Describe a nucleotide and a polynucleotide, and compare and contrast the general structure of DNA and RNA.

 b. Name the two purines and the three pyrimidines.

37. What is the function of RNA?

38. What is ATP, and what is its function in cells?

CASE FILE 2 PERSPECTIVE

Water is composed of two hydrogens covalently bonded to a single oxygen. The nature of the bonding makes water molecules polar; that is, having sides with opposite charges. In its liquid form, its polarity makes it an excellent solvent for dissolving ions and numerous biological compounds. Cells require molecules dissolved in solution for most chemical reactions, including membrane function, digestion, synthesis, and other metabolic activities.

It is now clear that Mars has abundant water but mainly in the ice and vapor forms. A major unknown is whether any significant liquid water exists. There are surface features that point to a past history of liquid water, but right now, it appears that the ice converts directly into water vapor when exposed to increasing temperatures. Some locations on the planet may have "hot spots" deeper in the crust where liquid water exists part of the time. It is here that any remnants of life are most likely to occur.

Carbon is a versatile element that can make bonds with numerous other atoms, including other carbon atoms. It forms compounds with elongate chains, side chains, and rings that make it possible to construct the complex macromolecules such as protein and DNA that are so crucial to functions and structures in cells.

Inorganic carbon in the form of carbon dioxide and carbonates are known to occur in places in the solar system that could not support life. So, even though these substances may be involved in cellular reactions such as photosynthesis and respiration, nonliving processes may also create them. With organic compounds, something as simple as methane may also be produced by nonliving reactions, and it is not a firm indicator of life. More complex compounds such as sugars and amino acids tend to be associated with life functions and are more reliable evidence. The discovery of intact proteins or nucleic acids would be a chemical signature that truly points to life.

Other bioelements that would ordinarily be a part of living chemistry are phosphorus, sulfur, nitrogen, magnesium, iron, sodium, potassium, calcium, and chloride. All of these have been tested for and found to be present in the Martian atmosphere or soil. So, chemically at least, the major participants for life as we know it exist on Mars.

For more information on the search for life, access the NASA Astrobiology Institute, NASA Mission to Mars, or NASA Exploration Program websites.

Chapter Summary with Key Terms

2.1 Atoms: Fundamental Building Blocks of All Matter in the Universe

A. Atomic Structure and Elements

1. All **matter** in the universe is composed of minute particles called **atoms**—the simplest form of matter not divisible into a simpler substance by chemical means. Atoms are composed of smaller particles called **protons, neutrons,** and **electrons.**

2. a. Protons are positively (+) charged, neutrons are without charge, and electrons are negatively (−) charged.

 b. Protons and neutrons form the **nucleus** of the atom.

 c. Electrons orbit the nucleus in energy shells.

3. Atoms that differ in numbers of the protons, neutrons, and electrons are elements. **Elements** can be described by **mass number (MN),** equal to the number of protons and neutrons it has, and **atomic number (AN),** the number of protons in the nucleus, and each is known by a distinct name and symbol. Elements may exist in variant forms called **isotopes.** The **atomic mass** or **weight** is equal to the average of the mass numbers.

2.2 Bonds and Molecules

A. Atoms interact to form **chemical bonds** and **molecules.** If the atoms combining to make a molecule are different elements, then the substance is termed a **compound.**

1. The type of bond is dictated by the electron makeup of the outer orbitals **(valence)** of the atoms. Bond types include:

2. **Covalent bonds,** with shared electrons. The molecule shares the electrons; the balance of charge will be **polar** if unequal or **nonpolar** if equally shared/electrically neutral.

3. **Ionic bonds,** where electrons are transferred to an atom that can come closer to filling up the outer orbital. Dissociation of these compounds leads to the formation of charged **cations** and **anions.**

4. **Hydrogen bonds** involve weak attractive forces between hydrogen and nearby oxygens and nitrogens.

5. **Van der Waals forces** are also weak interactions between polarized zones of molecules such as proteins.

6. Chemicals termed **reactants** can interact in a way to form different compounds termed **products.** Examples of reactions are synthesis and decomposition.

7. An **oxidation** is a loss of electrons and a **reduction** is a gain of electrons. A substance that causes an oxidation by taking electrons is called an **oxidizing agent,** and a substance that causes a reduction by giving electrons is called a **reducing agent.**

2.3 Chemical Reactions, Solutions, and pH

A. A **solution** is a combination of a solid, liquid, or gaseous chemical (the **solute**) dissolved in a liquid medium (the **solvent**). Water is the most common solvent in natural systems.

B. Ionization of water leads to the release of hydrogen ions (H^+) and hydroxyl (OH^-) ions. The **pH** scale expresses the concentration of H^+ such that a pH of less than 7.0 is considered **acidic,** and a pH of more than that, indicating fewer H^+, is considered **basic (alkaline).**

2.4 The Chemistry of Carbon and Organic Compounds

A. **Biochemistry** studies those molecules that are found in living things. These are based on **organic** compounds, which usually consist of carbon and hydrogen covalently bonded in various combinations. **Inorganic** compounds do not contain both carbon and hydrogen in combination.

B. **Macromolecules** are very large organic compounds and are generally assembled from single units called **monomers** by polymerization. Molecules of life fall into basic categories of carbohydrates, lipids, proteins, and nucleic acids.

2.5 Molecules of Life: Carbohydrates

A. **Carbohydrates** are composed of carbon, hydrogen, and oxygen and contain aldehyde or ketone groups.

1. **Monosaccharides** such as glucose are the simplest carbohydrates with 3 to 7 carbons; these are the monomers of carbohydrates.

2. **Disaccharides** such as lactose consist of two monosaccharides joined by **glycosidic bonds.**

3. **Polysaccharides** such as **starch** and **peptidoglycan** are chains of five or more monosaccharides.

2.6 Molecules of Life: Lipids

A. **Lipids** contain long hydrocarbon chains and are not soluble in polar solvents such as water due to their nonpolar, hydrophobic character. Common components of fats are **fatty acids,** elongate molecules with a carboxylic acid group. Examples are **triglycerides, phospholipids,** sterols, and waxes.

2.7 Molecules of Life: Proteins

A. **Proteins** are highly complex macromolecules that are crucial in most, if not all, life processes.

1. **Amino acids** are the basic building blocks of proteins. They all share a basic structure of an amino group, a carboxyl group, an R group, and hydrogen bonded to a carbon atom. There are 20 different R groups, which define the basic set of 20 amino acids, found in all life forms.

2. A **peptide** is a short chain of amino acids bound by **peptide bonds:** a **protein** contains at least 50 amino acids.

3. The structure of a protein is very important to the function it has. This is described by the **primary structure** (the chain of amino acids), the **secondary structure** (formation of α helixes and β-sheets due to hydrogen bonding within the chain), **tertiary structure** (cross-links, especially disulfide bonds, between secondary structures), and **quaternary structure** (formation of multisubunit proteins). The incredible variation in shapes is the basis for the diverse roles proteins play as **enzymes, antibodies,** receptors, and structural components.

2.8 Molecules of Life: Nucleic Acids—the basis for genetic functions.

A. **Nucleotides** are the building blocks of nucleic acids. They are composed of a **nitrogen base,** a pentose sugar, and a phosphate. Nitrogen bases are ringed compounds: **adenine (A), guanine (G), cytosine (C), thymine (T),** and **uracil (U).** Pentose sugars may be deoxyribose or ribose.

B. **Deoxyribonucleic acid (DNA)** is a polymer of nucleotides that occurs as a double-stranded helix with hydrogen bonding in pairs between the helices. It has all of the bases except uracil, and the pentose sugar is deoxyribose. DNA is the master code for a cell's life processes and must be transmitted to the offspring through **replication.**

C. **Ribonucleic acid (RNA)** is a polymer of nucleotides where the sugar is ribose and the **uracil** is used instead of thymine. It is almost always found single stranded and is used to express the DNA code into proteins.

D. **Adenosine triphosphate (ATP)** contains a nucleotide and is involved in the transfer and storage of energy in cells.

Multiple-Choice Questions

Select the correct answer from the answers provided. For questions with blanks, choose the combination of answers that most accurately completes the statement.

1. The smallest unit of matter with unique characteristics is
 a. an electron c. an atom
 b. a molecule d. a proton

2. The _____ charge of a proton is exactly balanced by the _____ charge of a (an) _____.
 a. negative, positive, electron c. positive, negative, electron
 b. positive, neutral, neutron d. neutral, negative, electron

3. Electrons move around the nucleus of an atom in pathways called
 a. shells c. circles
 b. orbitals d. rings

4. Which parts of an element do not vary in number?
 a. electrons c. protons
 b. neutrons d. All of these vary.

5. If a substance contains two or more elements of different types, it is considered
 a. a compound c. a molecule
 b. a monomer d. organic

6. Bonds in which atoms share electrons are defined as _____ bonds.
 a. hydrogen c. double
 b. ionic d. covalent

7. A hydrogen bond can form between _____ adjacent to each other.
 a. two hydrogen atoms
 b. two oxygen atoms
 c. a hydrogen atom and an oxygen atom
 d. negative charges

8. An atom that can donate electrons during a reaction is called
 a. an oxidizing agent c. an ionic agent
 b. a reducing agent d. an electrolyte

9. In a solution of NaCl and water, NaCl is the _____ and water is the _____.
 a. acid, base c. solute, solvent
 b. base, acid d. solvent, solute

10. A solution with a pH of 2 _____ than a solution with a pH of 8.
 a. has less H^+
 b. has more H^+
 c. has more OH^-
 d. is less concentrated

11. Fructose is a type of
 a. disaccharide
 b. monosaccharide
 c. polysaccharide
 d. amino acid

12. Bond formation in polysaccharides and polypeptides is accompanied by the removal of a
 a. hydrogen atom
 b. hydroxyl ion
 c. carbon atom
 d. water molecule

13. The monomer unit of polysaccharides such as starch and cellulose is
 a. fructose
 b. glucose
 c. ribose
 d. lactose

14. A phospholipid contains
 a. three fatty acids bound to glycerol
 b. three fatty acids, a glycerol, and a phosphate
 c. two fatty acids and a phosphate bound to glycerol
 d. three cholesterol molecules bound to glycerol

15. Proteins are synthesized by linking amino acids with _____ bonds.
 a. disulfide
 b. glycosidic
 c. peptide
 d. ester

16. The amino acid that accounts for disulfide bonds in the tertiary structure of proteins is
 a. tyrosine
 b. glycine
 c. cysteine
 d. serine

17. DNA is a hereditary molecule that is composed of
 a. deoxyribose, phosphate, and nitrogen bases
 b. deoxyribose, a pentose, and nucleic acids
 c. sugar, proteins, and thymine
 d. adenine, phosphate, and ribose

18. What is meant by the term *DNA replication?*
 a. synthesis of nucleotides
 b. cell division
 c. interpretation of the genetic code
 d. the exact copying of the DNA code into two new molecules

19. Proteins can function as
 a. enzymes
 b. receptors
 c. antibodies
 d. a, b, and c

20. RNA plays an important role in what biological process?
 a. replication
 b. protein synthesis
 c. lipid metabolism
 d. water transport

Case File Questions

1. Which of the following has not been an objective of Mars exploration?
 a. to identify plants
 b. to find fossils
 c. to detect organic compounds
 d. to detect CO_2

2. What was a significant result of the Mars *Phoenix* project?
 a. culturing bacteria
 b. verifying water
 c. finding organic matter
 d. discovering precious metals

Writing to Learn

These questions are suggested as a *writing-to-learn* experience. For each question, compose a one- or two-paragraph answer that includes the factual information needed to completely address the question. Note: The assess questions can also serve as a writing to learn assignment.

1. Explain this statement: "All compounds are molecules, but not all molecules are compounds." Give an example.

2. What causes atoms to form chemical bonds? Why do some elements not bond readily?

3. Why are some covalent molecules polar and others nonpolar?

4. Explain why some elements are diatomic.

5. Exactly what causes the charges to form on atoms in ionic bonds?

6. Why are hydrogen bonds relatively weak?

7. What kind of substances will be expected to be hydrophilic and hydrophobic, and what makes them so?

8. How can a neutral salt be formed from acids and bases?

9. a. Draw the atomic structure of magnesium and predict what kinds of bonds it will make.
 b. What kind of ion would you expect magnesium to make on the basis of its valence?

10. a. What characteristic of phospholipids makes them essential components of cell membranes?
 b. How are saturated and unsaturated fatty acids different?
 c. Why is the hydrophilic end of phospholipids attracted to water?

11. What makes the amino acids distinctive, and how many of them are there?

12. Why is DNA called a double helix?

13. Describe what occurs in a dehydration synthesis reaction.

14. How is our understanding of microbiology enhanced by a knowledge of chemistry?

Concept Mapping

Appendix E provides guidance for working with concept maps.

1. Supply your own linking words or phrases in this concept map, and provide the missing concepts in the empty boxes.

Critical Thinking Questions

Critical thinking is the ability to reason and solve problems using facts and concepts. These questions can be approached from a number of angles, and in most cases, they do not have a single correct answer.

1. a. The "octet rule" in chemistry helps predict the types of bonds that atoms will form. In general, an atom will be most stable if it fills its outer shell of 8 electrons. Atoms with fewer than 4 valence electrons tend to donate electrons and those with more than 4 valence electrons tend to accept additional electrons; those with exactly 4 can do both. Using this rule, determine what category each of the following elements falls into: N, S, C, P, O, H, Ca, Fe, and Mg. (You will need to work out the valence of the atoms.)
 b. Make simple drawings to show how atoms such as N and Cl form diatomic molecules.
 c. What type of chemical reaction is occurring between Na and Cl_2?

2. Predict the kinds of bonds that occur in ammonium (NH_3), phosphate (PO_4), disulfide (S—S), and magnesium chloride ($MgCl_2$). (Use simple models, such as those in figure 2.4.) *pg. 32*

3. Work out the following problems:
 a. What is the number of protons in helium? in iron?
 b. Will an H bond form between H_3C—CH=O and H_2O? Draw a simple figure to support your answer.
 c. Draw the following molecules and determine which are polar: Cl_2, NH_3, CH_4.
 d. What is the pH of a solution with a concentration of 0.00001 moles/ml (M) of H^+?
 e. What is the pH of a solution with a concentration of 0.00001 moles/ml (M) of OH^-?

4. a. Describe how hydration spheres are formed around cations and anions. *pg. 37*
 b. Distinguish between polar and ionic compounds. *pg 38*

5. In what way are carbon-based compounds like children's Tinker Toys or Lego blocks?

6. Is galactose an aldehyde or a ketone sugar? *pg. 42*

7. a. How many water molecules are released when a triglyceride is formed?
 b. How many peptide bonds are in a tetrapeptide?

8. Looking at figure 2.25, can you see why adenine forms hydrogen bonds with thymine and why cytosine forms them with guanine? *pg 52*

9. Saturated fats are solid at room temperature and unsaturated fats are not.
 a. Is butter an example of a saturated or an unsaturated fat?
 b. Is olive oil an example of a saturated or an unsaturated fat?
 c. Explain why sterols like cholesterol can add "stiffness" to membranes that contain them.

Visual Challenge

1. Figures 1 and 2 are both highly magnified views of biological substances. Using **figure 2.17** as your basis for comparison, speculate which molecules are shown and give the reasons for them having the microscopic appearance we see here.

(1) (200x) (2)

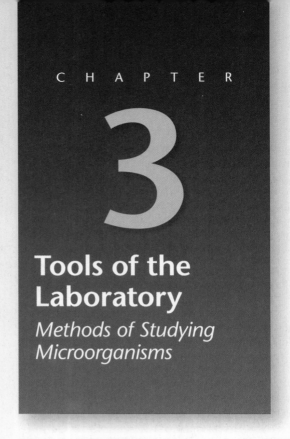

"A matter of life or death"

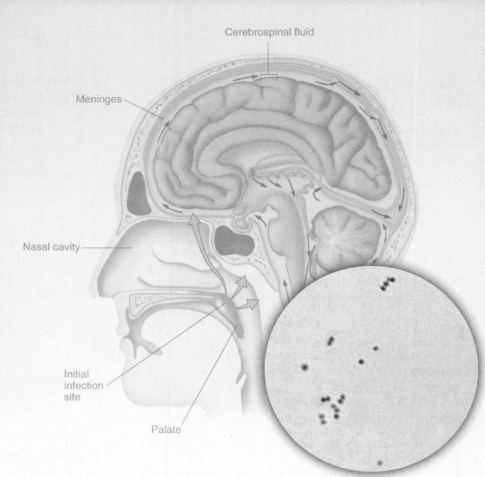

The meningococcus: A million of these tiny culprits could fit on the head of a pin, yet they can knock out a healthy adult in a few hours.

CASE FILE **3** **Battling a Brain Infection**

One Saturday evening in 2007, a 50-year-old woman began to suffer flulike symptoms, with fever, aching joints, sore throat, and a headache. Feeling miserable but not terribly concerned, she took some ibuprofen and went to bed. By the following morning, she began to feel increasingly ill and was unstable on her feet, confused, and complaining of light-headedness. Realizing this was more than just the flu, her husband rushed her immediately to the nearest emergency room.

An initial examination showed that most of her vital signs were normal. Conditions that may provide some clues were: rapid pulse and respiration, an inflamed throat, and a stiff neck. A chest X ray revealed no sign of pneumonia, and a blood test indicated an elevated white blood cell count. To rule out a possible brain infection, a puncture of the spinal canal was performed. As it turned out, the cerebrospinal fluid (CSF) the technician extracted appeared normal, microscopically and macroscopically.

Within an hour, she began to drift in and out of consciousness and was extremely lethargic. At one point, the medical team could not find a pulse and noticed dark brown spots developing on her legs. When her condition appeared to be deteriorating rapidly, she was immediately taken to the intensive care unit and placed on intravenous antibiotics. One of the emergency doctors was overheard saying, "Her medical situation was so critical that our intervention was truly a matter of life or death."

Because her symptoms pointed to a possible infection of the central nervous system, a second spinal puncture was performed. This time, the spinal fluid looked cloudy. A Gram stain was performed right away, and cultures were started.

▶ *What are signs and symptoms of disease? Give examples from the case that appear to be the most diagnostically significant.*

▶ *Why is so much importance placed on the CSF and its appearance?*

To continue the case, go to page 74.

3.1 Methods of Microbial Investigation

Expected Learning Outcomes

1. Explain what unique characteristics of microorganisms make them challenging subjects for study.

2. Briefly outline the processes and purposes of the six types of procedures that are used in handling, maintaining, and studying microorganisms.

Biologists studying large organisms such as animals and plants can, for the most part, immediately see and differentiate their experimental subjects from the surrounding environment and from one another. In fact, they can use their senses of sight, smell, hearing, and even touch to detect and evaluate identifying characteristics and to keep track of growth and developmental changes.

Because microbiologists cannot rely as much as other scientists on senses other than sight, they are confronted by some unique problems. First, most habitats (such as the soil and the human mouth) harbor microbes in complex associations. It is often necessary to separate the organisms from one another so they can be identified and studied. Second, to maintain and keep track of such small research subjects, microbiologists usually will need to grow them under artificial conditions. A third difficulty in working with microbes is that they are not visible to the naked eye. This, coupled with their wide distribution means that undesirable ones can be inconspicuously introduced into an experiment, where they may cause misleading results.

To deal with the challenges of their tiny and sometimes elusive targets, microbiologists have developed several types of procedures for investigating and characterizing microorganisms. These techniques can be summed up succinctly as the **six "I's": inoculation, incubation, isolation, inspection, information gathering,** and **identification,** so-called because they all begin with the letter "I". The main features of the 6 "I's" are laid out in **figure 3.1** and **table 3.1.** Depending on the purposes of the particular investigator, these may be performed in several combinations and orders, but together, they serve as the major tools of the microbiologist's trade. As novice microbiologists, most of you will be learning some basic microscope, inoculation, culturing, and identification techniques. Many professional researchers use more advanced investigation and identification techniques that may not even require growth or absolute isolation of the microbe in culture (see Insight 3.3). If you examine the example from the case file in chapter 1, you will notice that the researchers used only some of the 6 "I's", whereas the case file in this chapter includes all of them in some form.

The first three sections of this chapter cover some of the essential concepts that revolve around the 6 "I's", but not necessarily in the exact order presented in figure 3.1. Microscopes are so important to microbiological inquiry that we start out with the subject of microscopes, magnification, and staining techniques. This is followed by culturing procedures and media.

Check & Assess Section 3.1

✔ The small size and ubiquity of microorganisms make laboratory management and study of them difficult.

✔ The six "I's"—inoculation, incubation, isolation, inspection, information gathering, and identification—comprise the major kinds of laboratory procedures used by microbiologists.

1. Name the notable features of microorganisms that have created a need for the specialized tools of microbiology.

2. In one sentence, briefly define what is involved in each of the six "I's".

3.2 The Microscope: Window on an Invisible Realm

Expected Learning Outcomes

3. Describe the basic plan of an optical microscope, and differentiate between magnification and resolution.

4. Explain how the images are formed, along with the role of light and the different powers of lenses.

5. Indicate how the resolving power is determined and how resolution affects image visibility.

6. Differentiate between the major types of optical microscopes, their illumination sources, image appearance, and uses.

7. Describe the operating features of electron microscopes and how they differ from optical microscopes in illumination source, magnification, resolution, and image appearance.

8. Differentiate between transmission and scanning electron microsopes in image formation and appearance.

9. Explain the basic differences between fresh and fixed preparations for microscopy and how they are used.

10. Define dyes and describe the basic chemistry behind the process of staining.

11. Differentiate between negative and positive staining, giving examples.

12. Distinguish between simple, differential, and structural stains, including their applications.

13. Describe the process of Gram staining and how its results can aid the identification process.

Imagine **Leeuwenhoek's** excitement and wonder when he first viewed a drop of rainwater and glimpsed an amazing microscopic world teeming with unearthly creatures. Beginning microbiology students still experience this sensation, and even experienced microbiologists remember their first view. The microbial existence is indeed another world, but it would remain largely uncharted without an essential tool: the microscope. Your efforts in exploring

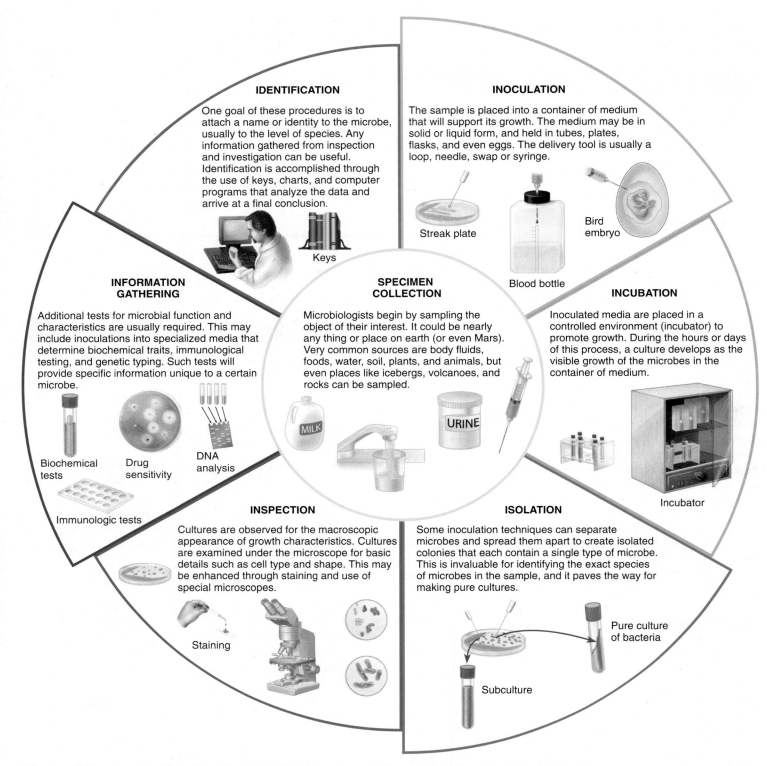

IDENTIFICATION

One goal of these procedures is to attach a name or identity to the microbe, usually to the level of species. Any information gathered from inspection and investigation can be useful. Identification is accomplished through the use of keys, charts, and computer programs that analyze the data and arrive at a final conclusion.

Keys

INOCULATION

The sample is placed into a container of medium that will support its growth. The medium may be in solid or liquid form, and held in tubes, plates, flasks, and even eggs. The delivery tool is usually a loop, needle, swap or syringe.

Streak plate

Blood bottle

Bird embryo

INFORMATION GATHERING

Additional tests for microbial function and characteristics are usually required. This may include inoculations into specialized media that determine biochemical traits, immunological testing, and genetic typing. Such tests will provide specific information unique to a certain microbe.

Biochemical tests

Drug sensitivity

DNA analysis

Immunologic tests

SPECIMEN COLLECTION

Microbiologists begin by sampling the object of their interest. It could be nearly any thing or place on earth (or even Mars). Very common sources are body fluids, foods, water, soil, plants, and animals, but even places like icebergs, volcanoes, and rocks can be sampled.

MILK

URINE

INCUBATION

Inoculated media are placed in a controlled environment (incubator) to promote growth. During the hours or days of this process, a culture develops as the visible growth of the microbes in the container of medium.

Incubator

INSPECTION

Cultures are observed for the macroscopic appearance of growth characteristics. Cultures are examined under the microscope for basic details such as cell type and shape. This may be enhanced through staining and use of special microscopes.

Staining

ISOLATION

Some inoculation techniques can separate microbes and spread them apart to create isolated colonies that each contain a single type of microbe. This is invaluable for identifying the exact species of microbes in the sample, and it paves the way for making pure cultures.

Pure culture of bacteria

Subculture

Figure 3.1 **An overview of some general laboratory techniques carried out by microbiologists.** "The six "I's". Procedures start at the central "hub" of specimen collection and flow from there to inoculation, incubation, and so on. But not all steps are always performed, nor do they necessarily proceed exactly in this order. Some investigators go right from sampling to microscopic inspection or from sampling to DNA testing. Others may require only inoculation and incubation on special media or test systems. *See Table 3.1 for a brief description of each procedure, its purpose, and intended outcome.

TABLE 3.1 An Overview of Microbiology Techniques

Technique	Process Involves	Purpose and Outcome	See Pages
Inoculation	Placing a sample into a container of medium that supplies nutrients for growth and is the first stage in culturing	To increase visibility; makes it possible to handle and manage microbes in an artificial environment and begin to analyze what the sample may contain	74
Incubation	Exposing the inoculated medium to optimal growth conditions, generally for a few hours to days	To promote multiplication and produce the actual culture. An increase in microbe numbers will provide the higher quantities needed for further testing.	76
Isolation	Methods for separating individual microbes and achieving isolated colonies that can be readily distinguished from one another macroscopically*	To make additional cultures from single colonies to ensure they are pure; that is, containing only a single species of microbe for further observation and testing	75–77
Inspection	Observing cultures macroscopically for appearance of growth and microscopically for appearance of cells	To analyze initial characteristics of microbes in samples; stains of cells may reveal information on cell type and morphology.	61–72
Information Gathering	Testing of cultures with procedures that analyze biochemical and enzyme characteristics, immunologic reactions, drug sensitivity, and genetic makeup	To provide much specific data and generate an overall profile of the microbes. These test results and descriptions will become key determinants in the last category, identification.	78
Identification	Analysis of collected data to help support a final determination of the types of microbes present in the original sample. This is accomplished by a variety of schemes.	This lays the groundwork for further research into the nature and roles of these microbes; it can also provide numerous applications in infection diagnosis, food safety, and potential biotechnology and bioremediation efforts.	76, 78

*Observable to the unaided or naked eye. This term usually applies to the cultural level of study.

microbes will be more meaningful if you understand some essentials of **microscopy*** and specimen preparation.

Magnification and Microscope Design

The two key characteristics of a reliable microscope are **magnification,*** the ability to make objects appear enlarged, and **resolving power,** the ability to show detail.

A discovery by early microscopists that spurred the advancement of microbiology was that a clear, glass sphere could act as a lens to magnify small objects. Magnification in most microscopes results from a complex interaction between visible light waves and the curvature of the lens. When a beam or ray of light transmitted through air strikes and passes through the convex surface of glass, it experiences some degree of **refraction,*** defined as the bending or change in the angle of the light ray as it passes through a medium such as a lens. The greater the difference in the composition of the two substances the light passes between, the more pronounced is the refraction. When an object is placed a certain distance from the spherical lens and illuminated with light, an optical replica, or **image,** of it is formed by the refracted light. Depending upon the size and curvature of the lens, the image appears enlarged to a particular degree, which is called its power of magnification and is usually identified with a number combined with × (read "times"). This behavior of light is evident if one looks through an everyday object such as a glass ball or a magnifying glass **(figure 3.2).** It is

Figure 3.2 Effects of magnification. Demonstration of the magnification and image-forming capacity of clear glass "lenses." Given a proper source of illumination, a magnifying glass can deliver 2 to 10 times magnification.

* *microscopy* (mye-kraw′-skuh-pee) Gr. The science that studies microscope techniques.
* *magnification* (mag′-nih-fih-kay″-shun) L. *magnus,* great, and *ficere,* to make.
* *refract, refraction* (ree-frakt′, ree-frak′-shun) L. *refringere,* to break apart.

Figure 3.3 **The parts of an optical microscope.** This microscope is a compound light microscope with two oculars (called binocular). It has four objective lenses, a mechanical stage to move the specimen, a condenser, an iris diaphragm, and a built-in lamp.

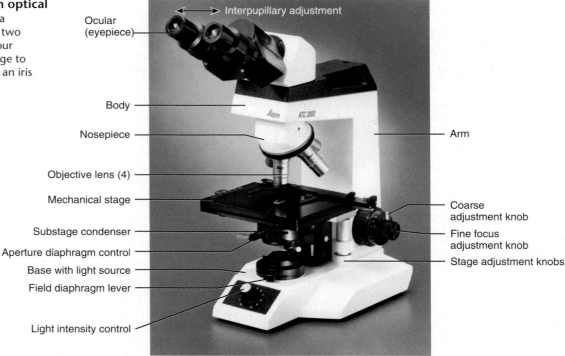

Interpupillary adjustment

Ocular (eyepiece)

Body

Nosepiece

Objective lens (4)

Mechanical stage

Substage condenser

Aperture diaphragm control

Base with light source

Field diaphragm lever

Light intensity control

Arm

Coarse adjustment knob

Fine focus adjustment knob

Stage adjustment knobs

basic to the function of all optical, or light, microscopes, though many of them have additional features that define, refine, and increase the size of the image.

The first microscopes were **simple,** meaning they contained just a single magnifying lens and a few working parts. Examples of this type of microscope are a magnifying glass, a hand lens, and Leeuwenhoek's basic little tool shown earlier in figure 1.9a. Among the refinements that led to the development of today's **compound** (two-lens) **microscope** were the addition of a second magnifying lens system, a lamp in the base to give off visible light and **illuminate*** the specimen, and a special lens called the **condenser** that converges or focuses the rays of light to a single point on the object. The fundamental parts of a modern compound light microscope are illustrated in **figure 3.3.**

Principles of Light Microscopy

To be most effective, a microscope should provide adequate magnification, resolution, and clarity of image. Magnification of the object or specimen by a compound microscope occurs in two phases. The first lens in this system (the one closest to the specimen) is the objective lens, and the second (the one closest to the eye) is the ocular lens, or eyepiece **(figure 3.4).** The objective forms the initial image of the specimen, called the **real image.** When the real image is projected to the plane of the eyepiece, the ocular lens magnifies it to produce a second image, the **virtual image.** The virtual image is the one that will be received by the eye and converted to a retinal and visual image. The magnifying power of the objective alone usually ranges from 4× to 100×, and the power of the ocular alone

ranges from 10× to 20×. The total power of magnification of the final image formed by the combined lenses is a product of the separate powers of the two lenses:

Power of Objective	×	Usual Power of Ocular	=	Total Magnification
4× scanning objective		10×	=	40×
10× low power objective		10×	=	100×
40× high dry objective		10×	=	400×
100× oil immersion objective		10×	=	1,000×

Microscopes are equipped with a nosepiece holding three or more objectives that can be rotated into position as needed. The power of the ocular usually remains constant for a given microscope. Depending on the power of the ocular, the total magnification of standard light microscopes can vary from 40× with the lowest power objective (called the scanning objective) to 2,000× with the highest power objective (the oil immersion objective).

Resolution: Distinguishing Magnified Objects Clearly In addition to magnification, a microscope must also have adequate **resolution,** or **resolving power.** Resolution defines the capacity of an optical system to distinguish two adjacent objects or points from one another. For example, at a distance of 25 cm (10 in), the lens in the human eye can resolve two small objects as separate points just as long as the two objects are no closer than 0.2 mm apart. The eye examination given by optometrists is in fact a test of the resolving power of the human eye for various-size letters read at a distance of 20 feet. Because microorganisms are extremely small and usually very close together, they will not be seen with clarity or any degree of detail unless the microscope's lenses can resolve them.

* *illuminate* (ill-oo′-mih-nayt) L. *illuminatus,* to light up.

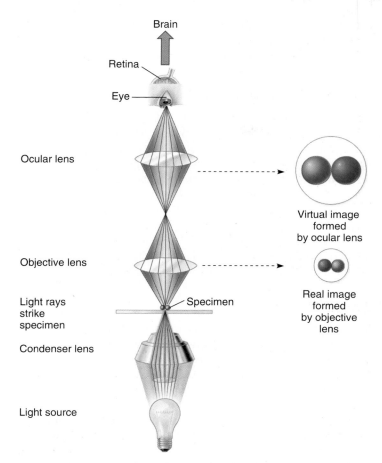

Figure 3.4 **The pathway of light and the two stages in magnification of a compound microscope.** As light passes through the condenser, it is gathered into a tight beam that is focused on the specimen. Light leaving the specimen enters the objective lens and is refracted so as to form an enlarged primary image, the real image. One does not see this image, but its degree of magnification is represented by the lower circle. The real image is projected through the ocular, and a second image, the virtual image, is formed by a similar process. The virtual image is the final magnified image that is received by the lens and retina of the eye and perceived by the brain. Notice that the lens systems reverse the image.

A simple equation in the form of a fraction expresses the main mathematical factors that influence the expression of resolving power.

$$\text{Resolving power (RP)} = \frac{\text{Wavelength of light in nm}}{2 \times \text{Numerical aperture of objective lens}}$$

From this equation, it is evident that the resolving power is a function of the wavelength of light that forms the image, along with certain characteristics of the objective. The light source for optical microscopes consists of a band of colored wavelengths in the visible spectrum. The shortest visible wavelengths are in the violet-blue portion of the spectrum (400 nm), and the longest are in the red portion (750 nm). Because the wavelength must pass between the objects that are being resolved, shorter wavelengths (in the 400–500 nm range) will provide better resolution (**figure 3.5**).

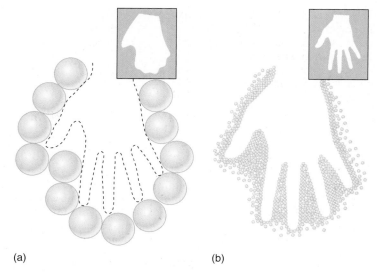

(a) (b)

Figure 3.5 **Effect of wavelength on resolution.** A simple model demonstrates how the wavelength influences the resolving power of a microscope. Here an outline of a hand represents the object being illuminated, and two different-size circles represent the wavelengths of light. In **(a)**, the longer waves are too large to penetrate between the finer spaces and produce a fuzzy, undetailed image. In **(b)**, shorter waves are small enough to enter small spaces and produce a much more detailed image that is recognizable as a hand.

The other factor influencing resolution is the **numerical aperture (NA),** a mathematical constant derived from the physical structure of the lens. This number represents the angle of light produced by refraction and is a measure of the quantity of light gathered by the lens. Each objective has a fixed numerical aperture reading ranging from 0.1 in the lowest power lens to approximately 1.25 in the highest power (oil immersion) lens. Lenses with higher NAs provide better resolving power because they increase the angle of refraction and widen the cone of light entering the lens. For the oil immersion lens to arrive at its maximum resolving capacity, a drop of oil must be inserted between the tip of the lens and the specimen on the glass slide. Because immersion oil has the same optical qualities as glass, it prevents refractive loss that normally occurs as peripheral light passes from the slide into the air; this property effectively increases the numerical aperture (**figure 3.6**). There is an absolute limitation to resolution in optical microscopes, which can be demonstrated by calculating the resolution of the oil immersion lens using a blue-green wavelength of light:

$$\text{RP} = \frac{500 \text{ nm}}{2 \times 1.25}$$

$$= 200 \text{ nm (or } 0.2 \text{ } \mu\text{m)}$$

In practical terms, this calculation means that the oil immersion lens can resolve any cell or cell part as long as it is at least 0.2 μm in diameter and that it can resolve two adjacent objects as long as they are no closer than 0.2 μm (**figure 3.7**). In general, organisms that are 0.5 μm or more in diameter are readily seen. This includes fungi and protozoa and some of their internal structures, and most bacteria. However, a few bacteria and most viruses are far too small to be resolved by the optical microscope and require electron

Figure 3.6 **Workings of an oil immersion lens.** To maximize its resolving power, an oil immersion lens (the one with highest magnification) must have a drop of oil placed at its tip. This transmits a continuous cone of light from the condenser to the objective, thereby increasing the amount of light and, consequently, the numerical aperture. Without oil, some of the peripheral light that passes through the specimen is scattered into the air or onto the glass slide; this scattering decreases resolution.

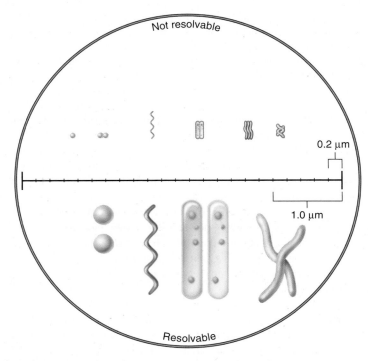

Figure 3.7 **Effect of resolution on image visibility.** Comparison of objects that would not be resolvable versus those that would be resolvable under oil immersion at 1,000× magnification. Note that in addition to differentiating two adjacent things, good resolution also means being able to observe an object clearly.

microscopy (see figure 1.7 and figure 3.13). The factor that most limits the clarity of a microscope's image is its resolving power. Even if a light microscope were designed to magnify several thousand times, its resolving power could not be increased, and the image it produced would be enlarged but blurry.

Despite this limit, small improvements to resolution are possible. One is to place a blue filter over the microscope lamp, keeping the wavelength at the shortest possible value. Another comes from

maximizing the numerical aperture. That is the effect of adding oil to the immersion lens, which effectively increases the NA from 1.25 to 1.4 and improves the resolving power to 0.17 μm.

Variations on the Optical Microscope

Optical microscopes that use visible light can be described by the nature of their *field,* meaning the circular area viewed through the ocular lens. With special adaptations in lenses, condensers, and light sources, four special types of microscopes can be described: bright-field, dark-field, phase-contrast, and interference. A fifth type of optical microscope, the fluorescence microscope, uses ultraviolet radiation as the illuminating source, and a sixth, the confocal microscope, uses a laser beam. Each of these microscopes is adapted for viewing specimens in a particular way, as described in the next sections and summarized in **table 3.2.** Refer back to figure 1.4 for size comparisons of microbes and molecules.

Bright-Field Microscopy

The bright-field microscope is the most widely used type of light microscope. Although we ordinarily view objects such as the words on this page with light reflected off the surface, a bright-field microscope forms its image when light is transmitted through the specimen. The specimen, being denser and more opaque than its surroundings, absorbs some of this light, and the rest of the light is transmitted directly up through the ocular into the field. As a result, the specimen will produce an image that is darker than the surrounding brightly illuminated field. The bright-field microscope is a multipurpose instrument that can be used for both live, unstained material and preserved, stained material. The bright-field image is compared with that of other microscopes in **figure 3.8.**

Dark-Field Microscopy

A bright-field microscope can be adapted as a dark-field microscope by adding a special disc called a *stop* to the condenser. The stop blocks all light from entering the objective lens except peripheral light that is reflected off the sides of the specimen itself. The resulting image is a particularly striking one: brightly illuminated specimens surrounded by a dark (black) field **(figure 3.8*b*).** Some of Leeuwenhoek's more successful microscopes probably operated with dark-field illumination. The most effective use of dark-field microscopy is to visualize living cells that would be distorted by drying or heat, or cannot be stained with the usual methods. It can outline the organism's shape and permit rapid recognition of swimming cells that may appear in fresh specimens, but it does not reveal fine internal details.

Phase-Contrast and Interference Microscopy

If similar objects made of clear glass, ice, and plastic are immersed in the same container of water, an observer would have difficulty telling them apart because they have similar optical properties. In the same way, internal components of a live, unstained cell also lack sufficient contrast to distinguish readily. But cell structures do differ slightly in density, enough that they can alter the light that

TABLE 3.2 **Comparisons of Types of Microscopy**

Microscope	Maximum Practical Magnification	Resolution	Important Features
Visible light as source of illumination			
Bright-field	2,000×	0.2 μm (200 nm)	Common multipurpose microscope for live and preserved stained specimens; specimen is dark, field is white; provides fair cellular detail
Dark-field	2,000×	0.2 μm	Best for observing live, unstained specimens; specimen is bright, field is black; provides outline of specimen with reduced internal cellular detail
Phase-contrast	2,000×	0.2 μm	Used for live specimens; specimen is contrasted against gray background; excellent for internal cellular detail
Differential interference	2,000×	0.2 μm	Provides brightly colored, highly contrasting, three-dimensional images of live specimens
Ultraviolet rays as source of illumination			
Fluorescent	2,000×	0.2 μm	Specimens stained with fluorescent dyes or combined with fluorescent antibodies emit visible light; specificity makes this microscope an excellent diagnostic tool.
Confocal	2,000×	0.2 μm	Specimens stained with fluorescent dyes are scanned by laser beam; multiple images (optical sections) are combined into three-dimensional image by a computer; unstained specimens can be viewed using light reflected from specimen.
Electron beam forms image of specimen			
Transmission electron microscope (TEM)	100,000×	0.5 nm	Sections of specimen are viewed under very high magnification; finest detailed structure of cells and viruses is shown; used only on preserved material
Scanning electron microscope (SEM)	650,000×	10 nm	Scans and magnifies external surface of specimen; produces striking three-dimensional image
Sharp tip probes atomic structure of specimen			
Atomic force microscope (AFM)	100,000,000×	0.01 Angstroms	Tip scans specimen and moves up and down with contour of surface; movement of tip is measured with laser and translated to image
Scanning tunneling microscope (STM)	100,000,000×	0.01 Angstroms	Tip moves over specimen while voltage is applied, generating current that is dependent on distance between tip and surface; atoms can be moved with tip.

passes through them in subtle ways. The phase-contrast microscope contains devices that transform the subtle changes in light waves passing through the specimen into differences in light intensity. For example, thicker cell parts such as organelles alter the pathway of light to a greater extent than thinner regions such as the cytoplasm. Light patterns coming from these regions will vary in contrast. The amount of internal detail visible by this method is greater than by either bright-field or dark-field methods. The phase-contrast microscope is most useful for observing intracellular structures such as bacterial spores, granules, and organelles, as well as the locomotor structures of eukaryotic cells (**figure 3.8c and figure 3.9a**).

Like the phase-contrast microscope, the differential interference contrast (DIC) microscope provides a detailed view of unstained, live specimens by manipulating the light. But this microscope has additional refinements, including two prisms that add contrasting colors to the image and two beams of light rather than a single one. DIC microscopes produce extremely well-defined images that are vividly colored and appear three-dimensional (**figure 3.9b**).

Fluorescence Microscopy

The fluorescence microscope is a specially modified compound microscope furnished with an ultraviolet (UV) radiation source and

(a)

(b)

(c)

Figure 3.8 **Comparing three versions of an image from optical microscopes.** A live cell of *Paramecium* viewed with **(a)** bright-field (400×), **(b)** dark-field (400×), and **(c)** phase-contrast (400×). Note the difference in the appearance of the field and the degree of detail shown by each method of microscopy. Only in phase-contrast are the cilia (fine hairs) on the cells noticeable.

Spores

(a)

(b)

Figure 3.9 **Visualizing internal structures. (a)** Phase-contrast micrograph of clostridial cells containing spores. The denser quality of the spores causes them to appear as bright, shiny objects against the darker cells (600×). **(b)** Differential interference micrograph of *Vorticella,* a common protozoan, showing outstanding internal detail, depth of field, and bright colors, which are not natural (400×). Notice the cilia at the top of the cell and the contractile fiber in the stalk.

filters that protect the viewer's eye from injury by these dangerous rays. The name for this type of microscopy is based on the use of certain dyes (acridine, fluorescein) and minerals that show **fluorescence.** This means that the dyes give off visible light when bombarded by shorter ultraviolet rays. For an image to be formed, the specimen must first be coated or placed in contact with a source of fluorescence. Subsequent illumination by ultraviolet radiation causes the specimen to emit visible light, producing an intense blue, yellow, orange, or red image against a black field.

Fluorescence microscopy has its most useful applications in diagnosing infections caused by specific bacteria, protozoans, and viruses. A staining technique with fluorescent dyes is commonly used to detect *Mycobacterium tuberculosis* (the agent of tuberculosis) in patients' specimens (see figure 19.20). In a number of diagnostic procedures, fluorescent dyes are bound to specific antibodies. These *fluorescent antibodies* can be used to detect the causative agents in such diseases as syphilis, chlamydiosis, trichomoniasis, herpes, and influenza. A newer technology using fluorescent nucleic acid stains can differentiate between live and dead cells in mixtures **(figure 3.10)** or detect uncultured cells (see Insight 3.3). A fluorescence microscope can be handy for locating microbes in complex mixtures because only those cells targeted by the technique will fluoresce.

Optical microscopes may be unable to form a clear image at higher magnifications, because samples are often too thick for conventional lenses to focus all levels of cells simultaneously. This is especially true of larger cells with complex internal structures. A newer type of microscope that overcomes this impediment is called the *scanning confocal microscope.* This microscope uses a laser beam of light to scan various depths in the specimen and deliver a sharp image focusing on just a single plane. It is thus able to capture a highly focused view at any level, ranging from the surface to the

Figure 3.10 **Fluorescent staining on a fresh sample of cheek scrapings from the oral cavity.** Cheek epithelial cells are the larger green cells with red nuclei. Bacteria are streptococci (red spheres in long chains) and tiny green rods. This particular staining technique also indicates whether cells are alive or dead; live cells fluoresce green, and dead cells fluoresce red (60×). Notice the size ratio between the human cell, its nucleus, and the bacterial cells.

Figure 3.11 **Confocal microscopic images.** This *Paramecium* is stained with fluorescent dyes and visualized by a scanning confocal microscope. Note the degree of detail that may be observed at different depths of focus: top shows surface cilia, bottom shows nucleus and organelles.

middle of the cell. It is most often used on fluorescently stained specimens, but it can also be used to visualize live unstained cells and tissues (**figure 3.11**).

Electron Microscopy

If conventional light microscopes are our windows on the microscopic world, then the electron microscope (EM) is our window on the tiniest details of that world. Although this microscope was originally conceived and developed for studying metals and small electronics parts, biologists immediately recognized the importance of the tool and began to use it in the early 1930s. One of the most impressive features of the electron microscope is the resolution it provides.

Unlike light microscopes, the electron microscope forms an image with a beam of electrons that can be made to travel in wavelike

patterns when accelerated to high speeds. These waves are 100,000 times shorter than the waves of visible light. Because resolving power is a function of wavelength, the resolving power fraction becomes very small. Indeed, it is possible to resolve atoms with an electron microscope, even though the practical resolution for most biological applications is approximately 0.5 nm. The degree of resolution allows magnification to be extremely high—usually between 5,000× and 1,000,000× for biological specimens and up to 5,000,000× in some applications. Its capacity for magnification and resolution makes the EM an invaluable tool for seeing the finest structure of cells and viruses. If not for electron microscopes, our understanding of biological structure and function would still be in its early theoretical stages.

In fundamental ways, the electron microscope is similar to the optical microscope. It employs components analogous to, but not necessarily the same as, those in light microscopy (**figure 3.12**). For

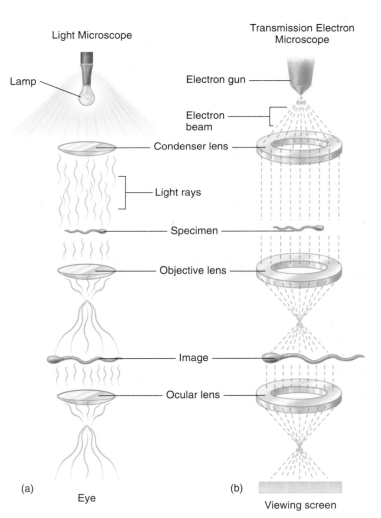

Figure 3.12 **Comparison of light and electron microscopes. (a)** The light microscope and **(b)** one type of electron microscope (EM; transmission type). These diagrams are highly simplified, especially for the electron microscope, to indicate the common components. Note that the EM's image pathway is actually upside down compared with that of a light microscope.

From Cell Ultrastructure 1st edition by Jensen/Park. © 1967. Reprinted with permission of Brooks/Cole, a division of Thomson Learning: www.thomsonrights.com. Fax 800 730-2215.

TABLE 3.3	Comparison of Light Microscopes and Electron Microscopes	
Characteristic	Light or Optical	Electron (Transmission)
Useful magnification	2,000×	1,000,000× or more
Maximum resolution	200 nm	0.5 nm
Image produced by	Light rays	Electron beam
Image focused by	Glass objective lens	Electromagnetic objective lenses
Image viewed through	Glass ocular lens	Fluorescent screen
Specimen placed on	Glass slide	Copper mesh
Specimen may be alive.	Yes	Usually not
Specimen requires special stains or treatment.	Depends on technique	Yes
Colored images formed	Yes	No

(a)

(b)

instance, it magnifies in stages by means of two lens systems, and it has a condensing lens, a specimen holder, and a focusing apparatus. Otherwise, the two types have numerous differences **(table 3.3).** An electron gun aims its beam through a vacuum to ring-shaped electromagnets that focus this beam on the specimen. Specimens must be pretreated with chemicals or dyes to increase contrast and usually cannot be observed in a live state. The enlarged image is displayed on a viewing screen or photographed for further study rather than being observed directly through an eyepiece. Because images produced by electrons lack color, electron micrographs (a micrograph is a photograph of a microscopic object) are always shades of black, gray, and white. The color-enhanced micrographs seen in this and other textbooks have computer-added color.

Two general forms of EM are the **transmission electron microscope (TEM)** and the **scanning electron microscope (SEM)** (see table 3.2). Transmission electron microscopes are the method of choice for viewing the detailed structure of cells and viruses. This microscope produces its image by transmitting electrons through the specimen. Because electrons cannot readily penetrate thick preparations, the specimen must be stained or coated with metals that will increase image contrast and sectioned into extremely thin slices (20–100 nm thick). The electrons passing through the specimen travel to the fluorescent screen and display a pattern or image. The darker and lighter areas of the image correspond to more and less dense parts on the specimen **(figure 3.13).** The TEM can also be used to produce negative images and shadow casts of whole microbes (see figure 6.3).

The scanning electron microscope provides some of the most dramatic and realistic images in existence. This instrument can create an extremely detailed three-dimensional view of all things biological—from dental plaque to tapeworm heads. To produce its

Figure 3.13 Transmission electron micrographs.
(a) Human parvoviruses (B–19) isolated from the serum of a patient with erythema infectiosum. This DNA virus and disease are covered in chapter 24. What is the average diameter in nanometers of a single virus? **(b)** A section through *Vorticella*, magnified 2150× (bar is 2 μm) emphasizes its complex ultrastructure. Labels indicate the micronucleus (mic), the macronucleus, the cilia around the oral cavity (pak), the contractile vacuole (cv), food vacuoles (fv), and the myoneme. Compare the detail with figure 3.9 of the same protozoan. You will learn more about eukaryotic cell structure in chapter 5.

images, the SEM bombards the surface of a whole, metal-coated specimen with electrons while scanning back and forth over it. A shower of electrons deflected from the surface is picked up with great fidelity by a sophisticated detector, and the electron pattern is

(a)

2 microns

(b)

Figure 3.14 Scanning electron micrographs—outstanding images in three dimensions. (a) An incredible view of the predatory ciliate *Didinium*. These little "dancers" use the central rows of cilia to swim, and the oral rows of cilia to trap their favorite food *Paramecium*, which they suck into their cone-shaped mouth. (b) A fantastic-looking algal cell called a coccolithophore displays an ornamental cell wall formed by layers of delicate calcium discs. These algae often form massive spring blooms in the world's oceans.

displayed as an image on a monitor screen. The contours of the specimens resolved with scanning electron micrography are very revealing and often surprising. Areas that look smooth and flat with the light microscope display intriguing surface features with the SEM **(figure 3.14)**. Improved technology has continued to refine

electron microscopes and to develop variations on the basic plan. Some inventive relatives of the EM are the scanning probe microscope and atomic force microscope **(Insight 3.1).**

Preparing Specimens for Optical Microscopes

A specimen for optical microscopy is generally prepared by mounting a sample on a suitable glass slide that sits on the stage between the condenser and the objective lens. The manner in which a slide specimen, or mount, is prepared depends upon: (1) the condition of the specimen, either in a living or preserved state; (2) the aims of the examiner, whether to observe overall structure, identify the microorganisms, or see movement; and (3) the type of microscopy available, whether it is bright-field, dark-field, phase-contrast, or fluorescence.

Fresh, Living Preparations

Live samples of microorganisms are used to prepare **wet mounts** so that they can be observed as near to their natural state as possible. The cells are suspended in a suitable fluid (water, broth, saline) that temporarily maintains viability and provides space and a medium for locomotion. A wet mount consists of a drop or two of the culture placed on a slide and overlaid with a cover glass. Although this preparation is quick and easy to make, it has certain disadvantages. The cover glass can damage larger cells, and the slide is very susceptible to drying and can contaminate the handler's fingers.

A more satisfactory alternative is the **hanging drop slide** (below) made with a special concave (depression) slide, an adhesive or sealant, and a coverslip from which a tiny drop of sample is suspended. These types of short-term mounts provide a true assessment of the size, shape, arrangement, color, and motility of cells. Greater cellular detail can be observed with phase-contrast or interference microscopy.

Coverslip

Hanging drop containing specimen

Vaseline

Depression slide

Fixed, Stained Smears

A more permanent mount for long-term study can be obtained by preparing fixed, stained specimens. The smear technique, developed by Robert Koch more than 100 years ago, consists of spreading a thin film made from a liquid suspension of cells on a slide and air-drying it. Next, the air-dried **smear** is usually heated gently by a process called **heat fixation** that simultaneously kills the specimen and secures it to the slide. Fixation also preserves various cellular components in a natural state with minimal distortion. Fixation of some microbial cells is performed with chemicals such as alcohol and formalin.

INSIGHT 3.1

The Evolution in Resolution: Probing Microscopes

In the past, chemists, physicists, and biologists had to rely on indirect methods to provide information on the structures of the smallest molecules. But technological advances have created a new generation of microscopes that "see" atomic structure by actually feeling it. *Scanning probe microscopes* operate with a minute needle tapered to a tip that can be as narrow as a single atom! This probe scans over the exposed surface of a material and records an image of its outer texture. These revolutionary microscopes have such profound resolution that they have the potential to image single atoms and to magnify 100 million times. The scanning tunneling microscope (STM) was the first of these microscopes. It uses a tungsten probe that hovers near the surface of an object and follows its topography while simultaneously giving off an electrical signal of its pathway, which is then imaged on a screen. The STM is used primarily for detecting defects on the surfaces of electrical conductors and computer chips but it has also provided the first incredible close-up views of DNA, the genetic material (see Insight 9.2).

Another variety, the atomic force microscope (AFM), gently forces a diamond and metal probe down onto the surface of a specimen like a needle on a record. As it moves along the surface, any deflection of the metal probe is detected by a sensitive device that relays the information to an imager. The AFM is very useful in viewing the detailed functions of biological molecules such as antibodies and enzymes.

The latest versions of these microscopes have recently increased the resolving power to around 0.5 Å, which allowed technicians to image a pair of electrons! Such powerful tools for observing and positioning atoms have spawned a field called *nanotechnology*—the science of the "small." Scientists in this area use physics, chemistry, biology, and engineering to manipulate small molecules and atoms.

Working at these dimensions, they are currently creating tiny molecular tools to miniaturize computers and other electronic devices. In the future, it may be possible to use microstructures to deliver drugs, analyze DNA, and treat disease.

Looking back at figure 2.1, name some other chemical features that may become visible using these high-power microscopes. *Answer available at http://www.mhhe.com/talaro8*

"Carbon monoxide man." This molecule was constructed from 28 single CO molecules (red spheres) and photographed by a scanning tunneling microscope. Each CO molecule is approximately 5 Å wide.

Like images on undeveloped photographic film, the unstained cells of a fixed smear are quite indistinct, no matter how great the magnification or how fine the resolving power of the microscope. The process of "developing" a smear to create contrast and make inconspicuous features stand out requires staining. Staining is any procedure that applies colored chemicals called dyes to specimens. Dyes impart a color to cells or cell parts by becoming affixed to them through a chemical reaction. In general, they are classified as **basic** (cationic) **dyes,** which have a positive charge, or **acidic** (anionic) **dyes,** which have a negative charge.

Negative versus Positive Staining Two basic types of staining technique are used, depending upon how a dye reacts with the specimen (summarized in **table 3.4**). Most procedures involve a **positive stain,** in which the dye actually sticks to cells and gives them color. A **negative stain,** on the other hand, is just the reverse (like a photographic negative). The dye does not stick to the specimen but dries around its outer boundary, forming a silhouette. Nigrosin (blue-black) and India ink (a black suspension of carbon particles) are the dyes most commonly used for negative staining. The cells themselves do not stain because these dyes are negatively charged and are repelled by the negatively charged surface of the cells. The value of negative staining is its relative simplicity and the reduced shrinkage or distortion of cells, as the smear is not heat fixed. A quick assessment can thus be made regarding

TAKE NOTE: DYES AND STAINING

Because many microbial cells lack contrast, it is necessary to use dyes to observe their detailed structure and identify them. Dyes are colored compounds related to or derived from the common organic solvent benzene. When certain double-bonded groups ($C{=}O$, $C{=}N$, $N{=}N$) are attached to complex molecules, the resultant compound gives off a specific color. Most dyes form ions when dissolved in a compatible solvent. The color-bearing ion, termed a **chromophore,** has a charge that attracts it to certain cell parts that have the opposite charge **(figure 3.15).**

Basic dyes carry a positively charged chromophore and are attracted to negatively charged cell components (nucleic acids and proteins). Because bacteria contain large amounts of negatively charged substances, they stain readily with basic dyes such as methylene blue, crystal violet, fuchsin, and safranin.

Acidic dyes with a negatively charged chromophore bind to positively charged molecules in some parts of cells. One example is eosin, a red dye used in staining blood cells. Because bacterial cells have numerous acidic substances and carry a slightly negative charge on their surface, they tend to repel acidic dyes. Acidic dyes such as nigrosin and India ink can still be used successfully on bacteria using the negative stain method **(figure 3.15c).** With this technique, the dye settles around the cell and creates an outline of it.

TABLE 3.4	Comparison of Positive and Negative Stains	
	Positive Staining	**Negative Staining**
Appearance of cell	Colored by dye	Clear and colorless
Background	Not stained (generally white)	Stained (dark gray or black)
Dyes employed	Basic dyes: Crystal violet Methylene blue Safranin Malachite green	Acidic dyes: Nigrosin India ink
Subtypes of stains	Several types: Simple stain Differential stains Gram stain Acid-fast stain Spore stain Structural stains One type of capsule stain Flagella stain Spore stain Granules stain Nucleic acid stain	Few types: Capsule Spore Negative stain of *Bacillus anthracis*

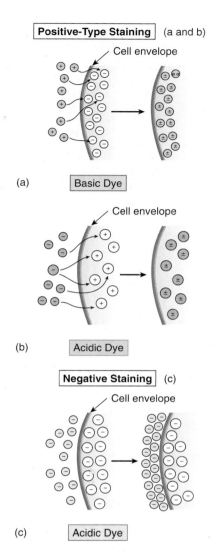

Figure 3.15 Staining reactions of dyes. (a) Basic dyes are positively charged and react with negatively charged cell areas. **(b)** Acidic dyes are negatively charged and react with positively charged cell areas. **(c)** Acidic dyes can be used with negative staining to create a background around a cell.

cellular size, shape, and arrangement. Negative staining is also used to accentuate the capsule that surrounds certain bacteria and yeasts **(figure 3.16).**

Simple versus Differential Staining Positive staining methods are classified as simple, differential, or structural (figure 3.16). **Simple stains** require only a single dye and an uncomplicated procedure, while **differential stains** use two different-colored dyes, called the *primary dye* and the *counterstain,* to distinguish between cell types or parts. These staining techniques tend to be more complex and sometimes require additional chemical reagents to produce the desired reaction.

Most simple staining techniques take advantage of the ready binding of bacterial cells to dyes like malachite green, crystal violet, basic fuchsin, and safranin. Simple stains cause all cells in a smear to appear more or less the same color, regardless of type, but they can still reveal bacterial characteristics such as shape, size, and arrangement. A simple stain with methylene blue is often used to stain granules in bacteria such as *Corynebacterium* **(figure 3.16a),** which can be a factor in identification.

Types of Differential Stains An effective differential stain uses dyes of contrasting color to clearly emphasize differences between two cell types or cell parts. Common combinations are red and purple, red and green, or pink and blue. Differential stains can also pinpoint other characteristics, such as the size, shape, and arrangement of cells. Typical examples include Gram, acid-fast, and endospore stains. Some staining techniques (spore, capsule) fall into more than one category.

Gram staining is a 130-year-old method named for its developer, **Hans Christian Gram.** Even today, it is an important diagnostic staining technique for bacteria. It permits ready differentiation of major categories based upon the color reaction of the cells: **gram-positive,** which stain purple, and **gram-negative,** which stain red **(figure 3.16c).** This difference in staining quality is due to structural variations found in the cell walls of bacteria. The Gram stain is the basis of several important bacteriologic topics, including bacterial taxonomy, cell wall structure, identification, and drug

(a) Simple Stains

Methylene blue
stain of *Corynebacterium*
(1,000×)

(b) Differential Stains

Gram stain
Purple cells are gram-positive.
Red cells are gram-negative
(1,000×).

Acid-fast stain
Red cells are acid-fast.
Blue cells are non-acid-fast
(750×).

(c) Structural Stains

India ink capsule stain of
Cryptococcus neoformans
(500×)

Flagellar stain of *Proteus vulgaris.*
Note the fine fringe of flagella
(1,500×).

**Figure 3.16 Types of microbiological
stains. (a)** Simple stains. **(b)** Differential stains:
Gram, acid-fast, and spore. **(c)** Structural stains:
capsule and flagellar. The spore stain (bottom) is
a method that fits two categories: differential and
special.

Spore stain, showing spores (green)
and vegetative cells (red)
(1,000×)

therapy. As we will see in the case file to continue, it is a critical step in diagnosing meningitis. Gram staining is discussed in greater detail in **Insight 3.2.**

The **acid-fast stain,** like the Gram stain, is an important diagnostic stain that differentiates acid-fast bacteria (pink) from non-acid-fast bacteria (blue). This stain originated as a specific method to detect *Mycobacterium tuberculosis* in specimens. It was determined that these bacterial cells have a particularly impervious outer wall that holds fast (tightly or tenaciously) to the dye (carbol fuchsin), even when washed with a solution containing acid or acid alcohol (**figure 3.16c**). This stain is used for other medically important mycobacteria such as the Hansen's disease (leprosy) bacillus and for *Nocardia,* an agent of lung or skin infections (see chapter 19).

The **endospore stain** (spore stain) is similar to the acid-fast method in that a dye is forced by heat into resistant survival cells called spores or endospores. These are formed in response to adverse conditions and are not reproductive (see chapter 4). This stain is designed to distinguish between spores and the vegetative cells that make them (**figure 3.16c**). Of significance in medical microbiology

INSIGHT 3.2

The Gram Stain: A Grand Stain

In 1884, Hans Christian Gram discovered a staining technique that could be used to make bacteria in infectious specimens more visible. His technique consisted of timed, sequential applications of crystal violet (the primary dye), Gram's iodine (IKI, the mordant), an alcohol rinse (decolorizer), and a contrasting counterstain. The initial counterstain used was yellow or brown and was later replaced by the red dye, safranin. Since that substitution, bacteria that stained purple are called gram-positive, and those that stained red are called gram-negative.

Although these staining reactions involve an attraction of the cell to a charged dye, it is important to note that the terms *gram-positive* and *gram-negative* are not used to indicate the electrical charge of cells or dyes but whether or not a cell retains the primary dye-iodine complex after decolorization. There is nothing specific in the reaction of gram-positive cells to the primary dye or in the reaction of gram-negative cells to the counterstain. The different results in the Gram stain are due to differences in the structure of the cell wall and how it reacts to the series of reagents applied to the cells.

In the first step, crystal violet stains cells in a smear all the same purple color. The second and key differentiating step is the mordant—Gram's iodine. The mordant is a stabilizer that causes the dye to form large crystals that get trapped by the thick meshwork of the cell wall. Because this layer in gram-positive cells is thicker, the entrapment of the dye is far more extensive in them than in gram-negative cells. Application of alcohol in the third step dissolves lipids in the outer membrane and removes the dye from the gram-negative cells. By contrast, the crystals of dye tightly embedded in the gram-positive bacteria are relatively inaccessible and resistant to removal. Because gram-negative bacteria are colorless after decolorization, their presence is demonstrated by applying the counterstain safranin in the final step.

This staining method remains an important basis for bacterial classification and identification. It permits differentiation of four major categories based upon color reaction and shape: gram-positive rods, gram-positive cocci, gram-negative rods, and gram-negative cocci (see table 4.4). The Gram stain can also be a practical aid in diagnosing infection and in guiding drug treatment. For example, Gram staining a fresh urine or throat specimen can help pinpoint the possible cause of infection, and in some cases, it is possible to begin drug therapy on the basis of this stain. Even in this day of elaborate and expensive medical technology, the Gram stain remains an important and unbeatable first tool in diagnosis.

What would be the primary concerns in selecting a counterstain dye for the Gram stain? *Answer available at http://www.mhhe.com/talaro8*

Step	Microscopic Appearance of Cell		Chemical Reaction in Cell (very magnified view)	
	Gram (+)	Gram (−)	Gram (+)	Gram (−)
1 Crystal violet (primary dye)			Both cell walls stain with the dye.	
2 Gram's iodine (mordant)			Dye crystals trapped in cell	No effect of iodine
3 Alcohol (decolorizer)			Crystals remain in cell.	Outer wall is weakened; cell loses dye.
4 Safranin (red dye counterstain)			Red dye has no effect.	Red dye stains the colorless cell.

are the gram-positive, spore-forming members of the genus *Bacillus* (the cause of anthrax) and *Clostridium* (the cause of botulism and tetanus)—dramatic diseases of universal fascination that we consider in later chapters.

Structural stains are used to emphasize special cell parts such as capsules, endospores, and flagella that are not revealed by conventional staining methods. Capsule staining is a method of observing the microbial capsule, an unstructured protective layer surrounding the cells of some bacteria and fungi. Because the capsule does not react with most stains, it is often negatively stained with India ink, or it may be demonstrated by special positive stains. The fact that not all microbes exhibit capsules is a useful feature for identifying pathogens. One example is *Cryptococcus,* which causes a serious fungal meningitis in AIDS patients **(figure 3.16b).**

Flagellar staining is a method of revealing flagella, the tiny, slender filaments used by bacteria for locomotion. Because the width of bacterial flagella lies beyond the resolving power of the light microscope, in order to be seen, they must be enlarged by depositing a coating on the outside of the filament and then staining it. The presence, number, and arrangement of flagella can be helpful in identification of bacteria **(figure 3.16c).**

CONTINUING CASE FILE 3

Lab results from the Gram stain and culture were conclusive: she was infected by the meningococcus, *Neisseria meningitidis**, which is an agent of both meningitis* and septicemia*. The microscopic examination yielded the classic appearance of tiny pairs of red cocci (diplococci) and white blood cells carrying the same cocci inside. A blood culture also showed growth, indicating that the bacteria had entered her bloodstream. Plates of agar inoculated with the CSF grew typical off-white, smooth, isolated colonies that tested out as *N. meningitidis*. The woman remained on the antibiotic regimen for 10 days and was released, fortunately without long term damage.

- What is the importance of a Gram stain in diagnosis of an infection like this?
- What kinds of media would be used to culture and identify this microbe?

For a wrap-up, see the Case File Perspective on page 85.

Check & Assess Section 3.2

✔ Magnification, resolving power, lens quality, and illumination source all influence the clarity of specimens viewed through the optical microscope.

✔ The maximum resolving power of the optical microscope is 200 nm, or 0.2 μm. This is sufficient to see the internal structures of eukaryotes and the morphology of most bacteria.

✔ There are six types of optical microscopes. Four types use visible light for illumination: bright-field, dark-field, phase-contrast, and interference microscopes. The fluorescence microscope uses UV light for illumination, but it has the same resolving power as the other optical microscopes. The confocal microscope can use UV light or visible light reflected from specimens.

✔ Electron microscopes (EMs) use electrons, not light waves, as an illumination source to provide high magnification (5,000× to 1,000,000×) and high resolution (0.5 nm). Electron microscopes can visualize cell ultrastructure (TEM) and three-dimensional images of cell and virus surface features (SEM).

✔ Specimens viewed through optical microscopes can be either alive or dead, depending on the type of specimen preparation, but most EM specimens have been killed because they must be viewed in a vacuum.

✔ Stains are important diagnostic tools in microbiology because they can be designed to differentiate cell shape, structure, and other features of microscopic morphology.

3. Differentiate between the concepts of magnification, refraction, and resolution.

4. Briefly explain how an image is made and magnified.

5. On the basis of the formula for resolving power, explain why a smaller RP value is preferred to a larger one and explain what it means in practical terms if the resolving power is 1.0 μm.

6. What does a value greater than 1.0 μm mean? (Is it better or worse?) and what does a value less than 1.0 μm mean?

7. What can be done to a microscope to improve resolution?

8. Compare bright-field, dark-field, phase-contrast, confocal, and fluorescence microscopy as to field appearance, specimen appearance, light source, and uses.

9. Compare and contrast the optical compound microscope with the electron microscope.

10. Why is the resolution so superior in the electron microscope?

11. Compare the way that the image is formed in the TEM and SEM.

12. Itemize the various staining methods, and briefly characterize each.

13. Explain what happens in positive staining to cause the reaction in the cell.

14. Explain what happens in negative staining that causes the final result.

15. For a stain to be considered differential, what must it do?

3.3 Additional Features of the Six "I's"

Expected Learning Outcomes

14. Define *inoculation*, *media*, and *culture*, and describe sampling methods and instruments, and what events must be controlled.

15. Describe three basic techniques for isolation, including tools, media, incubation, and outcome.

16. Explain what an isolated colony is and indicate how it forms.

17. Differentiate between a pure culture, subculture, mixed culture, and contaminated culture. Define *contaminant*.

18. What kinds of data are collected during information gathering?

19. Describe some of the processes involved in identifying microbes from samples.

* *Neisseria meningitidis* (ny-serr′ee-uh men′-in-jih′-tih-dis) From the German physician, *Neisser,* and Gr. *meninx,* a membrane.

* *meningitis* (men′-in-jy′-tis) Gr. *meninx,* a membrane, and *it is,* an inflammation; specifically, an inflammation of the meninges, the membranes that surround the brain.

* *septicemia* (sep′-tih-see′-mee-uh) Gr. *septikos,* to make putrid, and *haima,* blood; the presence of infectious agents or their toxins in the blood.

INSIGHT 3.3

The Uncultured

For some time, microbiologists have suspected that culture-based methods are unable to identify many kinds of bacteria. This was first confirmed by environmental researchers, who came to believe that only about 1% (and in some environments, it was 0.001%) of microbes present in lakes, soil, and saltwater environments could be grown in laboratories by the usual methods and, therefore, were unknown and unstudied. These microbes are termed **viable but nonculturable,** or **VBNC.**

Scientists had spent several decades concocting recipes for media and having great success in growing all kinds of bacteria from all kinds of environments. Just identifying and studying those kept them very occupied. But by the 1990s, a number of specific, nonculturing tools based on genetic testing had become widely available. When these methods were used to sample various environments, they revealed a vast "jungle" of new species that had never before been cultured. These were the techniques that Venter's team applied to ocean samples (Case File 1). This discovery seemed reasonable, because it may not yet be possible to exactly re-create the many correct media and conditions to grow such organisms in the lab.

Medical microbiologists have also missed a large proportion of microbes that are normal residents of the body. Stanford University scientists applied these techniques to subgingival plaque harvested from one of their own mouths. They used small, known fragments of DNA or probes that can highlight microbes in specimens. Oral biologists had previously recovered about 500 bacterial strains from this site; the Stanford scientists found 30 species that had never before been cultured or described.

A fluorescent micrograph of a biofilm of *Vibrio cholerae* (orange cells) taken from a water sample in India. After an attempt to culture these bacteria over 495 days, a few cells enlarged, but most remained inactive. This pathogen appears to remain dormant in water where it can serve as a source of infection (cholera).

This discovery energized the medical world and spurred the use of nonculture-based methods to find VBNCs in the human body.

The new realization that our bodies are hosts to a wide variety of unknown microbes has several implications. As evolutionary microbiologist Paul Ewald has asked, "What are all those microbes doing in there?" He points out that many oral microbes previously assumed to be innocuous are now associated with cancer and heart disease. Many of the diseases that we currently think of as noninfectious will likely be found to have an infectious cause once we continue to look for VBNCs.

Do you think that all of the VBNCs are really not culturable? Suggest some other possibilities. *Answer available at http://www.mhhe.com/talaro8*

Inoculation, Growth, and Identification of Cultures

To cultivate, or **culture,*** microorganisms, one introduces a tiny sample (the inoculum) into a container of nutrient **medium*** (pl. media), which provides an environment in which they multiply. This process is called **inoculation.*** The observable growth that later appears in or on the medium is known as a culture. The nature of the sample being cultured depends on the objectives of the analysis. Clinical specimens for determining the cause of an infectious disease are obtained from body fluids (blood, cerebrospinal fluid), discharges (sputum, urine, feces), or diseased tissue. Samples subject to microbiological analysis can include nearly any natural material. Some common ones are soil, water, sewage, foods, air, and inanimate objects. The important concept of media will be covered in more detail in section 3.4.

A past assumption has been that most microbes could be teased out of samples and cultured, given the proper media. But this view has had to be greatly revised **(Insight 3.3).**

TAKE NOTE: SPECIAL CONCERNS IN CULTURING

Inherent in these practices are the concepts of **sterile, aseptic,** and **pure culture**[1] techniques. Contamination is a constant problem, so sterile techniques (media, transfer equipment) help ensure that only microbes that came from the sample are present. Another concern is the possible release of infectious agents from cultures, which is prevented by aseptic techniques. Many of these techniques revolve around keeping species in the pure culture form for further study, identification, or biotechnology applications.

Isolation Techniques

Certain **isolation** techniques are based on the concept that if an individual bacterial cell is separated from other cells and provided adequate space on a nutrient surface, it will grow into a discrete

* *culture* (kul'-chur) Gr. *cultus,* to tend or cultivate. It can be used as a verb or a noun.
* *medium* (mee'-dee-um) pl. media; L., middle.
* *inoculation* (in-ok"-yoo-lay'-shun) L. *in,* and *oculus,* eye.

1. *Sterile* means the complete absence of viable microbes: *aseptic* refers to prevention of infection; *pure culture* refers to growth of a single species of microbe.

Mixture of cells in sample

Parent cells

Separation of cells by spreading or dilution on agar medium

Incubation

Growth increases the number of cells.

Microbes become visible as isolated colonies containing millions of cells.

Microscopic view Cellular level

Macroscopic view Colony level

Figure 3.17 Isolation technique. Stages in the formation of an isolated colony, showing the microscopic events and the macroscopic result. Separation techniques such as streaking can be used to isolate single cells. After numerous cell divisions, a macroscopic mound of cells, or a colony, will be formed. This is a relatively simple yet successful way to separate different types of bacteria in a mixed sample.

mound of cells called a **colony (figure 3.17).** Because it arises from a single cell or cluster of cells, an isolated colony consists of just one species. Proper isolation requires that a small number of cells be inoculated into a relatively large volume or over an expansive area of medium. It generally requires the following materials: a medium that has a relatively firm surface (see description of agar on page 80) contained in a clear, flat covered plate called a *Petri dish,* and inoculating tools.

In the **streak plate method,** a small droplet of sample is spread with a tool called an **inoculating loop** over the surface of the medium according to a pattern that gradually thins out the sample and separates the cells spatially over several sections of the plate **(figure 3.18a, b).** Because of its ease and effectiveness, the streak plate is the method of choice for most applications.

In the **loop dilution,** or **pour plate, technique,** the sample is inoculated, also with a loop, into a series of cooled but still liquid agar tubes so as to dilute the number of cells in each successive tube in the series **(figure 3.18c, d).** Inoculated tubes are then plated out (poured) into sterile Petri dishes and are allowed to solidify (harden). The number of cells per volume is so decreased that cells have ample space to form separate colonies in the second or third plate. One difference between this and the streak plate method is that in this technique, some of the colonies will develop deep in the medium itself and not just on the surface.

With the **spread plate technique,** a small volume of liquid, a diluted sample is pipetted onto the surface of the medium and spread around evenly by a sterile spreading tool (sometimes called a "hockey stick"). As with the streak plate, cells are spread over separate areas on the surface so that they can form individual colonies **(figure 3.18e, f).**

Once a container of medium has been inoculated, it is **incubated** in a temperature-controlled chamber (incubator) to encourage microbial growth. Although microbes have adapted to growth at temperatures ranging from freezing to boiling, the usual temperatures used in laboratory propagation fall between 20°C and 40°C. Incubators can also control the content of atmospheric gases such as oxygen and carbon dioxide that may be required for the growth of certain microbes. During the incubation period (ranging from a few hours to several weeks), the microbe multiplies and produces a culture with macroscopically observable growth. Microbial growth in a liquid medium materializes as cloudiness, sediment, a surface mat, or colored pigment. Growth on solid media may take the form of a spreading mat or separate colonies.

In some ways, culturing microbes is analogous to gardening. Cultures are formed by "seeding" tiny plots (media) with microbial cells. Extreme care is taken to exclude weeds (contaminants). A **pure culture** is a container of medium that grows only a single known species or type of microorganism **(figure 3.19a).** This type of culture is most frequently used for laboratory study, because it allows the precise examination and control of one microorganism by itself. Instead of the term *pure culture,* some microbiologists prefer the term **axenic,** meaning that the culture is free of other living things except for the one being studied. A standard method for preparing a pure culture is to use a **subculture** technique to make a second-level culture. A tiny bit of cells from a well-isolated colony is transferred into a separate container of media and incubated (see figure 3.1, isolation).

A **mixed culture (figure 3.19b)** is a container that holds two or more easily differentiated species of microorganisms, not unlike a garden plot containing both carrots and onions. A **contaminated culture (figure 3.19c)** has had **contaminants** (unwanted microbes of uncertain identity) introduced into it, like weeds into a garden. Because contaminants have the potential for disrupting experiments and tests, special procedures have been developed to control them, as you will no doubt witness in your own laboratory.

Identification Techniques

How does one determine what sorts of microorganisms have been isolated in cultures? Certainly microscopic appearance can be valuable in differentiating the smaller, simpler prokaryotic cells from the larger, more complex eukaryotic cells. Appearance can often be used to identify eukaryotic microorganisms to the level of genus or species because of their more distinctive appearance.

Bacteria are generally not as readily identifiable by these methods because very different species may appear quite similar. For them, we must include other techniques, some of which characterize their cellular metabolism. These methods, called **biochemical tests,** can determine fundamental chemical characteristics such as nutrient

Note: This method only works if the spreading tool (usually an **inoculating loop**) is resterilized (flamed) after each of steps 1–4.

Loop containing sample

1 2 3 4 5

(a) **Steps in a Streak Plate;** this one is a four-part or quadrant streak.

(b)

Loop containing sample

1 2 3

1 2 3

(c) **Steps in Loop Dilution;** also called a pour plate or serial dilution

(d)

"Hockey stick"

1 2

(e) **Steps in a Spread Plate**

(f)

Figure 3.18 Methods for isolating bacteria. (a) Steps in a quadrant streak plate and **(b)** resulting isolated colonies of bacteria. **(c)** Steps in the loop dilution method and **(d)** the appearance of plate 3. **(e)** Spread plate and **(f)** its result. Techniques in (a) and (b) use a loopful of culture, whereas (c) starts with a pre-diluted sample.

(a)

(b)

(c)

Figure 3.19 Various states of cultures. (a) A mixed culture of *M. luteus* and *E. coli* readily differentiated by their colors. **(b)** This plate of *S. marcescens* was overexposed to room air and has developed a large, white colony. Because this intruder is not desirable and not identified, the culture is now contaminated. **(c)** Tubes containing pure cultures of *Escherichia coli* (white), *Micrococcus luteus* (yellow), and *Serratia marcescens* (red).

(a)

Scheme for Differentating Gram-Negative Cocci and Coccobacilli

(b)

Figure 3.20 **Rapid mini testing with identification key for *Neisseria meningitidis*.** **(a)** A test system for common gram negative cocci uses 8 small wells of media to be inoculated with a pure culture and incubated. A certain combination of reactions will be consistent with this species. **(b)** A sample of a key that uses test data (positive or negative reactions) to guide identification.

requirements, products given off during growth, presence of enzymes, and mechanisms for deriving energy. **Figure 3.20***a* provides an example of a multiple-test, miniaturized system for obtaining physiological characteristics. Biochemical tests are discussed in more depth in chapter 17 and chapters that cover identification of pathogens.

Several modern diagnostic tools that analyze genetic characteristics can detect microbes based on their DNA. These DNA profiles can be extremely specific, even sufficient by themselves to identify some microbes (see chapters 10 and 17). Identification can be assisted by testing the isolate against known antibodies (immunologic testing). In the case of certain pathogens, further information is obtained by inoculating a suitable laboratory animal. By compiling physiological testing results with both macroscopic and microscopic traits, a complete picture of the microbe is developed. Expertise in final identification comes from specialists, manuals, and computerized programs. A traditional pathway in bacterial identification uses flow charts or keys that apply the results of tests

to a selection process (**figure 3.20***b*). The top of the key provides a general category from which to begin a series of branching points, usually into two pathways at each new junction. The points of separation are based on having a positive or negative result for each test. Following the pathway that fits the characteristics leads to an end point where a name for an organism is given. This process of "keying out" the organism can simplify the identification process.

Check & Assess Section 3.3

✔ During *inoculation*, a specimen is introduced into a container of medium. Inoculated media are *incubated* at a specified temperature to encourage growth and form a culture.

✔ *Isolation* occurs when *colonies* originate from single cells. Colonies are composed of large numbers of cells massed together in visible mounds.

✔ A culture may exist in one of the following forms: A *pure culture* contains only one species or type of microorganism. A *mixed culture* contains two or more known species. A *contaminated culture* contains both known and unknown (unwanted) microorganisms.

✔ During *inspection*, the cultures are examined and evaluated macroscopically for growth characteristics and microscopically for cellular appearance.

✔ Microorganisms are *identified* by a variety of *information* gathered through inspection and further tests, including biochemical reactions, immunologic reactions, and their genetic characteristics.

16. Provide the definitions of inoculation, growth, and contamination.
17. Name two ways that pure, mixed, and contaminated cultures are similar and two ways that they differ from each other.
18. What are some efforts for avoiding contamination?
19. Explain what is involved in isolating microorganisms and why it may be necessary to do this.
20. Compare and contrast three common laboratory techniques for separating bacteria in a mixed sample.
21. Describe how an isolated colony forms.
22. Explain why an isolated colony and a pure culture are not the same thing.

3.4 Media: The Foundations of Culturing

Expected Learning Outcomes

20. Explain the importance of media for culturing microbes in the laboratory.

21. Name the three general categories of media, based on their inherent properties and uses.

22. Compare and contrast liquid, solid, and semisolid media, giving examples.

23. Analyze chemically defined and complex media, describing their basic differences and content.

24. Describe functional media; list several different categories, and explain what characterizes each type of functional media.

25. Identify the qualities of enriched, selective, and differential media; use examples to explain their content and purposes.

26. Explain what it means when microorganisms are not culturable.

27. Describe live media and the circumstances that require it.

A major stimulus to the rise of microbiology in the late 1800s was the development of techniques for growing microbes out of their natural habitats and in pure form in the laboratory. This milestone enabled the close examination of a microbe and its morphology, physiology, and genetics. It was evident from the very first that for successful cultivation, the microorganisms being cultured had to be provided with all of their required nutrients in an artificial medium.

Some microbes require only a very few simple inorganic compounds for growth; others need a complex list of specific inorganic and organic compounds. This tremendous diversity is evident in the types of media that can be prepared. At least 500 different types of media are used in culturing and identifying microorganisms. Culture media are contained in test tubes, flasks, or Petri dishes, and they are inoculated by such tools as loops, needles, pipettes, and swabs. Media are extremely varied in nutrient content and consistency, and can be specially formulated for a particular purpose.

For an experiment to be properly controlled, sterile technique is necessary. This means that the inoculation must start with a sterile medium and inoculating tools with sterile tips must be used. Measures must be taken to prevent introduction of nonsterile materials, such as room air and fingers, directly into the media.

Types of Media

Most media discussed here are designed for bacteria and fungi, though algae and some protozoa can be propagated in media. Viruses can only be cultivated in live host cells.

Media fall into three general categories based on their properties: physical state, chemical composition, and functional type **(table 3.5).**

Physical States of Media

Liquid media are defined as water-based solutions that do not solidify at temperatures above freezing and that tend to flow freely when the container is tilted **(figure 3.21).** These media, termed **broths,** milks, or infusions, are made by dissolving various solutes in distilled water. Growth occurs throughout the container and can then present a dispersed, cloudy, or flaky appearance. A common laboratory medium, *nutrient broth,* contains beef extract and peptone dissolved in water. Other examples

of liquid media are methylene blue milk and litmus milk that contain whole milk and dyes. Fluid thioglycollate is a slightly viscous broth used for determining patterns of growth in oxygen (see figure 7.11).

At ordinary room temperature, **semisolid media** exhibit a clotlike consistency **(figure 3.22)** because they contain an amount of solidifying agent (agar or gelatin) that thickens them but does not produce a firm substrate. Semisolid media are used to determine the motility of bacteria and to localize a reaction at a specific site. Motility test medium and sulfur indole motility (SIM) medium both contain a small amount (0.3–0.5%) of agar. In both cases, the medium is stabbed carefully in the center with an inoculating

TABLE 3.5	Three Categories of Media Classification		
Physical State (Medium's Normal Consistency)	**Chemical Composition (Type of Chemicals Medium Contains)**	**Functional Type (Purpose of Medium)***	
1. Liquid	1. Synthetic (chemically defined)	1. General purpose	
2. Semisolid		2. Enriched	
3. Solid (can be converted to liquid)	2. Nonsynthetic (complex; not chemically defined)	3. Selective	
		4. Differential	
4. Solid (cannot be liquefied)		5. Anaerobic growth	
		6. Specimen transport	
		7. Assay	
		8. Enumeration	

*Some media can serve more than one function. For example, a medium such as brain-heart infusion is general purpose and enriched; mannitol salt agar is both selective and differential; and blood agar is both enriched and differential.

(a)

(b) (c)

Figure 3.21 **Sample liquid media. (a)** Liquid media tend to flow freely when the container is tilted. **(b)** Urea broth is used to show a biochemical reaction in which the enzyme urease digests urea and releases ammonium. This raises the pH of the solution and causes the dye to become increasingly pink. Left: uninoculated broth, pH 7; middle: growth with no change; right: positive, pH 8.0. **(c)** Presence-absence broth is for detecting the presence of coliform bacteria in water samples. It contains lactose and bromcresol purple dye. As coliforms use lactose, they release acidic substances. This lowers the pH and changes the dye from purple to yellow (right is *Escherichia coli*). Noncoliforms such as *Pseudomonas* (left) grow but do not change the pH (purple color indicates neutral pH).

(a)

1 2 3 4

(b)

Figure 3.22 Sample semisolid media. (a) Semisolid media have more body than liquid media but are softer than solid media. They do not flow freely and have a soft, clotlike consistency. **(b)** Sulfur indole motility (SIM) medium. (1) An uninoculated tube. The location of growth can be used to determine nonmotility (2) or motility (3). The medium reacts with any H_2S gas to produce a black precipitate (4).

needle and later observed for the pattern of growth around the stab line. In addition to motility, SIM can test for physiological characteristics used in identification (hydrogen sulfide production and indole reaction).

Solid media provide a firm surface on which cells can form discrete colonies (see figure 3.18) and are advantageous for isolating and culturing bacteria and fungi. They come in two forms: liquefiable and nonliquefiable. Liquefiable solid media, sometimes called reversible solid media, contain a solidifying agent that changes its physical properties in response to temperature. By far the most widely used and effective of these agents is **agar**,[2] a polysaccharide isolated from the red alga *Gelidium*. The benefits of agar are numerous. It is solid at room temperature, and it melts (liquefies) at the boiling temperature of water (100°C or 212°F). Once liquefied, agar does not resolidify until it cools to 42°C (108°F), so it can be inoculated and poured in liquid form at temperatures (45°C to 50°C) that will not harm the microbes or the handler (body temperature is about 37°C or 98.6°F). Agar is flexible and moldable, and it provides a basic framework to hold moisture and nutrients. Another useful property is that it is not readily digestible and thus not a nutrient for most microorganisms.

Any medium containing 1% to 5% agar usually has the word *agar* in its name. *Nutrient agar* is a common one. Like nutrient broth, it contains beef extract and peptone, as well as 1.5% agar by weight. Many of the examples covered in the section on functional categories of media contain agar.

Although gelatin is not nearly as satisfactory as agar, it will create a reasonably solid surface in concentrations of 10% to 15%. The main drawback for gelatin is that it can be digested by microbes and will melt at room and warmer temperatures, leaving a liquid. Agar and gelatin media are illustrated in **figure 3.23.**

Nonliquefiable solid media do not melt. They include materials such as rice grains (used to grow fungi), cooked meat media (good for anaerobes), and egg or serum media that are permanently coagulated or hardened by moist heat.

Chemical Content of Media

Media with a **chemically defined** composition are termed **synthetic.** Such media contain pure chemical nutrients that vary little from one source to another and have a molecular content specified by means of an exact formula. Synthetic media come in many forms. Some media, such as minimal media for fungi, contain nothing more than a few salts and amino acids dissolved in water. Others contain dozens of precisely measured ingredients **(table 3.6A).** Such standardized and reproducible media are most useful in research and cell culture. But they can only be used when the exact nutritional needs of the test organisms are known. Recently a defined medium that was developed to grow the parasitic protozoan *Leishmania* required 75 different chemicals.

If even one component of a given medium is not chemically definable, the medium is a **nonsynthetic,** or **complex,**[3] **medium.** The composition of this type of medium is not definable by an exact chemical formula. Substances that can make it nonsynthetic are extracts from animal or plant tissues including such materials as ground-up cells and secretions. Other examples are blood, serum, meat extracts, infusions, milk, soybean digests, and peptone. Peptone is a partially digested protein, rich in amino acids, that is often used as a carbon and nitrogen source. Nutrient broth, blood agar, and MacConkey agar, though different in function and appearance, are all complex nonsynthetic media. They present a rich mixture of nutrients for microbes with complex nutritional needs.

Table 3.6 provides a practical comparison of the two categories, using a medium to grow *Staphylococcus aureus*. Every substance in medium A is known to a very precise degree. The dominant substances in medium B are macromolecules that contain unknown (but potentially required) nutrients. Both A and B will satisfactorily grow the bacteria. (Which one would you rather make?)

2. This material was first employed by Dr. Hesse (see appendix A, 1881).

3. *Complex* means that the medium has large molecules such as proteins, polysaccharides, lipids, and other chemicals that can vary greatly in exact composition.

(a)

(b)

Figure 3.23 Solid media that are reversible to liquids.
(a) Media containing 1%–5% agar are solid enough to remain in place when containers are tilted or inverted. They are reversibly solid and can be liquefied with heat, poured into a different container, and resolidified. **(b)** Nutrient gelatin contains enough gelatin (12%) to take on a solid consistency. The top tube shows it as a solid. The bottom tube indicates what happens when it is warmed or when microbial enzymes digest the gelatin and liquefy it.

TABLE 3.6A	Chemically Defined Synthetic Medium for Growth and Maintenance of Pathogenic *Staphylococcus aureus*		
0.25 Grams Each of These Amino Acids		**0.5 Grams Each of These Amino Acids**	**0.12 Grams Each of These Amino Acids**
Cystine		Arginine	Aspartic acid
Histidine		Glycine	Glutamic acid
Leucine		Isoleucine	
Phenylalanine		Lysine	
Proline		Methionine	
Tryptophan		Serine	
Tyrosine		Threonine	
		Valine	

Additional ingredients

0.005 mole nicotinamide ⎤
0.005 mole thiamine ⎥ —Vitamins
0.005 mole pyridoxine ⎥
0.5 micrograms biotin ⎦

1.25 grams magnesium sulfate ⎤
1.25 grams dipotassium hydrogen phosphate ⎥ —Salts
1.25 grams sodium chloride ⎥
0.125 grams iron chloride ⎦

Ingredients dissolved in 1,000 milliliters of distilled water and buffered to a final pH of 7.0.

TABLE 3.6B	Brain-Heart Infusion Broth: A Complex, Nonsynthetic Medium for Growth and Maintenance of Pathogenic *Staphylococcus aureus*

27.5 grams brain-heart extract, peptone extract

2 grams glucose

5 grams sodium chloride

2.5 grams disodium hydrogen phosphate

Ingredients dissolved in 1,000 milliliters of distilled water and buffered to a final pH of 7.0.

Media to Suit Every Function

Microbiologists have many types of media at their disposal, with new ones being devised all the time. Depending upon what is added, a microbiologist can fine-tune a medium for nearly any purpose. Microbiologists have always been aware of microbes that could not be cultivated artificially, and now we can detect a single bacterium in its natural habitat without cultivation (see Insight 3.3). But even with this new technology, it is highly unlikely that microbiologists will abandon culturing, simply because it provides a constant source of microbes for detailed study, research, and diagnosis.

General-purpose media are designed to grow a broad spectrum of microbes that do not have special growth requirements. As a rule, these media are nonsynthetic (complex) and contain a mixture

TABLE 3.8	Examples of Differential Media	
Medium	**Substances That Facilitate Differentiation**	**Differentiates**
Blood agar (BAP)	Intact red blood cells	Types of RBC damage (hemolysis)
Mannitol salt agar (MSA)	Mannitol and phenol red	Pathogenic *Staphylococcus* species
Hektoen enteric (HE) agar*	Brom thymol blue, acid fuchsin, sucrose, salicin, thiosulfate, ferric ammonium citrate, and bile	*Salmonella, Shigella,* and other lactose nonfermenters from fermenters; H_2S reactions are also observable.
MacConkey agar (MAC)	Lactose, neutral red	Bacteria that ferment lactose (lowering the pH) from those that do not
Eosin-methylene blue (EMB)	Lactose, eosin, methylene blue	Same as MacConkey agar
Urea broth	Urea, phenol red	Bacteria that hydrolyze urea to ammonia and increase the pH
Sulfur indole motility (SIM)	Thiosulfate, iron	H_2S gas producers; motility; indole formation
Triple-sugar iron agar (TSIA)	Triple sugars, iron, and phenol red dye	Fermentation of sugars, H_2S production
XLD agar	Lysine, xylose, iron, thiosulfate, phenol red	Can differentiate *Enterobacter, Escherichia, Proteus, Providencia, Salmonella,* and *Shigella*

*Contains dyes and bile to inhibit gram-positive bacteria.

(a)

(b)

Figure 3.27 Media that differentiate multiple characteristics of bacteria. (a) Triple sugar iron agar (TSIA) inoculated on the surface and stabbed into the thicker region at the bottom (butt). This medium contains three sugars, phenol red dye to indicate pH changes (bright yellow is acid, various shades of red, basic), and iron salt to show H_2S gas production. Reactions are (1) no growth; (2) growth with no acid production (sugars not used); (3) acid production in the butt only; (4) acid production in all areas of the medium; (5) acid and H_2S production in butt (black precipitate). **(b)** A medium developed for culturing and identifying the most common urinary pathogens. CHROMagar Orientation™ uses color-forming reactions to distinguish at least seven species and permits rapid identification and treatment. In the example, the bacteria were streaked so as to spell their own names. Which bacterium was probably used to write the name at the top?

for growing anaerobic bacteria or for determining oxygen requirements of isolates (described in chapter 7). Carbohydrate fermentation media contain sugars that can be fermented (converted to acids) and a pH indicator to show this reaction (see figure 3.26a and figure 3.28). Media for other biochemical reactions that provide the basis for identifying bacteria and fungi are presented in several later chapters.

Transport media are used to maintain and preserve specimens that have to be held for a period of time before clinical analysis or to sustain delicate species that die rapidly if not held under stable conditions. Stuart's and Amie's transport media contain buffers and absorbants to prevent cell destruction but will not support growth. *Assay media* are used by technologists to test the effectiveness of antimicrobial drugs (see chapter 12) and by drug manufacturers to assess the effect of disinfectants, antiseptics, cosmetics,

and preservatives on the growth of microorganisms. Enumeration media are used by industrial and environmental microbiologists to count the numbers of organisms in milk, water, food, soil, and other samples.

A number of significant microbial groups (viruses, rickettsias, and a few bacteria) will only grow on live cells or animals. These obligate parasites have unique requirements that must be provided

Gas bubble

Outline of
Durham tube

Cloudiness
indicating
growth

Figure 3.28 Carbohydrate fermentation in broths. This medium is designed to show fermentation (acid production) using phenol red broth and gas formation by means of a small, inverted Durham tube for collecting gas bubbles. The tube on the left is an uninoculated negative control; the center tube is positive for acid (yellow) and gas (open space); the tube on the right shows growth but neither acid nor gas.

by living animals such as rabbits, guinea pigs, mice, chickens, and the early life stages (embryos) of birds. Such animals can be an indispensable aid for studying, growing, and identifying microorganisms. Animal inoculation is an essential step in testing the effects of drugs and the effectiveness of vaccines before they are administered to humans. Animals are an important source of antibodies, antisera, antitoxins, and other immune products that can be used in therapy or testing.

Check & Assess Section 3.4

✔ Microorganisms can be cultured on a variety of laboratory *media* that provide them with all of their required nutrients.

✔ Media can be classified by their *physical state* as liquid, semisolid, liquefiable solid, or nonliquefiable solid.

✔ Media can be classified by their *chemical composition* as either *synthetic* or *nonsynthetic*, depending on the precise content of their chemical composition.

✔ Media can be classified by their *function* as either general-purpose media or media with one or more specific purposes. Enriched, selective, differential, transport, assay, and enumerating media are all examples of media designed for specific purposes.

✔ Some microbes can be cultured only in living cells (animals, embryos, cell cultures).

23. Describe the main purposes of media, and compare the three categories based on physical state, chemical composition, and usage.
24. Differentiate among the ingredients and functions of enriched, selective, and differential media.
25. Explain the two principal functions of dyes in media.
26. Why are some bacteria difficult to grow in the laboratory? Relate this to what you know so far about metabolism.
27. What conditions are necessary to cultivate viruses in the laboratory?

CASE FILE 3 PERSPECTIVE

Signs and symptoms are noticeable manifestations in a patient that can direct diagnosis of a disease and help pinpoint which anatomical sites are affected. A sign is an objective observation that is measurable, such as a fever or white blood cell count. A symptom is something that a patient feels and reports; for instance, a headache or stiff neck. The most important initial diagnostic signs of infection in this case are an elevated white blood cell count, cloudy spinal fluid, and splotches on the legs. The most important symptoms are headache, stiff neck, and mental confusion.

Monitoring the cerebrospinal fluid (CSF) provides a means to quickly assess the presence of infectious agents in the spinal column and the brain. The CSF bathes the brain, spinal cord, and membranes and should be sterile. If it is cloudy macroscopically, this could indicate growth of bacteria or other infectious agents. A microscopic inspection can provide immediate feedback as to a possible cause.

A Gram stain is one of the key tests for getting quick feedback on the kind of microbes that might be present in a sample. It is routine in meningitis because it can differentiate among several bacteria and certain other infectious agents such as fungi, but it will not detect viruses. Within moments, a lab technician can tell if there are bacterial cells and can identify their gram reaction and shape. For example, gram-negative cocci were the finding in this case, but this disease is also commonly caused by *Streptococcus pneumoniae*, which would have shown gram-positive cocci, and *Haemophilus influenzae*, which has gram-negative rods. Having a good clue as to the microorganism also helps to instruct the types of drugs that will be given. Since meningitis can cause death in 5% to 10% of people in a few hours, early drug treatment is critical.

Neisseria meningitidis is routinely isolated and grown on blood agar, chocolate agar, and Thayer Martin medium. Its identity can be confirmed by a series of biochemical tests that differentiate it from close relatives that may look microscopically similar. See figure 3.20 for an example.

For more information on the nature of this agent and its disease, see chapter 18 and log on to *http://www.meningitis.org/symptoms*

Chapter Summary with Key Terms

3.1 Methods of Microbial Investigation

A. Microbiology as a science is very dependent on a number of specialized laboratory techniques.
 1. Initially, a specimen must be collected from a source, whether environmental or patient.
 2. **Inoculation** of a **medium** is the first step in obtaining a **culture** of the microorganisms present.
 3. **Isolation** of the microorganisms, so that each microbial cell present is separated from the others and forms discrete **colonies.**
 4. **Incubation** of the medium with the microbes under the right conditions allows growth of visible colonies.
 5. **Inspection** begins with macroscopic characteristics of the colonies and continues with microscopic analysis.
 6. **Information gathering** involves acquiring additional data from physiological, immune, and genetic tests.
 7. **Identification** correlates the key characteristics that can pinpoint the actual species of microbe.

3.2 The Microscope: Window on an Invisible Realm

A. Optical, or light, microscopy depends on lenses that **refract** light rays, drawing the rays to a focus to produce a magnified image.
 1. A simple microscope consists of a single magnifying lens, whereas a compound microscope relies on two lenses: the ocular lens and the objective lens.
 2. The total power of magnification is calculated from the product of the ocular and objective magnifying powers.
 3. **Resolution,** or the **resolving power,** is a measure of a microscope's capacity to make clear images of very small objects. Resolution is improved with shorter wavelengths of illumination and with a higher **numerical aperture** of the lens. Light microscopes are limited by resolution to magnifications around 2,000×.
 4. Modifications in the lighting or the lens system give rise to the bright-field, dark-field, phase-contrast, interference, fluorescence, and confocal microscopes.

B. Electron microscopy depends on electromagnets that serve as lenses to focus electron beams. A **transmission electron microscope (TEM)** projects the electrons through prepared sections of the specimen, providing detailed structural images of cells, cell parts, and viruses. A **scanning electron microscope (SEM)** is more like dark-field microscopy, bouncing the electrons off the surface of the specimen to detectors.

C. Specimen preparation in optical microscopy varies according to the specimen, the purpose of the inspection, and the type of microscope being used.
 1. Wet mounts and hanging drop mounts permit examination of the characteristics of live cells, such as motility, shape, and arrangement.
 2. Fixed mounts are made by drying and heating a film of the specimen called a smear. This is then stained using dyes to permit visualization of cells or cell parts.

D. Staining uses either **basic** (cationic) **dyes** with positive charges or **acidic** (anionic) **dyes** with negative charges. The surfaces of microbes are negatively charged and attract basic dyes. This is the basis of **positive staining.** In **negative staining,** the microbe repels the dye and it stains the background. Dyes may be used alone and in combination.
 1. **Simple stains** use just one dye and highlight cell morphology.
 2. **Differential stains** require a primary dye and a contrasting counterstain in order to distinguish cell types or parts. Important differential stains include the **Gram stain, acid-fast stain,** and the endospore stain.
 3. **Structural stains** are designed to bring out distinctive characteristics. Examples include capsule stains and flagellar stains.

3.3 Additional Features of the Six "I's"

A. Inoculation of media followed by incubation produces visible growth in the form of **cultures**.

B. Techniques with solid media in Petri dishes provide a means of separating individual microbes by producing **isolated colonies**.
 1. With the **streak plate,** a loop is used to thin out the sample over the surface of the medium.
 2. With the **loop dilution** (pour plate), a series of tubes of media is used to dilute the sample.
 3. The **spread plate** method evenly distributes a tiny drop of inoculum with a special stick.
 4. Isolated colonies can be **subcultured** for further testing at this point. The goal is a **pure culture,** in most cases, or a **mixed culture. Contaminated cultures** can ruin correct analysis and study.

3.4 Media: The Foundations of Culturing

 1. Artificial media allow the growth and isolation of microorganisms in the laboratory and can be classified by their physical state, chemical composition, and functional types. The nutritional requirements of microorganisms in the laboratory may be simple or complex.
 2. Physical types of media include those that are **liquid,** such as **broths** and milk, those that are **semisolid,** and those that are solid. Solid media may be liquefiable, containing a solidifying agent such as **agar** or gelatin.
 3. Chemical composition of a medium may be completely **chemically defined,** thus **synthetic. Nonsynthetic,** or **complex,** media contain ingredients that are not completely definable.
 4. Functional types of media serve different purposes, often allowing biochemical tests to be performed at the same time. Types include **general-purpose, enriched, selective,** and **differential media.**
 a. Enriched media contain **growth factors** required by microbes.
 b. Selective media permit the growth of desired microbes while inhibiting unwanted ones.
 c. Differential media bring out visible variations in microbial growth.
 d. Others include anaerobic (reducing), assay, and enumeration media. Transport media are important for conveying certain clinical specimens to the laboratory.
 5. In certain instances, microorganisms have to be grown in cell cultures or host animals.

Multiple-Choice Questions

Select the correct answer from the answers provided. For questions with blanks, choose the combination of answers that most accurately completes the statement.

1. Which of the following is not one of the six "I's"?
 a. inspection
 d. incubation
 b. identification
 e. inoculation
 c. illumination

2. The term *culture* refers to the _____ growth of microorganisms in _____.
 a. rapid, an incubator
 c. microscopic, the body
 b. macroscopic, media
 d. artificial, colonies

3. A mixed culture is
 a. the same as a contaminated culture
 b. one that has been adequately stirred
 c. one that contains two or more known species
 d. a pond sample containing algae and protozoa

4. Agar is superior to gelatin as a solidifying agent because agar
 a. does not melt at room temperature
 b. solidifies at 75°C
 c. is not usually decomposed by microorganisms
 d. both a and c

5. The process that most accounts for magnification is
 a. a condenser
 c. illumination
 b. refraction of light rays
 d. resolution

6. A subculture is a
 a. colony growing beneath the media surface
 b. culture made from a contaminant
 c. culture made in an embryo
 d. culture made from an isolated colony

7. Resolution is _____ with a longer wavelength of light.
 a. improved
 c. not changed
 b. worsened
 d. not possible

8. A real image is produced by the
 a. ocular
 c. condenser
 b. objective
 d. eye

9. A microscope that has a total magnification of 1,500× when using the oil immersion objective has an ocular of what power?
 a. 150×
 c. 15×
 b. 1.5×
 d. 30×

10. The specimen for an electron microscope is always
 a. stained with dyes
 c. killed
 b. sliced into thin sections
 d. viewed directly

11. Motility is best observed with a
 a. hanging drop preparation
 c. streak plate
 b. negative stain
 d. flagellar stain

12. Bacteria tend to stain more readily with cationic (positively charged) dyes because bacteria
 a. contain large amounts of alkaline substances
 b. contain large amounts of acidic substances
 c. are neutral
 d. have thick cell walls

13. The primary difference between a TEM and SEM is in
 a. magnification capability
 b. colored versus black-and-white images
 c. preparation of the specimen
 d. type of lenses

14. A fastidious organism must be grown on what type of medium?
 a. general-purpose medium
 c. synthetic medium
 b. differential medium
 d. enriched medium

15. What type of medium is used to maintain and preserve specimens before clinical analysis?
 a. selective medium
 c. enriched medium
 b. transport medium
 d. differential medium

16. Which of the following is NOT an optical microscope?
 a. dark-field
 c. atomic force
 b. confocal
 d. fluorescent

17. **Multiple Matching.** For each type of medium, select all descriptions that fit. For media that fit more than one description, briefly explain why this is the case.

 a,b,f mannitol salt agar
 d,f chocolate agar
 a,b,f MacConkey agar
 b,f nutrient broth
 d,f brain-heart infusion broth
 a,f Sabouraud's agar
 b,f triple-sugar iron agar
 b,f SIM medium

 a. selective medium
 b. differential medium
 c. chemically defined (synthetic) medium
 d. enriched medium
 e. general-purpose medium
 f. complex medium
 g. transport medium

Case File Questions

1. Which of the following techniques from the six "I's" was not used in the diagnosis?
 a. incubation
 c. inspection
 b. inoculation
 d. All of them were used.

2. Which of these would not be a *sign* of meningitis?
 a. brown blotches on the skin
 c. cloudy spinal fluid
 b. stiff neck
 d. gram-negative cocci in specimens

Writing to Learn

These questions are suggested as a *writing-to-learn* experience. For each question, compose a one- or two-paragraph answer that includes the factual information needed to completely address the question.

1. a. When buying a microscope, what features are most important to check for? pg. 61
 b. What is probably true of a $20 microscope that claims to magnify 1,000×?

2. How can one obtain 2,000× magnification with a 100× objective?

3. a. In what ways are dark-field microscopy and negative staining alike?
 b. How is the dark-field microscope like the scanning electron microscope?

4. Differentiate between microscopic and macroscopic methods of observing microorganisms, citing a specific example of each method.

5. Describe the steps of the Gram stain, and explain how it can be an important diagnostic tool for infections. pg. 73

6. Describe the steps you would take to isolate, cultivate, and identify a microbial pathogen from a urine sample. isolate cultivate identify

7. Trace the pathway of light from its source to the eye, explaining what happens as it passes through the major parts of the microscope.

Concept Mapping

Appendix E provides guidance for working with concept maps.

1. Supply your own linking words or phrases in the concept map, and provide the missing concepts in the empty boxes.

Critical Thinking Questions

Critical thinking is the ability to reason and solve problems using facts and concepts. These questions can be approached from a number of angles, and in most cases, they do not have a single correct answer.

1. A certain medium has the following composition:

Glucose	15 g
Yeast extract	5 g
Peptone	5 g
KH_2PO_4	2 g
Distilled water	1,000 ml

 a. Tell what chemical category this medium belongs to, and explain why this is true.
 b. Both A and B media in table 3.6 have the necessary nutrients for *S. aureus,* yet they are very different. Suggest the sources of most required chemicals in B.
 c. How could you convert *Staphylococcus* medium (table 3.6A) into a nonsynthetic medium?

2. a. Name four categories that blood agar fits.
 b. Name four differential reactions that TSIA shows.
 c. Observe figure 3.22. Suggest what causes the difference in growth pattern between nonmotile and motile bacteria.
 d. Explain what a medium that is both selective and differential does, using figure 3.25.

3. a. What kind of medium might you make to selectively grow a bacterium that lives in the ocean?
 b. One that lives in the human stomach?
 c. What characteristic of dyes makes them useful in differential media?
 d. Why are intestinal bacteria able to grow on media containing bile?

4. Go back to page 6 and observe the six micrographs in figure 1.3. See if you can tell what kind of microscope was used to make the photograph based on magnification and appearance.

5. Could the Gram stain be used to diagnose the flu? Why or why not?

6. Go to figure 3.20 and trace what information was used to "key out" the bacterial species that caused the meningitis in the case file.

7. Evaluate the following preparations in terms of showing microbial size, shape, motility, and differentiation: spore stain, negative stain, simple stain, hanging drop slide, and Gram stain.

8. Biotechnology companies have engineered hundreds of different types of mice, rats, pigs, goats, cattle, and rabbits to have genetic diseases similar to diseases of humans or to synthesize drugs and other biochemical products. They have patented these animals and are selling them to researchers for study and experimentation. Comment on the benefits, safety, and ethics of creating new animals just for experimentation.

9. This is a test of your living optical system's resolving power. Prop your book against a wall about 20 inches away and determine the line in the illustration below that is no longer resolvable by your eye. See if you can determine your actual resolving power, using a millimeter ruler.

So, Naturalists observe,

a flea has smaller

fleas that on him prey;

and these have smaller still

to bite 'em; and so proceed,

ad infinitum.

Poem by Jonathan Swift.

10. Some human pathogenic bacteria are resistant to most antibiotics. How would you prove a bacterium is resistant to antibiotics using laboratory culture techniques?

Visual Challenge

1. The image shown here is a Gram stain of spinal fluid. What can you conclude about its appearance and how it came to look this way?

2. Look at figure 3.1 on page 60. Precisely which of the six "I's" were used in case file 3? Which ones were used in case file 1? Were any used in case file 2?

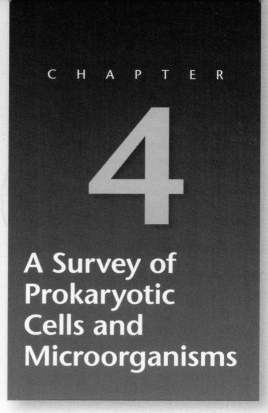

CHAPTER

4

A Survey of Prokaryotic Cells and Microorganisms

"Over 65% of chronic infections are caused by microbial biofilms."

Section of a prosthetic heart valve has a patch of MRSA biofilm attached (purple).

Looking as harmless as clusters of tiny purple grapes, the gram-positive pathogen *Staphylococcus aureus* is anything but.

CASE FILE 4 **Heart Valves and Biofilms**

On a summer morning in 2008, Mr. Maxwell Jones, a 65-year-old man, woke up complaining of abnormal fatigue and a scratchy throat. His wife said he felt hot and took his temperature. It was slightly elevated at 100°F. He dismissed his condition, saying he was probably tired from working in his garden and suffering one of his regular allergy attacks. Over the next few days, his symptom list grew. He lost his appetite, his joints and muscles were sore, and he woke up wringing wet from night sweats. He continued to have a fever, and his wife was worried over how pale he looked. She insisted he see a physician, who performed a physical and took a throat culture. Mr. Jones was sent home with instructions to take oral penicillin and Tylenol, and to come back in a week.

At the next appointment, the patient reported that he still had some of the same symptoms, including the fever, and that now he had begun to have headaches, rapid breathing, and coughing. The physician recorded a rapid heart rate and slight heart murmur. When the lab report indicated that the throat culture was negative, he had to look for other causes.

He began to wonder if the patient had a prior medical history of possible risk factors. From interviewing Mr. Jones, he learned that an artificial valve had been implanted in his heart 10 years before, a fact that had been omitted from his current chart. This finding immediately caused alarm, and Mr. Jones was admitted to the intensive care unit and placed on a mixture of intravenous antibiotics. Tests for blood cultures and a white blood cell count were ordered as backup. By that evening, Mr. Jones had become confused and lost consciousness. He was rushed to the operating room but died during open heart surgery.

▶ *What appear to be the most important facts in this case?*

▶ *What information would have helped in earlier diagnosis?*

To continue the case, go to page 96.

4.1 Basic Characteristics of Cells and Life Forms

Expected Learning Outcomes

1. Describe the fundamental characteristics of cells.
2. Identify the primary properties that define life and living things.

There is a universal biological truth that the basic unit of life is the cell, whether the organism is a bacterium whose entire body is just a single cell or an elephant made up of trillions of cells. Regardless of their origins, all cells share a few common characteristics. They tend to assume cubical, spherical, or cylindrical shapes, and have a **cell membrane** that encases an internal matrix called the **cytoplasm.** All cells have one or more chromosomes containing DNA, ribosomes for protein synthesis, and they exhibit highly complex chemical reactions. As we learned in chapter 1, all cells discovered thus far fall into one of two fundamentally different lines: the small, seemingly simple **prokaryotic cells** and the larger, structurally more complicated **eukaryotic cells.**

Eukaryotic cells are found in animals, plants, fungi, and protists. They contain a number of complex internal parts called **organelles** that perform useful functions for the cell. By convention, organelles are defined as cell components enclosed by membranes that carry out specific activities involving metabolism, nutrition, and synthesis. Organelles also partition the eukaryotic cell into smaller compartments. The most visible organelle is the nucleus, a roughly ball-shaped mass surrounded by a double membrane that contains the DNA of the cell. Other organelles include the Golgi apparatus, endoplasmic reticulum, vacuoles, and mitochondria (discussed in chapter 5).

Prokaryotic cells are present only in the bacteria and archaea. Sometimes it may seem that prokaryotes are the microbial "havenots" because, for the sake of comparison, they are described by what they lack. They have no nucleus or other organelles. This apparent simplicity is misleading, because the fine structure of prokaryotes can be complex. Overall, prokaryotic cells can engage in nearly every activity that eukaryotic cells can, and many can function in ways that eukaryotes cannot.

We start this chapter with a description of the characteristics that give living things the essence of life, followed by a look at prokaryotic cell anatomy and a survey of major groups of prokaryotes. In chapter 5, we will similarly survey the eukaryotic world.

After you have studied the cells as presented in this chapter and chapter 5, refer to table 5.2 (page 135), which summarizes the major differences between prokaryotic and eukaryotic cells.

What Is Life?

Biologists have long debated over universal characteristics we see in organisms that are solid indicators of life or of being alive. What does a cell do that sets it apart from a nonliving rock? One of the first things that often comes to mind is the ability to move or to grow. Unfortunately, taken individually, these are not life signs. After all, inanimate objects can move and crystals can grow. There is probably no single property that we can hold up as the ultimate indicator of life. In fact, defining life requires a whole collection of behaviors and properties that even the simplest organisms will have. First and foremost on this list would be a self-contained staging unit to carry out the activities of life, namely a cell. It is here that the life-supporting events of heredity, reproduction, growth, development, metabolism, responsiveness, and transport happen. Additional qualities that are often included in a "life list" are self-regulation and evolutionary change.

Heredity is the transmission of genetic material to the next generation in the form of chromosomes carrying DNA. It is tied to **reproduction,** which is the generation of offspring that keeps an organism's line going. Some organisms reproduce asexually, meaning they simply divide into two new cells by fission or mitosis. Other organisms produce new individuals sexually through the union of sex cells from two parents. **Growth** has two general meanings in microbiology. In one sense, it can mean the increase in the size of a population due to reproduction. It can also mean the increase in overall size of a single organism by an increase in bulk. **Development** is the process of cell change that leads to the full expression of an organism's genetic traits. This group of life properties are the forces behind long-term, gradual changes associated with evolution.

Cells undergo thousands of chemical reactions that form the basis of **metabolism.** Some reactions are synthetic and give rise to new cell components. Other reactions release energy that can drive cell activities. Both types of metabolism are made possible and regulated by special molecules of life called enzymes.

To say that cells are **responsive** means that they can react to external conditions. Some of the ways that they respond are through irritability, communication, and movement. Cells demonstrate irritability by receiving and reacting to stimuli such as light and chemicals from their environment. Part of this response may involve communicating with other cells by giving or receiving signals. Some cells are capable of self-propulsion or motility by moving themselves around in their environment with specialized locomotor structures, often in response to stimuli.

As part of their life scheme, cells must **transport** chemical nutrients from the external environment across their outer boundaries to the cell interior. Without these raw materials, metabolism would cease. Cells also secrete substances or expel waste in the reverse direction. Most transport of this type is largely the work of the cell membrane, which is the gatekeeper for many cell activities.

One of the main reasons that viruses are usually considered nonliving is that they do not have cells. Even if one argues that they are tiny units somewhat like cells and that they do show some traits of life such as heredity, development, and evolution, they do not display these characteristics independent of their living host cell. Outside of their host, they lack most other features of life we just described and are inactive and inert. We will return to this discussion about viruses in chapter 6.

4.2 Prokaryotic Profiles: The Bacteria and Archaea

3. Characterize the organization of a prokaryotic cell.
4. Describe the generalized anatomy of bacterial cells.
5. Distinguish among the types of external cell appendages.
6. Describe the structure and position of bacterial flagella and axial filaments, and their attachment patterns.
7. Explain how flagella influence motility and motile behavior.
8. Discuss the structure and functions of pili and fimbriae.
9. Define glycocalyx, and describe its different forms and functions.

The evolutionary history of prokaryotic cells extends back over 3.5 billion years. It is now generally thought that the very first cells to appear on the earth were similar to archaea that live on sulfur compounds in geothermal ocean vents (see figure 4.33). The fact that these organisms have endured for so long in such a variety of habitats indicates a cellular structure and function that are amazingly versatile and adaptable. The general cellular organization of a prokaryotic cell can be represented with this flowchart:

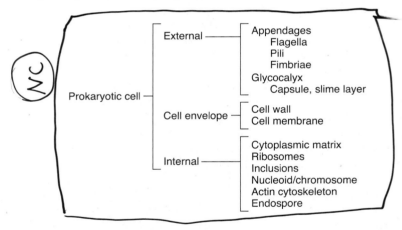

Structures that are essential to the functions of all prokaryotic cells are a cell membrane, cytoplasm, ribosomes, and one or a few chromosomes. The majority also have a cell wall and some form of surface coating or glycocalyx. Specific structures that are found in some, but not all, bacteria are flagella, pili, fimbriae, capsules, slime layers, inclusions, an actin cytoskeleton, and endospores.

The Structure of a Generalized Bacterial Cell

Prokaryotic cells appear featureless and two-dimensional when viewed with an ordinary microscope, but this is only because of their small size. Higher magnification provides increased insight into their intricate and often complex structure (see figures 4.17 and 4.28). The descriptions of prokaryotic structure, except where otherwise noted, refer to the **bacteria,** a category of prokaryotes with peptidoglycan

in their cell walls. **Figure 4.1** shows a three-dimensional illustration of a generalized (rod-shaped) bacterial cell with most of the structures from the flowchart. As we survey the principal anatomical features of this cell, we begin with the outer cell structures and proceed to the internal contents. Archaea—the other major group of prokaryotes—are discussed in section 4.7.

Cell Extensions and Surface Structures

Bacteria often bear accessory **appendages** sprouting from their surfaces. Appendages can be divided into two major groups: those that provide motility (flagella and axial filaments) and those that provide attachments or channels (fimbriae and pili).

Flagella—Bacterial Propellers

The prokaryotic **flagellum*** is an appendage of truly amazing construction and is certainly unique in the biological world. Flagella provide the power of motility or self-propulsion. This allows a cell to swim freely through an aqueous habitat. The bacterial flagellum when viewed under high magnification displays three distinct parts: the filament, the hook (sheath), and the basal body (**figure 4.2**). The **filament** is a helical structure composed of a protein called flagellin. It is approximately 20 nm in diameter and varies from 1 to 70 µm in length. It is inserted into a curved, tubular hook. The hook is anchored to the cell by the **basal body,** a stack of rings firmly anchored through the cell wall to the cell membrane. The hook and its filament are free to rotate 360°—like a tiny propeller. This is in contrast to the flagella of eukaryotic cells, which undulate back and forth.

One can generalize that all spirilla, about half of the bacilli, and a small number of cocci are flagellated (these bacterial shapes are shown in figure 4.23). Flagella vary both in number and arrangement according to two general patterns: (1) In a **polar** arrangement, the flagella are attached at one or both ends of the cell. Three subtypes of this pattern are: **monotrichous*** with a single flagellum; **lophotrichous*** with small bunches or tufts of flagella emerging from the same site; and **amphitrichous*** with flagella at both poles of the cell. (2) In a **peritrichous*** arrangement, flagella are dispersed randomly over the surface of the cell (**figure 4.3**).

The presence of motility is one piece of information used in the laboratory identification of various groups of bacteria. Special stains or electron microscope preparations must be used to see arrangement, because flagella are too minute to be seen in live preparations with a light microscope. Often it is sufficient to know simply whether a bacterial species is motile. One way to detect motility is to stab a tiny mass of cells into a soft (semisolid) medium in a test tube (see figure 3.5). Growth spreading rapidly through the entire medium is indicative of motility. Alternatively, cells can be observed microscopically with a hanging drop slide. A truly motile cell will flit, dart, or wobble around the field, making some progress, whereas one that is nonmotile jiggles about in one place but makes no progress.

* *flagellum* (flah-jel′-em) pl. flagella; L., a whip.
* *monotrichous* (mah″-noh-trik′-us) Gr. *mono*, one, and *tricho*, hair.
* *lophotrichous* (lo″-foh-trik′-us) Gr. *lopho*, tuft or ridge.
* *amphitrichous* (am″-fee-trik′-us) Gr. *amphi*, on both sides.
* *peritrichous* (per″-ee-trik′-us) Gr. *peri*, around.

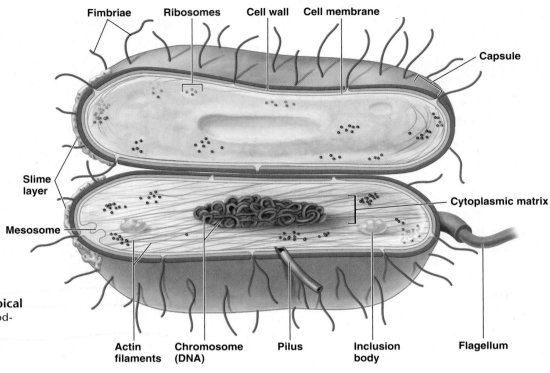

Figure 4.1 Structure of a typical bacterial cell. Cutaway view of a rod-shaped bacterium, showing major structural features. Note that not all components are found in all cells.

Figure 4.2 Details of the flagellar basal body and its position in the cell wall. The hook, rings, and rod function together as a tiny device that rotates the filament 360°. **(a)** Structure in gram-negative cells **(b)** Structure in gram-positive cells.

Flagellar Responses Flagellated bacteria can perform some rather sophisticated feats. They can detect and move in response to chemical signals—a type of behavior called **chemotaxis.*** Positive chemotaxis is movement of a cell in the direction of a favorable chemical stimulus (usually a nutrient); negative chemotaxis is movement away from a repellent (potentially harmful) compound.

The flagellum can guide bacteria in a certain direction because the system for detecting chemicals is linked to the mechanisms that drive the flagellum. Located in the cell membrane are clusters of receptors[1] that bind specific molecules coming from the immediate environment. The attachment of sufficient numbers of these molecules transmits signals to the flagellum and sets it into rotary motion. If several flagella are present, they become aligned and rotate

* *chemotaxis* (ke"-moh-tak′-sis) Gr. *chemo,* chemicals, and *taxis,* an ordering or arrangement.

1. Cell surface molecules that bind specifically with other molecules.

as a group (**figure 4.4**). As a flagellum rotates counterclockwise, the cell itself swims in a smooth linear direction toward the stimulus; this action is called a *run*. Runs are interrupted at various intervals by *tumbles* caused by the flagellum reversing its direction. This makes the cell stop and change its course. It is believed that attractant molecules inhibit tumbles, increase runs, and permit progress toward the stimulus (**figure 4.5**). Repellents cause numerous tumbles, allowing the bacterium to redirect itself away from the stimulus. Some photosynthetic bacteria exhibit *phototaxis*, a type of movement in response to light rather than chemicals.

Periplasmic Flagella

Corkscrew-shaped bacteria called **spirochetes*** show a worm-like or serpentine mode of locomotion caused by two or more long, coiled threads, the periplasmic flagella or *axial filaments*. A periplasmic flagellum is a type of internal flagellum that is enclosed in the space between the outer sheath and the cell wall peptidoglycan (**figure 4.6**). The filaments curl closely around the spirochete coils yet are free to contract and impart a twisting or flexing motion to the cell. This form of locomotion must be seen in live cells such as the spirochete of syphilis to be truly appreciated (this can be viewed on youtube: http://www.youtube.com/watch?v=i7BV4IOVKyg&feature=related)

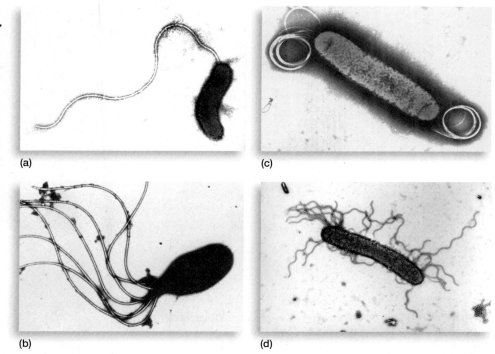

(a) (b) (c) (d)

Figure 4.3 **Electron micrographs depicting types of flagellar arrangements.** (a) Monotrichous flagellum on the predatory bacterium *Bdellovibrio*. (b) Lophotrichous flagella on *Vibrio fischeri,* a common marine bacterium (23,000×). (c) Unusual flagella on *Aquaspirillum* are amphitrichous (and lophotrichous) in arrangement and coil up into tight loops. (d) An unidentified bacterium discovered inside *Paramecium* cells exhibits peritrichous flagella.

* *spirochete* (spy´-roh-keet) Gr. *speira,* coil, and *chaite,* hair.

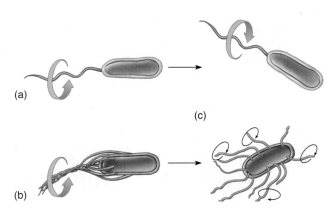

(a)

(c)

(b)

Figure 4.4 **The operation of flagella and the mode of locomotion in bacteria with polar and peritrichous flagella.** (a) In general, when a polar flagellum rotates in a counterclockwise direction, the cell swims forward in runs. When the flagellum reverses direction and rotates clockwise, the cell stops and tumbles. (b) In peritrichous forms, all flagella sweep toward one end of the cell and rotate as a single group. (c) During tumbles, the flagella rotate in the opposite direction and cause the cells to lose coordination.

Key

Tumble (T) Run (R) Tumble (T)

(a) **No attractant or repellent** (b) **Gradient of attractant concentration**

Figure 4.5 **Chemotaxis in bacteria.** (a) A cell moves via a random series of short runs and tumbles when there is no attractant or repellent. (b) The cell spends more time on runs as it gets closer to an attractant.

(a)

(b)

Figure 4.6 **The orientation of periplasmic flagella on the spirochete cell. (a)** Longitudinal section. **(b)** Cross section. Contraction of the filaments imparts a spinning and undulating pattern of locomotion.

Other Appendages: Fimbriae and Pili

The structures termed **fimbria*** and **pilus*** both refer to bacterial surface appendages that are involved in interactions with other cells but do not provide locomotion, except for some specialized pili.

Fimbriae are small, bristlelike fibers emerging from the surface of many bacterial cells **(figure 4.7).** Their exact composition varies, but most of them contain protein. Fimbriae have an inherent tendency to stick to each other and to surfaces. They may be responsible for the mutual clinging of cells that leads to biofilms and other thick aggregates of cells on the surface of liquids and for the microbial colonization of inanimate solids such as rocks and glass **(Insight 4.1).** Some pathogens can colonize and infect host tissues because of a tight adhesion between their fimbriae and epithelial cells **(figure 4.7b).** For example, the gonococcus (agent of gonorrhea) colonizes the genitourinary tract, and *Escherichia coli* colonizes the intestine by this means. Mutant forms of these pathogens that lack fimbriae are unable to cause infections.

A pilus (also called a *sex pilus*) is an elongate, rigid tubular structure made of a special protein, *pilin*. So far, true pili have been found only on gram-negative bacteria, where they are utilized primarily in a "mating" process between cells called **conjugation,**[2] which involves a transfer of DNA from one cell to another **(figure 4.8).** A pilus from a donor cell unites with a recipient cell, thereby providing a connection for making the transfer. Production of pili is controlled genetically, and conjugation using pili takes place only between compatible gram-negative cells. Conjugation in

(a)

(b)

Figure 4.7 **Form and function of bacterial fimbriae.**
(a) Several cells of pathogenic *Escherichia coli* covered with numerous stiff fibers called fimbriae (30,000×). Note also the dark blue masses, which are chromosomes. **(b)** A row of *E. coli* cells tightly adheres by their fimbriae to the surface of intestinal cells (12,000×). This is how the bacterium clings and gains access to the inside of cells during an infection. (G = glycocalyx)

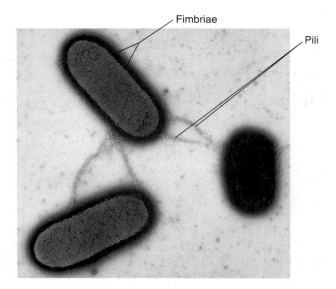

Figure 4.8 **Three bacteria in the process of conjugating.**
Clearly evident are the sex pili forming mutual conjugation bridges between a donor (upper cell) and two recipients (two lower cells). Fimbriae can also be seen on the donor cell.

* *fimbria* (fim′-bree-ah) pl. fimbriae; L., a fringe.
* *pilus* (py′-lus) pl. pili; L., hair.

2. Although the term *mating* is sometimes used for this process, it is not a form of sexual reproduction.

INSIGHT 4.1

 ## Biofilms—The Glue of Life

Microbes rarely live a solitary existence. More often they cling together in complex masses called **biofilms.** The formation of these living layers is actually a universal phenomenon that all of us have observed. Consider the scum that builds up in toilet bowls and shower stalls in a short time if they are not cleaned; or the algae that collect on the walls of swimming pools; and, more intimately, the constant deposition of plaque on teeth.

Fossils from ancient deposits tell us that microbes have been making biofilms for billions of years. It is through this process that they have colonized most habitats on earth and created stable communities that provide access to nutrients and other essential factors. Biofilms are often cooperative associations among several microbial groups (bacteria, fungi, algae, and protozoa) as well as plants and animals.

Substrates favorable to biofilm development have a moist, thin layer of organic material such as polysaccharides or glycoproteins deposited on their exposed surface. This has a sticky texture that attracts primary colonists, usually bacteria. These early cells attach and begin to multiply on the surface. As they secrete their glycocalyx (receptors, fimbriae, slime layers, capsules), the cells bind to the substrate and thicken the biofilm.

As the biofilm evolves, it undergoes specific adaptations to the habitat in which it forms. In many cases, the earliest colonists contribute nutrients and create microhabitats that serve as a matrix for other microbes to attach and grow into the film, forming complete communities. The biofilm varies in thickness and complexity, depending upon where it occurs and how long it keeps developing. Complexity ranges from single cell layers to thick microbial mats with dozens of dynamic interactive layers.

Biofilms are a profoundly important force in the development of terrestrial and aquatic environments. They dwell permanently in bedrock and the earth's sediments, where they play essential roles in recycling elements, leaching minerals, and forming soil. Biofilms associated with plant roots promote the mutual exchange of nutrients between the microbes and roots. Invasive biofilms can wreak havoc with human-made structures such as cooling towers, storage tanks, air conditioners, and even stone buildings.

New evidence now points to biofilm formation by our own microbiota—microbes that live naturally on the body. These associations are common on the skin, the oral cavity, and large intestine. In these locations, microbes signal each other as well as human cells in ways that shape the conditions there.

First colonists

Organic surface coating

Surface

Cells stick to coating.

Glycocalyx

As cells divide, they form a dense mat bound together by sticky extracellular deposits.

Additional microbes are attracted to developing film and create a mature community with complex function.

Biofilms accumulate on damaged tissues (such as rheumatic heart valves), hard tissues (teeth), and foreign materials (catheters, IUDs, artificial hip joints). Microbes in a biofilm are extremely difficult to eradicate with antimicrobials. New evidence indicates that bacteria in biofilms turn on different genes when they are in a biofilm than when they are "free-floating." This altered gene expression gives the bacteria a different set of characteristics, often making them impervious to antibiotics and disinfectants. For additional discussions of biofilms, see chapters 7 and 12.

Describe some possible benefits of having biofilms form in or on the human body. *Answer available at http://www.mhhe.com/talaro8*

gram-positive bacteria does occur but involves aggregation proteins rather than sex pili. The roles of pili in conjugation are further explored in chapter 9.

Certain pili can serve another function besides being cell connectors and transfer agents. Type IV pili found in *Pseudomonas* bacteria carry out a remarkable sort of twitching motility. Cells can slide in jerking movements over moist surfaces by extending and then retracting pili in a repetitive motion. Microbiologists have found that this mechanism is every bit as complex and regulated as the flagellar mode of locomotion. To observe this motility, go to http://www.youtube.com/watch?v=mlvJKz_bV7U.

The Bacterial Surface Coating, or Glycocalyx

The bacterial cell surface is frequently exposed to severe environmental conditions. The **glycocalyx** develops as a coating of macromolecules to protect the cell and, in some cases, help it adhere to its environment. Glycocalyces differ among bacteria in thickness, organization, and chemical composition. Some bacteria are covered with a loose shield called a **slime layer** that evidently protects them from dehydration and loss of nutrients (**figure 4.9a**). Other bacteria produce **capsules** composed of repeating polysaccharide units, of protein, or of both. A capsule is

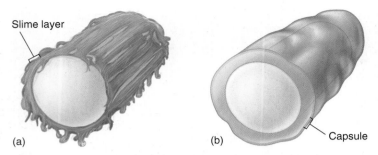

Slime layer

(a)

(b)

Capsule

Figure 4.9 Types of glycocalyces seen through cutaway views of cells. (a) The slime layer is a loose structure that is easily washed off. (b) The capsule is a thick, structured layer that is not readily removed.

bound more tightly to the cell than a slime layer is, and it has a thicker, gummy consistency that gives a prominently sticky (mucoid) character to the colonies of most encapsulated bacteria (**figure 4.10a**).

Specialized Functions of the Glycocalyx Capsules are formed by many pathogenic bacteria, such as *Streptococcus pneumoniae* (a cause of pneumonia, an infection of the lung), *Haemophilus influenzae* (one cause of meningitis), and *Bacillus anthracis* (the cause of anthrax). Encapsulated bacterial cells generally have greater pathogenicity because capsules protect the bacteria against white blood cells called phagocytes. Phagocytes are a natural body defense that can engulf and destroy foreign cells, which helps to prevent infection. A capsular coating blocks the mechanisms that phagocytes use to attach to and engulf bacteria. By escaping phagocytosis, the bacteria are free to multiply and infect body tissues. Encapsulated bacteria that mutate to nonencapsulated forms usually lose their pathogenicity.

Other types of glycocalyces can be important in formation of biofilms. The thick, white plaque that forms on teeth comes in

part from the surface slimes produced by certain streptococci in the oral cavity. This slime protects them from being dislodged from the teeth and provides a niche for other oral bacteria that, in time, can lead to dental disease. The glycocalyx of some bacteria is so highly adherent that it is responsible for persistent colonization of nonliving materials such as plastic catheters, intrauterine devices, and metal pacemakers that are in common medical use (**figure 4.11**).

CONTINUING CASE FILE 4

During an autopsy of Mr. Jones's body, the pathologist observed that the prosthetic valve was covered with small patches he called vegetations. The original blood cultures grew a strain of *Staphylococcus aureus,** known as MRSA. Microscopic examination of the valve revealed a thick biofilm coating containing that same bacterium. The pathologist concluded that the patient had infective endocarditis* and that vegetations on the valve lesions had broken loose and entered the circulation. This event created emboli that blocked arteries in his brain and gave rise to a massive stroke. Upon closer review of Mr. Jones's case, the physician discovered that he had suffered from a skin infection the previous spring that had been treated and cured by a different physician. It turned out to be caused by the MRSA type of *Staphylococcus aureus*.

- What is a biofilm and how did it form on the heart valve?
- What is the significance of MRSA?

For a wrap-up, see the Case File Perspective on page 119.

* *Staphylococcus aureus* (staf'-uh-loh-cok'-us ar-ee-us) Gr. *staphyle,* a bunch of grapes; *kokkus,* berry; and *aurum,* golden.
* *endocarditis* (en'-doh-car-dye'-tis) Gr. *endon,* within; *kardia,* heart; and *itis,* an inflammation.
An inflammation of the lining of the heart and its valves usually caused by infection.

Colony without a capsule

Colonies with a capsule

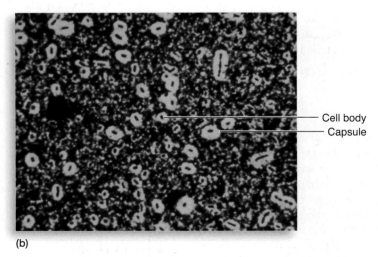

Cell body

Capsule

(a)

(b)

Figure 4.10 Appearance of encapsulated bacteria. (a) Close-up view of colonies of *Bacillus* species with and without capsules. Even at the macroscopic level, the moist slimy character of the capsule is evident. (b) Special staining reveals the microscopic appearance of a large, well-developed capsule (the clear "halo" around the cells) of *Klebsiella.*

Catheter
surface

Fungal
cells

Staphylococci

Figure 4.11 **Magnified view of a biofilm.** Scanning electron micrograph of a mixed biofilm of staphylococci and fungal cells attached to a catheter.

Check & Assess Sections 4.1 and 4.2

✔ Cells demonstrate a number of characteristics essential to life. Objects and macromolecules do not show these characteristics.

✔ Cells can be divided into two basic types: prokaryotes and eukaryotes. Prokaryotic cells include the bacteria and archaea.

✔ Prokaryotes are the oldest form of cellular life. They are also the most widely dispersed, occupying every conceivable microclimate on the planet.

✔ The external structures of bacteria include appendages (flagella, fimbriae, and pili) and a surface coating—the glycocalyx.

✔ Flagella vary in number and arrangement as well as in the type and rate of motion they produce.

✔ Bacterial cells have an outermost carbohydrate layer that functions in protection and adhesion termed a glycocalyx.

✔ Glycocalyces are thick clinging masses called capsules or thin, sticky surfaces called slime layers.

1. Name several general characteristics that could be used to define the prokaryotes.

2. What other microbial groups besides bacteria have prokaryotic cells?

3. Describe the structure of a flagellum and how it operates. What are the four main types of flagellar arrangement?

4. How does the flagellum dictate the behavior of a motile bacterium? Differentiate between flagella and periplasmic flagella.

5. List some direct and indirect ways that one can determine bacterial motility.

6. Differentiate between the structure and functions of pili and fimbriae.

7. Explain the position of the glycocalyx.

8. What are the functions of slime layers and capsules?

9. How is the presence of a slime layer evident even at the level of a colony?

10. Explain how the bacterial glycocalyx and certain surface appendages contribute to biofilm formation.

4.3 The Cell Envelope: The Boundary Layer of Bacteria

Expected Learning Outcomes

10. Explain the concept of the cell envelope, and describe its structure.

11. Outline the structure and functions of cell walls, and explain the role of peptidoglycan.

12. Contrast the major structure of gram-positive and gram-negative cell walls.

13. Summarize how gram-positive and gram-negative cells differ in their reactions.

14. Relate the characteristics of other types of cell walls and wall-free cells.

15. Describe the structure of the cell membrane, and explain several of its major roles in bacterial cells.

The majority of bacteria have a chemically complex external covering, termed the cell envelope, that lies outside of the cytoplasm. It is composed of two main layers: the cell wall and the cell membrane. These layers are stacked together and often tightly bound into a unit like the outer husk and casings of a coconut. Although each envelope layer performs a distinct function, together they act as a single unit that maintains cell integrity.

Basic Types of Cell Envelopes

Long before the detailed anatomy of bacteria was even remotely known, a Danish physician named **Hans Christian Gram** developed a staining technique, the **Gram stain,**[3] that delineates two generally different groups of bacteria. This stain was described in chapter 3. The two major groups shown by this technique are the gram-positive bacteria and the gram-negative bacteria. Because the Gram stain does not actually reveal the reasons that these two groups stain differently, we must turn to the electron microscope and to biochemical analysis of their structure.

3. This text follows the American Society of Microbiology style, which calls for capitalization of the terms *Gram stain* and *Gram staining* and lowercase treatment of gram-negative and gram-positive, except in headings.

The extent of the differences between gram-positive and gram-negative bacteria is evident in the physical appearance of their cell envelopes **(figure 4.12).** In **gram-positive** (see footnote 3) cells, a microscopic section resembles an open-faced sandwich with two layers: the thick cell wall, composed primarily of peptidoglycan (defined in the next section), and the cell membrane. A similar section of a gram-negative cell envelope shows a complete sandwich with three layers: an outer membrane, a thin peptidoglycan layer, and the cell membrane. **Table 4.1** provides a summary of the major similarities and differences between the wall types.

Structure of Cell Walls

The cell wall accounts for a number of important bacterial characteristics. In general, it helps determine the shape of a bacterium, and it also provides the kind of strong structural support necessary to keep a bacterium from bursting or collapsing because of changes in osmotic pressure. The cell walls of most bacteria gain their relative strength and stability from a unique macromolecule called **peptidoglycan (PG).** This compound is composed of a repeating framework of long **glycan*** chains cross-linked by short peptide fragments **(figure 4.13).** The amount and exact composition of peptidoglycan vary among the major bacterial groups.

Because many bacteria live in aqueous habitats with a low solute concentration, they are constantly absorbing excess water by osmosis. Were it not for the structural support of the peptidoglycan in the cell wall, they would rupture from internal pressure. Several types of drugs used to treat infection (penicillin, cephalosporins) are effective because they target the peptide cross-links in the peptidoglycan, thereby disrupting its integrity. With their cell walls incomplete or missing, such cells have very little protection from **lysis.*** Lysozyme, an enzyme contained in tears and saliva, provides a natural defense against certain bacteria by hydrolyzing the bonds in the glycan chains and causing the wall to break down. (Chapters 11 and 12 discuss the actions of antimicrobial chemical agents and drugs.)

The Gram-Positive Cell Wall

The bulk of the gram-positive cell wall is a thick, homogeneous sheath of peptidoglycan ranging from 20 to 80 nm in thickness. It also contains tightly bound acidic polysaccharides, including teichoic acid directly attached to the peptidoglycan and lipoteichoic acid **(figure 4.14).** Cell wall **teichoic acid** is a polymer of ribitol or glycerol and phosphate embedded in the peptidoglycan sheath. Lipoteichoic acid is similar in structure but is attached to the lipids in the plasma membrane. These molecules appear to function in cell wall maintenance and enlargement during cell division. They also move cations into and out of the cell and stimulate a specific immune response. The cell wall of gram-positive bacteria is often pressed tightly against the cell membrane with very little space between them, but in some cells, a thin **periplasmic*** **space** is evident between the cell membrane and cell wall.

Peptidoglycan
Cell membrane
Gram (+)
Cell membrane
Cell wall (peptidoglycan)
(a)

Cell membrane
Peptidoglycan
Outer membrane
Gram (−)
Cell membrane
Periplasmic space
Peptidoglycan — Cell wall
Outer membrane
(b)

Figure 4.12 **Comparative views of the envelopes of gram-positive and gram-negative cells. (a)** A section through a gram-positive cell wall/membrane with an interpretation of the main layers visible (85,000×). **(b)** A section through a gram-negative cell wall/membrane with an interpretation of its three sandwich-style layers (90,000×).

TABLE 4.1	Comparison of Gram-Positive and Gram-Negative Cell Walls	
Characteristic	**Gram-Positive**	**Gram-Negative**
Number of major layers	One	Two
Chemical composition	Peptidoglycan Teichoic acid Lipoteichoic acid Mycolic acids and polysaccharides*	Lipopolysaccharide (LPS) Lipoprotein Peptidoglycan Porin proteins
Overall thickness	Thicker (20–80 nm)	Thinner (8–11 nm)
Outer membrane	No	Yes
Periplasmic space	Narrow	Extensive
Permeability to molecules	More penetrable	Less penetrable

*In some cells.

* *glycan* (gly'-kan) Gr., sugar. These are large polymers of simple sugars.
* *lysis* (ly'-sis) Gr., to loosen. A process of cell destruction, as occurs in bursting.
* *periplasmic* (per"-ih-plaz'-mik) Gr. *peri,* around, and *plastos,* the fluid substances of a cell.

(a) The peptidoglycan of a cell wall is a huge, 3-dimensional lattice work that is actually one giant molecule to surround and support the cell.

(b) This shows the molecular pattern of peptidoglycan. It has alternating glycans (NAG and NAM) bound together in long strands. The NAG stands for *N*-acetyl glucosamine, and the NAM stands for *N*-acetyl muramic acid. Adjacent muramic acid molecules on parallel chains are bound by a cross-linkage of peptides (green spheres).

(c) An enlarged view of the links between the NAM molecules. Tetrapeptide chains branching off the muramic acids connect by amino acid interbridges. The amino acids in the interbridge can vary or may be lacking entirely. It is this linkage that provides rigid yet flexible support to the cell.

Figure 4.13 Structure of peptidoglycan component of the cell walls.

The Gram-Negative Cell Wall

The gram-negative cell wall is more complex in morphology because it is composed of an **outer membrane (OM)** and a thinner shell of peptidoglycan (see figures 4.12 and 4.14). The outer membrane is somewhat similar in construction to the cell membrane, except that it contains specialized types of **lipopolysaccharides (LPS)** and **lipoproteins.** Lipopolysaccharides are composed of lipid molecules bound to polysaccharides. The lipids form the matrix of the top layer of the OM, and the polysaccharide strands project from the lipid surface. The lipid portion may become toxic when it is released during infections. Its role as an endotoxin is described in chapters 13 and 20. The polysaccharides give rise to the somatic (O) antigen in gram-negative pathogens and can be used in identification. They may also function as receptors and in blocking host defenses.

Two types of proteins are located in the OM. The **porins** are inserted in the upper layer of the outer membrane. They have some regulatory control over molecules entering and leaving the cell. Many qualities of the selective permeability of gram-negative bacteria to bile, disinfectants, and drugs are due to the porins. Some structural proteins are also embedded in the upper layer of the OM. The bottom layer of the outer membrane is similar to the cell membrane in its overall structure and is composed of phospholipids and lipoproteins.

Gram-Positive **Gram-Negative**

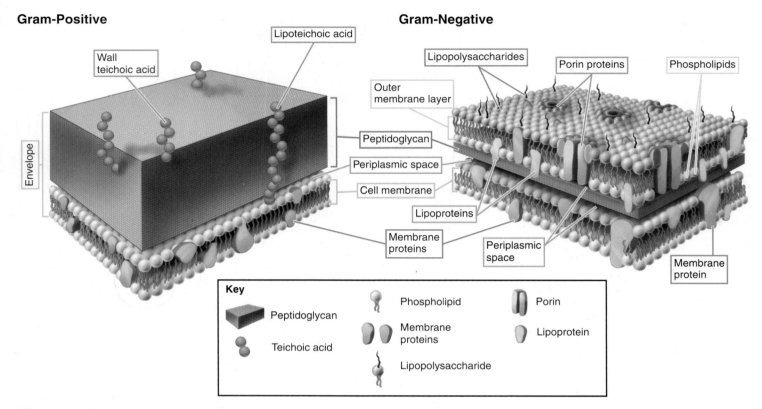

Figure 4.14 **A comparison of the detailed structure of gram-positive and gram-negative cell envelopes and walls.**

The bottom layer of the gram-negative wall is a single, thin (1–3 nm) sheet of peptidoglycan. Although it acts as a somewhat rigid protective structure, as previously described, its thinness gives gram-negative bacteria a relatively greater flexibility and sensitivity to lysis. There is a well-developed **periplasmic space** above and below the peptidoglycan. This space is an important reaction site for a large and varied pool of substances that enter and leave the cell.

Practical Considerations of Differences in Cell Wall Structure

Variations in cell wall anatomy contribute to several differences between the two cell types besides staining reactions. The outer membrane contributes an extra barrier in gram-negative bacteria that makes them more impervious to some antimicrobic chemicals such as dyes and disinfectants, so they are generally more difficult to inhibit or kill than are gram-positive bacteria. One exception is for alcohol-based compounds, which can dissolve the lipids in the outer membrane and disturb its integrity. Treating infections caused by gram-negative bacteria often requires different drugs than for gram-positive infections, especially drugs that can cross the outer membrane.

The cell wall or its parts can interact with human tissues and contribute to disease. The lipids have been referred to as *endotoxins* because they stimulate fever and shock reactions in gram-negative infections such as meningitis and typhoid fever. Proteins attached to the outer portion of the cell wall of several gram-positive species, including *Corynebacterium diphtheriae* (the agent of diphtheria) and *Streptococcus pyogenes* (the cause of strep throat), also have toxic properties. The lipids in the cell walls of certain *Mycobacterium* species are harmful to human cells as well. Because most macromolecules in the cell walls are foreign to humans, they stimulate antibody production by the immune system (see chapter 15).

Nontypical Cell Walls

Several bacterial groups lack the cell wall structure of gram-positive or gram-negative bacteria, and some bacteria have no cell wall at all. Although these exceptional forms can stain positive or negative in the Gram stain, examination of their fine structure and chemistry shows that they do not fit the descriptions for typical gram-negative or positive cells. For example, the cells of *Mycobacterium* and *Nocardia* contain peptidoglycan and stain gram-positive, but the bulk of their cell wall is composed of unique types of lipids. One of these is a very-long-chain fatty acid called *mycolic acid,* or cord factor, that contributes to the pathogenicity of this group (see chapter 19). The thick, waxy nature imparted to the cell wall by these lipids is also responsible for a high degree of resistance to certain chemicals and dyes. Such resistance is the basis for the **acid-fast stain** used to diagnose tuberculosis and leprosy. In this stain, hot carbol fuchsin dye becomes tenaciously attached (is held fast) to these cells so that an acid-alcohol solution will not remove the dye.

Because they are from a more ancient and primitive line of prokaryotes, the archaea exhibit unusual and chemically distinct cell walls. In some, the walls are composed almost entirely of polysaccharides, and in others, the walls are pure protein; but as a group, they all lack the true peptidoglycan structure described previously. Because a few archaea and all mycoplasmas (next section) lack a cell wall entirely, their cell membrane must serve the dual functions of support as well as transport.

Figure 4.15 **Scanning electron micrograph of** *Mycoplasma pneumoniae* **(100,000×).** Cells exhibit extreme variation in shape due to the lack of a cell wall.

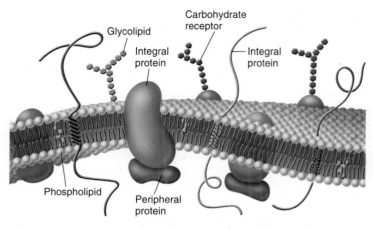

Figure 4.16 **Cell membrane structure.** A generalized version of the fluid mosaic model of a cell membrane indicates a bilayer of lipids with globular proteins embedded to some degree in the lipid matrix. This structure explains many characteristics of membranes, including flexibility, solubility, permeability, and transport.

Mycoplasmas and Other Cell-Wall-Deficient Bacteria

Mycoplasmas are bacteria that naturally lack a cell wall. Although other bacteria require an intact cell wall to prevent the bursting of the cell, the mycoplasma cell membrane contains sterols that make it resistant to lysis. These extremely tiny bacteria range from 0.1 to 0.5 μm in size. They range in shape from filamentous to coccus or doughnut-shaped. This property of extreme variations in shape is a type of **pleomorphism.*** They can be grown on artificial media, although added sterols are required for the cell membranes of some species. Mycoplasmas are found in many habitats, including plants, soil, and animals. The most important medical species is *Mycoplasma pneumoniae* **(figure 4.15),** which adheres to the epithelial cells in the lung and causes an atypical form of pneumonia in humans.

Some bacteria that ordinarily have a cell wall can lose it during part of their life cycle. These wall-deficient forms are referred to as **L forms** or L-phase variants (for the Lister Institute, where they were discovered). L forms arise naturally from a mutation in the wall-forming genes, or they can be induced artificially by treatment with a chemical such as lysozyme or penicillin that disrupts the cell wall. When a gram-positive cell is exposed to either of these two chemicals, it will lose the cell wall completely and become a **protoplast,*** a fragile cell bounded only by a membrane that is highly susceptible to lysis. A gram-negative cell exposed to these same substances loses its peptidoglycan but retains its outer membrane, leaving a less fragile but nevertheless weakened **spheroplast.*** Evidence points to a role for L forms in certain chronic infections.

Cell Membrane Structure

Appearing just beneath the cell wall is the **cell,** or **cytoplasmic, membrane,** a very thin (5–10 nm), flexible sheet molded completely around the cytoplasm. In general composition, it is a lipid bilayer with proteins embedded to varying degrees **(figure 4.16).**

The structure, first proposed by S. J. Singer and G. L. Nicolson, is called the **fluid mosaic model.** The model describes a membrane as a continuous bilayer formed by lipids with the polar heads oriented toward the outside and the nonpolar heads toward the center of the membrane. Embedded at numerous sites in this bilayer are various-size globular proteins. Some proteins are situated only at the surface; others extend fully through the entire membrane. The configuration of the inner and outer sides of the membrane can be quite different because of the variations in protein shape and position.

Membranes are dynamic and constantly changing because the lipid phase is in motion and many proteins can migrate freely about. This fluidity is essential to such activities as engulfment of food and discharge or secretion by cells. The structure of the lipid phase provides an impenetrable barrier to many substances. This property accounts for the selective permeability and capacity to regulate transport of molecules. Bacterial cell membranes have this typical structure, containing primarily phospholipids (making up about 30%–40% of the membrane mass) and proteins (contributing 60%–70%). Major exceptions to this description are the membranes of mycoplasmas, which contain high amounts of sterols—rigid lipids that stabilize and reinforce the membrane—and the membranes of archaea, which contain unique branched hydrocarbons rather than fatty acids.

In some locations, the cell membrane forms internal folds in the cytoplasm called **mesosomes*** (see figure 4.1). These are prominent in gram-positive bacteria but are harder to see in gram-negative bacteria because of their relatively small size. Mesosomes

* *pleomorphism* (plee″-oh-mor′-fizm) Gr. *pleon,* more, and *morph,* form or shape. The tendency for cells of the same species to vary to some extent in shape and size.
* *protoplast* (proh′-toh-plast) Gr. *proto,* first, and *plastos,* formed.
* *spheroplast* (sfer′-oh-plast) Gr. *sphaira,* sphere.

* *mesosome* (mes′-oh-sohm) Gr. *mesos,* middle, and *soma,* body.

presumably increase the internal surface area available for membrane activities. It has been proposed that mesosomes participate in cell wall synthesis and guiding the duplicated bacterial chromosomes into the two daughter cells during cell division (see figure 7.14). Some scientists are not convinced that mesosomes exist in live cells and believe they are artifacts that appear in bacteria fixed for electron microscopy. Support for the existence of functioning internal membranes such as mesosomes comes from specialized prokaryotes such as cyanobacteria and even mitochondria, considered as a type of prokaryotic cell. Some photosynthetic prokaryotes contain dense stacks of internal membranes that carry the photosynthetic pigments (see figure 4.28a).

Functions of the Cell Membrane

Because bacteria have none of the eukaryotic organelles, the cell membrane provides a site for energy reactions, nutrient processing, and synthesis. A major action of the cell membrane is to regulate *transport,* that is, the passage of nutrients into the cell and the discharge of wastes. Although water and small uncharged molecules can diffuse across the membrane unaided, the membrane is a *selectively permeable* structure with special carrier mechanisms for passage of most molecules (see chapter 7). The glycocalyx and cell wall can bar the passage of large molecules, but they are not the primary transport apparatus. The cell membrane is also involved in *secretion,* or the release of a metabolic product into the extracellular environment.

The membranes of bacteria are an important site for a number of metabolic activities. For example, most enzymes that handle the energy reactions of respiration reside in the cell membrane (see chapter 8). Enzyme structures located in the cell membrane also help synthesize structural macromolecules to be incorporated into the cell envelope and appendages. Other products (enzymes and toxins) are secreted by the membrane into the extracellular environment.

Check & Assess Section 4.3

✔ The cell envelope is the complex boundary structure surrounding a bacterial cell. It consists of the cell wall and membrane. In gram-negative bacteria, the cell envelope has three layers—the outer membrane, peptidoglycan, and cell membrane. In gram-positive bacteria, there are two—a thick layer of peptidoglycan and the cell membrane.

✔ In a Gram stain, gram-positive bacteria retain the crystal violet and stain purple. Gram-negative bacteria lose the crystal violet and stain red from the safranin counterstain.

✔ Gram-positive bacteria have thick cell walls of peptidoglycan and acidic polysaccharides such as teichoic acid, and they have a thin periplasmic space. The cell walls of gram-negative bacteria are thinner and have a wide periplasmic space.

✔ The outer membrane of gram-negative cells contains lipopolysaccharide (LPS). LPS is a source of substances called endotoxins.

✔ The bacterial cell membrane is typically composed of phospholipids and proteins, and it performs many metabolic functions as well as transport activities.

11. Compare the cell envelopes of gram-positive and gram-negative bacteria.
12. Explain the function of peptidoglycan and give a simple description of its structure.
13. What happens to a cell that has its peptidoglycan disrupted or removed?
14. What functions does the LPS layer serve?
15. How does the precise structure of the cell walls differ in gram-positive and gram-negative bacteria?
16. What other properties besides staining are different in gram-positive and gram-negative bacteria?
17. What is the periplasmic space, and how does it function?
18. What characteristics does the outer membrane confer on gram-negative bacteria?
19. Describe the structure of the cell membrane.
20. Why is it considered selectively permeable?
21. What are mesosomes and some proposed roles they play?
22. List five essential functions that the cell membrane performs in bacteria.

4.4 Bacterial Internal Structure

Expected Learning Outcomes

16. List the contents of the cell cytoplasm.
17. Describe features of the bacterial chromosome and plasmids.
18. Characterize the bacterial ribosomes and cytoskeleton.
19. Describe inclusion bodies and granules, and explain their importance to cells.
20. Describe the life cycle of endospore-forming bacteria, including the formation and germination of endospores.
21. Discuss the resistance and significance of endospores.

Contents of the Cell Cytoplasm

The cell membrane surrounds a complex solution referred to as **cytoplasm,** or cytoplasmic matrix. This chemical "pool" is a prominent site for many of the cell's biochemical and synthetic activities. Its major component is water (70%–80%), which serves as a solvent for a complex mixture of nutrients including sugars, amino acids, and other organic molecules and salts. The components of this pool serve as building blocks for cell synthesis or as sources of energy. The cytoplasm also holds larger, discrete bodies such as the chromosome, ribosomes, granules, and actin strands.

Bacterial Chromosomes and Plasmids: The Sources of Genetic Information

The hereditary material of most bacteria exists in the form of a single circular strand of DNA designated as the **bacterial chromosome,** although a few bacteria have multiple or linear chromosomes. By definition, bacteria do not have a true nucleus. Their

DNA is not enclosed by a nuclear membrane but instead is aggregated in a central area of the cell called the **nucleoid.** The chromosome is actually an extremely long molecule of DNA that is tightly coiled to fit inside the cell compartment. Arranged along its length are genetic units (genes) that carry information required for bacterial maintenance and growth. When exposed to special stains or observed with an electron microscope, chromosomes have a granular or fibrous appearance **(figure 4.17).**

Although the chromosome is the minimal genetic requirement for bacterial survival, many bacteria contain other, nonessential pieces of DNA called **plasmids.** These tiny strands exist apart from the chromosome, although at times they can become integrated into it. During bacterial reproduction, they are duplicated and passed on to offspring. They are not essential to bacterial growth and metabolism, but they often confer protective traits such as resisting drugs and producing toxins and enzymes (see chapter 9). Because they can be readily manipulated in the laboratory and transferred from one bacterial cell to another, plasmids are an important agent in modern genetic engineering techniques.

Ribosomes: Sites of Protein Synthesis

A bacterial cell contains thousands of **ribosomes,** which are made of RNA and protein. When viewed even by very high magnification, ribosomes show up as fine, spherical specks dispersed throughout the cytoplasm that often occur in chains (polysomes). Many are also attached to the cell membrane. Chemically, a ribosome is a combination of a special type of RNA called ribosomal RNA, or rRNA (about 60%), and protein (40%).

One method of characterizing ribosomes is by S, or Svedberg,[4] units, which rate the molecular sizes of various cell parts that have been spun down and separated by molecular weight and shape in a centrifuge. Heavier, more compact structures sediment faster and are assigned a higher S rating. Combining this method of analysis with high-resolution electron micrography has revealed that the prokaryotic ribosome, which has an overall rating of 70S, is actually composed of two smaller subunits **(figure 4.18).** They fit together to form a miniature "factory" where protein synthesis occurs. We examine the more detailed functions of ribosomes in chapter 9.

Inclusions, or Granules: Storage Bodies

Most bacteria are exposed to severe shifts in the availability of food. During periods of nutrient abundance, some can compensate by storing nutrients as **inclusion bodies,** or **inclusions,** of varying size, number, and content. As the environmental source of these nutrients becomes depleted, the bacterial cell can mobilize its own storehouse as required. Some inclusion bodies contain condensed, energy-rich organic substances, such as glycogen and poly β-hydroxybutyrate (PHB), within special single-layered membranes **(figure 4.19a).** A unique type of inclusion found in some aquatic bacteria is gas vesicles that provide buoyancy and flotation. Other inclusions, also called **granules,** contain crystals of inorganic compounds and are not enclosed by membranes. Sulfur granules of photosynthetic bacteria and polyphosphate granules of *Corynebacterium* and *Mycobacterium* are of this type. The latter represent an important source of building blocks for nucleic acid and ATP synthesis. They have been termed **metachromatic granules** because they stain a contrasting color (red, purple) in the presence of methylene blue dye.

4. Named in honor of T. Svedberg, the Swedish chemist who developed the ultracentrifuge in 1926.

Figure 4.17 Bacterial cells stained to highlight their chromosomes. The individual cocci of *Deinococcus radiodurans* each contains one or more prominent bodies that are the chromosomes or nucleoids (1200×). Some cells are in the process of dividing. Notice how much of the cell's space the chromosomes (light blue bodies) occupy.

Ribosome (70S)

Large subunit (50S)

Small subunit (30S)

Figure 4.18 A model of a prokaryotic ribosome, showing the small (30S) and large (50S) subunits, both separate and joined.

(a)　(b)

Figure 4.19 **Bacterial inclusion bodies.** **(a)** Large particles (pink) of polyhydroxybutyrate are deposited in an insoluble, concentrated form that provides an ample, long-term supply of that nutrient (32,500×). **(b)** A section through *Aquaspirillum* reveals a chain of tiny iron magnets (magnetosomes = MP). These unusual bacteria use these inclusions to orient within their habitat (123,000×).

Perhaps the most remarkable cell granule is involved not in nutrition but rather in navigation. Magnetotactic bacteria contain crystalline particles of iron oxide (magnetosomes) that have magnetic properties **(figure 4.19b).** These granules occur in a variety of bacteria living in oceans and swamps. Their primary function is to orient the cells in the earth's magnetic field, somewhat like a compass. It is thought that magnetosomes direct these bacteria into locations with favorable oxygen levels or nutrient-rich sediments.

The Bacterial Cytoskeleton

Until very recently, bacteriologists thought bacteria lacked any real form of cytoskeleton.[5] The cell wall was considered to be the sole framework involved in support and shape. Research into the fine structure of certain rod- and spiral-shaped bacteria has provided several new insights. It seems that many of them possess an internal network of protein polymers that is closely associated with the wall **(figure 4.20).** The proteins are chemically similar to the actin fila-

5. An intracellular framework of fibers and tubules that bind and support eukaryotic cells.

ments universal in the cytoskeleton of eukaryotic cells. Presently this **bacterial actin** is thought to help stabilize shape. It may also influence cell wall formation by providing sites for synthesis when the wall is being repaired or enlarged.

Bacterial Endospores: An Extremely Resistant Life Form

Ample evidence indicates that the anatomy of bacteria helps them adjust rather well to adverse habitats. But of all microbial structures, nothing can compare to the bacterial **endospore** (or simply spore) for withstanding hostile conditions and facilitating survival. Endospores are dormant bodies produced by the bacteria *Bacillus, Clostridium,* and *Sporosarcina.* These bacteria have a two-phase life cycle that shifts between a **vegetative cell** and an endospore **(figure 4.21).** The vegetative cell is the metabolically active and growing phase. When exposed to certain environmental signals, it forms an endospore by a process termed **sporulation.** The spore exists in an inert, resting condition that is capable of high resistance and very long-term survival.

Actin
filaments

Figure 4.20 **Bacterial cytoskeleton of *Bacillus*.** Fluorescent stain of actin fibers appears as fine helical ribbons wound inside the cell.

Figure 4.21 **Development of endospores.** These biological "safety pins" are actually stages in endospore formation of *Bacillus subtilis,* stained with fluorescent proteins. The large red and blue cell is a vegetative cell in the early stages of sporulating. The developing spores are shown in green and orange.

Process Figure 4.22 **A typical sporulation cycle in *Bacillus* species from the active vegetative cell to release and germination.** This process takes, on average, about 10 hours. Inset is a high magnification (10,000×) cross section of a single spore showing the dense protective layers that surround the core with its chromosome.

Endospore Formation and Resistance

The major stimulus for endospore formation is the depletion of nutrients, especially amino acids. Once this stimulus has been received by the vegetative cell, it converts to a committed sporulating cell called a **sporangium.** Complete transformation of a vegetative cell into a sporangium and then into an endospore requires 6 to 12 hours in most spore-forming species. **Figure 4.22** illustrates some major physical and chemical events in this process.

Bacterial endospores are the hardiest of all life forms, capable of withstanding extremes in heat, drying, freezing, radiation, and chemicals that would readily kill ordinary cells. Their survival under such harsh conditions is due to several factors. The heat resistance of spores has been linked to their high content of calcium and *dipicolinic acid,* although the exact role of these chemicals is not yet clear. We know, for instance, that heat destroys cells by inactivating proteins and DNA and that this process requires a certain amount of water in the protoplasm. Because the deposition of calcium dipicolinate in the endospore removes water and leaves the endospore very dehydrated, it is less vulnerable to the effects of heat. It is also metabolically inactive and highly resistant to damage from further drying. The thick, impervious cortex and spore coats also protect against radiation and chemicals.

The longevity of bacterial spores verges on immortality! One record describes the isolation of viable endospores from a fossilized bee that was 25 million years old. More recently, microbiologists unearthed a viable endospore from a 250-million-year-old salt crystal. Initial analysis of this ancient microbe indicates it is a species of *Bacillus* that is genetically different from known species.

...ge in microbiol-
...y compact cells
...ive structures of
... in origin, form,
...here is called an
...ll. It functions in
...ase in cell num-
...e fungi produce
...l and reproduc-

The Germination of Endospores

After lying in a state of inactivity, endospores can be revitalized when favorable conditions arise. The breaking of dormancy, or **germination,** happens in the presence of water and a specific germination agent. Once initiated, it proceeds to completion quite rapidly (1½ hours). Although the specific germination agent varies among species, it is generally a small organic molecule such as an amino acid or an inorganic salt. This agent stimulates the formation of hydrolytic (digestive) enzymes by the endospore membranes. These enzymes digest the cortex and expose the core to water. As the core rehydrates and takes up nutrients, it begins to grow out of the endospore coats. In time, it reverts to a fully active vegetative cell, resuming the vegetative cycle (figure 4.22).

Medical Significance of Bacterial Spores

Although the majority of spore-forming bacteria are relatively harmless, several bacterial pathogens are sporeformers. In fact, some aspects of the diseases they cause are related to the persistence and resistance of their spores. *Bacillus anthracis,* the agent of anthrax, has been a frequent candidate for bioterrorism. The genus *Clostridium* includes even more pathogens, including *C. tetani,* the cause of tetanus (lockjaw), and *C. perfringens,* the cause of gas gangrene. When the spores of these species are embedded in a wound that contains dead tissue, they can germinate, grow, and release potent toxins. Another toxin-forming species, *C. botulinum,* is the agent of botulism, a deadly form of food poisoning. These diseases are discussed further in chapter 19.

Because they inhabit the soil and dust, endospores are a constant intruder where sterility and cleanliness are important. They resist ordinary cleaning methods that use boiling water, soaps, and disinfectants, and they frequently contaminate cultures and media. Hospitals and clinics must take precautions to guard against the potential harmful effects of endospores in wounds. Endospore destruction is a particular concern of the food-canning industry. Several endospore-forming species cause food spoilage or poisoning. Ordinary boiling (100°C) will usually not destroy such spores, so canning is carried out in pressurized steam at 120°C for 20 to 30 minutes. Such rigorous conditions will ensure that the food is sterile and free from viable bacteria.

Check & Assess Section 4.4

✔ The cytoplasm of bacterial cells serves as a solvent for materials used in all cell functions.

✔ The genetic material of bacteria is DNA. Genes are arranged on large, circular chromosomes. Additional genes are carried on plasmids.

✔ Bacterial ribosomes are dispersed in the cytoplasm in chains (polysomes) and are also embedded in the cell membrane.

✔ Bacteria may store nutrients in their cytoplasm in structures called inclusions. Inclusions vary in structure and the materials that are stored.

✔ Some bacteria manufacture long actin filaments that help stabilize their cellular shape.

✔ A few families of bacteria produce dormant bodies called endospores, which are the hardiest of all life forms, surviving for thousands and even millions of years.

✔ Examples of sporeformers are the genera *Bacillus* and *Clostridium,* both of which contain deadly pathogens.

23. Compare the functions of the bacterial chromosome (nucleoid) and plasmids.
24. What is unique about the structure of bacterial ribosomes, what is their function, and where are they located?
25. Compare and contrast the structure and function of inclusions and granules.
26. What are metachromatic granules, and what do they contain?
27. Describe the formation of bacterial endospores.
28. Describe the structure of an endospore, and explain its function.
29. Explain why an endospore is not considered a reproductive body.
30. Why are spores so difficult to destroy?

4.5 Bacterial Shapes, Arrangements, and Sizes

Expected Learning Outcomes

22. Describe the shapes of bacteria and their possible variants.
23. Identify several arrangements of bacteria and how they are formed.
24. Indicate the size ranges in bacteria in comparison to other organisms.

For the most part, bacteria function as independent single-celled, or unicellular, organisms. Although it is true that an individual bacterial cell can live attached to others in colonies or other such groupings, each one is fully capable of carrying out all necessary life activities, such as reproduction, metabolism, and nutrient processing unlike the more specialized cells of a multicellular organism.

Key to Micrographs

(a) *Micrococcus luteus* (22,000×) (b) *Legionella pneumophila* (6500×) (c) *Vibrio cholerae* (13,000×) (d) *Aquaspirillum* (7,500×)
(e) Spirochetes on a filter (14,000×) (f) *Streptomyces species* (6500×)

Figure 4.23 Common bacterial shapes. Drawings show examples of shape variations for cocci, rods, vibrios, spirilla, spirochetes, and branching filaments. Below each shape is a micrograph of a representative example.

Bacteria exhibit considerable variety in shape, size, and colonial arrangement. It is convenient to describe most bacteria by one of three general shapes as dictated by the configuration of the cell wall **(figure 4.23).** If the cell is spherical or ball-shaped, the bacterium is described as a **coccus*.** Cocci can be perfect spheres, but they also can exist as oval, bean-shaped, or even pointed variants.

A cell that is cylindrical (longer than wide) is termed a rod, or **bacillus*.** There is also a genus named *Bacillus.* As might be expected, rods are also quite varied in their actual form. Depending on the bacterial species, they can be blocky, spindle-shaped, round-ended, long and threadlike (filamentous), or even clubbed or drumstick-shaped. When a rod is short and plump, it is called a **coccobacillus;** if it is gently curved, it is a **vibrio*.**

* *coccus* (kok´-us) pl. cocci (kok´-seye) Gr.*Kokkos,* berry.

* *bacillus* (bah-sil´-lus) pl. bacilli (bah-sil´-eye) L. *bacill,* small staff or rod.
* *vibrio* (vib´-ree-oh) L. *vibrare,* to shake.

TABLE 4.2	Comparison of the Two Spiral-Shaped Bacteria				
	Overall Appearance	**Mode of Locomotion**	**Number of Helical Turns**	**Gram Reaction (Cell Wall Type)**	**Examples of Important Types**
Spirilla	Rigid helix Spirilla	Polar flagella; cells swim by rotating around like corkscrews; do not flex; have one to several flagella; can be in tufts	Varies from 1 to 20	Gram-negative	Most are harmless; one species, *Spirillum minor,* causes rat bite fever.
Spirochetes	Flexible helix Curved or spiral forms: Spirillum/Spirochete	Periplasmic flagella within sheath; cells flex; can swim by rotation or by creeping on surfaces; have 2 to 100 periplasmic flagella	Varies from 3 to 70	Gram-negative	*Treponema pallidum,* cause of syphilis; *Borrelia* and *Leptospira,* important pathogens

A bacterium having the shape of a curviform or spiral-shaped cylinder is called a **spirillum***, a rigid helix, twisted twice or more along its axis (like a corkscrew). Another spiral cell mentioned earlier in conjunction with periplasmic flagella is the **spirochete,** a more flexible form that resembles a spring. Refer to **table 4.2** for a comparison of other features of the two helical bacterial forms. Because bacterial cells look two-dimensional and flat with traditional staining and microscope techniques, they are seen to best advantage using a scanning electron microscope to emphasize their striking three-dimensional forms (figure 4.23).

It is common for cells of a single species to show **pleomorphism* (figure 4.24).** This is due to individual variations in cell

wall structure caused by nutritional or slight hereditary differences. For example, although the cells of *Corynebacterium diphtheriae* are generally considered rod-shaped, in culture they display club-shaped, swollen, curved, filamentous, and coccoid variations. Pleomorphism reaches an extreme in the mycoplasmas, which entirely lack cell walls and thus display extreme variations in shape (see figure 4.15).

Bacterial cells can also be categorized according to **arrangement,** or style of grouping. The main factors influencing the arrangement of a particular cell type are its pattern of division and how the cells remain attached afterward. The greatest variety in arrangement occurs in cocci **(figure 4.25).** They may exist as singles, in pairs **(diplococci*),** in **tetrads** (groups of four), in irregular clusters (both **staphylococci*** and **micrococci),** or in chains of a few to hundreds of cells **(streptococci).** An even more complex grouping is a cubical packet of 8, 16, or more cells called a **sarcina*.** These different coccal groupings are the result of the division of a coccus in a single plane, in two perpendicular planes, or in several intersecting planes; after division, the resultant daughter cells remain attached.

Bacilli are less varied in arrangement because they divide only in the transverse plane (perpendicular to the axis). They occur either as single cells, as a pair of cells with their ends attached **(diplobacilli),** or as a chain of several cells **(streptobacilli).** A **palisades*** arrangement, typical of the corynebacteria, is formed when the cells of a chain remain partially attached by a small hinge region at the ends. The cells tend to fold (snap) back upon each other, forming a row of cells oriented side by side (see figure 4.24). The reaction can be compared to the behavior of boxcars on a jackknifed train, and the result looks superficially like an irregular picket fence. Spirilla are occasionally found in short chains, but spirochetes rarely remain attached after division. Comparative sizes of typical cells are presented in **figure 4.26.**

* *spirillum* (spy-ril′-em) pl. spirilla; L. *spira,* a coil.
* *pleomorphism* (plee″-oh-mor′-fizm) Gr.*Pleon,* more, and *morph,* form or shape.

Metachromatic granules

Metachromatic granules

Palisades arrangement

Palisades arrangement

Figure 4.24 **Pleomorphism and other morphological features of *Corynebacterium.*** Cells are irregular in shape and size (800×). This genus typically exhibits a palisades arrangement, with cells in parallel array (inset). Close examination will also reveal darkly stained granules called metachromatic granules.

* *diplococci* (dih-plo-kok-seye) Gr. *diplo,* double**.**
* *staphylococci* (staf″-ih-loh-kok′-seye) Gr. *staphyle,* a bunch of grapes.
* *sarcina* (sar′-sin-uh) L. *sarcina,* a packet.
* *palisades* (pal′-ih-saydz) L. *pale,* a stake. A fence made of a row of stakes.

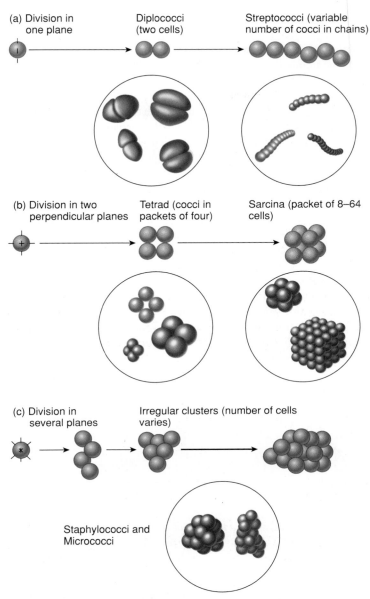

Figure 4.25 **Arrangement of cocci resulting from different planes of cell division.** (a) Division in one plane produces diplococci and streptococci. (b) Division in two planes at right angles produces tetrads and packets. (c) Division in several planes produces irregular clusters.

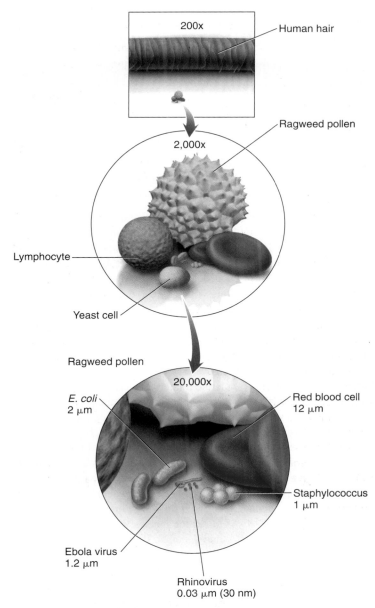

Figure 4.26 **The dimensions of bacteria.** The sizes of bacteria range from those just barely visible with light microscopy (0.2 μm) to those measuring a thousand times that size. Cocci measure anywhere from 0.5 to 3.0 μm in diameter; bacilli range from 0.2 to 2.0 μm in diameter and from 0.5 to 20 μm in length. Note the range of sizes as compared with eukaryotic cells and viruses. Comparisons are given as average sizes.

Check & Assess Section 4.5

✔ Most bacteria have one of three general shapes: coccus (sphere), bacillus (rod), or spiral, based on the configuration of the cell wall. Two types of spiral cells are spirochetes and spirilla.

✔ Shape and arrangement of cells are key means of describing bacteria. Arrangements of cells are based on the number of planes in which a given cell type divides.

✔ Cocci can divide in many planes to form pairs (diplococci), chains (streptococci), packets (sarcinae), or clusters (micrococci or staphylococci). Bacilli divide only in the transverse plane. If they remain attached, they form pairs, chains, or palisades arrangements.

31. How are spirochetes and spirilla different?
32. What is a vibrio? A coccobacillus?
33. What is pleomorphism?
34. What is the difference between the use of the shape bacillus and the name *Bacillus*? *Staphylococcus* and staphylococcus?
35. Rank the size ranges in bacteria according to shape.
36. Rank the bacteria in relationship to viruses and eukaryotic cell size.

4.6 Classification Systems of Prokaryotic Domains: Archaea and Bacteria

Expected Learning Outcomes

25. Describe the purposes of classification and taxonomy in the study of prokaryotes.

26. Overview characteristics used to classify bacteria.

27. Outline a basic system of bacterial taxonomy.

28. Explain the species and subspecies levels for bacteria.

Classification systems serve both practical and academic purposes. They aid in differentiating and identifying unknown species in medical and applied microbiology. They are also useful in organizing bacteria and as a means of studying their relationships and evolutionary origins. Since classification was started around 200 years ago, several thousand species of bacteria and archaea have been identified, named, and catalogued.

Tracing the origins of and evolutionary relationships among bacteria has not been an easy task. As a rule, tiny, relatively soft organisms do not form fossils very readily. Several times since the 1960s, however, scientists have discovered billion-year-old fossils of prokaryotes that look very much like modern bacteria (see figure 4.28d).

One of the questions that has plagued taxonomists is, What characteristics are reliable indicators of closeness in ancestry? Early bacteriologists found it convenient to classify bacteria according to shape, variations in arrangement, growth characteristics, and habitat. However, as more species were discovered and as techniques for studying their biochemistry were developed, it soon became clear that similarities in cell shape, arrangement, and staining reactions do not automatically indicate relatedness. Even though gram-negative rods look alike, there are hundreds of different species with highly significant differences in biochemistry and genetics. If we attempted to classify them on the basis of Gram stain and shape alone, we could not assign them to a more specific level than phylum. Newer classification schemes have turned to more specific genetic and molecular data.

The methods that a microbiologist uses to identify bacteria to the level of genus and species fall into the main categories of morphology (microscopic and macroscopic), bacterial physiology or biochemistry, serological analysis, and genetic techniques (chapter 17 and **appendix table D.8**). Data from a cross section of such tests can produce a unique profile of each bacterium. Final differentiation of any unknown species is accomplished by comparing its profile with the characteristics of known bacteria in tables, charts, and keys (see figure 17.5). Many identification systems are automated and computerized to process data and provide a "best fit" identification. However, not all methods are used on all bacteria. A few bacteria can be identified by a special technology that analyzes only the kind of fatty acids they contain; in contrast, some are identifiable by a Gram stain and a few physiological tests; others may require a diverse spectrum of morphological, biochemical, and genetic tests.

Bacterial Taxonomy Based on *Bergey's Manual*

There is no single official system for classifying the prokaryotes. Indeed, most plans are in a state of flux as new information and

methods of analysis become available. One widely used reference is *Bergey's Manual of Systematic Bacteriology,* a major resource that covers all known prokaryotes. In the past, the classification scheme in this guide has been based mostly on characteristics such as Gram stain and metabolic reactions. This is called a **phenetic,** or phenotypic, method of classification.

The second edition of this large collection now includes genetic information that clarifies the **phylogenetic** (evolutionary) history and relationships of the thousands of known species (**figure 4.27**). This change has somewhat complicated the presentation of their classification, because prokaryotes are now placed in five major subgroups and 25 different phyla instead of two domains split into four divisions. The Domains Archaea and Bacteria are based on genetic characteristics and have been retained. But the bacteria of clinical importance are no longer as closely aligned, and the 250 or so species that cause disease in humans are found within seven or eight of the revised phyla. The grouping by Gram reaction remains significant but primarily at the lower taxonomic levels.

The major headings for this revised taxonomic scheme are presented in **table 4.3,** and a complete version of all major genera can be located in **appendix table D.9.** The following section provides a brief survey of characteristics of the major taxonomic groups, along with examples of representative members (A–L)

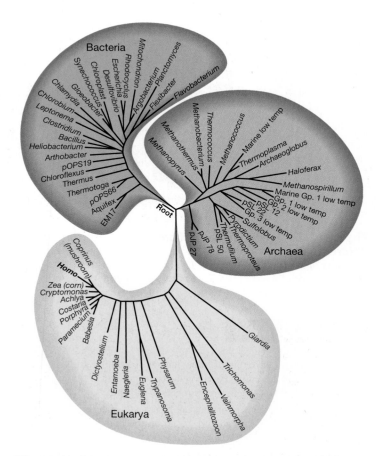

Figure 4.27 **A universal phylogenetic tree as proposed by Norman Pace.** This pattern is based on ribosomal RNA sequences. Balloons compare the two prokaryotic domains (Bacteria and Archaea) with the Domain Eukarya. Branches indicate the origins of major taxonomic groups, and their positions show degrees of relatedness.

| TABLE 4.3 | The General Classification Scheme of *Bergey's Manual* (2nd Ed.) (see appendix table D.9) |

Taxonomic Rank

Volume 1. The Archaea and the Deeply Branching and Phototrophic Bacteria

■■■■■■ Domain Archaea ■■■■■■ Domain Bacteria

1A. Domain Archaea

Phylum Crenarchaeota
Phylum Euryarchaeota
 Class I. Methanobacteria
 Class II. Methanococci
 Class III. Halobacteria
 Class IV. Thermoplasmata
 Class V. Thermococci
 Class VI. Archaeoglobi
 Class VII. Methanopyri

1B. Domain Bacteria

Phylum Aquificae
Phylum Thermotogae
Phylum Thermodesulfobacteria
Phylum "Deinococcus-Thermus"
Phylum Chrysiogenetes
Phylum Chloroflexi
Phylum Thermomicrobia
Phylum Nitrospira
Phylum Deferribacteres
Phylum Cyanobacteria
Phylum Chlorobi

Volume 3. The Low G + C Gram-Positive Bacteria*

Phylum Firmicutes
 Class I. Clostridia
 Class II. Mollicutes
 Class III. Bacilli

Volume 4. The High G + C Gram-Positive Bacteria*

Phylum Actinobacteria
 Class Actinobacteria

Volume 5. The Planctomycetes, Spirochaetes, Fibrobacteres, Bacteriodetes, and Fusobacteria

Phylum Planctomycetes Phylum Bacteroidetes
Phylum Chlamydiae Phylum Fusobacteria
Phylum Spirochaetes Phylum Verrucomicrobia
Phylum Fibrobacteres Phylum Dictyoglomus
Phylum Acidobacteria

Domain Bacteria

Volume 2. The Proteobacteria

Phylum Proteobacteria
 Class I. Alphaproteobacteria Class IV. Deltaproteobacteria
 Class II. Betaproteobacteria Class V. Epsilonproteobacteria
 Class III. Gammaproteobacteria

***G + C base composition** The overall percentage of guanine and cytosine in DNA is a general indicator of relatedness because it is a trait that does not change rapidly. Bacteria with a significant difference in G + C percentage are less likely to be genetically related. This classification scheme is partly based on this percentage.

Survey of Major Procaryotic Groups

- **Volume 1** This volume includes all of the **Domain Archaea** and the most ancient members of **Domain Bacteria.**

1A The Archaea are unusual primitive prokaryotes adapted to extreme habitats and modes of nutrition. There are two phyla and seven classes in this group. More information on Archaea can be found in Section 4.7.

A. An example of archaeal methanogens from a deep Antarctic lake.

2 microns

(A) Methanogens

1B Domain Bacteria. The section of deeply branching and phototropic bacteria contains 11 phyla. These are among the earliest inhabitants of earth. It contains members with a wide variety of adaptations. Many of them are photosynthetic (bacteria called cyanobacteria (figure B) and green sulfur bacteria), others are inhabitants of extreme environments such as thermal vents (*Aquifex*), and one genus is highly radiation resistant (*Deinococcus*) (figure C).

B. A cyanobacterium, *Trichodesmium.*
C. *Deinococcus radians* tetrad.

(B)

(C)

- **Volume 2** This includes representatives of the **Phylum Proteobacteria,** a group with five classes and an extremely complex and diverse cross section of over 2,000 species of bacteria. One characteristic they all share is a gram-negative cell wall. This group includes several medically significant members, including the obligate intracellular parasites called **rickettsias** (figure D) and gram-negative enteric rods such as *Escherichia* (figure E) and *Salmonella*. Vibrios (*Desulfvibrio* [figure F] and *Campylobacter*), photosynthetic bacteria with brilliant pigments (see figure 4.29), and unusual gliding bacteria with complex fruiting bodies (see figure 4.30) are additional examples.

(D) (E) (F)

D. A *Rickettsia* cell is being engulfed by a host cell.
E. *Escherichia coli* 0157:H7, a food-borne pathogen.

F. *Desulfvibrio* in a biofilm from an acidic pond.

- **Volume 3** This collection represents the **Phylum Firmicutes,** whose members are characterized as being mostly gram-positive and having low G + C content (less than 50%) (see table 4.3 note). The three classes in the phylum contain a wide diversity of members, many of which are medically important. The endospore formers *Clostridium* and *Bacillus* (figure G) account for a significant portion of species. Gram-positive cocci such as *Staphylococcus* and *Streptococcus* (figure H) are included as well. Significantly, the mycoplasmas are genetically allied with this group even though they have lost their cell wall at some point in time (see figure 4.15).

(G) (H)

G. *Bacillus anthracis* cell and endospore (B).
H. *Streptococcus pneumoniae* in pairs.

- **Volume 4** This includes the **Phylum Actinobacteria,** the taxonomic category that includes the high G + C (over 50%) gram-positive bacteria. The single class represents bacteria of many different shapes and life cycles. Prominent members are the branching filamentous actinomycetes, the spore-producing streptomycetes (figure I), and the common genera *Corynebacterium, Mycobacterium* (figure J), and *Micrococcus.*

I. *Streptomyces* sp., a source of antibiotics.
J. *Mycobacterium tuberculosis,* the cause of TB.

(I) (J)

- **Volume 5** The final volume contains a loose assemblage of nine phyla that may or may not be related, but all of them are gram-negative. Subgroups include the Planctomyces, Spirochaetes, Fibrobacteres, Bacteriodetes, and Fusobacteria. Once again, there is extreme variety in the members included, and many of them are (figure K) medically important.*Chlamydia* (figure K) are tiny obligate parasites that live inside their host cells like rickettsias. Other significant members are spirochetes such as *Treponema,* the cause of syphilis, and *Leptospira* (figure L), the cause of leptospirosis.

K. *Chlamydia* cells attached to a host cell.
L. *Leptospira,* a tightly coiled spirochete.

(K) (L)

Diagnostic Scheme for Medical Use

Many medical microbiologists prefer an informal working system that outlines the major families and genera (**table 4.4**). This scheme uses the phenotypic qualities of bacteria in identification. It is restricted to bacterial disease agents and depends less on nomenclature. It also divides the bacteria into gram-positive, gram-negative, and those without cell walls and then subgroups them according to cell shape, arrangement, and certain physiological traits such as oxygen usage: *Aerobic* bacteria use oxygen in metabolism; *anaerobic* bacteria do not use oxygen in metabolism; and facultative bacteria may or may not use oxygen. Further tests not listed on the table would be required to separate closely related genera and species. Many of these are included in later chapters on specific bacterial groups.

Species and Subspecies in Bacteria

Among most organisms, the **species** level is a distinct, readily defined, and natural taxonomic category. In animals, for instance, a species is a distinct type of organism that can produce viable offspring only when it mates with others of its own kind. This definition does not work for bacteria primarily because they do not exhibit a typical mode of sexual reproduction. They can accept genetic information from unrelated forms, and they can also alter their genetic makeup by a variety of mechanisms. Thus, it is necessary to use a modified definition for a bacterial species. Theoretically, it is a collection of bacterial cells, all of which share an overall similar pattern of traits, in contrast to other groups whose pattern differs significantly. Although the boundaries that separate two closely related species in a genus can be somewhat arbitrary, this definition still serves as a method to separate the bacteria into various kinds that can be cultured and studied. As additional information on bacterial genomes is discovered, it may be possible to define species according to specific combinations of genetic codes found only in a particular isolate.

Because the individual members of given species can show variations, we must also define levels within species (subspecies) called **strains** and **types.** A strain or variety of bacteria is a culture derived from a single parent that differs in structure.or metabolism from other cultures of that species (also called biovars or morphovars). For example, there are pigmented and nonpigmented strains of *Serratia marcescens* and flagellated and nonflagellated strains of *Pseudomonas fluorescens.* A type is a subspecies that can show differences in antigenic makeup (serotype or serovar), in susceptibility to bacterial viruses (phage type), and in pathogenicity (pathotype).

Check & Assess Section 4.6

✔ Key traits that are used to identify a bacterial species include (1) cell morphology, (2) Gram and other staining characteristics, (3) presence of specialized structures, (4) macroscopic appearance, (5) biochemical reactions, and (6) unique composition of both DNA and rRNA.

✔ Bacteria are formally classified by phylogenetic relationships and phenotypic characteristics.

✔ Medical identification of pathogens uses a more informal system of classification based on Gram stain, morphology, biochemical reactions, and metabolic requirements.

✔ A bacterial species is loosely defined as a collection of bacterial cells that shares an overall similar pattern of traits different from other groups of bacteria.

✔ Variant forms within a species (subspecies) include strains and types.

37. What general characteristics are used to classify bacteria?
38. What are the most useful characteristics for categorizing bacteria into phyla?
39. List 10 major phyla of bacteria and include an example of a species or genus that is a member.
40. Explain how the species level in bacteria is defined and name at least three ways bacteria are classified below the species level.

4.7 Survey of Prokaryotic Groups with Unusual Characteristics

Expected Learning Outcomes

29. Differentiate various groups of photosynthetic bacteria.

30. Characterize the types of obligate intracellular bacteria.

31. Describe bacteria with extremes in size.

32. Summarize the basic characteristics of archaea.

33. Compare Domain Archaea with Domains Bacteria and Eukarya.

34. Explain archaeal adaptations that place them in the category of extremophiles.

The bacterial world is so diverse that we cannot do complete justice to it in this introductory chapter. This variety extends into all areas of bacterial biology, including nutrition, mode of life, and behavior. Certain types of bacteria exhibit such unusual qualities that they deserve special mention. In this minisurvey, we consider some medically important groups and some more remarkable representatives of bacteria living free in the environment that are ecologically important. Many of the bacteria mentioned here do not have the morphology typical of bacteria discussed previously, and in a few cases, they are vividly different (**Insight 4.2**).

Free-Living Nonpathogenic Bacteria

Photosynthetic Bacteria

The nutrition of most bacteria is heterotrophic, meaning that they derive their nutrients from other organisms. Photosynthetic bacteria, however, are independent cells that contain special light-trapping pigments and can use the energy of sunlight to synthesize all

TABLE 4.4 Medically Important Families and Genera of Bacteria, with Notes on Some Diseases*

I. Bacteria with gram-positive cell wall structure (Firmicutes)

Cocci in clusters or packets that are aerobic or facultative

Family Micrococcaceae: *Staphylococcus* (members cause boils, skin infections)

Cocci in pairs and chains that are facultative

Family Streptococcaceae: *Streptococcus* (species cause strep throat, dental caries)

Anaerobic cocci in pairs, tetrads, irregular clusters
Family Peptococcaceae: *Peptococcus, Peptostreptococcus* (involved in wound infections)

Spore-forming rods

Family Bacillaceae: *Bacillus* (anthrax), *Clostridium* (tetanus, gas gangrene, botulism)

Non-spore-forming rods
Family Lactobacillaceae: *Lactobacillus, Listeria* (milk-borne disease), *Erysipelothrix* (erysipeloid)

Family Propionibacteriaceae: *Propionibacterium* (involved in acne)

Family Corynebacteriaceae: *Corynebacterium* (diphtheria)

Family Mycobacteriaceae: *Mycobacterium* (tuberculosis, leprosy)

Family Nocardiaceae: *Nocardia* (lung abscesses)

Family Actinomycetaceae: *Actinomyces* (lumpy jaw), *Bifidobacterium*

Family Streptomycetaceae: *Streptomyces* (important source of antibiotics)

II. Bacteria with gram-negative cell wall structure (Gracilicutes)

Aerobic cocci
Neisseria (gonorrhea, meningitis), *Branhamella*

Aerobic coccobacilli
Moraxella, Acinetobacter

Anaerobic cocci
Family Veillonellaceae
Veillonella (dental disease)

Miscellaneous rods
Brucella (undulant fever), *Bordetella* (whooping cough), *Francisella* (tularemia)

Aerobic rods
Family Pseudomonadaceae: *Pseudomonas* (pneumonia, burn infections)
Miscellaneous: *Legionella* (Legionnaires' disease)

Facultative or anaerobic rods and vibrios
Family Enterobacteriaceae: *Escherichia, Edwardsiella, Citrobacter, Salmonella* (typhoid fever), *Shigella* (dysentery),
Klebsiella, Enterobacter, Serratia, Proteus, Yersinia (one species causes plague)

Family Vibronaceae: *Vibrio* (cholera, food infection), *Campylobacter, Aeromonas*

Miscellaneous genera: *Chromobacterium, Flavobacterium, Haemophilus* (meningitis),
Pasteurella, Cardiobacterium, Streptobacillus

Anaerobic rods
Family Bacteroidaceae: *Bacteroides, Fusobacterium* (anaerobic wound and dental infections)

Helical and curviform bacteria
Family Spirochaetaceae: *Treponema* (syphilis), *Borrelia* (Lyme disease), *Leptospira* (kidney infection)

Obligate intracellular bacteria
Family Rickettsiaceae: *Rickettsia* (Rocky Mountain spotted fever), *Coxiella* (Q fever)

Family Bartonellaceae: *Bartonella* (trench fever, cat scratch disease)
Family Chlamydiaceae: *Chlamydia* (sexually transmitted infection)

III. Bacteria with no cell walls (Tenericutes)

Family Mycoplasmataceae: *Mycoplasma* (pneumonia), *Ureaplasma* (urinary infection)

*Details of pathogens and diseases in later chapters.

INSIGHT 4.2

Redefining Bacterial Size

Microbiologists keep being reminded how far we are from having a complete assessment of the bacterial world, mostly because the world is so large and bacteria are so small. There seem to be frequent reports of exceptional bacteria discovered in places like the deep ocean volcanoes or Antarctic ice. Among the most remarkable are giant and dwarf bacteria.

Big Bacteria Break Records

In 1985, biologists discovered a new bacterium living in the intestine of surgeonfish that at the time was a candidate for the *Guinness Book of World Records*. The large cells, named *Epulopiscium fishelsoni* ("guest at a banquet of fish"), measure around 100 μm in length, although some specimens were as large as 300 μm. This record was broken when a marine microbiologist discovered an even larger species of bacteria living in ocean sediments near the African country of Namibia. These gigantic cocci are arranged in strands that look like pearls and contain hundreds of golden sulfur granules, inspiring their name, *Thiomargarita namibia* ("sulfur pearl of Namibia") (see photo). The size of the individual cells ranges from 100 up to 750 μm (3/4 mm), and many are large enough to see with the naked eye. By way of comparison, if the average bacterium were the size of a mouse, *Thiomargarita* would be as large as a blue whale!

Closer study revealed that they are indeed prokaryotic and have bacterial ribosomes and DNA but that they also have some unusual adaptations to their life cycle. They live an attached existence embedded in sulfide sediments (H_2S) that are free of gaseous oxygen. They obtain energy through oxidizing these sulfides using dissolved nitrates (NO_3). These bacteria are found in such large numbers in the sediments that it is thought that they are essential to the ecological cycling of H_2S and other substances in this region, converting them to less toxic substances.

Miniature Microbes—The Smallest of the Small

At the other extreme, microbiologists are being asked to reevaluate the lower limits of bacterial size. Up until now it has been generally accepted that the smallest cells on the planet are some form of mycoplasma with dimensions of 0.2 to 0.3 μm, which is right at the limit of resolution with light microscopes. A new controversy is brewing over the discovery of tiny cells that look like dwarf bacteria but are 10 times smaller than mycoplasmas and 100 times smaller than the average bacterial cell. These

Thiomargarita namibia—giant cocci.

Section through sandstone shows tiny blobs that some scientists think are nanobes.

minute cells have been given the name **nanobacteria** or **nanobes** (Gr. *nanos,* one-billionth).

Nanobacterialike forms were first isolated from blood and serum samples. The tiny cells appear to grow in culture, have cell walls, and contain protein and nucleic acids, but their size range is only from 0.05 to 0.2 μm. Similar nanobes have been extracted by minerologists studying deposits in the ocean at temperatures of 100°C to 170°C. They have isolated minute filaments that can grow and deposit minerals in a test tube. Many geologists are convinced that these nanobes are real cells, that they are probably similar to the first microbes on earth, and that they play a strategic role in the evolution of the earth's crust.

Microbiologists tend to be more skeptical. It has been postulated that the minimum cell size to contain a functioning genome and reproductive and synthetic machinery is approximately 0.14 μm. They believe that the nanobes are really just artifacts or bits of larger cells that have broken free. It is partly for this reason that most bacteriologists rejected the idea that small objects found in a Martian meteor were microbes but were more likely caused by chemical reactions (see chapter 2). Additional studies are needed to test this curious question of nanobes and possibly to answer some questions about the origins of life on earth and even on other planets.

List additional qualities a nanobe would require to be considered a living cell. *Answer available at http://www.mhhe.com/talaro8*

required nutrients from simple inorganic compounds. The two general types of photosynthetic bacteria are those that produce oxygen during photosynthesis and those that produce some other substance, such as sulfur granules or sulfates.

Cyanobacteria: Blue-Green Bacteria

The cyanobacteria were called blue-green algae for many years and were grouped with the eukaryotic algae. However, further study verified that they are indeed bacteria with a gram-negative cell wall and general prokaryotic structure. These bacteria range in size from 1 μm to 10 μm, and they can be unicellular or can occur in colonial or filamentous groupings **(figure 4.28*a*).**

Cyanobacteria are among the oldest types of bacteria on earth. Fossil forms have been isolated in rocks that are over 3 billion years old. Interesting remnants of these microbes exist in stromatolites—fossil biofilms in oceanic deposits **(figure 4.28*b*).** When magnified, they reveal ancient cells that look remarkably like modern ones **(figure 4.28*c*).** A specialized adaptation of cyanobacteria is extensive internal membranes called **thylakoids,** which contain granules of chlorophyll *a* and other photosynthetic pigments **(figure 4.28*d*).** They also have gas inclusions, which permit them to float on the water surface and increase their light exposure, and cysts that convert gaseous nitrogen (N_2) into a form usable by plants. This group is sometimes called the blue-green

(a) 20 microns (a2) (a3)

Stromatolite

(b) (c) 10 μm

(d) Thylakoid membranes

Figure 4.28 Structure and examples of cyanobacteria.
(a) Striking cyanobacteria: (1000×) display a variety of colors and shapes (1000×). *Oscillatoria* (left), *Trichodesmium* (center) and *Merismopedia* (right) (b) stromatolite sectioned to show layers of biofilm laid down over a billion years by cyanobacteria. (c) A microfossil of a cyanobacterial filament from Siberia. You probably notice how similar it looks to some of the examples in (a). (d) Electron micrograph of a cyanobacterial cell (80,000×) reveals folded stacks of membranes that contain the photosynthetic pigments and increase surface area for photosynthesis.

bacteria in reference to their content of phycocyanin pigment that tints some members a shade of blue, although other members are colored yellow and orange. Some representatives glide or sway gently in the water from the action of filaments in the cell envelope that cause wavelike contractions.

Cyanobacteria are very widely distributed in nature. They grow profusely in freshwater and seawater and are thought to be responsible for periodic blooms that kill off fish. Some members are so pollution-resistant that they serve as biological indicators of polluted water. Cyanobacteria inhabit and flourish in hot springs and have even exploited a niche in dry desert soils and rock surfaces.

Green and Purple Sulfur Bacteria

The green and purple bacteria are also photosynthetic and contain pigments. They differ from the cyanobacteria in having a different type of chlorophyll called *bacteriochlorophyll* and by not giving off

oxygen as a product of photosynthesis. They live in sulfur springs, freshwater lakes, and swamps that are deep enough for the anaerobic conditions they require yet where their pigment can still absorb wavelengths of light **(figure 4.29)**. These bacteria are named for their predominant colors, but they can also develop brown, pink, purple, blue, and orange coloration. Both groups utilize sulfur compounds (H_2S, S) in their metabolism.

Gliding, Fruiting Bacteria

The gliding bacteria are a mixed collection of gram-negative bacteria that live in water and soil. The name is derived from the tendency of members to glide over moist surfaces. The gliding property evidently involves rotation of filaments or fibers just under the outer membrane of the cell wall. They do not have flagella. Several morphological forms exist, including slender rods, long filaments, cocci, and some miniature, tree-shaped fruiting bodies. Probably the most

Figure 4.29 **Photosynthetic bacteria.** Purple colored masses in a fall pond contain a concentrated bloom of purple sulfur bacteria (inset 1,500×). The pond also harbors a mixed population of algae (green clumps).

Rickettsial cells

Nucleus

Vacuole

Figure 4.31 **Transmission electron micrograph of the rickettsia *Coxiella burnetii* (20,000×), the cause of Q fever.** Its mass growth inside a host cell has filled a vacuole and displaced the nucleus to one side.

Figure 4.30 *Myxobacterium.* A photograph of the mature fruiting body and its cluster of myxospores held by a stalk.

Rickettsias[6]

Rickettsias are distinctive, very tiny, gram-negative bacteria **(figure 4.31).** Although they are somewhat typical in morphology, they are atypical in their life cycle and other adaptations. Most are pathogens that alternate between a mammalian host and blood-sucking arthropods,[7] such as fleas, lice, or ticks. Rickettsias cannot survive or multiply outside a host cell and cannot carry out metabolism completely on their own, so they are closely attached to their hosts. Several important human diseases are caused by rickettsias. Among these are Rocky Mountain spotted fever, caused by *Rickettsia rickettsii* (transmitted by ticks), and endemic typhus, caused by *Rickettsia typhi* (transmitted by lice).

Chlamydias

Bacteria of the genera *Chlamydia* and *Chlamydophila,* termed **chlamydias,** are similar to the rickettsias in that they require host cells for growth and metabolism; but they are not closely related to them and are not transmitted by arthropods. Because of their tiny size and obligately parasitic lifestyle, they were at one time considered a type of virus. Species that carry the greatest medical impact are *Chlamydia trachomatis,* the cause of both a severe eye infection (trachoma) that can lead to blindness and one of the most common sexually transmitted diseases, and *Chlamydophila pneumoniae,* an agent in lung infections.

Diseases caused by rickettsias and chlamydias are described in more detail in the infectious disease chapters.

intriguing and exceptional members of this group are the slime bacteria, or myxobacteria. What sets the myxobacteria apart from other bacteria is the complexity and advancement of their life cycle. During this cycle, the vegetative cells respond to chemotactic signals by swarming together and differentiating into a many-celled, colored structure called the fruiting body **(figure 4.30).** The fruiting body is a survival structure that makes spores by a method very similar to that of certain fungi. These fruiting structures are often large enough to be seen with the unaided eye on tree bark and plant debris.

Unusual Forms of Medically Significant Bacteria

Most bacteria are free-living or parasitic forms that can metabolize and reproduce by independent means. Two groups of bacteria—the rickettsias and chlamydias—have adapted to life inside their host cells, where they are considered **obligate intracellular parasites.**

6. Named for Howard Ricketts, a physician who first worked with these organisms and later lost his life to typhus.

7. An arthropod is an invertebrate with jointed legs, such as an insect, tick, or spider.

TABLE 4.5 **Comparison of Three Cellular Domains**

Characteristic	Bacteria	Archaea	Eukarya
Cell type	Prokaryotic	Prokaryotic	Eukaryotic
Chromosomes	Single, or few, circular	Single, circular	Several, linear
Types of ribosomes	70S	70S but structure is similar to 80S	80S
Contains unique ribosomal RNA signature sequences	+	+	+
Number of RNA sequences shared with Eukarya	One	Three	
Protein synthesis similar to Eukarya	−	+	
Presence of peptidoglycan in cell wall	+	−	−
Cell membrane lipids	Fatty acids with ester linkages	Long-chain, branched hydrocarbons with ether linkages	Fatty acids with ester linkages
Sterols in membrane	− (some exceptions)	−	+

Archaea: The Other Prokaryotes

The discovery and characterization of novel prokaryotic cells that have unusual anatomy, physiology, and genetics changed our views of microbial taxonomy and classification (see chapter 1 and table 4.3). These single-celled, simple organisms, called **archaea,** or **archaeons,** are considered a third cell type in a separate super-kingdom (the Domain Archaea). We include them in this chapter because they are prokaryotic in general structure and they do share many bacterial characteristics. But evidence is accumulating that they are actually more closely related to Domain Eukarya than to bacteria. For example, archaea and eukaryotes share a number of ribosomal RNA sequences that are not found in bacteria, and their protein synthesis and ribosomal subunit structures are similar. Table 4.5 outlines selected points of comparison of the three domains.

Among the ways that the archaea differ significantly from other cell types are that certain genetic sequences are found only in their ribosomal RNA and that they have unique membrane lipids and cell wall construction. It is clear that the archaea are the most primitive of all life forms and have retained characteristics of the first cells that originated on the earth nearly 4 billion years ago. The early earth is thought to have contained a hot, anaerobic "soup" with sulfuric gases and salts in abundance. The modern archaea still live in the remaining habitats on the earth that have some of the same ancient conditions—the most extreme habitats in nature. It is for this reason that they are considered **extremophiles,** meaning that they "love" extreme conditions in the environment.

Metabolically, the archaea exhibit nearly incredible adaptations to what would be deadly conditions for other organisms. These hardy microbes have adapted to multiple combinations of temperature, salt, acid, pH, pressure, and atmosphere. Included in this group are methane producers, hyperthermophiles, extreme halophiles, and sulfur reducers.

Members of the group called **methanogens** can convert CO_2 and H_2 into methane gas (CH_4) through unusual and complex pathways. These archaea are common inhabitants of anaerobic mud and the bottom sediments of lakes and oceans. The gas they produce collects in swamps and may become a source of fuel. Methane may also contribute to the "greenhouse effect," which maintains the earth's temperature and can contribute to global warming. Not all methanogens live in extreme environments. Some are commonly found in the oral cavity and large intestine of humans.

Other types of archaea—the **extreme halophiles**—require salt to grow and may have such a high salt tolerance that they can multiply in sodium chloride solutions (36% NaCl) that would destroy most cells. They exist in the saltiest places on the earth—inland seas, salt lakes, and salt mines. They are not particularly common in the ocean because the salt content is not high enough to support them. Many of the "halobacteria" use a red pigment to synthesize ATP in the presence of light. These pigments are responsible for "red herrings," the color of the Red Sea, and the red color of salt ponds **(figure 4.32).**

Archaea adapted to growth at very low temperatures are called *psychrophilic* (loving cold temperatures); those growing at very high temperatures are *hyperthermophilic* (loving high temperatures). **Hyperthermophiles** flourish at temperatures between 80°C and 121°C (boiling temperature) and cannot grow at 50°C. They live in volcanic waters and soils and submarine vents, and are often salt- and acid-tolerant as well **(figure 4.33).** Researchers sampling sulfur vents in the deep ocean discovered thermophilic archaea flourishing at temperatures up to 250°C—150° above the temperature of boiling water! Not only were these archaea growing prolifically at this high temperature, but they were also living at 265 atmospheres of pressure. (On the earth's surface, pressure is about 1 atmosphere.) For additional discussion of the unusual adaptations of archaea, see chapter 7.

(a)

5 µm

(b)

Figure 4.32 Halophiles around the world. (a) An aerial view of a salt pond at San Francisco Bay, California. The archaea that thrive in this warm, highly saline habitat produce brilliant red, pink and orange pigments. **(b)** A sample taken from a saltern in Australia viewed by fluorescent microscopy (1,000×). Note the range of cell shapes (cocci, rods, and squares) found in this community.

Figure 4.33 An electron micrograph of the "hottest" microbe on earth. This tiny archaea (bar is 1 µm) thrives deep in hydrothermal vents that regularly reach 121°C—the working temperature of an autoclave. (S = slime layer, CM = cell membrane)

Check & Assess Section 4.7

✔ Bacteria exist in a tremendous variety of structure and lifestyles. Most of them are free-living rather than parasitic. Notable examples are the photosynthetic bacteria and gliding bacteria, which have unique adaptations in physiology and development.

✔ The rickettsias are a group of intracellular parasitic bacteria dependent on their eukaryote host for energy and nutrients. Most are pathogens that alternate between arthropods and mammalian hosts.

✔ The chlamydias are also small, intracellular parasites that infect humans, mammals, and birds. They do not require arthropod vectors.

✔ Archaea are another type of prokaryotic cell that constitute the third domain of life. They exhibit unusual biochemistry and genetics that make them different from bacteria. Many members are adapted to extreme habitats with low or high temperature, salt, pressure, or acid.

41. Discuss several ways in which bacteria are medically and ecologically important.

42. Name two main groups of obligate intracellular parasitic bacteria and explain why these groups can't live independently.

43. Explain the characteristics of archaea that indicate that they constitute a unique domain of living things that is neither bacterial nor eukaryotic.

44. What is meant by the term *extremophile*? Describe some archaeal adaptations to extreme habitats.

CASE FILE **4** PERSPECTIVE

All of the symptoms and signs in this case can be linked to endocarditis, but most of them are too nonspecific to be helpful in diagnosis. The most important considerations are the constant fever, the history of a replacement heart valve, and the respiratory and circulatory symptoms. A fever of unknown origin (FUO) is often traced to a chronic infection. The artificial heart valve provides an opportunity for microorganisms to grow in the body because it readily supports a biofilm that will be resistant to immune reactions. The dysfunctional valve causes blood to back up in the heart, leading to respiratory distress and abnormal heart function.

The patient might have survived if the attending physician had known about the heart valve and a prior infection early on. If he had taken blood cultures earlier, the infectious agent could have been isolated and identified, and the correct treatment started.

Most bacteria can form structured multicellular communities or biofilms on objects in a moist environment. This is even true of bacterial pathogens in the body. The CDC estimates that over 65% of chronic infections are caused by microbial biofilms. In this case, the MRSA bacteria in the patient's skin infection must have entered the circulation and colonized the artificial valve over several weeks to months. Most cases of chronic endocarditis are caused by biofilms on valves. When the biofilm grows into larger vegetations, portions of it break loose into the circulation. These infect the blood and are spread into organs,

causing fever and other signs and symptoms, including the ones that were fatal for Mr. Jones.

Staphylococcus aureus is the most common cause of endocarditis, especially associated with artificial heart valves. MRSA stands for methicillin resistant *Staphylococcus aureus*, which is a shorthand way of saying that the microbe is resistant to numerous drugs used in treatment. The reason that the penicillin Mr. Jones was given to take at home would not have worked is that most strains of *S. aureus* are resistant to it. Treatment with a mixture of drugs is indicated, but it needs to start early in the infection. MRSA is an emerging pathogen that started as a problem in hospitals but is now prominent in nonhospital settings as well.

For more background on MRSA and endocarditis, see chapter 18.

Chapter Summary with Key Terms

4.1 Basic Characteristics of Cells and Life Forms
A. All living things are composed of **cells,** which are complex collections of macromolecules that carry out living processes. All cells must have the minimum structure of an outer **cell membrane, cytoplasm,** a **chromosome,** and **ribosomes.**
B. Cells can be divided into two basic types: prokaryotes and eukaryotes.
 1. Prokaryotic cells are the basic structural unit of bacteria and archaea. They lack a nucleus or **organelles.** They are highly successful and adaptable single-cell life forms.
 2. Eukaryotic cells contain a membrane-surrounded nucleus and a number of organelles that function in specific ways. A wide variety of organisms, from single-celled protozoans to humans, are composed of eukaryotic cells.
 3. Viruses are not generally considered living or cells, and rely on host cells to replicate.
C. Cells show the basic essential characteristics of life. Parts of cells and macromolecules do not show these characteristics independently.
 1. **Heredity** and **Reproduction:** Cells must pass on genetic information to their offspring, whether **asexually** (with one parent) or **sexually** (with two parents).
 2. **Growth:** Living entities are able to grow and increase in size, often renewing and rebuilding themselves over time.
 3. **Metabolism:** This refers to the chemical reactions in the cell, including the synthesis of proteins on ribosomes and the capture and release of energy using ATP.
 4. **Responsiveness: Irritability** is responsiveness to external stimuli, **communication** is interaction with other cells, and **motility** originates from special locomotor structures such as flagella and cilia.
 5. **Transport:** Nutrients must be brought into the cell through the membrane and wastes expelled from the cell.

4.2 Prokaryotic Profiles: The Bacteria and Archaea
A. Prokaryotes consist of two major groups, the bacteria and the archaea. Life on earth would not be possible without them.

B. Prokaryotic cells lack the membrane-surrounded organelles and nuclear compartment of eukaryotic cells but are still complex in their structure and function. All prokaryotes have a cell membrane, cytoplasm, ribosomes, and a chromosome.
C. **Appendages:** Some bacteria have projections that extend from the cell.
 1. **Flagella** (and internal **axial filaments** found in spirochetes) are used for motility.
 2. **Fimbriae** function in adhering to the environment; **pili** provide a means for genetic exchange.
 3. The **glycocalyx** may be a slime layer or a capsule.

4.3 The Cell Envelope: The Boundary Layer of Bacteria
A. Most prokaryotes are surrounded by a protective envelope that consists of the cell wall and the **cell membrane.**
B. The wall is relatively rigid due to **peptidoglycan.**
C. Structural differences give rise to **gram-positive** and **gram-negative** cells, as differentiated by the **Gram stain.**
 1. Gram-positive bacteria contain a thick wall composed of peptidoglycan and teichoic acid in a single layer.
 2. Gram-negative bacteria have a thinner two-layer cell wall with an **outer membrane,** thin layer of peptidoglycan, and a well-developed **periplasmic space.**
 3. Wall structure gives rise to differences in staining, toxicity, and effects of drugs and disinfectants.

4.4 Bacterial Internal Structure
The cell cytoplasm is a watery substance that holds some or all of the following internal structures in bacteria: the **chromosome**(s) condensed in the **nucleoid; ribosomes,** which serve as the sites of protein synthesis and are 70S in size; extra genetic information in the form of **plasmids;** storage structures known as **inclusions;** a **cytoskeleton** of **bacterial actin,** which helps give the bacterium its shape; and in some bacteria an **endospore,** which is a highly resistant structure for survival, not reproduction.

4.5 Bacterial Shapes, Arrangements, and Sizes
A. Most bacteria are unicellular and are found in a great variety of shapes, arrangements, and sizes. General shapes include **cocci, bacilli,** and helical forms such as **spirilla** and **spirochetes.** Some show great variation within the species in shape and size and are **pleomorphic.** Other variations include **coccobacilli, vibrios,** and filamentous forms.
B. Prokaryotes divide by binary fission and do not utilize mitosis. Various **arrangements** result from cell division and are termed **diplococci,** streptococci, **staphylococci, tetrads,** and **sarcina** for cocci; bacilli may form pairs, chains, or **palisades.**
C. Variant members of bacterial species are called **strains** and **types.**

4.6 Classification Systems of the Prokaryotic Domains: Archaea and Bacteria
A. An important taxonomic system is standardized by *Bergey's Manual of Determinative Bacteriology,* which presents the prokaryotes in five major volumes.

Volume 1 Domain Archaea
 Domain Bacteria: Deeply Branching and Phototropic Bacteria
Domain Bacteria
Volume 2 Proteobacteria (gram-negative cell walls)
Volume 3 Low G + C gram-positive Bacteria
Volume 4 High G + C gram-positive Bacteria
Volume 5 Planctomyces, Spirochaetes, Fibrobacteres, Bacteriodetes, and Fusobacteria (gram-negative cell walls)

4.7 Survey of Prokaryotic Groups with Unusual Characteristics

A. Several groups of bacteria have unusual adaptations and life cycles.
1. Medically important bacteria: **Rickettsias** and **chlamydias** are within the gram-negative group but are small **obligate intracellular parasites** that replicate within cells of the hosts they invade.
2. Nonpathogenic bacterial groups: The majority of bacterial species are free-living and not involved in disease. Unusual groups include photosynthetic bacteria such as cyanobacteria, which provide oxygen to the environment, and the green and purple bacteria.

B. **Archaea** share many characteristics with bacteria but vary in certain genetic aspects and structures such as the cell wall and ribosomes.
1. Many are adapted to extreme environments similar to the earliest of earth's inhabitants.
2. They are not considered medically important but are of ecological and potential economic importance.

Multiple-Choice Questions

Select the correct answer from the answers provided. For questions with blanks, choose the combination of answers that most accurately completes the statement.

1. Which structure is not a component of all cells?
 a. cell wall
 b. cell membrane
 c. genetic material
 d. ribosomes

2. Viruses are not considered living things because
 a. they are not cells
 b. they cannot reproduce by themselves
 c. they lack metabolism
 d. All of these are correct.

3. Which of the following is not found in all bacterial cells?
 a. cell membrane
 b. a nucleoid
 c. ribosomes
 d. actin cytoskeleton

4. The major locomotor structures in bacteria are
 a. flagella
 b. pili
 c. fimbriae
 d. cilia

5. Pili are tubular shafts in _____ bacteria that serve as a means of _____.
 a. gram-positive, genetic exchange
 b. gram-positive, attachment
 c. gram-negative, genetic exchange
 d. gram-negative, protection

6. An example of a glycocalyx is
 a. a capsule
 b. pili
 c. outer membrane
 d. a cell wall

7. Which of the following is a primary bacterial cell wall function?
 a. transport
 b. motility
 c. support
 d. adhesion

8. Which of the following is present in both gram-positive and gram-negative cell walls?
 a. an outer membrane
 b. peptidoglycan
 c. teichoic acid
 d. lipopolysaccharides

9. Metachromatic granules are concentrated _____ found in _____.
 a. fat, *Mycobacterium*
 b. dipicolinic acid, *Bacillus*
 c. sulfur, *Thiobacillus*
 d. PO$_4$, *Corynebacterium*

10. Bacterial endospores function in
 a. reproduction
 b. survival
 c. protein synthesis
 d. storage

11. An arrangement in packets of eight cells is described as a _____.
 a. micrococcus
 b. diplococcus
 c. tetrad
 d. sarcina

12. The major difference between a spirochete and a spirillum is
 a. presence of flagella
 b. the presence of twists
 c. the nature of motility
 d. size

13. Which phylum contains bacteria with a gram-positive cell wall?
 a. Proteobacteria
 b. Chlorobi
 c. Firmicutes
 d. Spirochetes

14. To which taxonomic group do cyanobacteria belong?
 a. Domain Archaea
 b. Phylum Actinobacteria
 c. Domain Bacteria
 d. Phylum Fusobacteria

15. Which stain is used to distinguish differences between the cell walls of medically important bacteria?
 a. simple stain
 b. acridine orange stain
 c. Gram stain
 d. negative stain

16. The first living cell on earth was most similar to
 a. a cyanobacterium
 b. an endospore former
 c. a gram-positive cell
 d. an archaea

Case File Questions

1. What is true of the condition endocarditis?
 a. It occurs in the heart muscle.
 b. It is caused by microbes growing in the internal organs.
 c. It is an infection of the heart valves and lining.
 d. It can be transmitted to others.

2. Where did the MRSA pathogen that made the biofilm originate?
 a. from the artificial valve itself
 b. from an earlier skin infection
 c. from the surgery
 d. from the patient's wife

Writing to Learn

These questions are suggested as a *writing-to-learn* experience. For each question, compose a one- or two-paragraph answer that includes the factual information needed to completely address the question. Any question listed in a section's Check and Assess may be considered as a writing-to-learn exercise.

1. Label the parts on the bacterial cell featured here and write a brief description of its function.

2. Discuss the collection of properties that are used to define life and the prokaryotic cell structures that are involved in carrying out these life processes.

3. Describe the basic process of biofilm formation. pg. 95

4. What leads microbiologists to believe the archaea are more closely related to eukaryotes than to bacteria?

Concept Mapping

Appendix E provides guidance for working with concept maps.

1. Construct your own concept map using the following words as the *concepts*. Supply the linking words between each pair of concepts.

genus	species
serotype	domain
Borrelia	*burgdorferi*
Spirochaetes	phylum

Critical Thinking Questions

Critical thinking is the ability to reason and solve problems using facts and concepts. These questions can be approached from a number of angles, and in most cases, they do not have a single correct answer.

1. What is required to kill endospores? How do you suppose archaeologists were able to date some spores as being thousands (or millions) of years old?

2. Using clay, demonstrate how cocci can divide in several planes and show the outcome of this division. Show how the arrangements of bacilli occur, including palisades.

3. Using a corkscrew and a spring to compare the flexibility and locomotion of spirilla and spirochetes, explain which cell type is represented by each object.

4. Under the microscope, you see a rod-shaped cell that is swimming rapidly forward.
 a. What do you automatically know about that bacterium's structure?
 b. How would a bacterium use its flagellum for phototaxis?
 c. Propose another function of flagella besides locomotion.

5. a. Name a bacterium that has no cell walls.
 b. How is it protected from osmotic destruction?

6. a. Name a bacterium that is aerobic, gram-positive, and spore-forming.
 b. What habitat would you expect this species to occupy?

7. a. Name an acid-fast bacterium.
 b. What characteristics make this bacterium different from other gram-positive bacteria?

8. a. Name a bacterium that uses chlorophyll to photosynthesize.
 b. Describe the two major groups of photosynthetic bacteria.
 c. How are they similar?
 d. How are they different?

9. a. What are some possible adaptations that the giant bacterium *Thiomargarita* has had to make because of its large size?
 b. If a regular bacterium were the size of an elephant, estimate the size of a nanobe at that scale.

10. Propose a hypothesis to explain how bacteria and archaea could have, together, given rise to eukaryotes.

11. Explain or illustrate exactly what will happen to the cell wall if the synthesis of the interbridge is blocked by penicillin. What if the glycan is hydrolyzed by lysozyme?

12. Ask your lab instructor to help you make a biofilm and examine it under the microscope. One possible technique is to suspend a glass slide in an aquarium for a few weeks, then carefully air-dry, fix, and Gram stain it. Observe the diversity of cell types.

13. a. Use the size bars to measure the cells in figure 4.32, 4.33, and Insight 4.2.
 b. Describe the shapes and arrangements of bacteria in figure 4.23a, b, e, and f.

Visual Challenge

1. From **chapter 3, figure 3.27b.** Which bacteria has a well-developed capsule: "*Klebsiella*" or "*S. aureus*"? Defend your answer.

2. Using **figure 4.22** as a guide, label the stages in the endospore cycle shown in the figure, and explain the events depicted.

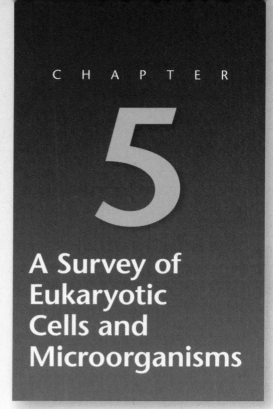

CHAPTER

5

A Survey of Eukaryotic Cells and Microorganisms

Trypanosoma, a flagellated protozoan

Larval schistosome, a parasitic roundworm

Hookworm larvae burrowing into the intestine

Three of the neglected microorganisms distributed across the earth's tropical zone.

"Almost everyone in the bottom billion has at least one of these diseases."

CASE FILE 5 **Neglected Tropical Diseases**

Some of the greatest suffering in the world originates from a group of ancient infectious diseases that exist primarily in the tropical and subtropical regions of Africa, India, Latin America, and Asia. Nearly 1.4 billion people worldwide, sometimes referred to as the "bottom billion," experience disability or death as a result of 13 common diseases.

These neglected tropical diseases, or NTDs, have received less attention than highly publicized diseases such as AIDS, malaria, and tuberculosis, and they have been largely ignored by the medical establishment in many countries. Unfortunately, they also tend to occur in the poorest rural areas, where access to treatment is often severely limited. "Almost everyone in the bottom billion has at least one of these diseases," said Dr. Peter Hotez, a parasitologist and medical

doctor at George Washington University. Taken together, NTDs rank second only to HIV/AIDS in their medical, social, and economic impact.

Eleven of the 13 neglected disease pathogens are eukaryotic parasites, either parasitic helminth worms or protozoans. The leading parasitic worm infections are ascariasis*, trichiuriasis, hookworm, and schistosomiasis, and the leading protozoan infections are Chagas disease, African trypanosomiasis, and leishmaniasis. The chronic, progressive actions of the infectious agents can cause debility and disfigurement. Consider Chagas disease, caused by a protozoan *Trypanosoma cruzi*, which over time lodges in the heart and other organs, destroys health, and shortens life. Some forms of leishmaniasis affect the skin and give rise to growths that deform the face and other body parts.

The growth of hookworms causes blood loss, saps strength, and impairs development. Some parasites cause blindness and others disfigure the limbs. In addition to medical effects, people often lose their abilities to go to school and work, and become victims of harsh ostracizing by their communities.

▶ *What subjects are studied by the science of parasitology?*

▶ *What does Dr. Hotez imply by his comment that everyone in the bottom billion has at least one infection?*

To continue the case, go to page 147.

*Some diseases are named by adding the suffix *-iasis* to the name of the organism that causes it.

5.1 The History of Eukaryotes

Expected Learning Outcomes

1. Describe the evolutionary history of eukaryotic cells.
2. Provide a substantial theory regarding how eukaryotic cells originated and how multicellularity came to be.
3. List the eukaryotic groups and their body plans.

Evidence from paleontology indicates that the first eukaryotic cells appeared on the earth approximately 2 billion years ago. Some fossilized cells that look remarkably like modern-day algae or protozoa appear in shale sediments from China, Russia, and Australia that date from 850 million to 950 million years ago (**figure 5.1**). Biologists have discovered convincing evidence to suggest that the eukaryotic cell evolved from prokaryotic organisms by a process of intracellular **symbiosis*** (**Insight 5.1**). Biologists have clear cut evidence that the first eukaryotic cells were the result of two prokaryotic cells meeting and merging. The starting event probably occurred when a larger prokaryotic cell engulfed smaller prokaryotic cells and kept them alive. Over millions of years, this combination evolved into a stable, beneficial partnership. Some of the small cells trapped inside these evolving cells become **organelles*** that are the distinguishing feature of eukaryotic cells. The structure of these early cells was so versatile that eukaryotic microorganisms soon spread out into available habitats and adopted greatly diverse styles of living.

The first primitive eukaryotes were probably single-celled, independent microorganisms, but, over time, some forms began to cluster in permanent groupings called colonies. With further evolution, some of the cells within colonies became *specialized,* or adapted to perform a particular function advantageous to the whole colony, such as locomotion, feeding, or reproduction. Complex multicellular organisms evolved as individual cells in the organism lost the ability to survive apart from the intact colony. Although a multicellular organism is composed of many cells, it is more than just a disorganized assemblage of cells like a colony. Rather, it is composed of distinct groups of cells that cannot exist independently of the rest of the body. The cell groupings of multicellular organisms that have a specific function are termed *tissues,* and groups of tissues make up *organs.*

Looking at modern eukaryotic organisms, we find examples of many levels of cellular complexity (**table 5.1**). All protozoa, as well as numerous algae and fungi, are unicellular. Truly multicellular organisms are found only among plants and animals and some of the fungi (mushrooms) and algae (seaweeds). Only certain eukaryotes are traditionally studied by microbiologists—primarily the protozoa, the microscopic algae and fungi, and animal parasites, or helminths.

(a) (b)

Figure 5.1 Ancient eukaryotic protists discovered in fossilized rocks. (a) A cell preserved in Siberian shale deposits dates from 850 million to 950 million years ago. **(b)** A disclike cell was recovered from a Chinese rock dating 590 million to 610 million years ago. Both cells are relatively simple, with (a) showing chloroplastlike bodies like algae. Example (b) has a cell wall with spines, very similar to the modern algae *Pediastrum.*

TABLE 5.1	Eukaryotic Organisms Studied in Microbiology	
Always Unicellular	**May Be Unicellular or Multicellular**	**Multicellular except Reproductive Stages**
Protozoa	Fungi Algae	Helminths (animals with unicellular egg or larval forms)

5.2 Form and Function of the Eukaryotic Cell: External Structures

Expected Learning Outcomes

4. Describe the plan of a basic eukaryotic cell and organelles, and indicate the structures all cells possess and those found only in some groups.
5. Describe the types of eukaryotic locomotor appendages.
6. Differentiate the structure and functions of flagella and cilia, and the types of cells that possess them.
7. Define the glycocalyx for eukaryotic cells and list its basic functions.
8. Characterize the cell wall and membrane of eukaryotic cells.

The cells of eukaryotic organisms are so varied that no one member can serve as a typical example. The composite structure of a eukaryotic cell is depicted in **figure 5.2.** Due to the organelles, it has much greater complexity and compartmentalization than a prokaryotic cell. No single type of microbial cell would contain all structures represented. Differences among fungi, protozoa, algae, and animal cells are introduced in later sections.

* *symbiosis* (sim-beye-oh′-sis) Gr. *sym,* together, and *bios,* to live. A close association between two organisms.
* *organelles* (or-gan′-elz) Gr. *organa,* tool, and *ella,* little. Structures inside cells that perform specific functions.

The Extraordinary Evolution of Eukaryotic Cells

For years, biologists have grappled with the problem of how a cell as complex as the eukaryotic cell originated. One of the most fascinating explanations is that of **endosymbiosis.** This theory proposes that eukaryotic cells arose when a much larger prokaryotic cell engulfed smaller prokaryotic cells that began to live and reproduce inside the prokaryotic cell rather than being destroyed. As the smaller cells took up permanent residence, they came to perform specialized functions for the larger cell, such as food synthesis and oxygen utilization, that enhanced the cell's versatility and survival. In time, the cells evolved into a single functioning entity, and the relationship became obligatory.

At first, the idea of endosymbiosis was greeted with some controversy; however, we now know that associations of this sort are rather common in the microbial world. Hundreds of protozoa have been discovered harboring living microbes internally. In some cases, this is a temporary symbiosis and not obligatory, but certain species of *Paramecium* contain algae and bacteria that they need to stay alive.

The biologist most responsible for validation of the theory of endosymbiosis is *Dr. Lynn Margulis.* Using modern molecular techniques, she has accumulated convincing evidence of the relationships between the organelles of modern eukaryotic cells and the structure of bacteria. In many ways, the mitochondrion of eukaryotic cells behaves as a tiny cell within a cell. It is capable of independent division, contains a circular chromosome with bacterial DNA sequences, and has ribosomes that are clearly prokaryotic. Mitochondria also have bacterial membranes and can be inhibited by drugs that affect only bacteria.

One possible origin of chloroplasts could have been endosymbiotic cyanobacteria that provided their host cells with a built-in feeding mechanism. Evidence is seen in a modern flagellated protist that harbors specialized chloroplasts with cyanobacterial chlorophyll and thylakoids. Dr. Margulis also has convincing evidence that eukaryotic cilia and flagella have arisen from endosymbiosis between spiral bacteria and the cell membrane of early eukaryotic cells. Most notably is a myxotrich found in termites that has spirochetes functioning as flagella.

The accompanying figure shows a possible sequence of events. A large archaeal cell with its flexible envelope could engulf a smaller bacterial cell, probably similar to a modern purple bacterium. The archaea would contribute its cytoplasmic ribosomes and some unique aspects of protein synthesis, as predicted by known characteristics (see table 4.5). Folds in the cell membrane could wrap around the chromosome to form a nuclear envelope. For its part, the bacterial cell would forge a metabolic relationship with the archaea, making use of archaeal molecules and contributing energy through aerobic respiration. Evolution of a stable mutualistic existence would maintain both cell types and create the eukaryotic

cell and its organelles. This scenario also has the advantage of explaining the relationships among the major domains. So well accepted is the theory that bacteriologists have placed both mitochondria and chloroplasts on the family tree of bacteria (see figure 4.27). In fact, the closest relative of mitochondria are rickettsias, which are also obligate intracellular bacteria!

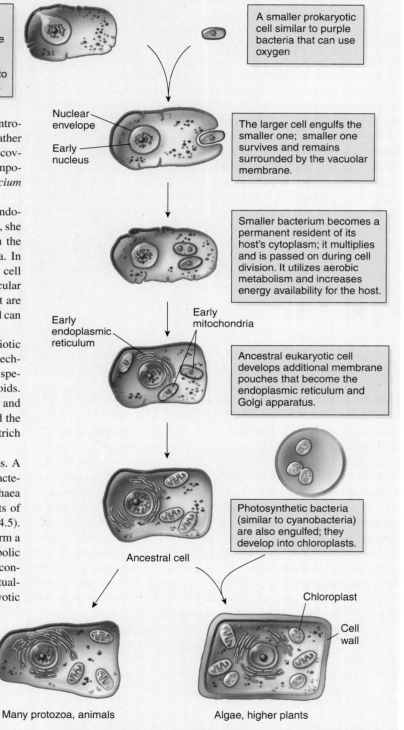

A larger prokaryotic cell such as an archaea has a flexible outer envelope and mesosomelike internal membranes to enclose the nucleoid.

A smaller prokaryotic cell similar to purple bacteria that can use oxygen

Nuclear envelope

Early nucleus

The larger cell engulfs the smaller one; smaller one survives and remains surrounded by the vacuolar membrane.

Smaller bacterium becomes a permanent resident of its host's cytoplasm; it multiplies and is passed on during cell division. It utilizes aerobic metabolism and increases energy availability for the host.

Early endoplasmic reticulum

Early mitochondria

Ancestral eukaryotic cell develops additional membrane pouches that become the endoplasmic reticulum and Golgi apparatus.

Photosynthetic bacteria (similar to cyanobacteria) are also engulfed; they develop into chloroplasts.

Ancestral cell

Chloroplast

Cell wall

Many protozoa, animals

Algae, higher plants

Paramecium labeled with fluorescent dyes. Orange spheres are endosymbiotic *Chlorella* algae; nuclei, are stained green.

Explain some of the ways the early mitochondria and chloroplasts could have assisted the nutrition of the larger cell. *Answer available at http://www.mhhe.com/talaro8*

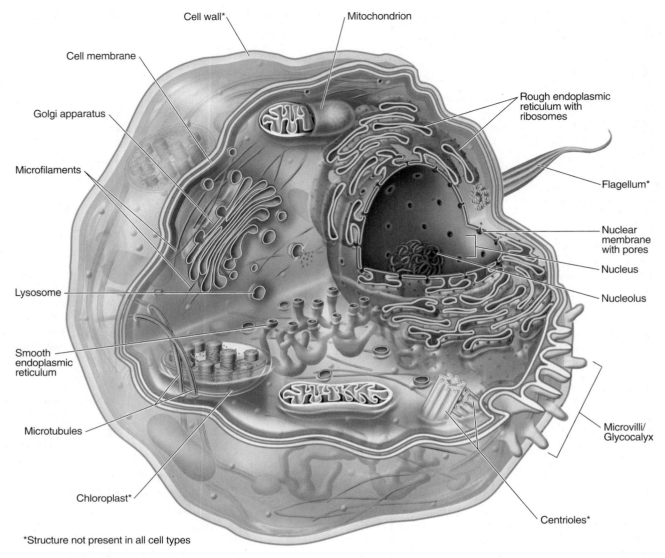

Cell wall*

Mitochondrion

Cell membrane

Golgi apparatus

Rough endoplasmic reticulum with ribosomes

Microfilaments

Flagellum*

Nuclear membrane with pores

Nucleus

Lysosome

Nucleolus

Smooth endoplasmic reticulum

Microtubules

Microvilli/ Glycocalyx

Chloroplast*

Centrioles*

*Structure not present in all cell types

Figure 5.2 Overview of composite eukaryotic cell. This drawing represents all structures associated with eukaryotic cells, but no microbial cell possesses all of them. See figures 5.16, 5.26, and 5.28 for examples of individual cell types.

This flowchart maps the organization of a eukaryotic cell. Compare this flowchart to the one found on page 91 in chapter 4.

Eukaryotic cell

External organelles and other structures
- Appendages
 - Flagella
 - Cilia
- Glycocalyx
 - Capsules
 - Slimes

Boundary of cell
- Cell wall
- Cell/cytoplasmic membrane

Internal organelles and other contents
- Cytoplasmic matrix
- Nucleus
 - Nuclear envelope
 - Nucleolus
 - Chromosomes
- Organelles
 - Endoplasmic reticulum
 - Golgi complex
 - Mitochondria
 - Chloroplasts
- Ribosomes
- Cytoskeleton
 - Microtubules
 - Microfilaments

In general, eukaryotic microbial cells have a cytoplasmic membrane, nucleus, mitochondria, endoplasmic reticulum, Golgi apparatus, vacuoles, cytoskeleton, and glycocalyx. A cell wall, locomotor appendages, and chloroplasts are found only in some groups. In the following sections, we cover the microscopic structure and functions of the eukaryotic cell. As with the prokaryotes, we begin on the outside and proceed inward through the cell.

Locomotor Appendages: Cilia and Flagella

Motility allows a microorganism to locate life-sustaining nutrients and to migrate toward positive stimuli such as sunlight; it also permits avoidance of harmful substances and stimuli. Locomotion by means of flagella is common in protozoa, algae, and a few fungal and animal cells. Cilia are found only in protozoa and animal cells.

Although they share the same name, eukaryotic flagella are much different from those of prokaryotes. The eukaryotic flagellum is thicker (by a factor of 10), has a much different construction, and

(c) Whips back and forth and pushes in snakelike pattern Twiddles the tip Lashes, grabs the substrate, and pulls

Figure 5.3 The structure of cilia and flagella. (a) A cross section through a protozoan cilium reveals the typical 9 + 2 arrangement of microtubules seen in both cilia and flagella. **(b)** Longitudinal section through a cilium, showing the lengthwise orientation of the microtubules and the basal body (bb) from which they arise. Note the membrane that surrounds the cilium that is an extension of the cell membrane and shows that it is indeed an organelle. **(c)** Locomotor patterns seen in flagellates.

is covered by an extension of the cell membrane. A flagellum is a long, sheathed cylinder containing regularly spaced hollow tubules—microtubules—that extend along its entire length **(figure 5.3b).** A cross section reveals nine pairs of closely attached microtubules surrounding a single central pair. This scheme, called the 9 + 2 arrangement, is a typical pattern of flagella and cilia **(figure 5.3a).** The nine pairs are linked together and anchored to the pair in the center. This architecture permits the microtubules to "walk" by sliding past each other, whipping the flagellum back and forth. Although details of this process are too complex to discuss here, it involves expenditure of energy and a coordinating mechanism in the cell membrane. Flagella can move the cell by pushing it forward like a fishtail or by pulling it by a lashing or twirling motion **(figure 5.3c).** The placement and number of flagella can be useful in identifying flagellated protozoa and certain algae.

Cilia are very similar in overall architecture to flagella, but they are shorter and more numerous (some cells have several thousand). They are found only in certain protozoa and animal cells. In the ciliated protozoa, the cilia occur in rows over the cell surface, where they beat back and forth in regular oarlike strokes **(figure 5.4)** and provide rapid motility. The fastest ciliated protozoan can swim up to 2,500 μm/s—a meter and a half per minute! On some cells, cilia also function as feeding and filtering structures.

The Glycocalyx

Most eukaryotic microbes have a **glycocalyx,** an outermost boundary that comes into direct contact with the environment. This structure is usually composed of polysaccharides and appears as a network of fibers, a slime layer, or a capsule much like the glycocalyx of prokaryotes. From its position as the exposed cell layer, the glycocalyx serves a variety of functions. Most prominently, it promotes adherence to environmental surfaces and the development of biofilms and mats. It also serves important receptor and communication functions and offers some protection against environmental changes.

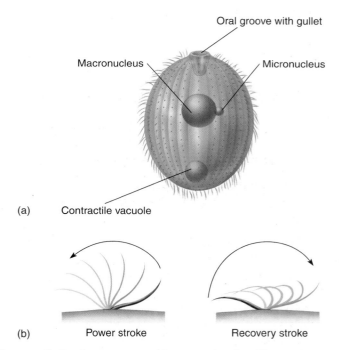

Figure 5.4 Structure and locomotion in ciliates. (a) The structure of a simple representative, *Holophrya,* with a regular pattern of cilia in rows over the cell surface. **(b)** Cilia beat in coordinated waves, driving the cell forward and backward. The pattern of ciliary movement is like a swimmer's arms, with a power forward stroke and a repositioning stroke. Ciliary moves are made possible by the microtubules they contain.

The nature of the layer beneath the glycocalyx varies among the several eukaryotic groups. Fungi and most algae have a thick, rigid cell wall surrounding a cell membrane. Protozoa, a few algae, and all animal cells lack a cell wall and are encased primarily by a cell membrane.

Form and Function of the Eukaryotic Cell: Boundary Structures

The Cell Wall

The cell walls of the fungi and algae are rigid and provide structural support and shape, but they are different in chemical composition from prokaryotic cell walls. Fungal cell walls have a thick, inner layer of polysaccharide fibers composed of chitin or cellulose and a thin outer layer of mixed glycans. The cell walls of algae are quite varied in chemical composition. Substances commonly found among various algal groups are cellulose, pectin,[1] mannans,[2] and minerals such as silicon dioxide and calcium carbonate.

The Cytoplasmic Membrane

The cytoplasmic (cell) membrane of eukaryotic cells is a typical bilayer of phospholipids in which protein molecules are embedded (see figure 4.16). In addition to phospholipids, eukaryotic membranes also contain *sterols* of various kinds. Sterols are different from phospholipids in both structure and behavior, as you may recall from chapter 2. Their relative rigidity confers stability on eukaryotic membranes. This strengthening feature is extremely important in cells that lack a cell wall. Cytoplasmic membranes of eukaryotes are functionally similar to those of prokaryotes, serving as selectively permeable barriers in transport. Unlike prokaryotes, eukaryotes have extensive membrane-bound organelles that can account for 60% to 80% of their volume.

Check & Assess Sections 5.1 and 5.2

✔ Eukaryotes have cells with a nucleus and organelles compartmentalized by membranes. Evidence shows that they originated from prokaryote ancestors about 2 billion years ago. Eukaryotic cell structure enabled eukaryotes to diversify from single cells into a huge variety of complex multicellular forms.

✔ The cell structures common to most eukaryotes are the cell membrane, nucleus, vacuoles, mitochondria, endoplasmic reticulum, Golgi apparatus, and a cytoskeleton. Cell walls, chloroplasts, and locomotor organs are present in some eukaryote groups.

✔ Microscopic eukaryotes use locomotor organs such as flagella or cilia for moving themselves or their food.

✔ The glycocalyx is the outermost boundary of most eukaryotic cells. Its functions are adherence, protection, and reception of chemical signals from the environment or from other organisms. The glycocalyx is supported by either a cell wall or a cell membrane.

✔ The cytoplasmic (cell) membrane of eukaryotes is similar in function and structure to that of prokaryotes, but it differs in containing sterols as additional stabilizing agents.

1. Briefly explain how the eukaryotic cell might have evolved from prokaryotic ones.
2. Which kingdoms of the five-kingdom system contain eukaryotic microorganisms?

3. How do unicellular, colonial, and multicellular organisms differ from each other? Give examples of each type.
4. How are flagella and cilia similar? How are they different?
5. Which eukaryotic cells have a cell wall?
6. What are the functions of the glycocalyx, cell wall, and membrane?

5.3 Form and Function of the Eukaryotic Cell: Internal Structures

Expected Learning Outcomes

9. Describe the structure of the nucleus and its outstanding features.
10. Outline the stages in cell division and mitosis.
11. Describe the structure of the two types of endoplasmic reticulum and their functions.
12. Identify the parts of the Golgi apparatus, and explain its basic actions and uses in the cell.
13. Summarize the stages in processing by the nucleus, endoplasmic reticulum, and Golgi apparatus involved in synthesis, packaging, and export.
14. Describe the structure of a mitochondrion, and explain its importance and functions.
15. Describe the structure of chloroplasts, and explain their importance and functions.
16. Discuss features of eukaryotic ribosomes.
17. Indicate the basic structure of the cytoskeleton, and explain its main features and functions.

The Nucleus: The Control Center

The nucleus is a compact sphere that is the most prominent organelle of eukaryotic cells. It is separated from the cell cytoplasm by an external boundary called a nuclear envelope. The envelope has a unique architecture. It is composed of two parallel membranes separated by a narrow space, and it is perforated with small, regularly spaced openings, or pores, at sites where the two membranes unite (**figure 5.5**). The pores are structured to serve as selective passageways for molecules to migrate between the nucleus and cytoplasm. The main body of the nucleus consists of a matrix called the nucleoplasm and a granular mass, the **nucleolus.** The nucleolus is the site for ribosomal RNA synthesis and a collection area for ribosomal subunits. The subunits are transported through the nuclear pores into the cytoplasm for final assembly into ribosomes.

A prominent feature of the nucleoplasm in stained preparations is a network of dark fibers known as **chromatin** because of its attraction for dyes. Analysis has shown that chromatin actually comprises the eukaryotic **chromosomes,** large units of genetic information in the cell. The chromosomes in the nucleus of nondividing cells are not readily visible because they are long, linear DNA molecules bound in varying degrees to **histone** proteins, and they are far too fine to be

1. A polysaccharide composed of galacturonic acid subunits.

2. A polymer of the sugar known as mannose.

Endoplasmic reticulum

Chromatin

Nuclear pore

Nuclear envelope

Nucleolus

(a)

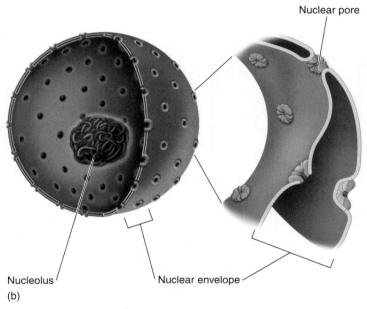

Nuclear pore

Nucleolus

Nuclear envelope

(b)

Figure 5.5 **The nucleus.** **(a)** Electron micrograph section of an interphase nucleus, showing its most prominent features. **(b)** Cutaway three-dimensional view of the relationships of the nuclear envelope and pores.

resolved as distinct structures without extremely high magnification. During **mitosis,** however, when the duplicated chromosomes are separated equally into daughter cells, the chromosomes themselves become readily visible as discrete bodies **(figure 5.6).** This appearance arises when the DNA becomes highly condensed by forming coils and supercoils around the histones to prevent the chromosomes from tangling as they are separated into new cells. This process is described in more detail in chapter 9.

Although we correctly view the nucleus as the primary genetic center, it does not function in isolation. As we show in the next three sections, it is closely tied to cytoplasmic organelles that perform elaborate cell functions.

Endoplasmic Reticulum: A Passageway in the Cell

The **endoplasmic reticulum (ER)** is a microscopic series of tunnels used in transport synthesis and storage. Two kinds of endoplasmic reticulum are the **rough endoplasmic reticulum (RER) (figure 5.7)** and the **smooth endoplasmic reticulum (SER).** Electron micrographs show that the RER originates from the outer membrane of the nuclear envelope and extends in a continuous network through the cytoplasm, even out to the cell membrane. This architecture permits the spaces in the RER, or **cisternae,** to serve as a passageway from the nucleus to the cytoplasm and ultimately to the cell's exterior. The RER appears rough because of large numbers of ribosomes attached to its membrane surface that synthesize proteins and transport them into the cavity of the reticulum. Additional functions of the RER are to process and modify proteins by adding essential molecules and cofactors prior to final packaging and transport. The SER has similar functions in synthesis, storage, and transport, but mainly of non-protein molecules such as lipids. It also serves as a site for detoxification of a variety of metabolic products.

Golgi Apparatus: A Packaging Machine

The **Golgi**[3] **apparatus,** also called the **Golgi complex** or body, is the site in the cell in which proteins are collected and packaged for transport to their final destinations. It is a discrete organelle consisting of a stack of flattened, disc-shaped sacs with spaces or cisternae, giving an appearance of pita breads. These sacs have outer limiting membranes and cavities like those of the endoplasmic reticulum, but they do not form a continuous network **(figure 5.8).** This organelle is always closely associated with the endoplasmic reticulum both in its location and function. At a site where it borders on the Golgi apparatus, the endoplasmic reticulum buds off tiny membrane-bound packets of protein called **transport vesicles** that are picked up by the forming face of the Golgi apparatus. Once in the complex itself, the proteins are often modified by the addition of polysaccharides and lipids. The final action of this apparatus is to pinch off finished **condensing vesicles** that will be conveyed to organelles such as lysosomes or transported outside the cell as **secretory vesicles (figure 5.9).**

Nucleus, Endoplasmic Reticulum, and Golgi Apparatus: Nature's Assembly Line

As the keeper of the eukaryotic genetic code, the nucleus ultimately governs and regulates all cell activities. But because the nucleus remains fixed in a specific cellular site, it must direct these activities through a structural and chemical network (figure 5.9). This network includes ribosomes, which originate in the nucleus, and the rough endoplasmic reticulum, which is continuously connected with the nuclear envelope.

Initially, a segment of the genetic code of DNA containing the instructions for producing a protein is copied into RNA and passed out through the nuclear pores directly to the ribosomes on

3. (gol′-jee) Named for C. Golgi, an Italian histologist who first described the apparatus in 1898.

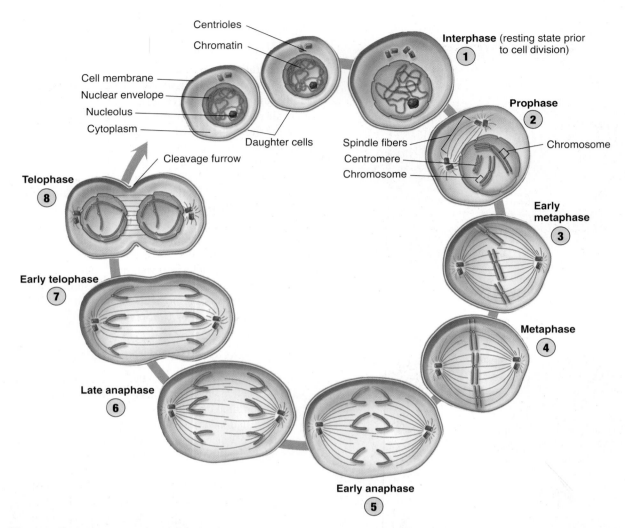

Process Figure 5.6 **Changes in the cell and nucleus during mitosis of a eukaryotic cell** **(1)** Before mitosis (at interphase), chromosomes are visible only as chromatin. **(2)** As mitosis proceeds (early prophase), chromosomes take on a fine, threadlike appearance as they condense, and the nuclear membrane and nucleolus are temporarily disrupted. **(3)–(4)** By metaphase, the chromosomes are fully visible as X-shaped structures. The shape is due to duplicated chromosomes attached at a central point, the centromere. **(5)–(6)** Spindle fibers attach to these and facilitate the separation of individual chromosomes during anaphase. **(7)–(8)** Telophase completes chromosomal separation and division of the cell proper into daughter cells. Note: This mechanism is how wall-free cells divide. Cells with walls will have a different pattern.

the endoplasmic reticulum. Here, specific proteins are synthesized from the RNA code and deposited in the lumen (space) of the endoplasmic reticulum. Details of this process are covered in chapter 9. After being transported to the Golgi apparatus, the protein products are chemically modified and packaged into **vesicles*** that can be used by the cell in a variety of ways. Some of the vesicles contain enzymes to digest food inside the cell; other vesicles are secreted to digest materials outside the cell; and yet others are important in the enlargement and repair of the cell wall and membrane.

A **lysosome*** is one type of vesicle originating from the Golgi apparatus that contains a variety of enzymes. Lysosomes are involved in intracellular digestion of food particles and in protection against invading microorganisms. They also participate in digestion and removal of cell debris in damaged tissue.

Other types of vesicles include **vacuoles,*** which are membrane-bound sacs containing fluids or solid particles to be digested, excreted, or stored. They are formed in phagocytic cells (certain white blood cells and protozoa) in response to food and other substances that have been engulfed. The contents of a food vacuole are digested through the merger of the vacuole with a lysosome. This merged structure is called a phagosome **(figure 5.10).** Other types of vacuoles are used in storing reserve food such as fats and glycogen. Protozoa living in freshwater habitats regulate osmotic pressure by means of contractile vacuoles, which regularly expel excess water that has diffused into the cell (described later).

* *vesicle* (ves′-ik-l) L. *vesios,* bladder. A small sac containing fluid.
* *lysosome* (ly′-soh-sohm) Gr. *Lysis,* dissolution, and *soma,* body.

* *vacuole* (vak′-yoo-ohl) L. *vacuus,* empty. Any membranous space in the cytoplasm.

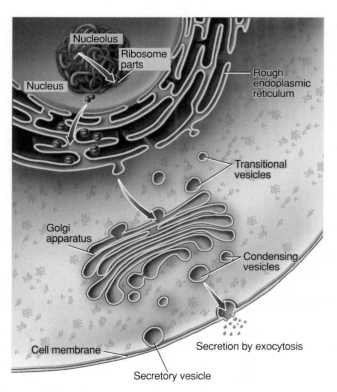

Figure 5.7 **The origin and detailed structure of the rough endoplasmic reticulum (RER).** **(a)** Schematic view of the origin of the RER from the outer membrane of the nuclear envelope. **(b)** Three-dimensional projection of the RER. **(c)** Detail of the orientation of a ribosome on the RER membrane. Proteins are collected within the RER cisternae and distributed through its network to other destinations.

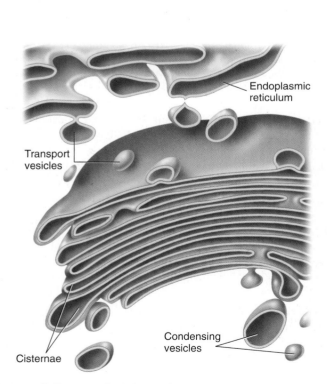

Figure 5.8 **Detail of the Golgi apparatus.** The flattened layers are cisternae. Specialized vesicles enter the upper surface and leave the lower surface.

Figure 5.9 **The transport process.** The cooperation of organelles in protein synthesis and transport: nucleus → RER → Golgi apparatus → vesicles → secretion.

Figure 5.10 **The origin and action of lysosomes in phagocytosis.**

Mitochondria: Energy Generators of the Cell

None of the cellular activities of the genetic assembly line could proceed without a constant supply of energy, the bulk of which is generated in most eukaryotes by **mitochondria.*** When viewed with light microscopy, mitochondria appear as round or elongated particles

Figure 5.11 **General structure of a mitochondrion.** (a) A three-dimensional projection. (b) An electron micrograph of a longitudinal section. In most cells, mitochondria are elliptical or spherical, although in certain fungi, algae, and protozoa, they are long and filamentlike.

scattered throughout the cytoplasm. Higher magnification reveals that a mitochondrion consists of a smooth, continuous outer membrane that forms the external contour and an inner, folded membrane nestled neatly within the outer membrane **(figure 5.11a).** The folds on the inner membrane, called **cristae,*** vary in exact structure among eukaryotic cell types. Plant, animal, and fungal mitochondria have lamellar cristae folded into shelflike layers. Those of algae and protozoa are tubular, fingerlike projections or flattened discs.

The cristae membranes hold the enzymes and electron carriers of aerobic respiration. This is an oxygen-using process that extracts chemical energy contained in nutrient molecules and stores it in the form of high-energy molecules, or ATP. More detailed functions of mitochondria are covered in chapter 8. The spaces around the cristae are filled with a chemically complex fluid called the **matrix,*** which holds ribosomes, DNA, and the pool of enzymes and other compounds involved in the metabolic cycle. Mitochondria (along with chloroplasts) are unique among organelles in that they divide

* *mitochondria* (my″-toh-kon′-dree-uh) sing. Mitochondrion; Gr. *Mitos,* thread, and *chondrion,* granule.

* *cristae* (kris′-te) sing. Crista; L. *crista,* a comb.
* *matrix* (may′-triks) L. *mater,* mother or origin.

independently of the cell, contain circular strands of DNA, and have prokaryote-sized 70S ribosomes. These findings have given support to the endosymbiotic theory of their evolutionary origins discussed in Insight 5.1.

Chloroplasts: Photosynthesis Machines

Chloroplasts are remarkable organelles found in algae and plant cells that are capable of converting the energy of sunlight into chemical energy through photosynthesis. The photosynthetic role of chloroplasts makes them the primary producers of organic nutrients upon which all other organisms (except certain bacteria) ultimately depend. Another important photosynthetic product of chloroplasts is oxygen gas. Although chloroplasts resemble mitochondria, chloroplasts are larger, contain special pigments, and are much more varied in shape.

There are differences among various algal chloroplasts, but most are generally composed of two membranes, one enclosing the other. The smooth, outer membrane completely covers an inner membrane folded into small, disclike sacs called **thylakoids** that are stacked upon one another into **grana.** These structures carry the green pigment chlorophyll and sometimes additional pigments as well. Surrounding the thylakoids is a ground substance called the **stroma* (figure 5.12).** The role of the photosynthetic pigments is to absorb and transform solar energy into chemical energy, which is then used during reactions in the stroma to synthesize carbohydrates. We further explore some important aspects of photosynthesis in chapters 7 and 8.

Ribosomes: Protein Synthesizers

In an electron micrograph of a eukaryotic cell, ribosomes are numerous, tiny particles that give a "dotted" appearance to the cytoplasm (see figure 5.31*b*). Ribosomes are distributed in two ways: Some are

* *stroma* (stroh′-mah) Gr. *Stroma,* mattress or bed.

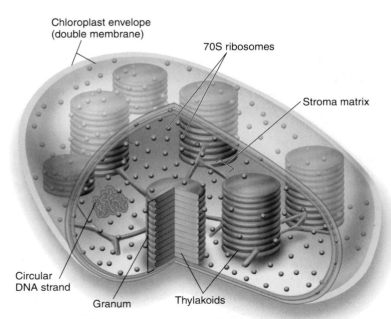

Chloroplast envelope (double membrane)

70S ribosomes

Stroma matrix

Circular DNA strand

Granum

Thylakoids

Figure 5.12 Detail of an algal chloroplast.

scattered freely in the cytoplasm and cytoskeleton; others are intimately associated with the rough endoplasmic reticulum, as previously described. Multiple ribosomes are often found arranged in short chains called polyribosomes (polysomes). The basic structure of eukaryotic ribosomes is similar to that of prokaryotic ribosomes, described in chapter 4. Both are composed of large and small subunits of ribonucleoprotein (see figure 5.7). By contrast, however, the eukaryotic ribosome (except in the mitochondrion) is the larger 80S variety that is a combination of 60S and 40S subunits. As in the prokaryotes, eukaryotic ribosomes are the staging areas for protein synthesis.

The Cytoskeleton: A Support Network

All cells share a generalized region encased by the cell membrane called the cytoplasm or, in eukaryotic cells, the **cytoplasmic matrix.** This complex area houses the organelles and sustains major metabolic and synthetic activities. It also contains the framework of support in cells lacking walls. The matrix is interwoven by a flexible framework of molecules called the **cytoskeleton (figure 5.13).** This framework appears to have several functions, such as anchoring organelles, providing support, and permitting shape changes and movement in some cells. The two main types of cytoskeletal elements are *microfilaments* and *microtubules.*

Microfilaments are thin strands composed of the protein actin that attach to the cell membrane and form a network through the cytoplasm. Some microfilaments are responsible for movements of the cytoplasm, often made evident by the streaming of organelles around the cell in a cyclic pattern. Other microfilaments are active in *amoeboid motion,* a type of movement typical of cells such as amoebas and phagocytes that produces extensions of the cell membrane (pseudopods) into which the cytoplasm flows.

Microtubules are long, hollow tubes that maintain the shape of eukaryotic cells such as protozoa that lack cell walls. They may also serve as an alternative transport system for molecules. The spindle fibers that play an essential role in mitosis are actually microtubules that attach to chromosomes and separate them into daughter cells. As indicated earlier, microtubules are also responsible for the movement of cilia and flagella.

Check & Assess Section 5.3

- ✔ The genome of eukaryotes is located in the nucleus, a spherical structure surrounded by a double membrane. The nucleus contains the nucleolus, the site of ribosome synthesis. DNA is organized into chromosomes in the nucleus.
- ✔ The endoplasmic reticulum (ER) is an internal network of membranous passageways extending throughout the cell.
- ✔ The Golgi apparatus is a packaging center that receives materials from the ER and then forms vesicles around them for storage or for transport to the cell membrane for secretion.
- ✔ The mitochondria generate energy in the form of ATP to be used in numerous cellular activities.
- ✔ Chloroplasts, membranous packets found in plants and algae, are used in photosynthesis.
- ✔ Ribosomes are the sites for protein synthesis present in both eukaryotes and prokaryotes.

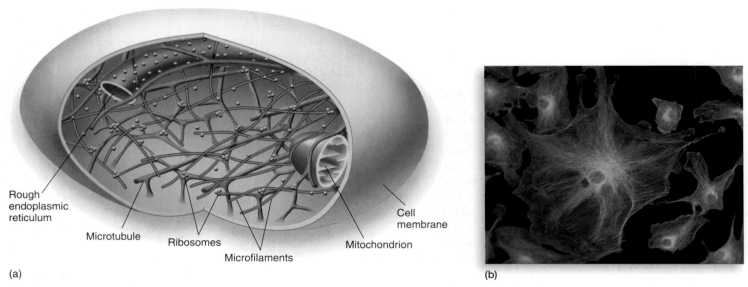

(a)

(b)

Figure 5.13 A model of the cytoskeleton. (a) Depicted is the relationship between microtubules, microfilaments, and organelles. (Not to scale.) **(b)** The cytoskeleton is highlighted by fluorescent dyes. Microtubules are stained green, microfilaments are red, and the nucleus is blue.

✔ The cytoskeleton, consisting of microfilaments and microtubules, maintains the shape of cells and produces movement of cytoplasm within the cell, movement of chromosomes at cell division, and, in some groups, movement of the cell as a unit.

7. In what ways does the nucleus function like the "brain" of the cell?

8. Explain how ribosomes and the nucleolus are related.

9. How does the nucleus communicate with the cytoplasm?

10. Compare and contrast the smooth ER, the rough ER, and the Golgi apparatus in structure and function.

11. How could a larger cell benefit from having an endoplasmic reticulum and a Golgi apparatus?

12. Compare the structure of the mitochondrion and the chloroplast.

13. What makes the mitochondrion and chloroplast unique among the organelles?

14. How are eukaryotic ribosomes different from prokaryotic ribosomes?

15. Describe some of the ways that organisms use lysosomes.

16. For what reasons would a cell need a "skeleton"?

5.4 Eukaryotic-Prokaryotic Comparisons and Taxonomy of Eukaryotes

Expected Learning Outcomes

18. Compare and contrast prokaryotic cells, eukaryotic cells, and viruses.

19. Outline the basics of eukaryotic taxonomy.

20. Explain what is meant by the term *protist,* and outline the types of organisms belonging to this designation.

At this point, you have been introduced to the major features of eukaryotic cells. **Table 5.2** provides a general summary of these characteristics. It also presents an opportunity to compare and contrast eukaryotic with prokaryotic cells (chapter 4) on the basis of significant anatomical and physiological traits. We have included viruses for the purposes of summarizing the biological characteristics of all microbes and to show you just how much viruses differ from cells. We will explore this fascinating group of microbes in chapter 6.

Overview of Taxonomy

Exploring the origins of eukaryotic cells with molecular technology has significantly clarified our understanding of relationships among the organisms in Domain Eukarya. The phenetic characteristics traditionally used for placing plants, animals, and fungi into separate kingdoms are general cell type, level of organization or body plan, and nutritional type. It now appears that these criteria really do reflect accurate differences among these organisms and give rise to the same basic kingdoms as do genetic tests using ribosomal RNA (see **figures** 1.15 and **5.14***a***).** The phylogenetic tree based on RNA relationships presents the Domain Eukarya spreading over a wide range, with the simplest protists (*Giardia*) at one end and fungi (*Coprinus*) at the other. *Homo* represents the animal branch and *Zea* (corn), the plants. The rest of the branches represent primarily groups that are traditionally labeled algae and protozoa.

Because our understanding of the phylogenetic relationships is still in development, there is not yet a single official system of taxonomy for presenting all of the eukaryotes. One of the proposed alternate plans of classification is shown in **figure 5.14***b***.** This simplified system, also based on data from ribosomal RNA analysis, retains some of the traditional kingdoms.

The one group that has created the greatest challenge in establishing reliable relationships is **Kingdom Protista**—the protists. Any simple eukaryotic cell that lacked multicellular structure or cell specialization has been placed into the Kingdom Protista generally

TABLE 5.2 A General Comparison of Prokaryotic and Eukaryotic Cells and Viruses

Function or Structure	Characteristic*	Prokaryotic Cells	Eukaryotic Cells	Viruses**
Genetics	Nucleic acids	+	+	+
	True nucleus	−	+	−
	Nuclear envelope	−	+	−
	Nucleoid	+	−	−
Reproduction	Mitosis	−	+	−
	Production of sex cells	+/−	+	−
	Binary fission	+	+	−
Biosynthesis	Independent	+	+	−
	Golgi apparatus	−	+	−
	Endoplasmic reticulum	−	+	−
	Ribosomes	+***	+	−
Respiration	Enzymes	+	+	−
	Mitochondria	−	+	−
Photosynthesis	Pigments	+/−	+/−	−
	Chloroplasts	−	+/−	−
Motility/locomotor structures	Flagella	+/−***	+/−	−
	Cilia	−	+/−	−
Shape/protection	Cell membrane	+	+	+/−
	Cell wall	+/−***	+/−	− (have capsids instead)
	Capsule	+/−	+/−	−
Size (in general)		0.5–3 µm****	2–100 µm	< 0.2 µm

*+ means most members of the group exhibit this characteristic; − means most lack it; +/− means some members have it and some do not.
**Viruses cannot participate in metabolic or genetic activity outside their host cells.
The prokaryotic type is structurally very different. *Much smaller and much larger bacteria exist; see Insight 4.2.

as an alga (photosynthetic) or protozoan (non-photosynthetic). Although exceptions arise with this classification scheme—notably multicellular algae and the photosynthetic protozoans—most eukaryotic microbes have been readily accommodated in the Kingdom Protista as traditionally presented.

Newer genetic data reveal that organisms placed in Kingdom Protista may be as different from each other as plants are different from animals. They are highly complex organisms that are likely to have evolved from a number of separate ancestors.

Microbiologists prefer to set up taxonomic systems that reflect true relationships based on all valid scientific data, including genotypic, molecular, morphological, physiological, and ecological. As a result, several alternative taxonomic strategies are being proposed to deal with this changing picture. Suggestions have ranged from distributing the protists among five new kingdoms to 25 kingdoms or more.

Note that figure 5.14b has arrayed a newer system alongside the original Kingdom Protista with its assorted phyla and divisions. One advantage of this plan is that it continues to use most of the original taxons at the phylum, division, and lower levels, especially as they relate to medically significant microbial groups. We feel the term **protist** is still a useful shorthand reference and will continue to use it for any eukaryote that is not a fungus, animal, or plant.

In the interest of space and ease in presentation, we have adopted a system of taxonomy that emphasizes natural groupings with the least complexity. A final note about taxonomy: no one system is perfect or permanent. Let your instructor guide you as to which one is satisfactory for your course.

With the general structure of the eukaryotic cell in mind, let us next examine the range of adaptations that this cell type has undergone. The following sections contain a general survey of the principal eukaryotic microorganisms—fungi, algae, protozoa, and parasitic worms—while also introducing elements of their structure, life history, classification, identification, and importance. A survey of diseases associated with some of these microbes is found in chapters 22 and 23.

5.5 The Kingdom of the Fungi

Expected Learning Outcomes

21. Describe the basic characteristics of the Kingdom Fungi in terms of general types of cells and organisms, structure, and nutrition.

22. Differentiate between characteristics of yeasts and molds, and define fungal spores.

23. Classify types of fungal spores and explain their functions.

24. Discuss the main features of fungal classification and representative examples of each group.

25. Explain how fungi are identified.

26. Discuss the importance of fungi in ecology, agriculture, commerce, and medicine.

(a)

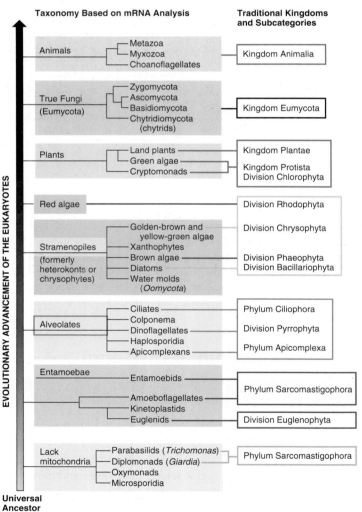

(b)

Figure 5.14 RNA-based phylogenetic tree and taxonomy of Domain Eukarya. (a) The branches are labeled with a genus that represents a particular rRNA profile. All branches after genus *Zea* (plants) represent various protist groups with many separate lines of evolution. **(b)** A comparison of two possible systems for presenting the taxonomy of major eukaryotic groups.

The position of the **fungi*** in the biological world has been debated for many years. Although they were originally classified with the green plants (along with algae and bacteria), they were separated from plants and placed in a group with algae and protozoa (the Protista). Even at that time, however, many microbiologists were struck by several unique qualities of fungi that warranted placement into a separate kingdom. Confirmation of their status by genetic testing eliminated any question that they belong in a kingdom of their own.

The **Kingdom Fungi,** or **Eumycota,** is filled with organisms of great variety and complexity that have survived on earth for approximately 650 million years. About 100,000 species are known, although experts estimate a count much higher than this—perhaps even 1.5 million different types. For practical purposes, mycologists divide the fungi into two groups: the *macroscopic fungi* (mushrooms, puffballs, gill fungi) and the *microscopic fungi* (molds, yeasts). Although the majority of fungi are either unicellular or colonial, a few complex forms such as mushrooms and puffballs are considered multicellular.

Chemical traits of fungal cells include the possession of a polysaccharide, **chitin,** in their cell walls and the sterol, ergosterol, in their cell membranes. Cells of the microscopic fungi exist in two basic morphological types: hyphae and yeasts. **Hyphae*** are long, threadlike cells that make up the bodies of filamentous fungi, or **molds (figure 5.15).** A **yeast** cell is distinguished by its round to oval shape and by its mode of asexual reproduction. It grows swellings on its surface called buds, which then become separate cells **(figure 5.16a).** Some species form a **pseudohypha,*** a chain of yeasts formed when buds remain attached in a row **(figure 5.16c).** Because of its manner of formation, it is not a true hypha like that of molds. While some fungal cells exist only in a yeast form and others occur primarily as hyphae, a few, called **dimorphic,*** can take either form, depending upon growth conditions, such as changing temperature. This variability in growth form is particularly characteristic of some pathogenic molds.

Fungal Nutrition

All fungi are **heterotrophic.*** They acquire nutrients from a wide variety of organic sources or **substrates (figure 5.17).** Most fungi are **saprobes,*** meaning that they obtain these substrates from the remnants of dead plants and animals in soil or aquatic habitats. Fungi can also be parasites on the bodies of living animals or plants, although very few fungi absolutely require a living host. In general, the fungus penetrates the substrate and secretes enzymes that reduce it to small molecules that can be absorbed. Fungi have enzymes for digesting an incredible array of substances, including feathers, hair, cellulose, petroleum products, wood, even rubber. It has been said that every naturally occurring organic material on the earth can be attacked by some type of fungus.

* *fungi* (fun′-jy) sing. fungus; Gr. *fungos,* mushroom.
* *hypha* (hy′-fuh) pl. hyphae (hy′-fee); Gr. *hyphe,* a web.
* *pseudohypha* (soo″-doh-hy′-fuh) pl. pseudohyphae; Gr. *pseudo,* false, and *hyphe.*
* *dimorphic* (dy-mor′-fik) Gr. *di,* two, and *morphe,* form.
* *heterotrophic* (het-ur-oh-tro′-fik) Gr. *hetero,* other, and *troph,* to feed. A type of nutrition that relies on an organic nutrient source.
* *saprobe* (sap′-rohb) Gr. *sapros,* rotten, and *bios,* to live. Also called saprotroph or saprophyte.

(a)

(b)

Septum

Figure 5.15 Macroscopic and microscopic views of molds.
(a) Plate of selective media growing a mixed culture of mold colonies (fuzzy texture) and bacterial colonies (slimy texture) isolated from soil. Note the various textures of mycelia and the array of color differences due to spores.
(b) Close-up of hyphal structure (1,200×).
(c) Basic structural types of hyphae. The septate hyphae develop small pores that allow communication between cells. Nonseptate hyphae lack septa and are single, long multinucleate cells.

Septa

Septate hyphae
Septum with pores
Nucleus
As in *Penicillium*

(c)

Nonseptate hyphae
Nuclei
As in *Rhizopus*

Bud Bud scar

Ribosomes
Mitochondrion
Endoplasmic reticulum
Nucleus
Nucleolus
Cell wall
Cell membrane
Golgi apparatus
Storage vacuole

(a) **Fungal (Yeast) Cell**

(b)

Bud
Nucleus
Bud scars

(c) Pseudohypha

Figure 5.16 Microscopic morphology of yeasts.
(a) General structure of a yeast cell, representing major organelles. Note the presence of a cell wall and lack of locomotor organelles.
(b) Scanning electron micrograph of *Malassezia furfur*, a yeast that causes a type of superficial skin infection (25,000×). (c) Formation and release of yeast buds and pseudohypha (a chain of budding yeast cells).

(a)

(b)

Figure 5.17 **Nutritional sources (substrates) for fungi.** **(a)** A mixture of fungal growth on bread. The fine hyphal filaments and black sporangia are typical of *Rhizopus*. **(b)** The skin of the foot with athlete's foot, an infection caused by *Trichophyton rubrum*.

Fungi are often found in nutritionally poor or adverse environments. Various fungal types thrive in substrates with higher salt, sugar, or acid content, at relatively high temperatures, and even in snow and glaciers.

The medical and agricultural impact of fungi is extensive. A number of species cause **mycoses** (fungal infections) in animals, and thousands of species are important plant pathogens. Fungal toxins may cause disease in humans, and airborne fungi are a frequent cause of allergies and other medical conditions **(Insight 5.2).**

Organization of Microscopic Fungi

The cells of most microscopic fungi grow in loose associations or colonies. The colonies of yeasts are much like those of bacteria in that they have a soft, uniform texture and appearance. The colonies of filamentous fungi are noted for the striking cottony, hairy, or velvety textures that arise from their microscopic organization and morphology. The woven, intertwining mass of hyphae that makes up the body or colony of a mold is called a **mycelium.***

Although hyphae contain the usual eukaryotic organelles, they also have some unique organizational features. In most fungi, the hyphae are divided into segments by cross walls, or **septa,** a condition called septate (see figure 5.15c). The nature of the septa varies from solid partitions with no communication between the compartments to partial walls with small pores that allow the flow of organelles and nutrients between adjacent compartments. Nonseptate hyphae consist of one long, continuous cell *not* divided into individual compartments by cross walls. With this construction, the cytoplasm and organelles move freely from one region to another, and each hyphal element can have several nuclei (see figure 5.15c).

Hyphae can also be classified according to their particular function. Vegetative hyphae (mycelia) are responsible for the visible mass of growth that appears on the surface of a food source and penetrates it to digest and absorb nutrients. During the development of a fungal colony, the vegetative hyphae give rise to structures called reproductive, or fertile, hyphae that branch off vegetative mycelium. These hyphae are responsible for the production of fungal reproductive bodies called spores discussed next. Other specializations of hyphae are illustrated in **figure 5.18.**

Reproductive Strategies and Spore Formation

Fungi have many complex and successful reproductive strategies. Most can propagate by the simple outward growth of existing hyphae or by fragmentation, in which a separated piece of mycelium can generate a whole new colony. But the primary reproductive mode of fungi involves the production of various types of **spores.** Do not confuse fungal spores with the more resistant, nonreproductive bacterial spores. Fungal spores are responsible not only for multiplication but also for survival, producing genetic variation, and dissemination. Because of their compactness and relatively light weight, spores are dispersed widely through the environment by air, water, and living things. Upon encountering a favorable substrate, a spore will germinate and produce a new fungus colony in a very short time (figure 5.18).

The fungi exhibit such a marked diversity in spores that they are largely classified and identified by their spores and spore-forming structures. Although there are some elaborate systems for naming and classifying spores, we present only a basic overview of the principal types. The most general subdivision is based on the way the spores arise. Asexual spores are the products of mitotic division of a single parent cell, and sexual spores are formed through a process involving the fusing of two parental nuclei followed by meiosis.

Asexual Spore Formation

On the basis of the nature of the reproductive hypha and the manner in which the spores originate, there are two subtypes of asexual spore **(figure 5.19):**

1. **Sporangiospores (figure 5.19a)** are formed by successive cleavages within a saclike head called a **sporangium,*** which is attached to a stalk, the sporangiophore. These spores are initially enclosed but are released when the sporangium ruptures.
2. **Conidia*** (conidiospores) are free spores not enclosed by a spore-bearing sac **(figure 5.19b).** They develop either by pinching off the tip of a special fertile hypha or by segmentation of a preexisting vegetative hypha. Conidia are the most common asexual spores, and they occur in these forms:

 arthrospore (ar'-thro-spor) Gr. *arthron,* joint. A rectangular spore formed when a septate hypha fragments at the cross walls.
 chlamydospore (klam-ih'-doh-spor) Gr. *chlamys,* cloak. A spherical conidium formed by the thickening of a hyphal cell. It is released when the surrounding hypha fractures, and it serves as a survival or resting cell.
 blastospore. A spore produced by budding from a parent cell that is a yeast or another conidium; also called a bud.

* *mycelium* (my'-see-lee-yum) pl. mycelia; Gr. *mykes,* root word for fungus.

* *sporangium* (spo-ran'-jee-um) pl. sporangia; Gr. *sporos,* seed, and *angeion,* vessel.
* *conidia* (koh-nid'-ee-uh) sing. conidium; Gr. *konidion,* a particle of dust.

INSIGHT 5.2

Fungi: A Force of Nature

Life in the Fungal Jungle

Fungi are among the most prolific microbes living in soil. So dominant are they that, by one estimate, the average topsoil contains nearly 9 tons of fungal mycelia per acre. In this habitat, they associate with a wide variety of other organisms, from bacteria to animals to plants.

The activities of fungi under natural conditions include nutritional and protective interactions with plants and algae, biological control of soil organisms through competition and antagonism, decomposition of organic matter and recycling of the minerals it contains, and contributions to biofilm networks that are needed to successfully colonize new habitats.

The mold *Verticillium* parasitizing a nematode (coiled into a circle).

Mycelium and spores in the ascomycete *Stachybotrys*. This black mold (400×) is implicated in building contamination that leads to toxic diseases.

Some fungi have developed into normal residents of plants called endophytes. These fungi live inside the tissues of the plant itself and protect it against disease, presumably by secreting chemicals that deter infections by other pathogenic fungi. Understanding the roles of endophytes promises to open up a new field of biological control to inoculate plants with harmless fungi rather than using more toxic fungicides.

Lichens are unusual hybrid organisms that form when a fungus combines with a photosynthetic microbe, either an alga or cyanobacterium. The fungus acquires nutrients from its smaller partner and probably provides a protective haven that favors survival of the lichen body. Lichens occur in most habitats on earth, but they are especially important as early invaders of rock and in soil formation. Fungi also interact with animals in their habitats, most notably with small roundworms, called nematodes, and insects. Several species of molds prey upon or parasitize nematodes. The fungi have inspired a new type of biocontrol agent called a *mycoinsecticide* that is being tested on Colorado potato beetles and other pests.

Is There a Fungus in the House?

The fact that fungi are so widespread also means that they frequently share human living quarters, especially in locations that provide ample moisture and nutrients. Often their presence is harmless and limited to a film of mildew on shower stalls or in other moist environments.

In some cases, depending on the amount of contamination and the type of mold, these indoor fungi can also give rise to various medical problems. Such common air contaminants as *Aspergillus, Cladosporium,* and *Stachybotrys* all have the capacity to give off airborne spores and toxins that, when inhaled, cause a whole spectrum of symptoms sometimes referred to as "sick building syndrome." The usual source of harmful fungi is the presence of chronically water-damaged walls, ceilings, and other building materials that have come to harbor these fungi. People exposed to these houses or buildings report symptoms that range from skin rash, flulike reactions, sore throat, and headaches to fatigue, diarrhea, allergies, and immune suppression.

Reports of sick buildings have been on the rise, affecting thousands of people, and some deaths have been reported in small children. The control of indoor fungi requires correcting the moisture problem, removing the contaminated materials, and decontaminating the living spaces. Mycologists are currently studying the mechanisms of toxic effects with an aim to develop better diagnosis and treatment.

What characteristics of fungi allow them to spread into such a wide variety of habitats? *Answer available at http://www.mhhe.com/talaro8*

phialospore (fy'-ah-lo-spor) Gr. *phialos,* a vessel. A conidium that is budded from the mouth of a vase-shaped spore-bearing cell called a phialide or *sterigma,* leaving a small collar.

microconidium and *macroconidium.* The smaller and larger conidia formed by the same fungus under varying conditions. Microconidia are one-celled, and macroconidia have two or more cells.

porospore. A conidium that grows out through small pores in the spore-bearing cell; some are composed of several cells.

Sexual Spore Formation

Since fungi can propagate successfully by producing millions of asexual spores, it is natural to wonder why they also make sexual spores. The answer lies in important variations that occur when fungi of different genetic makeup combine their genetic material. As in plants and animals, a union of genes from two parents creates offspring with combinations of genes different from that of either parent. The offspring from such a union can have slight variations in form and function that are potentially advantageous to the adaptation and evolution of their species.

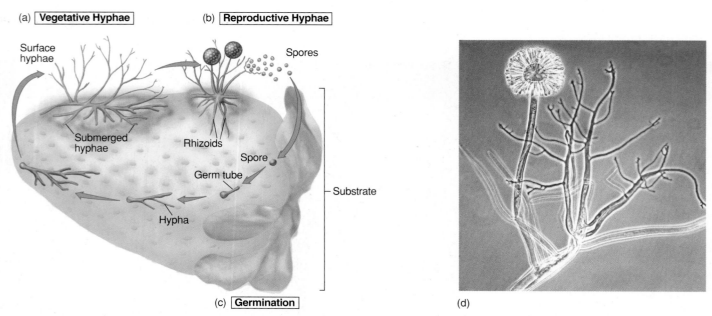

Figure 5.18 **Functional types of hyphae using the mold *Rhizopus* as an example. (a)** Vegetative hyphae are those surface and submerged filaments that digest, absorb, and distribute nutrients from the substrate. This species also has special anchoring structures called rhizoids. **(b)** Later, as the mold matures, it sprouts reproductive hyphae that produce asexual spores. **(c)** During the asexual life cycle, the free mold spores settle on a substrate and send out germ tubes that elongate into hyphae. Through continued growth and branching, an extensive mycelium is produced. So prolific are the fungi that a single colony of mold can easily contain 5,000 spore-bearing structures. If each of these released 2,000 single spores and if every spore were able to germinate, we would soon find ourselves in a sea of mycelia. Most spores do not germinate, but enough are successful to keep the numbers of fungi and their spores very high in most habitats. **(d)** *Syncephalastrum* depicts all major stages in the life cycle of a zygomycota.

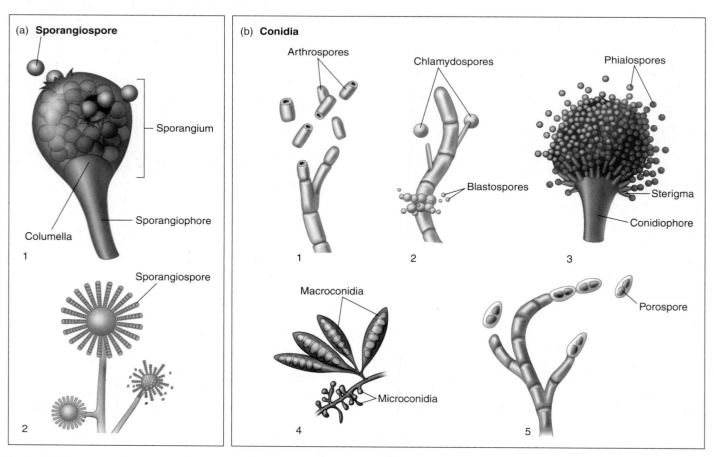

Figure 5.19 **Types of asexual mold spores. (a)** Sporangiospores: (1) *Absidia*, (2) *Syncephalastrum*, **(b)** Conidia: (1) arthrospores (e.g., *Coccidioides*), (2) chlamydospores and blastospores (e.g., *Candida albicans*), (3) phialospores (e.g., *Aspergillus*), (4) macroconidia and microconidia (e.g., *Microsporum*), and (5) porospores (e.g., *Alternaria*).

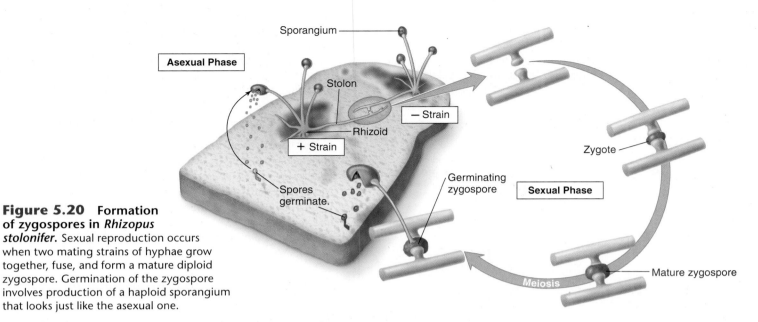

Figure 5.20 **Formation of zygospores in *Rhizopus stolonifer*.** Sexual reproduction occurs when two mating strains of hyphae grow together, fuse, and form a mature diploid zygospore. Germination of the zygospore involves production of a haploid sporangium that looks just like the asexual one.

The majority of fungi produce sexual spores at some point. The nature of this process varies from the simple fusion of fertile hyphae of two different strains to a complex union of differentiated male and female structures and the development of special fruiting structures. We consider the three most common sexual spores: zygospores, ascospores, and basidiospores. These spore types provide an important basis for classifying the major fungal divisions.

Zygospores* are sturdy diploid spores formed when hyphae of two opposite strains (called the plus and minus strains) fuse and create a diploid zygote that swells and becomes covered by strong, spiny walls **(figure 5.20).** When its wall is disrupted and moisture and nutrient conditions are suitable, the zygospore germinates and forms a mycelium that gives rise to a sporangium. Meiosis of diploid cells of the sporangium results in haploid nuclei that develop into sporangiospores. Both the sporangia and the sporangiospores that arise from sexual processes are outwardly identical to the asexual type, but because the spores arose from the union of two separate fungal parents, they are not genetically identical.

In general, haploid spores called **ascospores*** are created inside a special fungal sac, or **ascus** (pl. asci) **(figure 5.21).** Although details can vary among types of fungi, the ascus and ascospores are formed when two different strains or sexes join together to produce offspring. In many species, the male sexual organ fuses with the female sexual organ. The end result is a number of terminal cells called *dikaryons*, each containing a diploid nucleus. Through differentiation, each of these cells enlarges to form an ascus, and its diploid nucleus undergoes meiosis (often followed by mitosis) to form four to eight haploid nuclei that will mature into ascospores. A ripe ascus breaks open and releases the ascospores. Some species form an elaborate fruiting body to hold the asci (inset, figure 5.21).

Basidiospores* are haploid sexual spores formed on the outside of a club-shaped cell called a **basidium (figure 5.22).** In general, spore formation follows the same pattern of two mating types coming together, fusing, and forming terminal cells with diploid nuclei. Each of these cells becomes a basidium, and its nucleus produces, through meiosis, four haploid nuclei. These nuclei are extruded through the top of the basidium, where they develop into basidiospores. Notice the location of the basidia along the gills in mushrooms, which are often dark from the spores they contain. It may be a surprise to discover that the fleshy part of a mushroom is actually a fruiting body designed to protect and help disseminate its sexual spores.

Fungal Classification

It is often difficult for microbiologists to assign logical and useful classification schemes to microorganisms that also reflect their

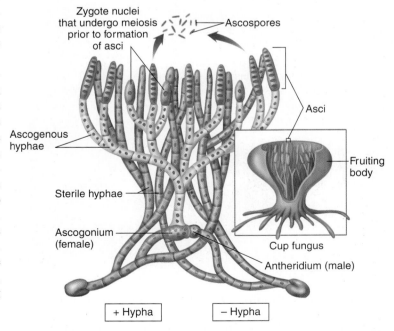

Figure 5.21 **Production of ascospores in a cup fungus.** Inset shows the cup-shaped fruiting body that houses the asci.

* *zygospores* (zy'-goh-sporz) Gr. *zygon*, yoke, to join.
* *ascospores* (as'-koh-sporz) Gr. *ascos*, a sac.
* *basidiospores* (bah-sid'-ee-oh-sporz) Gr. *basidi*, a pedestal.

evolutionary relationships. This difficulty is due to the fact that the organisms do not always perfectly fit the neat categories made for them, and even experts cannot always agree on the nature of the categories. The fungi are no exception, and there are several ways to classify them. For our purposes, we adopt a classification scheme with a medical mycology emphasis, in which the Kingdom Eumycota is subdivided into several phyla based upon the type of sexual reproduction, hyphal structure, and genetic profile. The next section outlines four phyla, including major characteristics and important members.

Phylum I—Zygomycota (also Zygomycetes)

- Sexual spores: zygospores.
- Asexual spores: mostly sporangiospores, some conidia.
- Hyphae are usually nonseptate. If septate, the septa are complete.
- Most species are free-living saprobes; some are animal parasites.
- Can be obnoxious contaminants in the laboratory and on food and vegetables.
- Examples of common molds: *Rhizopus,* a black bread mold; *Mucor; Absidia; Circinella* (**figure 5.23** and see figure 22.29).

Phylum II—Ascomycota (also Ascomycetes)

- Sexual spores: most produce ascospores in asci.
- Asexual spores: many types of conidia, formed at the tips of conidiophores.
- Hyphae with porous septa.
- Examples: This is by far the largest phylum. Most of the species are either molds or yeasts.
 - *Penicillium* is one source of antibiotics (**figure 5.24).**
 - *Saccharomyces* is a yeast used in making bread and beer.
 - Includes many human and plant pathogens, such as *Pneumocystis (carinii) jiroveci,* a pathogen of AIDS patients.
 - *Histoplasma* is the cause of Ohio Valley fever (see figure 22.7).
 - *Microsporum* is one cause of ringworm which is a common name for certain fungal skin infections that often grow in a ringed pattern (see figure 22.18).
 - *Coccidioides immitis* is the cause of Valley fever; (see figure 22.8) *Candida albicans* is the cause of various yeast infections; *Cladosporium* is a common mildew fungus; and *Stachybotrys* is a toxic mold (Insight 5.2).

Phylum III—Basidiomycota (also Basidiomycetes)

- Sexual reproduction by means of basidia and basidiospores.
- Asexual spores: conidia.
- Incompletely septate hyphae.

(a)

(b)

Figure 5.22 Examples of basidiomycota. (a) Formation of basidiospores in a mushroom. **(b)** *Armillaria* (honey) mushrooms sprouting from the base of a tree in Oregon provide only a tiny hint of the vast underground mycellium from which they arose. The massive mycelium covers 2,200 acres and stretches 3.5 miles across.

Figure 5.23 A representative Zygomycota, *Circinella.* Note the sporangia, sporangiospores, and nonseptate hyphae.

(a) (b)

Figure 5.24 Common members of Ascomycota, *Aspergillus* and *Penicillium*. (a) Macroscopic view of colonies. *Aspergillus* displays a granular texture due to large yellow conidial heads. (b) Microscopic view shows the brush arrangement of *Penicillium* phialospores (220×), the asexual phase. *Penicillium* has a velvety texture with typical blue-green color.

- Some plant parasites and one human pathogen.
- Fleshy fruiting bodies are common.
- Examples: mushrooms, puffballs, bracket fungi, and plant pathogens called rusts and smuts.
- The one human pathogen, the yeast *Cryptococcus neoformans,* causes an invasive systemic infection in several organs, including the skin, brain, and lungs (see figure 22.24).

Phylum IV Chytridomycota

Members of this phylum are unusual, primitive fungi commonly called **chytrids.*** Their cellular morphology ranges from single cells to clusters and colonies. They generally do not form hyphae or yeast-type cells. The one feature that most characterizes them is the presence of special flagellated spores, called zoospores and gametes. Most chytrids are saprobic and free-living in soil, water, and decaying matter. A significant number are parasites of plants, animals, and other microbes **(figure 5.25),** but they are not known to cause human disease. They are known to be serious frog pathogens, often responsible for destroying whole populations in some habitats.

Fungi That Produce Only Asexual Spores (Imperfect)

From the beginnings of fungal classification, any fungus that lacked a sexual state was called "imperfect" and was placed in a catchall category, the Fungi Imperfecti. A species would remain classified in that category until its sexual state was described. Gradually, many species were found to make sexual spores, and they were assigned to the taxonomic grouping that best fit those spores. Most of these species have been reassigned to one of the original phyla, particularly to Ascomycota.

* *chytrid* (kit'-rid) Gr. *chytridion,* little pot.

Fungal Identification and Cultivation

Fungi are identified in medical specimens by first being isolated on special types of media and then being observed macroscopically and microscopically. Examples of media for cultivating fungi are cornmeal, blood, and Sabouraud's agar. The latter medium is useful in isolating fungi from mixed samples because of its low pH, which inhibits the growth of bacteria but not of most fungi. Because the fungi are classified into general groups by the presence and type of sexual spores, it would seem logical to identify them in the same way, but sexual spores are rarely if ever demonstrated in the laboratory setting. As a result, the asexual spore-forming structures and spores are usually used to identify organisms to the level of genus and species. Other characteristics that contribute to identification are hyphal type, colony texture and pigmentation, physiological characteristics, and genetic makeup.

Fungi in Medicine, Nature, and Industry

Nearly all fungi are free-living and do not require a host to complete their life cycles. Even among those fungi that are pathogenic, most human infection occurs through accidental contact with an environmental source such as soil, water, or dust. Humans are generally quite resistant to fungal infection, except for two main types of fungal pathogens: the primary pathogens, which can infect even healthy persons, and the opportunistic pathogens, which attack persons who are already weakened in some way.

Mycoses (fungal infections) vary in the way the agent enters the body and the degree of tissue involvement **(table 5.3).** The list of opportunistic fungal pathogens has been increasing in the past few years because of newer medical techniques that keep immunocompromised patients alive. Even so-called harmless species found in the air and dust around us may be able to cause opportunistic infections in patients who already have AIDS, cancer, or diabetes.

Figure 5.25 Parasitic chytrid fungi. The filamentous diatom *Chaetoceros* is attacked by flagellated chytrids that dissolve and destroy their host (400×). Chytrids can infect a wide variety of animal, plant, and microbial cells.

TABLE 5.3 **Major Fungal Infections of Humans**

Degree of Tissue Involvement and Area Affected	Name of Infection	Name of Causative Fungus
Superficial (not deeply invasive)		
Outer epidermis	Tinea versicolor	*Malassezia furfur*
Epidermis, hair, and dermis can be attacked	Dermatophytosis, also called tinea or ringworm of the scalp, body, feet (athlete's foot), toenails	*Microsporum, Trichophyton*, and *Epidermophyton*
Mucous membranes, skin, nail	Candidiasis, or yeast infection	*Candida albicans*
Systemic (deep; organism enters lungs; can invade other organs)		
Lung	Coccidioidomycosis (San Joaquin Valley fever)	*Coccidioides immitis*
	North American blastomycosis (Chicago disease)	*Blastomyces dermatitidis*
	Histoplasmosis (Ohio Valley fever)	*Histoplasma capsulatum*
	Cryptococcosis (torulosis)	*Cryptococcus neoformans*
Lung, skin	Paracoccidioidomycosis (South American blastomycosis)	*Paracoccidioides brasiliensis*

Fungi are involved in other medical conditions besides infections (see Insight 5.2). Fungal cell walls give off chemical substances that can cause allergies. The toxins produced by poisonous mushrooms can induce neurological disturbances and even death. The mold *Aspergillus flavus* synthesizes a potentially lethal poison called aflatoxin,[4] which is the cause of a disease in domestic animals that have eaten grain contaminated with the mold and is also a cause of liver cancer in humans.

Fungi pose an ever-present economic hindrance to the agricultural industry. A number of species are pathogenic to field plants such as corn and grain, and fungi also rot fresh produce during shipping and storage. It has been estimated that as much as 40% of the yearly fruit crop is consumed not by humans but by fungi.

On the beneficial side, however, fungi play an essential role in decomposing organic matter and returning essential minerals to the soil. They form stable associations with plant roots (mycorrhizae) that increase the ability of the roots to absorb water and nutrients. Industry has tapped the biochemical potential of fungi to produce large quantities of antibiotics, alcohol, organic acids, and vitamins. Some fungi are eaten or used to impart flavorings to food. The yeast *Saccharomyces* produces the alcohol in beer and wine and the gas that causes bread to rise. Blue cheese, soy sauce, and cured meats derive their unique flavors from the actions of fungi.

4. From **aspergillus, flavus, toxin.**

Check & Assess Sections 5.4 and 5.5

✔ Eukaryotic cells differ from prokaryotic cells in major structural and physiological characteristics.

✔ Taxonomy of the Domain Eukarya is based on cell structure, body plan, and nutritional type, and results from genetic and ribosomal RNA analysis.

✔ The eukaryotic microorganisms include the Fungi, the Protista (algae and protozoa), and the Helminths (Kingdom Animalia).

✔ The Kingdom Fungi (Eumycota) is composed of nonphotosynthetic heterotrophic organisms with cell walls.

✔ The fungi are either saprobes or parasites and may be unicellular, colonial, or multicellular.

✔ Forms include yeasts (unicellular budding cells) and molds (filamentous cells called hyphae).

✔ Their primary means of reproduction involves asexual and sexual spores.

✔ Classification of the four phyla of fungi is based on the type of reproductive spores, hyphae, and genetic similarities.

✔ Fungi play exceedingly important roles in ecology, plant health, industry, agriculture, and medicine.

17. Differentiate between the yeast and hypha types of fungal cells. Describe the functional types of hyphae. What is a mold?

18. What does it mean if a fungus is dimorphic?

19. How does a fungus feed, and in what habitats would one expect to find fungi?

20. Describe the two main types of asexual fungal spores and how they are formed. What are some types of conidia?

21. Explain the importance of sexual spores in fungal life history.

22. Describe the three main types of sexual spores, and construct a simple diagram to show how each is formed.

23. How are fungi classified? Give an example of a member of each fungus phylum and describe its structure and importance.

24. What is a mycosis? What kind of mycosis is athlete's foot? What kind is coccidioidomycosis?

5.6 Survey of Protists: Algae

Expected Learning Outcomes

27. Discuss the major characteristics of algae, and explain how they are classified.

28. Describe several ways that algae are important microorganisms.

Even though the terms *algae* and *protozoa* do not have taxonomic status, they are still scientifically useful. These are terms, like *protist*, that provide a shorthand label for certain eukaryotes. Microbiologists use such general terms to reference organisms that possess a collection of predictable characteristics. For example, protozoa are considered unicellular eukaryotic protists that lack tissues and share similarities in cell structure, nutrition, life cycles, and

biochemistry. They are all microorganisms, and most of them are motile. Algae are eukaryotic protists, usually unicellular or colonial, that photosynthesize with chlorophyll *a*. They lack vascular systems for transport and have simple reproductive structures.

The Algae: Photosynthetic Protists

The **algae** are a group of photosynthetic organisms most readily recognized by their larger members, such as seaweeds and kelps. In addition to being beautifully colored and diverse in appearance, they vary in length from a few micrometers to 100 meters. Algae occur in unicellular, colonial, and filamentous forms, and the larger forms can possess tissues and simple organs. Examples of algal forms are shown in **figure 5.26** and **table 5.4.** An algal cell exhibits most of the organelles (**figure 5.26a**). The most noticeable of these are the chloroplasts, which contain, in addition to the green pigment chlorophyll, a number of other pigments that create the yellow, red, and brown coloration of some groups.

Algae are widespread inhabitants of fresh and marine waters. They are one of the main components of the large floating community of microscopic organisms called **plankton.** In this capacity, they play an essential role in the aquatic food web and produce most of the earth's oxygen. One of the most prevalent groups on earth are single-celled chrysophyta called *diatoms* (**figure 5.26b**). These beautiful algae have silicate cell walls and golden pigment in their chloroplasts.

Other algal habitats include the surface of soil, rocks, and plants; and several species are even hardy enough to live in hot springs or snowbanks. A desert-adapted green alga blooms during spring rains and survives by forming resistant spores (**figure 5.26c, d**).

Animal tissues would be rather inhospitable to algae, so algae are rarely infectious. One exception, *Prototheca*, is an unusual non-photosynthetic alga associated with skin and subcutaneous infections in humans and animals.

The primary medical threat from algae is due to a type of food poisoning caused by the toxins of marine algae such as dinoflagellates. During particular seasons of the year, the overgrowth of these

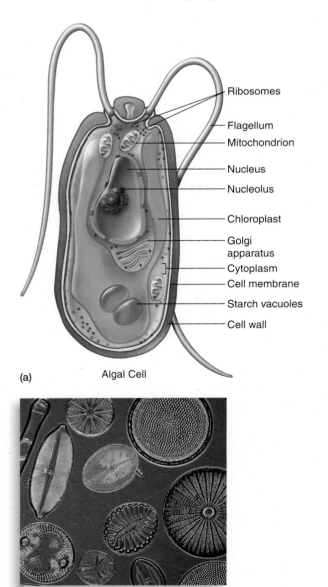

(a) Algal Cell

- Ribosomes
- Flagellum
- Mitochondrion
- Nucleus
- Nucleolus
- Chloroplast
- Golgi apparatus
- Cytoplasm
- Cell membrane
- Starch vacuoles
- Cell wall

(b)

(c)

(d)

Figure 5.26 Representative microscopic algae.
(a) Structure of *Chlamydomonas*, a motile green alga, indicating major organelles. **(b)** A strew of beautiful algae called diatoms shows the intricate and varied structure of their silica cell wall. **(c)** A member of the Chlorophyta, *Oedogonium,* displaying its green filaments and reproductive structures, including eggs (clear spheres) and zygotes (dark spheres) (200×). **(d)** Spring pond with mixed algal bloom. Active photosynthesis and oxygen release are evident in surface bubbles.

TABLE 5.4	Summary of Algal Characteristics			
Group/Common Name	Organization	Cell Wall	Pigmentation	Ecology/Importance
Euglenophyta (euglenids)	Mainly unicellular; motile by flagella	None; pellicle instead	Chlorophyll, carotenoids, xanthophyll	Some are close relatives of Mastigophora.
Pyrrophyta (dinoflagellates)	Unicellular, dual flagella	Cellulose or atypical wall	Chlorophyll, carotenoids	Cause of "red tide"
Chrysophyta (diatoms or golden-brown algae)	Mainly unicellular, some filamentous forms, unusual form of motility	Silicon dioxide	Chlorophyll, fucoxanthin	Diatomaceous earth, major component of plankton
Phaeophyta (brown algae—kelps)	Multicellular, vascular system, holdfasts	Cellulose, alginic acid	Chlorophyll, carotenoids, fucoxanthin	Source of an emulsifier, alginate
Rhodophyta (red seaweeds)	Multicellular	Cellulose	Chlorophyll, carotenoids, xanthophyll, phycobilin	Source of agar and carrageenan, a food additive
Chlorophyta (green algae, grouped with plants)	Varies from unicellular, colonial, filamentous, to multicellular	Cellulose	Chlorophyll, carotenoids, xanthophyll	Precursor of higher plants

motile algae imparts a brilliant red color to the water, which is referred to as a "red tide" (see figure 26.16). When intertidal animals feed, their bodies accumulate toxins given off by the algae that can persist for several months. Paralytic shellfish poisoning is caused by eating exposed clams or other invertebrates. It is marked by severe neurological symptoms and can be fatal. Ciguatera is a serious intoxication caused by algal toxins that have accumulated in fish such as bass and mackerel. Cooking does not destroy the toxin, and there is no antidote.

Several episodes of a severe infection caused by *Pfiesteria piscicida,* a toxic algal form, have been reported over the past several years in the United States. The disease was first reported in fish and was later transmitted to humans. This newly identified species occurs in at least 20 forms, including spores, cysts, and amoebas, that can release potent toxins. Both fish and humans develop neurological symptoms and bloody skin lesions. The cause of the epidemic has been traced to nutrient-rich agricultural runoff water that promoted the sudden "bloom" of *Pfiesteria.* These microbes first attacked and killed millions of fish and later people whose occupations exposed them to fish and contaminated water.

5.7 Survey of Protists: Protozoa

Expected Learning Outcomes

29. Summarize the main characteristics of protozoan form, nutrition, and locomotion.

30. Describe the general life cycle and mode of reproduction in protozoans.

31. Explain how protozoans are identified and classified.

32. Outline a classification scheme for protozoans, and provide examples of important members of each group.

33. Explain some biological properties of parasites, and list some important protozoan pathogens.

If a poll were taken to choose the most engrossing and vivid group of microorganisms, many biologists would choose the protozoa. Although their name comes from the Greek for "first animals," they are far from being simple, primitive organisms. The protozoa constitute a diverse large group (about 65,000 species) of single-celled creatures that display startling properties when it comes to movement, feeding, and behavior. Although most members of this group are harmless, free-living inhabitants of water and soil, a few species are parasites collectively responsible for hundreds of millions of infections of humans each year.

Protozoan Form and Function

Most protozoan cells are single cells containing the major eukaryotic organelles except chloroplasts. Their organelles can be highly specialized for feeding, reproduction, and locomotion. The cytoplasm is usually divided into a clear outer layer called the **ectoplasm** and a granular inner region called the *endoplasm.* Ectoplasm is involved in locomotion, feeding, and protection. Endoplasm houses the nucleus, mitochondria, and food and contractile vacuoles. Some ciliates and flagellates [5] even have organelles that work somewhat like a primitive nervous system to coordinate movement. Because protozoa lack a cell wall, they have a certain amount of flexibility. Their outer boundary is a cell membrane that regulates the movement of food, wastes, and secretions. Cell shape can remain constant (as in most ciliates) or can change constantly (as in amoebas). Certain amoebas (foraminiferans) encase themselves in hard shells made of calcium carbonate. The size of most protozoan cells falls within the range of 3 to 300 μm. Some notable exceptions are giant amoebas and ciliates that are large enough (3–4 mm in length) to be seen swimming in pond water.

5. The terms *ciliate* and *flagellate* are common names of protozoan groups that move by means of cilia and flagella.

Nutritional and Habitat Range

Protozoa are heterotrophic and usually require their food in a complex organic form. Free-living species graze on live cells of bacteria and algae, and even scavenge dead plant or animal debris. Some species have special feeding structures such as oral grooves, which carry food particles into a passageway or gullet that packages the captured food into vacuoles for digestion. A remarkable feeding adaptation can be seen in the ciliate *Coleps* (see figure 5.30*b*), which can easily devour another microbe that is nearly its size. Some protozoa absorb food directly through the cell membrane. Parasitic species live on the fluids of their host, such as plasma and digestive juices, or they can actively feed on tissues.

CONTINUING CASE FILE 5

There is much more to the story of neglected diseases of poverty. Most victims first become infected as children and continue to harbor the parasite for years, even for a lifetime. The chronic nature of the infection adds enormously to the accumulated damage that occurs. Infected people may become involved in an ongoing cycle that continues to produce more parasites. These parasites have adaptations that increase their survival and transmission. Many protozoans produce resistant survival cells called cysts, and worms go through complex reproductive phases with eggs and larvae. Some parasites are spread by direct contact with an infected person, some burrow into the skin, and others are ingested with contaminated soil or water. A significant factor in transmission of several parasites is the involvement of arthropod vectors such as mosquitoes and flies.

For most of these diseases, there is effective drug therapy, but getting the drugs to the poorest of the poor has been difficult. Many of the countries are not only deeply impoverished but are involved in civil wars and other conflicts that create upheaval in an already overburdened system. Controlling these NTDs is going to require a global partnership that brings together resources from many countries.

- What is the definition of a vector, and how are vectors important?
- In what ways are these diseases "neglected"?

For a wrap-up, see the Case File Perspective on page 155.

Although protozoa have adapted to a wide range of habitats, their main limiting factor is the availability of moisture. Their predominant habitats are freshwater and marine water, soil, plants, and animals. Even extremes in temperature and pH are not a barrier to their existence; hardy species are found in hot springs, ice, and habitats with low or high pH.

Styles of Locomotion

Except for one group (the Apicomplexa), protozoa are motile by means of **pseudopods** ("false foot"), **flagella,** or **cilia.** A few species have both pseudopods (also called pseudopodia) and flagella. Some unusual protozoa move by a gliding or twisting movement that does not appear to involve any of these locomotor structures. Pseudopods are blunt, branched, or long and pointed, depending on the particular species. The flowing action of the pseudopods results in amoeboid motion, and pseudopods also serve as feeding structures in many amoebas.

The structure and behavior of flagella and cilia were discussed in section 5.1. Flagella vary in number from one to several, and in certain species, they are attached along the length of the cell by an extension of the cytoplasmic membrane called the *undulating membrane* (see figure 5.28*b*). In most ciliates, the cilia are distributed over the surface of the cell in characteristic patterns. Because of the tremendous variety in ciliary arrangements and functions, ciliates are among the most diverse and awesome cells in the biological world. In certain protozoa, cilia line the oral groove and function in feeding; in others, they fuse together to form stiff props that serve as primitive rows of walking legs.

Life Cycles and Reproduction

Most protozoa are recognized by a motile feeding stage called the **trophozoite*** that requires ample food and moisture to remain active. A large number of species are also capable of entering into a dormant, resting stage called a **cyst** when conditions in the environment become unfavorable for growth and feeding. During *encystment*, the trophozoite cell rounds up into a sphere, and its ectoplasm secretes a tough, thick cuticle around the cell membrane **(figure 5.27).** Because cysts are more resistant than ordinary cells to heat, drying, and chemicals, they can survive adverse periods. They can be dispersed by air currents and may even be an important factor in the spread of diseases such as amebic dysentery. If provided with moisture and nutrients, a cyst breaks open and releases the active trophozoite.

The life cycles of protozoans vary from simple to complex. Several protozoan groups exist only in the trophozoite state. Many alternate between a trophozoite and a cyst stage, depending on the conditions of the habitat. The life cycle of a parasitic protozoan dictates its mode of transmission to other hosts. For example, the flagellate *Trichomonas vaginalis* causes a common sexually transmitted disease. Because it does not form cysts, it is more delicate and must be transmitted by intimate contact between sexual partners. In contrast, intestinal pathogens such as *Entamoeba histolytica* and *Giardia lamblia* form cysts and are readily transmitted in contaminated water and foods.

All protozoa reproduce by relatively simple, asexual methods, usually mitotic cell division. Several parasitic species, including the agents of malaria and toxoplasmosis, reproduce asexually inside a host cell by multiple fission. Sexual reproduction also occurs during the life cycle of most protozoa. Ciliates participate in **conjugation,** a form of genetic exchange in which members of two different mating types fuse temporarily and exchange nuclei. This process of sexual recombination yields new and different genetic combinations that can be advantageous in evolution.

Protozoan Identification and Cultivation

Taxonomists have not escaped problems classifying protozoa. They, too, are very diverse and frequently frustrate attempts to

* *trophozoite* (trof´-oh-zoh´-yte) Gr. *trophonikos*, to nourish, and *zoon*, animal.

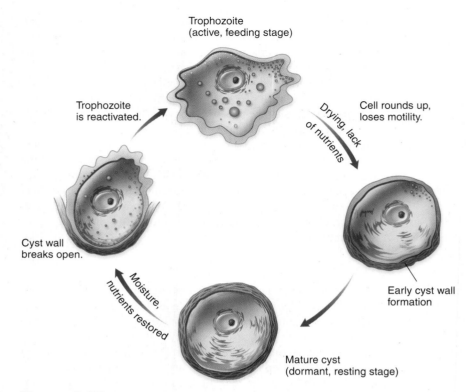

Trophozoite
(active, feeding stage)

Trophozoite
is reactivated.

Drying, lack
of nutrients

Cell rounds up,
loses motility.

Cyst wall
breaks open.

Moisture,
nutrients restored

Early cyst wall
formation

Mature cyst
(dormant, resting stage)

Figure 5.27 The general life cycle exhibited by many protozoa. All protozoa have a trophozoite form, but not all produce cysts.

generalize or place them in neat groupings. We will use a simple system of four groups, based on method of motility, mode of reproduction, and stages in the life cycle.

The unique appearance of most protozoa makes it possible for a knowledgeable person to identify them to the level of genus and often species by microscopic morphology alone. Characteristics to consider in identification include the shape and size of the cell; the type, number, and distribution of locomotor structures; the presence of special organelles or cysts; and the number of nuclei. Medical specimens taken from blood, sputum, cerebrospinal fluid, feces, or the vagina are smeared directly onto a slide and observed with or without special stains. Occasionally, protozoa are cultivated on artificial media or in laboratory animals for further identification or study.

Classification of Selected Medically Important Protozoa

The most recent system for classifying protozoa places them into 13 separate categories on the Eukarya tree (see figure 5.14a), but this method may be more complex than is necessary for our survey. We will simplify this system by presenting four groups, summarized as follows:

The Mastigophora (Also Called Zoomastigophora)

- Motility is primarily by flagella alone or by both flagellar and amoeboid motion.
- Single nucleus.

- Sexual reproduction, when present, by syngamy; division by longitudinal fission.
- Several parasitic forms lack mitochondria and Golgi apparatus.
- Most species form cysts and are free-living; the group also includes several parasites.
- Some species are found in loose aggregates or colonies, but most are solitary.
- Members include: *Trypanosoma* and *Leishmania,* important blood pathogens spread by insect vectors (see figures 23.8 and 23.9). *Giardia,* an intestinal parasite spread in water contaminated with feces; *Trichomonas,* a parasite of the reproductive tract of humans spread by sexual contact (see figure 23.6).
- Other members can be viewed in **figure 5.28.**

The Sarcodina[6] (Amoebas)

- Cell form is primarily an amoeba **(figure 5.29).**
- Major locomotor organelles are pseudopods, although some species have flagellated reproductive states.
- Asexual reproduction by fission.
- Mostly uninucleate; usually encyst.
- Most amoebas are free-living and not infectious.
- *Entamoeba* is a parasite of humans (see figure 23.1).
- Shelled amoebas called foraminifera and radiolarians are responsible for chalk deposits in the ocean.

The Ciliophora (Ciliates) (See Figure 5.30a)

- Trophozoites are motile by cilia.
- Some have cilia in tufts for feeding and attachment; most develop cysts.
- Have both macronuclei and micronuclei; division by transverse fission.
- Most have a definite mouth and feeding organelle.
- Show relatively advanced behavior **(figure 5.30).**
- The majority of ciliates are free-living and harmless.
- One pathogen is *Balantidium* (see figure 23.5).

The Apicomplexa (Sporozoa)

- Motility is absent in most cells except male gametes.
- Life cycles are complex, with well-developed asexual and sexual stages.
- Sporozoa produce special sporelike cells called **sporozoites*** **(figure 5.31)** following sexual reproduction, which are important in transmission of infections.

6. Some biologists prefer to combine Mastigophora and Sarcodina into the phylum Sarcomastigophora. Because the algal group Euglenophyta has flagella, it may also be included in the Mastigophora.

* *sporozoite* (spor″-oh-zoh′-yte) Gr. *sporos,* seed, and *zoon,* animal.

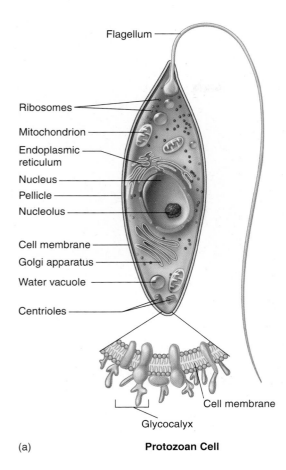

Flagellum

Ribosomes

Mitochondrion

Endoplasmic reticulum

Nucleus

Pellicle

Nucleolus

Cell membrane

Golgi apparatus

Water vacuole

Centrioles

Cell membrane

Glycocalyx

(a)

Protozoan Cell

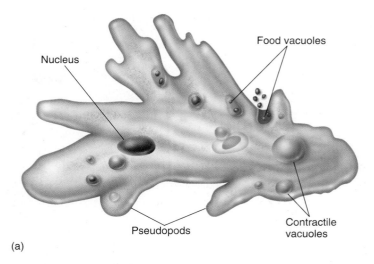

Nucleus

Food vacuoles

Pseudopods

Contractile vacuoles

(a)

(b)

Figure 5.29 **Examples of sarcodinians. (a)** The structure of an amoeba. **(b)** A freshwater amoeba has captured a diatom with its pseudopods.

- Most form thick-walled zygotes called oocysts; entire group is parasitic.
- *Plasmodium*, the most prevalent protozoan parasite, causes 100 million to 300 million cases of malaria each year worldwide. It is an intracellular parasite with a complex cycle alternating between humans and mosquitoes (see figure 23.13).
- *Toxoplasma gondii* causes an acute infection (toxoplasmosis) in humans, which is acquired from cats and other animals (see figure 23.16).
- *Cryptosporidium* is an emerging intestinal pathogen transmitted by contaminated water.

Important Protozoan Pathogens

Although protozoan infections are very common, they are actually caused by only a small number of species often restricted geographically to the tropics and subtropics (**table 5.5**). In this survey, we look at examples from two protozoan groups that illustrate some of the main features of protozoan diseases.

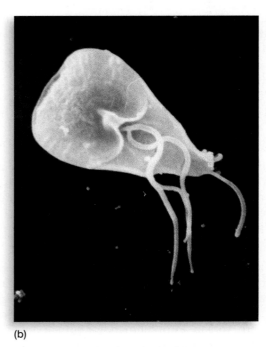

(b)

Figure 5.28 **Examples of mastigophorans. (a)** General structure of a mastigophoran such as *Peranema*. The outer pellicle is flexible and allows some shape changes in this group. **(b)** Scanning electron micrograph of *Giardia lamblia,* a cause of an intestinal infection (giardiasis). This trophozoite stage has 3 pairs of flagella.

Figure 5.30 Selected ciliate representatives. (a) Interference contrast image of the ciliate *Blepharisma,* a common inhabitant of pond water. Note the pattern of ciliary movement evident on the left side of the cell and the details of its internal structure. **(b)** Stages in the process of *Coleps* feeding on an alga (round cell). The predaceous ciliate sucks its prey into a large oral groove. See figure 5.4 for an additional view of a ciliate.

Figure 5.31 Apicomplexa protozoan. (a) General cell structure. Note the lack of specialized locomotor organelles. **(b)** Scanning electron micrograph of the sporozoite of *Cryptosporidium,* an intestinal parasite of humans and other mammals, often acquired through fecally contaminated water.

TAKE NOTE: THE PARASITIC LIFE

Protozoa are traditionally studied along with the helminths in the science of parasitology. Although a parasite is usually defined as an organism that obtains food and other requirements at the expense of a host, the term *parasite* is also used to denote protozoan and helminth pathogens. The range of host-parasite relationships can be very broad. At one extreme are the so-called "good" parasites, which occupy their host with little harm. An example is certain amebas that live in the human intestine and feed off organic matter there. At the other extreme are parasites (*Plasmodium, Trypanosoma*) that multiply in host tissues such as the blood or brain, causing severe damage and disease. Between these two extremes are parasites of varying pathogenicity, depending on their particular adaptations.

Most human parasites go through three general stages:

- The microbe is transmitted to the human host from a source such as soil, water, food, other humans, or animals.
- The microbe invades and multiplies in the host, producing more parasites that can infect other suitable hosts.
- The microbe leaves the host in large numbers by a specific means and must enter a new host to survive.

There are numerous variations on this theme. For instance, the microbe can invade more than one host species (alternate hosts) and undergo several changes as it cycles through these hosts, such as sexual reproduction or encystment. Some microbes are spread from human to human by means of **vectors,*** defined as animals such as insects that carry diseases. Others can be spread through bodily fluids and feces.

* *vector* (vek'-tur) L. *vectur,* one who carries.

| TABLE 5.5 | Major Pathogenic Protozoa, Infections, and Primary Sources | |
|---|---|
| **Protozoan/Disease** | **Reservoir/Source** |
| **Amoeboid Protozoa** | |
| Amoebiasis: *Entamoeba histolytica* | Human/water and food |
| Brain infection: *Naegleria, Acanthamoeba* | Free-living in water |
| **Ciliated Protozoa** | |
| Balantidiosis: *Balantidium coli* | Zoonotic in pigs |
| **Flagellated Protozoa** | |
| Giardiasis: *Giardia lamblia* | Zoonotic/water and food |
| Trichomoniasis: *T. hominis, T. vaginalis* | Human |
| Hemoflagellates | |
| Trypanosomiasis: *Trypanosoma brucei, T. cruzi* | Zoonotic/vector-borne |
| Leishmaniasis: *Leishmania donovani, L. tropica, L. brasiliensis* | Zoonotic/vector-borne |
| **Apicomplexan Protozoa** | |
| Malaria: *Plasmodium vivax, P. falciparum, P. malariae* | Human/vector-borne |
| Toxoplasmosis: *Toxoplasma gondii* | Zoonotic/vector-borne |
| Cryptosporidiosis: *Cryptosporidium* | Free-living/water and food |
| Cyclosporiasis: *Cyclospora cayetanensis* | Water/fresh produce |

Pathogenic Flagellates: Trypanosomes

Trypanosomes are protozoa belonging to the genus *Trypanosoma**. The two most important representatives are *T. brucei* and *T. cruzi,* species that are closely related but geographically restricted. *Trypanosoma brucei* occurs in Africa, where it causes approximately 35,000 new cases of sleeping sickness each year. *Trypanosoma cruzi,* the cause of Chagas disease,[7] is endemic to South and Central America, where it infects several million people a year. Both species have long, crescent-shaped cells with a single flagellum that is sometimes attached to the cell body by an undulating membrane. Both occur in the blood during infection and are transmitted by blood-sucking vectors. We use *T. cruzi* to illustrate the phases of a trypanosomal life cycle and to demonstrate the complexity of parasitic relationships.

The trypanosome of Chagas disease relies on the close relationship of a warm-blooded mammal and an insect that feeds on mammalian blood. The mammalian hosts are numerous, including dogs, cats, opossums, and armadillos. The vector is the *reduviid** bug, an insect that is sometimes called the "kissing bug" because of its habit of biting its host at the corner of the mouth. Transmission occurs from bug to mammal and from mammal to bug but usually not from mammal to mammal, except across the placenta

* *Trypanosoma* (try″-pan-oh-soh′-mah) Gr. *Trypanon,* borer, and *soma,* body.

7. Named for Carlos Chagas, the discoverer of *T. cruzi.*

* *reduviid* (ree-doo′-vee-id) A member of a large family of flying insects with sucking, beaklike mouths.

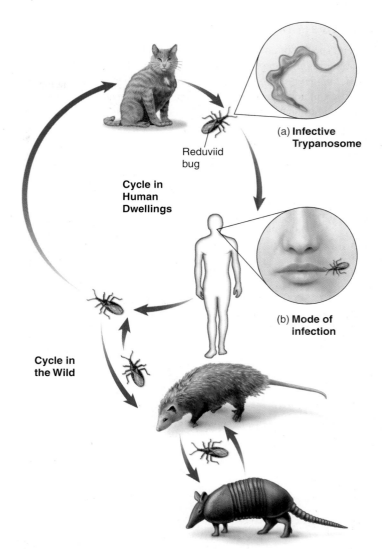

Figure 5.32 **Cycle of transmission in Chagas disease.** Trypanosomes (**inset a**) are transmitted among domestic and wild mammalian hosts and human hosts by means of a bite from the kissing bug (**inset b**).

during pregnancy. The general phases of this cycle are presented in **figure 5.32.**

The trypanosome trophozoite multiplies in the intestinal tract of the reduviid bug and is harbored in the feces. The bug seeks a host and bites the mucous membranes, usually of the eye, nose, or lips, releasing the trypanosome in feces near the bite. Ironically, the victims themselves inadvertently contribute to the entry of the microbe by scratching the bite wound. Once in the body, the trypanosomes become established and multiply in muscle and white blood cells. Periodically, these parasitized cells rupture, releasing large numbers of new trophozoites into the blood. Eventually, the trypanosomes can spread to many systems, including the lymphoid organs, heart, liver, and brain. Manifestations of the resultant disease range from mild to very severe and include fever, inflammation, and heart and brain damage. In many cases, the disease has an extended course and can cause death.

Infective Amoebas: *Entamoeba*

Several species of amoebas cause disease in humans, but probably the most common disease is amoebiasis, or amebic **dysentery,*** caused by *Entamoeba histolytica*. This microbe is widely distributed in the world, from northern zones to the tropics, and is nearly always associated with humans. Amebic dysentery is the fourth most common protozoan infection in the world. This microbe has a life cycle quite different from the trypanosomes in that it does not involve multiple hosts and a blood-sucking vector. It lives part of its cycle as a trophozoite and part as a cyst. Because the cyst is the more resistant form and can survive in water and soil for several weeks, it is the more important stage for transmission. The primary way that people become infected is by ingesting food or water contaminated with human feces.

* *dysentery* (dis′-en-ter″-ee) Any inflammation of the intestine accompanied by bloody stools. It can be caused by a number of factors, both microbial and nonmicrobial.

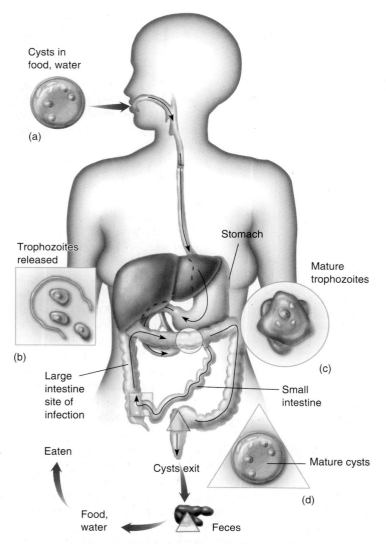

Cysts in food, water

(a)

Trophozoites released

(b)

Stomach

Mature trophozoites

Large intestine site of infection

Small intestine

Eaten

Cysts exit

Mature cysts

(c)

(d)

Food, water

Feces

Figure 5.33 Stages in the infection and transmission of amebic dysentery. Arrows show the route of infection; insets show the appearance of *Entamoeba histolytica*. **(a)** Cysts are eaten. **(b)** Trophozoites (amoebas) emerge from cysts. **(c)** Trophozoites invade the large intestinal wall. **(d)** Mature cysts are released in the feces and may be spread through contaminated food and water.

Figure 5.33 shows the major features of the amebic dysentery cycle, starting with the ingestion of cysts. The viable, heavy-walled cyst passes through the stomach unharmed. Once inside the small intestine, the cyst germinates into a large multinucleate amoeba that subsequently divides to form small amoebas (the trophozoite stage). These trophozoites migrate to the large intestine and begin to feed and grow. From this site, they can penetrate the lining of the intestine and invade the liver, lungs, and skin. Common symptoms include gastrointestinal disturbances such as nausea, vomiting, and diarrhea, leading to weight loss and dehydration. The cycle is completed in the infected human when certain trophozoites in the feces begin to form cysts, which then pass out of the body with fecal matter. Knowledge of the amoebic cycle and role of cysts has been helpful in controlling the disease. Important preventive measures include sewage treatment, curtailing the use of human feces as fertilizer, and adequate sanitation of food and water.

For additional information on protozoan parasites, see chapter 23.

5.8 Parasitic Helminths

Expected Learning Outcomes

34. Describe the major groups of helminths and their basic morphology and classification.

35. Explain the elements of helminth biology, life cycles, and reproduction.

36. Discuss the importance of the helminth parasites.

Tapeworms, flukes, and roundworms are collectively called *helminths,* from the Greek word meaning "worm." Adult animals are usually large enough to be seen with the naked eye, and they range from the longest tapeworms, measuring up to about 25 m in length, to roundworms less than 1 mm in length. Traditionally, they have been included among microorganisms because of their infective abilities and because the microscope is necessary to identify their eggs and larvae.

On the basis of morphological form, the two major groups of parasitic helminths are the **flatworms** (Phylum Platyhelminthes), with a very thin, often segmented body plan **(figure 5.34),** and the **roundworms** (Phylum Aschelminthes, also called **nematodes),*** with an elongate, cylindrical, unsegmented body **(figure 5.35).** The flatworm group is subdivided into the **cestodes,*** or tapeworms, named for their long, ribbonlike arrangement, and the **trematodes,*** or flukes, characterized by flat, ovoid bodies. Not all flatworms and roundworms are parasites by nature; many live free in soil and water.

General Worm Morphology

All helminths are multicellular animals equipped to some degree with organs and organ systems. In parasitic helminths, the most

* *nematode* (neem′-ah-tohd) Gr. *nemato,* thread, and *eidos,* form.
* *cestode* (sess′-tohd) L. *cestus,* a belt, and *ode,* like.
* *trematode* (treem′-a-tohd) Gr. *trema,* hole. Named for the appearance of having tiny holes.

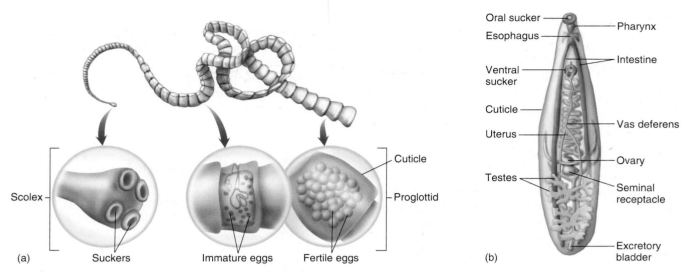

Figure 5.34 Parasitic flatworms. (a) A cestode (beef tapeworm), showing the scolex; long, tapelike body; and magnified views of immature and mature proglottids (body segments). **(b)** The structure of a trematode (liver fluke). Note the suckers that attach to host tissue and the dominance of reproductive and digestive organs.

developed organs are those of the reproductive tract, with some degree of reduction in the digestive, excretory, nervous, and muscular systems. In particular groups, such as the cestodes, reproduction is so dominant that the worms are reduced to little more than a series of flattened sacs filled with ovaries, testes, and eggs **(figure 5.34a).** Not all worms have such extreme adaptations as cestodes, but most have a highly developed reproductive potential, thick cuticles for protection, and mouth glands for breaking down the host's tissue.

Life Cycles and Reproduction

The complete life cycle of helminths includes the fertilized egg (embryo), larval, and adult stages. In the majority of helminths, adults derive nutrients and reproduce sexually in a host's body. In nematodes, the sexes are separate and usually different in appearance; in trematodes, the sexes can be either separate or **hermaphroditic,** meaning that male and female sex organs are in the same worm; cestodes are generally hermaphroditic. For a parasite's continued survival as a species, it must complete the life cycle by transmitting an infective form, usually an egg or larva, to the body of another host, either of the same or a different species. By convention, the host in which larval development occurs is the intermediate (secondary) host, and adulthood and mating occur in the **definitive (final) host.** A transport host is an intermediate host that experiences no parasitic development but is an essential link in the completion of the cycle.

In general, sources for human infection are contaminated food, soil, and water or infected animals, and routes of infection are by oral intake or penetration of unbroken skin. Humans are the definitive hosts for many of the parasites, and in about half the diseases, they are also the sole biological reservoir. In other cases, animals or insect vectors serve as reservoirs or are required to complete worm development. In the majority of helminth infections, the worms must leave their host to complete the entire life cycle.

Fertilized eggs are usually released to the environment and are provided with a protective shell and extra food to aid their development into larvae. Even so, most eggs and larvae are vulnerable to heat, cold, drying, and predators, and are destroyed or unable to

Figure 5.35 The life cycle of the pinworm, a roundworm. Eggs are the infective stage and are transmitted by unclean hands. Children frequently reinfect themselves and also pass the parasite on to others.

reach a new host. To counteract this formidable mortality rate, certain worms have adapted a reproductive capacity that borders on the incredible: A single female *Ascaris*[8] can lay 200,000 eggs a day, and a large female can contain over 25 million eggs at varying stages of development! If only a tiny number of these eggs makes it to another host, the parasite will have been successful in completing its life cycle.

A Helminth Cycle: The Pinworm

To illustrate a helminth cycle in humans, we use the example of a roundworm, *Enterobius vermicularis,* the pinworm or seatworm. This worm causes a very common infestation of the large intestine (**figure 5.35**). Worms range from 2 to 12 mm long and have a tapered, curved cylinder shape. The condition they cause, enterobiasis, is usually a simple, uncomplicated infection that does not spread beyond the intestine.

A cycle starts when a person swallows microscopic eggs picked up from another infected person by direct contact or by touching contaminated surfaces. The eggs hatch in the intestine and then release larvae that mature into adult worms within about one month. Male and female worms mate, and the female migrates out to the anus to deposit eggs, which cause intense itchiness that is relieved by scratching. Herein lies a significant means of dispersal: Scratching contaminates the fingers, which, in turn, transfers eggs to bedclothes and other inanimate objects. This person becomes a host and a source of eggs, and can spread them to others in addition to reinfesting himself. Enterobiasis occurs most often among families and in other close living situations. Its distribution is worldwide among all socioeconomic groups, but it seems to attack younger people more frequently than older ones.

Helminth Classification and Identification

The helminths are classified according to their shape; their size; the degree of development of various organs; the presence of hooks, suckers, or other special structures; the mode of reproduction; the kinds of hosts; and the appearance of eggs and larvae. They are identified in the laboratory by microscopic detection of the adult worm or its larvae and eggs, which often have distinctive shapes or external and internal structures. Occasionally, they are cultured in order to verify all of the life stages.

Distribution and Importance of Parasitic Worms

About 50 species of helminths parasitize humans (see table 23.4). They are distributed in all areas of the world that support human life. Some worms are restricted to a given geographic region, and many have a higher incidence in tropical areas. This knowledge must be tempered with the realization that jet-age travel, along with

human migration, is gradually changing the patterns of worm infections, especially of those species that do not require alternate hosts or special climatic conditions for development. The yearly estimate of worldwide cases numbers in the billions, and these are not confined to developing countries. A conservative estimate places 50 million helminth infections in North America alone. The primary targets are malnourished children.

Check & Assess Sections 5.6–5.8

✔ The protists are mostly unicellular or colonial eukaryotes that lack specialized tissues.

✔ There are two major organism types: the algae and the protozoa.

✔ Algae are photosynthetic organisms that contain chloroplasts with chlorophyll and other pigments and have cell walls.

✔ Algae are planktonic organisms that form the basis for aquatic food chains and produce most of the earth's oxygen through photosynthesis.

✔ Protozoa are heterotrophs that often display some form of locomotion and lack cell walls.

✔ Protozoa have single-celled trophozoites, and many produce a resistant stage, or cyst.

✔ Classification of protozoans uses type of locomotion and life cycle for placement into one of the general categories: Sarcodina, Mastigophora, Ciliophora, or Apicomplexa.

✔ Most protozoa are harmless and free-living, but some are medically important parasites.

✔ The Helminth worms belong to an animal group that is included in studies by microbiologists. Parasitic members include flatworms and roundworms that can invade and reproduce in human tissues and cause various diseases.

25. What is a working definition of a "protist"?

26. Describe the principal characteristics of algae that separate them from protozoa.

27. What causes the many colors in the algae?

28. How are algae important? Are there any algae of medical importance?

29. Explain the general characteristics of the protozoan life cycle.

30. Describe the protozoan adaptations for feeding.

31. Describe protozoan reproductive processes.

32. Briefly outline the characteristics of the four protozoan groups. What is an important pathogen in each group?

33. Which protozoan group is the most complex in structure, behavior, and life cycle?

34. What characteristics set the apicomplexa apart from the other protozoan groups?

35. Discuss the adaptations of parasitic worms to their lifestyles, and explain why these adaptations are necessary or advantageous to the worms' survival.

36. How are helminths similar to and different from microscopic eukaryotes?

8. *Ascaris* is a genus of parasitic intestinal roundworms.

CASE FILE **5** PERSPECTIVE

Parasitology is the study of eukaryotic organisms called parasites that invade the body and cause disease. It includes two major groups: the protozoans—primarily single-celled microorganisms, and the helminth worms—multicellular invertebrate animals that may or may not be microscopic. Parasitologists study the morphology, life cycles, epidemiology, and pathology of these parasites. A list of the most common parasitic diseases in humans would total around 35, distributed all around the globe.

Dr. Hotez's quote about the estimated worldwide case rate is meant to emphasize the enormity of the situation. In those regions where neglected tropical parasites prevail, every person is exposed to several different NTDs simultaneously, so the disease burden could easily top 3 billion or 4 billion cases overall.

A vector is any animal that passes an infectious microbe to humans. Vectors may be invertebrates or vertebrates. The most common vectors are arthropods such as insects (mosquitoes, fleas, bugs, biting flies) and arachnids (ticks and mites).

Vectors of this type usually have a unique relationship with their parasite, serving as an alternate host where it undergoes a phase of development. Many vectors infect humans by taking a blood meal, thereby providing a route of entry for the parasite.

These diseases are neglected out of economic necessity. One billion of the world's poorest live on less than $1 per day, and 2.7 billion live on less than $2 per day. In most cases, NTDs are "out of sight and out of mind"—often occurring in remote rural areas far removed from medical facilities. They are not given high priority for funding due to the fact that they are not rapidly fatal like tuberculosis or malaria.

Dr. Hotez has been looking for ways to educate people and find solutions for the terrible legacy of these diseases of poverty. Recently, he helped found a March of Dimes-like drive called "Just 50 Cents," which proclaims that a donation of only 50 cents will provide the necessary medication to save a child's life from one of these NTDs.

More information on this program can be found at http://www.globalnetwork.org/just50cents

Chapter Summary with Key Terms

5.1 The History of Eukaryotes

5.2 and 5.3 Form and Function of the Eukaryotic Cell: External and Internal Structures

A. Eukaryotic cells are complex and compartmentalized into individual organelles.

B. Major organelles and other structural features include: appendages (cilia, flagella), **glycocalyx,** cell wall, cytoplasmic (or cell) membrane, **organelles** (nucleus, **nucleolus, endoplasmic reticulum, Golgi complex, mitochondria, chloroplasts),** ribosomes, **cytoskeleton (microfilaments, microtubules).**

5.4 Eukaryotic-Prokaryotic Comparisons and Taxonomy of Eukaryotes

A. The eukaryotic cell can be compared with the prokaryotic cell in structure, size, metabolism, motility, and shape.

B. Taxonomic groups of the Domain Eukarya are based on level of organization, body plan, cell structure, nutrition, metabolism, and certain genetic characteristics.

5.5 The Kingdom of the Fungi

Common names of the macroscopic fungi are mushrooms, bracket fungi, and puffballs. Microscopic fungi are known as yeasts and molds.

A. *Overall Morphology:* At the cellular (microscopic) level, fungi are typical eukaryotic cells, with thick cell walls. Yeasts are single cells that form buds and **pseudohyphae. Hyphae** are long, tubular filaments that can be septate or nonseptate and grow in a network called a **mycelium;** hyphae are characteristic of the filamentous fungi called molds.

B. *Nutritional Mode/Distribution:* All are **heterotrophic.** The majority are harmless **saprobes** living off organic **substrates** such as dead animal and plant tissues. A few are **parasites,** living on the tissues of other organisms, but none is obligate. Distribution is extremely widespread in many habitats.

C. *Reproduction:* Primarily through **spores** formed on special reproductive hyphae. In asexual reproduction, spores are formed through budding, partitioning of a hypha, or in special sporogenous structures; examples are **conidia** and **sporangiospores.** In sexual reproduction, spores are formed following fusion of male and female strains and the formation of a sexual structure; sexual spores are one basis for classification.

D. *Major Groups:* The four main phyla among the terrestrial fungi, given with sexual spore type, are Zygomycota **(zygospores),** Ascomycota **(ascospores),** Basidiomycota **(basidiospores),** and Chytridiomycota (motile zoospores).

E. *Importance:* Fungi are essential decomposers of plant and animal detritis in the environment. Economically beneficial as sources of antibiotics; used in making foods and in genetic studies. Adverse impacts include: decomposition of fruits and vegetables, and human infections, or **mycoses;** some produce substances that are toxic if eaten.

5.6 Survey of Protists: Algae

General group that traditionally includes single-celled and colonial eukaryotic microbes that lack organization into tissues.

A. *Overall Morphology:* Are unicellular, colonial, filamentous or larger forms such as seaweeds.

B. *Nutritional Mode/Distribution:* Photosynthetic; freshwater and marine water habitats; main component of **plankton.**

C. *Importance:* Provide the basis of the food web in most aquatic habitats. Certain algae produce neurotoxins that are harmful to humans and animals.

5.7 Survey of Protists: Protozoa

Include large single-celled organisms; a few are pathogens.

A. *Overall Morphology:* Most are unicellular; lack a cell wall. The cytoplasm is divided into ectoplasm and endoplasm. Many convert to a resistant, dormant stage called a **cyst.**

B. *Nutritional Mode/Distribution:* All are heterotrophic. Most are free-living in a moist habitat (water, soil); feed by engulfing other microorganisms and organic matter.

C. *Reproduction:* Asexual by binary fission and **mitosis,** budding; sexual by fusion of free-swimming gametes, conjugation.

D. *Major Groups:* Protozoa are subdivided into four groups based upon mode of locomotion and type of reproduction: **Mastigophora,** the flagellates, motile by flagella; **Sarcodina,** the amoebas, motile by pseudopods; **Ciliophora,** the ciliates, motile by cilia; **Apicomplexa,** motility not well developed; produce unique reproductive structures.

E. *Importance:* Ecologically important in food webs and decomposing organic matter. Medical significance: hundreds of millions of people are afflicted with one of the many protozoan infections (malaria, trypanosomiasis, amoebiasis). Can be spread from host to host by insect **vectors.**

5.8 The Parasitic Helminths

Includes three categories: roundworms, tapeworms, and flukes.

A. *Overall Morphology:* Animal cells; multicellular; individual organs specialized for reproduction, digestion, movement, protection, though some of these are reduced.

B. *Reproductive Mode:* Includes embryo, larval, and adult stages. Majority reproduce sexually. Sexes may be hermaphroditic.

C. *Epidemiology:* Developing countries in the tropics are hardest hit by helminth infections; transmitted via ingestion, vectors, and direct contact with infectious stages. They afflict billions of humans.

Multiple-Choice Questions

Select the correct answer from the answers provided. For questions with blanks, choose the combination of answers that most accurately completes the statement.

1. Both flagella and cilia are found primarily in
 a. algae c. fungi
 b. protozoa d. both b and c

2. Features of the nuclear envelope include
 a. ribosomes
 b. a double membrane structure
 c. pores that allow communication with the cytoplasm
 d. b and c
 e. all of these

3. The cell wall is usually found in which eukaryotes?
 a. fungi c. protozoa
 b. algae d. a and b

4. What is embedded in rough endoplasmic reticulum?
 a. ribosomes c. chromatin
 b. Golgi apparatus d. vesicles

5. Yeasts are _____ fungi, and molds are _____ fungi.
 a. macroscopic, microscopic c. motile, nonmotile
 b. unicellular, filamentous d. water, terrestrial

6. In general, fungi derive nutrients through
 a. photosynthesis c. digesting organic substrates
 b. engulfing bacteria d. parasitism

7. A hypha divided into compartments by cross walls is called
 a. nonseptate c. septate
 b. imperfect d. perfect

8. Algae generally contain some type of
 a. spore c. locomotor organelle
 b. chlorophyll d. toxin

9. Which characteristic(s) is/are not typical of protozoan cells?
 a. locomotor organelle c. spore
 b. cyst d. trophozoite

10. The protozoan trophozoite is the
 a. active feeding stage c. infective stage
 b. inactive dormant stage d. spore-forming stage

11. All mature sporozoa are
 a. parasitic c. carried by an arthropod vector
 b. nonmotile d. both a and b

12. Parasitic helminths reproduce with
 a. spores d. cysts
 b. eggs and sperm e. all of these
 c. mitosis

13. Mitochondria likely originated from
 a. archaea c. purple bacteria
 b. invaginations of the d. cyanobacteria
 cell membrane

14. Human fungal infections involve and affect what areas of the human body?
 a. skin c. lungs
 b. mucous membranes d. all of these

15. Most helminth infections
 a. are localized to one site in the body
 b. spread through major systems of the body
 c. develop within the spleen
 d. develop within the liver

Multiple Matching. Select the description that best fits the word in the left column.

16. ___b___ diatom
17. ___e___ *Rhizopus*
18. ___c___ *Histoplasma*
19. ___h___ *Cryptococcus*
20. ___g___ euglenid
21. ___j___ dinoflagellate
22. ___i___ *Trichomonas*
23. ___d___ *Entamoeba*
24. ___a___ *Plasmodium*
25. ___f___ *Enterobius*

a. the cause of malaria
b. single-celled alga with silica in its cell wall
c. fungal cause of Ohio Valley fever
d. the cause of amebic dysentery
e. genus of black bread mold
f. helminth worm involved in pinworm infection
g. motile flagellated alga with eyespots
h. a yeast that infects the lungs
i. flagellated protozoan genus that causes an STD
j. alga that causes red tides

Case File Questions

1. Which of these is/are an example(s) of neglected tropical protozoan diseases?
 a. hookworm d. a and b
 b. Chagas disease e. b and c
 c. leishmaniasis f. all of these

2. Which is a possible vector of a tropical eukaryotic parasite?
 a. contaminated drinking water c. biting fly
 b. contaminated food d. dog tick

Writing to Learn

These questions are suggested as a *writing-to-learn* experience. For each question, compose a one- or two-paragraph answer that includes the factual information needed to completely address the question. Any question listed in a section's Check and Assess may be considered as a writing-to-learn exercise.

1. Describe the anatomy and functions of each of the major eukaryotic organelles.

2. Review the major similarities and differences between prokaryotic and eukaryotic cells.

3. Trace the synthesis of cell products, their processing, and their packaging through the organelle network.

4. a. What is the reproductive potential of molds in terms of spore production?
 b. How do mold spores differ from prokaryotic spores?

5. Provide some explanations for why the eukaryotic parasites are so widespread and successful.

6. a. Fill in the following summary table for defining, comparing, and contrasting eukaryotic cells.
 b. Briefly describe the manner of nutrition and body plan (unicellular, colonial, filamentous, or multicellular) for each group.

Organelle/ Structure	Briefly Describe Functions in Cell	Commonly Present In (Check)		
		Fungi	Algae	Protozoa
Flagella				
Cilia				
Glycocalyx				
Cell wall				
Cell membrane				
Nucleus				
Mitochondria				
Chloroplasts				
Endoplasmic reticulum				
Ribosomes				
Cytoskeleton				
Lysosomes				
Microvilli				
Centrioles				

Concept Mapping

Appendix E provides guidance for working with concept maps.

1. Construct your own concept map using the following words as the *concepts*. Supply the linking words between each pair of concepts.

Golgi apparatus	ribosomes
chloroplasts	flagella
cytoplasm	nucleolus
endoplasmic reticulum	cell membrane

Critical Thinking Questions

Critical thinking is the ability to reason and solve problems using facts and concepts. These questions can be approached from a number of angles, and in most cases, they do not have a single correct answer.

1. Suggest some ways that one would go about determining if ~132-133~ mitochondria and chloroplasts are a modified prokaryotic cell.

2. Give the common name of a eukaryotic microbe that is unicellular, ~138-39~ walled, nonphotosynthetic, nonmotile, and bud-forming. *Yeast*

3. Give the common name of a microbe that is unicellular, nonwalled, motile with flagella, and has chloroplasts. *protozoa*

4. What general type of multicellular parasite is composed primarily of thin sacs of reproductive organs? *Flatworm*

5. a. Name two parasites that are transmitted in the cyst form. *amoeba*
 b. How must a non-cyst-forming pathogenic protozoan be transmitted? Why? *sexually*

6. Explain what factors could cause opportunistic mycoses to be a growing medical problem. *alfatoxins*

7. a. How are bacterial endospores and cysts of protozoa alike? *dormant*
 b. How do they differ? *protect DNA*

8. You have gone camping in the mountains and plan to rely on water present in forest pools and creeks for drinking. Certain encysted pathogens often live in this type of water, but you do not discover this until you arrive at the campground and read your camping handbook. How might you treat the water to prevent becoming infected? *Boiling*

9. Can you think of a way to determine if a child is suffering from pinworms? Hint: Scotch tape is involved. *tape bottom*

Visual Challenge

1. Based on the rRNA tree (figure 5.14*a*), which organismic groups are closely related to animals (genus *Homo*)? Is any pattern evident among the protozoans on this tree?

2. Label the major structures you can observe in figures 5.18*d*, 5.23, 5.24*b*, and 5.29*b*.

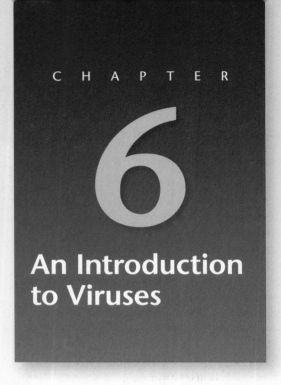

6

An Introduction to Viruses

Model of an influenza virus highlights the hemagglutinin spikes in red and the neuraminidase spikes in blue.

Humans, birds, and swine—the incubators for new strains of influenza.

"Beginning with the first diagnoses in March 2009, the influenza outbreak exploded into a pandemic in only six weeks."

CASE FILE 6 The Mystery Flu of 2009

In the middle of March 2009, the infectious disease monitoring system of Mexico began registering an increase in influenza (flu) cases. These reports caused concern because of the time frame and the age group affected. Influenza cases would normally have been in decline by this time in the season, and most of the cases occurred in healthy young adults, rather than very young or old patients, as is the usual pattern.

By mid-April, nearly 1,000 Mexican cases had been reported, with around 70 deaths. At about the same time, a few cases of this same type of flu cropped up in the United States, primarily in California and Texas. When the Centers for Disease Control and Prevention (CDC) received reports of these new cases, virologists immediately tested samples from American and Mexican patients to identify the strain of virus.

Major features used to identify the influenza virus are protein molecules on the surface of the virus called receptors or spikes. One of these receptors, called hemagglutinin (H), is essential for the virus to hook itself to respiratory cells. The other receptor is neuraminidase (N), which the virus uses to fuse with the cell. These receptors are genetically controlled and subject to frequent changes in structure. The spikes are assigned numbers to keep track of the several different forms, and from this practice comes the shorthand method of naming strains of a virus.

In both groups of patients, investigators discovered identical influenza viruses that typed out to a strain known as H1N1. Because a virus with this designation had appeared before in 1918 and 1976 as being associated with pigs, for a time, this outbreak was initially attributed to a "swine flu virus." But it turns out that this influenza virus was unlike any that had appeared in the past. It had genetic characteristics of a swine virus, but it also carried some genes from human and bird viruses. To allay unfounded fears about acquiring this disease from swine, public health groups urged that it be given the more accurate title of 2009 H1N1 type A influenza rather than swine flu. Determining the exact origin of this virus has proved difficult, and it may never be known. Viruses similar to it have been circulating in the United States since 1998, but this one was different in having genes and characteristics of European and Asian swine influenza viruses. It now appears that this virus underwent a genetic mutation that allowed it to be transmitted to and by humans.

▶ How do the receptor spikes on viruses play a role in infection?

▶ How is it possible for the same influenza virus to be able to infect both swine and humans?

To continue the case, go to page 169.

6.1 Overview of Viruses

Expected Learning Outcomes

1. Indicate how viruses were discovered and characterized.
2. Describe the unique characteristics of viruses.
3. Discuss the origin and importance of viruses.

Early Searches for the Tiniest Microbes

The discovery of the light microscope made it possible to see first-hand the agents of many bacterial, fungal, and protozoan diseases. But the techniques for observing and cultivating these relatively large microorganisms were useless for viruses. For many years, the cause of viral infections such as smallpox and polio was unknown, even though it was clear that the diseases were transmitted from person to person. The French bacteriologist Louis Pasteur was certainly on the right track when he postulated that rabies was caused by a "living thing" smaller than bacteria, and in 1884, he was able to develop the first vaccine for rabies. Pasteur also proposed the term *virus* (L. poison) to denote this special group of infectious agents.

The first substantial revelations about the unique characteristics of viruses occurred in the 1890s. First, D. Ivanovski and M. Beijerinck showed that a disease in tobacco was caused by a virus (tobacco mosaic virus). Then, Friedrich Loeffler and Paul Frosch discovered a virus that causes foot-and-mouth disease in cattle. These early researchers found that when infectious fluids from host organisms were passed through porcelain filters designed to trap bacteria, the filtrate remained infectious.

Over the succeeding decades, a remarkable picture of the physical, chemical, and biological nature of viruses began to take form. Years of experimentation were required to show that viruses were noncellular particles with a definite size, shape, and chemical composition. Using special techniques, they could be cultured in the laboratory. By the 1950s, virology had grown into a multifaceted discipline that promised to provide much information on disease, genetics, and even life itself **(Insight 6.1).**

The Position of Viruses in the Biological Spectrum

Viruses are a unique group of biological entities known to infect every type of cell, including bacteria, algae, fungi, protozoa, plants, and animals. Although the emphasis in this chapter is on animal viruses, much credit for our knowledge must be given to experiments with bacterial and plant viruses. The exceptional and curious nature of viruses prompts numerous questions, including:

1. How did viruses originate?
2. Are they organisms; that is, are they alive?
3. What are their distinctive biological characteristics?
4. How can particles so small, simple, and seemingly insignificant be capable of causing disease and death?
5. What is the connection between viruses and cancer?

In this chapter, we address these ideas and many others.

There is no universal agreement on how and when viruses originated. But there is little disagreement that they have been in existence for billions of years, probably ever since the early history of cells. One theory explains that viruses arose from loose strands of genetic material released by cells. These then developed a protective coating and a capacity to reenter a cell and use its machinery to reproduce. An alternate explanation for their origins suggests that they were once cells that have regressed to a highly parasitic existence inside other cells. Both of these viewpoints have some supportive evidence, and it is entirely possible that viruses have originated in more than one way.

In terms of numbers, viruses are considered the most abundant microbes on earth. Based on a decade of research on the viral populations in the ocean, virologists now have evidence that there are hundreds of thousands of distinct virus types that have never been described. Many of them are parasites of bacteria, which are also abundant in most ecosystems. By one estimate, viruses outnumber bacteria by a factor of 10. Because viruses tend to interact with the genetic material of their host cells and can carry genes from one host cell to another (chapter 9), they have played an important part in the evolution of Bacteria, Archaea, and Eukarya.

The unusual structure and behavior of viruses have led to debates about their connection to the rest of the microbial world. One viewpoint holds that viruses are unable to exist independently from the host cell, so they are not living things but are more akin to large, infectious molecules. Another viewpoint proposes that even though viruses do not exhibit most of the life processes of cells (discussed in section 4.1 of chapter 4), they can direct them and thus are certainly more than inert and lifeless molecules. Depending upon the circumstances, both views are defensible. This debate has greater philosophical than practical importance because viruses are agents of disease and must be dealt with through control, therapy, and prevention, whether we regard them as living or not. In keeping with their special position in the biological spectrum, it is best to describe viruses as *infectious particles* (rather than organisms) and as either *active* or *inactive* (rather than alive or dead).

Viruses are different from their host cells in size, structure, behavior, and physiology. They are a type of **obligate intracellular parasite** that cannot multiply unless it invades a specific host cell and instructs its genetic and metabolic machinery to make and release quantities of new viruses. Because of this characteristic, viruses are capable of causing serious damage and disease. Other unique properties of viruses are summarized in **table 6.1.**

Check & Assess Section 6.1

✔ Viruses are a unique group of tiny infectious particles that are obligate parasites of cells.

✔ Viruses do not exhibit the characteristics of life but can regulate the functions of host cells.

✔ They infect all groups of living things and produce a variety of diseases.

✔ They are not cells but resemble complex molecules.

✔ Viral replication inside a cell usually causes death or loss of function of that cell.

INSIGHT 6.1

An Alternate View of Viruses

Looking at this beautiful tulip, one would never guess that it derives part of its beauty from a viral infection. It contains tulip mosaic virus, which alters the development of the plant cells and causes varying patterns of colors in the petals. Aside from this, the virus does not cause severe harm to the plants. Despite the reputation of viruses as destructive pathogens, there is another side to viruses—that of being benign and, in some cases, even beneficial.

Recent research in extreme habitats in Antarctica uncovered a surprisingly rich diversity of viruses inhabiting a frigid lake. When the DNA of the samples was analyzed, it revealed over 10,000 unique viruses gathered from just a small area. Other studies of the ocean have documented the presence of 10 million viruses per milliliter of water—10 times the number of cellular microbes. This hints at the dominant role viruses have in the world's ecosystems, not only because they infect and can control populations of eukaryotic and prokaryotic cells, but because they appear to influence a number of biological activities, including photosynthesis and the cycling of nutrients.

Over the past several years, biomedical experts have been looking at viruses as vehicles to treat infections and disease. Vaccine experts have engineered new types of vaccines by combining a less harmful virus such as vaccinia or adenovirus with genetic material from a pathogen such as HIV and herpes simplex. This technique creates a vaccine that provides immunity but does not expose the person to the intact pathogen.

Several of these types of vaccines are currently in development.

The "harmless virus" approach is also being used to treat genetic diseases such as cystic fibrosis and sickle-cell anemia. With gene therapy, the normal gene is inserted into a virus vector, such as an adenovirus, and the patient is infected with this altered virus. It is hoped that the virus will introduce the needed gene into the cells and correct the defect. Dozens of experimental trials are currently underway to develop potential cures for diseases, with mixed success (see chapter 10).

An older therapy getting a second look involves use of bacteriophages to treat bacterial infections. The basis behind the therapy is that bacterial viruses can selectively attack and destroy their host bacteria without damaging human cells. Experiments with animals indicate that this method controls some infections as well as traditional drugs can. Potential applications include adding phage suspensions to skin grafts and to intravenous fluids to control infections.

The US Department of Agriculture has recently approved bacteriophage treatments to prevent food infections. One type is designed to spray on livestock to reduce the incidence of pathogenic *Escherichia coli*. The other is intended for use on processed meats and poultry to control *Listeria monocytogenes*, another prominent food-borne pathogen.

Explain why bacterial viruses would be harmless to humans.
Answer available at http://www.mhhe.com/talaro8

TABLE 6.1 Properties of Viruses

- Obligate intracellular parasites of bacteria, protozoa, fungi, algae, plants, and animals
- Ultramicroscopic size, ranging from 20 nm up to 450 nm (diameter)
- Not cellular in nature; structure is very compact and economical.
- Do not independently fulfill the characteristics of life
- Inactive macromolecules outside the host cell and active only inside host cells
- Basic structure consists of protein shell (capsid) surrounding nucleic acid core.
- Nucleic acid can be either DNA or RNA but not both.
- Nucleic acid can be double-stranded DNA, single-stranded DNA, single-stranded RNA, or double-stranded RNA.
- Molecules on virus surface impart high specificity for attachment to host cell.
- Multiply by taking control of host cell's genetic material and regulating the synthesis and assembly of new viruses
- Lack enzymes for most metabolic processes
- Lack machinery for synthesizing proteins

1. Describe 10 *unique* characteristics of viruses (can include structure, behavior, multiplication).
2. After consulting table 6.1, what additional statements can you make about viruses, especially as compared with cells?
3. Explain what it means to be an obligate intracellular parasite.
4. What is another way to describe the sort of parasitism exhibited by viruses?

6.2 The General Structure of Viruses

Expected Learning Outcomes

4. Describe the general structure and size range of viruses.
5. Distinguish among types of capsids and nucleocapsids.
6. Describe envelopes and spikes, and discuss their origins.
7. Explain the functions of capsids, nucleocapsids, envelopes, and spikes.
8. Summarize the different viral groups based on their basic structure.

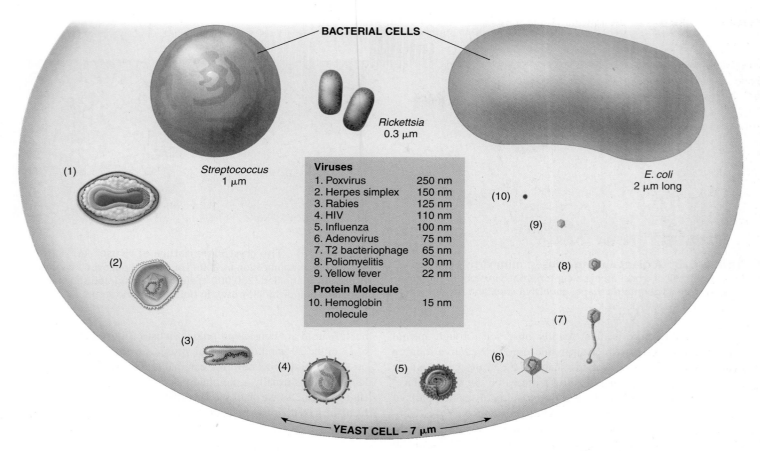

Figure 6.1 **Size comparison of viruses with a eukaryotic cell (yeast) and bacteria.** Viruses range from largest (1) to smallest (9). A molecule of a large protein (10) is included to indicate proportion of macromolecules.

The following labels appear within the figure:

BACTERIAL CELLS

Rickettsia
0.3 μm

Streptococcus
1 μm

E. coli
2 μm long

YEAST CELL – 7 μm

Viruses
1. Poxvirus	250 nm
2. Herpes simplex	150 nm
3. Rabies	125 nm
4. HIV	110 nm
5. Influenza	100 nm
6. Adenovirus	75 nm
7. T2 bacteriophage	65 nm
8. Poliomyelitis	30 nm
9. Yellow fever	22 nm

Protein Molecule
10. Hemoglobin molecule	15 nm

Size Range

As a group, viruses represent the smallest infectious agents (with some unusual exceptions to be discussed in section 6.8). Their size places them in the realm of the *ultramicroscopic*. This term means that most of them are so minute (< 0.2 μm) that an electron microscope is necessary to detect them or to examine their fine structure. They are dwarfed by their host cells: More than 2,000 bacterial viruses could fit into an average bacterial cell, and more than 50 million polioviruses could be accommodated by an average human cell. Animal viruses range in size from the small parvoviruses[1] (around 20 nm in diameter) to poxviruses[2] that are as large as small bacteria (up to 450 nm in length) (**figure 6.1**). Some cylindrical viruses are relatively long (800 nm in length) but so narrow in diameter (15 nm) that their visibility is still limited without the high magnification and resolution of an electron microscope. Figure 6.1 compares the sizes of several viruses with prokaryotic and eukaryotic cells and molecules.

TAKE NOTE: MIMIVIRUSES

A Transitional Form between Viruses and Cells?

In chapter 4, we featured recent discoveries of unusual types of bacteria (see Insight 4.2). Viruses are also known for their exceptional members. Probably one of the most outstanding examples is the mimivirus. These are giants in the viral world, averaging about 500 nm in diameter and being readily visible with light microscopy—as large as small bacteria such as rickettsias and mycoplasmas. They were first isolated as parasites of amoebas (*Acanthamoeba*) living in aquatic habitats (**figure 6.2**). Their name is derived from the word *mimic*, meaning that their characteristics, at least on the surface, give them the appearance of simple bacteria.

In addition to their large size, mimiviruses also contain a large number of genes for a virus (around 900). Extensive microscopic and molecular analysis revealed that they lack the major genes that all bacteria possess, and they do not have a true cellular structure. Analysis shows that they are related to other large virus families such as poxviruses, with a capsid, a complex membrane structure and a molecule of DNA in their core. They fit more traditional characteristics of viruses such as being obligate intracellular parasites, lacking ribosomes and metabolic enzymes, and not undergoing binary fission.

Where mimiviruses fit on the "tree of life," if at all, is still the subject of debate. Some microbiologists suggest that they may belong to a separate domain of microbes that is different in origin from other viral groups. Another possibility is that these viruses are related to ancient forms that evolved into the first cells.

1. DNA viruses that cause respiratory infections in humans.
2. A group of large, complex viruses, including smallpox, that cause raised skin swellings.

Figure 6.2 A giant among viruses: mimivirus. A section through the cell of a host amoeba captures one virus particle inside a vacuole. Note its geometric shape, dark DNA core, and the fine surface fibers (4,500×).

(a) (b)

Figure 6.3 The crystalline nature of viruses. (a) Light microscope magnification (1,200×) of purified poliovirus crystals. (b) Highly magnified (200,000×) electron micrograph showing hundreds of individual viruses, in tight geometric arrays.

Viral architecture is most readily observed through special stains in combination with electron microscopy. Negative staining uses an opaque salt to outline the shape of the virus against a dark background and to enhance textural features on the viral surface. Internal details are revealed by positive staining of specific parts of the virus such as protein or nucleic acid. The *shadowcasting* technique showers a virus preparation with a dense metallic vapor directed from a certain angle. These techniques are featured in several figures throughout this chapter.

Viral Components: Capsids, Nucleic Acids, and Envelopes

It is important to realize that viruses bear no real resemblance to cells and that they lack any of the synthetic machinery found in even the simplest cells. Their molecular structure consists of regular, repeating molecules that give rise to their crystalline appearance. Indeed, many purified viruses can form large aggregates or crystals if subjected to special treatments **(figure 6.3)**. The general plan of virus organization is very simple and compact. Viruses contain only those parts needed to invade and control a host cell: an external coating and a core containing one or more nucleic acid strands of either DNA or RNA. This pattern can be represented with a flowchart:

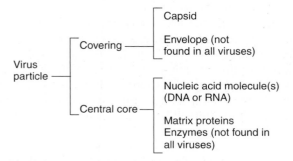

All viruses have a protein **capsid,*** or shell, that surrounds the nucleic acid in the central core. Together, the capsid and the nucleic acid are referred to as the **nucleocapsid.** Viruses that consist of only a

nucleocapsid are considered **naked viruses (figure 6.4a).** Members of 13 of the 20 families of animal viruses are **enveloped,** that is, they possess an additional covering external to the capsid called an envelope, which is usually a modified piece of the host's cell membrane **(figure 6.4b).** As we shall see later, the enveloped viruses also differ from the naked viruses in the way that they enter and leave a host cell.

(a) **Naked Nucleocapsid Virus**

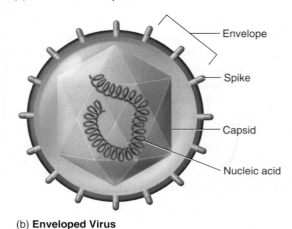

(b) **Enveloped Virus**

Figure 6.4 Generalized structure of viruses. (a) The simplest virus is a naked nucleocapsid consisting of a geometric capsid assembled around a nucleic acid strand or strands. (b) An enveloped virus is composed of a nucleocapsid surrounded by a flexible membrane called an envelope. The envelope usually has special receptor spikes inserted into it.

* *capsid* (kap'-sid) L. *capsa*, box.

The Viral Capsid: The Protective Outer Shell

When a virus particle is magnified several hundred thousand times, the capsid appears as the most prominent geometric feature (figure 6.4). In general, the capsid of any virus is constructed from a number of identical protein subunits called **capsomers.*** The capsomers can spontaneously self-assemble into the finished capsid. Depending on how the capsomers are shaped and arranged, this assembly results in two different types: helical and icosahedral.

The simpler **helical capsids** have rod-shaped capsomers that bind together to form a series of hollow discs resembling a bracelet. During the formation of the nucleocapsid, these discs link together and form a continuous helix into which the nucleic acid strand is coiled (**figure 6.5**). In electron micrographs, the appearance of a helical capsid varies with the type of virus. The nucleocapsids of naked helical viruses are very rigid and tightly wound into a cylinder-shaped package (**figure 6.6a**). Several viruses that infect plants are of this type (**figure 6.6b**). Enveloped helical nucleocapsids are more flexible and tend to be arranged as a looser helix within the envelope (**figure 6.6c, d**). This type of capsid is found in several enveloped human viruses, including those of influenza, measles, and rabies.

Several major virus families have capsids arranged in an **icosahedron***—a three-dimensional, 20-sided figure with 12 evenly spaced corners. The arrangements of the capsomers vary from one virus to another. The capsid of some viruses contains a single type of

capsomer, whereas others may contain several types of capsomers (**figure 6.7**). Although the capsids of all icosahedral viruses have this sort of symmetry, they can vary in the number of capsomers; for example, a poliovirus has 32 and an adenovirus has 242 capsomers. During assembly of the virus, the nucleic acid is packed into the center of this icosahedron, forming a nucleocapsid. Another factor that

* *capsomer* (kap′-soh-meer) L. *capsa,* box, and *mer,* part.
* *icosahedron* (eye″-koh-suh-hee′-drun) Gr. *eikosi,* twenty, and *hedra,* side. A type of polygon.

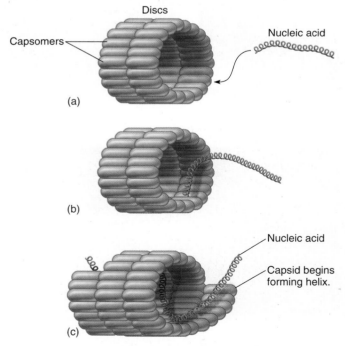

Figure 6.5 **Assembly of helical nucleocapsids.**
(a) Capsomers assemble into hollow discs. (b) The nucleic acid is inserted into the center of the disc. (c) Elongation of the nucleocapsid progresses from both ends, as the nucleic acid is wound "within" the lengthening helix.

Figure 6.6 **Typical variations of viruses with helical nucleocapsids. Naked helical virus (tobacco mosaic virus)** (a) a schematic view and (b) a greatly magnified micrograph. Note the overall cylindrical morphology. **Enveloped helical virus (influenza virus): (c)** a schematic view and **(d)** a colorized micrograph featuring a positive stain of the avian influenza virus (300,000×). This virus has a well-developed envelope with prominent spikes termed H5N1 type.

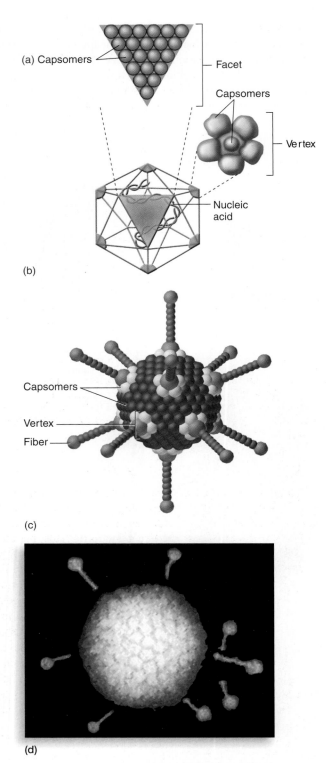

Figure 6.7 **The structure and formation of an icosahedral virus (adenovirus is the model). (a)** A facet or "face" of the capsid is composed of 21 identical capsomers arranged in a triangular shape. A vertex or "point" consists of five capsomers arranged with a single penton in the center. Other viruses can vary in the number, types, and arrangement of capsomers. **(b)** An assembled virus shows how the facets and vertices come together to form a shell around the nucleic acid. **(c)** A three-dimensional model (640,000×) of this virus shows fibers attached to the pentons. **(d)** A negative stain of this virus highlights its texture and fibers that have fallen off.

alters the appearance of icosahedral viruses is whether or not they have an outer envelope. Inspect **figure 6.8** to compare a rotavirus and its naked nucleocapsid with herpes simplex (cold sores) and its enveloped nucleocapsid.

The Viral Envelope

When **enveloped viruses** (mostly animal) are released from the host cell, they take with them a bit of its membrane system in the form of an envelope. Some viruses bud off the cell membrane; others leave via the nuclear envelope or the endoplasmic reticulum. Although the envelope is derived from the host, it is different because some or all of the regular membrane proteins are replaced with special viral proteins during the virus assembly process (see figure 6.11). Some proteins form a binding layer between the envelope and capsid of the virus, and glycoproteins (proteins bound to a carbohydrate) remain exposed on the outside of the envelope. These protruding molecules, called **spikes** or peplomers, are essential for the attachment of viruses to the next host cell. Because the envelope is more supple than the capsid, enveloped viruses are pleomorphic and range from spherical to filamentous in shape.

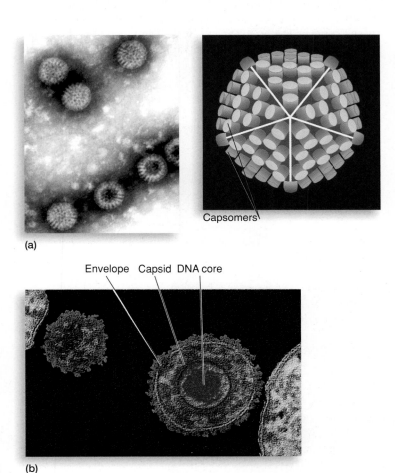

Figure 6.8 **Two types of icosahedral viruses, highly magnified. (a)** Left view: A negative stain of rotaviruses with unusual capsomers that look like spokes on a wheel (150,000×); right view is a three-dimensional model of this virus. **(b)** Herpes simplex virus, a type of enveloped icosahedral virus (300,000×).

(a)

(b)

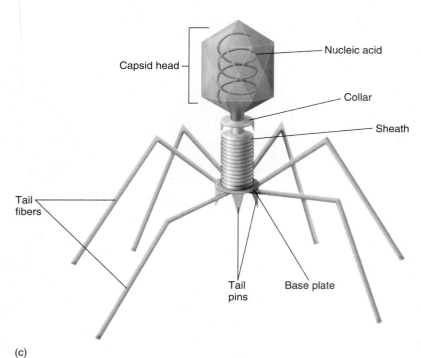

(c)

Figure 6.9 **Detailed structure of complex viruses.**
(a) Section through the vaccinia virus, a poxvirus, shows its internal components. (b) Photomicrograph of a bacteriophage from cyanobacteria (c) Diagram of a T4 bacteriophage of *E. coli*, (280,000×).

Functions of the Viral Capsid/Envelope

The outermost covering of a virus is indispensable to viral function because it protects the nucleic acid from the effects of various enzymes and chemicals when the virus is outside the host cell. We see this in the capsids of enteric (intestinal) viruses such as polio and hepatitis A, which are resistant to the acid- and protein-digesting enzymes of the gastrointestinal tract. Capsids and envelopes are also responsible for helping to introduce the viral DNA or RNA into a suitable host cell, first by binding to the cell surface and then by assisting in penetration of the viral nucleic acid (discussed in later sections). In addition, parts of viral capsids and envelopes stimulate the immune system to produce antibodies that can neutralize viruses and protect the host's cells against future infections (see chapter 15).

Complex Viruses: Atypical Viruses

Two special groups of viruses, termed complex viruses (**figure 6.9**), are more intricate in structure than the helical, icosahedral, naked, or enveloped viruses just described. The **poxviruses** (including the agent of smallpox) are very large DNA viruses that lack a typical capsid and are covered by a dense layer of lipoproteins and coarse fibrils on their outer surface. Some members of another group of very complex viruses, the **bacteriophages,*** have a polyhedral capsid head as well as a helical tail and fibers for attachment to the host cell. Their mode of multiplication is covered in section 6.5. **Figure 6.10** summarizes the primary morphological types found among the viruses.

* *bacteriophage* (bak-teer′-ee-oh-fayj″) From *bacteria,* and Gr. *phagein,* to eat. These viruses parasitize bacteria.

Nucleic Acids: At the Core of a Virus

The sum total of the genetic information carried by an organism is known as its **genome.** So far, one biological constant is that the genome of organisms is carried and expressed by nucleic acids (DNA, RNA). Even viruses are no exception to this rule, but there is a significant difference. Unlike cells, which contain both DNA and RNA, **viruses contain either DNA or RNA *but not both.*** Because viruses must pack into a tiny space all of the genes necessary to instruct the host cell to make new viruses, the number of viral genes is quite small compared with that of a cell. It varies from nine genes in human immunodeficiency virus (HIV) to hundreds of genes in some herpesviruses. By comparison, the bacterium *Escherichia coli* has approximately 4,000 genes, and a human cell has about 25,000 genes. The larger genome allows cells to carry out the complex metabolic activity necessary for independent life. Viruses possess only the genes needed to invade host cells and redirect their activity to make new viruses.

In chapter 2, you learned that DNA usually exists as a double-stranded molecule and that RNA is single-stranded. Although most viruses follow this same pattern, a few exhibit distinctive and exceptional forms. Notable examples are the parvoviruses, which contain single-stranded DNA, and reoviruses (a cause of respiratory and intestinal tract infections), which contain double-stranded RNA. In fact, viruses exhibit variety in how their RNA or DNA is configured. DNA viruses can have single-stranded (ss) or double-stranded (ds) DNA; the dsDNA can be arranged linearly or in circles. RNA viruses can be double-stranded but are more often single-stranded. You will learn in chapter 9 that all proteins are

A. Complex Viruses

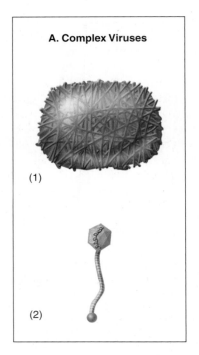

(1)

(2)

B. Enveloped Viruses

Helical Icosahedral

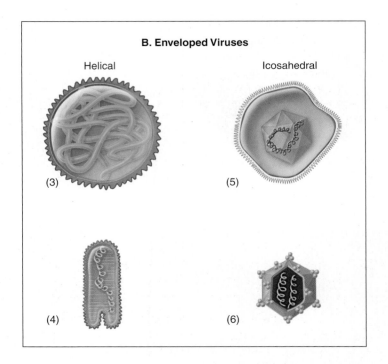

(3) (5)

(4) (6)

C. Nonenveloped Naked Viruses

Helical Icosahedral

(7) (8)

(9)

A. Complex viruses:
 (1) poxvirus, a large DNA virus
 (2) flexible-tailed bacteriophage

B. Enveloped viruses:
 With a helical nucleocapsid:
 (3) mumps virus
 (4) rhabdovirus
 With an icosahedral nucleocapsid:
 (5) herpesvirus
 (6) HIV (AIDS)

C. Naked viruses:
 Helical capsid:
 (7) plum poxvirus
 Icosahedral capsid:
 (8) poliovirus
 (9) papillomavirus

Figure 6.10 **Basic types of viral morphology.**

made by "translating" the nucleic acid code on a single strand of RNA into an amino acid sequence. Single-stranded RNA genomes that are ready for immediate translation into proteins are called *positive-strand* RNA. RNA genomes that have to be converted into the proper form for translation are called *negative-strand* RNA.

RNA genomes may also be *segmented,* meaning that the individual genes exist in separate RNA molecules. The influenza virus (an orthomyxovirus) is an example of this form. An RNA virus with some unusual features is a *retrovirus,* one of the few virus types that converts its nucleic acid from RNA to DNA inside its host cell.

Whatever the virus type, these tiny strands of genetic material carry the blueprint for viral structure and functions. In a very real sense, viruses are genetic parasites because they cannot multiply until their nucleic acid has reached the internal habitat of the host cell. At the minimum, they must carry genes for synthesizing the viral capsid and genetic material, for regulating the actions of the host, and for packaging the mature virus.

Other Substances in the Virus Particle

In addition to the protein of the capsid, the proteins and lipids of envelopes, and the nucleic acid of the core, viruses can contain enzymes for specific operations within their host cell. They may come with preformed enzymes that are required for viral **replication.*** Examples include **polymerases*** that synthesize DNA and RNA and replicases that copy RNA. The AIDS virus comes equipped with *reverse transcriptase* for synthesizing DNA

* *replication* (rep-lih-kay′-shun) L. *replicare,* to reply. To make an exact duplicate.
* *polymerase* (pol-im′-ur-ace) An enzyme that synthesizes a large molecule from smaller subunits.

from RNA. However, viruses completely lack the genes for synthesis of metabolic enzymes. As we shall see, this deficiency has little consequence, because viruses have adapted to assume total control over the cell's metabolic resources. Some viruses can actually carry away substances from their host cell. For instance, arenaviruses pack along host ribosomes, and retroviruses "borrow" the host's tRNA molecules.

6.3 How Viruses Are Classified and Named

Expected Learning Outcomes

9. Explain the classification scheme used for viruses.

10. Indicate the characteristics used in identifying and naming viruses.

Although viruses are not classified as members of the domains discussed in chapter 1, they are diverse enough to require their own classification scheme to aid in their study and identification. In an informal and general way, we have already begun classifying viruses—as animal, plant, or bacterial viruses; enveloped or naked viruses; DNA or RNA viruses; and helical or icosahedral viruses. These introductory categories are certainly useful in organization and description, but the study of specific viruses requires a more standardized method of nomenclature. For many years, the animal viruses were classified mainly on the basis of their hosts and the kind of diseases they caused. Newer systems for naming viruses also take into account the actual nature of the virus particles themselves, with only partial emphasis on host and disease. The main criteria presently used to group viruses are structure, chemical composition, and similarities in genetic makeup, which indicate evolutionary relatedness.

The International Committee on the Taxonomy of Viruses lists 3 orders, 63 families, and 263 genera of viruses. Note the naming conventions—that is, virus families are written with -*viridae* on the end of the name, and genera end with -*virus*.

Historically, some virologists had created an informal *species* naming system that mirrors the species names in higher organisms, using genus and species epithets such as *Measles morbillivirus*. The species category has created a lot of controversy within the virology community. Many scientists argue that nonorganisms such as viruses are too changeable and that fine distinctions used for deciding on species classifications will quickly disappear. Over the past decade, virologists have largely accepted the concept of viral species, defining them as consisting of members that have a number of properties in common but have some variations. In other words, a virus is placed in a species on the basis of a collection of properties such as host range, pathogenicity, and genetic makeup.

Because the use of standardized species names has not been widely accepted, the genus or common English vernacular names (for example, poliovirus and rabies virus) predominate in discussions of specific viruses in this text. **Table 6.2** illustrates a system of classification for important viruses and the diseases they cause. To-scale examples of each virus family are included.

Characteristics used for placement in a particular family include type of capsid, nucleic acid strand number, presence and type of envelope, overall viral size, and area of the host cell in which the virus multiplies. Some virus families are named for their microscopic appearance (shape and size). Examples include **rhabdoviruses,*** which have a bullet-shaped envelope, and **togaviruses,*** which have a cloaklike envelope. Anatomical or geographic areas have also been used in naming. For instance, **adenoviruses*** were first discovered in adenoids (one type of tonsil), and hantaviruses were originally isolated in the Korean Province of Hantaan. Viruses can also be named for their effects on the host. **Lentiviruses*** tend to cause slow, chronic infections. Acronyms made from blending several characteristics include **picornaviruses,*** which are tiny RNA viruses, and **hepadnaviruses** (for hepatitis and DNA).

Check & Assess Sections 6.2 and 6.3

✔ Virus size range is from 20 nm to 450 nm (diameter). Viruses consist of an outer capsid composed of protein units. It surrounds a nucleic acid core composed of DNA or RNA, along with essential enzymes. Some viruses also exhibit an envelope around the capsid containing spikes.

✔ Classification of viruses is based on nucleic acid type and structure, capsid type, presence of an envelope, and host organism, characteristics that are used to place them into families.

✔ Viruses that cause human infections are found in seven families of DNA viruses and 13 families of RNA viruses.

✔ Viruses have been assigned genus and species designations; in most contexts, they are known by their common or vernacular names.

5. Characterize viruses according to size range. What does it mean to say that they are ultramicroscopic?

6. Describe the general structure of viruses.

7. What is the capsid, and what is its function; how are the two types of capsids constructed?

8. What is a nucleocapsid? Give examples of viruses with the two capsid types.

9. What is an enveloped virus, and how does the envelope arise? What are spikes, how are they formed, and what is their function?

10. What are bacteriophages, and what is unique about their structure?

11. How are the poxviruses different from other animal viruses?

12. How are viruses classified? What are virus families?

13. How are generic and common names used?

14. Look at table 6.2 and count the total number of different viral diseases. How many are caused by DNA viruses? How many are RNA-virus diseases?

* *rhabdovirus* (rab″-doh-vy′-rus) Gr. *rhabdo,* little rod.
* *togavirus* (toh″-guh-vy′-rus) L. *toga,* covering or robe.
* *adenovirus* (ad″-uh-noh-vy′-rus) G. *aden,* gland.
* *lentivirus* (len″-tee-vy′-rus) Gr. *lente,* slow. HIV, the AIDS virus, belongs in this group.
* *picornavirus* (py-kor″-nah-vy′-rus) Sp. *pico,* small, plus RNA.

TABLE 6.2 Important Human Virus Families, Genera, Common Names, and Types of Diseases

Nucleic Acid Type	Family	Genus of Virus	Common Name of Genus Members	Name of Disease
DNA Viruses				
	Poxviridae	*Orthopoxvirus*	Variola and vaccinia	Smallpox, cowpox
	Herpesviridae	*Simplexvirus*	Herpes simplex 1 virus (HSV-1)	Fever blister, cold sores
			Herpes simplex 2 virus (HSV-2)	Genital herpes
		Varicellovirus	Varicella zoster virus (VZV)	Chicken pox, shingles
		Cytomegalovirus	Human cytomegalovirus (CMV)	CMV infections
	Adenoviridae	*Mastadenovirus*	Human adenoviruses	Adenovirus infection
	Papillomaviridae	*Papillomavirus*	Human papillomavirus (HPV)	Several types of warts
	Polyomaviridae	*Polyomavirus*	JC virus (JCV)	Progressive multifocal leukoencephalopathy
	Hepadnaviridae	*Hepadnavirus*	Hepatitis B virus (HBV or Dane particle)	Serum hepatitis
	Parvoviridae	*Erythrovirus*	Parvovirus B19	Erythema infectiosum
RNA Viruses				
	Picornaviridae	*Enterovirus*	Poliovirus	Poliomyelitis
			Coxsackievirus	Hand-foot-mouth disease
		Hepatovirus	Hepatitis A virus (HAV)	Short-term hepatitis
		Rhinovirus	Human rhinovirus	Common cold, bronchitis
	Caliciviridae	*Calicivirus*	Norwalk virus	Viral diarrhea, Norwalk virus syndrome
	Togaviridae	*Alphavirus*	Eastern equine encephalitis virus	Eastern equine encephalitis (EEE)
			Western equine encephalitis virus	Western equine encephalitis (WEE)
			Yellow fever virus	Yellow fever
			St. Louis encephalitis virus	St. Louis encephalitis
		Rubivirus	Rubella virus	Rubella (German measles)
	Flaviviridae	*Flavivirus*	Dengue fever virus	Dengue fever
			West Nile fever virus	West Nile fever
	Bunyaviridae	*Bunyavirus*	Bunyamwera viruses	California encephalitis
		Hantavirus	Sin Nombre virus	Respiratory syndrome
		Phlebovirus	Rift Valley fever virus	Rift Valley fever
		Nairovirus	Crimean–Congo hemorrhagic fever (CCHF) virus	Crimean–Congo hemorrhagic fever
	Filoviridae	*Filovirus*	Ebola, Marburg virus	Ebola fever
	Reoviridae	*Coltivirus*	Colorado tick fever virus	Colorado tick fever
		Rotavirus	Human rotavirus	Rotavirus gastroenteritis
	Orthomyxoviridae	*Influenza virus*	Influenza virus, type A (Asian, Hong Kong, and swine influenza viruses)	Influenza or "flu"
	Paramyxoviridae	*Paramyxovirus*	Parainfluenza virus	Parainfluenza
			Mumps virus	Mumps
		Morbillivirus	Measles virus	Measles (red)
		Pneumovirus	Respiratory syncytial virus (RSV)	Common cold syndrome

DNA Viruses illustrations: Poxviridae (Chordopoxvirinae), Herpesviridae, Adenoviridae, Papillomaviridae, Polyomaviridae, Hepadnaviridae, Parvoviridae (Parvovirinae)

RNA Viruses illustrations: Picornaviridae, Caliciviridae, Togaviridae, Flaviviridae, Bunyaviridae, Filoviridae, Reoviridae, Orthomyxoviridae, Paramyxoviridae

TABLE 6.2 (*Continued*)

Rhabdoviridae	*Retroviridae*	Rhabdoviridae	*Lyssavirus*	Rabies virus	Rabies (hydrophobia)
		Retroviridae	*Oncornavirus*	Human T cell leukemia virus (HTLV)	T-cell leukemia
			Lentivirus	HIV (human immunodeficiency viruses 1 and 2)	Acquired immunodeficiency syndrome (AIDS)
Arenaviridae	*Coronaviridae*	Arenaviridae	*Arenavirus*	Lassa virus	Lassa fever
		Coronaviridae	*Coronavirus*	Infectious bronchitis virus (IBV)	Bronchitis
100 nm				Enteric corona virus	Coronavirus enteritis
				SARS virus	Severe acute respiratory syndrome

6.4 Modes of Viral Multiplication

xpected Learning Outcomes

11. Describe the virus-host relationship.

12. Relate the stages in the multiplication cycle of animal viruses, and summarize what is happening in each stage.

13. Describe ways that animal viruses attach and enter a host cell.

14. Define *replication* as it relates to viruses.

15. Explain two ways that animal virus are released by a host cell.

16. Describe cytopathic effects of viruses and the possible results of persistent viral infections.

Viruses are closely associated with their hosts. In addition to providing the viral habitat, the host cell is absolutely necessary for viral multiplication, which is synonymous with infection. The way that viruses invade a host cell is an extraordinary biological phenomenon. Viruses have been aptly described as minute parasites that seize control of the synthetic and genetic machinery of cells. The nature of this cycle profoundly affects pathogenicity, transmission, the responses of the immune defenses, and human measures to control viral infections. From these perspectives, we cannot overemphasize the importance of a working knowledge of the relationship between viruses and their host cells.

Multiplication Cycles in Animal Viruses

The general phases in the life cycle of animal viruses are **adsorption,*** **penetration, synthesis, assembly,** and **release** from the host cell. Viruses vary in the exact mechanisms of these processes, but we will use a simple animal virus to illustrate the major events **(figure 6.11).** Other examples are given in chapters 9, 24, and 25.

Adsorption and Host Range

Invasion begins when the virus encounters a susceptible host cell and adsorbs specifically to receptor sites on the cell membrane. The membrane receptors that viruses attach to are usually glycoproteins the cell requires for its normal function. For example, the rabies virus affixes to receptors found on mammalian nerve cells, and the human immunodeficiency virus (HIV or AIDS virus) attaches to

the CD4 protein on certain white blood cells. The mode of attachment varies between the two general types of viruses. In enveloped forms such as influenza virus and HIV, glycoprotein spikes bind to the cell membrane receptors. Viruses with naked nucleocapsids (adenovirus, for example) use molecules on their capsids that adhere to cell membrane receptors **(figure 6.12).**

CONTINUING CASE FILE 6

Beginning with the first diagnoses in March 2009, the influenza outbreak exploded into a pandemic in only six weeks. Cases rapidly appeared in Canada, Central and South America, then Europe and Asia, and eventually more than 200 countries. By CDC estimates, from April to November 2009 in the United States alone, there were 50 million cases and close to 10,000 deaths. Deaths were particularly high among young children and pregnant women whose treatment had been delayed. Fortunately, the disease experienced by most people was milder than the usual seasonal flu, and it cleared up with few complications. The common symptoms are fever, muscle aches, and problems with breathing and coughing that subside in one or two weeks. The most serious complication is pneumonia. One group that seemed to be less susceptible to H1N1 influenza virus were people 60 years or older.

- What is a pandemic?
- Why would some people be more resistant to the virus?

For a wrap-up, see the Case File Perspective on page 181.

Because a virus can invade its host cell only through making an exact fit with a specific host molecule, the range of hosts it can infect in a natural setting is limited. This limitation, known as the **host range,** can vary from one virus to another. For example, hepatitis B infects only liver cells of humans; the poliovirus infects primarily intestinal and nerve cells of primates (humans, apes, and monkeys); and the rabies virus infects nerve cells of most mammals. Cells that lack compatible virus receptors are resistant to adsorption and invasion by that virus. This explains why, for example, human liver cells are not infected by the canine hepatitis virus and dog liver cells cannot host the human hepatitis A virus. It also explains why viruses usually have tissue specificities called **tropisms*** for certain

* *adsorption* (ad-sorp′-shun) L. ad, to, and *sorbere,* to suck. The attachment of one thing onto the surface of another.

* *tropism* (troh′-pizm) Gr. *trope,* a turn. Having a special affinity for an object or substance.

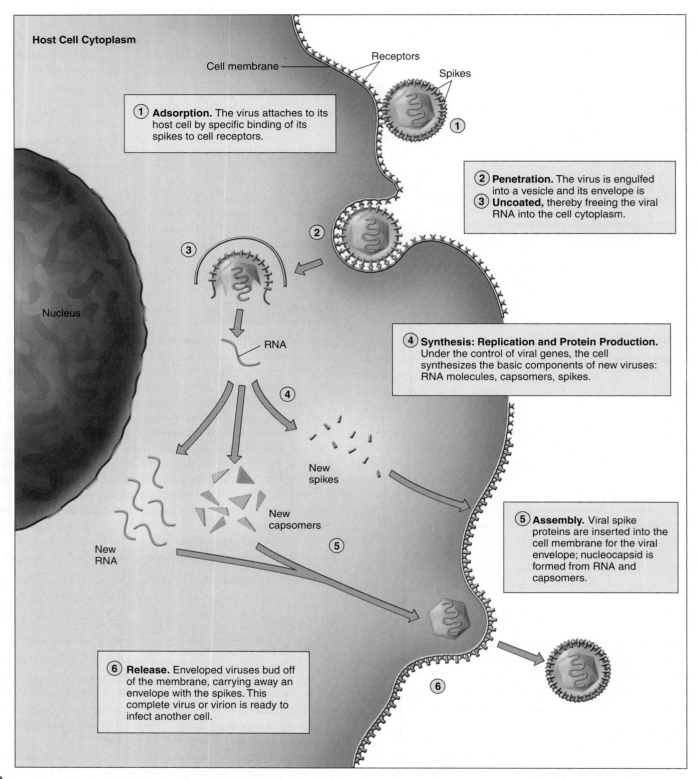

Host Cell Cytoplasm

Cell membrane

Receptors

Spikes

① **Adsorption.** The virus attaches to its host cell by specific binding of its spikes to cell receptors.

② **Penetration.** The virus is engulfed into a vesicle and its envelope is ③ **Uncoated,** thereby freeing the viral RNA into the cell cytoplasm.

Nucleus

RNA

④ **Synthesis: Replication and Protein Production.** Under the control of viral genes, the cell synthesizes the basic components of new viruses: RNA molecules, capsomers, spikes.

New spikes

New capsomers

New RNA

⑤ **Assembly.** Viral spike proteins are inserted into the cell membrane for the viral envelope; nucleocapsid is formed from RNA and capsomers.

⑥ **Release.** Enveloped viruses bud off of the membrane, carrying away an envelope with the spikes. This complete virus or virion is ready to infect another cell.

Process Figure 6.11 **General features in the multiplication cycle of an enveloped animal virus.** Using an RNA virus (rubella virus), the major events are outlined, although other viruses will vary in exact details of the cycle.

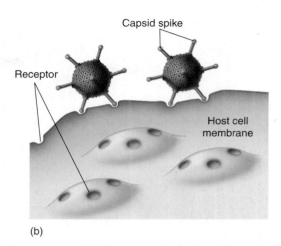

(a) (b)

Figure 6.12 **The mode by which animal viruses adsorb to the host cell membrane.** **(a)** An enveloped coronavirus with prominent spikes. The configuration of the spike has a complementary fit for cell receptors. The process in which the virus lands on the cell and plugs into receptors is termed docking. **(b)** An adenovirus has a naked capsid that adheres to its host cell by nestling surface molecules on its capsid into the receptors on the host cell's membrane.

cells in the body. The hepatitis B virus targets the liver, and the mumps virus targets salivary glands. Many viruses can be manipulated in the laboratory to infect cells that they do not infect naturally, thus making it possible to cultivate them.

Penetration/Uncoating of Animal Viruses

For an animal virus to successfully infect a cell, it must penetrate the cell membrane of the host cell and deliver the viral nucleic acid into the host cell's interior. How animal viruses do this varies with the type of virus and type of host cell, but most of them enter through some form of fusion or endocytosis **(figure 6.13).** In the case of **fusion,** the viral envelope fuses directly with the host cell membrane, so it can occur only in enveloped viruses **(figure 6.13*a*).** Following attachment of the virus to host cell receptors, the lipids within the adjacent membranes become rearranged so that the nucleocapsid can be translocated into the cytoplasm for the synthesis phase.

In the **endocytosis** version of penetration, the virus can be either enveloped **(figure 6.13*b*)** or naked **(figure 6.13*c*),** and it is engulfed entirely into a vesicle after its initial attachment. Once inside the cell, the virus is **uncoated;** that is, its nucleic acid or nucleocapsid is released by the actions of enzymes in the cytoplasm that dissolve the vesicle wall. There are numerous variations in the details of this process.

Synthesis: Replication and Protein Production

The synthetic and replicative phases of animal viruses are highly regulated and extremely complex at the molecular level. Free viral nucleic acid exerts control over the host's metabolism and synthetic machinery. How this control proceeds will vary, depending on whether the virus is a DNA or an RNA virus. In general, the DNA viruses (except poxviruses) enter the host cell's nucleus and are replicated and assembled there. RNA viruses are replicated and assembled in the cytoplasm, with some exceptions.

The details of animal virus replication are discussed in chapter 9. Here we provide a brief overview of the process, using an RNA virus as a model. Almost immediately upon entry, the viral nucleic

acid alters the genetic expression of the host and instructs it to synthesize the building blocks for new viruses. For example, it was recently shown that HIV depends on the expression of 250 human genes to complete its multiplication cycle. First, the RNA of the virus becomes a message for synthesizing viral proteins (translation). The viruses with positive-strand RNA molecules already contain the correct message for translation into proteins. Viruses with negative-strand RNA molecules must first be converted into a positive-strand message. Some viruses come equipped with the necessary enzymes for synthesis of viral components; others utilize those of the host. During the final phase, the host's replication and synthesis machinery produces new RNA, proteins for the capsid, spikes, and viral enzymes.

Assembly of Animal Viruses: Host Cell as Factory

Toward the end of the cycle, mature virus particles are constructed from the growing pool of parts. In most instances, the capsid is first laid down as an empty shell that will serve as a receptacle for the nucleic acid strand. Electron micrographs taken during this time show cells with masses of viruses, often enclosed in packets (see figure 6.15*a*). One important event leading to the release of enveloped viruses is the insertion of viral spikes into the host's cell membrane so they can be picked up as the virus buds off with its envelope, as discussed.

Release of Mature Viruses

To complete the cycle, assembled viruses leave their host in one of two ways. Nonenveloped and complex viruses that reach maturation in the cell nucleus or cytoplasm are released through cell **lysis** or rupturing. Enveloped viruses are liberated by **budding** or **exocytosis**[3] from the membranes of the cytoplasm, nucleus, endoplasmic reticulum, or vesicles. During this process, the nucleocapsid binds

3. For enveloped viruses, these terms are interchangeable. They mean the release of a virus from an animal cell by enclosing it in a portion of membrane derived from the cell.

Figure 6.13 Modes of Virus Penetration. Pictured in **(a)** is the process of fusion between the envelope of the virus and the host cell membrane, with release of the nucleocapsid into the cytoplasm. This is the mechanism for mumps and HIV viruses. Both (b) and (c) are examples of entry by endocytosis or engulfment, followed by uncoating. **(b)** In an enveloped virus such herpesvirus, the entire virus is taken into a vesicle with subsequent release of the DNA. **(c)** A naked virus such as polio virus is first taken into a vesicle and then releases its nucleic acid through a pore or rupture in the membrane.

to the membrane, which curves completely around it and forms a small pouch. Pinching off the pouch releases the virus with its envelope **(figure 6.14).** Budding of enveloped viruses causes them to be shed gradually, without the sudden destruction of the cell. Regardless of how the virus leaves, most active viral infections are ultimately lethal to the cell because of accumulated damage. Lethal damages include a permanent shutdown of metabolism and genetic expression, destruction of cell membrane and organelles, toxicity of virus components, and release of lysosomes.

The length of a multiplication cycle, from adsorption to lysis, varies to some extent, but it is usually measured in hours. A simple virus such as poliovirus takes about 8 hours; parvovirus takes 16 to 18 hours; and more complex viruses, such as herpesviruses, require 72 hours or more.

A fully formed, extracellular virus particle that is virulent and able to establish infection in a host is called a **virion.*** The number

of virions released by infected cells is variable, controlled by factors such as the size of the virus and the health of the host cell. About 3,000 to 4,000 virions are released from a single cell infected with poxviruses, whereas a poliovirus-infected cell can release over 100,000 virions. If even a small number of these virions happens to meet other susceptible cells and infect them, the potential for viral proliferation is immense.

Damage to the Host Cell and Persistent Infections

The short- and long-term effects of viral infections on animal cells are well documented. **Cytopathic*** **effects** (CPEs) are defined as virus-induced damage to the cell that alters its microscopic

* *virion* (vir′-ee-on) Gr. *ios,* poison.

* *cytopathic* (sy″-toh-path′-ik) Gr. *cyto,* cell, and *pathos,* disease.

Host cell membrane

Viral nucleocapsid

Viral glycoprotein spikes

Cytoplasm

Capsid

RNA

Viral matrix protein

Budding virion

Free infectious virion with envelope

(a)

(b)

Figure 6.14
Maturation and release of enveloped viruses.
(a) As parainfluenza virus is budded off the membrane, it simultaneously picks up an envelope and spikes.
(b) AIDS viruses (HIV) leave their host T cell by budding off its surface.

nucleus and cytoplasm (**figure 6.15b**). Examination of cells and tissues for cytopathic effects is an important part of the diagnosis of viral infections. **Table 6.3** summarizes some prominent cytopathic effects associated with specific viruses. One very common CPE is the fusion of multiple host cells into single large cells containing multiple nuclei. These **syncytia** are a result of some viruses' ability to fuse membranes. One virus (respiratory syncytial virus) is even named for this effect.

As a general rule, a virus infection kills its host cell, but some cells escape destruction by harboring the virus in some form. These so-called **persistent infections** can last from a few weeks to years and even for the life of the host. Take for example the measles virus, which occasionally remains hidden in brain cells for many years, eventually causing progressive damage and disease. Several viruses remain in a **latent state,** meaning that they are inactive over long periods. Examples of this are herpes simplex viruses (cold sores and genital herpes) and herpes zoster virus (chicken pox and shingles). Both viruses go into latency in nerve cells and later emerge under the influence of various stimuli to cause recurrent symptoms. Specific damage that occurs in viral diseases is covered more completely in chapters 24 and 25.

appearance. Individual cells can become disoriented, undergo gross changes in shape or size, or develop intracellular changes (**figure 6.15a**). It is common to note **inclusion bodies,** or compacted masses of viruses or damaged cell organelles, in the

Normal cell

Multiple nuclei

Giant cell

(a)

Inclusion bodies

(b)

Figure 6.15 **Cytopathic changes in cells and cell cultures infected by viruses. (a)** Human epithelial cells (400×) infected by herpes simplex virus demonstrate multinucleate giant cells; inset with a cluster of viruses from an intranuclear inclusion (100,000×) **(b)** Fluorescent-stained human cells infected with cytomegalovirus. Note the inclusion bodies (1,000×) and the loss of cohesive junctions between cells.

TABLE 6.3 Cytopathic Changes in Selected Virus-Infected Animal Cells

Virus	Response in Animal Cell
Smallpox virus	Cells round up; inclusions appear in cytoplasm.
Herpes simplex	Cells fuse to form multinucleated syncytia; nuclear inclusions (see figure 6.15)
Adenovirus	Clumping of cells; nuclear inclusions
Poliovirus	Cell lysis; no inclusions
Reovirus	Cell enlargement; vacuoles and inclusions in cytoplasm
Influenza virus	Cells round up; no inclusions
Rabies virus	No change in cell shape; cytoplasmic inclusions (Negri bodies)
Measles virus	Syncytia form (multinucleate).

Some animal viruses enter their host cell and permanently alter its genetic material, leading to cancer. These viruses are termed *oncogenic,* and their effect on the cell is called **transformation.** A startling feature of some of these viruses is that their nucleic acid becomes integrated into the host DNA. Transformed cells generally have an increased rate of growth, alterations in chromosomes, changes in the cell's surface molecules, and the capacity to divide for an indefinite period. Mammalian viruses capable of initiating tumors are called **oncoviruses.** Some of these are DNA viruses such as papillomavirus (associated with cervical cancer), herpesviruses (Epstein-Barr virus causes Burkitt's lymphoma), and hepatitis B virus (liver cancer). These findings have spurred a great deal of speculation on the possible involvement of viruses in cancers whose cause is still unknown. Additional information on the connection between viruses and cancer is found in chapter 24.

Check & Assess Section 6.4

✔ Viruses are genetic parasites that take over the host cell's metabolism and synthetic machinery.

✔ Animal viruses have a multiplication cycle in the host cell that goes through phases of adsorption, penetration (sometimes followed by uncoating), viral synthesis and assembly, and viral release.

✔ During adsorption, the virus attaches to specific host receptors.

✔ Penetration involves entry of the viral nucleic acid through engulfment or fusion.

✔ The host cell synthesizes and assembles viral molecules.

✔ Mature viruses are released by cell lysis or budding.

✔ These events turn the host cell into a factory solely for making and shedding new viruses and usually result in destruction of the cell.

✔ Animal viruses can cause acute infections or can persist in host tissues as latent infections that can reactivate periodically throughout the host's life. Some persistent animal viruses can be involved in cancer (oncogenic).

15. Write a narrative that describes the stages in the multiplication of an enveloped animal virus.
16. What are ways that animal viruses penetrate the host cell? What is uncoating?
17. Describe the two ways that animal viruses leave their host cell.
18. What is meant by the term *virion*?
19. Describe several cytopathic effects of viruses. What causes the appearance of the host cell, and how might it be used to diagnose viral infections?
20. What does it mean for a virus to be persistent or latent, and how are these events important?
21. Briefly describe the action of an oncogenic virus.

6.5 The Multiplication Cycle in Bacteriophages

Expected Learning Outcomes

17. Describe the stages in the multiplication cycle of bacteriophages.
18. Explain what is meant by lysogeny, prophage, and lysogenic induction and conversion.
19. Compare the major stages in multiplication of animal viruses and bacteriophage.

For the sake of comparison, we now turn to multiplication in bacterial viruses—the bacteriophages. When Frederick Twort and Felix d'Herelle discovered these viruses in 1915, it first appeared that the bacterial host cells were being eaten by some unseen parasite; hence, the name bacteriophage was used.

So far as is known, all bacteria are parasitized by various specific viruses. Most bacteriophages (often shortened to *phage*) contain double-stranded DNA, though single-stranded DNA and RNA types exist as well. Bacteriophages are of great interest to medical microbiologists because they often make the bacteria they infect more pathogenic for humans.

Probably the most widely studied bacteriophages are those of the intestinal bacterium *Escherichia coli*—especially the ones known as the T-even phages, such as T2 and T4. They are complex in structure, with an icosahedral capsid head containing DNA, a central tube (surrounded by a sheath), collar, base plate, tail pins, and fibers (see figure 6.9c). Momentarily setting aside a strictly scientific and objective tone, it is tempting to think of these extraordinary viruses as minute spacecrafts docking on an alien planet, ready to unload their genetic cargo.

T-even bacteriophages go through similar stages as the animal viruses described earlier **(figure 6.16).** First, they adsorb to host bacteria using specific receptors on the bacterial surface. Although the entire phage cannot enter the host cell, the nucleic acid is injected through a rigid tube the phage inserts through the bacterial membrane and wall **(figure 6.17).** The empty capsid remains attached to the cell surface. Entry of the nucleic acid stops host cell DNA

E. coli host

Bacteriophage

① Adsorption

⑦ Release of viruses

Lysogenic State

Bacterial DNA Viral DNA

② Penetration

Viral DNA becomes latent as prophage.

DNA splits

Spliced viral genome

■ Viral DNA ■ Bacterial DNA molecule

The lysogenic state in bacteria.
The viral DNA molecule is inserted at specific sites on the bacterial chromosome. The viral DNA is duplicated along with the regular genome and can provide adaptive genes for the host bacterium.

③ Duplication of phage components; replication of virus genetic material

④ Assembly of new virions

Lytic Cycle

⑥ Lysis of weakened cell

⑤ Maturation

Capsid DNA

Tail Tail fibers

Sheath

Bacteriophage

Bacteriophage assembly line.
First the capsomers are synthesized by the host cell. A strand of viral nucleic acid is inserted during capsid formation. In final assembly, the prefabricated components fit together into whole parts and finally into the finished viruses.

Process Figure 6.16 Events in the multiplication cycle of T-even bacteriophages. The lytic cycle (1–7) involves full completion of viral infection through lysis and release of virions. Occasionally the virus enters a reversible state of lysogeny (left) and is incorporated into the host's genetic material.

replication and protein synthesis, and it soon prepares the cell machinery for viral replication and synthesis of viral proteins. As the host cell produces new phage parts, the parts spontaneously assemble into bacteriophages.

An average-size *Escherichia coli* cell can contain up to 200 new phage units at the end of this period. Eventually, the host cell becomes so packed with viruses that it **lyses** and splits open, thereby releasing the mature virions (**figure 6.18**). This process is hastened by viral enzymes produced late in the infection cycle that weaken the cell envelope. Upon release, the virulent phages can spread to other susceptible bacterial cells and begin a new cycle of infection.

Lysogeny: The Silent Virus Infection

The lethal effects of a virulent phage on the host cell present a dramatic view of virus-host interaction. Not all bacteriophages complete the lytic cycle, however. Special DNA phages, called **temperate* phages,** undergo adsorption and penetration into the bacterial host but are not replicated or released immediately. Instead, the viral DNA enters an inactive **prophage*** state, during which it is inserted into the bacterial chromosome. This viral DNA

* *temperate* (tem'-pur-ut) A reduction in intensity.
* *prophage* (pro'-fayj) L. *pro,* before, plus phage.

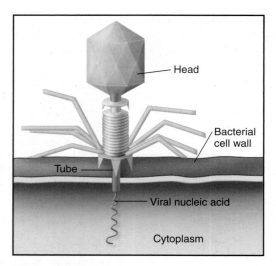

Figure 6.17 Penetration of a bacterial cell by a T-even bacteriophage. After adsorption, the phage plate becomes embedded in the cell wall, and the sheath contracts, pushing the tube through the cell wall and membrane and releasing the nucleic acid into the interior of the cell.

Figure 6.18 A weakened bacterial cell, crowded with viruses. The cell has ruptured and released numerous virions that can then attack nearby susceptible host cells. Note the empty heads of "spent" phages lined up around the ruptured wall.

will be retained by the bacterial cell and copied during its normal cell division so that the cell's progeny will also have the phage DNA. This condition, in which the host chromosome carries bacteriophage DNA, is termed **lysogeny.*** Because viral particles are not produced, the bacterial cells carrying temperate phages do not lyse and they appear entirely normal. On occasion, in a process called **induction,** the prophage in a lysogenic cell will be activated and progress directly into viral replication and the lytic cycle. The lysogenic phase is depicted in the green section of figure 6.16. Lysogeny is a less deadly form of infection than the full lytic cycle and is thought to be an advancement that allows the virus to spread without killing the host.

Because of the intimate association between the genetic material of the virus and host, phages occasionally serve as transporters of bacterial genes from one bacterium to another and consequently can

* *lysogeny* (ly-soj'-uhn-ee) The potential ability to produce phage.

TABLE 6.4	Comparison of Bacteriophage and Animal Virus Multiplication	
	Bacteriophage	**Animal Virus**
Adsorption	Precise attachment of special tail fibers to cell wall	Attachment of capsid or envelope to cell surface receptors
Penetration	Injection of nucleic acid through cell wall; no uncoating of nucleic acid	Whole virus is engulfed and uncoated, or virus surface fuses with cell membrane; nucleic acid is released.
Synthesis and Assembly	Occurs in cytoplasm Cessation of host synthesis Viral DNA or RNA replicated Viral components synthesized	Occurs in cytoplasm and nucleus Cessation of host synthesis Viral DNA or RNA replicated Viral components synthesized
Viral Persistence	Lysogeny	Latency, chronic infection, cancer
Release from Host Cell	Cell lyses when viral enzymes weaken it.	Some cells lyse; enveloped viruses bud off host cell membrane.
Cell Destruction	Immediate	Immediate or delayed

play a profound role in bacterial genetics. This phenomenon, called transduction, is one way that genes for toxin production and drug resistance are transferred between bacteria (see chapters 9 and 12).

Occasionally, phage genes carried by a bacterial chromosome code for toxins or enzymes that can cause pathology in the human. When a bacterium acquires genes from its temperate phage, it is called **lysogenic conversion** (see figure 6.16). The phenomenon was first discovered in the 1950s in the pathogen that causes diphtheria, *Corynebacterium diphtheriae.* The diphtheria toxin responsible for the severe nature of the disease is a bacteriophage product. *C. diphtheriae* without the phage are relatively harmless. Other bacteria that are made virulent by their prophages are *Vibrio cholerae,* the agent of cholera, and *Clostridium botulinum,* the cause of botulism. On page 173, we described a similar relationship that exists between certain animal viruses and human cells.

The cycles of bacterial and animal viruses illustrate different patterns of viral multiplication, which are summarized in **table 6.4.**

Check & Assess Section 6.5

✔ Bacteriophages vary significantly from animal viruses in their methods of adsorption, penetration, and exit from host cells.

✔ Lysogeny is a condition in which viral DNA is inserted into the bacterial chromosome and remains inactive for an extended period as a prophage. It is replicated right along with the chromosome every time the bacterium divides.

> ✔ Some bacteria express virulence traits that are coded for by the bacteriophage DNA in their chromosomes. This phenomenon is called lysogenic conversion.

22. In simple terms, what does the virus nucleic acid do once it gets into the cell?
23. What processes are involved in bacteriophage assembly?
24. What is a prophage or temperate phage? What is lysogeny?
25. Compare and contrast the main phases in the lytic multiplication cycle in bacteriophages and animal viruses.
26. What is necessary for adsorption?
27. Why is penetration so different in the two groups?

6.6 Techniques in Cultivating and Identifying Animal Viruses

Expected Learning Outcomes

20. Describe the general purposes of cultivating viruses.

21. Compare the methods and uses of cell culture, bird embryos, and live animals in growing viruses.

One problem hampering early animal virologists was their inability to propagate specific viruses routinely in pure culture and in sufficient quantities for their studies. Virtually all of the pioneering attempts at cultivation had to be performed in an organism that was the usual host for the virus. But this method had its limitations. How could researchers have ever traced the stages of viral multiplication if they had been restricted to the natural host, especially in the case of human viruses? Fortunately, systems of cultivation with broader applications were developed, including **in vitro*** cell (or tissue) culture methods and **in vivo*** inoculation of laboratory-bred animals and embryonic bird tissues. Such use of substitute host systems permits greater control, uniformity, and wide-scale harvesting of viruses.

The primary purposes of viral cultivation are: (1) to isolate and identify viruses in clinical specimens; (2) to prepare viruses for vaccines; and (3) to do detailed research on viral structure, multiplication cycles, genetics, and effects on host cells.

Recently, scientists succeeded in artificially creating viruses in the laboratory (**Insight 6.2**).

Using Cell (Tissue) Culture Techniques

The most important early discovery that led to easier cultivation of viruses in the laboratory was the development of a simple and effective way to grow populations of isolated animal cells in culture. These types of in vitro cultivation systems are termed **cell culture**

* *in vitro* (in vee'-troh) L. *vitros*, glass. Experiments performed in test tubes or other artificial environments.
* *in vivo* (in vee'-voh) L. *vivos*, life. Experiments performed in a living body.

or tissue culture. Although these terms are used interchangeably, cell culture is probably a more accurate description. This method makes it possible to propagate most viruses. Much of the virologist's work involves developing and maintaining these cultures. Animal cell cultures are grown in sterile chambers with special media that contain the correct nutrients required by animal cells to survive. The cultured cells grow in the form of a *mono-layer,* a single, confluent sheet of cells that supports viral multiplication and permits close inspection of the culture for signs of infection (**figure 6.19**).

Cultures of animal cells usually exist in the primary or continuous form. *Primary cell cultures* are prepared by placing freshly isolated animal tissue in a growth medium. Embryonic, fetal, adult, and even cancerous tissues have served as sources of primary cultures. A primary culture retains several characteristics of the

(a)

(b)

(c)

Figure 6.19 Microscopic views of normal and infected cell cultures. (a) A normal, undisturbed layer of cultured cells. **(b)** Plaques, which consist of open spaces where cells have been disrupted by viral infection. **(c)** Petri dish culture of *E. coli* bacteria shows macroscopic plaques (clear, round spaces) at points of infection by phages.

INSIGHT 6.2

Artificial Viruses Created!

The media are filled with stories of the debate over the ethics of creating life through cloning techniques. Dolly the cloned sheep and other animals created since then have raised ethical questions about scientists "playing God" when they harvest genetic material from an animal and create an identical organism from it.

Meanwhile, in a much less publicized event, scientists at the State University of New York at Stony Brook succeeded in artificially creating a virus that is virtually identical to natural poliovirus. They used DNA nucleotides they bought "off the shelf" and put them together according to the published poliovirus sequence. They then added an enzyme that would transcribe the DNA sequence into the RNA genome used by poliovirus. They ended up with a virus that was nearly identical to poliovirus. Its capsid, infectivity, and replication in host cells are similar to the natural virus.

The creation of the virus was greeted with controversy, particularly because poliovirus is potentially devastating to human health. The scientists, who were working on a biowarfare defense project funded by the US Department of Defense, argued that they were demonstrating

A duplicate of the 1918 flu virus created by the CDC.

what could be accomplished if information and chemicals fell into the wrong hands.

In yet another re-creation, in 2005 a team of virologists with the CDC resurrected the original H1N1 strain of influenza virus that erupted in 1918 (see photograph). This virus had been one of the most deadly pathogens of all time, possibly killing 50 million people in a single pandemic. Researchers wanted to study it in greater detail and to consider what characteristics led to the severity of the disease it caused. Having a sample of this virus would also make it possible to develop new flu vaccines to control future epidemics. Of course, the virus is held under the highest biosafety levels to prevent it from ever being released.

The prospects of harmful misuse of these technologies in virus manipulation have prompted scientific experts to team with national security and bioethics experts to discuss the pros and cons of the new technology and ways to ensure its acceptable uses.

What basic materials, molecules, and other components would be required to create viruses in a test tube? *Answer available at http://www.mhhe.com/talaro8*

original tissue from which it was derived, but this original line generally has a limited existence. Eventually, it will die out or mutate into a line of cells that can grow by continuous subculture in fresh nutrient medium. One very clear advantage of cell culture is that a specific cell line can be available for viruses with a very narrow host range. Strictly human viruses can be propagated in one of several primary or continuous human cell lines, such as embryonic kidney cells, fibroblasts, bone marrow, and heart cells.

One way to detect the growth of a virus in culture is to observe degeneration and lysis of infected cells in the monolayer of cells. The areas where virus-infected cells have been destroyed show up as clear, well-defined patches in the cell sheet called **plaques*** **(figure 6.19*b*).** Plaques are essentially the macroscopic manifestation of cytopathic effects (CPEs), discussed in section 6.4. This same technique is used to detect and count bacteriophages, because they also produce plaques when grown in soft agar cultures of their host cells **(figure 6.19*c*).** A plaque develops when the viruses released by an infected host cell radiate out to adjacent host cells. As new cells become infected, they die and release more viruses, and so on. As this process continues, the infection spreads gradually and symmetrically from the original point of infection, causing the macroscopic appearance of round, clear spaces that correspond to areas of lysed cells.

Using Bird Embryos

An embryo is an early developmental stage of animals marked by rapid differentiation of cells. Birds undergo their embryonic period within the closed protective case of an egg, which makes an incubating bird egg a nearly perfect system for viral propagation. It is an intact and self-supporting unit, complete with its own sterile environment and nourishment. Furthermore, it furnishes several embryonic tissues that readily support viral multiplication. Every year hundreds of millions of chicken embryos are inoculated to prepare influenza vaccines (chapter 25).

Chicken, duck, and turkey eggs are the most common choices for inoculation. The egg must be injected through the shell, necessitating rigorous sterile techniques to prevent contamination by bacteria and fungi from the air and the outer surface of the shell. The exact tissue inoculated is guided by the type of virus being cultivated and the goals of the experiment **(figure 6.20).**

Viruses multiplying in embryos may or may not cause effects visible to the naked eye. The signs of viral growth include death of the embryo, defects in embryonic development, and localized areas of damage in the membranes, resulting in discrete, opaque spots called pocks (a variant of *pox*). Embryonic fluids and tissues can be prepared for direct examination with an electron microscope. Certain viruses can also be detected by their ability to agglutinate red blood cells (form big clumps) or by their reaction with an antibody of known specificity that will affix to its corresponding virus, if it is present.

* *plaque* (plak) Fr. *placke*, patch or spot.

Using Live Animal Inoculation

Specially bred strains of white mice, rats, hamsters, guinea pigs, and rabbits are the usual choices for animal cultivation of viruses. Invertebrates or nonhuman primates are occasionally used as well. Because viruses can exhibit host specificity, certain animals can propagate a given virus more readily than others. Depending on the

(a)

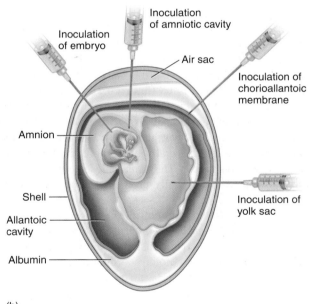

(b)

Figure 6.20 **Cultivating animal viruses in a developing bird embryo.** (a) A technician inoculates fertilized chicken eggs with viruses in the first stage of preparing vaccines. This process requires the highest levels of sterile and aseptic precautions. Influenza vaccine is prepared this way. (b) The shell is perforated using sterile techniques, and a virus preparation is injected into a site selected to grow the viruses. Targets include the allantoic cavity, a sac for embryonic waste removal; the amniotic cavity that cushions and protects the embryo; the chorioallantoic membrane, for embryonic gas exchange; the yolk sac, for the nourishment of the embryo; and the embryo itself.

particular experiment, tests can be performed on adult, juvenile, or newborn animals. The animal is exposed to the virus by injection of a viral preparation or specimen into the brain, blood, muscle, body cavity, skin, or footpads.

6.7 Viral Infection, Detection, and Treatment

Expected Learning Outcomes

22. Discuss the medical impact and importance of viruses.

23. Explain how animal viral infections are treated and detected.

The number of viral infections that occur on a worldwide basis is nearly impossible to measure accurately. Certainly, viruses are the most common cause of acute infections that do not result in hospitalization, especially when one considers widespread diseases such as colds, hepatitis, chicken pox, influenza, herpes, and warts. If one also takes into account prominent viral infections found only in certain regions of the world (dengue fever, Rift Valley fever, and yellow fever), the total could easily exceed several billion cases each year. Although most viral infections do not result in death, some, such as rabies, AIDS, and Ebola, have very high mortality rates, and others can lead to long-term debility (polio, hepatitis). Current research is focused on the possible connection of viruses to chronic afflictions of unknown cause, such as type I diabetes, multiple sclerosis, various cancers, and even conditions such as obesity **(Insight 6.3).**

Because some viral diseases can be life threatening, it is essential to have a correct diagnosis as soon as possible. Obtaining the overall clinical picture of the disease (specific signs) is often the first step in diagnosis. This may be followed by identification of the virus in clinical specimens by means of rapid tests that detect the virus or signs of cytopathic changes in cells or tissues (see CMV herpesvirus, figure 6.15). Immunofluorescence techniques or direct examination with an electron microscope are often used for this (see figure 6.8). Samples can also be screened for the presence of indicator molecules (antigens) from the virus itself. A standard procedure for many viruses is the polymerase chain reaction (PCR), which can detect and amplify even minute amounts of viral DNA or RNA in a sample. In certain infections, definitive diagnosis requires cell culture, embryos, or animals, but this method can be time-consuming and slow to give results. Screening tests can detect specific antibodies that indicate signs of virus infection in a patient's blood. This is the main test for HIV infection (see figure 17.16). Additional details of viral diagnosis are provided in chapter 17.

The nature of viruses has at times been a major impediment to effective therapy. Because viruses are not bacteria, antibiotics aimed at bacterial infections do not work. While there are increasing numbers of antiviral drugs, most of them block virus replication by targeting the function of host cells. This can cause severe side effects. Antiviral drugs are designed to target one of the steps in the viral life cycle you learned about earlier in this chapter. Azidothymide (AZT), a drug used to treat AIDS, targets the nucleic acid synthesis stage. A newer class of HIV drugs, the protease inhibitors, disrupts the final assembly phase of the viral life cycle.

INSIGHT 6.3

Seeking Your Inner Viruses

Would you be alarmed to hear that your cells carry around bits and pieces of fossil viruses? Well, we now know that they do. A fascinating aspect of the virus-host relationship is the extent to which viral genetic material becomes affixed to host chromosomes and is passed on, possibly even for millions of years. We know this from data obtained by the human genome project that sequenced all of the genetic codes on the 46 human chromosomes. While searching through the genome sequences, virologists began to find DNA they identified as viral in origin. So far they have found about 100,000 different fragments of viral DNA. In fact, over 8% of the DNA in human chromosomes comes from viruses!

These researchers are doing the work of molecular fossil hunters, locating and identifying these ancient viruses. Many of them are retroviruses that converted their RNA codes to DNA codes, inserted the DNA into a site in a host chromosome, and then became dormant and did not kill the cell. When this happened in an egg or sperm cell, the virus could be transmitted basically unchanged for hundreds of generations. One of the most tantalizing questions is what effect, if any, such retroviruses might have on modern humans. Some virologists contend that these virus genes would not have been maintained for thousands and even millions of years if they did not serve some function. Others argue that they are just genetic "garbage" that has accumulated over a long human history.

So far, we have only small glimpses of the possible roles of these viruses. One type of endogenous retrovirus has been shown to be intimately involved in forming the human placenta, leading microbiologists to conclude that some viruses have become an essential factor in evolution and development. Other retroviruses may be involved in diseases such as prostate cancer and chronic fatigue syndrome. We now have evidence that some types of viruses can even contribute to obesity in humans.

One of the latest findings is a human bornavirus that belongs to a family of animal viruses that are not retroviruses. Japanese scientists isolated this same virus from monkeys and apes as well, which allows them to fix a timeline for the age of the virus of about 40 million years. Bornaviruses are common among many animal groups, including ground squirrels, elephants, guinea pigs, and horses, where it causes a severe brain disease. Although we do not know what these agents do to humans, some researchers suggest they could be involved in psychoses such as schizophrenia. One thing is for sure: the discovery of this viral baggage will spur many years of research and provide greater understanding of the human genome and its tiny passengers.

Using information you have learned about viruses, explain how viruses could become a permanent component of an organism's genetic material. *Answer available at http://www.mhhe.com/talaro 8*

Another compound that shows some potential for treating and preventing viral infections is a naturally occurring human cell product called *interferon* (see chapters 12 and 14). Vaccines that stimulate immunity are an extremely valuable tool but are available for only a limited number of viral diseases (see chapter 15).

6.8 Prions and Other Nonviral Infectious Particles

Expected Learning Outcomes

24. Describe the properties of nonviral infectious particles.

25. Discuss the importance of prions and viroids and the diseases they cause.

Prions are a group of noncellular infectious agents that are not viruses and really belong in a category all by themselves. The term *prion* is derived from the words **pr**oteinaceous **in**fectious particle to suggest its primary structure—that of a naked protein molecule. They are quite remarkable in being the only biologically active agent that lacks any sort of nucleic acid (DNA or RNA). Up until their discovery about 30 years ago, it was considered impossible that an agent lacking genetic material could ever be infectious or transmissible.

The diseases associated with prions are known as **transmissible spongiform encephalopathies (TSEs).** This description recognizes that the diseases are spread from host to host by direct contact,

contaminated food, or other means. It also refers to the effects of the agent on nervous tissue, which develops a spongelike appearance due to loss of nerve and glial cells (see figure 25.28b). Another pathological effect observed in these diseases is the buildup of tiny protein fibrils in the brain tissue **(figure 6.21a).** Several forms of prion diseases are known in mammals, including scrapie in sheep, bovine spongiform encephalopathy (mad cow disease) in cattle, and wasting disease in elk, deer, and mink. These diseases have a long latent period (usually several years) before the first symptoms of brain degeneration appear. The animals lose coordination, have difficulty moving, and eventually progress to collapse and death **(figure 6.21b).**

Humans are host to similar chronic diseases, including *Creutzfeldt-Jakob syndrome* (CJS), kuru, fatal familial insomnia, and others. In all of these conditions, the brain progressively deteriorates and the patient loses motor coordination, along with sensory and cognitive abilities. There is no treatment and most cases so far have been fatal. Recently, a variant of CJS that appeared in Europe was traced to people eating meat from cows infected with the bovine form of encephalopathy. Several hundred people have developed the disease and over 100 died. This was the first indicator that prions of animals could cause infections in humans. It sparked a crisis in the beef industry and strict controls on imported beef. Although no cases of human disease have occurred in the United States, infected cattle have been reported.

The medical importance of these novel infectious agents has led to a great deal of research into how they function. Researchers have discovered that prion or prionlike proteins are very common in the cell membranes of plants, yeasts, and animals. One widely

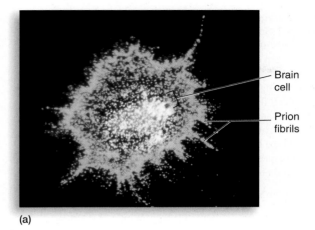

Brain cell

Prion fibrils

(a)

(b)

Figure 6.21 Some effects of prion-based diseases. (a) An isolated brain cell with prion fibrils on its surface. **(b)** Cow in the early phase of bovine spongiform encephalopathy. Symptoms are a staggering gait, weakness, and weight loss.

accepted theory suggests that the prion protein is an abnormal version of one of these proteins. When it comes in contact with a normal protein, the prion can induce spontaneous abnormal folding in the normal protein. Ultimately, the buildup of these abnormal proteins damages and kills the cell.

Another serious issue with prions is their extreme resistance. They are not destroyed by disinfectants, radiation, and the usual sterilization techniques. Even treatments with extreme high temperatures and concentrated chemicals are not always reliable methods to eliminate them. Additional information on prion diseases can be found in chapter 25.

Other fascinating viruslike agents in human disease are defective forms called satellite viruses that actually depend on other viruses for replication. Two remarkable examples are the adeno-associated virus (AAV), which can replicate only in cells infected with adenovirus, and the delta agent, a naked strand of RNA that is expressed only in the presence of the hepatitis B virus and can worsen the severity of liver damage.

Plants are also parasitized by viruslike agents called **viroids** that differ from ordinary viruses by being very small (about one-tenth the size of an average virus) and being composed of only naked strands of RNA, lacking a capsid or any other type of coating. The existence of these unusual agents is one bit of supportive evidence that viruses may have evolved from naked bits of nucleic acid. Viroids are significant pathogens in several economically important plants, including tomatoes, potatoes, cucumbers, citrus trees, and chrysanthemums.

CASE FILE 6 PERSPECTIVE

Viruses must attach to their host cells by their receptor molecules before they can enter the cells and complete the infection cycle. The shape of these receptors is complementary to specific molecules on the host cell. This specificity of attachment limits the kinds of cells they can invade. Once they are bound, viruses go to the next phase, which is entry into the interior of the cell and production of new viruses. The flu virus attaches by its H receptors to the epithelial cells of the upper respiratory tract. From here, it is poised with the assistance of the N receptors to fuse with the cell and continue its path of destruction.

Influenza viruses come in about 144 different subtypes and circulate within many vertebrate groups. As long as the viruses lack the correct receptors for a host, they cannot jump hosts. But influenza viruses are also notorious for altering the shapes of their receptors so that they can invade more than one host. This has happened several times with bird (avian) flu virus and swine flu viruses. In one possible circumstance, a single animal becomes infected with strains of viruses from two different hosts. The recombination of genes from these viruses can give rise to new viruses with receptors that fit all of the hosts. What evidently happened with the 2009 H1N1 strain is that a small change in a receptor was just enough to allow the virus to infect humans.

A pandemic is generally defined as an epidemic that spreads across continents. Health officials consider influenza at pandemic levels when a new virus emerges that many people are susceptible to, when there is sustained human-to-human transmission, and when infection becomes widespread. The rapid spread of a pandemic can be attributed to factors such as the prevalence and rate of travel, the ready spread of the virus through the air by means of respiratory secretions, and the lack of prior exposure and immunity in a large number of people.

One benefit of prior influenza infections is that a person usually develops long-term immunity to that strain of virus. Older members of the population are likely to have encountered viruses in the past that were similar enough to 2009 H1N1 that protective immunities from the earlier infection will also work to counteract the new virus. So, an immune person will be able to inactivate the virus before it can get a foothold. This is the same idea behind the flu vaccines: they prevent the infection by preparing the immune system to respond to the virus in the future.

See chapter 25 for additional details on influenza. Track the latest information on the flu pandemic at: http://www.cdc.gov/h1n1flu/update.htm

TAKE NOTE: WHAT COMES NEXT

We have completed our survey of prokaryotes, eukaryotes, and viruses and have described the significant characteristics of these three groups. Now that we know more about these microbes, we will use chapters 7, 8, and 9 to explore how they maintain themselves, beginning with microbial nutrition (chapter 7) followed by metabolism (chapter 8), and ending with genetics (chapter 9).

Check & Assess Sections 6.6–6.8

✔ Animal viruses must be studied in some type of host cell environment such as cell cultures, bird embryos, or laboratory animals.

✔ Cell and tissue cultures are cultures of host cells grown in special nutrients, using aseptic techniques to exclude unwanted microorganisms.

✔ Virus growth in cell culture is detected by the appearance of plaques.

✔ Eggs are used to cultivate viruses in large quantities for vaccine and research studies.

✔ Inoculation of animals is an alternate method for viruses that do not readily grow in cultures or embryos.

✔ Viruses are easily responsible for several billion infections each year. It is conceivable that many chronic diseases of unknown cause will eventually be connected to viral agents.

✔ Other noncellular agents of disease are: prions, which are not viruses at all but protein fibers that cause spongiform encephalopathies; viroids, extremely small lengths of protein-coated nucleic acid; and satellite viruses, which require larger viruses to cause disease.

✔ Viral infections are difficult to treat because the drugs that attack the viral replication cycle also cause serious side effects in the host.

28. Describe the three main techniques for cultivating viruses.
29. What are the advantages of using cell culture and what are the disadvantages?
30. What is a disadvantage of using live, intact animals or embryos?
31. What are cell lines and monolayers, and how are plaques formed?
32. What is the principal effect of the agent of Creutzfeldt-Jakob disease?
33. How are prions different from viruses, and what are viroids?

Chapter Summary with Key Terms

6.1 Overview of Viruses
A. Viruses, being much smaller than bacteria, fungi, and protozoa, had to be indirectly studied until the 20th century, when they were finally seen with an electron microscope.
B. Scientists don't agree about whether viruses are living or not. They are obligate intracellular parasites.

6.2 The General Structure of Viruses
A. Viruses are infectious particles and not cells; lack organelles and locomotion of any kind; are large, complex molecules; and can be crystalline in form. A virus particle is composed of a nucleic acid core (DNA or RNA, not both) surrounded by a geometric protein shell, or **capsid;** the combination is called a **nucleocapsid;** a capsid is **helical** or **icosahedral** in configuration; many are covered by a membranous **envelope** containing viral protein **spikes.** Complex viruses have additional external and internal structures.

B. *Shapes/Sizes:* Icosahedral, helical, spherical, and cylindrical shaped. Smallest infectious forms range from the largest poxvirus (0.45 mm or 450 nm) to the smallest viruses (0.02 mm or 20 nm).
C. *Nutritional and Other Requirements:* Lack enzymes for processing food or generating energy; are tied entirely to the host cell for all needs **(obligate intracellular parasites).**
D. Viruses are known to parasitize all types of cells, including bacteria, algae, fungi, protozoa, animals, and plants. Each viral type is limited in its host range to a single species or group, mostly due to specificity of adsorption of virus to specific host receptors.

6.3 How Viruses Are Classified and Named
A. The two major types of viruses are *DNA* and *RNA viruses.* These are further subdivided into families, depending on shape and size of capsid, presence or absence of an envelope, whether double- or single-stranded nucleic acid, antigenic similarities, and host cell.
B. Viruses are classified into orders, families, and genera. These groupings are based on virus structure, chemical composition, and genetic makeup.

6.4 Modes of Viral Multiplication
A. *Multiplication Cycle: Animal Cells*
1. The life cycle steps of an animal virus are **adsorption, penetration, synthesis** and **assembly,** and **release** from the host cell. A fully infective virus is a **virion.**
2. Some animal viruses cause chronic and **persistent infections.**
3. Viruses that alter host genetic material may cause *oncogenic* effects.

6.5 The Multiplication Cycle in Bacteriophages
A. Bacteriophages are viruses that attack bacteria. They penetrate by injecting their nucleic acid and are released as virulent phage upon **lysis** of the cell.
B. Some viruses go into a latent, or **lysogenic,** phase in which they integrate into the DNA of the host cell and later may be active and produce a lytic infection.

6.6 Techniques in Cultivating and Identifying Animal Viruses
A. The need for an intracellular habitat makes it necessary to grow viruses in living cells, either in isolated cultures of host cells **(cell culture),** in bird embryos, or in the intact host animal.
B. *Identification:* Viruses are identified by means of cytopathic effects (CPE) in host cells, direct examination of viruses or their components in samples, genetic analysis to detect virus nucleic acid, and growing viruses in culture.

6.7 Viral Infection, Detection, and Treatment
A. *Medical:* Viruses attach to specific target hosts or cells. They cause a variety of infectious diseases, ranging from mild respiratory illnesses (common cold) to destructive and potentially fatal conditions (rabies, AIDS). Some viruses can cause birth defects and cancer in humans and other animals.
B. *Research:* Viruses have become an invaluable tool for studying basic genetic principles. Current research is focused on the possible connection of viruses to afflictions of unknown causes, such as type I diabetes and multiple sclerosis.
C. Viral infections are detected by direct examination of specimens, genetic tests, testing patients' blood, and characteristic symptoms. Some viral infections can be treated with drugs that block viral replication.

6.8 Prions and Other Nonviral Infectious Particles

A. Spongiform encephalopathies are chronic persistent neurological diseases caused by **prions.**

B. Examples of neurological diseases include "mad cow disease" and Creutzfeldt-Jakob disease.

C. Other noncellular infectious agents include satellite viruses and **viroids.**

Multiple-Choice Questions

Select the correct answer from the answers provided. For questions with blanks, choose the combination of answers that most accurately completes the statement.

1. A virus is a tiny infectious
 a. cell
 b. living thing
 c. particle
 d. nucleic acid

2. Viruses are known to infect
 a. plants
 b. bacteria
 c. fungi
 d. all organisms

3. The capsid is composed of protein subunits called
 a. spikes
 b. protomers
 c. virions
 d. capsomers

4. The envelope of an animal virus is derived from the _____ of its host cell.
 a. cell wall
 b. membrane
 c. glycocalyx
 d. receptors

5. The nucleic acid of a virus is
 a. DNA only
 b. RNA only
 c. both DNA and RNA
 d. either DNA or RNA

6. The general steps in a viral multiplication cycle are
 a. adsorption, penetration, synthesis, assembly, and release
 b. endocytosis, replication, assembly, and budding
 c. adsorption, duplication, assembly, and lysis
 d. endocytosis, penetration, replication, maturation, and exocytosis

7. A prophage is a/an _____ stage in the cycle of _____.
 a. latent, bacterial viruses
 b. infective, RNA viruses
 c. early, poxviruses
 d. late, enveloped viruses

8. The nucleic acid of animal viruses enters the host cell through
 a. injection
 b. fusion
 c. endocytosis
 d. b and c

9. In general, RNA viruses multiply in the cell _____, and DNA viruses multiply in the cell _____.
 a. nucleus, cytoplasm
 b. cytoplasm, nucleus
 c. vesicles, ribosomes
 d. endoplasmic reticulum, nucleolus

10. Enveloped viruses carry surface receptors called
 a. buds
 b. spikes
 c. fibers
 d. sheaths

11. Viruses that persist in the cell and cause recurrent disease are considered
 a. oncogenic
 b. cytopathic
 c. latent
 d. resistant

12. Viruses cannot be cultivated in
 a. cell culture
 b. bird embryos
 c. live mammals
 d. blood agar

13. Clear patches in cell cultures that indicate sites of virus infection are called
 a. plaques
 b. pocks
 c. colonies
 d. prions

14. Which of these is *not* a general pattern of virus morphology?
 a. enveloped, helical
 b. naked, icosahedral
 c. enveloped, icosahedral
 d. complex helical

Case File Questions

1. Which receptor of the influenza virus most involved in binding to the respiratory cells?
 a. neuraminidase spikes
 b. hemagglutinin spikes
 c. the viral capsid
 d. the viral envelope

2. What features of the new H1N1 influenza classified it as a pandemic?
 a. It was transmitted from pigs to humans.
 b. It was transmitted from Mexico to the United States.
 c. It was readily transmitted from human to human.
 d. It was spread from the Americas to Europe.
 e. both c and d

Writing to Learn

These questions are suggested as a *writing-to-learn* experience. For each question, compose a one- or two-paragraph answer that includes the factual information needed to completely address the question. Any question listed in a section's Check and Assess may be considered as a writing-to-learn exercise.

1. a. What characteristics of viruses could be used to describe them as life forms?
 b. What makes them more similar to lifeless molecules?

2. HIV attacks only specific types of human cells, such as certain white blood cells and nerve cells. Can you explain why a virus can enter some types of human cells but not others?

3. a. Since viruses lack metabolic enzymes, how can they synthesize necessary components?
 b. What are some enzymes with which the virus is equipped?

4. What dictates the host range of animal viruses?

5. a. Outline the events that took place in the development of the 2009 pandemic H1N1 virus.
 b. How is the influenza virus different in its host range from most other viruses?

Concept Mapping

Appendix E provides guidance for working with concept maps.

1. Supply your own linking words or phrases in this concept map, and provide the missing concepts in the empty boxes.

Critical Thinking Questions

Critical thinking is the ability to reason and solve problems using facts and concepts. These questions can be approached from a number of angles, and in most cases, they do not have a single correct answer.

1. Comment on the possible origin of viruses. Is it not curious that the human cell welcomes a virus in and hospitably removes its coat as if it were an old acquaintance?

2. a. If viruses that normally form envelopes were prevented from budding, would they still be infectious? Why or why not?
 b. If only the RNA of an influenza virus were injected into a cell by itself, could it cause a lytic infection?

3. The end result of most viral infections is death of the host cell.
 a. If this is the case, how can we account for such differences in the damage that viruses do? (Compare the effects of the cold virus with those of the rabies virus.)
 b. Describe the adaptation of viruses that does not immediately kill the host cell and explain what its function may be.

4. a. Given that DNA viruses can actually be carried in the DNA of the host cell's chromosomes, comment on what this phenomenon means in terms of inheritance in the offspring.
 b. Discuss the connection between viruses and cancers, giving possible mechanisms for viruses that cause cancer.

5. One early problem in cultivating HIV was the lack of a cell line that would sustain indefinitely in vitro, but eventually, one was developed. What do you expect were the stages in developing this cell line?

6. a. If you were involved in developing an antiviral drug, what would be some important considerations? (Can a drug "kill" a virus?)
 b. How could multiplication be blocked?

7. Is there such a thing as a "good virus"? Explain why or why not. Consider both bacteriophages and viruses of eukaryotic organisms.

8. Why is an embryonic or fetal viral infection so harmful?

9. How are computer viruses analogous to real viruses?

10. Discuss some advantages and disadvantages of bacteriophage therapy in treating bacterial infections.

11. a. Consult table 6.2, page 168, right-hand column to determine which viral diseases you have had and which ones you have been vaccinated against.
 b. Which viruses would more likely be possible oncoviruses and why would they be?

12. Circle the viral infections on this list: cholera, rabies, plague, cold sores, whooping cough, tetanus, genital warts, gonorrhea, mumps, Rocky Mountain spotted fever, syphilis, rubella, rat bite fever.

Visual Challenge

1. Label the parts of viruses in figures 6.7*d*, 6.8*a*, and 6.10*b*, *c*.

2. Observe table 6.2 figures and determine which viruses are enveloped, which ones are naked, and which ones have an icosahedral capsid. Provide the sizes for the viruses illustrated there.

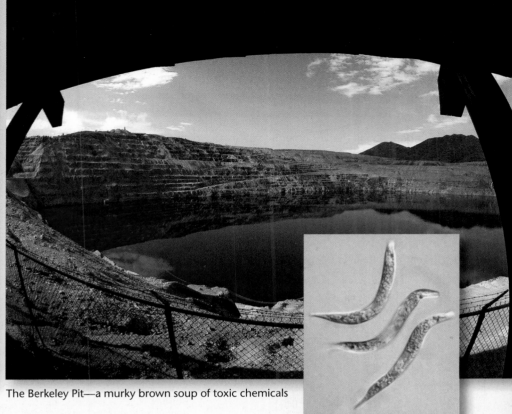

The Berkeley Pit—a murky brown soup of toxic chemicals

Euglena mutabilis—metal eaters extraordinaire

"In spite of the hostile conditions that were seemingly deadly, the water turned out to be teeming with microscopic life."

CASE FILE 7 Life Will Find a Way

Carved into a hillside near Butte, Montana, lies the Berkeley Pit, an industrial body of water that stretches about 1 mile across and contains a volume of close to 40 billion gallons. This site was formerly an open pit copper mine abandoned in 1982 and left to fill up with water seeping out of the local aquifer. At the bottom of the pit lay a massive deposit of mining waste that was like an accident waiting to happen.

A gradual buildup over 20 years transformed the pit into a lake-size cauldron of concentrated chemicals so toxic that it quickly killed any animals or plants that came in contact with it. Substances found in abundance are lead, cadmium, iron, copper, arsenic, and sulfides. Chemical reactions caused the pit to become 10,000 times more acidic than normal freshwater. There is serious concern that water from the pit will contaminate local groundwater and river drainages, creating one of the greatest ecological disasters on record. The federal Environmental Protection Agency designated it as a major superfund cleanup site. So far, the only actions taken have been to divert the drainage water, treat it, and remove some of the heavy metals. But this is a short-term solution to a very long-term problem.

Enter some curious scientists from nearby Montana Tech University. When they began examining samples of the water under a microscope, they were startled at what they found. In spite of the hostile conditions that were seemingly deadly, the water turned out to be teeming with microscopic life. It included an array of very hardy eukaryotes and prokaryotes—a rich community of microorganisms—that had established a foothold in this toxic soup. Rather than being doomed to death, these robust colonists had survived, grown, and spread into available habitats in a relatively short time.

▶ *What types of microbes would one expect to be living in such polluted water?*

▶ *Suggest some possible ways that microbes survive and even thrive under such conditions.*

To continue the case, go to page 201.

7.1 Microbial Nutrition

Expected Learning Outcomes

1. Describe the major environmental factors to which microbes must adapt for survival.
2. Define nutrition and nutrients and their subcategories based on need and quantity.
3. Differentiate between organic and inorganic nutrients.
4. Discuss the origins, types, and functions of bioelements and nutrients.
5. Describe the main categories of nutritional types among organisms.
6. Distinguish different types of autotrophs and their energy sources.
7. Distinguish different types of heterotrophs and their energy sources.

There are millions of habitats on earth, of both natural and human origin. In these settings, microorganisms are exposed to a tremendous variety of conditions that affect their survival. Environmental factors with the greatest impact on microorganisms are nutrient and energy sources, temperature, gas content, water, salt, pH, radiation, and other organisms (**figure 7.1**). Microbes survive in their habitats through the process of gradual adjustment of anatomy and

Gases: the atmosphere is a reservoir for nitrogen, oxygen, and carbon dioxide essential to living processes.

Sunlight supplies the basic source of energy on earth for most organisms. Photosynthesizers can use it directly to produce organic nutrients that feed other organisms. Non photosynthetic organisms extract the energy from chemical reactions to power cell processes.

Soil community

Aquatic microbes

Nutrients are constantly being formed by decomposition and synthesis and released into the environment. Many inorganic nutrients originate from non-living environments such as the air, water, and bedrock.

Complex communities of microbes exist in nearly every place on earth. Microbes residing in these communities must associate physically and share the habitat, often establishing biofilms and other interrelationships.

The temperature of habitats varies to a significant extent among all places on earth, and microbes exist at most points along this wide temperature scale.

pH 0 1 2 3 4 5 6 7 8 9 10 11 12 13 14

Acidic [H⁺] **Neutral** [OH⁻] **Basic (alkaline)**

Acid Base

The acid or base content (pH) can show extreme variations from habitat to habitat. Microbes are the most adaptable organisms with regard to pH.

Figure 7.1 Environmental conditions that influence microbial adaptations. The earth's habitats provide a constant supply of nutrients, gases, and energy; maintain a certain pH and temperature; and establish communities of other organisms to interact with.

INSIGHT 7.1

Dining with an Amoeba

An amoeba gorging itself on bacteria could be compared to a person eating a bowl of vegetable soup, because its nutrient needs are fundamentally similar to that of a human. Most food is a complex substance that contains many different types of nutrients. Some smaller molecules such as sugars can be absorbed directly by the cell; larger food debris and molecules must first be ingested and broken down into molecules that can be absorbed. As nutrients are taken in, they add to a dynamic pool of inorganic and organic compounds dissolved in the cytoplasm. This pool will provide raw materials to be assimilated into the organism's own specialized proteins, carbohydrates, lipids, and other macromolecules used in growth and metabolism.

In the case of an amoeba, food particles are phagocytosed into a vacuole that fuses with a lysosome containing digestive enzymes (E). Smaller subunits of digested macromolecules are transported out of the vacuole into the cell pool and are used in hundreds of cell activities. These range from cell division and synthesis to generating energy and locomotion.

In what general ways do an amoeba and a human differ in their digestion of nutrients? *Answer available at http://www.mhhe.com/talaro8*

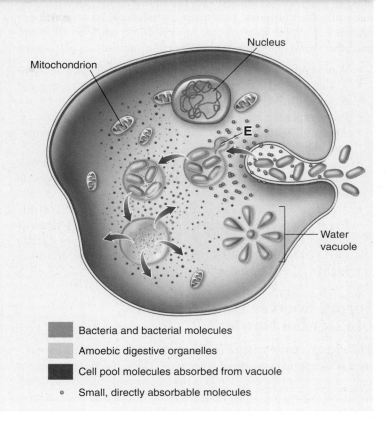

Nucleus

Mitochondrion

E

Water vacuole

- ■ Bacteria and bacterial molecules
- ■ Amoebic digestive organelles
- ■ Cell pool molecules absorbed from vacuole
- • Small, directly absorbable molecules

physiology, a process called **adaptation.*** It is this adaptability that allows microbes to inhabit all parts of the biosphere. The process that selects for favorable adaptations is also a major force behind the evolution of species. With these concepts as a theme for the chapter, we take a closer look at how microbes interact with their environment, how they transport materials, and how they grow.

Nutrition is a process by which chemical substances called **nutrients** are acquired from the environment and used in cellular activities such as metabolism and growth. With respect to nutrition, microbes are not really so different from humans **(Insight 7.1).** Bacteria living deep in a swamp on a diet of inorganic sulfur or protozoa digesting wood in a termite's intestine seem to show radical adaptations, but even these organisms require a constant influx of certain substances from their habitat.

In general, all living things have an absolute need for the bioelements, traditionally listed as carbon, hydrogen, oxygen, phosphorus, potassium, nitrogen, sulfur, calcium, iron, sodium, chlorine, magnesium, and certain other elements.[1] Beyond these basic requirements, microbes have significant differences in the source, chemical form, and amount of the elements they use. Any substance, whether an element or molecule, that must be provided to an

organism is called an **essential nutrient.** Once absorbed, nutrients are processed and transformed into the chemicals of the cell.

Two categories of essential nutrients are **macronutrients** and **micronutrients.** Macronutrients are required in relatively large quantities and play principal roles in cell structure and metabolism. Examples of macronutrients are compounds containing carbon, hydrogen, and oxygen. Micronutrients, or **trace elements,** such as manganese, zinc, and nickel are present in much smaller amounts and are involved in enzyme function and maintenance of protein structure. What constitutes a micronutrient can vary from one microbe to another and often must be determined in the laboratory. This determination is made by deliberately omitting the substance in question from a growth medium to see if the microbe can grow in its absence.

Another way to categorize nutrients is according to their carbon content. Most **organic** nutrients are molecules that contain a basic framework of carbon and hydrogen. Natural organic molecules are nearly always the products of living things. They range from the simplest organic molecule, methane (CH_4), to large polymers (carbohydrates, lipids, proteins, and nucleic acids).

In contrast, an **inorganic** nutrient is composed of an element or elements other than carbon and hydrogen. The natural reservoirs of many inorganic compounds are mineral deposits in the crust of the earth, bodies of water, and the atmosphere. Examples include metals and their salts (magnesium sulfate, ferric nitrate, sodium phosphate), gases (oxygen, carbon dioxide), and water.

* *adaptation* L. *adaptare,* to fit. Changes in structure and function that improve an organism's survival in a given environment.

1. See critical thinking question 1. c for a useful mnemonic device to recall the essential elements.

Chemical Analysis of Cell Contents

To gain insight into a cell's nutritional make up, it can be useful to analyze its chemical composition. The following list is a brief summary of some nutritional patterns in the intestinal bacterium *Escherichia coli*. Some nutrients are absorbed in a ready-to-use form, and others must be synthesized by the cell from simple compounds.

- Water content is the highest of all components (70%).
- About 97% of the dry cell weight is composed of organic compounds.
- Proteins are the most prevalent organic compound.
- About 96% of the cell is composed of six elements (represented by CHONPS).
- Chemical elements are needed in the overall scheme of cell growth, but most of them are available to the cell as compounds and not as pure elements.
- A cell as "simple" as *E. coli* contains on the order of 5,000 different compounds, yet it needs to absorb only a few types of nutrients to synthesize this great diversity. These include $(NH_4)_2SO_4$, $FeCl_2$, $NaCl$, trace elements, glucose, KH_2PO_4, $MgSO_4$, $CaHPO_4$, and water.

Forms, Sources, and Functions of Essential Nutrients

The elements that comprise nutrients ultimately exist in an environmental inorganic reservoir of some type. These reservoirs serve not only as a source of these elements but can be replenished by the activities of organisms. Thus, elements cycle in a pattern from an inorganic form in an environmental reservoir to an organic form in organisms. Organisms may, in turn, serve as a continuing source of that element for other organisms. Eventually, the element is recycled to the inorganic form, usually by the actions of microorganisms. All in all, a tremendous variety of microorganisms are involved in processing the elements. From a larger perspective, the overall cycle of life on this planet depends upon the combined action of several interacting nutritional schemes, each performing a necessary step. In fact, as we shall see in chapter 26 the framework of microbial nutrition underlies the nutrient cycles of all life on earth.

The source of nutrients is extremely varied: Microbes such as photosynthetic bacteria obtain their nutrients entirely in inorganic form from the environment. Others require a combination of organic and inorganic nutrients. For example, parasites that invade and live on the human body derive all essential nutrients from host tissues, tissue fluids, secretions, and wastes.

Refer to **table 7.1** to see an overview of the major bioelements, compounds, their sources, and their importance to microorganisms.

Carbon-based Nutritional Types

The element carbon is so key to the structure and metabolism of all life forms that the source of carbon defines two basic nutritional groups:

- A **heterotroph*** is an organism that must obtain its carbon in an organic form. Because organic carbon usually originates from organisms, heterotrophs are nutritionally dependent on other life forms. Among the common organic molecules that can satisfy this requirement are proteins, carbohydrates, lipids, and nucleic acids. In most cases, these nutrients provide several other elements as well. Some organic nutrients already exist in a form that is simple enough for absorption (for example, monosaccharides and amino acids), but many larger molecules must be digested by the cell before absorption. Not all heterotrophs can use the same organic carbon sources. Some are restricted to a few substrates, whereas others (certain *Pseudomonas* bacteria, for example) are so versatile that they can metabolize hundreds of different substrates.
- An **autotroph*** is an organism that uses inorganic CO_2 as its carbon source. Because autotrophs have the special capacity to convert CO_2 into organic compounds, they are not nutritionally dependent on other living things. We later enlarge on the topic of nutritional types based on carbon and energy sources.

TAKE NOTE: A CARBON CLARIFICATION

It seems worthwhile to emphasize a point about the *extracellular source* of carbon as opposed to the *intracellular function* of carbon compounds. Although a distinction is made between the type of carbon compound cells absorb as nutrients (inorganic or organic), the majority of carbon compounds involved in the normal structure and metabolism of all cells are organic.

Growth Factors: Essential Organic Nutrients

Many fastidious bacteria lack the genetic and metabolic mechanisms to synthesize every organic compound they need for survival. An organic compound such as an amino acid, nitrogenous base, or vitamin that cannot be synthesized by an organism and must be provided as a nutrient is a **growth factor.** For example, all cells require 20 different amino acids for proper assembly of proteins, but many cells cannot synthesize all of them. Those that must be obtained from food are called essential amino acids. A notable example of the need for growth factors occurs in *Haemophilus influenzae,* a bacterium that causes meningitis and respiratory infections in humans. It can grow only when hemin (factor X), NAD^+ (factor V), thiamine and pantothenic acid (vitamins), uracil, and cysteine are provided by another organism or a growth medium.

Classification of Nutritional Types

The earth's limitless habitats and microbial adaptations are matched by an elaborate menu of microbial nutritional schemes. Fortunately, most organisms show consistent trends and can be described by a few general categories (**table 7.2**) and a few selected terms (see Take Note: Terminology).

The main determinants of a microbe's nutritional type are its sources of carbon and energy. We saw earlier that microbes can be defined according to their carbon sources as autotrophs or

* *heterotroph* (het'-uhr-oh-trohf) Gr. *hetero*, other, and *troph*, to feed.

* *autotroph* (aw'-toh-trohf) Gr. *auto*, self, and *troph*, to feed.

TABLE 7.1 Sources and Biological Functions of Essential Elements and Nutrients

Element/Nutrient Forms Found in Nature		Sources/Reservoirs of Compounds	Significance to Cells
Carbon	CO_2 (carbon dioxide) gas CO_3^{2-} (carbonate) Organic compounds	Air (0.036%*) Sediments Living things	CO_2 is produced by respiration and used in photosynthesis; CO_3^{2-} is found in cell walls and skeletons; organic compounds are essential to the structure and function of all organisms and viruses.
Nitrogen	N_2 gas NO_3^- (nitrate) NO_2^- (nitrate) NH_3 (ammonium) Organic nitrogen (proteins, nucleic acids)	Air (79%*) Soil and water Soil and water Soil and water Organisms	Nitrogen gas is available only to certain microbes that fix it into other inorganic nitrogen compounds—nitrates, nitrites, and ammonium—the primary sources of nitrogen for algae, plants, and the majority of bacteria; animals and protozoa require organic nitrogen; all organisms use NH_3 to synthesize amino acids and nucleic acids.
Oxygen	O_2 gas Oxides H_2O	Air (20%*), a major product of photosynthesis	Oxygen gas is necessary for the metabolism of nutrients by aerobes. Oxygen is a significant element in organic compounds and inorganic compounds (see water, sulfates, phosphates, nitrates, carbon dioxide).
Hydrogen	H_2 gas H_2O H_2S (hydrogen sulfide) CH_4 (methane) Organic compounds	Waters, swamps, volcanoes, vents Organisms	Water is the most abundant compound in cells and a solvent for metabolic reactions; H_2, H_2S, and CH_4 gases are produced and used by bacteria and archaea; H^+ ions are the basis for transfers of cellular energy and help maintain the pH of cells.
Phosphorus	PO_4^{3-} (phosphate)	Rocks, Mineral deposits	Phosphate, a key component of DNA and RNA, is critical to the genetic makeup of cells and viruses; also found in ATP and NAD, where it takes part in numerous metabolic reactions; its presence in phospholipids provides stability to cell membranes.
Sulfur	S PO_4^{2-} (sulfate) SH (sulfhydryl)	Mineral deposits, volcanic sediments	Elemental sulfur is oxidized by some bacteria as an energy source; sulfur is found in vitamin B1; sulfhydryl groups are part of certain amino acids, where they form disulfide bonds that shape and stabilize proteins.
Potassium	K^+	Mineral deposits, ocean water	Plays a role in protein synthesis and membrane transport
Sodium	Na^+	Same as potassium	Major participant in membrane actions; maintains osmotic pressure in cells
Calcium	Ca^+	Oceanic sediments, rocks, and minerals	A component of protozoan shells (as $CaCO_3$); stabilizes cell walls; adds resistance to bacterial endospores
Magnesium	Mg^{2+}	Geologic sediments, rocks	A central atom in the chlorophyll molecule; required for function of membranes, ribosomes, and some enzymes
Chloride	Cl^-	Ocean water, salt lakes	May function in membrane transport; required by obligate halophiles to regulate osmotic pressure
Zinc	Zn^{2+}	Rocks, minerals	An enzyme cofactor; regulates eukaryotic genetics
Iron	Fe^{2+}	Rocks, minerals	Essential element for the structure of respiratory proteins (cytochromes)
Micronutrients: copper, cobalt, nickel, molybdenum manganese, iodine		Geologic sediments	Required in tiny amounts to serve as cofactors in specialized enzyme systems of some microbes but not all

*As a portion of the earth's atmosphere.

heterotrophs. Another system classifies them according to their energy source as **phototrophs** or **chemotrophs.** Microbes that photosynthesize are phototrophs, and those that gain energy from chemical compounds are chemotrophs. The terms for carbon and energy source are often merged into a single word for convenience (as in table 7.2). The categories described here are meant to describe only the major nutritional groups and do not include unusual exceptions.

Autotrophs and Their Energy Sources

Autotrophs derive energy from one of two possible nonliving sources: sunlight and chemical reactions involving simple chemicals.

TAKE NOTE: TERMINOLOGY

Much of the vocabulary for describing microbial adaptations is based on some common root words. These are combined in various ways that assist in discussing the types of nutritional or ecological adaptations, as shown in this partial list:

Root	Meaning	Example of Use
troph-	Food, nourishment	Trophozoite—the feeding stage of protozoa
-phile	To love	Extremophile—an organism that has adapted to ("loves") extreme environments
-obe	To live	Microbe—to "live small"
hetero-	Other	Heterotroph—an organism that requires nutrients from other organisms
auto-	Self	Autotroph—an organism that "feeds by itself"
photo-	Light	Phototroph—an organism that uses light as an energy source
chemo-	Chemical	Chemotroph—an organism that uses chemicals rather than light for energy
sapro-	Rotten	Saprobe—an organism that lives on dead organic matter
halo-	Salt	Halophile—an organism that can grow in high-salt environments
thermo-	Heat	Thermophile—an organism that grows at high temperatures
psychro-	Cold	Psychrophile—an organism that grows at cold temperatures
aero-	Air (O_2)	Aerobe—an organism that uses oxygen in metabolism

Modifier terms are also used to specify the nature of an organism's adaptations. **Obligate** or **strict** refers to being restricted to a narrow niche or habitat, such as an obligate thermophile that requires high temperatures to grow. By contrast, **facultative*** means not being so restricted but adapting to a wider range of environmental conditions. A facultative halophile can grow with or without high salt concentration. **Tolerance** is another term that can be used to describe the capacity to survive a range of conditions.

* *facultative* (fak′-uhl-tay-tiv) Gr. *facult*, capability or skill, and *tatos*, most.

TABLE 7.2 Nutritional Categories of Microbes by Energy and Carbon Source

Category/ Carbon Source	Energy Source	Example
Autotroph/CO₂	**Nonliving Environment**	
Photoautotroph	Sunlight	Photosynthetic organisms, such as algae, plants, cyanobacteria
Chemoautotroph	Simple inorganic chemicals	Only certain bacteria, such as methanogens, deep-sea vent bacteria
Heterotroph/ Organic	**Other Organisms or Sunlight**	
Chemoheterotroph	Metabolic conversion of the nutrients from other organisms	Protozoa, fungi, many bacteria, animals
1. Saprobe	Metabolizing the organic matter of dead organisms	Fungi, bacteria (decomposers)
2. Parasite	Utilizing the tissues, fluids of a live host	Various parasites and pathogens; can be bacteria, fungi, protozoa, animals
Photoheterotroph	Sunlight or organic matter	Purple and green photosynthetic bacteria

Photoautotrophs and Photosynthesis Photoautotrophs are photosynthetic; that is, they capture the energy of light rays and transform it into chemical energy that can be used in cell metabolism. In general, photosynthesis relies on special pigments to collect the light and uses the energy to convert CO_2 into simple organic compounds. There are variations in the mechanisms of photosynthesis.

Oxygenic (oxygen-producing) photosynthesis can be summed up by the equation:

$$CO_2 + H_2O \xrightarrow[\text{by chlorophyll}]{\text{Sunlight absorbed}} (CH_2O)_n + O_2$$

in which $(CH_2O)n$ is shorthand for a carbohydrate. This type of photosynthesis occurs in plants, algae, and cyanobacteria and uses chlorophyll as the primary pigment. Carbohydrates formed by the reaction can be used by the cell to synthesize other cell components. This topic is covered in more depth in chapter 8. Since these organisms are the primary producers in most ecosystems, they constitute the basis of food chains by providing nutrition for heterotrophs. This type of photosynthesis is also responsible for maintaining the level of oxygen gas in the atmosphere that is so vital to life.

The other form of photosynthesis is termed *anoxygenic* (no oxygen produced). It may be summarized by the equation:

$$CO_2 + H_2S \xrightarrow[\text{by bacteriochlorophyll}]{\text{Sunlight absorbed}} (CH_2O)_n + S^0 + H_2O$$

Note that this type of photosynthesis is different in several respects. It uses a unique pigment, bacteriochlorophyll; its hydrogen source is hydrogen sulfide gas; and it gives off elemental sulfur as one product. The reactions all occur in the absence of oxygen as well. Common groups of photosynthetic bacteria are the purple and green sulfur bacteria that live in various aquatic habitats, often in mixtures with other photosynthetic microbes (see figure 4.32).

Chemoautotrophy—Life on the Fringes Compared to common, familiar organisms, **chemoautotrophs** have adapted to the most stringent nutritional strategy on earth. These prokaryotes survive totally on inorganic substances such as minerals. They require neither light nor organic nutrients in any form, and they derive energy in diverse and sometimes surprising ways. In very simple terms, they remove electrons from inorganic substrates such as hydrogen gas, hydrogen sulfide, sulfur, or iron and combine them with carbon dioxide and hydrogen. This reaction gives off simple organic molecules and a modest amount of energy to drive the synthetic processes of the cell. Chemoautotrophic bacteria play an important part in recycling inorganic nutrients and elements. For an example of chemoautotrophy and its importance to deep-sea communities, see Insight 7.3.

The Methanogen World Methanogens* are a unique type of chemoautotroph widely distributed in the earth's habitats. Some of these tiny archaeons live in extreme places such as hot springs and frigid oceanic depths (**figure 7.2**). Others are common in soil, swamps, and even the intestines of humans and other animals.

They have a metabolism adapted to producing methane gas (CH_4 or "swamp gas") under anaerobic conditions as summarized by this equation:

$$4H_2 + CO_2 \rightarrow CH_4 + 2\,H_2O$$

Researchers sampling deep underneath the seafloor have uncovered massive deposits of methanogens. The immensity of this community has led one group of scientists to estimate that it comprises nearly one-third of all life on the planet! Evidence points to its extreme age: it has been embedded in the earth's crust for billions of years. Placed under tremendous pressures, the methane it releases becomes frozen into crystals. Some of it serves as a nutrient source for other extreme archaeons, and some of it escapes into the ocean. From this position, it may be a significant factor in the development of the earth's climate and atmosphere. Some microbial ecologists suggest that future rises in sea temperature could increase the melting of these methane deposits. Because methane is an important "greenhouse gas," this process could contribute considerably to ongoing global warming.

* *methanogen* (meth-an-oh-gen) From *methane*, a colorless, odorless gas, and *gennan*, to produce.

(a)

(b)

Figure 7.2 Methane-producing archaeons. Members of this group are primitive prokaryotes with unusual cell walls and membranes. **(a)** Fluorescent image of *Methanospirillum hungatei* (4,000×). This archaeon is found in anaerobic habitats of soil and water. **(b)** *Methanococcus jannaschii,* a motile archaeon that inhabits hot vents in the seafloor and uses hydrogen gas as a source of energy (180,000×).

Heterotrophs and Their Energy Sources

The majority of heterotrophic microorganisms are **chemoheterotrophs** that derive both carbon and energy from organic compounds. Processing these organic molecules by respiration or fermentation releases energy that is stored as ATP. We will explore these topics in chapter 8. An example of chemoheterotrophy is *aerobic respiration,* the principal energy-yielding pathway in animals, most protozoa and fungi, and aerobic bacteria. It can be simply represented by the equation:

$$\text{Glucose } [(CH_2O)_n] + O_2 \rightarrow CO_2 + H_2O + \text{Energy(ATP)}$$

Note that this reaction is complementary to photosynthesis. Here, glucose and oxygen are reactants and carbon dioxide is given off. Indeed, the earth's balance of both energy and metabolic gases greatly depends on this relationship. Chemoheterotrophic microorganisms belong to one of two main categories that differ in how they obtain their organic nutrients: **Saprobes**[2] are free-living microorganisms that feed primarily on organic detritus from dead organisms, and **parasites** ordinarily derive nutrients from the cells or tissues of a host.

Saprobic Microorganisms

Saprobes occupy a niche as decomposers of plant litter, animal matter, and dead microbes. If not for the work of decomposers, the earth would gradually fill up with organic material and the nutrients it contains would not be recycled. Most saprobes, notably bacteria and fungi, have a rigid cell wall and cannot engulf large particles of food. To compensate, they release enzymes to the extracellular environment and digest the food particles into smaller molecules that can be transported into the cell (**figure 7.3**).

Many saprobes exist strictly on dead organic matter in soil and water and are unable to adapt to the body of a live host. This group includes free-living protozoa, fungi, and certain bacteria. Apparently, there are fewer of these microbes than was once thought, and many supposedly nonpathogenic saprobes can adapt to and invade a susceptible host. When a saprobe does infect a host, it is considered a *facultative parasite*. Such an infection usually occurs when the host is compromised, and the microbe is considered an *opportunistic pathogen*. For example, although its natural habitat is soil and water, *Pseudomonas aeruginosa* frequently causes infections in patients when it is carried into the hospital environment.

Parasitic Microorganisms

Parasites grow in or on the body of a host, which they harm to some degree. Because parasites can damage tissues (disease) or even cause death, they are also called **pathogens.** Parasites range from viruses to helminth worms, and they can live on the body (ectoparasites), in the organs and tissues (endoparasites), or even within cells (intracellular parasites, the most extreme type). Although there are several degrees of parasitism, the more successful parasites generally have no fatal effects and may eventually evolve to a less harmful relationship with their host (see section 7.4).

Obligate parasites (for example, the leprosy bacillus and the syphilis spirochete) are so dependent that they are unable to grow outside of a living host. Less strict parasites such as the gonococcus and pneumococcus can be cultured artificially if provided with the correct nutrients and environmental conditions.

Obligate intracellular parasitism is an extreme but relatively common mode of life. Microorganisms that spend all or part of their life cycle inside a host cell include viruses, a few bacteria (rickettsias, chlamydias), and certain protozoa (apicomplexa). Contrary to what one may think, the cell interior is not completely without hazards, and microbes must overcome some difficult

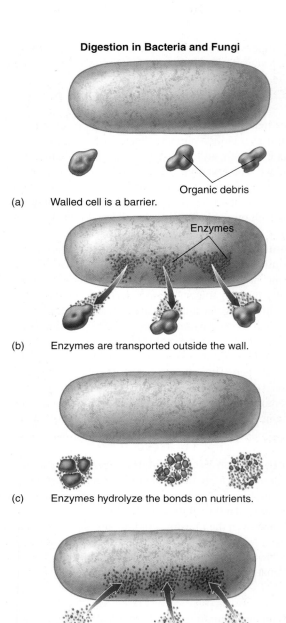

Digestion in Bacteria and Fungi

(a) Walled cell is a barrier.

(b) Enzymes are transported outside the wall.

(c) Enzymes hydrolyze the bonds on nutrients.

(d) Smaller molecules are transported across the wall and cell membrane into the cytoplasm.

Figure 7.3 Extracellular digestion in a saprobe with a cell wall (bacterium or fungus). (a) A walled cell is inflexible and cannot engulf large pieces of organic debris. **(b)** In response to a usable substrate, the cell synthesizes enzymes that are transported across the wall into the extracellular environment. **(c)** The enzymes hydrolyze the bonds in the debris molecules. **(d)** Digestion produces molecules small enough to be transported into the cytoplasm.

challenges. They must find a way into the cell, keep from being destroyed, not destroy the host cell too soon, multiply, and find a way to infect other cells. Intracellular parasites obtain different substances from the host cell, depending on the group. Viruses are the most extreme, parasitizing the host's genetic and metabolic machinery. Rickettsias are primarily energy parasites, and the malaria protozoan is a hemoglobin parasite.

2. Synonyms are *saprotroph* and *saprophyte*. We prefer to use the terms *saprobe* and *saprotroph* because they are more consistent with other terminology.

Check & Assess Section 7.1

✔ Microbes exist in every known natural habitat on earth.

✔ Microbes show enormous capacity to adapt to environmental factors.

✔ Nutrition is a process by which all living organisms obtain substances from their environment to convert to metabolic uses.

✔ Although the chemical form of nutrients varies widely, all organisms require six bioelements—carbon, hydrogen, oxygen, nitrogen, phosphorus, and sulfur—to survive, grow, and reproduce.

✔ Nutrients are categorized by the amount required (macronutrients or micronutrients), by chemical structure (organic or inorganic), and by their importance to the organism's survival (essential or nonessential).

✔ Microorganisms are classified both by the chemical form of their nutrients and the energy sources they utilize.

✔ Autotrophs can exist solely on inorganic nutrients, whereas heterotrophs require both inorganic and organic nutrients.

✔ Energy sources for microbes may come from light or chemicals.

✔ Heterotrophic microbes rely on an organic source for both energy and nutrients.

1. Differentiate between micronutrients, macronutrients, and essential nutrients.
2. List the general functions of the essential bioelements in the cell.
3. Define growth factors, and give examples of them.
4. Name some functions of metallic ions in cells.
5. Compare autotrophs and heterotrophs with respect to the form of carbon-based nutrients they require.
6. Describe the nutritional strategy of two types of chemoautotrophs described in the chapter.
7. Contrast chemoautotrophs and chemoheterotrophs as to carbon and energy sources and other unique strategies they may have.
8. What are the main differences between saprobes and parasites?

7.2 Transport: Movement of Substances across the Cell Membrane

Expected Learning Outcomes

8. Describe the basic factors in cell transport.

9. Discuss diffusion and passive transport systems.

10. Define *osmosis* and describe varying osmotic conditions.

11. Analyze adaptations microbes make in response to osmosis.

12. Describe the features of active transport and differentitate among its mechanisms.

A microorganism's habitat provides necessary substances—some abundant, others scarce—that must still be taken into the cell. For survival cells must transport waste materials out of the cell (and into the environment). Whatever the direction, transport occurs across the **cell membrane,** the structure specialized for this role. This is true even in organisms with cell walls (bacteria, algae, and fungi), because the cell wall is usually only a partial, nonselective barrier. In this section, we discuss some important physical forces in cell transport.

Diffusion and Molecular Motion

All atoms and molecules, regardless of being in a solid, liquid, or gas, are in continuous movement. As the temperature increases, the molecular movement becomes faster. This is called "thermal" movement. In any solution, including cytoplasm, these moving molecules cannot travel very far without having collisions with other molecules and, therefore, will bounce off each other like millions of pool balls every second **(figure 7.4).** As a result of each collision, the directions of the colliding molecules are altered and the direction of any one molecule is unpredictable and considered random. An example would be a situation in which molecules of a substance are more concentrated in one area than another. Just by random thermal movement, the molecules will become dispersed away from an area of higher concentration to an area of lower

How Molecules Diffuse in Aqueous Solutions

Figure 7.4 Diffusion of molecules in aqueous solutions. A high concentration of sugar exists in the cube at the bottom of the liquid. An imaginary molecular view of this area shows that sugar molecules are in a constant state of motion. Those at the edge of the cube diffuse from the concentrated area into more dilute regions. As diffusion continues, the sugar will spread evenly throughout the aqueous phase, and eventually there will be no gradient. At that point, the system is said to be in equilibrium.

concentration. In time, this will evenly distribute the molecules in the solution. This net movement of molecules down their concentration gradient by random thermal motion is known as **diffusion.** It can be demonstrated by a variety of simple observations. A drop of perfume released into one part of a room is soon smelled in another part, or a lump of sugar in a cup of tea spreads through the whole cup without stirring (figure 7.4).

Diffusion is a driving force in cell activities, but its effects are greatly controlled by membranes. Two special cases are osmosis and facilitated diffusion. Both processes are considered a type of **passive transport.** This means that the cell does not expend extra energy for them to function. The inherent energy of the molecules moving down a gradient does the work of transport.

The Diffusion of Water: Osmosis

Diffusion of water through a selectively permeable membrane, a process called **osmosis,** * is a physical phenomenon that is easily demonstrated in the laboratory with nonliving materials. Simple experiments provide a model of how cells deal with various solute concentrations in aqueous solutions **(figure 7.5).** In an osmotic system, the membrane is *selectively,* or *differentially, permeable,* having passageways that allow free diffusion of water but can block

certain dissolved molecules. When this type of membrane is placed between solutions of differing concentrations where the solute cannot pass across (protein, for example), then under the laws of diffusion, water will diffuse at a faster rate from the side that has more water to the side that has less water. As long as the concentrations of the solutions differ, one side will experience a net loss of water and the other a net gain of water until equilibrium is reached and the rate of diffusion is equalized.

Osmosis in living systems is similar to the model shown in figure 7.5. Living membranes generally block the entrance and exit of larger molecules and permit free diffusion of water. Because most cells contain and are surrounded by some sort of aqueous solution, osmosis can have far-reaching effects on cellular activities and survival. Depending on the water content of a cell as compared with its environment, a cell can gain or lose water, or it may remain unaffected. Terms that we use for describing these conditions are *isotonic, hypotonic,* and *hypertonic*[3] **(figure 7.6).**

Under **isotonic*** conditions, the environment is equal in solute concentration to the cell's internal environment; and because diffusion of water proceeds at the same rate in both directions, there is no net change in cell volume. Isotonic solutions are generally the most stable environments for cells, because they are already in an osmotic steady state with the cell. Parasites living in host tissues are most likely to be living in isotonic habitats.

* *osmosis* (oz-moh′-sis) Gr. *osmos,* impulsion, and *osis,* a process.

3. It will help you to recall these osmotic conditions if you remember that the prefixes *iso-, hypo-,* and *hyper-* refer to the environment *outside* of the cell.

* *isotonic* (eye-soh-tahn′-ik) Gr. *iso,* same, and *tonos,* tension.

Solute

Water

a. Inset shows a close-up of the osmotic process. The gradient goes from the outer container (higher concentration of H_2O) to the sac (lower concentration of H_2O). Some water will diffuse the opposite direction but the net gradient favors osmosis into the sac.

b. As the H_2O diffuses into the sac, the volume increases and forces the excess solution into the tube, which will rise continually.

c. Even as the solution becomes diluted, there will still be osmosis into the sac. Equilibrium will not occur because the solutions can never become equal. (Why?)

Figure 7.5 Model system to demonstrate osmosis. Here we have a solution enclosed in a membranous sac and attached to a hollow tube. The membrane is permeable to water (solvent) but not to solute. The sac is immersed in a container of pure water and observed over time.

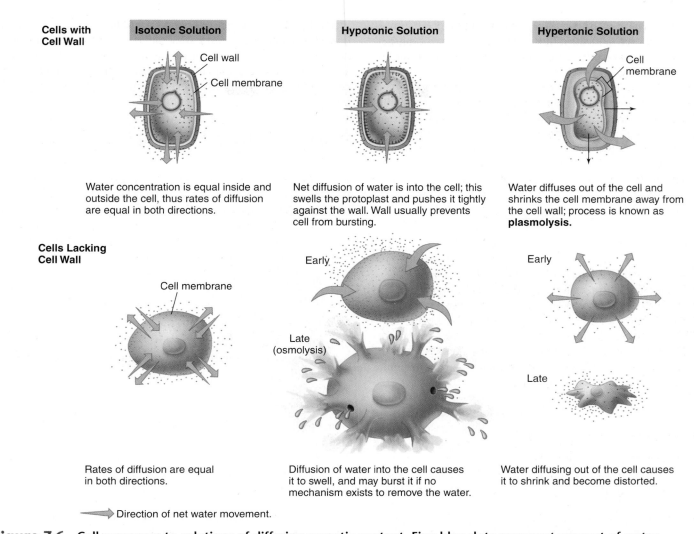

Figure 7.6 Cell responses to solutions of differing osmotic content. Fine blue dots represent amount of water.

Under **hypotonic*** conditions, the solute concentration of the external environment is lower than that of the cell's internal environment. Pure water provides the most hypotonic environment for cells because it has no solute. Because the net direction of osmosis is from the hypotonic solution into the cell, cells without walls swell and can burst when exposed to this condition.

Slight hypotonicity is tolerated quite well by most bacteria because of their rigid cell walls. The light flow of water into the cell keeps the cell membrane fully extended and the cytoplasm full. This is the optimum condition for the many processes occurring in and on the membrane.

A cell in a **hypertonic*** environment is exposed to a solution with higher solute concentration than its cytoplasm. Because hypertonicity will force water to diffuse out of a cell, it is said to create high *osmotic pressure* or potential. In cells with a wall, water loss causes shrinkage of the protoplast away from the wall, a condition called **plasmolysis.*** Although the whole cell does not collapse, this event

can still damage and even kill many kinds of cells. The effect on cells lacking a wall is to shrink down and usually to collapse (see figure 7.6). The growth-limiting effect of hypertonic solutions on microbes is the principle behind using concentrated salt and sugar solutions as preservatives for food, such as in salted hams and fish.

Adaptations to Osmotic Variations in the Environment

Let us now see how specific microbes have adapted osmotically to their environments. In general, isotonic conditions pose little stress on cells, so survival depends on counteracting the adverse effects of hypertonic and hypotonic environments.

An alga and an amoeba living in fresh pond water are examples of cells that live in constantly hypotonic conditions. The rate of water diffusing across the cell membrane into the cytoplasm is rapid and constant, and the cells would die without a way to adapt. As with bacteria, the majority of algae have a cell wall that protects them from bursting even as the cytoplasmic membrane becomes **turgid***

* *hypotonic* (hy-poh-tahn´-ik) Gr. *hypo,* under, and *tonos,* tension.
* *hypertonic* (hy-pur-tahn´-ik) Gr. *hyper,* above, and *tonos,* tension.
* *plasmolysis* (plaz´-moh-ly´-sis).

* *turgid* (ter´-jid) A condition of being swollen or congested.

from pressure. The amoeba has no cell wall to protect it, so it must expend energy to deal with the influx of water. This is accomplished with a water, or contractile, vacuole that siphons excess water out of the cell like a tiny pump.

A microbe living in a high-salt environment (hypertonic) has the opposite problem and must either restrict its loss of water to the environment or increase the salinity of its internal environment. Halobacteria living in the Great Salt Lake and the Dead Sea actually absorb salt to make their cells isotonic with the environment; thus, they have a physiological need for a high-salt concentration in their habitats (see halophiles on page 203).

The Movement of Solutes across Membranes

Simple diffusion works well for movement of small nonpolar molecules such as oxygen or lipid soluble molecules that readily pass through membranes. But many substances that cells require are polar and ionic chemicals with greatly reduced permeability. Simple diffusion alone would be unable to transport these substances. One way that cells have adapted to this limitation involves a process called **facilitated diffusion (figure 7.7).** This passive transport mechanism utilizes a carrier protein in the membrane that will bind a specific substance. This binding changes the conformation of the carrier proteins in a way that facilitates movement of the substance across the membrane. Once the substance is transported, the carrier protein resumes its original shape and is ready to transport again. These carrier proteins exhibit **specificity,** which means that they bind and transport only a single type of molecule. For example, a carrier protein that transports sodium will not bind glucose. A second characteristic exhibited by facilitated diffusion is *saturation.* The rate of transport of a substance is limited by the number of binding sites on the transport proteins. As the substance's concentration

increases, so does the rate of transport until the concentration of the transported substance causes all of the transporters' binding sites to be occupied. Then the rate of transport reaches a steady state and cannot move faster despite further increases in the substance's concentration. Facilitated diffusion is more important to eukaryotic than prokaryotic cells.

Active Transport: Bringing in Molecules against a Gradient

Free-living microbes exist under relatively nutrient-starved conditions and cannot rely completely on slow and rather inefficient passive transport mechanisms. To ensure a constant supply of nutrients and other required substances, microbes must capture those that are in low concentrations and actively transport them into the cell. Features inherent in **active transport** systems are

1. the transport of nutrients against the diffusion gradient or in the same direction as the natural gradient but at a rate faster than by diffusion alone,
2. the presence of specific membrane proteins (permeases and pumps; **figure 7.8a),** and
3. the expenditure of additional cellular energy in the form of ATP-driven uptake. Examples of substances transported actively are monosaccharides, amino acids, organic acids, phosphates, and metal ions.

Some freshwater algae have such efficient active transport systems that an essential nutrient can be found in intracellular concentrations that are 200 times greater than those in the habitat.

Carrier-mediated active transport functions with specific membrane proteins that bind both ATP and the molecules to be transported. Release of energy from the ATP drives the movement of the molecule through the protein carrier. This can occur in either direction. Some bacteria transport certain sugars, amino acids, vitamins, and phosphate into the cell by this mechanism. Other bacteria can actively pump drugs out of the cell, thereby providing them resistance to the drugs. Other active transport pumps can rapidly carry ions such as K^+, Na^+, and H^+ across the membrane. This behavior is particularly important in mitochondrial ATP formation, as described in chapter 8.

Another type of active transport, **group translocation,** couples the transport of a nutrient with its conversion to a substance that is immediately useful inside the cell **(figure 7.8b).** This method is used by certain bacteria to transport sugars (glucose, fructose) while simultaneously adding molecules such as phosphate that prepare them for the next stage in metabolism.

Endocytosis: Eating and Drinking by Cells

Some cells can transport large molecules, particles, liquids, or even other cells across the cell membrane. Because the cell usually expends energy to carry out this movement, it is also a form of active transport. The substances transported do not pass physically through the membrane but are carried into the cell by **endocytosis.** First the cell encloses the substance in its membrane, simultaneously forming a vacuole and engulfing it **(figure 7.8c).** Amoebas and certain white blood cells ingest whole cells or large solid matter by a type of

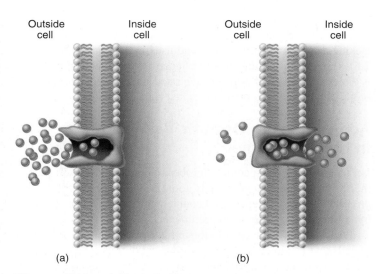

Outside cell Inside cell Outside cell Inside cell

(a) (b)

Figure 7.7 Facilitated diffusion. Facilitated diffusion involves the attachment of a molecule to a specific protein carrier. **(a)** Bonding of the molecule causes a conformational change in the protein that facilitates the molecule's passage across the membrane. **(b)** The membrane protein releases the molecule into the cell interior. The cell does not have to expend energy for transport.

(a) Carrier-mediated active transport. The membrane proteins (permeases) have attachment sites for essential nutrient molecules. As these molecules bind to the permease, energy from ATP pumps them into the cell's interior through special membrane protein channels. Microbes have these systems for transporting various ions (sodium, iron) and small organic molecules.

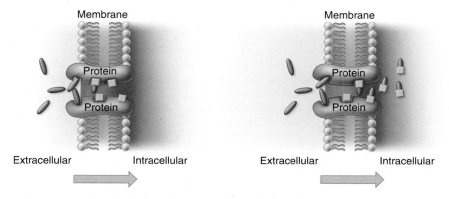

(b) In group translocation, the molecule is actively captured, but along the route of transport, it is chemically altered. By coupling transport with synthesis, the cell conserves energy.

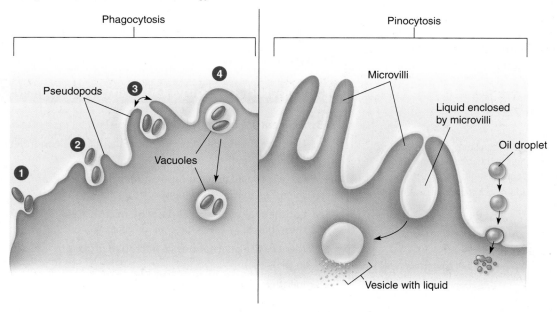

(c) Endocytosis. With phagocytosis, solid particles are engulfed by large cell extensions called pseudopods (1000×). With pinocytosis, fluids and/or dissolved substances are enclosed in vesicles by very fine cell protrusions called microvilli (3000×). Oil droplets fuse with the membrane and are released directly into the cell.

Figure 7.8 **Active transport.** In active transport mechanisms, energy is expended (ATP) to transport the molecule across the cell membrane.

TABLE 7.3	Summary of Transport Processes in Cells			
General Process	**Nature of Transport**	**Examples**	**Description**	**Qualities**
Passive	Energy expenditure by the cell is not required. Substances exist in a gradient and move from areas of higher concentration toward areas of lower concentration in the gradient.	Diffusion osmosis	A fundamental property of atoms and molecules that exist in a state of random motion	Nonspecific Brownian movement Movement of small uncharged molecules across membranes
		Facilitated diffusion	Molecule binds to a carrier protein in membrane and is carried across to other side.	Molecule specific; transports both ways Transports sugars, amino acids, water
Active	Energy expenditure is required. Molecules need not exist in a gradient. Rate of transport is increased. Transport may occur against a concentration gradient.	Carrier-mediated active transport	Atoms or molecules are pumped into or out of the cell by specialized receptors; driven by ATP or the proton motive force	Transports simple sugars, amino acids, inorganic ions (Na^+, K^+)
		Group translocation	Molecule is moved across membrane and simultaneously converted to a metabolically useful substance.	Alternate system for transporting nutrients (sugars, amino acids)
		Bulk transport	Mass transport of large particles, cells, and liquids by engulfment and vesicle formation	Includes endocytosis, phagocytosis, pinocytosis

endocytosis called **phagocytosis.** Liquids, such as oils or molecules in solution, enter the cell through **pinocytosis.** The mechanisms for transport of molecules into cells are summarized in **table 7.3.**

Check & Assess Section 7.2

✔ Substances are transported into microorganisms by two kinds of processes: active transport that expends cellular energy and passive transport that occurs without added energy input.

✔ The molecular size and concentration of a nutrient determine the method of transport.

9. Compare and contrast passive and active forms of transport, using examples of what is being transported and the requirements for each.

10. How are phagocytosis and pinocytosis similar? How are they different?

11. Explain the differences between facilitated diffusion and group translocation.

12. Compare the effects of isotonic, hypotonic, and hypertonic solutions on an ameba and on a bacterial cell. If a cell lives in a hypotonic environment, what will occur if it is placed in a hypertonic one? Answer for the opposite case as well.

7.3 Environmental Factors That Influence Microbes

Expected Learning Outcomes

13. Differentiate between habitat and niche.

14. Describe the range of temperatures a microbe can function within.

15. Explain the adaptive temperature groups with examples of microbes that exist in them.

16. List the major gases and describe microbial requirements for these gases.

17. Identify the adaptations of microbial groups to variations in pH.

18. Identify microbial adaptations to osmotic pressure.

Microbes are exposed to a wide variety of environmental factors that affect growth and survival. Microbial ecology focuses on ways that microorganisms deal with or adapt to such factors as heat, cold, gases, acid, radiation, osmotic and hydrostatic pressures, and even other microbes. Adaptation involves a complex adjustment in biochemistry or genetics that enables long-term survival and growth. The term biologists use to describe the totality of adaptations

organisms make to their habitats is **niche.*** For most microbes, environmental factors fundamentally affect the function of metabolic enzymes. Thus, survival is largely a matter of whether the enzyme systems of microorganisms can continue to function even in a changing environment. See **Insight 7.2** to discover the broad range of conditions that support microbial life.

Adaptations to Temperature

Microbial cells cannot control their temperature and therefore assume the ambient temperature of their natural habitats. To survive, they must adapt to whatever temperature variations are encountered in their habitat. The range of temperatures for microbial growth can be expressed as three **cardinal temperatures.** The **minimum temperature** is the lowest temperature that permits a microbe's continued growth and metabolism; below this temperature, its activities are inhibited. The **maximum temperature** is the highest temperature at which growth and metabolism can proceed. If the temperature rises slightly above maximum, growth will stop. If it continues to rise beyond that point, the enzymes and nucleic acids will eventually become permanently inactivated—a condition known as denaturation—and the cell will die. This is why heat works so well as an agent in microbial control. The **optimum temperature** covers a small range, intermediate between the minimum and maximum, which promotes the fastest rate of growth and metabolism. Only rarely is the optimum a single point.

Depending on their natural habitats, some microbes have a narrow cardinal range, others a broad one. Some strict parasites will not grow if the temperature varies more than a few degrees below or above the host's body temperature. For instance, rhinoviruses (one cause of the common cold) multiply successfully only in tissues that are slightly below normal body temperature (33°C to 35°C). Other microbes are not so limited. Strains of *Staphylococcus aureus* grow within the range of 6°C to 46°C, and the intestinal bacterium *Enterococcus faecalis* grows within the range of 0°C to 44°C.

Another way to express temperature adaptation is to describe whether an organism grows optimally in a cold, moderate, or hot temperature range. The terms used for these ecological groups are *psychrophile, mesophile,* and *thermophile* (**figure 7.9**).

A **psychrophile*** is a microorganism that has an optimum temperature below 15°C and is capable of growth at 0°C. It is obligate with respect to cold and generally cannot grow above 20°C. Laboratory work with true psychrophiles can be a real challenge. Inoculations have to be done in a cold room because ordinary room temperature can be lethal to these organisms. Unlike most laboratory cultures, storage in the refrigerator incubates rather than inhibits them. As one might predict, the habitats of psychrophilic bacteria, fungi, and algae are lakes, snowfields (**figure 7.10**), polar ice, and the deep ocean. Rarely, if ever, are they pathogenic. True psychrophiles must be distinguished from *psychrotrophs* or *facultative psychrophiles* that grow slowly in cold but have an optimum temperature above 20°C. Psychrotrophs such as *Staphylococcus aureus* and *Listeria monocytogenes* are a concern because they grow in refrigerated food and cause food-borne illness.

The majority of medically significant microorganisms are **mesophiles,*** organisms that grow at intermediate temperatures. Although an individual species can grow at the extremes of 10°C or 50°C, the optimum growth temperatures (optima) of most mesophiles fall into the range of 20°C to 40°C. Organisms in this group inhabit animals and plants as well as soil and water in temperate, subtropical, and tropical regions. Most human pathogens have optima somewhere between 30°C and 40°C (human body temperature is 37°C). *Thermoduric* microbes, which can survive short exposure to high temperatures but are normally mesophiles, are common contaminants of heated or pasteurized foods (see chapter 11). Examples include heat-resistant cysts such as *Giardia* or sporeformers such as *Bacillus* and *Clostridium.*

* *mesophiles* (mez'-oh-fylz)

Figure 7.9 **Ecological groups by temperature of adaptation.** Psychrophiles can grow at or near 0°C and have an optimum below 15°C. As a group, mesophiles can grow between 10°C and 50°C, but their optima usually fall between 20°C and 40°C. Generally speaking, thermophiles require temperatures above 45°C and grow optimally between this temperature and 80°C. Note that the extremes of the ranges can overlap to an extent.

Figure 7.10 **Red snow. (a)** The surface of an Alaskan glacier provides a perfect habitat for psychrophilic photosynthetic organisms such as *Chlamydomonas nivalis.* **(b)** Microscopic views of this snow alga, actually classified as a "green" alga although a red pigment dominates at this stage of its life cycle (600×).

* *niche* (nitch) Fr. *nichier,* to nest.
* *psychrophile* (sy'-kroh-fyl)

INSIGHT 7.2

Life in the Extremes

Any extreme habitat—whether hot, cold, salty, acidic, alkaline, high pressure, arid, oxygen-free, or toxic—is likely to harbor microorganisms with special adaptations to these conditions. Although in most instances the inhabitants are archaeons and bacteria, certain fungi, protozoans, algae, and even viruses are also capable of living in severe habitats. Microbiologists have termed such remarkable organisms **extremophiles.**

Extreme Temperatures

Some of the most extreme habitats are hot springs, geysers, volcanoes, and ocean vents, all of which support flourishing microbial populations. Temperatures in these regions range from 50°C to well above the boiling point of water, with some ocean vents even approaching 350°C. Many heat-adapted microbes are ancient archaeons whose genetics and metabolism are extremely modified for this mode of existence. A unique ecosystem based on hydrogen sulfide–oxidizing bacteria exists in the hydrothermal vents lying along deep oceanic ridges (see Insight 7.3). Microbiologists have recently uncovered large numbers of thermophilic viruses in the hot springs of Yellowstone National Park. Many of these viruses are bacteriophages of archaeons that can survive temperatures near boiling and a pH of 1 to 2.

A large part of the earth exists at cold temperatures. Microbes settle and grow throughout the Arctic and Antarctic and in the deepest parts of the ocean in near-freezing temperatures. Several species of algae and fungi thrive on the surfaces of snow and glacier ice (see figure 7.10). More surprising still is that some bacteria and algae flourish in the sea ice of Antarctica. Although the ice appears to be completely solid, it is honeycombed by various-size pores and tunnels filled with a solution that is −20°C. These frigid microhabitats harbor a microcosm of planktonic bacteria, algae, and predators that feed on them.

High Salt, Acidity, and Alkalinity

The growth of most microbial cells is inhibited by high amounts of salt; for this reason, salt is a common food preservative. But there are many salt lovers, including rich communities of bacteria and algae living in oceans, salt lakes, and inland seas, some of which are saturated with salt (30%). Microbiologists recently discovered salt pockets that supported archaeons living in water 100 times more concentrated than seawater.

Brave "Old" World

When samples of ancient Antarctic ice (several million years old) were brought into the lab, they yielded up living bacteria. Some of the oldest isolates were unusual, very slow-growing forms, but others were similar to modern bacteria.

"Conan the Bacterium"

A tiny gram-positive coccus, *Deinococcus radiodurans*, has been called "Conan the bacterium" because of its capacity to survive extreme drying and very high levels of radiation. This bacterium is widespread in many habitats, but its resistance to radiation sets it apart from every other organism ever known. The DNA and proteins of most organisms are very sensitive to X rays, gamma rays, and other high-energy radiation, which is why they are so damaging to cells. But we recently learned its secret. *Deinococcus* carries around several copies of its genome, making it easily replaceable, and it contains radiation-resistant enzymes that help it rapidly repair DNA damage. This microbe is being studied for its potential in bioremediation of radioactive waste and as a model organism for Mars studies.

Numerous species have carved a niche for themselves in the deepest layers of mud, swamps, and oceans, where oxygen gas and sunlight cannot penetrate. The predominant living things in the deepest part of the oceans (10,000 m or below) are pressure- and cold-loving microorganisms. Even parched zones in sand dunes and deserts harbor a hardy brand of microbes; and thriving bacterial populations can be found in petroleum, coal, and mineral deposits containing copper, zinc, gold, and uranium. Microbes that can adapt and survive under such lethal conditions are called toxophiles.

As a rule, a microbe that has adapted to an extreme habitat will die if placed in a moderate one. And, except for rare cases, none of the organisms living in these extremes is a pathogen because the human body is a rather hostile habitat for them.

Provide some examples of extreme habitats created by human technologies that would support microbes. *Answer available at http://www.mhhe.com/talaro8*

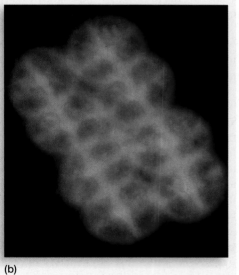

(a) A newly discovered thermal virus from a Yellowstone National Park hot spring. Bar is 100 nm. **(b)** A fluorescent micrograph of *Deinococcus radiodurans,* showing its packet arrangement, cell walls (red), and chromosomes (green).

The unique ecology of the pit has provided numerous topics for study by the Montana researchers. They conducted additional laboratory explorations and discovered around 130 different species of microbes, many of them never before identified. The dominant microbes appear to be algae. In their role as photosynthesizers, the algae provide nutritional support for the growth of heterotrophs such as fungi and protozoans. The researchers determined that some of these algae not only thrive in the presence of high levels of metals, but they also can remove them from solution. Algae also add carbonates to the water, which would raise its pH over time. These observations quickly pointed to some possible applications for future bioremediation of the pit. One line of research will attempt to stimulate the growth of the algae and create blooms of algae that could speed the remediation of the pit and lessen the threat to the watershed in the area.

- In what other ways might the algae impact the ecology of the pit?
- What is involved in the process of bioremediation?

For a wrapup, see the Case File Perspective on page 213.

A **thermophile*** is a microbe that grows optimally at temperatures greater than 45°C. Such heat-loving microbes live in soil and water associated with volcanic activity, in compost piles, and in habitats directly exposed to the sun. Thermophiles generally range in growth temperatures from 45°C to 80°C. Most eukaryotic forms cannot survive above 60°C, but a few bacteria, called hyperthermophiles, grow at between 80°C and 250°C (currently thought to be the temperature limit endured by enzymes and cell structures). Strict thermophiles are so heat tolerant that researchers may use a heat-sterilizing device to isolate them in culture.

Currently, there is intense interest in thermal microorganisms on the part of biotechnology companies. One of the most profitable discoveries so far was a strict thermophile *Thermus aquaticus,* which produces an enzyme that can make copies of DNA even at high temperatures. This enzyme—Taq polymerase—is now an essential component of the polymerase chain reaction or PCR, a process used in many areas of medicine, forensics, and biotechnology (see chapter 10).

Gas Requirements

The atmospheric gases that most influence microbial growth are oxygen (O_2) and carbon dioxide (CO_2). Oxygen comprises about 20% of the atmosphere and CO_2 about 0.03%. Oxygen gas has a great impact on microbial adaptation. It is an important respiratory gas, and it is also a powerful oxidizing agent that exists in many toxic forms. In general, microbes fall into one of three categories: those that use oxygen and can detoxify it; those that can neither use oxygen nor detoxify it; and those that do not use oxygen but can detoxify it.

** thermophile* (thur'-moh-fyl)

How Microbes Process Oxygen

As oxygen gas enters into cellular reactions, it can be transformed into several toxic products. Singlet oxygen (1O_2) is an extremely reactive molecule produced by both living and nonliving processes. Notably, it is one of the substances produced by phagocytes to kill invading bacteria (see chapter 14). The buildup of singlet oxygen and the oxidation of membrane lipids and other molecules can damage and destroy a cell. The highly reactive superoxide ion (O_2^-), peroxide (H_2O_2), and hydroxyl radicals (OH) are other destructive metabolic by-products of oxygen. Most cells have developed enzymes that go about the business of scavenging and neutralizing these chemicals. The complete conversion of superoxide ion into harmless oxygen involves a two-step process and at least two enzymes:

$$\text{Step 1.} \quad 2O_2^- + 2H^+ \xrightarrow{\text{Superoxide dismutase}} H_2O_2 \text{ (hydrogen peroxide)} + O_2$$

$$\text{Step 2.} \quad 2H_2O_2 \xrightarrow{\text{Catalase}} 2H_2O_2 + O_2$$

In this series of reactions essential for aerobic organisms, the superoxide ion is first converted to hydrogen peroxide and normal oxygen by the action of an enzyme called superoxide dismutase. Because hydrogen peroxide is also toxic to cells (it is used as a disinfectant and antiseptic), it must be degraded by an enzyme—either catalase or peroxidase—into water and oxygen. If a microbe is not capable of dealing with toxic oxygen by these or similar mechanisms, it will be restricted to habitats free of oxygen.

With respect to oxygen requirements, several general categories are recognized. An **aerobe*** (aerobic organism) can use gaseous oxygen in its metabolism and possesses the enzymes needed to process toxic oxygen products. An organism that cannot grow without oxygen is an **obligate aerobe.** Most fungi and protozoa, as well as many bacteria (genera *Micrococcus* and *Bacillus*), have strict requirements for oxygen in their metabolism.

A **facultative anaerobe** is an aerobe that does not require oxygen for its metabolism and is capable of growth in the absence of it. This type of organism metabolizes by aerobic respiration when oxygen is present, but in its absence, it adopts an anaerobic mode of metabolism such as fermentation. Facultative anaerobes usually possess catalase and superoxide dismutase. A number of bacterial pathogens fall into this group. This includes gram-negative intestinal bacteria and staphylococci. A **microaerophile*** does not grow at normal atmospheric concentrations of oxygen but requires a small amount of it (1–15%) in metabolism. Most organisms in this category live in a habitat such as soil, water, or the human body that provides small amounts of oxygen but is not directly exposed to the atmosphere.

A true **anaerobe** (anaerobic microorganism) lacks the metabolic enzyme systems for using oxygen gas in respiration. Because **strict, or obligate, anaerobes** also lack the enzymes for processing toxic oxygen, they cannot tolerate any free oxygen in the immediate environment and will die if exposed to it. Strict anaerobes live in highly reduced habitats, such as deep muds, lakes, oceans, and soil. Growing strictly anaerobic bacteria usually requires special media, methods of incubation, and handling chambers that exclude oxygen. **Figure 7.11** shows special systems for handling and growing anaerobes.

** aerobe* (air'-ohb) Although the prefix means air, it is used in the sense of oxygen.
** microaerophile* (myk"-roh-air'-oh-fyl)

(a) (b)

Figure 7.11 **Culturing techniques for anaerobes. (a)** An anaerobic environmental chamber is equipped with ports for handling strict anaerobes without exposing them to air. It also has provisions for incubation and inspection in a completely O_2-free system. **(b)** A small anaerobic or CO_2 incubator system.

Even though human cells use oxygen, and oxygen is found in the blood and tissues, some body sites present anaerobic pockets or microhabitats where colonization or infection can occur. Dental caries are partly due to the complex actions of aerobic and anaerobic bacteria in plaque. Most gingival infections consist of similar mixtures of oral bacteria that have invaded damaged gum tissues. Another common site for anaerobic infections is the large intestine, a relatively oxygen-free habitat that harbors a rich assortment of strictly anaerobic bacteria. Anaerobic infections can occur following abdominal surgery and traumatic injuries (gas gangrene and tetanus).

Aerotolerant anaerobes do not utilize oxygen gas but can survive and grow in its presence. These anaerobes are not harmed by oxygen, and some of them possess alternate mechanisms for breaking down peroxide and superoxide. For instance, lactobacilli, which are common residents of the intestine, inactivate these compounds with manganese ions.

Determining the oxygen requirements of a microbe from a biochemical standpoint can be a very time-consuming process. An initial clarification comes from cultures made with reducing media that contain an oxygen-removing chemical such as thioglycollate. The location of growth in a tube of fluid thioglycollate medium is a fair indicator of an organism's adaptation to oxygen use (**figure 7.12**).

Although all microbes require some carbon dioxide in their metabolism, **capnophiles** grow best at higher CO_2 tensions (3–10%) than are normally present in the atmosphere (0.033%). This becomes important in the initial isolation of some pathogens from clinical specimens, notably *Neisseria* (gonorrhea, meningitis), *Brucella* (undulant fever), and *Streptococcus pneumoniae*. Incubation is carried out in a CO_2 incubator that provides the correct range of CO_2 (**figure 7.11b**). Keep in mind that CO_2 is an essential nutrient for autotrophs, which use it to synthesize organic compounds.

Effects of pH

Microbial growth and survival are also influenced by the pH of the habitat. The pH was defined in chapter 2 as the degree of acidity or alkalinity of a solution. It is expressed by the pH scale, a series of numbers ranging from 0 to 14, pH 7 being neither acidic nor alkaline. As the pH value decreases toward 0, the acidity increases, and as the pH increases toward 14, the alkalinity increases. The majority of organisms live or grow in habitats between pH 6 and pH 8

because strong acids and bases can be highly damaging to enzymes and other cellular substances.

Although most microbes are **neutrophiles,** living around pH 7, a few microorganisms live at pH extremes. Highly acidic or alkaline habitats are not common, but acidic bogs, lakes, and alkaline soils contain a specialized microbial flora. Obligate **acidophiles** include *Euglena mutabilis,* an alga that grows in acid pools between 0 and 1.0 pH, and *Thermoplasma,* an archaeon that lacks a cell wall, lives in hot coal piles at a pH of 1 to 2, and will lyse if exposed to pH 7. A few species of algae and bacteria can actually survive at a pH near that of concentrated hydrochloric acid. Not

Figure 7.12 **Use of thioglycollate broth to demonstrate oxygen requirements.** Thioglycollate is a reducing medium that can establish a gradation in oxygen content. Oxygen concentration is highest at the top of the tube and absent in the deeper regions. When a series of tubes is inoculated with bacteria that differ in O_2 requirements, the relative position of growth provides some indication of their adaptations to oxygen use. Tube 1 (far left): aerobic (*Pseudomonas aeruginosa*); Tube 2: facultative (*Staphylococcus aureus*); Tube 3: facultative (*Escherichia coli*); Tube 4: obligate anaerobe (*Clostridium butyricum*).

only do they require such a low pH for growth but particular bacteria actually help maintain the low pH by releasing strong acid. Because many molds and yeasts tolerate moderate acidity, they are the most common spoilage agents of pickled foods.

Alkalinophiles live in hot pools and soils that contain high levels of basic minerals (up to pH 10.0). Bacteria that decompose urine create alkaline conditions, because ammonium (NH_4^+) can be produced when urea (a component of urine) is digested. Metabolism of urea is one way that *Proteus* spp. can neutralize the acidity of the urine to colonize and infect the urinary system.

Osmotic Pressure

Although most microbes exist under hypotonic or isotonic conditions, a few, called osmophiles, live in habitats with a high solute concentration. One common type of osmophile requires high concentrations of salt; these organisms are called **halophiles***

Obligate *halophiles* such as *Halobacterium* and *Halococcus* inhabit salt lakes, ponds, and other hypersaline habitats. They grow optimally in solutions of 25% NaCl but require at least 9% NaCl (combined with other salts) for growth. These archaeons have significant modifications in their cell walls and membranes, and will lyse in hypotonic habitats.

Some microbes adapt to wide concentrations in solutes. These are called **osmotolerant.** Such organisms are remarkably resistant to salt, even though they do not normally reside in high-salt environments. For example, *Staphylococcus aureus* can grow on NaCl media ranging from 0.1% up to 20%. Although it is common to use high concentrations of salt and sugar to preserve food (jellies, syrups, and brines), many bacteria and fungi actually thrive under these conditions and are common spoilage agents. A particularly hardy sugar-loving or *saccharophilic* yeast withstands the high sugar concentration of honey and candy.

Miscellaneous Environmental Factors

Descent into the ocean depths subjects organisms to increasing hydrostatic pressure. Deep-sea microbes called **barophiles** exist under pressures many times that of the atmosphere. Marine biologists sampling deep-sea trenches 7 miles below the surface isolated unusual eukaryotes called foraminifera that were being exposed to pressures 1,100 times normal. These microbes are so strictly adapted to high pressures that they will rupture when exposed to normal atmospheric pressure.

Because of the high water content of cytoplasm, all cells require water from their environment to sustain growth and metabolism. Water is the solvent for cell chemicals, and it is needed for enzyme function and digestion of macromolecules. A certain amount of water on the external surface of the cell is required for the diffusion of nutrients and wastes. Even in apparently dry habitats, such as sand or dry soil, the particles retain a thin layer of water usable by microorganisms. Only dormant, dehydrated cell stages (for example, spores and cysts) tolerate extreme drying because of the inactivity of their enzymes.

7.4 Ecological Associations among Microorganisms

19. Discuss the range of associations among microorganisms and their basic qualities.

20. Explain what occurs in symbiosis and coevolution.

21. Differentiate among mutual, commensal, and parasitic associations, providing examples.

22. Differentiate among synergism and antagonism with examples.

23. Describe some interactions between humans and their microbiota.

24. Describe the development and significance of biofilm associations.

Up to now, we have considered the importance of nonliving environmental influences on the growth of microorganisms. Another profound influence comes from other organisms that share (or sometimes are) their habitats. In all but the rarest instances, microbes live in shared habitats, which give rise to complex and fascinating associations. Some associations are between similar or dissimilar types of microbes; others involve multicellular organisms such as animals or plants. Interactions can have beneficial, harmful, or no particular effects on the organisms involved; they can be obligatory or nonobligatory to the members; and they often involve nutritional interactions. This outline provides an overview of some major types of microbial associations:

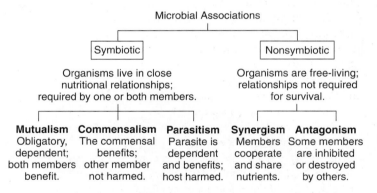

A general term used to denote a situation in which two organisms live together in a close partnership is **symbiosis,**[4]* and the members are termed *symbionts*. Three main types of symbiosis occur. **Mutualism** exists when organisms live in an obligatory but mutually beneficial relationship. This association is rather common in nature because of the survival value it has for the members involved. **Insight 7.3** gives several examples to illustrate this concept. In other symbiotic relationships, the relationship tends to be unequal, meaning it benefits one member and not the other, and it can be obligatory to one of the members but not both.

4. Note that *symbiosis* is a neutral term and does not by itself imply benefit or detriment.

* *symbiosis* (sim"-bye-oh'-sis) Gr. *syn*, together, and *bios*, to live.

* *halophiles* (hay'-loh-fylz).

TAKE NOTE: COEVOLUTION

The interrelationships of microbes run the gamut from very intimate, long-term, even essential partnerships to loose, temporary unions. Many of these associations are of ancient origins—the members have been evolving together for millions and even billions of years. We saw an example of this in the origin of the eukaryotic cell in chapter 5. The term that biologists use to specify this sort of occurrence is **coevolution.** When organisms live closely together, changes that occur in one member give rise to changes in the other as they adapt to one another. This can involve expression of genes that affect structure and function, so that the relationship even becomes imprinted in their genomes. Generally, these coadaptations improve the survival and success of both partners.

There are thousands of these partnerships in the natural world, most of them being mutualistic or cooperative. We see this in the tendency for certain insects (aphids) to harbor bacteria in their cells. These insects have come to rely on the bacteria to synthesize essential amino acids. Other bacteria can even manipulate the reproductive cycle of insects, ensuring their transmission to the next generation inside the insect's eggs.

One of the most remarkable cases of coevolution can be seen in the "photosynthetic worm." This small marine flatworm (*Convolvula*) has coevolved symbiotically with a green alga (*Tetraselmis*). Migration of the algal cells into the worm's tissues during development stimulates a dramatic conversion of its anatomy so that it loses its ability to feed independently. In time the algae embed themselves close to the surface of the worm's transparent body in a favorable position to receive sunlight. The algae make sugars and oxygen that feed the worm, and the worm provides a ready nitrogen source and shelter for the algae. It can also migrate to the locations on the beach to ensure optimum photosynthesis.

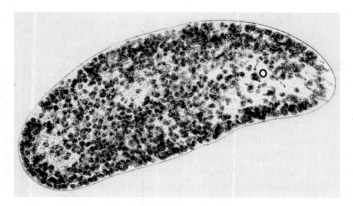

A blind marine flatworm containing its algae passengers.

In a relationship known as **commensalism,*** the member called the commensal receives benefits, while its coinhabitant is neither harmed nor benefited. A classic commensal interaction between microorganisms called **satellitism** arises when one member

* *commensalism* (kuh-men'-sul-izm) L. *com*, together, and *mensa*, table.

Staphylococcus aureus growth

Haemophilus satellite colonies

Figure 7.13 Satellitism, a type of commensalism between two microbes. In this example, *Staphylococcus aureus* provides growth factors to *Haemophilus influenzae*, which grows as tiny satellite colonies near the streak of *Staphylococcus*. By itself, *Haemophilus* could not grow on blood agar. The *Staphylococcus* gives off several nutrients such as vitamins and amino acids that diffuse out to the *Haemophilus*, thereby promoting its growth.

provides nutritional or protective factors needed by the other (**figure 7.13**). In others, microbes can break down a substance that would be toxic or inhibitory to the commensal partner. Relationships between humans and resident commensals that derive nutrients from the body are discussed under the next heading.

In section 7.1, we introduced the concept of **parasitism,** in which the host organism provides the parasitic microbe with nutrients and a habitat. Multiplication of the parasite usually harms the host to some extent. As this relationship evolves, the host may even develop tolerance for or dependence on a parasite, at which point we call the relationship commensalism or mutualism.

The partnership between fungi and algae or cyanobacteria in a lichen is an unusual form of parasitism. The two symbionts can live separately but begin coexisting when the nutrient supply is depleted. Together, they form a distinct organism different from the individual members. The fungus derives its nutrients from its photosynthetic host and may contribute some protection. It is considered parasitism because the fungus invades the cells of the algae.

Synergism or cooperation is an interrelationship between two or more organisms that benefits all members but is not necessary for their survival. Together, the participants cooperate to produce a result that none of them could do alone. One form of shared metabolism can be seen in certain soil bacteria involved in nutrient cycling:

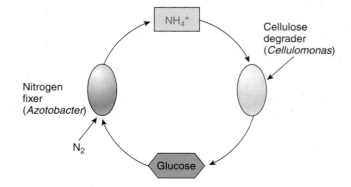

INSIGHT 7.3

A Mutual Attraction

A tremendous variety of mutualistic partnerships occur in nature. These associations gradually evolve a reciprocal life as participating members come to require some substance or habitat that the other members provide. In many cases, they will not survive outside of this association.

Protozoan cells often receive growth factors from symbiotic bacteria and algae that are mutually nurtured by the protozoan cell. One peculiar ciliate propels itself by affixing symbiotic bacteria to its cell membrane to act as "oars." Other ciliates and amoebas require specialized bacteria or algae inside their cells to provide nutrients.

This kind of relationship is especially striking in the complex mutualism of termites, which harbor 250 or more specialized microbes inside their gut. The termite's role is simply to gnaw off pieces of wood, but it cannot digest the wood. Endosymbiotic protozoans and bacteria do the final work of breaking down the wood particles, and all mutualists share in the nutrients provided (see photo).

Mutualism between Microbes and Animals

Microorganisms carry on mutual symbiotic relationships with animals as diverse as sponges, worms, and mammals. Bacteria and protozoa are essential in the operation of the rumen (a complex, four-chambered stomach) of cud-chewing mammals. These mammals produce no enzymes of their own to break down the cellulose that is a major part of their diet, but the microbial population harbored in their rumens does. The complex food materials are digested through several stages, during which time the animal regurgitates and chews the partially digested plant matter (the cud) and occasionally burps methane produced by the microbial symbionts (see chapter 8 case file).

Thermal Vent Symbionts

Another fascinating relationship has been found in the deep hydrothermal vents in the seafloor, where geologic forces spread the crustal plates and release heat and gas. These vents are a focus of tremendous biological and geologic activity. The source of energy in this community is not the sun, because the vents are too deep for light to penetrate (2,600 m). Instead, this ecosystem is based on a massive chemoautotrophic bacterial population that oxidizes the abundant hydrogen sulfide (H_2S) gas given off by the volcanic activity there. As the bottom of the food web, these bacteria serve as the primary producers of nutrients that service a broad spectrum of specialized animals.

How are these mutualistic relationships different from synergistic ones? *Answer available at http://www.mhhe.com/talaro8*

Upper: A termite displayed alongside its gut. Lower: Some gut microbes involved in wood digestion are protozoa (blue) and spirochete bacteria.

Cross Section of Worm

A view of a vent community based on mutualism and chemoautotrophy. The giant tube worm *Riftia* houses bacteria in its specialized feeding organ, the trophosome. Raw materials in the form of dissolved inorganic molecules are provided to the bacteria through the worm's circulation. With these, the bacteria produce usable organic nutrients absorbed by the worm.

In this relationship, *Cellulomonas* breaks down the cellulose left by dead plants and releases glucose. *Azotobacter* uses this glucose to derive energy and fixes atmospheric nitrogen into ammonium. This fixed nitrogen is taken up by *Cellulomonas* and feeds its metabolism and, thus, continues this cross feeding.

Another example of synergism is observed in the exchange between soil bacteria and plant roots (see chapter 26). The plant provides various growth factors, and the bacteria help fertilize the plant by supplying it with minerals. In synergistic infections, a combination of organisms can produce tissue damage that a single organism would not cause alone. Gum disease, dental caries, and gas gangrene involve mixed infections by bacteria interacting synergistically.

Antagonism, a type of competition, occurs when the actions of one organism affect the success or survival of others in the same

community. Often, what happens is that one microbe secretes chemical substances into the surrounding environment that inhibit or destroy other microbes in the same habitat. The first microbe may gain a competitive advantage by increasing the space and nutrients available to it. Interactions of this type are common in the soil, where mixed communities often compete for space and food. **Antibiosis** —the production of inhibitory compounds such as antibiotics—is actually a form of antagonism. Hundreds of naturally occurring antibiotics have been isolated from bacteria and fungi and used as drugs to control diseases (see chapter 12). Bacteriocins are another class of antimicrobial proteins produced mainly by gram-negative bacteria that are toxic to bacteria other than the ones that produced them.

Interrelationships between Microbes and Humans

The human body is a rich habitat for symbiotic bacteria, fungi, and a few protozoa. Microbes that normally live on the skin, in the alimentary tract, and in other sites are called the *normal resident microbiota* (see chapter 13). These residents participate in commensal, parasitic, and synergistic relationships with their human hosts. For example, *Escherichia coli* living symbiotically in the intestine produces vitamin K, and species of symbiotic *Lactobacillus* residing in the vagina help maintain an acidic environment that protects against infection by other microorganisms. Hundreds of commensal species "make a living" on the body without either harming or benefiting it. For example, many bacteria and yeasts feed on dead epidermal cells of the skin, and billions of bacteria live on the wastes in the large intestine. Commensal gut bacteria are considered to have coevolved with their human hosts. The hosts have developed mechanisms to prevent the disease effects of their bacterial passengers, and the bacteria have developed mechanisms to be less pathogenic to their hosts.

Because the normal residents and the body are in a constant state of change, these relationships are not absolute, and a commensal can convert to a parasite by invading body tissues and causing disease.

Microbial Biofilms—A Meeting Ground

Among the most common and prolific associations of microbes are **biofilms,** first described in Insight 4.1. Biofilms result when organisms attach to a substrate by some form of extracellular matrix that binds them together in complex organized layers. Biofilms are so prevalent that they dominate the structure of most natural environments on earth. This tendency of microbes to form biofilm communities is an ancient and effective adaptive strategy. Not only would biofilms favor microbial persistence in habitats, but they would also offer greater access to life-sustaining conditions. In a sense, these living networks operate as "superorganisms" that influence such microbial activities as adaptation to a particular habitat, content of soil and water, nutrient cycling, and the course of infections.

It is generally accepted that the individual cells in the bodies of multicellular organisms such as animals and plants have the capacity to produce, receive, and react to chemical signals such as hormones made by other cells. For many years, biologists regarded most single-celled microbes as simple individuals without compa-

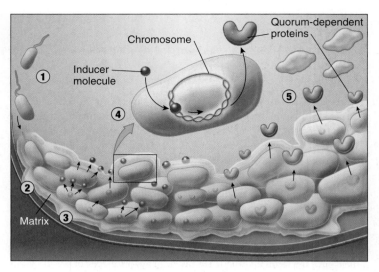

① Free-swimming cells settle on a surface and remain there.

② Cells synthesize a sticky matrix that holds them tightly to the substrate.

③ When biofilm grows to a certain density (quorum), the cells release inducer molecules that can coordinate a response.

④ Enlargement of one cell to show genetic induction. Inducer molecule stimulates expression of a particular gene and synthesis of a protein product, such as an enzyme.

⑤ Cells secrete their enzymes in unison to digest food particles.

Process Figure 7.14 Stages in biofilm formation quorum sensing, induction and expression, shown in progression from left to right.

rable properties, other than to cling together in colonies. But these assumptions have turned out to be incorrect. It is now evident that microbes show a well-developed capacity to communicate and cooperate in the formation and function of biofilms. This is especially true of bacteria, although fungi and other microorganisms can participate in these activities.

One idea to explain the development and behavior of biofilms is termed **quorum sensing.** This process occurs in several stages, including self-monitoring of cell density, secretion of chemical signals, and genetic activation **(figure 7.14).** Early in biofilm formation, free-floating or swimming microbes—often described as planktonic—are attracted to a surface and come to rest or settle down. Settling stimulates the cells to secrete a slimy or sticky matrix, usually made of polysaccharide, that binds them to the substrate. Once attached, cells begin to release **inducer* molecules** that accumulate as the cell population grows. By this means, they can monitor the size of their own population. In time, a critical number of cells, termed a **quorum,*** accumulates and this ensures that there will be sufficient quantities of inducer molecules. These inducer molecules enter biofilm cells and stimulate specific genes on their chromosomes to begin expression.

* *inducer* (in-doos′-ur) L. *inducere,* to lead. Something that brings about or causes an effect.
* *quorum* (kwor′-uhm) Latin *qui,* who. Indicates the minimum number of group members necessary to conduct business.

The nature of this expression varies, but generally it allows the biofilm to react as a unit. For example, by coordinating the expression of genes that code for proteins, the biofilm can simultaneously produce large quantities of a digestive enzyme or toxin. This regulation of expression accounts for several observations made about microbial activities. It explains how saprobic microbes in soil and water rapidly break down complex substrates.

The effect of quorum sensing has also given greater insight into how pathogens invade their hosts and produce large quantities of substances that damage host defenses. It has been well studied in a number of pathogenic bacteria. Infections by *Pseudomonas* give rise to tenacious lung biofilms in some types of pneumonia. *Staphylococcus aureus* commonly forms biofilms on inanimate medical devices and in wounds. *Streptococcus* species are the initial colonists on teeth surfaces leading to dental plaque (see figure 21.30).

Although the best-studied biofilms involve just a single type of microorganism, most biofilms observed in nature are polymicrobial. In fact, many symbiotic and cooperative relationships are based on complex communication patterns among coexisting organisms. As each organism in the biofilm carries out its specific niche, signaling among the members sustains the overall partnership. Biofilms are known to be a rich ground for genetic transfers among neighboring cells that involve conjugation and transformation (see chapter 9). As our knowledge of biofilm patterns grows, it will likely lead to greater understanding of their involvement in infections and their contributions to disinfectant and drug resistance (see chapter 12).

Check & Assess Section 7.4

✔ The environmental factors that control microbial growth are temperature, pH, moisture, radiation, gases, and other microorganisms.

✔ Environmental factors control microbial growth by their influence on microbial enzymes.

✔ Three cardinal temperatures for a microorganism describe its temperature range and the temperature at which it grows best. These are the minimum temperature, the maximum temperature, and the optimum temperature.

✔ Microorganisms are classified by their temperature requirements as psychrophiles, mesophiles, or thermophiles.

✔ Most eukaryotic microorganisms are aerobic, whereas bacteria vary widely in their oxygen requirements from obligately aerobic to anaerobic.

✔ Microorganisms have symbiotic associations with other species that range from mutualism to commensalism to parasitism that are often the result of coevolution.

✔ Microbes established in biofilms often communicate through complex chemical messages called quorum sensing. This allows members to coordinate their reactions and favor colonization and survival.

13. Why are most pathogens mesophilic?
14. What are the ecological roles of psychrophiles and thermophiles?
15. Explain what it means to be an obligate intracellular parasite. Name three groups of obligate intracellular parasites.
16. Classify a human with respect to oxygen requirements.

17. Where in the body are anaerobic habitats apt to be found?
18. Where do superoxide ions and hydrogen peroxide originate? What are their toxic effects?
19. Define *symbiosis* and differentiate among mutualism, commensalism, synergism, parasitism, and antagonism, using examples.
20. Explain the process of quorum sensing by using a flow diagram to organize the stages.
21. Relate several advantages to communication within a biofilm.

7.5 The Study of Microbial Growth

Expected Learning Outcomes

25. Define *growth* and explain the process of binary fission.
26. Describe the process of population growth and how it is measured.
27. Explain the stages in the population growth curve and its practical importance.
28. Relate the various ways that microbes in cultures and samples are enumerated.

When microbes are provided with nutrients and the required environmental factors, they become metabolically active and grow. Growth takes place on two levels. On one level, a cell synthesizes new cell components and increases its size; on the other level, the number of cells in the population increases. This capacity for multiplication, increasing the size of the population by cell division, has tremendous importance in microbial control, infectious disease, and biotechnology. In the following sections, we focus primarily on the characteristics of bacterial growth that are generally representative of single-celled microorganisms.

The Basis of Population Growth: Binary Fission

The division of a bacterial cell occurs mainly through **binary,** or **transverse, fission.**[5] The term *binary* means that one cell becomes two, and *transverse* refers to the division plane forming across the width of the cell. During binary fission, the parent cell enlarges, duplicates its chromosome, and forms a central transverse septum that divides the cell into two daughter cells. This process is repeated at intervals by each new daughter cell in turn; and with each successive round of division, the population increases. The stages in this continuous process are shown in greater detail in **figures 7.15** and **7.16.**

The Rate of Population Growth

The time required for a complete fission cycle—from parent cell to two new daughter cells—is called the **generation,** or **doubling, time.** The term *generation* has a similar meaning as it does in

5. Research with bacteria that are not culturable indicates that not all cells divide this way. Some bud and others divide by multiple fission.

① A young cell at early phase of cycle

② A parent cell prepares for division by enlarging its cell wall, cell membrane, and overall volume. Midway in the cell, the wall develops notches that will eventually form the transverse septum, and the duplicated chromosome becomes affixed to a special membrane site.

③ The septum wall grows inward, and the chromosomes are pulled toward opposite cell ends as the membrane enlarges. Other cytoplasmic components are distributed (randomly) to the two developing cells.

④ The septum is synthesized completely through the cell center, and the cell membrane patches itself so that there are two separate cell chambers.

⑤ At this point, the daughter cells are divided. Some species will separate completely as shown here, while others will remain attached, forming chains or doublets, for example.

■ Cell wall ■ Cell membrane ◯ Chromosome 1 ◯ Chromosome 2 • Ribosomes

Process Figure 7.15 **Steps in binary fission of a rod-shaped bacterium.**

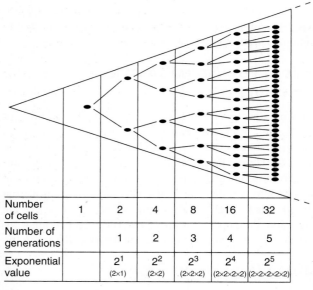

Number of cells	1	2	4	8	16	32
Number of generations		1	2	3	4	5
Exponential value		2^1 (2×1)	2^2 (2×2)	2^3 (2×2×2)	2^4 (2×2×2×2)	2^5 (2×2×2×2×2)

(a)

Figure 7.16 The mathematics of population growth. (a) Starting with a single cell, if each product of reproduction goes on to divide by binary fission, the population doubles with each new cell division or generation. This process can be represented by logarithms (2 raised to an exponent) or by simple numbers. **(b)** Plotting the logarithm of the cells produces a straight line indicative of exponential growth, whereas plotting the cell numbers arithmetically gives a curved slope. *Note that the left scale is logarithmic, and the right scale is arithmetic.

humans. It is the period between an individual's birth and the time of producing offspring. In bacteria, each new fission cycle or generation increases the population by a factor of 2, or doubles it. Thus, the initial parent stage consists of 1 cell, the first generation consists of 2 cells, the second 4, the third 8, then 16, 32, 64, and so on. As long as the environment remains favorable, this doubling effect can continue at a constant rate. With the passing of each generation, the population will double, over and over again.

The length of the generation time is a measure of the growth rate of an organism. Compared with the growth rates of most other living things, bacteria are notoriously rapid. The average generation time is 30 to 60 minutes under optimum conditions. The shortest generation times average 5 to 10 minutes, and longer generation times require days. For example, *Mycobacterium leprae,* the cause of Hansen's disease, has a generation time of 10 to 30 days—as long as some animals. Most pathogens have relatively short doubling times. *Salmonella enteritidis* and *Staphylococcus aureus,* bacteria that cause food-borne illness, double in 20 to 30 minutes, which explains why leaving food at room temperature even for a short period has caused many a person to be suddenly stricken with an attack of food-borne disease. In a few hours, a population of these bacteria can easily grow from a small number of cells to several million.

Figure 7.16 shows some quantitative characteristics of growth: (1) The cell population size can be represented by the number 2 with an exponent (2^1, 2^2, 2^3, 2^4); (2) the exponent increases by one in each generation; and (3) the number of the exponent also gives the number of the generation. This growth pattern is termed **exponential.** Because these populations often contain very large numbers of cells, it is useful to express them by means of exponents or logarithms (see appendix A). The data from a growing bacterial population are graphed by plotting the number of cells as a function of time. The cell number can be represented logarithmically or arithmetically. Plotting the logarithm number over time provides a straight line indicative of exponential growth. Plotting the data arithmetically gives a constantly curved slope. In general, logarithmic graphs are preferred because an accurate cell number is easier to read, especially during early growth phases.

Predicting the number of cells that will arise during a long growth period (yielding millions of cells) is based on a relatively simple concept. One could use the method of addition $2 + 2 = 4$; $4 + 4 = 8$; $8 + 8 = 16$; $16 + 16 = 32$, and so on, or a method of multiplication (for example, $2^5 = 2 \times 2 \times 2 \times 2 \times 2$), but it is easy to see that for 20 or 30 generations, this calculation could be very tedious. An easier way to calculate the size of a population over time is to use an equation such as:

$$N_f = (N_i)2^n$$

In this equation, N_f is the total number of cells in the population at some point in the growth phase, N_i is the starting number, the exponent n denotes the generation number, and 2^n represents the number of cells in that generation. If we know any two of the values, the other values can be calculated. Let us use the example of *Staphylococcus aureus* to calculate how many cells (N_f) will be present in an egg salad sandwich after it sits in a warm car for 4 hours. We will assume that N_i is 10 (number of cells deposited in the sandwich while it was being prepared). To derive n, we need to divide 4 hours (240 minutes) by the generation time (we will use 20 minutes). This calculation comes out to 12, so 2^n is equal to 2^{12}. Using a calculator or table,[6] we find that 2^{12} is 4,096.

Final number (N_f) = $10 \times 4,096$
 = $40,960$ bacterial cells in the sandwich

6. See appendix A for a table with powers of 2.

This same equation, with modifications, is used to determine the generation time, a more complex calculation that requires knowing the number of cells at the beginning and end of a growth period. Such data are obtained through actual testing by a method discussed in the following section.

Determinants of Population Growth

In reality, a population of bacteria does not maintain its potential growth rate and does not double endlessly, because in most systems numerous factors prevent the cells from continuously dividing at their maximum rate. Quantitative laboratory studies indicate that a population typically displays a predictable pattern, or **growth curve,** over time. The method traditionally used to observe the population growth pattern is a viable count technique, in which the live cells in a culture are sampled, grown, and counted during a growth period, as described in the next section.

The Viable Plate Count: Batch Culture Method

A growing population is established by inoculating a flask containing a known quantity of sterile liquid medium with a few cells of a pure culture. The flask is incubated at that bacterium's optimum temperature and timed. The population size at any point in the growth cycle is quantified by removing a tiny measured sample of the culture from the growth chamber and plating it out on a solid medium to develop isolated colonies. This procedure is repeated at evenly spaced intervals (i.e., every hour for 24 hours) (**Figure 7.17**).

Evaluating the samples involves a common and important principle in microbiology: One colony on the plate represents one cell or colony-forming unit (CFU) from the original sample. Because the CFU of some bacteria is actually composed of several cells (consider the clustered arrangement of *Staphylococcus,* for instance), using a colony count can underestimate the exact population size to an extent. This is not a serious problem because, in such bacteria, the CFU is the smallest unit of colony formation and dispersal. Multiplication of the number of colonies in a single sample by the container's volume gives a fair estimate of the total population size (number of cells) at any given point. The growth curve is determined by graphing the number for each sample in sequence for the whole incubation period (see **figure 7.18**).

Because of the scarcity of cells in the early stages of growth, some samples can give a zero reading even if there are viable cells in the culture. The sampling itself can remove enough viable cells to alter the tabulations, but since the purpose is to compare relative trends in growth, these factors do not significantly change the overall pattern.

Stages in the Normal Growth Curve

The system of batch culturing just described is *closed,* meaning that nutrients and space are finite and there is no mechanism for the removal of waste products. Data from an entire growth period of 3 to 4 days typically produce a curve with a series of phases termed the lag phase, the exponential growth (log) phase, the stationary phase, and the death phase (figure 7.18).

	60 min 0.1 ml	120 min	180 min	240 min	300 min	360 min	420 min	480 min	540 min	600 min
Sample is diluted in liquid agar medium and poured or spread over surface of solidified medium										
Plates are incubated, colonies are counted	None									
Number of colonies (CFU) per 0.1 ml	<1*	2	4	7	13	23	45	80	135	230
Total estimated cell population in flask	<10,000	5,000	20,000	35,000	65,000	115,000	225,000	400,000	675,000	1,150,000

*Only means that too few cells are present to be assayed.

Figure 7.17 The general steps in a viable plate count. 1. placing a few cells into a sterile liquid medium, **2.** incubating this culture over a period of several hours, **3.** sampling the broth at regular intervals during incubation, **4.** plating each sample onto solid media, and **5.** counting the number of colonies present after incubation.

Total cells in population, live and dead, at each phase
[] Few cells [] Live cells [] Dead cells (not part of count)

Figure 7.18 The growth curve in a bacterial culture. On this graph, the number of viable cells expressed as a logarithm (log) is plotted against time. See text for discussion of the various phases. Note that with a generation time of 30 minutes, the population has risen from 10 (10^1) cells to 1,000,000,000 (10^9) cells in only 16 hours.

The **lag phase** is an early "flat" period on the graph when the population appears not to be growing or is growing at less than the exponential rate. Growth lags primarily because

1. the newly inoculated cells require a period of adjustment, enlargement, and synthesis of DNA, enzymes, and ribosomes;

2. the cells are not yet multiplying at their maximum rate; and
3. the population of cells is so sparse or dilute that the sampling misses them.

The length of the lag period varies from one population to another, depending on the condition of the microbes and medium. It is

important to note that even though the population of cells is not increasing (growing), individual cells are metabolically active as they increase their contents and prepare to divide.

The cells reach the maximum rate of cell division during the **exponential growth (logarithmic or log) phase,** a period during which the curve increases geometrically. This phase will continue as long as cells have adequate nutrients and the environment is favorable. During this phase, the population fulfills its potential generation time, and growth is balanced and genetically coordinated.

At the **stationary growth phase,** the population enters a survival mode in which cells stop growing or grow slowly. The curve levels off because the rate of cell inhibition or death balances out the rate of multiplication. The number of viable cells has reached maximum and remains relatively constant during this period. The decline in the growth rate is caused by several factors. A common reason is the depletion of nutrients and oxygen. Another is that the increased cell density often causes an accumulation of organic acids and other toxic biochemicals.

As the limiting factors intensify, cells begin to die at an exponential rate (literally perishing in their own wastes), and most are unable to multiply. The curve now dips downward as the **death phase** begins. The speed with which death occurs depends on the relative resistance of the species and how toxic the conditions are, but it is usually slower than the exponential growth phase. Viable cells often remain many weeks and months after this phase has begun. In the laboratory, refrigeration is used to slow the progression of the death phase so that cultures will remain viable as long as possible.

Practical Importance of the Growth Curve The tendency for populations to exhibit phases of rapid growth, slow growth, and death has important implications in microbial control, infection, food microbiology, and culture technology. Antimicrobial agents such as heat and disinfectants rapidly accelerate the death phase in all populations, but microbes in the exponential growth phase are more vulnerable to these agents than are those that have entered the stationary phase. In general, actively growing cells are more vulnerable to conditions that disrupt cell metabolism and binary fission.

Growth patterns in microorganisms can account for the stages of infection (see chapter 13). A person shedding bacteria in the early and middle stages of an infection is more likely to spread it to others than is a person in the late stages. The course of an infection is also influenced by the relatively faster rate of multiplication of the microbe, which can overwhelm the slower growth rate of the host's own cellular defenses.

Understanding the stages of cell growth is crucial for work with cultures. Sometimes a culture that has reached the stationary phase is incubated under the mistaken impression that enough nutrients are present for the culture to continue to multiply. In most cases, it is unwise to incubate a culture beyond the stationary phase, because doing so will reduce the number of viable cells and the culture could die out completely. It is also preferable to do stains (an exception is the spore stain) and motility tests on young cultures, because the cells will show their natural size and correct reaction, and motile cells will have functioning flagella.

For certain research or industrial applications, closed-batch culturing with its four phases is inefficient. The alternative is an automatic growth chamber called the **chemostat,** or continuous culture system. This device can admit a steady stream of new nutrients and siphon off used media and old bacterial cells, thereby stabilizing the growth rate and cell number. The chemostat is very similar to industrial fermentors used to produce vitamins and antibiotics. It has the advantage of maintaining the culture in a biochemically active state and preventing it from entering the death phase.

Other Methods of Analyzing Population Growth

Microbiologists have developed several alternative ways of analyzing bacterial growth qualitatively and quantitatively. One of the simplest methods for estimating the size of a population is through turbidometry. This technique relies on the simple observation that a tube of clear nutrient solution becomes cloudy, or **turbid,** as microbes grow in it. In general, a greater turbidity indicates a higher microbial count. Differences in turbidity can be measured by means of sensitive instruments (**figure 7.19**).

Enumeration of Bacteria

Turbidity readings are useful for evaluating relative amounts of growth, but if a more quantitative evaluation is required, the viable colony count or some other enumeration (counting) procedure is necessary. The **direct,** or **total, cell count** involves counting the number of cells in a sample microscopically (**figure 7.20**). This technique, very similar to that used in blood cell counts, employs a special microscope slide (cytometer) calibrated to accept a tiny sample that is spread over a premeasured grid. The cell count from a cytometer can be used to estimate the total number of cells in a larger sample (for instance, of milk or water). One inherent inaccuracy in this method as well as in spectrophotometry is that no distinction can be made between dead and live cells, both of which are included in the count. Additional procedures that can assay the microbial content of water are the most probable number (MPN) and membrane filtration techniques. These are discussed in Chapter 26.

Counting can be automated by sensitive devices such as the *Coulter counter,* which electronically scans a culture as it passes through a tiny pipette. As each cell flows by, it is detected and registered on an electronic sensor (**figure 7.21**). A *flow cytometer* works on a similar principle, but in addition to counting, it can measure cell size and even differentiate between live and dead cells. When used in conjunction with fluorescent dyes and antibodies to tag cells, it has been used to differentiate between gram-positive and gram-negative bacteria. It is being adapted for use as a rapid method to identify pathogens in patient specimens and to differentiate blood cells. More sophisticated forms of the flow cytometer can actually sort cells of different types into separate compartments of a collecting device.

Although flow cytometry can be used to count bacteria in natural samples without the need for culturing them, it requires fluorescent labeling of the cells you are interested in detecting, which is not always possible.

A variation of the polymerase chain reaction (PCR) (see figure 10.7), called real-time PCR, allows scientists to quantify bacteria and other microorganisms that are present in environmental or tissue samples without isolating them and without culturing them.

Percent of light transmitted

(1)

(2)

Figure 7.19 **Turbidity measurements as indicators of growth.** **(a)** Holding a broth to the light is one method of checking for gross differences in cloudiness (turbidity). The broth on the left is transparent, indicating little or no growth; the broth on the right is cloudy and opaque, indicating heavy growth. **(b)** The eye is not sensitive enough to pick up fine degrees in turbidity; more sensitive measurements can be made with a spectrophotometer. (1) A tube with no growth will allow light to easily pass. Therefore, more light will reach the photodetector and give a higher transmittance value. (2) In a tube with growth, the cells scatter the light, resulting in less light reaching the photodetector and, therefore, giving a lower transmittance value.

Figure 7.20 **Direct microscopic count of bacteria.** A small sample is placed on the grid under a cover glass. Individual cells, both living and dead, are counted. This number can be used to calculate the total count of a sample.

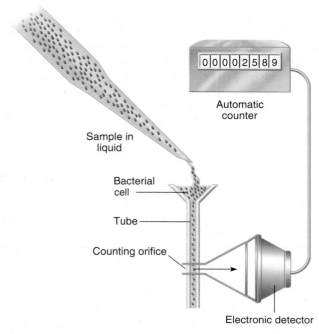

00002589

Automatic counter

Sample in liquid

Bacterial cell

Tube

Counting orifice

Electronic detector

Figure 7.21 **Coulter counter.** As cells pass through this device, they trigger an electronic sensor that tallies their numbers.

Check & Assess Section 7.5

✔ Microbial growth refers both to increase in cell size and increase in number of cells in a population.

✔ The generation time is a measure of the growth rate of a microbial population. It varies in length according to environmental conditions.

✔ The pattern of population growth in simple microbes is to double the number of cells in each generation.

✔ Growth rate is limited by availability of nutrients and buildup of waste products.

✔ Microbial cultures in a nutrient-limited batch environment exhibit four distinct stages of growth: the lag phase, the exponential growth (log) phase, the stationary phase, and the death phase.

✔ These distinct phases of growth occur in response to changing nutrient and waste conditions.

✔ Population growth can be quantified by measuring turbidity, colony counts, and direct cell counts. Other techniques can be used to count bacteria without growing them.

22. What is microbial growth? Explain the stages in binary fission.

23. Why is growth called exponential? What causes the pattern of the growth curve?

24. Contrast two different methods of detecting and counting microbial populations, and discuss their advantages and disadvantages.

25. Explain the relationship between colony counts and colony-forming units. Why can one use the number of colonies as an index of population size?

CASE FILE 7 PERSPECTIVE

In a word, the microbes found by the researchers are extremophiles. They "love" being in environmental conditions on earth that stretch the limits of hot, cold, salty, acidic, and other factors. Most investigations have shown a significant biodiversity in these environments, and we have probably just skimmed the surface of what is present. At one time, it was thought that primarily bacteria and archaeons would be able to colonize such extreme environments, but now we know that some algae, fungi, and protozoa can adapt to extremes almost as readily as these prokaryotes.

To understand how extremophiles survive such extremes, it may be helpful to remind you that these habitats are not extreme to the microbes—this is where they live. They are only extreme from the human perspective. Many extremophiles are the descendents of ancient microbes that evolved in the severe conditions of early earth.

The Berkeley Pit is an example of evolution in action. Only those microbes with traits to adapt to the conditions there (are fit enough) will survive to reproduce. These hardy pioneers must express hidden genetic traits or develop new ones for removing, modifying, or even utilizing the harsh chemicals of their habitat. Major adaptations include alterations of membrane and enzyme structure so they can function in the presence of heavy metals or acid. Some develop mechanisms for transporting the metal or acid out of their cells back into the environment. In many cases, they establish biofilms and other associations that provide a buffer against the conditions.

In being photosynthetic, the algae are a prolific source of oxygen gas that promotes survival and growth of aerobic organisms—namely, fungi, protozoa, and many types of bacteria. In addition, algae play an important role in the biogeochemical cycling of several elements—nitrogen, phosphorus, sulfur, and carbon—that all organisms use in metabolism.

Bioremediation is the introduction of microbes into polluted environments to reduce or degrade the toxic chemicals present there. Since the organisms that naturally detoxify the pit system seem to already be there, the next step would be to enrich the water with extra inorganic nutrients such as nitrates and phosphates that favor algal growth. It is expected that the increased population of algae would absorb the toxic metals and concentrate them in the sediments at the bottom of the pit, where they would be less accessible and less likely to leak out.

Additional coverage of these topics appears in chapter 26.

For more information on the biology of the Berkeley pit, go to http://www.pitwatch.org/.

Chapter Summary with Key Terms

7.1 Microbial Nutrition

Environmental factors that have the greatest impact on microbes are nutrients, temperature, pH, amount of available water, atmospheric gases, light, pressure, and other organisms.

Nutrition consists of taking in chemical substances (nutrients) and assimilating and extracting energy from them.

A. Substances required for survival are **essential nutrients**—usually containing the elements (C, H, N, O, P, S, Na, Cl, K, Ca, Fe, Mg). Essential nutrients are considered **macronutrients** (required in larger amounts) or **micronutrients** (trace elements required in smaller amounts—Zn, Mn, Cu). Nutrients are classed as either **inorganic** or **organic.**

B. A **growth factor** is an organic nutrient (amino acid and vitamin) that cannot be synthesized and must be provided.

C. *Nutritional Types*
 1. An **autotroph** depends on carbon dioxide for its carbon needs. An autotroph that derives energy from light is a **photoautotroph;** one that extracts energy from inorganic substances is a **chemoautotroph. Methanogens** are chemoautotrophs that produce methane.
 2. A **heterotroph** acquires carbon from organic molecules. A **saprobe** is a decomposer that feeds upon dead organic matter, and a **parasite** feeds from a live host and usually causes harm. Parasites that cause damage or death are called **pathogens.**

7.2 Transport: Movement of Substances Across The Cell Membrane

A. A microbial cell must take in nutrients from its surroundings by transporting them across the cell membrane.

B. **Osmosis** is diffusion of water through a selectively permeable membrane.

C. Osmotic changes that affect cells are **hypotonic** solutions, which contain a lower solute concentration, and **hypertonic** solutions, which contain a higher solute concentration. **Isotonic** solutions have the same solute concentration as the inside of the cell. A **halophile** thrives in hypertonic surroundings, and an **obligate halophile** requires a salt concentration of at least 15% but grows optimally in 25%.

D. Passive transport involves the natural movement of substances down a concentration gradient and requires no additional energy (**diffusion**). A form of passive transport that can move specific substances is **facilitated diffusion.**

E. In **active transport,** substances are taken into the cell by a process that consumes energy. In **group translocation,** molecules are altered during transport.

F. **Phagocytosis** and **pinocytosis** are forms of active transport in which bulk quantities of solid and fluid material are taken into the cell.

7.3 Environmental Factors That Influence Microbes

Every organism adapts to a particular habitat and niche.

A. *Temperature:* An organism exhibits **optimum, minimum, and maximum temperatures.** Organisms that cannot grow above 20°C but thrive below 15°C and continue to grow even at 0°C are known as **psychrophiles. Mesophiles** grow from 10°C to 50°C, having temperature optima from 20°C to 40°C. The growth range of **thermophiles** is 45°C to 80°C. Extreme thermophiles may grow to temperatures of 250°C.

B. *Oxygen Requirements:* The ecological need for free oxygen (O_2) is based on whether a cell can handle toxic by-products such as superoxide and peroxide.
 1. **Aerobes** grow in normal atmospheric oxygen and have enzymes to handle toxic oxygen by-products. An aerobic

organism capable of living without oxygen if necessary is a **facultative anaerobe.** An aerobe that requires a small amount of oxygen but does not grow under anaerobic conditions is a **microaerophile.**

2. **Strict (obligate) anaerobes** do not use free oxygen and cannot produce enzymes to dismantle reactive oxides. They are actually damaged or killed by oxygen. An aerotolerant anaerobe cannot use oxygen for respiration yet is not injured by it.

C. *Effects of pH:* Acidity and alkalinity affect the activity and integrity of enzymes and the structural components of a cell. Optimum pH for most microbes ranges from 6 to 8. **Acidophiles** prefer a pH below 7, **alkalinophiles** prefer a pH above 7, and **neutrophiles** grow best around pH 7.

D. *Other Environmental Factors:* Radiation and barometric pressure affect microbial growth. A **barophile** is adapted to life under high pressure (bottom dwellers in the ocean, for example).

7.4 Ecological Associations among Microorganisms

A. *Microbial Interactions:* Microbes coexist in varied relationships in nature.

1. Types of **symbiosis** are **mutualism,** a reciprocal, obligatory, and beneficial relationship between two organisms, and **commensalism,** an organism receiving benefits from another without harming the other organism in the relationship. One form of this is *satellitism.* **Parasitism** occurs between a host and an infectious agent; the host is harmed in the interaction.

2. **Synergism** is a mutually beneficial but not obligatory coexistence in which organisms cooperate to produce a reaction. **Antagonism** entails competition, inhibition, and injury directed against the opposing organism. A special case of antagonism is antibiotic production.

3. **Quorum sensing** is an interaction among members of a biofilm that results in a coordinated reaction, such as secretion of **inducer molecules** and enzymes. By behaving as a unit, the biofilm remains stable and organized. It can occupy and exploit a wide variety of habitats.

7.5 The Study of Microbial Growth

A. The splitting of a parent bacterial cell to form a pair of similar size daughter cells is known as **binary,** or **transverse, fission.**

B. The duration of each division is called the **generation,** or **doubling, time.** A population theoretically doubles with each generation, so the growth rate is **exponential,** and each cycle increases in geometric progression.

C. A **growth curve** is a graphic representation of a closed population over time. Plotting a curve requires an estimate of live cells, called a viable count. The initial flat period of the curve is called the **lag phase,** followed by the **exponential growth phase,** in which viable cells increase in logarithmic progression. Adverse environmental conditions combine to inhibit the growth rate, causing a plateau, or **stationary growth phase.** In the **death phase,** nutrient depletion and waste buildup cause increased cell death.

D. Cell numbers can be counted by a viable plate count, microscope counting chamber, Coulter counter, flow cytometer, or newer methods that detect organisms in their natural settings. Cell growth can also be determined by turbidometry and a total cell count.

Multiple-Choice Questions

Select the correct answer from the answers provided. For questions with blanks, choose the combination of answers that most accurately completes the statement.

1. An organic nutrient essential to an organism's metabolism that cannot be synthesized itself is termed a
 a. trace element c. growth factor
 b. micronutrient d. mineral

2. The source of the necessary elements of life is
 a. an inorganic environmental reservoir
 b. the sun
 c. rocks
 d. the air

3. An organism that can synthesize all its required organic components from CO_2 using energy from the sun is a
 a. photoautotroph c. chemoautotroph
 b. photoheterotroph d. chemoheterotroph

4. An obligate halophile requires high
 a. pH c. salt
 b. temperature d. pressure

5. Chemoautotrophs can survive on _____ alone.
 a. minerals c. minerals and CO_2
 b. CO_2 d. methane

6. Which of the following substances are required by *all* organisms?
 a. organic nutrients c. growth factors
 b. inorganic nutrients d. oxygen gas

7. A pathogen would most accurately be described as a
 a. parasite c. saprobe
 b. commensal d. symbiont

8. Which of the following is true of passive transport?
 a. it requires a gradient c. it includes endocytosis
 b. it uses the cell wall d. it only moves water

9. A cell exposed to a hypertonic environment will _____ by osmosis.
 a. gain water c. neither gain nor lose water
 b. lose water d. burst

10. Active transport of a substance across a membrane requires
 a. a gradient c. water
 b. the expenditure of ATP d. diffusion

11. Psychrophiles would be expected to grow
 a. in hot springs c. at refrigeration temperatures
 b. on the human body d. at low pH

12. Which of the following is *not* involved in quorum sensing?
 a. accumulation of inducer molecules
 b. a certain level of biofilm density
 c. self-monitoring
 d. release of genetic material

13. Superoxide ion is toxic to strict anaerobes because they lack
 a. catalase c. dismutase
 b. peroxidase d. oxidase

14. The time required for a cell to undergo binary fission is called the
 a. exponential growth rate
 b. growth curve
 c. generation time
 d. lag period

15. In a viable plate count, each _____ represents a _____ from the sample population.
 a. cell, colony
 b. colony, cell
 c. hour, generation
 d. cell, generation

16. The stage in population growth with the highest rate of cell division is the
 a. stationary phase
 b. lag phase
 c. log phase
 d. enlargement phase

Case File Questions

1. Which group are the dominant microorganisms in the Berkeley Pit?
 a. archaeons
 b. bacteria
 c. algae
 d. protozoa

2. How are the microbes in the Berkeley Pit detoxifying the water?
 a. They are removing heavy metals.
 b. They are neutralizing the acidity.
 c. They are recycling elements.
 d. a and b
 e. a, b, and c

Writing to Learn

These questions are suggested as a *writing-to-learn* experience. For each question (except 5), compose a one- or two-paragraph answer that includes the factual information needed to completely address the question. All Check & Assess questions can be used for writing-to-learn exercises.

1. Look at the following diagrams and predict in which direction osmosis will take place. Use arrows to show the net direction of osmosis. Is one of these microbes a halophile? Which one?

2. Describe the basic metabolism of methanogens.

3. Discuss extreme environments and types of extremophiles that live there.

4. Differentiate between mutualism and synergism with examples.

5. Write an essay on microbial associations that addresses such ideas as the possible sequence of events during coevolution, using examples in the chapter.

6. a. What biochemical events in quorum sensing ensure that a biofilm can function as a unit?
 b. Explain how biofilm self-monitoring would favor infections.

7. Explain what is happening to the population at points A, B, C, and D in the following diagram.

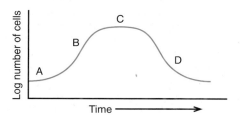

Concept Mapping

Appendix E provides guidance for working with concept maps.

1. Supply your own lines (linkers) and linking words or phrases in this concept map, and provide the missing concepts in the empty box.

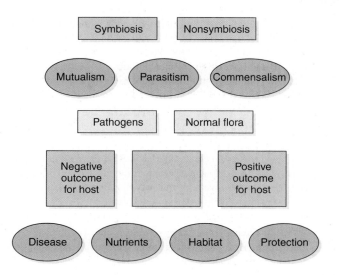

Critical Thinking Questions

Critical thinking is the ability to reason and solve problems using facts and concepts. These questions can be approached from a number of angles, and in most cases, they do not have a single correct answer.

1. a. Is there a microbe that could grow on a medium that contains only the following compounds dissolved in water: $CaCO_3$, $MgNO_3$, $FeCl_2$, $ZnSO_4$, and glucose? Defend your answer.
 b. Check the last entry in the cell analysis summary on page 188. Are all needed elements present here? Where is the carbon?
 c. A useful mnemonic device for keeping track of major elements required by most organisms is: **C HOPKINS Ca Fe, Na Cl Mg**ood. Read aloud, it becomes, See Hopkin's Cafe, NaCl (salt) M good! What does each letter stand for?

2. Describe how one might determine the nutrient requirements of a microbe from Mars. If, after exhausting various nutrient schemes, it still does not grow, what other factors might one take into account?

3. Patients with ketoacidosis associated with diabetes are especially susceptible to fungal infections. Propose why this might be the case.

4. Fill in the following table:

	Source of Carbon	Usual Source of Energy	Example
Photoautotroph			
Photoheterotroph			
Chemoautotroph			
Chemoheterotroph			
Saprobe			
Parasite			

5. a. Explain how osmotic pressure and pH are used in preserving foods.
 b. What effects do they have on microbes?

6. Provide some suggestions for treating anaerobic infections by manipulating the gas content.

7. Could anaerobes and aerobes coexist in the same habitat? Why or why not?

8. a. If an egg salad sandwich sitting in a warm car for 4 hours develops 40,960 bacterial cells, how many more cells would result with just one more hour of incubation? Use the same criteria that were stated in the sample problem on page 209.
 b. With 10 additional hours of incubation?
 c. What do your answers tell you about using clean techniques in food preparation and storage (other than aesthetic considerations)?

9. Looking at figure 7.5, how would you describe the solution in the container? Can you think of a biological setting where this type of osmosis occurs?

10. a. The noted microbial ecologist Martin W. Beijerinck has stated: "Everything is everywhere, the environment selects." Use this concept to explain what ultimately determines whether a microorganism can live in a certain habitat.
 b. Give two examples of ways that microbes become modified to survive.

11. a. Place appropriate points on the axes and draw a graph that fits an obligate thermophile.

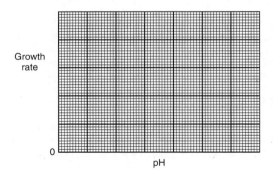

b. Enter points on the axes and show the growth results of an acidophilic microbe.

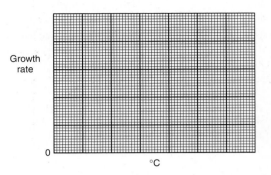

Visual Challenge

1. **Figure 7.9.** Draw a growth curve for a psychrotroph on the following diagram.

2. **Figure 7.21.** What effect will noncellular particulate material have on the automated cell counter shown below? Why?

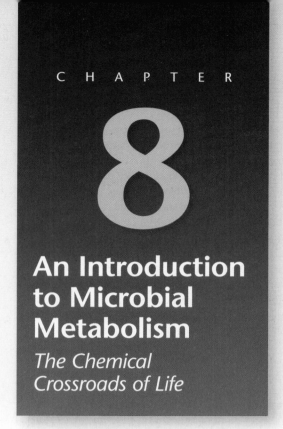

CHAPTER

8

An Introduction to Microbial Metabolism

The Chemical Crossroads of Life

The rumen—a living fermentation factory

Model of the enzyme cellulase—the lifeline for herbivores

An unusual protozoan isolated from a cow's rumen

"Rumen microbes should be considered a cow's best friend. Without microbes, a cow's digestive system would shut down and she would starve to death."

CASE FILE 8 **Symbiosis That Helps Feed The World**

Watching a cow grazing in a summer pasture, you'll notice that it spends a great deal of time chewing and belching—behaviors that are a natural and familiar part of feeding. In reality, these activities provide hints of the workings of a highly complex digestive system that is shared by many herbivorous animals called ruminants. Cattle, bison, goats, sheep, deer, antelopes, and even giraffes and camels all possess a digestive chamber called the rumen—a four-part stomach that is one of the most dynamic ecosystems in biology.

Ruminants feed on grasses and other forage that would ordinarily be very poor nutrition for most animals. While it is true that cattle and other ruminants cannot independently break down the complex carbohydrates in plant fibers, having a rumen makes all the difference. It supports

an incredible mutualistic community with trillions of microorganisms that produce the necessary enzymes for digesting and processing their feed. According to Dr. Paul Weimar, a microbiologist at the US Department of Agriculture: "Rumen microbes should be considered a cow's best friend. Without microbes, a cow's digestive system would shut down and she would starve to death."

As the animals graze on hay and grass, they moisten the feed with saliva and swallow it. The mixture enters the rumen first, where it is churned by muscular action and forms a liquid pulp. The rumen provides constant and carefully controlled conditions. It is maintained at about 39° C, its pH ranges between 5.5 and 7, and it is strictly anaerobic.

Specialized bacteria, protozoa, and a few fungi contribute a variety of enzymes

to the mix, and this begins the digestion process. After a short time, some of the rumen contents are regurgitated as a packet called a cud, which is chewed or ruminated. Chewing the cud pulverizes the plant fibers into smaller particles, which exposes them to increased enzymatic action. The cud is swallowed back into the rumen accompanied by further mixing and digestion. It is at this point that microbes go into fermentation mode and convert the digestive products into a form usable by the cow. From there, the nutrient-rich juices are transferred to the rest of the digestive tract, where they are absorbed by the cow's small intestine.

▶ *What is happening during digestion and fermentation of food molecules?*

To continue the case, go to page 242.

8.1 The Metabolism of Microbes

Expected Learning Outcomes

1. Define *metabolism* and explain two primary categories it contains.
2. Describe the basic functions of enzymes in cells.
3. Outline the outstanding characteristics of enzymes.
4. Explain how enzymes lower the energy required for a reaction to occur.
5. Discuss enzyme structure and interactions between enzymes and substrates.
6. Describe the types of enzyme functions and methods of naming them.
7. Summarize key features of enzyme regulation.

Metabolism, from the Greek term *metaballein,* meaning change, pertains to all chemical and physical workings of cells. It is a very useful term that can be used to designate almost any physiological process. Although metabolism entails thousands of different reactions, most of them can be placed into one of two general categories. In the case of **catabolism,** larger molecules are degraded or broken down into smaller molecules, usually with the release of energy. In the case of **anabolism,** also called biosynthesis, larger molecules are built from smaller ones, which results in the formation of cell structures. It is usually driven by energy derived from catabolism. Even though they have opposite effects, these two forms of metabolism are tightly linked and highly complementary. As we will see, the result of them working together is that all metabolism is coordinated for the cell to carry out essential life functions **(figure 8.1).** The major metabolic functions in cells and the sections of the chapter that cover them are listed here:

1. Enzymes serve as catalysts for both catabolic and anabolic activities. Because of their importance in all areas of microbiology, the remainder of section 8.1 is devoted to characteristics of enzymes.
2. Cells acquire energy from an external source (outside the cell) such as nutrients or the sun. This idea was first discussed in chapter 7 and continues in sections 8.2 and 8.5.
3. Cells extract the energy and transfer it to a high-energy molecule, adenosine triphosphate (ATP), a universal compound for managing energy. This is accomplished by several different types of catabolic pathways, covered in section 8.3.
4. Cells use ATP to power anabolic processes. Simple building blocks are bonded into larger molecules and cell structures, thereby creating greater complexity through the input of this energy. Anabolic pathways are covered in section 8.4.

We suggest you review relevant sections in Chapter 2 covering organic molecules and biochemistry.

Enzymes: Catalyzing the Chemical Reactions of Life

A microbial cell could be viewed as a microscopic factory, complete with basic building materials, a source of energy, and a genetic "blueprint" for running its extensive network of metabolic reactions. But the chemical reactions of life would never proceed without a special class of proteins called **enzymes.** Enzymes are a remarkable example of **catalysts,** chemicals that increase the rate of a chemical reaction without becoming part of the products or being consumed in the reaction. Do not make the mistake of thinking that an enzyme creates a reaction. Because of the free energy inherent in molecules, a reaction could occur spontaneously at some point even without an enzyme but at a very slow rate. A study of the enzyme urease shows that it increases the rate of the breakdown of urea by a factor of 100 trillion as compared to an uncatalyzed reaction. Because most uncatalyzed metabolic reactions do not occur fast enough to sustain

Figure 8.1 Simplified model of metabolism. Cellular reactions fall into two major categories. Catabolism involves the breakdown of complex organic molecules to extract energy and form simpler end products. Anabolism uses the energy to synthesize necessary macromolecules and cell structures from simple precursors, increasing chemical complexity. The two systems interact in a continuous cycle that maintains the metabolic balance of cells.

INSIGHT 8.1

Enzymes as Biochemical Levers

An analogy will allow us to envision the relationship of enzymes to the energy of activation. A large boulder sitting precariously on a cliff's edge contains a great deal of potential energy, but it will not fall to the ground and release this energy unless something disturbs it. It might eventually be disturbed spontaneously as the cliff erodes below it, but that could take a very long time. Moving it with a crowbar (to overcome the resistance of the boulder's weight) will cause it to tumble down freely by its own momentum. If we relate this analogy to chemicals, the reactants are the boulder on the cliff, the activation energy is the energy needed to move the boulder, the enzyme is the crowbar, and the product is the boulder at the bottom of the cliff. Like the crowbar, an enzyme permits a reaction to occur rapidly by getting over the energy "hump." Although this analogy concretely illustrates the concept of the energy of activation, it is imperfect in that the enzyme, unlike the person with the crowbar, does not actually add any energy to the system.

This phenomenon can be represented by plotting the energy of the reaction against the direction of the reaction. The curve of reaction is shown in the accompanying graph. Reactants must overcome the energy of activation (E_{act}) for the reaction to proceed. When a catalyst is present, it lowers the E_{act}. It does this by providing the reactants with an alternate pathway, requiring less energy, by which they can progress to the final state. Note that the products reach the same final energy state with or without an enzyme.

What is the actual physical process performed by the enzyme that speeds up a reaction? *Answer available at http://www.mhhe.com/ talaro8*

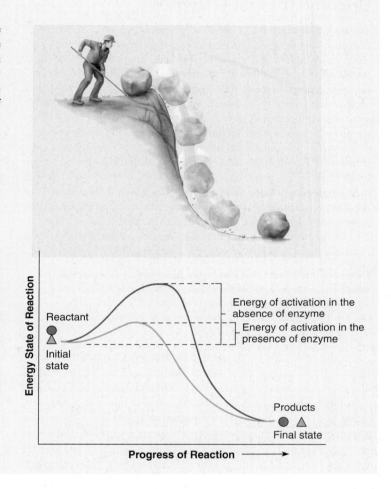

cell processes, enzymes are indispensable to life. Other major characteristics of enzymes are summarized in **table 8.1.**

TABLE 8.1	Checklist of Enzyme Characteristics

- Most composed of protein; may require cofactors
- Act as organic catalysts to speed up the rate of cellular reactions
- Lower the activation energy required for a chemical reaction to proceed (Insight 8.1)
- Enable metabolic reactions to proceed at a speed compatible with life
- Have unique characteristics such as shape, specificity, and function
- Provide an active site for target molecules called substrates
- Are much larger in size than their substrates
- Associate closely with substrates but do not become integrated into the reaction products
- Are not used up or permanently changed by the reaction
- Can be recycled, thus function in extremely low concentrations
- Are greatly affected by temperature and pH
- Can be regulated by feedback and genetic mechanisms

How Do Enzymes Work?

We have said that an enzyme speeds up the rate of a metabolic reaction, but just how does it do this? During a chemical reaction, reactants are converted to products by bond formation or breakage. A certain amount of energy is required to initiate every such reaction, which limits its rate. This resistance to a reaction, which must be overcome for a reaction to proceed, is measurable and is called the **energy of activation** or activation energy (**Insight 8.1**). In the laboratory, overcoming this initial resistance can be achieved by

- increasing thermal energy (heating) to increase molecular velocity,
- increasing the concentration of reactants to increase the rate of molecular collisions, or
- adding a catalyst.

In most living systems, the first two alternatives are not feasible, because elevating the temperature is potentially harmful and higher concentrations of reactants are not practical. This leaves only the action of catalysts, and enzymes fill this need efficiently and potently.

INSIGHT 8.2

Unconventional Enzymes

The molecular reactions of cells are still an active and rich source of discovery, and new findings come along on a regular basis that break older "rules" and adjust our understanding. Originally, proteins were considered the only biological molecules that act as catalysts; then biologists found a novel type of RNA termed **ribozymes.** Ribozymes are associated with many types of cells and a few viruses, and they display some of the properties of protein catalysts, such as having a specific active site and interacting with a substrate. But these molecules are remarkable because their substrate is other RNA. Ribozymes are thought to be remnants of the earliest molecules on earth that could have served as both catalysts and genetic material. In natural systems, ribozymes are involved in self-splicing or cutting of RNA molecules during final processing of the genetic code (see chapter 9). Further research has shown that ribozymes can be designed to handle other kinds of activities, such as inhibiting gene expression. Several companies have developed and are testing ribozyme-based therapies for treating cancer and AIDS.

Propose some effects that ribozymes have on microorganisms that would make them suitable for therapy. *Answer available at* http://www.mhhe.com/talaro8

A "hammerhead" ribozyme consisting of a single RNA strand curved around to form an active site (indentation between tetraloop and stem I).

At the molecular level, an enzyme promotes a reaction by serving as a physical site upon which the reactant molecules, called **substrates,** can be positioned for various interactions. The enzyme is much larger in size than its substrate, and it presents a uniquely shaped pocket that fits only that particular substrate. Although an enzyme binds to the substrate and participates directly in changes to the substrate, it does not become a part of the products, is not used up by the reaction, and can function over and over again. Enzyme speed, defined as the number of substrate molecules converted per enzyme per second, is well documented. Speeds range from several million for catalase to a thousand for lactate dehydrogenase. A key to understanding the roles of enzymes in metabolism is their structure.

Enzyme Structure

The primary structure of all enzymes is protein (with some exceptions—**Insight 8.2),** and enzymes can be classified as simple or conjugated. Simple enzymes consist of protein alone, whereas conjugated enzymes **(figure 8.2)** contain protein and nonprotein molecules. A conjugated enzyme, sometimes referred to as a **holoenzyme,** is a combination of a protein, now called the **apoenzyme,** and one or more **cofactors (table 8.2).** Cofactors are either organic molecules, called **coenzymes,** or inorganic elements (metal ions) that many enzymes require to become functional. In some enzymes, the cofactor is loosely associated with the apoenzyme by noncovalent bonds; in others, it is linked by covalent bonds.

Apoenzymes: Specificity and the Active Site

Apoenzymes range in size from small polypeptides with about 100 amino acids and a molecular weight of 12,000 to large polypeptide

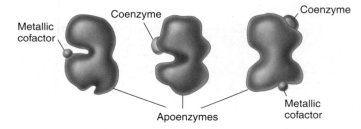

Figure 8.2 **Conjugated enzyme structure.** All examples have an apoenzyme (polypeptide or protein) component and one or more cofactors.

TABLE 8.2	Selected Enzymes, Catalytic Actions, and Cofactors	
Enzyme	**Action**	**Metallic Cofactor Required**
Catalase	Breaks down hydrogen peroxide	Iron (Fe)
Oxidase	Adds electrons to oxygen	Iron, copper (Cu)
Hexokinase	Transfers phosphate to glucose	Magnesium (Mg)
Urease	Splits urea into an ammonium ion	Nickel (Ni)
Nitrate reductase	Reduces nitrate to nitrite	Molybdenum (Mo)
DNA polymerase complex	Synthesis of DNA	Zinc (Zn) and Mg

Levels of Structure

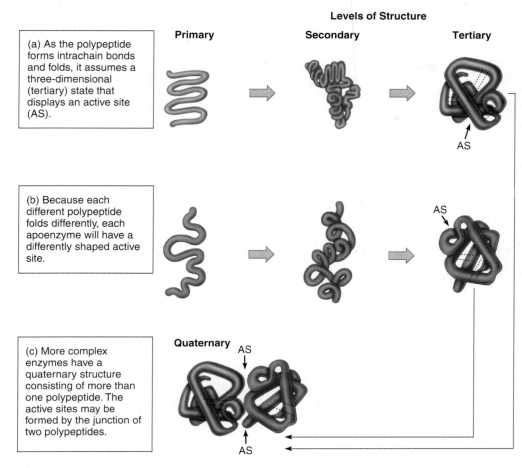

Primary **Secondary** **Tertiary**

(a) As the polypeptide forms intrachain bonds and folds, it assumes a three-dimensional (tertiary) state that displays an active site (AS).

AS

(b) Because each different polypeptide folds differently, each apoenzyme will have a differently shaped active site.

AS

(c) More complex enzymes have a quaternary structure consisting of more than one polypeptide. The active sites may be formed by the junction of two polypeptides.

Quaternary AS

AS

Figure 8.3 How the active site and specificity of the apoenzyme arise. Review figure 2.22 for details of this process.

conglomerates with thousands of amino acids and a molecular weight of over 1 million. Like all proteins, an apoenzyme exhibits levels of molecular complexity called the primary, secondary, tertiary, and, in larger enzymes, quaternary organization (**figure 8.3**). As we saw in chapter 2 (see figure 2.22), the first three levels of structure arise when a single polypeptide chain undergoes an automatic folding process and achieves stability by forming disulfide and other types of bonds. Folding causes the surface of the apoenzyme to acquire three-dimensional features that result in the enzyme's specificity for substrates. The actual pocket where the substrate binds is called the **active site,** or **catalytic site,** and there can be from one to several such sites (figure 8.3). Because each type of enzyme has a different primary structure (type and sequence of amino acids), this gives rise to variations in folding and unique active sites.

Enzyme-Substrate Interactions

For a reaction to take place, a temporary **enzyme-substrate** union must occur at the active site. There are several explanations for the way this happens (**figure 8.4**). The specificity is often

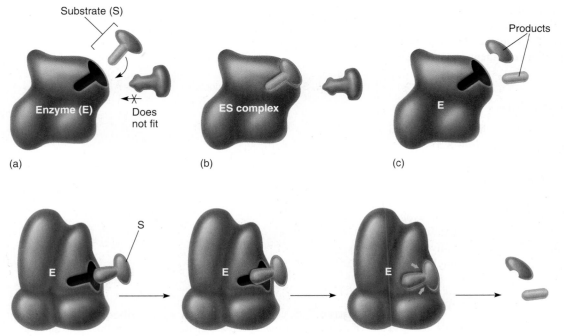

Substrate (S)

Products

Enzyme (E)

Does not fit

ES complex

E

(a) (b) (c)

S

E E E

(d)

Figure 8.4 Enzyme-substrate reactions. **(a)** When the enzyme and substrate come together, the substrate (S) must show the correct fit and position with respect to the enzyme (E). **(b)** When the ES complex is formed, it enters a transition state. During this temporary but tight interlocking union, the enzyme participates directly in breaking or making bonds. **(c)** Once the reaction is complete, the enzyme releases the products. **(d)** The induced-fit model proposes that the enzyme recognizes its substrate and adapts slightly to it so that the final binding is even more precise.

described as a "lock-and-key" fit in which the substrate is inserted into the active site's pocket. This is a useful analogy, but an enzyme is not a rigid lock mechanism. It is likely that the enzyme actually helps the substrate move into the active site through slight changes in its shape. This is called an *induced fit* (**figure 8.4d**).

The bonds formed between the substrate and enzyme are weak and, of necessity, easily reversible. Once the enzyme-substrate complex has formed, appropriate reactions occur on the substrate, often with the aid of a cofactor, and a product is formed and released. The enzyme can then attach to another substrate molecule and repeat this action. Although enzymes can potentially catalyze reactions in both directions, most examples in this chapter depict them working in one direction only.

Cofactors: Supporting the Work of Enzymes

In chapter 7, you learned that microorganisms require specific metal ions called trace elements and certain organic growth factors. In many cases, the need for these substances arises from their roles as **cofactors.** The **metallic cofactors,** including iron, copper, magnesium, manganese, zinc, cobalt, selenium, and many others, participate in precise functions between the enzyme and its substrate. In general, metals activate enzymes, help bring the active site and substrate close together, and participate directly in chemical reactions with the enzyme-substrate complex.

Coenzymes are organic cofactors that work in conjunction with the apoenzyme to perform a necessary alteration of a substrate. The general function of a coenzyme is to remove a chemical group from one substrate molecule and add it to another substrate, thereby serving as a temporary carrier of this group (**figure 8.5**). The specific activities of coenzymes are many and varied. In section 8.3, we shall see that coenzymes carry and transfer hydrogen atoms, electrons, carbon dioxide, and amino groups. One of the most common components of coenzymes are **vitamins,** which explains why vitamins are important to nutrition and may be required as growth factors for living things. Vitamin deficiencies prevent the complete holoenzyme from forming. Consequently, both the chemical reaction and the structure or function dependent upon that reaction are compromised.

Classification of Enzyme Functions

Enzymes are classified and named according to characteristics such as site of action, type of action, and substrate (**Insight 8.3**).

Location and Regularity of Enzyme Action

Enzymes perform their tasks either inside or outside of the cell in which they were produced. After initial synthesis in the cell, **exoenzymes** are transported extracellularly, where they break down large food molecules or harmful chemicals (see discussion of hydrolysis reactions). Examples of exoenzymes are cellulase, amylase, and penicillinase. By contrast, **endoenzymes** are retained intracellularly and function there (**figure 8.6**). Most enzymes of the metabolic pathways are of this variety.

In terms of their presence in the cell, enzymes are not all produced in equal amounts or at equal rates. Some, called

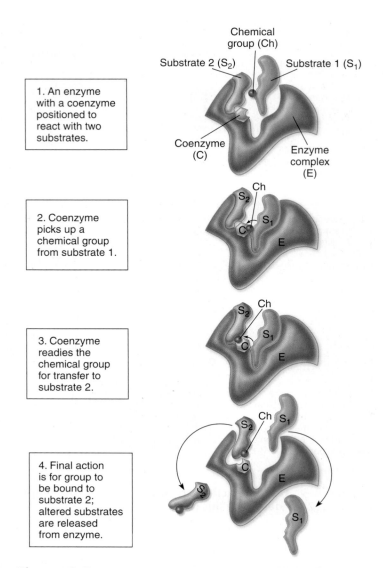

1. An enzyme with a coenzyme positioned to react with two substrates.

2. Coenzyme picks up a chemical group from substrate 1.

3. Coenzyme readies the chemical group for transfer to substrate 2.

4. Final action is for group to be bound to substrate 2; altered substrates are released from enzyme.

Figure 8.5 **The carrier functions of coenzymes.** Many enzymes use coenzymes to transfer chemical groups from one substrate to another.

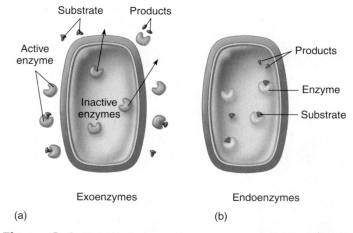

(a) Exoenzymes

(b) Endoenzymes

Figure 8.6 **Types of enzymes, as described by their location of action.** **(a)** Exoenzymes are transported outside the cell to become active. **(b)** Endoenzymes remain in the cell and function there.

INSIGHT 8.3

The Enzyme Name Game

Most metabolic reactions require a separate and unique enzyme. Up to the present time, researchers have discovered and named over 5,000 of them. Given the complex chemistry of the cell and the extraordinary diversity of living things, many more enzymes probably remain to be discovered. A standardized system of nomenclature and classification was developed to prevent discrepancies.

In general, an enzyme name is composed of two parts: a prefix or stem word derived from a certain characteristic—usually the substrate acted upon or the type of reaction catalyzed, or both—followed by the ending *-ase.*

The system classifies the enzyme in one of these six classes, on the basis of its general biochemical action:

1. *Oxidoreductases* transfer electrons from one substrate to another, and *dehydrogenases* transfer a hydrogen from one compound to another.
2. *Transferases* transfer functional groups from one substrate to another.
3. *Hydrolases* cleave bonds on molecules with the addition of water.
4. *Lyases* add groups to or remove groups from double-bonded substrates.
5. *Isomerases* change a substrate into its isomeric form. An **isomer** is a compound that has the same molecular formula as another compound but differs in arrangement of the atoms.
6. *Ligases* catalyze the formation of bonds with the input of ATP and the removal of water.

Each enzyme is also assigned a technical name that indicates the specific reaction it catalyses. These names are useful for biochemists but are often too lengthy for routine use. More often, a common name is derived from some notable feature such as the substrate or enzyme action. For example, *hydrogen peroxide oxidoreductase* is more commonly known as *catalase.* With this system, an enzyme that digests a carbohydrate substrate is a *carbohydrase;* a specific carbohydrase, *amylase,* acts on starch (amylose is a major component of starch). An enzyme that hydrolyzes peptide bonds of a protein is a *proteinase, protease,* or *peptidase,* depending on the size of the protein substrate. Some fats and other lipids are digested by *lipases.* DNA is hydrolyzed by *deoxyribonuclease,* generally shortened to *DNase.* A *synthetase* or *polymerase* bonds together many small molecules into large molecules. Other examples of enzymes are presented in **table 8.A.** (See also table 8.2.)

What would we call an enzyme that clots blood? One that digests collagen? *Answer available at http://www.mhhe.com/talaro8*

TABLE 8.A A Sampling of Enzymes, Their Substrates, and Their Reactions

Common Name	Systematic Name	Enzyme Class	Substrate	Action
Lactase	β-D-galactosidase	Hydrolase	Lactose	Breaks lactose down into glucose and galactose
Penicillinase	Beta-lactamase	Hydrolase	Penicillin	Hydrolyzes beta-lactam ring
DNA polymerase	DNA nucleotidyl-transferase	Transferase	DNA nucleosides	Synthesizes a strand of DNA using the complementary strand as a model
Lactate dehydrogenase	Same as common name	Oxidoreductase	Pyruvic acid	Catalyzes the conversion of pyruvic acid to lactic acid
Oxidase	Cytochrome oxidase	Oxidoreductase	Molecular oxygen	Catalyzes the reduction (addition of electrons and hydrogen) to O_2

constitutive enzymes (figure 8.7a), are always present and in relatively constant amounts, regardless of the amount of substrate. The enzymes involved in utilizing glucose, for example, are very important in metabolism and thus are constitutive. Other enzymes are **regulated enzymes (figure 8.7b),** the production of which is either turned on (induced) or turned off (repressed) in response to changes in concentration of the substrate or product. The level of inducible and repressible enzymes is controlled by the degree to which the genes for these proteins are expressed, as discussed in chapter 9.

Synthesis and Hydrolysis Reactions

A growing cell is in a constant state of synthesis. Such anabolic reactions require enzymes (ligases) to form covalent bonds between smaller substrate molecules. Also known as **condensation reactions,** synthesis reactions typically require ATP and release one water molecule for each bond made **(figure 8.8a).** Catabolic reactions involving energy transactions are equally active in a growing cell and require their own set of enzymes. For example, digestion of large substrates requires enzymes to break them down into smaller molecules so they can be used. Because the breaking of bonds usually requires the input of water, digestion is also termed a **hydrolysis* reaction (figure 8.8b).**

The Role of Microbial Enzymes in Disease

Many pathogens secrete unique exoenzymes that help them avoid host defenses or promote their multiplication in tissues. Because these enzymes contribute to pathogenicity, they are referred to as virulence factors, or toxins in some cases (see chapter 13). *Streptococcus pyogenes* (a cause of throat and skin infections) produces a streptokinase that

* *hydrolysis* (hy-drol'-uh-sis) Gr. *hydro,* water, and *lysis,* setting free.

digests blood clots and apparently assists in invasion of wounds. Another exoenzyme from this bacterium is called streptolysin.[1] In mammalian hosts, streptolysin damages blood cells and tissues. It is also responsible for lysing red blood cells used in blood agar media, and this trait is used for identifying the bacteria growing in culture (see figure 3.7).

Pseudomonas aeruginosa, a respiratory and skin pathogen, produces elastase and collagenase, which digest elastin and collagen, two proteins found in connective tissue. These increase the severity of certain lung diseases and burn infections. *Clostridium perfringens,* an agent of gas gangrene, synthesizes lecithinase C, a lipase that profoundly damages cell membranes and accounts for

1. Even though most enzyme names end in *-ase* (see Insight 8.3), not all do.

(a)

(b)

Figure 8.7 **Constitutive and regulated enzymes.**
(a) Constitutive enzymes are present in constant amounts in a cell. The addition of more substrate does not increase the numbers of these enzymes. **(b)** The concentration of some regulated enzymes in a cell increases or decreases in response to substrate levels.

Glycosidic bond

(a) Condensation Reaction. Forming a glycosidic bond between two glucose molecules to generate maltose requires the removal of a water molecule and energy from ATP.

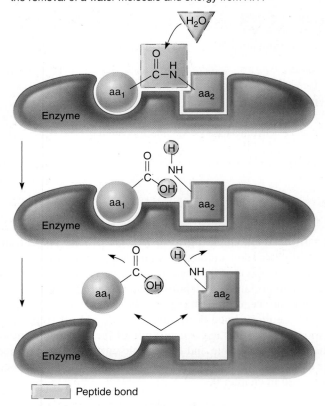

Peptide bond

(b) Hydrolysis Reaction. Breaking a peptide bond between two amino acids requires a water molecule that adds OH to one amino acid and H to the other.

Figure 8.8 **Examples of enzyme-catalyzed synthesis and hydrolysis reactions.**

the tissue death associated with this disease. Not all enzymes digest tissues; some, such as penicillinase, inactivate penicillin and thereby protect a microbe from this drug.

The Sensitivity of Enzymes to Their Environment

The activity of an enzyme is highly influenced by the cell's environment. In general, enzymes operate only under the natural temperature, pH, and osmotic pressure of an organism's habitat. When enzymes are subjected to changes in these normal conditions, they tend to be chemically unstable, or **labile.** Low temperatures inhibit catalysis, and high temperatures denature the apoenzyme. **Denaturation** is a process by which the bonds that collectively maintain the native shape of the apoenzyme are disrupted. This loss causes extreme distortion of the enzyme's shape and prevents the substrate from attaching to the active site (described in chapter 11). Such nonfunctional enzymes block metabolic reactions and thereby can lead to cell death. Low or high pH and certain chemicals (heavy metals, alcohol) are also denaturing agents.

Regulation of Enzymatic Activity and Metabolic Pathways

Metabolic reactions proceed in a systematic, highly regulated manner that maximizes the use of available nutrients and energy. The cell responds to environmental conditions by adopting those metabolic reactions that most favor growth and survival. Because enzymes are critical to these reactions, the regulation of metabolism is largely the regulation of enzymes by an elaborate system of checks and balances. Let us take a look at some general features of metabolic pathways.

Patterns of Metabolic Pathways

Metabolic reactions rarely consist of a single action or step. More often, they occur in a multistep series or pathway, with each step catalyzed by an enzyme. An individual reaction is shown in various ways, depending on the purpose at hand **(figure 8.9).** The product of one reaction is often the reactant (substrate) for the next, forming a linear chain of reactions. Many pathways have branches that provide alternate methods for nutrient processing. Others take a cyclic form, in which the starting molecule is regenerated to initiate another turn of the cycle. Pathways generally do not stand alone; they are interconnected and merge at many sites.

Every pathway has one or more enzyme pacemakers that set the rate of a pathway's progression. Regulation of pacemaker enzymes proceeds on two fundamental levels. Either the enzyme itself is directly inhibited or activated, or the amount of the enzyme in the system is altered (decreased or increased). When factors affect the enzyme directly, the system can be finely controlled or "tuned." When regulation is at the genetic level (enzyme synthesis), control is slower and less sensitive.

Direct Controls on the Action of Enzymes

The bacterial cell has many ways of directly influencing the activity of its enzymes. It can inhibit enzyme activity by supplying a molecule that resembles the enzyme's normal substrate. The "mimic"

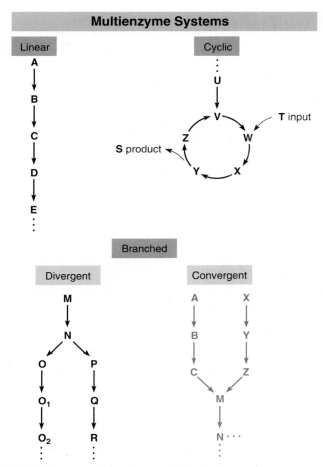

Figure 8.9 Patterns of metabolic pathways. In general, metabolic pathways consist of a linked series of individual chemical reactions that produce intermediary metabolites and lead to a final product. These pathways occur in several patterns, including linear, cyclic, and branched. Anabolic pathways involved in biosynthesis result in a more complex molecule, each step adding on a functional group, whereas catabolic pathways involve the dismantling of molecules and can generate energy. Virtually every reaction in a series involves a specific enzyme.

can then occupy the enzyme's active site, preventing the actual substrate from binding there. Because the mimic cannot actually be acted on by the enzyme and a product is not released, the enzyme is effectively shut down. This form of inhibition is called **competitive inhibition,** because the mimic is competing with the substrate for the binding site **(figure 8.10).** In chapter 12, you will see that some antibiotics have the same strategy of competing with enzymatic active sites to shut down metabolic processes.

Another form of inhibition can occur with special types of enzymes that have two binding sites: the active site and another area called the **regulatory** or **allosteric* site** (figure 8.10). The action of these enzymes is controlled by the binding of molecules other than the substrate in the regulatory site. Often the regulatory molecule is the product of the enzymatic reaction itself. This provides a negative

* *allosteric* (al-oh-stair´-ik) Gr. *allos,* other, and *steros,* solid. Literally, "another space."

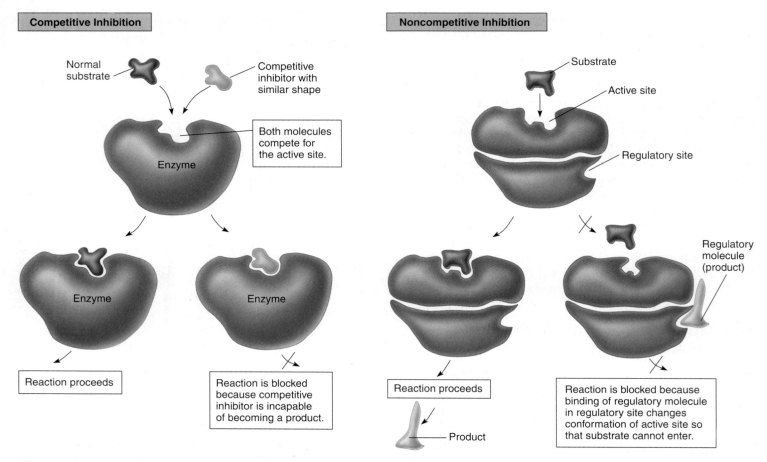

Figure 8.10 **Examples of two common mechanisms for regulation of enzyme action.**

feedback mechanism that can slow down enzymatic activity once a certain concentration of product is produced. This is **noncompetitive inhibition,** because the regulator molecule does not bind in the same site as the substrate.

Controls on Enzyme Synthesis

Controlling enzymes by controlling their synthesis is another effective mechanism. Because enzymes do not last indefinitely, they eventually degrade and must be replaced. This cycle works into the scheme of the cell, where replacement of enzymes can be regulated according to cell demand. The mechanisms of this system are genetic in nature; that is, they require regulation of DNA and protein synthesis, a topic we cover in more depth in chapter 9.

Enzyme repression is a means to stop further synthesis of an enzyme somewhere along its pathway. As the level of the end product from a given enzymatic reaction has built to excess, the genetic apparatus responsible for replacing these enzymes is automatically suppressed **(figure 8.11).** The response time is longer than for feedback inhibition, but its effects are more enduring.

A response that resembles the inverse of enzyme repression is **enzyme induction.** In this process, enzymes appear (are induced) only when suitable substrates are present; that is, the synthesis of enzyme is induced by its substrate. This response enables the organism to adapt to a variety of nutrients, and it also prevents a microbe from wasting of energy by making enzymes for which substrates are lacking.

Check & Assess Section 8.1

✔ Metabolism includes all the biochemical reactions that occur in the cell. It is a self-regulating complex of interdependent processes that encompass many thousands of chemical reactions.

✔ Anabolism is the energy-requiring subset of metabolic reactions, which synthesize large molecules from smaller ones.

✔ Catabolism is the energy-releasing subset of metabolic reactions, which degrade or break down large molecules into smaller ones.

✔ Enzymes are proteins or RNA that catalyze most biochemical reactions by forming enzyme-substrate complexes.

✔ The binding of the substrate by an enzyme makes possible both bond-forming and bond-breaking reactions, depending on the pathway involved. Enzymes may utilize cofactors as carriers and activators.

✔ Enzymes are classified and named according to the kinds of reactions they catalyze.

✔ To function effectively, enzymes require specific conditions of temperature, pH, and osmotic pressure.

✔ Enzyme activity is regulated by processes that affect the enzymes directly, or by altering the amount of enzyme produced during protein synthesis.

Process Figure 8.11 **One type of genetic control of enzyme synthesis: enzyme repression.** (1), (2), (3), (4), (5) Genetic controls are active and the enzyme is synthesized continuously until enough product has been made. (6), (7) Excess product reacts with a site on DNA that regulates the enzyme's synthesis, thereby inhibiting further enzyme production.

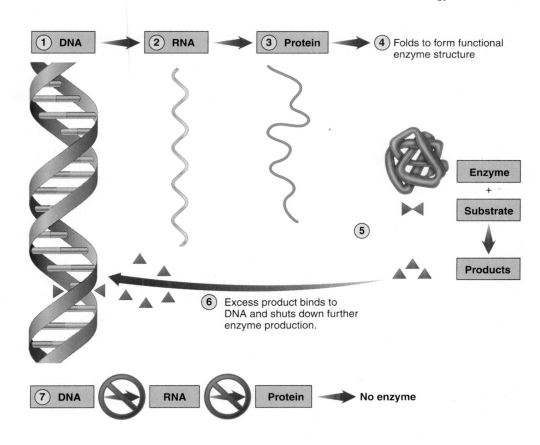

1. Describe 10 important biochemical properties of enzymes.
2. Describe the chemistry of enzymes, and explain how the apoenzyme forms.
3. Show diagrammatically the interaction of holoenzyme and its substrate and general products that can be formed from a reaction.
4. Differentiate among the chemical composition and functions of various cofactors. Provide examples of each type.
5. Summarize the direct and indirect controls that regulate enzymes.
6. Explain what denaturation is, and describe the conditions that denature enzymes.

8.2 The Pursuit and Utilization of Energy

Expected Learning Outcomes

8. Relate the diverse sources of energy and the cellular management of energy.

9. Describe biological oxidation-reduction and energy transfer.

10. Discuss the structure, basic formation, and roles of ATP in cells.

Energy is defined as the capacity to do work or to cause change. In order to carry out the work of an array of metabolic processes, cells require constant input and expenditure of some form of usable energy. Energy commonly exists in these forms:

- thermal (heat) energy from molecular motion,
- radiant (wave) energy from visible light or other rays,
- electrical energy from a flow of electrons,
- mechanical energy from a physical change in position,
- atomic energy from reactions in the nucleus of an atom, and
- chemical energy present in the bonds of molecules.

Cells are far too fragile to rely in any constant way on thermal or atomic energy for cell transactions. Except for certain chemoautotrophic bacteria, the ultimate source of energy is the sun, but only photosynthetic autotrophs can tap this source directly. This capacity of photosynthesis to convert solar energy into chemical energy provides both a nutritional and an energy basis for all heterotrophic living things. We cover this unique form of metabolism in section 8.5. For the most part, only chemical energy and reactions can fuel cellular energetics.

In this section, we further explore the role of energy as it is related to the use of nutrients, redox carriers, and the generation of ATP.

Cell Energetics

Cells manage energy in the form of chemical reactions that change molecules. This often involves activities such as the making or breaking of bonds and the transfer of electrons. Not all cellular reactions have the same relation to energy. Some release energy and others require it to proceed. For example, a reaction with this equation:

$$X + Y \xrightarrow{\text{Enzyme}} Z + \text{Energy}$$

releases energy as it goes forward. This type of reaction is termed **exergonic.*** Energy of this type is considered free: it is available for doing cellular work. Energy transactions such as the following:

$$\text{Energy} + A + B \xrightarrow{\text{Enzyme}} C$$

are called **endergonic*** because they are driven forward with the addition of energy. Exergonic and endergonic reactions are coupled, so that released energy is immediately put to use.

Summaries of metabolism might make it seem that cells "create" energy from nutrients, but they do not. What they actually do is extract chemical energy already present in nutrient fuels and apply that energy toward useful work in the cell, much like a gasoline engine releases energy as it burns fuel. The engine does not actually produce energy, but it converts the potential energy of the fuel to the free energy of work.

The basis for cell energetics lies in chemical bonds, electron transfers, and special pathways and their enzymes. During exergonic reactions, electrons are pulled out of atoms, which releases their inherent energy. The harvested energy is transferred to specialized molecules, mainly ATP. Newly formed ATP immediately releases its energy

* *exergonic* (ex-er-gon′-ik) Gr. *exo,* without, and *ergon,* work. In general, catabolic reactions are exergonic.
* *endergonic* (en-der-gon′-ik) Gr. *endo,* within, and *ergon,* work. Anabolic reactions tend to be endergonic.

to support endergonic reactions such as synthesis **(figure 8.12).** These reactions are balanced and cyclic, so that ATP is constantly being made, used, regenerated, and so on. The actual process of electron transfer is a type of redox reaction that was first described in chapter 2. Let us revisit this topic and see how it is involved in metabolism.

A Closer Look at Biological Oxidation and Reduction

Many compounds readily give or receive electrons and participate in oxidation (the loss of electrons) or reduction (the gain of electrons). The compound that loses the electrons is **oxidized,** and the compound that receives the electrons is **reduced.** Such oxidation-reduction or **redox reactions** are common in the cell and indispensable to energy transactions (see also figure 2.9). Such reactions always occur in pairs, with an electron donor and an electron acceptor, which constitute a *conjugate pair,* or *redox pair.* The reaction can be represented as follows:

Electron donor + Electron acceptor ⟶ Electron donor + Electron acceptor
•e⁻ •e⁻
(Reduced) (Oxidized) (Oxidized) (Reduced)

Electron Transfer Is Energy Transfer In cells, a redox reaction involves not only electrons but also the hydrogen atoms from which

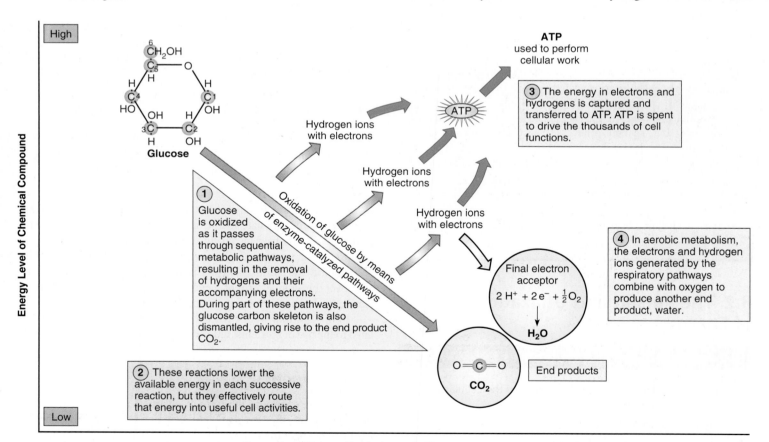

Process Figure 8.12 A simplified model for harvesting energy. The central events of cell energetics include the release of energy during the systematic dismantling of an energy-rich fuel such as glucose. This is achieved by the shuttling of hydrogens and electrons to sites in the cell where their energy can be transferred to ATP. The final products are low-energy CO_2 and H_2O molecules.

they come. Recall that a hydrogen atom consists of a single proton and a single electron. Hydrogen readily gives up its single electron to certain polar molecules, leaving a hydrogen ion or proton (H^+). So, the hydrogen ion is an important part of cellular metabolism right along with its electron. For simplicity's sake, we use the term *electron transfer* in our discussions, but keep in mind that hydrogen is directly involved in these transactions.

TAKE NOTE: ELECTRON AND PROTON CARRIERS: MOLECULAR SHUTTLES

Carrier molecules resemble shuttles that are alternately loaded and unloaded, repeatedly accepting and releasing electrons and hydrogens to facilitate the transfer of redox energy. The most common carrier is the coenzyme **nicotinamide adenine dinucleotide (NAD$^+$)**, which carries hydrogens and a pair of electrons **(figure 8.13)**. Other common redox carriers are FAD, NADP (NAD phosphate), coenzyme A, and the compounds of the respiratory chain, which are fixed into membranes.

A molecule such as NAD$^+$ carries the electrons between substrates. During catabolic reactions discussed in section 8.3, some energy given off by electron transfer is used to synthesize ATP. To complete the reaction, the electrons are passed to a final electron acceptor.

In aerobic metabolism, this acceptor is molecular oxygen; in anaerobic metabolism, it is some other inorganic or organic compound. Reduced NAD$^+$ can be represented in various ways. Because 2 hydrogens are removed, the actual carrier state is NADH + H$^+$, but this is somewhat cumbersome, so we will represent it with the shorter NADH.

With the transfer process being cyclic, there is a built-in efficiency. Oxidized NAD$^+$ is automatically regenerated and can be ready to accept more electrons.

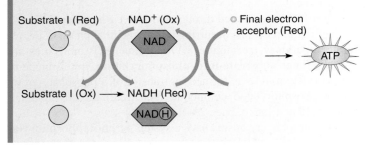

One example of a typical redox reaction is seen in this overall equation associated with aerobic respiration (compare with figure 8.12):

$$\overbrace{C_6H_{12}O_6 + 6O_2 \rightarrow 6CO_2}^{\text{Oxidation}} \underbrace{+ H_2O}_{\text{Reduction}} + \text{Energy}$$

Glucose

Note that the electron donor is hydrogen and the electron acceptor is oxygen. The electrons are transported and energy is released through a stepwise series of reactions.

For it to be useful, the energy is ultimately captured through **phosphorylation,** a process that adds an inorganic phosphate P_i to adenosine diphosphate, converting it to ATP.

$$\text{ADP} + P_i + \text{Energy} \leftrightarrows \text{ATP}$$

Most of the time, there are enzyme-bound carriers for handling and transporting electrons and hydrogens until they can be transferred to a final electron acceptor. Such redox reactions that remove hydrogen from organic compounds are a constant supplier of electrons for the main powerhouse in cells—the electron transport system (see figure 8.21).

Adenosine Triphosphate: Metabolic Money

ATP has been described as metabolic currency because it can be earned, banked, saved, spent, and exchanged. You can gain some appreciation for the enormous need for ATP if you consider that even one tiny cell of *Escherichia coli* consumes around 2.5 million ATP molecules per second for synthesis alone! As a temporary energy repository, ATP provides a connection between energy-yielding catabolism and all other cellular activities that require energy. Some clues to its energy-storing properties lie in its unique molecular structure.

The Molecular Structure of ATP ATP is a three-part molecule consisting of a nitrogen base (adenine) linked to a 5-carbon sugar

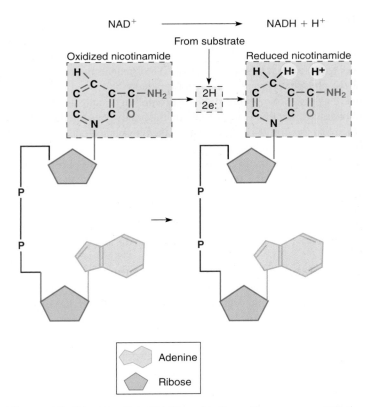

Figure 8.13 Details of NAD reduction. The coenzyme NAD$^+$ contains the vitamin nicotinamide (niacin) and the purine adenine attached to a pair of ribose phosphate molecules (a dinucleotide). The principal site of action is on the nicotinamide (boxed area). Hydrogens and electrons donated by a substrate interact with a carbon on the top of the ring. One hydrogen bonds there, carrying two electrons (H:), and the other hydrogen is carried in solution as H$^+$ (a proton).

Figure 8.14 The structure of adenosine triphosphate (ATP) and its partner compounds, ADP and AMP.

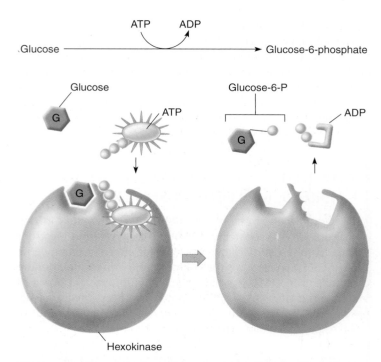

Figure 8.15 The phosphorylation of glucose by ATP. The first step in catabolizing glucose is the addition of a phosphate from ATP by an enzyme called hexokinase. This use of high-energy phosphate as an activator is a recurring feature of many metabolic pathways.

(ribose), with a chain of three phosphate groups bonded to the ribose (**figure 8.14**). The type, arrangement, and especially the proximity of atoms in ATP combine to form a compatible but unstable high-energy molecule. The high energy of ATP originates in the orientation of the phosphate groups, which are relatively bulky and carry negative charges. The proximity of these repelling electrostatic charges imposes a strain that is most acute on the bonds between the last two phosphate groups. When the strain on the phosphate bonds is released by removal of the terminal phosphates, free energy is released.

Breaking the bonds between the second and third phosphates of ATP yields adenosine diphosphate (ADP), which can then be converted to adenosine monophosphate (AMP). It is worthwhile noting how economically a cell manages its pool of adenosine nucleotides. AMP derivatives help form the backbone of RNA and are also a major component of certain coenzymes (NAD^+, FAD, and coenzyme A).

The Metabolic Role of ATP ATP is at the center of a cell's energy cycle, regardless of whether it is a heterotroph using respiration, a photosynthetic organism, or a chemoautotroph. Expenditure of ATP powers all processes that are part of cellular work. This energy cycle can be depicted as shown here:

Respiration
Fermentation
Photosynthesis
Chemoautotrophy

ADP + P_i

ATP

Chemical work
Transport work
Mechanical work

In many instances, the energy released from the high energy bond in ATP activates individual subunits before they are enzymatically bonded during synthesis. ATP is also used to prepare a

molecule for catabolism such as the phosphorylation of a 6-carbon sugar during the early stages of glycolysis (**figure 8.15**).

Because ATP needs to be constantly replenished, reactions are constantly continuing in the reverse order. The ADPs are situated to react immediately with phosphate and a high energy source. The energy to regenerate a high-energy phosphate bond comes from catabolic pathways where substrates or electron carriers are releasing their energy.

Most ATPs are formed through **oxidative phosphorylation,** a series of redox reactions occurring during the final phase of the respiratory pathway (see figure 8.17a). Some ATP molecules are formed through **substrate-level phosphorylation.** In this mode of phosphorylation, ATP is formed by transfer of a phosphate group from a phosphorylated compound (substrate) directly to ADP (**figure 8.18** and step 6, page 234).

Phototrophic organisms have a system of **photophosphorylation,** in which the ATP is formed through a series of sunlight-driven reactions discussed in more detail in section 8.5.

Figure 8.16 ATP formation by substrate-level phosphorylation. The substrate has been bound to a phosphate, creating a bond with high potential energy. Transfer of the high-energy phosphate to ADP gives rise to ATP and returns the substrate to its original form.

Check & Assess Section 8.2

✔ All metabolic processes require the constant expenditure of some form of usable energy. Chemical energy is the currency that runs the metabolic processes of the cell.

✔ Chemical energy is obtained from the electrons of nutrient molecules through catabolism. It is used to perform the cellular "work" of biosynthesis, movement, membrane transport, and growth.

✔ Energy is extracted from nutrient molecules by redox reactions. A redox pair of substances passes electrons and hydrogens between them. The donor substance loses electrons, becoming oxidized. The acceptor substance gains electrons, becoming reduced.

✔ The redox reaction is often facilitated by specialized carrier molecules such as nicotinamide adenine dinucleotide (NAD^+).

✔ ATP is an important energy molecule of the cell. It donates free energy to anabolic reactions and is continuously regenerated by three phosphorylation processes: substrate-level phosphorylation, oxidative phosphorylation, and photophosphorylation in certain organisms.

7. Explain how oxidation of a substrate proceeds without oxygen.
8. Refer to the redox equation on page 229 for glucose, and name which substance is reduced, which is oxidized, the oxidizing agent, and the reducing agent.
9. In the following redox pairs, which compound is reduced and which is oxidized?
 a. NAD^+ and NADH
 b. $FADH_2$ and FAD
 c. lactic acid and pyruvic acid
 d. NO_3^- and NO_2^-
 e. Ethanol and acetaldehyde
10. a. Describe the roles played by ATP and NAD^+ in metabolism.
 b. What particular features of their structure lend them to these functions?
11. Name the major ways that substrate-level phosphorylation is different from oxidative phosphorylation.
12. What is meant by the concept of the "final electron acceptor"?

8.3 Pathways of Bioenergetics

Expected Learning Outcomes

11. Relate the main points of bioenergetics and energy strategies in microorganisms.
12. Describe the main catabolic pathways and where they occur in aerobic respiration.
13. Define *glycolysis* and explain its input and output, its basic steps, and how it is linked to the Krebs cycle and oxidative phosphorylation.
14. Explain the basic steps in the Krebs cycle, its input and output, and how it is linked to oxidative phosphorylation.
15. Describe the components of the respiratory chain and their functions.

16. Explain the chemiosmotic mechanism of ATP formation, its relationship to electron transport, and the final step in electron transfer.
17. Summarize the results of aerobic respiration.
18. Define and describe some reactions of anaerobic respiration.
19. Define *fermentation* and explain several examples of the process in microorganisms.

One of the scientific community's greatest achievements was deciphering the biochemical pathways of cells. Initial work with bacteria and yeasts, followed by studies with animal and plant cells, clearly demonstrated metabolic similarities and strongly supported the presence of metabolic patterns that are universal. **Bioenergetics** is a study of the mechanisms of cellular energy release, including catabolic and anabolic routes. Although these pathways are interconnected and interdependent, anabolic pathways are not simply reversals of catabolic ones. At first glance, it may seem more economical to use identical pathways, but having different enzymes and divided pathways allows anabolism and catabolism to proceed simultaneously without interference.

In the following sections, we focus our discussion on the most common catabolic pathways that illustrate general principles of other pathways as well.

Catabolism: An Overview of Nutrient Breakdown and Energy Release

The primary catabolism of fuels (such as glucose) that results in energy release in many organisms proceeds through a series of three coupled pathways:

1. **glycolysis,*** also called the Embden-Meyerhof-Parnas (EMP) pathway;[2]
2. the **Krebs cycle,** also known as the citric acid or tricarboxylic acid cycle;[2] and
3. the **respiratory chain** (electron transport and oxidative phosphorylation).

Each segment of the pathway is responsible for a specific set of actions on various products of glucose. The major interconnections of these pathways in aerobic respiration are represented in **figure 8.17,** and their reactions are summarized in **table 8.3.** In the following sections, we observe each of these pathways in greater detail.

Energy Strategies in Microorganisms

Nutrient processing is extremely varied, especially in bacteria, yet in many cases it is based on three basic catabolic pathways. In previous discussions, microorganisms were categorized according to their requirement for oxygen gas, and this requirement is related directly to their mechanisms of energy release. As we shall see, **aerobic respiration** is a series of reactions (glycolysis, Krebs cycle, and the

* *glycolysis* (gly-kol'-ih-sis) Gr. *glykys,* sweet, and *lysis,* to split.

2. The EMP pathway is named for the biochemists who first outlined its steps. Krebs is in honor of Sir Hans Krebs who, with F. A. Lipmann, delineated this pathway, an achievement for which they won the Nobel Prize in Physiology or Medicine in 1953. TCA refers to the involvement of several organic acids containing three carboxylic acid groups, citric acid being the first tricarboxylic acid formed.

Figure 8.17 **Side-by-side comparison of the stages, reactions, and major products of pathways for aerobic respiration, anaerobic respiration, and fermentation.** Note that all begin with glycolysis but vary in use of other pathways, electron acceptors, and ATP yield.

respiratory chain) that converts glucose to CO_2, produces H_2O, and generates energy (**figure 8.17a**). It relies on free oxygen as the final acceptor for electrons and hydrogens, and generates a relatively large amount of ATP. Aerobic respiration is characteristic of many bacteria, fungi, protozoa, and animals, and it is the system we emphasize here.

Facultative and aerotolerant anaerobes may use only the glycolysis scheme to incompletely oxidize or **ferment** glucose. In this

case, oxygen is not required, organic compounds are the final electron acceptors, and a relatively small amount of ATP is synthesized (**figure 8.17c**).

Some strictly anaerobic microorganisms metabolize by means of **anaerobic respiration.** This system may involve the same three pathways as aerobic respiration, but it does not use molecular oxygen as the final electron acceptor (**figure 8.17b**). Instead, it uses

TABLE 8.3	Metabolic Strategies among Heterotrophic Microorganisms Growing on Glucose			
Scheme	**Pathways Involved**	**Final Electron Acceptor**	**Products**	**Primary Pathway Found In**
Aerobic respiration	Glycolysis, Krebs cycle, electron transport	O_2	ATP, CO_2, H_2O	Aerobes; facultative anaerobes
Anaerobic metabolism				
Fermentative	Glycolysis	Organic molecules	ATP, CO_2, ethanol, lactic acid*	Facultative, aerotolerant, strict anaerobes
Respiration	Glycolysis, Krebs cycle, electron transport	Various inorganic ions $(NO_3^-, SO_4^{2-}, CO_3^{3-})$	CO_2, ATP, organic acids, H_2S, CH_4, N_2	Anaerobes; some facultatives

*The products of microbial fermentations are extremely varied and include organic acids, alcohols, and gases.

other oxidized ions such as NO_3^-, SO_4^{2-}, or CO_3^{3-} for the final electron acceptor. Aspects of fermentation and anaerobic respiration are covered in later parts of section 8.3.

Aerobic Respiration

Aerobic respiration is a series of enzyme-catalyzed reactions in which electrons are transferred from fuel molecules such as glucose to oxygen as a final electron acceptor. This pathway is the principal energy-yielding scheme for aerobic heterotrophs, and it provides both ATP and metabolic intermediates for many other pathways in the cell, including those of protein, lipid, and carbohydrate synthesis (see figure 8.26).

Aerobic respiration in microorganisms can be summarized by an equation:

$$\text{Glucose (C}_6\text{H}_{12}\text{O}_6) + 6\text{ O}_2 + 38\text{ ADP} + 38\text{ P}_i \rightarrow$$
$$6\text{ CO}_2 + 6\text{ H}_2\text{O} + 38\text{ ATP}$$

The seeming simplicity of the equation for aerobic respiration conceals its complexity. Fortunately, we do not have to present all of the details to address some important concepts concerning its reactants and products, as follows:

1. the steps in the oxidation of glucose,
2. the involvement of coenzyme carriers and the final electron acceptor,
3. where and how ATP originates,
4. where carbon dioxide originates, and
5. where oxygen is required.

Glucose: The Starting Compound

Carbohydrates such as glucose are good fuels because these compounds are readily oxidized; that is, they are superior hydrogen and electron donors. The hydrogen and electrons removed from them can be used in energy transfers. The end products of these carbon compounds are energy-rich ATP and energy-poor carbon dioxide and water. Polysaccharides (starch, glycogen) and disaccharides (maltose, lactose) are stored sources of glucose for the respiratory pathways. Although we use glucose as the main starting compound, other hexoses (fructose, galactose) and fatty acid subunits can enter the pathways of aerobic respiration as well, as we will see in section 8.4.

Glycolysis: The Starting Lineup

The process called glycolysis (EMP) is an anaerobic pathway that converts glucose through several steps into pyruvic acid. Depending on the organism and the conditions, it may be only the first phase of aerobic respiration, or it may serve as the primary metabolic pathway (fermentation). Glycolysis provides a significant means to synthesize a small amount of ATP in the absence of oxygen and also to generate pyruvic acid, an essential intermediary metabolite.

Steps in the Glycolytic Pathway Glycolysis proceeds along nine linear steps. The first portion of glycolysis involves activation of glucose, which is followed by oxidation reactions of the glucose fragments, the synthesis of ATP, and the formation of pyruvic acid. **Although each step of metabolism is catalyzed by a specific enzyme, we will mention only certain key ones.** The following outline lists the principal steps of glycolysis. To see an overview of glycolysis, go to **figure 8.18.** Detailed illustrations appear in **appendix B.**

1. **Glucose** is phosphorylated by an **ATP** to produce glucose-6 phosphate. This is a way of "priming" the system and keeping the glucose inside the cell.
2. Glucose-6-phosphate is converted to its isomer, fructose-6-phosphate.
3. Another ATP is spent in phosphorylating the first carbon of fructose-6-phosphate, which yields fructose-1,6-diphosphate.

Up to this point, no energy has been released, no oxidation-reduction has occurred, and, in fact, 2 ATPs have been used and the molecules remain in the 6-carbon state.

4. Fructose-1,6-diphosphate is split into two 3-carbon fragments: glyceraldehyde-3-phosphate (G-3-P) and dihydroxyacetone phosphate (DHAP). These molecules are isomers, and DHAP is converted to G-3-P, which is the more reactive form for the following reactions.

The effect of the splitting of fructose diphosphate is to double every subsequent reaction, because where there was once a single molecule, there are now two to be fed into the remaining pathways.

5. Each molecule of glyceraldehyde-3-phosphate becomes involved in the single oxidation-reduction reaction of glycolysis. This sets the scene for ATP synthesis. The **NAD$^+$** coenzyme complex picks up hydrogens from G-3-P, forming **NADH.** This step is accompanied by the addition of an inorganic phosphate (PO_4^{3-}) to form an unstable bond on the third carbon of the G-3-P substrate. The product of these reactions is diphosphoglyceric acid (DPGA).
6. One of the high-energy phosphates of diphosphoglyceric acid is donated to ADP via substrate-level phosphorylation resulting in a molecule of ATP. The other product of this reaction is 3-phosphoglyceric acid (3 PGA).
7, 8. During this phase, the 3-phosphoglyceric acid is converted to 2-phosphoglyceric acid (2 PGA) through the shift of a phosphate from the third to the second carbon. Then, the removal of a water molecule from 2-phosphoglyceric acid converts it to phosphoenolpyruvic acid (PEPA). The result gives rise to a high-energy phosphate bond.
9. In the final reaction of glycolysis, phosphoenolpyruvic acid gives up its high-energy phosphate to form a second ATP, again via substrate-level phosphorylation. This reaction also produces **pyruvic acid** (pyruvate), a compound with many roles in metabolism.

The reactions of glycolysis thus give rise to two pyruvic acids per glucose. The two substrate-level phosphorylations produce a total of **4** ATPs. However, 2 ATPs were expended for steps 1 and 3, so the net number of ATPs available to the cell from these reactions is **2.**

In aerobic organisms, the NADH formed during step 5 will be shuttled to the electron transport system, where the final H acceptor will be oxygen, and additional ATPs will be generated per NADH. In organisms that ferment glucose, the NADH will be oxidized back to NAD$^+$, and the hydrogen acceptor will be an organic compound (see figure 8.17*b*).

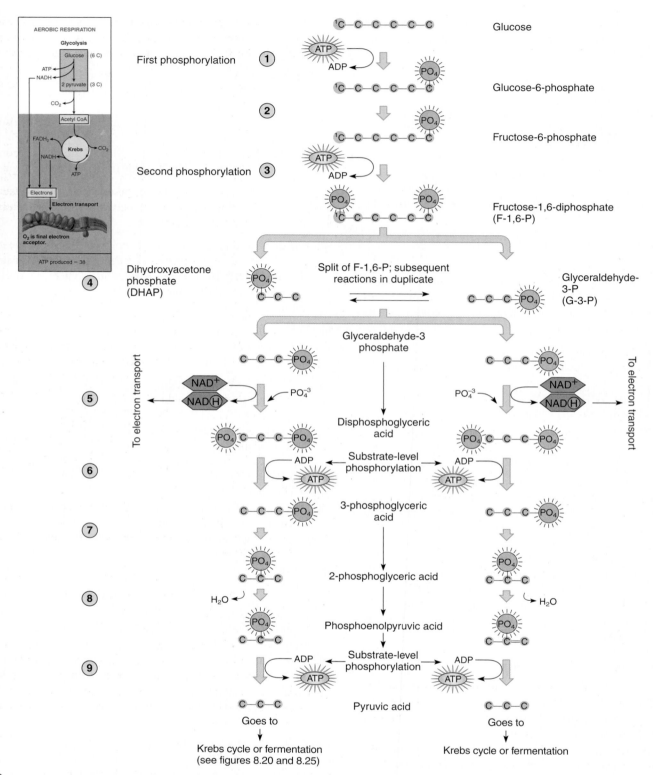

Process Figure 8.18 **Summary of the steps in glycolysis.** The primary reaction steps are described in the accompanying text. These reactions all take place in the cytoplasm, regardless of the cell type. Inset in upper corner has a gold box to show the position of this pathway in the scheme of aerobic respiration. A detailed analysis of specific chemical structures and reactions can be found in Appendix B.

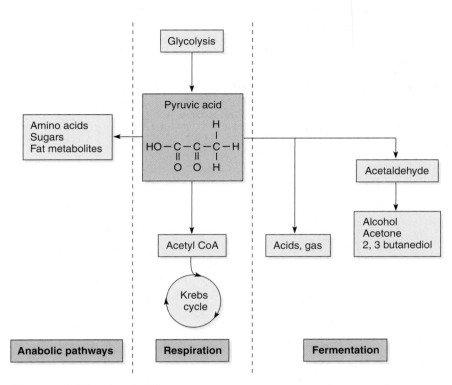

Figure 8.19 **The fates of pyruvic acid (pyruvate).** This metabolite is an important hub in the processing of nutrients by microbes. It may be fermented anaerobically to several end products or oxidized completely to CO_2 and H_2O through the Krebs cycle and the electron transport system. It can also serve as a source of raw material for synthesizing amino acids and carbohydrates.

Pyruvic Acid—A Central Metabolite

Pyruvic acid occupies an important position in several pathways, and different organisms handle it in different ways **(figure 8.19)**. In strictly aerobic organisms and some anaerobes, pyruvic acid is channeled into the Krebs cycle for processing and energy release. Facultative anaerobes can adopt a fermentative metabolism, in which pyruvic acid is further reduced into acids or other products.

The Preparation of Pyruvic Acid for the Krebs Cycle

The oxidation of glucose by glycolysis yields a comparatively small amount of energy. Pyruvic acid is still energy-rich, containing a number of extractable hydrogens and electrons to power ATP synthesis, and this will be achieved through the work of the second and third phases of aerobic respiration, during which pyruvic acid is converted to CO_2 and H_2O. In the following section, we examine the second phase—the Krebs cycle—that takes place in the mitochondrial matrix in eukaryotes and in the cytoplasm of bacteria.

To connect the glycolysis pathway to the Krebs cycle, the pyruvic acid is first converted to a starting compound for that cycle **(figure 8.20,** orange oval at top). This step involves another redox reaction, and it also releases the first carbon dioxide molecule. It brings in a cluster of enzymes and coenzyme A to participate in the oxidation of pyruvic acid, the reduction of NAD^+ to NADH, and the decarboxylation of pyruvic acid to a 2-carbon acetyl group. The

acetyl group remains attached to coenzyme A, forming **acetyl coenzyme A** (acetyl CoA) that feeds into the Krebs cycle. Pyruvate dehydrogenase, the enzyme complex that makes this reaction possible, is huge. In *E. coli,* it contains four vitamins and has a molecular weight of 6 million—larger than a ribosome!

The NADH produced in the formation of acetyl CoA will be shuttled into electron transport and used to generate ATP via oxidative phosphorylation in the third phase of aerobic respiration.

Keep in mind that all reactions during these subsequent phases happen twice for each glucose because of the two pyruvates that are released during glycolysis.

The Krebs Cycle—A Carbon and Energy Wheel

As you learned earlier, a cyclic pathway is one in which the starting compound is regenerated at the end. The Krebs cycle has eight steps, beginning with citric acid formation and ending with oxaloacetic acid (figure 8.20). The 2-carbon acetyl groups from coenzyme A are combined with a 4-carbon oxaloacetate molecule to form citric acid, a 6-carbon molecule. Although this seems a cumbersome way to dismantle such a small molecule, it is necessary in biological systems for extracting larger amounts of energy from the remaining acetyl fragment.

As we take a single spin around the Krebs cycle, it will be helpful to keep track of

1. the numbers of carbons (#C) of each substrate and product,
2. reactions where CO_2 is generated,
3. the involvement of the electron carriers NAD^+ and FAD, and
4. the site of ATP synthesis.

Be aware that the terms used for organic acids can be shown as either the acid form (oxaloacetic acid) or its salt (oxaloacetate). We use the acid form in this text for the most part.

The eight reactions in the Krebs cycle are:

1. Oxaloacetic acid (oxaloacetate; 4C) reacts with the acetyl group (2C) on acetyl CoA, thereby forming citric acid (citrate; 6C) and releasing coenzyme A so it can join with another acetyl group.
2. Citric acid is converted to its isomer, isocitric acid (isocitrate; 6C), to prepare this substrate for the decarboxylation and redox reaction of the next step.
3. Isocitric acid is acted upon by an enzyme complex including NAD^+ or NADP (depending on the organism) in a reaction that generates NADH or NADPH, splits off a carbon dioxide, and leaves α-ketoglutaric acid (α-ketoglutarate; 5C).
4. Alpha-ketoglutaric acid serves as a substrate for the final decarboxylation reaction and yet another redox reaction, involving coenzyme A and yielding NADH. The product is the high-energy compound succinyl CoA (4C).

At this point, the cycle has completed the formation of 3 CO_2 molecules that balance out the 3original 3-carbon pyruvic acid released

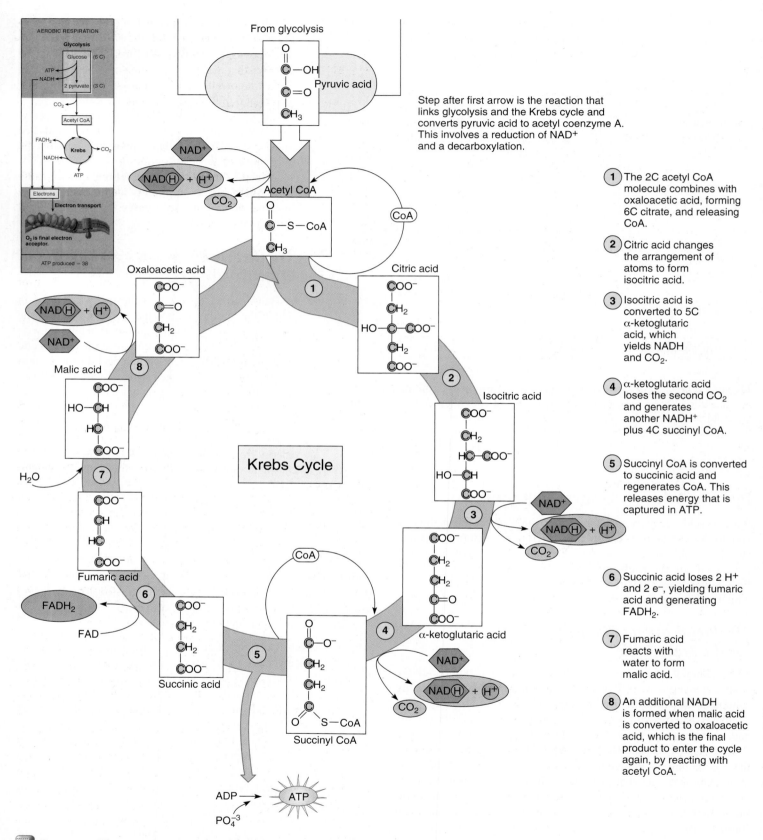

Process Figure 8.20 **The reactions of a single turn of the Krebs cycle.** Each glucose will produce two spins of this pathway. Note that this is an enlarged, more detailed view of the middle phase depicted in figure 8.17a (inset). It occurs in the cytoplasm of prokaryotes and the mitochondrial matrix of eukaryotes.

Step after first arrow is the reaction that links glycolysis and the Krebs cycle and converts pyruvic acid to acetyl coenzyme A. This involves a reduction of NAD⁺ and a decarboxylation.

1. The 2C acetyl CoA molecule combines with oxaloacetic acid, forming 6C citrate, and releasing CoA.

2. Citric acid changes the arrangement of atoms to form isocitric acid.

3. Isocitric acid is converted to 5C α-ketoglutaric acid, which yields NADH and CO₂.

4. α-ketoglutaric acid loses the second CO₂ and generates another NADH⁺ plus 4C succinyl CoA.

5. Succinyl CoA is converted to succinic acid and regenerates CoA. This releases energy that is captured in ATP.

6. Succinic acid loses 2 H⁺ and 2 e⁻, yielding fumaric acid and generating FADH₂.

7. Fumaric acid reacts with water to form malic acid.

8. An additional NADH is formed when malic acid is converted to oxaloacetic acid, which is the final product to enter the cycle again, by reacting with acetyl CoA.

by glycolysis. The remaining steps are needed not only to regenerate the oxaloacetic acid to start the cycle again but also to extract more energy from the intermediate compounds leading to oxaloacetic acid.

5. Succinyl CoA is the source of the one substrate level phosphorylation in the Krebs cycle. In most bacteria, it proceeds with the formation of ATP, although eukaryotes produce guanosine triphosphate (GTP), a similar source of energy. The other product at this step is succinic acid (succinate; 4C).

6. Succinic acid undergoes a redox reaction, but in this case, the electron and H^+ acceptor is flavin adenine dinucleotide (FAD). The enzyme that catalyzes this reaction, succinyl dehydrogenase, is found in the bacterial cell membrane and mitochondrial crista of eukaryotic cells. The $FADH_2$ generated directly enters the electron transport system. Fumaric acid (fumarate; 4C) is the product of this reaction.

7. The addition of water to fumaric acid results in malic acid (malate; 4C). This is one of the few reactions in respiration that directly incorporates water.

8. Malic acid is dehydrogenated (with formation of a final NADH), and oxaloacetic acid is formed. This step brings the cycle back to its original starting position, where oxaloacetic acid can react with acetyl coenzyme A.

An important feature of this pathway is that acetyl groups from the breakdown of certain fats can enter the pathway and be used as an energy source (discussed in section 8.4).

The Krebs cycle is not present in all cells, and it may not function under all metabolic conditions. But even in cells with alternate metabolic schemes, it is essential for generating small organic molecules that microbes require for synthesis (see figure 8.26).

The Respiratory Chain: Electron Transport and Oxidative Phosphorylation

We now come to the energy chain, which is the final "processing mill" for electrons and hydrogen and the major generator of ATP. Overall, the electron transport system (ETS) consists of a chain of special redox carriers that receive electrons from reduced carriers (NADH, $FADH_2$) generated by glycolysis and the Krebs cycle. The ETS shuttles the electrons in a sequential and orderly fashion (**figure 8.21**). The flow of electrons down this chain is highly energetic and gives off ATP at various points. The step that finalizes the transport process is the acceptance of electrons and hydrogen by oxygen, with the production of water.

Some variability exists from one organism to another, but the principal compounds that carry out these complex reactions are NADH dehydrogenase, flavoproteins, coenzyme Q (**ubiquinone**),* and **cytochromes**.* The cytochromes contain a tightly bound metal atom at their center that is actively involved in accepting electrons and donating them to the next carrier in the series. The highly compartmentalized structure of the respiratory chain is an important

* *ubiquinone* (yoo-bik′-wih-nohn) L. *ubique,* everywhere. A type of chemical, similar to vitamin K, that is very common in cells.
* *cytochrome* (sy′-toh-krohm) Gr. *cyto,* cell, and *kroma,* color. Cytochromes are pigmented, iron-containing molecules similar to hemoglobin.

factor in its function. Note in figure 8.21 that the electron transport carriers and enzymes are embedded in the inner mitochondrial membranes in eukaryotes. The cell membrane is the equivalent structure for housing them in bacteria.

The Carriers of Electron Transport: The Energy Cascade

The principal questions about the electron transport system are: How are the electrons passed from one carrier to another in the series? How is this progression coupled to ATP synthesis? and, Where and how is oxygen utilized? Although the biochemical details of this process are rather complicated, the basic reactions consist of a number of redox reactions now familiar to us. In general, the seven carrier compounds and their enzymes are arranged in linear sequence and are reduced and oxidized in turn (figure 8.21, lower view).

The sequence of electron carriers in the respiratory chain of most aerobic organisms is

1. NADH dehydrogenase, which is closely associated in a complex with the adjacent carrier, which is
2. flavin mononucleotide (FMN);
3. coenzyme Q;
4. cytochrome b;
5. cytochrome c_1;
6. cytochrome c; and
7. cytochromes a and a_3, which are complexed together.

The NADHs formed during glycolysis and the Krebs cycle are conveyed to the first carrier in the ETS. This sets in motion the remaining six steps. With each redox exchange, the energy level of the reactants is lessened. The released energy is captured and used by **ATP synthase** complexes, stationed along the cristae in close association with the ETS carriers. Each NADH that enters electron transport can give rise to a maximum of 3 ATPs. This coupling of ATP synthesis to electron transport is termed **oxidative phosphorylation.** Because $FADH_2$ from the Krebs cycle enters the cycle after the NAD^+ and FMN complex reactions, its reactions result in a maximum yield of about 2 ATPs.

The Formation of ATP and Chemiosmosis What biochemical processes are involved in coupling electron transport to the production of ATP? We first look at the system in eukaryotes, which have the components of electron transport embedded in a precise sequence on mitochondrial membranes. They are stationed between the inner mitochondrial matrix and the outer intermembrane space (figure 8.21). According to a widely accepted concept called **chemiosmosis,** as the electron transport carriers shuttle electrons, they actively pump hydrogen ions (protons) into the outer compartment of the mitochondrion. This process sets up a concentration gradient of hydrogen ions called the **proton motive force (PMF).** The PMF consists of a difference in charge between the outer membrane compartment (+) and the inner membrane compartment (−) (**figure 8.22a**).

Separating the charge has the effect of a battery, which can temporarily store potential energy. This charge will be maintained by the impermeability of the inner cristae membranes to H^+.

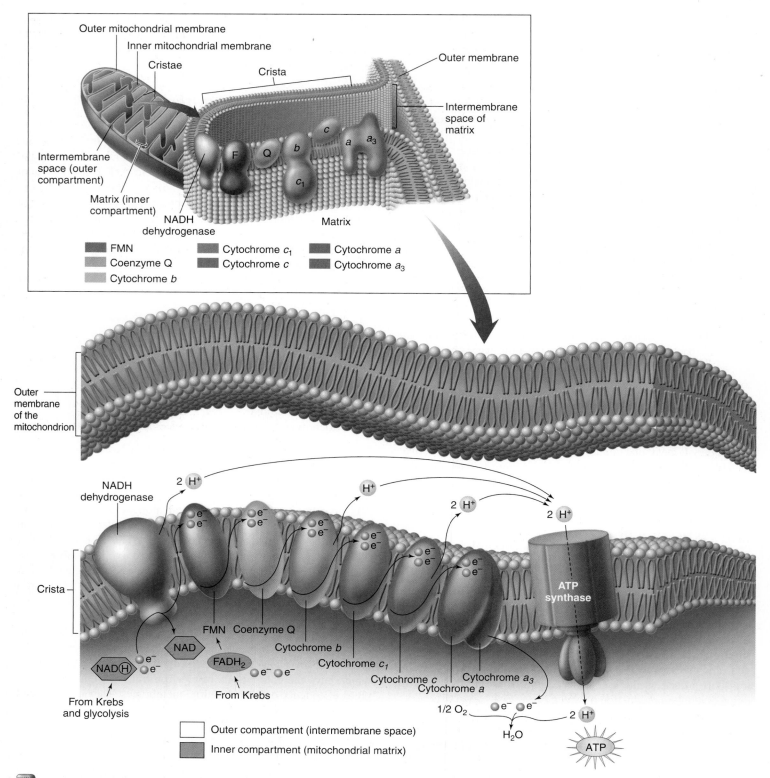

Figure 8.21 **The electron transport system and oxidative phosphorylation on the mitochondrial crista.** Starting at NADH dehydrogenase, electrons brought in from the Krebs cycle by NADH are passed along the chain of electron transport carriers. Each adjacent pair of transport molecules undergoes a redox reaction. Coupled to the transport of electrons is the simultaneous active transport of H^+ into the outer compartment by specific carriers. These processes set the scene for ATP synthesis and final H^+ and e^- acceptance by oxygen.

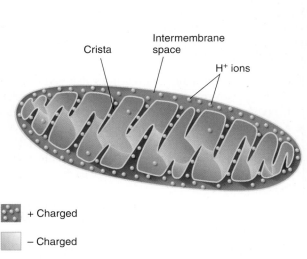

Crista

Intermembrane space

H+ ions

+ Charged

− Charged

(a) As the carriers in the mitochondrial cristae transport electrons, they also actively pump H+ ions (protons) to the intermembrane space, producing a chemical and charge gradient between the outer and inner mitochondrial compartments.

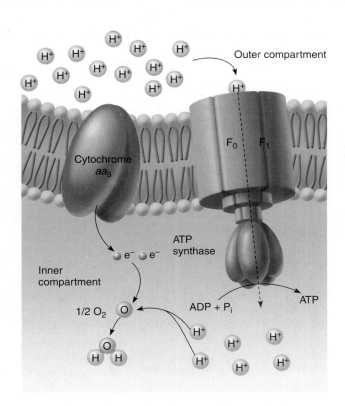

Outer compartment

Cytochrome aa_3

F_0 F_1

ATP synthase

Inner compartment

e^- e^-

ATP

ADP + P_i

1/2 O_2 O

O

H H

(b) The distribution of electric potential and the concentration gradient of protons across the membrane drive the synthesis of ATP by ATP synthase. The rotation of this enzyme couples diffusion of H+ to the inner compartment with the bonding of ADP and P_i. The final event of electron transport is the reaction of the electrons with the H+ and O_2 to form metabolic H_2O. This step is catalyzed by cytochrome oxidase (cytochrome aa_3).

Cell wall

Cell membrane with ETS

Cytoplasm

ATP synthase

ADP

ATP

(c) Enlarged view of bacterial cell envelope to show the relationship of electron transport and ATP synthesis. Bacteria have the ETS and ATP synthase stationed in the cell membrane. ETS carriers transport H+ and electrons from the cytoplasm to the exterior of the membrane. Here, it is collected to create a gradient just as it occurs in mitochondria.

Figure 8.22 Chemiosmosis—the force behind ATP synthesis.

The only site where H^+ can diffuse into the inner compartment is at the ATP synthase complex, which sets the stage for the final processing of H^+ and ATP synthesis.

ATP synthase is a complex enzyme composed of two large units, F_0 and F_1 **(figure 8.22b).** It is embedded in the membrane, but part of it rotates like a motor and traps chemical energy. As the H^+ ions flow through the F_0 center of the enzyme by diffusion, the F_1 compartments pull in ADP and P_i. Rotation induces a change in the three-dimensional structure of ATP synthase. This adjustment creates a high energy bond between ADP and P_i. The ATP that forms is released into the inner compartment (figure 8.22b). The enzyme is then rotated back to the start position and will continue the process.

Bacterial ATP synthesis occurs by means of this same overall process. However, bacteria have the ETS stationed in the cell membrane, and the direction of the proton movement is from the cytoplasm to the periplasmic space between the membrane and cell wall. Bacteria may also lack one or more of the electron carriers **(figure 8.22c).** As we will see, these differences will affect the amount of ATP produced. In both cell types, the chemiosmotic theory has been supported by tests showing that oxidative phosphorylation is blocked if the mitochondrial or bacterial cell membranes are disrupted.

The Terminal Step of Electron Transport

The terminal step, during which oxygen accepts the electrons, is catalyzed by cytochrome aa_3, also called cytochrome oxidase. This large enzyme complex is specifically adapted to receive electrons from cytochrome c, pick up hydrogens from the solution, and react with oxygen to form a molecule of water (figure 8.22b). This reaction, though in actuality more complex, is summarized as follows:

$$2\,H^+ + 2\,e^- + \tfrac{1}{2}O_2 \rightarrow H_2O$$

Most eukaryotic aerobes have a fully functioning cytochrome system, but bacteria exhibit wide-ranging variations in this part of the system. Some species lack one or more of the redox steps; others have several alternative electron transport schemes. Because many bacteria lack cytochrome oxidase, this variation can be used to differentiate among certain genera of bacteria. An oxidase detection test can be used to help identify members of the genera *Neisseria* and *Pseudomonas* and some species of *Bacillus*. Another variation in the cytochrome system is evident in certain bacteria (*Klebsiella, Enterobacter*) that can grow even in the presence of cyanide because they lack cytochrome oxidase. Cyanide will cause rapid death in humans and other eukaryotes because it blocks cytochrome oxidase, thereby completely terminating aerobic respiration, but it is harmless to these bacteria.

A potential side reaction of the respiratory chain in aerobic organisms is the incomplete reduction of oxygen to superoxide ion (O_2^-) and hydrogen peroxide (H_2O_2). As mentioned in chapter 7, these toxic oxygen products can be very damaging to cells. Aerobes have neutralizing enzymes to deal with these products, including *superoxide dismutase* and *catalase*. One exception is the genus *Streptococcus,* which can grow well in oxygen yet lacks both cytochromes and catalase. The tolerance of these organisms to oxygen can be explained by the neutralizing effects of a special peroxidase. The lack of cytochromes, catalase, and peroxidases in strict anaerobes

as a rule limits their ability to process free oxygen and contributes to its toxic effects on them.

Summary of Aerobic Respiration

We presented a summary equation for respiration on page 233. We are now in a position to estimate the theoretical number of ATP molecules that could arise from a combination of substrate-level phosphorylation, NADH, and $FADH_2$ generated from a single glucose molecule:

- There are 3 substrate-level reactions that give rise directly to ATP—2 in glycolysis and 1 in the Krebs cycle. Since these are happening twice for each event, this provides **6 ATPs** in total.
- There are 5 NADHs—1 from glycolysis, 1 from pyruvic acid oxidation, and 3 from the Krebs cycle. Each NADH could potentially power the synthesis of 3 ATPs, yielding $5 \times 3 = 15$ ATPs. Again, this happens twice for each glucose and could generate **30 ATPs.**
- There is a single reaction in the Krebs cycle that yields $FADH_2$, a carrier that can power synthesis of 2 ATPs, which multiplied by 2 provides **4 ATPs.**
- The grand total for ATP synthesis comes out to $6 + 30 + 4 = 40$; however, because the first and second reactions of glycolysis use up 2 ATPs, the final net number is **38 ATPs.**

We have emphasized that these totals are theoretical because the yields for NADH and $FADH_2$ are lower when the actual conditions in cells are taken into account. It is likely that oxidation of each $NADH^+$ provides closer to 2.5 ATPs and oxidation of $FADH_2$, about 1.5. This will decrease the practical total to around **30 ATPs.** Another reason this total will be reduced for eukaryotic cells is that they must use active transport (and ATP) to carry the NADH from glycolysis into the mitochondrial cristae. Most prokaryotic cells probably produce even less ATP because they lack one or more of the carriers in the electron transport system or do not generate as much NADH. **Figure 8.23** charts the course of these calculations as they occur in the three phases of aerobic respiration.

No summary would be complete without considering three other key participants:

- Six carbon dioxide molecules are generated by reactions just prior to and during the Krebs cycle. These represent all of the carbons and oxygens from the original glucose molecule.
- Six oxygen molecules are consumed during the final transfer of electrons and hydrogens to cytochrome oxidase.
- Six water molecules are produced in electron transport and 2 in glycolysis, but because 2 are used in the Krebs cycle, this leaves a net number of 6.

Alternate Catabolic Pathways

Certain bacteria follow a different pathway in carbohydrate catabolism. The phosphogluconate pathway (also called the hexose monophosphate shunt) provides ways to anaerobically oxidize glucose and other hexoses, to release ATP, to produce large amounts of NADPH, and to process pentoses (5-carbon sugars). This pathway, common in heterolactic fermentative bacteria, yields various end products, including lactic acid, ethanol, and carbon dioxide. Furthermore, it is a significant intermediate source of pentoses for nucleic acid synthesis.

Total aerobic yield 38 ATP maximum

Figure 8.23 Theoretic ATP yield from aerobic respiration. To attain the theoretic maximum yield of ATP, one must assume a ratio of 3 for the oxidation of NADH and 2 for $FADH_2$. The actual yield is generally lower and varies between eukaryotes and prokaryotes and among prokaryotic species.

Anaerobic Respiration

Some bacteria have evolved an anaerobic respiratory system that functions like the aerobic cytochrome system except that it utilizes oxygen-containing ions rather than free oxygen as the final electron acceptor. Of these, the nitrate (NO_3^-) and nitrite (NO_2^-) reduction systems are best known. The reaction in species such as *E. coli* is represented as:

$$\text{NO}_3^- + \text{NADH} + \text{H}^+ \xrightarrow{\text{Nitrate reductase}} \text{NO}_2^- + \text{H}_2\text{O} + \text{NAD}^+$$

The enzyme nitrate reductase catalyzes the removal of oxygen from nitrate, leaving nitrite and water as products. This reaction is the basis for a physiological test used in identifying certain bacteria.

Some species of *Pseudomonas* and *Bacillus* possess enzymes that can further reduce nitrite to nitric oxide (NO), nitrous oxide (N_2O), and even nitrogen gas (N_2). This process, called **denitrification,** is a very important step in recycling nitrogen in the biosphere. Other oxygen-containing nutrients reduced anaerobically by various bacteria are carbonates and sulfates.

Most obligate anaerobes use the H^+ they generate during glycolysis and the Krebs cycle to reduce some compound other than O_2 but still use the oxidative phosphorylation pathway. Methanogens (described in chapter 7) reduce CO_2 or CO_3^- (carbonate) to CH_4 (methane gas), and certain sulfate bacteria reduce SO_4^{2-} to S sulfide or hydrogen sulfide gas (H_2S). Some use metals or organic compounds as the final H^+ acceptor. These microbes live in subterranean habitats such as swamps and deep oceanic vents that are devoid of O_2 gas.

The Importance of Fermentation

Of all the results of pyruvate metabolism, probably the most varied is fermentation. Technically speaking, **fermentation*** is the incomplete oxidation of glucose or other carbohydrates in the absence of oxygen. This process uses organic compounds as the terminal electron acceptors and yields a small amount of ATP.

Over time, the term *fermentation* has acquired several looser meanings. Originally, Pasteur called the microbial action of yeast during wine production *ferments,* and to this day, biochemists use the term in reference to the production of ethyl alcohol by yeasts acting on glucose and other carbohydrates. Fermentation is also what bacteriologists call the formation of acid, gas, and other products by the action of various bacteria on pyruvic acid. The process is a common metabolic strategy among bacteria. Industrial processes that produce chemicals on a massive scale through the actions of microbes are also called fermentations. Each of these usages is acceptable for one application or another.

It may seem that fermentation would yield only meager amounts of energy (2 ATPs maximum per glucose) and that would slow down growth. What actually happens, however, is that many bacteria can grow as fast as they would in the presence of oxygen. This rapid growth is made possible by an increase in the rate of glycolysis. From another standpoint, fermentation permits survival and growth in the absence of molecular oxygen and allows colonization of anaerobic environments. It also enables microorganisms with a versatile metabolism to adapt to variations in the availability of oxygen. For these facultative microbes, fermentation provides a means to function even when oxygen levels are too low for aerobic respiration.

Bacteria that digest cellulose in the rumens of cattle are largely fermentative. After initially hydrolyzing cellulose to glucose, they ferment the glucose to organic acids, which are then absorbed as the bovine's principal energy source. Even human muscle cells can undergo a form of fermentation that permits short periods of activity after the oxygen supply in the muscle has been exhausted. Muscle cells convert pyruvic acid into lactic acid, which allows anaerobic production of ATP to proceed for a time. But this cannot go on indefinitely, and after a few minutes, the accumulated lactic acid causes muscle fatigue.

* *fermentation* (fur-men-tay'-shun) L. *fervere,* to boil, or *fermentatum,* leaven or yeast.

CONTINUING CASE FILE 8

The rumen microbiota include around 70 known types of organisms. Bacterial genera such as *Bacteroides, Ruminococcus,* and *Streptococcus* are present in the highest numbers. Second in prominence are unique types of protozoa. Other microbes present in smaller but significant numbers are yeasts, molds, and archaeons. Experts are certain that eventually they will identify thousands of species of rumen microbes. Together, these microbes release a combination of enzymes that hydrolyze the bonds of the two main carbohydrates in ruminant diets—cellulose and starch—releasing the simple sugars they contain. The sugars go on to a complex series of fermentations that release nutrients absorbable by the cow and gases that pass back into the air.

Another benefit of ruminant physiology and nutrition is the massive load of rumen microbes itself. Once they reach a chamber called the abomasum, they are destroyed by the extreme acidity there. Rather than just being flushed out, they actually become another source of nutrients, especially protein and amino acids the cow needs for its growth and metabolism. This is accomplished when a different set of enzymes from the animal's digestive organs dismantle the dead microbial cells.

- What types of enzymes do microbes produce for digesting the ruminant's feed?

- What types of chemical products are given off by rumen fermentation?

For a wrap-up, see the Case File Perspective on page 249.

Figure 8.24 The fermentation systems that produce acid and alcohol. In both cases, the final electron acceptor is an organic compound. In yeasts, pyruvic acid is decarboxylated to acetaldehyde, and the NADH given off in the glycolytic pathway reduces acetaldehyde to ethyl alcohol. In homolactic fermentative bacteria, pyruvic acid is reduced by NADH to lactic acid. Both systems regenerate NAD^+ to feed back into glycolysis or other cycles.

Products of Fermentation in Microorganisms

Alcoholic beverages (wine, beer, whiskey) are perhaps the most prominent among fermentation products; others are solvents (acetone, butanol), organic acids (lactic, acetic), dairy products, and many other foods. Derivatives of proteins, nucleic acids, and other organic compounds are fermented to produce vitamins, antibiotics, and even hormones such as hydrocortisone.

Fermentation products can be grouped into two general categories: alcoholic fermentation products and acidic fermentation products **(figure 8.24). Alcoholic fermentation** occurs in yeast or bacterial species that have metabolic pathways for converting pyruvic acid to ethanol. This process involves a decarboxylation of pyruvic acid to acetaldehyde, followed by a reduction of the acetaldehyde to ethanol. In oxidizing the NADH formed during glycolysis, NAD^+ is regenerated, thereby allowing the glycolytic pathway to continue. These processes are crucial in the production of beer and wine, though the actual techniques for arriving at the desired amount of ethanol and the prevention of unwanted side reactions are important tricks of the brewer's trade **(Insight 8.4).** Note that the products of alcoholic fermentation include both ethanol *and* CO_2, a gas that accounts for the bubbles in champagne and beer and the rising of bread dough.

Alcohols other than ethanol can be produced during bacterial fermentation pathways. Certain clostridia produce butanol and isopropanol through a complex series of reactions. Although this process was once an important source of alcohols for industrial use, it has been largely replaced by a nonmicrobial petroleum process.

The pathways of **acidic fermentation** are extremely varied. Lactic acid bacteria ferment pyruvate in the same way that humans do—by reducing it to lactic acid. If the product of this fermentation is mainly lactic acid, as in certain species of *Streptococcus* and *Lactobacillus,* it is termed *homolactic.* The souring of milk is due largely to the production of this acid by bacteria. When glucose is fermented to a mixture of lactic acid, acetic acid, and carbon dioxide, as is the case with *Leuconostoc* and other species of *Lactobacillus,* the process is termed *heterolactic fermentation.*

Many members of the family Enterobacteriaceae (*Escherichia, Shigella,* and *Salmonella*) possess enzyme systems for converting pyruvic acid to several acids simultaneously **(figure 8.25). Mixed acid fermentation** produces a combination of acetic, lactic, succinic, and formic acids, and it lowers the pH of a medium to about 4.0. *Propionibacterium* produces primarily propionic acid, which gives the characteristic flavor to Swiss cheese while fermentation gas (CO_2) produces the holes. Some members also further decompose formic acid completely to carbon dioxide and hydrogen gases. Because enteric bacteria commonly occupy the intestine, this fermentative activity accounts for the accumulation of some types of gas—primarily CO_2 and H_2—in the intestine. Some bacteria reduce the organic acids and produce the neutral end

INSIGHT 8.4

Pasteur and the Wine-to-Vinegar Connection

The microbiology of alcoholic fermentation was greatly clarified by Louis Pasteur after French winemakers hired him to uncover the causes of periodic spoilage in wines. Especially troublesome was the conversion of wine to vinegar and the resultant sour flavor. Up to that time, wine formation had been considered strictly a chemical process. After extensively studying beer making and wine grapes, Pasteur concluded that wine, both fine and not so fine, was the result of microbial action on the juices of the grape and that wine "disease" was caused by contaminating organisms that produced undesirable products such as acid. Although he did not know it at the time, the bacterial contaminants responsible for the acidity of the spoiled wines were likely to be *Acetobacter* or *Gluconobacter* introduced by the grapes, air, or wine-making apparatus. These common gram-negative genera further oxidized ethanol to acetic acid and are presently used in commercial vinegar production. The following formula shows how this is accomplished:

$$\underset{\text{Ethanol}}{\text{H}-\overset{\text{H}}{\underset{\text{H}}{\text{C}}}-\overset{\text{H}}{\underset{\text{H}}{\text{C}}}-\text{OH}} \xrightarrow{\text{O}_2} \underset{\text{Acetic acid}}{\text{H}-\overset{\text{H}}{\underset{\text{H}}{\text{C}}}-\text{C}\overset{\text{O}}{\underset{\text{OH}}{}}} + 2\,\text{H}_2\text{O}$$

Pasteur's far-reaching solution to the problem is still with us today—mild heating, or *pasteurization,* of the grape juice to destroy the contaminants, followed by inoculation of the juice with a pure yeast culture. The topic of pasteurization is explored further in chapter 11.

An analysis of wine shows that it contains thousands of different organic compounds (see http://www.vinterus.com/pdffiles/Wine.pdf.). What does this tell you about the metabolism of yeasts during alcoholic fermentation? *Answer available at http://www.mhhe.com/talaro8*

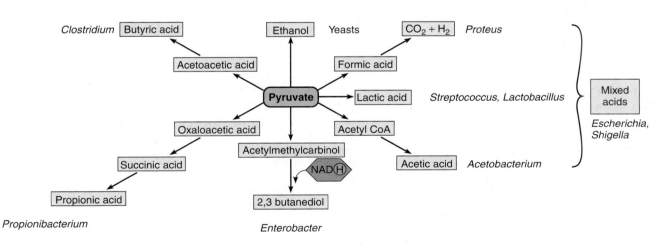

Figure 8.25 **Miscellaneous products of pyruvate fermentation and some of the bacteria involved in their production.**

product 2,3-butanediol. Several of these end products are the basis of biochemical tests discussed in chapters 17 and 20.

We have provided only a brief survey of fermentation products, but it is worth noting that microbes can be harnessed to synthesize a variety of other substances by varying the raw materials provided

them. In fact, so broad is the meaning of the word *fermentation* that the large-scale industrial syntheses by microorganisms often utilize entirely different mechanisms from those described here, and they even occur aerobically, particularly in antibiotic, hormone, vitamin, and amino acid production (see chapter 27).

Check & Assess Section 8.3

- ✔ Bioenergetics describes metabolism in terms of release, utilization, and transfer of energy by cells.
- ✔ Catabolic pathways release energy through three pathways: glycolysis, the Krebs cycle, and the respiratory electron transport system.
- ✔ Cellular respiration is described by the nature of the final electron acceptor. Aerobic respiration implies that O_2 is the final electron acceptor. Anaerobic respiration implies that some other molecule is the final electron acceptor. In some instances such as fermentation, the acceptor is an organic molecule.
- ✔ Carbohydrates are often used as cellular energy sources because they are superior hydrogen (electron) donors.
- ✔ Glycolysis is the catabolic process by which glucose is oxidized and converted into two molecules of pyruvic acid, with a net gain of 2 ATPs. The formation of ATP is via substrate-level phosphorylation.
- ✔ The Krebs cycle processes the 3-carbon pyruvic acid and generates three CO_2 molecules. The electrons it releases are transferred to electron transport carriers for energy harvesting.
- ✔ The electron transport chain generates free energy through sequential redox reactions collectively called oxidative phosphorylation.
- ✔ ATP synthase enzymes stationed in membranes generate the ATP molecules through chemiosmosis.
- ✔ For each glucose molecule catabolized, a potential maximum of 36 to 38 ATPs are released.
- ✔ Many microbes have metabolic systems based on fermentation for generating a small yield of ATP anaerobically.
- ✔ The starting compound for fermentation is an organic molecule such as pyruvic acid, and the final electron acceptor is also an organic molecule.
- ✔ Depending on the microbe, fermentation products can include alcohols, organic acids, and gases. Wide-scale fermentation is used in industry to produce a variety of drugs, vitamins, and even foods.

13. Compare and contrast the location of glycolysis, Krebs cycle, and electron transport in prokaryotic and eukaryotic cells.
14. Outline the basic steps in glycolysis, indicating where ATP is used and given off.
15. Where does NADH originate, and what is its fate in an aerobe?
16. What is the fate of NADH in a fermentative organism?
17. What is the source of ATP directly from the Krebs cycle?
18. How many ATPs could be formed from the original glucose molecule carried through aerobic respiration?
19. How does the total of ATPs generated differ between bacteria and many eukaryotes? What causes this difference?
20. Summarize the chemiosmotic theory of ATP formation.
21. Name the sources of oxygen in bacteria that use anaerobic respiration.
22. What are the final electron acceptors in aerobic, anaerobic, and fermentative metabolism?
23. Describe the reaction between H^+ and oxygen at the final step of electron transport.
24. What adaptive advantages does a fermentative metabolism confer on a microbe?
25. Describe three patterns of fermentation typical of microbes.

8.4 Biosynthesis and the Crossing Pathways of Metabolism

Expected Learning Outcomes

20. Explain how cells perform anabolic functions and what is involved in amphibolism.
21. Identify several pathways where molecules can be redirected or converted into useful cell products.
22. Briefly describe several mechanisms in biosynthesis of macromolecules.

Our discussion now turns from catabolism and energy extraction to anabolic functions and biosynthesis. In this section, we present aspects of intermediary metabolism, including amphibolic pathways, the synthesis of simple molecules, and the synthesis of macromolecules.

The Frugality of the Cell—Waste Not, Want Not

It must be obvious by now that cells have mechanisms for careful management of carbon compounds. Rather than being dead ends, most catabolic pathways contain strategic molecular intermediates (metabolites) that can be diverted into anabolic pathways. In this way, a given molecule can serve multiple purposes, and the maximum benefit can be derived from all nutrients and metabolites of the cell pool. The property of a system to integrate catabolic and anabolic pathways to improve cell efficiency is termed **amphibolism.***

At this point in the chapter, you can appreciate a more complex view of metabolism than that presented at the beginning in figure 8.1. **Figure 8.26** demonstrates the amphibolic nature of intermediary metabolism. The pathways of glucose catabolism are an especially rich "metabolic marketplace." The principal sites of amphibolic interaction occur during glycolysis (glyceraldehyde-3-phosphate and pyruvic acid) and the Krebs cycle (acetyl coenzyme A and various organic acids).

Amphibolic Sources of Cellular Building Blocks

Glyceraldehyde-3-phosphate can be diverted away from glycolysis and converted into precursors for amino acid, carbohydrate, and triglyceride (fat) synthesis. (A precursor molecule is one that gives rise to a different compound.) Earlier, we noted the numerous directions that pyruvic acid catabolism can take. In terms of synthesis, pyruvate also plays a pivotal role in providing intermediates for amino acids. In the event of an inadequate glucose supply, pyruvate serves as the starting point in glucose synthesis from various metabolic intermediates, a process called **gluconeogenesis.***

The acetyl group that starts the Krebs cycle is another extremely versatile metabolite that can be fed into a number of synthetic pathways. This 2-carbon fragment can be converted as a single unit into

* *amphibolism* (am-fee-bol′-izm) Gr. *amphi*, two-sided, and *bole*, a throw.
* *gluconeogenesis* (gloo″-koh-nee″-oh-gen′-uh-sis) Literally, the formation of "new" glucose.

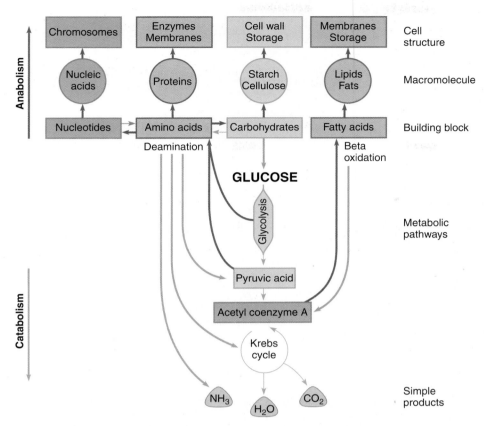

Figure 8.26 An amphibolic view of metabolism. Intermediate compounds such as pyruvic acid and acetyl coenzyme A serve multiple functions. With comparatively small modifications, these compounds can be converted into other compounds and enter a different pathway. Note that catabolism of glucose (center) furnishes numerous intermediates for anabolic pathways that synthesize amino acids, fats, nucleic acids, and carbohydrates. These building blocks can serve in further synthesis of larger molecules to construct various cell components.

one of several amino acids, or a number of these fragments can be condensed into hydrocarbon chains that are important building blocks for fatty acid and lipid synthesis. Note that the reverse is also true: fats can be degraded to acetyl and thereby enter the Krebs cycle via acetyl coenzyme A. This aerobic process, called **beta oxidation,** releases significant energy. Oxidation of a 6-carbon fatty acid potentially yields 50 ATPs, compared with 38 for a 6-carbon sugar.

Two carbohydrate metabolites from the Krebs cycle are oxalo-acetic acid and α-ketoglutaric acid—essential intermediates in the synthesis of certain amino acids. This occurs through **amination,** the addition of an amino group to a carbon skeleton **(figure 8.27a).** A certain core group of amino acids can then be used to synthesize others. Amino acids and carbohydrates can be interchanged through **transamination** as well **(figure 8.27b).**

Pathways that synthesize the nitrogen bases (purines, pyrimidines), which are components of DNA and RNA, originate in amino acids and so can be dependent on intermediates from the Krebs cycle as well. Because the coenzymes NAD^+, NADP, FAD, and others contain purines and pyrimidines similar to the nucleic acids, their synthetic pathways are also dependent on amino acids. During times of carbohydrate deprivation, organisms can likewise convert amino acids to intermediates of the Krebs cycle by **deamination**

(removal of an amino group) and thereby derive energy from proteins. Deamination results in the formation of nitrogen waste products such as ammonium ions or urea **(figure 8.27c).**

Formation of Macromolecules

Monosaccharides, amino acids, fatty acids, nitrogen bases, and vitamins—the building blocks that make up the various macromolecules and organelles of the cell—come from two possible sources. They can enter the cell from the outside as nutrients, or they can be synthesized through various cellular pathways. The degree to which an organism can synthesize its own building blocks (simple molecules) is determined by its genetic makeup, a factor that varies tremendously from group to group. In chapter 7, you learned that autotrophs require only CO_2 as a carbon source, a few minerals to synthesize all cell substances, and no organic nutrients. Some heterotrophic organisms (*E. coli,* yeasts) are so efficient that they can synthesize all cellular substances from minerals and one organic carbon source such as glucose. Compare this with a strict parasite having few synthetic abilities of its own, obtaining most of its nutrient molecules from the host.

Whatever their source, once these building blocks are added to the metabolic pool, they are available for synthesis of polymers by the cell. The details of synthesis vary among the types of macromolecules, but all of them involve the formation of bonds by specialized enzymes and the expenditure of ATP.

Carbohydrate Biosynthesis

The role of glucose in bioenergetics is so crucial that its biosynthesis is ensured by several alternative pathways. Certain structures in the cell depend on an adequate supply of glucose as well. It is the major component of the cellulose cell walls of some eukaryotes and of certain storage granules (starch, glycogen). One of the intermediaries in glycolysis, glucose-6-P, is used to form glycogen. Monosaccharides other than glucose are important in the synthesis of bacterial cell walls. Peptidoglycan contains a linked polymer of muramic acid and glucosamine. Fructose-6-P from glycolysis is used to form these two sugars. Carbohydrates (deoxyribose, ribose) are also essential building blocks in nucleic acids. Polysaccharides are the predominant components of cell surface structures such as capsules and the glycocalyx, and they are commonly found in slime layers (dextran). Remember that most polymerization reactions occur via loss of a water molecule (see figure 2.16) and the input of energy (see figure 8.8).

Assembly of the Cell

The component parts of a bacterial cell are being synthesized on a continuous basis, and catabolism is also taking place, as long as

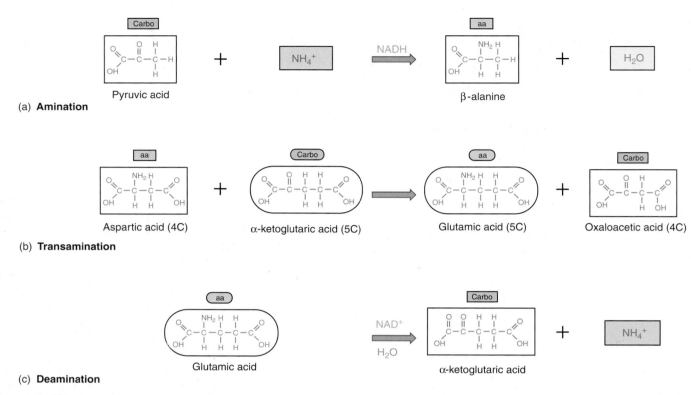

(a) **Amination**

(b) **Transamination**

(c) **Deamination**

Figure 8.27 Reactions that produce and convert amino acids. All of the reactions require energy as ATP or NADH and specialized enzymes. **(a)** Through amination (the addition of an ammonium molecule [amino group]), a carbohydrate can be converted to an amino acid. **(b)** Through transamination (transfer of an amino group from an amino acid to a carbohydrate), metabolic intermediates can be converted to amino acids that are in low supply. **(c)** Through deamination (removal of an amino group), an amino acid can be converted to a useful intermediate of carbohydrate catabolism. This is how proteins are used to derive energy. Ammonium is one waste product. Note that carbohydrate structures are in brown and amino acids are in blue.

nutrients are present and the cell is in a nondormant state. When anabolism produces enough macromolecules to serve two cells, and when DNA replication produces duplicate copies of the cell's genetic material, the cell undergoes binary fission, which results in two cells from one parent cell. The two cells will need twice as many ribosomes, twice as many enzymes, and so on. The cell has created these during the initial anabolic phases we have described. Before cell division, the membrane(s) and the cell wall will have increased in size to create a cell that is almost twice as big as a "newborn" cell. Once synthesized, the phospholipid bilayer components of the membranes assemble themselves spontaneously with no energy input. But proteins and other components must be added to the membranes. Growth of the cell wall, accomplished by the addition and coupling of sugars and peptides, requires energy input. The catabolic processes provide all the energy for these complex building reactions.

✔ Amphibolic compounds are the "crossroads compounds" of metabolism. They not only participate in catabolic pathways but also are precursor molecules to biosynthetic pathways.

✔ Biosynthetic pathways utilize building-block molecules from two sources: the environment and the cell's own catabolic pathways. Microorganisms construct macromolecules from these monomers using ATP and specialized enzymes.

✔ Carbohydrates are crucial as energy sources, cell wall constituents, and components of nucleotides.

✔ Proteins are essential macromolecules in all cells because they function as structural constituents, enzymes, and cell appendages.

✔ Final assembly of a new cell occurs via binary fission and requires an input of energy.

Check & Assess Section 8.4

✔ "Intermediary metabolism" refers to the metabolic pathways that use intermediate compounds and connect anabolic and catabolic reactions.

26. What is meant by amphibolism, and what are its most important functions?

27. Identify major ways that amino acids are processed.

28. Which macromolecules are synthesized by specialized anabolic pathways?

8.5 Photosynthesis: The Earth's Lifeline

Expected Learning Outcomes

23. Outline the general reactions of photosynthesis and the importance it holds for the energetics of life.

24. Describe the pigment systems and how they function.

25. Describe the main events in the light-dependent reactions and the products that result.

26. Describe the main events in the light-independent reactions and the products that result.

As mentioned earlier, the ultimate source of most of the chemical energy in cells is the sun. Because this source is directly available only to the cells of photosynthesizers, most organisms are either directly or indirectly dependent on photosynthesis except for a few chemoautotrophs that derive their energy and nutrients solely from inorganic substrates. The other major products of photosynthesis are organic carbon compounds, which are produced from carbon dioxide through a process called carbon fixation. On land, green plants are the primary photosynthesizers; and in aquatic ecosystems, where 80% to 90% of all photosynthesis occurs, this role is filled by algae, cyanobacteria, and green sulfur, purple sulfur, and purple nonsulfur bacteria.

The summary equation for the main reactants and products of photosynthesis is

$$6CO_2 + 6H_2O \xrightarrow{\text{Light and pigment}} \underset{\text{Glucose}}{C_6H_{12}O_6} + 6O_2 \uparrow$$

The anatomy of photosynthetic cells is adapted to trapping sunlight, and their physiology effectively uses this solar energy to produce high-energy glucose from low-energy CO_2 and water. Photosynthetic organisms achieve this remarkable feat through a series of reactions involving light, pigment, CO_2, and water, which is used as a source for electrons.

Photosynthesis proceeds in two phases: the **light-dependent reactions,** which proceed only in the presence of sunlight, and the **light-independent reactions,** which proceed regardless of the lighting conditions (light or dark) **(figure 8.28).**

Solar energy is delivered in discrete energy packets called **photons** (also called quanta) that travel as waves. The wavelengths of light operating in photosynthesis occur in the visible spectrum between 400 (violet) and 700 nanometers (red). As this light strikes photosynthetic pigments, some wavelengths are absorbed, some pass through, and some are reflected. The activity that has the greatest impact on photosynthesis is the absorbance of light by photosynthetic pigments. These include the **chlorophylls,** which are green; **carotenoids,** which are yellow, orange, or red; and **phycobilins,** which are red or blue-green.[3] By far the most important of these pigments are the chlorophylls, which contain a *photocenter* that consists of a magnesium atom held in the center of a complex ringed molecule called a *porphyrin* **(figure 8.29b).** As we will see, the chlorophyll molecule harvests the energy of photons and

3. The color of the pigment corresponds to the wavelength of light it reflects.

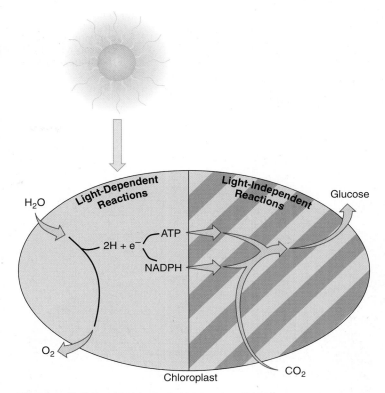

Figure 8.28 Overview of photosynthesis. The general reactions of photosynthesis, divided into two phases called light-dependent reactions and light-independent reactions. The dependent reactions require light to activate chlorophyll pigment and use the energy given off during activation to split an H_2O molecule into oxygen and hydrogen, producing ATP and NADPH. The independent reactions, which occur either with or without light, utilize ATP and NADPH produced during the light reactions to fix CO_2 into organic compounds such as glucose.

converts it to electron (chemical) energy. Accessory photosynthetic pigments such as carotenes trap light energy and shuttle it to chlorophyll, thereby functioning like antennae. These light-dependent reactions are catabolic (energy-producing) reactions, which pave the way for the next set of reactions, the light-independent reactions, which use the extracted energy for synthesis. During this phase, carbon atoms from CO_2 are fixed to the carbon backbones of organic molecules.

The detailed biochemistry of photosynthesis is beyond the scope of this text, but we will provide an overview of the general process as it occurs in green plants and algae. Many of the basic activities (electron transport and phosphorylation) are biochemically similar to certain pathways of respiration.

Light-Dependent Reactions

The same systems that carry the photosynthetic pigments are also the sites for the light reactions. They occur in the **thylakoid** membranes of compartments called grana (singular, granum) in chloroplasts **(figure 8.29a)** and in specialized parts of the cell membranes in prokaryotes (see figure 4.28a). These systems exist as two separate complexes called *photosystem I* (P700) and *photosystem*

(a) **A cell of the motile alga *Chlamydomonas*, with a single large chloroplast (magnified cutaway view). The chloroplast contains membranous compartments called grana where chlorophyll molecules and the photosystems for the light reactions are located.**

(b) **A chlorophyll molecule, with a central magnesium atom held by a porphyrin ring.**

(c) **The main events of the light reactions shown as an exploded view in one granum.**

① When light activates photosystem II, it sets up a chain reaction, in which electrons are released from chlorophyll.

② These electrons are transported along a chain of carriers to photosystem I.

③ The empty position in photosystem II is replenished by photolysis of H_2O. Other products of photolysis are O_2 and H^+.

④ Pumping of H^+ into the interior of the granum produces conditions for ATP to be synthesized.

⑤ The final electron and H^+ acceptor is NADP, which receives these from photosystem I.

⑥ Both NADPH and ATP are fed into the stroma for the Calvin cycle.

Process Figure 8.29 **The reactions of photosynthesis.**

II (P680)[4] (**figure 8.29c**). Both systems contain chlorophyll and they are simultaneously activated by light, but the reactions in photosystem II help drive photosystem I. Together the systems are activated by light, transport electrons, pump hydrogen ions, and form ATP and NADPH.

When photons enter the photocenter of the P680 system (PS II), the magnesium atom in chlorophyll becomes excited and releases 2 electrons. The loss of electrons from the photocenter has two major effects:

1. It creates a vacancy in the chlorophyll molecule forceful enough to split an H_2O molecule into hydrogens (H^+) (electrons and hydrogen ions) and oxygen (O_2). This splitting of water, termed **photolysis,** is the ultimate source of the O_2 gas that is an important product of photosynthesis. The electrons released from the lysed water regenerate photosystem II for its next reaction with light.

2. Electrons generated by the first photoevent are immediately boosted through a series of carriers (cytochromes) to the P700 system. At this same time, hydrogen ions accumulate in the internal space of the thylakoid complex, thereby producing an electrochemical gradient.

The P700 system (PS I) has been activated by light so that it is ready to accept electrons generated by the PS II. The electrons it receives are passed along a second transport chain to a complex that uses electrons and hydrogen ions to reduce NADP to NADPH. (Recall that reduction in this sense entails the addition of electrons and hydrogens to a substrate.)

A second energy reaction involves synthesis of ATP by a chemiosmotic mechanism similar to that shown in figure 8.22. Channels in the thylakoids of the granum actively pump H^+ into the inner chamber, producing a charge gradient. ATP synthase located in this same thylakoid uses the energy from H^+ transport to phosphorylate ADP to ATP. Because it occurs in light, this process is termed **photophosphorylation.** Both NADPH and ATP are released into the stroma of the chloroplast, where they drive the reactions of the **Calvin cycle.**

Light-Independent Reactions

The subsequent photosynthetic reactions that do not require light occur in the chloroplast stroma or the cytoplasm of cyanobacteria. These reactions use energy produced by the light phase to synthesize glucose by means of the Calvin cycle (**figure 8.30**).

4. The numbers refer to the wavelength of light to which each system is most sensitive.

Figure 8.30 **The Calvin cycle.** The main events of the reactions in photosynthesis that do not require light. It is during this cycle that carbon is fixed into organic form using the energy (ATP and NADPH) released by the light reactions. The end product, glucose, can be stored as complex carbohydrates, or it can be used in various amphibolic pathways to produce other carbohydrate intermediates or amino acids.

The cycle begins at the point where CO_2 is combined with a doubly phosphorylated 5-carbon acceptor molecule called ribulose-1,5-bisphosphate**(RuBP).** This process, called **carbon fixation,** generates a 6-carbon intermediate compound that immediately splits into two 3-carbon molecules of 3-phosphoglyceric acid (PGA). The subsequent steps use the ATP and NADPH generated by the photosystems to form high-energy intermediates. First, ATP adds a second phosphate to 3-PGA and produces 1,3-bisphosphoglyceric acid (BPG). Then, during the same step, NADPH contributes its hydrogen to BPG, and one high-energy phosphate is removed. These events give rise to glyceraldehyde-3-phosphate (PGAL). This molecule and its isomer dihydroxyacetone phosphate (DHAP) are key molecules in hexose synthesis leading to fructose and glucose. You may notice that this pathway is very similar to glycolysis, except that it runs in reverse (see figure 8.18). Bringing the cycle back to regenerate RuBP requires PGAL and several steps not depicted in figure 8.30.

Other Mechanisms of Photosynthesis

The **oxygenic,** or oxygen-releasing, photosynthesis that occurs in plants, algae, and cyanobacteria is the dominant type on the earth. Other photosynthesizers such as green and purple bacteria possess bacteriochlorophyll, which is more versatile in capturing light. They have only a cyclic photosystem I, which routes the electrons from the photocenter to the electron carriers and back to the photosystem

again. This pathway generates a relatively small amount of ATP, and it may not produce NADPH. As photolithotrophs, these bacteria use H_2, H_2S, or elemental sulfur rather than H_2O as a source of electrons and reducing power. As a consequence, they are **anoxygenic** (non-oxygen-producing), and many are strict anaerobes.

Check & Assess Section 8.5

✔ The sun is the primary energy source for most surface ecosystems. Photosynthesis captures this energy and utilizes it for carbon fixation by producer populations.

✔ Producers include plants, algae, cyanobacteria, and certain bacterial species.

✔ Photosynthesis proceeds in two stages: (1) photophosphorylation, in which light is trapped, energy is extracted to make ATP, and oxygen is evolved; (2) Calvin cycle, in which ATP is used to fix CO_2 into organic compounds such as glucose.

29. Tell whether each of the following is produced during the light-independent or light-dependent reactions of photosynthesis: O_2, ATP, NADPH, and glucose.
30. When are water and CO_2 consumed?
31. What is the function of chlorophyll and the photosystems?
32. What is the fate of the ATP and NADH?
33. Compare oxygenic with nonoxygenic photosynthesis.

CASE FILE **8** PERSPECTIVE

During digestion, larger nutrient molecules are broken down into smaller individual units that can be used in cellular metabolism. It requires a specific enzyme that fits the substrate in an exact way to assist in splitting the bonds between the units. As a rule, digestive enzymes require a water molecule as part of the bond-breaking event. Because of this, digestion is considered a type of hydrolysis (water splitting). This is the first step in the catabolism of a ruminant's feed.

Cellulose, which forms the cell walls of grasses and other forage plants, makes up 70% of the plant's weight. It is composed of glucose molecules arranged tightly in parallel chains. Bacteria, protozoa, and fungi synthesize a variety of cellulases that digest cellulose and free the glucose. Starch is also composed of glucose units but with a different bonding pattern that requires microbial amylases for hydrolysis. Pectin and lipids are some other dietary molecules that will be digested.

From a biochemical perspective, fermentation refers to various catabolic reactions carried out on organic molecules such as sugars that proceed in the absence of oxygen. The output of most fermentation reactions yields 2-and 3-carbon products and a small amount of energy. Because the microbes have coevolved in the oxygen-free environment of the rumen, almost all of the known members are strict anaerobes (even the protozoa and fungi), and many of them carry out fermentative metabolism.

The sugars from digestion of carbohydrates are not directly usable by the cow. They will be rapidly fermented by several different microbial groups into organic acids such as acetic acid and propionic acid. These substances along with fatty acids from lipid digestion are the most significant energy sources for the animal and greatly affect the quality of its milk and meat. Gaseous products of rumen fermentation are carbon dioxide, methane, and smaller amounts of hydrogen that are eliminated by constant belching. Carbon dioxide is a common product of fermentation pathways, and methane is produced by the archaeons living in the rumen. This release of greenhouse gases explains why cows are being blamed for contributing to climate change. Their actual impact is still a source of debate.

Because they can survive on a limited diet of dry grass, ruminants are among the most successful mammals on earth, residing in most land habitats. In these places, they are an important food source, especially for communities in harsh climates with limited agriculture.

For more coverage of this topic, visit http://ars.usda.gov/mwa/madison/dfrc.

Chapter Summary with Key Terms

8.1 The Metabolism of Microbes
A. *Metabolism*
 1. Metabolism is the sum of cellular chemical and physical activities; it involves chemical changes to reactants and the release of products using well-established pathways.
 2. Metabolism is a complementary process consisting of **anabolism,** synthetic reactions that convert small molecules into large molecules, and **catabolism,** in which large molecules are degraded. Together, they generate thousands of intermediate molecular states, called metabolites, which are regulated at many levels.
B. *Enzymes: Metabolic Catalysts*
 1. Metabolism is made possible by organic **catalysts,** or **enzymes,** that speed up reactions by lowering the **energy of activation.** Enzymes are not consumed and can be reused. Each enzyme acts specifically upon its matching molecule or **substrate.**
 2. *Enzyme Structure:* Depending upon its composition, an enzyme is either conjugated or simple. A conjugated enzyme consists of a protein component called the **apoenzyme** and one or more activators called **cofactors.**
 3. *Enzyme Specificity:* Substrate attachment occurs in the special pocket called the **active, or catalytic, site.** In order to fit, a substrate must conform to the active site of the enzyme. This three-dimensional state is determined by the amino acid content, sequence, and folding of the apoenzyme. Thus, enzymes are usually **substrate-specific.**
 4. *Cofactors:* Metallic cofactors impart greater reactivity to the enzyme-substrate complex. Coenzymes such as

NAD$^+$ (nicotinamide adenine dinucleotide) are transfer agents that pass functional groups from one substrate to another. Coenzymes usually contain **vitamins.**
 5. *Enzyme Classification:* Enzyme names often consist of a prefix derived from the type of reaction or the substrate and the ending *-ase.*
 6. *Microbial Enzymes and Disease:* Many pathogens secrete enzymes or toxins, which are referred to as virulence factors, that enable them to avoid host defenses.
 7. *Types of Enzyme Function:* The release of water that comes with formation of new covalent bonds is a condensation reaction. Hydrolysis reactions involve addition of water to break bonds. Functional groups may be added, removed, or traded in many reactions. Coupled **redox reactions** transfer electrons and protons (H$^+$) from one substrate to another.
 8. *Enzyme Sensitivity:* Enzymes are **labile** (unstable) and function only within narrow operating ranges of temperature, osmotic pressure, and pH; extreme conditions can cause **denaturation.** Enzymes are frequently the targets for physical and chemical agents used in control.
C. *Regulation of Enzymatic Activity*
 Regulatory controls can act on enzymes directly or on the process that gives rise to the enzymes.
 1. A substance that resembles the normal substrate and can occupy the same active site is said to exert **competitive inhibition.**
 2. **Noncompetitive inhibition** occurs when the regulator molecule does not bind to the same site as the substrate.
 3. The mechanisms of **enzyme repression** and **enzyme induction** exert control at the genetic level by controlling the synthesis of key enzymes.

8.2 The Pursuit and Utilization of Energy
A. *Cell Energetics*
 Energy is the capacity of a system to perform work. It is consumed in **endergonic reactions** and is released in **exergonic reactions.**
B. *A Closer Look at Biological Oxidation and Reduction*
 1. Extracting energy requires a series of electron carriers arrayed in a downhill redox chain between electron donors and electron acceptors.
 2. NAD$^+$ is the primary carrier of electrons from one pathway to another.
 3. In **substrate-level** and **oxidative phosphorylation,** energy is transferred to high-energy compounds such as ATP.

8.3 Pathways of Bioenergetics
A. *Principal Pathways in Oxidation of Glucose:* Carbohydrates, such as glucose, are energy-rich because they can yield a large number of electrons per molecule. Glucose is dismantled in stages.
 1. **Glycolysis** is a pathway that degrades glucose to **pyruvic acid** without requiring oxygen.
 2. Pyruvic acid is an important key molecule in several pathways.
B. *Fate of Pyruvic Acid in Krebs Cycle and Electron Transport*
 1. Pyruvic acid is processed in **aerobic respiration** via the **Krebs cycle** and its associated electron transport chain.
 2. Acetyl coenzyme A is the product of pyruvic acid processing that undergoes further oxidation and

decarboxylation in the Krebs cycle, which generates ATP, CO_2, and H_2O.

3. The respiratory chain completes energy extraction.
4. The chemiosmotic theory is a conceptual model that explains the origin and maintenance of electropotential gradients across a membrane that leads to ATP synthesis by **ATP synthase.**
5. The final electron acceptor in aerobic respiration is oxygen. In **anaerobic respiration,** compounds such as sulfate, nitrate, or nitrite serve this function. Bacteria serve as important agents in the nitrogen cycle (**denitrification**).

C. **Fermentation** is anaerobic respiration in which both the electron donor and final electron acceptors are organic compounds.
1. Fermentation enables anaerobic and facultative microbes to survive in environments devoid of oxygen. Production of alcohol, vinegar, and certain industrial solvents relies upon fermentation.
2. The phosphogluconate pathway is an alternative anaerobic pathway.

8.4 Biosynthesis and the Crossing Pathways of Metabolism
A. *Amphibolism*
The ability of a cell or system to integrate catabolic and anabolic pathways to improve efficiency is called **amphibolism.**
B. *Anabolism: Formation of Macromolecules*
Macromolecules, such as proteins, carbohydrates, and nucleic acids, are made of building blocks from two possible sources: from outside the cell (preformed) or via synthesis in one of the anabolic pathways.
C. *Versatility of Glycolysis and Krebs Cycle:* Glycolysis and the Krebs cycle are central pathways that link catabolic and anabolic pathways, allowing cells to break down different classes of molecules in order to synthesize compounds required by the cell.
1. Metabolites of these pathways double as building blocks and sources of energy. Intermediates such as pyruvic acid are convertible into amino acids through **amination.** Amino acids can be **deaminated** and used as a source of glucose and other carbohydrates (**gluconeogenesis**).
2. Two-carbon acetyl molecules from pyruvate **decarboxylation** can be used in fatty acid synthesis.

8.5 Photosynthesis: The Earth's Lifeline
Photosynthesis takes place in two stages—**light-dependent** and **light-independent reactions.**
A. During the light-dependent reactions, **photons** of solar energy are absorbed by **chlorophyll, carotenoid,** and **phycobilin** pigments in **thylakoid** membranes. Light energy captured by two photosystems splits water by **photolysis** and releases oxygen gas; it provides electrons to drive **photophosphorylation;** released light energy is used to synthesize ATP and NADPH.
B. The **Calvin cycle** (light-independent) uses ATP to **fix** carbon dioxide to a carrier molecule, ribulose-1,5-bisphosphate, and convert it to glucose in a multistep process.
C. The type of photosynthesis commonly found in plants, algae, and cyanobacteria is called **oxygenic** because it liberates oxygen. **Anoxygenic** photosynthesis occurs in photolithotrophs, which do not produce oxygen.

Multiple-Choice Questions

Select the correct answer from the answers provided. For questions with blanks, choose the combination of answers that most accurately completes the statement.

1. _____ is another term for biosynthesis.
 a. catabolism c. metabolism
 b. anabolism d. catalyst

2. Catabolism is a form of metabolism in which _____ molecules are converted into _____ molecules.
 a. large, small c. amino acid, protein
 b. small, large d. food, storage

3. An enzyme _____ the activation energy required for a chemical reaction.
 a. increases c. lowers
 b. converts d. catalyzes

4. An enzyme
 a. becomes part of the final products
 b. is nonspecific for substrate
 c. is consumed by the reaction
 d. is heat and pH labile

5. An apoenzyme is where the _____ is located.
 a. cofactor c. redox reaction
 b. coenzyme d. active site

6. Many coenzymes are
 a. metals c. proteins
 b. vitamins d. substrates

7. To digest cellulose in its environment, a fungus produces a/an
 a. endoenzyme c. catalase
 b. exoenzyme d. polymerase

8. Energy in biological systems is primarily
 a. electrical c. radiant
 b. chemical d. mechanical

9. Energy is carried from catabolic to anabolic reactions in the form of _____.
 a. ADP c. coenzymes
 b. high-energy ATP bonds d. inorganic phosphate

10. Exergonic reactions
 a. release potential energy c. form bonds
 b. consume energy d. occur only outside the cell

11. A reduced compound is
 a. NAD^+ c. NADH
 b. FAD d. ADP

12. Most oxidation reactions in microbial bioenergetics involve the
 a. removal of electrons and hydrogens
 b. addition of electrons and hydrogens
 c. addition of oxygen
 d. removal of oxygen

13. A product or products of glycolysis is/are
 a. ATP c. CO_2
 b. pyruvic acid d. both a and b

14. Fermentation of a glucose molecule has the potential to produce a net number of _____ ATPs.
 a. 4 c. 40
 b. 2 d. 0

15. Complete oxidation of glucose in aerobic respiration could potentially yield a net output of _____ ATPs.
 a. 40
 c. 38
 b. 32
 d. 2

16. The compound that enters the Krebs cycle is
 a. citric acid
 b. oxaloacetic acid
 c. pyruvic acid
 d. acetyl coenzyme A

17. The $FADH_2$ formed during the Krebs cycle enters the electron transport system at which site?
 a. NADH dehydrogenase
 c. coenzyme Q
 b. cytochrome
 d. ATP synthase

18. ATP synthase complexes can potentially generate _____ ATP/ATPs for each NADH that enters electron transport.
 a. 1
 c. 3
 b. 2
 d. 4

19. Photosynthetic organisms convert the energy of _____ into chemical energy.
 a. electrons
 c. photons
 b. protons
 d. hydrogen atoms

20. The Calvin cycle operates during which part of photosynthesis?
 a. only in the light
 c. in both light and dark
 b. only in the dark
 d. only during photosystem I

21. The oxygen produced by photosynthesis comes from:
 a. CO_2
 c. H_2O
 b. NO_3
 d. glucose

22. **Multiple Matching.** Match the process a, b, or c with the metabolic events in the list.
 a. glycolysis
 b. Krebs cycle
 c. electron transport/oxidative phosphorylation
 c H^+ and e^- are delivered to O_2 as the final acceptor.
 a Pyruvic acid is formed.
 b GTP is formed.
 c H_2O is produced.
 b CO_2 is formed.
 a Fructose diphosphate is split into two 3-carbon fragments.
 a NADH is oxidized.
 c ATP synthase is active.

Case File Questions

1. Which microbial group dominates in the rumen?
 a. bacteria
 c. protozoa
 b. fungi
 d. methanogens

2. What condition in the rumen makes fermentation possible?
 a. low pH
 b. incubation temperature of 39°C
 c. lack of oxygen
 d. presence of carbon dioxide

Writing to Learn

These questions are suggested as a *writing-to-learn* experience. For each question, compose a one- or two-paragraph answer that includes the factual information needed to completely address the question. All Check & Assess questions can be used for writing-to-learn exercises.

1. Discuss the relationship of
 a. anabolism to catabolism
 b. ATP to ADP
 c. glycolysis to fermentation
 d. electron transport to oxidative phosphorylation

2. Why does the Krebs cycle require oxygen?

3. What is unique about the actions of ATP synthase?

4. Compare the general equation for aerobic metabolism with figure 8.23 and verify that all figures balance.

5. How are aerobic and anaerobic respiration different?

6. Is fermentation catabolic or anabolic? Explain your choice.

7. Explain why cows do so much chewing and belching.

8. Explain the various ways that ruminant microbes contribute to cow nutrition.

Concept Mapping

Appendix E provides guidance for working with concept maps.

1. Supply your own linking words or phrases in this concept map, and provide the missing concepts in the empty box.

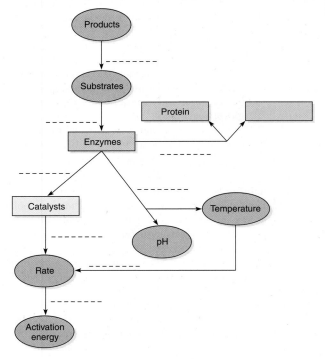

2. Construct your own concept map using the following words as the *concepts.* Supply the linking words between each pair of concepts.

anabolism nucleotides
catabolism DNA
precursor molecules ATP
bacterial cell

Critical Thinking Questions

Critical thinking is the ability to reason and solve problems using facts and concepts. These questions can be approached from a number of angles, and in most cases, they do not have a single correct answer.

1. Use the following graph to diagram the energetics of a chemical reaction, with and without an enzyme. Be sure to position reactants and products at appropriate points and to indicate the stages in the reaction and the energy levels.

Energy Involved in Reaction (vertical axis)

Progress of Reaction ⟶

2. Using the following simplified chart, fill in a summary of the major starting compounds required and the products given off by each phase of metabolism. Use arrows to pinpoint approximately where the reactions take place on the pathways.

Input	Start	Output
	Glycolysis	
	Krebs	
	ETS	

3. Describe the relationship between photosynthesis and respiration with regard to CO_2 and O_2.

4. Explain how it is possible for certain microbes to survive and grow in the presence of cyanide, which would kill many other organisms.

5. Suggest the advantages of having metabolic pathways staged in specific membrane or organelle locations rather than being free in the cytoplasm.

6. Two steps in glycolysis are catalyzed by allosteric enzymes. These are: (1) step 2, catabolized by phosphoglucoisomerase, and (2) step 9, catabolized by pyruvate kinase. Suggest what metabolic products might regulate these enzymes. How might one place these regulators in figure 8.18?

7. Give the general name of the enzyme that
 a. synthesizes ATP; digests RNA
 b. phosphorylates glucose
 c. reduces pyruvic acid to lactic acid
 d. reduces nitrate to nitrite

8. Beer production requires an early period of rapid aerobic metabolism of glucose by yeast. Given that anaerobic conditions are necessary to produce alcohol, can you explain why this step is necessary?

9. Draw a model of ATP synthase in three dimensions, showing how it works. Where in the mitochondrion does the ATP supply collect?

10. Microorganisms are being developed to control human-made pollutants and oil spills that are metabolic poisons to animal cells. A promising approach has been to genetically engineer bacteria to degrade these chemicals. What is actually being manipulated in these microbes?

11. Explain the relationship between enzyme sensitivity and the adaptations microbes make to their environments.

12. Using terminology introduced in chapter 7, describe ruminant microbes according to their nutritional, temperature, pH, and oxygen adaptations.

Visual Challenge

1. From **chapter 7, figure 7.12.** Describe the likely metabolic strategies of the microbes growing in tubes 1 through 4. Specify the pathways, final electron acceptor, and end products.

2. Find examples of reactions that are exergonic and some that are endergonic, using figures 8.18, 8.26, and 8.30.

CHAPTER 9

Microbial Genetics

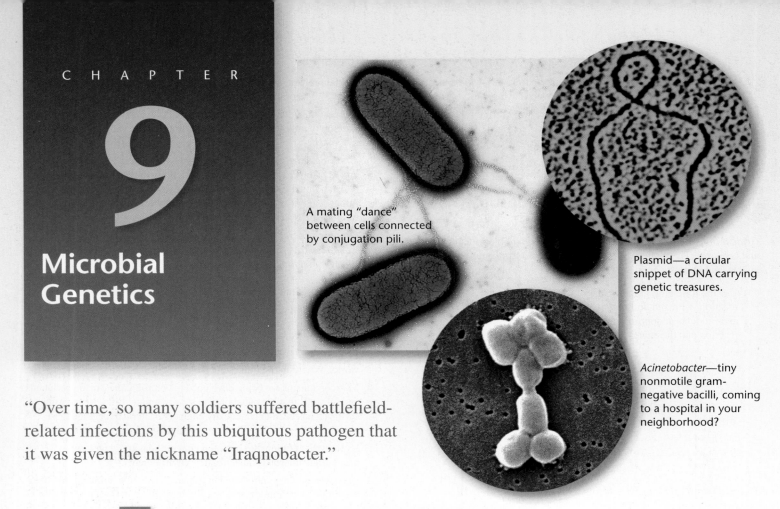

A mating "dance" between cells connected by conjugation pili.

Plasmid—a circular snippet of DNA carrying genetic treasures.

Acinetobacter—tiny nonmotile gram-negative bacilli, coming to a hospital in your neighborhood?

"Over time, so many soldiers suffered battlefield-related infections by this ubiquitous pathogen that it was given the nickname "Iraqnobacter.""

CASE FILE 9 At War with an Emerging Pathogen

A 55-year-old woman being treated in intensive care for a lung transplant developed a case of pneumonia from a ventilator she needed for breathing. The bacteria isolated from her lungs turned out to be *Acinetobacter baumannii*, a widespread resident of soil and water that has found its way into hospitals and long-term care facilities. Tests showed that all of the drugs available to treat the infection were ineffective against the pathogen—except for colistin. Early in the treatment with this drug, the lung infection appeared to be clearing up, but after a few days, it returned and the patient died of respiratory failure. Cultures taken late in the case detected a new strain of the infectious agent that had developed resistance to all known drugs, including colistin.

During the same period, several other patients in the ICU developed drug-resistant *A. baumannii* infections. This sudden outbreak of nosocomial infections spurred an investigation by an infection control nurse, who took samples from

multiple locations and cultured them. The bacteria were isolated from nearly every possible source, including medical devices, patient skin samples, bedding, hospital workers, the air, and room surfaces.

A world away in the Iraq and Afghanistan wars, this same microbe became one of the most fearsome pathogens to infect soldiers with battle injuries. In one case, a 55-year-old man sustained a grenade injury while transporting soldiers. Shrapnel penetrated his thigh and fractured the femur. When X rays revealed the presence of an infection in his broken bone, he was transported to a trauma hospital. Cultures of the wound isolated a gram-negative coccobacillus identified as *Acinetobacter baumannii*. Testing indicated that it was resistant to all drugs except for an experimental drug, tigecycline. After a several weeks' course of tigecycline, the infection was resolved and the patient was released. Over time, so many soldiers suffered battlefield-related infections by this ubiquitous pathogen that it was given the nickname "Iraqnobacter."

For many years, the rise of "super-bugs" has become a major concern of the medical community. The working definition of this term includes any bacterial species that have developed significant resistance to several drugs that would ordinarily have been given in therapy. Two of the *A. baumannii* strains in these cases are defined as multiple drug resistant (MDR), in that they resist a fairly broad range of drugs. One strain fits the category of extreme drug resistance (XDR), meaning that it is resistant to all possible drugs. Both versions of this pathogen have been added to the list of emerging diseases.

▶ *What is the probable origin of the pathogen, and what is meant by the term nosocomial?*

▶ *What does drug resistance mean from a genetic standpoint?*

To continue the case, go to page 285.

254

9.1 Introduction to Genetics and Genes: Unlocking the Secrets of Heredity

Expected Learning Outcomes

1. Define *heredity, genetics, genome, gene, phenotype,* and *genotype.*

2. Discuss the basic nature of genetic material in eukaryotes, prokaryotes, and viruses.

3. Explain how DNA is organized and packaged.

4. Describe the chemical structure of DNA and its significance.

5. List the nitrogen bases and explain their bonding patterns.

6. Describe the process of DNA replication as it occurs in prokaryotic cells.

Genetics is the science that studies the inheritance of biological characteristics by living things. This subject, also known as heredity, is a wide-ranging science that examines:

- the transmission of biological properties (traits) from parent to offspring,
- the expression and variation of those traits,
- the structure and function of the genetic material, and
- how this material changes or evolves.

The study of genetics takes place on several levels **(figure 9.1)**. Organismal genetics observes the transmission and expression of genetic factors in the whole organism or cell; chromosomal genetics examines the characteristics and actions of chromosomes; and molecular genetics deals with the biochemistry of gene function. All of these levels provide insight into microbial structure, physiology, evolution, and pathogenicity. It should also be noted that the study of microbial genetics will emphasize universal themes in the genetics of all organisms, and it will increase appreciation for the astounding advances in genetic engineering covered in chapter 10.

The Nature of the Genetic Material

Deoxyribonucleic acid or **DNA** is the central molecule of genetics, so it is essential that you develop a working knowledge of its characteristics to understand a broad range of genetic principles. With this in mind, we will use the first section of this chapter to lay some groundwork on DNA organization, structure, and duplication. This will be followed by coverage of related concepts such as RNA and protein synthesis, genetic control, mutations, and genetic transfers in the remaining five sections.

The Levels of Structure and Function of the Genome

The **genome** is the sum total of genetic material carried within a cell. Although most of the genome exists in the form of chromosomes, genetic material can appear in nonchromosomal sites as well **(figure 9.2)**. For example, bacteria and some fungi contain tiny extra pieces of DNA (**plasmids**), and the mitochondria and chloroplasts of eukaryotes are equipped with their own functional chromosomes. Genomes of cells are composed exclusively of DNA, but viruses contain either DNA or RNA as the principal genetic material. Although the specific genome of an individual organism is unique, the general pattern of nucleic acid structure and function is similar among all organisms.

In general, a **chromosome** is a discrete cellular structure composed of a neatly packaged DNA molecule. The chromosomes of eukaryotes and bacterial cells differ in several respects. The structure of eukaryotic chromosomes consists of a DNA molecule tightly wound around histone proteins, whereas a bacterial chromosome is condensed and secured into a packet by means of a different type of protein. Eukaryotic chromosomes are located in the nucleus; they vary in number from a few to hundreds; they can occur in pairs (diploid) or singles (haploid); and they are linear in format. In contrast, most bacteria have a single, circular chromosome, although some have multiple chromosomes and a few have linear chromosomes.

Organism level Cell level Chromosome level Molecular level

Figure 9.1 Levels of genetic study. The operations of genetics can be observed at the levels of organism, cell, chromosome, and DNA sequence (molecular level). Although levels shown here are for a eukaryotic organism they would follow a similar pattern in prokaryotes.

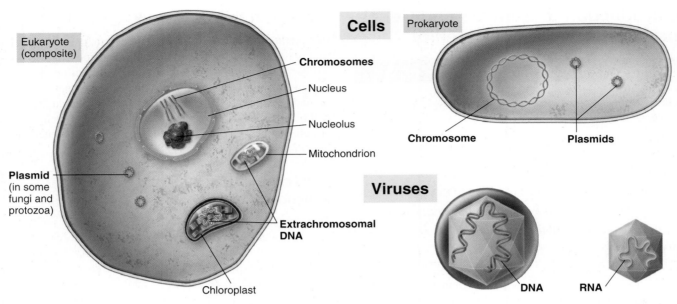

Figure 9.2 **The general location and forms of the genome in two cell types and selected viruses (not to scale).**

All chromosomes contain a series of basic informational "packets" called genes. A **gene** can be defined from more than one perspective. In classical genetics, the term refers to the fundamental unit of heredity responsible for a given trait in an organism. In the molecular and biochemical sense, it is a portion of the chromosome that provides information for a given cell function. More specifically still, it is a specific segment of DNA that contains the necessary codes to make a **protein** or RNA molecule. This last definition of a gene is emphasized in this chapter.

Genes fall into three basic categories: *structural genes* that code for proteins, genes that code for RNA, and *regulatory genes* that control gene expression. The sum of all of these types of genes constitutes an organism's distinctive genetic makeup, or **genotype.*** The expression of the genotype creates traits (certain structures or functions) referred to as the **phenotype.*** For example a person inherits a combination of genes (genotype) that gives a certain eye color or height (phenotype), a bacterium inherits genes that direct the formation of a flagellum or the ability to metabolize a certain substrate, and a virus has genes that dictate its capsid structure. All organisms contain more genes in their genotypes than are being seen as a phenotype at any given time. In other words, the phenotype can change depending on which genes are "turned on" or expressed.

The Size and Packaging of Genomes

Genomes vary greatly in size. Viruses have anywhere from a few to several hundred genes; the bacterium *Escherichia coli* has a single chromosome containing 4,288 genes, and a human cell packs about five times that many into 46 chromosomes. The chromosome of *E. coli* would measure about 1 mm if unwound and stretched out linearly, and yet this fits within a cell that measures just over 1 μm

across, making the stretched-out DNA 1,000 times longer than the cell **(figure 9.3).** Still, the bacterial chromosome takes up only about one-third to one-half of the cell's volume. Likewise, if the sum of all DNA contained in the 46 human chromosomes were unraveled and laid end to end, it would measure about 6 feet. How can such elongated genomes fit into the minuscule volume of a cell and, in the case of eukaryotes, into an even smaller compartment, the nucleus? The answer lies in the complex coiling of the DNA chain **(Insight 9.1).**

The Structure of DNA: A Double Helix with Its Own Language

To analyze the structure of DNA at the molecular level, imagine being able to magnify a small piece of a gene about 5 million

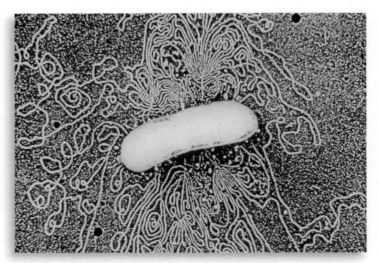

Figure 9.3 **A disrupted *Escherichia coli* cell has spewed out its chromosome in a single, uncoiled DNA strand.**

* *genotype* (jee′-noh-typ) Gr. *gennan*, to produce, and *typos*, type.
* *phenotype* (fee′-noh-typ) Gr. *phainein*, to show. The physical manifestation of gene expression.

INSIGHT 9.1

The Packaging of DNA: Winding, Twisting, and Coiling

Packing the mass of DNA into the cell involves compacting the DNA molecule by means of supercoils or superhelices. In the simpler system of prokaryotes, the circular chromosome is packaged by the action of a special enzyme called a **topoisomerase** (specifically, DNA **gyrase**). This enzyme coils the chromosome into a tight bundle by introducing a reversible series of twists into the DNA molecule.

The system in eukaryotes is more complex, with three or more levels of coiling. First, the DNA molecule of a chromosome, which is linear, is wound twice around the histone proteins, creating a chain of **nucleosomes.** The nucleosomes fold in a spiral formation upon one another. An even greater supercoiling occurs when this spiral arrangement further twists on its radius into a giant spiral with loops radiating from the outside. This extreme degree of compactness is what makes the eukaryotic chromosome visible during mitosis.

The condensation of DNA and protein into chromatin was once thought to function mainly as a way for such a long molecule to fit into a cell. However, new research has shown that the coiling of DNA also serves to make certain segments of the genetic program either more or less available to the cell. This is a major control on the expression of many traits, especially in eukaryotes. Having an additional regulatory function creates an entirely new set of information to guide phenotypic expression. The realization that the phenotype of a cell is due to more than just the order of nucleotides in the DNA has led to a rethinking of some of the most basic aspects of genetics (see Insight 9.3).

Because genetic codes cannot be properly accessed when DNA is supercoiled, what other system would a cell need to manage DNA? *Answer available at http://www.mhhe.com/talaro8*

The packaging of DNA. Eukaryotic DNA is intimately associated with a variety of proteins and small molecules. This relationship modulates the expression of many genes while at the same time allowing for the ordered condensation of DNA into a compact molecule.

times. What you would see is one of the great marvels of life. The pieces of this molecular puzzle were finally worked out by James Watson and Francis Crick in 1953 **(Insight 9.2).** They discovered that DNA is a giant molecule with two strands forming a double helix. Extensive evidence shows that the general structure of DNA is universal, except in some viruses that contain single-stranded DNA.

The basic unit of DNA structure is a **nucleotide,** composed of **phosphate, deoxyribose sugar,** and a **nitrogen base,** as shown in this simplified model (see figure 2.23):

INSIGHT 9.2

Deciphering the Structure of DNA

The search for the primary molecules of heredity was a serious focus throughout the first half of the twentieth century. At first, many biologists thought that protein was the genetic material. An important milestone occurred in 1944 when Oswald Avery, Colin MacLeod, and Maclyn McCarty purified DNA and demonstrated at last that it was indeed the blueprint for life. This was followed by an avalanche of research, which continues today.

"We Have Discovered the Secret of Life."

One area of extreme interest concerned the molecular structure of DNA. In 1951, American biologist James Watson and English physicist Francis Crick collaborated on solving the DNA puzzle. Although they did little of the original research, they were intrigued by several findings from other scientists. It had been determined by Erwin Chargaff that any model of DNA structure would have to contain deoxyribose, phosphate, purines, and pyrimidines arranged in a way that would provide variation and a simple way of copying itself. Watson and Crick spent long hours constructing models with cardboard cutouts and kept alert for any and every bit of information that might give them an edge.

Two English biophysicists, Maurice Wilkins and Rosalind Franklin, had been painstakingly collecting data on X-ray crystallographs of DNA for several years. With this technique, molecules of DNA bombarded by

The men who cracked the code of life. Dr. James Watson (left) and Dr. Francis Crick (right) stand next to their model that finally explained the structure of DNA in 1953.

X rays produce a photographic image that can predict the three-dimensional structure of the molecule. After being allowed to view certain X-ray data, Watson and Crick noticed an unmistakable pattern: The molecule appeared to be a double helix. Gradually, the pieces of the puzzle fell into place, and a final model was assembled—a model that explained all of the qualities of DNA, including how it is copied. Although Watson and Crick were rightly hailed for the clarity of their solution, it must be emphasized that their success was due to the considerable efforts of a number of English and American scientists. This historic discovery showed that the tools of physics and chemistry have useful applications in biological systems, and it also spawned ingenious research in all areas of molecular genetics.

Since the discovery of the double helix in 1953, an extensive body of biochemical, microscopic, and crystallographic analysis has left little doubt that the model first proposed by Watson and Crick is correct. Scanning tunneling microscopes produce three-dimensional images of DNA that verify the helical shape and twists of DNA represented by models.

What particular requirements did Watson and Crick's model of DNA have to fit to be accurate? *Answer available at http://www. mhhe.com/talaro8*

Each deoxyribose sugar bonds covalently in a repeating pattern with two phosphates. One of the bonds is to the number 5′ (read "five prime") carbon on deoxyribose, and the other is to the 3′ carbon, which specifies the order and direction of each strand (**figure 9.4**). This formation results in an elongate strand with a sugar-phosphate backbone.

The nitrogen bases, **purines** and **pyrimidines,** attach by covalent bonds at the 1′ position of the sugar (**figure 9.4a**). They span the center of the molecule and pair with appropriate complementary bases from the other strand, thereby forming a double-stranded helix. The **paired bases** are so aligned as to be joined by hydrogen bonds. Such weak bonds are easily broken, allowing the molecule to be "unzipped" into its complementary strands. This feature is of great importance in gaining access to the information encoded in the nitrogen base sequence. Pairing of purines and pyrimidines is not random; it is dictated by the formation of hydrogen bonds between certain bases. Thus, in DNA, the purine **adenine** (A) pairs with the pyrimidine **thymine** (T), and the purine **guanine** (G) pairs with the pyrimidine **cytosine** (C). Note that adenine forms two hydrogen bonds with thymine, and cytosine forms three hydrogen

bonds with guanine. This difference will influence a number of DNA functions. New research also indicates that the bases are attracted to each other in this pattern because each has a complementary three-dimensional shape that matches its pair. Although the base-pairing partners generally do not vary, the sequence of base pairs along the DNA molecule can assume any order, resulting in an infinite number of possible nucleotide sequences.

Other important considerations of DNA structure concern the nature of the double helix itself. The two strands are not oriented in the same direction. One side of the helix runs in the opposite direction of the other, in what is called an *antiparallel arrangement* (**figure 9.4b**). The order of the bond between the carbon on deoxyribose and the phosphates is used to keep track of the direction of the two sides of the helix. Thus, one helix runs from the 5′ to 3′ direction, and the other runs from the 3′ to 5′ direction. This characteristic is a significant factor in DNA synthesis and translation. As apparently perfect and regular as the DNA molecule may seem, it is not exactly symmetrical. The torsion in the helix and the stepwise stacking of the nitrogen bases produce two different-size surface features, the major and minor grooves (**figure 9.4c**).

Figure 9.4 **Three views of DNA structure.** **(a)** A schematic nonhelical model shows the basic layout of the two strands. Note that the order of phosphate and sugar bonds goes in the opposite direction for the two strands, going from the 5′ carbon to the 3′ carbon on one strand and from the 3′ carbon to the 5′ carbon on the other strand. Insets show details of the nitrogen base pairs. **(b)** Simplified model that highlights the antiparallel arrangement and the major and minor grooves. **(c)** Space-filling model that more accurately depicts the three-dimensional structure of DNA.

One of the significant findings that has emerged from sequencing the DNA of chromosomes is knowing how long it can be. An average bacterial chromosome consists of 5 million to 6 million nucleotides; the 46 chromosomes of humans amount to about 3 billion nucleotides.

DNA Replication: Preserving the Code and Passing It On

For a species to survive, it must reproduce. In bacteria, this involves division of the cell by means of binary fission or budding, but it also involves the accurate duplication and separation of the genetic material into each daughter cell to ensure normal function.

It is the sequence of bases along the length of DNA that constitutes its "language." For this language to be preserved for hundreds of generations, the codes it contains will need to be duplicated with high fidelity. This process of duplication is called **DNA replication.** In the following example, we will show replication in bacteria, but with some exceptions, it also applies to the process as it works in eukaryotes and some viruses. Recall from chapter 7 that early in binary fission, the metabolic machinery of a

bacterium is programmed to initiate the duplication of the chromosome. Despite its complexity, DNA replication can be very rapid: This must be completed in a single generation time (around 20 minutes in *E. coli*).

TAKE NOTE: THE SIGNIFICANCE OF DNA STRUCTURE

The arrangement of nitrogen bases in DNA has two essential effects.

1. **Maintenance of the genetic codes during reproduction.** The constancy of base-pairing guarantees that the codes will be retained during cell growth and division. When the two strands are separated, each one provides a **template** (pattern or model) for the replication (exact copying) of a new molecule **(figure 9.5)**. Because the sequence of one strand automatically gives the sequence of its partner, the codes can be duplicated with fidelity.
2. **Providing variety.** The order of bases along the length of the DNA strand provides the information needed to produce RNA and protein molecules, which in turn are responsible for the phenotype of the cell. As we will see, changing the identity of the bases or their order in the DNA molecule can have a dramatic effect on the phenotype of the organism.

It is tempting to consider how such a seemingly simple set of codes can account for the extreme differences among forms as diverse as a virus, *E. coli,* and a human. The English language, based on 26 letters, can create an infinite variety of words, but how can an apparently complex genetic language such as DNA be based on just four nitrogen base "letters"? The answer lies in the fact that the four bases of DNA are arranged into much longer sequences. An average bacterial gene may contain 1,000 nucleotides: so while DNA contains fewer "letters," they are arranged into much longer "words." Mathematically speaking, 1,000 nucleotides can be arranged in $4^{1,000}$ different sequences, a number so large (1.5×10^{602}) that it provides nearly endless degrees of variation.

The Overall Replication Process

What features allow the DNA molecule to be exactly duplicated, and how is its integrity retained? DNA replication requires a careful orchestration of the actions of 30 different enzymes (partial list in **table 9.1**), which separate the strands of the existing DNA molecule, copy its template, and produce two complete daughter molecules. A simplified version of replication is shown in **figure 9.5** and includes the following:

1. uncoiling the parent DNA molecule, beginning at a predetermined point of origin;
2. unzipping the hydrogen bonds between the base pairs, thus separating the two strands and exposing the nucleotide sequence of each strand (which is normally buried in the center of the helix) to serve as templates; and

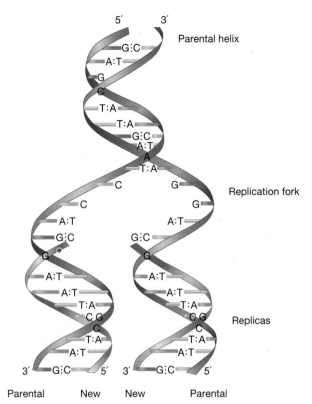

Figure 9.5 **Simplified steps to show the semiconservative replication of DNA.** The two strands of the parental double helix are unwound and separated. Each single strand will serve as a template (blue) to synthesize a new strand of DNA (purple), starting at the replication fork. As synthesis proceeds, the correct nucleotides are added according to the pattern of the template. An A on the template will pair with a T on the new molecule, and a C will pair with a G. The resultant new DNA molecules contain one strand of the newly synthesized DNA and the original template (parental) strand. This keeps the code intact because the same linear arrangement of the bases is maintained during this process. Note that the fine details of the process are presented in figure 9.6.

TABLE 9.1	Some Enzymes Involved in DNA Replication and Their Functions
Enzyme	**Function**
Helicase	Unzipping the DNA helix
Primase	Synthesizing an RNA primer
DNA polymerase III	Adding bases to the new DNA chain; proofreading the chain for mistakes
DNA polymerase I	Removing RNA primer, closing gaps, repairing mismatches
Ligase	Final binding of nicks in DNA during synthesis and repair
Gyrase	Supercoiling

3. synthesizing two new strands by attachment of the correct **complementary** nucleotides to each single-stranded template.

A critical feature of DNA replication is that each daughter molecule will be identical to the parent in composition, but neither one is

completely new; the strand that serves as a template is an original parental DNA strand. The preservation of the parent molecule in this way, termed **semiconservative replication,** helps explain the reliability and fidelity of replication (figure 9.5).

Events in Replication

All chromosomes have a specific **origin of replication** site that serves as the place where replication will be initiated. It is recognized by a short sequence rich in adenine and thymine that, you will recall, are held together by only two hydrogen bonds rather than three. Because the origin of replication is AT-rich, less energy is required to separate the two strands than would be required if the origin were rich in guanine and cytosine.

The process of synthesizing a new daughter strand of DNA using the parental strand as a template is carried out by a giant enzyme complex that brings in the primary replication enzyme—**DNA polymerase III**—along with numerous accessory enzymes (see

table 9.1). The DNA polymerase III is responsible for deciphering and duplicating DNA codes, but it has some major restrictions that will affect the overall process **(figure 9.6):**

- DNA polymerase cannot add nucleotides to a DNA strand without a preexisting unbound nucleotide.
- Before replication can begin, a piece of RNA called a primer is inserted at the origin of replication to serve as the starting point for bases to be added.
- The polymerase can add nucleotides only in the direction of 5′ to 3′, which means that synthesis of the two template strands will be somewhat different.

Prior to the start of replication, enzymes called **helicases** (unzipping enzymes) bind to the DNA at the origin. These enzymes untwist the helix and break the hydrogen bonds holding the two strands together, resulting in two separate strands, each of which will be used as a template for the synthesis of a new strand **(figure 9.6, step 1).**

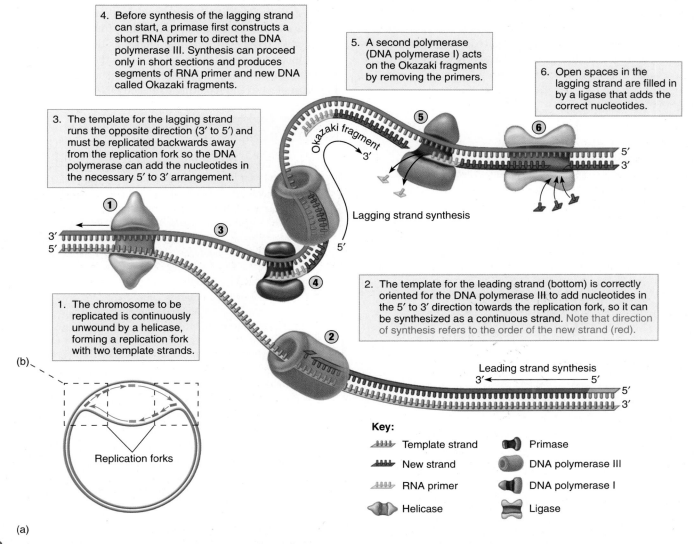

4. Before synthesis of the lagging strand can start, a primase first constructs a short RNA primer to direct the DNA polymerase III. Synthesis can proceed only in short sections and produces segments of RNA primer and new DNA called Okazaki fragments.

5. A second polymerase (DNA polymerase I) acts on the Okazaki fragments by removing the primers.

6. Open spaces in the lagging strand are filled in by a ligase that adds the correct nucleotides.

3. The template for the lagging strand runs the opposite direction (3′ to 5′) and must be replicated backwards away from the replication fork so the DNA polymerase can add the nucleotides in the necessary 5′ to 3′ arrangement.

Okazaki fragment

Lagging strand synthesis

1. The chromosome to be replicated is continuously unwound by a helicase, forming a replication fork with two template strands.

2. The template for the leading strand (bottom) is correctly oriented for the DNA polymerase III to add nucleotides in the 5′ to 3′ direction towards the replication fork, so it can be synthesized as a continuous strand. Note that direction of synthesis refers to the order of the new strand (red).

Replication forks

Leading strand synthesis

Key:

- Template strand
- New strand
- RNA primer
- Helicase
- Primase
- DNA polymerase III
- DNA polymerase I
- Ligase

Process Figure 9.6 **The Assembly line of DNA replication in a circular bacterial chromosome. (a)** A bacterial chromosome showing the overall pattern of replication. There are two replication forks where new DNA is being synthesized. **(b)** An enlarged view of the left replication fork to show the details of replication.

Replication begins when an RNA **primer** is synthesized by a primase at the origin of replication (**figure 9.6a**). DNA polymerase III must have this short strand of RNA to serve as a starting point for adding nucleotides. Because the bacterial DNA molecule is circular, opening of the circle forms two **replication forks,** each containing its own set of replication enzymes. Each fork contains two active DNA polymerases along with several other proteins and enzymes whose main functions are to stabilize the polymerase and provide a means of removing improperly incorporated nucleotides. The enzyme complex encircles the DNA near the replication fork and synthesizes new DNA using the parent strand as a template. As synthesis proceeds, the forks continually open up to expose the template for replication.

Because DNA polymerase III can only synthesize new DNA in the 5′ to 3′ direction, just one strand, called the **leading strand,** can be synthesized continuously (**figure 9.6, step 2**). One enzyme does this as it moves along the template strand toward the replication fork. The other strand, which is oriented 3′ to 5′ with respect to the polymerase, is termed the **lagging strand (figure 9.6, steps 3 and 4).** Because this strand cannot be synthesized continuously, an RNA primer is placed ahead of the polymerase to initiate synthesis. It then adds nucleotides a few at a time in the direction away from the fork (5′ to 3′). As the fork opens up a bit, the next segment is synthesized backward near the point of the previous segment. In this way, the two new strands can be synthesized simultaneously. This manner of synthesis produces short fragments of DNA (100 to 1,000 bases long) called **Okazaki fragments** on the side of the lagging strand.

The lagging strand will be completed when the enzyme, **DNA polymerase I,** removes the RNA primers from the Okazaki fragments and a **ligase** follows behind and inserts the correct DNA nucleotides into the open spaces (**figure 9.6, steps 5 and 6**). The final result should be identical to the leading strand.

The addition of nucleotides proceeds at an astonishing pace, estimated in some bacteria to be 750 bases per second at each fork! As replication proceeds, one newly synthesized strand loops down (**figure 9.7a**). When the forks have gone full circle, a termination site shuts replication down. The two circular daughter molecules remain connected briefly but are nicked and separated by a helicase (**figure 9.7b**).

Like any language, DNA is occasionally "misspelled" when an incorrect base is added to the growing chain. Studies have shown that such mistakes are made once in approximately 10^8 to 10^9 bases, but most of these are corrected. If not corrected, they are referred to as mutations and can lead to serious cell dysfunction. Because continued cellular integrity is very dependent on accurate replication, cells have evolved their own proofreading function for DNA. DNA polymerase III, the enzyme that elongates the molecule, can also detect incorrect, unmatching bases; excise them; and replace them with the correct base. DNA polymerase I is involved in proofreading the molecule and repairing damaged DNA.

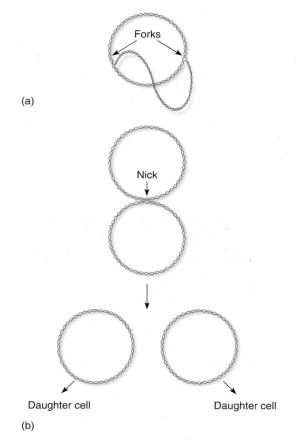

(a)

Nick

Daughter cell Daughter cell

(b)

Figure 9.7 Completion of chromosome replication in bacteria. (a) As replication proceeds, one strand loops down. **(b)** Final separation is achieved when an enzyme cuts and releases the two completed molecules. The daughter cells receive these during binary fission.

Check & Assess Section 9.1

✔ Genetics is the study of the expression of biological information and its transfer between organisms.

✔ Nucleic acids are molecules that contain the blueprints of life in the form of genes. DNA is the blueprint molecule for all cellular organisms. The blueprints of viruses, however, can be either DNA or RNA.

✔ The total amount of DNA in an organism's chromosomes is termed its genome. The genotype of each species contains a unique arrangement of genes that, when expressed, result in its traits or phenotype, including metabolic activities and pattern of reproduction.

✔ The genome of prokaryotes is quite small compared with the genome of eukaryotes. A bacterial chromosome usually consists of a few hundreds to thousands of genes. Eukaryotic genomes range from *thousands* to *tens of thousands* of genes. Their DNA is packaged in tightly wound spirals arranged in discrete chromosomes.

✔ DNA is a very long molecule composed of small subunits called nucleotides. The sequence of the nucleotides contains information needed to direct the synthesis of all proteins in the cell.

✔ The DNA molecule must be copied so that the genetic material can be transferred to offspring.

✔ DNA's chemical structure makes its replication possible.

✔ DNA copies itself just before cellular division by the process of semiconservative replication. Semiconservative replication means that each "old" DNA strand is the template upon which each "new" strand is synthesized.

✔ The circular bacterial chromosome is replicated at two forks as directed by DNA polymerase III. At each fork, two new strands are synthesized—one continuously and one in short fragments, and mistakes are proofread and removed.

1. Compare the genetic material of eukaryotes, bacteria, and viruses in terms of general structure, size, and mode of replication.
2. Characterize the organization of genetic material, going from the genome down to the gene.
3. Briefly describe how DNA is packaged to fit inside a cell.
4. What are the fundamental building blocks of DNA and how are they bonded together?
5. Describe what is meant by the antiparallel arrangement of DNA.
6. Explain the synthesis of the leading and lagging strands during DNA replication and identify the enzymes required for producing a finished product.
7. Name several characteristics of DNA structure that enable it to be replicated with such great fidelity generation after generation.

9.2 Applications of the DNA Code: Transcription and Translation

Expected Learning Outcomes

7. Present an overview of the main aspects of the flow of genetic information in cells.
8. Explain the relationship between the structure of DNA and the structure of proteins.
9. Describe the different types of RNA and their basic functions in genetic expression.
10. Explain what happens during transcription.
11. Describe the genetic code, codons, and anticodons, and how they relate to one another.
12. Relate the participants and steps in translation (protein synthesis).
13. Distinguish major points of difference between prokaryotic and eukaryotic transcription and translation.

Figure 9.8 Summary of the flow of genetic information in cells. DNA is the ultimate storehouse and distributor of genetic information. **(a)** DNA must be deciphered into a usable cell language. It does this by transcribing its code into RNA helper molecules that translate that code into protein. **(b)** Other sections of the DNA produce additional types of RNA molecules that regulate genes and their products.

We have explored how the genetic language of the DNA molecule is conserved through replication. Next, we consider exactly how DNA is expressed. Given that the sequence of bases in DNA is a genetic code, just what is the nature of this code and how is it utilized? Although the genome is full of critical information, the DNA molecule itself cannot perform cell processes directly. Instead, its information is conveyed to RNA molecules, which carry out the instructions it contains. The concept that genetic information flows from DNA to RNA to protein has long been a central theme of molecular biology (**figure 9.8a**). It states that the master code of DNA is first used to synthesize RNA via a process called **transcription,** and the information contained in the RNA is then used to produce proteins in a process known as **translation.** The principal exceptions to this pattern are found in RNA viruses, which convert RNA to other RNA, and in retroviruses, which convert RNA to DNA.

We now know that this so-called "central dogma," though accurate, is somewhat incomplete. In addition to the RNA used to produce proteins, a wide variety of specialized RNAs act by

regulating gene function (**figure 9.8b**). We will introduce some of the expanded functions for RNA in Insight 9.4, but we will start with its direct participation in protein synthesis.

The Gene-Protein Connection

The Triplet Code and the Relationship to Proteins

Several questions invariably arise concerning the relationship between genes and cell function. For instance, how does gene structure lead to the expression of traits in the individual, and what features of gene expression cause one organism to be so distinctly different from another? For answers, we must turn to the correlation

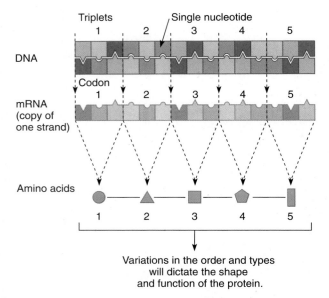

Figure 9.9 **Simplified view of the DNA-protein relationship.** The DNA molecule is a continuous chain of base pairs, but the sequence must be interpreted in groups of three base pairs (a triplet). Each triplet as copied into mRNA codons will translate into one amino acid; consequently, the ratio of base pairs to amino acids is 3:1.

between gene and protein structure. We know that each structural gene is an ordered sequence of nucleotides that codes for a protein. Because each protein is different, each gene must also differ somehow in its composition. In fact, the language of DNA exists in the order of groups of three consecutive bases, or **triplets,** on one DNA strand **(figure 9.9).** In short, one gene differs from another in its order of triplets. An equally important part of this concept is that each triplet represents a code for a particular amino acid. When the triplet code is transcribed and translated, it dictates the type and order of amino acids in a polypeptide (protein) chain.

The final key points that connect DNA and protein function are:

1. DNA is a blueprint that indicates which kinds of proteins to make and how to make them. This blueprint exists in the order of triplets along the DNA strands.
2. The order of triplets directs a protein's primary structure—the order and type of amino acids in the chain—which determines its characteristic shape and function.
3. Proteins contribute significantly to the phenotype by functioning as enzymes and structural molecules.

The Major Participants in Transcription and Translation

Transcription, the formation of RNA using DNA as a template, and translation, the synthesis of proteins using RNA as a template, are highly complex. A number of components participate: most prominently, messenger RNA, transfer RNA, ribosomes, several types of enzymes, and a storehouse of raw materials. After first examining each of these components, we shall see how they come together in the assembly line of the cell.

RNAs: Tools in the Cell's Assembly Line

Ribonucleic acid (**RNA**) is an encoded molecule like DNA, but its general structure is different in several ways:

1. It is a single-stranded molecule that can assume secondary and tertiary levels of complexity due to bonds within the molecule, leading to specialized forms of RNA (mRNA, tRNA, and rRNA— see figure 9.8).
2. RNA contains **uracil,** instead of thymine, as the complementary base-pairing mate for adenine. This does not change the inherent DNA code in any way because the uracil still follows the pairing rules.
3. Although RNA is a single helix that consists of alternating sugar and phosphate molecules, the sugar in RNA is **ribose** rather than deoxyribose (see figure 2.24).

The many functional types of RNA range from small regulatory pieces to large structural ones. All types of RNA are formed through transcription of a DNA gene, but only mRNA is further translated into protein. See **table 9.2** for a comparison of major types of RNA and its functions. Other forms of RNA are introduced in section 9.3.

Messenger RNA: Carrying DNA's Message

Messenger RNA (mRNA) is a transcribed version of a structural gene or genes in DNA. It is synthesized by a process similar to synthesis of the leading strand during DNA replication, and the complementary base-pairing rules ensure that the code will be faithfully copied in the mRNA transcript. The message of this transcribed strand is later read as a series of triplets called **codons** **(figure 9.10a),** and the length of the mRNA molecule varies from about 100 nucleotides to several thousand. The details of transcription and the function of mRNA in translation are covered shortly.

TABLE 9.2	Major Types of Ribonucleic Acid Involved in Protein Synthesis		
RNA Type	**Contains Codes For**	**Function in Cell**	**Translated**
Messenger (mRNA)	Sequence of amino acids in protein	Carries the DNA master code to the ribosome	Yes
Transfer (tRNA)	A cloverleaf tRNA to carry amino acids	Brings amino acids to ribosome during translation	No
Ribosomal (rRNA)	Several large structural rRNA molecules	Forms the major part of a ribosome and participates in protein synthesis	No
Primer	An RNA that can begin DNA replication	Primes DNA	No

(a) Messenger RNA (mRNA)
A short piece of messenger RNA (mRNA) illustrates the general structure of RNA: single strandedness, repeating phosphate-ribose sugar backbone attached to single nitrogen bases; use of uracil instead of thymine.

Codon 1 Codon 2 Codon 3

℗ = Phosphate ℞ = Ribose **U** = Uracil

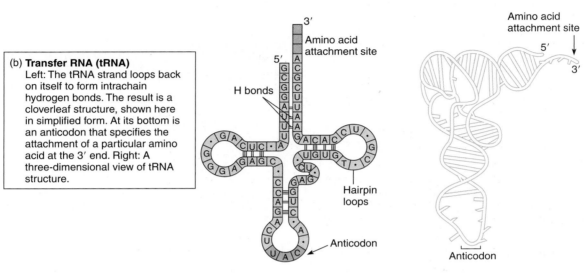

(b) Transfer RNA (tRNA)
Left: The tRNA strand loops back on itself to form intrachain hydrogen bonds. The result is a cloverleaf structure, shown here in simplified form. At its bottom is an anticodon that specifies the attachment of a particular amino acid at the 3′ end. Right: A three-dimensional view of tRNA structure.

Figure 9.10 Characteristics of messenger and transfer RNA.

Transfer RNA: The Key to Translation

Transfer RNAs (tRNA) are also complementary copies of specific regions of DNA; however, they differ from mRNA. TRNA is uniform in length, being 75 to 95 nucleotides long, and it contains sequences of bases that form hydrogen bonds with complementary sections of the same tRNA strand. At these points, the molecule bends back upon itself into several *hairpin loops,* giving the molecule a secondary *cloverleaf* structure that folds even further into a complex, three-dimensional helix **(figure 9.10b).** This compact molecule acts as a translator that converts RNA language into protein language. The bottom loop of the cloverleaf exposes a triplet, the **anticodon,** that both designates the specificity of the tRNA and complements mRNA's codons. At the opposite end of the molecule is a binding site for the amino acid that is specific for that tRNA's anticodon. For each of the 20 amino acids, there is at least one specialized type of tRNA to carry it. Binding of an amino acid to its specific tRNA requires a specific enzyme that can correctly match each tRNA with its amino acid.

The Ribosome: A Mobile Molecular Factory for Translation

The prokaryotic (70S) ribosome is a particle composed of tightly packaged **ribosomal RNA (rRNA)** and protein. The rRNA component of the ribosome is also a long polynucleotide molecule. It forms complex three-dimensional figures that contribute to the structure and function of ribosomes. The interactions of proteins and rRNA create the two subunits of the ribosome that engage in final translation of

the genetic code (see figure 9.12). A metabolically active bacterial cell can accommodate up to 20,000 of these minuscule factories—all actively engaged in reading the genetic program, taking in raw materials, and producing proteins at an impressive rate.

Transcription: The First Stage of Gene Expression

Like so many other genetic events, transcription is a marvel of efficiency. It is highly organized and guided by special codes on DNA itself. To describe it in more detail, we use the formation of mRNA in bacteria as a model. It is important to keep in mind that all RNA (tRNA, rRNA) is synthesized by a similar process. The other RNAs have important functions; they just do not provide the code for the amino acids in a protein as mRNA does.

During transcription, an RNA molecule is synthesized using the codes on DNA as a guide or template. A large enzyme complex, **RNA polymerase,** is responsible for this process. This polymerase is more multipurpose than DNA polymerase, because the RNA enzyme works alone and does not require a helicase. It can bind to DNA and unwind it, as well as synthesize RNA **(figure 9.11).**

Transcription proceeds in three stages: *initiation, elongation,* and *termination.* Initiation requires the RNA polymerase to recognize a region on a gene called the **promoter region (figure 9.11, step 1).** This region consists of two sets of DNA sequences located just before the *initiation site.* The first sequence is about 35 nucleotides from the initiation site, and the second is about 10 nucleotides from it. The primary function of the promoter is to provide a position for initial binding of

① Each gene contains a specific promoter region and a leader sequence for guiding the beginning of transcription. Next is the region of the gene that codes for a polypeptide and ends with a series of terminal sequences that stop translation.

RNA polymerase binding site

Leader sequence

Initiation codon

Termination sequences

Promoter region

| T | A | C | G | A | C | T | G | A | T | G | C |
| A | T | G | C | T | G | A | C | T | A | C | G |

Intervening sequence of variable size

Termination sequence

RNA polymerase

Template strand

3′ 5′

② DNA is unwound at the promoter by RNA polymerase. Only one strand of DNA, called the template strand, is copied by the RNA polymerase. This strand runs in the 3′ to 5′ direction.

5′ 3′

Nontemplate strand

Unwinding of DNA

Direction of transcription

③ The RNA polymerase moves along the strand, adding complementary nucleotides as dictated by the DNA template. The mRNA strand reads in the 5′ to 3′ direction.

3′

Nucleotide pool

5′

Early mRNA transcript

Elongation

④ The polymerase continues transcribing until it reaches a termination site, and the mRNA transcript is released to be translated. Note that the section of the transcribed DNA is rewound into its original configuration.

3′

5′

Late mRNA transcript

Process Figure 9.11 **The major events in transcription, using mRNA as a model.**

the RNA polymerase. The promoter sequences are not highly varied, and they tend to be rich in adenine and thymine base pairs, which have one less hydrogen bond, thus facilitating separation of the DNA strands.

Prior to the first synthesis step of transcription, the RNA polymerase begins to separate the two strands of the DNA helix and forms an open "bubble" of sorts **(figure 9.11, steps 2 and 3).** This bubble serves as the space where the bonds between the nucleotides of mRNA will actually be made. Only one strand of DNA—the **template strand**—is transcribed. This is the one that carries a message that can be translated into a protein. The other strand—the **nontemplate strand**—can serve a genetic function but not as a message for protein structure. Which strand becomes the template varies from one gene to another. Note that the promotor sequences are not transcribed as part of the final mRNA molecule.

The first DNA triplet to be transcribed is almost invariably TAC. Recall that, since RNA will have uracil, not thymine, as the complement of adenine, this means that the first codon on the

mRNA will be AUG (see start codon in figure 9.13). As elongation proceeds, the polymerase moves the transcription bubble forward, exposing subsequent sections of DNA. It simultaneously brings in nucleotides that are complementary to the DNA template and continues to assemble the mRNA strand. As with replication, this occurs in the 5′ to 3′ direction with regard to the growing mRNA. DNA that has already been transcribed rewinds back into its double helix structure **(figure 9.11, step 4).** The part of the mRNA strand that is already assembled remains attached to the enzyme complex but hangs out of the way of the processing machinery.

At termination, the polymerase recognizes another code on DNA near the end of the gene that signals the separation and release of the completed mRNA. It will go immediately to the translation phase on the ribosome. As we will see, this mRNA delivers its message in codons that are the master code for protein synthesis. The length of the final mRNA molecule depends upon the polypeptide that it encodes. The smallest one may consist of 100 nucleotides, an average one of 1,200 nucleotides, and the largest several thousand.

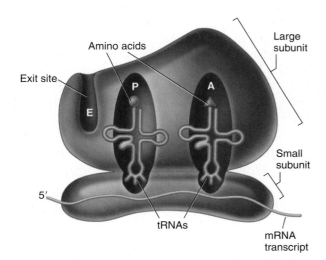

Figure 9.12 **The "players" in translation.** A ribosome serves as the stage for protein synthesis. Assembly of the small and large subunits results in specific sites for holding the mRNA and two tRNAs with their amino acids.

Translation: The Second Stage of Gene Expression

In translation, all of the elements needed to synthesize a protein, from the mRNA to the tRNAs with amino acids, are brought together on the ribosomes (**figure 9.12**). The process occurs in five stages: initiation, elongation, termination, and protein folding and processing.

Initiation of Translation

The mRNA molecule leaves the DNA transcription site and is transported directly to ribosomes. Ribosomal subunits are specifically assembled in a way that forms sites to hold the mRNA and tRNAs. The ribosome thus recognizes these molecules and stabilizes reactions between them. The small subunit binds to the 5′ end of the mRNA, and the large subunit supplies enzymes for making peptide bonds on the protein. The ribosome begins to scan the mRNA by moving in the 5′ to 3′ direction along the mRNA. Translation begins when the ribosome encounters the **start codon,** which is almost always AUG (and rarely, GUG).

With the mRNA message in place on the assembled ribosome, the next step in translation involves entrance of tRNAs with their amino acids. The pool of cytoplasm around the region contains a complete array of tRNAs, previously charged by having the correct amino acid attached. The step in which the complementary tRNA meets with the mRNA code is guided by the two sites on the large subunit of the ribosome called the P site (left) and the A site (right).[1] Think of these sites as recessed spaces tucked within the two subunits of the ribosome, with each site accommodating a tRNA. The ribosome also has an exit or E site where used tRNAs are released.

The Master Genetic Code: The Message in Messenger RNA

By convention, the master genetic code is represented by the mRNA codons and the amino acids they specify (**figure 9.13**).

1. P stands for peptide site; A stands for aminoacyl (amino acid) site; E stands for exit site.

		Second Base Position							
		U		**C**		**A**		**G**	
U	UUU } Phenylalanine, UUC	UCU, UCC, UCA, UCG Serine	UAU } Tyrosine, UAC	UGU } Cysteine, UGC	U, C				
	UUA } Leucine, UUG		UAA } STOP**, UAG	UGA STOP**, UGG Tryptophan	A, G				
C	CUU, CUC, CUA, CUG Leucine	CCU, CCC, CCA, CCG Proline	CAU } Histidine, CAC	CGU, CGC, CGA, CGG Arginine	U, C				
			CAA } Glutamine, CAG		A, G				
A	AUU, AUC Isoleucine, AUA	ACU, ACC, ACA, ACG Threonine	AAU } Asparagine, AAC	AGU } Serine, AGC	U, C				
	AUG* START Methionine		AAA } Lysine, AAG	AGA } Arginine, AGG	A, G				
G	GUU, GUC Valine, GUA, GUG	GCU, GCC, GCA, GCG Alanine	GAU } Aspartic acid, GAC	GGU, GGC, GGA, GGG Glycine	U, C				
			GAA } Glutamic acid, GAG		A, G				

First Base Position (left side) / Third Base Position (right side)

* This codon initiates translation.
**For these codons, which give the orders to stop translation, there are no corresponding tRNAs with amino acids.

Figure 9.13 **The genetic code: codons of mRNA that specify a given amino acid.** The master code for translation is provided by mRNA codons.

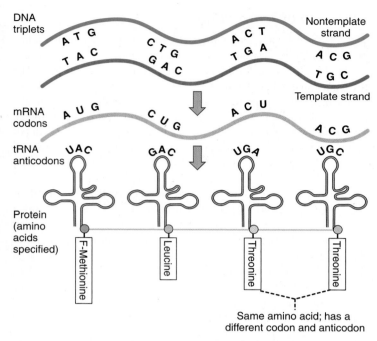

Figure 9.14 Interpreting the DNA code. If the DNA sequence is known, the mRNA codon can be surmised. If a codon is known, the anticodon and, finally, the amino acid sequence can be determined. The reverse is not possible (determining the exact codon or anticodon from amino acid sequence) due to the redundancy of the code.

Except in a very few cases (the genes of mitochondria and chloroplasts, for example), this code is universal, whether for prokaryotes, eukaryotes, or viruses. It is worth noting that once the triplet code on mRNA is known, the original DNA sequence, the complementary tRNA code, and the types of amino acids in the protein are automatically known (**figure 9.14**). However, one cannot predict (backward) from protein structure what the exact mRNA codons are because of a factor called **redundancy** or **degeneracy,** meaning that a particular amino acid can be coded by more than one codon.

In figure 9.13, the mRNA codons and their corresponding amino acid specificities are given. Because there are 64 different triplet codes[2] and only 20 different amino acids, it is not surprising that some amino acids are represented by several codons. For example, leucine and serine can each be represented by any of six different triplets, and only tryptophan and methionine are represented by a single codon. In such codons as leucine, only the first two nucleotides are required to encode the correct amino acid, and the third nucleotide does not change its sense. This property, called *wobble,* is thought to permit some variation or mutation without changing the message.

The Beginning of Protein Synthesis

With mRNA serving as the guide, the stage is finally set for actual protein assembly. The correct tRNA (labeled 1 on **figure 9.15**)

2. $64 = 4^3$ (the 4 different nucleotides in all possible combinations of 3).

enters the P site and binds to the **start codon (AUG)** presented by the mRNA. Rules of pairing dictate that the anticodon of this tRNA will complement the mRNA codon AUG; thus, the tRNA with anticodon UAC will first occupy site P. It happens that the amino acid carried by the initiator tRNA in bacteria is *formyl methionine* ($_f$Met; see figure 9.13), though in many cases, it may not remain a permanent part of the finished protein.

Continuation and Completion of Protein Synthesis: Elongation and Termination

The mRNA formed during transcription provides a specific order of nucleotides (codons) called the *reading frame* that must be translated in that exact order. In general, the ribosome shifts its reading frame along the mRNA from one codon to the next. This brings each succeeding codon into place on the ribosome and makes a space for the next tRNA to enter. Then, a peptide bond is formed between the amino acids on the adjacent tRNAs, and the polypeptide grows in length. The process is outlined here:

- Elongation begins with the filling of the A site by a second tRNA (**figure 9.15, step 1**). The identity of this tRNA and its amino acid is dictated by the second mRNA codon.
- The entry of tRNA 2 into the A site brings the two adjacent tRNAs in favorable proximity for a peptide bond to form between the amino acids (aa) they carry. The $_f$Met is transferred from the first tRNA to aa 2, resulting in two coupled amino acids called a dipeptide (**figure 9.15, step 2).**
- For the next step to proceed, some room must be made on the ribosome, and the next codon in sequence must be brought into position for reading. This process is accomplished by *translocation,* the enzyme-directed shifting of the ribosome to the next position on the mRNA strand, which causes the blank tRNA (1) to be discharged from the ribosome (**figure 9.15, step 3**) at the E site.
- This shifts the tRNA holding the dipeptide into P position. Site A is temporarily left empty. The tRNA that has been released is now free to drift off into the cytoplasm and become recharged with an amino acid for later additions to this or another protein.
- The stage is now set for the insertion of tRNA 3 at site A as directed by the third mRNA codon (**figure 9.15, step 4**). This insertion is followed once again by peptide bond formation between the dipeptide and aa 3 (making a tripeptide), splitting of the peptide from tRNA 2, and translocation (**figure 9.15, step 5**).
- This releases tRNA 2, shifts mRNA to the next position, moves tRNA 3 to position P, and opens position A for tRNA 4 (**step 6**). From this point on, peptide elongation proceeds repetitively by this same series of actions out to the end of the mRNA (**steps 7 and 8**).

The termination of protein synthesis is not simply a matter of reaching the last codon on mRNA. It is brought about by the presence of at least one special codon occurring just after the codon for the last amino acid. Termination codons—UAA, UAG, and UGA—are codons for which there is no corresponding tRNA. Also termed **nonsense** or **stop codons,** they carry a necessary message: *Stop here.* When this codon is reached, a special enzyme breaks the bond

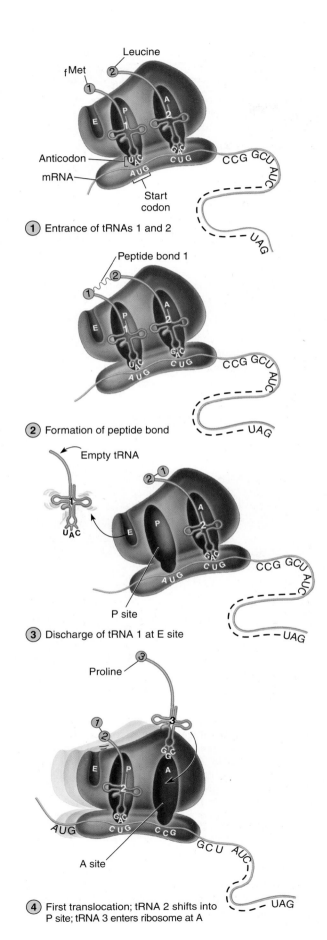

① Entrance of tRNAs 1 and 2

② Formation of peptide bond

③ Discharge of tRNA 1 at E site

④ First translocation; tRNA 2 shifts into P site; tRNA 3 enters ribosome at A

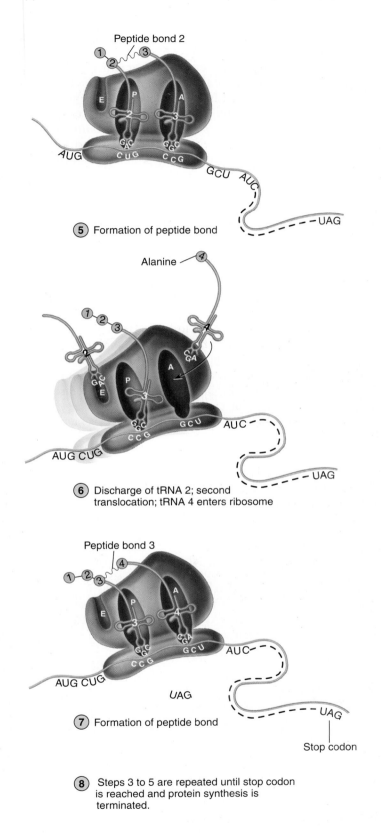

⑤ Formation of peptide bond

⑥ Discharge of tRNA 2; second translocation; tRNA 4 enters ribosome

⑦ Formation of peptide bond

⑧ Steps 3 to 5 are repeated until stop codon is reached and protein synthesis is terminated.

Process Figure 9.15 The events in protein synthesis.

between the final tRNA and the finished polypeptide chain, releasing it from the ribosome.

Before newly made proteins can carry out their structural or enzymatic roles, they often require finishing touches. Even before the peptide chain is released from the ribosome, it begins folding upon itself to achieve its biologically active tertiary conformation. Other alterations, called *posttranslational modifications,* may be necessary. Some proteins must have the starting amino acid (formyl methionine) clipped off; proteins destined to become complex enzymes have cofactors added; and some join with other completed proteins to form quaternary levels of structure.

The operation of transcription and translation is machine-like in its precision. Protein synthesis in bacteria is both efficient and rapid. At 37°C, 12 to 17 amino acids per second are added to a growing peptide chain. An average protein consisting of about 400 amino acids requires less than half a minute for complete synthesis. Further efficiency is gained when the translation of mRNA starts while transcription is still occurring (**figure 9.16**). A single mRNA is long enough to be fed through more than one ribosome simultaneously. This permits the synthesis of hundreds of protein molecules from the same mRNA transcript arrayed along a chain of ribosomes. This **polyribosomal complex** is indeed an assembly line for mass production of proteins. Protein synthesis consumes an enormous amount of energy. Nearly 1,200 ATPs are required just for synthesis of an average-size protein.

Eukaryotic Transcription and Translation: Similar yet Different

Eukaryotes share many aspects of protein synthesis with prokaryotes. They show similarities in the operation of the ribosomes, the functions of start and stop codons, and the mass production of proteins by means of polyribosomes. But they do differ in significant ways. The start codon in eukaryotes is also AUG, but it codes for an alternate form of methionine. Another difference is that eukaryotic mRNAs code for just one protein, unlike bacterial mRNAs, which often contain information from several genes in series.

Prokaryotic and eukaryotic cells also vary in location and structure of genes. The presence of DNA in the nucleus of eukaryotic cells means that eukaryotic transcription and translation cannot be simultaneous as it is in prokaryotes. The mRNA transcript must pass through pores in the nuclear membrane and be carried to the ribosomes in the cytoplasm or on the endoplasmic reticulum for translation.

We have given the simplified definition of a gene that works well for most prokaryotes,[3] but eukaryotic genes are generally not colinear[4]—meaning that they do *not* exist as an uninterrupted series of triplets coding for a protein. A eukaryotic gene contains the code for a protein, but located along the gene are one to several intervening sequences of bases, called **introns,** that do not code for protein. Introns are interspersed between coding regions, called **exons,** that will be translated into protein (**figure 9.17**). We can use words as examples. A short section of colinear prokaryotic gene might read

(a)

(b)

(c)

Figure 9.16 **Speeding up the protein assembly line in bacteria.** **(a)** The mRNA transcript encounters ribosomal parts immediately as it leaves the DNA. **(b)** The ribosomal factories assemble along the mRNA in a chain, each ribosome reading the message and translating it into protein. Many products will thus be well along the synthetic pathway before transcription has even terminated. **(c)** Photomicrograph of a polyribosomal complex in action. Note that the protein "tails" vary in length depending on the stage of translation (30,000×).

3. Introns have been found in cyanobacteria and archaea.

4. Colinearity means that the base sequence can be read directly into a series of amino acids.

Revising Some Rules of Genetics

What was true for E. coli *would be true for the elephant.*

Jacques Monod, Nobel laureate and discover of the *lac* operon

Bacteria—with their small genomes and short generation times—were the ideal model for research into molecular genetics. From this research came elegant and universal rules governing the workings of DNA. This led to the Central Dogma of Biology, which says that DNA is the repository of genetic information, RNA is a molecule that ferried information between DNA and the ribosome, and proteins produced from the RNA codes are responsible for carrying out the structural and metabolic needs of the cell. And all of this is still accurate.

It is just that there is more to the story. Recent discoveries in eukaryotic organisms have revealed that certain aspects of their genetics have turned out to be somewhat more complex and varied. This does not change our introduction to microbial genetics for two reasons. First, the laws of genetics covered in this chapter do hold true for the vast majority of prokaryotic organisms. Second, this is an *introduction* to genetics and it is certainly fair to leave the exceptions, the unusual, and the overly complex for future courses. We would like to use this Insight to introduce some of these revolutionary new findings.

The Genetic Code Is Universal with Some Exceptions

Based on early studies, it was assumed that the genetic code was the same in all cells and viruses, as no exceptions had ever been seen. Even more convincing, eukaryotic mRNAs were appropriately translated by bacterial cells, arguing for a single genetic dictionary. But later work with mitochondria isolated from yeast and humans found a slightly different code; for example, the stop codon UGA specifies the amino acid tryptophan in mitochondria. This discovery was followed by the detection of other exceptions to the genetic code in a few prokaryotic and eukaryotic organisms. This difference in the genetic code of mitochondria is one of the most compelling pieces of evidence supporting the endosymbiotic theory (see Insight 5.1).

Genetic Expression Can Be Modified by Epigenetic Mechanisms

For years, geneticists have debated the effects of nature (genes) versus nurture (environment) on the development of organisms. Though both have profound influence, we now have greater understanding about the direct effects of nongenetic factors.

The term *epigenetics* is used to designate factors that effect genetic expression other than the specific sequence of nucleotides in genes. One major epigenetic effect is in the manner that DNA coils around proteins during chromatin formation (see Insight 9.1). This layer of information is only now being fully appreciated and is still far from being understood. Epigenetic effects seem to work by increasing or decreasing the expression of certain genes by making them more or less accessible for transcription. For example, having methyl groups ($-CH_3$) attached to numerous cytosine nucleotides reduces the level of transcription of a section of DNA that is methylated.

A recent study in mice determined that the mother's diet during gestation could greatly alter gene expression in her offspring. Genetically identical mothers (clones) were placed into two groups—one that was fed a diet high in the soy chemical genistein and the other a normal diet. The soy group had children with significant differences in color coat and weight. They were larger and yellower than the mice from the normal group. This was a permanent effect, since the soy group remained unchanged even when placed on a normal diet. The effect appears to be due to methylation of the embryonic genes by genistein during fetal development (see figure).

90,000 Proteins . . . 25,000 Genes?

Humans have about 90,000 different proteins, and, going by our earlier understanding, we would have expected at least 90,000 different genes. Imagine the surprise most biologists felt when the Human Genome Project revealed that we have fewer than 25,000 protein-coding sequences, about 15,000 fewer than a simple corn plant.

For some time, it has been known that eukaryotic genes consist of expressed sequences (exons) interspersed with intervening sequences (introns) and that the final mRNA need not contain every exon. Until our stunningly low gene count was revealed, however, no one realized how often a single gene could be used to make several different proteins. Utilizing *alternative splicing,* a cell can select which exons will be retained and spliced into a final mRNA. Like a Scrabble player deciding which lettered tiles to use to make a word, a cell can place any combination of the exons in a gene to produce many alternate mRNAs (and resulting proteins). It is currently estimated that three of every four human genes are edited in this way and that the prevalence of alternative splicing appears to increase with the complexity of an organism.

Can you give some reasons for such a large proportion of a genome being apparently silenced or unused? *Answer available at http:// www.mhhe.com/talaro8*

Normal cytosine Methylated cytosine

Section of gene with methylation that blocks its activation

One example of epigenetic effects due to methylation of bases during early development.

Mice from mothers fed a high-soy diet were larger and different-colored from mice whose mothers were fed a normal diet.

TOM SAW OUR DOG DIG OUT; a eukaryotic gene that codes for the same portion would read TOM SAW XZKP FPL OUR DOG QZWVP DIG OUT. The recognizable words are the exons, and the nonsense letters represent the introns.

This unusual genetic architecture, also called a split gene, requires further processing before translation. Transcription of the entire gene with both exons and introns occurs first, producing a pre-mRNA. Next, a type of RNA and protein called a *spliceosome* recognizes the exon-intron junctions and enzymatically cuts through them in a process called RNA splicing. The action of this splicer enzyme loops the introns into lariat-shaped pieces, excises them, and joins the exons end to end. By this means, a strand of mRNA with no intron material is produced. This completed mRNA strand can then proceed to the cytoplasm to be translated.

Several different types of introns have been discovered, some of which do code for cell substances. A great deal of non-protein-coding DNA is proving to be vital for cell function. In humans, this intron material represents 98% of the DNA found in chromosomes. Another surprising discovery is that some proteins have sections that are in reverse order from the DNA sequence that encoded them. Previously, it was considered a hard and fast rule that DNA sequence determined the mRNA and consequently amino acid sequence after the introns were removed. This new data revealed that cellular machinery can reverse the order of DNA, essentially creating new proteins from the same gene. Many introns have been found to code for an enzyme called reverse transcriptase, which can convert RNA into DNA. Other introns are translated into endonucleases, enzymes that can snip DNA and allow insertions and deletions into the sequence. See **Insight 9.3** for other examples of exceptional eukaryotic genetics.

Figure 9.17 **The split gene of eukaryotes.** Eukaryotic genes have an additional complicating factor in their translation. Their coding sequences, or exons (E), are interrupted at intervals by segments called introns (I) that are not part of that protein's code. Introns are transcribed but not translated, which necessitates their removal by RNA splicing enzymes before translation.

Check & Assess Section 9.2

✔ Information in DNA is converted to proteins by the process of transcription and translation. Structural proteins contribute to the architecture of the cell, while functional proteins (enzymes) control an organism's metabolic activities.

✔ The codes on DNA are transferred to various RNA molecules, which carry out numerous functions, including protein synthesis and genetic regulation.

✔ The DNA code occurs in groups of three bases called codons; codes are transferred to RNA through transcription.

✔ Messenger (m) RNA carries the master message that instructs which amino acids are added during translation.

✔ Transfer (t) RNA carries the amino acid that will match the correct codon on mRNA.

✔ Ribosomes are sites of protein assembly by mRNA and tRNA.

✔ The processes of transcription and translation are similar but not identical for prokaryotes and eukaryotes.

8. How is the language of the gene expressed?

9. If a bacterial protein is 3,300 amino acids long, how many nucleotide pairs long is the gene sequence that codes for it?

10. Construct a table that compares the structure and functions of DNA and RNA.

11. Where does transcription begin?

12. What are the template and nontemplate strands of DNA?

13. Why is only one strand transcribed, and is the same strand of DNA always transcribed?

14. Distinguish between a codon and an anticodon.

15. Briefly describe the events in translation.

16. What are the functions of start and stop codons? Give examples.

17. Summarize how bacterial and eukaryotic cells differ in gene structure, transcription, and translation.

18. Discuss the roles of exons and introns.

9.3 Genetic Regulation of Protein Synthesis and Metabolism

14. Explain the functions of operons in bacterial genetic control.
15. Describe the main features of the lactose operon and inducible operons.
16. Describe the main features of repressible operons.
17. Summarize some aspects of genetic control by RNA.

In chapter 8, we surveyed the metabolic reactions in cells and the enzymes involved in those reactions. At that time, we mentioned that some enzymes are regulated and that one form of regulation occurs at the genetic level. This control mechanism ensures that genes are active only when their products are required. In this way, enzymes will be produced as they are needed and the waste of energy and materials in dead-end synthesis is prevented. A major form of gene regulation in prokaryotes is through systems called **operons.** An operon is a section of DNA that contains one or more structural genes along with a corresponding operator gene that controls transcription. Having this kind of set-up will permit genes for a particular metabolic pathway to be induced or repressed in unison by the same regulatory element.

Operons are described as either inducible or repressible. The category of an operon is determined by how it is affected by the environment within the cell. Many catabolic operons are **inducible,** meaning that the operon is turned on (induced) by the substrate of the enzyme for which the structural genes code. In this way, the enzymes needed to metabolize a nutrient (lactose, for example) are only produced when that nutrient is present. **Repressible operons** often contain genes coding for anabolic enzymes, such as those used to synthesize amino acids. In the case of these operons, several genes in series are turned off (repressed) by the product synthesized by the enzyme.

The Lactose Operon: A Model for Inducible Gene Regulation in Bacteria

The best understood cell system for explaining control through genetic induction is the **lactose (lac) operon.** The system, first described in 1961 by François Jacob and Jacques Monod, accounts for the regulation of lactose metabolism in *Escherichia coli.* Many other operons with similar modes of action have since been identified, and together they furnish convincing evidence that a cell's metabolic actions can have great impact on gene expression.

The lactose operon has three important features **(figure 9.18):**

1. the **regulator,** composed of the gene that codes for a protein capable of repressing the operon (a **repressor**);
2. the *control locus,* composed of two areas, the promoter recognized by RNA polymerase (see figure 9.11) and the **operator,** a sequence that acts as an on/off switch for transcription; and

3. the *structural locus,* made up of three genes, each coding for a different enzyme needed to catabolize lactose. One of the enzymes, β-galactosidase, hydrolyzes the lactose into its monosaccharides; another, permease, brings lactose across the cell membrane.

The promoter, operator, and structural components lie adjacent to one another, but the regulator can be at a distant site.

In inducible systems such as the *lac* operon, the operon is normally in an *off* mode and does not initiate enzyme synthesis when the appropriate substrate is absent **(figure 9.18, step 1).** How is the operon maintained in this mode? The key is in the repressor protein that is coded by the regulatory gene. This relatively large molecule is **allosteric,** meaning it has two binding sites, one for the operator and another for lactose. In the absence of lactose, this repressor binds with the operator locus. This has the effect of blocking the attachment of RNA polymerase and, thus, transcription of the structural genes lying downstream. Think of the repressor as a lock on the operator, and if the operator is locked, the structural genes cannot be transcribed. Importantly, the regulator gene is located away from the operator region and is not controlled by this block on the operon.

If lactose is added to the cell's environment, it triggers several events that turn the operon *on.* First, the binding of lactose to the repressor protein causes a conformational change in the repressor that dislodges it from the operator segment **(figure 9.18, step 2).** Then with the operator opened up, RNA polymerase can now bind to the promoter. The structural genes are transcribed in a single unbroken transcript coding for all three enzymes. During translation, however, each protein is synthesized separately. Because lactose is ultimately responsible for stimulating protein synthesis, it is called an **inducer.**

As lactose is depleted, further enzyme synthesis is not necessary, so the order of events reverses. At this point, there is no longer sufficient lactose to inhibit the repressor; hence, the repressor regains its shape for attachment to the operator. The operator is locked, and transcription of the structural genes and enzyme synthesis related to lactose both stop.

A fine but important point about the *lac* operon is that it functions only in the absence of glucose or if the cell's energy needs are not being met by the available glucose. Glucose is the preferred carbon source because it can be used immediately in growth and does not require induction of an operon. When glucose is present, a second regulatory system ensures that the *lac* operon is inactive, regardless of lactose levels in the environment.

A Repressible Operon

Bacterial systems for synthesis of amino acids, purines and pyrimidines, and many other processes work on a slightly different principle—that of repression. Similar factors such as repressor proteins, operators, and a series of structural genes exist for this operon, but with some important differences. Unlike the *lac* operon, this operon is normally in the *on* mode and will be turned *off* only when the product of the pathway is no longer required. Excess product plays a role as a **corepressor** that slows the transcription of the operon.

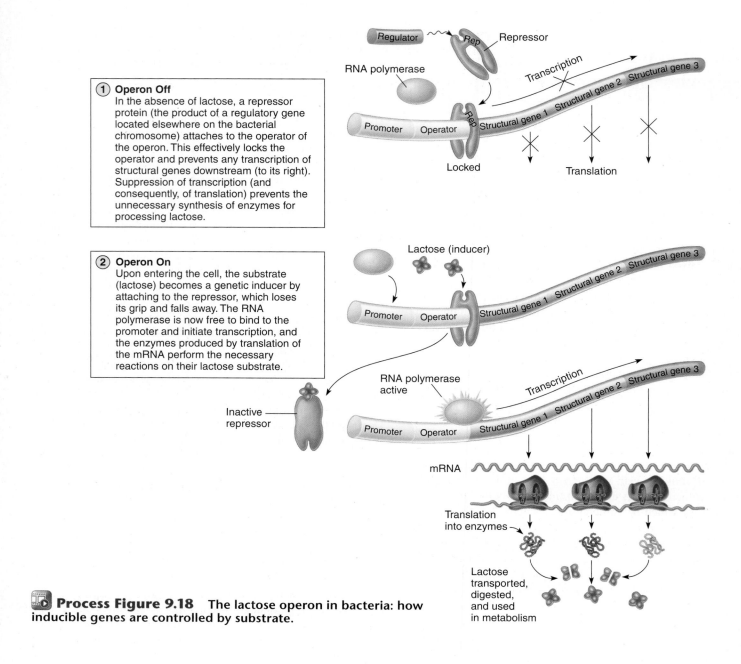

① Operon Off
In the absence of lactose, a repressor protein (the product of a regulatory gene located elsewhere on the bacterial chromosome) attaches to the operator of the operon. This effectively locks the operator and prevents any transcription of structural genes downstream (to its right). Suppression of transcription (and consequently, of translation) prevents the unnecessary synthesis of enzymes for processing lactose.

② Operon On
Upon entering the cell, the substrate (lactose) becomes a genetic inducer by attaching to the repressor, which loses its grip and falls away. The RNA polymerase is now free to bind to the promoter and initiate transcription, and the enzymes produced by translation of the mRNA perform the necessary reactions on their lactose substrate.

Process Figure 9.18 The lactose operon in bacteria: how inducible genes are controlled by substrate.

A growing cell that needs the amino acid arginine (arg) effectively illustrates the operation of a repressible operon. Under these conditions, the *arg* operon is set to *on,* and arginine is being actively synthesized through the action of the operon's enzymatic products (**figure 9.19, step 1**). In an active cell, the arginine will be used immediately, and the repressor will remain inactive (unable to bind the operator) because there is too little free arginine to activate it. As the cell's metabolism begins to slow down, however, the synthesized arginine is no longer used up and accumulates. The free arginine is then available to act as a corepressor by attaching to the repressor. This reaction changes the shape of the repressor, making it capable of binding to the operator and stopping transcription. Arginine will cease to be synthesized until the cell once again requires it in metabolism (**figure 9.19, step 2/2**).

Non-Operon Control Mechanisms

Analogous gene control mechanisms in eukaryotic cells are not as well understood, but it is known that gene function can be altered by intrinsic regulatory segments similar to operons. Some molecules, called transcription factors, insert on the grooves of the DNA molecule and enhance transcription of specific genes. These transcription factors can regulate gene expression in response to environmental stimuli such as nutrients, toxin levels, or even temperature. Eukaryotic genes are also regulated during growth and development, leading to the hundreds of different tissue types found in higher multicellular organisms. Considerable interest has been generated in the functions of regulatory RNAs that exert control on several levels in both prokaryotic and eukaryotic cells (**Insight 9.4**).

① Operon On
A repressible operon remains on when its nutrient products (here, arginine) are in great demand by the cell. The repressor cannot bind to the operator at low nutrient levels.

RNA polymerase

Transcription

Structural gene 1 Structural gene 2 Structural gene 3

Promoter Operator

mRNA

Repressor is inactive (wrong shape to attach to operator)

Rep

Enzymes synthesize arginine.

Arginine is immediately used in metabolism

② Operon Off
The operon is repressed when (1) arginine builds up and, serving as a corepressor, activates the repressor. (2) The repressor complex affixes to the operator and blocks the RNA polymerase and further transcription of genes for arginine synthesis.

Structural gene 1 Structural gene 2 Structural gene 3

Promoter Operator

Arginine accumulates

(1) Repressor is active (correct shape achieved)

RNA polymerase

Transcription blocked

Structural gene 1 Structural gene 2 Structural gene 3

Arginine synthesis arrested

(2) Promoter Operator

Process Figure 9.19 Repressible operon: genetic control through excess nutrient.

Check & Assess Section 9.3

✔ Gene expression must be orchestrated to coordinate the organism's needs with nutritional resources. Genes can be turned "on" and "off" by specific molecules, which expose or hide their nucleotide codes for transcribing proteins. Most, but not all, of these proteins are enzymes.

✔ Operons are collections of genes in bacteria that code for products with a coordinated function. They include genes for operational and structural components of the cell.

✔ Nutrients can combine with regulator gene products to turn a set of structural genes on (inducible genes) or off (repressible genes). The *lac* (lactose) operon is an example of an inducible operon. The *arg* (arginine) operon is an example of a repressible operon.

✔ Several roles in gene regulation have been tied to special types of RNA, including micro RNA, small, interfering RNA, antisense RNA, and riboswitches.

19. What is an operon? Describe the functions of regulators, promoters, and operators.

20. Compare and contrast the *lac* operon and repressible operon systems.

21. For both figures 9.18 and 9.19, write a simple stepwise diagram indicating the events that would reverse the operon's action.

22. At which levels of DNA regulation do small RNAs function?

INSIGHT 9.4

Unraveling the Secrets of RNA

Evidence on the roles of RNAs in gene expression is accumulating rapidly. We have learned how the introns of eukaryotic mRNAs are removed by the catalytic RNA of the spliceosome complex, and the peptide bonds of proteins are catalyzed by ribosomal RNA. Recent research has revealed that short lengths of RNA also have the ability to control the expression of certain genes. Some of these are called *micro RNAs* and others are called *small interfering (si) RNAs*. One of their actions is to activate a system inside cells that degrades double-stranded (ds) RNA of the type that would exist in certain invading viruses. The micro RNAs also degrade mRNA of regular genes, which shuts down translation of that gene.

The interfering nature of siRNA was discovered quite accidentally when researchers tried to induce expression of genes by providing them in dsRNA form. Instead of being activated, the genes matching these RNA sequences were turned off entirely.

Bacteria such as *E. coli* have about 40 small RNAs, and eukaryotes likely have hundreds and perhaps thousands.

A second type of regulation also occurs when small RNAs alter the structure of chromosomes. While DNA and proteins are coiling together to form chromatin, small RNAs direct how tightly or loosely the chromatin is constructed (see Insight 9.1). Just as a closed book cannot be read, DNA sequences contained within tightly coiled chromatin are generally inaccessible to the cell and silenced.

Antisense RNA provides another level of regulation common in cells. It is formed by transcription of the nontemplate strand of a gene that produces mRNA. This antisense molecule has the ability to base pair with the "sense," or messenger, RNA and thus keep it from being translated (see figure 10.15).

Riboswitches are RNA-based controls on both transcription and translation. One type of riboswitch causes a change in the folding of mRNA to favor or block transcription. Many bacteria use riboswitches to control genes of metabolic enzymes. Other riboswitches can alter binding of mRNA on the ribosome itself to regulate protein synthesis.

These newly discovered RNA molecules have answered some vexing questions that came out of genome sequencing studies. Most of the DNA in organisms was found *not* to code for functional proteins. This DNA was once considered "junk," but its percentage in humans was 98%! Obviously, in bacteria as well as in humans, it is far from junk. In fact, this DNA was preserved in the same form for many millions of years of evolution, suggesting it has a very important function. We now know that much of it codes for these important RNA regulatory molecules.

9.4 Mutations: Changes in the Genetic Code

Expected Learning Outcomes

18. Define a mutation and explain how mutant forms are identified.
19. Explain the causes and types of mutations and their effects.
20. Describe some mutation repair mechanisms.
21. Compare beneficial and detrimental effects of mutations.

As precise and predictable as the rules of genetic expression seem, permanent changes do occur in the genetic code. Indeed, genetic change is the driving force of evolution by adding variation to populations of organisms. Changes may become evident in altered gene expression, such as in the appearance or disappearance of anatomical or physiological traits. For example, a pigmented bacterium can lose its ability to form pigment, or a strain of the malarial parasite can develop resistance to a drug. A phenotypic change that is due to an alteration in the genotype is called a **mutation.** On a strictly molecular level, a mutation is an alteration in the nitrogen base sequence of DNA. It can involve the loss of base pairs, the addition of base pairs, or a rearrangement in the order of base pairs. Do not confuse this with genetic recombination, in which microbes transfer whole segments of genetic information between themselves.

A microorganism that exhibits a natural, nonmutated characteristic is known as a **wild type,** or wild strain. If a microorganism bears a mutation, it is called a **mutant strain.** Mutant strains can show variance in morphology, nutritional characteristics, genetic control mechanisms, resistance to chemicals, temperature preference, and nearly any type of enzymatic function. Mutant strains are very useful for tracking genetic events, unraveling genetic organization, and pinpointing genetic markers.

The simplest way to detect mutant bacteria is to inoculate solid media containing differential or selective agents such as metabolic substrates or antibiotics. Growing a pure culture on a plate of media and observing the appearance of colonies can reveal strains of a species that express genetic variations. For instance, in cultures of wild-type *E. coli* grown on differential MacConkey agar, most of the isolated colonies will indicate a positive result for lactose fermentation, but invariably, there will be a small number of colonies that are negative and have lost the genes for using lactose. This colony can be readily pulled out for further study. Antibiotics are frequently added to media as selective agents for separating out strains of bacteria that are drug resistant. Figure 10.11 shows how this method works for detecting recombinant cells used in genetic engineering.

Another method for detecting and isolating mutant strains of microbes that grow in colonies is the *replica plating* technique **(figure 9.20).** This method is based on a variation between wild strains that can synthesize a nutrient, such as an amino acid, and mutant strains that cannot synthesize it. First a culture of test microbes is exposed to a chemical agent known to cause mutations and then plated on a complete medium that contains the particular nutrient. It is important for this to be done in a way that individual colonies are formed. Colony position on the plate is

retained by pressing a pickup or replica carrier onto the colonies and then using the carrier to inoculate two plates in the same orientation. These two plates contain: (1) a complete medium containing the nutrient being tested for and (2) an incomplete medium that lacks it. Colonies that grow on the complete medium could be of either the wild or mutant type, but only the wild strains will grow on the incomplete medium. By comparing the two plates side by side, it is possible to locate the position of the colony on the complete medium that does not grow on the incomplete one. This mutant colony is now detectable and can be isolated. This method provides a way to test the effects of various mutagenic agents as well as providing data on the rate of mutation for this particular characteristic.

Causes of Mutations

A mutation is described as spontaneous or induced, depending upon its origin. A **spontaneous mutation** is a random change in the DNA arising from errors in replication that occur without a known cause. The frequency of spontaneous mutations has been measured for a number of organisms. Mutation rates vary from one mutation in 10^5 replications (a high rate) to one mutation in 10^{10} replications (a low rate). The rapid rate of bacterial reproduction allows these mutations to be observed more readily in bacteria than in most eukaryotes.

Induced mutations result from exposure to known **mutagens,** which are primarily physical or chemical agents that damage DNA and interfere with its functioning **(table 9.3).** The carefully controlled use of mutagens has proved a useful way to induce mutant strains of microorganisms for study.

Chemical mutagenic agents act in a variety of ways to change the DNA. Agents such as acridine dyes insert completely across the DNA helices between adjacent bases. This results in distortion of the helix and can cause frameshift mutations. Analogs[5] of the nitrogen bases (5-bromodeoxyuridine and 2-aminopurine, for example) are chemical mimics of natural bases that are incorporated into DNA during replication. Addition of these abnormal bases leads to mistakes in base-pairing. Many chemical mutagens are also carcinogens, or cancer-causing agents (see the discussion of the Ames test on page 279).

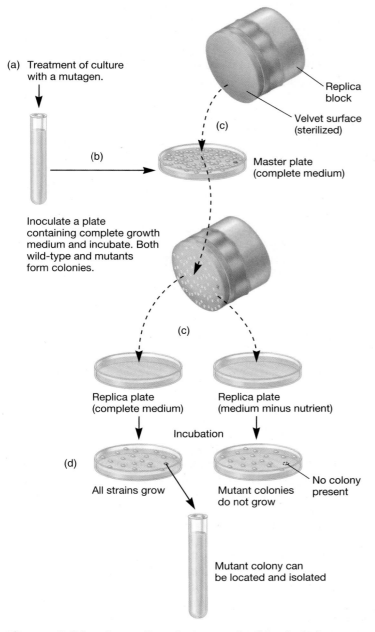

(a) Treatment of culture with a mutagen.

(b) Inoculate a plate containing complete growth medium and incubate. Both wild-type and mutants form colonies.

Replica block

Velvet surface (sterilized)

(c)

Master plate (complete medium)

(c)

Replica plate (complete medium)

Replica plate (medium minus nutrient)

Incubation

(d) All strains grow

Mutant colonies do not grow

No colony present

Mutant colony can be located and isolated

Figure 9.20 The replica plating method for isolating mutant bacteria. This technique was developed by Joshua Lederberg. It is based on **(a)** exposing a culture to a mutagen, **(b)** plating the culture for isolated colonies on a master plate with complete nutrients, **(c)** pressing a sterile carrier onto the surface of the colonies to pick up a tiny clump of cells and pressing it again onto a plate of complete and then onto one of incomplete medium. **(d)** After growth, the two plates are compared. The colonies that occupy a specific position on the complete medium but are missing from that same position on the incomplete medium are mutant colonies that can be subcultured.

5. An analog is a chemical structured very similarly to another chemical except for minor differences in functional groups.

TABLE 9.3	Selected Mutagenic Agents and Their Effects
Agent	**Effect**
Chemical	
Nitrous acid, bisulfite	Removes an amino group from some bases
Ethidium bromide	Inserts between the paired bases
Acridine dyes	Cause frameshifts due to insertion between base pairs
Nitrogen base analogs	Compete with natural bases for sites on replicating DNA
Radiation	
Ionizing (gamma rays, X rays)	Form free radicals that cause single or double breaks in DNA
Ultraviolet	Causes cross-links between adjacent pyrimidines

Physical agents that alter DNA are primarily types of radiation. High-energy gamma rays and X rays introduce major physical changes into DNA, and it accumulates breaks that may not be repairable. Ultraviolet (UV) radiation induces abnormal bonds between adjacent pyrimidines so that normal replication of that region is blocked. Exposure to large doses of radiation can be fatal, which is why radiation is so effective in microbial control; it can also be carcinogenic in animals. The use of radiation to control microorganisms is described further in chapter 11.

Categories of Mutations

Mutations range from large mutations, in which large genetic sequences are gained or lost, to small ones that affect only a single base on a gene. These latter mutations, which involve addition, deletion, or substitution of single bases, are called **point mutations.**

To understand how a change in DNA influences the cell, remember that the DNA code appears in a particular order of triplets (three bases) that is transcribed into mRNA codons, each of which specifies an amino acid. A permanent alteration in the DNA that is copied faithfully into mRNA and translated can change the structure of the protein. A change in a protein can likewise change the morphology and physiology of a cell. Most mutations have a harmful effect on the cell, leading to cell dysfunction or death; these are called lethal mutations. Neutral mutations produce neither adverse nor helpful changes. A small number of mutations are beneficial in that they provide the cell with a useful change in structure or physiology.

Any change in the code that leads to placement of a different amino acid is called a **missense mutation.** A missense mutation can do one of the following:

- create a faulty, nonfunctional (or less functional) protein,
- produce a protein that functions in a different manner, or
- cause no significant alteration in protein function.

A **nonsense mutation,** on the other hand, changes a normal codon into a stop codon that does not code for an amino acid and stops the production of the protein wherever it occurs. A nonsense mutation almost always results in a nonfunctional protein. A **silent mutation** alters a base but does not change the amino acid and thus has no effect. For example, because of the redundancy of the code, ACU, ACC, ACG, and ACA all code for threonine, so a mutation that changes only the last base will not alter the sense of the message in any way. A **back-mutation** occurs when a gene that has undergone mutation reverses (mutates back) to its original base composition.

Mutations also occur when one or more bases are inserted into or deleted from a newly synthesized DNA strand. This type of mutation, known as a **frameshift,** is so named because the reading frame of the mRNA has been changed. Frameshift mutations nearly always result in a nonfunctional protein because every amino acid after the mutation is different from what was coded for in the original DNA. Also note that insertion or deletion of bases in multiples of three (3, 6, 9, etc.) results in the addition or deletion of amino acids but does not disturb the reading frame. The effects of all of these types of mutations can be seen in **table 9.4.**

TABLE 9.4	Classification of Major Types of Mutations
	Example
I. Wild-type (original, nonmutated) sequence	THE BIG BAD DOG ATE THE FAT RED CAT

The wild-type sequence of a gene is the DNA sequence found in most organisms and is generally considered the "normal" sequence.

II. Categories of mutations based on type of DNA alteration

A. Substitution mutations

1. Missense	THE BIG BAD DOG ATE THE FIT RED CAT

A missense mutation causes a different amino acid to be incorporated into a protein. Effects range from unnoticeable to severe, based on how different the two amino acids are.

2. Nonsense	THE BIG BAD (stop)

A nonsense mutation converts a codon to a stop codon, resulting in premature termination of protein synthesis. Effects of this type of mutation are almost always severe.

B. Frameshift mutations

1. Insertion	B↓ THE BIG BAB DDO GAT ETH EFA TRE DCA T
2. Deletion	THE BIG BDD OGA TET HEF ATR EDC AT ↘A

Insertion and deletion mutations cause a change in the reading frame of the mRNA, resulting in a protein in which every amino acid after the mutation is affected. Because of this, frameshift mutations almost always result in a nonfunctional protein.

Repair of Mutations

Earlier, we indicated that DNA has a proofreading mechanism to repair mistakes in replication that might otherwise become permanent (see page 262). Because mutations are potentially life-threatening, the cell has additional systems for finding and repairing DNA that has been damaged by various mutagenic agents and processes. Most ordinary DNA damage is resolved by enzymatic systems specialized for finding and fixing such defects.

DNA that has been damaged by ultraviolet radiation can be restored by photoactivation or light repair. This repair mechanism requires visible light and a light-sensitive enzyme, DNA photolyase, which can attach to sites of abnormal pyrimidine bonding and restore the original DNA structure. Ultraviolet repair mechanisms are successful only for a relatively small number of UV mutations. Mutations can be excised by a series of enzymes that remove the incorrect bases and add the correct ones. This process is known as *excision repair.* First, enzymes break the bonds between the bases and the sugar-phosphate strand at the site of the error. A different enzyme subsequently removes the defective bases one at a time, leaving a gap that will be filled in by DNA polymerase I and ligase **(figure 9.21).** A repair system can also locate mismatched bases that were missed during proofreading: for example, C mistakenly paired with A, or G with T. The base must be replaced soon after the mismatch is made, or it will not be recognized by the repair enzymes.

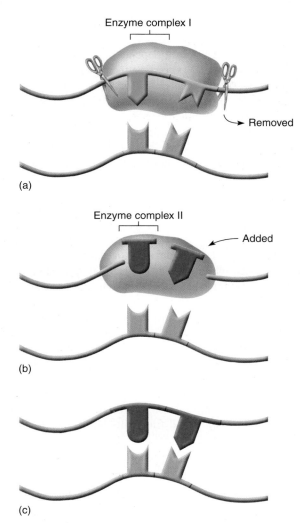

Figure 9.21 **Excision repair of mutation by enzymes.**
(a) The first enzyme complex recognizes one or several incorrect bases and removes them. **(b)** The second complex (DNA polymerase I and ligase) places correct bases and seals the gaps. **(c)** Repaired DNA.

The Ames Test

New agricultural, industrial, and medicinal chemicals are constantly being added to the environment, and exposure to them is widespread. The discovery that many such compounds are mutagenic and that up to 83% of these mutagens are linked to cancer is significant. Although animal testing has been a standard method of detecting chemicals with carcinogenic potential, a more rapid screening system called the **Ames test**[6] is also commonly used. In this ingenious test, the experimental subjects are bacteria whose gene expression and mutation rate can be readily observed and monitored. The premise is that any chemical capable of mutating bacterial DNA can similarly mutate mammalian (and thus human) DNA and is therefore potentially hazardous.

One indicator organism in the Ames test is a mutant strain of *Salmonella enterica*[7] that has lost the ability to synthesize the amino acid histidine, a defect highly susceptible to back-mutation because

6. Named for its creator, Bruce Ames.

7. *S. enterica* inhabits the intestine of poultry and causes food poisoning in humans. It is used extensively in genetic studies of bacteria.

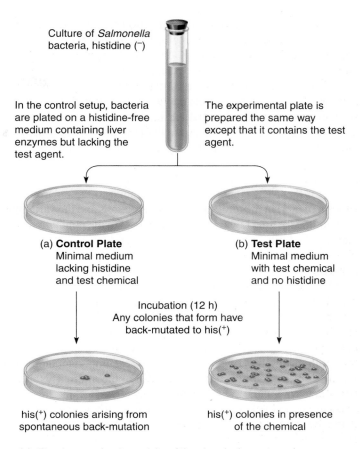

(a) **Control Plate** Minimal medium lacking histidine and test chemical

(b) **Test Plate** Minimal medium with test chemical and no histidine

Incubation (12 h) Any colonies that form have back-mutated to his(+)

his(+) colonies arising from spontaneous back-mutation

his(+) colonies in presence of the chemical

(c) The degree of mutagenicity of the chemical agent can be calculated by comparing the number of colonies growing on the control plate with the number on the test plate. Chemicals that induce an increased incidence of back-mutation (right side) are considered carcinogens.

Figure 9.22 **The Ames test.** This test is based on a strain of *Salmonella enterica* that cannot synthesize histidine his(−). It lacks the enzymes to repair DNA so that mutations show up readily and has leaky cell walls that permit the ready entrance of chemicals. Many potential carcinogens (benzanthracene and aflatoxin, for example) are mutagenic agents only after being acted on by mammalian liver enzymes, so an extract of these enzymes is added to the test medium.

the strain also lacks DNA repair mechanisms. Mutations that cause reversion to the wild strain, which is capable of synthesizing histidine, occur spontaneously at a low rate. A test agent is considered a mutagen if it enhances the rate of back-mutation beyond levels that would occur spontaneously. One variation on this testing procedure is outlined in **figure 9.22.** The Ames test has proved invaluable for screening an assortment of environmental and dietary chemicals for mutagenicity and carcinogenicity without resorting to animal studies.

Positive and Negative Effects of Mutations

Many mutations are not repaired. How the cell copes with them depends on the nature of the mutation and the strategies available to that organism. Mutations are passed on to the offspring of organisms during reproduction and to new viruses during replication. They become a long-term part of the gene pool. Most mutations are harmful to organisms, but some can provide adaptive advantages.

If a mutation leading to a nonfunctional protein occurs in a gene for which there is only a single copy, as in haploid or simple organisms, the cell will probably die. This happens when certain mutant strains of *E. coli* acquire mutations in the genes needed to repair damage by UV radiation. Mutations of the human genome affecting the action of a single protein (mostly enzymes) are responsible for more than 3,500 diseases.

Although most spontaneous mutations are not beneficial, a small number contribute to the success of the individual and the population by creating variant strains with alternate ways of expressing a trait. Microbes are not "aware" of this advantage and do not direct these changes; they simply respond to the environment they encounter. In the long-range view, mutations and the variations they produce are the raw materials for change in the population and, thus, for evolution.

Mutations that create variants occur frequently enough that any population contains mutant strains for a number of characteristics, but as long as the environment is stable, these mutants will never comprise more than a tiny percentage of the population. When the environment changes, however, it can become hostile for the survival of certain individuals, and only those microbes bearing protective mutations will be equipped to survive in the new environment.

Such observations can be explained by the important biological principles of natural selection and survival of the fittest. This theory states that environmental pressure selects for organisms that have greater adaptability due to beneficial mutations. They are more likely to survive, reproduce, and pass the changed genes on to succeeding generations. In time, these survivors become the dominant members of the population as long as the mutation continues to favor their survival. One of the clearest models for natural selection in action is acquired drug resistance in bacteria (discussed in chapter 12).

As we will see in section 9.5, bacteria have additional mechanisms for changing their genetic makeup by sharing genes with other bacteria through recombination.

9.5 DNA Recombination Events

Expected Learning Outcomes

22. Explain recombination in bacteria and what it involves.

23. Describe the main features of conjugation and its outcomes.

24. Discuss what is required for bacterial transformation and its outcomes.

25. Identify the basic processes involved in transduction and the end results.

26. Discuss transposons and their importance to microorganisms.

Genetic recombination through sexual reproduction is an important means of genetic variation in eukaryotes. Although bacteria have no exact equivalent to sexual reproduction, they exhibit a primitive means for sharing or recombining parts of their genome. An event in which one bacterium donates DNA to another bacterium is a type of genetic transfer termed **recombination,** the end result of which is a new strain different from both the donor and the original

recipient strain. In general, any organism that has acquired genes that originated in another organism is called a **recombinant.**

Recombination in bacteria depends largely on their extreme versatility in acquiring and expressing the genetic material of other bacteria and even other organisms. This capacity for genetic sharing has tremendous effects on the diversity and evolution of bacteria. Unlike spontaneous mutations, recombinations are generally beneficial. They can provide additional genes for resistance to drugs and metabolic poisons, new nutritional and metabolic schemes, increased virulence, and adaptations to changing environmental conditions.

Transmission of Genetic Material in Bacteria

DNA transfer between bacterial cells typically involves small pieces of DNA in the form of **plasmids** or chromosomal fragments. Plasmids are small, circular pieces of DNA that contain their own origin of replication and therefore can replicate independently of the bacterial chromosome. Plasmids are found in many bacteria (as well as some fungi) and typically contain a few dozen genes. Although plasmids are not necessary for bacterial survival, they often carry adaptive genes, such as those for drug resistance.

Case file 9 details a pathogen that has received dozens of resistance genes from other species. Chromosomal fragments that have escaped from a lysed bacterial cell are also commonly involved in the transfer of genetic information between cells. An important difference between plasmids and fragments is that, while a plasmid can be stably replicated and inherited, chromosomal fragments must integrate themselves into the bacterial chromosome in order to be replicated and eventually passed to progeny cells. The process of genetic recombination is relatively rare in nature, but its frequency can be increased in the laboratory, where the ability to transfer genes between organisms has become an important tool of research and technology.

Depending upon the mode of transmission, the means of genetic recombination in bacteria is called conjugation, transformation, or transduction. **Conjugation** requires the attachment of two related species and the formation of a bridge that can transport DNA. **Transformation** entails the transfer of naked DNA and requires no special vehicle. **Transduction** is DNA transfer mediated through the action of a bacterial virus **(table 9.5).**

Conjugation: Genetic Transmission through Direct Contact

Conjugation is a mode of genetic recombination in which a plasmid or fragment of DNA is transferred from a donor cell to a recipient cell via a direct connection. Both gram-negative and gram-positive bacteria can conjugate, but only gram-negative cells operate with a specialized plasmid called a **fertility,** or **F factor.** This plasmid directs the synthesis of a unique **pilus,** also called a sex pilus[8], that functions in most conjugation transfers. The recipient cell is usually a related bacterium with a recognition site on its surface for interacting with the pilus **(figure 9.23, step 1).**

8. This term refers to its function in the mating of bacteria, but it has nothing to do with sexual reproduction, which is absent in bacteria.

TABLE 9.5	Types of Intermicrobial Exchange		
Mode	**Factors Involved**	**Direct or Indirect***	**Examples of Genes Transferred**
Conjugation	Donor cell with pilus Fertility plasmid in donor Both donor and recipient alive Bridge forms between cells to transfer DNA.	Direct	Drug resistance; resistance to metals; toxin production; enzymes; adherence molecules; degradation of toxic substances; uptake of iron
Transformation	Free donor DNA (fragment or plasmid) Live, competent recipient cell	Indirect	Polysaccharide capsule; unlimited with cloning techniques; metabolic enzymes
Transduction	Donor is lysed bacterial cell. Defective bacteriophage is carrier of donor DNA. Live recipient cell of same species as donor	Indirect	Exotoxins; enzymes for sugar fermentation; drug resistance

*Direct means the donor and recipient are in contact during exchange; indirect means they are not.

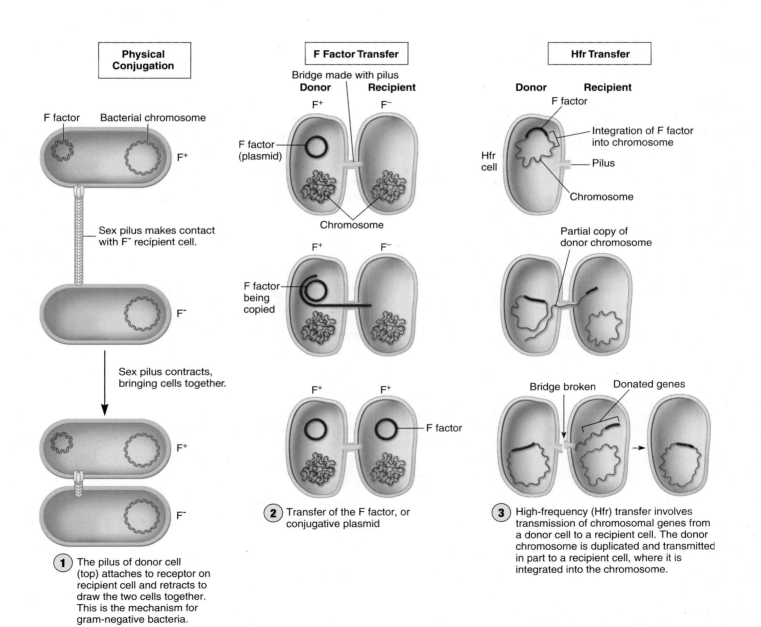

Process Figure 9.23 Conjugation: genetic transmission through direct contact between two cells.

A cell's role in conjugation is denoted by F$^+$ for the cell that has the F plasmid and by F$^-$ for the cell that lacks it. Contact is made when a pilus grows out from the F$^+$ cell, attaches to the surface of the F$^-$ cell, contracts, and draws the two cells together (**figure 9.23, step 1**). At the site where the pilus attaches the two cells, a mating or conjugative bridge is formed that serves as a transfer system for the plasmid. The latest reliable information indicates that the pilus becomes part of a complex secretory system that opens a gateway between the cell walls and membranes of the two cells and serves as a passageway for the plasmid.

There are hundreds of conjugative plasmids with some variations in their properties. One of the best understood plasmids is the F factor in *E. coli,* which exhibits these patterns of transfer:

1. Because bacteria tend to conserve plasmids, the donor cell first makes a copy of the F factor, which is simultaneously transferred across the bridge (**figure 9.23, step 2**). This way, it retains the original plasmid, remains F$^+$, and can continue to conjugate with other bacteria. The F$^-$ cell has been changed into an F$^+$ cell capable of producing a pilus and conjugating with other cells. No additional donor genes are transferred during this transfer.
2. In high-frequency recombination (Hfr) donors, the fertility factor has been integrated into the F$^+$ donor chromosome.

The term *high-frequency recombination* denotes that a cell with an integrated F factor transmits its chromosomal genes at a higher frequency than other cells. This is because the F factor can direct a more comprehensive transfer of part of the donor chromosome to a recipient cell. This transfer occurs through replication of the donor chromosome along with the F factor. One strand of DNA is retained by the donor, and the other strand is transported across to the recipient cell (**figure 9.23, step 3**). The F factor may not be transferred during this process. The transfer of an entire chromosome takes about 100 minutes, but the pilus bridge between cells is ordinarily broken before this time, and rarely is the entire genome of the donor cell transferred.

Conjugation has great biomedical importance. Special **resistance (R) plasmids,** or **factors,** that bear genes for resisting antibiotics and other drugs are commonly shared among bacteria through conjugation. Transfer of R factors can confer multiple resistance to antibiotics such as tetracycline, chloramphenicol, sulfonamides, and penicillin. This phenomenon is discussed further in chapter 12. Other types of R factors carry genetic codes for resistance to heavy metals (nickel and mercury) or for synthesizing virulence factors (toxins, enzymes, and adhesion molecules) that increase the pathogenicity of the bacterial strain. Conjugation studies have also provided a means to map the bacterial chromosome.

Gram-positive bacteria such as *Bacillus* and *Streptococcus* are known to recombine through conjugation. Since they lack pili, they must transport the plasmids via specialized proteins activated during adhesion of the two cells. The mechanisms involved are yet to be fully worked out.

Transformation: Capturing Free DNA from Solution

One of the cornerstone discoveries in microbial genetics was made in the late 1920s by the English biochemist Frederick Griffith working with *Streptococcus pneumoniae* and laboratory mice. The pneumococcus exists in two major strains based on the presence of the capsule, colonial morphology, and pathogenicity. Encapsulated strains bear a smooth (S) colonial appearance and are virulent; strains lacking a capsule have a rough (R) appearance and are nonvirulent. (Recall from chapter 4 that the capsule protects a bacterium from the phagocytic host defenses.) Griffith first showed that when mice were injected with a live, virulent (S) strain, they soon died (**figure 9.24a**). Mice injected with a live, nonvirulent (R) strain remained alive and healthy (**figure 9.24b**). Next he tried a variation on this theme. He heat-killed an S strain and injected it into mice, which remained healthy (**figure 9.24c**). Then came the ultimate test: Griffith injected both dead S cells and live R cells into mice, with the result that the mice died from pneumococcal blood infection (**figure 9.24d**). If killed bacterial cells do not come back to life and the nonvirulent live strain was harmless, why did the mice die? Although he did not know it at the time, Griffith had demonstrated that dead S cells, while passing through the body of the mouse, broke open and released some of their DNA (by chance, that part containing the genes for making a capsule). A few of the live R cells subsequently picked up this loose DNA and were transformed by it into virulent, capsule-forming strains.

Later studies supported the concept that a chromosome released by a lysed cell breaks into fragments small enough to be accepted by a recipient cell and that DNA, even from a dead cell, retains its genetic code. This nonspecific acceptance by a bacterial cell of small fragments of soluble DNA from the surrounding environment is termed **transformation (figure 9.25).** Transformation is facilitated by special DNA-binding proteins on the cell wall that capture DNA from the surrounding medium. Cells that are capable of accepting genetic material through this means are termed **competent.** The new DNA is passed through a DNA uptake system in the cell wall and membrane. Within the cytoplasm, the donated strand is inserted into the chromosome of the recipient cell, now termed *transformed*. Natural transformation is found in several groups of gram-positive and gram-negative bacterial species. In addition to genes coding for the capsule, bacteria also exchange genes for antibiotic resistance and bacteriocin synthesis in this way.

Because transformation requires no special appendages, and the donor and recipient cells do not have to be in direct contact, the process is useful for certain types of recombinant DNA technology. With this technique, foreign genes from a completely unrelated organism are inserted into a plasmid, which is then introduced into a competent bacterial cell through transformation. These recombinations can be carried out in a test tube, and human genes can be experimented upon and even expressed outside the human body by placing them in a microbial cell. Transformation is often used in the production of genetically engineered organisms such as yeast plants and mice and has been proposed as a future technique for curing genetic diseases in humans. These topics are covered in more detail in chapter 10.

Transduction: Viruses as Vectors

Bacteriophages (bacterial viruses) have been previously described as destructive bacterial parasites. Viruses can in fact serve as genetic vectors (an entity that can bring foreign DNA into a cell). The process by which a bacteriophage serves as the carrier of DNA

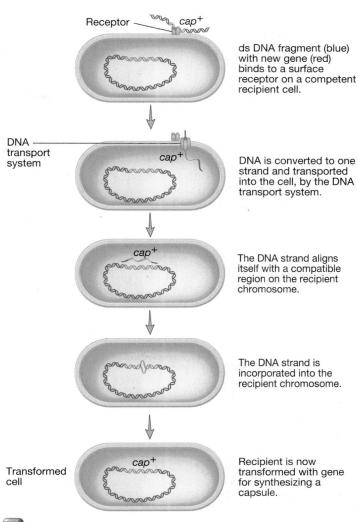

Figure 9.25 **Steps in bacterial transformation as seen in _S. pneumoniae._** This shows how genes coding for capsule formation are brought into the cell and inserted into the chromosome. _Cap_⁺ indicates genes for forming a capsule.

from a donor cell to a recipient cell is **transduction.** Although it occurs naturally in a broad spectrum of bacteria, the participating bacteria in a single transduction event must be the same species because of the specificity of viruses for host cells.

There are two versions of transduction. In _generalized transduction_ (**figure 9.26**), random fragments of disintegrating host DNA are taken up by the phage during assembly. Virtually any gene from the bacterium can be transmitted through this means. In _specialized transduction_ (**figure 9.27**), a highly specific part of the host genome is regularly incorporated into the virus. This specificity is explained by the prior existence of a temperate prophage inserted in a fixed site on the bacterial chromosome. When activated, the prophage DNA separates from the bacterial chromosome, carrying a small segment of host genes with it. During a lytic cycle, these specific viral-host gene combinations are incorporated into the viral particles and carried to another bacterial cell. Instances of transduction include the transfer of drug resistance seen in _Staphylococcus_ spp. and the transmission of gene regulators in gram-negative rods such as _Escherichia_ and _Salmonella._

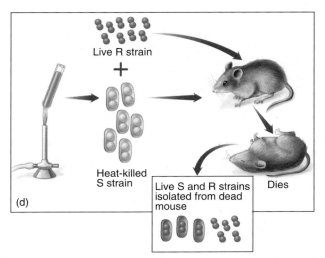

Figure 9.24 **Griffith's classic experiment in transformation.** In essence, this experiment proved that DNA released from a killed cell can be acquired by a live cell. The cell receiving this new DNA is genetically transformed—in this case, from a nonvirulent strain to a virulent one.

Recombination results in two possible outcomes.

Figure 9.26 **Generalized transduction: genetic transfer by means of a virus carrier.** (1) A phage infects cell A (the donor cell) by normal means. (2) During replication and assembly, a phage particle incorporates a segment of bacterial DNA by mistake. (3) Cell A then lyses and releases the mature phages, including the genetically altered one. (4) The altered phage adsorbs to and penetrates another host cell (cell B), injecting the DNA from cell A rather than viral nucleic acid. (5) Cell B receives this donated DNA, which recombines with its own DNA. Because the virus is defective (biologically inactive as a virus), it is unable to complete a lytic cycle. The transduced cell survives and can use this new genetic material.

Figure 9.27 **Specialized transduction: transfer of specific genetic material by means of a virus carrier.** (1) Specialized transduction begins with a cell that contains a prophage (red). (2) When the virus enters a lytic cycle, it excises itself from its host cell, carrying along some host DNA. (3) Replication and assembly results in production of a chimeric virus containing viral and bacterial DNA. (4) Release of the recombinant virus and infection of a new host result in transfer of bacterial DNA between cells. (5) Recombination can occur between the new host chromosome and the virus DNA, resulting in either bacterial DNA or a combination of viral and bacterial DNA being incorporated into the bacterial chromosome.

Transposons: "This Gene Is Jumping" One type of genetic transferral of great interest involves transposable elements, or **transposons.** Transposons have the distinction of shifting from one part of the genome to another and so are termed "jumping genes." When the idea of their existence in corn plants was first postulated by the geneticist Barbara McClintock, it was greeted with wide skepticism, because it had long been believed that the location of a given gene was set and that genes did not or could not move around. Now it is evident that jumping genes are widespread among prokaryotic and eukaryotic cells and viruses. Results from analysis of human DNA indicates that they make up nearly 45% of the human genome.

The first cases of *A. baumannii* infections were reported in the 1970s. At that time, it was not considered a serious pathogen, and it was sensitive to most available drugs. How did it go from being an uncommon infectious agent to one that accounts for around 6% of nosocomial infections, and how did the number of MDR strains increase from 3% to 40%? It had not developed greater virulence*. It is actually relatively harmless for most healthy people with intact immune systems. It is an opportunist that attacks mainly immunocompromised patients with various underlying medical problems. This pathogen causes primarily respiratory, skin, burn, urinary tract, and blood infections. Medical specialists suggest that its rapid evolution can be explained by the combination of four factors: the widespread distribution of the bacteria in the environment, the drug-rich environment of hospitals and clinics, the increase in severely compromised patients, and the pathogen's extreme genetic adaptability to challenges from new drugs.

- What is meant by the term *opportunist*?
- Which genetic events have allowed *Acinetobacter* to become so resistant?

For a wrap-up, see the Case File Perspectives on page 287.

All transposons share the general characteristic of traveling from one location to another on the genome—from one chromosomal site to another, from a chromosome to a plasmid, or from a plasmid to a chromosome (**figure 9.28**). Because transposons can occur in plasmids, they can also be transmitted from one cell to another in bacteria and a few eukaryotes. Some transposons replicate themselves before jumping to the next location, and others simply move without replicating first.

Transposons contain DNA that codes for the enzymes needed to remove and reintegrate the transposon at another site in the genome. Flanking the coding region of the DNA are sequences called inverted repeats, which mark the point at which the transposon is removed or reinserted into the genome. The smallest transposons consist of only these two genetic sequences and are often referred to as *insertion elements*. A type of transposon called *retrotransposon* can transcribe DNA into RNA and then back into DNA for insertion in a new location. Another example is the *integron,* which can carry large blocks of genetic material that are especially prominent in transmitting drug resistance.

The overall effect of transposons—to scramble the genetic language—can be beneficial or adverse, depending upon such variables as where insertion occurs in a chromosome, what kinds of genes are relocated, and the type of cell involved. In bacteria, transposons are known to be involved in

- changes in traits such as colony morphology, pigmentation, and antigenic characteristics;
- replacement of damaged DNA; and
- the transfer of drug resistance in bacteria.

* *virulence* (veer'-yoo-lents) ability of a microbe to invade and damage a host.

Figure 9.28 Transposons: shifting segments of the genome. (1) A transposon exists as a small piece of DNA integrated into the host cell chromosome (red). **(2)** The transposon may excise itself and move from one location to another in the genome, maintaining itself at single copy per cell. **(3)** It may also replicate prior to moving, leading to an increase in the copy number and a greater effect on the genome of the host. **(4)** Finally, the transposon may jump to a plasmid, which can then be transferred to another bacterial cell.

Check & Assess Sections 9.4 and 9.5

✔ Changes in the genetic code can occur by two means: mutation and recombination. Mutation means a change in the nucleotide sequence of the organism's genome.

✔ Recombination means the addition of genes from an outside source, such as a virus or another cell.

✔ Mutations can be either spontaneous or induced by exposure to some external mutagenic agent.

✔ All cells have enzymes that repair damaged DNA. When the degree of damage exceeds the ability of the enzymes to make repairs, mutations occur.

✔ Mutation-induced changes in DNA nucleotide sequencing range from a single nucleotide to addition or deletion of large sections of genetic material.

✔ The Ames test is used to identify potential carcinogens on the basis of their ability to cause back-mutations in bacteria.

✔ Because they lack sexual reproduction, genetic recombination in bacteria occurs through the processes of transformation, conjugation, and transduction.

✔ In all cases, a portion of one cell's genetic material is transferred to another cell.

✔ Transposons are genes that can relocate from one part of the genome to another, causing rearrangement of genetic material. Such rearrangements have either beneficial or harmful consequences for the organism involved.

23. Describe the principal types of mutations. Give an example of a mutation that is beneficial and one that is lethal or harmful.
24. Explain the purposes behind replica plating and the Ames test.
25. Compare conjugation, transformation, and transduction on the basis of general method, donor, and recipient.
26. Explain the differences between general and specialized transduction, using drawings.
27. By means of a flowchart, show the possible jumps that a transposon can make. Show the involvement of viruses in its movement.

9.6 The Genetics of Animal Viruses

Expected Learning Outcomes

27. Explain the major elements of viral genetics.
28. Compare aspects of the genetics of DNA and RNA viruses, using examples.

Viruses essentially consist of one or more pieces of DNA or RNA enclosed in a protective coating. They are genetic parasites that require access to their host cell's genetic and metabolic machinery to be replicated, transcribed, and translated, and they also have the potential for genetically changing the cells. Because they contain only those genes needed for the production of new viruses, the genomes of viruses tend to be very compact and economical.

Viruses show significant variety in genetic patterns. In many viruses, the nucleic acid is linear in form; in others, it is circular. The genome of most viruses exists in a single molecule, though in a few, it is segmented into several smaller molecules. Most viruses contain normal double-stranded (ds) DNA or single-stranded (ss) RNA, but other patterns exist. There are ssDNA viruses, dsRNA viruses, and retroviruses, which work backward by making dsDNA from ssRNA. In some instances, viral genes overlap one another, and in a few DNA viruses, both strands contain a translatable message.

With few exceptions, replication of the DNA molecule of DNA animal viruses occurs in the nucleus, where the cell's DNA replication machinery lies, and the genome of RNA viruses is replicated in the cytoplasm. In all viruses, viral mRNA is translated into viral proteins on host cell ribosomes using host tRNA.

Replication Strategies in Animal Viruses

Replication, Transcription, and Translation of dsDNA Viruses

Replication of dsDNA viruses is divided into phases (**figure 9.29**). During the early phase, viral DNA enters the nucleus, where several genes are transcribed into a messenger RNA. Next, the mRNA transcript then moves into the cytoplasm to be translated into viral proteins (enzymes) needed to replicate the viral DNA. This replication usually occurs in the nucleus with the host cell's own DNA polymerase, though some viruses (herpes, for example) have their own. During the late phase, other viral genes are transcribed and translated into proteins required to form the capsid and other

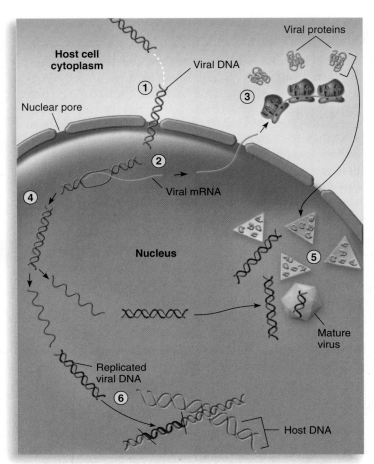

Figure 9.29 Stages in the multiplication of double-stranded DNA viruses. The virus penetrates the host cell and releases DNA, which, **(1)** enters the nucleus and **(2)** is transcribed by host cell enzymes into mRNA **(3)** Viral mRNA leaves the nucleus and is translated into structural proteins; these proteins are transported into the nucleus. **(4)** Viral DNA is replicated repeatedly in the nucleus. **(5)** Viral DNA and proteins are assembled into a mature virus in the nucleus. **(6)** Because it is double-stranded, some viral DNA can insert itself into host DNA (latency).

structures. The new viral genomes and capsids are assembled, and the mature viruses are released by budding or cell disintegration.

Double-stranded DNA viruses can interact directly with the DNA of their host cell. In some viruses, the viral DNA becomes silently *integrated* into the host's genome by insertion at a particular site on the host genome. Persistence of viral DNA may also lead to the transformation of the host cell into a cancer cell. Several DNA viruses, including hepatitis B (HBV), the herpesviruses, and papillomaviruses (warts), are known to be initiators of cancers and are thus termed **oncogenic.** The mechanisms of oncogenic **transformation** involve genes that can regulate cellular genomes and control the onset of cell division.

Replication, Transcription, and Translation of RNA Viruses

RNA viruses exhibit several differences from DNA viruses. Their genomes are smaller, they enter the host cell already in an RNA form,

and the virus cycle occurs entirely in the cytoplasm for most viruses. RNA viruses can have one of the following genetic messages:

1. a positive-strand (+) genome that comes ready to be translated into proteins,
2. a negative-strand (−) genome that must be converted to a positive-strand before translation, and
3. a positive-strand (+) genome that can be converted to DNA or a dsRNA genome.

Positive-Strand Single-Stranded RNA Viruses (Figure 9.30)

The ssRNA of positive-strand viruses such as polio come ready to be translated, so it is immediately translated into a large protein and cleaved into individual functional units. One of these is an RNA polymerase that initiates the replication of the viral strand. Replication of a positive-message strand happens in two steps. First, a negative strand is synthesized using the positive strand as a template by the usual base-pairing mechanism. The resultant negative strand

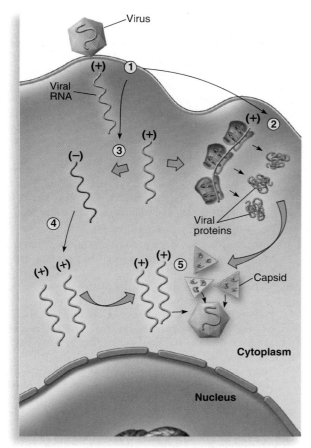

Figure 9.30 Replication of positive-strand, single-stranded RNA viruses. In general, these viruses do not enter the nucleus. **(1)** Penetration of viral RNA **(2)** Because it is positive in message and single-stranded, the RNA can be directly translated on host cell ribosomes into various necessary viral proteins. **(3)** A negative genome is synthesized against the positive template to produce large numbers of positive genomes for final assembly. **(4)** The negative template is then used to synthesize a series of positive replicates. **(5)** RNA strands and proteins assemble into mature viruses.

becomes a master template against which new positive daughter strands are made. Further translation of the viral genome produces large numbers of structural proteins for final assembly and maturation of the virus.

RNA Viruses with Reverse Transcriptase: Retroviruses

A most unusual class of viruses can reverse the order of the flow of genetic information. Thus far in our discussion, all genetic entities have shown the patterns DNA → DNA, DNA → RNA, or RNA → RNA. Retroviruses, including HIV, the cause of AIDS, and HTLV I, a cause of one type of human leukemia, synthesize DNA using their RNA genome as a template. They accomplish this by means of a **reverse transcriptase** that comes packaged with each virus particle. This enzyme synthesizes a single-stranded DNA against the viral RNA template and then directs the formation of a complementary strand of this ssDNA, resulting in a double strand of viral DNA. The dsDNA strand enters the nucleus, where it can be transcribed by the usual mechanisms into new viral ssRNA and integrated into the host genome. When the DNA of some retroviruses becomes inserted into the host's DNA as a provirus, cells may be transformed and produce tumors. Insertion allows HIV to remain latent in an infected cell for several years (see figure 25.16).

Check & Assess Section 9.6

✔ The genetic material of viruses can be either DNA or RNA occurring in single, double, positive-, or negative-message strands. Viruses contain genes primarily for taking over the cell machinery and synthesizing new virus components.

28. Explain why some viruses must enter the nucleus to replicate whereas others do not.
29. Explain the difference between positive strand and negative strand viral genomes.

CASE FILE 9 PERSPECTIVE

The source of the infection in the first case was most likely the ventilator that controlled the woman's breathing. Medical devices are readily contaminated by patients and healthcare workers. Any case acquired during a stay in a hospital is defined as a nosocomial infection. These infections are most problematic in very compromised patients, but stringent disinfection procedures can greatly reduce their incidence. The second case probably came from the soiled piece of shrapnel rather than the field hospital.

To say that a microbe has resistance to a drug means that it naturally possesses or has acquired a genetic mechanism to avoid the effects of the drug. Usually, its genome carries one or more genes that can eliminate the drug or prevent it from acting on the cells of the microbe. This leaves the microbe free to grow and infect even in the presence of that drug.

An opportunist or opportunistic pathogen is one that invades people whose defenses and immunities have been compromised. For various reasons they have lost a protective barrier or lack an effective immune response. Infections caused by *A. baumannii* dominate among patients with preexisting medical conditions such as cancer, surgery, organic diseases, or traumatic injury.

Most bacteria are rather efficient in transferring and acquiring genes from each other, and the genes for drug resistance provide an adaptive advantage wherever these drugs are being used, whether it be various hospital environments or even the patient's body.

Results from studies indicate that *A. baumannii* has an impressive repertoire for acquiring and expressing drug resistance. It demonstrates all of the properties for picking up genes from other bacteria, including pili for conjugation, DNA capture proteins for transformation, and bacteriophages for transduction. This pathogen acts like a magnet for plasmids, transposons, and integrons that encode multiple resistance genes. In one example, A. *baumannii* acquired an integron from a species of *Pseudomonas* carrying 45 genes that conferred some type of drug resistance.

Once the pathogen has obtained these genes, actual contact with the drug will exert a selective pressure on gene expression and favor the survival of a drug resistant strain. This may be repeated with every drug used in therapy until the pathogen has complete resistance. This is apparently what happened with the patient in the first case. Medical experts are alarmed that we may soon be facing a crisis with this pathogen, since there are few viable drugs and none in current development.

Many of these topics are further discussed in chapter 12.

For more detailed coverage, go to http://www.medscape.com/viewarticle/575837.

Chapter Summary with Key Terms

9.1 Introduction to Genetics and Genes: Unlocking the Secrets of Heredity

A. **Genetics** is the study of **heredity** and can be studied at the level of the organism, **genome, chromosome, gene,** and **DNA.** Genes provide the information needed to construct **proteins,** which have structural or catalytic functions in the cell.

B. **DNA** is a long molecule in the form of a double helix. Each strand of the helix consists of a string of **nucleotides.**

1. A nucleotide consists of a **phosphate, deoxyribose sugar,** and **nitrogen base.** The bases form hydrogen bonds with their counterparts on the other strand.

2. **Adenine** base-pairs with **thymine** while **guanine** base-pairs with **cytosine.**

3. The order of the nucleotides specifies which amino acids will be used to construct proteins during the process of translation.

C. DNA **replication** is **semiconservative** and requires the participation of several enzymes.

9.2 Applications of the DNA Code: Transcription and Translation

A. DNA is used to produce RNA (**transcription**) and RNA is then used to produce protein (**translation**).

B. RNA: Unlike DNA, RNA is single stranded, contains **uracil** instead of thymine and **ribose** instead of deoxyribose.

1. Major forms of RNA found in the cell include **mRNA, tRNA,** and **rRNA.**

2. The genetic information contained in DNA is copied to produce an RNA molecule. **Codons** in the mRNA pair with **anticodons** in the tRNA to specify what amino acids to assemble on the ribosome during translation.

C. *Transcription and Translation:*

1. Transcription occurs when **RNA polymerase** copies the template strand of a segment of DNA. RNA is always made in the 5′ to 3′ direction.

2. Translation occurs when the mRNA is used to direct the synthesis of proteins on the ribosome.

3. Codons in the mRNA pair with anticodons in the tRNA to assemble a string of amino acids. This occurs until a stop codon is reached.

D. *Eukaryotic Gene Expression:* Eukaryotic genes are composed of **exons** (expressed sequences) and **introns** (intervening sequences). The introns must be removed and the exons spliced together to create the final mRNA.

9.3 Genetic Regulation of Protein Synthesis and Metabolism

Protein synthesis is regulated through gene induction or repression, as controlled by an **operon.** Operons consist of several structural genes controlled by a common regulatory element.

A. *Inducible operons* such as the lactose operon are normally off but can be turned on by a lactose inducer.

B. *Repressible operons* are usually on but can be turned off when their end product is no longer needed.

C. Newly described types of RNA can serve in gene regulation.

1. Micro and small interfering RNAs can block the action of other RNAs and shut down gene expression.

2. Antisense RNA and riboswitches can also control RNA functions related to transcription and translation.

9.4 Mutations: Changes in the Genetic Code

A. Permanent changes in the genome of a microorganism are known as **mutations.** Mutations may be **spontaneous** or **induced.**

B. Mutations are detected by the use of selective media and the replica plating technique.

C. **Point mutations** entail a change in one or a few bases and are categorized as **missense, nonsense, or silent mutations** and **back-mutations,** based on the effect of the change in nucleotide(s).

D. Many mutations, particularly those involving mismatched bases or damage from ultraviolet light, can be corrected using enzymes found in the cell.

E. The **Ames test** measures the mutagenicity of chemicals by their ability to induce mutations in bacteria.

9.5 DNA Recombination Events

A. Intermicrobial transfer and genetic recombination permit gene sharing between bacteria. Major types of recombination include **conjugation, transformation,** and **transduction.**

B. **Transposons** are DNA sequences that regularly move to different places within the genome of a cell, generating mutations and variations in chromosome structure.

9.6 The Genetics of Animal Viruses

A. Viruses contain various forms of DNA and RNA that are transcribed and translated by the genetic machinery of their host cells to form functioning viral particles.

1. Genomes of viruses are found in double-stranded (ds) DNA and single-stranded (ss) RNA forms. A few contain unusual ssDNA and dsRNA not found in cells.
2. DNA viruses tend to replicate in the nucleus while RNA viruses replicate in the cytoplasm. Retroviruses synthesize dsDNA using ssRNA as a template.

Multiple-Choice Questions

Select the correct answer from the answers provided. For questions with blanks, choose the combination of answers that most accurately completes the statement.

1. What is the smallest unit of heredity (genotype)?
 a. chromosome
 b. gene
 c. codon
 d. nucleotide

2. A nucleotide contains which of the following?
 a. 5-carbon sugar
 b. nitrogen base
 c. phosphate
 d. b and c only
 e. all of these

3. The nitrogen bases in DNA are bonded to the
 a. phosphate
 b. deoxyribose
 c. ribose
 d. hydrogen

4. DNA replication is semiconservative because the _____ strand will become half of the _____ molecule.
 a. RNA, DNA
 b. template, finished
 c. sense, mRNA
 d. codon, anticodon

5. In DNA, adenine is the complementary base for _____, and cytosine is the complement for _____.
 a. guanine, thymine
 b. uracil, guanine
 c. thymine, guanine
 d. thymine, uracil

6. The base pairs are held together primarily by
 a. covalent bonds
 b. hydrogen bonds
 c. ionic bonds
 d. gyrases

7. Why must the lagging strand of DNA be replicated in short pieces?
 a. because of limited space
 b. otherwise, the helix will become distorted
 c. the DNA polymerase can synthesize in only one direction
 d. to make proofreading of code easier

8. Messenger RNA is formed by _____ of a gene on the DNA template strand.
 a. transcription
 b. replication
 c. translation
 d. transformation

9. Transfer RNA is the molecule that
 a. contributes to the structure of ribosomes
 b. adapts the genetic code to protein structure
 c. transfers the DNA code to mRNA
 d. provides the master code for amino acids

10. As a general rule, the template strand on DNA will always begin with
 a. TAC
 b. AUG
 c. ATG
 d. UAC

11. The *lac* operon is usually in the _____ position and is activated by a/an _____ molecule.
 a. on, repressor
 b. off, inducer
 c. on, inducer
 d. off, repressor

12. For mutations to have an effect on populations of microbes, they must be
 a. inheritable
 b. permanent
 c. beneficial
 d. a and b
 e. all of these

13. Which of the following characteristics is *not* true of a plasmid?
 a. It is a circular piece of DNA.
 b. It is required for normal cell function.
 c. It is found in bacteria.
 d. It can be transferred from cell to cell.

14. Which genes can be transferred by all three methods of intermicrobial transfer?
 a. capsule production
 b. toxin production
 c. F factor
 d. drug resistance

15. Which of these features is found in eukaryotes but not prokaryotes?
 a. polygene mRNAs
 b. introns
 c. stop codon
 d. AUG codon

16. Which of the following is present in prokaryotes but not eukaryotes?
 a. exons
 b. polyribosomes
 c. spliceosomes
 d. simultaneous transcription and translation

17. **Multiple Matching.** Fill in the blanks with all the letters of the words below that apply.

 __d__ site of protein synthesis
 __f__ carries the codon
 __b__ carries the anticodon
 __g__ a process synonymous with mRNA synthesis
 _____ bacteriophages participate in this transfer
 __e__ duplication of the DNA molecule
 __a__ process in which transcribed DNA code is deciphered into a polypeptide
 __i__ involves plasmids

 a. replication
 b. tRNA
 c. conjugation
 d. ribosome
 e. transduction
 f. mRNA
 g. transcription
 h. transformation
 i. translation
 j. none of these

Case File Questions

1. Which of the following is not a factor in the rise of drug-resistant *Acinetobacter baumannii*?
 a. It is very widespread in many habitats.
 b. It has acquired greater toxicity.
 c. It readily acquires genetic material from other species.
 d. The hospital environment brings together vulnerable patients and drugs.

2. Drug-resistance plasmids could be transferred in *A. baumannii* by which of the following mechanisms:
 a. conjugation pilus
 b. bacteriophages
 c. capture by cell wall
 d. DNA pumps
 e. a, b, and c
 f. all of these

Writing to Learn

These questions are suggested as a *writing-to-learn* experience. For each question, compose a one- or two-paragraph answer that includes the factual information needed to completely address the question. All Check & Assess questions can be used for Writing-to-Learn exercises.

1. Explain what is meant by the following diagram.

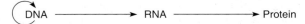

 DNA ⟶ RNA ⟶ Protein

2. On paper, replicate the following segment of DNA:

 5′ A T C G G C T A C G T T C A C 3′
 3′ T A G C C G A T G C A A G T G 5′

 a. Show the direction of replication of the new strands and explain what causes lagging and leading strands to show different patterns of replication.
 b. Explain how this is semiconservative replication. Are the new strands identical to the original segment of DNA?

3. The following sequence represents triplets on DNA:

 TAC CAG ATA CAC TCC CCT GCG ACT

 a. Give the mRNA codons and tRNA anticodons that correspond with this sequence, and then give the sequence of amino acids in the polypeptide.
 b. Provide a different mRNA strand that can be used to synthesize this same protein.
 c. Looking at figure 9.15, give the type and order of the amino acids in the peptide that is encoded.

4. Give an overview of the types of drug-resistance mechanisms that *A. baumannii* is likely to use.

5. Write an essay that discusses the four factors that have led to the rise of *A. baumannii* as a hospital pathogen.

6. Describe the actions of all of the enzymes involved in replication of a DNA molecule.

Concept Mapping

Appendix E provides guidance for working with concept maps.

1. Supply your own linking words or phrases in this concept map, and provide the missing concepts in the empty boxes.

Critical Thinking Questions

Critical thinking is the ability to reason and solve problems using facts and concepts. These questions can be approached from a number of angles, and in most cases, they do not have a single correct answer.

1. Knowing that retroviruses operate on the principle of reversing the direction of transcription from RNA to DNA, propose a drug that might possibly interfere with their replication.

2. Using the piece of DNA in writing-to-learn question 3, show a deletion, an insertion, a substitution, and nonsense mutations. Which ones are frameshift mutations? Are any of your mutations nonsense? Missense? (Use the universal code to determine this.)

3. Using figure 9.13 and table 9.4, go through the steps in mutation of a codon followed by its transcription and translation that will give the end result in silent, missense, and nonsense mutations. Why is a change in the RNA code alone not really a mutation?

4. The enzymes required to carry out transcription and translation are themselves produced through these same processes. Speculate which may have come first in evolution—proteins or nucleic acids—and explain your choice.

5. Why can one not reliably predict the sequence of nucleotides on mRNA or DNA by observing the amino acid sequence of proteins?

6. Activation of the lac operon will release which nutrients?

7. Explain the various mechanisms by which RNA can control transcription, translation, and gene expression.

8. Explain how epigenetics is related to the expression:
 phenotype = genotype + environment.

9. Compare and contrast the actions of DNA polymerase and RNA polymerase.

Visual Challenge

1. From **figure 9.15, step 3.** Label each part of the illustration.

2. Explain how replication shown in these steps from figure 6.17 would differ from those in figure 9.29.

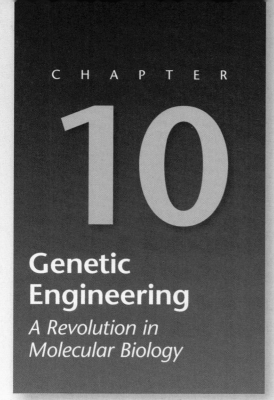

CHAPTER 10

Genetic Engineering

A Revolution in Molecular Biology

A representation of single nucleotide polymorphisms in humans (yellow)

Biochemists setting up automated systems for identifying DNA taken from World Trade Center samples

"It is a stunning testament to the extreme power and precision of DNA technology that it was able to extract the genetic profiles of the terrorists who hijacked the airplanes!"

CASE FILE 10 A Tragic Trial For DNA Technology

The devastating attack on the World Trade Center towers on September 11, 2001 has had repercussions that continue to reverberate throughout the world even today. But there is nothing to compare with the emotional horror of families and friends left to wonder: What was the fate of our loved one? Without tangible evidence of death, many families spent untold hours and days anguishing over the possibility that their missing family member could have survived and might return home. Gradually, a grim reality sank in: Nearly 3,000 people had perished and were not going home. This left their survivors with a hope that, somehow, authorities could salvage at least a small fragment of remains they could lay to rest.

The responsibility for identifying the victims fell to the office of the medical examiner of New York City, who organized teams to search for human remains buried under 1.2 million tons of rubble spread over 16 acres. They painstakingly sifted through masses of debris for any piece of bone or tissue to be preserved for later identification. Over 21,000 pieces were isolated, most of which were not identifiable by the usual forensic methods. It became clear that such an enormous endeavor would require a much more specific method—the powerful technology of DNA identification, sometimes referred to as DNA fingerprinting.

Several problems are inherent in using DNA to identify a person after a mass fatality such as this. One is that the high temperatures of the disaster site, which had burned for several days, would have severe effects on organic material. DNA would likely be degraded into small fragments and in some cases even incinerated. Another complication is not having an immediate way of knowing how an isolated bit of DNA belonged to a particular person. And last, it is expected that the genetic material from many people and even non-human organisms would have become cross-contaminated, or commingled.

Hundreds of consultants from dozens of fields were brought in to oversee the DNA analysis and to prepare a database to keep track of the DNA records as they were gathered. The specimens were stored in refrigerated mobile trailers on site. Because of the huge number of individual tests and the need for rapid results, much of this operation was carried out by automated robots. In the end, the endeavor was the largest mass screening of DNA on record and took 3 years for the completion of the first phase. By 2009, the World Trade Center ID Project had positively identified 1,626 people, or 59% of the 2,819 known victims. Once identified, their remains were finally restored to the families.

▶ *Why is DNA such a specific way to identify a person?*

▶ *How did the DNA investigators deal with the problems of identifying DNA as outlined here?*

To continue the case, go to page 313.

10.1 Basic Elements and Applications of Genetic Engineering

Expected Learning Outcomes

1. Define *genetic engineering,* and describe some of its subcategories and aims.
2. Explain the properties of DNA that lend it to manipulation in the laboratory.
3. Summarize the major methods of analyzing DNA and their results.
4. Describe the technology behind identifying, sequencing, synthesizing, and amplifying DNA.

In chapter 9, we looked at the ways in which microorganisms duplicate, exchange, and use their genetic information. In scientific parlance, this is called *basic science* because no product or application is directly derived from it. Human beings being what they are, however, it is never long before basic knowledge is used to derive *applied science* or useful products and applications that owe their invention to the basic research that preceded them. As an example, basic research into the workings of the electron has led to the development of television, computers, and cell phones. None of these staples of modern life was envisioned when early physicists were deciphering the nature of subatomic particles; but without the knowledge of how electrons worked, our ability to harness them for our own uses would never have materialized.

The knowledge of DNA that emerged from basic science over the past 30 years has spawned a revolution in a wide variety of human endeavors. In many ways, it has changed how we look at the world and, in fact, has profoundly changed the world. It now dominates research and development in hundreds of areas of biotechnology, medicine, forensics, diagnosis of disease, drug development and treatment, agriculture, industrial production, and ecology. It is now possible to accomplish feats that were once only imagined. The new wave of DNA scientists have gone back in history to diagnose the ailments of Abraham Lincoln and King Tutankhamen and to sequence DNA from a wooly mammoth. They have changed the face of agriculture by developing plants that resist disease and animals that can synthesize human hormones, and they have created entirely new viruses and bacteria. This technology has diagnosed and cured genetic diseases, identified the mass victims of disasters, and now it is a key part of law enforcement. Because of it, we are able to fully decode the DNA of thousands of organisms, including humans, and this provides the basic tools for deep understanding of all living things.

The primary goals of this science are to isolate, identify, and manipulate DNA in various ways. A major part of its work falls into the category of **genetic engineering** or **bioengineering,** defined as the direct, deliberate modification of an organism's genome. It also belongs to the larger scheme of **biotechnology,** which includes any applications of DNA, genes, or genetically altered organisms in commercial production.

Information on genetic engineering and its biotechnological applications is growing at such an expanding rate that some new discovery or product is disclosed almost daily. To keep this subject somewhat manageable, we present essential concepts and applications, organized under the following six topics:

- Basic Elements and Applications of Genetic Engineering
- Recombinant DNA Technology
- Biochemical Products of Recombinant DNA Technology
- Genetically Modified Organisms
- Genetic Treatments
- Genome Analysis

Tools and Techniques of DNA Technology

All of the intrinsic properties of DNA hold true whether the DNA is in a bacterium or a test tube. For example, the enzyme helicase is able to unwind the two strands of the double helix just as easily inside or outside a bacterial cell. But in the laboratory, we can take advantage of our knowledge of DNA chemistry to make helicase unnecessary. It turns out that when DNA is heated to just below boiling (90°C to 95°C), the two strands separate, revealing the information contained in their bases. With the nucleotides exposed, DNA can be easily identified, replicated, or transcribed. If heat-denatured DNA is then slowly cooled, complementary nucleotides will hydrogen bond with one another, and the strands will renature or regain their familiar double-stranded form **(figure 10.1a).** As we shall see, this process is a necessary feature of the polymerase chain reaction and nucleic acid probes described later.

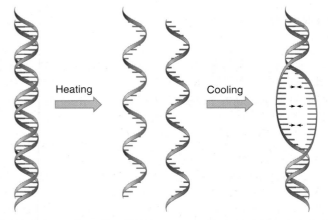

(a) **DNA heating and cooling.** DNA responds to heat by denaturing—losing its hydrogen bonding and thereby separating into its two strands. When cooled, the two strands rejoin at complementary regions. The two strands need not be from the same organisms as long as they have matching sites.

Endonuclease	*Eco*RI	*Hind*III	*Hae*III
Cutting pattern	G A A T T C C T T A A G	A A G C T T T T C G A A	G G C C C C G G

(b) **Examples of endonucleases, palindromes and cutting patterns.** The first two are staggered cuts, and the third is a blunt cut.

Figure 10.1 **Some properties of DNA that are an integral part of DNA manipulation.** The method in **(a)** can be used to separate the DNA strands lengthwise and make them single stranded. The method in **(b)** uses enzymes called endonucleases to make cuts across the DNA molecule at specific sites called palindromes.

Enzymes for Dicing, Splicing, and Reversing Nucleic Acids

The polynucleotide strands of DNA can also be clipped crosswise at selected positions by means of **restriction endonucleases.**[1] These enzymes originate in bacterial cells. They recognize foreign DNA and are capable of breaking the phosphodiester bonds between adjacent nucleotides on both strands of DNA, leading to a break in the DNA strand. In the bacterial cell, this ability protects against the incompatible DNA from bacteriophages or plasmids. In the biotechnologist's lab, the enzymes can be used to cleave DNA at desired sites and are a must for the techniques of recombinant DNA technology.

So far, about 3,000 restriction endonucleases have been discovered in bacteria. Each type has a known sequence of 4 to 10 base pairs as its target, so sites of cutting can be finely controlled. These enzymes have the unique property of recognizing and clipping at sites called **palindromes**[2] **(figure 10.1*b*).** Palindromes are sequences of DNA bases that are identical when read from the 5′ to 3′ direction on one strand and the opposite direction on the complementary strand.

Endonucleases are named by combining the first letter of the bacterial genus, the first two letters of the species, and the endonuclease number. Thus, *Eco* RI is the first endonuclease found in *Escherichia coli's* R strain, and *Hind*III is the third endonuclease discovered in *Haemophilus influenzae* Type d (figure 10.1*b*).

Endonucleases are used in the laboratory to cut DNA into smaller pieces for further study as well as to remove and insert it during recombinant DNA techniques, described in section 10.2. Usually, the enzymes make staggered symmetrical cuts that leave short, single-stranded tails called "sticky ends." Such adhesive tails will base-pair with complementary tails on other DNA fragments or plasmids. This effect makes it possible to splice genes into specific sites **(figure 10.2).** A few endonucleases such as *Hae*III make straight, blunt cuts on DNA.

The pieces of DNA produced by restriction endonucleases are termed **restriction fragments.** Because DNA sequences vary, even among members of the same species, differences in the cutting pattern of specific restriction endonucleases give rise to restriction fragments of differing lengths, known as **restriction fragment length polymorphisms (RFLPs).** RFLPs allow the direct comparison of the DNA of two different organisms at a specific site, which, as we will see, has many uses.

Another enzyme, called a **ligase,** is necessary to seal the sticky ends together by rejoining the phosphate-sugar bonds cut by endonucleases. Its main application is in final splicing of genes into plasmids and chromosomes.

An enzyme called **reverse transcriptase** is best known for its role in the replication of the AIDS virus and other retroviruses. It also provides geneticists with a valuable tool for converting RNA into DNA. Copies called **complementary DNA,** or **cDNA,** can be made from messenger, transfer, ribosomal, and other forms of RNA.

The technique provides a valuable means of synthesizing eukaryotic genes from mRNA transcripts **(figure 10.3).** The bacteria often used to clone eukaryotic genes in genetic engineering are unable to work with unmodified versions of these genes. The advantage is that the synthesized gene will be free of the intervening sequences (introns) that are present in eukaryotic genes.

Because the rules of nucleotide pairing hold true even between hybrid molecules consisting of one strand of RNA and one of DNA, complementary DNA can be used to analyze the nucleotide sequence of RNAs, such as those found in ribosomes and transfer RNA.

Analysis of DNA

One way to produce a readable pattern of DNA fragments is through **electrophoresis.** In this technique, samples are placed in compartments (wells) in a soft agar gel or similar soft medium and subjected to an electrical current. The phosphate groups in DNA give the entire molecule an overall negative charge, which causes the DNA to move toward the positive pole in the gel. The rate of movement is based primarily on the size of the fragments. The larger fragments move more slowly and remain nearer the wells, whereas the smaller fragments migrate faster and are positioned farther from

Action of restriction endonucleases. (1) A restriction endonuclease recognizes and cleaves DNA at the site of a specific palindromic sequence. Cleavage usually produces staggered tails called sticky ends that accept complementary tails for gene splicing. This palindrome is cut by *Aci* I. (2) The sticky ends can be used to join DNA from different organisms by cutting it with the same restriction enzyme, ensuring that both fragments have two complementary ends. This palindrome is cut by *Taq* I.

Figure 10.2 **Examples of cutting patterns by restriction endonucleases and how they are used in gene splicing.**

1. The meaning of restriction is that the enzymes do not act upon the bacterium's own DNA; an endonuclease nicks DNA internally, not at the ends.

2. This term originated from words or phrases with letters that read the same backward or forward, as in *radar,* "Madam, I'm Adam," and "Can I attain a C?"

(a)

Eukaryotic DNA

Transcription of eukaryotic DNA generates precursor mRNA.

Precursor mRNA

Splicing removes the introns from the precursor mRNA, generating the final mRNA for translation.

mRNA (within eukaryotic cells)

mRNA extracted from cells

(b)

mRNA (isolated)

Reverse transcriptase synthesizes DNA from mRNA.

Single strand of cDNA

DNA polymerase completes the cDNA, free from introns.

Double-stranded cDNA

the wells. The positions of DNA fragments are visualized by one of several methods **(figure 10.4)**. Electrophoresis patterns can be quite distinctive and are very useful in characterizing DNA fragments and comparing the degree of genetic similarities among samples as in a Southern blot or genetic fingerprint (discussed later).

Nucleic Acid Hybridization and Probes

Two different nucleic acids can **hybridize** by uniting at their complementary sites. All different combinations are possible: Single-stranded DNA can unite with other single-stranded DNA or RNA, and RNA can hybridize with other RNA. This property has been the inspiration for specially formulated oligonucleotide tags called

Figure 10.3 Synthesizing cDNA from eukaryotic mRNA. Bacterial cloning hosts cannot process intact eukaryotic genes. To provide a gene that will be transcribable and translatable by a bacterium, **(a)** first the modified mRNA, lacking introns, is isolated from the eukaryotic cell. Then **(b)** this strand is converted in a test tube to a single-stranded DNA by reverse transcriptase and double-stranded DNA by a polymerase. This copy of DNA (cDNA) can be recognized by bacteria.

Figure 10.4 Separating DNA restriction fragments with electrophoresis. (a) A restriction endonuclease is applied to a sample (using sample 3 as an example). The sample with DNA fragments is loaded into a well at the top of the agarose gel. Passage of an electrical current through the gel (in the direction of negative to positive) causes the DNA to migrate toward the positive pole. Larger fragments migrate more slowly than smaller (shorter) fragments, thus they are arrayed according to size from largest to smallest. The final position of the fragments will be visualized by staining or the method shown in (b). **(b)** A developed gel reveals separated bands of the DNA fragments. The size of a given DNA band can be determined by comparing the distance traveled to the distance traveled by a set of DNA fragments of known size called a size ladder (lane 5), measured in number of base pairs. This gel contained DNA with radioactive bases that can expose a photographic negative.

gene probes. These probes consist of a short stretch of DNA of a known sequence that will base-pair with a stretch of DNA that has a complementary sequence, if one exists in the test sample. Hybridization probes have practical value because they can detect specific nucleotide sequences in unknown samples. So that areas of hybridization can be visualized, the probes carry reporter molecules, such as luminescent or radioactive labels, which are revealed by placing photographic film in contact with the test reaction (see figure 10.4*b*). Fluorescent probes contain dyes that can be visualized with ultraviolet radiation (see figure 10.6).

When probes hybridize with an unknown sample of DNA or RNA, they tag the precise area and degree of hybridization, and help determine the nature of nucleic acid present in a sample. In a method called the **Southern blot,**[3] DNA fragments are first separated by electrophoresis and then denatured and transferred to a special filter. A DNA probe is then incubated with the sample, and wherever this probe encounters the segment for which it is complementary, it will attach and form a hybrid. Development of the hybridization pattern will show up as one or more bands **(figure 10.5).** This method is a sensitive and specific way to isolate fragments from a complex mixture and to find specific gene sequences on DNA. Southern blot is also one of the important first steps for preparing isolated genes.

Probes are commonly used for diagnosing the cause of an infection from a patient's specimen and identifying a culture of an unknown bacterium or virus (see chapter 17). A simple and rapid method called a hybridization test does not use electrophoresis. DNA from a test sample is isolated, denatured, placed on an absorbent filter, and combined with a microbe-specific probe. The blot is then developed and observed for areas of hybridization. Commercially available diagnostic kits are now on the market for identifying intestinal pathogens such as *E. coli, Salmonella, Campylobacter, Shigella, Clostridium difficile,* rotaviruses, and adenoviruses. Other bacterial probes exist for *Mycobacterium, Staphylococcus, Mycoplasma,* and *Chlamydia*; viral probes are available for herpes simplex and zoster, papilloma (genital warts), hepatitis A and B, and HIV. DNA probes have also been developed for human genetic markers and some types of cancer.

With another method, called *fluorescent in situ hybridization* (FISH), probes are applied to intact cells and observed microscopically for the presence and location of specific genetic marker sequences on genes. It is a very effective way to locate genes on chromosomes **(figure 10.6).** In situ techniques can also be used to identify unknown bacteria living in natural habitats without having to culture them, and they can be used to detect RNA in cells and tissues.

① DNA samples are cut with restriction enzymes and loaded on agarose gel for electrophoresis.

Lane 1: Labeled size markers
Lane 2: DNA cut with restriction enzyme A
Lane 3: DNA with restriction enzyme B

Gel electrophoresis

DNA is denatured; gel is placed on sponge wick.

Weight
Paper towels
DNA-binding filter
Gel
Wick (sponge)
Buffer

② DNA is separated by electrophoresis and visualized by staining and photography in UV light. (When DNA travels, it appears as a smear rather than distinct bands.)

③ DNA-binding filter, paper towels, and weight are placed on gel. Buffer passes upward by capillary action transferring DNA fragments to filter.

④ Filter is placed in heat-sealed food bag with solution containing radioactive probe.

Overlay filter with X-ray film

Developed X-ray film with DNA bands

⑤ Filter is washed to remove excess probe, then dried. Film is then exposed to produce photographic image of DNA bands that reacted with the probes.

Process Figure 10.5 **Conducting a Southern blot hybridization test.** The five-step process reveals the hybridized fragments on the X-ray film.

3. Named for its developer, E. M. Southern. The "northern" blot is a similar method used to analyze RNA, while the "western" blot detects proteins. There is a reason scientists became scientists and not comedians.

Figure 10.6 **Fluorescent in-situ hybridization (FISH) with papillomavirus in host cells.** The virus integrates into chromosomes (orange) in specific locations (yellow dots) that glow under UV light. This technique can be used to detect viral infection of cells.

Methods Used to Size, Synthesize, and Sequence DNA

Whether one is studying a bacterial cell or the DNA from that cell, an accurate description is invaluable. To that end, analysis of DNA is aided by our ability to cut it into shorter fragments and determine its precise nucleotide sequence. The relative sizes of nucleic acids are usually denoted by the number of base pairs (bp) or nucleotides they contain. For example, the palindromic sequences recognized by endonucleases are usually 4 to 10 bp in length; an average gene in *E. coli* is approximately 1,300 bp, or 1.3 kilobases (kb); and its entire genome is approximately 4,700,000 bp, 4,700 kb, or 4.7 megabases (mb). The DNA of the human mitochondrion contains 16 kb, and the Epstein-Barr virus (a cause of infectious mononucleosis) has 172 kb. Humans have approximately 3.1 billion base pairs (bb) arrayed along 46 chromosomes. The Human Genome Project, finished in 2001, produced a map containing every base in the entire human genome **(Insight 10.1).**

DNA Sequencing: Determining the Exact Genetic Code

Testing DNA for size, restriction patterns, and hybridization characteristics has greatly expanded our knowledge, but the most detailed information comes from determining the actual order and types of bases in DNA. This process, called **DNA sequencing,** provides the identity and order of nucleotides for all types of DNA, including genomic, cDNA, chromosomes, plasmids, and cloned genes. A common sequencing technique developed by Frederick Sanger is based on the synthesis and analysis of a complementary strand of DNA **(figure 10.7).** Even though mostly automated techniques are used today, this sequencer will serve as an example of the basic principles involved.

Because most DNA being sequenced is very long, it is made more manageable by cutting it into a number of shorter fragments, which are then separated. The test strands, typically several hundred nucleotides long, are denatured to expose single strands that will serve as templates to synthesize complementary strands. Usually, these fragments must be increased in number by PCR (see page 300). The fragments are divided into four separate tubes that contain primers to initiate the synthesis on one of the strands. The nucleotides will attach at the 3′ end of the primer, using the template strands as a guide, essentially the same as regular DNA synthesis as performed in a cell.

The tubes are incubated with the necessary DNA polymerase and all four of the regular nucleotides needed to carry out the process of elongating the complementary strand. Each tube also contains a single type (A, T, G, or C) of dideoxynucleotide (dd), each of which carries a fluorescent label in a unique color. Dideoxynucleotides have no hydroxyl group (OH⁻) bound to the 3′ carbon of ribose. This hydroxyl is needed for the chemical attachment of the next nucleotide in the growing DNA strand. As the reaction proceeds, strands elongate by adding normal nucleotides, but a small percentage of fragments will randomly incorporate the complementary dd nucleotide and be terminated. Eventually, all possible positions in the sequence will incorporate a terminal dideoxynucleotide, thus producing a series of strands that reflect the correct sequence.

During analysis, the reaction products can be combined and subjected to electrophoresis with a special medium that separates strands differing by only a single nucleotide in length. Electrophoresis arrays the fragments in order of size, and only the fragments carrying the labeled nucleotides will be readable (figure 10.7).

Automatic sequencing machines scan the electrophoresis pattern and record the DNA sequence by detecting the color of each band passing by a sensing device. In this way, the DNA sequence of the complementary strand is found directly and that of the template strand can easily be inferred. Most sequencers have computers that read the sequence and display it as a graphic figure. This method of sequencing is remarkably accurate, with only about one mistake in every 1,000 bases. All of these steps have been automated so that most researchers need only insert a sample into a machine and wait for the printout of the results. This sort of automation was absolutely necessary for sequencing the entire genomes of humans, mice, and other organisms (see Insight 10.1).

Genome sequencing projects have been highly successful. As of 2009, the genome sequences of several thousand viruses, prokaryotes, and eukaryotes have been published. The eukaryotic genomes include human, mouse, a nematode worm, yeast, fruit fly, cats, dogs, cows, and corn. One of the remarkable discoveries in this huge enterprise has been how similar the genomes of relatively unrelated organisms are. Humans share around 80% of their DNA codes with mice, about 50% with corn, and even 30% with the roundworm *Caenorhabditis elegans*.

Although sequencing provides the ultimate genetic map, it does not automatically identify the exact genes and alleles. Analysis and storage of this massive amount of new data require specialized computers. Because the data consist of information of both a biological and mathematical content, whole new disciplines have grown up around managing these data: **genomics** (see Take Note: vocabulary of the New "-Omics") and **bioinformatics.** The job of

(1) Original DNA to be sequenced.

(2) DNA is denatured to produce single template strand.

(3) Strand is labeled with specific primer molecule.

(4) Template strands, DNA polymerase, and all four nucleotides (ATP, CTP, GTP, TTP) are added to each of four tubes. Each tube also contains a small amount of a single type (A or C or T or G) of dideoxy nucleotide to stop the chain lengthening reaction when it is incorporated into the growing DNA strand. The dideoxy nucleotides each have a unique fluorescent color which allows them to be differentiated.

(5) During replication each new strand is terminated (↓) at the point of addition of a dd nucleotide.

(6) Left: A schematic view illustrating how finished fragments will end with a labeled dideoxy nucleotide. Right: All four reaction tubes are mixed and electrophoresed in one well, producing the correct sequence of bases from bottom to top.

(7) Sequencing display of a short segment of human DNA. The computer substitutes black for G for ease in reading.

Figure 10.7 **The basic techniques in DNA sequencing.**
The sequencing of a 9 base strand and an example of an automated read-out (7).

genomics and bioinformatics is to analyze and classify DNA, determine protein sequences, and ultimately understand the functions of genes. This fact is likely to bring another golden era in biology and microbiology, that of unraveling the complexity of relationships of genes and gene regulation. In time, it will provide a complete understanding of such phenomena as normal cell function, disease, development, aging, and many other issues. In addition, it will allow us to characterize the exact genetic mechanisms behind pathogens and to develop new treatments against them.

One of the fastest growing areas of the new "-omics" is designed to survey the scope of organisms that live in various habitats. This science is called environmental genomics, or **metagenomics.** In the past, it has been difficult to study complex communities where millions of organisms coexist, and it has not been possible to isolate and culture most of them. Metagenomics provides microbiologists a window through which to view the diversity of such habitats and makes it possible to pull out the microbial genomes, study them, and begin to understand their interactions. We already had an introduction to this technique in the chapter 1 case file on the large-scale genetic sampling of the oceans. Other research efforts on dozens of habitats, including the human body, are under way.

TAKE NOTE: VOCABULARY OF THE NEW "-OMICS"

The ability to obtain the entire genetic sequences of organisms has spawned new terms that refer to the "total picture" of some aspect of a cell or organism.

genomics The systematic study of an organism's genes and their functions.

proteomics The study of an organism's complement of proteins (its "proteome") and functions mediated by the proteins.

metagenomics (also called "community genomics") The study of all the genomes in a particular ecological niche, as opposed to individual genomes from single species.

metabolomics The study of the complete complement of small chemicals present in a cell at any given time. Provides a snapshot of the physiological state of the cell and the end products of its metabolism.

INSIGHT 10.1

OK, the Genome's Sequenced—What's Next?

In early 2001, a news conference was held to announce that the human genome had been sequenced. The 3.1 billion base pairs that make up the DNA found in (nearly) every human cell had been identified and put in the proper order. Champagne corks popped, balloons fell, bands played, reporters reported on the significance of the occasion, and . . . nothing else seemed to change. What, you may ask, has the Human Genome Project done for me lately? Here, for your perusal, are a few FAQs.

Q: Who sequenced the genome?

A: Francis Collins and his team from the Human Genome Project (HGP), and Craig Venter and his company Celera Genomics. Venter developed a new, more powerful method of DNA sequencing that competed with the HGP. A compromise was finally reached whereby both groups took credit for sequencing the genome.

Q: How big was this project?

A: The Human Genome Project was first discussed in the mid-1980s and got under way in 1990. Although the project was to have taken at least 15 years, advances in technology led to its being completed in just over 10. The 3.1 billion base pairs of DNA code for only about 25,000 genes, not the 100,000 or so that the genome was thought to contain only a few years ago.

Q: Will I ever see any benefits from this project?

A: Absolutely. Sequencing the genome was only the first step. Knowing what proteins are produced in the body, what they do, and how they interact with one another is essential to understanding the workings of the human body. By knowing the genetic signatures of different diseases, we will be able to design extraordinarily specific diagnostic tests that detect the thousands of genetic diseases and their many forms. For example, the cancer drugs Iressa and Erbitux are so specific in their actions that they are only prescribed after a genetic test of the cancer has found it to be a suitable target. In this way, more sensitive diagnostic tests are leading to more effective treatments with fewer side effects.

Q: What's next?

A: While everyone was still digesting the human genome (being the equivalent of a million-page book, it's a long read), the genomes of over 5,000 organisms and viruses have been sequenced. Some astounding results show that humans share nearly 99% of their genome with chimpanzees, 80% with mice, and even 60% with rice plants.

Studying genomes of microbes has also given us insight into evolutionary origins and relatedness of pathogens that infect humans such as HIV and the agent of syphilis.

Most recently, researchers have examined DNA samples from around the world and found that the human genome varies only about 0.1% among individuals. But even this 0.1% amounts to at least 1 million different haplotypes (a block of DNA inherited as a unit). Studying "hap maps" promises to give new insights into genetic diseases and evolution.

The search for novel genes in other species is growing rapidly. The Genomes to Life Project has already sequenced several hundred microbial genomes in the hope of identifying genes that could be of use to humans. For more information, go to http://www.ornl.gov/hgmis/.

Once the sequence of bases in an organism's genome is known, what are the next logical steps in deciphering the map? *Answer available at http://www.mhhe.com/talaro8*

Human Genome Project

Pharmaceuticals
Agriculture
Forensics
Anthropology
Bioremediation
Biofuels
Industrial Processes

The project was the first major genomic and bioinformatics endeavor. It laid the groundwork for hundreds of new projects in applied genetics.

Cartoon by John Chase. Reprinted by permission.

"We finished the genome map, now we can't figure out how to fold it!"

Polymerase Chain Reaction: A Molecular Copy Machine for DNA

Some of the techniques used to analyze DNA and RNA are limited by the small amounts of test nucleic acid available. This problem was largely solved by the invention of a simple, versatile way to amplify DNA called the **polymerase chain reaction (PCR).** This technique rapidly increases the amount of DNA in a sample without the need for making cultures or carrying out complex purification

techniques. It is so sensitive that it holds the potential to detect cancer from a single cell or to diagnose an infection from a single gene copy. It is comparable to being able to pluck a single DNA "needle" out of a "haystack" of other molecules and make unlimited copies of the DNA. The rapid rate of PCR makes it possible to replicate a target DNA from a few copies to billions of copies in a few hours. It can amplify DNA fragments that consist of a few base pairs to whole genomes containing several million base pairs.

To understand the idea behind PCR, it will be instructive to review figure 9.5 and figure 9.6, which models DNA synthesis as it occurs naturally in cells. The PCR method uses essentially the same events, with the opening up of the double strand, using the exposed strands as templates, the addition of primers, and the actions of a DNA polymerase.

Initiating the reaction requires a few specialized ingredients (**figure 10.8**). The **primers** are synthetic oligonucleotides (short DNA strands) of a known sequence of 15 to 30 bases that serve as landmarks to indicate where DNA amplification will begin. These take the place of RNA primers that would normally be provided by the cell. Depending upon the purposes and what is known about the DNA being replicated, the primers can be random, attaching to any sequence they may fit, or they may be highly specific and chosen to amplify a known gene. To keep the DNA strands separated, processing must be carried out at a relatively high temperature. This necessitates the use of special **DNA polymerases** isolated from thermophilic bacteria. Examples of these unique enzymes are Taq polymerase obtained from *Thermus aquaticus* and Vent polymerase from *Thermococcus litoralis*. Enzymes isolated from these thermophilic organisms remain active at the elevated temperatures used in PCR. Another useful component of PCR is a machine called a thermal cycler that automatically initiates the cyclic temperature changes.

The PCR technique operates by repetitive cycling of three basic steps: denaturation, priming, and extension.

1. **Denaturation.** The first step involves heating target DNA to 94°C to separate it into two strands. Each strand will become a template DNA, or *amplicon.*
2. **Priming.** As the reaction is cooled to between 50°C and 65°C (depending on the exact sequence of the primers), the oligonucleotide primers base-pair with specific sites on the amplicons.
3. **Extension.** In the third phase, which proceeds at 72°C, DNA polymerase and nucleotides are added. Beginning at the free end of the primers on both strands, the polymerases extend the molecule by adding appropriate nucleotides and produce two complete strands of DNA.

It is through cyclic repetition of these steps that DNA becomes amplified. When the DNAs formed in the first cycle are denatured, they become amplicons to be primed and extended in the second cycle. Each subsequent cycle converts the new DNAs to amplicons and doubles the number of copies. The number of cycles required to produce a million molecules is 20, but the process is usually carried out to 30 or 40 cycles. One significant advantage of this technique has been its natural adaptability to automation. A PCR machine can perform 20 cycles on nearly 100 samples in 2 or 3 hours.

DNA amplified by the PCR can be analyzed by any of the techniques discussed earlier. PCR can also be adapted to analyze RNA by initially converting an RNA sample to DNA with reverse transcriptase. This cDNA can then be amplified by PCR in the usual manner. It is by such means that ribosomal RNA and messenger RNA are readied for sequencing, and RNA viruses are sequenced and identified. The polymerase chain reaction has found prominence as a powerful workhorse of molecular biology, medicine, and biotechnology. It often plays an essential role in gene mapping, the study of genetic defects and cancer, forensics, taxonomy, and evolutionary studies.

For all of its advantages, PCR has some problems. One concern is the introduction and amplification of nontarget DNA from the surrounding environment, such as a skin cell from the technician carrying out the PCR reaction, rather than material from the sample DNA that was supposed to be amplified. Such contamination can be minimized by using equipment and rooms dedicated for DNA analysis and maintained with the utmost degree of cleanliness. Problems with contaminants can also be reduced by using gene-specific primers and treating samples with special enzymes that can degrade the contaminating DNA before it is amplified.

Check & Assess Section 10.1

✔ The genetic revolution has produced a wide variety of industrial technologies that translate and radically alter the blueprints of life.

✔ Increased public understanding is essential for developing appropriate guidelines for responsible use of these revolutionary techniques.

✔ Genetic engineering utilizes a wide range of methods that physically manipulate DNA for purposes of visualization, sequencing, hybridizing, and identifying specific sequences.

✔ The tools of genetic engineering include specialized enzymes, gel electrophoresis, DNA sequencing machines, and gene probes.

✔ The Human Genome Project and other genome sequencing projects have revolutionized our understanding of organisms and led to two new biological disciplines, genomic and bioinformatics.

✔ The polymerase chain reaction (PCR) technique amplifies small amounts of DNA into larger quantities for further analysis.

1. Define *genetic engineering* and *biotechnology*, and summarize the important purposes of these fields. Review the use of the terms *genome, chromosome, gene, DNA,* and *RNA* from chapter 9.
2. Describe the processes involved in denaturing and renaturing of DNA, and explain what is useful about this procedure.
3. Define *restriction endonuclease* and explain what it does.
4. Define *palindrome* and draw three different palindromic sequences.
5. Explain how electrophoresis works and the general way that DNA is sized. Estimate the size of the DNA fragment in base pairs in the first lane of the gel in figure 10.4*b*.
6. How would you make a copy of DNA from an mRNA transcript and what are this process and molecule called?
7. Briefly describe the functions of DNA synthesizers and sequencers.
8. Describe the effects of temperature change in PCR.
9. Exactly what are the functions of the primer and Taq polymerase?
10. Explain how a Southern blot would be combined with PCR and cDNA in an analysis.

(a) In cycle 1, the DNA to be amplified is denatured, primed, and replicated by a polymerase that can function at high temperature. The two resulting strands then serve as templates for a second cycle of denaturation, priming, and synthesis.

Cycle 1

DNA Sample — Segment of DNA to be amplified

Heat to 94°C

Denaturation
Strands separate.

50°C to 65°C

Priming
Oligonucleotide primers attach at ends of strands to promote replication of amplicons.

Amplicons Primer Primer

72°C

Extension
Heat-stable DNA polymerase synthesizes complementary strand.

Polymerase

Cycle 1

Cycle 2

Denaturation

Completed copies New strand

New strand Original strands

Heat to 94°C

Priming

50°C–65°C

Cycle 2

Extension

72°C

4 copies

Cycles 3, 4, . . . repeat same steps, with the copies of DNA

(b) A view of the process after 6 cycles, with 64 copies of amplified DNA. Continuing this process for 20 to 40 cycles can produce millions of identical DNA molecules.

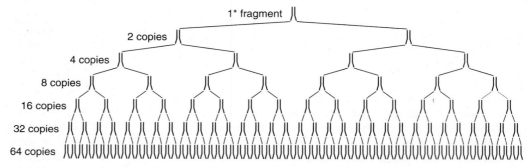

1* fragment

2 copies

4 copies

8 copies

16 copies

32 copies

64 copies

 Figure 10.8 Diagram of the polymerase chain reaction. (a) Details of the process over two cycles. **(b)** Yield after the cycle continues.

10.2 Recombinant DNA Technology: How to Imitate Nature

5. Explain what is involved in recombinant DNA technology.

6. Characterize the events in cloning, using an example.

7. List and discuss some protein products of recombinant DNA and their applications.

The primary intent of **recombinant DNA (rDNA) technology** is to deliberately remove genetic material from one organism and combine it with that of a different organism. Its origins can be traced to 1970, when microbiologists first began to duplicate the clever tricks bacteria do naturally with bits of extra DNA such as plasmids, transposons, and proviruses. As mentioned earlier, humans have been trying to artificially influence genetic transmission of traits for centuries. The discovery that bacteria can readily accept, replicate, and express foreign DNA made them powerful agents for studying the genes of other organisms in isolation. The practical applications of this work were soon realized by biotechnologists. Bacteria could be genetically engineered to mass-produce substances such as hormones, enzymes, and vaccines that were difficult to synthesize by the usual industrial methods.

Figure 10.9 provides an overview of the recombinant DNA procedure. An important objective of this technique is to form genetic **clones.** This type of cloning involves the removal of a selected gene from an animal, plant, or microorganism (the genetic donor) followed by its propagation in a different host organism. Cloning requires that the desired donor gene first be selected, excised by restriction endonucleases, and isolated. The gene is next inserted into a **vector** (usually a plasmid or a virus) that will insert the DNA into a **cloning host.** The cloning host is usually a bacterium or a yeast that can replicate the gene and translate it into the protein product for which it codes. In the next section, we examine the elements of gene isolation, vectors, and cloning hosts and show how they participate in a complete recombinant DNA procedure.

Technical Aspects of Recombinant DNA and Gene Cloning

The first hurdles in cloning a target gene are to locate its exact site on the genetic donor's chromosome and to isolate it. Some of the first isolated genes came from viruses, which have extremely small and manageable genomes, and later genes were isolated from bacteria and yeasts. At first, the complexity of the human genome made locating specific genes very difficult. However, this process has been greatly improved by tools for excising, mapping, and identifying genes. Among the most common strategies for obtaining genes in an isolated state are:

1. The DNA is removed from cells and separated into fragments by endonucleases. Each fragment is then inserted into a vector and cloned. The cloned fragments undergo Southern blotting and are probed to identify desired sequences. This is a long and tedious process, because each fragment of DNA must be examined for the cloned gene. For example, if the human genome

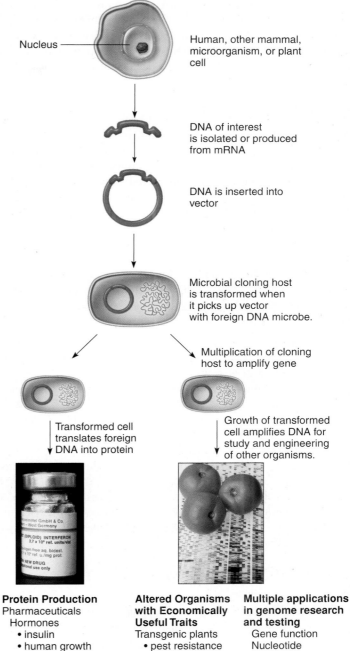

Protein Production
Pharmaceuticals
 Hormones
 • insulin
 • human growth
 hormone
 Vaccines
 • hepatitis B
Industrial
 chemicals

Altered Organisms with Economically Useful Traits
Transgenic plants
 • pest resistance
 • herbicide
 resistance
 • improved
 nutritional value

Multiple applications in genome research and testing
 Gene function
 Nucleotide
 sequencing
 Identification
 Gene libraries

Figure 10.9 **Overview of methods and applications of genetic engineering.** Practical applications of genetic engineering include the development of pharmaceuticals, genetically modified organisms, and identification techniques. See figure 10.11 for details of these steps used in recombinant DNA.

were separated into 20 kb fragments, there would be at least 150,000 clones to test. Alternatively,

2. A gene can be synthesized from isolated mRNA transcripts using reverse transcriptase (cDNA), or

3. A gene can be amplified using PCR in many cases.

Although gene cloning and isolation can be very laborious, a fortunate outcome is that, once isolated, genes can be maintained in a cloning host and vector just like a microbial pure culture. Genomic libraries, which contain fragments of DNA representing all the genes in a particular organism, have been created for thousands of life forms, from bacteria to humans.

Characteristics of Cloning Vectors

A good recombinant vector has two indispensable qualities: It must be capable of carrying a significant piece of the donor DNA, and it must be readily accepted by the cloning host. Plasmids are excellent vectors because they are small, well characterized, easy to manipulate, and can be transferred into appropriate host cells through transformation. Bacteriophages also serve well because they have the natural ability to inject DNA into bacterial hosts through transduction. A modified phage vector, the *Charon*[4] phage, has had sections of its genome deleted so it can carry a fairly large segment of foreign DNA.

Today, thousands of unique cloning vectors are available commercially. Although every vector has characteristics that make it ideal for a specific project, all vectors can be thought of as having three important attributes to consider **(figure 10.10):**

1. An origin of replication (ORI) is needed somewhere on the vector so that it will be replicated by the DNA polymerase of the cloning host.
2. The vector must accept DNA of the desired size. Early plasmids were limited to an insert size of less than 10 kb of DNA, far too small for most eukaryotic genes, with their sizable introns. Newer vectors called cosmids can hold 45 kb, whereas complex *bacterial artificial chromosomes* (BACs) and *yeast artificial chromosomes* (YACs) can hold as much as 300 kb and 1,000 kb, respectively.
3. Vectors typically contain a gene that confers drug resistance to their cloning host. In this way, cells can be grown on drug-containing media, and only those cells that harbor a plasmid will be selected for growth.

Characteristics of Cloning Hosts

The best cloning hosts possess several key characteristics **(table 10.1).** The traditional cloning host—and the one still used in the majority of experiments—is *Escherichia coli.* Because this bacterium was the original recombinant host, the protocols using it are well established, relatively easy, and reliable. Hundreds of specialized cloning vectors have been developed for it. The main disadvantages with this species are that the splicing of mRNA and the modification of proteins that would normally occur in the eukaryotic endoplasmic reticulum and Golgi apparatus are unavailable in this prokaryotic cloning host. One alternative host for certain industrial processes and research is the yeast *Saccharomyces cerevisiae,* which, being eukaryotic, already possesses mechanisms for processing and modifying eukaryotic gene products. Certain techniques may also employ different bacteria (*Bacillus subtilis*), animal cell culture, and even live animals and plants to serve as cloning hosts. In our coverage, we present the recombinant process as it is performed in bacteria and yeasts.

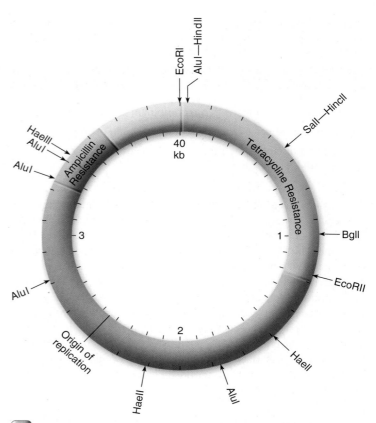

Figure 10.10 **Partial map of the pBR322 plasmid of** *Escherichia coli.* Arrows delineate sites cleaved by various restriction endonucleases. Numbers indicate the size of regions in kilobases (kb). Other important components are genes for ampicillin and tetracycline resistance and the origin of replication.

TABLE 10.1	Desirable Features in a Microbial Cloning Host
Rapid turnover, fast growth rate	
Can be grown in large quantities using ordinary culture methods	
Nonpathogenic	
Genome that is well delineated (mapped)	
Capable of accepting plasmid or bacteriophage vectors	
Maintains foreign genes through multiple generations	
Will secrete a high yield of proteins from expressed foreign genes	

Construction of a Recombinant, Insertion into a Cloning Host, and Genetic Expression

This section emphasizes the main steps in recombinant DNA technology needed to produce a protein. A drug called alpha-2a interferon (Roferon-A) will serve as an example, but it should be remembered that the steps are essentially the same regardless of the protein being produced. This form of interferon is used to treat cancers such as hairy-cell leukemia and Kaposi sarcoma in AIDS patients (see chapter 14). The human alpha interferon gene is a DNA molecule of approximately 500 bp that codes for a polypeptide of 166 amino acids. It was

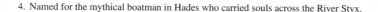

4. Named for the mythical boatman in Hades who carried souls across the River Styx.

originally isolated and identified from human blood cells and prepared from processed mRNA transcripts that are free of introns. This step is necessary because the bacterial cloning host has none of the machinery needed to excise this nontranslated part of a gene.

The first step in cloning is to prepare the isolated interferon gene for splicing into an *E. coli* plasmid (**figure 10.11**). One way this is accomplished is to digest both the gene and the plasmid with the same restriction enzyme, resulting in complementary sticky ends on both the vector and insert DNA. In the presence of the endonuclease, the plasmid's circular molecule is nicked open, and its sticky ends are exposed. When the gene and plasmid are placed together, their free ends base-pair, and a ligase makes the final covalent seals. The resultant gene and plasmid combination is called a **recombinant.**

Process Figure 10.11 **Using recombinant DNA for gene cloning.**

Following this procedure, the recombinant is introduced by transformation into the cloning host, a special laboratory strain of *E. coli* that lacks any extra plasmids that could complicate the expression of the gene. Because the recombinant plasmid enters only some of the cloning host cells, it is necessary to locate these recombinant clones. Cultures are plated out on medium containing ampicillin, and only those clones that carry the plasmid with ampicillin resistance can form colonies (**figure 10.12**). These recombinant colonies are selected from the plates and cultured. As the cells multiply, the plasmid is replicated along with the cell's chromosome. In a few hours of growth, there can be billions of cells, each containing the interferon gene. Once the gene has been successfully cloned and tested, this step does not have to be repeated; the recombinant strain can be maintained in culture indefinitely.

The bacteria's ability to express the eukaryotic gene is ensured because the plasmid has been modified with the necessary transcription and translation recognition sequences. As the *E. coli* culture grows, it transcribes and translates the interferon gene, synthesizes the peptide, and secretes it into the growth medium. Final processing to excise a terminal amino acid from the peptide yields the interferon product in a relatively pure form (see figure 10.11). The scale of this procedure can range from test tube size to gigantic industrial vats that can manufacture thousands of gallons of product.

Although the process we have presented here produces interferon, some variation of it can be used to mass-produce a variety of hormones, enzymes, and agricultural products such as pesticides, covered next.

Protein Products of Recombinant DNA Technology

Recombinant DNA technology is used by pharmaceutical companies to manufacture medications that cannot be manufactured by any other means. Diseases such as diabetes and dwarfism, caused by the lack of an essential hormone, are treated by replacing the missing hormone. Insulin of animal origin was once the only form available to treat diabetes, even though such animal products can cause allergic reactions in sensitive individuals.

In contrast, dwarfism cannot be treated with animal growth hormones, so originally the only source of human growth hormone (HGH) was the pituitaries of cadavers. At one time, not enough HGH was available to treat the thousands of children in need.

Another serious problem with using natural human products is the potential for infection. For example, infectious agents such as the prion responsible for Creutzfeld-Jakob disease, described in chapter 6, can be transmitted in this manner. Similarly, clotting factor VIII, a protein needed for blood to clot properly, is missing in persons suffering from hemophilia A. In the early 1980s, a large percentage of the patients receiving plasma-derived factor VIII contracted HIV infections as a result of being exposed to the virus through the donated plasma.

Recombinant technology changed the outcome of these and many other conditions by enabling large-scale manufacture of hormones and enzymes derived from human genes. Recombinant human insulin can now be prescribed for diabetics, and recombinant HGH can now be administered to children with dwarfism and Turner syndrome. HGH is also used to prevent the wasting syndrome that occurs in AIDS and cancer patients. In all of these applications, recombinant DNA technology has led to both a safer product and one that can be manufactured in quantities previously unfathomable. Other protein-based hormones, enzymes, and vaccines produced through recombinant DNA are summarized in **table 10.2.**

Nucleic acid products also have a number of medical applications. A new development in vaccine formulation involves using microbial DNA as a stimulus for the immune system. So far, animal tests using DNA vaccines for AIDS and influenza indicate that this may be a breakthrough in vaccine design. Recombinant DNA could also be used to produce DNA-based drugs for the types of gene and antisense therapy discussed in section 10.4.

Culture of cloning host after incubation with recombinant plasmid

Bacteria lack plasmid

Bacteria with recombinant plasmid

Bacteria carrying plasmid

(1) (2)

Ampicillin-resistance gene

Regular nonselective medium with two types of colonies

Selective medium with ampicillin

Pure culture of bacteria containing cloned gene

Figure 10.12 One method for screening clones of bacteria that have been transformed with the donor gene. (a) Plating the culture on nonselective medium will not separate the transformed cells from normal cells, which lack the plasmid. (b) Plating on selective medium containing ampicillin will permit only cells containing the plasmid to multiply. Colonies growing on this medium that carry the cloned gene can be used to make a culture for gene libraries, industrial production, and other processes. (Plasmids are shown disproportionate to size of cell.)

TABLE 10.2	Examples of Protein Products from Recombinant DNA Technology

Immune Treatments

Interferons—peptides used to treat some types of cancer, multiple sclerosis, and viral infections such as hepatitis and genital warts
Interleukins—types of cytokines that regulate the immune function of white blood cells; used in cancer treatment
Orthoclone—an immune suppressant in transplant patients
Macrophage-colony-stimulating factor—used to stimulate bone marrow activity after bone marrow grafts
Tumor necrosis factor (TNF)—used to treat cancer
Granulocyte-colony-stimulating factor (Neupogen)—developed for treating cancer patients suffering from low neutrophil counts

Hormones

Insulin—a small protein that is used to treat some forms of diabetes
Erythropoietin (EPO)—a peptide that stimulates bone marrow used to treat some forms of anemia
Tissue plasminogen activator (tPA)—can dissolve potentially dangerous blood clots
Hemoglobin A—form of artificial blood to be used in place of real blood for transfusions
Factor VIII—needed as replacement blood-clotting factor in type A hemophilia
Human growth hormone (HGH)—stimulates growth in children with dwarfism; prevents wasting syndrome

Enzymes

rH DNase (pulmozyme)—a treatment that can break down the thick lung secretions of cystic fibrosis
Antitrypsin—replacement therapy to benefit emphysema patients
PEG-SOD—a form of superoxide dismutase that minimizes damage to brain tissue after severe trauma

Vaccines

Vaccines for hepatitis A and B, *Haemophilus influenzae* meningitis, influenza, and rotavirus
Experimental malaria and AIDS vaccines based on recombinant surface antigens

Miscellaneous

Bovine growth hormone or bovine somatotropin (BST)—given to cows to increase milk production
Apolipoprotein—to deter the development of fatty deposits in the arteries and to prevent strokes and heart attacks
Spider silk—a light, tough fabric for parachutes and bulletproof vests

Check & Assess Section 10.2

✔ Recombinant DNA techniques combine DNA from different sources to create microorganism "factories" that produce biochemical substances on an industrial scale.

✔ Cloning is the process by which genes are removed from the original host and transferred into a cloning host by means of cloning vectors for duplication.

✔ Plasmids, bacteriophages, and cosmids are types of cloning vectors used to transfer recombinant DNA into a cloning host.

✔ Cloning hosts are simply organisms that readily accept recombinant DNA, grow easily, and synthesize large quantities of specific gene product.

✔ Bioengineered hormones, enzymes, and vaccines are safer and more effective than similar substances derived from animals. Recombinant DNA drugs and vaccines are useful alternatives to traditional treatments for disease.

11. What characteristics of plasmids and bacteriophages make them good cloning vectors?
12. Name several types of vectors, and list the types of genes that they can contain.
13. Describe the principles behind recombinant DNA technology.
14. What characteristics of bacteria make them good cloning hosts?
15. Outline the main steps in cloning a gene, including how this gene is used.
16. What is one way to determine whether a bacterial culture has received a recombinant plasmid?
17. Characterize several products that have resulted from recombinant DNA technology.

10.3 Genetically Modified Organisms

Expected Learning Outcomes

8. Define what is meant by the term *transgenic* or *genetically modified organisms*.

9. Describe the uses of genetically modified bacteria and viruses.

10. Discuss genetic modifications in plants and animals, explaining how they are carried out and giving examples of applications.

Recombinant organisms produced through the introduction of foreign genes are called **genetically modified organisms (GMOs)** or

INSIGHT 10.2

A Moment to Think

"When we seed millions of acres of land with these plants, what happens to foraging birds, to insects, to microbes, to the other animals, when they come in contact with and digest plants that are producing materials ranging from plastics to vaccines to pharmaceutical products?"

Jeremy Rifkin, Foundation on Economic Trends

"Our interest is in selling as much of it (GMO food) as possible. Assuring its safety is the FDA's (Food and Drug Administration) job."

Phil Angell, the director of corporate communications at Monsanto, the largest biotechnology firm selling engineered seeds

"There is no evidence to indicate that biotechnology is dangerous. After all, mother nature has been doing this kind of thing for God knows how long."

Nobel Prize laureate Norman E. Borlaug

There is no denying that many aspects of biotechnology are highly controversial and emotionally charged, but few would deny that this technology is here to stay. There are always several sides to every issue, and the best we can hope for is that the people developing genetic technology are as well informed as possible. Those who will determine the limits of genetic technology include not only scientists and politicians but voters and consumers as well. History is rife with both knee-jerk rejections to new technologies that have later proven to be invaluable (vaccinations) and complacency while harmful products were foisted onto an unsuspecting public (for example, Fen-Phen, Vioxx).

Ethical choices can be properly made only from a standpoint of intellectual awareness; and in this era of advertising, polling, and focus groups, people with a stake in genetic technology know that the most effective way to drum up support (both for or against) is *not* by carefully educating the public as to the uses and limits of our newfound powers but, rather, by publicizing exaggerated claims of frightening scenarios. Enlightenment is often a casualty of these advertising campaigns, and you as a student, voter, and potential consumer of bioengineered products need to realize that the truth lies somewhere between the photograph seen here and the blissful utopia often portrayed by some proponents of biotechnology.

A demonstrator at a biotechnology conference protests the development and sale of genetically modified foods.

Throughout this chapter, a common theme has been that biotechnology, genetically modified organisms, and gene-based therapies are simply new ways of achieving long-held goals. This is of course a vast understatement. Society has never before had the ability to do even a fraction of the things we can do today, and at some point we must decide where we should stop or if we need to stop at all. Where do we as individuals and as a society draw the line? The answer will differ for each person, based on his or her unique experiences and personal philosophy, but should certainly include a factual knowledge of the probable benefits and possible risks. We live in serious times, and these are choices requiring serious thought. Nothing worthwhile is ever easy.

Humans have the power to change the biology of living things. Can you name some of the major limitations to this potential? *Answer available at http://www.mhhe.com/talaro8*

transgenic organisms. Although foreign genes have been inserted into a variety of life forms including microbes, plants, and animals, the techniques used to create them owe their existence to the early work that took place in bacterial systems.

Recombinant Microbes: Modified Bacteria and Viruses

One of the first practical applications of recombinant DNA in agriculture was to create a genetically altered strain of the bacterium *Pseudomonas syringae.* The wild strain ordinarily contains an ice formation gene that promotes ice or frost formation on moist plant surfaces. Genetic alteration of the frost gene using recombinant

plasmids created a different strain that could prevent ice crystals from forming. A commercial product called Frostban has been successfully applied to stop frost damage in strawberry and potato crops. A strain of *Pseudomonas fluorescens* has been engineered with the gene from a bacterium (*Bacillus thuringiensis*) that codes for an insecticide. These recombinant bacteria are released to colonize plant roots and help destroy invading insects. All releases of recombinant microbes must be approved by the US Environmental Protection Agency and are closely monitored.

Although a number of recombinant proteins are produced by transformed bacterial hosts, many of the enzymes, hormones, and antibodies being used in drug therapy are currently being manufactured using mammalian cell cultures as the cloning and expression

hosts. One of the primary advantages to this alternative procedure is that these cell cultures can modify the proteins (adding carbohydrates, for example) so that they are biologically more active. Some forms of reproductive hormones, human growth hormone, and interferon are products of cell culture.

Another very significant bioengineering interest has been to create microbes to bioremediate disturbed environments. Biotechnologists have already developed and tested several types of bacteria that clean up oil spills and degrade pesticides and toxic substances.

Although primarily known for its ability to cause colds, the adenovirus has been bioengineered to attack and destroy cancer cells. These *oncolytic adenoviruses* have been altered so that their host range consists exclusively of cells that produce cancer-specific proteins. By simply infecting a cancerous cell and replicating, the engineered virus will eventually lyse and kill the cell. While not yet approved for use in humans, this novel use of viral vectors speaks to the potential of genetically engineered viruses. This technology is currently used to engineer viruses to deliver genes into animals and humans (see figure 10.15).

In mid-2007, a new milestone was reached when Craig Venter announced he had artificially converted one bacterial species into another—by completely replacing the DNA of the original bacterium with the DNA from a donor bacterium. While the scale of this experiment is small—a single-celled organism—the implications are large. The growing power of biotechnology has led to a deeper examination of the **bioethics**[5] involved in wielding the genetics of organisms (**Insight 10.2;** also see Insight 27.2).

Recombination in Multicellular Organisms

While the term *genetically modified organism* applies to any living being whose genome has been deliberately altered, it is often used in reference to plants and animals that possess genes that were not acquired through traditional breeding. The basic techniques covered in this chapter have given scientists a limited ability to custom build an organism that expresses a certain trait, in a sense accelerating the pace of evolution. Because of the universal nature of the genetic code, however, scientists can go nature one step better and combine genes from more than a single species, creating plants or animals that would never have existed otherwise. Examples of genetically modified organisms include plants with added nutritional value, animals that excrete clot-dissolving proteins in their milk, and humans who have gained the ability to produce an enzyme crucial to the immune response.

(a) The large plasmid (Ti) of this bacterium can be used as a cloning vector for foreign genes that code for herbicide or disease resistance.

(b) The recombinant plasmids are taken up by the *Agrobacterium* cells, which multiply and copy the foreign gene.

(c) Genetically engineered *Agrobacterium* is inoculated into a culture of target plant cells and infects the cells.

(d) Bacterium fuses with the plant cell wall and the Ti plasmid enters. The T-DNA carries the herbicide gene into the plant chromosome. Mature plants can be grown from single cells, and these transgenic plants will express the new gene.

(e) Because the gene will be part of the plant's genome, it will be transmitted to offspring in seeds.

Figure 10.13 **Bioengineering of plants.** Most techniques employ a natural tumor-producing bacterium called *Agrobacterium tumefaciens.* The example here shows the insertion of a gene to create a herbicide-resistant plant.

Transgenic Plants

Engineering plants to produce new traits often revolves around *Agrobacterium tumefaciens,* a plant pathogen particularly adept at transferring small amounts of DNA to plant cells. These bacteria live in soil and can invade injured plant tissues. Inside the wounded tissue, the bacteria adhere to plant cells and introduce a large plasmid termed **Ti** (tumor-inducing). This plasmid inserts into the genome of the infected plant cell and transforms it (**figure 10.13**). In time, this causes development of a tumor called *crown gall disease,* a mass of undifferentiated tissue on the stem.

The primary factor that induces this transformation is a discrete region of the Ti plasmid called T-DNA. This series of genes is integrated into a plant cell's chromosome, where it directs the synthesis of nutrients to feed the bacteria and hormones to stimulate plant

5. A field that relates biological issues to human conduct and moral judgment.

TABLE 10.3	Examples of Engineered (Transgenic) Plants	
Plant	**Trait/s**	**Results**
Cherry tomatoes	High levels of anthocyanin, an antioxidant	Tomatoes were engineered with pigment gene from snapdragons, producing five times the purple pigment anthocyanin and greatly increasing the nutrition value (see figure).
Endless summer tomato	Longer shelf life, better flavor	Genes were manipulated to slow ripening and increase acidity.
Wheat	Resistance to head blight	Genes added to break down the cell wall of the fungus *Fusarium* protect wheat plants from being attacked (see figure).
Wheat	Improved nutrition	Engineered with genes to increase zinc and iron levels in the plant.
Golden rice	Improved nutrition	Genes added to produce beta-carotene, a precursor to vitamin A; aim is to prevent deficiencies in developing countries that lead to blindness.
Soybeans, corn cotton	Herbicide resistance	Plants were engineered to withstand chemical herbicides.
	Insect resistance	Genes for a biopesticide called Bt were added to kill insects that feed on the plants.

These purple tomatoes contain high levels of pigment that has numerous health benefits.

Researcher holds a genetically engineered wheat plant being tested for its resistance to a fungal infection.

growth. Even after the bacteria in a tumor are dead, the plasmid genes remain in the cell nucleus and the tumor continues to grow.

This plasmid is an effective vector for inserting foreign genes into plant genomes. The procedure involves removing the Ti plasmid, inserting a specially selected gene such as herbicide resistance into it, and returning the plasmid to *Agrobacterium*. Infection of the plant by the recombinant bacteria automatically transfers the plasmid with the foreign gene into the plant cells. Transformed cells can be grown into mature plants that express the donor gene and transmit it to their offspring. The insertion of genes with a Ti plasmid works primarily to engineer important dicot plants such as potatoes, tomatoes, cotton, and grapes.

For plants that cannot be transformed with *Agrobacterium*, seed embryos have been successfully implanted with additional genes using gene guns—small "shotguns" that shoot the genes into plant embryos with tiny gold bullets.

Agricultural officials had approved more than 800 genetically modified plants for field testing through the year 2009. Over 200 million acres of these plants were grown in the United States alone in 2009, with herbicide- and insecticide-resistant soybeans,

corn, cotton, and canola making up the vast majority. The principal concern about such plants is the potential for genetic recombination with nearby wild plants, leading to the possible creation of "super-weeds" that are resistant to most herbicides and disease. Because this could alter the ecology of natural environments, the US Department of Agriculture is carefully regulating all releases of transgenic plants. See **table 10.3** for examples of plants that have been engineered.

Transgenic Animals: Engineering Embryos

Animals such as mice, pigs, sheep, and cows are prominent candidates for genetic engineering. Thousands of animal strains now exist for the purposes of agriculture, industry, and research into genetic diseases.

The most effective way to insert genes into animals is to use a virus to transfect the fertilized egg or early embryo (germline engineering). Foreign genes can also be delivered into a fertilized egg by pulsing the egg with high voltage or by injection (**figure 10.14**). The success rate does not have to be high, because once created, a transgenic animal will pass the genes on to its offspring. The

Several embryos recovered from female

Pronuclei

Embryos transferred to a depression slide containing culture medium

Culture medium

Oil

As embryo is held in place, DNA is injected into one pronucleus.

Holding pipette

Pronucleus

DNA to be injected

Injection pipette

Several injected embryos are placed into oviduct of receptive female.

Figure 10.14 **Steps in the creation of a transgenic animal.** Hundreds of mice and other mammals with genetic modifications have been produced this way.

dominant animals in the genetic engineer's laboratory are transgenic forms of mice. Mice are so easy to engineer that in some ways they have become as prominent in genetic engineering as *E. coli.* In one pioneering experiment, fertilized mouse embryos were injected with a gene from a jellyfish that produces a green fluorescent protein. The cells of these mice fluoresce in ultraviolet radiation. One serious use has been to study the developmental stages in mice (see table 10.4).

This advantage has led to the design of animal models to study human genetic diseases and to use these natural systems to test new genetic therapies before they are used in humans. Animals such as sheep or goats can also be engineered to become "factories" capable of manufacturing proteins and excreting them in their milk or semen, a process often referred to (when done to produce medically useful proteins) as "pharming."

One benefit of transgenic animals is that they will automatically transcribe and translate eukaryotic genes and accurately carry out posttranslational modification of the protein. By ensuring that the protein product is produced in the milk or semen of animals, recovery and purification are made that much simpler. Alpha-1 antitrypsin (AAT), produced in sheep, is used to treat an inherited form of emphysema. The sheep produce AAT in their milk at concentrations of 35 grams per liter, an enormous yield and more than enough to meet the needs of the 20,000 people a year who suffer from this debilitating disease. See **table 10.4** to examine some other examples and applications of transgenic animals.

Check & Assess Section 10.3

✔ When foreign genes are introduced into organisms, the organisms are called genetically modified or transgenic.

✔ Genetically modified microorganisms have been used to improve agricultural yields and reduce pest damage. Bacteria serve as agents of bioremediation, and viruses serve as vectors for introducing genes.

✔ Transgenic plants receive genes from a recombinant bacterium. Numerous species have been modified to improve nutrition and resistance to insects and herbicides.

✔ Transgenic animals are genetically designed to model genetic therapies, improve meat yield, or synthesize specific biological products.

✔ Transformed bacterial and mammalian cells are used to produce proteins of economic and pharmaceutical importance.

✔ Biotechnology promises not only improved quality of life and enhanced economic opportunity but also brings serious bioethical dilemmas.

18. Describe several instances in which genetically modified bacteria and viruses are used in medicine or industry.

19. Explain how many plants are genetically modified.

20. Why must animals be modified in the embryo stage?

21. Describe any moral, ethical, or biological problems associated with eating tomatoes from an engineered plant or pork from a transgenic pig.

22. What are the moral considerations of using transgenic animals to manufacture various human products?

10.4 Genetic Treatments: Introducing DNA into the Body

Expected Learning Outcomes

11. Explain how DNA technology can be used to treat genetic diseases.

12. Compare gene therapy with antisense RNA as genetic medicines.

Gene Therapy

We have known for decades that for certain diseases, the disease phenotype is due to the lack of a single specific protein. For example, type I diabetes is caused by a lack of insulin, leaving those with the disease unable to properly regulate their blood sugar. The initial treatment for this disease has been to provide diabetics with insulin isolated from a different source, in most cases the pancreas of pigs or cows. While this treatment was adequate for most diabetics, our increasingly sophisticated genetic engineering abilities made loitering around the slaughterhouse seem a decidedly low-tech solution to a high-tech problem. We've already discussed the use of

TABLE 10.4 Selected Applications for Transgenic Animals

Animal/Product	Application
Production of Pharmaceuticals	
Pig/Factors VIII and IX	Treatment of blood-clotting disorders
Sheep/Alpha antitrypsin	Replacement protein to prevent emphysema
Rabbit/Human interleukin-2	Stimulates the production of T cells to fight cancer
Cow/Lactoferrin	Iron-binding protein inhibits bacterial and viral infections.
Cow/Human albumin	Used in surgery to maintain blood volume
Goat/Tissue plasminogen activating factor	Dissolves blood clots and reduces heart damage
Goat/Antithrombin	*Atryn* was the first drug from a genetically engineered animal approved by the FDA; used to prevent blood clots.
Research/Industry/Environment	
Goat/Silk fibers	Female animals secrete silk protein in their milk (see photograph). Silk has many uses for constructing strong, lightweight material.
Pig/Pork meat	*Enviropig* has enzymes in its saliva to process excess phosphate in its feed, reducing pollution.
Mice, cats, pigs, monkeys/research animals	Engineered with the green fluorescent protein from jellyfish that causes their cells to glow green in UV light (see photograph). Animals are used to trace development in embryos and the stages of certain diseases.
Cats and dogs/Hypoallergenic	Genes that code for allergens from animal's saliva and sebaceous glands have been silenced.

Baby mice with green fluorescent protein alongside normal brown littermates.

The female offspring of these little male goats will release silk in their milk.

recombinant proteins to treat diseases. In fact, recombinant human insulin was the first genetically engineered drug to be approved for use in humans. The next logical step is to see if we can correct or repair a faulty gene in humans suffering from a fatal or debilitating disease, a process known as **gene therapy.**

The inherent potential of gene therapy would be to permanently cure the physiological dysfunction by repairing the genetic defect. There are two strategies for this therapy. In ex vivo therapy, the normal gene is cloned in vectors such as adenoviruses or retroviruses that have been modified so they do not cause disease in humans. Tissues removed from the patient are incubated with these genetically modified viruses to transfect them with the normal gene. The transfected cells are then reintroduced into the patient's body by transfusion (**figure 10.15**). In contrast, the in vivo type of therapy skips the intermediate step of incubating excised patient tissue. Instead, the naked DNA or a virus vector is directly introduced into the patient's tissues.

Experimentation with various types of gene therapy, or clinical testing, is performed on human volunteers with the particular genetic condition. Over 2,000 of these trials have been or are being carried out in the United States and other countries. Most trials target cancer, single-gene defects, and infections.

The most dramatic example of this type of treatment was seen in the very first gene therapy trial conducted in 1990. In this case, two girls suffering from adenosine deaminase (ADA) deficiency, a fatal immunologic disorder caused by a missing enzyme, received infusions of their own hematopoietic stem cells that had been transformed to contain a healthy copy of the adenosine deaminase gene. Although repeated treatments are still periodically needed, the girls and other children with ADA given the same type of therapy have shown improvement and continue to be healthy.

Unfortunately, progress on gene therapy has been bedeviled by setbacks. All human gene therapy trials were temporarily halted after two boys participating in a successful trial for X-linked severe combined immunodeficiency disease contracted a form of leukemia, most likely as a result of the viral vector integrating into a sensitive area of the chromosome. Since this event, researchers have been more selective in the types of viruses used to deliver the genes.

In 2009, medical researchers reported major successes in curing nine babies born with the devastating genetic disease termed X-1-linked SCID (severe combined deficiency disease). These children lacked a functioning immune system from birth, and the

researchers restored it by infecting their stem cells with a recombinant virus. In another trial, a team of geneticists improved the vision of several children with a genetic form of progressive blindness by transferring the gene directly into the retina.

DNA Technology as Genetic Medicine

Up to now, we have considered the use of genetic technology to replace a missing or faulty protein that is needed for normal cell function. A different problem arises when a disease results from the inappropriate expression of a protein. For example, Alzheimer disease, most viral diseases, and many cancers occur when an unwanted gene is expressed rather than when a desirable gene is missing.

In chapter 9, you read about unusual strategies used by bacteria (as well as some eukaryotes) to regulate their own genes. Small interfering RNAs and riboswitches are potentially useful in treating humans with diseases caused by the inappropriate or excessive expression of a gene. Because they stop the expression of a gene, these new therapies are often termed *gene silencing.*

1. Normal gene is isolated from healthy subject.
2. Gene is cloned.
3. Gene is inserted into retrovirus vector.
4. Bone marrow sample is taken from patient with genetic defect.
5. Stem cells from marrow are infected with retrovirus, (enlarged view)
6. Transfected cells (red dots) are infused into patient.
7. Patient is observed for expression of normal gene.

Marrow stem cell

Process Figure 10.15 Protocol for the ex vivo type of gene therapy in humans.

Antisense DNA and RNA: Targeting Messenger RNA

Note that the term **antisense** as used in molecular genetics does *not* mean "no sense." It is used to describe any nucleic acid strand with a base sequence that is complementary to the "sense" or translatable strand. For example, DNA contains a template strand that is transcribed and a matching strand that is not usually transcribed. We can also apply this terminology to RNA. Messenger RNA is considered the translatable strand, and a strand with its complementary sequence of nucleotides would be the antisense strand. To illustrate: If an mRNA sequence read AUGCGAGAC, then an antisense RNA strand for it would read UACGCUCUG.

Some genes in bacteria are repressed by **antisense RNA** that binds to them and prevents translation at the ribosome. Antisense RNA has bases that are complementary to the sense strand of mRNA in the area surrounding the initiation site. When the antisense RNA binds to its particular mRNA, the now double-stranded RNA no longer fits on a ribosome, resulting in a loss of translation of that mRNA.

As this process was tested in the laboratory, two significant differences between prokaryotic and eukaryotic systems emerged. The first is that in eukaryotes single-stranded DNA is the typical antisense agent, as it is easier to manufacture than RNA. The second is that, for some genes, once the antisense strand was bound to the mRNA it was unable even to leave the nucleus. When an adequate dose of **antisense DNA** is delivered across the cell membrane into the cytoplasm and enters the nucleus, it binds to specific sites on any mRNAs that are the targets of therapy. As a result, the reading of that mRNA transcript on ribosomes will be blocked, the gene product will not be synthesized, and the disease associated with the gene will be suppressed (**figure 10.16**).

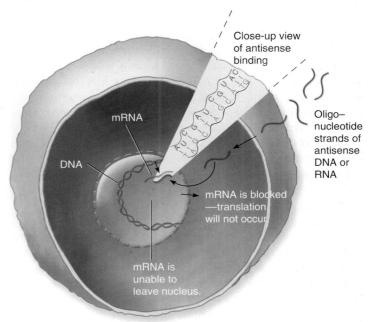

Close-up view of antisense binding

Oligo–nucleotide strands of antisense DNA or RNA

mRNA

DNA

mRNA is blocked—translation will not occur.

mRNA is unable to leave nucleus.

Figure 10.16 **The actions of antisense DNA.** When antisense strands are transported into the cell, they enter the nucleus and bind to complementary areas on an existing mRNA molecule (see enlarged view). Such bound RNA will be unavailable for translation on the cell's ribosomes, and no protein will be made.

Early clinical trials of antisense therapeutics have been promising, with most antisense therapies dramatically lowering the rate of synthesis of the targeted protein. Because the antisense agent can be targeted to a specific sequence of DNA, side effects are rare. Most concern has been centered on the difficulty of getting high enough levels of the drug into the nucleus of cells and the fact that these drugs would have to be taken for life.

10.5 Genome Analysis: Fingerprints and Genetic Testing

Expected Learning Outcomes

13. Discuss the concept of the use of DNA in genome identification.

14. Describe how a DNA fingerprint can be so precise, and explain variations on what the test identifies.

15. Describe several applications of DNA fingerprinting and microarray analysis.

An old geneticist's joke begins: "How would you describe a cat?" The answer, "GATCCT . . . ," is becoming more and more a reality. As was mentioned earlier, DNA technology has allowed us to accomplish many age-old goals by new and improved means. By remembering that phenotypes, whether human, bacterial, or even viral, are the result of specific sequences of DNA, we can easily see how DNA can be used to differentiate among organisms in the same way that observing any of these phenotypes can. Additionally, DNA can be used to "see" the phenotype of an organism no longer present, as when a criminal is identified by DNA extracted from a strand of hair left behind. Finally, possession of a particular sequence of DNA may indicate an increased risk of a genetic disease. Detection of this region of DNA, or marker, can identify a person as being at increased risk for cancer or Alzheimer disease long before symptoms arise. The ability to detect diseases before symptoms arise is especially important for diseases such as cancer, where early treatment is sometimes the difference between life and death. With examples like this in mind, let's look at several ways in which new DNA technology is allowing us to accomplish goals in ways that were only dreamed of a few years ago.

DNA Fingerprinting: A Unique Picture of a Genome

Although DNA is based on a structure of nucleotides, the exact way these nucleotides are combined is unique for each organism. It is now possible to apply DNA technology in a manner that emphasizes these differences and arrays the entire genome in a pattern for comparison. **DNA fingerprinting** (also called DNA profiling) is best known as a tool of forensic science first devised in the mid-1980s by Alex Jeffreys of Great Britain (**figure 10.17**).

Several of the methods discussed previously in this chapter are involved in the creation of a DNA fingerprint. Techniques such as the use of restriction endonucleases for cutting DNA pre-

cisely, PCR amplification for increasing the number of copies of a certain genome, electrophoresis to separate fragments, hybridization probes to locate specific loci and alleles, and the Southern blot technique for producing a visible record are all employed. Several different methods of DNA fingerprinting are available, but all depend on distinguishing one sequence of DNA from another by comparing the sequence of the strands at specific locations, or loci.

One type of analysis depends on the ability of a restriction enzyme to cut DNA at a specific recognition site. If a given strand of DNA possesses the recognition site for a particular restriction enzyme, the DNA strand is cut, resulting in two smaller pieces of DNA. If the same strand from a different person does not contain the recognition site (perhaps due to a mutation many generations ago), it is not cut by the restriction enzyme and remains as a single large piece of DNA. When each of these DNA samples is digested with restriction enzymes and separated on an electrophoretic gel, the first displays two small bands while the second displays one larger band. This is an example of a restriction fragment length polymorphism, which was discussed earlier. All methods of DNA fingerprinting depend on some variation of this strategy to ferret out differences in DNA sequence at the same location in the genome.

Tens of thousands of these differences at a single locus are known to exist throughout the genome. Other genetic **markers,** or observable variations in DNA structure, depend on small repeated sequences of nucleotides that vary in the number of repeated segments from person to person (and are hence polymorphic).

The most reliable methods of DNA fingerprinting presently available are: (1) use of **short tandem repeats (STRs)** to identify relatively short fragments of nuclear or chromosomal DNA; (2) identifying the DNA found in **mitochondria (mtDNA** or mitotyping); and (3) screening for **single nucleotide polymorphisms (SNPs).** All of these techniques rely on the polymerase chain reaction (PCR) to amplify the small amount of DNA that is present in most samples. The STR method locates and tags segments in a genome that have the same series of 3′ to 5′ bases repeated several times in a row. The bases can be readily detected because they are flanked by sites for restriction endonucleases. An example of this type of sequence might read:

AGCT|GATA|GATA|GATA|GATA|GATA|GGCC

The problem with this method is that it requires relatively longer pieces of DNA—200 to 400 nucleotides long, and generally will not work on DNA that is too degraded. This is the method used for most of the World Trade Center ID Project identifications. The miniSTR, a version that detects much shorter fragments, has been used for some types of forensic identification and is gaining in use. The mitochondrial DNA method isolates mitochondria from any cells present in the samples and runs specific tests to detect this type of DNA. One advantage of mtDNA is that it remains relatively intact over long periods and under harsh conditions. However, it is not as unique as nuclear DNA because in humans and other animals, it is inherited only from the mother's egg. This means that children of the same mother would have the same mtDNA profile. Single nucleotide polymorphisms are the most specific because they are the most

(b) Electrophoresis of DNA fragments separates them by size, with larger fragments nearer the wells. Although invisible to the unaided eye at this point, each lane contains a number of individual bands. Specific DNA sequences are identified by hybridization with fluorescently labeled nucleic acid probes or radioactive probes.

(a) Cells from different samples are processed to isolate their DNA. The DNA samples are exposed to endonucleases which snip them at specific sites into a series of different fragments.

(c) An actual DNA fingerprint used in a rape trial. Control lanes with known markers are in lanes 1, 5, 8, and 9. The second lane contains a sample of DNA from the victim's blood. Evidence samples 1 and 2 (lanes 3 and 4) contain semen samples taken from the victim. Suspects 1 and 2 (lanes 6 and 7) were tested. Can you tell by comparing evidence and suspect lanes which individual committed the rape?

Figure 10.17 **DNA fingerprints: the bar codes of life. Steps in the process of identifying DNA.**

variable of all polymorphisms, occurring 3 million to 10 million times among a person's 3 billion base pairs. They appear as one to two base differences at random sites throughout the chromosomes (see chapter opener figure).

CONTINUING CASE FILE 10

The method that was used for most of the identifications was the short tandem repeat (STR) analysis, a standard used by law enforcement that looks for variations in clusters of repeated bases in the DNA molecule unique to an individual. Some types of mitochondrial DNA were also used in cases that did not yield results using STR. These two methods alone could only clearly

identify about 1,500 specimens, probably because of the degraded nature of the DNA in many of the samples.

At this point, the medical examiner brought in two relatively new methods called the miniSTR and the single nucleotide polymorphism (SNP) that could potentially detect and identify much smaller pieces of DNA. These were tried out over a 2-year period, but they definitively identified only 130 more people. It is a stunning testament to the extreme power and precision of DNA technology that it was able to extract the genetic profiles of the terrorists who hijacked the airplanes!

■ How was it possible to manage millions of bits of DNA data?

■ What is meant by a positive identification or match?

For a wrap-up, see the Case File Perspective on page 315.

Applications of DNA testing

The most powerful uses of DNA fingerprinting are in forensics, detecting genetic diseases, determining parentage, analyzing the family trees of humans and animals, identifying microorganisms, and tracing the lineage of ancient organisms.

One of the first uses of the technique of DNA fingerprinting was in forensic medicine. Besides old-fashioned fingerprints, criminals often leave other evidence at the site of a crime—a hair, a piece of skin or fingernail, semen, blood, or saliva. Because DNA fingerprinting can be combined with amplifying techniques such as PCR, it provides a way to test even specimens that have only minute amounts of DNA, or even old specimens. Regardless of the method used, the primary function of the fingerprint is to provide a snapshot of the genome, displaying it as a unique picture of an individual organism. Markers such as those just described are often used in combination to increase the odds of a correct identification.

For example, imagine that a sample of blood found at a crime scene is thought to have come from a suspect and has tested positive for markers A, B, C, and D. If marker A is present in 1 person out of 20, marker B in 1 person out of 10, marker C in 1 person out of 25, and marker D in 1 person out of 50, the probability that a single person would have all four markers is $(1/20 \times 1/10 \times 1/25 \times 1/50)$ or 1 out of 250,000. When combined with traditional evidence, such as being seen leaving the crime scene, DNA fingerprinting is convincing indeed!

DNA fingerprinting is so widely accepted that it makes a significant contribution to trial evidence. It has been used to convict and exonerate thousands of accused defendants in criminal cases. In fact, the technology is so reliable that many older cases for which samples are available are being reopened and solved. The demand has been so high that many forensic units are overwhelmed by requests for testing. The FBI has established a national database of DNA fingerprints (CODIS), and several states allow DNA fingerprinting of felons in prison. In a stunning application of this profiling ability, forensic scientists were able to positively identify the Unabomber, Ted Kaczynski, by DNA recovered from the back of postage stamps he had licked. This same technology has also been effective in identifying the remains of the shuttle *Columbia* astronauts, some victims of Hurricane Katrina, and those in the World Trade Center (case file). For an example of its strength as a tool for determining paternity and maternity, see Visual Challenge question #1.

More and more, infectious disease laboratories are employing DNA test procedures to identify bacteria and viruses. DNA typing is used to identify several dozen pathogens, including *Neisseria gonorrhoeae, Chlamydia,* the syphilis spirochete, *Staphylococcus* MRSA, *M. tuberculosis* and a number of viruses. It is also an essential tool for determining genetic similarities between microbes involved in disease outbreaks.

Because even fossilized DNA can remain partially intact, ancient DNA is being studied for comparative and evolutionary purposes. Anthropologists have traced the migrations of ancient people by analyzing the mitochondrial DNA found in bone fragments. Mitochondrial DNA (mtDNA) is less subject to degradation than is chromosomal DNA and can be used as an evolutionary time clock. A group of evolutionary biologists have been able to recover Neanderthal mtDNA and use it to show that this group of ancient primates could not have been ancestors of modern humans. Biologists studying canine origins have also been able to show that the wolf is the ancestral form for modern canines.

Microarrays

Knowing the complete nucleotide sequence of the human genome is really only half the battle. With very few exceptions, all cells in an organism contain the same DNA, so knowing the sequence of that DNA, while certainly helpful, is of very little use when comparing two cells or tissues from the same organism. Recall that genes are expressed in response to both internal needs and external stimuli, and while the DNA content of a cell is static, the mRNA (and hence protein) content at any given time provides scientists with a profile of genes currently being expressed in the cell. What truly distinguishes a liver cell from a kidney cell or a healthy cell from a diseased cell is the genes expressed in each.

Twin advances in biology and electronics have allowed biologists to view the expression of genes in any given cell using a technique called DNA *microarray analysis.* Prior to the advent of this technology, scientists were able to track the expression of, at most, a few genes at a time. Given the complex interrelationship that exists between genes in a typical cell, tracking two or three genes was of very little practical use. Microarrays are able to track the expression of thousands of genes at once and are able to do so in a single efficient experiment. Microarrays consist of a glass "chip"[6] (occasionally a silicon chip or nylon membrane is used) which have bound sequences from tens of thousands of different genes. A solution containing fluorescently labeled cDNA, representing all of the mRNA molecules in a cell at a given time, is added to the chip. The labeled cDNA is allowed to bind to any complementary DNA bound to the chip. Bound cDNA is then detected by exciting the fluorescent tag on the cDNA with a laser and recording the fluorescence with a detector linked to a computer. The computer can then interpret this data to determine what mRNAs are present in the cell under a variety of conditions **(figure 10.18).**

Microarrays permit the development of extraordinarily sensitive diagnostic tests that search for a specific pattern of gene expression. As an example, being able to identify a patient's cancer as one of many subtypes (rather than just, for instance, breast cancer) will allow pharmacologists and doctors to treat each cancer with the drug that will be most effective. Again, we see that genetic technology can be a very effective way to reach long-held goals.

6. It's actually just a glass slide, but all microarray setups have come to be called chips.

Figure 10.18 Gene expression analysis using microarrays. The microarray seen here consists of oligonucleotides bound to a glass slide. The gene expression of two cells (for example, one healthy and one diseased) can be compared by labeling the cDNA from each cell with either a red or green fluorescent label and allowing the cDNAs to bind to the microarray. A laser can excite the bound cDNAs, while a detector records those spots that fluoresce. The color of each spot reveals whether the DNA on the microarray is present in the healthy sample, the diseased sample, both samples, or neither sample.

CASE FILE **10** PERSPECTIVE

We learned in chapter 9 that the precise sequence of base pairs provides a unique "blueprint" for each organism, except for clones and identical siblings. The less related individuals are, the more differences will be present in their genetic codes. The strength of DNA technology is that it can display and compare the patterns of these DNA codes in a way that makes differences observable and quantifiable.

It is true that the investigators in the World Trade Center ID Project were faced with huge problems, but fortunately, some of the solutions were available or already existed in the laboratory. For example, the minute amounts of DNA in some samples can be overcome by using the polymerase chain reaction (PCR), which amplifies even a single fragment of DNA to millions of copies, ready for identification.

You may have been wondering: How did they figure out that a certain DNA fingerprint matched a particular person? An essential part of DNA analysis is having a reference sample of DNA collected from a known source for comparison with an unknown DNA sample. Sources for a reference sample could include hair or personal items from the missing person, if available. Another reliable source is the family of the victim, who will share a similar DNA pattern through inheritance. The examiner's office requested cheek swabs from as many family members as possible, and these too, were tested and added to a database.

With respect to degraded DNA and commingled samples, there were no complete solutions. The medical examiner closed down the major part of the project in 2006, feeling that the technology had reached its limits. In a letter to families of victims, he said: "We may have paused in our work, but we didn't stop trying and we will never quit." The unidentified specimens remain preserved for future improvements in technology that might be successful with compromised DNA.

A second massive dimension to the project was managing all of the data compiled from the DNA tests. It started out with several separate computer databases that were eventually combined into a single one. Using computers in this way to organize, analyze, and track genetic data is a major part of the science of bioinformatics— a field that merges biology, computer science, mathematics, and statistics. It involves powerful search engines that can analyze and compare DNA STR sequences to locate matches. The investigation team decided to base the identification on the FBI's **C**ombined **D**NA **I**ndex **S**ystem (CODIS). The accepted criterion for a definitive identification is finding points of agreement between 13 different STR markers and comparable sequences in the reference samples. This number of markers gives an extremely high level of confidence: There is only a 1 in 10 billion chance that two unrelated persons would have the same profiles.

To read more about this remarkable project, go to http://www.massfatality.dna.gov/.

Check & Assess Sections 10.4 and 10.5

✔ Gene therapy is the replacement of faulty host genes with functional genes by use of cloning vectors such as viruses. This type of transfection can be used to treat genetic disorders and acquired disease.

✔ Antisense DNA and interfering RNAs are used to block expression of undesirable genes in plants, animals, and microbes.

✔ DNA technology has advanced understanding of basic genetic principles that have significant applications in a wide range of disciplines, particularly medicine, evolution, forensics, and anthropology.

✔ DNA fingerprinting is a technique by which organisms are identified for purposes of medical diagnosis, genetic ancestry, forensics, and parentage.

✔ Making a DNA fingerprint involves the separation of isolated DNA into a pattern of bands that is unique and identifiable.

✔ The bands used in DNA identification represent polymorphisms called short tandem repeats (STRs) and single nucleotide polymorphisms (SNPs).

✔ Microarray analysis can determine what genes are transcribed in a given tissue. It is used to identify and devise treatments for diseases based on the genetic profile of the disease.

23. Describe what a DNA fingerprint is and why and how polymorphic fragments can be used to form a unique DNA pattern.
24. Discuss briefly how DNA fingerprinting is being used routinely by medicine, the law, the military, and human biology.
25. Describe the uses of gene therapy and antisense therapy to control genetic diseases.
26. Explain the origins of mtDNA and how it is limited in its usefulness.
27. Explain the difference between a DNA fingerprint and a microarray.

Chapter Summary with Key Terms

10.1 Basic Elements and Applications of Genetic Engineering

Genetic engineering refers to the manipulation of an organism's genome and is often used in conjunction with biotechnology, the use of an organism's biochemical and metabolic pathways for industrial production of proteins.

A. *Practical Properties of DNA*
1. **Restriction endonucleases** are used to cut DNA at specific recognition sites, resulting in small segments of DNA that can be more easily managed in the laboratory.
2. The enzyme **ligase** is used to covalently join DNA molecules, whereas **reverse transcriptase** is used to obtain a DNA copy of a particular mRNA.
3. Short strands of DNA called oligonucleotides are used either as probes for specific DNA sequences or as primers for DNA polymerase.

B. *Nucleic Acid Hybridization and Probes*
1. The two strands of DNA can be separated (denatured) by heating. These single-stranded nucleic acids will bind, or **hybridize,** to their complementary sequence, allowing them to be used as **gene,** or **hybridization, probes.**
 a. In the Southern blot method, DNA isolated from an organism is probed to identify specific DNA sequences.

C. *Methods Used to Size, Synthesize, and Sequence DNA*
1. A technique called **DNA sequencing,** originally developed by Frederick Sanger, allows for the determination of the exact sequence of a stretch of DNA.
 a. Gene **mapping** is a method of identifying landmarks throughout the genome. Several types of maps are available, each with a different level of resolution.
 b. The ultimate map is a DNA sequence map, which provides the exact order of all of the nucleotides that make up the genome.
2. The **polymerase chain reaction** is a technique used to amplify a specific segment of DNA several million times in just a few hours.
3. The Human Genome Project has completed a sequence map for humans, identifying the sequence of the approximately 25,000 genes found in our DNA.

10.2 Recombinant DNA Technology: How to Imitate Nature

A. *Technical Aspects of Recombinant DNA Technology*
1. Recombinant DNA methods have as their goal the transfer of DNA from one organism to another. This commonly involves inserting the DNA into a **vector** (usually a plasmid or virus) and transferring the recombinant vector to a **cloning host** (usually a bacterium or yeast) for expression.
 a. Examples of cloning vectors include plasmids, bacteriophages, and cosmids, as well as the larger bacterial and yeast artificial chromosomes (BACs and YACs).
 b. Cloning hosts are fast-growing, nonpathogenic organisms that are genetically well understood. Examples include *E. coli* and *Saccharomyces cerevisiae.*
2. **Recombinant** organisms are created by cutting both the donor DNA and the plasmid DNA with restriction endonucleases and then joining them together with

ligase. The vector is then inserted into a cloning host by transformation, and those cells containing a vector are selected by their resistance to antibiotics.
3. When the appropriate recombinant cell has been identified, it is grown on an industrial scale. The cloning host expresses the gene it contains, and the resulting protein is excreted from the cell where it can be recovered from the media and purified.

B. *Protein Products of Recombinant DNA Technology*
In many instances, the final product of recombinant DNA technology is a protein.

Medical conditions such as diabetes, dwarfism, and blood-clotting disorders can be treated with recombinant insulin, human growth factor, and clotting factors, respectively.

10.3 Genetically Modified Organisms

A. Recombinant organisms (transgenics) include microbes, plants, and animals that have been engineered to express genes they did not formally possess.
1. Transformed bacteria produce a wide variety of drugs, vaccines, and other useful materials.
2. Modified viruses function as genetic vectors and vaccines.
3. Plants are **transfected** by a bacterium *Agrobacterium,* that can shuttle donor genes into cells.
4. Animals are genetically engineered by having selected genes inserted into the embryo.

10.4 Genetic Treatments: Introducing DNA into the Body Gene

therapy describes a collection of techniques in which defective genes are replaced or, more commonly, have their expression increased or decreased. Therapy is said to be in vivo if cells are altered within the body and ex vivo if the alteration is done on cells taken from the body.

10.5 Genome Analysis: Fingerprints and Genetic Testing

A. **DNA Fingerprinting**
1. Because the DNA sequence of an individual is unique, so too is the array of DNA fragments produced by cutting that DNA with a restriction endonuclease.
2. The identification of specific fragments produced by digestion of the DNA with a restriction endonucleases is commonly referred to as the DNA fingerprint of an organism, and serves as a unique identifier of any organism.
3. DNA analysis involves locating polymorphisms called STRs and SNPs and mitochondrial DNA testing.

B. *Microarray analysis* is a method of determining which genes are actively transcribed in a cell under a variety of conditions, such as health versus disease or growth versus differentiation.

Multiple-Choice Questions

Select the correct answer from the answers provided. For questions with blanks, choose the combination of answers that most accurately completes the statement.

1. Which gene is incorporated into plasmids to detect recombinant cells?
 a. restriction endonuclease
 b. Taq polymerase
 c. antibiotic resistance
 d. reverse transcriptase

2. Which of the following is *not* essential to carry out the polymerase chain reaction?
 a. primers
 b. DNA polymerase
 c. gel electrophoresis
 d. high temperature

3. Which of the following is *not* a part of the Sanger method to sequence DNA?
 a. dideoxynucleotides
 b. DNA polymerase
 c. electrophoresis
 d. reverse transcriptase

4. The function of ligase is to
 a. rejoin segments of DNA
 b. make longitudinal cuts in DNA
 c. synthesize cDNA
 d. break down ligaments

5. The pathogen of plant roots that is used as a cloning host is
 a. *Pseudomonas*
 b. *Agrobacterium*
 c. *Escherichia coli*
 d. *Saccharomyces cerevisiae*

6. Which of the following sequences, when combined with its complement, would be clipped by a restriction endonuclease?
 a. ATCGATCGTAGCTA
 b. AAGCTTCGAA
 c. GAATTC
 d. ACCATTGGA

7. Which DNA fragment will be closest to the top (negative pole) of an electrophoretic gel?
 a. 450 bp
 b. 3,560 bp
 c. 5 kb
 d. 1,500 bp

8. Which of the following is a primary participant in cloning an isolated gene?
 a. restriction endonuclease
 b. vector
 c. host organism
 d. all of these

9. For which of the following would a nucleic acid probe *not* be used?
 a. locating a gene on a chromosome
 b. developing a Southern blot
 c. identifying a microorganism
 d. constructing a recombinant plasmid

10. **Multiple Matching.** Match the term with its description.
 _____ nucleic acid probe
 _____ nontemplate strand
 _____ template strand
 _____ reverse transcriptase
 _____ Taq polymerase
 _____ antisense DNA
 _____ primer
 _____ restriction endonuclease
 a. enzyme that transcribes RNA into DNA
 b. complementary strand that blocks mRNA expression
 c. the nontranslated strand of DNA or RNA
 d. enzyme that snips DNA at palindromes
 e. oligonucleotide that initiates the PCR
 f. strand of nucleic acid that is transcribed or translated
 g. thermostable enzyme for synthesizing DNA
 h. oligonucleotide used in hybridization

Case File Questions

1. What was the main reason that the short tandem repeat method of analyzing DNA could not identify all victims?
 a. It is too nonspecific.
 b. It does not work on degraded DNA.

c. The amount of DNA in the specimen was too small.
 d. It is too slow to work on large numbers of samples.

2. What was a feasible and important part of accurately identifying the terrorists?
 a. a sample of their mitochondrial DNA
 b. a sample of their blood cells
 c. a known sample of DNA from their families
 d. identification papers discovered at ground zero

Writing to Learn

These questions are suggested as a *writing-to-learn* experience. For each question, compose a one- or two-paragraph answer that includes the factual information needed to completely address the question. All Check & Assess questions can be used for writing-to-learn exercises.

1. a. Why do bacteria make restriction endonucleases?
 b. What is it about the endonucleases that prevents bacteria from destroying their own DNA?

2. a. Using nucleotide letters, show how a piece of DNA can be cut to circularize it.
 b. What are restriction length polymorphisms, and how are they used?

3. Why would it be an advantage to synthesize eukaryotic genes by using restriction endonucleases to make cDNA?

4. a. Explain what is hybridization and how it is used in the Southern blot.
 b. How do gene probes work?

5. Look at figure 10.7, and determine the correct DNA sequence for the fragment's template strand. (Be careful of orientation.)

6. a. Explain why the PCR is unlikely to amplify contaminating bacterial DNA in a sample of human DNA.
 b. Explain how PCR could be used to pick a gene out of a complex genome and amplify it.

7. We now know that humans share 99.9% of their genome and vary in only about 0.1% of sequences. Explain how small differences could result in large differences in phenotype.

8. Explain the kinds of study involved in genomics, proteomics, and bioinformatics.

Concept Mapping

Appendix E provides guidance for working with concept maps.

1. Construct your own concept map using the following words as the *concepts.* Supply the linking words between each pair of concepts.

 DNA

 restriction endonuclease

 palindrome

 ligase

 plasmid

 vector

 origin of replication

 recombinant

Critical Thinking Questions

Critical thinking is the ability to reason and solve problems using facts and concepts. These questions can be approached from a number of angles, and in most cases, they do not have a single correct answer.

1. a. Give an example of a benefit of genetic engineering to society and a possible adverse outcome. Discuss.
 b. Give an example of an ecological benefit and a possible adverse side effect. Discuss.

2. a. In reference to Insight 10.2, how do you feel about the potential dangers of genetic engineering?
 b. Most of us would agree to growth hormone therapy for a child with dwarfism, but how do we deal with parents who want to give growth hormones to their 8-year-old-son so that he will be "better at sports"?

3. a. When gene probes, fingerprinting, and mapping make it possible for you to know of future genetic diseases in you or one of your children, would you wish to use this technology to find out?
 b. What if it were used as a screen for employment or insurance?

4. Which suspect was more likely the rapist, according to the fingerprint in figure 10.17*c*?

5. a. Describe how a virus might be genetically engineered to make it highly virulent.
 b. Can you trace the genetic steps in the development of a tomato plant that has become frost-free from the addition of a flounder's antifreeze genes? (Hint: You need to use *Agrobacterium*.)

6. You have obtained a blood sample in which only red blood cells are left to analyze.
 a. Can you conduct a DNA analysis of this blood?
 b. If no, explain why.
 c. If yes, explain what you would use to analyze it and how to do it.

7. The way that PCR amplifies DNA is similar to the doubling in a population of growing bacteria; a single DNA strand is used to synthesize 2 DNA strands, which become 4, then 8, then 16, etc. If a complete cycle takes 3 minutes, how many strands of DNA would theoretically be present after 10 minutes? After 1 hour?

8. You are on a jury to decide whether a person committed a homicide and you have to weigh DNA fingerprinting evidence. Two different sets of fingerprints were done: one that tested 5 markers and one that tested 10. Both sets match the defendant's profile. Which one is more reliable and why?

9. a. How do you suppose the fish and game department, using DNA evidence, could determine whether certain individuals had poached a deer or if a particular mountain lion had eaten part of a body?
 b. Can you think of some reasons that it would *not* be possible to re-create dinosaurs using the technology we have described in this chapter?

10. Recently a woman at high risk for breast cancer was directed by her physician to be tested for the breast cancer (BRCA) genes. When her blood was sent to the company that performs the tests, it denied her health insurance and told her she would have to cover the cost ($4,000). No other companies do this test, and she could not afford to have the tests done. It turns out that the company has the patent and thus "owns" the rights to these and several other genes. It can even

limit genetic researchers from studying them. In spring, 2010, a judge reversed these patents. Now that you know more about genetics and DNA, what are your opinions about the ownership of the genome? See the video about this subject at: http://www.youtube.com/watch?v=6h6X46-qz14.

11. What are some of the major impediments to genetically modifying human embryos?

12. Could actual fingerprints taken from human fingers be used to perform a DNA fingerprint? Explain.

Visual Challenge

1. These are two different DNA paternity tests: one indicates the man is the father and the other shows he is not. Can you tell which is which? C/AF = Child/Alleged father in one lane combined.

2. From **chapter 9, figure 9.24*d*.** Use the technology from figure 10.11 as your guide to arrive at the same result as shown here, but without the mouse.

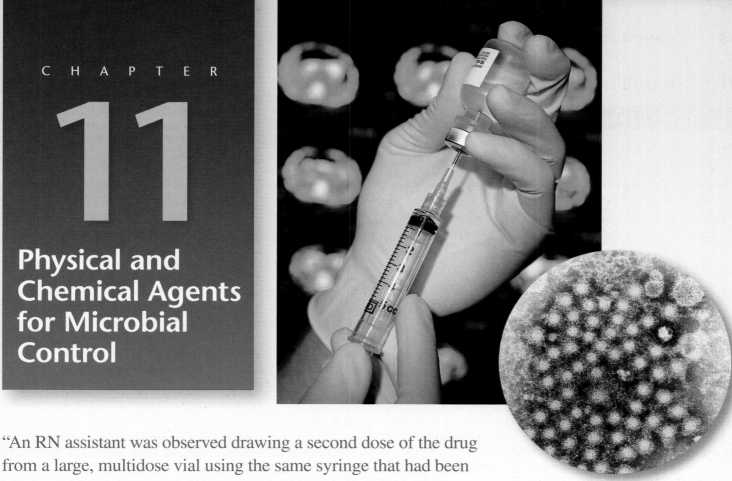

"An RN assistant was observed drawing a second dose of the drug from a large, multidose vial using the same syringe that had been used to take the first dose, although he did replace the needle."

Virions of Hepatitis C—a pathogen hiding in the blood

CASE FILE **11** **The Half Billion Dollar Medical Error**

Early in December 2007, a middle-aged man was hospitalized in Las Vegas with acute symptoms of abdominal pain, nausea, and vomiting. The attending physician also noticed that his eyes and skin were jaundiced (yellowed) and his urine was a dark color. Laboratory tests revealed an increased level of liver enzymes and antibodies to hepatitis C virus (HCV), and a nucleic acid test confirmed that RNA from HCV was present in his blood. Upon questioning, the patient denied having any of the risk factors for this infection such as IV drug abuse or exposure to blood or blood products. Further probing disclosed that he had recently undergone a dental procedure and had two endoscopic examinations of his gastrointestinal tract in September. This case was immediately reported to the Nevada Office of Epidemiology.

Within a month, two other cases of acute HCV infection were diagnosed in Las Vegas patients. These patients also claimed to lack the risk factors that caused this infection and they, too, had recently had endoscope procedures performed at the same clinic as the first patient. The seriousness of these findings made it necessary to request assistance in the investigation from the Centers for Disease Control (CDC). The connection of the three patients to the same endoscopy center made this facility the prime focus of the investigation.

The clinic had a large, thriving practice doing colonoscopies and other procedures on several thousand people each year. Because this raised the potential for other patients to have become infected with the virus, the CDC required that 50,000 persons who had a procedure in this clinic be immediately notified of their possible exposure to the virus.

Another crucial piece of evidence was to verify if these patients were infected by the same genetic strain of HCV. When it turned out that all three viruses were genetically linked to each other, authorities were certain that the patients were infected by the same source. Knowing that the three patients had never had any kind of contact, epidemiologists concluded that the clinic had to be the origin of the infection. This finding alone did not provide any clues as to what went wrong, but it had tremendous repercussions on the clinic, its patients, and the public health department.

▶ How are acute and chronic infections different, and why was it so important to differentiate them in this case?

▶ What sorts of procedures in a clinic might expose a person to blood-borne diseases?

To continue the case, go to page 342.

11.1 Controlling Microorganisms

Expected **Learning Outcomes**

1. Summarize the major categories of microbial control and its purposes.
2. Evaluate the relative resistance of different groups of infectious agents.
3. Define and differentiate among the major terms for microbial control, citing examples of each.
4. Identify the parameters of microbial death, and describe several factors that influence the rate of death.
5. Summarize what practices in microbial control influence the effectiveness of an antimicrobial agent.
6. Identify the targets of antimicrobial control agents, and explain what effects their modes of actions have.

Much of the time in our daily existence, we take for granted tap water that is drinkable, food that is not spoiled, shelves full of products to eradicate "germs," and drugs to treat infections. Controlling our degree of exposure to potentially harmful microbes is a monumental concern in our lives, and it has a long and eventful history **(Insight 11.1).**

General Considerations in Microbial Control

The methods of microbial control belong to the general category of *decontamination* procedures, in that they destroy or remove contaminants. In microbiology, contaminants are microbes present at a given place and time that are undesirable or unwanted. Most decontamination methods employ either physical agents, such as heat or radiation, or chemical agents, such as disinfectants and antiseptics. This separation is convenient, even though the categories overlap in some cases; for instance, radiation can cause damaging chemicals to form, and chemicals can generate heat. A flowchart **(figure 11.1)** summarizes the major applications and aims in microbial control.

Relative Resistance of Microbial Forms

The targets of microbial control processes are microorganisms capable of causing infection or spoilage that are constantly present in the external environment and on the human body. This targeted population is rarely simple or uniform; in fact, it often contains mixtures of microbes with extreme differences in resistance and harmfulness. Contaminants that can have far-reaching effects if not adequately controlled include bacterial vegetative cells and endospores, fungal hyphae and spores, yeasts, protozoan trophozoites and cysts, worms, viruses, and prions. This scheme compares the general resistance of these forms to physical and chemical methods of control:

Highest resistance
> Prions: proteinaceous infectious particles (see note on page 321); bacterial endospores, primarily from bacteria in the genera *Bacillus* and *Clostridium.*

Moderate resistance
> Protozoan cysts; some fungal sexual spores (zygospores); some viruses. In general, naked viruses are more resistant than enveloped forms.

Figure 11.1 **An overview of microbial control methods.**

INSIGHT 11.1

Microbial Control in Ancient Times

No one knows for sure when humans first applied methods that could control microorganisms, but perhaps the discovery and use of fire in prehistoric times were the starting points. We do know that records describing simple measures to control decay and disease appear from civilizations that existed several thousand years ago. We know, too, that these ancient people had no concept that germs caused disease, but they did have a mixture of beliefs; skills in observing natural phenomena; and, possibly, a bit of luck. This combination led them to carry out simple and sometimes rather hazardous measures that contributed to the control of microorganisms.

Salting, smoking, pickling, and drying foods, and exposing food, clothing, and bedding to sunlight were prevalent practices among early civilizations. The Egyptians showed surprising sophistication and understanding of decomposition by embalming the bodies of their dead with strong salts and pungent oils. They introduced filtration of wine and water as well. The Greeks and Romans burned clothing and corpses during epidemics, and they stored water in copper and silver containers. The armies of Alexander the Great reportedly boiled their drinking water and buried their wastes. Burning sulfur to fumigate houses and applying sulfur as a skin ointment also date approximately from this era.

Illustration of protective clothing used by doctors in the 1700s to avoid exposure to plague victims. The beaklike portion of the hood contained volatile perfumes to protect against foul odors and possibly inhaling "bad air."

During the great plague pandemic of the Middle Ages, it was commonplace to bury corpses in mass graves, burn the clothing of plague victims, and ignite aromatic woods in the houses of the sick in the belief that fumes would combat the disease. In a desperate search for some sort of protection, survivors wore peculiar garments and anointed their bodies with herbs, strong perfume, and vinegar. The children's rhyme "Ring-a-round a rosie" actually refers to many of the rituals related to the plague, such as filling the pockets of the victims with flowers ("pocket full of posies . . .") and burning the dead so as to reduce the spread of the disease ("Ashes! Ashes! We all fall down"). These attempts may sound foolish and antiquated, but it now appears that they may have had some benefits. Burning wood releases formaldehyde, which could have acted as a disinfectant; herbs, perfume, and vinegar contain mild antimicrobial substances. Each of these early methods, although somewhat crude, laid the foundations for microbial control methods that are still in use today.

What are some modern techniques that are equivalent to these early techniques described here? *Answer available at http://www.mhhe. com/talaro8*

Several groups of bacteria lack endospores but have vegetative cells with increased resistance to microbial control agents. These include *Mycobacterium,* whose cell walls have thick, waxy layers that can block the entrance of many disinfectants. The cell walls of *Pseudomonas, Acinetobacter,* and many other gram-negative rods have an outer membrane that also prevents penetration of chemical agents. *Staphylococcus* species are among the most heat- and chemical-resistant of all bacteria due to their thick, protective peptidoglycan cell walls.

Least resistance

Most bacterial vegetative cells; fungal spores (other than zygospores) and hyphae; enveloped viruses; yeasts; and protozoan trophozoites

Actual comparative figures on the requirements for destroying various groups of microorganisms are shown in **table 11.1.** Bacterial endospores have traditionally been considered the most resistant microbial entities, being as much as 18 times harder to destroy than vegetative cells. Because of endospore resistance to microbial control and their widespread distribution, destroying them is the goal of sterilization (see definition in following section). Any process that kills endospores will invariably kill all less resistant microbial forms. Other methods of control (disinfection, antisepsis) target primarily microbes that are less hardy than endospores.

TAKE NOTE: PRIONS

The role of prions as infectious particles in certain progressive brain diseases (Creutzfeldt-Jakob disease) was first introduced in chapter 6. Not only are prions highly unusual in their actions, but they are also in a class of their own when it comes to "sterilization" procedures. This chapter defines *sterile* as the absence of all viable microbial life. But prions are not microbes, and most of the procedures described here are by themselves not sufficient to destroy them. Prions are extraordinarily resistant to heat and chemicals. Instruments or other objects contaminated with these unique agents must be discarded as biohazards, incinerated, or decontaminated with a combination of chemicals and heat in accordance with CDC guidelines.

In terms of routine sterilization procedures, this chapter considers bacterial endospores as the most resistant form of microbial life that will be targeted by antimicrobial agents. When tissues, fluids, or instruments are suspected of containing prions, consultation with infection control experts, the CDC, or both is recommended to choose effective sterilization conditions. Prions are described in more detail in chapter 25.

TABLE 11.1	Relative Resistance of Bacterial Endospores and Vegetative Cells to Control Agents		
Method	Endospores*	Vegetative Forms*	Relative Resistance**
Heat (moist)	120°C	80°C	1.5×
Radiation (X-ray) dosage	4,000 grays	1,000 grays	4×
Sterilizing gas (ethylene oxide)	1,200 mg/1	700 mg/1	1.7×
Sporicidal liquid (2% glutaraldehyde)	3 hours	10 min	18×

*Values are based on methods (concentration, exposure time, intensity) that are required to destroy the most resistant pathogens in each group.

**This number represents the intensity of the treatment (temperature, radiation dosage, gas concentration, or time) needed to kill endospores versus vegetative cells and is an average of the most resistant pathogens in each group.

Terminology and Methods of Microbial Control

Through the years, a growing terminology has emerged for describing and defining measures that control microbes. To complicate matters, the everyday use of some of these terms can at times be vague and inexact. For example, occasionally one may be directed to "sterilize" or "disinfect" a patient's skin, even though this usage does not fit the technical definition of either term. To lay the groundwork for the concepts in microbial control to follow, we present here a series of concepts, definitions, and usages in antimicrobial control.

Sterilization

Sterilization is a process that destroys or removes all viable microorganisms, including viruses. Any material that has been subjected to this process is said to be **sterile.** These terms should be used only in the strictest sense for methods that have been proved to sterilize. An object cannot be slightly sterile or almost sterile—it is either sterile or not sterile. Control methods that sterilize are generally reserved for inanimate objects, because sterilizing parts of the human body would call for such harsh treatment that it would be highly dangerous and impractical. As we shall see in chapter 13, many parts of the body—the brain and other internal organs, for example—are naturally free of microbes.

Sterilized products—surgical instruments, syringes, and commercially packaged foods, just to name a few—are essential to human well-being. Although most sterilization is performed with a physical agent such as heat, a few chemicals called *sterilants* can be classified as sterilizing agents because of their ability to destroy spores.

At times, sterilization is neither practicable nor necessary, and only certain groups of microbes need to be controlled. Some antimicrobial agents eliminate only the susceptible vegetative states of microorganisms but do not destroy the more resistant endospore and cyst stages. Keep in mind that the destruction of spores is not always a necessity, because most of the infectious diseases of humans and animals are caused by non-spore-forming microbes.

Microbicidal Agents

The root -*cide,* meaning to kill, can be combined with other terms to define an antimicrobial agent aimed at destroying a certain group of microorganisms. For example, a **bactericide** is a chemical that destroys bacteria except for those in the endospore stage. It may or may not be effective on other microbial groups. A **fungicide** is a chemical that can kill fungal spores, hyphae, and yeasts. A **virucide** is any chemical known to inactivate viruses, especially on living tissue. A **sporicidal** agent can destroy bacterial endospores which makes it also a sterilizing agent.

Agents That Cause Microbistasis

The Greek words *stasis* and *static* mean to stand still. They can be used in combination with various prefixes to denote a condition in which microbes are temporarily prevented from multiplying but are not killed outright. Although killing or permanently inactivating microorganisms is the usual goal of microbial control, microbistasis does have meaningful applications. **Bacteriostatic** agents prevent the growth of bacteria on tissues or on objects in the environment, and *fungistatic* chemicals inhibit fungal growth. Materials used to control microorganisms in the body (antiseptics and drugs) have microbistatic effects because many microbicidal compounds can be highly toxic to human cells.

Germicides, Disinfection, Antisepsis

A **germicide,*** also called a *microbicide,* is any chemical agent that kills pathogenic microorganisms. A germicide can be used on inanimate (nonliving) materials or on living tissue, but it ordinarily cannot kill resistant microbial cells. Any physical or chemical agent that kills "germs" is said to have germicidal properties.

The related term **disinfection*** refers to the use of a physical process or a chemical agent (a disinfectant) to destroy vegetative pathogens but not bacterial endospores. It is important to note that disinfectants are normally used only on inanimate objects because, in the concentrations required to be effective, they can be toxic to living tissue. Disinfection processes can also remove the harmful products of microorganisms (toxins) from materials. Examples of disinfection include applying a solution of 5% bleach to an examining table, boiling food utensils used by a sick person, and immersing thermometers in an iodine solution between uses.

In modern usage, **sepsis** is defined as the growth of microorganisms in the blood and other tissues. The term **asepsis*** refers to any practice that prevents the entry of infectious agents into sterile tissues and thus prevents infection. Aseptic techniques commonly practiced in health care range from sterile methods that exclude all microbes to **antisepsis.*** In antisepsis, chemical agents called **antiseptics** are applied directly to exposed body surfaces (skin and mucous membranes), wounds, and surgical incisions to destroy or inhibit vegetative pathogens. Examples of antisepsis include preparing the skin before surgical incisions with iodine compounds, swabbing a wound with hydrogen peroxide, and ordinary hand washing with a germicidal soap.

* *germicide* (jer'-mih-syd) L. *germen,* germ, and *caedere,* to kill. Germ is a common term for a pathogenic microbe.
* *disinfection* (dis"-in-fek'-shun) L. *dis,* apart, and *inficere,* to corrupt.
* *asepsis* (ay-sep'-sis) Gr. *a,* no or none, and *sepsis,* decay.
* *antisepsis anti,* against.

Methods That Reduce the Numbers of Microorganisms

While medical applications quite often require sterility, especially with regard to invasive instruments such as scalpels, clamps, dental hand tools, and the like, this absolute level of microbial control is often unwarranted and perhaps even unwanted. In many cases, it is more important to focus on reducing the size of a microbial population, or its *microbial load*. **Sanitization*** refers to any cleansing technique that removes debris, microorganisms, and toxins, and in this way reduces the potential for infection and spoilage. Soaps and detergents are the most commonly employed sanitizers. It is important to note that sanitization is often preferable to sterilization. For example, in a restaurant you could be given a sterile fork with someone else's old food on it and a sterile glass with lipstick on the rim. Given that sanitization is in general far less expensive than sterilization, its cleansing effect has obvious advantages.

With living tissues the reduction in microbial load through mechanical means is referred to as **degermation.** This process usually involves scrubbing the skin or immersing it in chemicals, or both. Microorganisms are removed both mechanically and indirectly by emulsifying the oils that coat the skin and provide a habitat for their growth. Examples of degerming procedures include the surgical handscrub, the application of alcohol wipes, and the cleansing of a wound with germicidal soap and water. **Table 11.2** summarizes several of the important terms involved in microbial control.

What Is Microbial Death?

Death is a phenomenon that involves the permanent termination of an organism's vital processes. Signs of life in complex organisms such as animals are self-evident, and death is made clear by loss of nervous function, respiration, or heartbeat. In contrast, death in microscopic organisms that are composed of just one or a few cells is hard to detect, because they often reveal no conspicuous vital signs

* *sanitization* (san″ ih-tih-zay′-shun) L. *sanitas*, health.

to begin with. Lethal agents (such as radiation and chemicals) do not necessarily alter the overt appearance of microbial cells. Even the loss of movement in a motile microbe cannot be used to indicate death. This fact has made it necessary to develop special qualifications that define and delineate microbial death.

The destructive effects of chemical or physical agents occur at the cellular and molecular level. As a cell is exposed to an agent such as intense heat or toxic chemicals, various cell structures break down, and the entire cell can sustain irreversible damage. At present, the most practical way to detect this damage is to determine if a microbial cell can still reproduce when exposed to a suitable environment. If the microbe has sustained metabolic or structural damage to such an extent that it can no longer reproduce, even in an ideal environment, then it is no longer viable. **The permanent loss of reproductive capability, even under optimum growth conditions, has become the accepted microbiological definition of death.**

Factors That Affect Death Rate

The ability to define microbial death has tremendous theoretical and practical importance. It allows medicine and industry to test the conditions required to destroy microorganisms, to pinpoint the ways that antimicrobial agents kill cells, and to establish standards of sterilization and disinfection in these fields. Hundreds of testing procedures have been developed for evaluating physical and chemical agents. Some of these are given in **appendix C.**

The cells of a culture show marked variation in susceptibility to a given microbicidal agent. Death of the whole population is not instantaneous but begins when a certain threshold of the microbicidal agent (some combination of time and concentration) is met. Death continues in a logarithmic manner as the time of exposure is increased **(figure 11.2).** Because many microbicidal agents target the cell's metabolic processes, active cells (younger, rapidly dividing) tend to die more quickly than those that are less metabolically active (older, inactive). Eventually, a point is reached at which survival of any cells is highly unlikely; this point is equivalent to sterilization.

TABLE 11.2	Microbial Control Terminology	
Term	**Definition**	**Example**
Decontamination	The destruction, removal, or reduction in number of undesirable microbes	Asepsis, disinfection, sanitization, degermation
Sepsis	The growth of microorganisms in the tissues	Infected wounds, blood infection
Asepsis	Techniques that prevent the entry of microorganisms into sterile tissues	Cleansing the skin with iodine prior to surgery, using sterile needles
Antiseptic	Chemicals applied to body surfaces to destroy or inhibit vegetative pathogens	Iodophors, antibacterial soap, chlorhexidine
Disinfection	Destruction of vegetative pathogens on inanimate objects	5% bleach, boiling water
Sanitization	Cleansing technique that removes microorganisms and debris from inanimate surfaces	Dishwashing, laundering clothes
Degermation	Cleansing technique that removes microorganisms and debris from living tissue	Surgical handscrub, alcohol wipes
Sterilization	The removal or destruction of all viable microbes	Autoclaving, ionizing radiation (correctly applied)

Figure 11.2 **Factors that influence the rate at which microbes are killed by antimicrobial agents.** **(a)** Length of exposure to the agent. During exposure to a chemical or physical agent, all cells of a microbial population, even a pure culture, do not die simultaneously. Over time, the number of viable organisms remaining in the population decreases logarithmically, giving a straight-line relationship on a graph. The point at which the number of survivors is infinitesimally small is considered sterilization. **(b)** Effect of the microbial load. **(c)** Relative resistance of spores versus vegetative forms. **(d)** Action of the agent, whether microbicidal or microbistatic.

The effectiveness of a particular agent is governed by several factors besides time. These additional factors influence the action of antimicrobial agents:

1. The number of microorganisms (**figure 11.2b**). A higher load of contaminants requires more time to destroy.
2. The nature of the microorganisms in the population (**figure 11.2c**). In most actual circumstances of disinfection and sterilization, the target population is not a single species of microbe but a mixture of bacteria, fungi, spores, and viruses, presenting a broad spectrum of microbial resistance.
3. The temperature and pH of the environment.
4. The concentration (dosage, intensity) of the agent. For example, UV radiation is most effective at 260 nm and most disinfectants are more active at higher concentrations.
5. The mode of action of the agent (**figure 11.2d**). How does it kill or inhibit microorganisms?
6. The presence of solvents, interfering organic matter, and inhibitors. Saliva, blood, and feces can inhibit the actions of disinfectants and even of heat.

Some Practical Concerns in Selecting Antimicrobial Agents

A remarkable variety of substances and objects require sterilization, disinfection, and other forms of decontamination. They range from durable materials made of glass and rubber to highly sensitive liquids such as blood plasma and even human tissues such as bone. Sometimes entire rooms with their contents and large pieces of equipment such as space ships and satellites may need to be sterilized. No one method can possibly work for all of these examples.

Literally hundreds of situations occur in the health care setting alone where prevention of infection must be taken into consideration at all times. The CDC provides guidelines for recommended sterilization and disinfection procedures. The case file for this chapter provides a good example of some of the difficulties in ensuring that proper techniques are followed at all times.

One of the first questions that arises when choosing a method is, What will render materials free of potential pathogens? For example, reusable dental instruments are highly contaminated with saliva, blood, and other fluids and tissues. These are all potential sources of infection, so it is critical that they be sterilized between patients with a method that will kill spores.

The next concern is selecting the best method of sterilization. Because they are mostly made of metal, dental instruments can be sterilized with heat, although some forms of radiation and chemical treatments could also work. Clinics generally opt for a method such as steam sterilization that is simplest, fastest, and least expensive. This would not work on other types of delicate instruments such as endoscopes, which must be sterilized between patients using heat-free techniques.

On the other hand, certain widely used medical supplies such as plastic syringes and catheters used in hospitals and clinics are usually disposable and come already sterilized and packaged at the factory. Since they are usually made of heat-sensitive plastics, this is done with penetrating radiation or chemicals. These items are not meant to be reused, but they still must be decontaminated before being discarded to prevent the transmission of disease. Disposal of such medical wastes is regulated by various governmental agencies.

How Antimicrobial Agents Work: Their Modes of Action

An antimicrobial agent's adverse effect on cells is known as its **mode (or mechanism) of action.** Agents affect one or more cellular targets, inflicting damage progressively until the cell is no longer viable. Antimicrobials have a range of cellular targets, with the agents that are least selective in their targeting tending to be effective against the widest range of microbes (examples include heat and radiation). More selective agents (drugs, for example) tend to target only a single cellular component and are much more restricted as to the microbes they are effective against.

The cellular targets of physical and chemical agents fall into four general categories: (1) cell walls; (2) cell membranes; (3) protein and nucleic acid synthesis; and (4) protein structure and function.

The Effects of Agents on the Cell Wall

The cell wall maintains the structural integrity of bacterial and fungal cells. Several types of chemical agents damage the cell wall by blocking its synthesis, digesting it, or breaking down its surface. A cell deprived of a functioning cell wall becomes fragile and is lysed very easily. Examples of this mode of action include some antimicrobial drugs (penicillins) that interfere with the synthesis of the cell wall in bacteria (described in chapter 12). Detergents and alcohol can also disrupt cell walls, especially in gram-negative bacteria.

How Agents Affect the Cell Membrane

All microorganisms have a cell membrane composed of lipids and proteins, and even some viruses have an outer membranous envelope. As we learned in previous chapters, a cell's membrane provides a two-way system of transport. If this membrane is disrupted, a cell loses its selective permeability and can neither prevent the loss of vital molecules nor bar the entry of damaging chemicals. Loss of those functions usually leads to cell death. Chemicals called **surfactants*** work as microbicidal agents because they lower the surface tension of cell membranes. Surfactants are polar molecules such as detergents with hydrophilic and hydrophobic regions that can physically bind to the lipid layer and penetrate the internal hydrophobic region of membranes. In effect, this process "opens up" the once tight interface, leaving leaky spots that allow harmful chemicals to seep into the cell and important ions to seep out **(figure 11.3).** Alcohols exert a related effect by dissolving membrane lipids and stripping membranes away from cells.

* *surfactant* (sir-fak'-tunt) A word derived from **surf**ace-**act**ing ag**ent.**

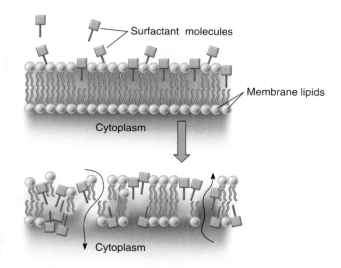

Figure 11.3 Mode of action of surfactants on the cell membrane. Surfactants inserting in the lipid bilayer disrupt it and create abnormal channels that alter permeability and cause leakage both into and out of the cell.

Agents That Affect Protein and Nucleic Acid Synthesis

Microbial life depends upon an orderly and continuous supply of proteins to function as enzymes and structural molecules. As we saw in chapter 9, these proteins are synthesized on the ribosomes during translation. Any chemical that blocks this process can affect function and survival. For example, certain drugs bind to the ribosomes of bacteria in a way that stops peptide bonds from forming. In its presence, many bacterial cells are inhibited from forming proteins required in growth and metabolism and are thus inhibited from multiplying. Many of the drugs used to treat infections are chemicals that block protein synthesis in microbes without adversely affecting human cells. This subject is discussed in greater detail in chapter 12.

The nucleic acids are likewise necessary for the continued functioning of microbes. DNA must be regularly replicated and transcribed in growing cells, and any agent that either impedes these processes or changes the genetic code is potentially antimicrobial. Some agents bind irreversibly to DNA, preventing both transcription and translation, whereas others are mutagenic agents. Gamma, ultraviolet, or X radiation causes mutations that result in permanent inactivation of DNA. Chemicals such as formaldehyde and ethylene oxide also interfere with DNA and RNA function.

Agents That Alter Protein Function

A microbial cell contains large quantities of proteins that function properly only if they remain in a normal three-dimensional configuration called the *native state*. The antimicrobial properties of some agents arise from their capacity to disrupt the structure of, or **denature,** proteins. In general, **denaturation** occurs when the bonds that maintain the secondary and tertiary structure of the protein are broken. Breaking these bonds will cause the protein to unfold or create random, irregular loops and coils **(figure 11.4).** One way that proteins can be denatured is through coagulation by moist heat (the same reaction seen in the irreversible solidification of the white of an egg

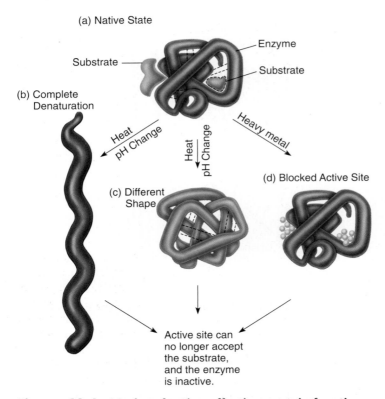

(a) Native State

Substrate

Enzyme

Substrate

(b) Complete Denaturation

Heat
pH Change

Heat
pH Change

Heavy metal

(d) Blocked Active Site

(c) Different Shape

Active site can no longer accept the substrate, and the enzyme is inactive.

Figure 11.4 Modes of action affecting protein function.
(a) The native (functional) state is maintained by bonds that create active sites to fit the substrate. Some agents denature the protein by breaking all or some secondary and tertiary bonds. Results are **(b)** complete unfolding or **(c)** random bonding and incorrect folding. **(d)** Some agents react with functional groups on the active site and interfere with bonding.

when boiled). Chemicals such as strong organic solvents (alcohols, acids) also coagulate proteins. A few antimicrobial agents, such as metallic ions, attach to the active site of the protein and prevent it from interacting with its correct substrate. Regardless of the exact mechanism, such losses in normal protein function can promptly arrest metabolism and exert broad effects on all types of microbes.

Check & Assess Section 11.1

✔ Microbial control methods consist of physical and chemical agents to eliminate or reduce the numbers of microorganisms in a specific environment. These measures prevent the spread of infectious agents, retard spoilage, and keep commercial products safe.

✔ The population of microbes that cause spoilage or infection varies widely in species composition, resistance, and harmfulness, so microbial control methods must be adjusted to fit individual situations.

✔ Several terms specify the purposes and targets of microbial control. Sterilization agents destroy all viable organisms, including viruses. Antisepsis, disinfection, and sanitization agents reduce the numbers of viable microbes to a specified level.

✔ Antimicrobial agents are described according to their ability to destroy or inhibit microbial growth. Microbicidal agents cause microbial death. They are described by what they are -*cidal* for: sporicides, bactericides, fungicides, virucides.

✔ Microbistatic agents are primarily inhibitors that slow microbial growth.

✔ An antiseptic agent is applied to living tissue to destroy or inhibit microbial growth.

✔ A disinfectant agent is used on inanimate objects to destroy vegetative pathogens but not bacterial endospores.

✔ Sanitization reduces microbial numbers on inanimate objects to safe levels by physical or chemical means.

✔ Degermation refers to the process of mechanically removing microbes from the skin.

✔ Microbial death is defined as the permanent loss of reproductive capability in microorganisms.

✔ Antimicrobial agents attack specific cell sites to cause microbial death or damage. Any given antimicrobial agent attacks one of four major cell targets: the cell wall, the cell membrane, biosynthesis pathways for DNA or RNA, or protein (enzyme) function.

1. Compare sterilization with disinfection and sanitization. Describe the relationship of the concepts of sepsis, asepsis, and antisepsis.

2. Contrast various microbes and rate their variations in resistance to control agents.

3. Briefly explain how the type of microorganisms present will influence the effectiveness of exposure to antimicrobial agents.

4. Explain how the numbers of contaminants can influence the measures used to control them.

5. Precisely what is microbial death?

6. Why does a population of microbes not die instantaneously when exposed to an antimicrobial agent?

7. Distinguish between microbicidal and microbistatic effects on microbes, and tell why it is important to understand the difference.

8. Why are antimicrobial processes inhibited in the presence of extraneous organic matter?

9. Describe four modes of action of antimicrobial agents, and provide examples of how they work.

11.2 Physical Methods of Control: Heat

Expected Learning Outcomes

7. List the major types of physical agents used in controlling microbes.

8. Describe the basic effects of temperature on microbes.

9. Define *moist* and *dry heat,* and explain their modes of action and relative effectiveness.

10. Differentiate between thermal death point and thermal death time, and provide examples of their uses.

11. Describe several moist heat methods and their chief applications for sterilization and disinfection.

12. Describe some methods using dry heat and its chief applications.

13. Explain how cold temperatures and desiccation are used in microbial control.

Microorganisms have adapted to a tremendous diversity of habitats on earth, even severe conditions of temperature, moisture, pressure, and light. For microbes that normally withstand such extreme physical conditions, our usual attempts at control would probably have little effect. Fortunately for us, we are most interested in controlling microbes that flourish in the same environment in which humans live. The vast majority of these microbes are readily controlled by abrupt changes in their environment. Most prominent among antimicrobial physical agents is heat. Other less widely used agents include radiation and filtration. The following sections examine these and other methods, and explore their practical applications in medicine, commerce, and the home.

Effects of Temperature on Microbial Activities

A sudden departure from a microbe's temperature of adaptation is likely to have a detrimental effect on it. As a rule, elevated temperatures that exceed the maximum growth temperature are microbicidal, whereas lower temperatures below the minimum growth temperature are microbistatic. **Figure 11.5** shows graphically how changing temperatures affect growth and survival of a typical mesophile.

The two physical states of heat used in microbial control are moist and dry. **Moist heat** occurs in the form of hot water, boiling water, or steam (vaporized water). In practice, the temperature of moist heat usually ranges from 60°C to 135°C. As we shall see, the temperature of steam can be regulated by adjusting its pressure in a closed container. The expression **dry heat** denotes air with a low moisture content that has been heated by a flame or electric heating coil. In practice, the temperature of dry heat ranges from 160°C to several thousand degrees Celsius.

Mode of Action and Relative Effectiveness of Heat

Moist heat and dry heat differ in their modes of action as well as in their efficiency. Moist heat operates at lower temperatures and shorter exposure times to achieve the same effectiveness as dry heat (**table 11.3**). Although many cellular structures are damaged by moist heat, its

TABLE 11.3	Comparison of Times and Temperatures to Achieve Sterilization with Moist and Dry Heat	
	Temperature	**Time to Sterilize**
Moist Heat	121°C	15 min
	125°C	10 min
	134°C	3 min
Dry Heat	121°C	600 min
	140°C	180 min
	160°C	120 min
	170°C	60 min

most microbicidal effects are the coagulation and denaturation of proteins, which quickly and permanently halt cellular metabolism.

Dry heat with a moderate temperature dehydrates the cell, removing the water necessary for metabolic reactions, and it alters protein structure. However, the lack of water actually increases the stability of some protein conformations, necessitating the use of higher temperatures when dry heat is employed as a method of microbial control. At very high temperatures, dry heat oxidizes cells, burning them to ashes. This method is the one used in the laboratory when a loop is flamed or in industry when medical waste is incinerated.

Heat Resistance and Thermal Death of Spores and Vegetative Cells

Bacterial endospores exhibit the greatest resistance, and vegetative states of bacteria and fungi are the least resistant to both moist and dry heat. Destruction of spores usually requires temperatures above boiling, although resistance varies widely. The following table compares some times and temperatures required to kill sporeformers using moist versus dry heat:

Moist Heat	Temp/Time	Dry Heat	Temp/Time
Bacillus subtilis	121°C/1 min	*Bacillus subtilis*	121°C/120 min
B. stearothermophilis	121°C/12 min	*B. stearothermophilis*	140°C/5 min
Clostridium botulinum	120°C/10 min	*Clostridium botulinum*	120°C/120 min
C. tetani	105°C/10 min	*C. tetani*	100°C/60 min

In general, reliable killing of the most heat-resistant species of sporeformers requires 121°C for 20 minutes in moist heat.

Vegetative cells also vary in their sensitivity to heat, though not to the same extent as spores. Among bacteria, the death conditions for moist heat range from 50°C for 3 minutes (*Neisseria gonorrhoeae*) to 60°C for 60 minutes (*Staphylococcus aureus*). It is worth noting that vegetative cells of sporeformers are just as susceptible as vegetative cells of non-sporeformers and that pathogens are neither more nor less susceptible than nonpathogens. Other microbes, including fungi, protozoa, and worms, are rather similar in their sensitivity to heat. Viruses can be surprisingly resistant to heat, with a tolerance range extending from 55°C for 2 to 5 minutes (adenoviruses) to 60°C for 600 minutes (hepatitis A virus). For practical purposes, all non-heat-resistant forms of bacteria, yeasts, molds, protozoa, worms, and viruses are destroyed by exposure to 80°C for 20 minutes.

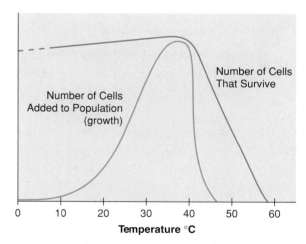

Figure 11.5 Comparative effects of temperature on growth rate and survival of a mesophilic microbe. (a) At temperatures below minimum (10°C), the microbe doesn't grow (brown) but can survive (blue). **(b)** At temperatures above maximum (45°C), the microbe rapidly stops growing and dies at a logarithmic rate with increasing temperature.

Practical Concerns in the Use of Heat: Thermal Death Measurements

Adequate sterilization requires that both temperature and length of exposure be considered. As a general rule, higher temperatures allow shorter exposure times, and lower temperatures require longer exposure times. A combination of these two variables constitutes the **thermal death time,** or TDT, defined as the shortest length of time required to kill all test microbes at a specified temperature. The TDT has been experimentally determined for the microbial species that are common or important contaminants in various heat-treated materials. Another way to compare the susceptibility of microbes to heat is the **thermal death point** (TDP), defined as the lowest temperature required to kill all microbes in a sample in 10 minutes.

Many perishable substances are processed with moist heat. Some of these products are intended to remain on the shelf at room temperature for several months or even years. The chosen heat treatment must render the product free of agents of spoilage or disease. At the same time, the quality of the product and the speed and cost of processing must be considered. For example, in the commercial preparation of canned green beans, one of the cannery's greatest concerns is to prevent growth of the agent of botulism. From several possible TDTs (that is, combinations of time and temperature) for *Clostridium botulinum* spores, the cannery must choose one that kills all spores but does not turn the beans to mush. These many considerations produce an optimal TDT for a given processing method. Commercial canneries heat low-acid foods at 121°C for 30 minutes, a treatment that sterilizes these foods. Because of such strict controls in canneries, cases of botulism from commercially canned foods are rare.

Common Methods of Moist Heat Control

The four ways that moist heat is employed to control microbes are

- steam under pressure;
- nonpressurized steam;
- boiling water; and
- pasteurization.

Sterilization with Steam under Pressure At sea level, normal atmospheric pressure is 15 pounds per square inch (psi), or 1 atmosphere. At this pressure, water will boil (change from a liquid to a gas) at 100°C, and the resultant steam will remain at exactly that temperature, which is unfortunately too low to reliably kill all microbes. In order to raise the temperature of steam, the pressure at which it is generated must be increased. As the pressure is increased, the temperature at which water boils and the temperature of the steam produced both rise. For example, at a pressure of 20 psi (5 psi above normal), the temperature of steam is 109°C. As the temperature is increased to 10 psi above normal, the steam's temperature rises to 115°C, and at 15 psi above normal (a total of 2 atmospheres), it will be 121°C. It is not the pressure by itself that is killing microbes but the increased temperature it produces.

Such pressure-temperature combinations can be achieved only with a special device that can subject pure steam to pressures greater than 1 atmosphere. Health and commercial industries use an **autoclave** for this purpose, and a comparable home appliance is the

pressure cooker. Autoclaves have a fundamentally similar plan: a cylindrical metal chamber with an airtight door on one end and racks to hold materials **(figure 11.6).** Its construction includes a complex network of valves, pressure and temperature gauges, and ducts for

(a)

(b)

Figure 11.6 Sterilization using steam under pressure.
(a) An autoclave for sterilizing media and cultures. **(b)** Cutaway section, showing autoclave components and the flow of steam through the autoclave. From John J. Perkens, *Principles and Methods of Sterilization in Health Science,* 2nd ed., 1969. Courtesy of Charles C. Thomas, Publisher, Springfield, Illinois.

regulating and measuring pressure and conducting the steam into the chamber. Sterilization is achieved when the steam condenses against the objects in the chamber and gradually raises their temperature.

Experience has shown that the most efficient pressure-temperature combination for achieving sterilization is 15 psi, which yields 121°C. It is possible to use higher pressure to reach higher temperatures (for instance, increasing the pressure to 30 psi raises the temperature to 132°C), but doing so will not significantly reduce the exposure time and can harm the items being sterilized.

The duration of the process is adjusted according to the bulkiness of the items in the load (thick bundles of material or large flasks of liquid) and how full the chamber is. The range of holding times varies from 10 minutes for light loads to 40 minutes for heavy or bulky ones; the average time is 20 minutes.

Nonpressurized Steam Selected substances that cannot withstand the high temperature of the autoclave can be subjected to *intermittent sterilization,* also called **tyndallization.**[1] This technique requires a chamber to hold the materials and a reservoir for boiling water. Items in the chamber are exposed to free-flowing steam for 30 to 60 minutes. This temperature is not sufficient to reliably kill spores, so a single exposure will not suffice. On the assumption that surviving spores will germinate into less resistant vegetative cells, the items are incubated at appropriate temperatures for 23 to 24 hours, and then again subjected to steam treatment. This cycle is repeated for 3 days in a row. Because the temperature never gets above 100°C, highly resistant spores that do not germinate could survive even after 3 days of this treatment.

Boiling Water: Disinfection A simple boiling water bath or chamber can quickly decontaminate items in the clinic and home. Because a single processing at 100°C will not kill all resistant cells, this method can be relied on only for disinfection and not for sterilization. Exposing materials to boiling water for 30 minutes will kill most non-spore-forming pathogens, including resistant species such as the tubercle bacillus and staphylococci. Probably the greatest disadvantage with this method is that the items can be easily recontaminated when removed from the water.

Pasteurization: Disinfection of Beverages Fresh beverages such as milk, fruit juices, beer, and wine are easily contaminated during collection and processing. Because microbes have the potential for spoiling these foods or causing illness, heat is frequently used to reduce the microbial load and destroy pathogens. **Pasteurization** is a technique in which heat is applied to liquids to kill potential agents of infection and spoilage, while at the same time retaining the liquid's flavor and food value.

A widely used pasteurization technique is the **flash method,** that exposes the liquid to heat exchangers at 71.6°C for 15 seconds. Some industries use the batch method which holds milk at 63°C to 66°C for 30 minutes. The first method is preferable because it is less likely to change flavor and nutrient content, and it is more effective against certain resistant pathogens such as *Coxiella* and *Mycobacterium.* Although these treatments inactivate most viruses

and destroy the vegetative stages of 97% to 99% of bacteria and fungi, they do not kill endospores or **thermoduric** microbes (mostly nonpathogenic lactobacilli, micrococci, and yeasts). Milk is not sterile after regular pasteurization. In fact, it can contain 20,000 microbes per milliliter or more, which explains why even an unopened carton of milk will eventually spoil. Newer techniques can also produce *sterile milk* that has a longer storage life. This milk is processed with ultrahigh temperature (UHT)—134°C for 2 to 5 seconds.

One important aim in pasteurization is to prevent the transmission of milk-borne diseases from infected cows or milk handlers. The primary targets of pasteurization are non-spore-forming pathogens: *Salmonella* species (a common cause of food infection), *Campylobacter jejuni* (acute intestinal infection), *Listeria monocytogenes* (listeriosis), *Brucella* species (undulant fever), *Coxiella burnetii* (Q fever), *Mycobacterium bovis, M. tuberculosis,* and several enteric viruses.

Dry Heat: Hot Air and Incineration

Dry heat is not as versatile or as widely used as moist heat, but it has several important sterilization applications. The temperatures and times employed in dry heat vary according to the particular method, but in general, they are greater than with moist heat. **Incineration** in a flame or electric heating coil is perhaps the most rigorous of all heat treatments. The flame of a Bunsen burner reaches 1,870°C at its hottest point, and furnaces/incinerators operate at temperatures of 800°C to 6,500°C. Direct exposure to such intense heat ignites and reduces microbes and other substances to ashes and gas.

Incineration of microbial samples on inoculating loops and needles using a Bunsen burner is a very common practice in the microbiology laboratory. This method is fast and effective, but it is also limited to metals and heat-resistant glass materials. Tabletop infrared incinerators **(table 11.4),** an alternative to Bunsen burners, are safer and prevent splatter of the inoculum.

The hot-air oven provides another means of dry-heat sterilization. The so-called *dry oven* is usually electric (occasionally gas) and has coils that radiate heat within an enclosed compartment. Heated, circulated air transfers its heat to the materials in the oven. Depending on the type of oven and the material being decontaminated, a cycle takes 12 minutes to 4 hours to complete and involves temperatures of 150°C to 180°C.

Table 11.4 presents an illustrated summary of the application of heat to control microbes.

The Effects of Cold and Desiccation

The principal benefit of cold treatment is to slow growth of cultures and microbes in food during processing and storage. *It must be emphasized that cold merely retards the activities of most microbes.* Although it is true that some sensitive microbes are killed by cold temperatures, most are not adversely affected by gradual cooling, long-term refrigeration, or deep-freezing. In fact, freezing temperatures, ranging from −70°C to −135°C, provide an environment that can preserve cultures of bacteria, viruses, and fungi for long periods. Recall from chapter 7 that some psychrotrophs grow very slowly even at freezing temperatures and can continue to secrete toxic products. Pathogens able to survive several months in the refrigerator are *Staphylococcus aureus; Clostridium* species (sporeformers); *Streptococcus*

1. Named for the British physicist John Tyndall, who did early experiments with sterilizing procedures.

TABLE 11.4 Applications Using Heat for Sterilization and Disinfection

Form of Heat	Equipment Temperature/Time	Medical and Commercial Uses and Limitations
Moist heat: Employs steam or boiling water; mode of action coagulates proteins		
Steam under pressure	Autoclave, pressure cooker; 121°C, pressure of 15 psi for 10 to 40 minutes, depending on load size; kills endospores	Sterilization of heat-resistant materials made of glass, cloth, rubber, metal; media; a few plastics; canned food, spices Decontamination of cultures Not appropriate for moisture-repellent oil or powder
Unpressurized steam (Tyndallization)	Arnold's Sterilizer—100°C for 30 minutes on 3 days; items are incubated between treatments to allow endospores to germinate.	Limited uses in sterilizing some media and foods that cannot be autoclaved; can be used to disinfect but not sterilize some medical supplies
Pasteurization	Pasteurizer; flash method works at 71.6°C for 15 seconds; ultrahigh temperature (UHT) uses 134°C for a few seconds.	Disinfection of milk and dairy products to destroy milk-borne pathogens; UHT can sterilize for longer shelf life. Fruit juices, beer, and wine may be pasteurized to destroy contaminants and preserve the products.
Boiling water	Water bath or pan heated to 100°C; items may be placed in water for 30 minutes to kill vegetative pathogens.	Limited to disinfection and sanitization of heat-resistant household objects such as eating utensils, clothing, sick room supplies, baby supplies, bedding, and water in emergencies; problem with recontamination while handling
Dry heat: Employs dried hot air, flame, or heat coil; works by dehydration or combustion		
Dry oven	Electric or gas chamber heated to 150°C to 180°C for 2 to 4 hours	Materials that can withstand hot temperatures and dehydration; glassware, metals, powders, oils Poor choice for liquids, rubber, plastics Time-consuming
Incineration	Bunsen burner flame up to 1,800°C; small infrared incinerators to 800°C Furnaces, incinerators up to 6,500°C	For sterilizing tips of inoculation instruments; limited to high-temperature metals Permanent disposal of waste material from laboratories; burning reduces bulky packages and garbage to ashes. For permanent disposal only

Figure A A small flash pasteurizer used by dairies for calve's milk.

Figure B **Dry heat incineration.** Infrared incinerator with shield to prevent spattering of microbial samples during flaming.

species; and several types of yeasts, molds, and viruses. Outbreaks of *Salmonella* food infection traced back to refrigerated foods such as ice cream and eggs are testimony to the inability of cold temperatures to reliably kill pathogens.

Most vegetative cells directly exposed to normal room air gradually become dehydrated, or **desiccated.*** Delicate pathogens such as *Streptococcus pneumoniae,* the spirochete of syphilis, and *Neisseria*

gonorrhoeae can die after a few hours of air-drying, but many others are not killed and some are even preserved. Staphylococci and streptococci in dried secretions and the tubercle bacillus surrounded by sputum can remain viable in air and dust for lengthy periods. Many viruses (especially nonenveloped) and fungal spores can also withstand long periods of desiccation. Endospores of *Bacillus* and *Clostridium* are viable for millions of years under extremely arid conditions. Desiccation can be a valuable way to preserve foods because it greatly reduces the amount of water available to support microbial growth.

* *desiccate* (des'-ih-kayt) To dry at normal environmental temperatures.

It is interesting to note that a combination of freezing and drying—**lyophilization***—is a common method of preserving microorganisms and other cells in a viable state for many years. Pure cultures are frozen instantaneously and exposed to a vacuum that rapidly removes the water. This method avoids the formation of ice crystals that would damage the cells. Although not all cells survive this process, enough of them do to permit future reconstitution of that culture.

As a general rule, chilling, freezing, and desiccation should not be construed as methods of disinfection or sterilization because their antimicrobial effects are erratic and uncertain, and one cannot be sure that pathogens subjected to them have been killed.

Check & Assess Section 11.2

✔ Physical methods of microbial control include heat, cold, radiation, drying, and filtration.

✔ Heat is the most widely used method of microbial control. It is used in combination with water (moist heat) or as dry heat (oven, flames).

✔ The thermal death time (TDT) is the shortest length of time required to kill all microbes at a specific temperature. The TDT is longest for spore-forming bacteria and certain viruses.

✔ The thermal death point (TDP) is the lowest temperature at which all microbes are killed in a specified length of time (10 minutes).

✔ Autoclaving, or steam sterilization, is the process by which steam is heated under pressure to sterilize a wide range of materials in a comparatively short time (minutes to hours). It is effective for most materials except water-resistant substances such as oils, waxes, and powders.

✔ Boiling water and pasteurization of beverages disinfect but do not sterilize materials.

✔ Dry heat is microbicidal under specified times and temperatures. Flame heat, or incineration, is microbicidal. It is used when total destruction of microbes and materials is desired.

✔ Chilling, freezing, and desiccation are microbistatic but not microbicidal. They are not considered true methods of disinfection because they are not consistent in their effectiveness.

10. Observe figure 11.5 and use it to explain the relationship between temperature versus growth and survival of microbes. What happens to cell structure as the temperature rises above maximum?

11. Summarize the nature, mode of action, and effectiveness of moist and dry heat on spores and vegetative cells.

12. Explain the concepts of TDT and TDP, using examples.

13. What are the minimum TDTs for vegetative cells and endospores?

14. How can the temperature of steam be raised above 100°C? Which instrument is based on this technique and how is it used?

15. What are the primary purposes of pasteurization, and what is ultrapasteurization?

16. What are several microbial targets of pasteurization?

17. Explain why desiccation and cold are not reliable methods of disinfection but how they can still be effective in microbial control.

* *lyophilization* (ly-off″-il-ih-za′-shun) Gr. *lyein,* to dissolve, and *philein,* to love.

11.3 Physical Methods of Control: Radiation

Expected Learning Outcomes

14. Define *radiation* and assess the use of irradiation as a method of microbial control.

15. Differentiate between ionizing and nonionizing radiation in their modes of action.

16. Summarize the primary uses of ionizing and nonionizing radiation in sterilization and disinfection.

Radiation as a Microbial Control Agent

Another way in which energy can serve as an antimicrobial agent is through the use of radiation. **Radiation** is defined as energy emitted from atomic activities and dispersed at high velocity through matter or space. It is characterized by a range of wavelengths known as the electromagnetic spectrum (**figure 11.7**). Although radiation exists in many energy states, we will consider only those types suitable for microbial control: gamma rays, X rays, and ultraviolet radiation.

Modes of Action of Ionizing versus Nonionizing Radiation

The actual physical effects of radiation on microbes can be understood by visualizing the process of **irradiation,** or bombardment

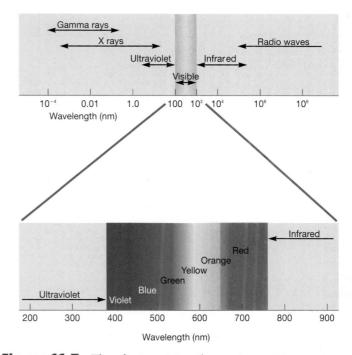

Figure 11.7 The electromagnetic spectrum. Waves range from the shortest gamma rays to the longest radiowaves. Inset shows the spectrum of visible light between ultraviolet (UV) and infrared. Only ionizing radiation, UV, and infrared have applications in microbial control.

with radiation, at the cellular level **(figure 11.8).** When a cell is bombarded by certain waves or particles, its molecules absorb some of the available energy, leading to one of two consequences: (1) If the radiation ejects orbital electrons from an atom, it causes ions to form; this type of radiation is termed **ionizing radiation.** One of the most sensitive targets for ionizing radiation is DNA, which will undergo mutations on a broad scale. Secondary lethal effects appear to be chemical changes in organelles and the production of toxic substances. Gamma rays, X rays, and high-speed electrons are all ionizing in their effects. (2) **Nonionizing radiation,** best exemplified by UV, excites atoms by raising them to a higher energy state, but it does not ionize them. This atomic excitation, in turn, leads to the formation of abnormal bonds within molecules such as DNA and is thus a source of mutations (see figure 11.10).

Figure 11.8 Cellular effects of irradiation. (a) Ionizing radiation can penetrate a solid barrier, bombard a cell, enter it, and dislodge electrons from molecules. Breakage of DNA creates massive mutations. **(b)** Nonionizing radiation enters a cell, strikes molecules, and excites them. The effect on DNA is mutation by formation of abnormal bonds. **(c)** A solid barrier cannot be penetrated by nonionizing radiation.

Ionizing Radiation: Gamma Rays, X Rays, and Cathode Rays

Over the past several years, ionizing radiation has become safer and more economical to use, and its applications have expanded. It is a highly effective alternative for sterilizing materials that are sensitive to heat or chemicals. Because it sterilizes in the absence of heat, irradiation is a type of **cold** (or low-temperature) **sterilization.**[2] Devices that emit ionizing rays include gamma-ray machines containing radioactive cobalt, X-ray machines similar to those used in medical diagnosis, and cathode-ray machines that operate like the vacuum tube in a television set. Items are placed in these machines and irradiated for a short time with a carefully chosen dosage. The dosage of radiation is measured in grays (which has replaced the older term, *rads*). Depending on the application, exposure ranges from 5 to 50 kilograys (kGy; a kilogray is equal to 1,000 grays). Although all ionizing radiations can penetrate liquids and most solid materials, gamma rays are most penetrating, X rays are intermediate, and cathode rays least penetrating.

Irradiation as a Method of Microbial Control

Foods have been subject to irradiation in limited circumstances for more than 50 years. From flour to pork and ground beef, to fruits and vegetables, radiation is used to kill not only bacterial pathogens but also insects and worms and even to inhibit the sprouting of white potatoes. **Figure 11.9** demonstrates how irradiation of fruit can greatly reduce spoilage and improve shelf life.

As soon as radiation is involved, however, consumer concern arises that food may be made less nutritious, unpalatable, or even unsafe by subjecting it to ionizing radiation. But irradiated food has been extensively studied, and each of these concerns has been addressed.

2. To clarify the use of the word *cold*, in this context, it only means the absence of heat. Beer manufacturers have sometimes used this terminology as well. When they say that their product is "cold-filtered," they mean that it has been freed of contaminants via filtration, without heating.

Raspberries exposed to X rays, 5 days later

Raspberries not treated with X rays, 5 days later

Figure 11.9 Sterilization with ionizing radiation. Many foods can be effectively sterilized by utilizing the penetrating power of ionizing radiation. The radura symbols applied to fresh foods signifies that the food has been irradiated and is required to be displayed on all food so treated.

Irradiation may lead to a small decrease in the amount of thiamine (vitamin B$_1$) in food, but this change is small enough to be inconsequential. The irradiation process does produce short-lived free radical oxidants, which disappear almost immediately (this same type of chemical intermediate is produced through cooking as well). Certain foods do not irradiate well and are not good candidates for this type of antimicrobial control. The white of eggs becomes milky and alfalfa seeds do not germinate properly. It should be emphasized that food is not made radioactive by the irradiation process, and many studies, in both animals and humans, have concluded that there are no ill effects from eating irradiated food. In fact, NASA relies on irradiated meat for its astronauts.

One of the potential benefits of irradiated foods has to do with infection control. It has been estimated that irradiation of 50% of the meat and poultry in the United States would result in 900,000 fewer cases of infection, 8,500 fewer hospitalizations, and 350 fewer deaths each year. Radiation is currently approved in the United States for the reduction of bacterial pathogens such as *E. coli* and *Salmonella* in beef and chicken, reduction of *Trichinella* worms in pork, and the reduction of pathogens and pests on fruits and vegetables. An additional benefit of irradiation is that microbes responsible for food spoilage are killed along with pathogens, leading to an increased shelf life. In any event, no irradiated food can be sold to consumers without clear labeling that this method has been used (figure 11.9).

Sterilizing medical products with ionizing radiation is a rapidly expanding field. Drugs, vaccines, medical instruments (especially plastics), and other delicate materials can be irradiated without harming them. Its main advantages include speed, high penetrating power (it can sterilize materials through outer packages and wrappings), and the absence of heat. Its main disadvantages are potential dangers from factory exposure to radiation and possible damage to some materials. See **table 11.5** for a summary of the applications of ionizing radiation.

Nonionizing Radiation: Ultraviolet Rays

Ultraviolet (UV) radiation ranges in wavelength from approximately 100 nm to 400 nm. It is most lethal from 240 nm to 280 nm (with a peak at 260 nm). In everyday practice, the source of UV radiation is the germicidal lamp, which generates radiation at 254 nm. Owing to its lower energy state, UV radiation is not as penetrating as ionizing radiation. Because UV radiation passes readily through air, slightly through liquids, and only poorly through solids, the object to be disinfected must be directly exposed to it for full effect.

As UV radiation passes through a cell, it is initially absorbed by DNA. Specific molecular damage occurs on the pyrimidine bases (thymine and cytosine), which form abnormal linkages with each other called **pyrimidine dimers (figure 11.10).** These bonds occur between adjacent bases on the same DNA strand and interfere with normal DNA replication and transcription. The results are inhibition of growth and cellular death. In addition to altering DNA directly, UV radiation also disrupts cells by generating toxic photochemical products called free radicals. These highly reactive molecules interfere with essential cell processes by binding to DNA,

Figure 11.10 Formation of pyrimidine dimers by the action of ultraviolet (UV) radiation. This shows what occurs when two adjacent thymine bases on one strand of DNA are induced by UV rays to bond laterally with each other. The result is a thymine dimer shown in greater detail. Dimers can also occur between adjacent cytosines and thymine and cytosine bases. If they are not repaired, dimers can prevent that segment of DNA from being correctly replicated or transcribed. Massive dimerization is lethal to cells.

RNA, and proteins. Ultraviolet rays are a powerful tool for destroying fungal cells and spores, bacterial vegetative cells, protozoa, and viruses. Bacterial spores are about 10 times more resistant to radiation than are vegetative cells, but they can be killed by increasing the time of exposure.

Ultraviolet radiation is usually directed at disinfection rather than sterilization. Germicidal lamps can cut down on the concentration of airborne microbes by as much as 99%. Ultraviolet disinfection of air has proved effective in reducing postoperative infections, preventing the transmission of infections by respiratory droplets, and curtailing the growth of microbes in food-processing plants and slaughterhouses.

Ultraviolet irradiation of liquids requires special equipment to spread the liquid into a thin, flowing film that is exposed directly to a lamp. This method can be used to treat water (**figure 11.11**) and to purify other liquids (milk and fruit juices) as an alternative to heat. The surfaces of solid, nonporous materials such as walls and floors, as well as meat, nuts, tissues for grafting, and drugs, can be successfully disinfected with UV. Several types of small UV disinfection appliances have been marketed for personal items.

One major disadvantage of UV is its poor powers of penetration through solid materials such as glass, metal, cloth, plastic, and even paper. Another drawback to UV is the damaging effect of overexposure on human tissues, including sunburn, retinal damage, cancer, and skin wrinkling. These harmful results occur only if a person is directly exposed to the UV rays, but it necessitates placement of barriers to cut down on this possibility. Table 11.5 summarizes UV applications.

TABLE 11.5 Applications Using Radiation for Sterilization and Disinfection

Form of Radiation	Equipment/Mode of Action	Medical and Commercial Uses and Limitations
Ionizing radiation		
Sporicidal and penetrating	Cobalt radiation gives off gamma rays; others are X rays, cathode rays. Rays fragment DNA and create ionized bases; highly mutagenic.	Cold sterilization of packaged products such as medical supplies and instruments that cannot be heat sterilized
		Sterilization of foods, including fruits, lettuce and spinach, grains, spices, meats, and packaged meals
		Sterilization of drugs and vaccines
		Rapid and penetrating, prevents infection and spoilage
Nonionizing ultraviolet (UV) radiation		
Sporicidal and nonpenetrating	Germicidal lamps or generators set to deliver UV rays from 240 to 280 nm in wavelength; action creates abnormal bonds between adjacent thymines and cytosines and produces deletion mutations on DNA.	Mostly disinfection of air, some liquids, a few solids; reduces airborne pathogens in medical and dental offices, hospital rooms, and operating rooms
		Commercial use for clean rooms, food processing areas
		Works on liquids such as water, milk, drugs, vaccines if exposed to UV in thin layers
		Solid materials such as room surfaces, foods, tissue grafts, drugs, and instruments
		Lack of penetration through solids limits applications; direct exposure can be dangerous to eyes and skin.

Figure A Pallets of apples are being irradiated with a cobalt 60 machine that emits gamma rays.

Figure B Table top chamber generates ultraviolet rays to disinfect a toothbrush.

Figure 11.11 An ultraviolet (UV) treatment system for disinfection of water. Water is directed through channels at a water treatment plant, past racks of UV lamps (blue glow). This system has a capacity of several million gallons per day and can be used as an alternative to chlorination. Home systems that fit under the sink are also available.

11.4 Using Filtration to Remove Microbes

Expected Learning Outcomes

17. Describe the concept behind sterilization by filtration, including examples of applications.

Filtration is an effective method to remove microbes from air and liquids. In practice, a fluid is strained through a filter with openings large enough for the fluid to pass through but too small for microorganisms to pass through (**figure 11.12a**).

Most modern microbiological filters are thin membranes of cellulose acetate, polycarbonate, and a variety of plastic materials (Teflon, nylon) whose pore size can be carefully controlled and standardized. Ordinary substances such as charcoal, diatomaceous earth, or unglazed porcelain are also used in some applications. Viewed microscopically, most filters are perforated by very precise, uniform pores (**figure 11.12b**). The pore diameters vary from coarse (8 µm) to ultrafine (0.02 µm), permitting selection of the minimum particle size to be trapped. Those with even smaller pore diameters permit true sterilization by removing viruses, and some will even remove large proteins. A sterile liquid filtrate is typically produced by suctioning the liquid through a sterile filter into a presterilized container. These filters are also used to separate mixtures of microorganisms and to enumerate bacteria in water analysis.

Applications of Filtration Sterilization

Filtration sterilization is used to prepare liquids that cannot withstand heat, including serum and other blood products, vaccines, drugs, IV fluids, enzymes, and media. Filtration has been employed as an alternative method for sterilizing milk and beer without altering their flavor. It is also an important step in water purification. Its use extends to filtering out particulate impurities (crystals, fibers) that can cause severe reactions in the body. It has the disadvantage of not removing soluble molecules (toxins) that can cause disease.

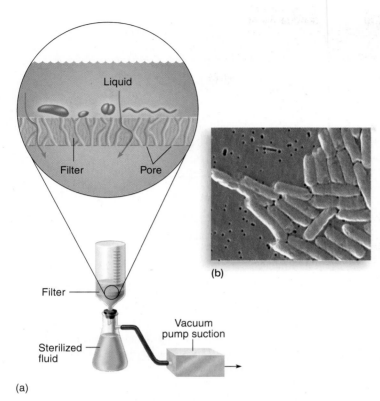

Figure 11.12 Membrane filtration. (a) Vacuum assembly for achieving filtration of liquids through suction. Inset shows filter as seen in cross section, with tiny passageways (pores) too small for the microbial cells to enter but large enough for liquid to pass through. **(b)** Scanning electron micrograph of *Salmonella* trapped on the surface of a membrane filter (10,400×). Average cells measure about 0.5 μm by 2 μm, and average pore size is about 0.1 μm.

Filtration is also an efficient means of removing airborne contaminants that are a common source of infection and spoilage. High-efficiency particulate air (HEPA) filters are widely used to provide a flow of sterile air to hospital rooms and sterile rooms.

Check & Assess Sections 11.3 and 11.4

✔ Radiation composed of high energy electromagnetic rays are highly microbicidal.

✔ Irradiation with ionizing radiation (cold sterilization) by gamma rays and X rays damages DNA and cell organelles by producing disruptive ions.

✔ Irradiation has numerous applications in sterilizing heat-sensitive items such as medical supplies, fruits, meats, and spices.

✔ Ultraviolet light, or nonionizing radiation, has limited penetrating ability. It is therefore restricted to disinfecting air and certain liquids and solids.

✔ Sterilization by filtration removes microbes from heat-sensitive liquids and circulating air. The pore size of the filter determines what kinds of microbes are removed.

18. What are some advantages and disadvantages of ionizing radiation as a method of control?
19. Describe some applications of ionizing radiation.
20. What is the precise mode of action of ultraviolet radiation?
21. How are its applications limited?
22. What are the advantages and disadvantages of using filtration as a method of control?

11.5 Chemical Agents in Microbial Control

Expected Learning Outcomes

18. Define the terms that are used to describe different methods of chemical control of microbes.

19. Explain the desirable features of antimicrobial chemicals and what factors influence their effectiveness.

20. Describe the types of halogens, their modes of action, and their clinical and commercial applications.

21. Identify the characteristics of phenolic compounds and alcohols, explain how they act, and evaluate their primary uses.

22. Describe the characteristics of oxidizing agents and aldehydes, and summarize their main applications.

23. Differentiate among the sterilizing gases and aerosols, and describe how they are used.

24. Explain how detergents, soaps, and heavy metals can be employed in clinical, commercial, and household microbial control.

Chemical control of microbes probably emerged as a serious science in the early 1800s, when physicians used chloride of lime and iodine solutions to treat wounds and to wash their hands before surgery. At the present time, approximately 10,000 different antimicrobial chemical agents are manufactured; probably 1,000 of them are used routinely in the health care and home environments. A genuine need exists to avoid infection and spoilage, but the abundance of products available to "kill germs," "disinfect," "antisepticize," "clean and sanitize," "deodorize," "fight plaque," and "purify the air" indicates a preoccupation with eliminating microbes from the environment that, at times, seems excessive **(Insight 11.2).**

Antimicrobial chemicals occur in the liquid, gaseous, or even solid state. They serve as **disinfectants, antiseptics, sterilants** (chemicals that sterilize), degermers, or preservatives (chemicals that inhibit the deterioration of substances). In most cases, solid or gaseous antimicrobial chemicals are dissolved in water, alcohol, or a mixture of the two to produce a liquid solution. Solutions containing pure water as the solvent are termed **aqueous,** whereas those dissolved in pure alcohol or water-alcohol mixtures are termed **tinctures.**

INSIGHT 11.2

Pathogen Paranoia: "The Only Good Microbe Is a Dead Microbe"

The sensational publicity over influenza, MRSA skin infections, and other outbreaks has had a dramatic impact on the public awareness of microorganisms. This glut of coverage has also had certain benefits, such as greater emphasis on hand washing, food handling, and personal hygiene. But sometimes a little knowledge can be dangerous. The trend also seems to have escalated into an obsessive fear of "germs" lurking around every corner and a fixation on eliminating microbes from the environment and the human body.

As might be expected, commercial industries have found a way to capitalize on those fears. Every year, a number of new products are released that incorporate antibacterial or germicidal "protection." A widespread array of cleansers and commonplace materials now contain antimicrobial chemicals. First, it was hand soaps and dishwashing detergents, and eventually, the list grew to include shampoos, laundry aids, hand lotions, foot pads for shoes, deodorants, sponges, kitty litter, acne medication, cutting boards, garbage bags, toys, and toothpaste.

One chemical agent routinely added to these products is a phenolic called *triclosan* (Irgasan). This substance is fairly mild and nontoxic and does indeed kill most pathogenic bacteria. However, it does not reliably destroy viruses or fungi and has been linked to cases of skin rashes due to hypersensitivity.

Medical experts are concerned that the widespread overuse of these antibacterial chemicals could favor the survival and growth of resistant strains of bacteria. One study reported that pathogens such as *Mycobacterium tuberculosis* and *Clostridium* are naturally resistant to triclosan and that *E. coli* and *Staphylococcus aureus* have developed decreased sensitivity to it. One fear has been that overuse of this chemical could select for "super bugs" that can survive all attempts at disinfection. Evidently, this has already happened. New research uncovered mutated species of *Pseudomonas* and *Acinetobacter,* both serious hospital pathogens, that are 12 times more resistant than other bacteria to triclosan.

But that's not all. More and more studies are showing that when some bacteria become resistant to antibacterial agents, including triclosan,

The molecular structure of triclosan, also known as Irgasan and Ster-Zac, a phenol-based chemical that destroys bacteria by disrupting cell walls and membranes.

they simultaneously become resistant to antibiotics, such as tetracycline and erythromycin.

Another outcome of overuse of environmental germicides is to reduce the natural contact with microbes that is required to maintain the normal microbial residents of the body and stimulate immunities. Constant use of these agents could shift the balance in the microbiota by killing off harmless or beneficial microbes.

An unfortunate side effect of negative views of microbes is how they foster the feeling that all microbes are harmful. We must not forget that most human beings manage to remain healthy despite the fact that they live in continual intimate contact with microorganisms. We really do not have to be preoccupied with microbes every minute or feel overly concerned that the things we touch, drink, or eat are sterile, as long as they are somewhat clean and free of pathogens. For most of us, resistance to infection is well maintained by our numerous host defenses.

Infectious disease specialists urge a happy medium approach. Instead of filling the home with questionable germicidal products, they encourage cleaning with traditional soaps and detergents, reserving more potent products to reduce the spread of infection in clinical settings.

Can you name other side effects of mass usage of antimicrobial chemicals in addition to those mentioned here? *Answer available at http://www.mhhe.com/talaro8*

Choosing a Microbicidal Chemical

The choice and appropriate use of antimicrobial chemical agents are of constant concern in medicine and dentistry. Although actual clinical practices of chemical decontamination vary widely, some desirable qualities in a germicide have been identified, including:

- rapid action in low concentrations,
- solubility in water or alcohol and long-term stability,
- broad-spectrum microbicidal action without being toxic to human and animal tissues,
- penetration of inanimate surfaces to sustain a cumulative or persistent action,
- resistance to becoming inactivated by organic matter,
- noncorrosive or nonstaining properties,
- sanitizing and deodorizing properties,
- affordability and ready availability, and
- nonoffensive odor.

As yet, no chemical can completely fulfill all of those requirements, but chlorhexidine and hydrogen peroxide approach this ideal. At the same time, we should question the rather overinflated claims made about certain commercial agents such as mouthwashes and disinfectant air sprays.

Germicides are evaluated in terms of their effectiveness in destroying microbes in medical and dental settings. The three levels of chemical decontamination procedures are *high, intermediate,* and *low* **(table 11.6).**

- High-level germicides kill endospores and, if properly used, are sterilants. Materials that necessitate high-level control are medical devices—for example, catheters, heart-lung equipment, and implants—that are not heat-sterilizable and are intended to enter body tissues during medical procedures.
- Intermediate-level germicides kill fungal (but not bacterial) spores, resistant pathogens such as the tubercle bacillus, and

TABLE 11.6 **Qualities of Chemical Agents Used in Health Care**

Agent	Target Microbes	Level of Activity	Toxicity	Comments
Chlorine	Sporicidal (slowly)	Intermediate	Gas is highly toxic; solution irritates skin	Inactivated by organics; unstable in sunlight
Iodine	Sporicidal (slowly)	Intermediate	Can irritate tissue; toxic if ingested	Iodophors* are milder forms
Phenolics	Some bacteria, viruses, fungi	Low to intermediate	Can be absorbed by skin; can cause CNS damage	Poor solubility; expensive
Chlorhexidine*	Most bacteria, some viruses, fungi	Low to intermediate	Low toxicity	Fast-acting, mild, has residual effects
Alcohols	Most bacteria, viruses, fungi	Intermediate	Toxic if ingested; a mild irritant; dries skin	Flammable, fast-acting
Hydrogen peroxide,* stabilized	Sporicidal	High	Toxic to eyes; toxic if ingested	Improved stability; works well in organic matter
Quaternary ammonium compounds	Some bactericidal, virucidal, fungicidal activity	Low	Irritating to mucous membranes; poisonous if taken internally	Weak solutions can support microbial growth; easily inactivated
Soaps	Certain very sensitive species	Very low	Nontoxic; few if any toxic effects	Used for removing soil, oils, debris
Mercurials	Weakly microbistatic	Low	Highly toxic if ingested, inhaled, absorbed	Easily inactivated
Silver nitrate	Bactericidal	Low	Toxic, irritating	Discolors skin
Glutaraldehyde*	Sporicidal	High	Can irritate skin; toxic if absorbed	Not inactivated by organic matter; unstable
Formaldehyde	Sporicidal	Intermediate to high	Very irritating; fumes damaging, carcinogenic	Slow rate of action; limited applications
Ethylene oxide gas*	Sporicidal	High	Very dangerous to eyes, lungs; carcinogenic	Explosive in pure state; good penetration; materials must be aerated
Dyes	Weakly bactericidal, fungicidal	Low	Low toxicity	Stains materials, skin

*These chemicals approach the ideal by having many of the following characteristics: broad spectrum, low toxicity, fast action, penetrating abilities, residual effects, stability, potency in organic matter, and solubility.

viruses. They are used to disinfect items (respiratory equipment, thermometers) that come into intimate contact with the mucous membranes but are noninvasive.

- Low levels of disinfection eliminate only vegetative bacteria, vegetative fungal cells, and some viruses. They are used to clean materials such as electrodes, straps, and furniture that touch the skin surfaces but not the mucous membranes.

Factors That Affect the Germicidal Activities of Chemical Agents

Several concepts of microbial control introduced in section 11.1 apply to germicides and other chemical agents. For instance, they target the same cellular structures and functions previously discussed: cell walls and membranes, proteins, and nucleic acids. The effectiveness of the agents depends on several factors. Among these are (1) the numbers and kinds of microbes that are present, (2) the kinds of

materials that are being treated, (3) the time of exposure required, and (4) the strength and mode of action of the agent. Here, we focus on the latter two because of their importance to chemical agents.

Length of exposure Most compounds require adequate contact time for the chemical to penetrate and to act on the microbes present. The composition of the material being treated can greatly affect this process. Smooth, solid objects are more rapidly disinfected than are those with pores or pockets that can trap soil. An item contaminated with common biological matter such as serum, blood, saliva, pus, fecal material, or urine presents an additional barrier to effective disinfection. Large amounts of this organic material can hinder penetration and, in some cases, can form bonds that reduce a chemical's activity. Adequate cleaning of instruments and other reusable materials will ensure that the germicide or sterilant will better accomplish the job for which it was chosen.

Concentration of the Chemical Agent A chemical's strength or concentration is expressed in various ways, depending upon convention and the method of preparation. The content of many chemical agents can be formulated in more than one way. For dilutions, a small volume of the liquid chemical (solute) is diluted in a larger volume of solvent to achieve a certain ratio. For example, a common laboratory phenolic disinfectant such as Lysol is usually diluted 1:200; that is, one part of chemical (solute) has been added to 200 parts of water (solvent) by volume. Solutions such as chlorine that are effective in very diluted concentrations are expressed in parts per million (ppm). In percentage solutions, the solute is added to solvent by weight or volume to achieve a certain percentage in the solution. Alcohol, for instance, is used in percentages ranging from 50% to 95%. In general, solutions of low dilution or high percentage have more of the active chemical (are more concentrated) and tend to be more germicidal, but expense and potential toxicity can necessitate using the minimum strength that is effective.

Standardized procedures for testing the effectiveness of germicides are summarized in **appendix C.**

Categories of Chemical Agents

Several general groups of chemical compounds are widely used for antimicrobial purposes in medicine and commerce (see table 11.6). Prominent agents include halogens, phenolic compounds, alcohols, oxidizers, aldehydes, gases, detergents, and heavy metals. These groups are surveyed in the following section from the standpoint of each agent's specific forms and modes of action. Applications and methods of delivery are described in tables.

The Halogen Antimicrobial Chemicals

The **halogens*** are fluorine, bromine, chlorine, and iodine, a group of nonmetallic elements with similar chemical properties. Although they can exist in either the ionic (halide) or nonionic state, most halogens exert their antimicrobial effect primarily in the nonionic state, not the halide state (chloride, iodide, for example). Because fluorine and bromine are difficult and dangerous to handle and no more effective than chlorine and iodine, only the latter two are used routinely in germicidal preparations. These elements are highly effective components of disinfectants and antiseptics because they are microbicidal and not just microbistatic, and they are sporicidal with longer exposure. For these reasons, halogens are the active ingredients in nearly one-third of all antimicrobial chemicals currently marketed.

Chlorine and Its Compounds Chlorine has been used for disinfection and antisepsis for approximately 200 years. The major forms used in microbial control are liquid and gaseous chlorine (Cl_2), hypochlorites (OCl), and chloramines (NH_2Cl). In solution, these compounds combine with water and release hypochlorous acid (HOCl):

$$Cl_2 + H_2O \rightarrow HC_1 + HClO$$

$$Ca(OCl)_2 + 2H_2O \rightarrow Ca(OH)_2 + 2HClO$$

$$HClO \rightarrow HCl + O$$

This reaction oxidizes the sulfhydryl (S—H) group on the amino acid cysteine and breaks disulfide (S—S) bridges on numerous enzymes. The resulting denaturation of the enzymes is permanent and suspends metabolic reactions.

Death of almost all microorganisms usually occurs within 30 minutes, although endospores may require several hours. Chlorine compounds are less effective and relatively unstable if exposed to light, alkaline pH, and excess organic matter.

Iodine and Its Compounds Iodine (I_2) is a pungent blue-black element that forms a brown-colored solution when dissolved in water. It rapidly penetrates the cells of microorganisms, where it disturbs a variety of metabolic functions by interfering with the disulfide bonds of proteins. This mode of action is similar to that seen in chlorine, although iodine is less susceptible to inactivation by organic material. All classes of microorganisms can be killed if proper exposure times and iodine concentrations are used.

See **table 11.7** for information on the main uses and forms of halogens.

Phenol and Its Derivatives

Phenol (carbolic acid) is an acrid, poisonous compound derived from the distillation of coal tar. First adopted by Joseph Lister in 1867 as a surgical germicide, phenol was the major antimicrobial chemical until other phenolics with fewer toxic and irritating effects were developed. Solutions of phenol are now used only in certain limited cases, but it remains one standard against which other phenolic disinfectants are rated. The *phenol coefficient* quantitatively compares a chemical's antimicrobic properties to those of phenol. Substances chemically related to phenol are often referred to as phenolics. Hundreds of these chemicals are now available.

Phenolics consist of one or more aromatic carbon rings with added functional groups **(figure 11.13).** Among the most important

Phenol
(basic aromatic
ring structure) *o*-cresol *p*-cresol

Chlorophene
(a chlorinated phenol) Hexachlorophene
(a bisphenol)

Figure 11.13 Examples of phenolic structures. All contain a basic aromatic ring, but they differ in the types of additional compounds such as Cl and CH_3.

* *halogens* (hay'-loh-jenz) Gr. *halos,* salt, and *gennan,* to produce.

TABLE 11.7	Applications of Halogen Compounds	
Form of Chlorine	**Primary Applications**	**How Delivered/Concerns**
Chlorine Gas (Cl$_2$)	Large-scale disinfection of drinking water, sewage, and wastewater	Chlorination to a concentration of 0.6 to 1.0 parts of chlorine per million parts of water destroys most vegetative pathogens.*
Hypochlorites (HClO) Bleach	Used extensively in sanitization and disinfection of food equipment, treatment of swimming pools, spas, drinking water, and fresh foods; for wound antisepsis and routine medical and household disinfection, deodorizing, and stain removal	Common household bleach is a 5% solution of sodium hypochlorite; dilutions of 1:10–1:1000 are highly effective germicides.
Chloramines (Dichloramine, Halazone)	An alternative to pure chlorine in treating water supplies; also as sanitizers and disinfectants; for treating wounds and skin surfaces	Because standard gas chlorination of water is now believed to produce unsafe levels of (trihalomethanes), some water districts are required to use chloramine treatment of water supplies.
Form of Iodine	**Primary Applications**	**How Delivered/Concerns**
Iodophors	Most common iodine for skin and mucous membranes; antiseptic prep for surgery and injections; for surgical handscrubs; to disinfect equipment and surfaces; possibly for burns; and may be an alternative preventive for eye infections in newborns	A complex of iodine and a neutral protein polymer provides slow release and reduced toxicity or irritation of tissues; less prone to staining. Common products are Betadine and Povidone (PVP), which contain 2% to 10% of available iodine.
Aqueous or Tinctures	Topical antiseptic prior to surgery; sometimes for burned or injured skin	Weak solutions of 1% to 3% in water or in alcohol tinctures
	Medium-level disinfection for plastic instruments, thermometers; tablet form available for disinfection of contaminated water	Aqueous solutions or tinctures of 5% to 10%; somewhat limited by their toxicity and tendency to stain

Figure A Water is passed through a chlorination tank during final processing .

Figure B Use of povidone-iodine sponge to prepare skin for a surgical incision.

*Some cyst-forming protozoans such as *Giardia* and *Cryptosporidium* can survive this level of chlorination.

are alkylated phenols (cresols), chlorinated phenols, and bisphenols. In high concentrations, they are cellular poisons, rapidly disrupting cell walls and membranes and precipitating proteins; in lower concentrations, they inactivate certain critical enzyme systems. The phenolics are strongly microbicidal and will destroy vegetative bacteria (including the tuberculosis bacillus), fungi, and most viruses (not hepatitis B), but they are not reliably sporicidal. Their continued activity in the presence of organic matter and their detergent actions contribute to their usefulness. Unfortunately, the toxicity of many of the phenolics makes them too dangerous to use as antiseptics.

Chlorhexidine

The compound chlorhexidine (Hibiclens, Hibitane) is a complex organic base containing chlorine and two phenolic rings. Its mode of action targets both cell membranes by lowering surface tension and protein structure by causing denaturation. At moderate to high concentrations, it is bactericidal for both gram-positive and gram-negative bacteria but inactive against spores. Its effects on viruses and fungi vary. It possesses distinct advantages over many other antiseptics because of its mildness, low toxicity, and rapid action, and it is not absorbed into deeper tissues to any extent. It is the antiseptic cleanser of choice to control *Staphylococcus* MRSA and *Acinetobacter* outbreaks in hospitals.

Applications for phenolic compounds and chlorhexidine are given in **table 11.8**.

Alcohols as Antimicrobial Agents

Alcohols are colorless hydrocarbons with one or more —OH functional groups. Of several alcohols available, only ethyl and isopropyl are suitable for microbial control. Methyl alcohol is not particularly microbicidal, and more complex alcohols are either poorly soluble in water or too expensive for routine use. Alcohols are employed alone in aqueous solutions or as solvents for tinctures (with iodine, for example).

TABLE 11.8	Applications of Phenolics and Chlorhexidene	
Form of Phenolic	**Primary Applications**	**How Delivered/Concerns**
Lysol and Creolin	Common household versions of phenol; for low or intermediate levels of disinfection in the hospital	1% to 3% emulsions that are combined with soap; can be too toxic for antisepsis; tend to be absorbed by membranes into the blood
Bisphenols	Widely employed commercially, clinically, and in the home; Lysol spray often used in hospital and laboratory disinfection	Orthophenyl phenol is the major ingredient in disinfectant aerosol sprays. This same phenolic is also found in some compounds.
Hexachlorophene	Was once a widespread additive of cleansing soaps (pHisoHex) for hospital and home use; now, occasionally used to control outbreaks of skin infections	When hexachlorophene was found to be absorbed through the skin and a cause of neurological damage, it was no longer available without a prescription.
Triclosan (Dichlorophenoxyphenol)	Widely used antibacterial compound added to soaps, cosmetics, and many other household products	It acts as both disinfectant and antiseptic and is broad-spectrum in its effects.
Form of Chlorhexidine	**Primary Applications**	**How Delivered/Concerns**
Chlorhexidine (Hibiclens, Hibitane)	Alcoholic or aqueous solutions are now commonly used for hand scrubbing, preparing skin sites for surgical incisions and injections, and whole-body washing. Solutions also serve as an obstetric and neonatal wash, a wound degermer, a mucous membrane irrigant, and a preservative for eye solutions.	Complex base containing chlorine and two phenolic rings; mode of action targets cell membranes and protein structure. At moderate to high concentrations, it destroys both gram-positive and gram-negative bacteria but not spores. It may be fungicidal and virucidal. It is milder and less toxic than phenolics and is not absorbed into the skin.

Alcohol's mechanism of action depends in part upon its concentration. Concentrations of 50% and higher dissolve membrane lipids, disrupt cell surface tension, and compromise membrane integrity. Alcohol that has entered the protoplasm denatures proteins through coagulation but only in alcohol-water solutions of 50% to 95%. Alcohol is the exception to the rule that higher concentrations of an antimicrobial chemical have greater microbicidal activity. Because water is needed for proteins to coagulate, alcohol shows a greater microbicidal activity at 70% concentration (that is, 30% water) than at 100% (0% water). Absolute alcohol (100%) dehydrates cells and inhibits their growth but is generally not a protein coagulant.

Although useful in intermediate- to low-level germicidal applications, alcohol does not destroy bacterial spores at room temperature. Alcohol can, however, destroy resistant vegetative forms, including tuberculosis bacteria and fungal spores, provided the time of exposure is adequate. Alcohol tends to inactivate enveloped viruses more readily than nonenveloped viruses such as poliovirus and hepatitis A virus.

Hydrogen Peroxide and Related Germicides

Hydrogen peroxide (H_2O_2) is a colorless, caustic liquid that decomposes in the presence of light, metals, or catalase into water and oxygen gas. Early formulations were unstable and inhibited by organic matter, but manufacturing methods now permit synthesis of H_2O_2 so stable that even dilute solutions retain activity through several months of storage.

The germicidal effects of hydrogen peroxide are due to the direct and indirect actions of oxygen. Oxygen forms hydroxyl free radicals (—OH), which, like the superoxide radical (see chapter 7),

are highly toxic and reactive to cells. Although most microbial cells produce catalase to inactivate the metabolic hydrogen peroxide, it cannot neutralize the amount of hydrogen peroxide entering the cell during disinfection and antisepsis. Hydrogen peroxide is bactericidal, virucidal, fungicidal, and, in higher concentrations, sporicidal.

A number of clinical procedures involve delicate reusable instruments such as endoscopes and dental handpieces. Because these devices can become heavily contaminated by tissues and fluids, they need to undergo sterilization, not just disinfection, between patients to prevent transmission of infections such as hepatitis, tuberculosis, and genital warts. These very effective and costly diagnostic tools (a colonoscope may cost up to $30,000) have created another dilemma. They may trap infectious agents where they cannot be easily removed; and they are delicate, complex, and difficult to clean. Traditional methods are either too damaging to the instruments (heat) or too slow to sterilize them in a timely fashion between patients (ethylene oxide gas). The need for effective rapid sterilization has led to the development of low-temperature sterilizing cabinets that contain liquid chemical sterilants (**figure 11.14**).

Coverage on the applications of alcohols and peroxides is provided in **table 11.9**.

Aldehyde Sterilants and Disinfectants

Organic substances bearing a —CHO functional group (a strong reducing group) on the terminal carbon are called aldehydes. The two aldehydes used most often in microbial control are *glutaraldehyde* and *formaldehyde*.

Glutaraldehyde is a yellow liquid with a mild odor. The molecule's two aldehyde groups favor the formation of polymers. The

Figure 11.14 Processing cabinet for sterilizing medical equipment with peroxide. Endoscope is coiled in the chamber and treated for 30 minutes.

mechanism of activity involves cross-linking protein molecules on the cell surface. In this process, amino acids are alkylated, meaning that a hydrogen atom on an amino acid is replaced by the glutaraldehyde molecule itself **(figure 11.15).** It can also irreversibly disrupt the activity of enzymes within cells. Glutaraldehyde is rapid and broad-spectrum, and is one of the few chemicals officially accepted as a sterilant and high-level disinfectant. It kills spores in 3 hours and fungi and vegetative bacteria (even *Mycobacterium* and *Pseudomonas*) in a few minutes. Viruses, including the most resistant forms, appear to be inactivated after relatively short exposure times. A special device known as a chemiclave (similar to an autoclave) vaporizes the solution under pressure to sterilize delicate instruments without heat. Glutaraldehyde retains its potency even in the presence of organic matter, is noncorrosive, does not damage plastics, and is less toxic or irritating than formaldehyde. Its principal disadvantage is that it is somewhat unstable, especially with increased pH and temperature.

TABLE 11.9	Applications of Alcohols and Peroxides (and Other Oxidizing Agents)	
Form of Alcohol*	**Primary Applications**	**How Delivered/Concerns**
Ethyl Alcohol (Ethanol, Grain Alcohol)	Skin degerming agent and antiseptic; surfactant action removes skin oil, soil, and some microbes sheltered in deeper skin layers; occasionally used to disinfect electrodes and face masks	Solutions of 70% to 95% are germicidal, inexpensive, nonirritating; one limitation is its rate of evaporation. Items need to be first cleaned and then soaked in alcohol for 15 to 20 minutes.
Isopropyl Alcohol (Rubbing Alcohol)	Some disinfection of objects, surfaces; limited skin cleansing	More microbicidal and less expensive than ethanol, but these benefits must be weighed against its toxicity; inhaling its vapors can adversely affect the nervous system.
Form of Oxidizing Agent	**Primary Applications**	**How Delivered/Concerns**
Hydrogen Peroxide (H_2O_2)	Versatile uses as an antiseptic, including skin and wound cleansing, bedsore care, and mouthwashing; disinfectant for soft contact lenses, surgical implants, plastics, bedding, and room interiors	3% hydrogen peroxide—most common form—is useful in treating infections by anaerobic bacteria because of the lethal effects of the oxygen released.
Sterilizing H_2O_2	Vaporized H_2O_2 is a major type of sterilant; hydrogen peroxide plasma sterilizers used for industrial parts or medical items; for isolators, clean rooms, and space vehicles	Hydrogen peroxide (35%) penetrates into delicate medical instruments, kills the most resistant microbes, and does not corrode small parts; vapors are sporicidal.
Peracetic Acid	Used to sterilize rooms, space shuttle; decontamination of large areas	Oxidizing agent—similar actions to hydrogen peroxide
Ozone (O_3)	For disinfection of air, water, industrial air conditioners, and cooling towers	Difficult to handle

Figure A A nurse takes advantage of an alcohol gel hand cleansing station. These are very common in hospitals and clinics to encourage hand antisepsis.

Figure B Disinfection of contact lenses with peroxide-based cleaner. Note the bubbles produced, indicating the release of oxygen gas.

*Both forms of alcohol are in alcohol gels used for universal precautions and hand degerming in hospitals.

Figure 11.15 Actions of glutaraldehyde on cell walls of gram-negative and gram-positive cells. The molecule polymerizes easily. When these alkylating polymers react with the amino groups (N) on amino acids, they cross-link and inactivate proteins.

CONTINUING CASE FILE 11

By mid-2008, investigators had linked 9 cases and 106 possible cases to the facility. It was the largest outbreak of health care-associated hepatitis ever reported. Not only were 50,000 clinic patients tested for HCV, all of them were directed to have additional tests for hepatitis B virus (HBV) and human immunodeficiency virus (HIV).

In the meantime, the big question was: What went wrong at the endoscopy center? Teams of epidemiologists reviewed the records on the days the patients had procedures. Essentially, their investigation identified four potential problem areas and possible modes of transmission: staff to patient, contaminated equipment, IV technique, and method of preparing syringes for sedation drugs. First, they tested all of the health care workers at the clinic for HCV. Endoscopes are frequently suspected in outbreaks because they have to be reused and disinfected between patients. They found that the endoscopes were scrubbed with an enzymatic cleaner, soaked in glutaraldehyde for 30 minutes, and flushed with alcohol. Other equipment was either new and sterile or disinfected in the clinic. The IV lines were inserted in a standard fashion with sterile needles and syringes. Protocols for preparing drugs for sedation were not standardized and varied among the nurse assistants who did the procedures.

- Why was it necessary to also test the patients for HBV and HIV?
- Which of the clinic's practices was responsible for the outbreak?

For a wrap-up, see the Case File Perspective on page 347.

Formaldehyde is a sharp, irritating gas that readily dissolves in water to form an aqueous solution called **formalin.** Pure formalin is a 37% solution of formaldehyde gas dissolved in water. The chemical is microbicidal through its attachment to nucleic acids and functional groups of amino acids. Formalin is an intermediate- to high-level disinfectant, but its extreme toxicity (it is classified as a carcinogen) and irritating effects on the skin and mucous membranes greatly limit its clinical usefulness.

Another aldehyde, ortho-phthalaldehyde (OPA), has recently been registered by the EPA as a high-level disinfectant. OPA is a pale blue liquid with a barely detectable odor. It has a mechanism of action similar to glutaraldehyde, is stable and nonirritating, and,

for most uses, is much faster acting than glutaraldehyde. It is effective against vegetative bacteria, including *Mycobacterium* and *Pseudomonas,* fungi, and viruses. Chief among its disadvantages are an inability to reliably destroy spores and a tendency to stain proteins, including those in human skin.

Gaseous Sterilants and Disinfectants

Processing inanimate substances with chemical vapors, gases, and aerosols provides a versatile alternative to heat or liquid chemicals. Currently, those vapors and aerosols having the broadest applications are ethylene oxide (ETO), propylene oxide, and chlorine dioxide.

Ethylene oxide is a colorless substance that exists as a gas at room temperature. Its structure is shown here:

$$CH_2 - CH_2$$
$$O$$

It is very explosive in air, a feature that can be eliminated by combining it with a high percentage of carbon dioxide or fluorocarbon. Like the aldehydes, ETO is a very strong alkylating agent, and it reacts vigorously with functional groups of DNA and proteins. Through these actions, it blocks both DNA replication and enzymatic actions. Ethylene oxide is one of a very few gases generally accepted for chemical sterilization because, when employed according to strict procedures, it is a sporicide. A specially designed ETO sterilizer is equipped with a chamber, gas ports, and temperature, pressure, and humidity controls (**figure 11.16**). Ethylene oxide is rather penetrating but relatively slow-acting, requiring from 90 minutes to 3 hours. Some items absorb ETO residues and must be aerated with sterile air for several hours after exposure to remove as much residual gas as possible. For all of its effectiveness, ETO has some unfortunate features. Its explosiveness makes it dangerous to handle; it can damage the lungs, eyes, and mucous membranes if contacted directly; and it is rated as a carcinogen by the government.

Figure 11.16 Sterilization using an ethylene oxide sterilizer. The machine is equipped with gas canisters containing ethylene oxide (ETO) and carbon dioxide, a chamber to hold items, and mechanisms that evacuate gas and introduce air.

Chlorine dioxide is another gas used for sterilization. Despite the name, chlorine dioxide works in a completely different way from the chlorine compounds discussed earlier in the chapter. It is a strong alkylating agent, which disrupts proteins and is effective against vegetative bacteria, fungi, viruses, and endospores. Although chlorine dioxide is used for the treatment of drinking water, wastewater, food-processing equipment, and medical waste, its best known use was in the decontamination of several government offices after the anthrax attack of 2001.

Betapropiolactone (BPL) is a substance somewhat similar to ETO in applications. It is rapidly microbicidal when used as an aerosol or liquid to disinfect whole rooms and instruments, sterilize bone and arterial grafts, and inactivate viruses in vaccines.

A summary of applications for aldehydes and sterilizing gases is presented in **table 11.10.**

Chemicals with Surface Action: Detergents and Soaps

Detergents are polar molecules that act as surfactants. Most anionic detergents have limited microbicidal power. This includes most soaps. Much more effective are positively charged (cationic) detergents, particularly the **quaternary ammonium compounds** (usually shortened to *quats*).

The activity of cationic detergents arises from the amphipathic (two-headed) nature of the molecule. The positively charged end binds well with the predominantly negatively charged bacterial surface proteins, while the long, uncharged hydrocarbon chain allows the detergent to disrupt the cell membrane (**figure 11.17**). Eventually, the cell bursts and dies.

The effects of detergents are varied. When used at high enough concentrations, quaternary ammonium compounds are effective

TABLE 11.10	Applications of Aldehydes and Sterilant Gases	
Form of Aldehyde	**Primary Applications**	**How Delivered/Concerns**
Glutaraldehyde	A sterilant for materials usually damaged by heat. Examples include respiratory therapy equipment, hemostats, fiberoptic endoscopes, kidney dialysis equipment, and dental instruments; an alternate vaccine preservative, and sanitizer for poultry carcasses.	Solutions diluted to 2% are relatively mild but require 2–4 hours soaking. Usual commercial products are Cidex, Sporocidin. Instrument sterilization involves precleaning along with autoclaving to inactivate hepatitis B and other blood-borne viruses.
Formalin	Limited uses as a disinfectant for surgical instruments; used in aquaculture to kill fish parasites and control growth of algae and fungi; it is one active ingredient in embalming fluid (with an alcohol and a phenolic).	May be used as 8% tincture; objects intended for contact with the body must be thoroughly rinsed to remove the formalin residue, which is toxic and carcinogenic.
Ortho-phthalaldehyde (OPA)	High-level disinfectant similar in uses and effects to glutaraldehyde	Faster acting and more stable than glutaraldehyde but less sporicidal
Form of Gas	**Primary Applications**	**How Delivered/Concerns**
Ethylene Oxide (ETO)	An official sterilant for heat-sensitive plastics and delicate instruments in hospitals and industries— prepackaged medical supplies and disposable Petri dishes; used extensively to disinfect foods, spices, dried fruits, and drugs	Carboxide and cryoxide are commercial products; gas is explosive and must be used with a stabilizer; ETO is rather toxic to humans; penetrating but slow, requiring 1–3 hours of exposure inside a special chamber
Propylene Oxide	Sterilization of foods such as nuts, powders, starches, and spices	Similar physical properties and mode of action to ETO but safer because it breaks down into a relatively harmless substance
Chlorine Dioxide	Air and surface sterilant; for treatment of water, food processing equipment, and medical waste; the decontamination of whole rooms and space probes	Similar in action to ETO but does not require a chamber; thus, effective for large spaces or objects

Figure A
Technician treats surgical instruments with glutaraldehyde prior to heat sterilization.

Figure B
Medical supplies packaged for sterilization in an ethylene oxide chamber.

(a)

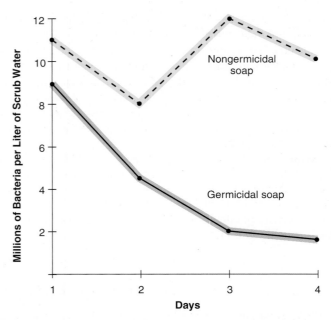

(b)

Benzalkonium chloride

Figure 11.17 The structure of detergents. (a) In general, detergents are polar molecules with a positively charged head and at least one long, uncharged hydrocarbon chain. The head contains a central nitrogen nucleus with various alkyl (R) groups attached. **(b)** A common quaternary ammonium detergent, benzalkonium chloride.

against some gram-positive bacteria, viruses, fungi, and algae. In low concentrations, they exhibit only microbistatic effects. Drawbacks to the quats include their ineffectiveness against the tuberculosis bacterium, hepatitis virus, *Pseudomonas,* and spores at any concentration. Furthermore, their activity is greatly reduced in the presence of organic matter, and they require alkaline conditions to function. As a result of these limitations, quats are rated only for low-level disinfection in the clinical setting.

Soaps are alkaline compounds made by combining the fatty acids in oils with sodium or potassium salts. In usual practice, soaps are weak microbicides, and they destroy only highly sensitive forms such as the agents of gonorrhea, meningitis, and syphilis. The common hospital pathogen *Pseudomonas* is so resistant to soap that various species grow abundantly in soap dishes.

Soaps function primarily as cleansing agents and sanitizers in industry, the home, and the medical setting. Vigorously brushing the hands with germicidal soap over a 15-second period is an effective way to remove dirt, oil, and surface contaminants as well as some resident microbes, but it is unable to remove the skin microbiota (**Insight 11.3** and **figure 11.18**).

See **table 11.11** for coverage on the main uses of detergents and soaps.

Heavy Metal Compounds

Various forms of the metallic elements mercury, silver, gold, copper, arsenic, and zinc have been applied in microbial control over several centuries. These are often referred to as heavy metals because of their relatively high atomic weight. However, from this list, only preparations containing mercury and silver still have any significance as germicides. Although some metals (zinc, iron) are actually needed in small concentrations as cofactors on enzymes, the higher molecular weight metals (mercury, silver, gold) can be very toxic, even in minute quantities (parts per million). This property of having antimicrobial effects in exceedingly small amounts is called an

Figure 11.18 Graph showing effects of hand scrubbing. Comparison of scrubbing over several days with a nongermicidal soap versus a germicidal soap. Germicidal soap has persistent effects on skin over time, keeping the microbial count low. Without germicide, soap does not show this sustained effect.

oligodynamic* action (**figure 11.19**). Heavy metal germicides contain either an inorganic or an organic metallic salt, and they come in the form of aqueous solutions, tinctures, ointments, or soaps.

Mercury, silver, and most other metals exert microbicidal effects by binding onto functional groups of proteins and inactivating them, rapidly bringing metabolism to a standstill (see figure 11.4c). This mode of action can destroy many types of microbes, including

* *oligodynamic* (ol″-ih-goh-dy-nam′-ik) Gr. *oligos,* little, and *dynamis,* power.

Silver amalgam Gold foil

Figure 11.19 Demonstration of the oligodynamic action of heavy metals. A pour plate inoculated with saliva has small fragments of heavy metals pressed lightly into it. During incubation, clear zones indicating growth inhibition developed around both fragments. The slightly larger zone surrounding the amalgam (used in tooth fillings) probably reflects the synergistic effect of the silver and mercury it contains.

INSIGHT 11.3

The Quest for Sterile Skin

About 130 years ago, before sterile gloves were a routine part of medical procedures, the hands remained bare during surgery. At this time, medical practitioners did understand the risk for infections and attempted to "sterilize" the hands of surgeons and their assistants. Several stringent (and probably very painful) techniques involving strong chemical germicides and vigorous scrubbing were practiced. Here are a few examples.

In Schatz's method, the hands and forearms were first cleansed by brisk scrubbing with liquid soap for 3 to 5 minutes, then soaked in a saturated solution of permanganate at a temperature of 110°F until they turned a deep mahogany brown. Next, the limbs were immersed in saturated oxalic acid until the skin became decolorized. Then, as if this were not enough, the hands and arms were rinsed with sterile limewater and washed in warm bichloride of mercury for 1 minute.

Or, there was Park's method (more like a torture). First, the surfaces of the hands and arms were rubbed completely with a mixture of cornmeal and green soap to remove loose dirt and superficial skin. Next, a paste of water and mustard flour was applied to the skin until it began to sting. This potion was rinsed off in sterile water, and the hands and arms were then soaked in hot bichloride of mercury for a few minutes, during which the solution was rubbed into the skin.

Another method once earnestly suggested for getting rid of microorganisms was to expose the hands to a hot-air cabinet to "sweat the germs" out of skin glands. Pasteur himself advocated a quick flaming of the hands to maintain asepsis.

The old dream of sterilizing the skin was finally reduced to some basic realities: The microbes entrenched in the epidermis and skin glands cannot be completely eradicated even with the most intense efforts, and

Heavy growth of microbial colonies on a plate of blood agar prepared by having a classroom of 30 students touch its surface. The result was a mixed population of bacteria and fungi, many of which could be pathogens (clear zones of hemolysis).

the skin cannot be sterilized without also seriously damaging it. Because this is true for both medical personnel and their patients, the chance always exists that infectious agents can be introduced during invasive medical procedures. Safe surgery had to wait until 1890, when rubber gloves were first made available for placing a sterile barrier around the hands. Of course, this did not mean that skin cleansing and antiseptic procedures were abandoned or downplayed. A thorough scrubbing of the skin, followed by application of an antiseptic, is still needed to remove the most dangerous source of infections—the superficial contaminants constantly picked up from whatever we touch.

What would be some of the more serious problems associated with sterilizing or trying to sterilize the skin? *Answer available at http://www.mhhe.com/talaro8*

TABLE 11.11	Applications of Detergents and Soaps	
Form of Detergents or Soaps	**Primary Applications**	**How Delivered/Concerns**
Quaternary Ammonium Compounds (Quats)	Mixed with cleaning agents to disinfect and sanitize floors, equipment surfaces, and restrooms; for sanitizing restaurant utensils, food-processing equipment, and clothing; common preservatives in ophthalmic solutions and cosmetics	Names include benzalkonium chloride, Zephiran, and cetylpyridinium chloride (Ceepryn); in dilutions ranging from 1:100 to 1:1,000; the level of disinfection is too low for disinfecting medical instruments.
Soaps	Cleansing agents and sanitizers in industry and the home; preparing instruments for heat sterilization; degerming patients' skin, routine hand washing by medical and dental personnel, surgical and preoperative hand scrubbing	Alkaline salts of fatty acids; weak germicides with superior sudsing and wetting properties; can remove large amounts of surface soil, grease, and other debris; antimicrobial chemicals added to make germicidal soaps with greater disinfection power

Worker vigorously scrubs a lettuce-processing machine with detergent to prevent contamination with food-borne pathogens.

vegetative bacteria, fungal cells and spores, algae, protozoa, and viruses (but not endospores).

Unfortunately, there are several drawbacks to using metals in microbial control:

- Metals are very toxic to humans if ingested, inhaled, or absorbed through the skin, even in small quantities, for the same reasons that they are toxic to microbial cells;
- they commonly cause allergic reactions;
- large quantities of biological fluids and wastes neutralize their actions; and
- microbes can develop resistance to metals.

Health and environmental considerations have dramatically reduced the use of metallic antimicrobial compounds in medicine, dentistry, commerce, and agriculture. **Table 11.12** provides an overview of applications for heavy metals.

Dyes as Antimicrobial Agents

Dyes are important in staining techniques and as selective and differential agents in media; they are also a primary source of certain drugs used in chemotherapy. Because aniline dyes such as crystal violet and malachite green are very active against gram-positive species of bacteria and various fungi, they are incorporated into solutions and ointments to treat skin infections (ringworm, for example). The yellow acridine dyes, acriflavine and proflavine, are sometimes utilized for antisepsis and wound treatment in medical and veterinary clinics. For the most part, dyes will continue to have limited applications because they stain and have a narrow spectrum of activity.

Acids and Alkalis

Conditions of very low or high pH can destroy or inhibit microbial cells, but they are limited in applications due to their corrosive, caustic, and hazardous nature. Aqueous solutions of ammonium hydroxide remain a common component of detergents, cleansers, and deodorizers. Organic acids are widely used in food preservation because they prevent spore germination and bacterial and fungal growth and because they are generally regarded as safe to eat. Acetic acid (in the form of vinegar) is a pickling agent that inhibits bacterial growth; propionic acid is commonly incorporated into breads and cakes to retard molds; lactic acid is added to sauerkraut and olives to prevent growth of anaerobic bacteria (especially the clostridia); and benzoic and sorbic acids are added to beverages, syrups, and margarine to inhibit yeasts.

For a look at the antimicrobial chemicals found in some common household products, see **table 11.13.**

TABLE 11.12	Applications of Heavy Metals	
Form of Metal	**Primary Applications**	**How Delivered/Concerns**
Organic Mercury	Thimerosal (Merthiolate) and nitromersol (Metaphen) are weak antiseptics and infection preventives; may be preservatives in cosmetics and ophthalmic solutions	Tinctures (0.001% to 0.2%) are fairly effective but a poor choice for broken skin because they are toxic and can delay healing. Mercurochrome is now considered among the poorest of antiseptics.
Silver Sulfadiazine Ointment	Added to dressings, it effectively prevents infection in second- and third-degree burn patients.	A yellow ointment containing silver salts and a sulfa drug
Silver Nitrate (AgNO$_3$)	Primarily as a topical germicide for mouth ulcers and occasionally root canals	1% to 2% solutions were first used as a prevention for gonococcal eye disease in newborns, now replaced by antibiotics.
Metallic Silver	Incorporated into catheters to prevent urinary tract infections in the hospital; added to paint, plastics, and steel as a way to control microbial survival on items such as toilet seats, stethoscopes, and even walls and floors in homes	
Colloidal Silver	Mild germicidal ointments or rinses for the mouth, nose, eyes, and vagina	

TABLE 11.13	Active Ingredients of Various Commercial Antimicrobial Products	
Product	**Specific Chemical Agent**	**Antimicrobial Category**
Lysol and Clorox Sanitizing Wipes	Dimethyl benzyl ammonium chloride	Detergent (quat)
Tilex and Lysol Mildew Remover	Sodium hypochlorites	Halogen
Ajax and Dial Antibacterial Hand Soap	Triclosan	Phenolic
Lysol Disinfecting Spray	Alkyl dimethyl benzyl ammonium/ethanol	Detergent (quats)/alcohol
ReNu Contact Lens Solution	Polyaminopropyl biguanide	Chlorhexidine
Wet Ones Antibacterial Moist Towelettes	Benzethonium chloride	Detergents (quat)
Noxzema Triple Clean	Triclosan	Phenolic
Scope Mouthwash	Ethanol	Alcohol
Purell Instant Hand Sanitizer	Ethanol	Alcohol
Pine-Sol	Pine oil and surfactant	Aldehydes, alcohols
Allergan Eye Drops	Sodium chlorite	Halogen

Check & Assess Section 11.5

✔ Chemical agents of microbial control are classified by their physical state and chemical nature.

✔ Chemical agents can be either microbicidal or microbistatic. They are also classified as high-, medium-, or low-level germicides.

✔ Factors that determine the effectiveness of a chemical agent include the type and numbers of microbes involved, the material involved, the strength of the agent, and the exposure time.

✔ Halogens are effective chemical agents at both microbicidal and microbistatic levels. Chlorine compounds disinfect water, food, and industrial equipment. Iodine is used as either free iodine or iodophor to disinfect water and equipment. Iodophors are also used as antiseptic agents.

✔ Phenols are strongly microbicidal agents used in general disinfection. Milder phenol compounds, the bisphenols, are also used as antiseptics.

✔ Alcohols dissolve membrane lipids and destroy cell proteins. Their action depends upon their concentration, but they are generally only microbistatic.

✔ Hydrogen peroxide is a versatile microbicide that can be used as an antiseptic for wounds and a disinfectant for utensils. A high concentration is an effective sporicide.

✔ Aldehydes are potent sterilizing agents and high-level disinfectants that irreversibly disrupt microbial enzymes.

✔ Ethylene oxide and chlorine dioxide are gaseous sterilants that work by alkylating protein and DNA.

✔ Surfactants are of two types: detergents and soaps. They reduce cell membrane surface tension, causing membrane rupture. Cationic detergents, or quats, are low-level germicides limited by the amount of organic matter present and the microbial load.

✔ Products containing the heavy metals silver and mercury have limited applications as antiseptics and preservatives.

23. Describe situations that require high-level disinfection and those for which medium- or low-level disinfection is adequate.

24. What is the difference between a tincture and an aqueous solution?

25. Name one chemical for which the general rule that a higher concentration is more effective is *not* true.

26. Define *sterilant*, and name the principal sporicidal chemical agents.

27. Summarize the chief forms and uses of chlorine compounds.

28. What are the superior characteristics of iodophors over free iodine solutions?

29. What are the modes of action of alcohols and phenolics?

30. Why is hydrogen peroxide solution so effective against anaerobes?

31. Give the uses and disadvantages of the heavy metal chemical agents, glutaraldehyde, and the sterilizing gases.

32. What does it mean to say that a chemical has an oligodynamic action, and which compounds have this effect?

CASE FILE 11 PERSPECTIVE

Hepatitis C is the most common blood-borne infection in the United States, with an estimated 3.2 million persons currently infected. This estimate is higher than the reported numbers of cases because only about 10% of infected persons present with an acute case of the infection. These are individuals who suddenly develop visible signs and symptoms indicative of a recent, diagnosable infection. The remainder have a chronic infection, meaning that they have had it for a longer period, generally lack symptoms, do not seek medical care, and are usually carriers of HCV.

The initial cases in this file were acute infections, implying that the HCV had been recently acquired, most likely from a chronic carrier who had also been treated in the clinic. What made it even more serious was that the circumstances of contact involved thousands of patients, 90% of whom were at risk for developing a disease for which there are no cures. Finding and alerting these patients was critical, not only because they may become chronic carriers, but also because they are prone to liver diseases such as cirrhosis and cancer.

Most diseases of medical origin involve exposure to the body fluids or tissues of another patient, although occasionally clinic workers are a source. Contact may occur when errors are made during injections, biopsies, and other invasive procedures that increase the chance of cross contamination. Most of the time, the infection arises from contaminated equipment that is not properly sterilized or handled. It is unlikely that one could get HCV from casual contact or environmental sources.

Patients had to be screened for HBV and HIV because these viruses are also transmitted by contact with blood. It is reasoned that people who are infected by one of these viruses have an increased risk of harboring other viruses in their blood. Fortunately, there were no cases of HBV or HIV infections traced to the clinic among the 50,000 patients tested.

A detailed study ruled out all but one possible mode of transmission. None of the clinic personnel carried HCV. Not only were the endoscopes properly sterilized, but the same ones had not been reused on any of the infected patients. It all narrowed down to the method used by some nurse assistants to draw up the anesthesia drugs from a vial into a syringe. An RN assistant was observed taking out a second dose of the drug from a large, multidose vial using the same syringe that had been used to take the first dose, although he did replace the needle. Often, these partly used vials were stored and reused on other patients. Some workers also neglected to observe proper hand washing and to wipe the top of the vial with alcohol before inserting the needle. In the end, lack of due diligence allowed a single patient to become the source for 106 cases of hepatitis, forced the clinic into bankruptcy, and cost the state of Nevada as much as $21 million. In 2010, a judge awarded the claimants in a lawsuit the huge sum of $500 million to be paid by the drug manufacturer and distributor for negligence in packaging of the drug.

Additional coverage of these topics can be found in chapters 13 and 25.

A summary of this case can be found at http://www.cdc.gov/mmwr/preview/mmwrhtml/mm5719a2.htm.

Chapter Summary with Key Terms

11.1 Controlling Microorganisms
- A. *Decontamination* procedures involve the destruction or removal of contaminants. Contaminants are defined as microbes present at a given place and time that are undesirable or unwanted.
- B. **Sterilization** is a process that destroys or removes all microbes, including viruses.
 1. A **bactericide** is a chemical that destroys bacteria.
 2. **Bacteriostatic** agents inhibit or prevent the growth of bacteria on tissues or on other objects in the environment.
 3. A **germicide** is a chemical that will kill any pathogenic microorganisms.
- C. **Disinfection** refers to a physical or chemical process that destroys vegetative pathogens but not bacterial endospores.
- D. **Sanitization** is any chemical technique that removes microorganisms to "safe levels or standards."
- E. Several factors influence the rate at which antimicrobial agents work. These factors are:
 1. Exposure time to the agent
 2. Numbers of microbes present
 3. Relative resistance of microbes (for example, endospores versus vegetative forms)
 4. Activity of the agent (microbicidal versus microbistatic)
- F. How antimicrobial agents work: their *modes of action.* Agents affect cell wall synthesis, membrane permeability, and protein and nucleic acid synthesis and function.

11.2 Physical Methods of Control: Heat
- A. Moist heat denatures proteins and DNA while destroying membranes.
 1. *Sterilization:* **Autoclaves** utilize steam under pressure to sterilize heat-resistant materials, whereas intermittent sterilization can be used to sterilize more delicate items.
 2. *Disinfection:* **Pasteurization** subjects liquids to temperatures below 100°C and is used to lower the microbial load in liquids. Boiling water can be used to destroy vegetative pathogens in the home.
- B. Dry heat, using higher temperatures than moist heat, can also be used to sterilize.
 1. **Incineration** can be carried out using a Bunsen burner or incinerator. Temperatures range between 800°C and 6,500°C.
 2. *Dry ovens* operate at temperatures of 150°C to 180°C.
- C. Cold temperatures (refrigeration and freezing) are microbistatic and are commonly used to preserve food, media, and cultures.
- D. Drying and **desiccation** lead to (often temporary) metabolic inhibition by reducing water in the cell.

11.3 Physical Methods of Control: Radiation
- A. *Radiation:* Energy in the form of **radiation** can be used for **cold sterilization,** which works by killing microbes without heat.
 1. **Ionizing radiation,** such as gamma rays and X rays, has deep penetrating power and works by causing breaks in the DNA of target organisms.
 2. **Nonionizing radiation** uses **ultraviolet** waves with very little penetrating power and works by creating dimers between adjacent pyrimidines, which interferes with replication.

11.4 Using Filtration to Remove Microbes
Filtration involves the physical removal of microbes by passing a gas or liquid through a fine filter, and can be used to sterilize air as well as heat-sensitive liquids.

11.5 Chemical Agents in Microbial Control
- A. Chemicals are divided into disinfectants, **antiseptics,** sterilants, **sanitizers,** and **degermers** based on their level of effectiveness and the surfaces to which they are applied.
- B. Antimicrobial chemicals are found as solids, gases, and liquids. Liquids can be either **aqueous** (water based) or **tinctures** (alcohol based).
- C. **Halogens** are chemicals based on elements from group VII of the periodic table.
 1. Chlorine is used as chlorine gas, hypochlorites, and chloramines. All work by disrupting disulfide bonds and, given adequate time, are sporicidal.
 2. Iodine is found both as free iodine (I_2) and as **iodophors** (iodine bound to organic polymers such as soaps). Iodine has a mode of action similar to chlorine and is also sporicidal, given enough time.
- D. Phenolics have phenol as a major component; they work by disrupting cell membranes and precipitating proteins. They are bactericidal, fungicidal, and virucidal but not sporicidal. Bisphenols are mild forms for disinfection and antisepsis.
- E. Chlorhexidine (Hibiclens, Hibitane) is a surfactant and protein denaturant with broad microbicidal properties, although it is not sporicidal. Solutions of chlorhexidine are used as skin degerming agents for preoperative scrubs, skin cleaning, and burns.
- F. Ethyl and isopropyl alcohol, in concentrations of 50% to 90%, are useful for microbial control. Alcohols act as **surfactants,** dissolving membrane lipids and coagulating proteins of vegetative bacterial cells and fungi. They are not sporicidal.
- G. Hydrogen peroxide produces highly reactive hydroxyl free radicals that damage protein and DNA while also decomposing to O_2 gas, which is toxic to anaerobes. Strong solutions of H_2O_2 are sporicidal.
- H. Aldehydes such as glutaraldehyde and formaldehyde kill microbes by alkylating protein and DNA molecules.
- I. Gases and aerosols such as **ethylene oxide (ETO),** propylene oxide, and chlorine dioxide are strong alkylating agents, all of which are sporicidal.
- J. Detergents and soaps
 1. Cationic detergents known as **quaternary ammonium compounds (quats)** act as surfactants that alter the membrane permeability of some bacteria and fungi. They are not sporicidal.
 2. Soaps have little microbicidal activity but, rather, function by removing grease and soil that contain microbes.
- K. Heavy metals: Solutions of silver and mercury kill vegetative cells (but not spores) in exceedingly low concentrations **(oligodynamic action)** by inactivating proteins.
- L. Dyes, acids, and alkalis can also inhibit or destroy microbes.

Multiple-Choice Questions

Select the correct answer form the answers provided. For questions with blanks, choose the combination of answers that most accurately completes the statement.

1. A microbicidal agent has what effect?
 a. sterilizes
 b. inhibits microorganisms
 c. is toxic to human cells
 d. destroys microorganisms

2. Microbial control methods that kill _____ are able to sterilize.
 a. viruses
 b. the tubercle bacillus
 c. endospores
 d. cysts

3. Any process that destroys the non-spore-forming contaminants on inanimate objects is
 a. antisepsis
 b. disinfection
 c. sterilization
 d. degermation

4. Sanitization is a process by which
 a. the microbial load on objects is reduced
 b. objects are made sterile with chemicals
 c. utensils are scrubbed
 d. skin is debrided

5. An example of an agent that lowers the surface tension of cells is
 a. phenol
 b. chlorine
 c. alcohol
 d. formalin

6. High temperatures _____ and low temperatures _____.
 a. sterilize, disinfect
 b. kill cells, inhibit cell growth
 c. denature proteins, burst cells
 d. speed up metabolism, slow down metabolism

7. The primary action of _____ heat is to _____.
 a. dry, destroy cell wall
 b. moist, kill vegetative cells
 c. dry, dissolve lipids
 d. moist, denature proteins

8. The temperature-pressure combination for an autoclave is
 a. 100°C and 4 psi
 b. 121°C and 15 psi
 c. 131°C and 9 psi
 d. 115°C and 3 psi

9. Microbe(s) that is/are the target(s) of pasteurization include(s)
 a. *Clostridium botulinum*
 b. *Mycobacterium* species
 c. *Salmonella* species
 d. both b and c

10. Ionizing radiation removes _____ from atoms.
 a. protons
 b. waves
 c. electrons
 d. ions

11. The primary mode of action of nonionizing radiation is to
 a. produce superoxide ions
 b. make pyrimidine dimers
 c. denature proteins
 d. break disulfide bonds

12. The most versatile method of sterilizing heat-sensitive liquids is
 a. UV radiation
 b. exposure to ozone
 c. beta propiolactone
 d. filtration

13. _____ is the iodine antiseptic of choice for wound treatment.
 a. Eight percent tincture
 b. Five percent aqueous
 c. Iodophor
 d. Potassium iodide solution

14. A chemical with sporicidal properties is
 a. phenol
 b. alcohol
 c. quaternary ammonium compound
 d. glutaraldehyde

15. Silver sulfadiazine is used
 a. in antisepsis of burns
 b. as a mouthwash
 c. to treat genital gonorrhea
 d. to disinfect water

16. Detergents are
 a. high-level germicides
 b. low-level germicides
 c. excellent antiseptics
 d. used in disinfecting surgical instruments

17. Which of the following is an approved sterilant?
 a. chlorhexidine
 b. betadyne
 c. ethylene oxide
 d. ethyl alcohol

Case File Questions

1. How would one best describe the state of being a *carrier* of HCV?
 a. It is on the skin and transmitted through direct contact.
 b. The person shows symptoms and is jaundiced.
 c. The person has a chronic infection of the liver and harbors the virus.
 d. The person has had contact with blood in a medical setting.

2. The main issue in the case of clinic-acquired HCV infections was:
 a. use of unclean equipment
 b. a high number of carriers
 c. contamination of drugs
 d. health workers who were IV drug abusers

Writing to Learn

These questions are suggested as a *writing-to-learn* experience. For each question, compose a one- or two-paragraph answer that includes the factual information needed to completely address the question. All Check & Assess questions can be used for writing-to-learn exercises.

1. What is wrong with this statement: "Prior to vaccination, the patient's skin was sterilized with alcohol"? What would be the more correct wording?

2. Would it be proper to say that a patient has been disinfected? Why or why not?

3. Think of three situations in which the same microbe would be considered a serious contaminant in one case and completely harmless in another.

4. From the case file, cite instances where the assistants used poor aseptic techniques and universal precautions. How would a greater awareness of microbiology have prevented these mistakes?

5. Was the source patient in the case file most likely acutely or chronically infected? Explain.

6. Explain some of the problems involved in sterilizing delicate and heat-sensitive instruments such as endoscopes, syringes, needles, catheters, and other instruments.

7. The shelf life and keeping qualities of fruit and other perishable foods are greatly enhanced through irradiation, saving industry and consumers billions of dollars. How would you personally feel about eating irradiated fruit? How about spices that had been sterilized with ETO?

Concept Mapping

Appendix E provides guidance for working with concept maps.

1. Construct your own concept map using the following words as the *concepts*. Supply the linking words between each pair of concepts.

halogens	sporicidal
oligodynamic	chemical
surfactants	physical
alcohol	silver
phenolics	

Critical Thinking Questions

Critical thinking is the ability to reason and solve problems using facts and concepts. These questions can be approached from a number of angles, and in most cases, they do not have a single correct answer.

1. For each item on the following list, give a reasonable method of sterilization. You cannot use the same method more than three times for the whole list. The method must sterilize, not just disinfect; and the method must not destroy the item or render it useless unless there is no other choice. After considering a workable method, think of a method that would not work. Note: Where an object containing something is given, you must sterilize everything (for example, both the jar and the Vaseline in it). Some examples of methods are autoclave, ethylene oxide gas, dry oven, and ionizing radiation.

room air	disposable syringes
blood in a syringe	carcasses of cows with "mad cow" disease
serum	
a pot of soil	inside of a refrigerator
plastic Petri dishes	wine
heat-sensitive drugs	a jar of Vaseline
cloth dressings	fruit in plastic bags
leather shoes from a thrift shop	talcum powder
a cheese sandwich	milk
human hair (for wigs)	orchid seeds
a flask of nutrient agar	metal instruments
an entire room (walls, floor, etc.)	mail contaminated with anthrax spores
rubber gloves	

2. Using pasteurization techniques as a model, compare the TDTs and explain the relationships between temperature and length of exposure.

3. It may seem contradictory that lyophilization can be used to preserve foods yet also be used to keep cultures of microbes viable. Explain how these applications can be reconciled.

4. A supermarket/drugstore assignment: Look at the labels of 10 different products used to control microbes, and make a list of their active ingredients, their suggested uses, and information on toxicity and precautions.

5. Devise an experiment that will differentiate between bactericidal and bacteriostatic effects.

6. There is quite a bit of concern that chlorine used as a water purification chemical presents serious dangers. What alternative methods covered by this chapter could be used to purify water supplies and keep them safe from contamination?

Visual Challenge

1. From **chapter 2, figure 2.20.** Study this illustration of a cell membrane. In what ways could alcohol (Hint: a solvent) damage this membrane? How would that harm the cell?

2. From **chapter 4, figure 4.22.** Explain the main features this microbe has that contribute to its resistance to control agents.

12

Drugs, Microbes, Host—The Elements of Chemotherapy

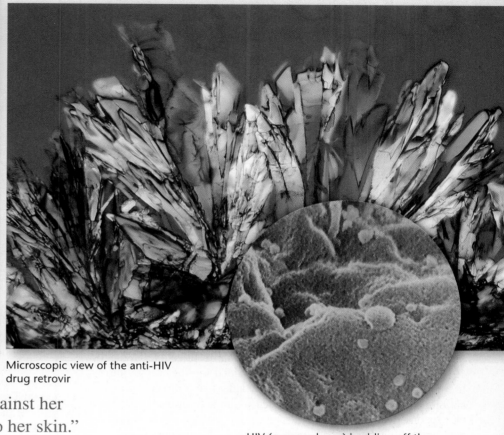

Microscopic view of the anti-HIV drug retrovir

"The metal needle accidently fell against her arm and jabbed with some force into her skin."

HIV (green spheres) budding off the surface of an infected white blood cell

CASE FILE 12 An Ounce of Prevention and a Pound of Cure

The 48-year-old registered nurse had safely removed needles from intravenous catheters hundreds of times, yet all of those years of care and experience seemed to evaporate in an instant. The needle accidently fell against her arm and jabbed with some force into her skin. The injury bled slightly, and she immediately reported it and began the standard precautions for a needle stick. She washed the site of the laceration with water and betadyne, as indicated by hospital protocol.

Needle sticks are a regular occurrence in hospitals and clinics, and usually the first concern that comes to mind is the type of infectious agents that might be carried in the offending needle. Unfortunately for the nurse, the patient had been suffering from acquired immune deficiency syndrome (AIDS) caused by human immunodeficiency virus (HIV). Within two hours, it was necessary for her to begin a regimen of anti-HIV drugs that are routinely given for postexposure prophylaxis. The drug chosen was combivir, a pill containing two different reverse transcriptase inhibitors. She was tested right away for HIV to establish a baseline for the possibility of a prior infection, and it came back negative.

Despite treatment, the AIDS patient died unexpectedly of respiratory failure due to *Pneumocystis* pneumonia. His chart did not show other infections transmitted by blood except HIV, but this did not stop the next several weeks from turning into a nightmare for the nurse. She was supposed to take the combivir for 6 weeks, but it caused severe side effects, including nausea, headache, muscle pain, weakness, abdominal pain, and rash. She discontinued taking it after 2 weeks and hoped that it had succeeded in preventing HIV infection.

During the next month, the site of her injury became abscessed and painful, and she suffered from a fever up to 102°F. Since the wound was obviously infected, her doctor biopsied it and had routine bacterial and fungal cultures made, but they came back negative. The course of dicloxacillin he prescribed had not im-

proved her condition. Signs that the infection was spreading came in the form of an enlarged lymph node in her armpit.

Eventually, she had a routine skin test for tuberculosis, and rather surprisingly, it developed a raised red patch that indicated TB positive. A new specimen taken from the wound verified *Mycobacterium tuberculosis*. Unknown to the medical staff was the fact that the AIDS patient had been suffering from both pneumocystis **and** tuberculosis lung infections. This now explained his sudden death, because tuberculosis is a deadly risk factor for AIDS patients, and he had not received any therapy for TB. The blood in the needle must have contained a high level of tubercle bacilli that were inoculated into the nurse's arm when it penetrated the skin.

▶ *What is meant by prophylactic drug treatment, and what are its purposes?*

▶ *What are some of the reasons that the cause of her infection was missed?*

To continue the case, go to page 363.

351

12.1 Principles of Antimicrobial Therapy

Expected Learning Outcomes

1. State the goals of antimicrobial therapy.

2. Discuss the various origins of antimicrobial drugs.

3. Differentiate between antibiotics, synthetic drugs, and semisynthetic drugs.

4. Describe some of the special strategies of antimicrobial therapy.

5. Explain what is meant by a drug's spectrum and how narrow- and broad-spectrum drugs differ.

6. Summarize the five modes of action of antimicrobial drugs, with specific examples of their effects.

TABLE 12.1	Characteristics of the Ideal Antimicrobial Drug

- Selectively toxic to the microbe but nontoxic to host cells
- Microbicidal rather than microbistatic
- Relatively soluble; functions even when highly diluted in body fluids
- Remains potent long enough to act and is not broken down or excreted prematurely
- Doesn't lead to the development of antimicrobial resistance
- Complements or assists the activities of the host's defenses
- Remains active in tissues and body fluids
- Readily delivered to the site of infection
- Reasonably priced
- Does not disrupt the host's health by causing allergies or predisposing the host to other infections

A hundred years ago in the United States, one out of three children was expected to die of an infectious disease before the age of 5. Early death or severe lifelong debilitation from scarlet fever, diphtheria, tuberculosis, meningitis, and many other bacterial diseases was a fearsome yet undeniable fact of life to most of the world's population.

The introduction of modern drugs to control infections in the 1930s was a medical revolution that has added significantly to the life span and health of humans. It is no wonder that for many years, antibiotics in particular were regarded as miracle drugs. In later discussions, we evaluate this misconception in light of the shortcomings of drug therapy. Although antimicrobial drugs have greatly reduced the incidence of certain infections, they have definitely not eradicated infectious disease and probably never will. In fact, in some parts of the world, mortality rates from infectious diseases are as high as before the arrival of antimicrobial drugs. Nevertheless, humans have been taking medicines to try to control diseases for thousands of years (**Insight 12.1**).

The goal of **antimicrobial chemotherapy**[1] is deceptively simple: administer a drug to an infected person that destroys the infective agent without harming the host's cells. In actuality, this goal is rather difficult to achieve, because many often contradictory factors must be taken into account. The ideal drug should be easily administered yet be able to reach the infectious agent anywhere in the body, be able to kill the infectious agent rather than merely inhibit its growth, and remain active in the body as long as needed yet be safely and easily broken down and excreted. In short, the perfect drug does not exist, but by balancing drug characteristics against one another, a satisfactory compromise can be achieved (**table 12.1**).

The Origins of Antimicrobial Drugs

The origin of modern antimicrobial drugs is varied. Some, called **antibiotics,** are substances produced by the natural metabolic processes of some microorganisms that can inhibit or destroy other microorganisms. Nature is a prolific producer of antimicrobial drugs. Antibiotics, come primarily from aerobic spore-forming bacteria and fungi. By inhibiting the growth of other microorganisms

in the same habitat (antagonism), antibiotic producers presumably enjoy less competition for nutrients and space. The greatest numbers of antibiotics are derived from bacteria in the genera *Streptomyces* and *Bacillus* and molds in the genera *Penicillium* and *Cephalosporium* (**figure 12.1** and **table 12.3**). Species of *Streptomyces* are the most prolific producers of all. They are the source of two-thirds of the antibiotics; about 9,000 different bioactive compounds have been isolated from them.

Chemotherapeutic agents are described with regard to their origin, range of effectiveness, and whether they are naturally produced or chemically synthesized. A few of the more important terms you will encounter are found in **table 12.2**.

In this chapter, we describe different types of antibiotic drugs, their mechanism of action, and the types of microbes on which they are effective.

Figure 12.1 A Colony of *Streptomyces coelicolor*. The surface droplets contain actinorhodin, an antibiotic secreted by the microorganism. *S. coelicolor* belongs to a group of bacteria called streptomycetes that produce a large number of useful antibiotics.

1. The word *chemotherapy,* meaning the use of chemicals in therapy, is often associated with the treatment of cancer, but its official meaning also includes antimicrobial treatment.

INSIGHT 12.1

From Witchcraft to Wonder Drugs

Early human cultures relied on various types of primitive medications such as potions, poultices, and mud plasters. Many were concocted from plant, animal, and mineral products that had been found, usually through trial and error or accident, to have some curative effect upon ailments and complaints. In one ancient Chinese folk remedy, a fermented soybean curd was applied to skin infections. The Greeks used wine and plant resins (myrrh and frankincense), rotting wood, and various mineral salts to treat diseases. Some folk medicines were effective, but others either had no effect or were even harmful. It is interesting that the Greek word *pharmakeutikos* originally meant the practice of witchcraft. These ancient remedies were handed down from generation to generation, but it was not until the Middle Ages that a specific disease was first treated with a specific chemical. Dosing syphilitic patients with inorganic arsenic and mercury compounds may have proved the ancient axiom: *Graviora quaedum sunt remedia periculus* ("Some remedies are worse than the disease").

Paul Ehrlich was the first scientist to formulate basic theoretical concepts in chemotherapy in the late 1800s. Ehrlich had observed that certain dyes affixed themselves to specific microorganisms and not to animal tissues. This observation led to the profound idea that if a drug was properly selective in its actions, it would zero in on and destroy a microbial target and leave human cells unaffected. His first discovery was an arsenic-based drug that was very toxic to the spirochete of syphilis but, unfortunately, to humans as well. Ehrlich systematically altered this parent molecule, creating numerous derivatives. Finally, on the 606th try, he arrived at a compound he called *salvarsan*. This drug had some therapeutic merit and was used for a few years, but it eventually had to be discontinued because it was still not selective enough in its toxicity. Ehrlich's work had laid important foundations for many of the developments to come.

Another pathfinder in early drug research was Gerhard Domagk, whose discoveries in the 1930s launched a breakthrough in therapy that marked the true beginning of broad-scale usage of antimicrobial drugs. Domagk showed that the red dye prontosil was chemically changed by the body into a compound with specific activity against bacteria. This substance was sulfonamide—the first *sulfa* drug. In a short time, the structure of this drug was determined, and it became possible to synthesize it on a wide scale and to develop scores of other sulfonamide drugs. Although these drugs had immediate applications in therapy (and still do), still another fortunate discovery ushered in the "age of antibiotics."

The discovery of antibiotics dramatically demonstrates how developments in science and medicine often occur through a combination of accident, persistence, collaboration, and vision. In 1928 in the London laboratory of Sir Alexander Fleming, a plate of *Staphylococcus aureus* became contaminated with the mold *Penicillium notatum*. Observing these plates, Fleming noted that the colonies of *Staphylococcus* were

A pioneer in the discovery of antibiotics. A Scottish physician, Sir Alexander Fleming, accidently discovered penicillin when his keen eye noticed that colonies of bacteria were being lysed by a fungal contaminant. He studied the active ingredient and set the scene for the development of the drug 10 years later.

evidently being destroyed by some activity of the nearby mold colonies. Struck by this curious phenomenon, he extracted from the fungus a compound he called penicillin and showed that it was responsible for the inhibitory effects.

Although Fleming understood the potential for penicillin, he was unable to develop it. A decade after his discovery, chemists Howard Florey and Ernst Chain worked out methods for industrial production of penicillin to help in the war effort. Clinical trials conducted in 1941 ultimately proved its effectiveness, and penicillin was made available to the world's population. At first, it seemed a godsend. By the 1950s, the pharmaceutical industry had entered an era of drug research and development that soon made penicillin only one of a large assortment of antimicrobial drugs. But in time, because of extreme overuse and misunderstanding of its capabilities, it also became the model for one of the most serious drug problems—namely, drug resistance.

Name some potential "therapies" you could employ to prevent and treat infections if you were cast away on a deserted island. *Answer available at http://www.mhhe.com/talaro8*

TABLE 12.2	Terminology of Chemotherapy
Chemotherapeutic drug	Any chemical used in the treatment, relief, or prophylaxis of a disease
Prophylaxis*	Use of a drug to prevent potential for infection of a person at risk
Antimicrobial chemotherapy*	The use of chemotherapeutic drugs to control infection
Antimicrobials	All-inclusive term for any antimicrobial drug, regardless of its origin
Antibiotics*	Substances produced by the natural metabolic processes of some microorganisms that can inhibit or destroy other microorganisms
Semisynthetic drugs	Drugs that are chemically modified in the laboratory after being isolated from natural sources
Synthetic drugs	Antimicrobial compounds synthesized in the laboratory through chemical reactions
Narrow spectrum (limited spectrum)	Antimicrobials effective against a limited array of microbial types; for example, a drug effective mainly on gram-positive bacteria
Broad spectrum (extended spectrum)	Antimicrobials effective against a wide variety of microbial types; for example, a drug effective against both gram-positive and gram-negative bacteria

* *prophylaxis* (proh″-fih-lak′-sis) Gr. *prophylassein*, to keep guard before. A process that prevents infection or disease in a person at risk.
* *chemotherapy* (kee″-moh-ther′-uh-pee) Gr. *chemieia*, chemistry, and *therapeia*, service to the sick. Use of drugs to treat disease.
* *antibiotic* (an-tee′-by-aw″-tik) Gr. *anti*, against, and *bios*, life.

TAKE NOTE: STRATEGIES TO OUTWIT INFECTIOUS AGENTS

You will encounter the terms **prophylaxis, combined therapy,** and drug **synergy** in several areas of the chapter. Each idea has its own special niche in drug therapy. With prophylaxis, drugs are administered to *prevent* an infection from happening. This is brought into play in cases of known or possible exposure. For example, certain dental patients are often pre-treated with antibiotics prior to instrument procedures that could leak saliva into the blood.

Combined therapy, also known as combination therapy, is a protocol for taking more than two drugs at a time to treat an infection. This is the usual method of treatment for tuberculosis, HIV/AIDS, and gonorrhea. Its main rationale is to prevent the survival of drug-resistant microbes, but it is also used for treating mixed infections by microbes that differ in the correct drug of choice.

Another advantage comes from the synergistic effect that two drugs working together can often provide. Synergy can produce an improved result over either drug used alone, and this enhanced effectiveness means that smaller quantities of the drugs can be used. Figure 12.18*b* shows a test result for this effect.

TABLE 12.3	Selected Microbial Sources of Antibiotics	
Producer Genus	**Type of Microbe**	**Drugs Formed**
Penicillium sp.	Mold	Penicillins Griseofulvin
Cephalosporium	Mold	Cephalosporins
Micromonospora	Bacteria	Gentamicin
Bacillus sp.	Bacteria	Bacitracin Polymixin B
Chromobacterium	Bacteria	Aztreonam
Streptomyces sp.	Filamentous bacteria	Streptomycin Erythromycin Tetracycline Vancomycin Chloramphenicol Amphotericin B

Interactions between Drugs and Microbes

The goal of antimicrobial drugs is either to disrupt the cell processes or structures of bacteria, fungi, and protozoa or to inhibit the virus multiplication cycle. Most of the drugs used in chemotherapy interfere with the function of enzymes required to synthesize or assemble macromolecules, but a few destroy structures already formed in the cell. Above all, drugs should be **selectively toxic,** which means they should kill or inhibit the growth of microbes without simultaneously damaging host tissues. This concept of selective toxicity is central to chemotherapy. The best drugs are those that act specifically on microbial structures or functions not found in vertebrate cells. Examples of drugs with excellent selective toxicity are those that block the synthesis of the cell wall in bacteria (penicillins). They have low toxicity and few direct effects on human cells because human cells lack a cell wall and are thus unaffected by this action of the antibiotic. Among the most toxic to human cells are drugs that act upon a structure common to both the infective agent and the host cell, such as the cell membrane (for example, amphotericin B used to treat fungal infections). As the characteristics of the infectious agent become more similar to those of the host cell, selective toxicity becomes more difficult to achieve and undesirable side effects are more likely to occur. We examine the subject of toxicity in a later section.

Mechanisms of Drug Action

The primary basis of chemotherapy is to target an infectious agent with a chemical that irreversibly damages or inhibits it. This knowledge guides drug development as well. Most classes of antimicrobic drugs are active against some unique feature of a microbe structure or metabolism. Observe **table 12.4** and take note of the diversity of cell types and possible effects of drug action. Overall, there are five major components that are useful drug targets in an actively dividing cell. These are the cell wall and membrane, the

TABLE 12.4	Mechanisms of Drug Effects on Major Categories of Pathogens		
Infectious Agent	Drug Groups/ Examples	General Targets	Outcome of Drug Action on the Microbe
Bacteria	Penicillins Penicillin, ampicillin	Cell wall synthesis	Lysis of cells
	Cephalosporins Keflex, cefotaxime	Cell wall synthesis	Lysis of cells
	Bacitracin	Cell wall	Lysis of cells
	Aminoglycosides Streptomycin Gentamicin	Prokaryotic ribosomes	Inhibit protein synthesis
	Macrolides	Various targets	
	Erythromycin	Ribosomes	Inhibit protein synthesis
	Vancomycin	Cell wall	Lysis of cells
	Tetracyclines Doxycycline	Prokaryotic ribosomes	Inhibit protein synthesis
	Chloramphenicol	Ribosomes	Inhibits protein synthesis
	Fluoroquinolones Ciprofloxacin Levofloxacin	DNA gyrase	Block replication of DNA
	Rifampin	RNA polymerase	Stops mRNA synthesis
	Sulfa drugs Sulfasoxazole	Metabolic pathway	Inhibit folic acid formation
	Trimethoprim	Metabolic pathway	Inhibits folic acid formation
Fungi	Macrolides Amphotericin B	Fungal cell membrane	Loss of selective permeability
	Azoles Miconazole Fluconazole	Fungal cell membrane	Loss of selective permeability
	Flucytosine	Fungal DNA and RNA synthesis	Stops cell reproduction
Protozoa	Quinines Chloroquine Mefloquine	Nutrition of the malaria parasite	Buildup of toxic wastes in the parasite's cells
	Metronidazole	Anaerobic cells	Buildup of toxic free radicals
Helminths	Bendazoles	Microtubules	Inhibit glucose metabolism
	Diethylcarbamide	Unknown	Kills larval forms
	Piperazine	Worm muscles	Worm is expelled
	Niclosamide	ATP formation	Loosens worm hold
	Ivermectin	Nerve transmission	Worm is expelled
Viruses	Amantidine	Host cell membrane	Blocks entry, fusion
	Cyclovirs	DNA synthesis	Stop virus replication
	Azidothymidine	Reverse transcriptase	Blocks DNA formation

genetic material, protein synthesis, and metabolic pathways. We cover the effects of drugs on these cell components based on this outline and **figure 12.2:**

1. inhibition of cell wall synthesis,
2. breakdown of the cell membrane structure or function,
3. inhibition of structures and functions of DNA and RNA,
4. inhibition of protein synthesis, and
5. blocks on key metabolic pathways.

These targets are not completely discrete, and some have overlapping effects. For example, if DNA transcription is inhibited, protein synthesis will be as well. Viruses are a special case and have very specialized targets for drugs (see table 12.6).

The Spectrum of an Antimicrobial Drug One result of drugs acting upon a particular microbial structure or function is that each drug has a particular range of activity. This has been termed the drug's *spectrum.* Traditionally, **narrow-spectrum** drugs are effective on a small range of cell types. This is usually because they target a specific component that is found only in certain bacteria. For instance, bacitracin blocks the elongation of the peptidoglycan in gram-positive bacteria but has no effect on gram-negatives. Polymyxin breaks down the outer membrane on gram-negative cells but does not target gram-positive ones.

Drugs that are effective against a wider range of microbes are termed medium or broad spectrum, depending on the particular drug. For example, the **medium-spectrum** drug ampicillin is effective on both gram-positive and gram-negative bacteria, but not every type. **Broad-spectrum** drugs such as tetracycline have the greatest range of activity. They work on most gram-negative and gram-positive bacteria, rickettsias, mycoplasmas, and spirochetes. The broader-spectrum drugs usually exert their effects on common cell components such as ribosomes, which are found in all cells and mitochondria.

1. Antimicrobial Drugs That Affect the Bacterial Cell Wall The cell walls of most bacteria contain a rigid girdle of peptidoglycan, which protects the cell against rupture from hypotonic environments. Active cells must constantly synthesize new peptidoglycan and transport it to its proper place in the cell envelope. Drugs such as penicillins and cephalosporins react with one or more of the enzymes required to complete this process, causing the cell to develop weak points at growth sites and to become osmotically fragile (**figure 12.3**). Antibiotics that produce this effect are considered bactericidal, because the weakened cell is subject to lysis. It is essential to note that most of these antibiotics are active only in young, growing cells, because old, inactive, or dormant cells do not synthesize peptidoglycan. (One exception is a new class of antibiotics termed "-penems.")

Figure 12.2 **Major targets of drugs acting on bacterial cells.** Each mode of action correlates with the numbered section where it is discussed in the text.

Penicillins and cephalosporins bind and block peptidases that cross-link the glycan molecules, thereby interrupting the completion of the cell wall **(figure 12.3b).** Penicillins that do not penetrate the outer membrane are less effective against gram-negative bacteria. Broad-spectrum penicillins (carbenicillin) and cephalosporins (ceftriaxone) can pass into the cell walls of gram-negative species. Cycloserine inhibits the formation of the basic peptidoglycan subunits, and vancomycin hinders the elongation of the peptidoglycan.

2. Antimicrobial Drugs That Disrupt Cell Membrane Function A cell with a damaged membrane invariably dies from disruption in metabolism or lysis. It does not even have to be actively dividing to be destroyed. The antibiotic classes that damage cell membranes usually have specificity for particular microbial groups, based on differences in the types of lipids in their cell membranes.

Polymyxins interact with membrane phospholipids and cause leakage of proteins and nitrogen bases, particularly in gram-negative bacteria **(figure 12.4).** The polyene antifungal antibiotics (amphotericin B and nystatin) form complexes with the sterols on fungal membranes, which causes abnormal openings and seepage of small ions. Unfortunately, this selectivity is not exact, and the similarities in membranes of microbial and animal cells means that most of these antibiotics can be quite toxic to humans cells.

3. Antimicrobial Drugs That Affect Nucleic Acid Synthesis
As you learned in chapter 9, the metabolic pathway that generates DNA and RNA is a long, enzyme-catalyzed series of reactions.

Like any complicated process, it is subject to breakdown at many different points along the way, and inhibition at any point in the sequence can block subsequent events. Antimicrobial drugs interfere with nucleic acid synthesis by blocking synthesis of nucleotides, inhibiting replication, or stopping transcription. Because functioning DNA and RNA are required for proper translation as well, the effects on protein metabolism can be far-reaching.

Several antimicrobials inhibit DNA synthesis. Chloroquine (an antimalarial drug) binds and cross-links the double helix. The broad-spectrum quinolones inhibit DNA unwinding enzymes or helicases, thereby stopping DNA replication and repair. Certain antiviral drugs are **analogs*** of purines and pyrimidines which will insert in the viral nucleic acid like a normal base. When synthesis can no longer continue, viral replication is blocked (see table 12.6).

4. Antimicrobial Drugs That Block Protein Synthesis Most drugs that inhibit protein synthesis react with the ribosome-mRNA complex. Although human cells also have ribosomes, the ribosomes of eukaryotes are different in size and structure from those of prokaryotes, so these antimicrobials usually have a selective action against bacteria. One potential therapeutic consequence of drugs that bind to the prokaryotic ribosome is the damage they can do to eukaryotic mitochondria, which contain a prokaryotic type of ribosome.

* *analog* (an'-uh-log) A compound whose configuration closely resembles another compound required for cellular reactions.

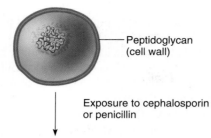

Peptidoglycan (cell wall)

Exposure to cephalosporin or penicillin

(a) Normal untreated cells. Left and center views are *Staphylococcus* cells; right view is the molecular structure of peptidoglycan, with chains of NAM (N-acetyl muramic acid) and NAG (N-acetyl glucosamine) linked together by cross bridges (green).

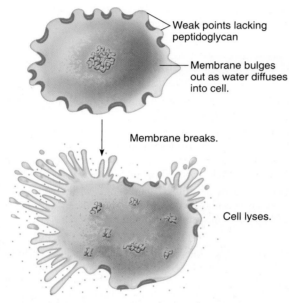

Weak points lacking peptidoglycan

Membrane bulges out as water diffuses into cell.

Membrane breaks.

Cell lyses.

(b) Effects of treatment with penicillin. Left side depicts the stages in the breakdown of a *Staphylococcus* cell exposed to penicillin. Center view shows actual cells undergoing lysis and cell death. Right is a molecular view of how penicillin blocks the formation of the cross bridges, which removes the support that holds the cell wall together.

10,000×

10,000×

Figure 12.3 **Effects of penicillin on bacterial cell walls.**

Two possible targets of ribosomal inhibition are the 30S subunit and the 50S subunit (**figure 12.5**). Aminoglycosides (streptomycin, gentamicin, for example) insert on sites on the 30S subunit and cause the misreading of the mRNA, leading to abnormal proteins. Tetracyclines block the attachment of tRNA on the A acceptor site and effectively stop further synthesis of that protein. Other antibiotics attach to sites on the 50S subunit in a way that prevents the formation of peptide bonds (chloramphenicol) or inhibits translocation of the subunit during translation (erythromycin). The oxazolidinones are a newer drug class that affect the assembly of the completed ribosome by blocking the attachment site for the 50S and 30S subunits.

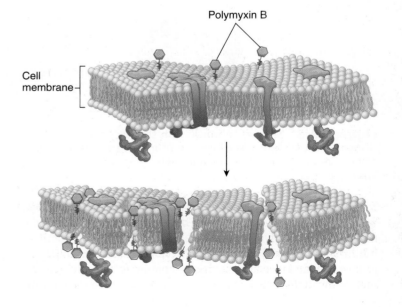

Polymyxin B

Cell membrane

Figure 12.4 **Effects of drugs on the bacterial cell envelope.** The detergent action of polymyxin. Polymyxin binds to the outer and cell membranes and forms abnormal openings that cause them to leak their contents and lyse.

mRNA is misread, protein is incorrect

Aminoglycosides → 50S aa aa 30S → mRNA

Formation of peptide bonds is blocked

Chloramphenicol → 50S aa X aa 30S → mRNA

Prevent initiation and block ribosome assembly

Oxazolidinones → 50S aa aa 30S → mRNA

tRNA is blocked, no protein is synthesized

Tetracyclines → 50S 30S → mRNA

Ribosome is prevented from translocating

Erythromycin → 50S aa aa 30S → mRNA

Figure 12.5 Sites of inhibition on the prokaryotic ribosome and major antibiotics that act on these sites. All have the general effect of blocking protein synthesis. Blockage actions are indicated by ×.

5. Antimicrobial Drugs That Affect Metabolic Pathways

Sulfonamides and trimethoprim are drugs that act by mimicking the normal substrate of an enzyme in a process called **competitive inhibition.** They are supplied to the cell in high concentrations to ensure that a needed enzyme is constantly occupied by the **metabolic analog** rather than by the true substrate of the enzyme.

As the enzyme is no longer able to produce a needed product, cellular metabolism slows or stops.

Sulfonamides and trimethoprim interfere with folate metabolism by blocking enzymes required for the synthesis of tetrahydrofolate, which is needed by bacterial cells for the synthesis of folic acid and the eventual production of DNA and RNA and amino acids. Trimethoprim and sulfonamides are often given simultaneously to achieve a synergistic effect so that a lower dosage of each drug can be used. **Figure 12.6** illustrates sulfonamides' competition with PABA for the active site of the enzyme that synthesizes the folic acid precursor.

The selective toxicity of these compounds is explained by the fact that mammals derive folic acid from their diet and so do not possess this enzyme system. This makes it possible to inhibit bacterial and protozoan parasites, which must synthesize folic acid, while leaving the human host unaffected.

Check & Assess Section 12.1

✔ Antimicrobial chemotherapy involves the use of drugs to control infection on or in the body.

✔ Antimicrobial drugs are produced either synthetically or from natural sources. They inhibit or destroy microbial growth in the infected host. Antibiotics are the subset of antimicrobials produced by the natural metabolic processes of microorganisms.

✔ Bacteria and fungi are the primary sources of most antibiotics. The molecular structures of these compounds can be chemically altered to form additional semisynthetic antimicrobials.

✔ Antimicrobial therapy involves three interacting factors: the microbe's sensitivity, the drug's toxicity, and the health of the patient.

✔ Modes of action of antimicrobial drugs fall into five categories: they inhibit or kill by acting on cell walls, cell membranes, DNA and RNA function, protein synthesis, or metabolic enzymes.

✔ Drugs should be selectively toxic to microbes and not cause disruptions in the host.

1. Differentiate between antibiotics and synthetic drugs.
2. Differentiate between narrow-spectrum and broad-spectrum antibiotics.
3. Can you determine why some drugs have narrower spectra than others? (Hint: Look at their mode of action.)
4. How might one determine whether a particular antimicrobial is broad or narrow spectrum?
5. What is the major source of antibiotics and what appears to be the natural function of antibiotics in microbes that produce them?
6. What drug characteristics will make treatment most effective?
7. Why is it better for a drug to be microbicidal than microbistatic?
8. Explain the major modes of action of antimicrobial drugs, and give an example of each.
9. What is competitive inhibition?
10. What is the basic reason that a metabolic analog molecule can inhibit metabolism?
11. What are the long-term effects of drugs that block transcription?
12. Why would a drug that blocks translation on the ribosomes of bacteria also affect human cells?

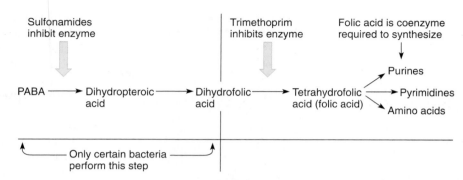

(a) The metabolic pathway needed to synthesize tetrahydrofolic acid (THFA) contains two enzymes that are chemotherapeutic targets. Under normal circumstances, PABA acts as a substrate for the enzyme pteridine synthetase, the product of which is needed for the eventual production of folic acid.

(b) Left: Formula of a sulfanilamide molecule compared alongside one of PABA. Note that, despite a similarity in overall shape, sulfa drugs cannot be used to make folic acid. Right: The sulfa drug molecules can insert into the active site in the enzyme that would ordinarily fit the PABA. This will work only when the sulfa molecules are more prevalent and can outcompete with the PABA and keep the enzyme sites filled. Since little or no PABA can bind, the synthesis of folic acid will be inhibited.

Figure 12.6 Competitive inhibition as a mode of action. This example shows how sulfa drugs block a metabolic pathway bacteria use to synthesize folic acid.

12.2 Survey of Major Antimicrobial Drug Groups

Expected Learning Outcomes

7. Describe the classes of antibacterial drugs that act on the cell wall and membrane, and list some of their applications.

8. Describe the classes of antibacterial drugs that act on DNA or RNA, and characterize their uses.

9. Indicate which types of antibacterial drugs block protein synthesis, and explain how they are used.

10. Describe some new classes of antibacterial drugs and their applications.

Scores of antimicrobial drugs are marketed in the United States. Although the medical and pharmaceutical literature contains a wide array of names for antimicrobials, most of them are variants of a small number of drug families. About 260 different antimicrobial drugs are currently classified in around 20 drug families. Drug reference books may give the impression that there are 10 times that many because various drug companies assign different trade names

to the very same generic drug. Ampicillin, for instance, is available under 50 different names. Most antibiotics are useful in controlling bacterial infections, although we shall also consider a number of antifungal, antiviral, and antiprotozoan drugs. Summaries of drug choices are located in later chapters on disease.

Antibacterial Drugs That Act on the Cell Wall

The **beta-lactam** (bey'-tuh-lak'-tam) group of antibiotics all contain a 3-carbon, 1-nitrogen ring that is highly reactive **(figure 12.7)**. Its primary mode of action is to interfere with proteins involved in synthesis of the cell wall, leading to lysis and cell death. More than half of all antimicrobic drugs are beta-lactams, with the penicillins and cephalosporins being the most prominent representatives.

Penicillin and Its Relatives

The **penicillin** group of antibiotics, named for the parent compound, is a large, diverse group of compounds, most of which end in the suffix -*cillin*. Although penicillins could be completely synthesized in the laboratory from simple raw materials, it is more practical and economical to obtain natural penicillin through microbial fermentation. The natural product can then be used either in unmodified form

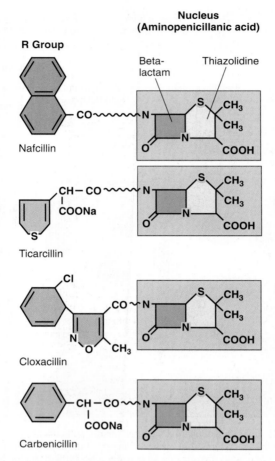

Figure 12.7 Chemical structure of penicillins. The basic nucleus of penicillin drugs (blue box) consists of a thiazolidine ring (yellow) and a beta-lactam ring (red), but specific types differ in the nature of the side chain (R group), which is also responsible for differences in each compound's activity against gram-negative cells.

or to make semisynthetic derivatives. *Penicillium chrysogenum* is the major source of the drug. All penicillins consist of three parts: a beta-lactam ring, a thiazolidine ring, and a variable side chain that dictates its microbicidal activity (figure 12.7 and **Insight 12.2**).

Subgroups and Uses of Penicillins The characteristics of certain penicillin drugs are shown in **table 12.5.** Penicillin G was the first antibiotic and is the parent compound for all "-cillin" drugs. It is narrow spectrum and considered the drug of choice for infections by sensitive gram-positive bacteria (streptococci) and some gram-negative bacteria (meningococci and the syphilis spirochete). A close relative, penicillin V, has similar uses but has been modified to be stable in the stomach acid and so can be taken orally.

Semisynthetic penicillins such as ampicillin, carbenicillin, and amoxicillin have been chemically altered with side chains that help them move across the outer membrane of gram-negative cell walls. This increases their spectrum and makes them useful in treating many types of gram-negative infections.

Many bacteria produce enzymes that are capable of destroying the beta-lactam ring of penicillin. The enzymes are referred to as **penicillinases** or **beta-lactamases,** and they make the bacteria that possess them resistant to many penicillins. Penicillinase-resistant penicillins such as methicillin, nafcillin, and cloxacillin are useful in treating infections caused by some penicillinase-producing bacteria. Mezlocillin and azlocillin have such an extended spectrum that they can be substituted for combinations of antibiotics. All of the "-cillin" drugs are relatively mild and well tolerated because of their specific mode of action on cell walls (which humans lack).

The primary problems in therapy include allergy, which is altogether different than toxicity, and resistant strains of pathogens. Clavulanic acid is a chemical that inhibits beta-lactamase enzymes, thereby increasing the longevity of beta-lactam antibiotics in the presence of penicillinase-producing bacteria. For this reason, clavulanic acid is often added to semisynthetic penicillins to augment

TABLE 12.5	Characteristics of Selected Penicillin Drugs		
Name	**Spectrum of Action**	**Uses, Advantages**	**Disadvantages**
Penicillin G	Narrow	Best drug of choice when bacteria are sensitive; low cost; low toxicity	Can be hydrolyzed by penicillinase; allergies occur; requires injection
Penicillin V	Narrow	Good absorption from intestine; otherwise, similar to penicillin G	Hydrolysis by penicillinase; allergies
Oxacillin, dicloxacillin	Narrow	Not susceptible to penicillinase; good absorption	Allergies; expensive
Methicillin, nafcillin	Narrow	Not usually susceptible to penicillinase	Poor absorption; allergies; growing resistance
Ampicillin	Broad	Works on gram-negative bacilli	Can be hydrolyzed by penicillinase; allergies; only fair absorption
Amoxicillin	Broad	Gram-negative infections; good absorption	Hydrolysis by penicillinase; allergies
Carbenicillin	Broad	Same as ampicillin	Poor absorption; used only parenterally
Piperacillin, mezlocillin, ticarcillin	Very broad	Effective against *Pseudomonas* species; low toxicity compared with aminoglycosides	Allergies, susceptible to many beta-lactamases

INSIGHT 12.2

A Modern Quest for Designer Drugs

Once the first significant antibiotic was developed, the world immediately witnessed a scientific scramble to find more. This search was advanced on several fronts. Hundreds of investigators began the laborious task of screening samples from soil, dust, muddy lake sediments, rivers, estuaries, oceans, plant surfaces, compost heaps, sewage, skin, and even the hair and skin of animals for antibiotic-producing bacteria and fungi. This intense effort has paid off over the past 50 years, because more than 10,000 antibiotics have been discovered (although surprisingly, only a relatively small number have actually been used clinically). Finding a new antimicrobial substance is only a first step. The complete pathway of drug development from discovery to therapy takes between 12 and 24 years at a cost of billions of dollars.

Antibiotics are products of fermentation pathways in many microorganisms. Among the greatest producers of antibiotics are the fungi, members of the soil-dwelling bacterial genera *Streptomyces* and the spore-forming bacteria. The role of antibiotics in the lives of these microbes must be important because the genes for antibiotic production are preserved in evolution. Some experts theorize that antibiotic-releasing microorganisms can inhibit or destroy nearby competitors or predators; others propose that antibiotics play a part in spore formation. Whatever benefit the microbes derive, these compounds have been extremely profitable for humans. Every year, the pharmaceutical industry farms vast quantities of microorganisms and harvests their products to treat diseases caused by other microorganisms. Researchers have facilitated the work of nature by selecting mutant species that yield more abundant or useful products, by varying the growth medium, or by altering the procedures for large-scale industrial production.

Another approach in the drug quest is to chemically manipulate molecules by adding or removing functional groups. Drugs produced in this way are designed to have advantages over related drugs. Using this **semisynthetic** method, a natural product of the microorganism is joined with various preselected functional groups. The antibiotic is reduced to its basic molecular framework (called the nucleus), and to this nucleus specially selected side chains (R groups) are added.

A case in point is the metamorphosis of the semisynthetic penicillins. The nucleus is an inactive penicillin derivative called aminopenicillanic acid, which has a special bonding site for addition of R groups. A particular carboxylic acid (R group) added to this nucleus can "fine-tune" the penicillin, giving it special characteristics. For instance, some R groups will make the product resistant to penicillinase (methicillin), some confer a broader activity spectrum (ampicillin), and others make the product acid resistant (penicillin V). Cephalosporins and tetracyclines also exist in several semisynthetic versions. The potential for using bioengineering techniques to design drugs seems almost limitless, and, indeed, several drugs have already been produced by manipulating the genes of antibiotic producers.

Explain why development of effective drugs takes so many years.

Answer available at http://www.mhhe.com/talaro8

R Groups

(a) Aminopenicillanic acid

(b) Methicillin

(c) Ampicillin

(d) Penicillin V

A plate with several discrete colonies of soil bacteria was sprayed with a culture of *Escherichia coli* and incubated. Zones of inhibition (clear areas with no growth) surrounding several colonies indicate species that produce antibiotics.

their effectiveness. For example, clavamox is a combination of amoxicillin and clavulanate and is marketed under the trade name Augmentin. Zosyn, a similar combination of the beta-lactamase inhibitor tazobactam and piperacillin, is used for a wide variety of systemic infections.

The Cephalosporin Group of Drugs

The **cephalosporin** antibiotics currently account for one-third of all antibiotics administered. Cephalosporins are similar to penicillins; they have a beta-lactam structure that can be synthetically

altered (**figure 12.8**) and have a similar mode of action. The generic names of these compounds are often recognized by the presence of the root *cef, ceph,* or *kef* in their names.

Subgroups and Uses of Cephalosporins The cephalosporins are versatile. They are relatively broad spectrum, resistant to most penicillinases, and cause fewer allergic reactions than penicillins. Although some cephalosporins are given orally, many are poorly absorbed from the intestine and must be administered **parenterally,*** by injection into a muscle or a vein.

Four generations of cephalosporins exist, with each group being more effective against gram-negative organisms than the generation before it. In addition, succeeding generations typically have improved dosing schedules and fewer side effects. First-generation cephalosporins (cephalothin and cefazolin) are most effective against gram-positive cocci and a few gram-negative bacteria. Second-generation forms (cefaclor and cefonacid) are more effective than the first-generation forms in treating infections by gram-negative

bacteria such as *Enterobacter, Proteus,* and *Haemophilus.* Third-generation cephalosporins, such as *cephalexin* (Keflex) and *ceftri-axone* (Rocephin), are broad spectrum with especially well-developed activity against enteric bacteria that produce beta-lactamases. Fourth-generation drugs, such as cefepime, have the widest range of antimicrobial properties. They are effective with both gram-negative and gram-positive bacterial infections and are rapidly microbicidal.

Carbapenems and monobactams Other beta-lactam drugs are imipenem and aztreonam. *Imipenem* is a broad-spectrum carbapenem used for infections with aerobic and anaerobic pathogens. It is active in very low concentrations and can be taken by mouth with few side effects except for allergies. *Aztreonam,* a monobactam, is a narrow-spectrum antibiotic for treating pneumonia, septicemia, and urinary tract infections by gram-negative aerobic bacilli. Aztreonam is especially useful when treating persons who are allergic to penicillin, because it is chemically different and does not cross-react with antibodies made to penicillin.

Miscellaneous Non-Beta-Lactam Cell Wall Inhibitors *Vancomycin* is a narrow-spectrum antibiotic most effective in treating staphylococcal infections in cases of penicillin and methicillin resistance or in patients with an allergy to penicillins. It has also been chosen to treat *Clostridium* infections in children and endocarditis (infection of the lining of the heart) caused by *Enterococcus faecalis.* Because it is very toxic and hard to administer, vancomycin is usually restricted to the most serious, life-threatening conditions.

Bacitracin is a narrow-spectrum peptide antibiotic produced by a strain of *Bacillus subtilis.* Its primary effect is to block the elongation of the peptidoglycan in gram-positive bacteria. Since it was first isolated, its greatest claim to fame has been as a major ingredient in a common drugstore antibiotic ointment (Neosporin) for combating superficial skin infections by streptococci and staphylococci. For this purpose, it is usually combined with neomycin (an aminoglycoside) and polymyxin.

Isoniazid (INH) works by interfering with the synthesis of mycolic acid, a necessary component of the cell wall of acid-fast organisms. It is used to treat infections with *Mycobacterium tuberculosis* but is effective only against growing cells. It is often prescribed with rifampin (see section on drugs that act on RNA) and other drugs to treat tuberculosis. Oral doses are indicated for both active tuberculosis and prophylaxis in cases of a positive TB test. Ethambutol, a closely related compound, is another anti tuberculosis drug added to treatment regimens (see continuing case file 12).

* *parenterally* (par-ehn′-tur-ah-lee) Gr. *para,* beyond, and *enteron,* intestine. A route of drug administration other than the gastrointestinal tract.

R Group 1	Basic Nucleus	R Group 2
	S or O	Cephalothin (first generation*)
	2nd	Cefotiam (second generation)
	3rd	Moxalactam (third generation)
	4th	Cefepime (fourth generation)

*New improved versions of drugs are referred to as new "generations."

Figure 12.8 **The structure of cephalosporins.** Like penicillin, their main nucleus consists of a beta-lactam ring (red) and a second ring (yellow). However, unlike penicillins, they have two sites for placement of R groups (at positions 3 and 7). This makes possible several generations of molecules with greater versatility in function and complexity in structure.

The discovery of the nurse's tuberculosis skin infection called for an entirely different drug regimen. The dicloxacillin given early on was ineffective because this relative of penicillin does not have a mode of action that works on *M. tuberculosis.* Major problems with drug treatment for tuberculosis are (1) the pathogen can already have or can develop drug resistance, and (2) it takes longer than most pathogenic bacteria to be eliminated.

This means that it is necessary to test for drug sensitivity, to choose drugs that will prevent resistance from happening, and to administer the drugs for several months. When tests indicated that the culture was sensitive to isoniazid, rifampin, and pyrazinamide, the nurse was started on this three-drug combined pill for 2 months, followed by 4 additional months on isoniazid and rifampin alone. By the end of this treatment, the infection was cured, and the nurse was relieved of her worst fears when a final HIV test came out negative.

- What is combined therapy, and what are the reasons for using it?
- Summarize some of the major problems with drug therapy presented in this case.

For a wrap-up, see the Case File Perspective on page 382.

Antibiotics That Damage Bacterial Cell Membranes

Bacillus polymyxa is the source of the **polymyxins,** narrow-spectrum peptide antibiotics with a unique fatty acid component that contributes to their detergent activity (see figure 12.4). Only two polymyxins—B and E (also known as colistin)—have any routine applications, and even these are limited by their toxicity to the kidney. Either drug can be indicated to treat drug-resistant infections caused by gram-negative rods such as *Pseudomonas aeruginosa* and *Acinetobacter baumannii.* It is also one of the topical agents in ointments to prevent skin infections, such as Neosporin.

Drugs That Act on DNA or RNA

Much excitement was generated by a class of synthetic drugs chemically related to quinine called **fluoroquinolones.** These drugs work by binding to DNA gyrase and a related enzyme, topoisomerase IV—both of which are essential for replication of the bacterial DNA. While DNA gyrase tends to be the primary target in gram-negative organisms, topoisomerase is targeted in gram-positive cells. It is not known how these drugs target different enzymes in different cell types, but this mode of action ensures that fluoroquinolones provide broad-spectrum effectiveness. In addition to being broad spectrum and highly potent, the drugs are readily absorbed from the intestine.

The principal quinolones, *norfloxacin* and *ciprofloxacin,* have been successful in therapy for urinary tract infections, sexually transmitted diseases, gastrointestinal infections, osteomyelitis, respiratory infections, and soft tissue infections. Newer drugs in this category are sparfloxacin and levofloxacin. These agents are especially recommended for pneumonia, bronchitis, and sinusitis. Now that concerns have arisen regarding the overuse of quinolone drugs, the CDC is recommending careful monitoring of their use to prevent the emergence of ciprofloxacin-resistant bacteria (see Insight 12.4). Side effects that can limit the use of quinolones include seizures and other brain disturbances.

Another drug essential to tuberculosis therapy is *rifampin.* The primary action of this drug is in blocking the action of RNA polymerase, thereby preventing transcription. It is somewhat limited in spectrum because the molecule cannot pass through the cell envelope of many gram-negative bacilli. It is mainly used to treat infections by selected gram-positive rods and cocci and a few gram-negative bacteria. Rifampin figures most prominently in treating mycobacterial infections, especially tuberculosis and leprosy, but it is usually given in combination with other drugs to prevent development of resistance. Rifampin is also recommended for prophylaxis in *Neisseria meningitidis* carriers and their contacts, and it is occasionally used to treat *Legionella, Brucella,* and *Staphylococcus* infections.

Drugs That Interfere with Protein Synthesis

The Aminoglycoside Drugs

Antibiotics composed of one or more amino sugars and an aminocyclitol (6-carbon) ring are referred to as **aminoglycosides (figure 12.9).** These complex compounds are exclusively the products of various species of soil **actinomycetes** in the genera *Streptomyces* and *Micromonospora.*

Subgroups and Uses of Aminoglycosides The aminoglycosides have a relatively broad antimicrobial spectrum because they inhibit protein synthesis by binding to one of the ribosomal subunits. They are especially useful in treating infections caused by aerobic

Figure 12.9 The structure of an aminoglycoside: streptomycin. Colored portions of the molecule are found in all numbers of this drug class.

gram-negative rods and certain gram-positive bacteria. *Streptomycin* is among the oldest of the drugs but has gradually been replaced by newer forms with less mammalian toxicity. It is still the antibiotic of choice for treating bubonic plague and tularemia, and is considered an effective antituberculosis agent. *Gentamicin* is less toxic and is widely administered for infections caused by gram-negative rods (*Escherichia, Pseudomonas, Salmonella,* and *Shigella*). Two relatively new aminoglycosides, tobramycin and amikacin, are also used for gram-negative infections, with tobramycin especially useful for treating *Pseudomonas* infections in cystic fibrosis patients.

Tetracycline Antibiotics

The first antibiotic in this class was aureomycin. It was used to synthesize terramycin, tetracycline, and several semisynthetic derivatives, commonly known as the **tetracyclines (figure 12.10a).** Their action of binding to ribosomes and blocking protein synthesis accounts for the broad-spectrum effects in the group.

Subgroups and Uses of Tetracyclines The scope of microorganisms inhibited by tetracyclines is very broad. It includes gram-positive and gram-negative rods and cocci, aerobic and anaerobic bacteria, mycoplasmas, rickettsias, and spirochetes. Tetracycline compounds such as *doxycycline* and *minocycline* are administered orally to treat several sexually transmitted diseases, Rocky Mountain spotted fever, Lyme disease, typhus, *Mycoplasma* pneumonia, cholera, leptospirosis, acne, and even some protozoan infections. Although generic tetracycline is low in cost and easy to administer, its use can be limited by its side effects. In addition to gastrointestinal disruption due to changes in the normal flora and staining of the teeth, ingestion during pregnancy can interfere with fetal bone development (see table 12.8).

Chloramphenicol

Chloramphenicol is a potent broad-spectrum antibiotic with a unique nitrobenzene structure (**figure 12.10b**). Its primary effect on cells is to block peptide bond formation and protein synthesis. It is one type of antibiotic that is no longer derived from the natural source but is entirely synthesized through chemical processes. Although this drug is fully as broad spectrum as the tetracyclines, it is so toxic to human cells that its uses are restricted. A small number of people undergoing long-term therapy with this drug incur irreversible damage to the bone marrow that usually results in a fatal form of aplastic anemia.[2] Its administration is now limited to typhoid fever, brain abscesses, and rickettsial and chlamydial infections for which an alternative therapy is not available. Chloramphenicol should never be given in large doses repeatedly over a long time period, and the patient's blood must be monitored during therapy.

Macrolides and Related Antibiotics

Erythromycin is a representative of antibiotics termed macrolides.[3] Its structure consists of a large lactone ring with sugars attached

(a) **Tetracyclines**

(b) **Chloramphenicol**

(c) **Erythromycin**

Figure 12.10 Structures of three broad-spectrum antibiotics. (a) Tetracyclines. These are named for their regular group of four rings. The several types vary in structure and activity by substitution at the four R groups. **(b)** Chloramphenicol. **(c)** Erythromycin, an example of a macrolide drug. Its central feature is a large lactone ring to which two hexose sugars are attached.

(**figure 12.10c**). This drug has a moderate spectrum and fairly low toxicity. Its mode of action is to block protein synthesis by attaching to the 50S subunit of the ribosome. It is administered orally for *Mycoplasma* pneumonia, legionellosis, *Chlamydia* infections, pertussis, and diphtheria, and as a prophylactic drug prior to intestinal surgery. It also offers a useful substitute for dealing with penicillin-resistant streptococci and gonococci and for treating syphilis and acne. Newer semisynthetic macrolides include *clarithromycin* and *azithromycin*. Both drugs are useful for middle ear, respiratory, and skin infections and have been approved for *Mycobacterium* (MAC) infections in AIDS patients. Clarithromycin has additional applications in controlling infections and gastric ulcers caused by *Helicobacter pylori*.

Clindamycin is a broad-spectrum antibiotic derived from lincomycin. The tendency of clindamycin to cause adverse reactions in the gastrointestinal tract limits its applications to (1) serious infections in the large intestine and abdomen due to anaerobic bacteria (*Bacteroides* and *Clostridium*), (2) infections with penicillin-resistant staphylococci, and (3) acne medications applied to the skin.

2. A failure of the blood-producing tissue that results in very low levels of blood cells.

3. Macrolide antibiotics get their name from the type of chemical ring structure (macrolide ring) they all possess.

Drugs That Block Metabolic Pathways

Many of the drugs used to interfere with metabolism are synthetic. These antimicrobials as a group do not originate from bacterial or fungal fermentations. Some were synthesized from aniline dyes, and others were originally isolated from plants. Although they have been largely replaced by antibiotics, several types remain essential to chemotherapy.

The most important group of drugs in this class is the **sulfonamides,** or **sulfa drugs.** Their name is derived from sulfanilamide, an early form of the drug **(figure 12.11).** These drugs function as metabolic analogs to block the synthesis of folic acid by bacteria (see figure 12.6). Dozens of these drugs have been developed, most of them narrow spectrum. Because of its solubility, sulfisoxazole is an effective choice for treating shigellosis, urinary tract infections, and certain protozoan infections. Silver sulfadiazine ointment and solution are prescribed for treatment of burns and eye infections. Another drug, *trimethoprim,* inhibits a second enzymatic step in the synthesis of folic acid. Because of this, trimethoprim is often given in combination with sulfamethoxazole to take advantage of the synergistic effect of the two drugs (Septra, Bactrim). This combination is one of the primary treatments for *Pneumocystis (carinii) jiroveci* pneumonia (PCP) in AIDS patients, urinary tract infections, and otitis media.

Sulfones are compounds chemically related to the sulfonamides but lacking their broad-spectrum effects. They still remain important as key drugs in treating Hansen's disease (leprosy). The most active form is dapsone, usually given in combination with rifampin and clofazamine (an antibacterial dye) over long periods.

Newly Developed Classes of Antibiotics

Although most new antibiotics are formulated from the drug classes that already exist, a few classes of drugs with novel actions against bacteria have recently been added to drug regimens.

Fosfomycin trimethamine is a phosphoric acid agent effective as alternate treatment for urinary tract infections caused by enteric bacteria. It works by inhibiting an enzyme necessary for cell wall synthesis.

Synercid is an antibiotic from the streptogramin group of drugs that contains two active chemicals. It is effective against *Staphylococcus* and *Enterococcus* species that cause endocarditis and surgical infections, and against resistant strains of *Streptococcus.* It is one of the main choices when other drugs are ineffective due to resistance. Synercid works by binding to sites on the 50S ribosome, inhibiting peptide transfer and elongation.

Daptomycin is a lipopeptide directed mainly against gram-positive bacteria, acting to disrupt multiple aspects of membrane function. Many experts are urging physicians to use these medications only when no other drugs are available to slow the development of drug resistance.

Another new class of drugs, the ketolides, is similar to macrolides like erythromycin but has a different ring structure. The main representative, *telithromycin* (Ketek), is used for respiratory tract infections caused by macrolide resistant bacteria.

The *oxazolidinone* class of drugs operates using a unique mechanism, namely to inhibit the *initiation* of protein synthesis. It does this by interfering with the interaction between the mRNA and the two ribosomal subunits necessary for the initiation of translation. The sole drug in this class, linezolid (sold under the trade name Zyvox), has been used to treat infections caused by two of the most difficult clinical pathogens, methicillin-resistant *Staphylococcus aureus* (MRSA) and vancomycin-resistant *Enterococcus* (VRE). Because this drug is synthetic and not found in nature, it is hoped that drug resistance will be slow to develop.

Figure 12.11 **The structures of some sulfonamides.** Boxed side chains attach at the —ONH group of the nucleus. **(a)** Sulfacetamide, **(b)** sulfadiazine, and **(c)** sulfisoxazole.

Check & Assess Section 12.2

✔ Antimicrobials are classified into around 20 major drug families, based on their chemical composition, source or origin, and their site of action.

✔ Antimicrobial drugs are classified by their range of effectiveness. Broad-spectrum antimicrobials are effective against many types of microbes. Narrow-spectrum antimicrobials are effective against a limited group of microbes.

✔ The majority of antimicrobials are effective against bacteria, but a limited number are effective against protozoa, helminths, fungi, and viruses.

✔ Penicillins, cephalosporins, bacitracin, vancomycin, and cycloserines block cell wall synthesis, primarily in gram-positive bacteria.

✔ Aminoglycosides and tetracyclines block protein synthesis in prokaryotes.

✔ Sulfonamides, trimethoprim, isoniazid, nitrofurantoin, and the fluoroquinolones are synthetic antimicrobials effective against a broad range of microorganisms. They block steps in the synthesis of nucleic acids.

13. Construct a chart that summarizes the modes of action and applications of the major groups of antibacterial drugs (antibiotics and synthetics), antifungal drugs, antiparasitic drugs, and antiviral drugs.

14. Why do the penicillin and cephalosporin groups of drugs have milder toxicity than other antibiotics? What are their primary side effects?

12.3 Drugs to Treat Fungal, Parasitic, and Viral Infections

Antifungal Drugs

Because the cells of fungi are eukaryotic, they present special problems in chemotherapy. For one, the great majority of chemotherapeutic drugs are designed to act on bacteria and are generally ineffective in combating fungal infections. For another, the similarities between fungal and human cells often mean that drugs toxic to fungal cells are also capable of harming human tissues. A few agents with special antifungal properties have been developed for treating systemic and superficial fungal infections. Five main drug groups currently in use are the macrolide polyene antibiotics, griseofulvin, synthetic azoles, flucytosine, and echinocandins (**figure 12.12**). See table 22.4 for a complete outline of antifungal drugs.

Polyenes bind to fungal membranes and cause loss of selective permeability. They are specific for fungal membranes because fungal membranes contain a particular sterol component called ergosterol, while human membranes do not. The toxicity of polyenes is not completely selective, however, because mammalian cell membranes contain compounds similar to ergosterol that bind polyenes to a small extent.

Macrolide polyenes, represented by *amphotericin B* (Fungizone) (named for its acidic and basic—amphoteric—properties) and nystatin (for New York State, where it was discovered), have a structure that mimics the lipids in some cell membranes. Amphotericin B is by far the most versatile and effective of all antifungals. Not only does it work on most fungal infections, including skin and mucous membrane lesions caused by *Candida albicans*, but it is one of the few drugs that can be injected to treat systemic fungal infections such as histoplasmosis and cryptococcus meningitis. A significant drawback to the widespread use of amphotericin centers on its serious side effects, which can include fatigue and irregular heartbeat. *Nystatin* (Mycostatin) is used only topically or orally to treat candidiasis of the skin and mucous membranes, but it is not useful for subcutaneous or systemic fungal infections or for ringworm.

Griseofulvin is an antifungal product especially active in certain dermatophyte infections such as athlete's foot. The drug is deposited in the epidermis, nails, and hair, where it inhibits fungal growth. Because complete eradication requires several months and griseofulvin is relatively nephrotoxic, this therapy is given only for the most stubborn cases.

The **azoles** are broad-spectrum antifungal agents with a complex ringed structure that also act on fungal membrane structure. The most effective drugs are ketoconazole (Nizoral), fluconazole (Diflucan), clotrimazole (Gyne-Lotrimin), miconazole (Monistat), and itraconazole (Sporanox). Ketoconazole is used orally and topically for cutaneous mycoses, vaginal and oral candidiasis, and some systemic mycoses. Fluconazole can be used in selected patients for AIDS-related mycoses such as aspergillosis and cryptococcus meningitis. Clotrimazole and miconazole are used mainly as topical ointments for infections in the skin, mouth, and vagina. Itraconazole is generally used orally for fungal infections of the nails and systemic candidiasis.

Flucytosine is an analog of the nucleotide cytosine that has antifungal properties. It is rapidly absorbed after oral therapy, and it is readily dissolved in the blood and cerebrospinal fluid. Alone, it can be used to treat certain cutaneous mycoses. Many fungi are resistant to flucytosine, so it is usually combined with amphotericin B to effectively treat systemic mycoses.

After many years of testing, a new antifungal drug category has been introduced. *Echinocandins,* such as caspofungin, damage the cell walls of several types of fungi, thereby making them sensitive to lysis. A significant advantage of this mode of action is that it cannot target human cells.

Figure 12.12 Some antifungal drug structures.
(a) Polyenes. The example shown is amphotericin B, a complex steroidal antibiotic that inserts into fungal cell membranes.
(b) Clotrimazole, an azole that inhibits synthesis of ergesterol, a component of the fungal cell membrane. **(c)** Flucytosine, a structural analog of cytosine that inhibits DNA and protein synthesis.

Antiparasitic Chemotherapy

The enormous diversity among protozoan and helminth parasites and their corresponding therapies reaches far beyond the scope of this textbook; however, a few of the more common drugs are surveyed here and described again for particular diseases in chapter 23. Presently, a small number of approved and experimental drugs are used to treat malaria, leishmaniasis, trypanosomiasis, amebic dysentery, and helminth infections, but the need for new and better drugs has spurred considerable research in this area.

Antimalarial Drugs: Quinine and Its Relatives

Quinine, extracted from the bark of the cinchona tree, was the principal treatment for malaria for hundreds of years, but it has been replaced by the synthesized quinolines, mainly *chloroquine* and *primaquine*, which have less toxicity to humans. Because there are several species of *Plasmodium* (the malaria parasite) and many stages in its life cycle, no single drug is universally effective for every species and stage, and each drug is restricted in application. For instance, primaquine eliminates the liver phase of infection, and chloroquine suppresses acute attacks associated with infection of red blood cells. Chloroquine is taken alone for prophylaxis and suppression of acute forms of malaria. Primiquine is administered to patients with relapsing cases of malaria. Mefloquine is a semisynthetic analog of quinine used to treat infections caused by chloroquine-resistant strains of *Plasmodium*, an increasing problem in Southeast Asia.

Chemotherapy for Other Protozoan Infections

A widely used amoebicide, metronidazole (Flagyl), is effective in treating mild and severe intestinal infections and hepatic disease caused by *Entamoeba histolytica*. Given orally, it also has applications for infections by *Giardia lamblia* and *Trichomonas vaginalis*. Other drugs with antiprotozoan activities are quinicrine (a quinine-based drug), sulfonamides, and tetracyclines.

Antihelminthic Drug Therapy

Treating helminthic infections has been one of the most challenging of all chemotherapeutic tasks. Flukes, tapeworms, and roundworms are much larger parasites than other microorganisms and, being animals, have greater similarities to human physiology. Also, the usual strategy of using drugs to block their reproduction is usually not successful in eradicating the adult worms. The most effective drugs immobilize, disintegrate, or inhibit the metabolism of all stages of the life cycle.

Mebendazole and *thiabendazole* are broad-spectrum antiparasitic drugs used to treat several types of roundworm infections. These drugs work locally in the intestine to inhibit the function of the microtubules of worms, eggs, and larvae, which interferes with their glucose utilization and disables them. The compounds *pyrantel* and *piperazine* paralyze the muscles of intestinal roundworms. Niclosamide destroys the scolex and the adjoining proglottids of tapeworms, thereby loosening the worm's holdfast. In these forms of therapy, the worms are unable to maintain their grip on the intestinal wall and are expelled along with the feces by the normal peristaltic action of the bowel. A drawback to the use of the drug is the severe abdominal cramps that often accompany treatment. Two newer antihelminthic drugs are *praziquantel*, a treatment for various tapeworm and fluke infections, and *ivermectin*, a veterinary drug now used for strongyloidiasis and oncocercosis in humans.

Antiviral Chemotherapeutic Agents

The chemotherapeutic treatment of viral infections presents unique problems. Throughout our discussion of infections caused by bacteria, fungi, protozoa, and helminths, we've emphasized differences in structure and metabolism that guide drug choice. With viruses, we are dealing with an infectious agent that relies upon the host cell for the vast majority of its metabolic functions. Disrupting viral function may require that we disrupt the metabolism of the host cell to a much greater extent than is desirable. Put another way, selective toxicity with regard to viral infection is almost impossible to achieve because a single metabolic system is responsible for the well-being of both virus and host. A few viral diseases such as measles, mumps, and hepatitis are routinely prevented by effective vaccinations. Unfortunately, vaccines are unavailable for many serious viral diseases, creating a real need for antiviral medications.

Over the past few years, several antiviral drugs have been developed that target specific points in the infectious cycle of viruses. But due to marked differences among the viruses, most antiviral compounds are quite limited in their spectrum.

Most antiviral drugs are designed to block a step in completion of the virus cycle. Major modes of action include

1. barring penetration of the virus into the host cell;
2. blocking replication, transcription, and/or translation of viral genetic material; and
3. preventing the normal maturation of viral particles.

Although antiviral drugs protect uninfected cells by keeping viruses from being synthesized and released, most are unable to eliminate extracellular viruses or those in a latent state. A summary covering the actions of several antiviral drugs appears in **Table 12.6.**

Drugs for Treating Influenza *Amantadine* and its relative, rimantidine, are restricted almost exclusively to treating infections by influenza A virus. They function by blocking the hemagglutinin receptors. In this way, they prevent fusion of the virus with cell membranes and consequently interfere with entry, uncoating, and release of the virus. *Relenza* and *Tamiflu* are slightly broader-spectrum antiviral medications because they block neuraminidase enzymes in both influenza A and influenza B. Both types of drugs may be used prophylactically and must be given early in an infection to be most effective (table 12.6).

Antiherpes Drugs Many antiviral agents mimic the structure of nucleotides (i.e., act as *analogs*) and compete for sites on replicating DNA. Once one of these nucleotide analogs is incorporated into the growing chain of DNA, replication of that strand stops, interrupting the viral life cycle. **Acyclovir** (Zovirax) and the related compounds *valacyclovir* (Valtrex), *famciclovir* (Famvir), and *penciclovir* (Denavir) all work in this manner and (with small exceptions from drug to drug) can be used either orally or topically to treat common herpesvirus infections such as oral and genital herpes, chickenpox, and shingles. A related drug, *ganciclovir* (Cytovene) is used parenterally to treat cytomegalovirus infection in patients with compromised immune systems.

An interesting aspect of some of these antiviral agents (specifically valacyclovir and famciclovir) is that they are activated by an enzyme encoded by the virus itself, making the drug active only in virally infected cells. The enzyme thymidine kinase is used by the virus to process nucleosides before incorporating them into viral RNA or DNA. When the inactive drug enters a virally infected cell, the virus's own thymidine kinase converts it to a working antiviral agent. In cells without viruses, the drug is never activated and normal DNA replication is allowed to continue.

TABLE 12.6 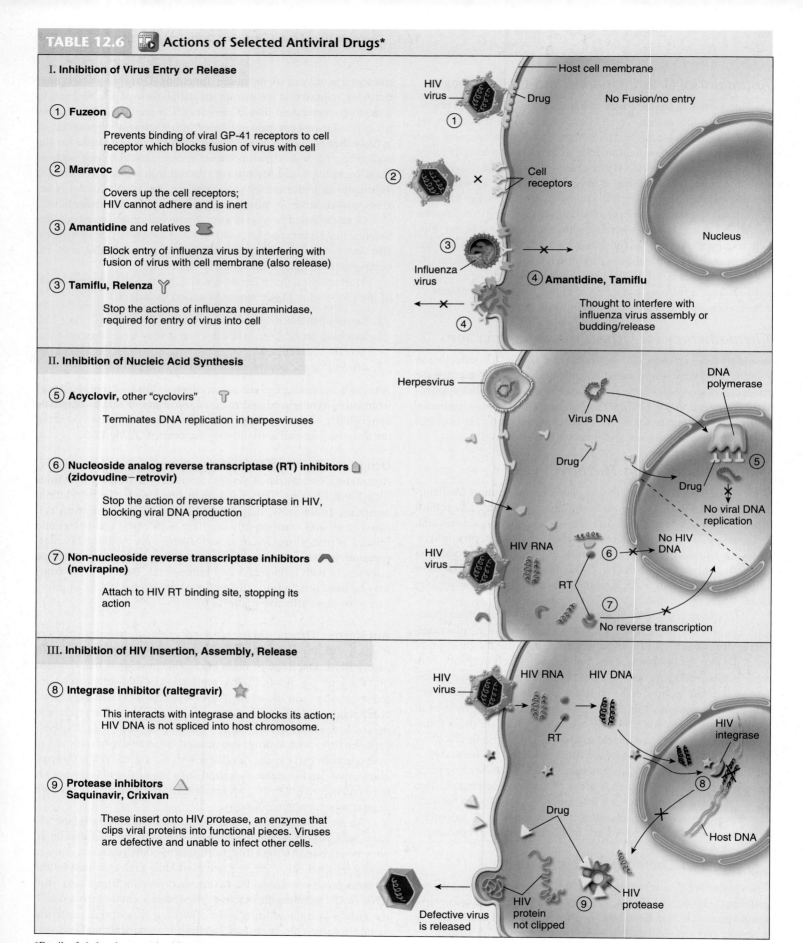 **Actions of Selected Antiviral Drugs***

I. Inhibition of Virus Entry or Release

① **Fuzeon**

Prevents binding of viral GP-41 receptors to cell receptor which blocks fusion of virus with cell

② **Maravoc**

Covers up the cell receptors; HIV cannot adhere and is inert

③ **Amantidine** and relatives

Block entry of influenza virus by interfering with fusion of virus with cell membrane (also release)

③ **Tamiflu, Relenza**

Stop the actions of influenza neuraminidase, required for entry of virus into cell

Host cell membrane

HIV virus — Drug

No Fusion/no entry

Cell receptors

Nucleus

Influenza virus

④ **Amantidine, Tamiflu**

Thought to interfere with influenza virus assembly or budding/release

II. Inhibition of Nucleic Acid Synthesis

⑤ **Acyclovir**, other "cyclovirs"

Terminates DNA replication in herpesviruses

⑥ **Nucleoside analog reverse transcriptase (RT) inhibitors (zidovudine–retrovir)**

Stop the action of reverse transcriptase in HIV, blocking viral DNA production

⑦ **Non-nucleoside reverse transcriptase inhibitors (nevirapine)**

Attach to HIV RT binding site, stopping its action

Herpesvirus

Virus DNA

DNA polymerase

Drug

Drug

No viral DNA replication

HIV virus

HIV RNA

No HIV DNA

RT

No reverse transcription

III. Inhibition of HIV Insertion, Assembly, Release

⑧ **Integrase inhibitor (raltegravir)**

This interacts with integrase and blocks its action; HIV DNA is not spliced into host chromosome.

⑨ **Protease inhibitors Saquinavir, Crixivan**

These insert onto HIV protease, an enzyme that clips viral proteins into functional pieces. Viruses are defective and unable to infect other cells.

HIV virus

HIV RNA

HIV DNA

RT

HIV integrase

Host DNA

Drug

HIV protease

Defective virus is released

HIV protein not clipped

*Details of viral cycles are omitted for ease in observing drug effects.

Drugs for Treating HIV Infection and AIDS The global epidemic of AIDS has been a powerful stimulus for drug discovery. Most of the newer drugs being introduced are designed for HIV and AIDS. Although there is now a considerable array of drugs and drug combinations to manage HIV, we cover just the basic ones here. All of the drugs target a specific part of the multiplication cycle of HIV (table 12.6). There are currently drugs that can inhibit or block every strategic step in the virus cycle. One of the most important steps is the action of the **reverse transcriptase (RT)** enzyme that HIV brings with it. Recall from chapter 9 that HIV is a retrovirus with an RNA genome. Upon entering a host, the RNA is used as a template by reverse transcriptase to produce a DNA copy of this RNA. This actually offers two types of targets for anti-HIV drugs. The first involves *nucleoside reverse transcriptase analogs* (NRTIs) that directly stop DNA synthesis, and the second comes from *nonnucleoside reverse transcriptase inhibitors* (NNRTIs) that interfere with the function of the RT enzyme itself.

Zidovudine (retrovir, AZT) was the first NRTI and the first drug developed to treat HIV and AIDS. It is an analog of thymine that becomes incorporated into the DNA strand of HIV but terminates synthesis because it does not offer the correct binding sites. Retrovir can be used at all stages of HIV infection, including as prophylaxis for accidental exposure. Other NRTIs are epivir, abacavir, didanosine, and stavudine.

NNRTIs accomplish a similar goal in preventing reverse transcription of the HIV RNA, but they bind to the enzyme itself and prevent its reactive site from synthesizing DNA. They are usually given together or in combination with NRTIs. Drugs with this mode of action include nevirapine, efavirenz, and delaviridine.

Another enzyme that is essential to the virus's release is HIV protease, which is involved in cutting elongate protein strands of the virus before they are assembled into the finished virion. The class of drugs called *protease inhibitors* has the effect of blocking this enzyme and causing the viruses to be defective and not infective. Examples are indinavir, crixivan, and saquinavir.

Newer classes of anti-HIV drugs are two that block entry or fusion of the virus at the beginning of its invasion and one that prevents viral integration into the host genome. The drug *maravoc* keeps HIV out of cells by covering up the CD4 receptors on the white blood cells that HIV usually infects. This has the effect of preventing the viruses from ever adhering to host cells, and without this step, they cannot proceed. *Fuzeon* prevents infection when it stops the envelope of the virus from fusing with the host cell after it has attached. The *integrase inhibitor* called raltegravir is effective in preventing the HIV DNA, once it is formed, from being incorporated or integrated into the host's chromosome. Since this step is required for full viral multiplication, virus replication cannot proceed.

Most therapies use several drugs that work together to keep viral activity to a minimum. This therapeutic approach has been shown to reduce the HIV load in the blood to undetectable levels and to improve the survival of AIDS patients. Combined therapy also decreases levels of drug resistance, which has been a serious problem with therapy. More on these topics can be found in chapter 25.

Interferons A sensible alternative to artificial drugs has been a human-based substance, **interferon (IFN).** Interferon is a glycoprotein produced primarily by fibroblasts and leukocytes in response to various immune stimuli. It has numerous biological activities, including antiviral and anticancer properties. Studies have shown that it is a versatile part of animal host defenses, having a major role in natural immunities. Its mechanism is discussed in chapter 14.

The first investigations of interferon's antiviral activity were limited by the extremely minute quantities that could be extracted from human blood. Several types of interferon (Betaseron, Roferon) are currently produced by the recombinant DNA technology techniques outlined in chapter 10. Extensive clinical trials have tested interferon's effectiveness in viral infections and cancer. Some of the known therapeutic benefits of interferon include

1. reducing the time of healing and some of the complications in certain infections (mainly of herpesviruses);
2. preventing or reducing some symptoms of cold and papillomaviruses (warts);
3. slowing the progress of certain cancers, including bone cancer and cervical cancer, and certain leukemias and lymphomas; and
4. treating a rare cancer called hairy-cell leukemia, hepatitis C (a viral liver infection), genital warts, and Kaposi's sarcoma in AIDS patients.

Medical science is giving close scrutiny to alternate methods of preventing or treating infections beyond the usual antibiotics and synthesized drugs **(Insight 12.3).**

Check & Assess Section 12.3

✔ Fungal antimicrobials, such as macrolide polyenes, griseofulvin, azoles, and flucytosine, are used for systemic and superficial infections.

✔ Most antifungal drugs are selective for cell membranes, causing lysis of the fungus, but their action on membranes makes them potentially toxic.

✔ There are fewer antiparasitic drugs than antibacterial drugs because parasites are eukaryotes like their human hosts and they have several life stages, some of which can be resistant to the drugs.

✔ Antihelminthic drugs immobilize or disintegrate infesting helminths or inhibit their metabolism in some manner.

✔ Antiviral drugs interfere with viral replication by blocking viral entry into cells, blocking the replication process, or preventing the assembly of viral subunits into complete virions.

✔ Many antiviral agents are analogs of nucleotides. They inactivate the replication process when incorporated into viral nucleic acids. HIV antivirals interfere with reverse transcriptase, proteases, and integrases to prevent the production of new viral particles.

✔ Interferon is effective in vivo against certain viral infections; commercial interferon is mainly restricted to certain cancers and hepatitis B and C.

15. Construct a chart to outline the applications of major groups of antifungal drugs, antiparasitic drugs, and antiviral drugs.
16. Explain why there are fewer antifungal, antiparasitic, and antiviral drugs than antibacterial drugs.
17. What is the strategy behind taking several different types of drugs to treat HIV?
18. Summarize the origins and biological actions of interferon.

INSIGHT 12.3

Alternative Antimicrobial Therapies

"The enemy of my enemy is my friend."

This old saying might well describe a novel way to kill bacteria using viruses. This technique, called phage therapy, was first attempted over a hundred years ago after the discovery of bacteriophages. The concept behind it is rather simple: When you expose bacterial cells to their specific viruses, the viruses will invade and destroy them. If you apply the viruses in large quantity to treat a human or animal infection, they could lyse the bacteria and cure the infection. Phage therapy went out of favor for many years except in Russia, but now it is making a comeback worldwide. An important push for renewed interest is the search for solutions to the high levels of drug resistance among bacteria. Bacteria do not become resistant to viruses. Other significant advantages are that viruses are very narrow spectrum—they can target just one species of bacteria—and they would not be toxic to humans as are many drugs.

For now, phage therapy is restricted in the United States to removing contaminants from meats, reducing the incidence of *Salmonella* and *Escherichia coli* in young farm animals, treating skin infections, and sanitizing food. In some clinics, it is in experimental trials for treatment of skin ulcers, wounds, and urinary tract infections. It will require extensive testing before it becomes a first-line therapy.

Other novel approaches to controlling infections include the use of **probiotics*** and **prebiotics.** Probiotics are preparations of live microorganisms that are fed to animals and humans to modify the intestinal flora. These microorganisms can replace microbes lost during antimicrobial therapy or simply augment the flora that is already there. Culture-enriched dairy products are the latest offerings in the market for probiotics. Several new types of yogurt now claim specific health benefits from eating high numbers of special strains of bacteria. According to some published research findings, the ingestion of *Bifidobacterium lactis regularis* can help to maintain regular bowel function, and *Lactobacillus casei immunitas* may boost the immune function of the gut. Purposely eating such high numbers of bacterial cells—estimated at from 1 billion to 10 billion per serving—is quite a departure from the usual view of microbes as disease agents. Bon appetit!

A probiotic approach is frequently used in female patients with recurring urinary tract infections. There is evidence that vaginal inserts or perineal swabs of *Lactobacillus* can restore a healthy acidic environment to the region and displace disease-causing microorganisms. Probiotics also seem to inhibit the development of food allergies.

Prebiotics are nutrients that encourage the growth of beneficial microbes in the intestine. For instance, fruit sugars such as fructans are thought to encourage the growth of *Bifidobacterium* in the large intestine and to discourage the growth of potential pathogens. Clearly, the use of these agents is a different type of antimicrobial strategy than we are used to, but it may have its place in a future in which traditional antibiotics are more problematic.

Another category of antimicrobial is lantibiotics—short peptides produced by bacteria that inhibit the growth of other bacteria. They are distinguished by the presence of unusual amino acids that are not seen elsewhere in the cell. They exert their antimicrobial activity either by puncturing cell membranes or by inhibiting bacterial enzymes. The most well-known lantibiotic is *nisin*. Lantibiotics have long been used in food preservation, in veterinary medicine, and more recently in personal care products such as deodorants. Because lantibiotics are apparently effective against human pathogens such as *Propionibacterium acnes* (acne), methicillin-resistant *S. aureus* (MRSA), and *Helicobacter pylori* (ulcers), they may soon be an alternate choice for these infections.

Describe several benefits of using nondrug regimens and foods to treat or prevent infections. *Answer available at http://www.mhhe.com/talaro8*

* *probiotic* (proh′-by-ah-tik′) Gr. *pro,* for, and *bios,* life.

12.4 Interactions between Microbes and Drugs: The Acquisition of Drug Resistance

Expected Learning Outcomes

15. Describe the development of drug resistance, and explain how it is acquired.

16. Explain the major mechanisms by which microbes become resistant to drugs.

17. Discuss some of the ways for reducing levels of drug resistance.

18. Discuss how drug resistance spreads through a population of microbes and the long-term effects of this.

One unfortunate outcome of the use of antimicrobials is the development of microbial **drug resistance,** an adaptive response in which microorganisms begin to tolerate an amount of drug that would ordinarily be inhibitory. The development of mechanisms for resisting antimicrobial drugs is a result of the genetic versatility and adaptability of microbial populations. The property of drug resistance can be intrinsic as well as acquired. Intrinsic drug resistance can best be exemplified by the fact that bacteria must, of course, be resistant to any antibiotic that they produce. This type of resistance is limited, however, to a small group of organisms and is generally not a problem with regard to antimicrobial chemotherapy. Of much greater importance is the acquisition of resistance to a drug by a microbe that was previously sensitive to the drug. In our context, the term *drug resistance* will refer to this last type of acquired resistance.

How Does Drug Resistance Develop?

Contrary to popular belief, antibiotic resistance is not a recent phenomenon. Resistance to penicillin developed in some bacteria as early as 1940, 3 years before the drug was even approved for public use. The scope of the problem became apparent in the 1980s and 1990s when scientists and physicians observed treatment failures on a large scale.

Microbes become newly resistant to a drug after one of the following events occurs: (1) spontaneous mutations in critical chromosomal genes, or (2) acquisition of entire new genes or sets of genes via transfer from another species. Chromosomal drug resistance usually results from spontaneous random mutations in bacterial populations. The chance that such a mutation will be advantageous is minimal, and the chance that it will confer resistance to a specific drug is lower still. Nevertheless, given the huge numbers of microorganisms in any population and the constant rate of mutation, such mutations do occur. The end result varies from slight changes in microbial sensitivity, which can be overcome by larger doses of the drug, to complete loss of sensitivity.

Resistance through intermicrobial transfer originates from chromosomal genes and plasmids called **resistance (R) factors** being transferred through the processes we considered in chapter 9: conjugation, transformation, or transduction (**figure 12.13**). Studies have shown that plasmids encoded with drug resistance are naturally present in microorganisms before they have been exposed to the drug. Such traits are "lying in wait" for an opportunity to be expressed and to confer adaptability on the species. Many bacteria also maintain transposable drug-resistance sequences (transposons) that are duplicated and inserted from one plasmid to another or from a plasmid to the chromosome. Chromosomal genes and plasmids containing codes for drug resistance are faithfully replicated and inherited by all subsequent progeny. This sharing of resistance genes accounts for the rapid proliferation of drug-resistant species. A growing body of evidence points to the ease and frequency of gene transfers in biofilms between totally unrelated bacteria living in the body's normal flora and the environment.

Specific Mechanisms of Drug Resistance

Inside a bacterial cell, the net effect of these genetic transfers is one of the following, which actually causes the bacterium to be resistant (note that the numbers that follow correspond to the numbers seen in **figure 12.14**):

1. Inducement of alternate enzymes can inactivate the drug (only occurs when new genes are acquired).
2. Permeability or uptake of drug into bacterium is decreased or eliminated.
3. The microbe engages special drug transport pumps that remove the drug.
4. Binding sites for drug are decreased in number or affinity (can occur via mutation or acquisition of new genes).
5. An affected metabolic pathway is shut down or an alternate pathway is used (occurs due to mutation of original enzyme(s)).

Some bacteria can become resistant indirectly by lapsing into dormancy or, in the case of penicillin, by converting to a cell-wall-deficient form (L form) that penicillin cannot affect.

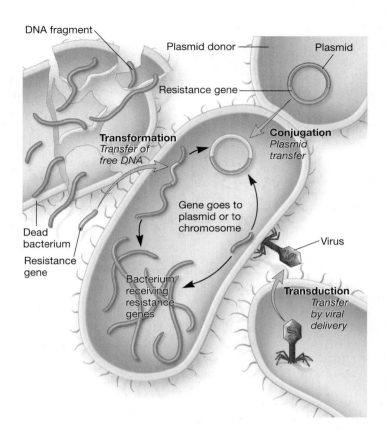

Figure 12.13 **Transfer of drug resistance.** Bacteria transmit R factors, often containing numerous genes for drug resistance by means of conjugation, transduction, and transformation. The widespread occurrence of such exchanges accounts for the rapid spread of drug resistance.

Drug Inactivation Mechanisms

Microbes inactivate drugs by producing enzymes that permanently alter drug structure. One example, bacterial exoenzymes called *beta-lactamases,* hydrolyze the beta-lactam ring structure of some penicillins and cephalosporins, rendering the drugs inactive. Two beta-lactamases—penicillinase and cephalosporinase—disrupt the structure of certain penicillin or cephalosporin molecules so their activity is lost (**figure 12.14, part 1**). So many strains of *Staphylococcus aureus* produce penicillinase that regular penicillin is rarely a possible therapeutic choice. Now that some strains of *Neisseria gonorrhoeae,* called PPNG,[4] have also acquired penicillinase, alternative drugs are required to treat gonorrhea. A large number of other gram-negative species are inherently resistant to some of the penicillins and cephalosporins because of naturally occurring beta-lactamases.

Decreased Drug Permeability or Increased Drug Elimination

The resistance of some bacteria can be due to a mechanism that prevents the drug from entering the cell and acting on its target. For example, the outer membrane of the cell wall of certain gram-negative bacteria is a natural blockade for some of the penicillin drugs.

4. Penicillinase-producing *Neisseria gonorrhoeae.*

Figure 12.14 **Examples of mechanisms of acquired drug resistance.**

Resistance to the tetracyclines can arise from plasmid-encoded proteins that pump the drug out of the cell. Aminoglycoside resistance is known to develop through changes in drug permeability caused by point mutations in proteins of the transport system or outer membrane (**figure 12.14, parts 2 and 3**).

Many bacteria possess multidrug-resistant (MDR) pumps that actively transport drugs and other chemicals out of cells. These pumps are proteins encoded by plasmids or chromosomes. They are stationed in the cell membrane and expel molecules by a proton-motive force similar to ATP synthesis (**figure 12.14, part 3**). They confer drug resistance on many gram-positive pathogens (*Staphylococcus, Streptococcus*) and gram-negative pathogens (*Pseudomonas, E. coli*). Because they lack selectivity, one type of pump can expel a broad array of antimicrobial drugs, detergents, and other toxic substances.

Change of Drug Receptors

Because most drugs act on a specific target such as protein, RNA, DNA, or membrane structure, microbes can circumvent drugs by altering the nature of this target. Bacteria can become resistant to aminoglycosides when point mutations in ribosomal proteins arise (**figure 12.14, part 4**). Erythromycin and clindamycin resistance is associated with an alteration on the 50S ribosomal binding site. Penicillin resistance in *Streptococcus pneumoniae* and methicillin resistance in *Staphylococcus aureus* are related to an alteration in the binding proteins in the cell wall. Several species of enterococci have acquired resistance to vancomycin through a similar alteration of cell wall proteins. Fungi can become resistant by decreasing their synthesis of ergosterol, the principal receptor for certain antifungal drugs.

TAKE NOTE: BIOFILMS AND DRUG RESISTANCE

In chapters 4 and 7, we described the development of microbial biofilms in various habitats. Recall that in these complex communities, certain "pioneer" microbes anchor themselves to a substrate by secreting a sticky matrix. This initial colonization stimulates a network of other microbes to migrate into the film. Most habitats support microbes that are adapted to these biofilm partnerships.

It is estimated that around 60% of infections also involve biofilms containing single or multiple species of microbes. Often the infections occur on natural tissues—for example, bacterial infections of the heart valves, middle ear, and teeth. The increased use of indwelling medical devices has created yet another habitat in which biofilms may become tenaciously attached. Catheters, artificial valves, and prosthetic devices can harbor persistent biofilm colonizations that are very difficult to treat and remove.

Microbes in infectious biofilms may be hundreds of times more drug resistant than the same free, unattached microbes. This resistance does not come from genetic mutations; rather, it is due to the character of the biofilm itself. For one thing, microbes are protected by the impenetrable nature of the extracellular matrix—drugs may be blocked from entry. We know that microbes also communicate and regulate resistance mechanisms such as drug pumps. These factors alone or in combination allow the persistence of the biofilm and a long-term source of infection for the patient.

With knowledge of biofilms, drug companies are researching newer drug strategies that can prevent or disrupt their formation. One concept already in use is to incorporate antimicrobic chemicals into implants, catheters, and other metal or plastic devices to inhibit microbial growth.

Changes in Metabolic Patterns

The action of antimetabolites can be circumvented if a microbe develops an alternative metabolic pathway or enzyme (**figure 12.14, part 5**). Sulfonamide and trimethoprim resistance develops when microbes deviate from the usual patterns of folic acid synthesis. Fungi can acquire resistance to flucytosine by completely shutting off certain metabolic activities.

Natural Selection and Drug Resistance

So far, we have been considering drug resistance at the cellular and molecular levels, but its full impact is felt only if this resistance occurs throughout the cell population. Let us examine how this might happen and its long-term therapeutic consequences.

Any large population of microbes is likely to contain a few individual cells that are already drug resistant because of prior mutations or transfer of plasmids (**figure 12.15a**). As long as the drug is not present in the habitat, the numbers of these resistant forms will remain low because they have no particular growth advantage. But if the population is subsequently exposed to this drug (**figure 12.15b**), sensitive individuals are inhibited or destroyed, and resistant forms survive and proliferate. As the population continues to grow, offspring of these resistant microbes will inherit this drug resistance. In time, the replacement population will have more of the drug-resistant forms and can eventually become completely resistant (**figure 12.15c**). In ecological and evolutionary terms, the environmental factor (in this case, the drug) has put selection pressure on the population, allowing the more "fit" microbe (the drug-resistant one) to survive and grow. The end result is a population that has evolved to a condition of drug resistance.

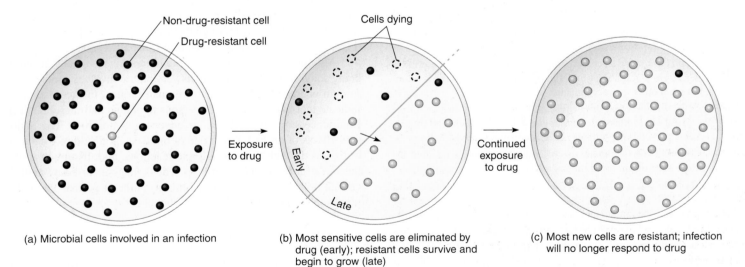

(a) Microbial cells involved in an infection

(b) Most sensitive cells are eliminated by drug (early); resistant cells survive and begin to grow (late)

(c) Most new cells are resistant; infection will no longer respond to drug

Figure 12.15 **A model of natural selection for drug resistance.** (**a**) Generally, any population of microbes will harbor a few cells carrying a prior mutation that confers drug resistance. (**b**) Environmental pressure (here, exposure to the drug) selects for survival of these cells. (**c**) In time, the resistant cells become the dominant cells.

TABLE 12.7	**Strategies to Limit Drug Resistance of Microorganisms**

Drug Usage

- Physicians have the responsibility for making an accurate diagnosis and prescribing the correct drug therapy.
- Compliance with physicians' guidelines: The patient should take the correct dosage, by the best route, for the appropriate period. This diminishes the selection for mutants that can resist low drug levels and ensures elimination of the pathogen.
- Combined therapy with two or more drugs together increases the chances that at least one of the drugs will be effective and that a resistant strain of either drug will not be able to persist.

Drug Research

- Research focuses on developing shorter-term, higher-dose antimicrobials that are more effective, less expensive, and have fewer side effects.
- Pharmaceutical companies continue to seek new antimicrobial drugs with structures that are not readily inactivated by microbial enzymes or drugs with modes of action that are not readily circumvented.

Long-Term Strategies

- Proposals to reduce the abuse of antibiotics range from educational programs for health workers to requiring justification for prescribing certain types of antibiotics.
- Especially valuable antimicrobials may be restricted in their use to only one or two types of infections.
- The addition of antimicrobials to animal feeds must be curtailed worldwide.
- Government programs that make effective therapy available to low-income populations should be increased.
- Vaccines should be used whenever possible to provide alternative protection.

Natural selection for drug-resistant forms is apparently a common phenomenon. It takes place most frequently in various natural habitats, medical environments, and it also can occur within the bodies of humans and animals during drug therapy. See **Insight 12.4** to find out some of the long-term implications of drug resistance. **Table 12.7** summarizes a number of actions that are being considered to slow its development.

Check & Assess Section 12.4

✔ Microorganisms are termed drug resistant when they are no longer inhibited by an antimicrobial to which they were previously sensitive.

✔ Drug resistance is genetic; microbes acquire genes that code for methods of inactivating or escaping the antimicrobial or acquire mutations that affect the drug's impact. Resistance is selected by environments where antimicrobials are present in high concentrations, such as hospitals.

✔ Microbial drug resistance develops through the selection of preexisting random mutations and through acquisition of resistance genes from other microorganisms.

✔ Varieties of microbial drug resistance include drug inactivation, decreased drug uptake, decreased drug receptor sites, and modification of metabolic pathways formerly attacked by the drug.

✔ Widespread indiscriminate use of antimicrobials has resulted in an explosion of microorganisms resistant to all common drugs.

19. Explain the phenomenon of drug resistance from the standpoint of microbial genetics (include a description of R factors).
20. Describe the five basic mechanisms of drug resistance.
21. Explain why biofilm infections are often resistant to drugs.
22. Name several ways that drug resistance can be prevented by medical personnel and patients (see table 12.7).

12.5 Interactions between Drugs and Hosts

Expected Learning Outcomes

19. Summarize the adverse effects of drugs on the human body.
20. Describe some of the toxic effects of drugs on organs and tissues.
21. Discuss how drugs can cause damaging immune responses.
22. Explain the ways that drugs alter the normal microbiota and the results.

Until now, this chapter has focused on the interactions between antimicrobials and the microorganisms they target. During an infection, the microbe is living in or on a host; therefore, the drug is administered to the host though its target is the microbe. Because of this, the effect of the drug on the host must always be considered.

Although selective antimicrobial toxicity is the ideal constantly being sought, chemotherapy by its very nature involves contact with foreign chemicals that can harm human tissues. In fact, estimates indicate that at least 5% of all persons taking an antimicrobial drug experience some type of serious adverse reaction to it. The major side effects of drugs fall into one of three categories: direct damage to tissues through toxicity, allergic reactions, and disruption in the balance of normal microbial flora. The damage incurred by antimicrobial drugs can be short term and reversible or permanent, and it ranges in severity from cosmetic to lethal.

Toxicity to Organs

Drugs can adversely affect the following organs: the liver (hepatotoxic), kidneys (nephrotoxic), gastrointestinal tract, cardiovascular system and blood-forming tissue (hemotoxic), nervous system (neurotoxic), respiratory tract, skin, bones, and teeth.

The Rise of Drug Resistance

"We may have finally arrived at the era of the untreatable bacterial infection."

Laura Stokowski, a nurse at Inova Fairfax Hospital

It is a fact of nature that if a large number of microbes are exposed to a variety of drugs, there will always be some genetically favored strains or types that survive and thrive. One could even say that microbes are merely practicing their livelihoods—to adapt and survive. Our very use of drugs is indeed a catalyst that speeds their evolution into drug resistance.

Ironically, thousands of patients die every year in the United States from infections that lack effective drugs, and 60% of hospital infections are caused by drug-resistant microbes. For many years, concerned observers reported the gradual development of drug resistance in a number of pathogens, but during the past decade, the scope of the problem has escalated. It is now a common event to discover microbes that have become resistant to relatively new drugs in a very short time.

The clinical setting is a prolific source of drug-resistant strains of bacteria. This environment continually exposes pathogens to a variety of drugs. The hospital also maintains patients with weakened defenses, making them highly susceptible to pathogens. A classic example occurred with *Staphylococcus aureus* and penicillin. In the 1950s, hospital strains began to show resistance to this drug, and because of indiscriminate use, these strains became nearly 100% resistant to penicillin in 30 years. In short order, MRSA (methicillin-resistant *S. aureus*) strains appeared, which can tolerate nearly all antibiotics. To complicate this problem, strains of MRSA have been spread from the clinic into the community. Up until recently, MRSA has been sensitive to the drug vancomycin. Then, in 2002, the first cases of complete resistance to this drug were reported (VRSA).

Drugs in Animal Feeds

Another practice that has contributed significantly to growing drug resistance is the addition of antimicrobials to livestock feed, with the idea of decreasing infections and thereby improving animal health and size. This practice has had serious impact in both the United States and Europe. Enteric bacteria such as *Salmonella, E. coli,* and enterococci that live as normal intestinal flora of these animals readily share resistance plasmids and are constantly selected and amplified by exposure to drugs. These pathogens subsequently "jump" to humans and cause drug-resistant infections, oftentimes at epidemic proportions. In a deadly outbreak of *Salmonella* infection in Denmark, the pathogen was found to be resistant to seven different antimicrobials. The opportunistic pathogen called VRE (vancomycin-resistant enterococcus) has been traced to the use of a vancomycinlike drug in cattle feed. It is now one of the most tenacious of hospital-acquired infections for which there are few drug choices.

To attempt to curb this source of resistance, Europe and the United States have begun to ban the use of human drugs in animal feeds. The move seems to be working. Denmark banned all agricultural antibiotic use for growth promotion in 1998. Resistance to the drugs declined dramatically among bacteria isolated from the farm animals without significant reductions in animal size or health.

Worldwide Drug Resistance

The drug dilemma has become a widespread problem, affecting all countries and socioeconomic groups. In general, the majority of infectious diseases, whether bacterial, fungal, protozoan, or viral, are showing increases in drug resistance. In parts of India, the main drugs used to treat cholera (furazolidone, ampicillin) have gone from being highly effective to essentially useless in 10 years. In Southeast Asia, 98% of gonococcus infections are multidrug resistant. Malaria, tuberculosis, and typhoid fever pathogens are gaining in resistance, with few alternative drugs to control them. To add to the problem, global travel and globalization of food products mean that drug resistance can be spread around the world in a short time.

In the United States alone, the extra cost for treating drug-resistant infections is around $10 billion per year. In many developing countries, drugs are mishandled by overuse and underuse, either of which can contribute to drug resistance. Many countries that do not regulate the sale of prescription drugs make them readily available to purchase over the counter. For example, the antituberculosis drug INH (isoniazid) is sometimes used as a "lung vitamin" to improve health, and antibiotics are often taken in the wrong dose for undiagnosed conditions. These countries serve as breeding grounds for drug resistance that can eventually be carried to other countries.

"Bad Bugs–No Drugs"

The development of new antimicrobial drugs is considered the only viable way to gain some advantage in this situation, often referred to as the "bad bugs–no drugs" problem. A case in point is the alarming trend of pathogens known collectively as ESKAPE. This acronym comes from the first letters of their generic names, but it also tells us that these six pathogens are very successful at escaping the effects of drugs and are getting better at it all the time. The microbes—*Enterococcus, Staphylococcus aureus, Klebsiella, Acinetobacter, Pseudomonas,* and *Enterobacter*—make up the majority of hospital-acquired infections and are also the most drug-resistant bacteria in the hospital environment. It is not unusual to find extremely drug-resistant strains or "superbugs" for which there are only two, one, and sometimes no drugs available. Even more problematic is that right now, there are no new drugs that could control these pathogens making their way through the pipeline of development. To avert a public health crisis, a number of governments are actively working with pharmaceutical companies on drug discovery and development, sometimes called the "new bugs–new drugs" approach. It may be our only hope.

What are some of the reasons that microbes develop drug resistance so rapidly? *Answer available at http://www.mhhe.com/talaro8*

Because the liver is responsible for metabolizing and detoxifying foreign chemicals in the blood, it can be damaged by a drug or its metabolic products. Injury to liver cells can result in enzymatic abnormalities, fatty liver deposits, hepatitis, and liver failure. The kidney is involved in excreting drugs and their metabolites. Some drugs irritate the nephron tubules, creating changes that interfere with their filtration abilities. Drugs such as sulfonamides can crystallize in the kidney and form stones that can obstruct the flow of urine.

The most common complaint associated with oral antimicrobial therapy is diarrhea, which can progress to severe intestinal irritation or colitis. Although some drugs directly irritate the intestinal lining, the usual gastrointestinal complaints are caused by disruption of the intestinal microflora (discussed in a subsequent section).

Many drugs given for parasitic infections are toxic to the heart, causing irregular heartbeats and even cardiac arrest in extreme cases. Chloramphenicol can severely depress blood-forming cells in the bone marrow, resulting in either a reversible or a permanent (fatal) anemia. Some drugs hemolyze the red blood cells, others reduce white blood cell counts, and still others damage platelets or interfere with their formation, thereby inhibiting blood clotting.

Certain antimicrobials act directly on the brain and cause seizures. Others, such as aminoglycosides, damage nerves (very commonly, the eighth cranial nerve), leading to dizziness, deafness, or motor and sensory disturbances. When drugs block the transmission of impulses to the diaphragm, respiratory failure can result.

The skin is a frequent target of drug-induced side effects. The skin response can be a symptom of drug allergy or a direct toxic effect. Some drugs interact with sunlight to cause photodermatitis, a skin inflammation. Tetracyclines are contraindicated (not advisable) for children from birth to 8 years of age because they bind to the enamel of the teeth, creating a permanent gray to brown discoloration (**figure 12.16**). Pregnant women should avoid tetracyclines because they cross the placenta and can be deposited in the developing fetal bones and teeth.

Allergic Responses to Drugs

One of the most frequent drug reactions is heightened sensitivity, or **allergy.** This reaction occurs because the drug acts as an antigen (a foreign material capable of stimulating the immune system) and stimulates an allergic response. This response can be provoked by the intact drug molecule or by substances that develop from the body's metabolic alteration of the drug. In the case of penicillin, for instance, it is not the penicillin molecule itself that causes the allergic response but a product, *benzylpenicilloyl.* Allergic reactions have been reported for every major type of antimicrobial drug, but the penicillins account for the greatest number of antimicrobial allergies, followed by the sulfonamides.

People who are allergic to a drug become sensitized to it during the first contact, usually without symptoms. Once the immune system is sensitized, a second exposure to the drug can lead to a reaction such as a skin rash (hives); respiratory inflammation; and, rarely, anaphylaxis,

an acute, overwhelming allergic response that develops rapidly and can be fatal. This topic is discussed in greater detail in chapter 16.

Suppression and Alteration of the Microflora by Antimicrobials

Most normal, healthy body surfaces, such as the skin, large intestine, outer openings of the urogenital tract, and oral cavity, provide numerous habitats for a virtual "garden" of microorganisms. These normal colonists or residents, called the microbiota, consist mostly of harmless or beneficial bacteria, but a small number can potentially be pathogens. Although we defer a more detailed discussion of this topic to chapter 13 and later chapters, here we focus on the general effects of drugs on this population.

If a broad-spectrum antimicrobial is administered to treat infection, it will destroy microbes regardless of their roles in maintaining a balance, affecting not only the targeted infectious agent but also many others in sites far removed from the original infection (**figure 12.17**). When this therapy destroys beneficial resident species, resistant microbes that were once in small numbers begin to overgrow and cause disease. This complication is called a **superinfection.**

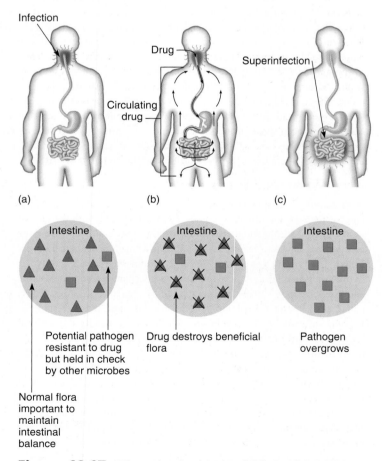

Figure 12.17 **The role of antimicrobials in disrupting microbial flora and causing superinfections.** **(a)** A primary infection in the throat is treated with an oral antibiotic. **(b)** The drug is carried to the intestine and is absorbed into the circulation. **(c)** The primary infection is cured, but drug-resistant pathogens have survived and create an intestinal superinfection.

Figure 12.16 **Drug-induced side effect.** An adverse effect of tetracycline given to young children is the permanent discoloration of tooth enamel.

Some common examples demonstrate how a disturbance in microbial flora leads to replacement flora and superinfection. A broad-spectrum cephalosporin used to treat a urinary tract infection by *E. coli* will cure the infection, but it will also destroy the lactobacilli in the vagina that normally maintain a protective acidic environment there. The drug has no effect, however, on *Candida albicans,* a yeast that also resides in normal vaginas. Released from the inhibitory environment provided by lactobacilli, the yeasts proliferate and cause symptoms. *Candida* can cause similar superinfections of the oropharynx (thrush) and the large intestine.

Oral therapy with tetracyclines, clindamycin, and broad-spectrum penicillins and cephalosporins is associated with a serious and potentially fatal condition known as *antibiotic-associated colitis* (pseudomembranous colitis). This condition is due to the overgrowth in the bowel of *Clostridium difficile,* an endospore-forming bacterium that is resistant to the antibiotic. It invades the intestinal lining and releases toxins that induce diarrhea, fever, and abdominal pain. You'll learn more about *C. difficile* in chapter 19.

Refer to **table 12.8** for a general summary of major drug groups and their side effects.

Check & Assess Section 12.5

✔ The three medical side effects of antimicrobials are toxicity to organs, allergic reactions, and problems resulting from suppression or alteration of resident microbes.

✔ Toxicity may be mild and short-lived or severe and very damaging.

✔ Drugs can stimulate an overreaction by the immune system that causes various allergic symptoms.

✔ Antimicrobials that destroy most but not all normal residents allow the unaffected survivors to overgrow, causing a superinfection.

23. Why do drugs that act on bacterial and fungal membranes generally have high toxicity?
24. Describe the series of events that gives rise to a superinfection, using an example.
25. What characteristic of antibiotics and other drugs causes hypersensitivity and allergic reactions?
26. Explain why it is better to use a narrow-spectrum drug instead of a broad-spectrum one, if possible.

12.6 Considerations in Selecting an Antimicrobial Drug

Expected Learning Outcomes

23. Outline some important factors to consider in drug therapy.
24. Describe the principles behind sensitivity testing for infectious agents.
25. Explain the uses of the minimum inhibitory concentration and the therapeutic index in drug selection.

26. Discuss medical considerations in final drug selection.
27. Summarize the problems in managing antimicrobial drugs.

Before actual antimicrobial therapy can begin, consideration must be given to at least three factors:

1. the nature of the microorganism causing the infection,
2. the degree of the microorganism's susceptibility (also called sensitivity) to various drugs, and
3. the overall medical condition of the patient.

Identifying the Agent

Identification of infectious agents from body specimens should be attempted as soon as possible. It is especially important that such specimens be taken before any antimicrobial drug is given to prevent loss of the infectious agent. Direct examination of body fluids, sputum, or stool can provide a rapid detection method for bacteria or fungi. A doctor often begins the therapy on the basis of such immediate findings. The choice of drug will be based on experience with drugs that are known to be effective against the microbe. For instance, if a sore throat appears to be caused by *Streptococcus pyogenes,* the physician may prescribe penicillin, because this species seems to be almost universally sensitive to it so far. If the infectious agent is not or cannot be isolated, epidemiological statistics may be required to predict the most likely agent in a given infection. For example, *Streptococcus pneumoniae* accounts for the majority of cases of meningitis in children, followed by *Neisseria meningitidis* (discussed in detail in chapter 18).

Testing for the Drug Susceptibility of Microorganisms

Testing is essential in those groups of bacteria commonly showing resistance, primarily *Staphylococcus* species, *Neisseria gonorrhoeae, Streptococcus pneumoniae, Enterococcus faecalis,* and the aerobic gram-negative enteric bacilli. However, not all infectious agents require antimicrobial sensitivity testing. When certain groups, such as group A streptococci and all anaerobes (except *Bacteroides*), are known to be uniformly susceptible to penicillin G, testing may not be necessary unless the patient is allergic to penicillin. Testing methods are available for fungi (see figure 12.21), protozoa, and viruses, although testing may not be done as often for these groups.

Selection of a proper antimicrobial agent begins by demonstrating the in vitro activity of several drugs against the infectious agent by means of standardized methods. In general, these tests involve exposing a pure culture of the bacterium to several different drugs and observing the effects of the drugs on growth.

The *Kirby-Bauer* technique is an agar diffusion test that provides useful data on antimicrobial susceptibility. In this test, the surface of a plate of special medium is spread with the test bacterium, and small discs containing a premeasured amount of antimicrobial are dispensed onto the bacterial lawn. After incubation, the zone of inhibition surrounding the discs is measured and compared with a standard for each drug **(figure 12.18).** The profile of antimicrobial sensitivity, or *antibiogram,* provides data for drug selection.

TABLE 12.8	Major Adverse Toxic Reactions to Common Drug Groups	
Antimicrobial Drug	**Primary Tissue Affected**	**Primary Damage or Abnormality Produced**
Antibacterials		
Penicillin G	Skin	Rash
Carbenicillin	Platelets	Abnormal bleeding
Ampicillin	GI tract	Diarrhea and enterocolitis
Cephalosporins	Platelet function	Inhibition of prothrombin synthesis
	White blood cells	Decreased circulation
	Kidney	Nephritis
Tetracyclines	GI tract	Diarrhea and enterocolitis
	Teeth, bones	Discoloration of tooth enamel; damage to fetal skeleton
	Skin	Reactions to sunlight (photosensitization)
Chloramphenicol	Bone marrow	Injury to red and white blood cell precursors
Aminoglycosides (streptomycin, gentamicin, amikacin)	GI tract, hair cells in cochlea, vestibular cells, neuromuscular cells, kidney tubules	Diarrhea and enterocolitis; malabsorption; loss of hearing, dizziness, kidney damage
Isoniazid	Liver	Hepatitis
	Brain	Seizures
	Skin	Dermatitis
Sulfonamides	Kidney	Formation of crystals; blockage of urine flow
	Red blood cells	Hemolysis
	Platelets	Reduction in number
Polymyxin	Kidney	Damage to membranes of tubule cells
	Neuromuscular system	Weakened muscular responses
Quinolones (ciprofloxacin, norfloxacin)	Nervous system, bones, GI tract	Headache, dizziness, tremors, GI distress
Rifampin	Liver	Damage to hepatic cells
	Skin	Dermatitis
Antifungals		
Amphotericin B	Kidney	Disruption of tubular filtration
Flucytosine	White blood cells	Decreased number
Antiprotozoan Drugs		
Metronidazole	GI tract	Nausea, vomiting
Chloroquine	GI tract	Vomiting
	Brain	Headache
	Skin	Itching
Antihelminthics		
Niclosamide	GI tract	Nausea, abdominal pain
Pyrantel	GI tract	Irritation
	Brain	Headache, dizziness
Antivirals		
Acyclovir	Brain	Seizures, confusion
	Skin	Rash
Amantadine	Brain	Nervousness, light-headedness
	GI tract	Nausea
AZT	Bone marrow	Immunosuppression, anemia
Protease inhibitors	Liver	Hepatic damage
	Lipid metabolism	Abnormal deposition of fats
	Pancreas	Diabetes

The Kirby-Bauer procedure is less effective for bacteria that are anaerobic, highly fastidious, or slow-growing (*Mycobacterium*).

An alternative quantitative system that provides a quantitative rating of drug effectiveness is the *Etest*® (**figure 12.19**). In addition to providing a precise numerical rating (the MIC), this method allows testing for a wide variety of drugs and microbial types, including anaerobes, mycobacteria, and fungi.

More sensitive and quantitative results can also be obtained with tube dilution tests. First the antimicrobial is diluted serially in tubes of broth, and then each tube is inoculated with a small uniform sample of pure culture, incubated, and examined for growth (turbidity). The smallest concentration (highest dilution) of drug that visibly inhibits growth is called the **minimum inhibitory concentration,** or **MIC.** The MIC is useful in determining the smallest effective dosage

Kirby-Bauer Disk Diffusion Test*

Enrofloxacin 5 µg
(R < 17 mm; S ≥ 22 mm)

Oxytetracycline 30 µg
(R < 17 mm; S ≥ 22 mm)

Gentamicin 10 µg
(R < 17 mm; S ≥ 21 mm)

Cefotaxime 30 µg
(R < 14 mm; S ≥ 23 mm)

Ampicillin 10 µg
(R < 14 mm; S ≥ 22 mm)

Chloramphenicol 30 µg
(R < 21 mm; S ≥ 21 mm)

◯ = Zone of Inhibition

◯ = Region of bacterial growth

(ENR 5) = Antibiotic carrier (disc) imprinted with abbreviation and concentration

Disk Diffusion Test (schematic).
Example and evaluation of a sensitivity test, agar diffusion method.
R = resistant, I = intermediate, S = sensitive

(a) If the test bacterium is sensitive to a drug, a zone of inhibition develops around its disk. The larger the size of this zone, the greater is the bacterium's sensitivity to the drug. The diameter of each zone is measured in millimeters and evaluated for susceptibility or resistance by means of a comparative standard (R, I, or S).

CTX AMC

(b) Result for gram negative *Klebsiella pneumoniae,* with zones of inhibition indicating sensitivity to cefotaxime (CTX) and ampicillin and clavulanic acid (AMC). Observe the expanded zone between these two drugs that indicates their enhanced action, or synergy, when combined.

(c) Result for gram positive *Staphylococcus aureus* using gentamicin (GM, left) and aztreonam (ATM, right), Although the bacteria are sensitive to both drugs, the GM has a few colonies growing in the zone of inhibition, indicating some cells that have developed resistance to that concentration of the drug.

Figure 12.18 Techniques for preparation and interpretation of disc diffusion tests.

of a drug and in providing a comparative index against other antimicrobials (**figures 12.20 and 12.21,** and **table 12.9**). In many clinical laboratories, these antimicrobial testing procedures are performed in automated machines that can test dozens of drugs simultaneously.

The MIC and the Therapeutic Index

The results of antimicrobial sensitivity tests guide the physician's choice of a suitable drug. If therapy has already commenced, it is imperative to determine if the tests bear out the use of that particular drug. Once therapy has begun, it is important to observe the patient's clinical response, because the in vitro activity of the drug is not always correlated with its in vivo effect. When antimicrobial treatment fails, the failure is due to

1. the inability of the drug to diffuse into that body compartment (the brain, joints, skin);
2. a few resistant cells in the culture that did not appear in the sensitivity test; or
3. an infection caused by more than one pathogen (mixed), some of which are resistant to the drug.

If therapy does fail, a different drug, combined therapy, or a different method of administration must be considered.

Figure 12.19 Etest®: a quantitative method for determining sensitivity. This test uses a plastic strip that contains a premeasured gradient of a drug and is labeled with a scale that indicates increasing concentrations of the drug in µg. When the strips are placed on a plate swabbed with the test microbe and incubated, an oval area called the ellipse of inhibition develops adjacent to the strip. The lowest point on the scale (lowest concentrations are towards the middle of the plate) where the ellipse begins to form corresponds with the MIC. (IP = imipenem and TZ = ceftazidime). Etest® and the Etest gradient strip are registered trademarks of AB BIODISK.

Same inoculum size of test bacteria added

0 0.2 0.4 0.8 1.6 3.2 6.4 12.8

Negative
control

μg/ml

Increasing concentration of drug

☐ Growth ☐ No growth

Figure 12.20 Tube dilution test for determining the minimum inhibitory concentration (MIC). The antibiotic is diluted serially through tubes of liquid nutrient from right to left. All tubes are inoculated with an identical amount of a test bacterium and then incubated. The first tube on the left is a control that lacks the drug and shows maximum growth. The dilution of the first tube in the series that shows no growth (no turbidity) is the MIC.

+ve growth control
AmB = 0.125 ug/ml
Fluc = 256 ug/ml
Itra = 0.125 ug/ml
Keto = 1.0 ug/ml
5FC = 1.0 ug/ml
Vori = 0.06 ug/ml
Caspo = 0.125 ug/ml

Figure 12.21 Microbroth dilution in a multiwell plate adapted for a pathogenic yeast. Drugs being tested are amphotericin B (AmB), flucytosine (5FC), and several azole drugs (Itra, Keto, Vori). Pink indicates growth and blue, no growth. Reading across numbers 2-12 are increasing concentrations of the drugs in μg/ml. Row 1 has no drug. The first dilution with no growth is crossed out, indicating the MIC for that drug. Results are given in black box.

Many factors influence the choice of an antimicrobial drug besides microbial sensitivity to it. The nature and spectrum of the drug, its potential adverse effects, and the condition of the patient can be critically important. When several antimicrobial drugs are available for treating an infection, final drug selection advances to a new series of considerations. In general, it is better to choose the narrowest-spectrum drug of those that are effective if the causative agent is known. This decreases the potential for superinfections and other adverse reactions.

Because drug toxicity is of concern, it is best to choose the drug with high selective toxicity for the infectious agent and low human toxicity. The **therapeutic index (TI)** is defined as the ratio of the dose of the drug that is toxic to humans as compared to its minimum effective (therapeutic) dose. The closer these two figures are (the smaller the ratio), the greater is the potential for toxic drug reactions. For example, a drug that has a therapeutic index of

$$\frac{10\,\mu g/ml: toxic\,dose}{9\,\mu g/ml\,(MIC)} \quad \boxed{TI = 1.1}$$

is a riskier choice than one with a therapeutic index of

$$\frac{10\,\mu g/ml}{1\,\mu g/ml} \quad \boxed{TI = 10}$$

Drug companies recommend dosages that will inhibit the microbes but not adversely affect patient cells. When a series of drugs being considered for therapy have similar MICs, the drug with the highest therapeutic index usually has the widest margin of safety.

Patient Factors in Choosing an Antimicrobial Drug

The physician must also take a careful history of the patient to discover any preexisting medical conditions that will influence the activity of the drug or the response of the patient. A history of allergy to a certain class of drugs should preclude the administration of that drug and any drugs related to it. Underlying liver or kidney disease will ordinarily necessitate the modification of drug therapy, because these organs play such an important part in metabolizing or excreting the drug. Infants, the elderly, and pregnant women require special precautions. For example, age can diminish gastrointestinal absorption and organ function, and most

TABLE 12.9	Comparative MICs (μg/ml) for Common Drugs and Pathogens				
Bacterium	**Penicillin G**	**Ampicillin**	**Sulfamethoxazole**	**Tetracycline**	**Cefaclor**
Staphylococcus aureus	4.0	0.05	3.0	0.3	4.0
Enterococcus faecalis	3.6	1.6	100.0	0.3	60.0
Neisseria gonorrhoeae	0.5	0.5	5.0	0.8	2.0
Escherichia coli	100.0	12.0	3.0	1–4.0	3.0
Pseudomonas aeruginosa	>500.0	>200.0	>100	>100.0	4.0
Salmonella species	12.0	6.0	10.0	1.0	0.8
Clostridium perfringens	0.16	NA	NA	3.0	12.0

NA = not available

antimicrobial drugs cross the placenta and could affect fetal development.

The intake of other drugs must be carefully scrutinized, because incompatibilities can result in increased toxicity or failure of one or more of the drugs. For example, the combination of aminoglycosides and cephalosporins increases nephrotoxic effects; antacids reduce the absorption of isoniazid; and the interaction of tetracycline or rifampin with oral contraceptives can abolish the contraceptive's effect. Some drugs (penicillin with certain aminoglycosides, or amphotericin B with flucytosine) act synergistically, so that reduced doses of each can be used.

Other concerns in choosing drugs include any genetic or metabolic abnormalities in the patient, the site of infection, the route of administration, and the cost of the drug.

Even when all the information is in, the final choice of a drug is not always straightforward. Consider the case of an elderly alcoholic patient with pneumonia caused by *Klebsiella* and complicated by diminished liver and kidney function. All drugs must be given parenterally because of prior damage to the gastrointestinal lining and poor absorption. Drug tests show that the infectious agent is sensitive to fourth-generation cephalosporins, gentamicin, imipenem, and azlocillin. The patient's history shows previous allergy to the penicillins, so these would be ruled out. Cephalosporins are associated with serious bleeding in elderly patients, so this may not be a good choice. Aminoglycosides such as gentamicin are nephrotoxic and poorly cleared by damaged kidneys. Imipenem is probably the best choice because of its broad spectrum and low toxicity.

TAKE NOTE: THE ANTIMICROBIAL DRUG DILEMMA

We began this chapter with a view of the exciting strides made in antimicrobial therapy, but we must end it on a note of qualification and caution. There is now a worldwide problem in the management of antimicrobial drugs, which rank second only to nervous system drugs in overall usage. The remarkable progress in treating many infectious diseases has spawned a view of antimicrobials as a "cure-all" for infections as diverse as the common cold and acne. And, although it is true that nothing is as dramatic as curing an infectious disease with the correct antimicrobial drug, in many instances, drugs have no effect or can be harmful. The depth of this problem can perhaps be appreciated better with a few facts:

1. Roughly 200 million prescriptions for antimicrobials are written in the United States every year. A recent study disclosed that 75% of antimicrobial prescriptions are for pharyngeal, sinus, lung, and upper respiratory infections. A fairly high percentage of these are viral in origin, and antibacterial drugs will have little or no effect.
2. Drugs are often prescribed without benefit of culture or susceptibility testing, even when such testing is clearly warranted.

3. Therapy is often applied in a "shotgun" approach for minor infections, which involves administering a broad-spectrum drug instead of a more specific narrow-spectrum one. This practice can lead to superinfections and other adverse reactions.
4. More expensive newer drugs are chosen when a less costly older one would be just as effective.
5. Tons of excess antimicrobial drugs produced in this country are exported to other countries, where controls are not as strict. Nearly 200 different antibiotics are sold over the counter in Latin American and Asian countries. Self-medication without understanding the correct medical indication is largely ineffectual; but, worse yet, it is partly responsible for emergence of drug-resistant bacteria that subsequently cause epidemics.

The medical community recognizes the motivation for immediate therapy to protect a sick patient and for defensive medicine to provide the very best care possible. But in the final analysis, every allied health professional should be critically aware not only of the admirable and utilitarian nature of antimicrobials but also of their limitations.

Check & Assess Section 12.6

✔ The three major considerations necessary to choose an effective antimicrobial are the nature of the infecting microbe, the microbe's sensitivity to available drugs, and the overall medical status of the infected host.

✔ The Kirby-Bauer test identifies antimicrobials that are effective against a specific infectious bacterial isolate.

✔ The MIC (minimum inhibitory concentration) identifies the smallest effective dose of an antimicrobial toxic to the infecting microbe.

✔ The therapeutic index is a ratio of the amount of drug toxic to the infected host and the MIC. The smaller the ratio, the greater the potential for toxic host-drug reactions.

✔ The effectiveness of antimicrobial drugs is being compromised by several alarming trends: inappropriate prescription, use of broad-spectrum instead of narrow-spectrum drugs, use of higher-cost drugs, sale of over-the-counter antimicrobials in other countries, and lack of sufficient testing before prescription.

27. How can dilution susceptibility tests and disc diffusion tests be used to determine microbial drug sensitivity?
28. Briefly describe the Kirby-Bauer test and its purpose.
29. What would you surmise if you examined a Kirby-Bauer assay and found individual colonies of the plated microbe growing within the zone of inhibition?
30. Explain the relationship between the minimum inhibitory concentration and the therapeutic index and what is optimum for each.
31. Discuss the general factors to consider in selecting an appropriate antimicrobial drug.

Health authorities estimate that between 600,000 and 800,000 needle sticks occur in medical facilities each year. Exposure to HIV by this route has a potential of causing about three infections in 1,000 people. The risk is high enough to warrant prophylaxis—an early preventive therapy to eliminate the virus before it gets established. In this case, the prophylaxis was given postexposure, although prophylaxis can also provide preexposure protection before certain dental procedures or surgery. If given early enough after HIV exposure, the drugs can act upon the virus before it gets a chance to replicate inside host cells. Both of the drugs in combivir (retrovir and epivir) have the action of preventing the HIV RNA from being converted to DNA and thus block further virus activity.

This case was complicated by the AIDS patient having undiagnosed tuberculosis. His pneumocystis pneumonia would have masked the symptoms of TB and interfered with analyzing a chest X ray. Tuberculosis is notoriously slow to culture, and a few days or weeks are not enough time to detect identifiable growth. Another contributing factor is that cutaneous tuberculosis is a relatively rare infection and would not ordinarily have been suspected even from a skin puncture.

The idea behind combined therapy applies to both of the pathogens in this case. Both HIV and *M. tuberculosis* have relatively high rates of drug resistance. Single drug therapy would select out a resistant strain and allow it to flourish. But two or more drugs given together increases the chance that any microbes resistant to one of the drugs will be wiped out by another drug. This effectively prevents the survival of any microbes that are drug resistant. Often the drugs are administered in a single daily dose that can serve for either prophylaxis or treatment of infections. For HIV infection or AIDS, these drugs are selected from one of the classes of reverse transcriptase inhibitors, protease inhibitors, and others (described in table 12.6). For tuberculosis, the choices are narrowed down to isoniazid, rifampin, ethambutol, pyrazinamide, and streptomycin.

This case provides insights into several aspects of antimicrobial therapy. It shows that each infectious disease must be treated individually with the drugs that have a specific mode of action for the pathogen. Testing for sensitivity is especially important in selecting the best drug and in detecting instances of drug resistance. Even then, more than one drug will often be required to slow the development of resistance. We have also touched on the issue of adverse side effects of drugs, which can prevent the compliance with therapy that is a major factor in the failure of HIV and TB treatment. Correct, rapid diagnosis of an infection is also essential to effective therapy. As we saw with the AIDS patient, the lack of correct diagnosis was probably fatal.

For more information on infections from needle sticks, go to http://emedicine.medscape.com/article/784812-overview.

For additional coverage of tuberculosis, go to chapter 19; for HIV and AIDS, go to chapter 25.

Chapter Summary with Key Terms

12.1 Principles of Antimicrobial Therapy

A. **Chemotherapeutic drugs** are used to control microorganisms in the body. Depending on their source, these drugs are described as **antibiotics, semisynthetic** or synthetic.

B. Based on their mode and spectrum of action, they are described as **broad spectrum** or **narrow spectrum** and microbistatic or microbicidal.

C. Strategic approaches to the use of chemotherapeutics include
 1. **Prophylaxis,** where drugs are administered to *prevent* infection in susceptible people.
 2. **Combined therapy,** where two or more drugs are given simultaneously, either to prevent the emergence of resistant species or achieve **synergism.**

D. *Interactions between Drugs and Microbes*
 1. The ideal antimicrobial is **selectively toxic,** highly potent, stable, and soluble in the body's tissues and fluids.
 2. It does not disrupt the immune system or microflora of the host and is exempt from drug resistance.

12.2 Survey of Major Antimicrobial Drug Groups

A. *Drugs That Act on the Cell Envelope*
 1. **Penicillins** are beta-lactam-based drugs originally isolated from the mold *Penicillium chrysogenum.*

The natural form is penicillin G, although various semisynthetic forms, such as ampicillin and methicillin, vary in their spectrum and applications.
 a. Penicillin is bactericidal, blocking completion of the cell wall, which causes eventual rupture of the cell.
 b. Major problems encountered in penicillin therapy include allergic reactions and bacterial resistance to the drug through **beta-lactamase.**

 2. **Cephalosporins** include both natural and semisynthetic drugs initially isolated from the mold *Cephalosporium.* Cephalosporins inhibit peptidoglycan synthesis (like penicillin) but have a much broader spectrum.

 3. Bacitracin and **polymyxin** are narrow-spectrum antibiotics.
 a. Bacitracin prevents cell wall synthesis in gram-positive organisms and is used in antibacterial skin ointments.
 b. Polymyxin interferes with the cell membrane. It is added to skin ointments and can be used to treat *Pseudomonas* infections.

 4. Vancomycin interferes with the early stages of cell wall synthesis and is used for life-threatening, methicillin-resistant staphylococcal infection.

 5. Isoniazid (INH) blocks the synthesis of cell wall components in *Mycobacteria.* It is used in the treatment of tuberculosis.

B. *Drugs That Affect Nucleic Acid Synthesis*
 1. Fluoroquinolones (ciprofloxacin) represent a new class of broad-spectrum synthetic drugs that have proven useful in a variety of infections.

2. Rifampin interferes with RNA polymerase (thereby affecting transcription) and is primarily used for tuberculosis and leprosy infections.

C. *Drugs That Inhibit Protein Synthesis*

1. **Aminoglycosides** include several narrow-spectrum drugs isolated from unique bacteria found in the genera *Streptomyces.* Examples include streptomycin, gentamicin, tobramycin, and amikacin.

2. **Tetracyclines** and chloramphenicol are very broad-spectrum drugs isolated from *Streptomyces.* Both interfere with translation but their use is limited by adverse effects.

3. Erythromycin and clindamycin both affect protein synthesis. Erythromycin is a broad-spectrum alternative for use with penicillin-resistant bacteria, while clindamycin is used primarily for intestinal infection by anaerobes.

D. *Antimetabolite Drugs*

1. **Sulfonamides** are synthetic drugs that act as **metabolic analogs, competitively inhibiting** enzymes needed for nucleic acid synthesis.

2. Trimethoprim is often used in combination with sulfa drugs.

3. Dapsone is a narrow-spectrum drug used (often in combination with rifampin) to treat leprosy.

4. Ketek is a new ketolide antibiotic with broad spectrum effects.

5. Oxazolidinones are another new class of synthetic drug that work by inhibiting the start of protein synthesis. They are useful for treatment of staphylococci that are resistant to other drugs (MRSA and VRE).

E. *Newly Developed Antibiotics* are useful when bacteria have developed resistance to more traditional drugs.

1. Fosfomycin inhibits cell wall synthesis and is effective against gram-negative organisms.

2. Synercid inhibits translation and is most effective against gram-positive cocci.

12.3 Drugs to Treat Fungal, Parasitic, and Viral Infections

A. *Antifungal Drugs*

1. Amphoteracin B and nystatin disrupt fungal membranes by detergent action.

2. **Azoles** are synthetic drugs that interfere with membrane synthesis. They include ketoconazole, fluconazole, and miconazole.

3. Flucytosine inhibits DNA synthesis. Due to fungal resistance, flucytosine must usually be used in conjunction with amphotericin.

B. *Drugs for Protozoan Infections*

1. Quinine or the related compounds chloraquine, primaquine, or mefloquine are used to treat infections by the malarial parasite *Plasmodium.*

2. Other drugs including metronidazole, suramin, melarsopral, and nitrifurimox are used for other protozoan infections.

C. *Drugs for Helminth Infections* include mebendazole, thiabendazole, praziquantel, pyrantel, piperizine, and niclosamide, which are all used for antihelminthic chemotherapy.

D. *Drugs for Viral Infections* inhibit viral penetration, multiplication, or assembly. Because viral and host metabolism are so closely related, toxicity is a potential adverse reaction.

1. **Acyclovir,** valacyclovir, famciclovir, and ribavirin act as nucleoside analogs, inhibiting viral DNA replication, especially in herpesviruses.

2. Amantadine acts to stop uncoating of the influenza A virus.

3. Reverse transcriptase inhibitors (zidovudine) and protease inhibitors are used as anti-HIV drugs, usually in combinations of two or three.

4. Interferon is a naturally occurring protein with both antiviral and anticancer properties.

12.4 Interactions between Microbes and Drugs: The Acquisition of Drug Resistance

A. Microbes can lose their sensitivity to a drug through the acquisition of **resistance (R) factors.** Drug resistance takes the form of drug inactivation, decreased permeability to drug or increased elimination of drug from the cell, change in drug receptors, or change of metabolic patterns.

B. New approaches to antimicrobial therapy include the use of **probiotics** and **prebiotics.**

12.5 Interactions between Drugs and Hosts

Side effects of chemotherapy include organ toxicity, allergic responses, alteration of microflora.

12.6 Considerations in Selecting an Antimicrobial Drug

A. Rapid identification of the infectious agent is important.

B. The pathogenic microbe should be tested for its susceptibility to different antimicrobial agents. This involves using standardized methods such as the *Kirby-Bauer* and **minimum inhibitory concentration (MIC)** techniques. The **therapeutic index** factors in the toxicity of a drug as compared with its MIC so as to guide proper drug selection.

C. Medical conditions of the patient such as age, allergy, disease, and pregnancy are also important guides.

D. The inappropriate use of drugs on a worldwide basis has led to numerous medical and economic problems.

Multiple-Choice Questions

Select the correct answer from the answers provided. For questions with blanks, choose the combination of answers that most accurately completes the statement.

1. A compound synthesized by bacteria or fungi that destroys or inhibits the growth of other microbes is a/an
 a. synthetic drug c. antimicrobial drug
 b. antibiotic d. competitive inhibitor

2. Which statement is *not* an aim in the use of drugs in antimicrobial chemotherapy? The drug should
 a. have selective toxicity
 b. be active even in high dilutions
 c. be broken down and excreted rapidly
 d. be microbicidal

3. Microbial resistance to drugs is acquired through
 a. conjugation c. transduction
 b. transformation d. all of these

4. R factors are _____ that contain a code for _____.
 a. genes, replication c. transposons, interferon
 b. plasmids, drug resistance d. plasmids, conjugation

5. When a patient's immune system becomes reactive to a drug, this is an example of
 a. superinfection c. allergy
 b. drug resistance d. toxicity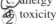

6. An antibiotic that disrupts the normal flora can cause
 a. the teeth to turn brown
 c. a superinfection
 b. aplastic anemia
 d. hepatotoxicity

7. Most antihelminthic drugs function by
 a. weakening the worms so they can be flushed out by the intestine
 b. inhibiting worm metabolism
 c. blocking the absorption of nutrients
 d. inhibiting egg production
 e. all of these

8. Select a drug or drugs that can prevent a viral nucleic acid from being replicated.
 a. azidothymidine
 c. amantadine
 b. acyclovir
 d. both a and b

9. Which of the following effects do antiviral drugs *not* have?
 a. killing extracellular viruses
 b. stopping virus synthesis
 c. inhibiting virus maturation
 d. blocking virus receptors

10. Which of the following modes of action would be most selectively toxic?
 a. interrupting ribosomal function
 b. dissolving the cell membrane
 c. preventing cell wall synthesis
 d. inhibiting DNA replication

11. The MIC is the _____ of a drug that is required to inhibit growth of a microbe.
 a. largest concentration
 c. smallest concentration
 b. standard dose
 d. lowest dilution

12. An antimicrobial drug with a _____ therapeutic index is a better choice than one with a _____ therapeutic index.
 a. low, high
 b. high, low

Case File Questions

1. What was the purpose of administering combivir to the nurse?
 a. to treat AIDS
 b. to prevent pneumocystis pneumonia
 c. to prevent HIV infection
 d. to block virus entry into cells

2. What is a probable reason that tuberculosis therapy must continue for several months?
 a. to prevent drug resistance
 b. to keep the infection from spreading
 c. the TB bacillus grows very slowly
 d. the drugs are slow acting

Matching. Select the mode of action for each drug in the left column; all but one choice is used.

_____ sulfonamides	a. reverse transcriptase inhibitor
_____ penicillin	b. blocks the attachment of tRNA on the ribosome
_____ tetracycline	c. interferes with fusion of virus and host cell
_____ erythromycin	d. interferes with synthesis of folic acid
_____ quinolone	e. breaks down cell membrane integrity
_____ amantidine	f. prevents the ribosome from translocating
_____ polymixin	g. blocks synthesis of peptidoglycan
	h. inhibits DNA gyrase

Writing to Learn

These questions are suggested as a *writing-to-learn* experience. For each question, compose a one- or two-paragraph answer that includes the factual information needed to completely address the question. Check & Assess questions can be used for writing-to-learn exercises.

1. Using the diagram below as a guide, briefly explain how the three factors in drug therapy interact.

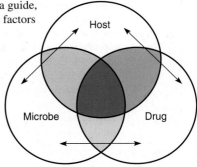

2. Observe table 12.4 with regard to type of microbe and structures or functions that are affected by drugs. Write a short paragraph that explains the different ways in which drugs can affect microbes and how this affects selectivity of drugs and their spectrum.

3. Drugs are often given to surgical patients, to dental patients with heart disease, or to healthy family members exposed to contagious infections.
 a. What word would you use to describe this use of drugs?
 b. What is the purpose of this form of treatment?
 c. Explain some potential undesired effects of this form of therapy.
 d. Define probiotics and give some ways they are used.

4. Write an essay summarizing some of the main concerns in antimicrobial drug therapy, including resistance, allergies, superinfections, and other adverse effects.

5. a. Explain the basis for combined therapy.
 b. Give reasons why it could be helpful to use combined therapy in treating HIV infection.

6. Explain the whys and hows of creating a semisynthetic antibiotic.

7. Summarize the primary reasons that we find ourselves in a "drug dilemma" with regard to antimicrobial drugs.

Concept Mapping

Appendix E provides guidance for working with concept maps.

1. Supply your own linking lines and phrases in this concept map.

Penicillin

Amphotericin B Metronidazole Sulfamethoxazole–trimethoprim

Protozoa Fungi Bacteria

2. Use 6 to 10 bolded words of your choice from the Chapter Summary to create a concept map. Finish it by providing linking words.

Critical Thinking Questions

Critical thinking is the ability to reason and solve problems using facts and concepts. These questions can be approached from a number of angles, and in most cases, they do not have a single correct answer.

1. Occasionally, one will hear the expression that a microbe has become "immune" to a drug.
 a. What is a better way to explain what is happening?
 b. Explain a simple test one could do to determine if drug resistance was developing in a culture.

2. a. Your pregnant neighbor has been prescribed a daily dose of oral tetracycline for acne. Do you think this therapy is advisable for her? Why or why not?
 b. A woman has been prescribed a broad-spectrum oral cephalosporin for a strep throat. What are some possible consequences in addition to cure of the infected throat?
 c. A man has a severe case of sinusitis that is negative for bacterial pathogens. A physician prescribes an oral antibacterial drug in treatment. What is right or wrong with this therapy?

3. You have been directed to take a sample from a growth-free portion of the zone of inhibition in the Kirby-Bauer test and inoculate it onto a plate of nonselective medium.
 a. What does it mean if growth occurs on the new plate?
 b. What if there is no growth?

4. Using table 12.8 and case file 12 as sources, list some of the possible side effects for a patient taking drugs for both HIV infection and tuberculosis simultaneously.

5. From the results shown in figure 12.21, which drug is likely to be the best choice to treat the yeast infection?

6. Explain the differences between the uses of prophylaxis and combined therapy.

7. In cases in which it is not possible to culture or drug test an infectious agent (such as middle ear infection), how would the appropriate drug be chosen?

8. Reviewing drug characteristics, choose an antimicrobial for each of the following situations (explain your choice):
 a. for an adult patient suffering from *Mycoplasma* pneumonia
 b. for a child with meningitis (drug must enter into cerebrospinal fluid)
 c. for a patient who has chlamydiosis and an allergy to erythromycin
 d. for a urinary tract infection by *Enterococcus*
 e. for PPNG drug-resistant gonorrhea

9. How would you personally feel about being told by a physician that your infection cannot be cured by an antibiotic and that the best thing to do is go home, drink a lot of fluids, and take aspirin or other symptom-relieving drugs?

10. a. Using table 12.9 as a reference, find and explain the differences between the results for *Staphylococcus* and *Pseudomonas*. Are any of these drugs broad spectrum?
 b. Explain the difference in the MIC of *E. coli* for penicillin versus ampicillin.
 c. Refer to figure 12.18a, take measurements, and interpret the results.

Visual Challenge

For the following figures 1–5, research the chapter to find an appropriate drug to treat an infection with the microbe shown, and explain what its effects will be.

(a) Figure 18.1

(b) Figure 6.12a

Envelope spike
Host cell membrane
Receptor

(c) Figure 5.16c

Nucleus
Bud scars

(d) Figure 5.33c

(e) Figure 5.35

Female
Anus
Eggs
Male

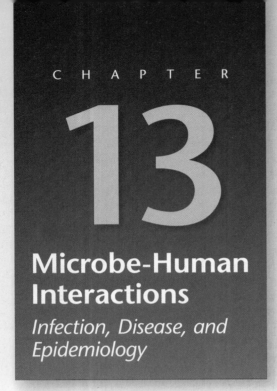

CHAPTER

13

Microbe-Human Interactions

Infection, Disease, and Epidemiology

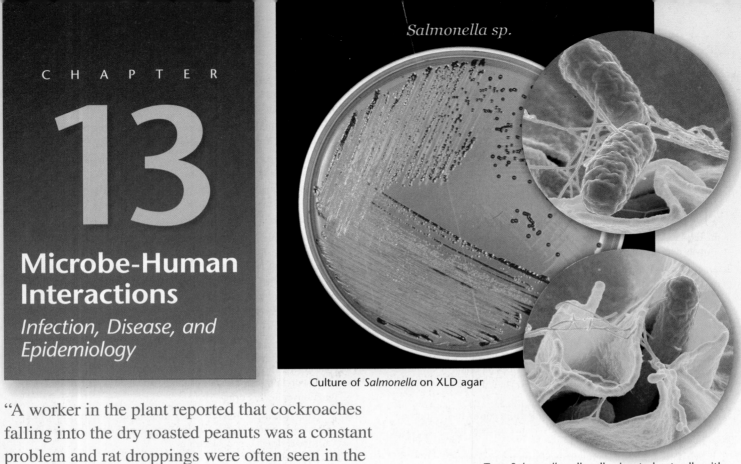

Salmonella sp.

Culture of *Salmonella* on XLD agar

"A worker in the plant reported that cockroaches falling into the dry roasted peanuts was a constant problem and rat droppings were often seen in the processing area."

Top: *Salmonella* cells adhering to host cells with flagella (pink strands); bottom: cells being engulfed by host cell membrane

CASE FILE 13 A Recipe for *Salmonella* Surprise

This case file contains numerous boldfaced concepts and terms that will provide a context for linking the actual events in a recent epidemic to chapter contents.

Starting in September, 2008, reports of *Salmonella enterica* **infections** began to trickle into the **CDC's Reportable and Notifiable Diseases** division, increasing steadily to a peak in late December, and tapering back to a small number in April 2009. The final tally was 714 **cases** reported from 46 states. The victims ranged in age from 2 months to 98 years, and 9 people died. What was responsible for this wave of gastrointestinal **disease**? The first suspicions would fall on the usual suspects—contaminated beef, eggs, or chicken—but by January, it was clear that all cases originated from an unsuspected **vehicle**: peanut butter products.

Infected people interviewed about their food history all told a similar story. About 12 to 48 hours after ingesting peanut butter or snacks, they developed **acute symptoms** of **enteritis**, including nausea and vomiting, followed by severe abdominal

cramping and diarrhea. At least half had a fever of 103° F or higher. Over 100 patients had to be hospitalized and some were gravely ill. The deaths occurred primarily in elderly and young people

Salmonella enterica, serotype Typhimurium was isolated from all of the stool samples tested. In addition, laboratories in several states isolated the same serotype from unopened containers of peanut butter. A serotype is an isolate with a small variation in the molecules (antigens) that stimulate an immune reaction, and it is one way to categorize the strains within a species. There are over 2300 different serotypes in this species, Typhimurium being a major one. This serotype has additional genetic subtypes that can narrow down the origin of the **pathogen** even more.

Epidemiologists from the CDC, Food and Drug Administration (FDA), and local public health agencies teamed up to identify the origin of the **outbreak**. Some of the questions they asked were: What was the specific subtype of the *Salmonella* Typhimurium, and was there a **common**

source for all cases? What was the pathogen's **reservoir** and how did it get into the peanut products?

The specific subtype was analyzed by PulseNet, a network that uses genetic fingerprinting to determine if isolated cultures from different sources are an exact match. It uses these data to track the **incidence** of new cases of food-borne infections and is an early warning system to indicate if new cases correspond to a **sporadic** or **epidemic** pattern. Tests determined that all of the peanut butter-associated cases in the reports originated from a single brand of peanut butter or paste from King Nut, a small company in Georgia that sold products in bulk form to other companies for making peanut butter snacks. Ultimately, thousands of products that contained this company's goods had to be recalled.

▶ *What term refers to the 12 to 48 hours between eating and symptoms?*

▶ *How would one distinguish if the outbreak were sporadic or epidemic?*

386

To continue the case, go to page 409.

13.1 We Are Not Alone

Expected Learning Outcomes

1. Describe some of the major interactions between humans and the microbes that share our habitats.

2. Identify and define the terms associated with infectious diseases.

3. Discuss the characteristics of the normal microbiota and the types of functions they serve.

4. Briefly relate the sources and conditions that influence the development of microbiota in the major body systems.

5. Identify which bodily sites remain free of living organisms, and explain why this is necessary.

6. Explain the germ-free condition and how it is important to microbiologists.

The human body exists in a state of dynamic equilibrium with microorganisms. The healthy individual is able to maintain a balanced coexistence with them. But on occasion, the balance is disrupted, allowing microorganisms to cause an infection or disease. In this chapter, we explore each component of the human-microbe relationship, beginning with the nature and function of normal microbial residents, moving to the stages of infection and disease, and closing with a study of epidemiology and the patterns of disease in populations. These topics will set the scene for chapters 14 and 15, which deal with host defenses, and for later chapters, which examine the agents of infectious disease.

Many of the interactions between the human body and microorganisms involve the development of biofilms (see Insight 4.1). These associations result from the binding of microbial cells to our cells and often to those of other microorganisms in various ways. As we discuss topics in this chapter, keep in mind that colonization of the body involves a constant "give and take" between us and them. The most significant of these interactions can be summarized as follows:

- Microbes provide a protective and stabilization effect on body surfaces by establishing themselves as normal residents;
- Microbes are involved in maturation of host defenses and development of the immune system.
- Microbes can invade and grow in sterile tissues where they cause disease by damaging tissues and organs.

Contact, Colonization, Infection, Disease

In chapter 7, we first considered several of the basic interrelationships between humans and microorganisms. Most of the microbial inhabitants benefit from the nutrients and protective habitat the body provides. From the human point of view, these relationships run the gamut from mutualism to commensalism to parasitism and can have beneficial, neutral, or harmful effects.

Microbes that engage in mutual or commensal associations with humans belong to the **normal resident microbiota.** Other terms sometimes applied are **indigenous* microflora** and commensals.

* *indigenous* (in-dih′-juh-nus) Belonging or native to.

Even though these resident microbes have colonized the body, they remain in the outer surfaces without penetrating into sterile tissues or fluids. When a microbe has penetrated the host defenses, invaded sterile tissues, and multiplied, the result is an **infection,** and the microbe is considered an infectious agent or **pathogen.** If the infection causes damage or disruption to tissues and organs, it is now considered an **infectious disease.** Note that a disease is defined as any deviation from health. There are hundreds of different diseases caused by such factors as infections, diet, genetics, and aging. In this chapter, however, we discuss only diseases arising from the disruption of a tissue or organ caused by microbes or their products.

Figure 13.1 provides a diagrammatic view of the potential phases in infection and disease. Because of numerous factors relating to host resistance and degree of pathogenicity, not all contacts lead to infection and not all infections lead to disease. In fact, contact without infection and infection without disease are the rule.

TAKE NOTE: SYMBIOSIS AND THE SUPERORGANISM

The intriguing idea that the human body is actually 90% microbial inspired Dr. Bruce Findley of the University of British Columbia to describe our shared life with microbiota as a "superorganism". It is expected that such an intimate coexistence with trillions of organisms exerts a profound effect on many aspects of human health and disease. However, the scientific details of *how* we interact with our microbial partners is incomplete. To fill in these details, the National Institutes of Health launched a massive study called the Human Microbiome Project (HGP) in 2007. Before this study, most information on our microbial residents had been obtained through culturing various bodily sites, but we now know that many of our symbiotic microbes are not readily cultured. Another breakthrough driving this research is the availability of metagenomic technology to identify the residents of a complex community by extracting and analyzing its DNA. Because the basis of these results will be the collective genome of these microbial residents, it has been given the name *microbiome.*

The HGP was designed to identify and catalogue the complete array of microbial residents, starting with the oral cavity, skin, nose, digestive tract, and vagina. Now that the first data are beginning to be published, it is immediately evident that the diversity and numbers of microbiota are far greater than originally thought. They include all categories of microorganisms, including bacteria, archaea, protozoa, fungi, and viruses. So far, researchers have isolated DNA from over 600 species of oral microbes, 1000 intestinal microbes, and at least that many inhabitants of the skin. It has been known for some time that the normal microbiota play important roles in human development, immunity, and nutrition. But this new study promises an even deeper understanding of these interactions, as well as the roles microbiota play in conditions such as acne, obesity, eczema, irritable bowel syndrome, dental infections, and cancer.

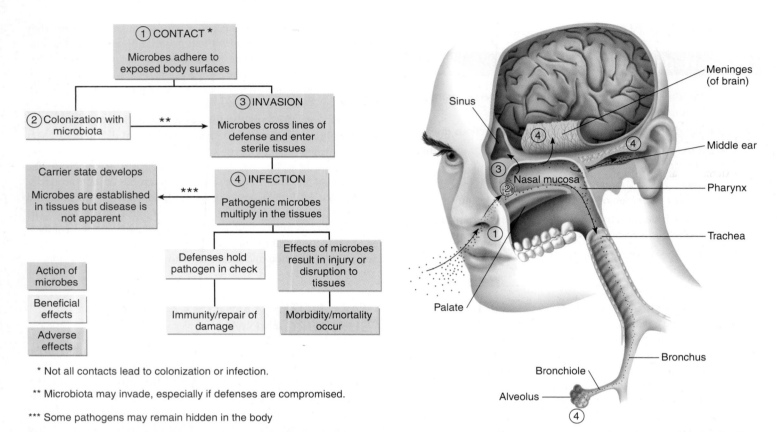

Figure 13.1 **Associations between microbes and humans.** Effects of contact with microbes can progress in a variety of directions, ranging from no effect to colonization, and from infection to disease and immunity. The example shown here follows the possible events in the case of contact with a pathogen such as *Streptococcus pneumoniae* (the pneumococcus). This bacterium can be harbored harmlessly in the upper respiratory tract, but it may also invade and infect the ear, cranium, and respiratory tract.

Before we consider further details of infection and disease, let us examine what happens when microbes colonize the body, thereby establishing a long-term, usually beneficial, relationship.

Resident Microbiota: The Human as a Habitat

The human body offers a seemingly endless variety of microenvironments and niches, with wide variations in temperature, pH, nutrients, moisture, and oxygen tension from one area to another. With such a wide range of habitats, it should not be surprising that the body supports an abundance of microbes. In fact, it is so favorable that, cell for cell, microbes on the human body outnumber human cells 10 to one!

As shown in **table 13.1,** most areas of the body in contact with the outside environment harbor resident microorganisms. Mucosal surfaces provide a particularly attractive surface for attachment. Most other body sites, including internal organs, tissue, and the fluids they contain, are generally microbe-free **(table 13.2).** In fact, the presence of microbes in these areas is usually indicative of infection.

The vast majority of microbes that come in contact with the body are removed or destroyed by the host's defenses long before they are able to colonize a particular area. Those organisms that occupy the body for only short periods are known as **transients.** The remaining microbes that do become established more permanently are considered **residents.** One of their most important

adaptations is to avoid the attention of the body's defenses. The resident organisms have coevolved along with their human hosts toward a complex relationship in which the effects of normal microflora are generally not harmful to the host, and vice versa.

Although generally stable, the microbiota can fluctuate to a limited extent with general health, age, variations in diet, hygiene, hormones, and drug therapy. In many cases, the microbiota actually benefit the human host by preventing the overgrowth of harmful microorganisms. A common example is the fermentation of glycogen by lactobacilli, which keep the pH in the vagina acidic enough to prevent the overgrowth of the yeast *Candida albicans*. A second example is seen in the large intestine, where a protein produced by *Escherichia coli* can prevent the growth of pathogenic bacteria such as *Salmonella* and *Shigella*.

The generally antagonistic effect "good" microbes have against intruder microorganisms is called **microbial antagonism.** Normal flora that exist in an established biofilm are unlikely to be displaced by incoming microbes. This antagonistic protection may simply be a result of a limited number of attachment sites in the host site, all of which are stably occupied by normal flora. Antagonism may also result from the chemical or physiological environment created by the resident microbiota, which is hostile to other microbes.

In experiments performed with mice, scientists discovered that a species of the intestinal bacterium *Bacteroides* controlled the host's production of a defense compound that suppressed

TABLE 13.1	**Sites That Harbor Normal Resident Microbes**

- Skin and its contiguous mucous membranes
- Upper respiratory tract
- Gastrointestinal tract (various parts)
- Outer opening of urethra
- External genitalia
- Vagina
- External ear and canal
- External eye (lids, lash follicles)

TABLE 13.2	**Sterile (Microbe-Free) Anatomical Sites and Fluids**

All Internal Tissues and Organs

Heart and circulatory system	Bones
Liver	Ovaries/testes
Kidneys and bladder	Glands (pancreas, salivary)
Lungs	Sinuses
Brain and spinal cord	Middle and inner ear
Muscles	Internal eye

Fluids within an Organ or Tissue

Blood
Urine in kidneys, ureters, bladder
Cerebrospinal fluid
Saliva prior to entering the oral cavity
Semen prior to entering the urethra
Amniotic fluid surrounding the embryo and fetus

Figure 13.2 **The origins of microbiota in newborns.** A newborn presents a rich and varied collection of habitats. Exposure to the environment through the birthing process, parents, health care workers, and visitors leads to colonization of the newborn by numerous microbes.

other microbes in the area. This is an extreme case of microbial antagonism, but the general phenomenon is of great importance to human health.

Generally, the normal residents will have beneficial or benign effects only if the host is in good health, with a fully functioning immune system, and if the microbiota remain in their natural microhabitat within the body. Hosts with compromised immune systems could very easily be infected by these residents (see table 13.4). We see this outcome when AIDS patients contract pneumonia caused by *Streptococcus pneumoniae*, often carried as normal flora in the nasopharynx. Other microbiota-associated infections can occur when residents are introduced to a site that was previously sterile, as when *E. coli* from the large intestine gets into the bladder, resulting in a urinary tract infection.

Initial Colonization of the Newborn

The uterus and its contents are normally sterile during embryonic and fetal development and remain essentially germ-free until just before birth. The event that first exposes the infant to microbes is the breaking of the fetal membranes, at which time microbes from the mother's vagina can enter the womb. Comprehensive exposure occurs during the birth process itself, when the baby unavoidably comes into intimate contact with the birth canal **(figure 13.2).** Within 8 to 12 hours after delivery, the newborn typically has been colonized by bacteria such as streptococci, staphylococci, and lactobacilli, acquired primarily from its mother. The skin, gastrointestinal tract, and portions of the respiratory and genitourinary tracts all continue to be colonized as contact continues with family members, health care personnel, the environment, and food.

The nature of the microbiota initially colonizing the large intestine depends upon whether the baby is bottle- or breast-fed. Bottle-fed infants (receiving milk or a milk-based formula) tend to acquire a mixed population of coliforms, lactobacilli, enteric streptococci, and staphylococci. In contrast, the intestinal residents of breast-fed infants consists primarily of *Bifidobacterium* species whose growth is favored by a growth factor from mother's milk. This bacterium metabolizes sugars into acids that protect the infant from infection by certain intestinal pathogens.

Milestones that contribute to further development of the microbiota are eruption of teeth, weaning, and introduction of the first solid food. Although exposure to microbes is unavoidable and even necessary for the maturation of the infant's microbial residents, contact with pathogens is dangerous. The neonate does not yet have the full protection they provide and, owing to its immature immune defenses, is extremely susceptible to infection.

TABLE 13.3 Life on Humans: Sites Containing Well-Established Microbiota and Representative Examples

Anatomic Sites	Common Genera	Remarks
Skin	**Bacteria:** *Staphylococcus, Micrococcus, Corynebacterium, Propionibacterium, Streptococcus*	Microbes live only in upper dead layers of epidermis, glands, and follicles; dermis and layers below are sterile.
	Fungi: *Candida, Malassezia*	Dependent on skin lipids for growth
	Arthropods: *Demodix* mite	Present in sebaceous glands and hair follicles
Gastrointestinal Tract		
Oral cavity	**Bacteria:** *Streptococcus, Neisseria, Veillonella, Fusobacterium, Lactobacillus, Bacteroides, Actinomyces, Eikenella, Treponema, Haemophilus*	Colonize the epidermal layer of cheeks, gingiva, pharynx; surface of teeth; found in saliva in huge numbers
	Fungi: *Candida* sp.	Can cause thrush
	Protozoa: *Entamoeba gingivalis*	Inhabit the gingiva of persons with poor oral hygiene
Large intestine and rectum	**Bacteria:** *Bacteroides, Fusobacterium, Bifidobacterium, Clostridium,* fecal streptococci and staphylococci, *Lactobacillus,* coliforms (*Escherichia, Enterobacter*), *Proteus* sp.	Areas of lower gastrointestinal tract other than large intestine and rectum have sparse or nonexistent residents. Microbiota consist predominantly of strict anaerobes; other microbes are aerotolerant or facultative.
	Fungi: *Candida*	Yeast can survive this habitat, but not molds.
	Protozoa: *Entamoeba coli, Trichomonas hominis*	Feed on waste materials in the large intestine
Upper Respiratory Tract	Microbial population exists in the nasal passages, throat, and pharynx; owing to proximity, microbes are similar to those of oral cavity.	Trachea may harbor a sparse population; bronchi, bronchioles, and alveoli are essentially sterile due to local host defenses.
Genital Tract	**Bacteria:** *Lactobacillus, Streptococcus,* diphtheroids (*Corynebacterium* and relatives) *Escherichia, Gardnerella*	In females, microbes occupy the external genitalia and vaginal and cervical surfaces; internal reproductive structures normally remain sterile. Vaginal colonists respond to hormonal changes during life.
	Fungi: *Candida*	Cause of yeast infections
Urinary Tract	**Bacteria:** *Staphylococcus, Streptococcus, Corynebacterium, Lactobacillus*	In females, microbiota exist only in the first portion of the urethral mucosa; the remainder of the tract is sterile. In males, the entire reproductive and urinary tract is sterile except for a short portion of the anterior urethra.
Eye	**Bacteria:** coagulase-negative staphylococci, *Streptococcus, Neisseria*	The lids and follicles harbor similar microbes as skin; the conjunctiva has a transient population; deep tissues are sterile.
Ear	**Bacteria:** staphylococci, diphtheroids	The external ear is similar to the skin in content; areas internal to the tympanum are generally sterile.
	Fungi: *Aspergillus, Penicillium, Candida,* yeasts	

Indigenous Microbiota of Specific Regions

Although we tend to speak of the microbiota as a single unit, it is a complex mixture of hundreds of species, differing somewhat in quality and quantity from one individual to another. Studies have shown that most people harbor certain specially adapted bacteria, fungi, and protozoa and, occasionally, viruses and arthropods. **Table 13.3** provides an overview of the primary locations and types of normal residents.

Colonizers of the Human Skin

The skin is the largest and most accessible of all organs. Its major layers are the epidermis, an outer layer of dead cells continually being sloughed off and replaced, and the dermis, which lies atop the subcutaneous layer of tissue (**figure 13.3a**). Depending on its location, skin also contains hair follicles and several types of glands; and the outermost surface is covered with a protective, waxy cuticle that can help microbes adhere. The normal microbiota reside only in or on the dead cell layers, and except for in follicles and glands, it does not extend into the dermis or subcutaneous levels. The nature of the population varies according to site. Oily, moist skin supports a more prolific flora than dry skin. Humidity, occupational exposure, and clothing also influence its character. Transition zones where the skin joins with the mucous membranes of the nose, mouth, and external genitalia harbor a particularly rich set of microbial colonists.

Ordinarily, there are two cutaneous populations. The **transients** cling to the skin surface but do not ordinarily grow there. They are acquired by routine contact and vary markedly from person to

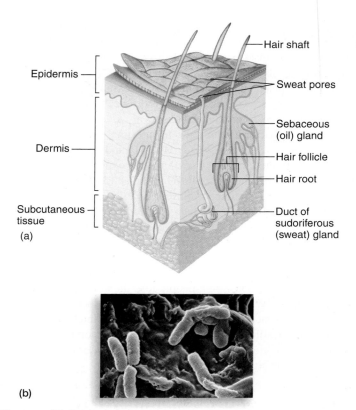

Hair shaft

Epidermis

Sweat pores

Dermis

Sebaceous
(oil) gland

Hair follicle

Hair root

Subcutaneous
tissue

Duct of
sudoriferous
(sweat) gland

(a)

(b)

Figure 13.3 **The landscape of the skin. (a)** The epidermis, along with associated glands and follicles (colored), provides rich and diverse habitats. Noncolored regions (dermis and subcutaneous layers) are free of microbes. **(b)** A highly magnified view (7500×) of skin layers reveals rod- and coccus-shaped bacteria adhering to skin cells (blue).

person and over time. The transients include any microbe a person has picked up, often species that do not ordinarily live on the body, and are greatly influenced by the hygiene of the individual.

The resident populations live and multiply in deeper layers of the epidermis and in glands and follicles (**figure 13.3b**). The composition of the resident microbiota is more stable, predictable, and less influenced by hygiene than are the transients. The normal skin residents consist primarily of bacteria (notably *Staphylococcus, Corynebacterium,* and *Propionibacterium*) and yeasts. Moist skin folds, especially between the toes, tend to harbor fungi, whereas lipophilic mycobacteria and staphylococci are prominent in sebaceous[1] secretions of the axilla, external genitalia, and external ear canal. One species, *Mycobacterium smegmatis,* lives in the waxy secretion, or *smegma,* on the external genitalia of men and women.

TAKE NOTE: VIRUSES AS NORMAL MICROBIOTA

The role of viruses as normal residents is complicated by their usual reputation as infectious agents. In addition, they are harbored inside cells, which makes them very different from most resident microbes. But we now know that they can be

harmless and even beneficial inhabitants of the body. Discoveries from the human genome sequencing project have shown that about 8% to 10% of the genetic material consists of endogenous retroviruses (ERVs). It has been postulated that these ERVs started out as ancient infections, but through co-evolution, the virus established a mutualistic existence with the human genetic material. In time, the viruses may have become an important factor in development and gene expression of their hosts.

Microbial Residents of the Gastrointestinal Tract

The gastrointestinal (GI) tract receives, moves, digests, and absorbs food and removes waste. It encompasses the oral cavity, esophagus, stomach, small intestine, large intestine, rectum, and anus. Stating that the GI tract harbors residents may seem to contradict our earlier statement that internal organs are generally sterile, but it is not an exception to the rule. How can this be true? In reality, the GI tract is a long, hollow tube (with numerous pockets and curves), bounded by the mucous membranes of the oral cavity on one extreme and those of the anus on the other. Because the innermost surface of this tube is exposed to the environment, it is topographically outside the body, so to speak.

The shifting conditions of pH and oxygen tension and differences in the microscopic anatomy of the GI tract are reflected by the variations in or distribution of the microbiota (**figure 13.4**). Some microbes remain attached to the mucous epithelium or its associated structures, and others dwell in the **lumen.*** Although the abundance of nutrients invites microbial growth, the only areas that harbor appreciable permanent microbes are the oral cavity, large intestine, and rectum. The esophagus contains an extremely light load of flora, primarily bacteria swallowed with saliva. The stomach acid inhibits most microbes, although small numbers of lactobacilli and *Helicobacter* can become established there. The small intestine has a sparse population of lactobacilli and streptococci except for its terminal segment, which has flora more similar to the adjacent large intestine.

Microbiota of the Mouth

The oral cavity has unique residents that are among the most diverse and abundant of the body. Microhabitats, including the cheek epithelium, gingiva, tongue, floor of the mouth, and tooth enamel, provide numerous adaptive niches. By one estimate, there are more than 600 species residing there. The most common residents are aerobic *Streptococcus* species—*S. sanguis, S. salivarius, S. mitis*—that colonize the smooth superficial epithelial surfaces. Two species, *S. mutans* and *S. sanguis,* make a major contribution to dental caries by forming sticky dextran slime layers in the presence of simple sugars. The adherence of dextrans to the tooth surface establishes the basis for a biofilm that attracts other bacteria.

1. From *sebum,* a lipid material secreted by glands in the hair follicles.

* *lumen* (loo′-men) The space within a tubular structure.

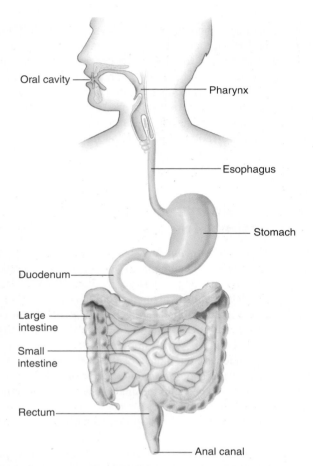

Figure 13.4 Distribution of microbes in the GI tract.
Areas of the gastrointestinal tract that shelter major communities of resident microbes are highlighted in color. Noncolored areas do not harbor residents in significant numbers.

Eruption of the teeth establishes an anaerobic habitat in the gingival crevice that favors the colonization by anaerobic bacteria that can be involved in dental caries and periodontal[2] infections. The instant that saliva is secreted from ducts into the oral cavity, it becomes laden with resident and transient flora. Saliva normally has a high bacterial count (up to 5×10^9 cells per milliliter), a fact that tends to make mouthwashes rather ineffective and a human bite quite dangerous.

Microbiota of the Large Intestine

Microbes inhabiting the intestinal tract have complex and profound interactions with the host. The large intestine (cecum and colon) and the rectum harbor a huge population of microbes ($10^8 - 10^{11}$ per gram of feces). So abundant and prolific are these microbes that they constitute 10% to 30% of the fecal volume. Even an individual on a long-term fast passes feces consisting primarily of bacteria.

The appendix has been considered an organ without much use, but recent research is verifying its importance in replenishment of the normal microbiota (**Insight 13.1**) The state of the intestinal environment favors strictly anaerobic bacteria (*Bacteroides, Bifidobacterium,*

Fusobacterium, and *Clostridium*). **Coliforms*** such as *E. coli, Enterobacter,* and *Citrobacter* are present in smaller numbers. Many species ferment waste materials in the feces, generating vitamins (B_{12}, K, pyridoxine, riboflavin, and thiamine) and acids (acetic, butyric, and propionic) of potential value to the host. Occasionally significant are bacterial digestive enzymes that convert disaccharides to monosaccharides or promote steroid metabolism.

Intestinal bacteria contribute to intestinal odor by producing **skatole,*** amines, and gases (CO_2, H_2, CH_4, and H_2S). Intestinal gas is known in polite circles as *flatus,* and the expulsion of it as *flatulence*. Some of the gas arises through the action of bacteria on dietary carbohydrate residues from vegetables such as cabbage, corn, and beans. The bacteria produce an average of 8.5 liters of gas daily, but only a small amount is ejected in flatus. Combustible gases occasionally form an explosive mixture in the presence of oxygen that has reportedly ignited during intestinal surgery and ruptured the colon!

Late in childhood, some people lose the ability to secrete the enzyme lactase. When they ingest milk or other lactose-containing dairy products, lactose is acted upon instead by intestinal bacteria, and severe intestinal distress can result. The recommended treatment for this deficiency is avoiding these foods or eating various lactose-free substitutes.

Inhabitants of the Respiratory Tract

The first microorganisms to colonize the upper respiratory tract (nasal passages and pharynx) are predominantly oral streptococci. *Staphylococcus aureus* preferentially resides in the nasal entrance, nasal vestibule, and anterior nasopharynx, and *Neisseria* species take up residence in the mucous membranes of the nasopharynx behind the soft palate (**figure 13.5**). Lower still are

* *coliform* (koh'-lih-form) L. *colum,* a sieve. Gram-negative, facultatively anaerobic, and lactose-fermenting microbes.
* *skatole* (skat'-ohl) Gr. *skatos,* dung. One chemical that gives feces their characteristic stench.

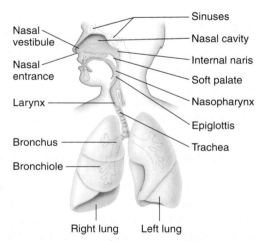

Figure 13.5 Colonized regions of the respiratory tract.
The moist mucous blanket of the nasopharynx has a well-entrenched resident microbes (indicated by color). Some colonization occurs in the upper trachea, but lower regions of bronchi, bronchioles, and lungs lack resident microbes.

2. Situated or occurring around the tooth.

INSIGHT 13.1

The Appendix: A Safe Haven for Microbiota?

The appendix, a 4-inch-long wormlike pouch that sprouts from the junction between the small and large intestine, has been written off for years as a lingering remnant of tissue without any real functions. Evidence offered for its uselessness is that it can be removed without real medical consequences. In recent times, the medical establishment has been reconsidering the accuracy of this claim. For one thing, immunologists have known for some time that the appendix contains a significant population of white blood cells, and they consider it a modest form of lymphatic organ that protects the intestines from pathogens in the same way that tonsils protect the respiratory tract.

Studies at Duke University have produced strong evidence that the appendix has an additional role to play—that of harboring and supporting the normal resident microbiota. Researchers uncovered an extensive biofilm of beneficial microbes on the internal surfaces of the organ that periodically reinoculates the large intestine and maintains the correct composition necessary for normal colon function (see figure). This way, the colon's own biofilms could be rapidly replaced after a case of diarrhea or other intestinal disruption. They also discovered that the immune system mediates and supports this biofilm formation in both systems.

These experts reason that humans can still function normally without an appendix because we don't need it as much in light of our improved medical care and sanitation, which greatly reduce the incidence of

Left: Section through appendix stained with acridine dye reveals a surface biofilm attached to mucosal cells (yellow nuclei). Right: Enlarged view of cells in biofilm. Left bar = 20μm; right bar = 5μm

intestinal infections. In fact, it appears that living in ultraclean environments offering less contact with microbes is a factor in infection and inflammation leading to appendicitis. In developing countries where levels of hygiene are lower, the rates of appendicitis and surgery to remove the appendix are far lower.

Why would living in a cleaner environment make the appendix more susceptible to infections? *Answer available at http://www.mhhe.com/ talaro8*

assorted streptococci and species of *Haemophilus* that colonize the tonsils and lower pharynx. Conditions lower in the respiratory tree (bronchi and lungs) are unfavorable habitats for permanent residents.

Microbiota of the Genitourinary Tract

The regions of the genitourinary tract that harbor microflora are the vagina and outer opening of the urethra in females and the anterior urethra in males **(figure 13.6)**. The internal reproductive organs are kept sterile through physical barriers such as the cervical plug and other host defenses. The kidney, ureter, bladder, and upper urethra are presumably kept sterile by urine flow and regular bladder emptying. The shortness of the urethra in women (about 3.5 cm long), frequently leads to urinary tract infections. The principal residents of the urethra are nonhemolytic streptococci, staphylococci, corynebacteria, and occasionally, coliforms.

The vagina presents a notable example of how changes in physiology can greatly influence the composition of the normal microbiota. An important factor influencing these changes in women is the hormone estrogen. Estrogen normally stimulates the vaginal mucosa to secrete glycogen, which certain bacteria (primarily *Lactobacillus* species) ferment, thus lowering the pH to

about 4.5. Before puberty, a girl produces little estrogen and little glycogen and has a vaginal pH of about 7. These conditions favor the establishment of diphtheroids,[3] staphylococci, streptococci, and some coliforms. As hormone levels rise at puberty, the vagina begins to deposit glycogen, and the microbiota shift to the acid-producing lactobacilli. It is thought that the acidic pH of the vagina during this time prevents the establishment and invasion of microbes with potential to harm a developing fetus. The estrogen-glycogen effect continues, with minor disruption, throughout the childbearing years until menopause, when the microbiota return to a mixed population similar to that of prepuberty. These transitions are not abrupt but occur over several months to years.

Maintenance of the Normal Microbiota

There is no question that the normal residents are essential to the health of humans and other animals. When living in balance with their host, the flora create an environment that may prevent infections and can enhance certain host defenses. In general, the microbes replace themselves naturally on a regular basis to maintain the types and numbers in their zones. However, because the exact content of

3. Any nonpathogenic species of *Corynebacterium.*

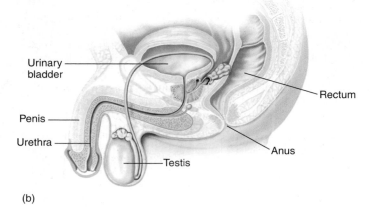

Figure 13.6 Microbiota of the reproductive tract.
(a) Female and **(b)** male genitourinary residents (location indicated by color).

the microbiota is not fixed, a number of changes can disrupt this balance. Use of broad-spectrum antibiotics, changes in diet, and underlying disease all have the potential of altering the makeup of the microbiota and tilting toward disease. A growing trend in therapy is the use of live cultures of known microbes in the form of **probiotics** (discussed in chapter 12). This essentially involves introducing known microbes back into the body through ingestion or inoculation. The microbes chosen for this process are known to be beneficial and are considered nonpathogenic.

For a look into laboratory studies that address the effects of microbiota, see **Insight 13.2.**

Check & Assess Section 13.1

✔ Humans are in contact with microorganisms from the moment of birth onward. An infection is a condition in which contaminating microorganisms overcome host defenses, multiply, and cause disease, damaging tissues and organs.

✔ The resident or normal microbiota include bacteria, fungi, protozoa, and even viruses.

✔ There are two types of cutaneous populations of microbes: the transients, which cling to but do not grow on the superficial layers of the skin, and the more permanent residents, which reside in the deeper layers of the epidermis and its glands.

✔ The inhabitants of the alimentary canal are confined primarily to the mouth, large intestine, and rectum.

✔ The residents of the respiratory tract extend from the nasal cavity to the lower pharynx.

✔ The microbiota of the genitourinary tract are restricted to the urethral opening in males and to the urethra and vagina in females.

1. Describe the significant relationships that humans have with microbes.
2. Differentiate between contamination, colonization, infection, and disease, and explain some possible outcomes in each.
3. How are infectious diseases different from other diseases?
4. Name the general body areas that are sterile and explain why the inside of the intestine is not sterile like other organs.
5. Differentiate between transient and resident microbes.
6. Explain what causes variations in the microbiota of the newborn intestine and the vaginal tract.

13.2 Major Factors in the Development of an Infection

Expected Learning Outcomes

7. Review the main stages in the development of an infection.
8. Distinguish between different states of pathogenicity and differentiate it from virulence.
9. Differentiate among the different portals of entry, and give examples of pathogens that invade by these means.
10. Explain what is meant by the infectious dose, using examples.
11. Describe the process of adhesion and various mechanisms by which microbes use it to gain entry.
12. Identify and discuss invasive factors and virulence factors.
13. Compare and contrast the major characteristics of exotoxins and endotoxins.

In this section, we trace the course of an infection, a sequence of events that is summarized in **figure 13.7.** Pathogens are parasitic microbes whose relationship with a host results in infection and disease. The type and severity of infection depend on both the pathogenicity of the organism and the condition of the host. Pathogenicity, you will recall, is a broad concept that describes an organism's potential to cause infection or disease and is used to divide pathogenic microbes into one of two general groups. **True pathogens** (primary pathogens) are capable of causing disease in healthy persons with normal immune defenses. They are generally associated with a specific, recognizable disease, which may vary in severity from mild (colds) to severe (malaria) to fatal (rabies). Examples of true pathogens include influenza virus, plague bacillus, and malaria protozoan.

INSIGHT 13.2

A Germ-Free Existence

For years, questions lingered about how essential the microbiota are to normal life and what functions they might serve. The need for animal models to further investigate these questions led eventually to development of laboratory strains of *germ-free,* or **axenic,** mammals and birds. After the young mammals are taken from the mother aseptically by cesarean section, they are immediately transferred to a sterile isolator or incubator. The newborns must be fed by hand through gloved ports in the isolator until they can eat on their own, and all materials entering their chamber must be sterile. Rats, mice, rabbits, guinea pigs, monkeys, and dogs are some of the mammals raised in the germ-free state.

Experiments with germ-free animals are of two basic varieties: (1) general studies on how the lack of normal microbial residents influences the nutrition, metabolism, and anatomy of the animal; and (2) **gnotobiotic*** studies, in which the germ-free subject is inoculated either with a single type of microbe to determine its individual effect or with several known microbes to determine interrelationships. Results are validated by comparing the germ-free group with a conventional, normal control group. **Table 13.A** summarizes some major conclusions arising from studies with germ-free animals.

A dramatic characteristic of germ-free animals is that they live longer and have fewer diseases than normal controls, as long as they remain in a sterile environment. From this standpoint, it is clear that the microbiota are not needed for survival and may even be the source of infectious agents. At the same time, it is also clear that axenic life is highly impractical.

Additional studies have revealed that microbiota contribute significantly to the development of the immune system. When germ-free animals are placed in contact with normal control animals, they gradually develop resident microbes similar to that of the controls. However, germ-free subjects are less tolerant of microorganisms and can die from infections by relatively harmless species. This susceptibility is due to the immature character of the immune system of germ-free animals. These animals have a reduced number of certain types of white blood cells and slower antibody response.

* (noh″-toh-by-ah′-tik)

Sterile enclosure for rearing and handling germ-free animals.

TABLE 13.A Effects of the Germ-Free State

Germ-Free Animals Display	Significance
Enlargement of the cecum; other degenerative diseases of the intestinal tract of rats, rabbits, chickens	Microbes are needed for normal intestinal development.
Vitamin deficiency in rats	Microbes are a significant nutritional source of vitamins.
Underdevelopment of immune system in most animals	Microbes are needed to stimulate development of certain host defenses.
Absence of dental caries and periodontal disease in dogs, rats, hamsters	Microbes are essential in caries formation and gum disease.
Heightened sensitivity to enteric pathogens (*Shigella, Salmonella, Vibrio cholerae*) and to fungal infections	Normal bacterial residents are antagonistic against pathogens.
Lessened susceptibility to amebic dysentery	Normal resident microbes facilitate the completion of the life cycle of the amoeba in the gut.
Less body fat	Normal microbiota help to break down indigestible carbohydrates and increase fat storage in the body.

One puzzling observation revolved around the fact that germ-free mice had as much as 42% less body fat than their normal counterparts. Recent experiments revealed that indigenous bacteria acted to increase the body fat of mice in two ways. First, certain bacteria in the gut helped to digest complex carbohydrates, making extra calories available to the mice. Second, these same bacteria secreted a substance that, by acting on mouse hormones, caused the animal to store more fat than it otherwise would.

Gnotobiotic experiments have clarified the dynamics of several infectious diseases. For years, the precise involvement of microbes in dental caries had been ambiguous. Studies with germ-free rats, hamsters, and beagles confirmed that caries development is influenced by heredity, a diet high in sugars, and poor oral hygiene. Even when all these predisposing factors are present, however, germ-free animals still remain free of caries unless they have been inoculated with specific bacteria.

Why is a cesarean section necessary for developing axenic mammals?
Answer available at http://www.mhhe.com/talaro8

Portal of entry	→	Adhesion	→	Invasion	→	Multiplication	→	Infection of target	→	Disease	→	Portal of exit
Microbes evade barriers		Microbes attach to host cells		Microbes make pathway into cells		Microbes grow and spread		Microbes attack specific tissues		Microbes damage tissues		Microbes leave host

Figure 13.7 **A flow diagram that relates the events in entry, establishment, and exit of infectious agents.**

Opportunistic pathogens cause disease when the host's defenses are compromised[4] or when they become established in a part of the body that is not natural to them. Opportunists are not considered pathogenic to a normal healthy person and, unlike primary pathogens, do not generally possess well-developed virulence properties. Examples of opportunistic pathogens include *Pseudomonas* species and *Candida albicans.* Factors that greatly predispose a person to infections, both primary and opportunistic, are shown in **table 13.4.**

The relative severity of the disease caused by a particular microorganism depends on the **virulence*** of the microbe. Although the terms *pathogenicity* and *virulence* are often used interchangeably, virulence is the accurate term for describing the degree of pathogenicity. The virulence of a microbe is determined by its

4. People with weakened immunity are often termed *immunocompromised.*
* *virulence* (veer-yoo-lents) L. *virulentia,* virus, poison.

TABLE 13.4	Factors That Weaken Host Defenses and Increase Susceptibility to Infection*

- Old age and extreme youth (infancy, prematurity)
- Genetic defects in immunity and acquired defects in immunity
- Surgery and organ transplants
- Organic disease: cancer, liver malfunction, diabetes
- Chemotherapy/immunosuppressive drugs
- Physical and mental stress
- Other infections

*These conditions compromise defense barriers or immune responses.

ability to establish itself in the host and cause damage. There is much involved in both of these steps. To establish themselves in a host, microbes must enter the host, attach firmly to host tissues, and survive the host defenses. To cause damage, microbes invade tissues, produce toxins, or induce a host response that is actually injurious to the host. Any characteristic or structure of the microbe that contributes to the infection or disease state is called a **virulence factor.** Virulence can be due to a single factor or to multiple factors. In some microbes, the causes of virulence are clearly established, but in others, they are not. In the following section, we examine the effects of virulence factors and their roles in the progress of an infection.

As with many concepts, the idea of a pathogen being true or opportunistic is a convenience to simplify discussion. In reality, pathogenicity actually extends along a range from low to high. To recognize this, the CDC has adopted a system of biosafety categories for pathogens based on their degree of pathogenicity and the relative danger in handling them. This system assigns microbes to one of four levels or classes. Microbes not known to cause disease in humans are assigned to class 1, and highly contagious viruses that pose an extreme risk to humans are classified as class 4. Microbes of intermediate virulence are assigned to class 2 or 3. Details of this system are presented in more detail in **table 13.5.**

Becoming Established: Phase One—Portals of Entry

To initiate an infection, a microbe enters the tissues of the body by a characteristic route, the **portal of entry (figure 13.8),** usually a cutaneous or membranous boundary. The source of the infectious agent

TABLE 13.5	Primary Biosafety Levels and Agents of Disease

Personnel handling infectious agents in the laboratory must be protected from possible infection through special risk management or containment procedures.

Biosafety Level	Facilities and Practices	Risk of Infection and Class of Pathogens
1	Standard, open bench, no special facilities needed; typical of most microbiology teaching labs; access may be restricted.	Low infection hazard; class 1 microbes not generally considered pathogens and will not invade the bodies of healthy persons; *Micrococcus luteus, Bacillus megaterium, Lactobacillus, Saccharomyces.*
2	At least Level 1 facilities and practices; plus personnel must be trained in handling pathogens; lab coats and gloves required; safety cabinets may be needed; biohazard signs posted; access restricted.	Agents with moderate potential to infect; class 2 pathogens can cause disease in healthy people but can be contained with proper facilities; pathogens that belong to class 2 include *Staphylococcus aureus, Escherichia coli, Salmonella* sp.; pathogenic helminths; hepatitis A, B, and rabies viruses; *Cryptococcus* and *Blastomyces;* HIV
3	Minimum of Level 2 facilities and practices; manipulation performed in safety cabinets with special containment features; only personnel require protective clothing; no unsterilized materials can leave the lab; personnel are monitored and vaccinated.	Agents can cause severe or lethal disease especially when inhaled; class 3 microbes include *Mycobacterium tuberculosis, Francisella tularensis, Yersinia pestis, Brucella* sp., *Coxiella burnetii, Coccidioides immitis,* and yellow fever, and western equine encephalitis.
4	Minimum of Level 3 facilities and practices; facilities have highest levels of controlled access; clothing changes and showers required for all people entering and leaving; materials must be autoclaved or gas sterilized prior to entering and leaving lab.	Agents being handled are highly virulent microbes that pose extreme risk for morbidity and mortality when inhaled in droplet or aerosol form; class 4 microbes include flaviviruses; arenaviruses (Lassa fever virus) and filoviruses (Ebola and Marburg viruses).

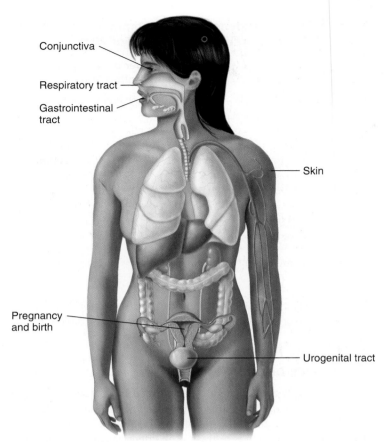

Figure 13.8 Portals of entry. Infectious microbes enter the body through characteristic routes. Most microbes have adapted to a specific portal.

Labels:
- Conjunctiva
- Respiratory tract
- Gastrointestinal tract
- Skin
- Pregnancy and birth
- Urogenital tract

can be **exogenous,** originating from a source outside the body such as the environment or another person or animal, or **endogenous,** already existing on or in the body from microbiota or a latent infection.

For the most part, the portals of entry are the same anatomical regions that also support resident microbes: the skin, gastrointestinal tract, respiratory tract, and urogenital tract. The majority of pathogens have adapted to a specific portal of entry, one that provides a habitat for further growth and spread. This adaptation can be so restrictive that if certain pathogens enter the "wrong" portal, they will not be infectious. For instance, inoculation of the nasal mucosa with the influenza virus invariably gives rise to the flu, but if this virus contacts only the skin, no infection will result. Likewise, contact with athlete's foot fungi in small cracks in the toe webs can induce an infection, but inhaling the fungus spores will not infect a healthy individual.

Occasionally, an infective agent can enter by more than one portal. For instance, *Mycobacterium tuberculosis* enters through both the respiratory and gastrointestinal tracts, and pathogens in the genera *Streptococcus* and *Staphylococcus* have adapted to invasion through several portals of entry such as the skin, urogenital tract, and respiratory tract. Understand that the same pathogen entering different portals of entry will likely lead to different infections or diseases. Infection of a skin follicle by *Staphylococcus aureus* will produce an abcess, but its invasion of the respiratory tract causes a serious form of pneumonia.

Infectious Agents That Enter the Skin

The skin is a very common portal of entry. The actual sites of entry are usually nicks, abrasions, and punctures (many of which are tiny and inapparent) rather than smooth, unbroken skin. *Staphylococcus aureus* (the cause of boils), *Streptococcus pyogenes* (an agent of impetigo), the fungal dermatophytes, and agents of gangrene and tetanus gain access through damaged skin. The cold sore virus (herpes simplex) enters through the mucous membranes near the lips.

Some infectious agents create their own passageways into the skin using digestive enzymes. For example, certain helminth worms burrow through the skin directly to gain access to the tissues. Other infectious agents enter through bites. The bites of insects, ticks, and other animals offer an avenue to a variety of viruses, rickettsias, and protozoa. An artificial means for breaching the skin barrier is contaminated hypodermic needles by intravenous drug abusers. Users who inject drugs are predisposed to a disturbing list of well-known diseases: hepatitis, HIV/AIDS, tetanus, tuberculosis, osteomyelitis, and malaria. A resurgence of some of these infections is directly traceable to drug use. Contaminated needles often contain bacteria from the skin or environment that induce heart disease (endocarditis), lung abscesses, and chronic infections at the injection site.

Although the conjunctiva, the outer protective covering of the eye, is a relatively good barrier to infection, bacteria such as *Chlamydia trachomatis* (trachoma), and *Neisseria gonorrhoeae* readily attach to this membrane.

The Gastrointestinal Tract as Portal

The gastrointestinal tract is the portal of entry for pathogens contained in food, drink, and other ingested substances. They are adapted to survive digestive enzymes and abrupt pH changes. Most enteric pathogens possess specialized mechanisms for entering and localizing in the mucosa of the small or large intestine. The best-known enteric agents of disease are gram-negative rods in the genera *Salmonella, Shigella, Vibrio,* and certain strains of *Escherichia coli.* Viruses that enter through the gut are poliovirus, hepatitis A virus, echovirus, and rotavirus. Important enteric protozoans are *Entamoeba histolytica* (amebiasis) and *Giardia lamblia* (giardiasis). Although the anus is not a typical portal of entry, it becomes one in people who practice anal sex.

The Respiratory Portal of Entry

The oral and nasal cavities are also the gateways to the respiratory tract, the portal of entry for the greatest number of pathogens. Because there is a continuous mucous membrane surface covering the upper respiratory tract, the sinuses, and the auditory tubes, microbes are often transferred from one site to another. The extent to which an agent is carried into the respiratory tree is based primarily on its size. In general, small cells and particles are inhaled more deeply than larger ones.

Infectious agents with this portal of entry include the bacteria of streptococcal sore throat, meningitis, diphtheria, and whooping cough and the viruses of influenza, measles, mumps, rubella, chickenpox, and the common cold. Pathogens that are inhaled into the lower regions of the respiratory tract (bronchioles and lungs) can cause pneumonia, an inflammatory condition of the lung. Bacteria (*Streptococcus*

pneumoniae, Klebsiella, Mycoplasma) and fungi (*Cryptococcus* and *Pneumocystis*) are a few of the agents involved in pneumonias. Other agents causing unique recognizable lung diseases are *Mycobacterium tuberculosis* and fungal pathogens such as *Histoplasma*.

Urogenital Portals of Entry

The urogenital tract is the portal of entry for pathogens that are contracted by sexual means (intercourse or intimate direct contact). In the past, a judgmental attitude toward women and a somewhat Victorian attitude toward sex led to these diseases being referred to as venereal, from Venus, the Latin name of the goddess of love. The more contemporary description, **sexually transmitted disease (STD),**[5] reflects the mode of transmission of these diseases more accurately. STDs account for an estimated 4% of infections worldwide, with approximately 19 million new cases occurring in the United States each year. The most recent available statistics (2008) for the estimated incidence of common STDs are provided in **table 13.6.**

The microbes of STDs enter the skin or mucosa of the penis, external genitalia, vagina, cervix, and urethra. Some can penetrate an unbroken surface; others require a cut or abrasion. Once-predominant STDs such as syphilis and gonorrhea have been surpassed by a growing list led by trichomoniasis, caused by a protozoan, genital warts (HPV), and chlamydia. Evolving sexual practices have increased the incidence of STDs that were once uncommon, and diseases that were not originally considered STDs are now so classified.[6] Other common sexually transmitted agents include HIV, herpes simplex virus, *Candida albicans* (a yeast), and hepatitis B virus.

Not all urogenital infections are STDs. Some of these infections are caused by displaced organisms (as when normal flora from

5. Also known as sexually transmitted infections (STIs).

6. Amebic dysentery, scabies, salmonellosis, and *Strongyloides* worms are examples.

TABLE 13.6	Incidence of Most Common Sexually Transmitted Diseases For 2008
STD	**Estimated Number of New Cases per Year in the United States**
Trichomoniasis*	7,400,000
Human papillomavirus* (HPV)	6,200,000
Herpes simplex*	1,000,000
Chlamydiosis	1,210,523
Gonorrhea	336,742
Syphilis	46,277
Hepatitis B	43,000
New HIV infections	41,269
AIDS diagnoses	37,151

*These STDs are not routinely reported; figures are estimates using clinical data.

the gastrointestinal tract cause urinary tract infections) or by opportunistic overgrowth of normal microbiota ("yeast infections").

Pathogens That Infect during Pregnancy and Birth

The placenta is an exchange organ formed by maternal and fetal tissues that separate the blood of the developing fetus from that of the mother yet permit diffusion of dissolved nutrients and gases to the fetus. The placenta is ordinarily an effective barrier against microorganisms in the maternal circulation. However, a few microbes, such as the syphilis spirochete, can cross the placenta, enter the umbilical vein, and spread by the fetal circulation into the fetal tissues **(figure 13.9).**

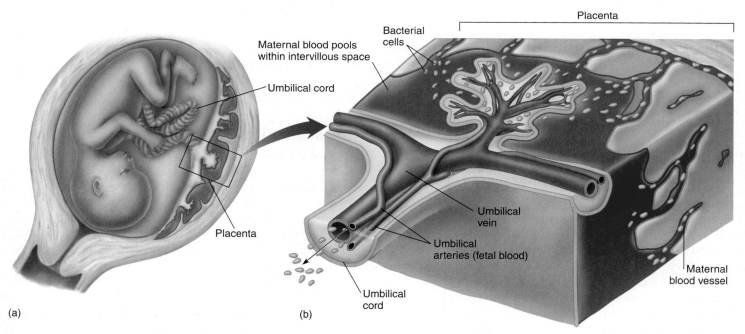

Figure 13.9 **Transplacental infection of the fetus.** **(a)** Fetus in the womb. **(b)** In a closer view, microbes are shown penetrating the maternal blood vessels and entering the blood pool of the placenta. They then invade the fetal circulation by way of the umbilical vein.

Other infections, such as herpes simplex, occur perinatally when the child is contaminated by the birth canal. The common infections of fetus and neonate are grouped together in a unified cluster known by the acronym **STORCH** that medical personnel must monitor. STORCH stands for **s**yphilis, **t**oxoplasmosis, **o**ther diseases (hepatitis B, AIDS, and chlamydia), **r**ubella, **c**ytomegalovirus, and **h**erpes simplex virus. The most serious complications of STORCH infections are spontaneous abortion, congenital abnormalities, brain damage, prematurity, and stillbirths.

The Requirement for an Infectious Dose

Another factor crucial to the course of an infection is the quantity of microbes that enters the portal of entry. For most agents, infection will proceed only if a minimum number, called the **infectious dose (ID),** is present. This number has been determined experimentally for many microbes. In general, microorganisms with smaller infectious doses have greater virulence. **Table 13.7** presents examples of dose sizes for a number of pathogens, ranging from small to large infectious doses. In addition, agents that must pass through the stomach (cholera, typhoid) require a higher ID due to their being exposed to the stomach acid. If the number of microbes entering a portal is below an infectious dose, infection and disease will generally not result. But if the quantity is far in excess of the ID, the onset of disease can be extremely rapid. Even weakly pathogenic species can be rendered more virulent with a large inoculum.

Much of our knowledge of modes of infection and infectious dose has come from human and animal experimentation **(Insight 13.3).**

Attaching to the Host: Phase Two

How Pathogens Attach

Adhesion is a process by which microbes gain a more stable foothold at the portal of entry. Because adhesion is dependent on binding between specific molecules on both the host and pathogen, a

TABLE 13.7	Estimated Infectious Doses of Selected Pathogens*	
Agent of	**Infectious Dose Estimate**	**Primary Route of Infection**
Measles	1 virus	Respiratory
Q fever	1–10 cells	Respiratory
Tularemia	10–50 cells	Various
Smallpox	10–100 viruses	Respiratory
Brucellosis	10–100 cells	Various
Viral encephalitis	10–100 viruses	Mosquito bite
Plague	100–500 cells	Flea bite
Gonorrhea	1,000 cells	Sexual contact
Anthrax	8,000–50,000 spores	Respiratory, cutaneous
Typhoid	10,000 cells	Ingestion
Cholera	100,000,000 cells	Ingestion

*Several of these agents are considered potential bioterror pathogens.

particular pathogen is limited to only those cells (and organisms) to which it can bind. Once attached, the pathogen is poised advantageously to invade the body compartments. Bacterial, fungal, and protozoal pathogens attach most often by appendages and surface structures such as fimbriae (pili), flagella, and adhesive slimes or capsules; viruses attach by means of specialized receptors **(figure 13.10).** In addition, parasitic worms are mechanically fastened to the portal of entry by suckers, hooks, and barbs. Adhesion methods of various microbes and the diseases they cause are shown in **table 13.8.** Firm attachment to host tissues is almost always a prerequisite for causing disease because the body has so many mechanisms for flushing microbes and foreign materials from its tissues.

Invading the Host and Becoming Established: Phase Three

Microbes have numerous properties that improve their invasiveness, that is, their capacity to evade host defenses and enter into deeper tissues where they will grow and become established. These properties all contribute to virulence and thus are considered some type of **virulence factor.** These same factors determine the degree of tissue damage that occurs and hence the severity of the disease. The effects of a pathogen's virulence factors on tissues vary greatly. Cold viruses, for example, invade and multiply but cause relatively little damage to

(a) **Fimbriae**

(b) **Capsules**

(c) **Spikes**

Figure 13.10 Mechanisms of adhesion by pathogens.
(a) Fimbriae, minute bristlelike appendages. **(b)** Adherent extracellular capsules made of slime or other sticky substances. **(c)** Viral envelope spikes. See table 13.8 for specific examples.

INSIGHT 13.3

Human Guinea Pigs

These days, human beings are not used as subjects for determining the effects of infectious disease, but in earlier times they were. The tradition of human experimentation dates well back into the eighteenth century, with the subject frequently the experimenter himself. In some studies, mycologists inoculated their own skin and even that of family members with scrapings from fungal lesions to demonstrate that the disease was transmissible. In the early days of parasitology, it was not uncommon for a brave researcher to swallow worm eggs in order to study the course of his infection and the life cycle of the worm.

In a sort of reverse test, a German colleague of Koch's named Max von Petenkofer believed so strongly that cholera was *not* caused by a bacterium that he and his assistant swallowed cultures of *Vibrio cholerae.* Fortunately for them, they acquired only a mild form of the disease. Many self-experimenters have not been so fortunate.

One of the most famous cases is that of Jesse Lazear, a Cuban physician who worked with Walter Reed on the etiology of yellow fever in 1900. Dr. Lazear was convinced that mosquitoes were directly involved in the spread of yellow fever; and by way of proof, he allowed himself and two volunteers to be bitten by mosquitoes infected with the blood of yellow fever patients. Although all three became ill as a result of this exposure, Dr. Lazear's sacrifice was the ultimate one: he died of yellow fever.

A remarkable, more recent use of human guinea pigs occurred from 1954 to 1973, when the US government staged 153 separate tests using biological warfare agents. The scientists in charge of so-called "Operation Whitecoat" wanted to determine the infectivity of certain agents and how readily they could be acquired through airborne or other means. A group of about 2,000 courageous volunteers agreed to be exposed to some highly virulent pathogens, including the agents of Q fever, tularemia, and leishmaniasis. None of the deadliest pathogens (anthrax, plague, smallpox) was tested on humans. The test subjects all shared some important characteristics: they were young, male draftees and they belonged to the Seventh Day Adventist religion.

What made them uniquely suited to this highly secret study was their status as conscientious objectors—opposed to service as combat soldiers on religious grounds. Many of them were assigned to other

A specially constructed chamber with vents for exposing test subjects to pathogens.

military positions of a medical or clerical type. One of the expectations was that they would consent to be subjects for experiments. The tests, run by the biological warfare program at Ft. Detrick, Maryland, involved direct exposure to airborne pathogens released inside special hoods or, in some cases, breathing aerosols of pure cultures released into the air. Frequently monkeys and actual guinea pigs were exposed simultaneously as a way to track the same infectious agent in animals for comparison.

A number of important discoveries were made. The infectious doses of several pathogens were determined by these studies, as was the effect of particle size. Other tests helped determine the effectiveness of vaccines. It is quite extraordinary that none of the men experienced long-term diseases as a result of these studies, given the high risks involved. It is very unlikely that this sort of testing would be sanctioned in the present era, but there is no doubt that the sacrifices of these men greatly advanced our understanding of the risks of biological warfare.

Give some possible reasons that epidemiologists needed to use human subjects to test for the effects of pathogens. *Answer available at http://www.mhhe.com/talaro8*

their host. At the other end of the spectrum, pathogens such as the tetanus bacillus or rabies virus can severely damage or even kill their host.

Most of the common virulence factors fall into one or more of three categories: antiphagocytic effects, exoenzymes, and toxins. **Figure 13.11** illustrates the workings of a few of these factors.

Antiphagocytic Factors

Microbes are likely to encounter resistance from host defenses when first entering the portal of entry. The initial response comes from white blood cells called **phagocytes.** These cells can ordinarily engulf pathogens and destroy them by means of enzymes and antimicrobial chemicals (see chapter 14).

Antiphagocytic factors are a type of virulence factor used by some pathogens to avoid phagocytes. The antiphagocytic factors of resistant microorganisms help them to circumvent some part of the phagocytic

process **(figure 13.11c).** The most aggressive strategy involves bacteria that kill phagocytes outright. Species of both *Streptococcus* and *Staphylococcus* produce **leukocidins,** substances that are toxic to white blood cells. Some microorganisms secrete an extracellular surface layer (slime or capsule) that makes it physically difficult for the phagocyte to engulf them. *Streptococcus pneumoniae, Salmonella typhi, Neisseria meningitidis,* and *Cryptococcus neoformans* are notable examples. Some bacteria are well adapted to survival inside phagocytes after ingestion. For instance, pathogenic species of *Legionella, Mycobacterium,* and many rickettsias are readily engulfed but are capable of avoiding further destruction. This survival inside phagocytes has special significance because it provides a place for the microbes to hide, grow, and be spread throughout the body.

Getting Inside Host Cells Certain pathogens have developed a secretion system for inserting specialized virulence proteins directly

TABLE 13.8	**Adhesion Properties of Microbes**	
Microbe	**Disease**	**Adhesion Mechanism**
Neisseria gonorrhoeae	Gonorrhea	Fimbriae attach to genital epithelium.
Escherichia coli	Diarrhea	Well-developed fimbrial adhesin
Shigella and *Salmonella*	Gastroenteritis	Fimbriae can attach to intestinal epithelium.
Vibrio	Cholera	Glycocalyx anchors microbe to intestinal epithelium.
Treponema	Syphilis	Tapered hook embeds in host cell.
Mycoplasma	Pneumonia	Specialized tip at ends of bacteria fuses tightly to lung epithelium.
Pseudomonas aeruginosa	Burn, lung infections	Fimbriae and slime layer
Streptococcus mutans, S. sobrinus	Dental caries	Dextran slime layer glues cocci to tooth surface.
Influenza virus	Influenza	Viral spikes react with receptor on respiratory surface.
Poliovirus	Polio	Capsid proteins attach to receptors on susceptible cells.
HIV	AIDS	Viral spikes adhere to white blood cell receptors.
Giardia lamblia (protozoan)	Giardiasis	Small suction disc on underside attaches to intestinal surface.
Trypanosoma (protozoan)	African and S. American trypanosomiasis	Flagellum is needed to penetrate and stay alive.

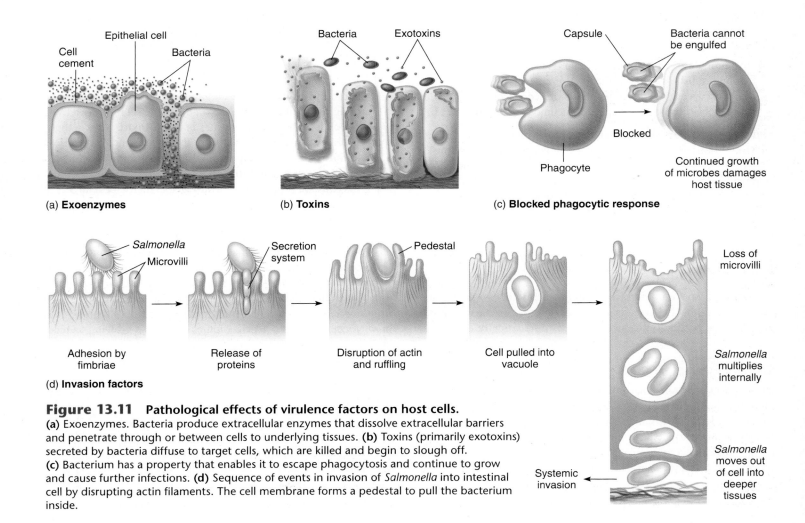

(a) **Exoenzymes**

(b) **Toxins**

(c) **Blocked phagocytic response**

(d) **Invasion factors**

Figure 13.11 **Pathological effects of virulence factors on host cells.**
(a) Exoenzymes. Bacteria produce extracellular enzymes that dissolve extracellular barriers and penetrate through or between cells to underlying tissues. **(b)** Toxins (primarily exotoxins) secreted by bacteria diffuse to target cells, which are killed and begin to slough off. **(c)** Bacterium has a property that enables it to escape phagocytosis and continue to grow and cause further infections. **(d)** Sequence of events in invasion of *Salmonella* into intestinal cell by disrupting actin filaments. The cell membrane forms a pedestal to pull the bacterium inside.

into their host cells (**figure 13.11d**). In the case of *Salmonella* and *E. coli,* the virulence proteins disrupt the actin cytoskeleton of an intestinal cell and force the cell surface to form ruffles that pull the pathogens into the cell interior. Once inside, they multiply within vacuoles and, in time, use the cell as a passageway to break out into the underlying layers of tissues. Other pathogens (*Listeria* and *Shigella*) collect actin filaments and use them as motility organelles to move about in the cells (see figure 19.10)

Extracellular Enzymes

Many pathogenic bacteria, fungi, protozoa, and worms secrete **exoenzymes** that disrupt the structure of tissues. Other enzymes dissolve the host's defense barriers and promote the spread of microbes to deeper regions of the body.

A few examples include:

1. mucinase, which digests the protective coating on mucous membranes and is a factor in amoebic dysentery;
2. keratinase, which digests the principal component of skin and hair, and is secreted by fungi that cause ringworm;
3. collagenase, which digests the principal fiber of connective tissue and is an invasive factor of *Clostridium* species and certain worms; and
4. hyaluronidase, which digests hyaluronic acid, the ground substance that cements animal cells together. This enzyme is an important virulence factor in staphylococci, clostridia, streptococci, and pneumococci (**Figure 13.11a**).

Some enzymes react with components of the blood. Coagulase, an enzyme produced by pathogenic staphylococci, causes clotting of blood or plasma. By contrast, the bacterial kinases (streptokinase, staphylokinase) do just the opposite, dissolving fibrin clots and expediting the invasion of damaged tissues. In fact, one form of streptokinase (streptase) is marketed as a therapy to dissolve blood clots in patients with problems with thrombi and emboli.[7]

Bacterial Toxins: A Potent Source of Cellular Damage

A **toxin** is a specific chemical product of microbes, plants, and some animals that has poisonous effects on other organisms. **Toxigenicity,** the power to produce toxins, is a genetically controlled characteristic of many bacterial species and is responsible for the adverse effects of a variety of diseases generally called **toxinoses.** Toxinoses in which the toxin is spread by the blood from the site of infection are called **toxemias** (tetanus and diphtheria, for example), whereas those caused by ingestion of toxins are **intoxications** (botulism). A toxin is named according to its specific target of action: Neurotoxins act on the nervous system; enterotoxins act on the intestine; hemotoxins lyse red blood cells; and nephrotoxins damage the kidneys.

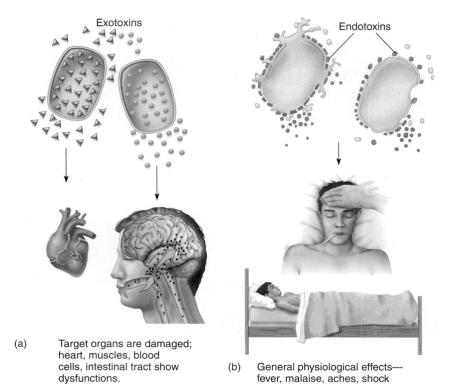

(a) Target organs are damaged; heart, muscles, blood cells, intestinal tract show dysfunctions.

(b) General physiological effects— fever, malaise, aches, shock

Figure 13.12 **The origins and effects of circulating exotoxins and endotoxin. (a)** Exotoxins, given off by live cells, have highly specific targets and physiological effects. **(b)** Endotoxin, given off when the cell wall of gram-negative bacteria disintegrates, has more generalized physiological effects.

A more traditional scheme classifies bacterial toxins according to their origins (**figure 13.12**). A toxin molecule secreted by a living bacterial cell into the infected tissues is an **exotoxin.** A toxin that is not secreted but is released only after the cell is damaged or lysed is an **endotoxin.** Other important differences between the two groups are summarized in **table 13.9.**

Exotoxins are proteins with a strong specificity for a target cell and extremely powerful, sometimes deadly, effects. They generally affect cells by damaging the cell membrane and initiating lysis or by disrupting intracellular function (**figure 13.11b**). **Hemolysins*** are a class of bacterial exotoxin that disrupts the cell membrane of red blood cells (and some other cells, too). This damage causes the red blood cells to **hemolyze**—to burst and release hemoglobin pigment. Hemolysins that increase pathogenicity include the streptolysins of *Streptococcus pyogenes* and the alpha (α) and beta (β) toxins of *Staphylococcus aureus.* When colonies of bacteria growing on blood agar produce hemolysin, distinct zones appear around the colony. The pattern of hemolysis is often used to identify bacteria and determine their degree of pathogenicity.

The exotoxins of diphtheria, tetanus, and botulism, among others, attach to a particular target cell, become internalized, and interrupt an essential cell pathway. The consequences of cell disruption depend upon the target.

One type of exotoxin is specified as an **A-B toxin (figure 13.13).** A-B toxins are so named because they consist of

7. These conditions are intravascular blood clots that can cause circulatory obstructions.

hemolysins (hee-mahl'-uh-sinz) Gt. *hemo,* blood, and *lysin,* to split.

Characteristic	Exotoxins	Endotoxin
TABLE 13.9	**Differential Characteristics of Bacterial Exotoxins and Endotoxin**	
Toxicity	Toxic in tiny amounts	Toxic in higher doses
Effects on the Body	Specific to a cell type (blood, liver, nerve)	Systemic: fever, inflammation, weakness, shock
Chemical Composition	Small proteins	Lipopolysaccharide of cell wall
Heat Denaturation at 60°C	Unstable	Stable
Toxoid Formation	Can be converted to toxoid*	Cannot be converted to toxoid
Immune Response	Stimulate antitoxins**	Does not stimulate antitoxins
Fever Stimulation	Usually not	Yes
Manner of Release	Secreted from live cell	Released from cell wall during lysis
Typical Sources	A few gram-positive and gram-negative	All gram-negative bacteria
Examples of Diseases	Tetanus, diphtheria, cholera, anthrax	Meningitis, endotoxic shock, salmonellosis.

*A toxoid is an inactivated toxin used in vaccines.

**An antitoxin is an antibody that reacts specifically with a toxin.

two parts, an A (active) component and a B (binding) component. The B component binds to a specific receptor molecule on the surface of the host cell. The next step is for the A component to be moved across the host cell membrane. The A components of most A-B toxins then catalyze a reaction that inactivates a normal cell protein. Loss of that protein's function damages a target cell or tissue specific to that toxin.

Examples include one toxin of *Clostridium tetani* that blocks the action of certain spinal neurons, the toxin of *Clostridium botulinum* that prevents the transmission of nerve-muscle stimuli, pertussis toxin inactivating the respiratory cilia, and cholera toxin provoking profuse salt and water loss from intestinal cells. More details of the pathology of exotoxins are found in later chapters on specific organisms (see figures 19.9 and 21.14).

Unlike exotoxins, which contain many specific examples, there is just a single type of endotoxin. It is chemically a substance called lipopolysaccharide (LPS), which is part of the outer membrane of gram-negative cell envelopes. When gram-negative bacteria cause infections, some of them eventually lyse and release these LPS molecules into the infection site or into the circulation. Endotoxin differs from exotoxins in having a variety of systemic effects on tissues and organs. Depending upon the amounts present, endotoxin can cause fever, inflammation, hemorrhage, and diarrhea. Blood infection by gram-negative bacteria such as *Salmonella, Shigella, Neisseria meningitidis,* and *Escherichia coli* is particularly dangerous in that it can lead to fatal endotoxic shock (see Insight 20.1).

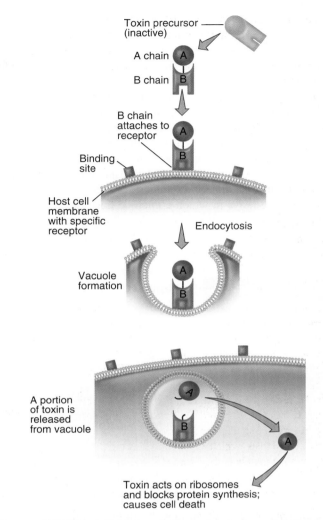

Figure 13.13 A-B toxin of *Corynebacterium diphtheriae*. The B chain attaches to a specific receptor on the cell membrane, then the toxin is engulfed into a vacuole. The two chains separate and the A chain enters the cytoplasm as an enzyme that inactivates a protein needed for protein synthesis. The lack of protein leads to cell dysfunction.

Effects of Infection on Organs and Body Systems

In addition to the adverse effects of enzymes, toxins, and other factors, multiplication by a pathogen frequently weakens host tissues. Pathogens can obstruct tubular structures such as blood vessels, lymphatic channels, fallopian tubes, and bile ducts. Accumulated damage can lead to cell and tissue death, a condition called **necrosis.** Although viruses do not produce toxins or destructive enzymes, they destroy cells by multiplying in and lysing them. Many of the cytopathic effects of viral infection arise from the impaired metabolism and death of cells (see chapter 6).

The direct effects of virulence factors such as enzymes and toxins can cause significant harm to cells, tissues, and organs. It is important to note that many microbial diseases are actually the result of indirect damage, or the host's excessive or inappropriate response to a microorganism. This means that pathogenicity is not a trait inherent in microorganisms alone but is really a consequence of the interplay between microbe and host. The capsule of *Streptococcus*

pneumoniae is a good example. Its presence prevents the bacterium from being cleared from the lungs by phagocytic cells, leading to a powerful inflammatory response by white blood cells and a continuous influx of fluids into the lung spaces—a condition we know as pneumonia.

Of course, it is easier to study and characterize the microbes that cause direct damage through toxins or enzymes. For this reason, these true pathogens were the first to be fully understood as the science of microbiology progressed. But in the last 15 to 20 years, microbiologists have come to appreciate exactly how important the relationship between microbe and host is, and this has greatly expanded our understanding of infectious diseases.

The various ways in which a microbe-host interaction causes damage are described in the chapters describing specific organisms (chapters 18–25).

Check & Assess Section 13.2

✔ Microbial infections result when a microorganism penetrates host defenses, multiplies, and damages host tissue. The *pathogenicity* of a microbe refers to its ability to cause infection or disease. The *virulence* of a pathogen refers to the degree of damage it inflicts on the host tissues.

✔ *True pathogens* cause infectious disease in healthy hosts, whereas *opportunistic pathogens* become infectious only when the host immune system is compromised in some way.

✔ The site at which a microorganism first contacts host tissue is called the *portal of entry*. Most pathogens have one preferred portal of entry, although some have more than one.

✔ The *infectious dose,* or ID, refers to the minimum number of microbial cells required to initiate infection in the host. The ID varies widely among microbial species.

✔ Fimbriae, hooks, and adhesive capsules are types of adherence factors by which pathogens physically attach to host tissues.

✔ Antiphagocytic factors produced by microorganisms include leukocidins, capsules, and factors that resist digestion by white blood cells.

✔ Exoenzymes, toxins, and the ability to induce injurious host responses are the three main types of *virulence factors* pathogens utilize to combat host defenses, invade, and damage host tissue.

✔ Exotoxins and endotoxins differ in their chemical composition and tissue specificity.

✔ Characteristics or structures of microbes that induce extreme host responses are a major factor in most infectious diseases.

7. Explain several ways that true pathogens differ from opportunistic pathogens.
8. Distinguish between pathogenicity and virulence.
9. Explain why most microbes are limited to a single portal of entry.
10. Explain how the portal of entry could differ from the site of infection.
11. Differentiate between exogenous and endogenous infections.
12. What factors possibly affect the size of the infectious dose?
13. Describe four factors involved in microbial adhesion.

14. Define *virulence factors* and give examples of them in gram-positive and gram-negative bacteria, viruses, and parasites.
15. Discuss the effects of several virulence factors that act on white blood cells.
16. Define *toxigenicity.*
17. Which body cells or tissues are affected by hemolysins, leukocidins, hyaluronidase, kinases, tetanus toxin, pertussis toxin, and enterotoxin?

13.3 The Outcomes of Infection and Disease

Expected Learning Outcomes

14. Describe the clinical stages of infection.

15. Use key terms to describe different patterns of infection.

16. Use correct terminology to explain the manifestations of infections and inflammation.

17. Discuss the major portals of exit and how they influence the end stages of infection and disease.

The Stages of Clinical Infections

As the body of the host responds to the invasive and toxigenic activities of a parasite, it passes through four distinct phases of infection and disease: the incubation period, the prodrome, the period of invasion, and the convalescent period (**figure 13.14**).

The **incubation period** is the time from initial contact with the infectious agent (at the portal of entry) to the appearance of the first symptoms. During the incubation period, the agent is multiplying at the portal of entry but has not yet caused enough damage to elicit symptoms. Although this period is relatively well defined and predictable for each microorganism, it does vary according to host resistance, degree of virulence, and distance between the target organ and the portal of entry (the farther apart, the longer the incubation period). Overall, an incubation period can range from several hours in pneumonic plague to several years in leprosy. The majority of infections, however, have incubation periods ranging between 2 and 30 days.

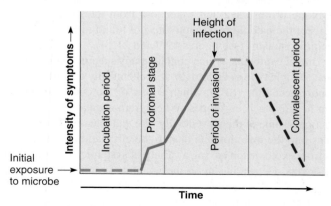

Figure 13.14 Stages in the course of infection and disease. Dashed lines represent periods with a variable length.

The earliest notable symptoms of infection appear as a vague feeling of discomfort, such as head and muscle aches, fatigue, upset stomach, and general malaise. This short period (1–2 days) is known as the **prodromal stage.** The infectious agent next enters a **period of invasion,** during which it multiplies at high levels, exhibits its greatest toxicity, and becomes well established in its target tissue. This period is often marked by fever and other prominent and more specific signs and symptoms, which can include cough, rashes, diarrhea, loss of muscle control, swelling, jaundice, discharge of exudates, or severe pain, depending on the particular infection. The length of this period is extremely variable.

As the patient begins to respond to the infection, the symptoms decline—sometimes dramatically, other times slowly. During the recovery that follows, called the **convalescent period,** the patient's strength and health gradually return owing to the healing nature of the immune response. An infection that results in death is called terminal.

The transmissibility of the microbe during these four stages must be considered on an individual basis. A few agents are released mostly during incubation (measles, for example); many are released during the invasive period (*Shigella*); and others can be transmitted during all of these periods (hepatitis B).

Patterns of Infection

Patterns of infection are many and varied. In the simplest situation, a **localized infection,** the microbe enters the body and remains confined to a specific tissue (**figure 13.15a**). Examples of localized infections are boils, fungal skin infections, and warts.

Many infectious agents do not remain localized but spread from the initial site of entry to other tissues. In fact, spreading is necessary for pathogens such as rabies and hepatitis A virus, whose target tissue is some distance from the site of entry. The rabies virus travels from a bite wound along nerve tracts to its target in the brain, and the hepatitis A virus moves from the intestine to the liver via the circulatory system. When an infection spreads to several sites and tissue fluids, usually in the bloodstream, it is called a **systemic infection (figure 13.15b).** Examples of systemic infections are viral diseases (measles, rubella, chickenpox, and AIDS); bacterial diseases (brucellosis, anthrax, typhoid fever, and syphilis); and fungal diseases (histoplasmosis and cryptococcosis). Infectious agents can also travel to their targets by means of nerves (as in rabies) or cerebrospinal fluid (as in meningitis).

A **focal infection** is said to exist when the infectious agent breaks loose from a local infection and is seeded or *disseminated* into other tissues (**figure 13.15c**). This pattern is exhibited by tuberculosis or by streptococcal pharyngitis, which gives rise to scarlet fever. In the condition called **toxemia,**[8] the infection itself remains localized at the portal of entry, but the toxins produced by the pathogens are carried by the blood to the actual target tissue. In this way, the target of the bacterial cells can be different from the target of their toxin.

An infection is not always caused by a single microbe. In a **mixed infection,** several agents establish themselves simultaneously at the infection site (**figure 13.15d**). In some mixed or synergistic

8. Not to be confused with toxemia of pregnancy, which is a metabolic disturbance and not an infection.

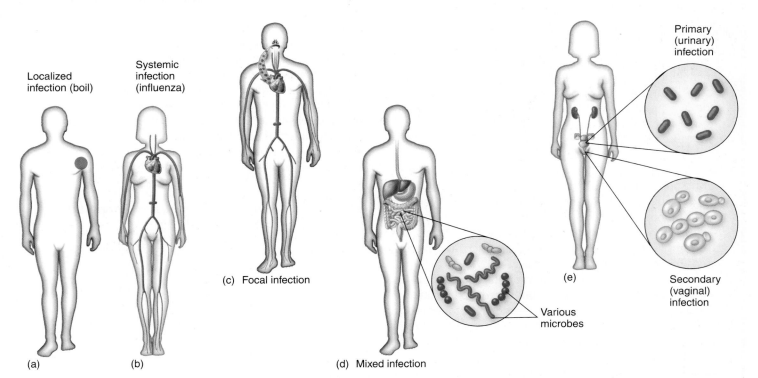

Figure 13.15 **The occurrence of infections with regard to location and sequence.** **(a)** A localized infection, in which the pathogen is restricted to one specific site. **(b)** Systemic infection, in which the pathogen spreads through circulation to many sites. **(c)** A focal infection occurs initially as a local infection, but circumstances cause the microbe to be carried to other sites systemically. **(d)** A mixed infection, in which the same site is infected with several microbes at the same time. **(e)** In a primary-secondary infection, an initial infection is complicated by a second one in the same or a different location and caused by a different microbe.

Localized infection (boil)

Systemic infection (influenza)

(c) Focal infection

(d) Mixed infection

Various microbes

Primary (urinary) infection

(e)

Secondary (vaginal) infection

(a)

(b)

infections, the microbes cooperate in breaking down a tissue. In other mixed infections, one microbe creates an environment that enables another microbe to invade. Gas gangrene, wound infections, dental caries, and human bite infections tend to be mixed. These are sometimes called **polymicrobial** diseases and may be the result of biofilm formation at the site of infection.

Some diseases are described according to a sequence of related infections. When an initial, or **primary, infection** is complicated by another infection caused by a different microbe, this subsequent infection is termed a **secondary infection (figure 13.15e).** This pattern often occurs in a child whose chickenpox (primary infection) become infected with *Staphylococcus aureus* (secondary infection). Another significant combination is seen in elderly patients who develop secondary pneumonia as a complication of influenza. The secondary infection need not be in the same site as the primary infection, and it usually indicates altered host defenses.

TAKE NOTE: A GUIDE TO THE TERMINOLOGY OF INFECTION AND DISEASE

Words in medicine have great power and economy. A single technical term can often replace a whole phrase or sentence, thereby saving time and space in patient charting. The beginning student may feel overwhelmed by what seems like a mountain of new words. However, having a grasp of a few root words and a fair amount of anatomy can help you learn many of these words and even deduce the meaning of unfamiliar ones. Some examples of medical shorthand follow.

- The suffix *-itis* means an inflammation and, when affixed to the end of an anatomical term, indicates an inflammatory condition in that location. Thus, meningitis is an inflammation of the meninges surrounding the brain; encephalitis is an inflammation of the brain itself; hepatitis involves the liver; gastroenteritis, the intestine; and otitis media, the middle ear. Although not all inflammatory conditions are caused by infections, many infectious diseases inflame their target organs.
- The suffix *-emia* is derived from the Greek word *haeima,* meaning blood. When added to a word, it means "associated with the blood." Thus, septicemia means sepsis (infection) of the blood; bacteremia, bacteria in the blood; viremia, viruses in the blood; and fungemia, fungi in the blood. It is also applicable to specific conditions such as toxemia, gonococcemia, and spirochetemia.
- The suffix *-osis* means "a disease or morbid process." It is frequently added to the names of pathogens to indicate the disease they cause: for example, listeriosis, histoplasmosis, toxoplasmosis, shigellosis, salmonellosis, and borreliosis. A variation of this suffix is *-iasis,* as in trichomoniasis and candidiasis.
- The suffix *-oma* comes from the Greek word *onkomas* (swelling) and means tumor. Although the root is often used to describe cancers (sarcoma, melanoma), it is also applied in some infectious diseases that cause masses or swellings (tuberculoma, leproma).

Infections that come on rapidly, with severe but short-lived effects, are called **acute infections.** Infections that progress and persist over a long period of time are **chronic infections.** The Take Note on terminology illustrates other common words used to describe infectious diseases.

Signs and Symptoms: Warning Signals of Disease

When an infection causes pathologic changes leading to disease, it is often accompanied by a variety of signs and symptoms. A **sign** is any objective evidence of disease as noted by an observer; a **symptom** is the subjective evidence of disease as sensed by the patient. Signs tend to be more precise than symptoms and are often measured. Both can be the result of the same underlying cause. For example, an infection of the brain may present with the sign of bacteria in the spinal fluid and symptom of headache. Or a streptococcal infection may produce a sore throat (symptom) and inflamed pharynx (sign). Disease indicators that can be sensed and observed can qualify as either a sign or a symptom depending upon how they are reported. When a disease can be identified or defined by a defined collection of signs and symptoms, it is termed a **syndrome.** Signs and symptoms with considerable importance in diagnosing infectious diseases are shown in **table 13.10.**

Signs and Symptoms of Inflammation

The earliest symptoms of disease result from the activation of the body defense process called **inflammation.*** The inflammatory response includes cells and chemicals that respond nonspecifically to disruptions in the tissue. This subject is discussed in greater detail in chapter 14, but as noted earlier, many signs and symptoms of infection are caused by the mobilization of this system. Some common symptoms of inflammation include fever, pain, soreness, and swelling. Signs of inflammation include **edema,*** the accumulation of fluid in an afflicted tissue; **granulomas** and **abscesses,** walled-off collections of inflammatory cells and microbes in the tissues; and **lymphadenitis,** swollen lymph nodes.

* *inflammation* (in-flam-aye'-shun) L. *inflammation,* to set on fire.
* *edema* (uh-dee'-muh) Gr. *oidema,* swelling.

| TABLE 13.10 | Common Signs and Symptoms of Infectious Diseases | |
|---|---|
| **Signs** | **Symptoms** |
| Fever | Chills |
| Septicemia | Pain, irritation |
| Microbes in tissue fluids | Nausea |
| Abnormal chest sounds | Malaise, fatigue |
| Skin eruptions | Chest tightness |
| Leukocytosis | Itching |
| Leukopenia | Headache |
| Swollen lymph nodes | Weakness |
| Abscesses | Abdominal cramps |
| Tachycardia (increased heart rate) | Anorexia (lack of appetite) |
| Antibodies in serum | Sore throat |

Rashes and other skin eruptions are common symptoms and signs in many diseases, and because they tend to mimic each other, it can be difficult to differentiate among diseases on this basis alone. The general term for the site of infection or disease is **lesion.*** Skin lesions can be restricted to the epidermis and its glands and follicles, or they can extend into the dermis and subcutaneous regions. The lesions of some infections undergo characteristic changes in appearance and location during the course of disease and thus fit more than one category.

Signs of Infection in the Blood

Changes in the number of circulating white blood cells, as determined by special counts, are considered to be signs of possible infection. **Leukocytosis*** is an increase in the level of white blood cells, whereas **leukopenia*** is a decrease. Other signs of infection revolve around the occurrence of a microbe or its products in the blood. The clinical term for blood infection, **septicemia,** refers to a general state in which microorganisms are multiplying in the blood and are present in large numbers. When small numbers of bacteria or viruses are found in the blood, the correct terminology is **bacteremia** or **viremia,** which means that these microbes are present in the blood but are not necessarily multiplying.

During infection, a normal host will invariably show signs of an immune response in the form of antibodies in the serum or some type of sensitivity to the microbe. This fact is the basis for several serological tests used in diagnosing infectious diseases such as AIDS or syphilis. Such specific immune reactions indicate the body's attempt to develop specific immunities against pathogens. We concentrate on this role of the host defenses in chapters 14 and 15.

Infections That Go Unnoticed

It is rather common for an infection to produce no noticeable symptoms, even though the microbe is active in the host tissue. In these cases the person does not show overt indications of his infection. Infections of this nature are known as **asymptomatic, subclinical,** or *inapparent* because the patient experiences no symptoms usually associated with disease and does not seek medical attention. However, it is important to note that most infections are attended by some sort of detectable sign. In section 13.4, we further address the significance of subclinical infections in the transmission of infectious agents.

The Portal of Exit: Vacating the Host

Earlier, we introduced the idea that a parasite is considered *unsuccessful* if it does not have a provision for leaving its host and moving to other susceptible hosts. With few exceptions, pathogens depart by a specific avenue called the **portal of exit (figure 13.16).** In most cases, the pathogen is shed or released from the body through secretion, excretion, discharge, or sloughed tissue. The usually very high number of infectious agents in these materials increases both virulence and the likelihood that the pathogen will reach other hosts. In

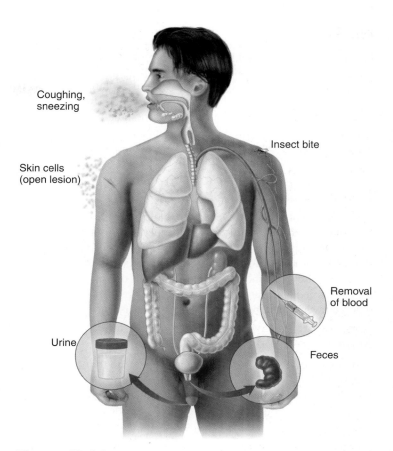

Figure 13.16 **Major portals of exit of infectious diseases.** Pathogens are released from the body through secretion, excretion, or discharge.

many cases, the portal of exit is the same as the portal of entry, but a few pathogens use a different route. As we see in the next section, the portal of exit concerns epidemiologists because it greatly influences the dissemination of infection in a population.

Respiratory and Salivary Portals

Mucus, sputum, nasal drainage, and other moist secretions are the media of escape for the pathogens that infect the lower or upper respiratory tract. The most effective means of releasing these secretions are coughing and sneezing (see figure 13.19), although they can also be released during talking and laughing. Tiny particles of liquid released into the air form aerosols or droplets that can spread the infectious agent to other people. The agents of tuberculosis, influenza, measles, and chickenpox most often leave the host through airborne droplets. Droplets of saliva are the exit route for several viruses, including those of mumps, rabies, and infectious mononucleosis.

Epithelial Cells

The outer layer of the skin and scalp are constantly being shed into the environment. A large proportion of household dust is actually composed of skin cells. A single person can shed several billion skin cells a day, and some persons, called shedders, disseminate massive numbers of bacteria into their immediate surroundings. Skin lesions and their exudates can serve as portals of exit in warts, fungal infections, boils, herpes simplex, smallpox, and syphilis.

* *lesion* (lee´-zhun) L. *laesio,* to hurt.
* *leukocytosis* (loo˝-koh´-sy-toh´-sis) From *leukocyte,* a white blood cell, and the suffix *-osis.*
* *leukopenia* (loo˝-koh´-pee´-nee-uh) From *leukocyte* and *penia,* a loss or lack of.

Fecal Exit

Feces are a very common portal of exit. Some intestinal pathogens grow in the intestinal mucosa and create an inflammation that increases the motility of the bowel. This increased motility speeds up peristalsis, resulting in diarrhea; and the more fluid stool provides a rapid exit for the pathogen. A number of helminth worms release cysts and eggs through the feces. Feces containing pathogens are a public health problem when allowed to contaminate drinking water or when used to fertilize crops.

Urogenital Tract

A number of agents involved in sexually transmitted infections leave the host in vaginal discharge or semen. This is also the source of neonatal infections such as herpes simplex, *Chlamydia,* and *Candida albicans,* which infect the infant as it passes through the birth canal. Less commonly, certain pathogens that infect the kidney are discharged in the urine: for instance, the agents of leptospirosis, typhoid fever, tuberculosis, and schistosomiasis.

Removal of Blood or Bleeding

Although the blood does not have a direct route to the outside, it can serve as a portal of exit when it is removed or released through a vascular puncture made by natural or artificial means. Blood-feeding animals such as ticks and fleas are common transmitters of pathogens (see chapter 23). Human immunodeficiency and hepatitis viruses are transmitted by shared needles or through small gashes in a mucous membrane caused by sexual intercourse.

Because many microbes can be transferred through blood donation, the monitoring of both donors and donated blood has to be scrupulous. In the mid-1980s, several thousand people suffering from hemophilia were infected with HIV as a result of receiving clotting factors contaminated with the virus. The current risk for transfusion-transmitted infections is about one case in 10 million transfusions. One factor that increases risk is the presence of emerging disease agents in the blood supply that have not yet been detected. This recently happened with West Nile virus and Chagas disease.

The Persistence of Microbes and Pathologic Conditions

The apparent recovery of the host does not always mean that the microbe has been completely removed or destroyed by the host defenses. After the initial symptoms in certain chronic infectious diseases, the infectious agent retreats into a state called persistence or **latency.** In some cases of latency, the microbe can periodically become active and produce a recurrent disease. The viral agents of herpes simplex, herpes zoster, hepatitis B, AIDS, and Epstein-Barr can persist in the host for long periods. The agents of syphilis, typhoid fever, tuberculosis, and malaria also enter into latent stages. The person harboring a persistent infectious agent may or may not shed it during the latent stage. If it is shed, such persons are chronic carriers who serve as sources of infection for the rest of the population.

Some diseases leave **sequelae*** in the form of long-term or permanent damage to tissues or organs. For example, meningitis can result in deafness, a strep throat can lead to rheumatic heart disease, Lyme disease can cause arthritis, and polio can produce paralysis.

Check & Assess Section 13.3

✔ Patterns of infection vary with the pathogen or pathogens involved. They range from local and focal to systemic.

✔ A mixed infection is caused by two or more microorganisms simultaneously.

✔ Infections can be characterized by their sequence as primary or secondary and by their duration as either acute or chronic.

✔ An infectious disease is characterized by both objective signs and subjective symptoms.

✔ Infectious diseases that are asymptomatic or subclinical may produce clinical signs but no noticeable symptoms.

✔ The portal of exit by which a pathogen leaves its host is often but not always the same as the portal of entry.

✔ The portals of exit and entry determine how pathogens spread in a population.

✔ Some pathogens persist in the body in a latent state; others cause long-term damages called *sequelae.*

18. Explain what is happening during each stage of infection. Compare and contrast: systemic versus local infections; primary versus secondary infections; infection versus intoxication.

19. What are the differences between signs and symptoms? (First put yourself in the place of a patient with an infection and then in the place of a physician examining you. Describe what you would feel and what the physician would detect upon examining the affected area.)

20. Name some examples of infections and their portals of exit.

21. Using terminology from the guide, give definitions for *urethritis, endotoxemia, chlamydiosis,* and *lymphoma.*

13.4 Origins and Transmission Patterns of Infectious Microbes

Expected Learning Outcomes

18. Relate the significant factors involved in the transmission of infectious diseases.

19. Explain the concept of carriers and describe several types.

20. Use examples to distinguish between the types of vectors, and discuss the importance of zoonoses.

21. Differentiate between communicable and noncommunicable infectious diseases.

22. Characterize the patterns of transmission for communicable diseases.

* *sequelae* (su-kwee′-lee) L. *sequi,* to follow.

23. Discuss nosocomial infections and the ways they impact the clinical environment.

24. Explain what is meant by universal precautions, and discuss how they are implemented.

Reservoirs: Where Pathogens Persist

For an infectious agent to continue to exist and be spread, it must have a permanent place to reside. The **reservoir** is the primary habitat in the natural world from which a pathogen originates. Often it is a human or animal carrier, although soil, water, and plants are also reservoirs. The reservoir can be distinguished from the infection **source,** which is the individual or object from which an infection is actually acquired. In diseases such as syphilis, the reservoir and the source are the same (the human body). In the case of hepatitis A, the reservoir (a human carrier) is usually different from the source of infection (contaminated food).

Living Reservoirs

Many pathogens continue to exist and spread because they are harbored by members of a host population. Persons or animals with frank symptomatic infection are obvious sources of infection, but a **carrier** is, by definition, an individual who *inconspicuously* shelters a pathogen and spreads it to others without any notice. Although human carriers are occasionally detected through routine screening (blood tests, cultures) and other epidemiological devices, they are unfortunately very difficult to discover and control. As long as a pathogenic reservoir is maintained by the carrier state, the disease will continue to exist in that population, and the potential for epidemics will be a constant threat. The duration of the carrier state can be short or long term, and the carrier may or may not have experienced disease due to the microbe.

Several situations can produce the carrier state. **Asymptomatic** (apparently healthy) **carriers** are infected, but as previously indicated, they show no symptoms (**figure 13.17a**). A few asymptomatic infections (gonorrhea and genital warts, for instance) can carry out their entire course without overt manifestations. **Figure 13.17b** demonstrates three types of carriers who have had or will have the disease but do not at the time they transmit the organism. *Incubation carriers* spread the infectious agent during the incubation period. For example, patients with HIV infection can harbor and spread the virus for months and years before their first symptoms appear. Recuperating patients without symptoms are considered *convalescent carriers* when they continue to shed viable microbes and convey the infection to others. Diphtheria patients, for example, spread the microbe for up to 30 days after the disease has subsided.

An individual who shelters the infectious agent for a long period after recovery is a *chronic carrier.* Patients who have recovered from tuberculosis, hepatitis, and herpes infections frequently carry the agent chronically. About one in 20 victims of typhoid fever continues to harbor *Salmonella typhi* in the gallbladder for several years and sometimes for life. The most infamous of these was "Typhoid Mary," a cook who spread the infection to hundreds of victims in the early 1900s.

The **passive carrier** state is of great concern during patient care (see a later section on nosocomial infections). Medical and dental personnel who must constantly handle materials that are heavily contaminated with patient secretions and blood risk picking up pathogens mechanically and accidently transferring them to other patients (**figure 13.17c**). Proper hand washing, handling of contaminated materials, and aseptic techniques greatly reduce this likelihood.

Animals as Reservoirs and Sources Up to now, we have lumped animals with humans in discussing living reservoirs or carriers, but animals deserve special consideration as vectors of infections. The word **vector** is used by epidemiologists to indicate a live animal that transmits an infectious agent from one host to another. (The term is sometimes misused to include any object that spreads disease.) The majority of vectors are arthropods such as fleas, mosquitoes, flies, and ticks, although larger animals can also spread infection; for example, mammals (rabies), birds (ornithosis), or lower vertebrates (salmonellosis).

CONTINUING CASE FILE **13**

The team of CDC investigators went to the plant in Georgia to observe the production facilities and to carry out **traceback** studies. They wished to determine how the peanut butter was originally contaminated with *Salmonella* and to what extent **vectors** were involved. A worker in the plant reported that cockroaches falling into the dry roasted peanuts had been a constant problem and rat droppings were often seen in the processing area. The team also found that the roof was a collecting place for hundreds of birds and holes in the roof that allowed contaminated rainwater to drip into the plant. Even though the peanuts had been roasted at about 90°C (190° F) for 30 minutes, which is considered long enough to kill most pathogens, they were not heated again, and contamination could occur during later stages of processing and packaging. *Salmonella* are relatively hardy bacteria, and although they would not grow in the low-moisture peanut butter, they could remain alive for several months, even years. How the large **infectious dose** of 100,000 cells could have developed under these circumstances is still a mystery, but it is obvious that the processed peanuts must have harbored high levels of the pathogen.

■ What terms would you use to describe the roles of cockroaches, rats, and birds in transmission?

■ What **pathologic** effect does *Salmonella* have on the intestinal tract?

For a wrap-up, see the Case File Perspective on page 418.

By tradition, vectors are placed into one of two categories, depending upon the animal's relationship with the microbe. A **biological vector** actively participates in a pathogen's life cycle, serving as a site in which it can multiply or complete its life cycle (see figure 23.28, page 713). A biological vector communicates the infectious agent to the human host by biting, aerosol formation, or touch. In the case of biting vectors, the animal can inject infected

Asymptomatic **Incubation** **Convalescent** **Chronic**

Stages of release during infection

Time

(a) (b)

Passive

(c) Transfer of infectious agent through contact ✳ Infectious agent

Figure 13.17 **Involvement of carriers in transmission of infectious agents.** (a) An asymptomatic carrier is infected without symptoms. (b) Incubation, convalescent, and chronic carriers can transmit the infection either before or after the period of symptoms. (c) A passive carrier is contaminated but not infected and spreads the pathogen through physical contact.

saliva into the blood (the mosquito), defecate around the bite wound (the flea), or regurgitate blood into the wound (the tsetse fly).

Mechanical vectors are not necessary to the life cycle of an infectious agent and merely transport it without being infected. The external body parts of these animals become contaminated when they come into physical contact with a source of pathogens. The agent is subsequently transferred to humans indirectly by an intermediate such as food or, occasionally, by direct contact (as in certain eye infections). Houseflies are noxious mechanical vectors. They feed on decaying garbage and feces, and while they are feeding, their feet and mouthparts easily become contaminated. They also regurgitate juices onto food to soften and digest it. Flies spread more than 20 bacterial, viral, protozoan, and worm infections. Other nonbiting flies transmit tropical ulcers, yaws, and trachoma. Cockroaches, which have similar unsavory habits, play a role in the mechanical transmission of fecal pathogens, as well as contributing to allergy attacks in asthmatic children.

Many vectors and animal reservoirs spread their own infections to humans. An infection indigenous to animals but naturally transmissible to humans is a **zoonosis.*** In these types of infections,

the human is essentially a dead-end host and does not contribute to the natural persistence of the microbe. Some zoonotic infections (rabies, for instance) can have multihost involvement, and others can have very complex cycles in the wild (plague). Zoonotic spread of disease is promoted by close associations of humans with animals, and people in animal-oriented or outdoor professions are at greatest risk. At least 150 zoonoses exist worldwide; the most common ones are listed in **table 13.11.** Zoonoses make up a full 70% of all new emerging diseases worldwide. It is worth noting that zoonotic infections are impossible to completely eradicate without also eradicating the animal reservoirs. Attempts have been made to eradicate mosquitoes and certain rodents. During the H1N1 flu epidemic of 2009, thousands of pigs were slaughtered in the wrongful attempt to control its spread.

One technique that can provide an early warning signal for the occurrence of certain mosquito-borne zoonoses has been the use of *sentinel animals.* These are usually domestic animals (most often chickens or horses) that can serve as hosts for diseases such as West Nile fever, various viral encephalitides, and malaria. Sentinel animals are placed at various sites throughout the community, and their blood is monitored periodically for antibodies to the infectious agents that

* *zoonosis* (zoh″-uh-noh′-sis) Gr. *zoion,* animal, and *nosos,* disease.

TABLE 13.11	Common Zoonotic Infections	
Disease/Agent	Primary Animal Reservoirs	Mode of Infection
Viruses		
Rabies	All mammals	Bite; aerosol
Yellow fever	Wild birds, mammals, mosquitoes	Mosquito bite
Viral fevers	Wild mammals	Various mosquitoes
Hantavirus	Rodents	Airborne wastes
Influenza	Chickens, swine	Airborne secretions
West Nile virus	Wild birds, mosquitoes	Mosquito bite
Bacteria		
Rocky Mountain spotted fever	Dogs, ticks	Tick bite
Psittacosis	Birds	Airborne secretions
Leptospirosis	Domestic animals	Animal wastes (urine)
Anthrax	Domestic animals	Airborne spores, cutaneous contact
Brucellosis	Cattle, sheep, pigs	Air- and food-borne
Plague	Rodents, fleas	Flea bite; aerosol
Salmonellosis	Variety of mammals, birds, and rodents	Food-borne
Tularemia	Rodents, birds, arthropods	Tick bite, airborne
Miscellaneous		
Ringworm	Domestic mammals	Cutaneous contact
Toxoplasmosis	Cats, rodents, birds	Food-borne, contact
Trypanosomiasis	Domestic and wild mammals	Tse-tse fly; reduviid bug
Trichinosis	Swine, bears	Ingestion of cysts
Tapeworm	Cattle, swine, fish	Ingestion of eggs
Scabies	Domestic animals	Cutaneous contact

would indicate a recent infection by means of a mosquito bite. The presence of infected animals provides useful data on the potential for human exposure, and it also helps establish the epidemiological pattern of the zoonosis including where it may have spread.

Nonliving Reservoirs

Clearly, microorganisms have adapted to nearly every habitat in the biosphere. They thrive in soil and water and often find their way into the air. Although most of these microbes are saprobic and cause little harm and considerable benefit to humans, some are opportunists and a few are regular pathogens. Because human hosts are in regular contact with these environmental sources, acquisition of pathogens from natural habitats is of diagnostical and epidemiological importance.

Soil harbors the vegetative forms of bacteria, protozoa, helminths, and fungi, as well as their resistant or developmental stages such as spores, cysts, ova, and larvae. Bacterial pathogens include the anthrax bacillus and species of *Clostridium* that are responsible for gas gangrene, botulism, and tetanus. Pathogenic fungi in the genera *Coccidioides* and *Blastomyces* are spread by spores in the soil and dust. The invasive stages of the hookworm *Necator* occur in the soil. Natural bodies of water carry fewer nutrients than soil does but still support pathogenic species such as *Legionella, Cryptosporidium,* and *Giardia.*

The Acquisition and Transmission of Infectious Agents

Infectious diseases can be categorized on the basis of how they are acquired. A disease is **communicable** when an infected host can transmit the infectious agent to another host and establish infection in that host. Although this terminology is standard, one must realize that it is not the disease that is communicated but the microbe. Also be aware that the word *infectious* is sometimes used interchangeably with the word *communicable,* but this is not precise usage. The transmission of the agent can be direct or indirect, and the ease with which the disease is transmitted varies considerably from one agent to another. If the agent is highly communicable, especially through direct contact, the disease is **contagious.** Influenza and measles move readily from host to host and thus are contagious, whereas leprosy is only weakly communicable. Because they can be spread through the population, communicable diseases are our main focus in the following sections.

In contrast, a **noncommunicable** infectious disease does *not* arise through transmission of the infectious agent from host to host. The infection and disease are acquired through some other, special circumstance. Noncommunicable infections occur primarily when a compromised person is invaded by his or her own microflora (as with certain pneumonias, for example) or when an individual has accidental contact with a facultative parasite that exists in a nonliving reservoir such as soil. Some examples are certain mycoses, acquired through inhalation of fungal spores, and tetanus, in which *Clostridium tetani* spores from a soiled object enter a cut or wound. Persons thus infected do not become a source of infection.

Patterns of Transmission in Communicable Diseases

The routes or patterns of disease transmission are many and varied. The spread of diseases is by direct or indirect contact with animate or inanimate objects and can be horizontal or vertical. The term *horizontal* means the disease is spread through a population from one infected individual to another; *vertical* signifies transmission from parent to offspring via the ovum, sperm, placenta, or milk. The extreme complexity of transmission patterns among microorganisms makes it very difficult to generalize. However, for easier organization, we divide microorganisms into two major groups, as shown in **figure 13.18:** transmission by direct routes or transmission by indirect routes, in which some sort of vehicle is involved.

Modes of Direct Transmission For microbes to be directly transferred, some type of contact must occur between the skin or mucous membranes of the infected person and those of the infectee. It may help to think of this route as the portal of exit meeting the portal of entry without the involvement of an intermediate object or

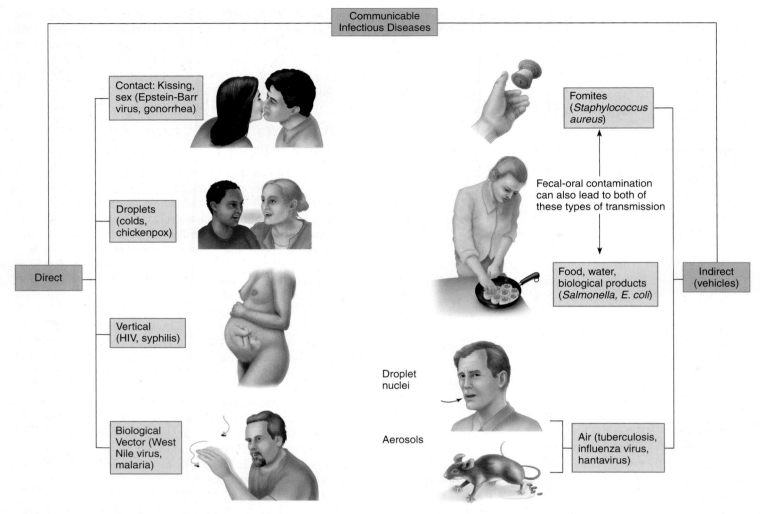

Figure 13.18 Summary of how communicable infectious diseases are acquired.

substance. Included in this category are fine droplets sprayed directly upon a person during sneezing or coughing (as distinguished from droplet nuclei that are transmitted some distance by air). Most sexually transmitted diseases are spread directly. In addition, infections that result from kissing or bites by biological vectors are direct. Most obligate parasites are far too sensitive to survive for long outside the host and can be transmitted only through direct contact. Diseases transmitted vertically from mother to baby fit in this contact category also.

Routes of Indirect Transmission For microbes to be indirectly transmitted, the infectious agent must pass from an infected host to an intermediate conveyor and from there to another host. This form of communication is especially pronounced when the infected individuals contaminate inanimate objects, food, or air through their activities. The transmitter of the infectious agent can be either openly infected or a carrier.

Indirect Spread by Vehicles: Contaminated Materials
The term **vehicle** specifies any inanimate material commonly used by humans that can transmit infectious agents. A **common vehicle** or **source** is a single material that serves as the source of infection

for many individuals. Some specific types of vehicles are food, water, various biological products (such as blood, serum, and tissue), and fomites.

A **fomite** is an inanimate object that harbors and transmits pathogens. The list of possible fomites is as long as your imagination allows. Probably highest on the list would be objects commonly in contact with the public such as doorknobs, telephones, push buttons, and faucet handles that are readily contaminated by touching. Shared bed linens, handkerchiefs, toilet seats, toys, eating utensils, clothing, personal articles, and syringes are other examples. Although paper money is impregnated with a disinfectant to inhibit microbes, pathogens are still isolated from bills as well as coins.

Outbreaks of food poisoning often result from the role of food as a common vehicle. The source of the agent can be soil, the handler, or a mechanical vector. Because milk provides a rich growth medium for microbes, it is a significant means of transmitting pathogens from diseased animals, infected milk handlers, and environmental sources of contamination. The agents of brucellosis, tuberculosis, Q fever, salmonellosis, and listeriosis are transmitted by contaminated milk. Water that has been contaminated by feces or urine can carry *Salmonella, Vibrio* (cholera), viruses (hepatitis A, polio), and pathogenic protozoans (*Giardia, Cryptosporidium*).

In the type of transmission termed the **fecal-oral route,** a fecal carrier contaminates food during handling, and an unsuspecting person ingests it. Hepatitis A, amebic dysentery, shigellosis, and typhoid fever are often transmitted this way. Fecal-oral transmission can also involve contaminated materials such as toys and diapers. It is really a special category of indirect transmission, which specifies that the vehicle became contaminated through contact with fecal material and has found its way to someone's mouth.

Indirect Spread by Vehicles: Air as a Vehicle Unlike soil and water, outdoor air cannot provide nutritional support for microbial growth and is less likely to transmit airborne pathogens. On the other hand, indoor air (especially in a closed space) can serve as an important medium for the suspension and dispersal of certain respiratory pathogens via droplet nuclei and aerosols.

Droplet nuclei are dried microscopic residues created when microscopic pellets of mucus and saliva are ejected from the mouth and nose. They are generated forcefully in an unstifled sneeze or cough (**figure 13.19**) or mildly during other vocalizations. Although the larger beads of moisture settle rapidly, smaller particles evaporate and remain suspended for longer periods. Droplet nuclei are implicated in the spread of hardier pathogens such as the tubercle bacillus and the influenza virus.

Aerosols are suspensions of fine dust or moisture particles in the air that contain live pathogens. Q fever is spread by dust from animal quarters, and psittacosis by aerosols from infected birds. Outbreaks of coccidioidomycosis (a lung infection) often occur in areas where soil is disturbed by wind or digging, releasing clouds of dust bearing the spores of *Coccidioides.*

Nosocomial Infections: The Hospital as a Source of Disease

Infectious diseases that are acquired or develop during a hospital stay are known as **nosocomial* infections.** This concept seems incongruous at first thought, because a hospital is regarded as a place to get treatment for a disease, not a place to acquire a disease. Yet it is not uncommon for a surgical patient's incision to become infected or a burn patient to develop a case of pneumonia in the clinical setting. The rate of nosocomial infections can be as low as 0.1% or as high as 20% of all admitted patients, depending on the clinical setting. The average rate is about 5%. In light of the number of admissions, this adds up to 2 million to 4 million cases a year, which result in nearly 99,000 deaths. Nosocomial infections cost time and money as well as suffering. By one estimate, they amount to 8 million additional days of hospitalization a year and an increased cost of $5 billion to $10 billion.

So many factors unique to the hospital environment are tied to nosocomial infections that a certain number of infections are virtually unavoidable. After all, the hospital both attracts and creates compromised patients, and it serves as a collection point for pathogens. Some patients become infected when surgical procedures or lowered defenses permit resident microbes to invade their bodies. Other patients acquire infections directly or indirectly from fomites, medical equipment, other patients, medical personnel, visitors, air, and water.

Figure 13.19 The explosiveness of a sneeze. Special photography dramatically captures droplet formation in an unstifled sneeze. Even the merest attempt to cover a sneeze with one's hand will reduce this effect considerably. When such droplets dry and remain suspended in air, they are droplet nuclei.

The health care process itself increases the likelihood that infectious agents will be transferred from one patient to another. Treatments using reusable instruments such as respirators and endoscopes constitute a possible source of infectious agents. Indwelling devices such as catheters, prosthetic heart valves, grafts, drainage tubes, and tracheostomy tubes form a ready portal of entry and habitat for infectious agents. Because such high numbers of the hospital population receive antimicrobial drugs during their stay, drug-resistant microbes are selected for at a much greater rate than is the case outside the hospital.

The most common nosocomial infections involve the urinary tract, the respiratory tract, surgical incisions, and the blood (sepsis) (**figure 13.20**). Gram-negative intestinal flora (*Escherichia coli, Klebsiella, Pseudomonas*) are cultured in more than half of patients with nosocomial infections. Gram-positive bacteria (staphylococci and streptococci) and yeasts make up most of the remainder. True pathogens such as *Mycobacterium tuberculosis, Salmonella,* hepatitis B, and influenza virus can be transmitted in the clinical setting as well.

The potential seriousness and impact of nosocomial infections have required hospitals to develop committees that monitor infectious outbreaks and develop guidelines for infection control and aseptic procedures.

Medical asepsis includes practices that lower the microbial load in patients, caregivers, and the hospital environment. These practices include proper hand washing, disinfection, and sanitization, as well as patient isolation. **Table 13.12** summarizes guidelines for the major types of isolation. The goal of these procedures is to limit the spread of infectious agents from person to person. An even higher level of stringency is seen with *surgical asepsis,* which involves all of the strategies listed previously plus ensuring that all surgical procedures are conducted under sterile conditions. This includes sterilization of surgical instruments, dressings, sponges and the like, as well as clothing personnel in sterile garments and scrupulously disinfecting the room surfaces and air.

Hospitals generally employ an *infection control officer* who not only implements proper practices and procedures throughout the hospital but also is charged with tracking potential outbreaks,

* *nosocomial* (nohz"-oh-koh'-mee-al) Gr. *nosos,* disease, and *komeion,* to take care of. Originating from a hospital or infirmary.

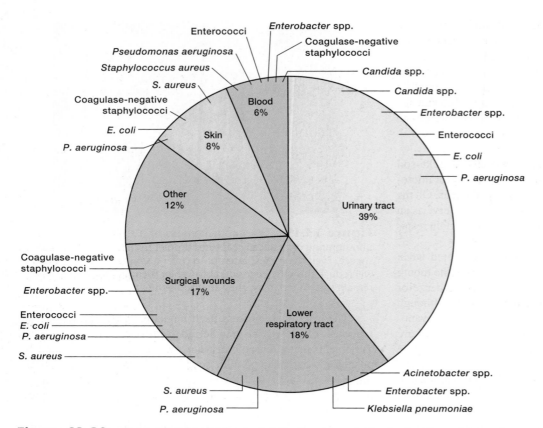

Figure 13.20 **Nosocomial Infections.** Relative frequency by body site. These data are from the National Nosocomial Infections Surveillance, which is conducted by the Centers for Disease Control and Prevention.

identifying breaches in asepsis, and training other health care workers in aseptic technique. Among those most in need of this training are nurses and other caregivers whose work, by its very nature, exposes them to needle-sticks, infectious secretions, blood, and physical contact with the patient. The same practices that interrupt the routes of infection in the patient can also protect the health care worker. It is for this reason that most hospitals have adopted universal precautions that recognize that all secretions from all persons in the clinical setting are potentially infectious and that transmission can occur in either direction.

Universal Blood and Body Fluid Precautions

Medical and dental settings require stringent measures to prevent the spread of nosocomial infections from patient to patient, from patient to worker, and from worker to patient. But even with precautions, the rate of such infections is rather high. Recent evidence indicates that more than one-third of nosocomial infections could be prevented by consistent and rigorous infection control methods.

TABLE 13.12	Levels of Isolation Used in Clinical Settings	
Type of Isolation*	**Protective Measures****	**To Prevent Spread of**
Enteric Precautions	Gowns and gloves must be worn by all persons having direct contact with patient; masks are not required; special precautions are taken for disposing of feces and urine.	Diarrheal diseases; *Shigella, Salmonella,* and *Escherichia coli* gastroenteritis; cholera; hepatitis A; rotavirus; and giardiasis
Respiratory Precautions	Private room with closed door is necessary; gowns and gloves are not required; masks are usually indicated; items contaminated with secretions must be disinfected.	Tuberculosis, measles, mumps, meningitis, pertussis, rubella, chickenpox
Drainage and Secretion Precautions	Gowns and gloves are required; masks are not needed; contaminated instruments and dressings require special precautions.	Staphylococcal and streptococcal infections; gas gangrene; herpes zoster; burn infections
Strict Isolation	Private room with closed door is required; gowns, masks, and gloves must be worn; contaminated items must be wrapped and decontaminated.	Mostly highly virulent or contagious microbes; includes tuberculosis, some types of pneumonia, extensive skin and burn infections, herpes simplex and zoster
Reverse Isolation (Also Called Protective Isolation)	Same guidelines as for strict isolation are required; room may be ventilated by laminar airflow filtered through a high-efficiency particulate (HEPA) filter that removes airborne pathogens; infected persons must be barred.	Used to protect patients extremely immunocompromised by cancer therapy, surgery, genetic defects, burns, prematurity, or AIDS and therefore vulnerable to opportunistic pathogens

*Precautions are based upon the primary portal of entry and communicability of the pathogen.

**In all cases, visitors to the patient's room must report to the nurses' station before entering the room; all visitors and personnel must wash their hands upon entering and leaving the room.

Previously, control guidelines were disease-specific, and clearly identified infections were managed with particular restrictions and techniques. With this arrangement, personnel tended to handle materials labeled *infectious* with much greater care than those that were not so labeled. The AIDS epidemic spurred a reexamination of that policy. Because of the potential for increased numbers of undiagnosed HIV-infected patients, the CDC laid down more-stringent guidelines for handling patients and body substances. These guidelines have been termed **universal precautions (UPs),** because they are based on the assumption that all patient specimens could harbor infectious agents and so must be treated with the same degree of care. They also include body substance isolation (BSI) techniques to be used in known cases of infection.

It is worth mentioning that these precautions are designed to protect all individuals in the clinical setting—patients, workers, and the public alike. In general, they include techniques designed to prevent contact with pathogens and contamination and, if prevention is not possible, to take purposeful measures to decontaminate potentially infectious materials. See appendix C for a more detailed list of UPs.

Check & Assess Section 13.4

- ✔ The primary habitat of a pathogen is called its reservoir. A human reservoir is also called a carrier.
- ✔ Animals can be either reservoirs or vectors of pathogens. An infected animal is a biological vector. Uninfected animals, especially insects, that transmit pathogens mechanically are called mechanical vectors.
- ✔ Soil and water are nonliving reservoirs for pathogenic bacteria, protozoa, fungi, and worms.
- ✔ A communicable disease can be transmitted from an infected host to others, but not all infectious diseases are communicable.
- ✔ The spread of infectious disease from person to person is called horizontal transmission. The spread of infectious disease from parent to offspring is called vertical transmission.
- ✔ Infectious diseases are spread by either contact or indirect routes of transmission. Vehicles of indirect transmission include soil, water, food, air, and fomites (inanimate objects).
- ✔ Nosocomial infections are acquired in a hospital from surgical procedures, equipment, personnel, and exposure to drug-resistant microorganisms.

22. Explain the precise difference between communicable and noncommunicable infectious diseases, and between contact and indirect modes of transmission.
23. What is a common vehicle or source?
24. Nosocomial infections can arise from what general sources?
25. Outline the major agents involved in nosocomial infections and determine whether they are true pathogens or opportunists.
26. Distinguish the primary differences between mechanical and biological vectors.
27. Explain what it means to be a carrier of infectious disease.
28. Describe four ways that humans can be carriers.
29. What is epidemiologically and medically important about carriers in the population?

13.5 Epidemiology: The Study of Disease in Populations

Expected Learning Outcomes

25. Define *epidemiology,* and summarize the major goals of its studies.
26. Explain the primary methods of tracking infections and diseases in a population.
27. Differentiate among the patterns of disease outbreaks according to frequency, number, and location.
28. Summarize the steps in Koch's postulates, and explain their importance to microbiologists.

So far, our discussion has revolved primarily around the impact of an infectious disease in individuals. Let us now turn our attention to the effects of diseases on the community—the realm of **epidemiology.***
By definition, this term involves the study of the frequency and distribution of disease and other health-related factors in defined human populations. It involves many disciplines—not only microbiology but also anatomy, physiology, immunology, medicine, psychology, sociology, ecology, and statistics—and it considers many diseases other than infectious ones, including heart disease, cancer, drug addiction, and mental illness.

The epidemiologist is a medical sleuth who collects clues on the causative agent, pathology, and sources and modes of transmission and tracks the numbers and distribution of cases of disease in the community. In fulfilling these demands, the epidemiologist asks who, when, where, how, why, and what about diseases. The outcome of these studies helps public health departments develop prevention and treatment programs and establish a basis for predictions.

Who, When, and Where? Tracking Disease in the Population

Epidemiologists are concerned with all of the factors covered earlier in this chapter: virulence, portals of entry and exit, and the course of disease. But they are also interested in surveillance; that is, collecting, analyzing, and reporting data on the rates of occurrence, mortality, morbidity, and transmission of infections. Surveillance involves keeping data for a large number of diseases seen by the medical community and reported to public health authorities. By law, certain **reportable,** or **notifiable, diseases** must be reported to authorities; others are reported on a voluntary basis.

A well-developed network of individuals and agencies at the local, district, state, national, and international levels keeps track of infectious diseases. Physicians and hospitals report all notifiable diseases that are brought to their attention. Case reporting can focus on a single individual or collectively on group data.

Local public health agencies first receive the case data and determine how they will be handled. In most cases, health officers investigate the history and movements of patients to trace their prior

* *epidemiology* (ep"-ih-dee-mee-alh'-uh-gee) Gr. *epidemios,* prevalent.

contacts and to control the further spread of the infection as soon as possible through drug therapy, immunization, and education. In sexually transmitted diseases, patients are asked to name their partners so that these persons can be notified, examined, and treated. It is very important to maintain the confidentiality of the persons in these reports. The principal government agency responsible for keeping track of infectious diseases nationwide is the Centers for Disease Control and Prevention (CDC) in Atlanta, Georgia, which is a part of the United States Public Health Service. The CDC publishes a weekly notice of diseases (the *Morbidity and Mortality Report*) that provides weekly and cumulative summaries of the case rates and deaths for about 50 notifiable and 65 reportable diseases, highlights important and unusual diseases, and presents data concerning disease occurrence in the major regions of the United States. This notice is available to anyone at http://www.cdc.gov/mmwr/. Ultimately, the CDC shares its statistics on disease with the World Health Organization (WHO) for worldwide tabulation and control.

Epidemiological Statistics: Frequency of Cases

The **prevalence** of a disease is the accumulated total of existing cases with respect to the entire population. It is usually reported as the percentage of the population having a particular disease and it provides a long-term record of a disease. Disease **incidence** measures the number of new cases over a certain time period, as compared with the general healthy population. This statistic, also called the case, or morbidity rate, indicates both the rate and the risk of infection at the time of reporting. The equations used to figure these rates follow.

$$\text{Prevalence} = \frac{\begin{array}{c}\text{Total number of} \\ \text{cases in population}\end{array}}{\begin{array}{c}\text{Total number of} \\ \text{persons in population}\end{array}} \times 100 = \%$$

$$\text{Incidence} = \frac{\begin{array}{c}\text{Number of} \\ \text{new cases}\end{array}}{\begin{array}{c}\text{Total number of} \\ \text{susceptible persons}\end{array}} \quad \begin{array}{c}\text{(Usually reported} \\ \text{per 100,000 persons)}\end{array}$$

As an example, let us use a classroom of 50 students exposed to a new strain of influenza. Before exposure, the prevalence and incidence in this population are both zero (0/50). If in one week, 5 out of the 50 people contract the disease, the prevalence is 5/50 = 10%, and the incidence is 1 in 10 (10 is used instead of 100,000 due to smaller population). If within a month 25 more people acquire the flu, the prevalence becomes 30/50 or 60%, and the incidence (the number of new infections) is 25/50, or 5 in 10.

Using reported cases of AIDS as a real-life example, the most recent statistics show the cumulative total of cases (prevalence) between 1981 and 2009 of 1,299,965 cases, or 0.43% of the population. The incidence of 37,151 cases for 2009 comes out to 0.081 cases per 100,000 people.

The changes in incidence and prevalence are usually followed over a seasonal, yearly, and long-term basis and are helpful in predicting trends **(figure 13.21).** Statistics of concern to the epidemiologist are the rates of disease with regard to sex, race, or geographic region. Also of importance is the **mortality rate,** which measures the total number of deaths in a population due to a certain disease.

Over the past century, the overall death rate from infectious diseases has dropped, although the **morbidity*** rate, or the number of persons afflicted with infectious diseases, has remained relatively high.

Monitoring statistics also makes it possible to define the frequency of a disease in the population. An infectious disease that exhibits a relatively steady frequency over a long time period in a particular geographic locale is **endemic (figure 13.22a).** Often, the reason for endemicity is the presence of a reservoir in that location. For example, Lyme disease is endemic to certain areas of the United States where the tick vector is found. A certain number of new cases are expected in these areas every year. When a disease is **sporadic,** occasional cases are reported at irregular intervals in random locales **(figure 13.22b).** Tetanus and typhoid fever are reported sporadically in the United States (fewer than 50 cases a year).

When statistics indicate that the number of new cases of an endemic or sporadic disease is increasing beyond what is expected for that population, the pattern is described as an **epidemic (figure 13.22c).** An epidemic exists when an increasing trend is observed in a particular population. The time period is not defined. It can range from hours in food poisoning to years in syphilis.

The term **outbreak** is often used interchangeably with *epidemic,* but an outbreak is usually on a smaller scale in a limited area. In this chapter's case file, the cases started as outbreaks, but as time went on, the numbers of cases and regions affected rose to the level of an epidemic. There is no exact percentage of increase required before an outbreak can qualify as an epidemic. Several diseases are at epidemic levels in the United States, most recent examples being chlamydia and pertussis. The spread of an epidemic across continents is a **pandemic,** as exemplified by AIDS and influenza **(figure 13.22d).** See case file 6 for an example of pandemic spread.

One important epidemiological truism might be called the "iceberg effect," which refers to the fact that only a small portion of an iceberg is visible above the surface of the ocean, with a much more massive part lingering unseen below the surface. Regardless of case reporting and public health screening, a large number of cases of infection in the community go undiagnosed and unreported. In the instance of salmonellosis, approximately 40,000 cases are reported each year. Epidemiologists estimate that the actual number is more likely somewhere between 1 million and 2 million. The iceberg effect can be even more lopsided for sexually transmitted diseases or for infections that are not reported to agencies. For a list of reportable diseases in the United States, see the tables in the back end papers of this book.

Investigative Strategies of the Epidemiologist

Initial evidence of a new disease or an epidemic in the community is fragmentary. A few sporadic cases are seen by physicians and are eventually reported to authorities. However, it can take several reports before any alarm is registered. Epidemiologists and public health departments must piece together odds and ends of data from similar cases and go through a **traceback** investigation. In essence, this means that they work backward to reconstruct the epidemic pattern, similar to this chapter's case file. A completely new disease requires even greater preliminary investigation, because the

* *morbidity* (mor-bih´-dih-tee) L. *morbidis,* sick. A condition of being diseased.

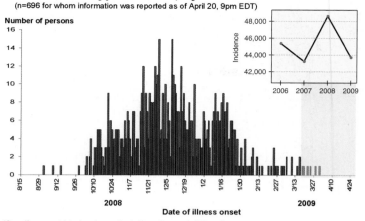

Infections with the outbreak strain of *Salmonella* Typhimurium, by date of illness onset
(n=696 for whom information was reported as of April 20, 9pm EDT)

(a)

*Some illness onset dates have been estimated from other reported information

(b)

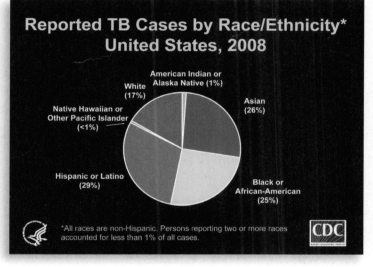

(c)

Figure 13.21 **Samples of epidemiological data as analyzed by the CDC.** **(a)** An epicurve tracks the incidence of salmonellosis cases discussed in the chapter case file; inset shows incidence for years 2006 to 2009. **(b)** The density of AIDS survivors depicted by state and territory. **(c)** Most recent statistics for tuberculosis: Left is a graph of cases according to age and sex; right is a piechart according to race/ethnicity.

infectious agent must be isolated and linked directly to the disease **(Insight 13.4).** All factors possibly impinging on the disease are scrutinized. Investigators search for *clusters* of cases indicating spread between persons or a public (common) source or vehicle. They also look at possible contact with animals, contaminated food, water, and public facilities and at human interrelationships or changes in community structure. Out of this maze of case information, the investigators hope to recognize a pattern that indicates the source of infection so that they can quickly move to control it.

Check & Assess Section 13.5

✔ Epidemiology is the study of the determinants and distribution of all diseases in populations. The study of infectious disease in populations is just one aspect of this field.

✔ Data on specific, reportable diseases are collected by local, national, and worldwide agencies.

✔ The *prevalence* of a disease is the percentage of existing cases in a given population. The disease *incidence,* or *morbidity rate,* is the ratio of newly infected members compared to the whole population during a specified time period.

✔ Disease frequency is described as sporadic, epidemic, pandemic, or endemic.

✔ Epidemiologists collect data on diseases and trace them back to their origins.

30. What types of information do epidemiologists require to differentiate between the incidence and prevalence of infectious diseases?

31. Outline the science of epidemiology and the work of an epidemiologist.

32. See questions 5a and 5b on page 422 to answer questions on patterns of disease in populations.

33. List the main features of Koch's postulates and explain why it is so difficult to prove them for some diseases.

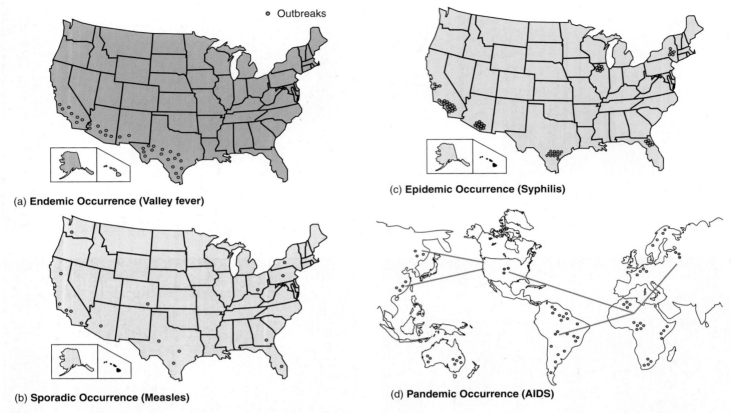

Figure 13.22 **Patterns of infectious disease occurrence.** **(a)** In endemic occurrence, cases are concentrated in one area at a relatively stable rate. **(b)** In sporadic occurrence, a few cases occur randomly over a wide area. **(c)** An epidemic is an increased number of cases that often appear in geographic clusters. The clusters may be local, as in the case of a restaurant-related food-borne epidemic, or nationwide, as is the case with *Chlamydia.* **(d)** Pandemic occurrence means that an epidemic ranges over more than one continent.

CASE FILE 13 PERSPECTIVE

Salmonellosis is the second most common **food infection,** with over 40,000 reported cases every year, but this number is considered only the tip of the iceberg. The CDC estimates an additional 1 million to 2 million cases go unreported. *Salmonella* Typhimurium, the most common serotype implicated in human disease, inhabits the large intestine of a variety of birds, rodents, cattle, and some reptiles. Outbreaks occur on a regular basis and are usually associated with foods such as beef, eggs, chicken, and produce. The time from the first contact with the pathogen to the onset of symptoms is called the **incubation period.** The earliest symptoms of nausea and vomiting occur in the **prodromal** phase and are followed by full-fledged cramping, diarrhea, and fever. In most cases, the manifestations subside within 1 to 2 weeks.

Evidence that differentiates sporadic from epidemic occurrence includes the number and location of cases and whether they are linked in any way. It became evident that this was an epidemic because it rose to a higher number of cases than was usually expected. This rise was sustained for several weeks, and the cases occurred in **clusters** associated with the exact same source. If the incidence had been sporadic, there would have been a small number of isolated cases that involved different foods and unrelated serotypes.

Transmission of salmonellosis frequently involves vectors. There were potentially two different types of vectors acting here.

In one instance, bird or rat feces would both be likely reservoirs of *Salmonella.* Because these animals harbor and nurture the microbe internally, they would be considered **biological vectors.** The tendency for cockroaches to transport the pathogen on the outside of their bodies and spread it by contact makes them **mechanical vectors.** The investigators were never able to make a final conclusion on the exact reservoir or which vector was more involved. It was probably a combination of factors. They did indicate that any one of the violations could have given rise to the contamination. The company did not adequately separate raw and finished products and failed to take seriously the 12 times its product tested positive for *Salmonella* by its own laboratory.

The infectious dose is relatively high because the microbe has to survive passage through the stomach acid. Once they reach the small intestine, the pathogens **adhere** to the intestinal cells with their pili and are engulfed. Being gram-negative, *Salmonella* contains **endotoxin** in its cell wall, which triggers inflammatory reactions that lead to the gastrointestinal disturbances and fever. In the more severe cases, the infection becomes **systemic** as the pathogen enters the blood and perforates the bowel.

For more information on this case, go to http://www.cdc.gov/salmonella/typhimurium/update.html.

Chapter 20 covers the important features of salmonellosis.

INSIGHT 13.4

Koch's Postulates: Solving the Puzzle of New Diseases

An essential aim in the study of infection and disease is determining the precise **etiologic,** or causative, **agent.** In our modern technological age, we take for granted that a certain infection is caused by a certain microbe, but such has not always been the case. More than 130 years ago, Robert Koch realized that in order to support the germ theory of disease, he would have to develop a standard for determining causation that would stand the test of scientific scrutiny. Out of his experimental observations on the transmission of anthrax in cows came a series of proofs, called **Koch's postulates,** that established the principal criteria for etiologic studies. A summary of these postulates is as follows:

1. find evidence of a particular microbe in every case of a disease,
2. isolate that microbe from an infected subject and cultivate it artificially in the laboratory,
3. inoculate a susceptible healthy subject with the laboratory isolate and observe the same resultant disease, and
4. reisolate the agent from this subject.

Valid application of Koch's postulates involves several critical details. Each culture must be pure, observed microscopically, and identified by means of characteristic tests; the first and second isolate must be identical; and the pathologic effects, signs, and symptoms of the disease in the first and second subject must be the same. Once established, these postulates were rapidly put to the test. Within a short time, they had helped determine the causative agents of tuberculosis, diphtheria, and plague. Today, most infectious diseases have been directly linked to a known agent.

Many infectious agents such as viruses and obligately parasitic bacteria can be difficult to culture or cannot be cultured. For these pathogens, the postulates have been modified to take this into account. It is now considered significant to repeatedly observe the infectious agent in tissue samples taken from people with the same disease. Likewise, the presence of specific genetic profiles or genes for virulence are viewed as highly valid evidence.

Koch's original postulates—or a modified version—continue to play an essential role in modern epidemiology. Nearly every year, new diseases arise or old ones spread out of their usual range. Notable examples include dengue fever, influenza, West Nile fever, and Whipple disease. Whipple disease is a chronic intestinal infection, which for many years lacked a known causative agent. Recently, investigators felt sure enough of their identification of a bacterium in infected tissues to announce they had finally identified the mystery pathogen, calling it *Tropheryma whipplei.*

Explain how we can be so certain that HIV causes AIDS even though there is no animal model for it. *Answer available at* http://www.mhhe.com/talaro8

1. Specimen from patient ill with infection of unknown etiology is carried through an isolation procedure.

2. A pure culture of the suspected agent is made.

Full microscopic and biological characterization

Inoculation of test subject

3. Observe animal for disease characteristics

Specimen taken

4. Pure culture and identification procedures

The microbe in the initial and second isolations and the disease in the patient and experimental animal must be identical for the postulates to be satisfied.

Chapter Summary with Key Terms

13.1 We Are Not Alone

A. The human body is in contact with microbes that are constantly present in the environment. Microbes are classified as either part of the normal residents of the body or as **pathogens,** depending on whether or not their colonization of the body results in infection.

B. *Resident Microbiota: The Human as Habitat*

1. **Normal resident microbiota** consist of a huge and varied population of microbes that reside permanently on all surfaces of the body exposed to the environment.

2. Colonization begins just prior to birth and continues throughout life. Variations in normal residents occur in response to changes in an individual's age, diet, hygiene, and health.

3. Normal residents often provide some benefit to the host but are also capable of causing disease, especially when the immune system of the host is compromised or when they invade a normally sterile area of the body.

4. An important benefit provided by microbiota is **microbial antagonism.**

C. *Indigenous Microbiota of Specific Regions*
Resident microbes include bacteria, fungi, and protozoa. They occupy the skin, mouth, gastrointestinal tract, large intestine, respiratory tract, and genitourinary tract.

13.2 Major Factors in the Development of an Infection

A. **Pathogenicity** is the ability of microorganisms to cause infection and disease; **virulence** refers to the degree to which a microbe can invade and damage host tissues.

1. **True pathogens** are able to cause disease in a normal healthy host with intact immune defenses.

2. **Opportunistic pathogens** can cause disease only in persons whose host defenses are compromised by *predisposing conditions.*

B. *Becoming Established: Phase One—Portals of Entry*

1. Microbes typically enter the body through a specific **portal of entry.** Portals of entry are generally the same as those areas of the body that harbor microflora.

2. The minimum number of microbes needed to produce an infection is termed the **infectious dose,** and these pathogens may be classified as **exogenous** or **endogenous,** based on their source.

C. *Attaching to the Host: Phase Two*
Microbes attach to the host cell, a process known as **adhesion,** by means of fimbriae, capsules, or receptors.

D. *Invading the Host and Becoming Established: Phase Three*
Microbes that persist in the human host have developed mechanisms of disabling early defensive strategies.

E. *How Virulence Factors Contribute to Tissue Damage and Disease*
Virulence factors are structures or properties of a microbe that lead to pathologic effects on the host. While some virulence factors are those that enable entry or adhesion, many virulence factors lead directly to damage. These fall in three categories: enzymes, toxins, and the inhibition or destruction of phagocytes.

1. **Exoenzymes** digest epithelial tissues and permit **invasion** of pathogens.

2. **Toxigenicity** refers to a microbe's capacity to produce **toxins** at the site of multiplication. Toxins are divided into **endotoxins** and **exotoxins,** and diseases caused by toxins are classified as **intoxications, toxemias, or toxinoses.**

13.3 The Outcomes of Infection and Disease

A. **Infectious diseases** follow a typical pattern consisting of an **incubation period, prodromal stage, period of invasion,** and **convalescent period.** Each stage is marked by symptoms of a specific intensity.

B. Infections are classified by

1. whether they remain localized at the site of inoculation (**localized infection, systemic infection, focal infection**);

2. the number of microbes involved in an infection, and the order in which they infect the body (**mixed infection, primary infection, secondary infection**); and

3. the persistence of the infection (**acute infection, chronic infection,** subacute infection).

C. **Signs** of infection refer to *objective* evidence of infection, and **symptoms** refer to *subjective* evidence. A **syndrome** is a disease that manifests as a predictable complex of symptoms.

D. After the initial infection, pathogens may remain in the body in a **latent** state and may later cause recurrent infections. Long-term damages to host cells are referred to as **sequelae.**

E. Microbes are released from the body through a specific **portal of exit** that allows them access to another host.

13.4 Origins and Transmission Patterns of Infectious Microbes

A. The **reservoir** of a microbe refers to its natural habitat; a pathogen's **source** refers to the immediate origin of an infectious agent.

B. **Carriers** are individuals who inconspicuously shelter a pathogen and spread it to others. Carriers are divided into **asymptomatic carriers,** incubation carriers, convalescent carriers, chronic carriers, and **passive carriers.**

C. **Vectors** refer to animals that transmit pathogens.

1. **Biological vectors** are animals that are involved in the life cycle of the pathogen they transmit; **mechanical vectors** transmit pathogens without being involved in the life cycle.

2. **Zoonoses** are diseases that can be transmitted to humans but are normally found in animal populations.

D. Agents of disease are either **communicable** or **noncommunicable,** with highly communicable diseases referred to as **contagious.**

1. Direct transmission of a disease occurs when a portal of exit meets a portal of entrance.

2. Indirect transmission occurs when an intermediary such as a **vehicle, fomite,** or **droplet nuclei** connects portals of entrance and exit.

E. **Nosocomial** infections are infectious diseases acquired in a hospital setting. Special vigilance is required to reduce their occurrence.

13.5 Epidemiology: The Study of Disease in Populations

A. Epidemiologists are involved in the surveillance of reportable diseases in populations. Diseases are tabulated with regard to their **prevalence** and **incidence,** while populations are analyzed with regard to **morbidity, mortality,** age, sex, and geographic region.

B. Diseases are described as being **endemic, sporadic, epidemic,** or **pandemic** based on the frequency of their occurrence in a population.

C. Epidemiologists look for changes in the occurrence of diseases such as clusters of new cases. They compile evidence and **trace back** to the original source or reservoir.

Multiple-Choice Questions

Select the correct answer from the answers provided. For questions with blanks, choose the combination of answers that most accurately completes the statement.

1. The best descriptive term for the resident microbes is
 a. commensals
 c. pathogens
 b. parasites
 d. contaminants

2. Resident microbiota is commonly found in the
 a. liver
 c. salivary glands
 b. kidney
 d. urethra

3. Normal resident microbes are absent from the
 a. pharynx
 c. intestine
 b. lungs
 d. hair follicles

4. Virulence factors include
 a. toxins
 c. capsules
 b. enzymes
 d. all of these

5. The specific action of hemolysins is to
 a. damage white blood cells
 c. damage red blood cells
 b. cause fever
 d. cause leukocytosis

6. The _____ is the time that lapses between encounter with a pathogen and the first symptoms.
 a. prodrome
 c. period of convalescence
 b. period of invasion
 d. period of incubation

7. A short period early in a disease that manifests with general malaise and achiness is the
 a. period of incubation
 c. sequela
 b. prodrome
 d. period of invasion

8. The presence of a few bacteria in the blood is termed
 a. septicemia
 c. bacteremia
 b. toxemia
 d. a secondary infection

9. A _____ infection is acquired in a hospital.
 a. subclinical
 c. nosocomial
 b. focal
 d. zoonosis

10. A/an _____ is a passive animal transporter of pathogens.
 a. zoonosis
 c. mechanical vector
 b. biological vector
 d. asymptomatic carrier

11. An example of a noncommunicable infection is
 a. measles
 c. tuberculosis
 b. leprosy
 d. tetanus

12. A general term that refers to an increased white blood cell count is
 a. leukopenia
 c. leukocytosis
 b. inflammation
 d. leukemia

13. The occurrence of Lyme disease mainly in areas where certain species of ticks live would define it as a/an _____ disease.
 a. epidemic
 c. endemic
 b. pandemic
 d. sporadic

14. A positive antibody test for HIV would be a _____ of infection.
 a. sign
 c. syndrome
 b. symptom
 d. sequela

15. Which of the following would *not* be a portal of entry?
 a. the meninges
 c. skin
 b. the placenta
 d. small intestine

16. Which of the following is *not* a condition of Koch's postulates?
 a. isolate the causative agent of a disease
 b. cultivate the microbe in a lab
 c. inoculate a test animal to observe the disease
 d. test the effects of a pathogen on humans

Case File Questions

1. This outbreak would be considered an epidemic because:
 a. it happened over a wide geographic region
 b. of a rise in the number of cases over a short time
 c. it was transmitted by a common vehicle
 d. it occurred in all age groups

2. Which of these is the most accurate description of the relationship between rats and *Salmonella*?
 a. Rats carry *Salmonella* on their feet and fur.
 b. Rats pick up *Salmonella* from infected cockroaches.
 c. Rats harbor *Salmonella* as intestinal microbiota.
 d. *Salmonella* lives in the soil around rat nests.

Writing to Learn

These questions are suggested as a *writing-to-learn* experience. For each question, compose a one- or two-paragraph answer that includes the factual information needed to completely address the question. All Check & Assess questions can be used for writing-to-learn exercises.

1. What are the important clinical implications of positive blood or cerebrospinal fluid?

2. What is the expected result if a compromised person is exposed to a true pathogen?

3. Explain how endotoxin gets into the bloodstream of a patient with endotoxic shock.

4. Use the following formula to explain the relationships among the several factors and what happens when they change.

$$\frac{\text{Infection}}{\text{(infectious disease)}} = \frac{\text{No. of organisms} \times \text{Virulence}}{\text{Host resistance}}$$

5. Give a brief description of how the A-B type of toxin works.

6. For each portal of entry, give a vehicle that carries the pathogen and the course it must travel to invade the tissues.

7. Describe the course of infection from contact with the pathogen to its exit from the host.

8. What are some important considerations about the portal of exit?

9. Using case file 13 as an example, explain the difference between a source and a reservoir.

10. List the signs and symptoms of the infections in case file 13.

11. What does it mean for salmonellosis to have an iceberg effect?

12. a. Outline the five types of clinical isolation.
 b. What is the work of an infection control officer?

Concept Mapping

Appendix E provides guidance for working with concept maps.

1. Supply your own linking words or phrases in this concept map, and provide the missing concepts in the empty boxes.

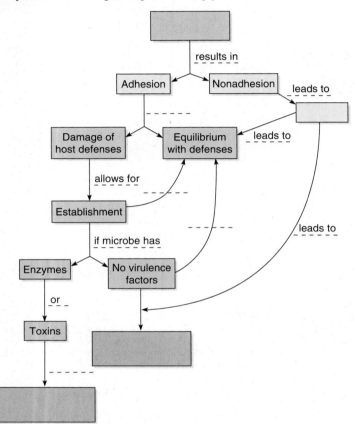

2. Use 6 to 10 bolded words of your choice from the Chapter Summary to create a concept map. Finish it by providing linking words.

Critical Thinking Questions

Critical thinking is the ability to reason and solve problems using facts and concepts. These questions can be approached from a number of angles, and in most cases, they do not have a single correct answer.

1. a. Discuss the relationship between the vaginal residents and the colonization of the newborn.
 b. Can you think of some serious medical consequences of this relationship?
 c. How could normal microbial residents cause some infections to be more severe and other infections to be less severe?

2. If the following patient specimens produced positive cultures when inoculated and grown on appropriate media, indicate whether this result indicates a disease state and why or why not:

Urine	Liver biopsy	Blood
Lung biopsy	Throat	Urine from bladder
Saliva	Feces	Semen
Cerebrospinal fluid		

3. Assume that you have been given the job of developing a colony of germ-free chickens.
 a. What will be the main steps in this process?
 b. What possible experiments can you do with these animals?

4. Complete the table.

	Exotoxins	Endotoxins
Chemical makeup		
General source		
Degree of toxicity		
Effects on cells		
Symptoms in disease		
Examples		

5. a. Use the following statistics to make simple maps and graphs of the following three diseases. Determine which show endemic, sporadic, or epidemic patterns. Explain how you can determine each type.

	Cases	
United States Region	**2008**	**2009**
Pertussis		
East Coast	2235	1673
Midwest	3390	4567
South	3316	5035
West	2229	1929
Total cases	11,170	13,204
Coccidioidomycosis		
East Coast	7	8
Midwest	45	4
South	6	5
West	6939	12,519
Total cases	7042	12,522
Typhoid fever		
East Coast	103	75
Midwest	117	83
South	85	62
West	144	104
Total cases	449	324

 b. Explain what would have to occur for these diseases to have a pandemic distribution.
 c. Look at the summary of notifiable diseases table inside the back cover and find examples of diseases that appear to be showing sporadic and epidemic patterns.

6. Describe each of the following infections using correct technical terminology. (Descriptions may fit more than one category.) Use terms such as *primary, secondary, nosocomial, STD, mixed, latent, toxemia, chronic, zoonotic, asymptomatic, local, systemic, -itis, -emia*.

 Caused by needlestick in dental office

 Pneumocystis pneumonia in AIDS patient

 Bubonic plague from rat flea bite

 Diphtheria

 (List continued on page 423)

Undiagnosed chlamydiosis

Acute necrotizing gingivitis

Syphilis of long duration

Large numbers of gram-negative rods in the blood

A boil on the back of the neck

An inflammation of the meninges

7. Name 10 fomites you contacted today.

8. a. Suggest several reasons why urinary tract, respiratory tract, and surgical infections are the most common nosocomial infections.

 b. Name several measures that health care providers must exercise at all times to prevent or reduce nosocomial infections.

Visual Challenge

1. From the case file for this chapter, explain the processes occurring between *Salmonella* (pink) and the surface of a host cell (yellow) in both views.

2. From **figure 13.20b.** Summarize the trends in TB incidence with regard to age, sex, and ethnicity.

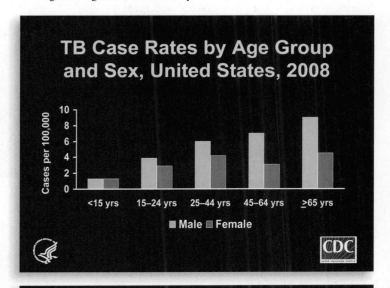

TB Case Rates by Age Group and Sex, United States, 2008

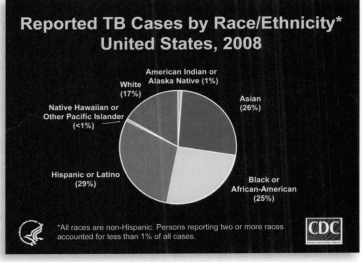

Reported TB Cases by Race/Ethnicity* United States, 2008

14

An Introduction to Host Defenses
and Innate Immunities

Child with fungal osteomyelitis of his ribcage, a sign of immune deficiency

"Systemic aspergillosis is one of the most common infections in children and adults with defective phagocytes."

The conidial head of *Aspergillus fumigatus,* a pathogenic mold that grows well in human tissues and organs

CASE FILE 14 **When Phagocytes Fail**

The 5-year-old boy's medical history included several episodes of bacterial pneumonia starting at 7 months of age that cleared up without complications following antibiotic therapy. But lately, his parents had begun to notice a troubling new set of symptoms and signs. He complained of constant pain in his wrist and jawbone, and the tissues overlying these bones were swollen and inflamed. They took him to the hospital, where he was admitted for lab tests. The results revealed no abnormalities, and the standard tests for infectious agents came out negative. When an X ray detected small spots in the bones of his wrist and jaw, a biopsy was performed that uncovered granulomalike lesions and signs of osteomyelitis. Because of his history of recurrent infections, he was checked for immune function, but his B and T cell activities were normal. Based on the assumption that he had an infection of unknown etiology, he was placed on a broad range of therapy consisting of antibacterial, antifungal, and antiviral medications and went home as soon as his condition improved.

After 8 months, he returned to the hospital in a state of severe distress and was readmitted. This time, he coughed, had difficulty in breathing, and complained of chest and abdominal pain. He was immediately given antibiotics and scanned by computerized tomography (CT), which revealed a mass in the liver that extended into the ribcage. Physicians drew a sample of fluid from the liver and sent it for testing. From the liver sample, technicians isolated a pure culture with greenish-grey, wooly colonies typical of the mold *Aspergillus fumigatus.* Microscopic preparations yielded branched, septate hyphae with brushlike conidial heads typical of this fungus.

It now became clear that the child had a serious systemic fungal infection that was likely responsible for his earlier episode as well. This finding required a different regimen using a powerful intravenous antifungal drug, amphotericin B, backed up with interferon and rifampin. He responded well to the medications and was able to return home.

After considering the nature of these new infections, his doctors began to search for an unknown loss of host defenses that could account for the child's recurring opportunistic infections. They were aware, for instance, that systemic aspergillosis is one of the most common infections in children and adults with defective phagocytes. They examined the boy's white blood cells using the nitroblue tetrazolium test, which monitors the effectiveness of phagocyte activity, and the neutrophil cytochrome assay, which can measure activities associated with the metabolism of oxygen. Both tests indicated a low rate of phagocyte responsiveness, which is highly diagnostic of a genetic disease known as chronic granulomatous disease (CGD).

▶ *What activities account for the child's inflammation and swelling?*

▶ *What is a granuloma, and what factors in CGD cause it?*

To continue the case, go to page 445.

14.1 Overview of Host Defense Mechanisms

Expected Learning Outcomes

1. Summarize the characteristics of basic host defenses.
2. Differentiate between the three lines of defense, describing examples of each.
3. Explain the nature of the different types of innate, nonspecific defenses.

The survival of the host depends upon an elaborate network of defenses that keep harmful microbes and other foreign materials from entering the body and causing damage. Should they penetrate, additional host defenses are immediately summoned to prevent them from becoming established in tissues. Defenses involve barriers, cells, and chemicals, and they range from nonspecific to specific and from inborn or innate to acquired. This chapter introduces the main lines of defense intrinsic to all humans. Topics included in this survey are the anatomical and physiological systems that detect, recognize, and destroy foreign substances and the major activities that account for an individual's nonspecific resistance to infection and disease.

In chapter 13, we explored the human-microbe relationship, with emphasis on the role of microorganisms in disease. In this chapter, we examine the other side of the relationship—that of the host defending itself against microorganisms. As previously stated, in light of the unrelenting contact and colonization of humans, it is something of a miracle that we are not constantly infected and diseased. Our ability to overcome this continual assault on our bodies is the result of a fascinating, highly complex system of defenses.

To protect the body against pathogens, the immune system relies on an overlapping network of **host defenses** that operate on several levels. These include **innate, natural defenses** present at birth that provide nonspecific resistance to infection and the **adaptive immunities** that are specific and must be acquired. These levels of protection are sometimes divided into categories called the *first, second,* and *third lines of defense* (**table 14.1**).

The *first line of defense* includes any barrier that blocks invasion at the portal of entry. This mostly nonspecific line of defense limits access to the internal tissues of the body. However, it is not considered a true immune response because it does not involve recognition of a specific foreign substance and is very general in its actions.

The *second line of defense* is a more internalized system of protective cells and fluids, which includes inflammation and phagocytosis. It acts rapidly at both the local and systemic levels once the first line of defense has been circumvented, and in most reactions, it is also nonspecific.

The highly specific *third line of defense* is adaptive and acquired on an individual basis as each foreign substance is encountered by white blood cells called lymphocytes. Responses directed toward each different microbe produce unique protective substances and memory cells that can come into play if that microbe is encountered again. Details of how this third line of defense provides long-term immunity is discussed in chapter 15, while this chapter focuses on the first and second lines of defense.

The various levels of defense do not operate in a completely separate fashion; most defenses overlap and are even redundant in some of their effects. This cooperative reaction targets invading microbes on several different fronts simultaneously, making their survival unlikely. To present the interactive nature of host defenses, we start with basic concepts that will build a foundation for understanding the more complex reactions of the immune system to come later.

Barriers at the Portal of Entry: An Inborn First Line of Defense

A number of defenses are a normal part of the body's anatomy and physiology. These inborn, nonspecific defenses can be divided into physical, chemical, and genetic barriers that impede the entry of not only microbes but any foreign agent, whether living or not (**figure 14.1**).

Physical or Anatomical Barriers at the Body's Surface

The skin and mucous membranes of the respiratory and digestive tracts have several built-in defenses. The outermost layer (stratum corneum) of the skin is composed of epithelial cells that have become compacted, cemented together, and impregnated with an insoluble protein, keratin. The result is a thick, tough layer that is highly impervious and waterproof. Few pathogens can penetrate this unbroken barrier, especially in regions such as the soles of the feet or the palms of the hands, where the stratum corneum is much thicker than on other parts of the body. Other cutaneous barriers include hair follicles and skin glands. The hair shaft is periodically

TABLE 14.1	General Features of Host Defenses			
Line of Defense	Innate/ Acquired	Specific or Nonspecific	Development of Immunologic Memory	Examples
First	Innate	Nonspecific	No	Physical barriers: skin, tears, coughing, sneezing
				Chemical barriers: low pH, lysozyme, digestive enzymes
				Genetic barriers: resistance inherent in genetic makeup of host (pathogen cannot invade)
Second	Innate	Mostly nonspecific	No	Phagocytosis, inflammation, fever, interferon
Third	Acquired	Specific	Yes	T lymphocytes, B lymphocytes, antibodies

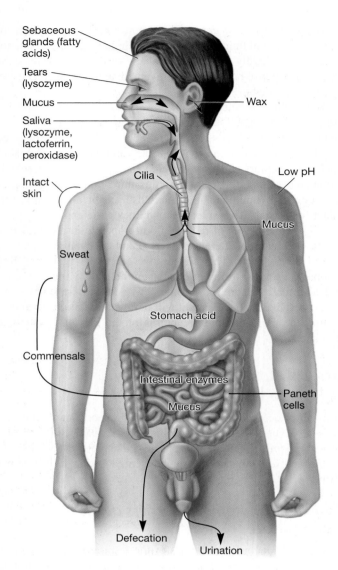

Figure 14.1 **Defenses.** Some nonspecific defense barriers that help prevent entry of microorganisms into the host's tissues.

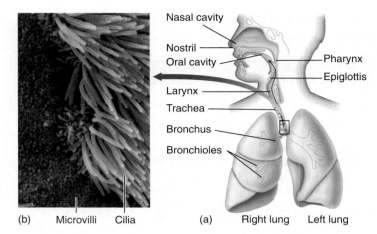

Figure 14.2 **The ciliary defense of the respiratory tree.** **(a)** The epithelial lining of the trachea and bronchi contains a brush border of cilia to entrap and propel particles upward toward the pharynx. **(b)** Detail of trachea (5,000×) features tufts of cilia and microvilli on the surface of cells.

trachea and bronchi), a ciliated epithelium (called the ciliary escalator) conveys foreign particles entrapped in mucus toward the pharynx to be removed **(figure 14.2).** Irritation of the nasal passage reflexively initiates a sneeze, which expels a large volume of air at high velocity. Similarly, the acute sensitivity of the bronchi, trachea, and larynx to foreign matter triggers coughing, which ejects irritants.

The genitourinary tract derives partial protection from the continuous trickle of urine through the ureters and from periodic bladder emptying that flushes the urethra.

The composition of microbiota and its protective effect were discussed in chapter 13. Even though the normal microbial residents do not constitute a physical or anatomical barrier, their presence can block the access of pathogens to epithelial surfaces and can create an unfavorable environment for pathogens by competing for limited nutrients or by altering the local pH. Some bacterial residents of the large intestine secrete bacteriocins, which can inhibit or kill other bacteria.

Nonspecific Chemical Defenses

The skin and mucous membranes offer a variety of chemical defenses. Sebaceous secretions exert an antimicrobial effect, and specialized glands such as the meibomian glands of the eyelids lubricate the conjunctiva with an antimicrobial secretion. Other defenses in tears and saliva include **lysozyme** and **defensins.** Lysozyme is an enzyme that hydrolyzes the peptidoglycan in the cell wall of bacteria. Defensins are peptides produced by various cells and tissues that damage cell membranes and lyse bacteria and fungi. Skin cells produce a defensin called dermicidin which helps eliminate bacteria. The high lactic acid and electrolyte concentrations of sweat and the skin's acidic pH and fatty acid content are also inhibitory to many microbes.

Internally, the hydrochloric acid content of the stomach renders protection against many pathogens that are swallowed. The intestine's digestive juices and bile are also potentially destructive to microbes. Even semen contains an antimicrobial chemical that inhibits bacteria, and the vagina has a protective acidic pH maintained by lactobacilli that are part of the microbiota.

extruded, and the follicle cells are **desquamated,*** or shed. The flushing effect of sweat glands also helps remove microbes.

The mucocutaneous membranes of the digestive, urinary, and respiratory tracts and of the eye are moist and permeable. Despite the normal wear and tear upon these epithelia, damaged cells are rapidly replaced. The mucous coat on the free surface of some membranes impedes the entry and attachment of bacteria. Blinking and tear production (lacrimation) flush the eye's surface with tears and rid it of irritants. The constant flow of saliva helps carry microbes into the harsh conditions of the stomach. Vomiting and defecation also evacuate noxious substances or microorganisms from the body.

The respiratory tract is constantly guarded from infection by elaborate and highly effective adaptations. Nasal hair traps larger particles. Rhinitis, the inflammation of the nasal mucosal surface marked by the copious flow of mucus and fluids, which occurs in allergy and colds, exerts a flushing action. In the respiratory tree (primarily the

** desquamated* (des'-kwuh-mayt) L. *desquamo,* to scale off. The casting off of epidermal scales.

Genetic Resistance to Infection

In certain cases, the genetic makeup of an individual is different enough to ensure protection from infection by some pathogens. One explanation for this phenomenon is that some pathogens have such great specificity for one host species that they are incapable of infecting other species. Another way of putting it is: "Humans can't acquire distemper from cats, and cats can't get mumps from humans." This specificity is particularly true of viruses, which can invade only by attaching to a specific host receptor. But it does not hold true for zoonotic infectious agents that attack a broad spectrum of animals. Genetic differences in susceptibility can also exist within members of one species. Humans carrying a gene or genes for sickle-cell anemia are resistant to malaria. Genetic differences also exist in susceptibility to tuberculosis, leprosy, and certain systemic fungal infections.

The vital contribution of barriers is clearly demonstrated in people who have lost them or never had them. Patients with severe skin damage due to burns are extremely susceptible to infections; those with blockages in the salivary glands, tear ducts, intestine, and urinary tract are also at greater risk for infection. But as important as it is, the first line of defense alone cannot provide adequate protection. Because many pathogens find a way to circumvent the barriers by using their virulence factors (discussed in chapter 13), a whole new set of defenses—inflammation, phagocytosis, specific immune responses—is brought into play.

Check & Assess Section 14.1

- ✔ The multilevel, interconnecting network of host protection against microbial invasion can be organized into three lines of defense.
- ✔ The first line consists of physical and chemical barricades.
- ✔ Examples of physical barriers are the skin, mucous membranes, and respiratory cilia.
- ✔ Examples of chemical barriers are stomach acid, lysozyme, and defensins in skin and saliva.
- ✔ The second line encompasses all the nonspecific defensive cells and chemicals found internally, including those of inflammation and phagocytosis.
- ✔ The third line, the specific immune response, is customized to react to specific characteristics of a microbial invader. This line of defense also creates memory cells that allow the immune system to immediately react if the same pathogen enters the body in the future.

1. Explain the functions of the three lines of defense and indicate which are the most essential to survival.
2. What is the difference between nonspecific host defenses and immune responses?
3. Differentiate innate and acquired immunity. What are the general characteristics of each? Discuss the advantages of having multiple defenses (innate and acquired) against the same microbe.
4. Give four innate immune responses present in the body, and explain how each type functions to combat infection.

14.2 Structure and Function of the Organs of Defense and Immunity

Expected Learning Outcomes

4. Describe the primary functions and organs of the immune system.
5. Describe several features of the recognition system of host defenses.
6. Characterize pattern recognition receptors and pathogen-associated molecular patterns.
7. Describe the microscopic anatomy of body compartments and how they interconnect.
8. Outline the anatomy and functions of the reticuloendothelial system.
9. Outline the origin and composition of the blood, and characterize the types of blood cells and their functions.
10. Explain the behavior of white blood cells with regard to the blood circulation and tissues.
11. Describe the characteristics and functions of the lymphatic system.

Immunology encompasses the study of all features of the body's second and third lines of defense. Although this chapter is concerned primarily with infectious microbial agents, be aware that immunology is central to the study of fields as diverse as cancer and allergy. In chapters 15, 16, and 17, you will see that many of the most powerful applications of immunology have been developed for use in health care, laboratory, or commercial settings.

The primary functions of a healthy functioning immune system can be summarized as

1. surveillance of the body,
2. recognition of self and foreign material such as pathogens, and
3. attack and destruction of foreign invaders (**figure 14.3**).

Because infectious agents could potentially enter through any number of portals, the cells of the immune system constantly move about the body, searching for potential pathogens. This process is carried out primarily by **white blood cells** (**WBCs**) or **leukocytes** which have an innate capacity to recognize and differentiate any foreign material in the body, such as invading bacterial cells. Such foreign matter is often called **nonself.** Normal cells of the body, called **self,** are monitored and recognized and are usually not attacked by the immune system. The ability to evaluate nonself and thus spare self from attack is central to the effectiveness of the immune system. In Chapter 16, we will see how autoimmune diseases can arise when this system goes awry. While foreign cells must be recognized as a potential threat and dealt with appropriately, normal cells must not come under attack by the immune defenses. Exceptions to this would be "abnormal" self such as cancer cells.

Figure 14.3 To search, to recognize, and to eliminate are the functions of the immune system. White blood cells are equipped with a very sensitive sense of "touch." As they sort through the tissues, they "feel" surface molecules that help them determine what is foreign or nonself. When nonself molecules (PAMPs) are detected, some type of reaction is mounted to eliminate the microbes that carry them. Self molecules do not trigger this reaction.* PAMPs are pathogen-associated molecular patterns that many pathogens display.

How Do White Blood Cells Carry Out Recognition and Surveillance?

White blood cells are motile, migratory cells that meander through tissues and body compartments, while evaluating the nature of molecules they encounter. Although WBCs have certain built-in functions such as phagocytosis or inflammation, they all have the capacity to discover and interact with microbes and other foreign substances. On their surface, they carry a variety of surface receptors known as **pattern recognition receptors (PRRs)** that serve as their "feelers" for foreign substances **(figure 14.3)**. One special collection of PRRs has been given the name **toll-like receptors (TLRs)**. These proteins reside in the membranes of phagocytes that are the early responders to invaders. These will be discussed again in section 14.4 on phagocytosis.

Although these PRRs and TLRs do not respond to a specific microbe, they can recognize and interact with molecules on the surface of many pathogens called **pathogen-associated molecular patterns** or **PAMPs.** PAMPs are molecules shared by many microorganisms that act like "red flags" for the white blood cells involved

in innate immunity. The detection of PAMPs on pathogens by PRRs and TLRs provides an early alert of invasion and triggers reactions that are necessary to control and eliminate pathogens. Examples of PAMPs include peptidoglycan, lipoteichoic acid, and lipopolysaccharide from bacterial cell walls, double-stranded RNA found in some viruses, zymosan from fungal cell walls, and bacterial flagellin.

Because these components are shared among many pathogens and not specific to a single cell type, only a small number of TLRs are needed to recognize an astonishing variety of microbes. In that this type of recognition is not specific to a certain microbe, this process is a type of nonspecific, innate immunity that should not be confused with the specific acquired immunity from the third line of defense.

Compartments and Connections of the Immune System

It will soon become clear in this and subsequent chapters that, unlike many body systems, the immune "system" does not exist in a single, well-defined site. Instead, it encompasses a large, complex, and diffuse network of cells and fluids that penetrate into every organ and tissue. It is this very arrangement that promotes the surveillance and recognition processes that help screen the body for harmful substances.

The body is partitioned into several fluid-filled spaces called the intracellular, extracellular, lymphatic, cerebrospinal, and circulatory compartments. Although these compartments are physically separated, they have numerous connections. Their structure and position permit extensive interchange and communication **(figure 14.4)**. Among the body compartments that participate in immune function are

1. the **reticuloendothelial*** system (RES),
2. the spaces surrounding tissue cells that contain **extracellular fluid (ECF),**
3. the **bloodstream,** and
4. the **lymphatic system.**

For effective immune responsiveness, the activities in one fluid compartment must be conveyed to other compartments. Let us see how this occurs by viewing tissue at the microscopic level **(figure 14.4a).** At this level, clusters of tissue cells are in direct contact with the reticuloendothelial system (RES) and the extracellular fluid (ECF), which bathes all cells and is an important site of exchange for nutrients and gases. Other compartments (vessels) that penetrate at this level are blood and lymphatic capillaries. This close association allows cells and chemicals that originate in the RES and ECF to seep or migrate into the blood and lymphatics; any products of a lymphatic reaction can be transmitted directly into the blood through the connection between these two

* *reticuloendothelial* (reh-tik"-yoo-loh-en"-doh-thee'-lee-al) L. *reticulum,* a small net, and *endothelium,* lining of the blood vessel.

Figure 14.4 **Connections between the body compartments. (a)** The meeting of the major fluid compartments at the microscopic level. The close proximity of these compartments at the cellular level allows for continuous exchange of cells, fluids, and molecules (arrows). **(b)** Schematic view of how main fluid compartments form a continuous cyclic system of exchange. Positioning of the compartments favors rapid circulation from one site to another, either through direct (→) or indirect (---→) connections.

systems; and certain cells and chemicals originating in the blood can move through the vessel walls into the extracellular spaces and migrate into the lymphatic system (**figure 14.4b**).

The flow of events among these systems depends on where an infectious agent or foreign substance first intrudes. A typical progression might begin in the extracellular spaces and RES, move to the lymphatic circulation, and ultimately end up in the bloodstream. Regardless of which compartment is first exposed, an immune reaction in any one of them will eventually be communicated to the others at the microscopic level. An obvious benefit of such an integrated system is that no cell of the body is far removed from competent protection, no matter how isolated. Let us take a closer look at each of these compartments separately.

Immune Functions of the Reticuloendothelial System

The tissues of the body are permeated by a network of connective tissue fibers, or a reticulum, which originates in the cellular basal lamina, interconnects nearby cells, and meshes with the massive connective tissue network surrounding all organs. This network, or reticuloendothelial system (**figure 14.5**), is intrinsic to immune function because it provides a passageway within and between tissues and organs. It also is heavily endowed with white blood cells of the **mononuclear phagocyte system.** This system is waiting to attack passing foreign intruders as they arrive in the skin, lungs, liver, lymph nodes, spleen, and bone marrow. The RES serves an additional role in carrying extracellular fluid throughout its network.

Origin, Composition, and Functions of the Blood

The circulatory system consists of the circulatory system proper, which includes the heart, arteries, veins, and capillaries that circulate the blood, and the lymphatic system, which includes lymphatic

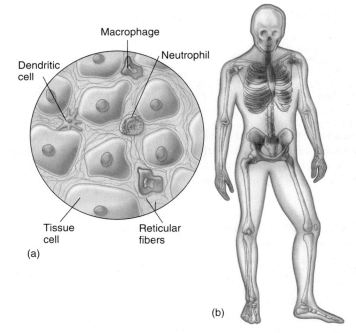

Figure 14.5 **The reticuloendothelial system occurs as a continuous connective tissue framework throughout the body. (a)** This system begins at the microscopic level with a fibrous support network (reticular fibers) enmeshing each cell. This web connects one cell to another within a tissue or organ and provides a niche for phagocytic white blood cells, which can crawl within and between tissues. **(b)** The degrees of shading in the body indicate variations in phagocyte concentration (darker = greater).

vessels and lymphatic organs (lymph nodes) that circulate lymph. As you will see, these two circulations parallel, interconnect with, and complement one another.

Figure 14.6 The macroscopic composition of whole blood. (a) When blood containing anticoagulants is allowed to sit for a period, it stratifies into a clear layer of plasma, a thin layer of off-white material called the buffy coat (which contains the white blood cells), and a layer of red blood cells in the bottom, thicker layer. **(b)** Serum is the clear fluid that separates from clotted blood. Clotting causes all cells to clump in the bottom of the tube in a single mass or clot.

The substance that courses through the arteries, veins, and capillaries is **whole blood,** a liquid connective tissue consisting of **blood cells** (formed elements) suspended in **plasma.** One can visualize these two components with the naked eye when a tube of unclotted blood is allowed to sit or is spun in a centrifuge. The cells' density causes them to settle into an opaque layer at the bottom of the tube, leaving the plasma, a clear, yellowish fluid, on top **(figure 14.6).** **Serum** is essentially the same as plasma, except it is the clear fluid from clotted blood so it lacks the clotting proteins that plasma contains. Serum is often used in immune testing and therapy (chapter 15).

Fundamental Characteristics of Plasma Plasma contains hundreds of different chemicals produced by the liver, white blood cells, endocrine glands, and nervous system and absorbed from the digestive tract. The main component of this fluid is water (92%), and the remainder consists of proteins such as albumin and globulins (including antibodies); other immunochemicals; fibrinogen and other clotting factors; hormones; nutrients; ions; dissolved gases (O_2 and CO_2); and waste products (urea). These substances support the normal physiological functions of nutrition, development, protection, homeostasis, and immunity. We return to the subject of plasma and its function in immune interactions later in this chapter and in chapter 15.

A Survey of Blood Cells The production of blood cells, known as **hemopoiesis,** or **hematopoiesis,*** begins early in embryonic development in the yolk sac (an embryonic membrane). Later, it is taken over by the liver and lymphatic organs, and it is finally assumed entirely and permanently by the red bone marrow. Although much of a newborn's red marrow is devoted to hematopoietic function, the active marrow sites gradually recede, and by the age of 4 years, only the ribs, sternum, pelvic girdle, flat bones of the skull and spinal column, and proximal portions of the humerus and femur are devoted to blood cell production.

The relatively short life of blood cells demands a rapid turnover that is continuous throughout a human life span. The primary precursor of new blood cells is a pool of undifferentiated cells called pluripotential **stem cells**[1] maintained in the marrow. During development, these stem cells proliferate and *differentiate*—meaning that immature or unspecialized cells develop the specialized form and function of mature cells. The primary lines of cells that arise from this process produce red blood cells (RBCs), white blood cells (WBCs, or leukocytes), and platelets (thrombocytes). The white blood cell lines are programmed to develop into several secondary lines of cells during the final process of differentiation **(figure 14.7).** These committed lines of WBCs are largely responsible for immune function.

The **leukocytes*** are traditionally evaluated by their reactions with a hematologic stain that contains a mixture of dyes and can differentiate cells by color and morphology. When this stain is used on blood smears and evaluated using the light microscope, the leukocytes appear either with or without noticeable colored granules in the cytoplasm and, on that basis, are divided into two groups: **granulocytes** and **agranulocytes.** Greater magnification reveals that even the agranulocytes have tiny granules in their cytoplasm, so some hematologists also use the appearance of the nucleus to distinguish them. Granulocytes have a lobed nucleus, and agranulocytes have an unlobed, rounded nucleus. Note both of these characteristics in circulating leukocytes shown in figure 14.7.

Granulocytes The types of granular leukocytes present in the bloodstream are neutrophils, eosinophils, and basophils. All three are known for prominent cytoplasmic granules that stain with some combination of acidic dye (eosin) or basic dye (methylene blue). Although these granules are useful diagnostically, they also function in numerous physiological events. Refer to figure 14.7 to view the cell types described.

Neutrophils* are distinguished from other leukocytes by their conspicuous lobed nuclei and by their fine, pale lavender granules. In cells newly released from the bone marrow, the nuclei are horseshoe-shaped, but as they age, they form multiple lobes (up to five). These cells, also called polymorphonuclear neutrophils (PMNs), make up 55% to 90% of the circulating leukocytes—about 25 billion cells in the circulation at any given moment. The main work of the neutrophils is in phagocytosis. Their high numbers in both the blood and tissues suggest a constant challenge from resident microbiota and environmental sources. Most of the cytoplasmic granules carry digestive enzymes and other chemicals that degrade the phagocytosed materials (see the discussion of phagocytosis in section 14.4). The average neutrophil lives only about 2 days, spending much of this time in the tissues and only about 4 to 10 hours in circulation.

Eosinophils* are readily distinguished in a stain preparation by their larger, orange to red (eosinophilic) granules and bilobed

1. Pluripotential stem cells can develop into several different types of blood cells; unipotential cells have already committed to a specific line of development.

 * *leukocyte* (loo′-koh-syte) Gr. *leukos,* white, and *kytos,* cell. The whiteness of unstained WBCs is best seen in the white layer, or buffy coat, of sedimented blood.
 * *neutrophil* (noo′-troh-fil) L. *neuter,* neither, and Gr. *philos,* to love. The granules are neutral and do not react markedly with either acidic or basic dyes. In clinical reports, they are often called "polys" or PMNs for short.
 * *eosinophil* (ee″-oh-sin′-oh-fil) Gr. *eos,* dawn, rosy, and *philos,* to love. Eosin is a red, acidic dye attracted to the granules.

* *hematopoiesis* (hem″-mat-o-poy-ee′-sis) Gr. *haima,* blood, and *poiesis,* a making.

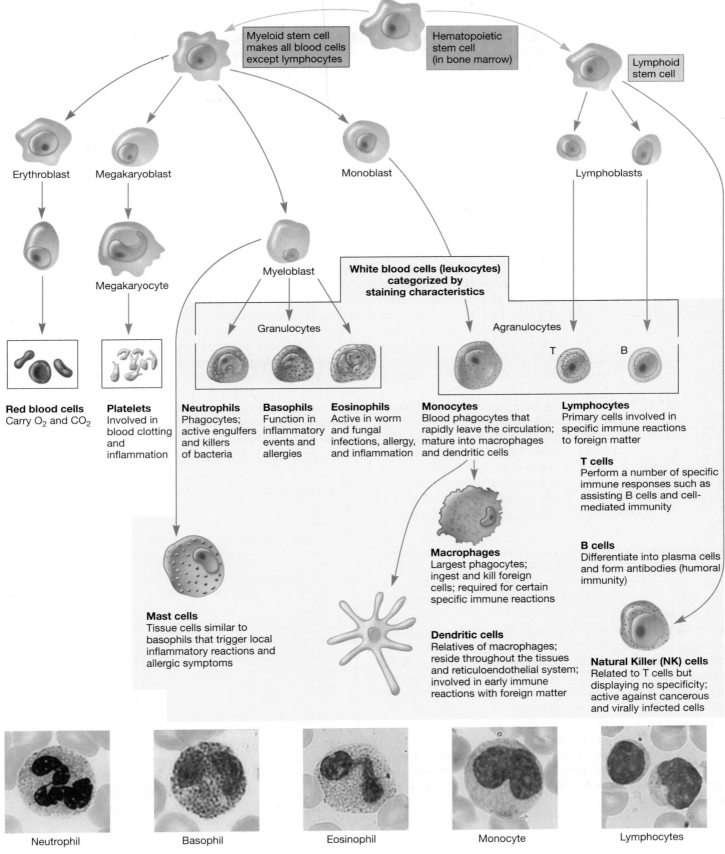

Myeloid stem cell makes all blood cells except lymphocytes

Hematopoietic stem cell (in bone marrow)

Lymphoid stem cell

Erythroblast

Megakaryoblast

Monoblast

Lymphoblasts

Megakaryocyte

Myeloblast

White blood cells (leukocytes) categorized by staining characteristics

Granulocytes

Agranulocytes

T

B

Red blood cells
Carry O_2 and CO_2

Platelets
Involved in blood clotting and inflammation

Neutrophils
Phagocytes; active engulfers and killers of bacteria

Basophils
Function in inflammatory events and allergies

Eosinophils
Active in worm and fungal infections, allergy, and inflammation

Monocytes
Blood phagocytes that rapidly leave the circulation; mature into macrophages and dendritic cells

Lymphocytes
Primary cells involved in specific immune reactions to foreign matter

T cells
Perform a number of specific immune responses such as assisting B cells and cell-mediated immunity

Macrophages
Largest phagocytes; ingest and kill foreign cells; required for certain specific immune reactions

B cells
Differentiate into plasma cells and form antibodies (humoral immunity)

Mast cells
Tissue cells similar to basophils that trigger local inflammatory reactions and allergic symptoms

Dendritic cells
Relatives of macrophages; reside throughout the tissues and reticuloendothelial system; involved in early immune reactions with foreign matter

Natural Killer (NK) cells
Related to T cells but displaying no specificity; active against cancerous and virally infected cells

Neutrophil

Basophil

Eosinophil

Monocyte

Lymphocytes

Figure 14.7 **Simplified diagram of blood cell and platelet development.** The details of several intermediate steps have been omitted. Undifferentiated stem cells in the red marrow give rise to several different cell lines that become increasingly specialized until mature cells are released into circulation. The shaded area indicates mature leukocytes (white blood cells). Section at bottom displays photomicrographs of major circulating leukocytes.

nucleus. They are much more numerous in the bone marrow and the spleen than in the circulation, contributing only 1% to 3% of the total WBC count. The role of the eosinophil in the immune system is not fully defined, although several functions have been suggested. Their granules contain peroxidase, lysozyme, and other digestive enzymes, as well as toxic proteins and inflammatory chemicals.

The main protective action of eosinophils is their ability to attack and destroy large eukaryotic pathogens. They are also involved in inflammation and allergic reactions. Among their most important targets are helminth worms and fungi. They physically bind to these parasites, release their toxic granules, and eliminate them through lysis. Eosinophils are among the earliest cells to accumulate near sites of inflammation and allergic reactions, where they attract other leukocytes and release chemical mediators.

Basophils* are characterized by pale-stained, constricted nuclei and very prominent dark blue to black granules. They are the scarcest type of leukocyte, making up less than 0.5% of the total circulating WBCs in a normal individual. Basophils share some morphological and functional similarities with widely distributed tissue cells called **mast* cells.** Although these two cell types were once regarded as identical, mast cells are nonmotile elements bound to connective tissue around blood vessels, nerves, and epithelia, and basophils are motile elements. Both cell types originate from the same bone marrow stem cell line.

Basophils parallel eosinophils in many of their actions, because they also contain granules with potent chemical mediators. These mediators act on other cells and tissues of the body. For example, they may attract white blood cells toward the site of an infection or cause blood vessels to dilate in response to an injury. Mast cells are first line defenders against the local invasion of pathogens; they recruit other inflammatory cells; and they are directly responsible for the release of histamine and other allergic stimulants during immediate allergies (see chapter 16).

Agranulocytes Agranular leukocytes have globular, nonlobed nuclei and lack prominent cytoplasmic granules when viewed with the light microscope. The two general types are monocytes and lymphocytes. Refer to figure 14.7 for the descriptions that follow.

Although lymphocytes are the cornerstone of the third line of defense, which is the subject of chapter 15, their origin and morphology are described here so their relationship to the other blood components is clear. **Lymphocytes** are the second most common WBC in the blood, comprising 20% to 35% of the total circulating leukocytes. The fact that their overall number throughout the body is among the highest of all cells indicates how important they are to immunity. One estimate suggests that about one-tenth of all adult body cells are lymphocytes, exceeded only by erythrocytes and fibroblasts.

In a stained blood smear, most lymphocytes are small, spherical cells with a uniformly dark, rounded nucleus surrounded by a thin fringe of clear cytoplasm, although in tissues, they can become much larger and can even mimic monocytes in appearance. Lymphocytes exist as two functional types—the **B lymphocytes (B cells,** for short), and the thymus-derived, or **T lymphocytes (T cells,** for short). In humans, B cells mature in special bone marrow sites. T cells mature in the thymus gland. Both populations of cells are transported by the bloodstream and lymph and move about freely between lymphoid organs and connective tissue. We return to this topic in chapter 15.

Lymphocytes are the key cells of the third line of defense and the specific immune response. When stimulated by foreign substances (antigens), lymphocytes are transformed into activated cells that neutralize and destroy those foreign substances. B cells contribute to **humoral immunity,** defined as protective molecules carried in the fluids of the body. Activated B cells form specialized **plasma cells,** which produce **antibodies.*** Antibodies are protein molecules that interlock with foreign cells or molecules and participate in their destruction. Activated T cells engage in a spectrum of immune functions characterized as **cell-mediated immunity** in which T cells modulate immune functions and kill foreign cells. The actions of both classes of lymphocytes account for the specificity and memory typical of immunity. Lymphocytes are so important to the defense of the body that most of chapter 15 is devoted to their reactions.

Monocytes* are generally the largest of all white blood cells and the third most common in the circulation (3% to 7%). As they mature, the nucleus becomes oval- or kidney-shaped—indented on one side, off center, and often contorted with fine wrinkles. The cytoplasm holds many fine vacuoles containing digestive enzymes. Monocytes are discharged by the bone marrow into the bloodstream, where they live as phagocytes for a few days. Later, they leave the circulation to undergo final differentiation into **macrophages.*** Macrophages are among the most versatile and important of cells. In general, they are responsible for

1. many types of specific and nonspecific phagocytic and killing functions (they assume the job of cellular housekeepers, "mopping up the messes" created by infection and inflammation);
2. processing foreign molecules and presenting them to lymphocytes; and
3. secreting biologically active compounds that assist, mediate, attract, and inhibit immune cells and reactions.

We touch upon these functions in several ensuing sections.

Another product of the monocyte cell line is **dendritic cells,** named for their long, thin cell processes. Immature dendritic cells move from the blood to the RES and lymphatic tissues, where they trap pathogens. Ingestion of bacteria and viruses stimulates dendritic cells to migrate to lymph nodes and the spleen to participate in reactions with lymphocytes. This part of adaptive immunity is covered in chapter 15.

Erythrocyte and Platelet Lines These elements stay in the circulatory system proper. Their development is also shown in figure 14.7.

Erythrocytes develop from stem cells in the bone marrow and lose their nucleus just prior to entering the circulation. The resultant red blood cells are simple, biconcave sacs of hemoglobin that transport oxygen and carbon dioxide to and from the tissues. These are the most numerous of circulating blood cells, appearing in stains as small pink circles. Red blood cells do not ordinarily have immune functions, though they can be the target of immune reactions.

* *basophil* (bay'-soh-fil) Gr. *basis,* foundation. The granules attract to basic dyes.
* *mast* From Ger. *mast,* food. Early cytologists thought these cells were filled with food vacuoles.

* *antibody* (an'-tih-bahd"-ee) Gr. *anti,* against, and O. E., *bodig,* body.
* *monocyte* (mon'-oh-syte) From *mono,* one, and *cytos,* cell.
* *macrophage* (mak'-roh-fayj) Gr. *macro,* large, and *phagein,* to eat. They are the "large eaters" of the tissues.

TABLE 14.2 **Characteristics of Leukocytes**

Cell Type	Prevalence in Circulation	Primary Function	Features	Appearance*
Neutrophils	55%–90% of white blood cells	General phagocytosis	Life span of 2 days, with only 4–10 hours spent in the circulation	Multilobed nuclei; small purple granules containing digestive enzymes
Eosinophils	1%–3% of white blood cells	Destruction of large eukaryotic pathogens such as worms and fungi	Found in much higher numbers in the spleen and bone marrow	Bilobed nucleus with large orange granules containing toxic proteins, inflammatory mediators, and digestive enzymes
Basophils	0.5% of white blood cells	Respond to large eukaryotic pathogens and recruit other leukocytes	Cytoplasmic granules contain histamines, prostaglandins, and other chemical mediators of the allergic response.	Pale-staining, constricted nuclei with dark blue to black granules
Monocytes	3%–7% of white blood cells	Phagocytosis, followed by final differentiation into macrophages and dendritic cells	Monocytes also secrete several chemicals that moderate the functions of the immune system.	Largest WBC; nuclei large, ovoid, and often indented—no cytoplasmic granules visible using a light microscope
Lymphocytes	20%–35% of white blood cells	Specific (acquired) immunity	Two types of lymphocytes exist. T cells are responsible for cell-mediated immunity, whereas B cells are responsible for humoral immunity.	Small spherical cells with uniformly staining dark, round nuclei

*See examples in figure 14.7.

Platelets are formed elements in circulating blood that are *not* whole cells. They are formed by the disintegration of a large, multinucleate cell, the megakaryocyte. In stains, platelets are blue-gray with fine red granules and are readily distinguished from cells by their small size. Platelets function primarily in hemostasis (plugging broken blood vessels to stop bleeding) and in releasing chemicals that act in blood clotting and inflammation. Some of the major characteristics of leukocytes are summarized in **table 14.2.**

Components and Functions of the Lymphatic System

The lymphatic part of the circulatory system is a compartmentalized network of vessels, cells, and specialized accessory organs **(figure 14.8).** It begins in the farthest reaches of the tissues as tiny capillaries that transport a special fluid (lymph) through an increasingly larger system of vessels and filters (lymph nodes). It eventually connects to major vessels that drain back into the regular circulatory system. Some major functions of the **lymphatic** or **lymphoid system** relating to immune defenses are

1. to provide an auxiliary route for the return of extracellular fluid to the circulatory system proper;
2. to act as a "drain-off" system for the inflammatory response; and
3. to render surveillance, recognition, and protection against foreign materials through a system of lymphocytes, phagocytes, and antibodies.

Lymphatic Fluid Lymph is a plasmalike liquid carried by the lymphatic circulation. It is formed when certain blood components move out of the blood vessels into the extracellular spaces and diffuse or migrate into the lymphatic capillaries. Because of this, the composition of lymph parallels that of plasma in many ways. It is made up of water, dissolved salts, and 2% to 5% protein (especially antibodies and albumin). Like blood, it also transports numerous white blood cells (especially lymphocytes) and miscellaneous materials such as fats, cellular debris, and infectious agents that have gained access to the tissue spaces. Unlike blood, red blood cells are not normally found in lymph.

Lymphatic Vessels The system of vessels that transports lymph is constructed along the lines of blood vessels **(figure 14.9).** The tiniest vessels, **lymphatic capillaries,** accompany the blood capillaries. The lymphatic capillaries permeate all parts of the body except the central nervous system and certain organs such as bone, placenta, and thymus. Their thin walls have a single layer of epithelial cells similar to blood capillaries, with loose junctions that allow free entry of ECF that collects from the circulatory system (see figure 4.4). Lymphatic vessels are found in particularly high numbers in the hands, feet, and around the areolae of the breast.

Two important differences between the bloodstream and the lymphatic system should be mentioned **(figure 14.10).** First, given that one of the main functions of the lymphatic system is returning lymph to the circulation, lymph flows only in one direction, moving from the extremities toward the heart. Eventually, it will be returned to the bloodstream through the thoracic duct and the right lymphatic duct near the base of the neck (figure 14.8). The second difference concerns how lymph travels through the vessels of the lymphatic system. While blood is transported through the body by means of a dedicated pump (the heart), lymph is moved only through the contraction of skeletal muscles that surround the lymphatic ducts. This dependence on muscle movement helps to explain the swelling of the hands and feet that sometimes occurs during the night (when muscles are inactive) yet goes away soon after waking.

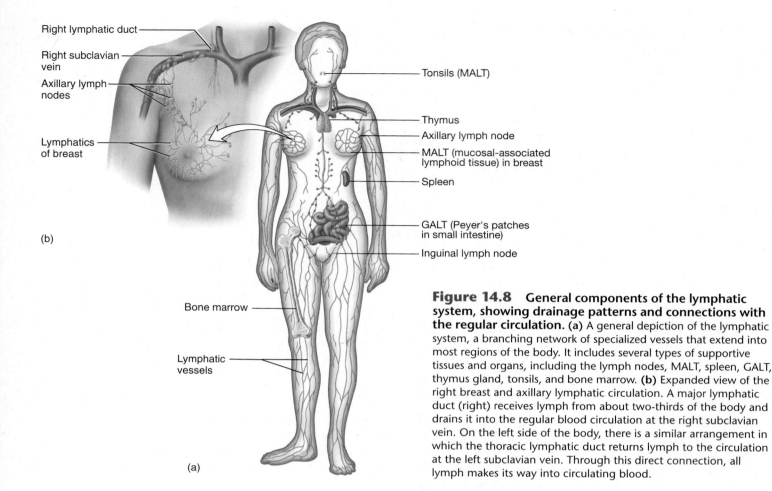

Right lymphatic duct

Right subclavian vein

Axillary lymph nodes

Lymphatics of breast

Tonsils (MALT)

Thymus

Axillary lymph node

MALT (mucosal-associated lymphoid tissue) in breast

Spleen

GALT (Peyer's patches in small intestine)

Inguinal lymph node

Bone marrow

Lymphatic vessels

(b)

(a)

Figure 14.8 **General components of the lymphatic system, showing drainage patterns and connections with the regular circulation.** **(a)** A general depiction of the lymphatic system, a branching network of specialized vessels that extend into most regions of the body. It includes several types of supportive tissues and organs, including the lymph nodes, MALT, spleen, GALT, thymus gland, tonsils, and bone marrow. **(b)** Expanded view of the right breast and axillary lymphatic circulation. A major lymphatic duct (right) receives lymph from about two-thirds of the body and drains it into the regular blood circulation at the right subclavian vein. On the left side of the body, there is a similar arrangement in which the thoracic lymphatic duct returns lymph to the circulation at the left subclavian vein. Through this direct connection, all lymph makes its way into circulating blood.

TAKE NOTE: A DISEASE OF CHRONIC EDEMA

The parasitic infection filariasis presents a dramatic example of what happens when the lymphatic drainage is blocked by an infectious agent (see figure 23.25). Filarial worms from the infection become caught in the lymph nodes and channels of the extremities and plug them up. This prevents lymph from flowing into the accessory ducts and back to the circulation. The buildup of lymph massively distorts limbs and other body parts such as the scrotum.

Lymphoid Organs Lymphatic organs and tissues with immune functions can be classified as primary and secondary, as summarized here:

Primary organs
 Thymus gland
 Bone marrow

Secondary organs and tissues
 Lymph nodes
 Spleen
 MALT—mucosal-associated lymphoid tissue
 SALT—skin-associated lymphoid tissue
 GALT—gut-associated lymphoid tissue (Peyer's patches)

The **primary lymphoid organs** include the thymus and bone marrow, because they are a source of immune cells and a "nursery" for developing white blood cells. They are the sites of origin and maturation of lymphocytes—white blood cells with specific immune function. They subsequently release these cells to populate the secondary lymphatic sites.

Secondary lymphoid organs such as the spleen and lymph nodes are circulatory-based locations where encounters with microbes and immune responses often take place. Associated lymphoid tissues are collections of cells widely dispersed throughout body tissues such as the skin and mucous membranes, ready to react with any locally entering infectious agents.

The Thymus Gland The **thymus** originates in the embryo as two lobes in the pharyngeal region that fuse into a triangular structure. It is located in the thoracic cavity near the tip of the sternum. The size of the thymus is greatest proportionately at birth **(figure 14.11),** and it continues to exhibit high rates of activity and growth until puberty, after which it begins to shrink gradually through adulthood.

Children born without a complete thymus (DiGeorge syndrome, see chapter 16) or who have had their thymus surgically removed are severely immunodeficient and fail to thrive. Adults have developed enough mature T cells that removal of the thymus or reduction in its function has milder effects. Do

(a) The finest level of lymphatic circulation begins with blind capillaries (green) that pick up foreign matter from the surrounding tissues and transport it in lymph away from the extremities via a system of small ducts.

(b) The ducts carry lymph into a circuit of larger ducts that ultimately flow into clusters of filtering organs, the lymph nodes.

(c) A section through a lymph node reveals the afferent ducts draining lymph into sinuses that house several types of white blood cells. Here, foreign material is filtered out and processed by lymphocytes, macrophages, and dendritic cells.

(d) and (e) Lymph continues to trickle from the lymph nodes via efferent ducts into a system of larger drainage vessels, which ultimately connect with large veins near the heart. In this way, cells and products of immunity continually enter the regular circulation.

Figure 14.9 Scheme of circulation in the lymphatic vessels and lymph nodes.

Lymph capillary
Tissue cells
Capillary bed
Venule
Lymph duct
Arteriole
a
b
c
Section of lymph node
Afferent lymphatic ducts
Parafollicle region
Medulla
Cortical follicles
Lymph node
Efferent lymphatic duct
d
Main lymphatic duct
Lymphatic duct
Lymphatic duct
Lymphatic duct
e
Blood circulation
Veins that feed into heart

At cellular level of anatomy
At regional level of anatomy
Major transport vessels

not confuse the thymus with the thyroid gland, which is located in the cervical region near the larynx and has an entirely different function.

Lymph Nodes Lymph nodes are small, encapsulated, bean-shaped organs stationed, usually in clusters, along lymphatic channels and large blood vessels of the thoracic and abdominal cavities (see figure 14.8). Major aggregations of nodes occur in the loose connective tissue of the armpit (axillary nodes), groin (inguinal nodes), and neck (cervical nodes). Both the location and architecture of these nodes clearly specialize them for filtering out materials that have entered the lymph and providing appropriate cells and niches for immune reactions.

Figure 14.10 **Comparison of the circulation patterns of the regular circulatory and lymphatic systems.** The movement of lymphatic flow is in one direction (green) from lymph capillaries to collecting vessels and ducts to large lymphatic trunks to subclavian veins to the heart. The flow of blood on the other hand, is cyclic, with blood continuously flowing through arteries to capillaries to veins to the heart and back around. With this combined system the lymphatics can collect excess tissue fluid and return it to the bloodstream. The two systems can also participate together in surveillance of the tissues for foreign invaders. Note: The lymphatic flow (green) is shown on only one side to keep the comparison uncluttered.

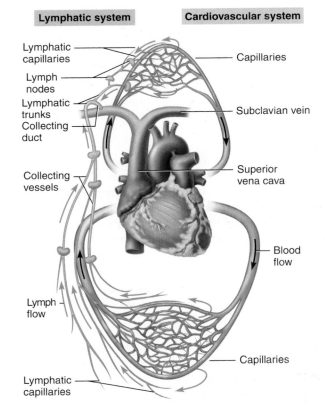

Lymphatic system
Cardiovascular system

Lymphatic capillaries
Capillaries
Lymph nodes
Lymphatic trunks
Subclavian vein
Collecting duct
Collecting vessels
Superior vena cava
Blood flow
Lymph flow
Capillaries
Lymphatic capillaries

Thymus gland

Blood vessels

Cortex

Medulla

Figure 14.11 The thymus gland. Immediately after birth, the thymus is a large organ that nearly fills the region over the midline of the upper thoracic region. In the adult, however, it is proportionately smaller (to compare, see figure 14.8*a*). Section shows the main anatomical regions of the thymus. Immature T cells enter through the cortex and migrate into the medulla as they mature.

Spleen The spleen is a lymphoid organ in the upper portion of the abdominal cavity nestled below the diaphragm and left of the stomach. It is somewhat similar to a lymph node except that it serves as a filter for blood instead of lymph. While the spleen's primary function is to remove worn-out red blood cells from circulation, its most important immunologic function centers on the filtering of pathogens from the blood and their subsequent phagocytosis by resident macrophages. Although adults whose spleens have been surgically removed can live a relatively normal life, asplenic children are severely immunocompromised.

Associated Lymphoid Tissue Embedded throughout systems lined with mucous membranes are discrete bundles of lymphocytes and other white blood cells termed the **mucosal-associated lymphoid tissue** or **MALT.** The positioning of this widespread system provides a local, rapid mechanism for responding to the constant influx of microbes entering via the gastrointestinal, respiratory, urinary, and other portals of entry. The pharynx houses a prominent source of MALT in the form of tonsils. The breasts of pregnant and lactating women also become temporary sites of lymphoid tissues that add protective antibodies to breast milk (see colostrum, Insight 15.3.)

The intestinal tract houses the best-developed collection of lymphoid tissue, called **gut-associated lymphoid tissue,** or **GALT.** Examples of GALT include the appendix, and **Peyer's patches,** compact aggregations of lymphocytes in the ileum of the small intestine. GALT provides immune functions against intestinal

pathogens and is a significant source of some types of antibodies. Other, less well-organized collections of secondary lymphoid tissue include *skin-associated lymphoid tissue* and *bronchial-associated lymphoid tissue.*

Check & Assess Section 14.2

- ✔ The science of immunology is concerned with reactions of the second and third lines of defense.
- ✔ White blood cells (leukocytes) are charged with surveillance recognition, and destruction or foreign invaders (nonself).
- ✔ Receptors on WBCs detect markers on microbes (PAMPs) that indicate infection.
- ✔ The immune system is a complex collection of fluids and cells that penetrate every organ, tissue space, fluid compartment, and vascular network of the body. The four major subdivisions of this system are the RES, the ECF, the blood vascular system, and the lymphatic system.
- ✔ The RES, or reticuloendothelial system, is a network of connective tissue fibers inhabited by macrophages ready to attack and ingest microbes that have managed to bypass the first line of defense.
- ✔ The ECF, or extracellular fluid, compartment surrounds all tissue cells and is penetrated by both blood and lymph vessels, which bring all components of the second and third lines of defense to attack infectious microbes.
- ✔ The blood contains both specific and nonspecific defenses. Nonspecific cellular defenses include the granulocytes, macrophages, and dendritic cells.
- ✔ The two components of the specific immune response are the T lymphocytes, which provide specific cell-mediated immunity, and the B lymphocytes, which produce specific antibody or humoral immunity.
- ✔ The lymphatic system (1) returns tissue fluid to general circulation; (2) carries away excess fluid in inflamed tissues; and (3) concentrates and processes foreign invaders and initiates the specific immune response. Important sites of lymphoid tissues are lymph nodes, spleen, thymus, tonsils, MALT, GALT, and bone marrow.

5. Describe the main elements of the process through which the immune system distinguishes self from foreign cells and particles.
6. How is surveillance of the tissues carried out and what is responsible for it?
7. What does the term *nonself* mean in reference to the immune system?
8. What are the main components of the reticuloendothelial system, and why is it also called the mononuclear phagocyte system?
9. How does it communicate with tissues and the vascular system?
10. Review the anatomical locations of hematopoiesis at various stages of development.
11. What are stem cells?
12. Differentiate between granulocytes and agranulocytes.
13. Describe the main cell types in granulocytes and agranulocytes, their functions, and their incidence in the circulation.

14. Give the principal function of the lymphocytes, and differentiate between the two lymphocyte types and between humoral and cell-mediated immunity.
15. What is lymph, and how is it formed?
16. Differentiate between primary and secondary lymphoid organs.
17. What are the functions of the lymphatic system?
18. Explain what is meant by MALT and GALT, and describe their functions.

14.3 Second Line Defenses: Inflammation

Expected Learning Outcomes

12. Review the basic components of inflammation.
13. Describe the main events in the inflammatory reaction, and explain what is occurring in each.
14. Summarize the functions of cytokines, chemokines, and other inflammatory mediators.
15. Describe the mechanism behind fever, and explain its beneficial and harmful effects.

Now that we have introduced the principal anatomical and physiological framework of the immune system, we address some mechanisms that play important roles in host defenses: (1) inflammation, (2) phagocytosis, (3) interferon, and (4) complement. Because of the generalized nature of these defenses, they are primarily nonspecific in their effects, but they also support and interact with the specific immune responses described in chapter 15.

The Inflammatory Response: A Complex Concert of Reactions to Injury

At its most general level, **inflammation** is a reaction to any traumatic event in the tissues that restores homeostasis: It is a normal and necessary process that helps to clear away cellular debris. Most of us manifest inflammation in some way every day. It appears in the flare of a cut, the blistering of a burn, the painful lesion of an infection, and the symptoms of allergy, but sometimes it goes unnoticed.

It is readily identifiable by a classic series of signs and symptoms characterized succinctly by four Latin terms: *rubor, calor, tumor,* and *dolor.* Rubor (redness) is caused by increased circulation and vasodilation in the injured tissues; calor (warmth) is the heat given off by the increased flow of blood; tumor (swelling) is caused by increased extracelluar fluid accumulating in the tissues; and dolor (pain) is caused by the stimulation of nerve endings from the pressure of swelling or by chemical mediators **(figure 14.12).** A fifth symptom, *loss of function,* is sometimes added for a complete picture of effects that can accompany inflammation. Although these manifestations are often unpleasant, they serve an important warning

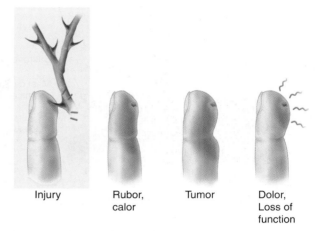

Figure 14.12 The response to injury. This classic series summarizes the reactions of the tissues to an assault. Each of the events is an indicator of one of the mechanisms of inflammation described in this chapter.

that injury has taken place and set in motion responses that can save the body from further injury.

Factors that can elicit an inflammatory response include trauma from infection (the primary emphasis here), tissue injury or death, and specific immune reactions. Although the details of inflammation are very complex, its chief functions can be summarized as follows:

1. to mobilize and attract immune components to the site of the injury,
2. to set in motion mechanisms to repair tissue damage and localize and clear away harmful substances, and
3. to destroy and block microbes from further invasion **(figure 14.13).**

The inflammatory response is a powerful defensive reaction, a means for the body to maintain stability and restore itself after an injury. But when it is chronic, it has the potential to actually *cause* tissue injury, destruction, and disease **(Insight 14.1).**

The Stages of Inflammation

The process leading to inflammation is a dynamic sequence of events that can be acute—lasting from a few minutes or hours—to chronic—lasting for days, weeks, or years. Once the initial injury has occurred, a chain reaction takes place at the site of damaged tissue, summoning beneficial cells and fluids into the injured area. As an example, we look at an injury at the microscopic level and observe the flow of major events (figure 14.13).

Vascular Changes: Early Inflammatory Events

Following an injury, some of the earliest changes occur in the vasculature (arterioles, capillaries, venules) in the vicinity of the damaged tissue. These changes are controlled by nervous stimulation, **chemical mediators,** and **cytokines*** released by blood cells,

* *cytokine* (sy'-toh-kyne) Gr. *cytos,* cell, and *kemein,* to move. A protein or polypeptide produced by WBCs that regulates host defenses.

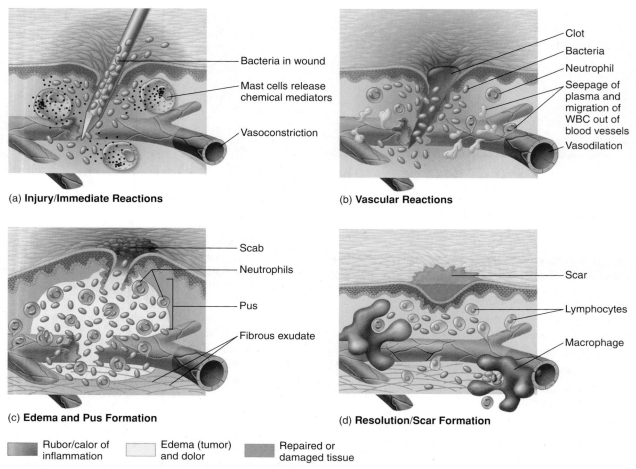

(a) **Injury/Immediate Reactions**
- Bacteria in wound
- Mast cells release chemical mediators
- Vasoconstriction

(b) **Vascular Reactions**
- Clot
- Bacteria
- Neutrophil
- Seepage of plasma and migration of WBC out of blood vessels
- Vasodilation

(c) **Edema and Pus Formation**
- Scab
- Neutrophils
- Pus
- Fibrous exudate

(d) **Resolution/Scar Formation**
- Scar
- Lymphocytes
- Macrophage

Rubor/calor of inflammation

Edema (tumor) and dolor

Repaired or damaged tissue

Figure 14.13 The major events in inflammation. (a) Injury → Narrowing of the blood vessels (vasoconstriction) lasting until clot formation **(b)** → Release of chemical mediators into area. **(b)** Increased diameter of blood vessels (vasodilation) → Increased blood flow → Increased vascular permeability → Leakage of fluid (plasma) from blood vessels into tissues (exudate formation). **(c)** Edema/swelling → Infiltration of site by neutrophils and accumulation of pus. **(d)** Macrophages and lymphocytes → Repair, either by complete resolution and return of tissue to normal state or by formation of scar tissue and possible loss of function.

tissue cells, and platelets in the injured area. Some mediators are *vasoactive*; that is, they affect the endothelial cells and smooth muscle cells of blood vessels. Others are **chemotactic factors,** also called **chemokines,** that affect white blood cells. Inflammatory mediators cause fever, stimulate lymphocytes, prevent infection spread, and cause allergic symptoms (**Insight 14.2** and **figure 14.14**). Although the constriction of arterioles is stimulated first, it lasts for only a few seconds or minutes. Once a clot has formed to prevent blood loss, it is followed in quick succession by the opposite reaction, **vasodilation.** The overall effect of vasodilation is to increase the flow of blood into the area, which facilitates the influx of immune components and also causes redness and warmth (rubor and calor).

Edema: Leakage of Vascular Fluid into Tissues

Postcapillary venules, small veins that drain blood from the capillary beds, are important in several aspects of inflammation. When they constrict, this causes a slowing of circulation and a pooling of extracellular fluid at the site. Some vasoactive substances cause the endothelial cells in the walls of these venules to separate and form gaps through which blood-borne components exude into the extracellular spaces. Capillaries also become leaky under the influence

of cytokines. The fluid part that escapes into the tissues is called the *exudate.* Accumulation of this fluid gives rise to local swelling and hardness known as **edema.** The edematous exudate contains varying amounts of plasma proteins, such as globulins, albumin, the clotting protein fibrinogen, blood cells, and cellular debris. Depending upon its content, the exudate may be clear (called *serous*), or it may contain red blood cells or **pus.** Pus is composed mainly of white blood cells, microbes, and the debris generated by phagocytosis. In some types of edema, the fibrinogen is converted to fibrin threads that enmesh the injury site (see figure 14.13*c*). Within an hour, multitudes of neutrophils responding chemotactically to special signaling molecules converge on the injured site.

Unique Dynamic Characteristics of White Blood Cells In order for WBCs to leave the blood vessels and enter the tissues, they adhere to the inner walls of the smaller blood vessels such as capillaries and venules. From this position, they are poised to migrate out of the blood into the tissue spaces by a process called **diapedesis.***

* *diapedesis* (dye"-ah-puh-dee'-sis) Gr. *dia*, through, and *pedan*, to leap.

INSIGHT 14.1

When Inflammation Gets Out of Hand

Not every aspect of inflammation is protective or results in the proficient resolution of tissue damage. As one looks over a list of diseases, it is rather striking how many of them are due in part or even completely to an overreactive or dysfunctional inflammatory response.

Some -*itis* reactions mentioned in chapter 13 are a case in point. Inflammatory exudates that build up in the brain in African trypanosomiasis, cryptococcosis, and other brain infections can be so injurious to the nervous system that impairment is permanent. Frequently, an inflammatory reaction leads to an abscess, a swollen mass of neutrophils and dead, liquefied tissue that can harbor live pathogens in the center. Abscesses are a prominent feature of staphylococcal, amebic, and enteric infections.

Other pathologic manifestations of chronic diseases—for example, the tubercles of tuberculosis, the lesions of late syphilis, the disfiguring nodules of leprosy, and the cutaneous ulcers of leishmaniasis—are due to an aberrant tissue response called *granuloma formation*. Granulomas develop not only in response to microbes but also in response to inanimate foreign bodies that are difficult to break down. This condition is initiated when neutrophils and macrophages ineffectively and incompletely phagocytose the pathogens or

materials involved in an inflammatory reaction. This failure to manage the debris of inflammation is sometimes referred to as "frustrated phagocytosis." Macrophages respond by storing ingested materials in vacuoles and becoming inactive. Over a given time period, macrophages fuse into giant, inactive multinucleate cells called foreign body giant cells. These sites are further infiltrated with lymphocytes. The resultant collections make the tissue appear granular—hence, the name. A granuloma can exist in the tissue for months, years, or even a lifetime. This chapter's case file provides a striking example of massive proliferation of failed phagocytes.

Medical science is rapidly searching for applications of the new information about inflammatory mediators. One highly promising area appears to be the use of chemokine inhibitors that could reduce chemotaxis and the massive, destructive influx of leukocytes. Such therapy could ultimately be used for certain cancers, hardening of arteries, and Alzheimer disease.

How could inflammatory mediators stimulate tissue injury and disease? *Answer available at http://www.mhhe.com/talaro8*

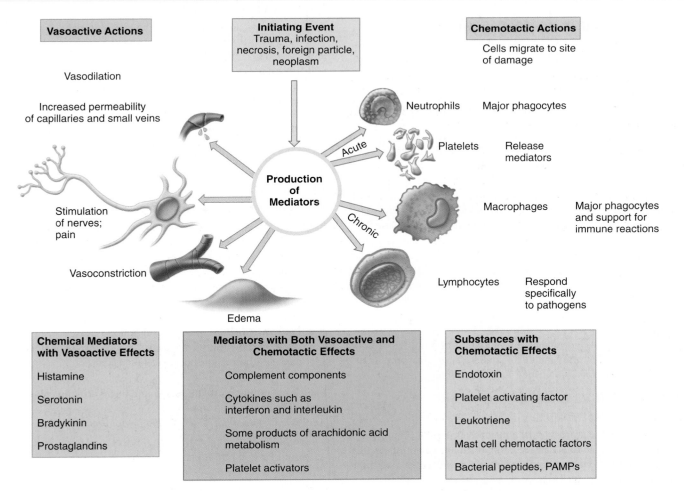

Figure 14.14 Major chemical mediators of the inflammatory response and their effects. See Insight 14.2 for more information on the effects of mediators.

INSIGHT 14.2

The Dynamics of Inflammatory Mediators

Just as the nervous system is coordinated by a complex communications network, so too is the immune system. Hundreds of small, active molecules are constantly being secreted to regulate, stimulate, suppress, and otherwise control the many aspects of cell development, inflammation, and immunity. These substances are the products of several types of cells, including monocytes, macrophages, lymphocytes, fibroblasts, mast cells, platelets, and endothelial cells of blood vessels. Their effects may be local or systemic, short term or long lasting, nonspecific or specific, protective or pathologic.

The field of cytokines has become so increasingly complex that we can include here only an overview of the major groups of important cytokines and other mediators. The major functional types can be categorized into

1. cytokines that mediate nonspecific immune reactions such as inflammation and phagocytosis,
2. cytokines that regulate the growth and activation of lymphocytes,
3. cytokines that activate immune reactions during inflammation,
4. hematopoiesis factors for white blood cells,
5. vasoactive mediators, and
6. miscellaneous inflammatory mediators.

Nonspecific Mediators of Inflammation and Immunity

- *Tumor necrosis factor* (*TNF*), a substance from macrophages, lymphocytes, and other cells. It increases chemotaxis and phagocytosis and stimulates other cells to secrete inflammatory cytokines. It also serves as an endogenous pyrogen, increases blood coagulation, suppresses bone marrow, and causes wasting of the body (cachexia).
- *Interferon alpha and interferon beta,* produced by leukocytes and fibroblasts, inhibit virus replication and cell division and increase the action of certain lymphocytes that kill other cells.
- *Interleukin* is a term that refers to a group of small peptides originally isolated from leukocytes. There are currently more than 20 known interleukins. We now know that other cells besides leukocytes can synthesize them and that they have a variety of biological activities.
 - *Interleukin-1* (*IL-1*), a product of macrophages and epithelial cells. It has many of the same biological activities as TNF, such as inducing fever and activating certain white blood cells.
 - *Interleukin-6,* secreted by macrophages, lymphocytes, and fibroblasts. Its primary effects are to stimulate the growth of B cells and to increase the synthesis of liver proteins.
- *Various chemokines.* By definition, chemokines are cytokines that stimulate the movement and migration of white blood cells (chemotactic factors). Included among these are complement C5A, interleukin-8, and platelet-activating factor.

Cytokines That Regulate Lymphocyte Growth and Activation

- *Interleukin-2,* the primary growth factor from T cells. Interestingly, it acts on the same cells that secrete it. It stimulates mitosis and secretion of other cytokines. In B cells, it is a growth factor and stimulus for antibody synthesis.

- *Interleukin-4,* a stimulus for the development of B cells and production of allergy antibodies. It inhibits macrophage actions.
- *Granulocyte colony-stimulating factor* (*G-CSF*), produced by T cells, macrophages, and neutrophils. It stimulates the activation and differentiation of neutrophils.
- *Macrophage colony-stimulating factor* (*M-CSF*), produced by a variety of cells. M-CSF promotes the growth and development of macrophages from undifferentiated precursor cells.

Cytokines That Activate Specific Immune Reactions

- *Interferon gamma,* a T-cell-derived mediator whose primary function is to activate macrophages. It also promotes the differentiation of T and B cells, activates neutrophils, and stimulates diapedesis.
- *Interleukin-5* activates eosinophils and B cells; *interleukin-10* inhibits macrophages and B cells; and *interleukin-12* activates T cells and killer cells.

Vasoactive Mediators

- *Histamine,* a vasoactive mediator produced by mast cells and basophils. It causes vasodilation, increased vascular permeability, and mucus production. It functions primarily in inflammation and allergy.
- *Serotonin,* a mediator produced by platelets and intestinal cells. It causes smooth muscle contraction, inhibits gastric secretion, and acts as a neurotransmitter.
- *Bradykinin,* a vasoactive amine from the blood or tissues. It stimulates smooth muscle contraction and increases vascular permeability, mucus production, and pain. It is particularly active in allergic reactions.

Miscellaneous Inflammatory Mediators

- *Prostaglandins,* produced by most body cells; complex chemical mediators that can have opposing effects (for example, dilation or constriction of blood vessels) and are powerful stimulants of inflammation and pain.
- *Leukotrienes* stimulate the contraction of smooth muscle and enhance vascular permeability. They are implicated in the more severe manifestations of immediate allergies (constriction of airways).
- *Platelet-activating factor,* a substance released from basophils, causes the aggregation of platelets and the release of other chemical mediators during immediate allergic reactions.

Which inflammatory mediators would be involved in rubor? In tumor? In dolor? In calor?
Answer available at http://www.mhhe.com/talaro8

Diapedesis, also known as transmigration, is aided by several related characteristics of WBCs. For example, they are actively motile and readily change shape. Their migration is assisted by the nature of the endothelial cells lining venules. Venules contain complex adhesive receptors that develop increased "stickiness" under the influence of inflammatory mediators. This causes WBCs to adhere or marginate at the endothelial cells. From this location, they can readily crawl into the extracellular spaces (**figure 14.15**).

Another factor in the migratory habits of these WBCs is **chemotaxis,*** defined as the tendency of cells to migrate in response to a specific chemical stimulus released at a site of injury or infection. Through this means, cells swarm from many compartments to the site of infection and remain there to perform general functions such as phagocytosis, repair, and specific immune reactions. These basic properties are absolutely essential for the sort of intercommunication and deployment of cells required for most immune reactions (figure 14.15).

The Benefits of Edema and Chemotaxis Both the formation of edematous exudate and the infiltration of neutrophils are physiologically beneficial activities. The influx of fluid dilutes toxic substances, and the fibrin clot can effectively trap microbes and prevent their further spread. The neutrophils that aggregate in the inflamed site are immediately involved in phagocytosing and destroying microbes, dead tissues, and particulate matter (by mechanisms discussed in section 14.4 on phagocytosis). In some types of

inflammation, accumulated phagocytes contribute to pus. Certain bacteria (streptococci, staphylococci, gonococci, and meningococci) are especially powerful attractants for neutrophils and are thus termed **pyogenic,** or pus-forming, bacteria.

Late Reactions of Inflammation Sometimes a mild inflammation can be resolved by edema and phagocytosis. Inflammatory reactions that last more than a few days attract a collection of monocytes, lymphocytes, and macrophages to the reaction site. Clearance of pus, cellular debris, dead neutrophils, and damaged tissue is performed by macrophages, the only cells that can engulf and dispose of such large masses. At the same time, specific immune reactions of acquired immunity are brought into play. B lymphocytes react with foreign molecules and cells by producing specific antimicrobial proteins (antibodies), and T lymphocytes kill intruders directly. Late in the process, the tissue is completely repaired, if possible, or replaced by connective tissue in the form of a scar (see figure 14.13*d*). If the inflammation cannot be relieved or resolved in this way, it can become chronic and create a long-term pathologic condition.

Fever: An Adjunct to Inflammation

An important systemic component of inflammation is fever, defined as an abnormally elevated body temperature. Although fever is a nearly universal symptom of infection, it is also associated with certain allergies, cancers, and other organic illnesses. Fevers whose causes are unknown are called fevers of unknown origin, or FUO.

The body temperature is normally maintained by a control center in the hypothalamus region of the brain. This thermostat regulates the body's heat production and heat loss and sets the core temperature at around 37°C (98.6°F), with slight fluctuations (1°F) during a daily cycle. Fever is initiated when a circulating substance called **pyrogen*** resets the hypothalamic thermostat to a higher setting. This change signals the musculature to increase heat production and peripheral arterioles to decrease heat loss through vasoconstriction (**Insight 14.3**). Fevers range in severity from low-grade (37.7°C to 38.3°C, or 100°F to 101°F) to moderate (38.8°C to 39.4°C, or 102°F to 103°F) to high (40.0°C to 41.1°C, or 104°F to 106°F). Pyrogens are described as *exogenous* (coming from outside the body) or *endogenous* (originating internally). Exogenous pyrogens are products of infectious agents such as viruses, bacteria, protozoans, and fungi. One well-characterized exogenous pyrogen is endotoxin, the lipopolysaccharide found in the cell walls of gram-negative bacteria. Blood, blood products, vaccines, or injectable solutions can also contain exogenous pyrogens. Endogenous pyrogens are liberated by monocytes, neutrophils, and macrophages during the process of phagocytosis and appear to be a natural part of the immune response. Two potent pyrogens released by macrophages are interleukin-1 (IL-1) and tumor necrosis factor (TNF).

*** *chemotaxis* (kee-moh-tak′-sis) NL. *chemo,* chemical, and *taxis,* arrangement.

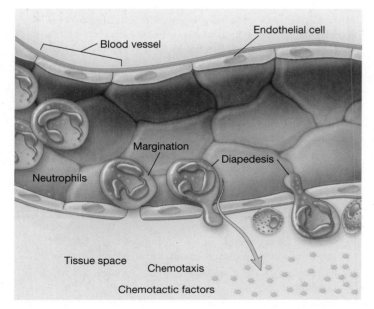

Figure 14.15 Diapedesis and chemotaxis of leukocytes.
Cross section of a venule depicts white blood cells squeezing themselves between spaces in the vessel wall through diapedesis. This process also indicates how the pool of leukocytes adheres to the endothelial wall. From this site, they are poised to migrate out of the vessel into the tissue space and respond to chemotactic factors released during various inflammatory reactions.

Image labels: Blood vessel; Endothelial cell; Neutrophils; Margination; Diapedesis; Tissue space; Chemotaxis; Chemotactic factors

Benefits of Fever The association of fever with infection strongly suggests that it serves a beneficial role, a view still being

*** *pyrogen* (py′-roh-jen) Gr. *pyr,* fire, and *gennan,* produce. Same origin as *funeral pyre* and *pyromaniac.*

INSIGHT 14.3

Some Facts about Fever

Fever is such a prevalent reaction that it is a prominent symptom of hundreds of diseases. For thousands of years, people believed fever was part of an innate protective response. Hippocrates offered the idea that it was the body's attempt to burn off a noxious agent. Sir Thomas Sydenham wrote over 400 years ago: "Why, fever itself is Nature's instrument!" So widely held was the view that fever could be therapeutic that pyretotherapy (treating disease by inducing an intermittent fever) was once used to treat syphilis, gonorrhea, leishmaniasis (a protozoan infection), and cancer. This attitude fell out of favor when drugs for relieving fever (aspirin) first came into use in the early 1900s, and an adverse view of fever began to dominate.

Changing Views of Fever

In recent times, the medical community has returned to the original concept of fever as more healthful than harmful. Experiments with vertebrates indicate that fever is a universal reaction, even in cold-blooded animals such as lizards and fish. A study with febrile (feverish) mice and frogs indicated that fever increases the rate of antibody synthesis. Work with tissue cultures showed that increased temperatures stimulate the activities of T cells and increase the effectiveness of interferon.

Artificially infected rabbits and pigs allowed to remain febrile survive at a higher rate than those given suppressant drugs. Fever appears to enhance phagocytosis of staphylococci by neutrophils in guinea pigs and humans. In recent years, malaria-induced fevers have been experimentally studied as a way to treat HIV infection.

Hot and Cold: Why Do Chills Accompany Fever?

Fever almost never occurs as a single response; it is usually accompanied by chills. What causes this oddity—that a person flushed with fever periodically feels cold and trembles uncontrollably? The explanation lies in the natural physiological interaction between the thermostat in the hypothalamus and the temperature of the blood. For example, if the thermostat has been set (by pyrogen) at 102°F but the blood temperature is 99°F, the muscles are stimulated to contract involuntarily (shivering) as a means of producing heat. In addition, the vessels in the skin constrict, creating a sensation of cold, and the piloerector muscles in the skin cause goose bumps to form.

Explain the physiological mechanism for feeling cold during a fever.
Answer available at http://www.mhhe.com/talaro8

debated but gaining acceptance. Aside from its practical and medical importance as a sign of a physiological disruption, increased body temperature has additional benefits:

- Fever inhibits multiplication of temperature-sensitive microorganisms such as the poliovirus, cold viruses, herpes zoster virus, systemic and subcutaneous fungal pathogens, *Mycobacterium* species, and the syphilis spirochete.
- Fever impedes the nutrition of bacteria by reducing the availability of iron. It has been demonstrated that during fever, the macrophages stop releasing their iron stores, which could retard several enzymatic reactions needed for bacterial growth.
- Fever increases metabolism and stimulates immune reactions and naturally protective physiological processes. It speeds up hematopoiesis, phagocytosis, and specific immune reactions.

Treatment of Fever With this revised perspective on fever, whether to suppress it or not can be a difficult decision. Some advocates feel that a slight to moderate fever in an otherwise healthy person should be allowed to run its course in light of its potential benefits and minimal side effects. All medical experts do agree that high and prolonged fevers or fevers in children or patients with cardiovascular disease, seizures, and respiratory ailments are risky and must be treated immediately with fever-suppressant drugs. The classic therapy for fever is an antipyretic drug such as aspirin or acetaminophen (Tylenol) that lowers the setting of the hypothalamic center and restores normal temperature. Any physical technique that increases heat loss (tepid baths, for example) can also help reduce the core temperature.

Check & Assess Section 14.3

✔ Communication between cells of the immune system is facilitated by the release of chemical messengers such as cytokines.

✔ These chemicals stimulate a variety of beneficial responses. They drive the inflammatory response, increase blood flow and the migration of white blood cells, initiate fever, or destroy infected cells.

✔ The inflammatory response is a complex reaction to infection that brings in protective defenses to fight foreign agents and limit further damage to the body.

✔ The four symptoms of inflammation are rubor (redness), calor (heat), tumor (edema), and dolor (pain). Loss of function often accompanies these.

✔ Fever is another component of nonspecific immunity. It is caused by both endogenous and exogenous pyrogens. Fever increases the rapidity of the host immune responses and reduces the viability of many microbial invaders.

19. Describe the major events in the inflammatory response, including the stimuli, physiological reactions, and signs and symptoms.
20. Of rubor, calor, dolor, and tumor, which are signs and which are symptoms?
21. What actions of the inflammatory and immune defenses account for swollen lymph nodes and leukocytosis?
22. What is pus, and what does it indicate?
23. In what ways can edema be beneficial and in what ways is it harmful?
24. Describe the physiological events leading to fever.

25. In what ways is fever naturally beneficial?
26. Explain the processes of diapedesis and chemotaxis, and show how they interrelate.
27. Briefly account for the origins and actions of the major types of inflammatory mediators and cytokines.

14.4 Second Line Defenses: Phagocytosis, Interferon, and Complement

Expected Learning Outcomes

16. Define *phagocytosis* and identify which cells serve this function.
17. Indicate the major stages of phagocytosis, and describe what is happening at each stage.
18. Explain how phagocytes kill pathogens.
19. Describe the origin of interferon, and explain its role in controlling viruses and other immune functions.
20. Characterize the complement system, its origins, and basic functions.

Phagocytosis: Partner to Inflammation and Immunity

By any standard, a phagocyte represents an impressive piece of living machinery, meandering through the tissues to seek, capture, and destroy a target. The general activities of phagocytes are

1. to survey the tissue compartments and discover microbes, particulate matter (dust, carbon particles, antigen-antibody complexes), and injured or dead cells;
2. to ingest and eliminate these materials; and
3. to extract immunogenic information (antigens) from foreign matter.

Major Categories of Phagocytes

It is generally accepted that all cells have some capacity to engulf materials, but professional phagocytes do it for a living. The three main types of phagocytes are neutrophils, monocytes, and macrophages.

Neutrophils and Eosinophils As previously stated, neutrophils are general-purpose phagocytes that react early in the inflammatory response to bacteria and other foreign materials and to damaged tissue. A common sign of bacterial infection is a high neutrophil count in the blood (neutrophilia), and neutrophils are also a primary component of pus. Eosinophils are attracted to sites of parasitic infections and antigen-antibody reactions, though they play only a minor phagocytic role.

TAKE NOTE: NEUTROPHILS GIVE THEIR ALL

Neutrophils are truly one of the "workhorses" of innate immunity. In addition to their phagocytic functions, a recent discovery revealed that they also have a distinct system of capturing pathogens called *neutrophil extracellular traps* or NETS. Neutrophils are programmed to die after they have gone through their regular engulfment and killing of bacteria and other pathogens. But they undergo one last aggressive act while they are dying, which is to throw out a fibrous matrix that consists of their lysed DNA, enzymes, histones, and other cell contents that works outside the cell even after it is dead. This NET works at several levels to trap bacteria and fungi, degrade their virulence factors, kill them with microbicidal chemicals, and, to top it off, immobilize microbes and prevent them from spreading.

Macrophages: Large-Tissue Phagocytes After emigrating out of the bloodstream into the tissues, monocytes are transformed by various inflammatory mediators into macrophages. This process is marked by an increase in size and by enhanced development of lysosomes and other organelles (**figure 14.16**). At one time, macrophages were classified as either fixed (adherent to tissue) or wandering, but this terminology can be misleading. All macrophages retain the capacity to move about. Whether they reside in a specific organ or wander depends upon their stage of development and

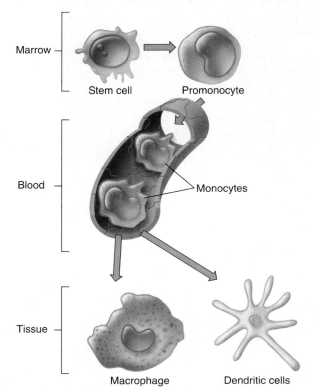

Marrow — Stem cell → Promonocyte

Blood — Monocytes

Tissue — Macrophage — Dendritic cells

Figure 14.16 The developmental stages of monocytes and macrophages. The cells progress through maturational stages in the bone marrow and peripheral blood. Once in the tissues, a macrophage can remain nomadic or take up residence in a specific organ.

the immune stimuli they receive. Specialized macrophages called *histiocytes* migrate to a certain tissue and remain there during their life span. Examples are alveolar (lung) macrophages; Kupffer cells in the liver; Langerhans cells in the skin; and macrophages in the spleen, lymph nodes, bone marrow, kidney, bone, and brain. Other macrophages do not reside permanently in a particular tissue and drift nomadically throughout the RES. Not only are macrophages dynamic scavengers, but they also process foreign substances and prepare them for reactions with certain lymphocytes (see chapter 15).

Mechanisms of Phagocytic Recognition, Engulfment, and Killing

The term **phagocytosis** literally means "eating cell process." But phagocytosis is more than just the physical act of engulfment, because phagocytes also actively attack and dismantle foreign cells with a wide array of antimicrobial substances. Phagocytosis can occur as an isolated event performed by a lone phagocytic cell responding to a minor irritant in its area or as part of the orchestrated events of inflammation described in section 14.3. The events in phagocytosis include chemotaxis, ingestion, phagolysosome formation, destruction, and elimination (**figure 14.17**).

Chemotaxis, Binding, and Ingestion Phagocytes migrate into a region of inflammation with a deliberate sense of direction, attracted by a gradient of stimulant products from the pathogen and host tissue at the site of injury.

Once a phagocyte encounters the pathogen, it uses its toll-like receptors (TLRs, discussed in section 14.2) to make contact with the pathogen (figure 14.17). Recall that TLRs are receptors that recognize and bind the PAMP receptors of various microbes. There are about 10 different TLRs in the membranes of phagocytes. The exposed end of a receptor hooks on to a PAMP and immediately dimerizes or joins with a second TLR to encase the molecule (**figure 14.18**). This relays a signal into the nucleus that stimulates the intracellular phagocytic processes and the release of chemical mediators.

On the scene of an inflammatory reaction, phagocytes often trap cells or debris against the fibrous network of connective

Process Figure 14.17 **The sequential events in phagocytosis.** (1) Phagocyte is attracted to bacteria. (2) Close-up view of process showing bacteria adhering to special receptors by their PAMPs. (3) Vacuole is formed around bacteria during engulfment. (4) Phagosome digestive vacuole results. (5) Lysosomes fuse with phagosome, forming a phagolysosome. (6) Enzymes and toxic oxygen products kill and digest bacteria. (7) Undigested particles are released. Inset: Scanning electron micrograph of a neutrophil actively engaged in devouring anthrax bacteria (10,000×).

(1) Chemotaxis by phagocyte

Bacterial cells

PAMPs

(2) Adhesion of bacteria

Receptor on host cell

(3) Engulfment into phagocytic vacuole

Lysosomes

Golgi apparatus

Rough endoplasmic reticulum

(4) Phagosome

(5) Phagolysosome formation

Nucleus

Enzymes
Lysozyme
DNase
RNase
Proteases
Peroxidase

(6) Killing and destruction of bacterial cells

Reactive oxygen products
Superoxide ($O_2^{-\bullet}$)
Hydrogen peroxide (H_2O_2)
Singlet oxygen (1O_2)
Hydroxyl ion (OH^-)
Hypochlorite ion ($HClO^-$)

(7) Release of residual debris

Figure 14.18 Phagocyte detection and signaling with toll-like receptors. Toll-like receptors (TLRs) span the membrane of phagocytes and other cells of the immune system. When a molecule such as a PAMP on a particular pathogen is recognized by this receptor, the TLRs merge and bind the foreign molecule. This induces production of chemicals and triggers engulfment.

tissue or the wall of blood and lymphatic vessels. Once the phagocyte has "caught" its prey, it extends pseudopods that enclose the cells or particles in a pocket and internalize them in a vacuole called a *phagosome.*

Phagolysosome Formation and Killing In a short time, **lysosomes** migrate to the scene of the phagosome and fuse with it to form a **phagolysosome.** Other granules containing antimicrobial chemicals are released into the phagolysosome, forming a potent brew designed to poison and then dismantle the ingested material (figure 14.17). The destructiveness of phagocytosis is evident by the death of bacteria within 30 minutes after contacting this battery of antimicrobial substances.

Destruction and Elimination Systems Destructive chemicals await the microbes in the phagolysosome. The oxygen-dependent system, known as the **respiratory burst** or oxidative burst, elaborates products of oxygen metabolism called **reactive oxygen intermediates** (ROI). Myeloperoxidase, an enzyme found in granulocytes, forms halogen ions (OCl^-) that are strong oxidizing agents. Other products of oxygen metabolism such as hydrogen peroxide, the superoxide anion (O_2^-), activated or singlet oxygen ($^1O^-$), and the hydroxyl free radical (HO) separately and together have formidable killing power. This series of reactive oxygen products delivers a "knockout" punch necessary to kill aerobic pathogens such as fungi and many bacteria (see this chapter's case file). Other chemicals that come into play are the lactic acid, lysozyme, and *nitric oxide* (NO), a powerful mediator that kills bacteria and inhibits viral replication. Cationic proteins that injure bacterial cell membranes and a number of proteolytic and other hydrolytic enzymes complete the job. The bits of undigestible debris are released from the macrophage by exocytosis. As we shall see in Chapter 15, macrophages and dendritic cells use this system to process microbes for specific immune responses with lymphocytes.

About one in 200,000 children in the United States are diagnosed yearly with chronic granulomatous disease (CGD); of these, around 90% are males. CGD is somewhat complex in its inheritance, but there are essentially two forms. The X-linked type occurs as a defective gene on the X chromosome and is more severe. Over half of all cases are X-linked. The autosomal defect is located on a different chromosome and tends to be milder. In both types, the genetic codes carry errors that prevent phagocytes from producing certain key metabolic enzymes.

The neutrophils and macrophages of CGD patients can carry out the usual phagocytic activities such as amoeboid motion, chemotaxis, engulfment, and digestion of microbes in phagocytic vesicles, but they are unable to actively kill certain types of aerobic pathogens. They lack the enzymes to synthesize the battery of reactive oxygen intermediates (ROIs)—superoxide ion, hydrogen peroxide, and hydroxyl radical—that are essential to kill these microbes. As a result, the ineffective phagocytes harbor live microbes that can break away and continue their infectious path. In time, these phagocytes shut down, die, and build up in the infected tissue in masses (granulomas).

- Why would children with CGD be more prone to infections from fungi and bacteria and not viruses?
- Why is interferon an effective treatment for CGD?

For a wrap-up, see the Case File Perspective on page 448.

Interferon: Antiviral Cytokines and Immune Stimulants

Interferon (IFN) was described in chapter 12 as a small protein produced naturally by certain white blood and tissue cells. It is used in therapy against certain viral infections and cancer and can be used as an immune enhancer, as we saw in this chapter's case file. Although the interferon system was originally thought to be directed exclusively against viruses, it is now known to be involved also in defenses against other microbes and in immune regulation and intercommunication. Three major types are *interferon alpha,* a product of lymphocytes and macrophages; *interferon beta,* a product of fibroblasts and epithelial cells; and *interferon gamma,* a product of T cells. See Insight 14.2 for a summary of interferon actions.

All three classes of interferon are produced in response to viruses, RNA, immune products, and various antigens. Their biological activities are extensive. In all cases, they bind to cell surfaces and induce changes in genetic expression, but the exact results vary. In addition to antiviral effects discussed in the next section, all three IFNs can inhibit the expression of cancer genes and have tumor suppressor effects. Interferon alpha and IFN beta stimulate phagocytes, and IFN gamma is an immune regulator of macrophages and T and B cells.

Characteristics of Antiviral Interferon

When a virus binds to the receptors on a host cell, a signal is sent to the nucleus that directs the cell to synthesize interferon. After transcribing and translating the interferon gene, newly synthesized interferon molecules are rapidly secreted by the cell into the extracellular space, where they bind to other host cells. The binding of interferon to a second cell induces the production of another class of proteins that inhibit viral multiplication by preventing the translation of viral proteins (**figure 14.19**). Interferon is not virus-specific, so its synthesis in response to one type of virus will also protect against other types. Because this protein is an inhibitor of viruses, it has been a valuable treatment for a number of virus infections.

Other Roles of Interferon

Interferons are also important immune regulatory cytokines that activate or instruct the development of white blood cells. For example, interferon alpha produced by T lymphocytes activates a subset of cells called natural killer (NK) cells. In addition, one type of interferon beta plays a role in the maturation of B and T lymphocytes and in inflammation. Interferon gamma inhibits cancer cells, stimulates B lymphocytes, activates macrophages, and enhances the effectiveness of phagocytosis.

Complement: A Versatile Backup System

Among its many overlapping functions, the immune system has another complex and multiple-duty system called **complement** that, like inflammation and phagocytosis, is brought into play at several levels. The complement system, named for its property of "complementing" immune reactions, consists of at least 26 blood proteins that work in concert to destroy a wide variety of bacteria, viruses, and parasites. The sources of complement factors are liver hepatocytes, lymphocytes, and monocytes.

Complement functions as a *positive feedback loop* or *cascade reaction*. Its primary action is a sequential physiological response like that of blood clotting, in which the first substance in a chemical series activates the next substance, which activates the next, and so on, until a desired end product is reached.

Three different versions of the complement pathways exist (**figure 14.20**). Their main distinguishing features are how they are activated, major participating factors, and specificity. The end stages of all three converge at the same point and yield a similar end result, that is, destruction of a pathogen. The *classical pathway* is the most specific, activated mainly by the presence of antibody bound to microorganisms. The *lectin pathway* is a nonspecific reaction of a host serum protein that binds a sugar called mannan present in the walls of fungi and other microbes. (Proteins that bind carbohydrates are called lectins.) The *alternative pathway* begins when complement proteins bind to normal cell wall and surface components of microbes. Note that because the complement numbers (C1 to C9) are based on the order of their discovery, some factors are not activated in numerical order.

Overall Stages in the Complement Cascade

In general, the complement cascade goes through four stages: initiation, amplification and cascade, polymerization, and membrane attack. The starting reaction requires some type of initiator molecule such as antibodies, lectins, or microbial surface receptors (**figure 14.20a**), depending on the pathway. The presence of this initiator on the pathogen's membrane propels the chain of actions involving complement chemicals C1 through C4, with each step serving a dual purpose. We are omitting the fine details of the amplification part of the complement system, but the end result is a molecule, **C3,** that is a key factor in subsequent polymerization and membrane attack for all three pathways. C3 is acted on by a convertase enzyme that gives rise to **C5$_b$,** which forms the recognition site and anchor for the final series of reactions (**figure 14.20b**). These amplification reactions also release cleavage products that serve other immune functions, including viral neutralization, allergic reactions, and immune regulation.

The final events in the complement series involve polymerization of C5$_b$, C6, C7, and C8 into a half circular formation within the membrane (**figure 14.20c**). This is followed by the insertion of several C9 molecules that complete a full ring-shaped polymer termed the **membrane attack complex or MAC (figure 14.20d).** The MAC is the primary destructive force of the complement system. It effectively perforates and lyses the membranes of gram-negative bacteria, fungi, parasitic protozoans, and enveloped viruses, Except for the classical pathway, these reactions are non-specific and can be active against a wide spectrum of microbes.

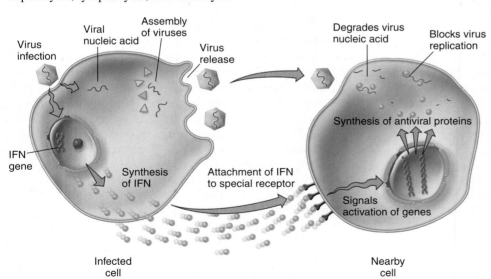

Figure 14.19 The antiviral activity of interferon. When a cell is infected by a virus, its nucleus is triggered to transcribe and translate the interferon (IFN) gene. Interferon diffuses out of the cell and binds to IFN receptors on nearby uninfected cells, where it induces production of proteins that eliminate viral genes and block viral replication. Note that the original cell is not protected by IFN and that IFN does not prevent viruses from invading the protected cells.

(a)

Classical Pathway	MB-Lectin Pathway	Alternative Pathway
Complement-fixing antibodies (discussed in chapter 15) rapid, specific	Mannose-binding lectin (MBL) binds mannose on pathogen surfaces. Nonspecific for bacteria and viruses	Molecules on surfaces of bacteria, fungi, viruses, and parasites. Nonspecific
C1q, C1r, C1s C4 C2 C3	MBL, MASP-1, MASP-2 C4 C2 C3	Factor B Factor D C3

C3 convertase enzyme converts C3 molecule into an activator—$C3_b$

(b) **Cascade and Amplification.** C5 factor is acted on by $C3_b$, which converts it to $C5_b$. $C5_b$ becomes bound to the membrane and serves as the starting molecule for the chain of events that assemble the complex in (c) and (d).

$C5_b$ $C3_b$ C3 Convertase

Other molecules given off: $C3_a$, $C5_a$ which are peptide mediators of inflammation, phagocyte recruitment

(c) **Polymerization.** $C5_b$ is a reactive site for the final assembly of an attack complex. In series, C6, C7, and C8 aggregate with $C5_b$ and become integrated into the membrane. They form a substrate upon which the final component, C9, can bind. Up to 15 of these C9 units ring the central core of the final membrane attack complex (MAC).

C8 Completed MAC C7 C6 $C5_b$

Terminal complement components
$C5_b$
C6
C7
C8
C9

(d) **Membrane Attack.** Insertion of MACs produces hundreds of tiny holes in the cell membrane. This can cause lysis and death of eukaryotic cells and many gram-negative bacteria.

Lysis C9

$\overline{C5_b}6789$ Membrane-attack complex, lysis of certain pathogens and cells

Figure 14.20 Overview of the complement pathways. (a) All three pathways have different triggers and starting points, but they all converge at the same place, C3 convertase. This enzyme begins the series of reactions **(b,c,d)** that give rise to the membrane attack complex characteristic of the complement "machinery." The final result is the formation of tiny openings in the cell membrane and the destruction of the target pathogen.

An Outline of Major Host Defenses

At this point, it should be apparent that the body's defenses often work in both parallel and redundant ways to fight pathogens. The result is a constant onslaught of defenses functioning at many levels. This can be illustrated by a final figure, organized by the three lines of defense that started out our discussion and that lead us naturally into chapter 15 (**figure 14.21).**

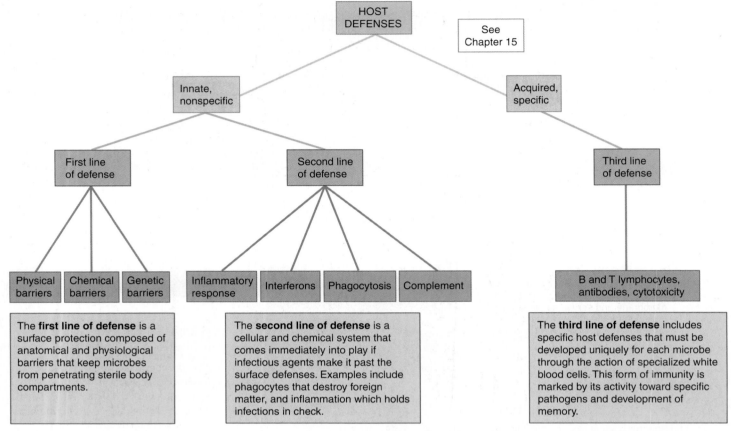

Figure 14.21 Flowchart summarizing the major components of the host defenses. Defenses are classified into one of two general categories: (1) innate and nonspecific or (2) acquired and specific. These can be further subdivided into the first, second, and third lines of defense, each being characterized by a different level and type of protection. The third line of defense is the most complex and is responsible for specific immunity. It is covered in greater detail in chapter 15.

Check & Assess Section 14.4

✔ White blood cells known as phagocytes help to clear microbes and foreign particles from the body.

✔ Macrophages are activated monocytes. Along with neutrophils, they are the key phagocytic agents of nonspecific response to disease.

✔ Phagocytes engulf and eliminate pathogens by producing highly reactive oxygen products such as peroxide and superoxide.

✔ Interferon are small proteins produced by several types of white blood and tissue cells that function against viruses and other infectious agents. They have a variety of other actions that regulate immune responses.

✔ The plasma contains complement, a nonspecific group of chemicals that works on its own or with the third line of defense to destroy foreign cells by lysing their membranes.

28. Describe the events that give rise to macrophages.

29. What types of macrophages are there, and what are their principal functions?

30. Outline the major phases of phagocytosis.

31. Briefly describe the three major types of interferon, their sources, and their biological effects.

32. Describe the mechanism by which interferon acts as an antiviral compound.

33. Describe the general complement reaction in terms of a feedback system or cascade.

34. What are some factors that trigger the complement cascade?

35. Using figure 14.21 as a guide, give examples for each category shaded orange, brown, and purple.

CASE FILE 14 PERSPECTIVE

This case is a striking example of how important every element of immunity can be. Although the components do overlap in some of their effects, each one carries out some essential function that cannot be totally duplicated. CGD individuals have a mostly intact immune response, but they lack a complete phagocyte defense that comes into play early in the infection process. Phagocytes (neutrophils, monocytes, and macrophages) are most active against bacteria and eukaryotic pathogens, but they also are important participants in many other immune reactions.

The swelling and inflammation of the child's bones (osteomyelitis) were triggered by the *Aspergillus* hyphae growing into

bone cavities. The presence of these foreign cells initiated a series of local events, beginning with infiltration of phagocytes and other white blood cells into the infected area. The white blood cells released biochemical mediators that give rise to the classic inflammatory signals of rubor, calor, tumor, and dolor. These mediators dilate blood vessels, creating rubor or redness, which also has the effect of increasing the local temperature (calor). The reactions flood the site with tissue fluid and cause swelling (tumor) and pain (dolor).

Even people with healthy phagocytes can produce granulomas, but CGD individuals produce them in pathological amounts. The granulomatous nature of this disease is triggered by chronic inflammation and buildup of dead neutrophils that could not completely finish the job of clearing the pathogen. When the backup team of macrophages enters the scene to "clean up" the dead neutrophils and infection debris, it also is unable to resolve the inflammation and leaves behind large inactive packets of cells that remain in tissues for long periods of time.

The major infectious agents that invade CGD patients are catalase-producing aerobic bacteria and fungi. These types of organisms would usually be kept in check by normal phagocytes that literally douse them in toxic oxygen intermediates such as superoxide ions, hydroxyl radicals, and hypochlorites. Because CGD phagocytes lack these intermediates, they have little or no effect on organisms such as *Staphylococcus, Klebsiella, Aspergillus,* and *Candida.* Viruses are not a significant cause of opportunistic infections in these patients because viruses are not controlled by oxygen-based chemicals. They are eliminated by other immune defenses such as interferon.

The immune stimulant gamma interferon has been beneficial in resolving infections in CGD patients. This may be due to several factors. One contribution of interferon is to activate macrophages and enhance the effectiveness of phagocytosis. Another is that it modulates T cells that can destroy infectious agents without the need for phagocyte-killing mechanisms. Interferon is often prescribed in combination with antimicrobial drugs that target the pathogen. Since this disease is genetic, it must be treated continuously with long-term prophylactic medications and other therapies. There is no cure, but several trials of gene therapy to correct the genetic defect are in progress.

For additional information on this disease, go to http://emedicine.medscape.com/article/1116022-print

Other immune disorders are covered in chapter 16.

Chapter Summary with Key Terms

14.1 Overview of Host Defense Mechanisms

The *first line of defense* is an inborn, nonspecific system composed of anatomical, chemical, and genetic barriers that block microbes at the portal of entry. The *second line of defense* is also inborn and nonspecific and includes protective cells and fluids in tissues, and the *third line of defense* is acquired and specific and is dependent on the function of T and B cells.

14.2 Structure and Function of the Organs of Defense and Immunity

A. **Immunology** is the study of **immunity,** which refers to the development of resistance to infectious agents by white blood cells (WBCs). WBCs conduct surveillance of the body, recognizing and differentiating **nonself** (foreign) cells by virtue of the **pathogen-associated molecular patterns (PAMPs)** they carry on their surface. Under normal conditions, **self** cells are left alone but nonself cells are destroyed.

B. For purposes of immunologic study, the body is divided into three compartments: the blood, lymphatics, and the **reticuloendothelial system.** A fourth component, the extracellular fluid, surrounds the first three and allows constant communication between all areas of the body.

C. **Circulatory System: Blood and Lymphatics**
 1. **Whole blood** consists of **blood cells** (formed by **hematopoiesis** in the bone marrow) dispersed in **plasma. Stem cells** in the bone marrow differentiate to produce white blood cells **(leukocytes),** red blood cells **(erythrocytes),** and megakaryocytes, which give rise to **platelets.**
 2. Leukocytes are the primary mediators of immune function and can be divided into **granulocytes (neutrophils, eosinophils,** and **basophils)** and **agranulocytes (monocytes** and **lymphocytes)** based on their appearance. Monocytes later differentiate to become **macrophages.** Lymphocytes are divided into **B cells,** which produce **antibodies** as part of **humoral immunity,** and **T cells,** which participate in **cell-mediated immunity.** Leukocytes display both **chemotaxis** and **diapedesis** in response to chemical mediators of the immune system.

D. *Lymphatic System*
 1. The **lymphatic system** parallels the circulatory system and transports lymph while also playing host to cells of the immune system. Lymphoid organs and tissues include **lymph nodes,** the spleen, and the **thymus,** as well as areas of less well-organized immune tissues such as **GALT** and **MALT.**

14.3 Second Line Defenses: Inflammation

A. The inflammatory response is a complex reaction to tissue injury marked by *redness, heat, swelling,* and *pain.*
 1. **Pattern recognition receptors (PRRs),** found on many cells of the immune system, bind pathogens and initiate an immune response.
 2. Blood vessels narrow and then dilate in response to **chemical mediators** and **cytokines.**
 3. **Edema** swells tissues, helping prevent the spread of infection.
 4. WBCs, microbes, debris, and fluid collect to form **pus.**
 5. Pyrogens may induce fever.

14.4 Second Line Defenses: Phagocytosis, Interferon, and Complement

A. Phagocytosis: Partner to Inflammation
 1. Macrophages and neutrophils engage in **phagocytosis,** engulfing microbes in a phagosome. Uniting the phagosome with a **lysosome** results in destruction of the phagosome contents.
 2. Phagocytes use PRRs called **toll-like receptors (TLRs)** to recognize and adhere to foreign markers on microbes.

B. **Interferon (IFN)** is a family of proteins produced by leukocytes and fibroblasts that inhibit the reproduction of viruses by degrading viral RNA or blocking the synthesis of viral proteins.

C. **Complement** is a complex defense system that results, by way of a cascade mechanism, in the formation of a membrane attack complex that kills cells by creating holes in their membranes.

14. Which of the following substances is *not* produced by phagocytes to destroy engulfed microorganisms?
 a. hydroxyl radical c. hydrogen peroxide
 b. superoxide anion d. bradykinin

15. Which of the following is the end product of the complement system?
 a. properdin c. membrane attack complex
 b. cascade reaction d. complement factor C9

Multiple-Choice Questions

Select the correct answer from the answers provided. For questions with blanks, choose the combination of answers that most accurately completes the statement.

1. An example/examples of a nonspecific chemical barrier to infection is/are
 a. unbroken skin c. cilia in respiratory tract
 b. lysozyme in saliva d. all of these

2. Which nonspecific host defense is associated with the trachea?
 a. lacrimation c. desquamation
 b. ciliary lining d. lactic acid

3. Which of the following blood cells function primarily as phagocytes?
 a. eosinophils c. lymphocytes
 b. basophils d. neutrophils

4. Which of the following is not a lymphoid tissue?
 a. spleen c. lymph node
 b. thyroid gland d. GALT

5. What is included in GALT?
 a. thymus c. tonsils
 b. Peyer's patches d. breast lymph nodes

6. A signaling molecule from microbes recognized by phagocytes is
 a. pyrogen c. complement
 b. PAMP d. lectin

7. Monocytes are _____ leukocytes that develop into _____.
 a. granular, phagocytes c. agranular, macrophages
 b. agranular, mast cells d. granular, T cells

8. Which of the following inflammatory signs specifies pain?
 a. tumor c. calor
 b. dolor d. rubor

9. Toll-like receptors are proteins on _____.
 a. phagocytes that recognize foreign molecules
 b. viruses that stimulate immune reactions
 c. skin cells that provide barriers to infection
 d. lymphocytes that damage parasitic worms

10. An example of an inflammatory mediator that stimulates vasodilation is
 a. histamine c. complement C5a
 b. collagen d. interferon

11. _____ is an example of an inflammatory mediator that stimulates chemotaxis.
 a. Endotoxin c. A fibrin clot
 b. Serotonin d. Interleukin-2

12. An example of an exogenous pyrogen is
 a. interleukin-1 c. interferon
 b. complement d. endotoxin

13. _____ interferon, produced by T lymphocytes, activates cells called _____ and is involved in destroying viruses.
 a. Gamma, fibroblasts c. Alpha, natural killer cells
 b. Beta, lymphocytes d. Beta, fibroblasts

Case File Questions

1. What are signs of chronic granulomatous disease?
 a. unresolved inflammation d. recurrent infections
 b. opportunistic fungal infections e. a, b, and c
 c. low levels of white blood cells f. a, b, and d

2. What feature of CGD phagocytes is defective?
 a. the ability to engulf pathogens
 b. production of toxic forms of oxygen
 c. mechanisms to eliminate viruses
 d. synthesis of interferon

Writing to Learn

These questions are suggested as a *writing-to-learn* experience. For each question, compose a one- or two-paragraph answer that includes the factual information needed to completely address the question. All Check & Assess questions can be used for writing-to-learn exercises.

1. a. Use the lines on the figure to describe the major components of the first line of defense.
 b. What effects do these defenses have on microbes?

2. a. Describe the initial response system that is found in the primary responders to foreign matter.

3. Explain what PPRs and TLRs are and the nature of their interaction with PAMPs.

4. Make a simple drawing of the functions of toll-like receptors.

5. Describe the mechanisms by which leucocytes migrate from the bloodstream into an area of infection. What chemical factors are they reacting to?

6. Prepare a simplified outline of the cell lines of hematopoiesis.

7. Discuss the multiple, overlapping functions of the immune system. Which tissues and organs are involved in the immune response, and how does each function?

8. In what ways is a phagocyte a tiny container of disinfectants?

9. Briefly outline what leads to the structure shown to the right. What is it called and what is its action?

10. Macrophages perform the final job of removing tissue debris and other products of infection. Indicate some of the possible effects when these scavengers cannot successfully complete the work of phagocytosis.

Concept Mapping

Appendix E provides guidance for working with concept maps.

1. Construct your own concept map using the following words as the *concepts.* Supply the linking words between each pair of concepts.

defenses	inflammation
leukocytes	antibodies
lymphocytes	neutrophils
monocytes	fever
macrophages	

Critical Thinking Questions

Critical thinking is the ability to reason and solve problems using facts and concepts. These questions can be approached from a number of angles, and in most cases, they do not have a single correct answer.

1. Suggest some reasons that there is so much redundancy of action and there are so many interacting aspects of immune responses.

2. a. What are some possible elements missing in children born without a functioning lymphocyte system?
 b. What is the most important component extracted in bone marrow transplants?

3. a. What is the likelihood that plants have some sort of immune protection?
 b. Explain your reasoning.

4. A patient's chart shows an increase in eosinophil levels.
 a. What does this cause you to suspect?

b. What does it mean if the basophil levels are very high?
c. What if the neutrophil levels are very high?

5. How can adults continue to function relatively normally after surgery to remove the thymus, tonsils, spleen, or lymph nodes?

6. Explain why the X-linked form of CGD is more common in boys and more severe than autosomal forms of the disease.

7. What elements of host defenses and immunity would remain intact in a person with CGD?

8. An obsolete treatment for syphilis involved inducing fever by deliberately infecting patients with the agent of relapsing fever. A recent experimental AIDS treatment involved infecting patients with malaria to induce high fevers. Can you provide some possible explanations behind these peculiar forms of treatment?

9. Patients with a history of tuberculosis often show scars and other lesions in the lungs and experience recurrent infection. Account for these effects on the basis of the inflammatory response.

10. *Shigella, Mycobacterium,* and numerous other pathogens have developed mechanisms that prevent them from being killed by phagocytes.
 a. Suggest two or three factors that help them avoid destruction by the powerful antiseptics in macrophages.
 b. Suggest the potential implications that these infected macrophages can have on the development of disease.

11. Account for the several inflammatory symptoms that occur in the injection site when one has been vaccinated against influenza and tetanus.

12. a. Knowing that fever is potentially both harmful and beneficial, what are some possible guidelines for deciding whether to suppress it or not?
 b. What is the specific target of fever-suppressing drugs?

Visual Challenge

1. Each numbered figure 1–4 represents a cell that provides an immune defense. Each lettered figure A–D shows a pathogenic microbe. Name the cell and match it with its primary target microbe and effects.

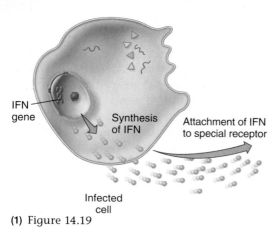

IFN gene
Synthesis of IFN
Attachment of IFN to special receptor
Infected cell
(1) Figure 14.19

(2) Figure 14.7

(3) Figure 14.7

(4) Figure 14.7

Spore

Vegetative cell

A Figure 19.1

B Figure 23.26

C Table 25.1

D Opener, Chapter 14

15

Adaptive, Specific Immunity and Immunization

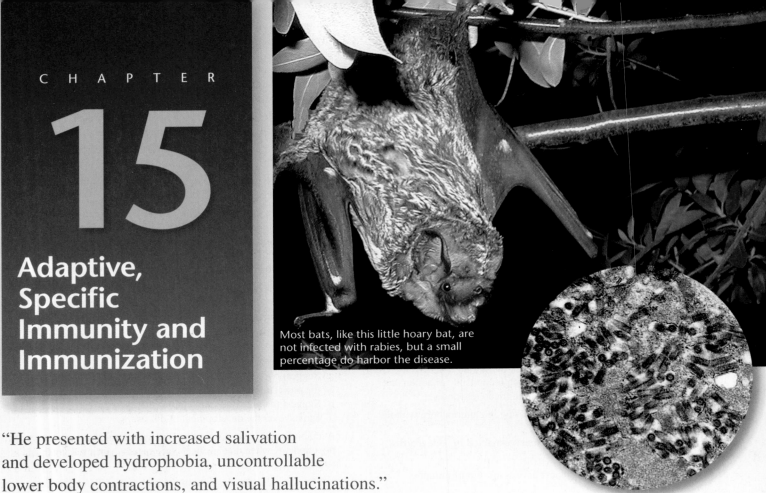

Most bats, like this little hoary bat, are not infected with rabies, but a small percentage do harbor the disease.

"He presented with increased salivation and developed hydrophobia, uncontrollable lower body contractions, and visual hallucinations."

The deadliest virus of them all. Masses of bullet-shaped rabies viruses inside an infected cell.

CASE FILE 15 A Tale of Two Bat Bites

Case #1. One morning in the fall of 2000, a 9-year-old Canadian boy awoke with a fever and shooting pain in one arm. Over the next several days, he became restless, had difficulty sleeping, and developed tremors in his arm and hand. After being hospitalized, he presented with increased salivation, hydrophobia, uncontrollable lower body contractions, and visual hallucinations. His respirations became so labored that he required a breathing tube. Medical personnel were forced to consider the dreadful diagnosis of rabies. The only definitive diagnostic result showed antibodies against the rabies virus. A week after he first developed symptoms, his medical team decided to intervene with postexposure prophylaxis (PEP) therapy. This involved administering both human rabies immune globulin (HRIG) and five spaced-out doses of rabies vaccine in an effort to stop the infection from spreading.

Digging deeper to discover how the child may have become infected, the

physicians learned that about a month before he showed symptoms, the father had captured and released a live bat trapped in the boy's bedroom. A day or so later, the child complained to his mother about a small laceration on his upper arm. It was cleaned and bandaged, but no further action was taken at the time.

Case #2. In early October 2008, a 15-year-old Brazilian youth was attacked by a bat, which bit him in several places. The bat escaped and was not available for testing. Clinical workers washed his wounds and applied peroxide and betadyne antiseptic to them. Even though he seemed fine, he was taken to a hospital for further evaluation. Physicians were concerned about the bat's behavior and the nature of the boy's wounds, so they treated him with a version of PEP that included five doses of rabies vaccine but not rabies immune globulin. This was started four days after the injuries.

Over the next two weeks, the young man began to show early symptoms of

rabies, including weakness, headache, fever, and a prickling sensation around the bites. Tests done on his blood and spinal fluid about this same time indicated antibodies specific to the rabies virus but no virus antigen. The medical staff had heard about a case of a young woman in Wisconsin who had survived rabies by being placed in a medically induced coma and given antiviral drugs for several weeks, and they decided to try this therapy on him. About three weeks from the time of the bite, it appeared that his condition had stabilized, and they removed him from sedation.

▶ *What are the basic immunological rationales behind PEP with rabies immune globulin and the rabies vaccine?*

To continue the case, go to page 474.

15.1 Specific Immunity: The Adaptive Line of Defense

Expected Learning Outcomes

1. Summarize the general requirements of adaptive, acquired immunity.
2. Define *immunocompetence, antigens, specificity,* and *memory* as they relate to the immune system.
3. Outline the overall phases in a specific immune response.
4. Explain the functions of immune receptors.
5. Discuss the major types of immune receptors, including where and how they develop.
6. Describe the major events in the origin of diversity in the immune system and features of the clonal selection theory.
7. Describe the development of antigen receptors on lymphocytes.

In chapter 14, we described the capacity of the immune system to survey, recognize, and react to foreign cells and molecules. Our main focus in that chapter was to overview innate host defenses such as anatomical and physiological barriers, phagocytosis, and inflammation. We also provided essential background in the main systems involved such as blood cells and lymphatic tissues. Refer to figure 14.21 and use the flowchart to refresh yourself on the relationships among various host defenses (see page 448). An essential system of defense that is not innate but must be acquired will be the main subject of this chapter.

Normal humans have an extremely specific and powerful system for resisting infectious agents called **adaptive** or **acquired immunity,** sometimes known as the third line of defense. It can be observed in the long-term protection provided by developing infections such as measles or mumps or being vaccinated for diphtheria. The absolute need for adaptive immunity is impressively documented in children with genetic defects in this system or in AIDS patients who have lost it. Even with heroic measures to isolate the patient, combat infection, or restore lymphoid tissue, the victim can be constantly vulnerable to life-threatening infections.

Acquired adaptive immunity is the product of a dual system of specialized leukocytes—the B and T lymphocytes. During fetal development, these lymphocytes undergo a selective process that prepares them for reacting only to one specific antigen. During this time, **immunocompetence,** the ability of the body to react with a wide spectrum of foreign substances begins to develop. An infant is born with the theoretical *potential* to produce an immune response to millions of different foreign molecules or antigens. But the completion of this immunocompetence takes many years, extending into late puberty.

Antigens figure very prominently in specific immunity. They are defined as any molecules that can stimulate a response by T and B cells. They consist of protein, polysaccharide, and other compounds from cells and viruses. Environmental chemicals can also be antigens, as related in the section of chapter 16 on allergy. In fact, any exposed or released substance is potentially an antigen, even those from our own cells. For reasons we discuss later, our own antigens do not usually evoke a response from our own immune systems, but they may do so in other persons.

In chapter 14, we discussed pathogen-associated molecular patterns (PAMPs) that stimulate responses by phagocytic cells during an innate defense response. PAMPs are molecules shared by many types of microbes that stimulate a nonspecific response. In contrast, antigens are unique molecules that stimulate specific immune responses. The two molecules do share two characteristics: (1) they are "parts" of foreign cells (microbes), and (2) they provoke a reaction by the white blood cells of the host.

Two features that most characterize adaptive immunity are **specificity** and **memory.** Unlike mechanisms such as anatomical barriers or phagocytosis, acquired immunity is selective. For example, the antibodies produced during an infection against the chickenpox virus will protect against that virus but not against the measles virus. The property of memory pertains to the rapid mobilization of lymphocytes that have been programmed to "recall" their first engagement with an invader and respond rapidly to it during subsequent exposures. These topics will be covered in some depth in several sections in this chapter.

An Overview of Specific Immune Responses

Immune responses are highly complex and regulated, and they represent one of the most elegant and coordinated networks of cells and chemicals in the body. To present this system in an organized and clear manner, we have found it helpful to divide it into separate but related sections, each detailing some event in development or the immune response. The sections and pages of coverage are as follows:

 I. Development and differentiation of the immune system (sections 15.1 and 15.2, pages 453–461)
 II. Lymphocytes, maturation, and the nature of antigens (section 15.2, pages 459–461)
III. Cooperation between lymphocytes during antigen presentation (section 15.3, pages 461–468)
 IV. B lymphocyte responses and the production and actions of antibodies (section 15.3, pages 461–468)
 V. T-lymphocyte responses (section 15.4, pages 468–472)

As we explore the information contained in these sections, we will be following a "map" of sorts. **Figure 15.1** lays out the flow of the main stages and will serve to coordinate the presentation. It also includes figure numbers that enlarge on the topics in that section.

Development of the Immune Response System

Before we examine lymphocyte development and function in greater detail, we must initially review concepts such as the unique structure of molecules (especially proteins), the characteristics of cell surfaces (membranes and envelopes), the ways that genes are expressed, immune recognition, and identification of self and nonself. Ultimately, the shape and function of protein receptors and markers protruding from the surfaces of certain white blood cells are the result of genetic expression, and these molecules are responsible for specific immune recognition and, thus, immune reactions.

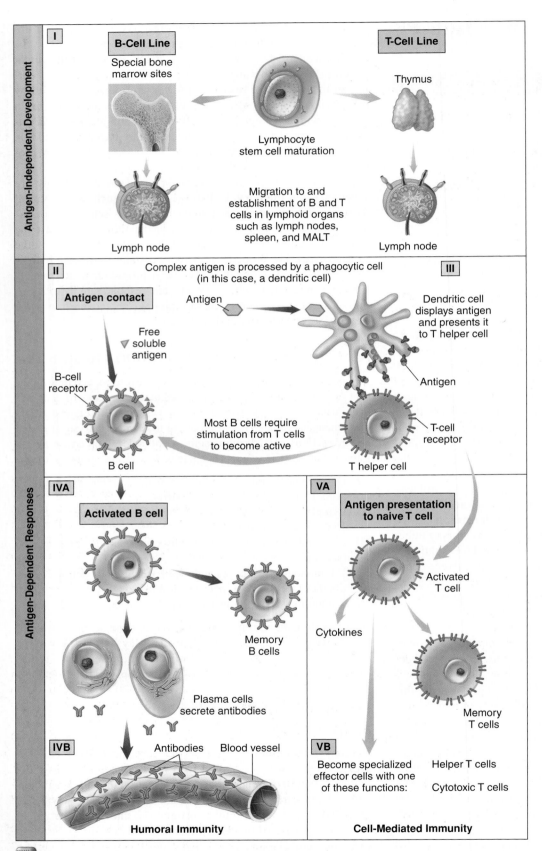

I. Development of the Lymphocyte System. Lymphocytes arise from the same stem cells but differentiate into two distinct cell types early on. B cells mature in specialized bone marrow sites, and T cells mature in the thymus gland. Mature cells settle in lymphoid organs and serve as a constant attack force for infectious agents. (See figures 15.3 and 15.4a.)

II. Contact with antigens and **III.** Presentation by antigen presenting cells. Foreign cells bear molecules (antigens) that are recognized and engulfed by dendritic cells, macrophages, or B cells. For most responses, specialized cells called T helper cells first receive the processed antigen from the phagocyte and stimulate B and other T cells. (See figures 15.4b, 15.9, and 15.10, 1–4).

IVA. B-Cell Responses. A B cell activated by T helper cells undergoes a spurt of cell division. Some new cells are memory cells that provide a rapid recall of the antigen, and some are plasma cells that secrete proteins called antibodies. (See figure 15.10, 5–7.)

IVB. Humoral Immunity*. Antibodies circulate in fluids (blood, ECF, and lymph) providing humoral* immunity. The antibodies react specifically with the antigen and mark it for an enhanced response. (See figures 15.11 to 15.15, table 15.2)

VA. Activation of T cells and **VB.** T-cell responses. Activated T cells form memory cells and differentiate into helper cells or cytotoxic cells. They synthesize cytokines that regulate immune reactions. T-cell immunity is termed cell mediated because the T cell acts directly against any cells bearing antigen. (See figures 15.16, 15.17, and table 15.3).

*A traditional term for such body fluids is the "humors."

Figure 15.1 **Overview of the origins and events of adaptive immune responses.** This summary lays out the plan of sections 15.2 to 15.4.

Markers on Cell Surfaces Involved in Recognition of Self and Nonself

Chapter 14 touched on the fundamental idea that cell surface receptors confer specificity and identity. A given cell can express several different receptors, each type playing a distinct and significant role in detection, recognition, and cell communication. Major functions of receptors are

- to perceive and attach to nonself or foreign antigens,
- to promote the recognition of self antigens,
- to receive and transmit chemical messages among other cells of the system, and
- to aid in cellular development.

Because of their importance in the immune response, we concentrate here on the major receptors of lymphocytes and macrophages.

Major Histocompatibility Complex One set of genes that codes for human cell receptors is the **major histocompatibility complex (MHC).** This gene complex gives rise to a series of glycoproteins (called MHC molecules) found on all cells except red blood cells. The MHC is also known as the **human leukocyte antigen (HLA)** system. This receptor complex plays a vital role in recognition of self by the immune system and in rejection of transplanted tissues.

The functions of the MHC groups have been identified. **Class I MHC** genes code for markers that display unique characteristics of self and allow for the recognition of self molecules and the regulation of immune reactions. Coverage in section 15.4 will outline how some T cells require interaction with class I MHC before they can react with foreign cells.

The system is rather complicated in its details, but in general, each person inherits a particular combination of class I MHC (HLA) genes in a relatively predictable fashion. Although millions of different combinations and variations of these genes are possible among humans, the closer the relationship, the greater the probability for similarity in MHC profile. Individual differences in the exact inheritance of MHC genes, however, make it highly unlikely that even closely related persons will express an identical MHC profile. This fact introduces an important recurring theme: Although humans are genetically the same species, the cells of each individual express molecules that are foreign (antigenic) to other humans. This is where the term *histo-* (tissue) *compatibility* (acceptance) originated. This fact necessitates testing for MHC and other antigens when blood is transfused and organs are transplanted (see graft rejection, chapter 16).

Class II MHC genes code for immune regulatory receptors. The cells that express class MHC II are grouped together by the term **antigen-presenting cells** (APCs). This includes macrophages, dendritic cells, and B cells. These are the only "professional" antigen presenters in that they possess both class I and II receptors necessary for interactions with T cells. See **figure 15.2** for depictions of the two MHC classes.

Lymphocyte Receptors and Specificity to Antigen The part lymphocytes play in immune surveillance and recognition is very much a function of their receptors. B-cell receptors bind

Class I MHC molecule found on all nucleated human cells

Class II MHC found on some types of white blood cells

Figure 15.2 Molecules of the human major histocompatibility complex.

free antigens, while T-cell receptors bind processed antigens together with the MHC molecules on the cells that present antigens to them. Antigen molecules exist in great diversity; there are potentially millions and even billions of unique types. The many sources of antigens include microorganisms as well as an awesome array of chemical compounds in the environment. One of the most fascinating questions in immunology is: How can the lymphocyte receptors be varied to react with such a large number of different antigens? After all, it is generally accepted that there will have to be a different lymphocyte receptor for each unique antigen. Some additional questions that naturally follow are: How can a cell accommodate enough genetic information to respond to millions or even billions of antigens? When, where, and how does the capacity to distinguish self from foreign tissue arise? To answer these questions, we must first introduce a central theory of immunity.

The Origin of Diversity and Specificity in the Immune Response

The Clonal Selection Theory and Lymphocyte Development Research findings have shown that lymphocytes use slightly more than 500 genes to produce the tremendous repertoire of specific receptors they must display for antigens. The most widely accepted explanation for how this diversity is generated is called the **clonal selection theory.** According to this theory, early undifferentiated lymphocytes in the embryo and fetus undergo a continuous series of divisions and genetic changes that generate hundreds of millions of different cell types, each carrying a particular receptor specificity.

The mechanism, generally true for both B and T cells, can be summarized as follows: In the bone marrow, blood stem cells can become some form of blood cell such as granulocytes, monocytes, or lymphocytes. The lymphocytic line of stem cells differentiates into either T cells or B cells. Cells destined to become B cells stay

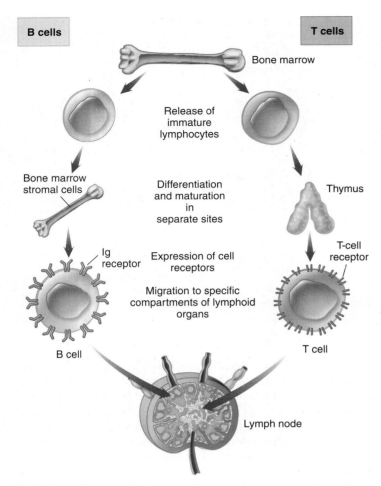

Figure 15.3 **Major stages in the development of B and T cells.**

clones of lymphocytes that can react against *self* MHC antigens. The presence of such "forbidden clones" could cause severe damage if the immune system mistakenly recognizes self molecules as foreign and mounts a response against the host's own tissues. Thus, part of the clonal selection theory says that some clones are eliminated during development through clonal *deletion*. The removal of such potentially harmful clones is one basis of **immune tolerance** or tolerance to self. Some diseases (autoimmunity) are thought to be caused by the loss of immune tolerance along with reactions directed at self (see chapter 16).

The second stage of development—**clonal selection and expansion**—does require stimulation by antigens such as those that come from microbes. When an antigen enters the immune surveillance system, it encounters specific lymphocytes ready to recognize it. Such contact stimulates that clone to undergo mitotic divisions and expands it into a larger population of lymphocytes all bearing the same specificity. This increases the capacity of the immune response for that antigen.

Two important generalities one can derive from the clonal selection theory are (1) lymphocyte specificity is preprogrammed, existing in the genetic makeup before an antigen has ever entered the tissues; and (2) each genetically distinct lymphocyte expresses only a single specificity and can react to only one type of antigen. Other important features of the lymphocyte response system are detailed in later sections.

The Specific B-Cell Receptor: An Immunoglobulin Molecule In the case of B lymphocytes, the receptor genes that undergo the recombination described are those governing **immunoglobulin* (Ig)** synthesis. Immunoglobulins are large glycoprotein molecules that serve as the specific receptors of B cells and as antibodies. The basic immunoglobulin molecule is a composite of four polypeptide chains: a pair of identical heavy (H) chains and a pair of identical light (L) chains (**figure 15.5a**). Each light chain binds to a heavy chain, and the two heavy chains bind to each other with disulfide bonds, creating a symmetrical, Y-shaped arrangement.

The ends of the forks formed by the light and heavy chains contain pockets called the **antigen binding sites.** These sites can be highly variable in shape to fit a wide range of antigens. This extreme versatility is due to **variable (V) regions** where amino acid composition is highly varied from one clone of B lymphocytes to another (discussed in section 15.3). The remainder of the light chains and heavy chains consist of constant (C) regions whose amino acid content does not vary greatly from one antibody to another. Figure 15.11 provides further details of antibody structure.

Development of Lymphocyte Receptors during Maturation
The genes that code for immunoglobulins lie on three different chromosomes. An undifferentiated lymphocyte has about 150 different genes that code for the variable region of light chains and a total of about 250 genes for the variable and **diversity (D) regions** of the heavy chains. It has only a few genes for coding for the constant regions and a few for the **joining (J) regions** that join segments of the molecule together. Owing to genetic recombination during

in the bone marrow; T cells "home" to the thymus. Mature B and T cells then migrate to secondary lymphoid tissues (**figure 15.3**). These secondary lymphoid tissues must constantly be resupplied with B and T cells through this series of activities in the primary lymphoid tissues.

By the time T and B cells reach the lymphoid tissues, each one is already equipped to respond to a single unique antigen. This amazing diversity is generated by extensive rearrangements of the gene segments that code for the antigen receptors on the T and B cells (**figure 15.4a**). The result of extensive genetic recombinations is a huge assortment of lymphocytes.[1] Each genetically unique line of lymphocytes arising from these recombinations is termed a **clone.** Keep in mind that the genetic code is expressed as a protein receptor of unique configuration on the surface of the lymphocyte, something like a "signpost" announcing its specificity and reactivity for an antigen.

This *proliferative* stage of lymphocyte development does not require the actual presence of foreign antigens. But it does require another important action—namely, the removal or "silencing" of

1. Estimates of the theoretical number of possible variations that may be created vary from 10^{14} to 10^{18} different specificities—clearly enough to react with all potential antigens that might be encountered.

* *immunoglobulin* (im″-yoo-noh-glahb′-yoo-lin)

(a) Antigen-Independent Period
1. During development of early lymphocytes from stem cells, a given stem cell undergoes rapid cell division to form numerous progeny.

During early cell differentiation, there are random rearrangements of the genes that code for cell surface protein receptors. The result is a large array of genetically distinct cells, called clones, each clone bearing a different receptor to react with only a single type of foreign molecule or antigen.

2. At this time, any lymphocyte clones that carry a specificity for self molecules and could be harmful are eliminated from the pool of diversity. This is a way to achieve immune tolerance.

3. The specificity for a single antigen molecule is programmed into the lymphocyte and is set for the life of a given clone. The end result is an enormous pool of mature but naive lymphocytes that are ready to further differentiate under the influence of certain organs and immune stimuli.

(b) Antigen-Dependent Period
4. Lymphocytes come to populate the lymphatic organs, where they are set to encounter antigens. These antigens will become the stimulus for the lymphocytes' final activation and immune function. Entry of a specific antigen selects only the lymphocyte clone or clones that carry matching surface receptors. This will trigger an immune response, which varies according to the type of lymphocyte involved.

Figure 15.4 Overview of the clonal selection theory of lymphocyte development and diversity. Receptors are stylized to show diversity in shape that clonal selection generates. See Figure 15.1 for the details of receptor development in B cells.

development, only the selected (V and D) receptor genes are active in the mature cell, and all the other V and D genes are deleted (**figure 15.5***b*). This is how the specificity of lymphocytes arises.

It is helpful to think of the immunoglobulin genes as blocks of DNA that code for polypeptides lying in sequence along a chromosome. During development of each lymphocyte, the genetic blocks are independently segregated, randomly selected, and assembled as follows:

- For a heavy chain, a variable region gene and diversity region gene are selected from among the hundreds available and spliced to one joining region gene and one constant region gene.
- For a light chain, one variable, one joining, and one constant gene are spliced together.
- After transcription and translation of each gene complex into a polypeptide, a heavy chain combines with a light chain to form half an immunoglobulin; two of these combine to form a completed protein (figure 15.5*b*).

Once synthesized, the immunoglobulin product is transported to the cell membrane and inserted there to act as a receptor that

expresses the specificity of that cell and to react with an antigen, as shown in figure 15.12. It is notable that for each lymphocyte, the genes that were selected for the variable region, and thus for its specificity, will be locked in for the rest of the life of that lymphocyte and its progeny. We discuss the different classes of Ig molecules further in section 15.3. It is worth noting that the constant region may be altered to provide different functional properties (see table 15.2).

T-Cell Receptors for Antigen The T-cell receptor for antigen belongs to the same protein family as the B-cell receptor. It is similar to B cells in being formed by genetic modification, having variable and constant regions, being inserted into the membrane, and having an antigen binding site formed from two parallel polypeptide chains (**figure 15.6**). Unlike the immunoglobulins, the T-cell receptor is relatively small. It is equivalent to one fork of the B-cell receptor and is not secreted. The other receptors of T cells are described in section 15.5.

Table 15.1 summarizes the main differences in the structure, functions, and actions of B cells and T cells.

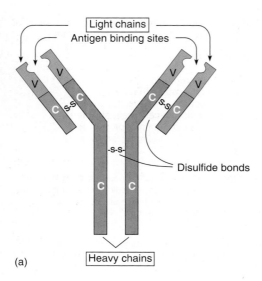

(a)

Figure 15.5 The basic structure and genetics of immunoglobulins. (a) Simple model of an immunoglobulin molecule. The main components are four polypeptide chains—two identical light chains and two identical heavy chains bound by disulfide bonds as shown. Each chain consists of a variable region (V) and a constant region (C). The variable regions of light and heavy chains form a binding site for antigen. **(b)** The final gene that codes for a heavy or light chain is assembled by splicing blocks of genetic material from several regions (1, 2, 3). These genes are transcribed and translated into the polypeptides that join to form the final molecule (4).

1. The heavy-chain genes are composed of four separate segments (V, D, J, and C) that are 2. transcribed and translated to form the heavy polypeptide chains.

3. The light-chain genes are put together like heavy ones, except that the final gene is spliced from three gene groups (V, J, and C), making smaller polypeptides.

Precisely shaped binding sites

4. During final assembly, first the heavy and light chains are bound, and then the heavy-light combinations are connected to form the immunoglobulin molecule.

(b)

Antigen binding site

Membrane

Figure 15.6 Proposed structure of the T-cell receptor (TCR) for antigen. The structure of this polypeptide is similar to that of an immunoglobulin. It consists of 2 polypeptides that mimic the structure of one "arm" of an immunoglobulin. Like this molecule, the TCR has variable (V) regions that can show high levels of diversity for antigens, and constant (C) regions that do not greatly vary.

Check & Assess Section 15.1

✔ Acquired specific immunity is an elegant but complex matrix of interrelationships between lymphocytes and macrophages consisting of several stages.

✔ The outstanding properties of adaptive immunity are immunocompetence, diversity of reactions, specificity, and memory.

✔ The surfaces of all cell membranes contain identity markers known as protein receptors. These cell markers function in identification, communication, and cell development. They are also self identity markers.

✔ Human cell markers are genetically determined by MHCs. MHC I codes for identity markers on all nucleated host cells and is important in antigen processing by APCs. MHC II codes for immune receptors on dendritic cells, macrophages, and B cells.

✔ The clonal selection theory explains that during development, both B and T cells develop millions of genetically different clones through independent segregation, random reassortment, and mutation.

✔ Together, these clones possess enough genetic variability to respond to many millions of different antigens. Each clone, however, can respond to only one specific antigen.

TABLE 15.1 Contrasting Properties of B Cells and T Cells

	B Cells	T Cells
Site of Maturation	Bone marrow	Thymus
Immune Surface Markers	Immunoglobulin MHC I and MHC II	T-cell receptor, CD molecules, MHC I
Circulation in Blood	Low numbers	High numbers
Receptors for Antigen	Immunoglobulins D and M	T-cell receptor (TCR)
Distribution in Lymphatic Organs	Cortex (in follicles)	Paracortical sites (interior to the follicles)
Require Antigen Presented with MHC	No	Yes
Product of Antigenic Stimulation	Plasma cells and memory cells	Several types of sensitized T cells and memory cells
General Functions	Production of antibodies to inactivate, neutralize, target antigens	Cells function in regulating immune functions, killing "foreign" cells; hypersensitivity; synthesize cytokines

1. Discuss what is meant by immunocompetence, immune specificity and memory, and adaptive immunity.
2. What function do receptors play in specific immune responses?
3. How can receptors be made to vary so widely?
4. Describe the major histocompatibility complex, and explain how it participates in immune reactions.
5. Explain the clonal selection theory of antibody specificity and diversity.
6. Why must the body develop tolerance to self?
7. Trace the origin and development of B lymphocytes and of T lymphocytes.

15.2 Lymphocyte Maturation and The Nature of Antigens

Expected Learning Outcomes

8. Discuss the events in B cell and T cell maturation.
9. Explain the characteristics of antigens, the property of antigenicity, and epitopes.
10. Discuss the main categories of antigens, based on function.

With a working knowledge of some factors in the development of immune specificity, we can now continue our coverage of the immune response as outlined in figure 15.1.

Specific Events in B-Cell Maturation

The site of B-cell maturation exists in certain bone marrow sites that harbor *stromal cells.* These huge cells nurture the lymphocyte stem cells and provide chemical signals that initiate B-cell development. The result of gene modification and selection is the development of hundreds of millions of distinct B cells (see figure 15.4*a*). These naive lymphocytes travel to specific sites in the lymph nodes, spleen, and mucosal-associated lymphoid tissue (MALT), where they adhere to specific binding molecules. Here, they will come into contact with antigens throughout life. B cells display one of several classes of immunoglobulins as surface receptors for antigens (see table 15.2).

Specific Events in T-Cell Maturation

The further development and maturation of T cells are directed by the thymus gland and its hormones. The complexity of T cell function discussed in section 15.4 is partly due to different classes of T-cell surface molecules termed **clusters of differentiation** or **CD receptors** added during maturation. CD molecules serve multiple roles as cell receptors and may be involved in cell adhesion and communication in a variety of cells. They are denoted with a number—CD1, CD2, etc. Our focus will be primarily on two CD groups found on T cells: CD4 receptors on T helper cells and CD8 receptors on T cytotoxic cells. These CD receptors interact with a particular MHC marker on other white blood cells during immune reactions. Like B cells, mature T cells also migrate to specific sites in lymphoid organs and constantly circulate from there. It is estimated that 25×10^9 T cells pass between the lymphatic and general circulation per day.

Characteristics of Antigens and Immunogens

Earlier, we encountered the concept of an **antigen,**[2] which can be more fully defined as any molecule or fragment of a molecule that has the *potential* to trigger a specific immune response by lymphocytes. This property is known as **antigenicity.** For a substance to be antigenic, it must meet several criteria related to foreignness, size, shape, and accessibility that are discussed next. A related set of more specific terms are **immunogen** and *immunogenicity.* An immunogen is a type of antigen that actually *does* induce a specific immune response when introduced into the body.

The difference between these terms is subtle and addresses the fact that some antigens may not fit all of the actual conditions to stimulate an immune reaction, whereas immunogens do. One example involves the small polysaccharides from the capsule of the bacterium *Hemophilus influenzae,* one cause of meningitis. These molecules are foreign and antigenic, but they are too small to trigger a lymphocyte response: They are antigenic but not immunogenic. To use this antigen to produce a vaccine, it must be rendered

2. *Antigen* (an'-tih-jen) means, literally, an **anti**body **gen**erator.

immunogenic by increasing its profile, and this is in fact what is done (see the discussion of haptens). In most of the succeeding discussions on immune responsiveness, we will use the more general term *antigen* because it encompasses all possibilities of eliciting immune reactions.

One important characteristic of an antigen is that the immune cells react to it as nonself or as foreign, meaning that it is not a "normal" constituent of the body. Whole microbes or their parts, cells, or substances that arise from other humans, animals, plants, and various molecules all possess this quality of foreignness and thus are potentially antigenic to the immune system of an individual **(figure 15.7)**. Molecules of complex composition such as proteins and protein-containing compounds prove to be more immunogenic than repetitious polymers composed of a single type of unit. Most materials that serve as antigens fall into these chemical categories with examples:

- Proteins and polypeptides (enzymes, cell surface structures, hormones, exotoxins)
- Lipoproteins (cell membranes)
- Glycoproteins (blood cell markers)

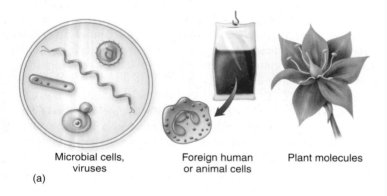

Microbial cells, viruses Foreign human or animal cells Plant molecules

(a)

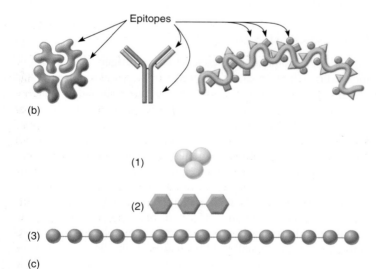

Epitopes

(b)

(1)

(2)

(3)

(c)

Figure 15.7 Characteristics of antigens. (a) Whole cells and viruses make good immunogens. **(b)** Complex molecules with several epitopes make good immunogens. **(c)** Poor immunogens include small molecules not attached to a carrier molecule (1), simple molecules (2), and large but repetitive molecules (3).

- Nucleoproteins (DNA complexed to proteins, but DNA alone is too regular and repetitive in structure)
- Polysaccharides (certain bacterial capsules) and lipopolysaccharides

Effects of Molecular Shape and Size on Antigenicity

To initiate an immune response, a substance must also be large enough to "catch the attention" of the surveillance cells. Molecules with a molecular weight (MW) of less than 1,000 are seldom complete antigens, and those between 1,000 MW and 10,000 MW are weakly so. Complex macromolecules of at least 100,000 MW are the most immunogenic, a category also dominated by large proteins. Note that large size alone is not sufficient for antigenicity; glycogen, a polymer of glucose with a highly repetitious structure, has an MW over 100,000 and is not normally antigenic, whereas insulin, a protein with an MW of 6,000, can be antigenic.

A lymphocyte's capacity to discriminate differences in molecular shape is so fine that it recognizes and responds to only a portion of the antigen molecule. This molecular fragment, called the **epitope** or antigenic determinant, is the primary signal that the molecule is foreign **(figure 15.7b)**. The shape of this determinant fits like a key to the receptor "lock" of the lymphocyte, which then recognizes and responds to it. Certain amino acids accessible at the surface of proteins or protruding carbohydrate side chains are typical examples. Many foreign cells and molecules are very complex antigenically, with numerous determinants, each of which will elicit a separate and different lymphocyte response. Examples of these multiple, or *mosaic,* antigens include bacterial cell wall, membrane, flagellar, capsular, and toxin antigens; and viruses, which express various surface and core antigens.

Small foreign molecules that consist only of a determinant group and are too small by themselves to elicit an immune response are termed **haptens.** However, if such an incomplete antigen is linked to a larger carrier molecule, the combination develops immunogenicity **(figure 15.8)**. The carrier group contributes to the size of the complex and enhances the orientation of the antigen, while the hapten serves as the epitope. Haptens include such molecules as vaccine antigens; drugs; metals; and household, industrial, and environmental chemicals. Many haptens develop immunogenicity in the body by combining with large carrier molecules such as serum proteins (see allergy in chapter 16).

Functional Categories of Antigens

Because each human being is genetically and biochemically unique (except for identical twins), the proteins and other molecules of one person can be antigenic to another. **Alloantigens** are cell surface markers and molecules that occur in some members of the same species but not in others. Alloantigens are the basis for an individual's blood group and major histocompatibility profile, and they are responsible for incompatibilities that can occur in blood transfusion or organ grafting.

Some bacterial proteins, called **superantigens,** are potent stimuli for T cells. Notable examples of these types of antigens are toxic shock toxin and enterotoxin of *Staphylococcus.* Reactions with superantigens may give rise to massive release of cytokines and cell death. The role of these compounds is covered in section 15.4.

Figure 15.8 **The hapten-carrier phenomenon. (a)** Haptens are antigens that are too small to be discovered by an animal's immune system and do not elicit a response. **(b)** When a hapten binds to a large molecule, the hapten serves as an epitope and stimulates a response and an antibody that is specific for it.

Antigens that evoke allergic reactions, called **allergens,** will be characterized in detail in chapter 16.

On occasion, even a part of the body can take on the character of an antigen. During lymphocyte differentiation, immune tolerance to self tissue occurs, but a few anatomical sites can contain sequestered (hidden) molecules that escape this assessment. Such molecules, called **autoantigens,** can occur in tissues (eye, thyroid gland, for example) that are walled off early in embryonic development before the surveillance system is in complete working order. Because tolerance to these substances has not yet been established, they can subsequently be mistaken as foreign; this mechanism appears to account for some types of autoimmune diseases such as rheumatoid arthritis (chapter 16).

Check & Assess Section 15.2

✔ Immature lymphocytes released from hematopoietic tissue undergo further development or maturation at different sites. B cells mature in the stromal cells of the bone marrow. T cells mature in the thymus.

✔ Antigens or immunogens are foreign proteins or other complex molecules of high molecular weight that trigger the immune response in the host.

✔ Lymphocytes respond to a specific portion of an antigen called the epitope that binds to their receptors. A given microorganism has many such epitopes, all of which stimulate individual specific immune responses.

✔ Haptens are molecules that are too small to trigger an immune response alone but can be immunogenic when they attach to a larger substance, such as host serum protein.

✔ Autoantigens and allergens are types of antigens that cause damage to host tissue as a consequence of the immune response.

8. What is happening during lymphocyte maturation?
9. Describe three ways in which B cells and T cells are similar and at least five major ways in which they are different.
10. What is an antigen, immunogen, and epitope?
11. How do foreignness, size, and complexity contribute to antigenicity?
12. Compare five unique types of antigens, and explain their functions or actions.

15.3 Cooperation in Immune Reactions to Antigens

Expected Learning Outcomes

11. Describe the cooperative interactions between antigen-presenting cells, T cells, and B cells.

12. Discuss the actions of interleukins in the early reactions of recognition and activation.

13. List and analyze the stages in activation of B lymphocytes, clonal expansion, and antibody formation.

14. Describe the structure and basic functions of the five major types of immunoglobulins.

15. Explain the actions of antibodies in protective immune reactions.

16. Analyze the primary and secondary responses to antigens and immunogens, including the importance of the anamnestic response and boosters.

17. Define monoclonal antibodies, and explain how they are produced and used.

The basis for most immune responses is the encounter between antigens and white blood cells. Microbes and other foreign substances enter most often through the respiratory or gastrointestinal mucosa and less frequently through other mucous membranes or the skin. Antigens introduced intravenously become localized in the liver, spleen, bone marrow, kidney, and lung. If introduced by some other route, antigens are carried in lymphatic fluid and concentrated by the lymph nodes. The lymph nodes and spleen are important in concentrating the antigens in areas where they will come in contact with antigen-presenting cells (APCs) and lymphocytes. APCs and lymphocytes can subsequently circulate into fluid compartments to seek out the antigens for which they are specific.

The Role of Antigen Processing and Presentation

In most immune reactions, the antigen is in a "raw" state and must be further acted upon by **antigen-presenting cells (APCs)** before it is

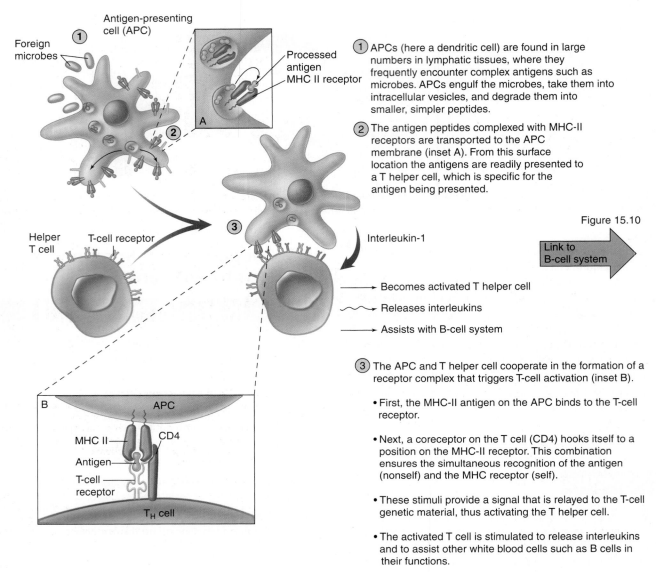

Antigen-presenting cell (APC)

Foreign microbes

Processed antigen
MHC II receptor

A

Helper T cell T-cell receptor

Interleukin-1

Becomes activated T helper cell

Releases interleukins

Assists with B-cell system

Figure 15.10

Link to B-cell system

① APCs (here a dendritic cell) are found in large numbers in lymphatic tissues, where they frequently encounter complex antigens such as microbes. APCs engulf the microbes, take them into intracellular vesicles, and degrade them into smaller, simpler peptides.

② The antigen peptides complexed with MHC-II receptors are transported to the APC membrane (inset A). From this surface location the antigens are readily presented to a T helper cell, which is specific for the antigen being presented.

③ The APC and T helper cell cooperate in the formation of a receptor complex that triggers T-cell activation (inset B).

- First, the MHC-II antigen on the APC binds to the T-cell receptor.

- Next, a coreceptor on the T cell (CD4) hooks itself to a position on the MHC-II receptor. This combination ensures the simultaneous recognition of the antigen (nonself) and the MHC receptor (self).

- These stimuli provide a signal that is relayed to the T-cell genetic material, thus activating the T helper cell.

- The activated T cell is stimulated to release interleukins and to assist other white blood cells such as B cells in their functions.

B APC

MHC II CD4

Antigen

T-cell receptor

T$_H$ cell

Process Figure 15.9 Interactions between antigen-presenting cells (APCs) and T helper (CD4) cells required for T-cell activation. For T cells to recognize foreign antigens, they must have the antigen processed and presented by a professional APC such as a dendritic cell. See Figure 15.10 for next phase.

presented to T cells. Three different cells can serve as APCs: macrophages, **dendritic* cells,** and B cells. Antigen-presenting cells modify the antigen so that it will be more immunogenic and recognizable. After processing is complete, the antigen is moved to the surface of the APC and bound to an MHC class II receptor to make it readily accessible to T cells during presentation **(figure 15.9).**

Before a T cell can respond to APC-bound antigens, certain conditions must be met. *T-cell-dependent antigens,* usually protein-based, require recognition steps between the APC, antigen, and lymphocytes. The first cells on the scene to assist in activating B cells and other T cells are a special class of **T helper cells (T$_H$).** The T cell receptor (TCR) of this class of T cell will bind simultaneously with the class II MHC receptor on the APC and with the antigen (figure 15.9). A second interaction involves the binding of the T-cell

CD4 receptor to the MHC of the APC. Once this identification has occurred, a cytokine, **interleukin-1 (IL-1),** produced by the APC, activates this T helper cell. The T$_H$ cell, in turn, produces a different cytokine, **interleukin-2 (IL-2),** that stimulates a general increase in activity of committed B and T cells. The manner in which B and T cells subsequently become activated by the APC–T helper cell complex and their individual responses to antigen are addressed in the next two sections.

A few antigens can trigger a response from B lymphocytes without the cooperation of APCs or T helper cells. These *T-cell-independent antigens* are usually simple molecules such as carbohydrates with many repeating and invariable determinant groups. Examples include lipopolysaccharide from the cell wall of *Escherichia coli,* polysaccharide from the capsule of *Streptococcus pneumoniae,* and molecules from rabies and Epstein-Barr virus. Because so few antigens are of this type, most B-cell reactions require assistance from T helper cells.

* *dendritic* (den′-drih-tik) Gr. *dendron,* branching like a tree.

1. Clonal Selection and Antigen Binding
B cells can independently recognize microbes (example here is a virus) and their foreign antigens, and can bind them with their Ig receptors. This is how the initial selection of the antigen-specific B-cell clone occurs.

2. Antigen Processing and Presentation
Once the microbe is attached, the B cell endocytoses it, processes it into smaller protein units, and displays these on the MHC II complex (similar to other APCs). This event readies the antigen for presentation to a specific T$_H$ cell.

3. B Cell/T$_H$ Cell Cooperation and Recognition
For most B cells to become functional, they must interact with a T helper cell that bears receptors for antigen from the same microbe. This T cell may have been activated by an APC (see figure 15.9). The two cells engage in linked recognition, in which the MHC II receptor bearing antigen on the B cell binds to both the T-cell antigen receptor and the CD4 molecule on the T cell (inset).

4. B-Cell Activation
The T cell gives off additional signals in the form of interleukins and B-cell growth factors. The linked receptors and the chemical stimuli serve to activate the B cell. Such activation signals an increase in cell metabolism, leading to cell enlargement, proliferation, and differentiation.

5–6. Clonal Expansion/Memory Cells
The activated B cell undergoes numerous mitotic divisions, which expand the clone of cells bearing this specificity and produce memory cells and plasma cells. The memory cells are persistent, long-term cells that can react with the same antigen on future exposures.

7. Plasma Cells/Antibody Synthesis
The plasma cells are short-lived, active secretory cells that synthesize and release antibodies. These antibodies (here IgM) have the same specificity as the Ig receptor and circulate in the fluid compartments of the body, where they react with the same antigens and microbes shown in step 1.

Process Figure 15.10 Events in B-cell activation and antibody synthesis. *Only key receptors for these reactions are shown.

B-Cell Responses

Activation of B Lymphocytes: Clonal Selection, Expansion, and Antibody Production

The immunologic activation of most B cells requires a series of events (**figure 15.10**).

1. **Clonal selection and binding of antigen.** In this case, a precommitted B cell of a particular clonal specificity picks up the antigen on its Ig receptors and processes it into small peptide determinants. The antigen is then bound to the MHC II receptors

on the B cell. The MHC/Ag receptor on the B cell is bound by a T$_H$ cell.

2. **Instruction by chemical mediators.** The B cell receives developmental signals from macrophages and T cells (interleukin-2 and interleukin-6) and various other growth factors, such as IL-4 and IL-5.

3. The combination of these stimuli on the membrane receptors causes a signal to be transmitted internally to the B-cell nucleus.

4. These events trigger B-cell activation. An activated B cell called a *lymphoblast* undergoes an increase in size and DNA and protein synthesis, in preparation for entering the cell cycle and mitosis.

Figure 15.11 **Working models of antibody structure.** **(a)** Diagrammatic view of IgG depicts the principal regions (Fabs and Fc) of the molecule. Note that the Fabs can swivel at the hinge region and provide flexibility in position. **(b)** Realistic model of immunoglobulin shows the tertiary and quaternary structure achieved by additional intrachain and interchain bonds and the position of the carbohydrate component. The individual spheres represent amino acids in the protein.

5–6. Clonal expansion. A stimulated B cell multiplies through successive mitotic divisions and produces a large population of genetically identical daughter cells. Some cells that stop short of becoming fully differentiated are **memory cells,** which remain for long periods to react with that same antigen at a later time. This reaction also expands the clone size, so that subsequent exposure to that antigen provides more cells with that specificity. This expansion of the clone size is one factor in the increased memory response. The most numerous progeny are large, specialized, terminally differentiated B cells called **plasma cells.**

7. Antibody production and secretion. The primary action of plasma cells is to secrete into the surrounding tissues copious amounts of antibodies with the same specificity as the original receptor (figure 15.10). Although an individual plasma cell can produce around 2,000 antibodies per second, production does not continue indefinitely. The plasma cells do not survive for long and deteriorate after they have synthesized antibodies.

Products of B Lymphocytes: Antibody Structure and Functions

The Structure of Immunoglobulins Earlier we saw that a basic **immunoglobulin (Ig)**[3] molecule contains four polypeptide chains connected by disulfide bonds. Because antibodies are a type of immunoglobulin, they will show this same basic structure. Let us review it using an IgG molecule as a model. Two functionally distinct segments called *fragments* can be differentiated. The two "arms" that bind antigen are termed **antigen binding fragments (Fabs),** and the rest of the molecule is the **crystallizable fragment (Fc),** so called because it can be crystallized in pure form. The amino-terminal end of each antigen binding fragment (consisting of the variable regions of the heavy and light chains) folds into a groove that will accommodate one epitope of the antigen. The presence of a special region at the site of attachment between the Fabs and the Fc allows swiveling of the Fabs. In this way, they can change their angle to accommodate nearby antigen sites that vary slightly in distance and position. The Fc portion is involved in binding to various cells and molecules of the immune system itself. **Figure 15.11** shows two views of antibody structure.

Antibody-Antigen Interactions and the Function of the Fab The site on the antibody where the antigenic determinant binds is composed of *hypervariable regions* with an extremely variable amino acid content. The groove of this antigen binding site has a specific three-dimensional fit for the antigen (**figure 15.12**). The specificity of antigen binding sites for antigens is very similar to enzymes and substrates, and some antibodies have been used as enzymes. So specific are some immunoglobulins for antigen

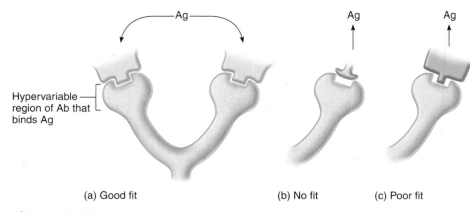

(a) Good fit (b) No fit (c) Poor fit

Figure 15.12 **Antigen-antibody binding.** The union of antibody (Ab) and antigen (Ag) is characterized by a certain degree of fit and is supported by a multitude of weak linkages, especially hydrogen bonds and electrostatic attraction. A better fit (i.e., antigen in **(a)** vs antigen in **(c)**,) provides greater lymphocyte stimulation during the activation stage.

3. *Ig* is a shorthand term. Immunoglobulins are proteins found in the globulin fraction of the serum and have immune function as antibodies.

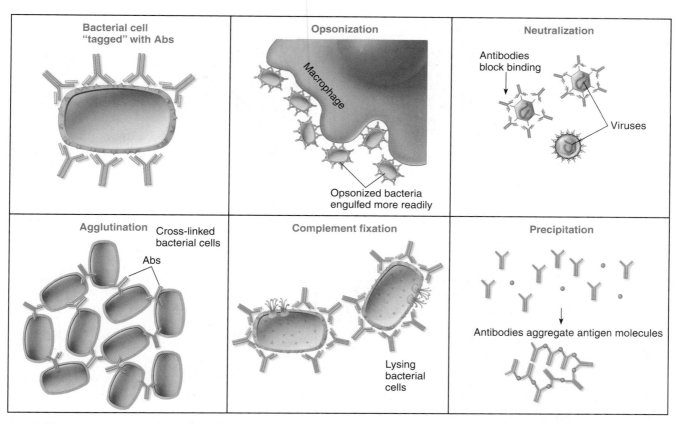

Figure 15.13 Summary of antibody functions. Complement fixation, agglutination, and precipitation are covered further in chapter 17.

that they can distinguish between a single functional group of a few atoms. Because the specificity of the Fab sites is identical, an Ig molecule can bind antigenic determinants on the same cell or on two separate cells and thereby link them.

The principal activity of an antibody is to unite with, immobilize, call attention to, or neutralize the antigen that initiated the antibody's formation **(figure 15.13).** Antibodies called opsonins stimulate **opsonization***, a process in which microorganisms or other particles are coated with specific antibodies so that they will be more readily recognized by phagocytes, which dispose of them. Opsonization has been likened to putting handles on a slippery object to provide a better grip. Antibodies can aggregate or **agglutinate** cells by cross-linking them into large clumps. Agglutination renders microbes immobile and enhances their phagocytosis. **Precipitation** is a similar reaction that occurs with small free antigen molecules. Both processes provide a basis for certain immune tests discussed in chapter 17. The interaction of an antibody with complement can result in the specific rupturing of cells and some viruses. This is seen in complement fixation, which refers to the classical pathway of complement activation (see figures 14.20 and 17.13). In **neutralization** reactions, antibodies fill the surface receptors on a virus or the active site on a bacterial protein, which prevents them from attaching to their target cells. **Antitoxins** are special types of antibodies that neutralize bacterial exotoxins. It should be noted that not all antibodies are protective; some neither benefit nor harm, and a few, such as autoantibodies and those originating from allergic reactions, can actually cause diseases.

Functions of the Crystallizable Fragment Although the Fabs bind antigen, the Fc has a different binding function. In most classes of immunoglobulin, the end of the Fc contains an effector that can bind to the membrane of cells, such as macrophages, neutrophils, eosinophils, mast cells, basophils, and lymphocytes. The effect of an antibody's Fc binding to a cell depends upon that cell's role. In the case of opsonization, the attachment of antibody to foreign cells and viruses exposes the Fcs to phagocytes.

Certain antibodies have regions on the Fc portion for fixing complement, and in some immune reactions, the binding of the Fc causes the release of cytokines. For example, the Fc end of the antibody of allergy (IgE) binds to basophils and mast cells, which causes the release of allergic mediators such as histamine. The size and amino acid composition of the Fc also determine an antibody's permeability, its distribution in the body, and its class.

Accessory Molecules on Immunoglobulins All antibodies contain molecules in addition to the basic polypeptides. Varying amounts of carbohydrates are affixed to the constant regions in most instances (see figure 15.11b and **table 15.2.)** Two additional accessory molecules are the *J chain,* which helps keep the monomers of IgA and IgM together, and the *secretory component,* which helps move IgA across mucous membranes. These proteins occur only in certain immunoglobulin classes.

The Classes of Immunoglobulins Immunoglobulins exist as structural and functional classes called *isotypes* (compared and contrasted in table 15.2). The differences in these classes are due

* *opsonization* (awp″-son-uh-zay′-shun) L., *opsonium,* food or provisions, and *izare,* to become.

TABLE 15.2 Characteristics of the Immunoglobulin (Ig) Classes

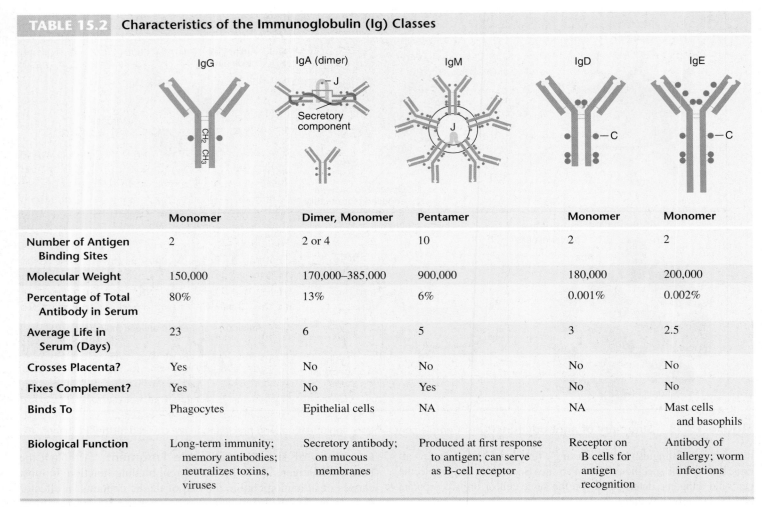

	IgG	IgA (dimer)	IgM	IgD	IgE
	Monomer	Dimer, Monomer	Pentamer	Monomer	Monomer
Number of Antigen Binding Sites	2	2 or 4	10	2	2
Molecular Weight	150,000	170,000–385,000	900,000	180,000	200,000
Percentage of Total Antibody in Serum	80%	13%	6%	0.001%	0.002%
Average Life in Serum (Days)	23	6	5	3	2.5
Crosses Placenta?	Yes	No	No	No	No
Fixes Complement?	Yes	No	Yes	No	No
Binds To	Phagocytes	Epithelial cells	NA	NA	Mast cells and basophils
Biological Function	Long-term immunity; memory antibodies; neutralizes toxins, viruses	Secretory antibody; on mucous membranes	Produced at first response to antigen; can serve as B-cell receptor	Receptor on B cells for antigen recognition	Antibody of allergy; worm infections

C = carbohydrate. J = J chain.

primarily to variations in the Fc. The classes are differentiated with shorthand names—Ig, followed by a letter: IgG, IgA, IgM, IgD, IgE.[4]

The structure of **IgG** has already been presented. It is a monomer[5] produced by a plasma cell late in a primary response and by memory cells responding the second time to a given antigenic stimulus. It is by far the most prevalent antibody circulating throughout the blood, lymph, and extracellular fluids. It has numerous functions: It neutralizes toxins, opsonizes, and fixes complement, and it is the only antibody capable of crossing the placenta.

The two forms of **IgA** are (1) a monomer that circulates in small amounts in the blood; and (2) a dimer that is a significant component of the mucous and serous secretions of the salivary glands, intestine, nasal membrane, breast, lung, and genitourinary tract. The dimer, called secretory IgA, is formed in a plasma cell by two monomers held together by a J chain. To facilitate the transport of IgA across membranes, a secretory piece is later added by the epithelial cells of the mucosa. IgA coats the surface of these membranes and appears free in saliva, tears, colostrum, and mucus. It

confers essential local immunity against enteric, respiratory, and genitourinary pathogens by excluding foreign intruders on mucosal membranes. Its contribution in protecting newborns who derive it passively from nursing is mentioned in Insight 15.2.

IgM is a huge molecule composed of five monomers (making it a pentamer) attached by the Fc portions to a central J chain. With its 10 binding sites, this molecule has tremendous capacity for binding antigen. IgM is the first class synthesized following the host's first encounter with antigen. Its ability to agglutinate and Fix *complement* make it an important antibody in many immune reactions. It circulates mainly in the blood and is far too large to cross the placental barrier.

IgD is a monomer found in minuscule amounts in the serum, and it does not fix complement, opsonize, or cross the placenta. Its main function is to serve as a receptor for antigen on B cells, usually along with IgM. It seems to be the triggering molecule for B-cell activation.

IgE is also an uncommon blood component unless one is allergic or has a parasitic worm infection. Its Fc region interacts with receptors on mast cells and basophils. Its biological significance is to stimulate an inflammatory response through the release of potent physiological substances by the basophils and mast cells. Because inflammation would enlist blood cells such as eosinophils and lymphocytes to the site of infection, it would certainly be one defense against parasites. Unfortunately, IgE has another, more insidious effect—that of mediating anaphylaxis, asthma, and certain other allergies (Chapter 16).

4. The letters correspond to Greek letters gamma, alpha, mu, delta, and epsilon, which are also used to designate the structure of their constant regions.

5. *Monomer* means "one unit" or "one part." Accordingly, *dimer* means "two units," *pentamer* means "five units," and *polymer* means "many units."

Evidence of Antibodies in Serum

Regardless of the site where antibodies are first secreted, a large quantity eventually ends up in the blood by way of the body's communicating networks. If one subjects a sample of **antiserum** (serum containing specific antibodies) to electrophoresis, the major groups of proteins migrate in a pattern consistent with their mobility and size **(figure 15.14)**. The albumins show up in one band, and the globulins in four bands called alpha-1 (α_1), alpha-2 (α_2), beta (β), and gamma (γ) globulins. Most of the globulins represent antibodies, which explains how the term *immunoglobulin* was derived. **Gamma globulin** is composed primarily of IgG, whereas β and α_2 globulins are a mixture of IgG, IgA, and IgM.

Monitoring Antibody Production over Time: Primary and Secondary Responses to Antigens

We can learn a great deal about how the immune system reacts to an antigen by studying the levels of antibodies in serum over time **(figure 15.15)**. This level is expressed quantitatively as the **titer,**[6] or concentration, of antibodies. Upon the first exposure to an antigen or immunogen, the system undergoes a **primary response.** The earliest part of this response, the *latent period,* is marked by a lack of antibodies for that antigen, but much activity is occurring. During this time, the antigen is being concentrated in lymphoid tissue and is being processed by the correct clones of B lymphocytes. As plasma cells synthesize antibodies, the serum titer increases to a certain plateau and then tapers off to a low level over a few weeks or months.

When the class of antibodies produced during this response is tested, an important characteristic of the response is uncovered. It turns out that, early in the primary response, most of the antibodies are the IgM type, which is the first class to be secreted by plasma cells. Later, the class of the antibodies (but not their specificity) is switched to IgG or some other class (IgA or IgE).

Figure 15.14 Pattern of human serum after electrophoresis. When antiserum is subjected to electrical current, the various proteinaceous components are separated into bands. This is a means of separating the different antibodies and serum proteins as well as quantifying them. By far the most significant immune component of serum is found in the gamma (γ) band, which is high in IgG and IgM.

6. *Titer* (ty'-tur) Fr. *titre,* standard. One method for determining titer is shown in figure 17.9.

Primary response. A latent period with no measurable antibody occurs early on. The first antibody to appear is IgM, followed later by IgG arising from activation of the first memory cells. Within weeks, the titer tapers back to low levels.

Secondary response. A latent period is lacking because other memory lymphocytes from the earlier response are immediately ready to react. A rapid rise in antibody titer, mainly of IgG, is sustained for several weeks. A smaller amount of IgM is also produced by naive B cells.

Figure 15.15 A graphic view of the primary and secondary immune responses to antigenic stimuli. Monitoring the antibody titer or concentration throughout the initial (primary) response and the second (secondary) response produces a distinct pattern. Note that the anamnestic response yields an amount of antibody that is nearly 1000 times that of the primary response (the scale is logarithmic). This in effect "boosts" this defense to even more effective levels.

When the immune system is exposed again to the same antigen or immunogen within weeks, months, or even years, a **secondary response** occurs. The rate of antibody synthesis, the peak titer, and the length of antibody persistence are greatly increased over the primary response. The rapidity and amplification seen in this response are attributable to the memory B cells that were formed during the primary response. Because of its association with recall, the secondary response is also called the **anamnestic* response.** The recall effect shortens the latent or lag period and yields a faster, stronger, and longer-lasting antibody response. Factors that account for the anamnestic effect are: the memory B cells do not have to go through the early steps in activation, and they do not require as many signals to form plasma cells.

The advantage of this response is evident: It provides a quick and potent strike against subsequent exposures to infectious agents. This memory effect is the fundamental basis for vaccine boosters, which we discuss in section 15.6.

Monoclonal Antibodies: Useful Products from Cancer Cells

The value of antibodies as tools for locating or identifying antigens is well established. For many years, antiserum extracted from human or animal blood was the main source of antibodies for tests and therapy, but most antiserum has a basic problem. It contains **polyclonal antibodies,** meaning that it is a mixture of different antibodies because it reflects dozens of immune reactions from a wide variety of B-cell clones. This characteristic is to be expected, because several immune reactions may be occurring simultaneously; and even a single species of microbe can induce several different types of antibodies. Certain applications in immunology require a pure preparation of **monoclonal antibodies (MABs)** that originate from a single clone and have a single specificity for antigen. The development of human MABs and other novel approaches using genetic engineering is currently under way **(Insight 15.1).**

Check & Assess Section 15.3

✔ Antigen-presenting cells (APCs) such as dendritic cells and macrophages bind foreign antigen to their cell surfaces for presentation to T lymphocytes. Physical binding between the APC, T helper cells, and B cells and stimulation by interleukins activate lymphocytes to proceed with their respective immune responses.

✔ Lymphocytes proliferate, producing clones of progeny that include groups of responder cells and memory cells.

✔ Activated B lymphocytes become plasma cells that produce and secrete antibodies.

✔ B cells produce five classes of immunoglobulins (antibodies): IgM, IgG, IgA, IgD, and IgE. IgG and IgM are important antibodies in the globulin fraction of plasma. IgA provides protection in body secretions. IgD is expressed on B cells as an antigen receptor. IgE binds to tissue cells, promoting inflammation.

✔ Antibodies bind physically to the specific antigen that induces their production, thereby immobilizing the antigen and enabling it to be destroyed by other components of the immune system.

* *anamnestic* (an-am-ness´-tik) Gr. *anamnesis,* a recalling.

✔ The memory response means that the second exposure to antigen calls forth a much faster and more vigorous response than the first.

✔ Monoclonal antibodies are immunoglobulins produced by technology for treating disease and immune research.

13. Describe the actions of an antigen-presenting cell.
14. What is the difference between a T-cell-dependent and T-cell-independent response?
15. Trace the immune response system, beginning with the entry of a T-cell-dependent antigen, antigen processing, presentation, the cooperative response among the macrophage and lymphocytes, and the reactions of activated B and T cells.
16. What are the actions of interleukins 1, 2, 4, and 12?
17. On what basis is a particular B-cell clone selected?
18. How are B cells activated, what events are involved in this process, and what happens after B cells are activated?
19. What are the functions of plasma cells, clonal expansion, and memory cells?
20. Describe the structure of immunoglobulin and the functions of the Fab and Fc portions.
21. Describe four or five ways that antibodies function in immunity.
22. Describe the attachment of Abs to Ags. What eventually happens to the Ags?
23. Contrast the primary and secondary response to Ag.
24. Explain the type, order of appearance, and amount of immunoglobulin in each response and what causes the differences.
25. What causes the latent period and the anamnestic responses?
26. Describe several possible applications of monoclonals in medicine.

15.4 T-Cell Responses

Expected Learning Outcomes

18. Describe the main activities of cell-mediated immunities.

19. Analyze the relationships between the types of T cells and their receptors, and how they are activated.

20. Describe the primary functions of helper T cells and cytotoxic T cells.

21. Explain the concepts of natural killer cells and superantigens.

Cell-Mediated Immunity (CMI)

During the time that B cells are actively responding to antigens, the T-cell portion of the immune system is equally active. The responses of T cells, however, are considered **cell-mediated immunities** (CMI). These reactions are among the most complex and diverse in the immune system. They involve several subsets of T cells that differ in the types of CD receptors and the precise actions they have against foreign antigens and cells. T cells are restricted. Before they can be activated, they must have the antigen offered by an MHC complex on an APC to ensure recognition of self.

INSIGHT 15.1

Monoclonal Antibodies: Variety without Limit

Imagine releasing millions of tiny homing pigeons into a molecular forest and having them navigate directly to their proper roost, and you have some sense of what monoclonal antibodies can do.

The technology for producing monoclonal antibodies is possible by hybridizing cancer cells and activated B cells in vitro. This technique began with the discovery that tumors isolated from multiple myelomas in mice consist of identical plasma cells. These monoclonal plasma cells secrete a strikingly pure form of antibodies with a single specificity and continue to divide indefinitely. Immunologists recognized the potential in these plasma cells and devised a **hybridoma** approach to creating **mono**clonal **anti**bodies or **MABs**. The basic idea behind this approach is to hybridize or fuse a **myeloma*** cell with a normal plasma cell from a mouse spleen to create an immortal cell that secretes a supply of functional antibodies with a single specificity.

Monoclonals can be hybridized with plant or bacterial toxins to form immunotoxin complexes that attach to a target cell and poison it. The most exciting prospect of this therapy is that it can target a specified cancer cell and not normal cells. This type of monoclonal is currently employed in treatment of breast and colon cancers. Drugs such as Rituxan and Herceptin bind to cancer cells and trigger their death.

This technology is also being used for a wide variety of biomedical applications. Monoclonal antibodies have provided immunologists with excellent standardized tools for studying the immune system and for expanding disease diagnosis and treatment. Many of the successful applications thus far use MABs in in vitro diagnostic testing and research.

Another approach with some success has been the development of "humanized" monoclonals to be used in therapeutic drugs. This process involves plants and mice genetically engineered to produce human MABs. About 20 drugs are currently on the market, mainly for treating cancer and preventing immune diseases. See **table 15.A** for examples of these drugs.

* *myeloma* (my-uh-loh′-muh) Gr. *myelos,* marrow, and *oma,* tumor. A malignancy of the bone marrow.

TABLE 15.A	Selected Monoclonal Antibody–Based Drugs	
Drug Names (Chemical* and Trade)		**Used in Therapy for**
Cancer Drugs		
Trastuzumab	Herceptin	Breast cancer
Retuximab	Rituxan	Non-Hodgkin's lymphoma
Bevacizumab	Avastin	Colorectal cancer/lung cancer
Gemtuzumab	Mylotarg	Acute myelogenous leukemia
Others		
Omalizumab	Xolair	Asthma
Infliximab	Remicade	Crohn's disease
Palivizumab	Synagis	Respiratory syncytial virus (RSV)

*Note that these names are coded to end in *mab*—short for monoclonal antibody.

What would be the major drawback to using purely mouse-derived monoclonal antibodies in therapy? *Answer available at http://www.mhhe.com/talaro8*

Summary of the technique for producing monoclonal antibodies by hybridizing myeloma tumor cells with normal plasma cells.

Antigen Isolated myeloma cells

(a) A normal mouse is inoculated with an antigen having the desired specificity, and activated cells are isolated from its spleen. A special strain of mouse provides the myeloma cells.

Mouse spleen cells producing antibody Myeloma cells in culture

Fusion into hybridoma

(b) The two cell populations are mixed with polyethylene glycol, which causes some cells in the mixture to fuse and form hybridomas.

Selection of Ab-producing clones

(c) Surviving cells are cultured and separated into individual wells. Tests are performed on each hybridoma to determine the specificity of the antibody (Ab) it secretes.

Maintenance of hybridoma Product

Hybridoma cells injected into mice for culture Purified monoclonal antibodies

(d) A hybridoma with the desired specificity is grown in tissue culture. The hybridoma is maintained in a susceptible mouse for future use. Antibody product is isolated and purified.

All produce cytokines that, working together, have a spectrum of biological effects and immune functions.

T cells have other notable differences in function from B cells. Rather than releasing antibodies to control foreign antigens, the whole T cell reacts directly in contact with other cells rather than by secretion of molecules into circulation. T cells also stimulate other T cells, B cells, and phagocytes.

The Activation of T Cells and Their Differentiation into Subsets

The mature T cells in lymphoid organs are primed to react with antigens that have been processed and presented to them by dendritic cells and macrophages. They recognize an antigen only when it is presented in association with an MHC carrier (**figure 15.16**). T cells with CD4 receptors recognize endocytosed peptides presented on MHC II, and T cells with CD8 receptors recognize peptides presented on MHC I.

A T cell is initially sensitized by the binding of antigen/MHC to its receptors. As with B cells, activated T cells transform in preparation for mitotic divisions, and they divide into one of the subsets of effector cells and memory cells that can interact with the antigen upon subsequent contact (**table 15.3).** Memory T cells are some of the longest-lived blood cells known (decades, rather than weeks or months for other lymphocytes).

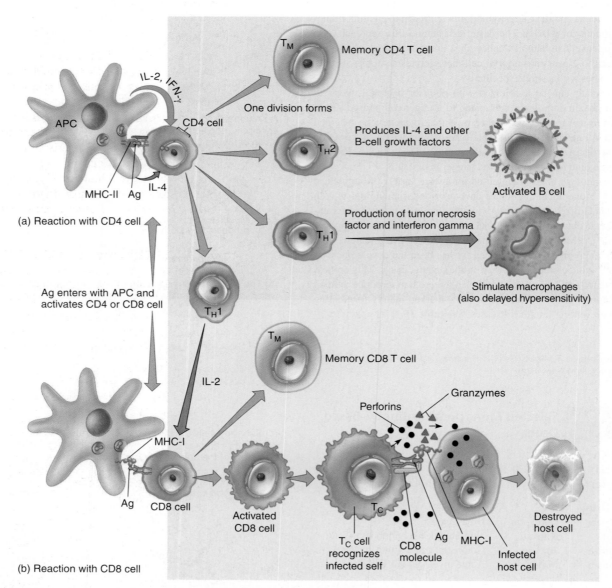

Figure 15.16 **Overall scheme of T-cell activation and differentiation into different types of T cells.** Antigen-presenting cells present antigenic peptides to T cells bearing either CD4 (a) or CD8 (b) markers. **(a)** CD4 cells bind antigen/MHC-II complexes on APCs and, depending on the type of cytokine released by the APC, become either T_H1 or T_H2 cells. T_H1 cells synthesize IL-2 that activates CD8 cells and causes macrophages to destroy ingested microbes or to become more active. T_H2 cells secrete cytokines that enhance B-cell activation. **(b)** Upon binding antigen/MHC-I complex, CD8 cells become cytotoxic T (T_C) cells, which bind infected host cells displaying antigen/MHC-I complexes and release perforins and granzymes. This leads to the apoptosis (programmed cell death) of the infected cell.

TABLE 15.3	Characteristics of Subsets of T Cells	
Types	**Primary Receptors on T Cell**	**Functions/Important Features**
T helper cell 1 (T_H1)	CD4	Activates other CD4 and CD8 cells; secretes IL-2, tumor necrosis factor, and interferon gamma; responsible for delayed hypersensitivity; interacts with MHC II receptors
T helper cell 2 (T_H2)	CD4	Drives B-cell proliferation; secretes IL-4, IL-5, IL-6, IL-10; can dampen T_H1 activity
T cytotoxic cell (T_C)	CD8	Destroys a target foreign cell by lysis; important in destruction of cancer cells, virus-infected cells; graft rejection; requires MHC I for function

T Helper (T_H) Cells: Activators of Specific Immune Responses **T helper (CD4) cells** play a central role in regulating immune reactions to antigens, including those of B cells and other T cells. They are also involved in activating macrophages and increasing phagocytosis. They do this directly by receptor contact and indirectly by releasing cytokines such as interleukin-2, which stimulates the primary growth and activation of B and T cells, and interleukins-4, -5, and -6, which stimulate various activities of B cells. T helper cells are the most prevalent type of T cell in the blood and lymphoid organs, making up about 65% of this population. The severe depression of the CD4 class of T cells by HIV is a major factor in the immunopathology of AIDS.

When T helper cells are activated by antigen/MHC II, they differentiate into either **T helper 1** (T_H1) cells, or **T helper 2** (T_H2) cells, depending on what type of cytokines the antigen-presenting cells secrete. It is thought that if the dendritic cell (APC) secretes IL-1 or IL-12, the T cell will become a T_H1 cell that can activate more T cells. It is also involved in delayed hypersensitivity **(figure 15.16a).** Delayed hypersensitivity is a type of response to allergens, distinct from immediate allergies such as hay fever and anaphylaxis. Both of these reactions are discussed in chapter 16.

If the APC secretes another set of cytokines (IL-4, IL-5, IL-6), the T cell will differentiate into a T_H2 cell. These cells have the function of secreting substances that influence B-cell differentiation and consequently affect the antibody response. They may also suppress the response of T_H1 cells when necessary.

Cytotoxic T (T_C) Cells: Cells That Kill Other Cells When CD8 cells are activated by antigen/MHC I, they differentiate into **T cytotoxic cells (killer T cells).**

Cytotoxicity is the capacity of certain T cells to kill a specific target cell. It is a fascinating and powerful property that accounts for much of our immunity to foreign cells and cancer; and yet, under some circumstances, it can lead to disease. For a killer T cell to become activated, it must recognize a foreign peptide carried by an MHC I receptor and mount a direct attack upon the target cell. After activation, the T_C cell severely injures the target cell **(figure 15.16b).** This process involves the secretion of **perforins**[7] and **granzymes.** Perforins are proteins that can punch holes in the membranes of target cells, and granzymes are enzymes that digest proteins. First the perforins cause ions to leak out of target cells and create a passageway for granzymes to enter. Granzymes induce the loss of selective permeability followed by target cell death through a process called *apoptosis**. The apoptosis is genetically programmed and results in destruction of the nucleus and complete cell lysis and death.

Target cells that T_C cells can destroy include the following:

- Virally infected cells (figure 15.16b). Cytotoxic cells recognize these because of virus peptide MHC combinations expressed on their surface. Cytotoxic defenses are an essential protection against viruses.
- Cancer cells. T cells constantly survey the tissues and immediately attack any abnormal cells they encounter **(figure 15.17).** The importance of this function is clearly demonstrated in the susceptibility of T-cell-deficient people to cancer (chapter 16).
- Cells from other animals and humans. Cytotoxic CMI is the most important factor in graft rejection. In this instance, the T_C cells attack the foreign tissues that have been implanted into a recipient's body.

Other Types of Killer Cells **Natural killer (NK)** cells are a type of lymphocyte related to T cells that are part of cytotoxic CMI. They circulate through the spleen, blood, and lungs, and are probably the first killer cells to attack cancer cells and virus-infected cells. They destroy such cells by similar mechanisms as cytotoxic T cells. They are considered part of cell-mediated immunity because their activities are moderated by cytokines such as interleukin-12 and interferon.

7. *Perforin* is from the term *perforate,* to penetrate with holes.

* *apoptosis* (ah-poh-toh′-sis)

Cytotoxic T cell Cancer cell Dead cancer cell Cytotoxic T cell

Figure 15.17 **Attack of the cytotoxic T cells.** Left: T cells recognize and bind to a cancer cell, which they perforate with many small holes. Right: The cancer cell collapses and dies by apoptosis.

T Cells and Superantigens Most of the T cell responses we have covered so far are primarily beneficial and protective. But there is one type of reaction that often has drastic consequences and may lead to serious diseases. Such is the case in exposure of T cells to **superantigens.** These antigens, found primarily in bacteria and viruses, are actually a form of virulence factor. Examples include enterotoxin given off by pathogenic staphylococci, certain toxins of Group A streptococci, and proteins of Epstein-Barr virus. They can provoke overwhelming immune responses by large numbers of T cells regardless of specificity. Superantigen molecules are structured so that they can span both MHC II receptors and some antigen receptors (TCR) on T cells. This can "trick" large numbers of T cells into releasing massive amounts of cytokines such as tumor necrosis factor and interleukins 1 and 6. The massive influx of such potent mediators can cause blood vessel damage, toxic shock, and multiorgan failure. This is thought to be the underlying mechanism of toxic shock syndrome (see chapter 18).

Check & Assess Section 15.4

✔ T cells do not produce antibodies. Instead, they produce different cytokines that play diverse roles in the immune response. Each subset of T cell produces a distinct set of cytokines that stimulate other lymphocytes.

✔ T cells provide cell-mediated immunity, meaning that the T cell reacts with a whole foreign or infected cell.

✔ After interacting with the appropriate MHC-II receptor, CD4 T cells become type 1 or type 2 T helper cells that regulate or facilitate immune functions.

✔ CD8 cells stimulated by MHC-I and antigen can differentiate into cytotoxic cells that destroy abnormal or foreign cells.

✔ Superantigens are microbial toxins and proteins that overstimulate T cells, leading to disease.

27. Why are the immunities involving T cells called cell-mediated?
28. Explain how the different types of T cells are activated and their responses.
29. What are the targets of cytotoxic cells and how do they destroy them?

15.5 A Classification Scheme for Specific, Acquired Immunities

Expected Learning Outcomes

22. Outline the major categories of adaptive immunity.
23. Differentiate between natural and artificial immunities and between active and passive immunities.
24. Expand on the four combinations of the defining categories, using examples.

Humans develop their adaptive immunity by several general routes, as summarized in **figure 15.18.** These categories are based on the source of the immunity and whether the protection is produced by one's own immune system or whether it is received from another person. This classification system is a useful medical guide that can rapidly identify and explain the sources of immunity.

Defining Categories by Mode of Acquisition

Active immunity occurs when an individual receives an immune stimulus (microbe) that activates specific lymphocytes, causing an immune response such as production of antibodies. Active immunity is marked by several characteristics: (1) It is an essential attribute of an immunocompetent individual; (2) it creates a memory that renders the person ready for quick action upon reexposure to that same microbe; (3) it requires several days to develop; and (4) it can last for a relatively long time, sometimes for life. Active immunity can be stimulated by natural or artificial means.

Passive immunity occurs when an individual receives immune substances (primarily antibodies) that were produced actively by the immune system of another human or animal donor. Recipients are protected for a time even though they have not had prior exposure to the microbe. It is characterized by (1) lack of memory for the original antigen, (2) no production of new antibodies against that disease, (3) immediate onset of protection, and (4) short-term effectiveness because antibodies have a limited period of function. Ultimately, the recipient's body disposes of them. Passive immunity can also be natural or artificial in origin.

Natural immunity encompasses immunity that is acquired during any normal biological experiences of an individual but not through medical intervention.

Artificial immunity is protection from infection obtained through medical procedures. This type of immunity is induced by immunization with vaccines and immune serum.

Putting together all combinations of these categories, we arrive at four possible examples of origins of immunity.

1. Natural Active Immunity: Getting an Infection

After recovering from infectious disease, a person may be actively resistant to reinfection for a period that varies according to the disease. In the case of childhood viral infections such as measles, mumps, and rubella, this natural active stimulus provides nearly lifelong immunity. Other diseases result in a less extended immunity of a few months to years (such as pneumococcal pneumonia and shigellosis), and reinfection is possible. Even a subclinical infection can stimulate natural active immunity. This probably accounts for the fact that some people are immune to an infectious agent without ever having been noticeably infected with or vaccinated for it.

2. Natural Passive Immunity: Mother to Child

Natural, passively acquired immunity occurs only as a result of the prenatal and postnatal, mother-child relationship. During fetal life, IgG antibodies circulating in the maternal bloodstream are small enough to pass or be actively transported across the placenta. Antibodies against tetanus, diphtheria, pertussis, and several viruses regularly cross the placenta. This natural mechanism provides an infant with a

Acquired Immunity

Natural Immunity
acquired through the normal life experiences
not induced through medical means.

Artificial Immunity
produced purposefully through
medical procedures (also called immunization).

Active Immunity
is the consequence of
a person developing his
own immune response
to a microbe.

Passive Immunity
is the consequence of
one person receiving
preformed immunity
made by another person.

Active Immunity (same as vaccination)
is the consequence of
a person developing his
own immune response
to a microbe.

Passive Immunity
is the consequence of
one person receiving
preformed immunity
made by another person.

Figure 15.18 **Categories of acquired immunities.** Natural immunities, which occur during the normal course of life, are either active (acquired from an infection and then recovering) or passive (antibodies donated by the mother to her child). Artificial immunities are acquired through medical practices and can be active (vaccinations with antigen, to stimulate an immune response) or passive (immune therapy with a serum containing antibodies).

mixture of many maternal antibodies that can protect it for the first few critical months outside the womb, while its own immune system is gradually developing active immunity. Depending upon the microbe, passive protection lasts anywhere from a few months to a year, but it eventually wanes. Most childhood vaccinations are timed so that there is no lapse in protection against common childhood infections.

Another source of natural passive immunity comes to the baby by way of mother's milk **(Insight 15.2)**. Although the human infant acquires most natural passive immunity in utero and some through nursing, the milk-borne antibodies provide IgA antibodies that react against microbes entering the intestine. This unique protection is not available from transplacental antibodies.

Artificial Immunity: Immunization

Immunization is any clinical process that produces immunity in a subject. Because it is often used to give advance protection against infection, it is also called *immunoprophylaxis*. The details of this form of immunity are presented in section 15.6.

3. Artificial Active Immunization: Vaccination

The term *vaccination* originated from the Latin word *vacca* (cow), because the cowpox virus was used in the first preparation for active immunization against smallpox (see Insight 15.3). Vaccination exposes a person to a specially prepared microbial (antigenic) stimulus, which then triggers the immune system to produce antibodies and lymphocytes to protect the person upon future exposure to that microbe. As with natural active immunity, the degree and length of protection vary. Commercial vaccines are currently available for many diseases.

4. Artificial Passive Immunization: Immunotherapy

In immunotherapy, a patient at risk for acquiring a particular infection is administered a preparation that contains specific antibodies against that infectious agent. Pooled human serum from donor blood (gamma globulin) and immune serum globulins containing high quantities of antibodies are the usual sources. Immune serum globulins are used to protect people who have been exposed to certain diseases such as hepatitis A, rabies, and tetanus.

Check & Assess Section 15.5

✔ Active immunity means that the body has an immune response caused by a disease agent. If you contract the disease, you can develop natural active immunity. If you are vaccinated, your body will produce artificial active immunity.

✔ In passive immunity, you receive antibodies from another person. Natural passive immunity comes from the mother. Artificial passive immunity involves medical therapy with globulins from blood donors.

30. Contrast active and passive immunity in terms of how each is acquired, how long it lasts, whether memory is triggered, how soon it becomes effective, and what immune cells and substances are involved.

31. Name at least two major ways in which natural and artificial immunities are different.

INSIGHT 15.2

Breast Feeding: The Gift of Antibodies

An advertising slogan from the past claims that cow's milk is "nature's most nearly perfect food." One could go a step further and assert that human milk is nature's *perfect* food for young humans. Clearly, it is loaded with essential nutrients, not to mention being available on demand from a readily portable, hygienic container that does not require refrigeration or warming. But there is another and perhaps even greater benefit. During lactation, the breast becomes a site for the proliferation of lymphocytes that produce IgA antibodies that protect the mucosal surfaces from local invasion by microbes. The very earliest secretion of the breast, a thin, yellow milk called **colostrum,** is very high in these antibodies. They form a protective coating in the gastrointestinal tract of a nursing infant that guards against infection by a number of enteric pathogens (*Escherichia coli, Salmonella,* poliovirus, rotavirus). Protection at this level is especially critical because an infant's own antibodies and natural intestinal barriers are not yet developed. As with immunity in utero, the necessary antibodies will be donated only if the mother herself has active immunity to the microbe through a prior infection or vaccination.

In recent times, the ready availability of artificial formulas and the changing lifestyles of women have reduced the incidence of breast feeding. Where adequate hygiene and medical care prevail, bottle-fed infants get through the critical period with few problems because the foods given them are relatively free of pathogens and they have received protection against some childhood infections in utero. Mothers in developing countries with untreated water supplies or poor medical services are strongly discouraged from using prepared formulas, because they can actually inoculate the baby's intestine with pathogens from the formula. Millions of neonates suffer from severe and life-threatening diarrhea that could have been prevented by the hygienic practice of nursing.

In the past, baby formula manufacturers have marketed prepared formula in developing countries. They provided free samples of formula, and once babies were weaned from breast milk, mothers felt pressured to buy formula. The health effects were so damaging that the World Health Organization issued an *International Code on the Marketing of Breast Milk Substitutes,* discouraging formula use in the developing world.

Explain the reasons that donated antibodies (either placental or breast milk) are only a temporary protection. *Answer available at http://www.mhhe.com/talaro8*

CONTINUING CASE FILE 15

Case #1. Despite intensive medical intervention on his behalf, the Canadian boy's condition deteriorated, and he died about 6 weeks after the initial bat bite. By this time, the laboratory had isolated rabies virus from the skin on the back of his neck and from his saliva and tears as well. Genetic testing of the virus revealed that it was a variant carried by silver-haired bats. Due to the boy's contact with his family and medical personnel, it was important that they also be given a course of PEP to prevent them from contracting the disease. This meant that around 60 people were injected with HRIG and rabies vaccine. Fortunately, none of his contacts became infected.

Case #2. The Brazilian youth was one of those rare cases in which medical intervention was successful. When he first came out of the coma, he could open his eyes and respond, and he continued to improve from that point. By November, he was well enough to return home. He was the first recorded case in Brazil to survive a case of rabies. Tests done from skin samples detected rabies virus RNA, which was found to be a variant of the virus carried by vampire bats. One finding that may have made a difference in his case was the early presence of neutralizing antibodies that continued to increase during his hospital stay.

- Provide some likely reasons that the first boy died and the second did not.
- Explain the strategy behind giving several injections of rabies vaccine.

For a wrap-up, see the Case File Perspective on page 482.

15.6 Immunization: Methods of Manipulating Immunity for Therapeutic Purposes

Expected Learning Outcomes

25. Explain the purposes of immunotherapy and immunization.
26. Describe the sources and uses of artificial passive immunization.
27. Describe the concepts behind artificial active immunization or vaccination.
28. Discuss which factors are involved in vaccine development.
29. Identify the major categories of vaccine antigens, citing examples.
30. Discuss some new strategies for developing vaccines.
31. Describe the medical guidelines for vaccination, its side effects, and how it relates to herd immunity.

Methods that actively or passively immunize people are widely used in disease prevention and treatment. The use of these terms is sometimes imprecise; thus, it should be stressed that active immunization, in which a person is administered antigen, is synonymous with vaccination. Passive immunization, in which a person is given antibodies, is a type of **immunotherapy.**

INSIGHT 15.3

The Lively History of Active Immunization

The basic notion of immunization has existed for thousands of years. It probably stemmed from the observation that persons who had recovered from certain communicable diseases rarely, if ever, got a second case. The earliest crude vaccination attempts involved bringing a susceptible person into contact with a diseased person or animal. The first recorded attempt occurred in sixth-century China. It consisted of drying and grinding up smallpox scabs and blowing them with a straw into the nostrils of vulnerable family members.

By the tenth century, this practice had changed to the deliberate inoculation of dried pus from the smallpox pustules of one patient into the arm of a healthy person, a technique later called **variolation** (variola is the smallpox virus). This method was used in parts of Asia for centuries before Lady Mary Montagu brought it to England in 1721. Although the principles of the technique had some merit, unfortunately many recipients and their contacts died of smallpox. This outcome vividly demonstrates a cardinal rule for a workable vaccine: It must contain an antigen that will provide protection but not cause the disease. Variolation was so controversial that any English practitioner caught doing it was charged with a felony.

Eventually, this human experimentation paved the way for the first really effective vaccine, developed by the English physician Edward Jenner in 1796. Jenner conducted the first scientific test of a vaccine, one that had a tremendous impact on the advance of medicine. His work gave rise to the words **vaccine** and *vaccination* (from L., *vacca,* cow), which now apply to any immunity obtained by inoculation with selected antigens. Jenner was inspired by the case of a dairymaid who had been infected by a pustular infection called cowpox. This related virus afflicts cattle but causes a milder condition in humans. She explained that she and other milkmaids had remained free of smallpox. To test the effectiveness of this cross protection, Jenner prepared material from human cowpox lesions and inoculated a young boy. When challenged 2 months later with an injection of crusts from a smallpox patient, the boy proved immune.

Jenner's discovery—that a less pathogenic agent could confer protection against a more pathogenic one—is especially remarkable in view

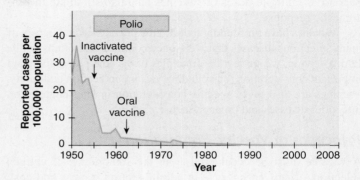

of the fact that microscopy was still in its infancy and the nature of viruses was unknown. At first, the use of the vaccine was regarded with some fear and skepticism. When Jenner's method proved successful and word of its significance spread, it was eventually adopted in many other countries. Eventually, the original virus mutated into a unique strain (*vaccinia* virus) that became the basis of the current vaccine. In 1973, the World Health Organization declared that smallpox had been eradicated. As a result, smallpox vaccination had been discontinued until recently, due to the threat of bioterrorism.

Other historical developments in vaccination included using heat-killed bacteria in vaccines for typhoid fever, cholera, and plague, and techniques for using neutralized toxins for diphtheria and tetanus. Throughout the history of vaccination, there have been vocal opponents and minimizers, but one fact seems clear: Whenever a vaccine has been introduced, the prevalence of that disease has declined dramatically (see graph).

Using principles of immunology, explain how a vaccine based on one type of microbe (vaccinia) could produce an effective immunity to a different microbe (variola). *Answer available at http://www.mhhe.com/talaro8*

Artificial Passive Immunization

The first attempts at passive immunization involved the transfusion of horse serum containing antitoxins to prevent tetanus and to treat patients exposed to diphtheria. Since then, antisera from animals have been largely replaced with products of human origin that function with various degrees of specificity. **Immune serum globulin (ISG),** sometimes called *gamma globulin,* contains immunoglobulin extracted from the pooled blood of at least 1,000 human donors. The method of processing ISG concentrates the antibodies to increase potency and eliminates potential pathogens (such as the hepatitis B and HIV viruses). It is a treatment of choice in preventing measles and hepatitis A and in replacing antibodies in immunodeficient patients. Most forms of ISG are injected intramuscularly to minimize adverse reactions, and the protection they provide lasts 2 to 3 months.

A preparation called **specific immune globulin (SIG)** is derived from a more defined group of donors. Companies that prepare SIG obtain serum from patients who are convalescing and in a hyperimmune state after such infections as pertussis, tetanus, chickenpox, and hepatitis B. This chapter's case file examines how a specific antirabies

immune globulin (HRIG) can be used in cases of rabies exposure. These globulins are preferable to ISG because they contain higher titers of specific antibodies obtained from a smaller pool of patients. Although useful for prophylaxis in persons who may be exposed to infectious agents, these sera are often limited in availability.

When a human immune globulin is not available, antisera and antitoxins of animal origin can be effective. Sera produced in horses are available for diphtheria, botulism, rabies, and spider and snake bites. Unfortunately, horse antigens often stimulate allergies such as serum sickness or anaphylaxis. Although donated immunities only last a relatively short time, they act immediately and can protect patients for whom no other useful medication or vaccine exists, such as immunocompromised patients who cannot be vaccinated.

Artificial Active Immunity: Vaccination

Active immunity can be conferred artificially by **vaccination**—exposing a person to material that is antigenic but not pathogenic. The discovery of vaccination was one of the farthest-reaching and most important developments in medical science (**Insight 15.3**).

The basic principle behind vaccination is to stimulate a primary response that primes the immune system for future exposure to a virulent pathogen. If this pathogen enters the body, the secondary immune response will be immediate, powerful, and sustained. Second or multiple doses of vaccines can also stimulate the anamnestic response.

Vaccines have profoundly reduced the prevalence and impact of many infectious diseases that were once common and often deadly. In this section, we survey the principles of vaccine preparation and important considerations surrounding vaccine indication and safety. Vaccines are also given specific consideration in later chapters on infectious diseases and organ systems.

Principles of Vaccine Preparation

A vaccine must be considered from the standpoints of antigen selection, effectiveness, ease in administration, safety, and cost. In natural immunity, an infectious agent stimulates a relatively long-term protective response. In artificial active immunity, the objective is to obtain this same response with a modified version of the microbe or its components. Qualities of an effective vaccine are listed in **table 15.4.** Most vaccine preparations are based on one of the following antigen preparations **(figure 15.19):**

1. killed whole cells or inactivated viruses;
2. live, attenuated bacterial cells or viruses;
3. antigenic molecules derived from bacterial cells or viruses; or
4. genetically engineered microbes or microbial antigens.

A survey of the major licensed vaccines and their indications is presented in **table 15.5.**

Large, complex antigens such as whole cells or viruses are very effective immunogens. Depending on the vaccine, these are either killed or attenuated. **Killed** or **inactivated vaccines** are prepared by cultivating the desired strain or strains of a bacterium or virus and treating them with formalin, radiation, heat, or some other agent that does not change the antigenic structure. One type of vaccine for the bacterial disease cholera is of this type. The Salk polio vaccine and one form of influenza vaccine contain inactivated viruses. Because the microbe does not multiply, killed vaccines often require a larger dose and more boosters to be effective.

A number of vaccines are prepared from **attenuated** microbes. Attenuation is any process that substantially lessens or negates the virulence of viruses or bacteria. It is usually achieved by modifying the growth conditions or manipulating microbial genes in a way

that eliminates virulence factors. Attenuation methods include long-term cultivation, selection of mutant strains that grow at colder temperatures (cold mutants), passage of the microbe through unnatural hosts or tissue culture, and removal of virulence genes. The vaccine for tuberculosis (BCG) was obtained after 13 years of subculturing the agent of bovine tuberculosis. Vaccines for measles, mumps, polio (Sabin), and rubella contain active, nonvirulent viruses. The primary advantages of these preparations are that: 1) Viable microorganisms can multiply and produce infection (but not disease) like the natural organism, 2) they confer long-lasting protection, and 3) they usually require fewer doses and boosters than other types of vaccines.

Disadvantages of using live microbes in vaccines are that they require special storage facilities, can be transmitted to other people, and can conceivably mutate back to a virulent strain.

If the exact antigenic determinants that stimulate immunity are known, these substances can be isolated from microorganisms and serve as the basis for a vaccine (figure 15.19b). Vaccines made from bacterial cell parts are called **acellular** or **subcellular vaccines.** If isolated from viruses, they are called **subunit vaccines.** The antigen used in these vaccines may be extracted from cultures of the microbes, produced by genetic engineering or synthesized chemically.

Examples of extracted antigens currently in use are the capsules of the pneumococcus and meningococcus, the protein surface antigen of anthrax, and the surface proteins of hepatitis B virus. A special type of vaccine is the **toxoid,** which consists of a purified fragment of bacterial exotoxin that has been inactivated. By eliciting the production of antitoxins that can neutralize the natural toxin, toxoid vaccines provide protection against toxinoses such as diphtheria and tetanus.

Development of New Vaccines

Despite considerable successes, dozens of bacterial, viral, protozoan, and fungal diseases still remain without a functional vaccine. At the present time, no reliable vaccines are available for malaria, HIV/AIDS, various diarrheal diseases (*E. coli, Shigella*), a number of respiratory diseases, and worm infections, which affect over 200 million people per year worldwide. Of all of the challenges facing vaccine specialists, probably the most difficult has been choosing a vaccine antigen that is safe and that properly stimulates immunity. Currently, much attention is being focused on newer strategies for vaccine preparation that employ antigen synthesis, recombinant DNA, and gene cloning technology.

When the exact composition of an antigenic determinant is known, it is sometimes possible to artificially synthesize it. This ability permits preservation of antigenicity while greatly increasing antigen purity and concentration. The malaria vaccine currently being tested in areas of South America and Africa is composed of three synthetic peptides from the parasite. Several biotechnology companies are using plants to mass produce vaccine antigens. Tests are underway to grow tomatoes, potatoes, and bananas that synthesize proteins from cholera, hepatitis, papillomavirus, and *E. coli* pathogens. This strategy aims to economically harvest vaccine antigens and make vaccines for populations that otherwise would not have access to them.

TABLE 15.4	**Checklist of Requirements for an Effective Vaccine**

- It should have a low level of adverse side effects or toxicity and not cause serious harm.
- It should protect against exposure to natural, wild forms of pathogen.
- It should stimulate both antibody (B-cell) response and cell-mediated (T-cell) response.
- It should have long-term, lasting effects (produce memory).
- It should not require numerous doses or boosters.
- It should be inexpensive, have a relatively long shelf life, and be easy to administer.

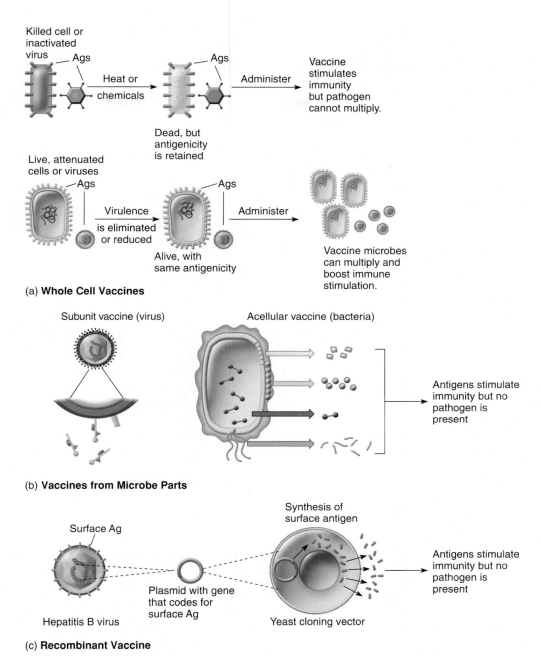

Killed cell or inactivated virus — Ags

Heat or chemicals

Dead, but antigenicity is retained — Ags

Administer

Vaccine stimulates immunity but pathogen cannot multiply.

Live, attenuated cells or viruses — Ags

Virulence is eliminated or reduced

Alive, with same antigenicity — Ags

Administer

Vaccine microbes can multiply and boost immune stimulation.

(a) **Whole Cell Vaccines**

Subunit vaccine (virus)

Acellular vaccine (bacteria)

Antigens stimulate immunity but no pathogen is present

(b) **Vaccines from Microbe Parts**

Surface Ag

Hepatitis B virus

Plasmid with gene that codes for surface Ag

Synthesis of surface antigen

Yeast cloning vector

Antigens stimulate immunity but no pathogen is present

(c) **Recombinant Vaccine**

Figure 15.19 Strategies in vaccine design. (a) Whole cells or viruses, killed or attenuated. **(b)** Acellular or subunit vaccines are made by disrupting the microbe to release various molecules or cell parts that can be isolated and purified. **(c)** Recombinant vaccines are made by isolating a gene for the antigen from the pathogen (here a hepatitis virus) and splicing it into a plasmid. Insertion of the recombinant plasmid into a cloning host (yeast) results in the production of large amounts of viral surface antigen to use in vaccine preparation.

Genetically Engineered Vaccines

Some of the genetic engineering concepts introduced in chapter 10 offer novel approaches to vaccine development. These methods are particularly effective in designing vaccines for obligate parasites that are difficult or expensive to culture, such as the syphilis spirochete or the malaria parasite. Recombinant DNA technology provides a means of isolating the genes that encode various microbial antigens, inserting them into plasmid vectors, and cloning them in appropriate hosts. The outcome of recombination can be as varied as desired. For instance, the cloning host can be stimulated to synthesize and secrete a protein product (antigen), which is then harvested and purified **(figure 15.19c).** Vaccines for hepatitis and human papillomavirus are currently being prepared in this way. Antigens from the agents of syphilis, *Schistosoma,* and influenza have been similarly isolated and cloned and are currently being considered as potential vaccine material.

Another ingenious technique using genetic recombination has been nicknamed the *Trojan horse* vaccine. The term derives from an ancient legend in which the Greeks sneaked soldiers into the fortress of their Trojan enemies by hiding them inside a large, mobile wooden horse. In the microbial equivalent, genetic material

TABLE 15.5 Currently Approved Vaccines

Disease/Vaccine Preparation	Route of Administration	Recommended Usage/Comments
Contain killed whole bacteria		
Cholera	Subcutaneous (SQ) injection	For travelers; effect not long term
Plague	SQ	For exposed individuals and animal workers; variable protection
Contain live, attenuated bacteria		
Tuberculosis (BCG)	Intradermal (ID) injection	For high-risk occupations only; protection variable
Typhoid	Oral	For travelers only; low rate of effectiveness
Acellular vaccines (capsular polysaccharides or proteins)		
Anthrax	SQ	For protection in military recruits, occupationally exposed
Meningitis (meningococcal)	SQ	For protection in high-risk infants, military recruits, college students; short duration
Meningitis (*Haemophilus influenzae*)	IM	For infants and children; may be administered with DTaP
Pneumococcal pneumonia	IM or SQ	Important for people at high risk: the young, elderly, and immunocompromised; moderate protection
Pertussis (aP)	IM	For newborns and children; contains recombinant protein antigens
Toxoids (formaldehyde-inactivated bacterial exotoxins)		
Diphtheria	IM	A routine childhood vaccination; highly effective in systemic protection
Tetanus	IM	A routine childhood vaccination; highly effective
Pertussis	IM	A routine childhood vaccination; highly effective
Botulism	IM	Only for exposed individuals such as laboratory personnel
Contain inactivated whole viruses		
Poliomyelitis (Salk)	IM	Routine childhood vaccine; now used as first choice
Rabies	IM	For victims of animal bites or otherwise exposed; effective
Influenza	IM	For high-risk populations; requires constant updating for new strains; immunity not durable
Japanese encephalitis	SQ	For those residing in endemic areas, lab workers
Hepatitis A	IM	Protection for travelers, institutionalized people
Contain live, attenuated viruses		
Adenovirus infection	Oral	For immunizing military recruits
Measles (rubeola)	SQ	Routine childhood vaccine; very effective
Mumps (parotitis)	SQ	Routine childhood vaccine; very effective
Poliomyelitis (Sabin)	Oral	Routine childhood vaccine; very effective but can cause polio
Rubella	SQ	Routine childhood vaccine; very effective
Chickenpox (varicella)	SQ	Routine childhood vaccine; immunity can diminish over time
Rotavirus (RotaTeq)	Oral	New live oral vaccine for infants at risk for rotavirus infection
Smallpox (live vaccinia virus, not attenuated variola)	Multiple punctures	Since 2003 offered on voluntary basis for health care workers, some military personnel
Yellow fever	SQ	Travelers, military personnel in endemic areas
Influenza	Inhaled	Same as for inactivated
Subunit viral vaccines		
Hepatitis B*	IM	Recommended for all children, starting at birth; also for health workers and others at risk
Influenza	IM	See influenza, inactivated, above
Human papillomavirus*	IM	To prevent HPV infection and uterine cancer in girls and women
Recombinant vaccines		
Hepatitis B	IM	Used more often than subunit but for same groups
Pertussis	IM	See acellular, above

*These vaccines are also made by recombinant methods.

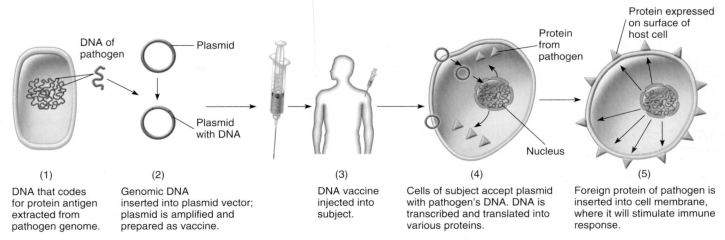

(1) DNA that codes for protein antigen extracted from pathogen genome.

(2) Genomic DNA inserted into plasmid vector; plasmid is amplified and prepared as vaccine.

(3) DNA vaccine injected into subject.

(4) Cells of subject accept plasmid with pathogen's DNA. DNA is transcribed and translated into various proteins.

(5) Foreign protein of pathogen is inserted into cell membrane, where it will stimulate immune response.

Figure 15.20 DNA vaccine preparation. DNA vaccines contain all or part of the pathogen's DNA, which is used to "infect" a recipient's cells. Processing of the DNA leads to production of an antigen protein that can stimulate a specific response against that pathogen.

from a selected infectious agent is inserted into a live carrier microbe that is nonpathogenic. In theory, the recombinant microbe will multiply and express the foreign genes, and the vaccine recipient will be immunized against the microbial antigens. Vaccinia, the virus originally used to vaccinate for smallpox, and adenoviruses have proved practical agents for this technique. Vaccinia is used as the carrier in experimental vaccines for AIDS, herpes simplex 2, leprosy, and tuberculosis.

DNA vaccines are another promising approach to immunization. The technique in these formulations is very similar to gene therapy as described in chapter 10, except in this case, microbial (not human) DNA is inserted into a plasmid vector and inoculated into a recipient **(figure 15.20).** The expectation is that the human cells will take up some of the plasmids and express the microbial DNA in the form of proteins. Because these proteins are foreign, they will be recognized during immune surveillance and cause B and T cells to be sensitized and form memory cells. Note: Even though the DNA itself is from a microbe, it will not cause an immune response because the structure of DNA is not highly immunogenic.

Experiments with animals have shown that these vaccines are very safe and that only a small amount of the foreign antigen need be expressed to produce effective immunity. Another advantage to this method is that any number of potential microbial proteins can be expressed, making the antigenic stimulus more complex and improving the likelihood that it will stimulate both antibody and cell-mediated immunity. At the present time, vaccines for Lyme disease, hepatitis C, herpes simplex, influenza, tuberculosis, papillomavirus, malaria, and SARS are undergoing animal trials, most with encouraging results.

Routes of Administration and Side Effects of Vaccines

Most vaccines are injected by subcutaneous, intramuscular, or intradermal routes. Oral vaccines are available for only a few diseases (see table 15.5), but they have some distinct advantages. An oral dose of a vaccine can stimulate protection (IgA) on the mucous membrane of the portal of entry. Oral vaccines are also easier to give, more readily accepted, and well tolerated. Another method that shows promise is an intranasal vaccine delivered into the nose by spray or drops (FluMist for influenza, for example).

Some vaccines are made more effective by adding an **adjuvant.*** An adjuvant is any compound that enhances immunogenicity and prolongs antigen retention at the injection site. The adjuvant precipitates the antigen and holds it in the tissues so that it will be released gradually. Its gradual release presumably facilitates contact with antigen-presenting cells and lymphocytes. Common adjuvants are alum (aluminum hydroxide salts), Freund's complete adjuvant (emulsion of mineral oil, water, and extracts of mycobacteria), and beeswax.

Vaccines must go through many years of trials in experimental animals and human volunteers before they are licensed for general use. Even after they have been approved, like all therapeutic products, they are not without complications. The most common of these are local reactions at the injection site, fever, allergies, and other adverse reactions. Relatively rare reactions (about one case out of 300,000 vaccinations) are panencephalitis (from measles vaccine), back-mutation to a virulent strain (from polio vaccine), disease due to contamination with dangerous viruses or chemicals, and neurological effects of unknown cause (from pertussis and swine flu vaccines). Some patients experience allergic reactions to the medium (eggs or tissue culture) rather than to vaccine antigens.

When known or suspected adverse effects have been detected, vaccines are altered or withdrawn. Recently, the whole-cell pertussis vaccine was replaced by the acellular capsule (aP) form when it was associated with adverse neurological effects. The Lyme disease vaccine was withdrawn when recipients developed arthritis and other side effects. Polio vaccine was switched from live, oral to inactivated when too many cases of paralytic disease occurred from back-mutated vaccine stocks. Vaccine companies have also phased out certain preservatives, such as thimerosal, that could cause allergies and other medical problems in some patients.

* *adjuvant* (ad′-joo-vunt) L., *adjuvans,* a thing that helps or aids.

Professionals involved in giving vaccinations must understand their inherent risks but also realize that the risks from the infectious disease almost always outweigh the chance of an adverse vaccine reaction. The greatest caution must be exercised in giving live vaccines to immunocompromised or pregnant patients, the latter because of possible risk to the fetus.

TAKE NOTE: HAS VACCINATION BECOME A VICTIM OF ITS OWN SUCCESS?

Vaccination has always been considered one of humanity's triumphs, one that has made an enormous contribution to health and well-being. Such life-threatening diseases as tetanus, diphtheria, polio, and others are no longer a reality for most people. But lately, vaccination's good reputation is being tarnished. It began with a poorly-designed 1998 English study that hinted at a link between the MMR vaccine and autism. The publicity from this report has spurred a decade-long controversy and a trend towards some parents choosing not to vaccinate their children against the usual childhood diseases. The study's results have since been refuted, but the vaccine phobia persists. Twelve separate epidemiologic studies have discounted any connection between the two, and additional studies show that the preservative thimerosal is also not correlated with autism (MMR is a live attenuated vaccine and would not contain this substance anyway). Despite the fact that routine childhood vaccines have not contained thimerosal since 2001, the numbers of new autism cases unfortunately continue to rise.

Many decisions not to vaccinate are based on misconceptions and emotion rather than science, yet consequences are already being felt in some populations. In the past two years, outbreaks of measles, mumps, meningitis, and pertussis have been tied to inadequate vaccinations. The reality is that these diseases are still around, and they pose a greater risk for young children than the possibility of a vaccine reaction. In fact, 1% of babies develop a severe brain condition from measles virus, just as mumps, meningitis, and pertussis all have their own serious complications. The reduction in vaccination will lead to less herd immunity and create unknown pockets where children lack protection. Now more than ever before, people in health care fields will need to educate and counsel patients about the importance of routine immunizations for both individuals and the community.

To Vaccinate: Why, Whom, and When?

Vaccination confers long-lasting, sometimes lifetime, protection in the individual, but an equally important effect is to protect public health. Vaccination is a proven method of establishing **herd immunity** in the population. According to this concept, individuals immune to a communicable infectious disease will not harbor it, thus reducing the occurrence of that pathogen. With a larger number of immune individuals in a population (herd), it will be less likely that an unimmunized member of the population will encounter the agent and become infected. In effect, collective immunity through mass immunization confers indirect protection on the nonimmune members (such as children). The most effective herd immunity outcome occurs when the healthiest members are vaccinated. Because most vaccine programs target the youngest and oldest (and thus weaker) people instead, herd immunity in many populations is not complete. Herd immunity maintained through immunization considered a valuable factor in preventing epidemics.

Table 15.6 contains the recommended childhood and adolescent immunization schedule, categorized by age group and by medical condition. A second table in the inside of the back cover lists the current recommended schedule for adult immunizations. Some vaccines are mixtures of antigens from several pathogens, notably Pediatrix (DTaP, IPV, and HB). It is also common for several vaccines to be given simultaneously, as occurs in military recruits who receive as many as 15 injections within a few minutes and children who receive boosters for DTaP and polio at the same time they receive the MMR vaccine. Experts doubt that immune interference (inhibition of one immune response by another) is a significant problem in these instances, and the mixed vaccines are carefully balanced to prevent this eventuality. The main problem with simultaneous administration is that side effects can be amplified.

Check & Assess Section 15.6

✔ Knowledge of the specific immune response has a practical application: commercial production of antisera and vaccines.

✔ Artificial passive immunity usually involves administration of antiserum. This serum, collected from donors, is high in antibodies that can confer immediate protection to infectious agents. Examples include ISG (immune serum globulin) and SIG (specific immune globulin).

✔ Artificial active agents are vaccines that provoke a protective immune response in the recipient but do not cause the actual disease. Vaccination is the process of challenging the immune system with a specially selected antigen. Examples are (1) killed or inactivated microbes; (2) live, attenuated microbes; (3) subunits of microbes; and (4) genetically engineered microbes or microbial parts.

✔ Vaccines can cause side effects such as fever, pain, allergies, and occasionally more serious diseases, but they provide more protection than risk.

✔ Vaccination programs seek to protect the individual directly through raising the antibody titer and indirectly through the development of herd immunity.

32. What are the advantages and disadvantages of a killed vaccine; a live, attenuated vaccine; a subunit vaccine; a recombinant vaccine; and a DNA vaccine?

33. What are boosters, adjuvants, Trojan horse vaccines, and toxoids?

34. Describe the concept of herd immunity, and explain how vaccination contributes to its development in a community.

35. Give some possible adverse side effects of vaccination.

TABLE 15.6 **Recommended Immunization Schedule United States 2010***

Persons Aged 0–6 Years

Vaccine ▼ / Age ▶	Birth	1 month	2 months	4 months	6 months	12 months	15 months	18 months	19–23 months	2–3 years	4–6 years
Hepatitis B	HepB	HepB			HepB						
Rotavirus			RV	RV	RV						
Diphtheria, Tetanus, Pertussis			DTaP	DTaP	DTaP		DTaP				DTaP
Haemophilus influenzae type b			Hib	Hib	Hib	Hib					
Pneumococcal			PCV	PCV	PCV	PCV				PPSV	
Inactivated Poliovirus			IPV	IPV		IPV					IPV
Influenza						Influenza (Yearly)					
Measles, Mumps, Rubella						MMR					MMR
Varicella						Varicella					Varicella
Hepatitis A						HepA (2 doses)				HepA Series	
Meningococcal										MCV	

Persons Aged 7–18 Years

Vaccine ▼ / Age ▶	7–10 years	11–12 years	13–18 years
Diphtheria, Tetanus, Pertussis		Tdap	Tdap
Human Papillomavirus		HPV (3 doses)	HPV Series
Meningococcal	MCV	MCV	MCV
Influenza	Influenza (Yearly)		
Pneumococcal	PPSV		
Hepatitis A	HepA Series		
Hepatitis B	HepB Series		
Inactivated Poliovirus	IPV Series		
Measles, Mumps, Rubella	MMR Series		
Varicella	Varicella Series		

Range of recommended ages for all children except certain high-risk groups

Range of recommended ages for certain high-risk groups

Range of recommended ages for all children except certain high-risk groups

Range of recommended ages for catch-up immunization

Range of recommended ages for certain high-risk groups

* For details for vaccine administration go to http://www.cdc.gov/vaccines/recs/schedules/child-schedule.htm

CASE FILE **15** PERSPECTIVE

Rabies has the distinction of being the most deadly of all infectious diseases, being considered 100% fatal once symptoms appear. Worldwide, only 10 people have ever been known to survive the disease, and most of these survivors received some treatment as in the cases presented here. The disease is rather rare in humans but not in certain wild mammals. About 2 to 3 human cases are reported per year in the United States, but 5,000 to 6,000 animal cases are reported. The most prevalent animal carriers are raccoons, skunks, coyotes, foxes, and bats, but the virus can infect all mammals. Infection is most likely to occur following an animal bite or scratch, although exposure to saliva and other infectious fluids has been implicated.

Rabies is a type of encephalitis in which the virus migrates by nerve cell processes from the portal of entry, enters the brain, multiplies, and causes severe inflammation and damage to brain cells. Postexposure prophylaxis is an immunization method designed to be administered as soon after contact as possible. As we see in the two cases, its timing is critical. First, the wound area is infused with human rabies immunoblobulin (HRIG) and HRIG is also injected into the muscle of the arm. It contains high levels of specific antibodies that adhere to the virus and neutralize it. This artificial passive immunity will deliver immediate antirabies antibodies to the wound as well as the circulation, hopefully before the virus can invade the brain.

The rabies vaccine contains whole inactivated viruses grown in cell culture. The usual protocol requires a series of four to five intramuscular injections that start on day 0 (first dose) and then proceed on days 3, 7, 14, and sometimes 28. This provides longer-term artificial active immunity that stimulates the patient to produce his own protective antibodies and T cells. It will also

trigger a buildup in the number of antibodies in the blood and spinal fluid to eliminate the virus.

The incubation period in rabies varies from 2 weeks to 3 months or more, depending on the amount of virus and the distance it must travel to the brain. It was relatively short in both of these cases—around 2 to 3 weeks. The different outcome in these cases comes down to knowledge and timing. For instance, the parents of the child in case #1 did not know that the injury came from a bat, whereas in case #2, it was clearly known. This situation caused the delay in treatment for child #1, who had developed symptoms before it was started. In child #2, it was started before any symptoms appeared. Another significant factor was that child #2 had antiviral antibodies fairly early and showed a rising antibody titer during a critical period needed to control the virus. Keeping him sedated during this time and slowing brain activity may also have been a factor. Child #1 already had such a high viral load in the brain that his immune response could not catch up.

The main purpose of giving so many injections of vaccine is to flood the immune system with antigen, which will expand the clones of lymphocytes that can react with this virus. This will result in more antibodies and more activated T cells to destroy infected cells. Three doses are known to provide 100% protection, but additional doses are given to boost the immune system. Every year in the US about 40,000 people report potential exposure to a rabid animal and are given PEP right away. The strategy has been so effective that not one healthy person has ever developed a case of rabies after having this treatment.

To read more about rabies immunization, go to http://www.cdc.gov/rabies/exposure/index.html.

Additional information on rabies is available in chapter 25.

Chapter Summary with Key Terms

15.1 Specific Immunity: The Adaptive Line of Defense

Development of Lymphocyte Specificity/Receptors Acquired immunity involves the reactions of B and T lymphocytes to foreign molecules, or **antigens.** Before they can react, each lymphocyte must undergo differentiation into its final functional type by developing protein receptors for antigen, the specificity of which is randomly generated and unique for each lymphocyte.

A. Development of the Immune Response System
 1. Genetic recombination and mutation during embryonic and fetal development produce billions of different lymphocytes, each bearing a different receptor. The process by which diversity is generated is **clonal selection.**
 2. **Immune tolerance** or tolerance to self occurs during this time.
 3. The receptors on B cells are **immunoglobulin (Ig)** molecules, and receptors on T cells are smaller unrelated glycoprotein molecules called **CD markers.**
 4. T-cell receptors bind antigenic determinants on **MHC** molecules, which in humans are called *human leukocyte antigens (HLA).*

15.2 Lymphocyte Maturation and the Nature of Antigens

An antigen (Ag) is any substance that stimulates an immune response.

A. Requirements for **antigenicity** include foreignness (recognition as nonself), large size, and complexity of cell or molecule.
B. Foreign molecules less than 1,000 MW (**haptens**) are not antigenic unless attached to a larger carrier molecule.
C. The epitope is the small molecular group of the foreign substance that is recognized by lymphocytes. Cells, viruses, and large molecules can have numerous antigenic determinants.
D. Other antigens are mosaic alloantigens, superantigens, allergens, and autoantigens.

15.3 Cooperation in Immune Reactions to Antigens

A. T-cell-dependent antigens must be processed by large phagocytes such as dendritic cells or macrophages called **antigen-presenting cells (APCs).**
B. The presentation of a single antigen involves a direct collaboration between an APC, a T helper (T_H), and an antigen-specific B or T cell.
C. Cytokines involved are **interleukin-1** from the APC, which activates the T_H cells, and **interleukin-2** produced by the T_H cell, which activates B and other T cells.

D. **B-Cell Responses**
1. Once B cells process the antigen, interact with T_H cells, and are stimulated by B-cell growth and differentiation factors, they enter the cell cycle in preparation for mitosis and clonal expansion.
2. Divisions give rise to **plasma cells** that secrete antibodies and **memory cells** that can react to that same antigen later.
 a. *Nature of antibodies* (immunoglobulins): A single immunoglobulin molecule (monomer) is a large Y-shaped protein molecule consisting of four polypeptide chains. It contains two identical regions with ends that form the **antigen binding fragments (Fabs)** that bind with a unique specificity to an antigen. The **crystallizable fragment (Fc)** determines the location and the function of the antibody molecule.
 b. *Antigen-antibody (Ag-Ab) reactions* include **opsonization, neutralization, agglutination,** and complement fixation.
 c. The five **antibody classes,** which differ in size and function, are **IgG, IgA** (secretory Ab), **IgM, IgD,** and **IgE.**
3. *Antibodies in serum* (**antiserum**)
 a. The first introduction of an Ag to the immune system produces a **primary response,** with a gradual increase in Ab titer.
 b. The second contact with the same Ag produces a **secondary,** or **anamnestic, response** due to memory cells produced during initial response.
 c. **Monoclonal antibodies** (MABs) are pure forms of immunoglobulins produced in hybridoma cells. They may be tailored to high specificities and have numerous applications in medicine and research.

15.4 T-Cell Responses
A. **T helper cells** (**T_H or CD4**) are the dominant T cells.
1. They regulate immune reactions to antigens and require interaction with MHC II receptors on host cells.
2. They activate macrophages, B cells, and other T cells by means of secreted interleukins.
 a. **T_H1** cells function as T-cell activators and stimulate delayed hypersensitivity.
 b. **T_H2** cells modulate B-cell functions and may suppress certain undesirable T cells' actions.
B. **Cytotoxic T cells** (**T_C or CD8**) are responsive to MHC I receptors on host cells and are sometimes called killer cells.
1. They recognize and react to foreign cells, cancer cells, and graft tissues and secrete chemicals that directly damage and kill these cells.
C. **Natural killer cells** (**NK**) are T-cell relatives responsible for early killing of infected cells and cancer cells in response to various cytokines.

15.5 A Classification Scheme for Specific, Acquired Immunities
Immunities acquired through B and T lymphocytes can be classified by a simple system.
A. **Active immunity** results when a person is challenged with antigen that stimulates production of antibodies. It creates memory, takes time, and is lasting.
B. In **passive immunity,** preformed antibodies are donated to an individual. It does not create memory, acts immediately, and is short term.

C. **Natural immunity** is acquired as part of normal life experiences.
D. **Artificial immunity** is acquired through medical procedures such as immunization.

15.6 Immunization: Methods of Manipulating Immunity for Therapeutic Purposes
A. Passive **immunotherapy** includes administering immune serum globulin and specific immune globulins pooled from donated serum to prevent infection and disease in those at risk; antisera and antitoxins from animals are occasionally used.
B. Active immunization is synonymous with **vaccination** and provides an antigenic stimulus that does not cause disease but can produce long-lasting, protective immunity. **Vaccines** are made with:
1. **Killed** whole cells or **inactivated viruses** that do not reproduce but are antigenic
2. Live, **attenuated** cells or viruses that are able to reproduce but have lost virulence
3. **Acellular** or **subunit** components of microbes such as surface antigens or neutralized toxins (**toxoids**)
4. Genetic engineering techniques, including cloning of antigens, recombinant attenuated microbes, and **DNA**-based **vaccines**
C. Boosters (additional doses) are often required.
D. Vaccination increases **herd immunity,** protection provided by mass immunity in a population.

Multiple-Choice Questions

Select the correct answer from the answers provided. For questions with blanks, choose the combination of answers that most accurately completes the statement.

1. Which of these characteristics is not a major factor in adaptive or acquired immunity?
 a. specificity c. recognition
 b. chemotaxis d. memory

2. The primary B-cell receptor is
 a. IgD c. IgE
 b. IgA d. IgG

3. In humans, B cells mature in the _____ and T cells mature in the _____.
 a. GALT, liver c. bone marrow, thymus
 b. bursa, thymus d. lymph nodes, spleen

4. Small, simple molecules are _____ antigens.
 a. poor c. effective
 b. never d. mosaic

5. Which type of cell actually secretes antibodies?
 a. T cell c. plasma cell
 b. macrophage d. monocyte

6. CD4 cells are _____ cells and CD8 cells are _____ cells.
 a. killer, suppressor c. cytotoxic, helper
 b. helper, cytotoxic d. B, T

7. Helper T cells receive antigen from _____, and cytotoxic T cells receive antigen from _____.
 a. macrophages, B cells c. viruses, bacteria
 b. class II MHC, d. class I MHC, class II MHC
 class I MHC

8. The cross-linkage of antigens by antibodies is known as
 a. opsonization
 c. agglutination
 b. a cross-reaction
 d. complement fixation

9. The greatest concentration of antibodies is found in the _____ fraction of the serum.
 a. gamma globulin
 c. beta globulin
 b. albumin
 d. alpha globulin

10. T_____ cells assist in the functions of certain B cells and other T cells.
 a. sensitized
 c. helper
 b. cytotoxic
 d. natural killer

11. Cytotoxic T cells are important in controlling
 a. virus infections
 c. autoimmunity
 b. allergy
 d. all of these

12. Vaccination is synonymous with _____ immunity.
 a. natural active
 c. artificial active
 b. artificial passive
 d. natural passive

13. Which of the following can serve as antigen-presenting cells (APCs)?
 a. T cells
 d. dendritic cells
 b. B cells
 e. b, c, and d
 c. macrophages

14. A living microbe with reduced virulence that is used for vaccination is considered
 a. a toxoid
 c. denatured
 b. attenuated
 d. an adjuvant

15. A vaccine that contains parts of viruses is called
 a. acellular
 c. subunit
 b. recombinant
 d. attenuated

16. Widespread immunity that protects the population from the spread of disease is called
 a. seropositivity
 b. cross-reactivity
 c. epidemic prophylaxis
 d. herd immunity

17. DNA vaccines contain _____ DNA that stimulates cells to make _____ antigens.
 a. human, RNA
 c. human, protein
 b. microbial, protein
 d. microbial, polysaccharide

18. What is the purpose of an adjuvant?
 a. to kill the microbe
 b. to stop allergic reactions
 c. to improve the contact between the antigen and lymphocytes
 d. to make the antigen more soluble in the tissues

19. **Multiple Matching.** Place all possible matches in the spaces at the left.

 _____ IgG _____ IgE
 _____ IgA _____ IgM
 _____ IgD

 a. found in mucous secretions
 b. a monomer
 c. a dimer
 d. has greatest number of Fabs
 e. first antibody made during the primary response
 f. first antibody made during the secondary response
 g. crosses the placenta
 h. fixes complement
 i. involved in allergic reactions
 j. primarily a surface receptor on B cells

20. **Multiple Matching.** An outline summarizing the general host defenses covered in chapters 14 and 15 was presented in figure 14.21 (page 448). You may want to use this resource to review major aspects of immunity and to guide you in answering this question. In the blanks on the left, place the letters of all of the host defenses and immune responses in the right column that can fit the description.

 _____ vaccination for tetanus
 _____ lysozyme in tears
 _____ immunization with horse serum
 _____ in utero transfer of antibodies
 _____ booster injection for diphtheria
 _____ colostrum
 _____ interferon
 _____ action of neutrophils
 _____ injection of gamma globulin
 _____ recovery from a case of mumps
 _____ edema
 _____ humans having protection from canine distemper virus
 _____ stomach acid
 _____ cilia in trachea

 a. active
 b. passive
 c. natural
 d. artificial
 e. acquired
 f. innate, inborn
 g. chemical barrier
 h. mechanical barrier
 i. genetic barrier
 j. specific
 k. nonspecific
 l. inflammatory response
 m. second line of defense

Case File Questions

1. What does the rabies vaccine contain that the rabies immunoglobulin does not?
 a. antibodies
 c. live virus
 b. inactive virus
 d. animal serum

2. What types of cells could be involved in specific immune reactions against the rabies virus?
 a. T cells
 d. dendritic cells
 b. B cells
 e. a, b, and d
 c. macrophages
 f. all of these

Writing to Learn

These questions are suggested as a *writing-to-learn* experience. For each question, compose a one- or two-paragraph answer that includes the factual information needed to completely address the question. All Check & Assess questions can be used for writing-to-learn exercises.

1. Using words and arrows, complete a flow outline of an immune response, beginning with entrance of antigen; include processing, cell interaction, involvement of cytokines, and the end results for B and T cells.

2. a. Explain why most immune reactions result in a polyclonal collection of antibodies.
 b. How do monoclonal antibodies differ from this?

3. Give an immunological explanation as to why people can get influenza every year even though they have been vaccinated.

4. How can the anamnestic response be explained in immunologic terms?

5. Differentiate between humoral and cell-mediated immunity.

6. a. Combine information on the functions of different classes of Ig to explain the exact mechanisms of natural passive immunity (both transplacental and colostrum-induced).
 b. Why are these sorts of immunity short term?

Concept Mapping

Appendix E provides guidance for working with concept maps.

1. Supply your own linking lines, as well as the linking words or phrases, in this concept map, and provide the missing concepts in the empty boxes.

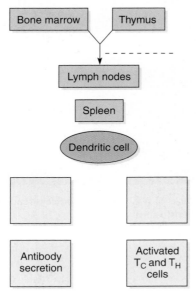

2. Construct your own concept map using the following words as the *concepts*. Supply the links and the linking words between each pair of concepts.

active immunity vaccines

passive immunity interferon

natural immunity inflammation

artificial immunity memory

innate immunity

Critical Thinking Questions

Critical thinking is the ability to reason and solve problems using facts and concepts. These questions can be approached from a number of angles, and in most cases, they do not have a single correct answer.

1. What is the advantage of having lymphatic organs screen the body fluids, directly and indirectly?

2. Cells contain built-in suicide genes to self-destruct by apoptosis under certain conditions. Can you explain why development of the immune system might depend in part on this sort of adaptation?

3. a. Describe what could be summarized as the three Rs of immunity: recognize, react, remember.
 b. Why would it be necessary for the T cells to bind both antigen and self (MHC) receptors?

4. Double-stranded DNA is a large, complex molecule, but it is not generally immunogenic unless it is associated with proteins or carbohydrates. Can you think why this might be so? (Hint: How universal is DNA?)

5. Explain how it is possible for people to give a false-positive reaction in blood tests for syphilis, AIDS, and infectious mononucleosis.

6. Describe the cellular/microscopic pathology in the immune system of AIDS patients that results in opportunistic infections and cancers.

7. Describe the relationship between an antitoxin, a toxin, and a toxoid.

8. It is often said that a vaccine does not prevent infection; rather, it primes the immune system to undergo an immediate response to prevent an infection from spreading. Explain what is meant by this statement, and outline what is happening at the cellular/molecular level from the time of vaccination until subsequent contact with the infectious agent actually occurs.

9. Describe the specific components of acquired immunity that would be necessary to control rabies virus.

10. A recent study by the CDC reported that only 73% of children in the United States are being adequately vaccinated. Several million children are at high risk for infections.
 a. Name some factors that account for this trend.
 b. Explain how the reduction in vaccination impacts herd immunity.

11. a. From table 15.6 and the back inside cover, determine the vaccines you have been given and those for which you will require periodic boosters.
 b. Suggest vaccines you may need in the future.

12. When traders and missionaries first went to the Hawaiian Islands, the natives there experienced severe disease and high mortality rates from smallpox, measles, and certain STDs.
 a. Explain what immunologic factors are involved in the devastating effects of new diseases in previously unexposed populations.
 b. Explain the ways in which vaccination has been responsible for the worldwide eradication of smallpox and near elimination of polio.

Visual Challenge

1. Using figure 15.10, label the lines and arrows, and explain what is happening at numbers 1, 2, and 3.

2. Examine figure 25.13, and determine which components of HIV would be epitopes or antigens.

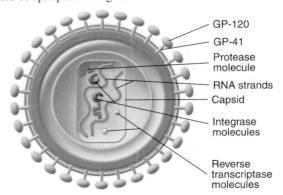

16

Disorders in Immunity

Antibodies (IgM)

Packs of washed red blood cells ready for transfusions

The wrong blood type: RBCs, clumped by antibodies

"Despite being surrounded by the best medical care available, none of the last minute interventions could save his life."

CASE FILE | 16 | A Deadly Transfusion Reaction

On December 29, 2007, a 67-year-old male patient at a medical center in Florida underwent a minor procedure that required a blood transfusion. In anticipation of this surgery, the patient's blood had been typed by a lab technician a few days earlier. The man was transfused with 1 pint of type A$^+$ blood during the surgery and was returned to his room to recover. In a few hours, nurses noticed he had a sudden onset of fever, chills, and difficulty in breathing. When a vital sign check noted tachycardia*, hypotension**, and urine with a pinkish tinge, an emergency team was called in to evaluate his condition and administer treatment. A blood sample taken at this time showed unusual levels of hemoglobin in the plasma and clumping of red blood cells (RBCs). Despite being surrounded by the best medical care available, none of the last minute interventions could save his life. The patient died just a few hours after his operation from heart failure, shock, and

disseminated intravascular coagulation (DIC)***. Attending physicians concluded that he had experienced an acute hemolytic transfusion reaction (AHTR).

The usual cause of a transfusion reaction is the presence of RBCs of the wrong blood type in the patient's body. A pathological reaction occurs when these donated RBCs carry antigens on their surface that react with IgM and IgG antibodies that already exist in the plasma of the recipient. When the antibodies combine with the antigens, the complement system in the blood is activated, which leads to the formation of membrane attack complexes and hemolysis of the donated red blood cells. Since such a reaction can occur even if only 5 to 20 ml of blood is transfused, the 500-ml transfusion was more than enough to be fatal.

What seemed most puzzling to investigators in this case was that the records showed that the donated blood (type A$^+$) had been matched to the patient's by cross-matching just minutes before the

transfusion. This procedure involves mixing a drop of the patient's blood with a drop of the blood to be transfused and observing the mixture macroscopically and microscopically for evidence of agglutination. Technicians had reported that the cross-match was successful and showed no evidence of a reaction.

▶ *What is responsible for the ABO and Rh blood types, and why does transfusion of an incompatible blood type lead to severe complications?*

▶ *What causes the abnormal signs indicative of a transfusion reaction?*

To continue the case, go to page 500.

*Tachycardia is an increased heart rate.

**Hypotension is decreased blood pressure.

***DIC is a clumping of RBCs in blood vessels throughout the body.

16.1 The Immune Response: A Two-Sided Coin

1. Summarize the main categories of immunopathology and their medical consequences.
2. Discuss the factors involved in allergies and hypersensitivities.

Humans possess a powerful and intricate system of defense, which by its very nature also carries the potential to cause injury and disease. In most instances, a malfunction in immunity shows itself in commonplace but miserable symptoms such as those of hay fever and dermatitis. But abnormal immune functions are also actively involved in debilitating or life-threatening diseases such as asthma, anaphylaxis, rheumatoid arthritis, and graft rejection.

With few exceptions, our previous discussions of the immune response have centered around its numerous beneficial effects. The precisely coordinated system that recognizes and destroys invading microbes is clearly protective, but it also presents another side—a side that promotes rather than prevents disease. In this chapter, we survey **immunopathology,** the study of disease states associated with overreactivity or underreactivity of the immune response **(figure 16.1).** In the cases of allergies and *autoimmunity,* the tissues are innocent bystanders attacked by immunologic functions that can't distinguish one's own tissues from those expressing foreign antigens. In **grafts** and **transfusions,** a recipient reacts to the foreign tissues and cells of another individual. In **immunodeficiency** diseases, immune function is incompletely developed, suppressed, or destroyed. Cancer falls into a special category, because it is both a cause and an effect of immune dysfunction. As we shall see, one fascinating by-product of studies of immune disorders has been our increased understanding of the basic workings of the immune system.

Overreactions to Antigens: Allergy/Hypersensitivity

The term **allergy** means a condition of altered reactivity or exaggerated immune responsiveness that is manifested by inflammatory symptoms. Although it is sometimes used interchangeably with **hypersensitivity,** some experts refer to immediate reactions such as hay fever as allergies and to delayed reactions as hypersensitivities.

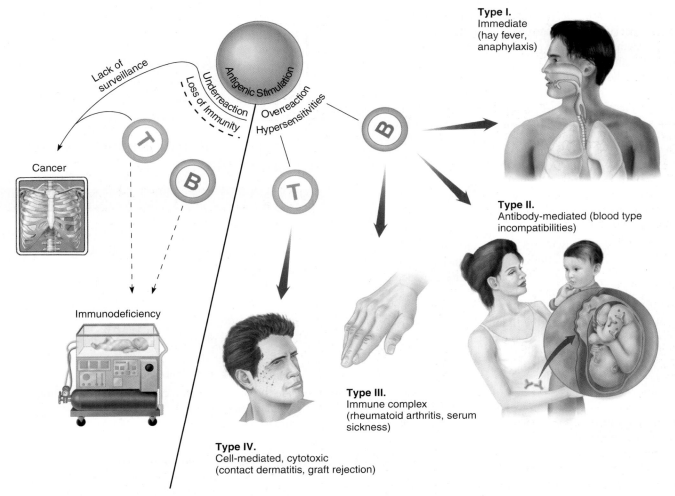

Figure 16.1 Overview of diseases of the immune system. Just as the system of T cells and B cells provides necessary protection against infection and disease, the same system can cause serious and debilitating conditions by overreacting or underreacting to immune stimuli.

TABLE 16.1 **Hypersensitivity States**

Type		Systems and Mechanisms Involved	Examples
I	Immediate hypersensitivity	IgE-mediated; involves mast cells, basophils, and allergic mediators	Anaphylaxis, allergies such as hay fever, asthma
II	Antibody-mediated	IgG, IgM antibodies act upon cells with complement and cause cell lysis; includes some autoimmune diseases	Blood group incompatibility, pernicious anemia; myasthenia gravis
III	Immune complex	Antibody-mediated inflammation; circulating IgG complexes deposited in basement membranes of target organs; includes some autoimmune diseases	Systemic lupus erythematosus; rheumatoid arthritis; serum sickness; rheumatic fever
IV	T-cell-mediated	Delayed hypersensitivity and cytotoxic reactions in tissues	Infection reactions; contact dermatitis; graft rejection; some types of autoimmunity

Allergic and hypersensitive individuals are acutely sensitive to repeated contact with antigens called **allergens** that do not noticeably affect nonallergic individuals. Although the general effects of these conditions are detrimental, we must be aware that they involve the very same types of immune reactions as those at work in protective immunities. These include humoral and cell-mediated actions, the inflammatory response, phagocytosis, and complement. Such an association means that all humans have the potential to develop allergy or hypersensitivity under particular circumstances.

Originally, allergies were defined as either immediate or delayed, depending upon the time lapse between contact with the allergen and onset of symptoms. Subsequently, they were differentiated as humoral versus cell-mediated. But as information on the nature of the allergic immune response accumulated, it became evident that, although useful, these schemes oversimplified what is really a very complex spectrum of reactions. The most widely accepted classification, first introduced by immunologists P. Gell and R. Coombs, includes four major categories: type I ("common" allergy and anaphylaxis), type II (IgG- and IgM-mediated cell damage), type III (immune complex), and type IV (delayed hypersensitivity) **(table 16.1).** In general, types I, II, and III involve a B-cell–immunoglobulin response, and type IV involves a T-cell response (figure 16.1). The antigens that elicit these reactions can be exogenous, originating from outside the body (microbes, pollen grains, and foreign cells and proteins), or endogenous, arising from self tissue (autoimmunities).

One of the reasons allergies are easily mistaken for infections is that both involve damage to the tissues and thus trigger the inflammatory response, as described in chapter 14. Many symptoms and signs of inflammation (redness, heat, skin eruptions, edema, and granuloma) are prominent features of allergies and are caused by some of the same chemical mediators.

Check & **Assess** Section 16.1

✔ Immunopathology is the study of diseases associated with excesses and deficiencies of the immune response. Such diseases include allergies, autoimmunity, grafts, transfusions, immunodeficiency disease, and cancer.

✔ An allergy or hypersensitivity is an exaggerated immune response that injures or inflames tissues.

✔ There are four categories of hypersensitivity reactions: type I (allergy and anaphylaxis), type II (IgG and IgM tissue destruction), type III (immune complex reactions), and type IV (delayed hypersensitivity reactions).

1. Define *allergy* and *hypersensitivity.*
2. What accounts for the reactions that occur in these conditions?
3. What does it mean when a reaction is immediate or delayed?
4. Give examples of each type.

16.2 Type I Allergic Reactions: Atopy and Anaphylaxis

Expected **Learning Outcomes**

3. Describe general characteristics of type I allergic reactions.
4. Outline the major allergen categories, giving examples.
5. Explain the mechanism of immediate allergies, including the concepts of IgE, mast cells, sensitization, and provocation.
6. Summarize the physiological effects of type I allergies that lead to symptoms.
7. Define the types of atopic allergies and describe their major features.
8. Explain the process of anaphylaxis.
9. Relate the primary methods of diagnosing, treating, and preventing allergies.

All type I allergies share a similar physiological mechanism, are immediate in onset, and are associated with exposure to specific antigens. However, it is convenient to recognize two levels of severity: **Atopy*** is any chronic local allergy such as hay fever or asthma;

* *atopy* (at′-oh-pee) Gr. *atop,* out of place.

and **anaphylaxis*** is a systemic, sometimes fatal reaction that involves airway obstruction and circulatory collapse. In the following sections, we consider the allergens, routes of inoculation, mechanisms of disease, and specific syndromes involved in type I allergies.

Modes of Contact with Allergens

Allergies exert profound medical and economic impact. Allergists (physicians who specialize in treating allergies) estimate that about 10% to 30% of the population is prone to atopic allergy. It is generally acknowledged that self-treatment with over-the-counter medicines accounts for significant underreporting of cases. The 35 million people afflicted by hay fever (15% to 20% of the population) spend about half a billion dollars annually for medical treatment. The monetary loss due to employee debilitation and absenteeism is immeasurable. The majority of type I allergies are relatively mild, but certain forms such as asthma and anaphylaxis may require hospitalization and can cause death.

The predisposition for type I allergies has a strong familial association. Be aware that what is hereditary is a generalized *susceptibility,* not the allergy to a specific substance. For example, a parent who is allergic to ragweed pollen can have a child who is allergic to cat hair. The prospect of a child's developing atopic allergy is at least 25% if one parent is atopic, increasing up to 50% if grandparents or siblings are also afflicted. The actual basis for atopy appears to be the inheritance of genes that favor allergic antibody (IgE) production, increased reactivity of mast cells, and increased susceptibility of target tissue to allergic mediators. Allergic persons often exhibit a combination of syndromes, such as hay fever, eczema, and asthma.

Other factors that affect the onset of allergy are age, infection, and geographic locale. New allergies tend to crop up throughout an allergic person's life, especially as new exposures occur after moving or changing lifestyle. In some persons, atopic allergies last for a lifetime; others "outgrow" them, and still others suddenly develop them later in life. Some features of allergy are not yet completely explained.

TAKE NOTE: ALLERGIES–THE PRICE OF BEING TOO CLEAN?

Allergists have been observing an increased incidence of allergies such as asthma, allergic rhinitis, and other atopic allergies that is not tied to genetic factors or better diagnosis. Most of these increases have occurred in affluent industrialized regions in the United States and Europe. Although this phenomenon cannot easily be explained, many medical experts think it is related to something called the **hygiene hypothesis.** This explanation basically says that ultraclean lifestyles and living environments prevent the kind of contact with microbes that is required for the normal development of the immune system. Because of modern housekeeping, vaccines, and antimicrobial drugs, children are not exposed as much to non-infectious environmental microbes and viral and parasitic infections. Some support for this exists in the lower levels of allergies seen in developing countries and rural areas, where children

have greater exposure to soil and infectious agents. Although the factors involved appear to be complex, it is thought that contact with certain types of microbes initiates protective responses and shapes the future direction of immunities. The immune systems of children who lack sufficient microbial contact become programmed to react with allergens such as household chemicals, animal and plant molecules, and even food, which sets the scene for longterm allergies. Since people are unlikely to return to less clean lifestyles, researchers are on the trail of finding out what kinds of stimuli can inoculate people against allergies. It may turn out that "a little dirt can't hurt," and it may even be beneficial.

The Nature of Allergens and Their Portals of Entry

As with other antigens, allergens have certain immunogenic characteristics. Not unexpectedly, proteins are more allergenic than carbohydrates, fats, or nucleic acids. Some allergens are haptens, nonproteinaceous substances with a molecular weight of less than 1,000 that can form complexes with carrier molecules in the body (shown in figure 15.9). Organic and inorganic chemicals found in industrial and household products, cosmetics, food, and drugs are commonly of this type. **Table 16.2** lists a number of common allergenic substances and their portal of entry.

Allergens typically enter through epithelial portals in the respiratory tract, gastrointestinal tract, and skin. The mucosal surfaces of the gut and respiratory system present a thin, moist surface that is normally quite penetrable. The dry, tough keratin coating of skin is less permeable, but access still occurs through tiny breaks, glands, and hair follicles. It is worth noting that the organ of allergic expression may or may not be the same as the portal of entry.

Airborne environmental allergens such as pollen, house dust, dander (shed skin scales), or fungal spores are termed *inhalants.* Each geographic region harbors a particular combination of airborne substances; this combination varies with the season and humidity (**figure 16.2a**). Pollen, the most common offender, is given off seasonally by the reproductive structures of pines and flowering plants (weeds, trees, and grasses). Unlike pollen, mold spores are released throughout the year and are especially profuse in moist areas of the home and garden. Airborne animal hair and dander (skin flakes), feathers, and the saliva of dogs and cats are common

TABLE 16.2	Common Allergens, Classified by Portal of Entry		
Inhalants	**Ingestants**	**Injectants**	**Contactants**
Pollen	Food (milk, peanuts,	Hymenopteran	Drugs
Dust	wheat, shellfish,	venom (bee,	Cosmetics
Mold spores	soybeans,	wasp)	Heavy metals
Dander	nuts, eggs,		Detergents
Animal hair	fruits)	Drugs	Formalin
Insect parts		Vaccines	Rubber
Formalin	Food additives	Serum	Solvents
Drugs	Drugs (aspirin,	Enzymes	Dyes
	penicillin)	Hormones	

* *anaphylaxis* (an"-uh-fih-lax'-us) Gr. *ana,* excessive, and *phylaxis,* protection.

National Allergy Bureau
Pollen and Mold Report*

Location: Flower Mound, TX Date: April 29, 2010
Counting Station: Allergy Medical Group of the North Area

Trees	High concentration	Total count: 163 / m³
Weeds	Low concentration	Total count: 2 / m³
Grass	Moderate concentration	Total count: 13 / m³
Mold	Low concentration	Total count: 3,259 / m³

(a) *Count standards vary with the type of pollen and mold spore

(b)

(c)

Figure 16.2 **Monitoring airborne allergens. (a)** The air in heavily vegetated places with a mild climate is especially laden with allergens such as pollen and mold spores. These counts vary seasonally. **(b)** The dust mite *Dermatophagoides* feeds primarily on human skin cells in house dust, and is found in abundance in bedding and carpets. Airborne mite feces and particles from their exoskeletons are an important source of allergies. **(c)** Scanning electron micrograph of a pollen grain from a morning glory flower (5,000×). Millions of these are released from a single flower.

sources of allergens. The component of house dust that appears to account for most dust allergies is not soil or other debris but the decomposed skeletons of tiny mites that commonly live in this dust **(figure 16.2b).** Some people are allergic to their work, in the sense that they are exposed to allergens on the job. Examples include florists, woodworkers, farmers, drug processors, and plastics manufacturers whose work can aggravate inhalant and contact allergies.

Allergens that enter by mouth, called *ingestants,* often cause food allergies. *Injectant* allergies are an important adverse side effect of drugs or other substances used in diagnosing, treating, or preventing disease. A natural source of injectants is venom from stings by hymenopterans, a family of insects that includes honeybees and wasps. *Contactants* are allergens that enter through the skin. Many contact allergies are of the type IV, delayed variety discussed in section 16.5.

Mechanisms of Type I Allergy: Sensitization and Provocation

What causes some people to sneeze and wheeze every time they step out into the spring air, while others suffer no ill effects? We can answer this question by examining what occurs in the tissues of the allergic individual that does not occur in the normal person. In general, type I allergies develop in stages **(figure 16.3).** The initial encounter with an allergen provides a **sensitizing dose** that primes the immune system for a subsequent encounter with that allergen but generally elicits no signs or symptoms. The memory cells and immunoglobulin are then ready to react with a subsequent **provocative dose** of the same allergen. It is this dose that precipitates the signs and symptoms of allergy. Despite numerous anecdotal reports of people showing an allergy upon first contact with an allergen, it is generally believed that these individuals unknowingly had contact at some previous time. Fetal exposure to allergens from the mother's bloodstream is one possibility, and foods can be a prime source of "hidden" allergens such as penicillin.

The Physiology of IgE-Mediated Allergies

During primary contact and sensitization, the allergen penetrates the portal of entry **(figure 16.3, steps 1 through 6).** When large particles such as pollen grains, hair, and spores encounter a moist membrane, they release molecules of allergen that pass into the tissue fluids and lymphatics. The lymphatics then carry the allergen to the lymph nodes, where specific clones of B cells recognize it, are activated, and proliferate into plasma cells. These plasma cells produce immunoglobulin E (IgE), the antibody of allergy. IgE is different from other immunoglobulins in having an Fc region with great affinity for mast cells and basophils. The binding of IgE to these cells in the tissues sets the scene for the reactions that occur upon repeated exposure to the same allergen **(figure 16.3, steps 7 through 10).**

The Role of Mast Cells and Basophils

The most important characteristics of mast cells and basophils relating to their roles in allergy are

1. Their ubiquitous location in tissues. Mast cells are located in the connective tissue of virtually all organs, but particularly high concentrations exist in the lungs, skin, gastrointestinal tract, and genitourinary tract. Basophils circulate in the blood but migrate readily into tissues.
2. Their capacity to bind IgE during sensitization (figure 16.3). Each cell carries 30,000 to 100,000 cell receptors that bind 10,000 to 40,000 IgE antibodies.
3. Their cytoplasmic granules (secretory vesicles), which contain physiologically active cytokines (histamine, serotonin—introduced in chapter 14).
4. Their tendency to **degranulate** (figures 16.3 and 16.4) or release the contents of the granules into the tissues when triggered by a specific allergen through the IgE bound to them.

Let us now see what occurs when sensitized cells are challenged with allergen a second time.

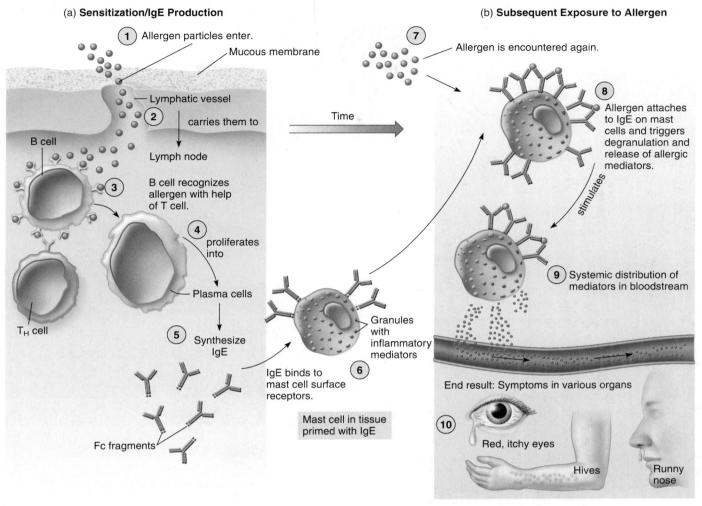

(a) Sensitization/IgE Production

① Allergen particles enter.

Mucous membrane

Lymphatic vessel

② carries them to

B cell

Lymph node

③ B cell recognizes allergen with help of T cell.

④ proliferates into

Plasma cells

T_H cell

⑤ Synthesize IgE

Granules with inflammatory mediators

⑥ IgE binds to mast cell surface receptors.

Fc fragments

Mast cell in tissue primed with IgE

Time

(b) Subsequent Exposure to Allergen

⑦ Allergen is encountered again.

⑧ Allergen attaches to IgE on mast cells and triggers degranulation and release of allergic mediators.

stimulates

⑨ Systemic distribution of mediators in bloodstream

End result: Symptoms in various organs

⑩ Red, itchy eyes

Hives

Runny nose

Process Figure 16.3 A schematic view of cellular reactions during the type I allergic response. (a) Sensitization (initial contact with sensitizing dose), 1–6. **(b)** Provocation (later contacts with provocative dose), 7–10.

The Second Contact with Allergen

After sensitization, the IgE-primed mast cells can remain in the tissues for years. Even after long periods without contact, a person can retain the capacity to react immediately upon reexposure. The next time allergen molecules contact these sensitized cells, they bind across adjacent receptors and stimulate degranulation. As chemical mediators are released, they diffuse into the tissues and bloodstream. Cytokines give rise to numerous local and systemic reactions, many of which appear quite rapidly (figure 16.3, step 10). Thus, the symptoms of allergy are not caused by the direct action of allergen on tissues but by the physiological effects of mast cell mediators on target organs.

Cytokines, Target Organs, and Allergic Symptoms

Numerous substances involved in mediating allergy (and inflammation) have been identified. The principal chemical mediators produced by mast cells and basophils are histamine, serotonin, leukotriene, platelet-activating factor, prostaglandins, and bradykinin

(figure 16.4). These chemicals, acting alone or in combination, account for the tremendous scope of allergic symptoms. Targets of these mediators include the skin, upper respiratory tract, gastrointestinal tract, and conjunctiva. The general responses of these organs include rashes, itching, redness, rhinitis, sneezing, diarrhea, and shedding of tears. Systemic targets include smooth muscle, mucous glands, and nervous tissue. Because smooth muscle is responsible for regulating the size of blood vessels and respiratory passageways, changes in its activity can profoundly alter blood flow, blood pressure, and respiration. Pain, anxiety, agitation, and lethargy are also attributable to the effects of mediators on the nervous system. For some theories pertaining to this function of the allergic response, see **Insight 16.1.**

Histamine* is the most profuse and fastest-acting allergic mediator. It is a potent stimulator of smooth muscle, glands, and eosinophils. Histamine's actions on smooth muscle vary with location. It *constricts* the smooth muscle layers of the bronchioles and intestine, thereby causing labored breathing and increased intestinal

* *histamine* (his'-tah-meen) Gr. *histio,* tissue, and amine.

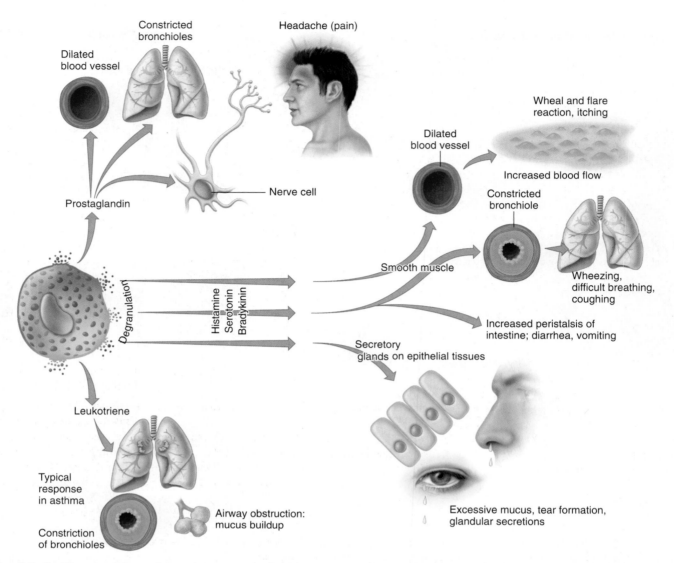

Figure 16.4 **The spectrum of reactions to inflammatory cytokines released by mast cells and the common symptoms they elicit in target tissues and organs.** Note the extensive overlapping effects.

motility. In contrast, histamine *relaxes* vascular smooth muscle and dilates arterioles and venules. It is responsible for the **wheal* and flare reaction** (see figure 16.6), pruritis (itching), and headache. More severe reactions such as anaphylaxis can be accompanied by edema and vascular dilation, which lead to hypotension, tachycardia, circulatory failure, and, frequently, shock. Salivary, lacrimal, mucous, and gastric glands are also histamine targets.

Although the role of **serotonin*** in human allergy is uncertain, its effects appear to complement those of histamine. In experimental animals, serotonin increases vascular permeability, capillary dilation, smooth muscle contraction, intestinal peristalsis, and respiratory rate, but it diminishes central nervous system activity.

A type of **leukotriene,*** known as the "slow-reacting substance of anaphylaxis," induces gradual contraction of smooth muscle.

This leukotriene is responsible for the prolonged bronchospasm, vascular permeability, and mucous secretion of the asthmatic individual. Other leukotrienes stimulate the activities of polymorphonuclear leukocytes.

Platelet-activating factor is a lipid released by basophils, neutrophils, monocytes, and macrophages. The physiological response to this factor is similar to that of histamine, including increased vascular permeability, pulmonary smooth muscle contraction, pulmonary edema, hypotension, and a wheal and flare response in the skin.

Prostaglandins* are a group of powerful inflammatory agents. Normally, these substances regulate smooth muscle contraction. For example, they stimulate uterine contractions during delivery. In allergic reactions, they are responsible for vasodilation, increased vascular permeability, increased sensitivity to pain, and bronchoconstriction. Certain anti-inflammatory drugs work by preventing the actions of prostaglandins.

* *wheal* (weel) A smooth, slightly elevated, temporary welt that is surrounded by a flushed patch of skin (flare).

* *serotonin* (ser″-oh-toh′-nin) L. *serum,* whey, and *tonin,* tone.

* *leukotriene* (loo″-koh-try′-een) Gr. *leukos,* white blood cell, and *triene,* a chemical suffix

* *prostaglandin* (pross″-tah-glan′-din) From prostate gland. The substance was originally isolated from semen.

INSIGHT 16.1

The Origins of Allergy

Why would humans and other mammals evolve an allergic response that is capable of doing so much harm and even causing death? It is unlikely that this limb of immunity exists merely to make people miserable; it must have a role in protection and survival. What are the underlying biological functions of IgE, mast cells, and the array of potent cytokines? Analysis has revealed that, although allergic persons have high levels of IgE, trace quantities are present even in the sera of nonallergic individuals, just as mast cells and inflammatory chemicals are also part of normal human physiology. In chapter 14, you learned that inflammatory mediators serve valuable functions, such as increasing blood flow and vascular permeability to summon essential immune components to an injured site. They are also responsible for increased mucus secretion, gastric motility, sneezing, and coughing, which help expel noxious agents. The difference is that, in allergic persons, the quantity and quality of these reactions are excessive and uncontrolled.

Some surprising insights have emerged from studies at the University of California, San Francisco. It seems that a significant source of many allergies is the polysaccharide *chitin*. This ubiquitous compound is found in insect exoskeletons, fungal cell walls, and parasitic worms. The digestion of chitin by enzymes present in the human body causes the release of smaller antigen molecules. These set off powerful immune and inflammatory reactions. There is even a specific pathway in lymphocytes and other white blood cells for responding to chitin and its antigens. It is theorized that this system was originally designed to aid in the attack against worms, fungi, and other pathogens containing chitin, but modern hygiene and therapy have greatly reduced these infections. The remnant of this formerly protective mechanism now actually causes "out of control" reactions such as asthma and other respiratory allergies. This appears to be one basis of allergies to dust (from mites) and the increase in childhood asthma.

List some other potential benefits of allergy. *Answer available at http://www.mhhe.com/talaro8*

Bradykinin* is related to a group of plasma and tissue peptides known as kinins that participate in blood clotting and chemotaxis. In allergy, it causes prolonged contraction of the bronchioles, dilation of peripheral arterioles, increased capillary permeability, and increased mucus secretion.

Specific Diseases Associated with IgE- and Mast-Cell-Mediated Allergy

The mechanisms just described are basic to hay fever, allergic asthma, food allergy, drug allergy, eczema, and anaphylaxis. In this section, we cover the main characteristics of these conditions, followed by methods of detection and treatment.

Atopic Diseases

Hay fever is a generic term for **allergic rhinitis,*** a seasonal reaction to inhaled plant pollen or molds or a chronic, year-round reaction to a wide spectrum of airborne allergens or inhalants (see table 16.2). The targets are typically respiratory membranes, and the symptoms include nasal congestion; sneezing; coughing; profuse mucus secretion; itchy, red, and teary eyes; and mild bronchoconstriction.

Asthma* is a respiratory disease characterized by episodes of impaired breathing due to severe bronchoconstriction. The airways of asthmatic people are exquisitely responsive to minute amounts of inhalant allergens, food, or other stimuli, such as infectious agents. The symptoms of asthma range from occasional, annoying bouts of difficult breathing to fatal suffocation. Labored breathing, shortness of breath, wheezing, cough, and ventilatory

rales* are present to one degree or another. The respiratory tract of an asthmatic person is chronically inflamed and severely overreactive to allergy chemicals, especially leukotrienes and serotonin from pulmonary mast cells. Other pathologic components are thick mucous plugs in the air sacs and lung damage that can result in long-term respiratory compromise. An imbalance in the nervous control of the respiratory smooth muscles is apparently involved in asthma, and the episodes are influenced by the psychological state of the person, which strongly supports a neurological connection.

The number of asthma sufferers in the United States is estimated at more than 10 million, with nearly one-third of them children. For reasons that are not completely understood, asthma is on the increase, and deaths from it have doubled since 1982, even though effective agents to control it are more available now than they have ever been before. It has been suggested that more highly insulated buildings, mandated by energy efficiency regulations, have created indoor air conditions that harbor higher concentrations of contaminants, including airborne allergens and ozone (see Insight 16.1).

Atopic dermatitis is an intensely itchy inflammatory condition of the skin, sometimes also called **eczema.*** Sensitization occurs through ingestion, inhalation, and, occasionally, skin contact with allergens. It usually begins in infancy with reddened, vesicular, weeping, encrusted skin lesions (**figure 16.5a**). It then progresses in childhood and adulthood to a dry, scaly, thickened skin condition. Lesions can occur on the face, scalp, neck, and inner surfaces of the limbs and trunk. The itchy, painful lesions cause considerable discomfort, and they are often predisposed to secondary bacterial infections. An anonymous writer once aptly described eczema as "the itch that rashes" or "one scratch is too many but one thousand is not enough."

* *bradykinin* (brad″-ee-kye′-nin) Gr. *bradys,* slow, and *kinein,* to move.
* *rhinitis* (rye-nye′-tis) Gr. *rhis,* nose, and *itis,* inflammation.
* *asthma* (az′-muh) The Greek word for gasping.

* *rales* (rails) Abnormal breathing sounds.
* *eczema* (eks′-uh-mah; also ek-zeem′-uh) Gr. *ekzeo,* to boil over.

(a) (b)

Figure 16.5 Skin manifestations in atopic allergies. (a) Atopic dermatitis, or eczema. Vesicular, encrusted lesions are typical in afflicted infants. This condition is prevalent enough to account for 1% of pediatric care. **(b)** Skin rash caused by an allergy to strawberries. The slightly raised red lesions, termed urticaria or hives, can sometimes merge to form a solid red rash.

Food Allergy

The ordinary diet contains a vast variety of compounds that are potentially allergenic. Although the mode of entry is intestinal, food allergies can affect the skin and respiratory tract as well. Gastrointestinal symptoms include vomiting, diarrhea, and abdominal pain. In severe cases, nutrients are poorly absorbed, leading to growth retardation and failure to thrive in young children. Other manifestations of food allergies include eczema, hives (**figure 16.5***b*), rhinitis, asthma, and occasionally, anaphylaxis. Classic food hypersensitivity involves IgE and degranulation of mast cells, but not all reactions involve this mechanism. The most common food allergens come from peanuts, fish, cow's milk, eggs, shellfish, and soybeans.[1]

Drug Allergy

Modern chemotherapy has been responsible for many medical advances. Unfortunately, it has also been hampered by the fact that drugs are foreign compounds capable of stimulating allergic reactions. In fact, allergy to drugs is one of the most common side effects of treatment (present in 5% to 10% of hospitalized patients). Depending upon the allergen, route of entry, and individual sensitivities, virtually any tissue of the body can be affected, and reactions range from mild atopy to fatal anaphylaxis. Compounds implicated most often are antibiotics (penicillin is number one in prevalence), synthetic antimicrobials (sulfa drugs), aspirin, opiates, and contrast dye used in X rays. The actual allergen is not the intact drug itself but a hapten given off when the liver processes the drug. Some forms of penicillin sensitivity are due to the presence of small amounts of the drug in meat, milk, and other foods and to exposure to *Penicillium* mold in the environment.

Anaphylaxis: A Powerful Systemic Reaction to Allergens

The term **anaphylaxis,** or **anaphylactic shock,** was first used to denote a reaction of animals injected with a foreign protein. Although the animals showed no response during the first contact, upon reinoculation with the same protein at a later time, they exhibited acute symptoms—itching, sneezing, difficult breathing, prostration, and convulsions—and many died in a few minutes. Two clinical types of anaphylaxis are seen in humans. **Cutaneous anaphylaxis** is the wheal and flare inflammatory reaction to the local injection of allergen. **Systemic anaphylaxis,** on the other hand, is characterized by sudden respiratory and circulatory disruption that can be fatal in a few minutes. In humans, the allergen and route of entry are variable, though bee stings and injections of antibiotics or serum are implicated most often. Bee venom is a complex material containing several allergens and enzymes that create a sensitivity that can last for decades after exposure.

The underlying physiological events in systemic anaphylaxis parallel those of atopy, but the concentration of chemical mediators and the strength of the response are greatly amplified. The immune system of a sensitized person exposed to a provocative dose of allergen responds with a sudden, massive release of chemicals into the tissues and blood, which act rapidly on the target organs. Anaphylactic persons have been known to die in 15 minutes from complete airway blockage.

Diagnosis of Allergy

Because allergy mimics infection and other conditions, it is important to determine if a person is actually allergic. If possible or necessary, it is also helpful to identify the specific allergen or allergens. Allergy diagnosis involves several levels of tests, including nonspecific, specific, in vitro, and in vivo methods.

A test that can distinguish whether a patient has experienced an allergic attack measures elevated blood levels of tryptase, an enzyme released by mast cells. The enzyme level increases during an allergic response. Several types of specific in vitro tests can determine significant signs of allergy from a patient's blood sample. A differential blood cell count can indicate the levels of basophils and eosinophils—a higher level of these indicates allergy. The leukocyte histamine-release test measures the amount of histamine released from the patient's basophils when exposed to a specific allergen. Serological tests that use radioimmune assays (see chapter 17) to reveal the quantity and type of IgE are also clinically helpful.

Skin Testing

Skin testing can precisely detect atopic or anaphylactic sensitivities. With this technique, a patient's skin is injected, scratched, or pricked with a small amount of a pure allergen extract. Drug companies can supply a large battery of standard allergen extracts containing common airborne allergens (plant and mold pollen) and more unusual allergens (mule dander, theater dust, bird feathers). Unfortunately, skin tests for food allergies using food extracts are unreliable in most cases.

1. Do not confuse food allergy with food intolerance. Many people are lactose intolerant, for example, due to a deficiency in the enzyme that degrades the milk sugar.

Figure 16.6 A method for conducting an allergy skin test. The forearm (or back) is mapped and then injected with a selection of allergen extracts. The allergist must be very aware of potential anaphylaxis attacks triggered by these injections. **(a)** Close-up of skin wheals showing a number of positive reactions (dark lines are measurer's marks). **(b)** An actual skin test record for some common environmental allergens in a patient with severe allergies.

Environmental Allergens

No. 1 Standard Series				No. 2 Airborne Particles		
	ID				ID	
	+++	1. Acacia gum			+++	1. Ant
	+++	2. Cat dander			+++++	2. Aphid
	++++	3. Chicken feathers			++++	3. Bee
	++++	4. Cotton lint			++++	4. Housefly
	++	5. Dog dander			✕	5. House mite
	+	6. Duck feathers			+++	6. Mosquito
	+	7. Glue, animal			++++	7. Moth
	++	8. Horse dander			+++	8. Roach
	✕	9. Horse serum			++	9. Wasp
	+++	10. House dust #1			0	10. Yellow jacket
	+	11. Kapok				Airborne mold spores
	+	12. Mohair (goat)			++	11. *Alternaria*
	+	13. Paper			+++	12. *Aspergillus*
	++++	14. Pyrethrum			++	13. *Cladosporium*
	+++	15. Rug pad, ozite			+++	14. *Fonsecaea*
	+	16. Silk dust			0	15. *Penicillium*
	+	17. Tobacco dust			+	16. *Phoma*
	+	18. Tragacanth gum			+++	17. *Rhizopus*
	+++++	19. Upholstery dust				18.
	+++	20. Wool				

✕ - not done ++ - mild reaction
0 - no reaction +++ - moderate reaction
+ - slight reaction ++++ - severe reaction

(b)

In patients with numerous allergies, the allergist maps the skin on the inner aspect of the forearms or back and injects the allergens intradermally according to this predetermined pattern **(figure 16.6a).** Approximately 20 minutes after antigenic challenge, each site is appraised for a wheal response indicative of histamine release. The diameter of the wheal is measured and rated on a scale of 0 (no reaction) to 4+ (greater than 15 mm). **Figure 16.6b** shows skin test results for a person with extreme inhalant allergies. For detecting contactant allergies, some specialists use a patch test. Small allergen-impregnated patches are applied to the skin and left for 48 hours. They are evaluated for size of reaction as with skin prick tests. A new rapid test provides an alternate system for testing that has greater accuracy than skin testing and does not require injection of allergen.

Treatment and Prevention of Allergy

In general, the methods of treating and preventing type I allergy involve: (1) avoiding the allergen, although this may be very difficult in many instances; (2) taking drugs that block the action of lymphocytes, mast cells, or chemical mediators; and (3) undergoing desensitization therapy.

It is not possible to completely prevent initial sensitization, because there is no way to tell in advance if a person will develop an allergy to a particular substance. The practice of delaying the introduction of solid foods apparently has some merit in preventing food allergies in children, although even breast milk can contain allergens ingested by the mother. Rigorous cleaning and air conditioning can reduce contact with airborne allergens, but it is not feasible to isolate a person from all allergens, which is the reason drugs are so important in control.

Therapy to Counteract Allergies

The aim of antiallergy medication is to block the progress of the allergic response somewhere along the route between IgE production and the appearance of symptoms **(figure 16.7).** Oral anti-inflammatory drugs such as corticosteroids inhibit the activity of lymphocytes and thereby reduce the production of IgE, but they also have dangerous side effects and should not be taken for prolonged periods. Some drugs block the degranulation of mast cells and reduce the levels of inflammatory cytokines. The most effective of these are diethylcarbamazine and cromolyn. Asthma and rhinitis sufferers can find relief with a drug that blocks synthesis of leukotriene (Singulair) and a monoclonal antibody that inactivates IgE (Xolair).

Widely used medications for preventing symptoms of atopic allergy are **antihistamines,** the active ingredients in most over-the-counter allergy-control drugs. Antihistamines interfere with histamine activity by binding to histamine receptors on target organs. Although many of them have major side effects such as drowsiness, some of the newer antihistamines lack this side effect because they do not cross the blood-brain barrier. Other drugs that relieve inflammatory symptoms are aspirin and acetaminophen, which reduce pain by interfering with prostaglandin, and theophylline, a bronchodilator that reverses spasms in the respiratory smooth muscles. Persons who suffer from anaphylactic attacks are urged to carry at all times injectable epinephrine (adrenaline) and an identification tag indicating their sensitivity. An aerosol inhaler containing epinephrine can also provide rapid relief. Epinephrine reverses constriction of the airways and slows the release of allergic mediators.

Approximately 70% of allergic patients benefit from controlled injections of specific allergens as determined by skin tests. This technique, called **desensitization** or **hyposensitization,** is a therapeutic way to

Figure 16.7 **Strategies for circumventing allergic attacks.**

prevent reactions between allergen, IgE, and mast cells. The allergen preparations contain pure, preserved suspensions of plant antigens, venoms, dust mites, dander, and molds (but so far, hyposensitization for foods has not proved very effective). The immunologic basis of this treatment is open to differences in interpretation. One theory suggests that injected allergens stimulate the formation of high levels of allergen-specific IgG (**figure 16.8**) instead of IgE. It has been proposed that these IgG **blocking antibodies** remove allergen from the system before it can bind to IgE, thus preventing the degranulation of mast cells. It is also possible that allergen delivered in this fashion combines with the IgE itself and takes it from circulation before it can react with the mast cells.

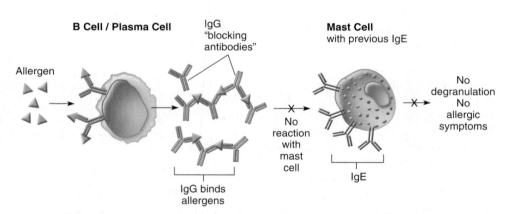

Figure 16.8 **The blocking antibody theory for allergic desensitization.** An injection of allergen causes IgG antibodies to be formed instead of IgE; these blocking antibodies cross-link and effectively remove the allergen before it can react with the IgE in the mast cell.

Check & Assess Section 16.2

✔ Antigens that trigger hypersensitivity reactions are allergens. They can be either exogenous (originate outside the host) or endogenous (involve the host's own tissue).

✔ Type I hypersensitivity reactions result from excessive IgE production in response to an exogenous antigen.

✔ The two kinds of type I hypersensitivities are atopy, a chronic, local allergy, and anaphylaxis, a systemic, potentially fatal allergic response.

✔ The predisposition to type I hypersensitivities is inherited, but age, geographic locale, and infection also influence allergic response.

✔ Type I allergens include inhalants, ingestants, injectants, and contactants.

✔ The portals of entry for type I antigens are the skin, respiratory tract, and gastrointestinal tract.

✔ Type I hypersensitivities are set up by a sensitizing dose of allergen and expressed when a second provocative dose triggers the allergic response. The time interval between the two can be many years.

✔ The primary participants in type I hypersensitivities are IgE, basophils, mast cells, and agents of the inflammatory response.

✔ Allergies are diagnosed by a variety of in vitro and in vivo tests that assay specific cells, IgE, and local reactions.

✔ Allergies are treated by medications that interrupt the allergic response at certain points.

✔ Allergic reactions can often be prevented by desensitization therapy that blocks the primary sensitization process.

5. Give some examples of allergens and explain their routes of access to the body.

6. Describe several factors that influence types and severity of allergic responses.

7. Describe the events in sensitization and provocation.

8. How are atopic allergies similar to anaphylaxis, and how are they different?

9. Trace the course of a pollen grain through sensitization and provocation in type I allergies. Include in the discussion the role of mast cells, basophils, IgE, and allergic mediators.

10. Relate the roles of major mediators to the signs and symptoms of allergy.

11. Outline the target organs and symptoms of the principal atopic diseases and their diagnosis and treatment.

12. Describe the allergic response that leads to anaphylaxis. Include its usual causes, how it is diagnosed and treated, and two effective physiological targets for treatment.

13. Explain how hyposensitization is achieved and suggest two mechanisms by which it might work.

16.3 Type II Hypersensitivities: Reactions That Lyse Foreign Cells

Expected Learning Outcomes

10. Explain the requirements for developing a type II hypersensitivity.

11. Define what is meant by blood groups, and explain how they are inherited and expressed.

12. Discuss the presence of blood group antibodies and the primary medical concerns of blood transfusions.

13. Discuss how blood is typed and notated.

14. Explain the special concerns with the Rh factor as a blood type.

15. Relate the causes of hemolytic disease of the newborn and the ways to prevent it.

The diseases termed type II hypersensitivities are a complex group of syndromes that involve complement-assisted destruction (lysis) of cells by antibodies (IgG and IgM) directed against those cells' surface antigens. This category includes transfusion reactions and some types of autoimmunities (discussed in section 16.6). The cells targeted for destruction are often red blood cells, but other cells can be involved.

Chapters 14 and 15 described the functions of unique surface markers on cell membranes. Ordinarily, these molecules play essential roles in transport, recognition, and development, but they become medically important when the tissues of one person are placed into the body of another person. Blood transfusions and organ donations introduce alloantigens (molecules that differ in the same species) on donor cells that are recognized by the lymphocytes of the recipient. These reactions are not really immune dysfunctions as allergy and autoimmunity are. The immune system is in fact working normally, but it is not equipped to distinguish between the desirable foreign cells of a transplanted tissue and the undesirable ones of a microbe.

The Basis of Human ABO Antigens and Blood Types

The existence of human blood types was first demonstrated by an Austrian pathologist, Karl Landsteiner, in 1904. While studying incompatibilities in blood transfusions, he found that the serum of one person could clump the red blood cells of another. Landsteiner identified four distinct types, subsequently called the **ABO blood groups.**

Like the MHC antigens on white blood cells, the ABO antigen markers on red blood cells are genetically determined and composed of glycoproteins. These ABO antigens are inherited as two (one from each parent) of three alternative **alleles:*** A, B, or O. A and B alleles are dominant over O and codominant with one another. As **table 16.3** indicates, this mode of inheritance gives rise to four blood types (phenotypes), depending on the particular combination of genes. Thus, a person with an *AA* or *AO* genotype has type A blood; genotype *BB* or *BO* gives type B; genotype *AB* produces

* *allele* (ah-leel′) Gr. *allelon,* of one another. An alternate form of a gene for a given trait.

TABLE 16.3	Characteristics of ABO Blood Groups					
					Incidence of Type in United States	
Genotype	Blood Type	Antigen Present on Erythrocyte Membranes	Antibody in Plasma	Among Whites (%)	Among Asians (%)	Among Those of African and Caribbean Descent (%)
AA, AO	A	A	Anti-b	41	28	27
BB, BO	B	B	Anti-a	10	27	20
AB	AB	A and B	Neither anti-a nor anti-b	4	5	7
OO	O	Neither A nor B	Anti-a and anti-b	45	40	46

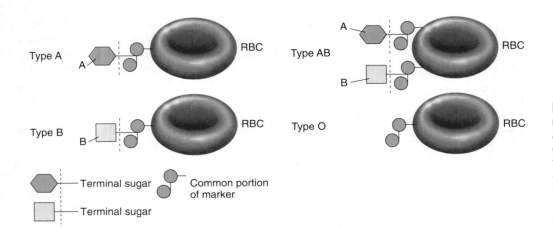

Figure 16.9 The genetic/molecular basis for the A and B antigens (receptors) on red blood cells. In general, persons with blood types A, B, and AB inherit a gene or genes for enzymes that add a terminal sugar to the basic RBC receptor. Type O persons do not have these enzymes and lack the terminal sugar.

type AB; and genotype *OO* produces type O. Some important points about the blood types are

1. they are named for the dominant antigen(s);
2. the RBCs of type O persons lack the A and B antigens, but they do have other types of antigens; and
3. tissues other than RBCs carry A and B antigens.

A diagram of the AB antigens and blood types is shown in **figure 16.9.** The A and B genes each code for an enzyme that adds a terminal carbohydrate to RBC surface molecules during maturation. RBCs of type A contain an enzyme that adds *N*-acetylgalactosamine to the molecule; RBCs of type B have an enzyme that adds D-galactose; RBCs of type AB contain both enzymes that add both carbohydrates; and RBCs of type O lack the genes and enzymes to add a terminal molecule.

Antibodies against A and B Antigens

Although an individual does not normally produce antibodies in response to his or her own RBC antigens, the serum can contain antibodies that react with blood of another antigenic type, even though contact with this other blood type has *never* occurred. These preformed antibodies account for the immediate and intense quality of transfusion reactions. As a rule, type A blood contains antibodies (anti-b) that react against the B antigens on type B and AB red blood cells. Type B blood contains antibodies (anti-a) that react with A antigen on type A and AB red blood cells. Type O blood contains antibodies against both A and B antigens. Type AB blood does not contain antibodies against either A or B antigens[2] (see table 16.3). What could be the source of these anti-a and anti-b antibodies? It appears that they develop in early infancy because of exposure to certain antigens that are widely distributed in nature. These antigens are surface molecules on bacteria and plant cells that mimic the structure of A and B antigens and stimulate the production of the antibodies that cause transfusion reactions.

Clinical Concerns in Transfusions

The presence of ABO antigens and a, b antibodies underlies several clinical concerns in giving blood transfusions. First, the individual blood types of donor and recipient must be determined. By use of a

standard technique, drops of blood are mixed with antisera that contain antibodies against the A and B antigens and are then observed for the evidence of agglutination **(figure 16.10).**

Knowing the blood types involved makes it possible to determine which transfusions are safe to do. The general rule of compatibility is that the RBC antigens of the donor must not be agglutinated by antibodies in the recipient's blood **(figure 16.11).** The ideal practice is to transfuse blood that is a perfect match (A to A, B to B). But even in this event, blood samples must be cross-matched before the transfusion because other blood group incompatibilities can exist. This test involves mixing the blood of the donor with the serum of the recipient to check for agglutination.

Under certain circumstances (medical emergencies, the battlefield), the concept of universal transfusions can be used. To appreciate how this works, we must apply the rule stated in the previous paragraph. Type O blood lacks A and B antigens and will not be agglutinated by other blood types, so it could theoretically be used in any transfusion. We call a person with this blood type a **universal donor.** Because type AB blood lacks agglutinating antibodies, an individual

Figure 16.10 Interpretation of blood typing. In this test, a drop of blood is mixed with a specially prepared antiserum (anti-A and anti-B) known to contain antibodies against the A or B antigens. If that particular antigen is not present, the red blood cells in that droplet do not agglutinate, and form a uniform suspension. If that antigen is present, agglutination occurs and the RBCs form visible clumps (small red particles). The patterns and interpretations for the four blood types are shown.

Type A

Type B

Type AB

Type O

2. Why would this be true? The answer lies in the first sentence of the paragraph.

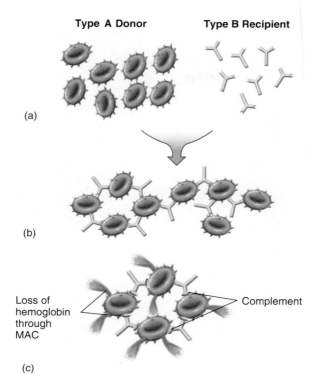

Type A Donor **Type B Recipient**

(a)

(b)

Loss of
hemoglobin
through
MAC Complement

(c)

Figure 16.11 Microscopic view of a transfusion reaction.
(a) Incompatible blood types. The red blood cells of the type A donor
contain antigen A, while the serum of the type B recipient contains
anti-a antibodies that can agglutinate donor cells. **(b)** Agglutination
complexes can block the circulation in vital organs. **(c)** Activation of the
complement by antibody on the RBCs causes hemolysis and anemia.
This sort of incorrect transfusion is very rare because of the great care
taken by blood banks to ensure a correct match.

with this blood could conceivably receive any type of blood. Type AB
persons are consequently called **universal recipients.** Although both
types of transfusions involve antibody incompatibilities, these are of
less concern because of the dilution of the donor's blood in the body
of the recipient. Additional RBC markers that can be significant in
transfusions are the Rh, MN, and Kell antigens (see next sections).

Transfusion of the wrong blood type causes differing degrees
of adverse reaction. The severest reaction is massive hemolysis
when the donated red blood cells react with recipient antibody and
trigger the complement cascade (figure 16.11). The resultant de-
struction of red cells leads to systemic shock and kidney failure
brought on by the blockage of glomeruli (blood-filtering apparatus)
by cell debris. Death is a common outcome. Other reactions caused
by RBC destruction are fever, anemia, and jaundice. A transfusion
reaction is managed by immediately halting the transfusion, admin-
istering drugs to remove hemoglobin from the blood, and beginning
another transfusion with red blood cells of the correct type. See this
chapter's Case File for a description of a transfusion reaction.

The Rh Factor and Its Clinical Importance

Another RBC antigen of major clinical concern is the **Rh factor** (or
D antigen). This factor was first discovered in experiments explor-
ing the genetic relationships among animals. Rabbits inoculated
with the RBCs of rhesus monkeys produced an antibody that also

reacted with human RBCs. Further tests showed that this monkey
antigen (termed Rh for rhesus) was present in about 85% of humans
and absent in the other 15%. The details of Rh inheritance are more
complicated than those of ABO; but in simplest terms, a person's
Rh type results from a combination of two possible alleles—a
dominant one that codes for the factor and a recessive one that does
not. A person inheriting at least one Rh gene will be Rh^+; only
those persons inheriting no Rh genes are Rh^-. The "+" or "−" ap-
pearing after a blood type refers to the Rh status of the person, as in
O^+ or AB^- (see figure 16.10). However, unlike the ABO antigens,
exposure to environmental antigens does not sensitize Rh^- persons
to the Rh factor. The only ways one can develop antibodies against
this factor are through placental sensitization or transfusion.

Hemolytic Disease of the Newborn and Rh Incompatibility

The potential for placental sensitization occurs when a mother is
Rh^- and her unborn child is Rh^+. The obvious intimacy between
mother and fetus makes it possible for fetal RBCs to leak into the
mother's circulation during childbirth, when the detachment of the
placenta creates avenues for fetal blood to enter the maternal circu-
lation. The mother's immune system detects the foreign Rh factors
on the fetal RBCs and is sensitized to them by producing antibodies
and memory B cells. The first Rh^+ child is usually not affected be-
cause the process begins so late in pregnancy that the child is born
before maternal sensitization is completed. However, the mother's
immune system has been strongly primed for a second contact with
this factor in a subsequent pregnancy (**figure 16.12a**).

In the next pregnancy with an Rh^+ fetus, fetal blood cells escape
into the maternal circulation late in pregnancy and elicit a memory
response. The fetus is at risk when the maternal anti-Rh antibodies
cross the placenta into the fetal circulation, where they affix to fetal
RBCs and cause complement-mediated lysis. The outcome is a po-
tentially fatal **hemolytic disease of the newborn (HDN)** called
*erythroblastosis fetalis.** This term is derived from the presence of
immature nucleated RBCs called erythroblasts in the blood. They are
released into the infant's circulation to compensate for the massive
destruction of RBCs stimulated by maternal antibodies. Additional
symptoms are severe anemia, jaundice, and enlarged spleen and liver.

Maternal-fetal incompatibilities are also possible in the ABO
blood group, but adverse reactions occur less frequently than with
Rh sensitization because the antibodies to these blood group anti-
gens are IgM rather than IgG and are unable to cross the placenta in
large numbers. The maternal-fetal relationship itself is a fascinating
instance of foreign tissue not being rejected, despite the extensive
potential for contact. It is thought that the placenta forms a barrier
to keep the fetus isolated in its own antigen-free environment. The
placenta is surrounded by a dense, many-layered envelope that pre-
vents the passage of maternal cells; and it actively absorbs, re-
moves, and inactivates circulating antigens.

Preventing Hemolytic Disease of the Newborn

Once sensitization of the mother to Rh factor has occurred, all other
Rh^+ fetuses will be at risk for hemolytic disease of the newborn.

erythroblastosis fetalis (eh-rith″-roh-blas-toh′-sis fee-tal′-is)

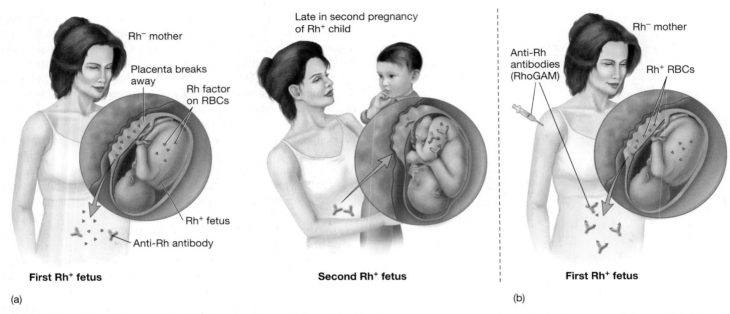

Figure 16.12 Rh factor incompatibility can result in RBC lysis. (a) A naturally occurring blood cell incompatibility results when a Rh⁺ fetus develops within a Rh⁻ mother. Initial sensitization of the maternal immune system occurs when fetal blood passes the placental barrier. In most cases, the fetus develops normally. However, a subsequent pregnancy with a Rh⁺ fetus results in a severe, fetal hemolysis. **(b)** Control of incompatibility: Anti-Rh antibody (RhoGAM) can be administered to Rh⁻ mothers during pregnancy to help bind, inactivate, and remove any Rh factor that may be transferred from the fetus. In some cases, RhoGAM is administered before sensitization occurs.

Prevention requires a careful family history of an Rh⁻ pregnant woman. It can predict the likelihood that she is already sensitized or is carrying an Rh⁺ fetus. It must take into account other children she has had, their Rh types, and the Rh status of the father. If the father is also Rh⁻, the child will be Rh⁻ and free of risk, but if the father is Rh⁺, the probability that the child will be Rh⁺ is 50% or 100%, depending on the exact genetic makeup of the father. If there is any possibility that the fetus is Rh⁺, the mother must be passively immunized with antiserum containing antibodies against the Rh factor (*Rh₀ [D] immune globulin,* or RhoGAM[3]). This antiserum, injected at 28 to 32 weeks and again immediately after delivery, keeps the maternal immune system from recognizing fetal RBCs that have escaped into the mother's circulation. This prevents the sensitization of her immune system to Rh factor (**figure 16.12b**). Anti-Rh antibody must be given with each pregnancy that involves an Rh⁺ fetus. It is ineffective if the mother has already been sensitized by a prior Rh⁺ fetus or an incorrect blood transfusion, which can be determined by a serological test.

As in ABO blood types, the Rh factor should be matched for a transfusion, although it is acceptable to transfuse Rh⁻ blood if the Rh type is not known.

CONTINUING CASE FILE **16**

Hospital investigators reviewing the events in the case were faced with explaining the fatal blood transfusion to the patient's family. What could have gone wrong? Going by the first typing of the patient's blood and the cross-match with the transfused

blood, nothing unusual would have been expected. But there was no question that he suffered an extreme transfusion reaction. When the patient's blood was retyped after death, it turned out that he was actually O⁺, the most common blood type among humans. One possibility might have been a reaction with a blood group other than ABO. But his blood had the same Rh factor as the transfused blood and was compatible for Rh factor. And in cases of the minor RBC antigens such as MN and Kell, the reactions are usually quite mild and never result in mass hemolysis and death. As the investigation continued, it became evident that the cause of this tragedy was not a mistake in the technology of the blood typing or some unusual characteristic of the blood, but rather it was placed squarely on human error in record keeping.

- Why do people carry antibodies against other blood types even when they have never been sensitized by prior contact with that blood?

- If the transfusion reaction had not progressed so rapidly, what blood could have been safely given to the patient and why?

For a wrap-up, see the Case File Perspective on page 513.

Other RBC Antigens

Although the ABO and Rh systems are of greatest medical significance, about 20 other red blood cell antigen groups have been discovered. Examples are the *MN, Ss, Kell,* and *P* blood groups. Because of incompatibilities that these blood groups present, transfused blood is screened to prevent possible cross-reactions. The study of these blood antigens (as well as ABO and Rh) has given

3. RhoGAM: Immunoglobulin fraction of human anti-Rh serum, prepared from pooled human sera.

rise to other useful applications. For example, they can be useful in forensic medicine (crime detection), studying ethnic ancestry, and tracing prehistoric migrations in anthropology. Many blood cell antigens are remarkably hardy and can be detected in dried blood stains, semen, and saliva. Even the 3,300-year-old mummy of King Tutankhamen has been typed A_2MN!

16.4 Type III Hypersensitivities: Immune Complex Reactions

Expected Learning Outcomes

16. Describe the background features of immune complex reactions.

17. Differentiate the major types of immune complex diseases, and discuss their physiological effects.

Type III hypersensitivity involves the reaction of soluble antigen with antibody and the deposition of the resulting complexes in basement membranes of epithelial tissue. It is similar to type II,

because it involves the production of IgG and IgM antibodies after repeated exposure to antigens and the activation of complement. Type III differs from type II because its antigens are not attached to the surface of a cell. The interaction of these antigens with antibodies produces free-floating complexes that can be deposited in the tissues, causing an **immune complex reaction** or disease. This category includes therapy-related disorders (serum sickness and the Arthus reaction) and a number of autoimmune diseases (such as glomerulonephritis and lupus erythematosus).

Mechanisms of Immune Complex Diseases

After initial exposure to a profuse amount of antigen, the immune system produces large quantities of antibodies that circulate in the fluid compartments. When this antigen enters the system a second time, it reacts with the antibodies to form antigen-antibody complexes (**figure 16.13**). These complexes summon various inflammatory components such as complement and neutrophils, which would ordinarily eliminate Ab-Ag complexes as part of the normal immune response. In an immune complex disease, however, these complexes are so abundant that they deposit in the **basement membranes**[4] of epithelial tissues and become inaccessible. In response to these events, neutrophils release lysosomal granules that digest tissues and

4. **Basement membranes** are basal partitions of epithelia that normally filter out circulating antigen-antibody complexes.

Phases:

Antibody combines with excess soluble antigen, forming large quantities of Ag-Ab complexes.

Circulating immune complexes become lodged in the basement membranes of epithelia in blood vessels, kidney, skin and other sites.

Complement factors trigger release of histamine and other inflammatory mediators.

Neutrophils migrate to sites of Ag-Ab complexes and release enzymes and chemokines that severely damage the target tissues and organs.

Major organs where immune complexes are deposited

Process Figure 16.13 Pathogenesis of immune complex disease.

cause a destructive inflammatory condition. The symptoms of type III hypersensitivities are due in great measure to this pathologic state.

Types of Immune Complex Disease

During the early tests of immunotherapy using animals, hypersensitivity reactions to serum and vaccines were common. In addition to anaphylaxis, two syndromes, the **Arthus reaction**[5] and **serum sickness,** were identified. These syndromes are associated with certain types of passive immunization (especially with animal serum).

Serum sickness and the Arthus reaction are like anaphylaxis in requiring sensitization and preformed antibodies. Characteristics that set them apart from anaphylaxis are (1) they depend upon IgG, IgM, or IgA (precipitating antibodies) rather than IgE; (2) they require large doses of antigen (not a minuscule dose as in anaphylaxis); and (3) their symptoms are delayed (a few hours to days).

The Arthus reaction and serum sickness differ from each other in some important ways. The Arthus reaction is a *localized* dermal injury due to inflamed blood vessels in the vicinity of any injected antigen. Serum sickness is a *systemic* injury initiated by antigen-antibody complexes that circulate in the blood and settle into membranes at various sites.

The Arthus Reaction

The Arthus reaction is usually an acute response to a second injection of vaccines (boosters) or drugs at the same site as the first injection. In a few hours, the area becomes red, hot to the touch, swollen, and very painful. These symptoms are mainly due to the destruction of tissues in and around the blood vessels and the release of histamine from mast cells and basophils. Although the reaction is usually self-limiting and rapidly cleared, intravascular blood clotting can occasionally cause necrosis and loss of tissue.

Serum Sickness

Serum sickness was named for a condition that appeared in soldiers after repeated injections of horse serum to treat tetanus. It can also be caused by injections of animal hormones and drugs. The immune complexes enter the circulation; are carried throughout the body; and are eventually deposited in blood vessels of the kidney, heart, skin, and joints (figure 16.13). The condition can become chronic, causing symptoms such as enlarged lymph nodes, rashes, painful joints, swelling, fever, and renal dysfunction.

Check & Assess Section 16.4

✔ Type III hypersensitivities are induced when a profuse amount of antigen enters the system and results in large quantities of antibody formation.

✔ Type III hypersensitivity reactions occur when large quantities of antigen react with host antibody to form small, soluble immune complexes that settle in tissue cell membranes, causing chronic destructive inflammation. The reactions appear hours or days after the antigen challenge.

✔ The mediators of type III hypersensitivity reactions include soluble IgA, IgG, or IgM and agents of the inflammatory response.

✔ Two kinds of type III hypersensitivities are localized (Arthus) reactions and systemic (serum sickness) reactions.

✔ Arthus reactions occur at the site of injected drugs or booster immunizations.

✔ Systemic reactions occur when repeated antigen challenges cause systemic distribution of the immune complexes and subsequent inflammation of joints, lymph nodes, and kidney tubules.

22. Contrast type II and type III hypersensitivities with respect to type of antigen, antibody, and manifestations of disease.
23. Explain what occurs in immune complex disease and give examples.

16.5 Immunopathologies Involving T Cells

Expected Learning Outcomes

18. Define what is meant by *type IV hypersensitivities.*
19. Describe the characteristics of allergies to infectious agents and contact dermatitis.
20. Outline the primary mechanisms in contact dermatitis.
21. Discuss the involvement of T cells in organ transplantation and the different ways that grafted tissues may be rejected.
22. Describe the categories of grafts and how rejection is prevented.

Type IV Delayed-Type Hypersensitivity

The adverse immune responses we have covered so far are explained primarily by B-cell involvement and antibodies. A notable difference exists in type IV hypersensitivity, which involves primarily the T-cell branch of the immune system. Type IV immune dysfunction has traditionally been known as delayed hypersensitivity because the symptoms arise one to several days following the second contact with an antigen. In general, type IV diseases result when T cells respond to antigens displayed on self tissues or transplanted foreign cells. Examples of type IV hypersensitivity include delayed allergic reactions to infectious agents, contact dermatitis, and graft rejection.

Infectious Allergy

A classic example of a delayed-type hypersensitivity occurs when a person sensitized by tuberculosis infection is injected with an extract (tuberculin) of the bacterium *Mycobacterium tuberculosis.* The so-called tuberculin reaction is an acute skin inflammation at the injection site appearing within 24 to 48 hours. So useful and diagnostic is this technique for detecting present or prior tuberculosis that it is the chosen screening device. Other infections that use similar skin testing are leprosy, syphilis, histoplasmosis, toxoplasmosis, and candidiasis. This form of hypersensitivity arises from a

5. Named after Maurice Arthus, the physiologist who first identified this localized inflammatory response.

INSIGHT 16.2

Pretty, Pesky, Poisonous Plants

As a cause of allergic contact dermatitis (affecting about 10 million people a year), nothing can compare with a single family of plants belonging to the genus *Toxicodendron*. At least one of these plants—poison ivy, poison oak, or poison sumac—flourishes in the forests, woodlands, or along the trails of most regions of America. The allergen in these plants, an oil called urushiol, has such extreme potency that a pinhead-sized amount could spur symptoms in 500 people, and it is so long-lasting that botanists must be careful when handling 100-year-old plant specimens. Although degrees of sensitivity vary among individuals, it is estimated that 85% of all Americans are potentially hypersensitive to this compound. Some people are so acutely sensitive that even the most minuscule contact, such as handling pets or clothes that have touched the plant or breathing vaporized urushiol, can trigger an attack.

Humans first become sensitized by contact during childhood. Individuals at great risk (firefighters, hikers) are advised to determine their degree of sensitivity using a skin test, so that they can be adequately cautious and prepared. Some odd remedies include skin potions containing bleach, buttermilk, ammonia, hair spray, and meat tenderizer. Commercial products are available for blocking or washing away the urushiol. Allergy researchers are currently testing oral vaccines containing a form of urushiol, which seem to desensitize experimental animals. An effective method using poison ivy desensitization injection is currently available to people with extreme sensitivity.

Provide an explanation for why occasionally a single, first-time exposure to poison oak could result in a dermatitis reaction.
Answer available at http://www.mhhe.com/talaro8

Poison oak

Poison sumac

Poison ivy

Learning to identify these common plants can prevent exposure and sensitivity. One old saying that might help warns, "Leaves of three, let it be; berries white, run with fright."

specific class of T cells (T$_H$1) that receive the processed allergens from dendritic cells. Activated T$_H$ cells release cytokines that recruit various inflammatory cells such as macrophages, neutrophils, and eosinophils. The buildup of fluid and cells at the site gives rise to a red papule **(figure 16.14).** In a chronic infection (tertiary syphilis, for example), this reaction can give rise to extensive damage to organs through granuloma formation.

Contact Dermatitis

The most common delayed allergic reaction, contact dermatitis, is caused by exposure to resins in poison ivy or poison oak **(Insight 16.2),** to simple haptens in household and personal articles (jewelry, cosmetics, elasticized undergarments), and to certain drugs. Like immediate atopic dermatitis, the reaction to these allergens usually requires a sensitizing and a provocative dose. The allergen first penetrates the outer skin layers, is processed by Langerhans cells (skin macrophages), and is presented to T cells. When subsequent exposures attract lymphocytes and macrophages to this area, these cells give off enzymes and inflammatory cytokines that severely

Figure 16.14 Positive tuberculin test. Intradermal injection of tuberculin extract in a person sensitized to tuberculosis yields a slightly raised red bump about 14 mm in diameter in this view.

damage the epidermis in the immediate vicinity **(figure 16.15a).** This response accounts for the intensely itchy papules and blisters that are the early symptoms **(figure 16.15b).** As healing progresses, the epidermis is replaced by a thick, horny layer. Depending upon the

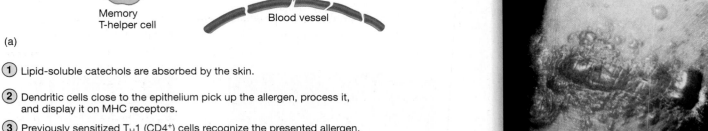

(a)

① Lipid-soluble catechols are absorbed by the skin.

② Dendritic cells close to the epithelium pick up the allergen, process it, and display it on MHC receptors.

③ Previously sensitized T$_H$1 (CD4$^+$) cells recognize the presented allergen.

④ Sensitized T$_H$1 cells are activated and secrete cytokines (IFN, TNF).

⑤ These cytokines attract macrophages and cytotoxic T cells to the site.

⑥ Macrophages release mediators that stimulate a strong, local inflammatory reaction. Cytotoxic T cells directly kill cells and damage the skin. Fluid-filled blisters result.

(b)

Process Figure 16.15 Contact dermatitis.
(a) Genesis of contact dermatitis. **(b)** Contact dermatitis from poison oak, showing various stages of involvement: blister, scales, and thickened patches.

dose and the sensitivity of the individual, the time from initial contact to healing can be a week to 10 days.

T Cells and Their Role in Organ Transplantation

Transplantation or grafting of organs and tissues is a common medical procedure. Although it is life-giving, this technique is plagued by the natural tendency of lymphocytes to seek out foreign antigens and mount a campaign to destroy them through **graft rejection.** The bulk of the damage that occurs in graft rejections can be attributed to expression of cytotoxic T cells and other killer cells. This section covers the mechanisms involved in graft rejection, tests for transplant compatibility, reactions against grafts, prevention of graft rejection, and types of grafts.

The Genetic and Biochemical Basis for Graft Rejection

In chapter 15, we discussed the role of major histocompatibility (MHC or HLA) genes and surface markers in immune function. In general, the genes and markers in MHC classes I and II are extremely important in recognizing self and in regulating the immune response. These molecules also set the events of graft rejection in motion. The MHC genes of humans are inherited from among a large pool of genes, so the cells of each person can exhibit

variability in the pattern of cell surface molecules. The pattern is identical in different cells of the same person and can be similar in related siblings and parents, but the more distant the relationship, the less likely that the MHC genes and markers will be similar. When donor tissue (a graft) displays surface molecules of a different MHC class, the T cells of the recipient (called the host) will recognize its foreignness and react against it.

T-Cell-Mediated Recognition of Foreign MHC Receptors

Host Rejection of Graft When certain T cells of a host recognize foreign class I MHC markers on the surface of grafted cells, they release interleukin-2 as part of a general immune mobilization. The effect is to expand the helper and cytotoxic T cells specific to the antigens displayed by the donated cells. The cytotoxic cells bind to the grafted tissue and secrete lymphokines that begin the rejection process within 2 weeks of transplantation (**figure 16.16***a*). Late in this process, antibodies formed against the graft tissue contribute to immune damage. A final blow is the destruction of the vascular supply, promoting death of the grafted tissue.

Graft Rejection of Host Some types of immunodeficiencies prevent the host from rejecting a graft. But this failure may not protect the host from serious damage, because graft incompatibility is a two-way phenomenon. Some grafted tissues (especially bone marrow)

(a)

(b)

Figure 16.16 **Potential reactions in transplantation. (a)** The host's immune system (primarily cytotoxic T cells) encounters the cells of the donated organ (heart) and mediates the rejection of the organ. **(b)** Grafted tissue (bone marrow) contains endogenous T cells that recognize the host's tissues as foreign and mount an attack on many tissues and organs. The recipient will develop symptoms of graft versus host disease.

contain an indigenous population called passenger lymphocytes. This makes it quite possible for the graft to reject the host, causing **graft versus host disease (GVHD) (figure 16.16***b***).** Because any host tissue bearing MHC markers foreign to the graft can be attacked, the effects of GVHD are widely systemic and toxic. A papular, peeling skin rash is the most common symptom. Other organs affected are the liver, intestine, muscles, and mucous membranes. Previously, GVHD occurred in approximately 30% of bone marrow transplants within 100 to 300 days of the graft. This percentage is declining with the development of better screening and selection of tissues.

Classes of Grafts Grafts are generally classified according to the genetic relationship between the donor and the recipient. Tissue transplanted from one site on an individual's body to another site on his body is known as an **autograft.** Typical examples are skin replacement in burn repair and the use of a vein to fashion a coronary artery bypass. In an **isograft,** tissue from an identical twin is used. Because isografts do not contain foreign antigens, they are not rejected, but this type of grafting has obvious limitations. **Allografts,** the most common type of grafts, are exchanges between genetically different individuals belonging to the same species (two humans). A close genetic correlation is sought for most allograft transplants (see next section). A **xenograft** is a tissue exchange between individuals of different species. Until rejection can be better controlled, most xenografts are experimental or for temporary therapy only.

Avoiding and Controlling Graft Incompatibility

Graft rejection can be averted or lessened by directly comparing the tissue of the recipient with that of potential donors. Several tissue matching procedures are used. In the *mixed lymphocyte reaction* *(MLR),* lymphocytes of the two individuals are mixed and incubated. If an incompatibility exists, some of the cells will become activated and proliferate. *Tissue typing* is similar to blood typing, except that specific antisera are used to disclose the HLA antigens

on the surface of lymphocytes. In most grafts (one exception is bone marrow transplants), the ABO blood type must also be matched. Although a small amount of incompatibility is tolerable in certain grafts (liver, heart, kidney), a closer match is more likely to be successful, so the closest match possible is sought.

Practical Examples in Transplantation

Today, transplantation is a recognized medical procedure whose benefit is reflected in the fact that more than 25,000 transplants take place each year in the United States. It has been performed on every major organ, including parts of the brain. The most frequent transplant operations involve skin, liver, heart, kidney, coronary artery, cornea, and bone marrow. The sources of organs and tissues are live donors (kidney, skin, bone marrow, liver), cadavers (heart, kidney, cornea), and fetal tissues. In the past decade, we have witnessed some unusual types of grafts. For instance, the fetal pancreas has been implanted as a potential treatment for diabetes and fetal brain tissues for Parkinson's disease.

Bone marrow transplantation is a rapidly growing medical procedure for patients with immune deficiencies, aplastic anemia, leukemia and other cancers, and radiation damage. This procedure is extremely expensive, costing up to $400,000 per patient. The process involves a search for a compatable donor with a closely matched MHC profile. This person is sedated, and a bone marrow/blood sample is aspirated by inserting a special needle into an accessible marrow cavity. The most favorable sites are the crest and spine of the ilium (major bone of the pelvis). During this procedure, between 500 and 800 mls of bone marrow are extracted with a special tool. In a few weeks, the donor's marrow will naturally replace itself. Implanting the harvested bone marrow is rather convenient, because it is not necessary to place it directly into the marrow cavities of the recipient. Instead, it is dripped intravenously into the circulation, and the new marrow cells automatically settle in the appropriate bone marrow regions.

To prepare the recipient for the infusion procedure, he must be treated with chemotherapy and whole-body irradiation, a procedure designed to destroy his own blood stem cells and thus prevent his rejection of the new marrow cells. Within 2 weeks to a month after infusion, the grafted cells become established in the host. Because donor lymphoid cells can still cause GVHD, antirejection drugs may be necessary. An amazing consequence of bone marrow transplantation is that a recipient's blood type may change to the blood type of the donor.

Recent advances in stem cell technology have made it possible to isolate stem cells directly from the blood of donors without bone marrow sampling. Another potential source is the umbilical cord blood from a newborn infant. These have expanded the possibilities for treatment and survival.

Check & Assess Section 16.5

✔ Type IV hypersensitivity reactions occur when cytotoxic T cells attack either self tissue or transplanted foreign cells. Type IV reactions are also termed delayed hypersensitivity reactions because they occur hours to days after the antigenic challenge.

✔ Type IV hypersensitivity reactions are mediated by T lymphocytes and are carried out against foreign cells that show both a foreign MHC and a nonself receptor site.

✔ Examples of type IV reactions include the tuberculin reaction, contact dermatitis, and mismatched organ transplants (host rejection and GVHD reactions).

✔ The four classes of transplants or grafts are determined by the degree of MHC similarity between graft and host. From most to least similar, these are: autografts, isografts, allografts, and xenografts.

✔ Rejection of or by a grafted organ occurs when T cells recognize and attack the foreign antigens the graft contains.

✔ Graft rejection can be minimized by tissue matching procedures, immunosuppressive drugs, and use of tissues that do not provoke a type IV response.

24. What is the molecular/cellular basis for a host rejecting the graft tissue?
25. Explain how a graft rejects the host.
26. Compare the four types of grafts.
27. What does it mean to say that two tissues constitute a close match?
28. Describe the procedure involved in a bone marrow transplant.

16.6 Autoimmune Diseases—An Attack on Self

Expected Learning Outcomes

23. Outline the pathology of autoimmune diseases.

24. Explain the origins of autoimmunity and which persons are most often targeted by it.

25. Review the major types of autoimmune diseases and the adverse medical effects they have.

The immune diseases we have covered so far are all caused by foreign antigens. In the case of autoimmunity, individuals actually develop hypersensitivity to their own cells. This pathologic process accounts for **autoimmune diseases,** in which **autoantibodies,** T cells, and in some cases, both, mount an abnormal attack against self antigens. The scope of autoimmune diseases is extremely varied. In general, they are either *systemic,* involving several major organs, or *organ-specific,* involving only one organ or tissue. They usually fall into the categories of type II or type III hypersensitivity, depending upon how the autoantibodies bring about injury. Some major autoimmune diseases, their targets, and basic pathology are presented in **table 16.4.** (For a reminder of hypersensitivity types, refer to table 16.1.)

Genetic and Gender Correlation in Autoimmune Disease

In most cases, the precipitating cause of autoimmune disease remains obscure, but we do know that susceptibility is determined by genetics and influenced by gender. Cases cluster in families, and even unaffected members tend to develop the autoantibodies for that disease. More direct evidence comes from studies of the major histocompatibility gene complexes. Particular genes in the class I and class II major histocompatibility complexes coincide with certain autoimmune diseases. For example, autoimmune joint diseases such as rheumatoid arthritis and ankylosing spondylitis are more common in persons with the B-27 human leukocyte antigen (HLA) type; systemic lupus erythematosus, Graves' disease, and myasthenia gravis are associated with the B-8 HLA. Why autoimmune diseases (except ankylosing spondylitis) afflict more females than males also remains a mystery. Females are more susceptible during childbearing years than before puberty or after menopause, suggesting a possible hormonal relationship.

The Origins of Autoimmune Disease

Because otherwise healthy individuals show very low levels of autoantibodies, it is suspected that there is a function for them. A moderate, regulated amount of autoimmunity is probably required to dispose of old cells and cellular debris. Disease apparently arises when this regulatory or recognition apparatus goes awry. Although there is not yet a complete explanation for autoimmune diseases, several theories have been proposed.

The **sequestered antigen theory** explains that during embryonic growth, some tissues are immunologically privileged; that is, they are sequestered behind anatomical barriers and cannot be scanned by the immune system. Examples of these sites are regions of the central nervous system, which are shielded by the meninges and blood-brain barrier; the lens of the eye, which is enclosed by a thick sheath; and antigens in the thyroid and testes, which are sequestered behind an epithelial barrier. Eventually, the antigen becomes exposed by means of infection, trauma, or deterioration and is perceived by the immune system as a foreign substance.

According to the clonal selection theory, the immune system of a fetus develops tolerance to self by eradicating or silencing all self-reacting lymphocyte clones, called **forbidden clones,** while retaining only those clones that react to foreign antigens. Some of these forbidden clones may survive; and because they have not been

TABLE 16.4	Selected Autoimmune Diseases		
Disease	Target	Type of Hypersensitivity	Characteristics
Systemic lupus erythematosus (SLE)	Systemic	II and III	Inflammation of many organs; antibodies against red and white blood cells, platelets, clotting factors, nucleus DNA
Rheumatoid arthritis and ankylosing spondylitis	Systemic	III and IV	Vasculitis; frequent target is joint lining; antibodies against other antibodies (rheumatoid factor)
Scleroderma	Systemic	II	Excess collagen deposition in organs; antibodies formed against many intracellular organelles
Hashimoto's thyroiditis	Thyroid	II	Destruction of the thyroid follicles
Graves disease	Thyroid	II	Antibodies against thyroid-stimulating hormone receptors
Pernicious anemia	Stomach lining	II	Antibodies against receptors prevent transport of vitamin B_{12}.
Myasthenia gravis	Muscle	II	Antibodies against the acetylcholine receptors on the nerve-muscle junction alter function.
Type I diabetes	Pancreas	II	Antibodies stimulate destruction of insulin-secreting cells.
Multiple sclerosis	Myelin	II and IV	T cells and antibodies sensitized to myelin sheath destroy neurons.
Goodpasture syndrome (glomerulonephritis)	Kidney	II	Antibodies to basement membrane of the glomerulus damage kidneys.
Rheumatic fever	Heart	II	Antibodies to group A streptococci cross-react with heart tissue.

subjected to this tolerance process, they can attack tissues carrying self molecules, which they mistake as antigens.

The **theory of immune deficiency** proposes that mutations in the receptor genes of some lymphocytes render them reactive to self or that a general breakdown in the normal T-suppressor function sets the scene for inappropriate immune responses. Inappropriate expression of MHC II markers on cells that don't normally express them has been found to cause abnormal immune reactions to self. In a related phenomenon, T-cell activation may incorrectly "turn on" B cells that can react with self antigens. This phenomenon is called the bystander effect.

Some autoimmune diseases appear to be caused by **molecular mimicry,** in which microbial antigens bear molecular determinants similar to normal human cells. An infection could cause formation of antibodies that can cross-react with tissues. This is one purported explanation for the pathology of rheumatic fever. Another probable example of mimicry leading to autoimmune disease is the skin condition psoriasis. Although the etiology of this condition is complex and involves the inheritance of certain types of MHC alleles, infection with group A streptococci also plays a role. Scientists report that T cells primed to react with streptococcal surface proteins also react with keratin cells in the skin, causing them to proliferate. For this reason, psoriasis patients often report flare-ups after a streptococcal throat infection.

Autoimmune disorders such as type I diabetes and multiple sclerosis are likely triggered by *viral infection*. Viruses can noticeably alter cell receptors, thereby causing immune cells to attack the tissues bearing viral receptors.

Some researchers believe that many, if not most, autoimmune diseases will someday be discovered to have an underlying microbial etiology, either through molecular mimicry or viral alteration of host antigens (just described) or due to the presence of as yet undetectable microbes in the sites affected by autoimmunity.

Examples of Autoimmune Disease
Systemic Autoimmunities

One of the most severe chronic autoimmune diseases is **systemic lupus erythematosus* (SLE)** or lupus. This name originated from the characteristic butterfly-shaped rash that drapes across the nose and cheeks **(figure 16.17a)**. Apparently, ancient physicians thought the rash resembled a wolf bite (*lupus* is Latin for "wolf"). Although the manifestations of the disease vary considerably, all patients produce autoantibodies against a great variety of organs and tissues. The organs most involved are the kidneys, bone marrow, skin, nervous system, joints, muscles, heart, and GI tract. Antibodies to intracellular materials such as the nucleoprotein of the nucleus and mitochondria are also common.

In SLE, autoantibody-autoantigen complexes appear to be deposited in the basement membranes of various organs. Kidney failure, blood abnormalities, lung inflammation, myocarditis, and skin lesions are the predominant symptoms. One form of chronic lupus (called discoid) is influenced by exposure to the sun and primarily afflicts the skin. The etiology of lupus is still a puzzle. It is not known how such a generalized loss of self-tolerance arises, though viral infection or loss of T-cell suppressor function is suspected. The fact that women of childbearing years account for 90% of cases indicates that hormones may be involved. The diagnosis of SLE can usually be made with blood tests. Antibodies against the nucleus and various tissues (detected by indirect fluorescent antibody or radioimmune assay techniques) are common, and a positive test for the lupus factor (an antinuclear factor) is also very indicative of the disease.

* *systemic lupus erythematosus* (sis-tem´-ik loo´-pis air″-uh-theem-uh-toh´-sis) L. *lupus,* wolf, and *erythema,* redness.

(a)

(b)

Figure 16.17 Common autoimmune diseases.
(a) Systemic lupus erythematosus. One symptom is a prominent rash across the bridge of the nose and on the cheeks. Papules and blotches can also occur on the chest and limbs. **(b)** Rheumatoid arthritis commonly targets the synovial membrane of joints. Chronic inflammation thickens this membrane, erodes the articular cartilage, and fuses the joint. These effects severely limit motion and can swell and distort the joints.

Rheumatoid arthritis,* another systemic autoimmune disease, incurs progressive, debilitating damage to the joints. In some patients, the lung, eye, skin, and nervous systems are also involved. In the joint form of the disease, autoantibodies form immune complexes that bind to the synovial membrane of the joints and activate phagocytes and stimulate release of cytokines. Chronic inflammation leads to scar tissue and joint destruction. The joints in the hands and feet are affected first, followed by the knee and hip joints **(figure 16.17b).** The precipitating cause in rheumatoid arthritis is not known, though infectious agents such as Epstein-Barr virus have been suspected. The most common feature of the disease is the presence of an IgM antibody, called rheumatoid factor (RF), directed against other antibodies. This does not cause the disease but is used mainly in diagnosis.

Autoimmunities of the Endocrine Glands

On occasion, the thyroid gland is the target of autoimmunity. The underlying cause of **Graves' disease** is the attachment of autoantibodies to receptors on the follicle cells that secrete the hormone thyroxin. The abnormal stimulation of these cells causes the overproduction of

* *rheumatoid arthritis* (roo'-muh-toyd ar-thry'-tis) Gr. *rheuma,* a moist discharge, and *arthron,* joint.

this hormone and the symptoms of hyperthyroidism. In **Hashimoto's thyroiditis,** both autoantibodies and T cells are reactive to the thyroid gland, but in this instance, they reduce the levels of thyroxin by destroying follicle cells and by inactivating the hormone. As a result of these reactions, the patient suffers from hypothyroidism.

The pancreas and its hormone, insulin, are other autoimmune targets. Type I **diabetes mellitus** is caused by a dysfunction in insulin production by beta cells in the pancreas or its utilization by cells. This form is associated with autoantibodies and sensitized T cells that damage the beta cells. A complex inflammatory reaction leading to lysis of these cells greatly reduces the amount of insulin secreted. A current theory links type I diabetes to infection by the Coxsackievirus, a common cause of colds and enteric infections in a possible case of molecular mimicry.

Neuromuscular Autoimmunities

Myasthenia gravis is named for the pronounced muscle weakness that is its principal symptom. Although the disease afflicts all skeletal muscle, the first effects are usually felt in the muscles of the eyes and throat. Eventually, it can progress to complete loss of muscle function and death. The classic syndrome is caused by autoantibodies binding to the receptors for acetylcholine, a chemical required to transmit a nerve impulse across the synaptic junction to a muscle **(figure 16.18).** The immune attack so severely damages the muscle cell membrane that transmission is blocked and paralysis ensues. Current treatment usually includes immunosuppressive drugs and therapy to remove the autoantibodies from the circulation. Experimental therapy using immunotoxins to destroy lymphocytes that produce autoantibodies shows some promise.

Multiple sclerosis (MS) is a paralyzing neuromuscular disease associated with lesions in the insulating myelin sheath that surrounds neurons in the white matter of the central nervous system. The underlying pathology involves damage to the sheath by both T cells and autoantibodies; this damage severely compromises the capacity of neurons to send impulses. The principal motor and sensory symptoms are muscular weakness and tremors, difficulties in

Figure 16.18 Mechanism for involvement of autoantibodies in myasthenia gravis. Left: Normal process. Right: Antibodies developed against receptors on the postsynaptic membrane block them so that acetylcholine cannot bind and muscle contraction is inhibited.

speech and vision, and some degree of paralysis. Most MS patients first experience symptoms as young adults, and they tend to experience remissions (periods of relief) alternating with recurrences of disease throughout their lives.

Convincing evidence from studies of the brain tissue of MS patients points to a strong connection between the disease and infection with viruses. Some research suggests that MS is associated with Epstein-Barr virus; newer evidence indicates that an endogenous retrovirus may be involved. The disease can be treated passively with monoclonal antibodies that target T cells, and a vaccine containing the myelin protein has shown beneficial effects. Immunosuppressants such as cortisone and interferon beta may also alleviate symptoms.

Check & Assess Section 16.6

- ✔ Autoimmune hypersensitivity reactions occur when autoantibodies or host T cells mount an abnormal attack against self antigens. Autoimmune antibody responses can be either local or systemic type II or type III hypersensitivity reactions. Autoimmune T-cell responses are type IV hypersensitivity reactions.
- ✔ Susceptibility to autoimmune disease appears to be influenced by gender and by genes in the MHC.
- ✔ Autoimmune disease may be an excessive response of a normal immune function, the appearance of sequestered antigens, "forbidden" clones of lymphocytes that react to self antigens, or the result of alterations in the immune response caused by infectious agents, particularly viruses.
- ✔ Examples of autoimmune diseases include systemic lupus erythematosus, rheumatoid arthritis, diabetes mellitus, myasthenia gravis, and multiple sclerosis.

29. Explain the pathologic process in autoimmunity.
30. Draw diagrams that explain five possible mechanisms for the development of autoimmunity.
31. Describe four major types of autoimmunity, comparing target organs and symptoms.

16.7 Immunodeficiency Diseases: Compromised Immune Responses

Expected Learning Outcomes

26. Outline the categories of immunodeficiency diseases.
27. Describe the origins and effects of primary immune diseases caused by B-cell, T-cell, or combined defects.
28. Relate examples of secondary immunodeficiencies.

It is a marvel that development and function of the immune system proceed as normally as they do. On occasion, however, an error occurs and a person is born with or develops weakened immune responses. In many cases, these "experiments of nature" have provided penetrating insights into the exact functions of certain cells, tissues, and organs because of the specific signs and symptoms

shown by the immunodeficient individuals. The predominant consequences of immunodeficiencies are recurrent, overwhelming infections, often with opportunistic microbes. Immunodeficiencies fall into two general categories: **primary diseases,** present at birth (congenital) and usually stemming from genetic errors, and **secondary diseases,** acquired after birth and caused by natural or artificial agents **(table 16.5).**

Primary Immunodeficiency Diseases

Deficiencies affect both specific immunities such as antibody production and less-specific ones such as phagocytosis. Consult **figure 16.19** to survey the places in the normal sequential development of lymphocytes where defects can occur and the possible

TABLE 16.5	**General Categories of Immunodeficiency Diseases with Selected Examples**

Primary Immune Deficiencies (Genetic)

B-Cell Defects (Low Levels of B Cells and Antibodies)
Agammaglobulinemia (X-linked, non-sex-linked)
Hypogammaglobulinemia
Selective immunoglobulin deficiencies

T-Cell Defects (Lack of All Classes of T Cells)
Thymic aplasia (DiGeorge syndrome)
Chronic mucocutaneous candidiasis

Combined B-Cell and T-Cell Defects
Severe combined immunodeficiency disease (SCID)
X-SCIDI due to an interleukin defect
Wiskott-Aldrich syndrome
Ataxia-telangiectasia

Phagocyte Defects
Chédiak-Higashi syndrome
Chronic granulomatous disease of children
(see Case File 14, chapter 14)
Lack of surface adhesion molecules

Complement Defects
Lacking one of C components
Hereditary angioedema
Associated with rheumatoid diseases

Secondary Immune Deficiencies (Acquired)

From Natural Causes
Infection: AIDS, leprosy, tuberculosis, measles
Other disease: cancer, diabetes
Nutrition deficiencies
Stress
Pregnancy
Aging

From Immunosuppressive Agents
Irradiation
Severe burns
Steroids (cortisones)
Drugs to treat graft rejection and cancer
Removal of spleen

Figure 16.19 **The stages of development and the functions of B cells and T cells, whose failure causes immunodeficiencies.** Dotted lines represent the phases in development where breakdown can occur.

consequences. In many cases, the deficiency is due to an inherited abnormality, though the exact nature of the abnormality is not known for a number of diseases. Because the development of B cells and T cells departs at some point, an individual can lack one or both cell lines. It must be emphasized, however, that some deficiencies affect other cell functions. For example a T-cell deficiency can affect B-cell function because of the role of T helper cells. In some deficiencies, the lymphocyte in question is completely absent or is present at very low levels, whereas in others, lymphocytes are present but do not function normally.

Clinical Deficiencies in B-Cell Development or Expression

Genetic deficiencies in B cells usually appear as an abnormality in immunoglobulin expression. In some instances, only certain immunoglobulin classes are absent; in others, the levels of all types of immunoglobulins (Ig) are reduced. A significant number of B-cell deficiencies are X-linked (also called sex-linked) recessive traits, meaning that the gene occurs on the X chromosome and the disease appears primarily in male children.

The term **agammaglobulinemia** literally means the absence of gamma globulin, the fraction of serum that contains immunoglobulins (Ig). Because it is very rare for Ig to be completely absent, some physicians prefer the term **hypogammaglobulinemia.*** T-cell function in these patients is usually normal. The symptoms of recurrent, serious bacterial infections usually appear about 6 months after birth. The bacteria most often implicated are pyogenic cocci,

Pseudomonas, and *Haemophilus influenzae;* and the most common infection sites are the lungs, sinuses, meninges, and blood. Many Ig-deficient patients can have recurrent infections with viruses and protozoa, as well. Patients often manifest a wasting syndrome and have a reduced life span, but modern therapy has improved their prognosis. The current treatment for this condition is passive immunotherapy with immune serum globulin and continuous antibiotic therapy.

The lack of a particular class of immunoglobulin is a relatively common condition. Although genetically controlled, its underlying mechanisms are not yet clear. IgA deficiency is the most prevalent, occurring in about 1 person in 600. Such persons have normal quantities of B cells and other immunoglobulins, but they are unable to synthesize IgA. Consequently, they lack protection against local microbial invasion of the mucous membranes and suffer recurrent respiratory and gastrointestinal infections. The usual treatment using Ig replacement does not work, because conventional preparations are high in IgG, not IgA.

Clinical Deficiencies in T-Cell Development or Expression

Due to T cells' critical role in immune defenses, a genetic defect in T cells results in a broad spectrum of disease, including severe opportunistic infections, wasting, and cancer. In fact, a dysfunctional T-cell line is usually more devastating than a defective B-cell line because T helper cells are required to assist in most specific immune reactions. The deficiency can occur anywhere along the developmental spectrum, from thymus to mature, circulating T cells.

Abnormal Development of the Thymus The most severe of the T-cell deficiencies involve the congenital absence or immaturity

* *hypogammaglobulinemia* (hy′-poh-gem-ah-glob-yoo-lin-ee′-mee-ah) Gr. *Hypo,* denoting a lowered level of Ig.

of the thymus gland. Thymic aplasia, or **DiGeorge syndrome,** results when the embryonic third and fourth pharyngeal pouches fail to develop (**figure 16.20**). Some cases are associated with a deletion in chromosome 22. The accompanying lack of cell-mediated immunity makes children highly susceptible to persistent infections by fungi, protozoa, and viruses. Common, usually benign childhood infections such as chickenpox, measles, or mumps can be overwhelming and fatal in these children. Even vaccinations using attenuated microbes pose a danger. Other symptoms of thymic failure are reduced growth, wasting of the body, unusual facial characteristics, and an increased incidence of lymphatic cancer. These children can have reduced antibody levels, and they are unable to reject transplants. The major therapy for them is a transplant of thymus tissue.

Severe Combined Immunodeficiencies: Dysfunction in B and T Cells

Severe combined immunodeficiencies (SCIDs) are the most dire and potentially lethal of the immunodeficiency diseases because they involve dysfunction in both lymphocyte systems. Some SCIDs are due to the complete absence of the lymphocyte stem cell in the marrow; others are attributable to the dysfunction of B cells and T cells later in development. Infants with SCIDs usually manifest the T-cell deficiencies within days after birth by developing candidiasis, sepsis, pneumonia, or systemic viral infections. This debilitating condition appears to have several forms. In the two most common forms, Swiss-type agammaglobulinemia and thymic alymphoplasia, the numbers of all types of lymphocytes are extremely low, the blood antibody content is greatly diminished, and the thymus and cell-mediated immunity are poorly developed. Both diseases are due to a genetic defect in the development of the lymphoid cell line.

A rarer form of SCID is **adenosine deaminase (ADA) deficiency,** which is caused by an autosomal recessive defect in the metabolism of adenosine. In this case, lymphocytes develop, but a metabolic product builds up to abnormal levels and selectively destroys the cells. Infants with ADA deficiency are subject to recurrent infections and severe wasting typical of severe deficiencies. A small number of SCID cases are due to a developmental defect in receptors for B and T cells. An X-linked deficiency in interleukin

Figure 16.20 Facial characteristics of a child with DiGeorge syndrome. Typical defects include low-set, deformed earlobes; wide-set, slanted eyes; a small, bowlike mouth; and the absence of a philtrum (the vertical furrow between the nose and upper lip).

receptors was responsible for the disease of David, the child in the "plastic bubble" (**Insight 16.3**). Another newly identified condition, *bare lymphocyte syndrome,* is caused by the lack of genes that code for class II MHC receptors.

Because of their profound lack of specific adaptive immunities, children with SCID require the most rigorous kinds of aseptic techniques to protect them from opportunistic infections. Aside from life in a sterile plastic bubble, the only serious option for their longtime survival is total replacement or correction of dysfunctional lymphoid cells. Some infants can benefit from fetal liver or stem cell grafts. Although transplanting compatible bone marrow has been about 50% successful in curing the disease, it is complicated by graft versus host disease. The condition of some ADA-deficient patients has been partly corrected by periodic transfusions of blood containing large amounts of the normal enzyme. A more lasting treatment for all types of SCID has been found in gene therapy. Recently, scientists cured children with this disease by inserting normal genes to replace the defective genes (see figure 10.13).

Secondary Immunodeficiency Diseases

Secondary acquired deficiencies in B cells and T cells are caused by one of four general agents: (1) infection, (2) organic disease, (3) chemotherapy, or (4) radiation.

The most recognized infection-induced immunodeficiency is **acquired immunodeficiency syndrome** or **AIDS.** This syndrome is caused when several types of immune cells, including T helper cells, monocytes, macrophages, and antigen-presenting cells, are infected by the human immunodeficiency virus (HIV). It is generally thought that the depletion of T helper cells and functional impairment of immune responses ultimately account for the cancers and opportunistic protozoan, fungal, and viral infections associated with this disease. (See chapter 25 for an extensive discussion of AIDS.) Other infections that can deplete immunities are tuberculosis, leprosy, and malaria.

Cancers that target the bone marrow or lymphoid organs can be responsible for extreme malfunction of both humoral and cellular immunity. In leukemia, the massive numbers of cancer cells compete for space and literally displace the normal cells of the bone marrow and blood. Plasma cell tumors produce large amounts of nonfunctional antibodies, and thymus gland tumors cause severe T-cell deficiencies.

16.8 The Function of the Immune System in Cancer

Expected Learning Outcomes

29. Describe the characteristics of cancer, and explain the mechanism of transformation.

30. Explain how immune function relates to the development of cancer.

The term **cancer** comes from the Latin word for crab and presumably refers to the appendagelike projections that a spreading tumor develops. A cancer is defined as the new growth of abnormal cells.

INSIGHT 16.3

An Answer to the Mystery of David*

David Vetter, the most famous SCID child, lived all but the last 2 weeks of his life in a sterile environment to isolate him from the microorganisms that could have quickly ended his life. David's brother had died from the disease, and David had a 50% chance of also inheriting it. To ensure his survival, he was delivered by cesarean section and immediately placed in a sterile isolette. When it was shown that he did have SCID, it was necessary to keep him in various sterile isolators—ranging from room size to a special suit that allowed him to walk outside. Throughout this time, his only physical contact with others was through rubber-glove ports.

When David was 12, his doctors decided to attempt a bone marrow transplant that might allow him to live free of his isolating bubbles. David was transplanted with his sister's bone marrow, but the marrow harbored a common herpesvirus called Epstein-Barr virus. Because he lacked any form of protective immunities against this oncogenic virus, a cancer spread rapidly through his body. No effective medical interventions were available, and David died a short time later from metastatic lymphoma.

After further analysis, researchers discovered the basis for David's immunodeficiency. He had inherited a form that arises from a defective genetic code in the receptors for interleukin-2, interleukin-4, and interleukin-7. The defect prevents the receptors on T cells and B cells

David Vetter, at age 10.

from receiving the appropriate interleukin signals for growth, development, and reactivity. The end result is that both cytotoxic immunities and antibody-producing systems are shut down, leaving the body defenseless against infections and cancer.

What would have been the major concern for David after the transplant other than infections and cancer? *Answer available at http://www.mhhe.com/talaro8*

* For powerful and heartbreaking insights into David's story, see http://www.pbs.org/wgbh/amex/bubble/filmmore/fd.html.

The disease is also known by the synonym **neoplasm,*** or by more specific terms that usually end in the suffix *-oma.* **Oncology*** is the field of medicine that specializes in cancer.

An abnormal growth or tumor is generally characterized as benign or malignant. A **benign tumor** is a self-contained mass within an organ that does not spread into adjacent tissues. The mass is usually slow-growing, rounded, and not greatly different from its tissue of origin. Benign tumors do not ordinarily cause death unless they grow into critical spaces such as the heart valves or brain ventricles. The feature that most distinguishes a **malignant tumor** (cancer) is uncontrolled growth of abnormal cells within normal tissue. As a general rule, cancer originates in cells such as skin and bone marrow cells that have retained the capacity to divide, whereas mature cells that have lost this power (neurons, for instance) do not commonly become cancerous.

Research findings link cancers to genetic damage or inherited genetic predispositions that alter the function of certain genes called oncogenes found normally in all cells. These genetic alterations disrupt the normal cell division cycle and **transform** a normal cell into a cancer cell. Cancer development involves many complex interactions between genes, their products, and external signals received by the cell. There is no single mechanism. Consider the fact that at least eight different genetic defects lead to colon cancer and nearly that many are implicated in breast cancers.

It is increasingly evident that susceptibility to certain cancers is inherited. At least 30 neoplastic syndromes that run in families have been described. A particularly stunning discovery is that some cancers in animals are actually due to endogenous viral genomes passed from parent to offspring in the gametes. Many cancers reflect an interplay of both hereditary and environmental factors. An example is lung cancer, which is associated mainly with tobacco smoking. But not all smokers develop lung cancer, and not all lung cancer victims smoke tobacco.

What role does the immune system play in controlling cancer? A concept that readily explains the detection and elimination of cancer cells is **immune surveillance.** Many experts postulate that cells with cancer-causing potential arise constantly in the body but that the immune system ordinarily discovers and destroys these cells, thus keeping cancer in check. Experiments with mammals have amply demonstrated that components of cell-mediated immunity interact with tumors and their antigens. The primary types of cells that operate in surveillance and destruction of tumor cells are cytotoxic T cells, natural killer (NK) cells, and macrophages. It appears that these cells recognize abnormal or foreign surface markers on the tumor cells and destroy them by mechanisms shown in figure 15.17. Antibodies help destroy tumors by interacting with macrophages and natural killer cells. This involvement of the immune system in destroying cancerous cells has inspired specific therapies.

How do we account for the commonness of cancer in light of this powerful range of defenses against tumors? To an extent,

* *neoplasm* (nee′-oh-plazm) Gr. *neo.*, new, and *plasm,* formation.
* *oncology* (awn-kaw′-luh-gee″) Gr. *onkos,* cancer.

the answer is simply that, as with infection, the immune system can and does fail. In some cases, the cancer may not be immunogenic enough; it may retain self markers and not be targeted by the surveillance system. In other cases, the tumor antigens may have mutated to escape detection. As we saw earlier, patients with immunodeficiencies such as AIDS and SCID are more susceptible to various cancers because they lack essential T-cell or cytotoxic functions. It may turn out that most cancers are associated with some sort of immunodeficiency, even a slight or transient one.

An ironic outcome of lifesaving medical procedures is the possible suppression of a patient's immune system. For instance, some immunosuppressive drugs that prevent graft rejection by T cells can likewise suppress beneficial immune responses. Although radiation and anticancer drugs are the first line of therapy for many types of cancer, both agents are extremely damaging to the bone marrow and other body cells.

Check & Assess Sections 16.7 and 16.8

✔ Immunodeficiency diseases occur when the immune response is reduced or absent.

✔ The impairments to the immune system result in major problems with recurrent infections and cancers.

✔ Primary immune diseases are genetically induced deficiencies of B cells, T cells, the thymus gland, or combinations of these.

✔ Secondary immune diseases are caused by infection, organic disease, chemotherapy, or radiation.

✔ The best-known infection-induced immunodeficiency is AIDS.

✔ Cancer is caused by genetic transformation of normal host cells into malignant cells. These transformed cells perform no useful function but, instead, grow unchecked and interfere with normal tissue function.

✔ Cytotoxic T cells, NK cells, and macrophages identify and destroy transformed cancerous cells by means of immune surveillance. Cancerous cells survive when these mechanisms fail.

32. In general, what causes primary immunodeficiencies and acquired immunodeficiencies?

33. Why do T-cell deficiencies have greater impact than B-cell deficiencies?

34. What kinds of symptoms accompany B-cell defects, T-cell defects, and combined defects?

35. Give examples of specific diseases that involve each type of defect.

36. Define *cancer,* and differentiate between a benign tumor and a malignant tumor, using examples.

37. Describe the relationship between cancer and the immune system.

CASE FILE 16 PERSPECTIVE

A person's ABO blood type is inherited as a combination of two alleles that code for glycoprotein antigens on red blood cells. Complications can arise when blood is transfused because most people carry antibodies that react with blood antigens that are different from theirs. (Who are the exceptions?) As long as a person does not receive blood of the incorrect antigenic type, these antibodies will not create a problem.

When an incompatible blood type is given, the donor RBCs release their hemoglobin into the blood stream. The red hemoglobin molecules will change the color of plasma and urine to a shade of pink. The disseminated intravascular coagulation is created by RBCs being agglutinated by IgM into large clumps, which clog the small arteries and capillaries, cutting off circulation to vital organs such as the heart, lungs, and kidney. In this case, the shutting down of the patient's circulation caused shock, low blood pressure, heart failure, and death.

It is generally thought that the plasma-borne antibodies against A and B antigens are produced through a natural, normal process that unfortunately collides with medical science. There are glycoproteins called heterophile antigens found in plants and bacteria that mimic the structure of human blood group molecules. During early contact with these substances from food and normal microbiota, the immune system is sensitized and makes antibodies that "lie in wait" in case of future contact with these antigens.

A transfusion reaction is a medical emergency. The treatment usually consists of flushing the bloodstream with saline and removing the hemoglobin with drugs. It is often necessary to give additional transfusions to replace the entire contents of the bloodstream. The decision under these circumstances is which blood to give safely when the actual blood type of the patient cannot be immediately clarified. This is an example of a circumstance where type O⁻ blood is considered for universal donation. It is safe to give because it contains none of the A, B, or Rh antigens and will not trigger further hemolysis and agglutination.

There are about 20,000 acute transfusion reactions per year in the United States, with about 50 deaths. The majority of these cases are attributable to staff errors in the hospital or bloodbank. The patient in this case was caught in a double worse-case scenario. First, he had the misfortune of being mistaken for his hospital roommate, and second, he was given type A blood, one of the most serious incompatibilities for type O. The blood vial was labeled with his correct name but was used to take blood from the man in the adjacent bed. This same blood was also used in the cross-match, so it did not react with the transfusion blood. A newer system used by many hospitals to prevent this sort of error is to have a microchip embedded in the patient's armband that reads out his blood type. It is compared to the blood being transfused and sounds an alarm if the bloods are incompatible.

To read more about the subject of transfusion reactions, go to http://emedicine.medscape.com/article/780074-overview.

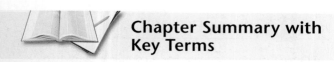

Chapter Summary with Key Terms

16.1 The Immune Response: A Two-Sided Coin
The study of disease states involving malfunctions of the immune system is called **immunopathology.**
 A. **Allergy,** or hypersensitivity, is an exaggerated, misdirected expression of certain immune responses.
 B. Autoimmunity involves abnormal responses to self antigens. A deficiency or loss in immune function is called **immunodeficiency.**
 C. Immediate-onset allergies involve contact with **allergens,** antigens that affect certain people; susceptibility is inherited.
 D. Allergens enter through four portals: Inhalants are breathed in (pollen, dust); ingestants are swallowed (food, drugs); injectants are inoculated (drugs, bee stings); contactants react on skin surface (cosmetics, glue).

16.2 Type I Allergic Reactions: Atopy and Anaphylaxis
 A. On first contact with allergen, specific B cells react with allergen and form a special antibody class called IgE, which affixes by its Fc receptor to mast cells and basophils.
 1. This **sensitizing dose** primes the allergic response system.
 2. Upon subsequent exposure with a **provocative dose,** the same allergen binds to the IgE–mast cell complex.
 3. This causes **degranulation,** release of intracellular granules containing mediators **(histamine, serotonin, leukotriene, prostaglandin),** with physiological effects such as vasodilation and bronchoconstriction.
 4. Conditions that are common types of atopy are **eczema,** food allergies, and **asthma;** severe systemic reaction is **anaphylaxis.**
 5. Other symptoms are rash, itching, redness, increased mucous discharge, pain, swelling, and difficulty in breathing.
 B. Diagnosis of allergy can be made by a histamine release test on basophils; serological assays for IgE; and skin testing, which injects allergen into the skin and mirrors the degree of reaction.
 C. Control of allergy involves drugs to interfere with the action of histamine, inflammation, and release of cytokines from mast cells. **Desensitization** therapy involves the administration of purified allergens.

16.3 Type II Hypersensitivities: Reactions That Lyse Foreign Cells
 A. Type II reactions involve the interaction of antibodies, foreign cells, and complement, leading to lysis of the foreign cells. In transfusion reactions, humans may become sensitized to special antigens on the surface of the red blood cells of other humans.
 B. The **ABO blood groups** are genetically controlled: Type A blood has A antigens on the RBCs; type B has B antigens; type AB has both A and B antigens; and type O has neither antigen. People produce antibodies against A or B antigens if they lack these antigens. Antibodies can react with antigens if the wrong blood type is transfused.
 C. **Rh factor** is another RBC antigen that becomes a problem if an Rh^- mother is sensitized by an Rh^+ fetus. A second fetus can receive antibodies she has made against the factor and develop hemolytic disease of the newborn. Prevention involves therapy with Rh immune globulin.

16.4 Type III Hypersensitivities: Immune Complex Reactions
Exposure to a large quantity of soluble foreign antigens (serum, drugs) stimulates antibodies that produce small, soluble Ab-Ag complexes. These **immune complexes** are trapped in various organs and tissues, which incites a damaging inflammatory response.

16.5 Immunopathologies Involving T Cells
 A. *Type IV delayed allergic response.*
 1. A delayed response to antigen involving the activation of and damage by T cells.
 2. Skin response to allergens, including infectious agents. Example is tuberculin reaction.
 3. Contact dermatitis is caused by exposure to plants (ivy, oak) and simple environmental molecules (metals, cosmetics).
 4. Cytotoxic T cells acting on allergen elicit a skin reaction.
 B. *Graft rejection:* Reaction of cytotoxic T cells directed against foreign cells of a grafted tissue; involves recognition of foreign human leukocyte antigens (HLA) by T cells and rejection of tissue.
 1. Host may reject graft; graft may reject host.
 2. Types of grafts include: **autograft,** from one part of body to another; **isograft,** between identical twins; **allograft,** between two members of same species; **xenograft,** between two different species.
 3. All major organs may be successfully transplanted.
 4. Allografts require tissue match (HLAs must correspond); rejection is controlled with drugs.

16.6 Autoimmune Diseases—An Attack on Self
 A. In certain type II and type III hypersensitivities, the immune system has lost tolerance to self molecules (autoantigens) and forms **autoantibodies** and sensitized T cells against them. Disruption of function can be systemic or organ-specific.
 B. **Autoimmune diseases** are genetically determined and more common in females.
 C. Examples include **systemic lupus erythematosus, rheumatoid arthritis, diabetes mellitus,** and diseases of the thyroid gland.

16.7 Immunodeficiency Diseases: Compromised Immune Responses
Components of the immune response system are absent. Deficiencies involve B and T cells, phagocytes, and complement.
 A. Primary immunodeficiency is genetically based, congenital; defect in inheritance leads to lack of B-cell activity, T-cell activity, or both.
 B. B-cell defect is called **agammaglobulinemia;** patient lacks antibodies; serious recurrent bacterial infections result. In Ig deficiency, one of the classes of antibodies is missing or deficient.
 C. In T-cell defects, the thymus is missing or abnormal. In **DiGeorge syndrome,** the thymus fails to develop; afflicted children experience recurrent infections with eukaryotic pathogens and viruses; immune response is generally underdeveloped.
 D. In **severe combined immunodeficiency (SCID),** both limbs of the lymphocyte system are missing or defective; no adaptive immune response exists; fatal without replacement of bone marrow or other therapies.
 E. Secondary (acquired) immunodeficiency is due to damage after birth (infections, drugs, radiation). **AIDS** is the most common of these; T helper cells are main target; deficiency manifests in numerous opportunistic infections and cancers.

16.8 The Function of the Immune System in Cancer

A. **Cancer** is characterized by overgrowth of abnormal tissue, also known as a **neoplasm.** It appears to arise from malfunction of **immune surveillance.**

 1. **Tumors** can be **benign** (a nonspreading local mass of tissue) or **malignant** (a cancer) that spreads (metastasizes) from the tissue of origin to other sites.

 2. Cancers occur in nearly every cell type (except mature, nondividing cells).

B. Cancer cells appear to share a common basic mechanism involving a genetic alteration that **transforms** a normally dividing cell into one that grows out of control.

C. Possible causes of cancerous transformation include irregularities in mitosis, genetic damage, activation of oncogenes, and infection by retroviruses.

Multiple-Choice Questions

Select the correct answer from the answers provided. For questions with blanks, choose the combination of answers that most accurately completes the statement.

1. Pollen is which type of allergen?
 a. contactant
 b. ingestant
 c. injectant
 d. inhalant

2. B cells are responsible for which allergies?
 a. asthma
 b. anaphylaxis
 c. tuberculin reactions
 d. both a and b

3. Which allergies are T-cell mediated?
 a. type I
 b. type II
 c. type III
 d. type IV

4. The contact with allergen that results in symptoms is called the
 a. sensitizing dose
 b. degranulation dose
 c. provocative dose
 d. desensitizing dose

5. Production of IgE and degranulation of mast cells are involved in
 a. contact dermatitis
 b. anaphylaxis
 c. Arthus reaction
 d. both a and b

6. The direct, immediate cause of allergic symptoms is the action of
 a. the allergen directly on smooth muscle
 b. the allergen on B lymphocytes
 c. allergic mediators released from mast cells and basophils
 d. IgE on smooth muscle

7. Theoretically, type _____ blood can be donated to all persons because it lacks _____.
 a. AB, antibodies
 b. O, antigens
 c. AB, antigens
 d. O, antibodies

8. An example of a type III immune complex disease is
 a. serum sickness
 b. contact dermatitis
 c. graft rejection
 d. atopy

9. Type II hypersensitivities are due to
 a. IgE reacting with mast cells
 b. activation of cytotoxic T cells
 c. IgG-allergen complexes that clog epithelial tissues
 d. complement-induced lysis of cells in the presence of antibodies

10. Production of autoantibodies may be due to
 a. emergence of forbidden clones of B cells
 b. production of antibodies against sequestered tissues
 c. infection-induced change in receptors
 d. All of these are possible.

11. Rheumatoid arthritis is an _____ that affects the _____.
 a. immunodeficiency disease, muscles
 b. autoimmune disease, nerves
 c. allergy, cartilage
 d. autoimmune disease, joints

12. A positive tuberculin skin test is an example of
 a. a delayed-type allergy
 b. autoimmunity
 c. acute contact dermatitis
 d. eczema

13. Contact dermatitis can be caused by
 a. pollen grains
 b. chemicals absorbed by the skin
 c. microbes
 d. proteins found in foods

14. Which disease would be most similar to AIDS in its pathology?
 a. X-linked agammaglobulinemia
 b. SCID
 c. ADA deficiency
 d. DiGeorge syndrome

15. The property whereby cancer cells display abnormal growth in normal tissue is called
 a. cytotoxicity
 b. mutation
 c. benign
 d. malignancy

16. In which of these ways is the immune system involved in development of cancer?
 a. failure in immune surveillance
 b. mutation in cytotoxic T cells
 c. autoantibody formation
 d. overreaction to environmental chemicals

Case File Questions

1. Which blood type antibodies would the patient have carried in his plasma? (Explain your answer.)
 a. anti-A alone
 b. anti-B alone
 c. anti-Rh alone
 d. anti-A and -B
 e. anti-A, -B, and -Rh
 f. none of these

2. If the patient had a blood type different from O^+, which one would have not resulted in his death? (Explain your answer.)
 a. type A^-
 b. type B^+
 c. type AB^-
 d. type O^-

Writing to Learn

These questions are suggested as a *writing-to-learn* experience. For each question, compose a one- or two-paragraph answer that includes the factual information needed to completely address the question. All Check & Assess questions can be used for writing-to-learn exercises.

1. Compare and contrast type I (atopic) and type IV (delayed) hypersensitivity as to mechanism, symptoms, eliciting factors, and allergens.

2. Why would it be necessary for an Rh^- woman who has had an abortion, miscarriage, or an ectopic pregnancy to be immunized against the Rh factor?

3. How does a baby inherit Rh⁺ blood from an Rh⁻ mother?

4. Why is a hemolytic transfusion reaction considered a type of hypersensitivity?

5. Given that it is the donated RBCs and not the patient's normal RBCs that are hemolyzed, why is a hemolytic transfusion reaction still fatal?

6. Describe three circumstances that might cause antibodies to develop against self tissues.

7. Explain how people with autoimmunity could develop antibodies against intracellular components (nucleus, mitochondria, and DNA).

7. How can a person prevent becoming allergic to poison oak? What is the basis for infectious allergy?

8. Why are primary immunodeficiencies considered "experiments of nature"?

9. a. Explain why babies with agammaglobulinemia do not develop opportunistic infections until about 6 months after birth.
 b. Explain why people with B-cell deficiencies can benefit from artificial passive immunotherapy. Explain whether vaccination would work for them.
 c. Explain why the lack of a specific enzyme (ADA, tyrosine kinase) would lead to the death of lymphocytes.

10. In what ways can cancer be both a cause and a symptom of immunodeficiency?

Concept Mapping

Appendix E provides guidance for working with concept maps.

1. Construct your own concept map using the following terms as the *concepts*. Supply the linking words between each pair of concepts.

 lysed cells

 degranulation release of mediators

 immune complexes

 damage by T cells

 allergens

 cell-bound antibody

 processed antigen

 soluble antigen

Critical Thinking Questions

Critical thinking is the ability to reason and solve problems using facts and concepts. These questions can be approached from a number of angles, and in most cases, they do not have a single correct answer.

1. a. Discuss the reasons that the immune system sometimes behaves like a double-edged sword.
 b. Suggest a possible function of allergy.

2. A 3-week-old neonate develops severe eczema after being given penicillin therapy for the first time. Explain what has happened.

3. Why would a person be allergic to strawberries when he eats them but show a negative skin test to them?

4. a. Where in the course of type I allergies do antihistamine drugs, cortisone, and desensitization work? Exactly what do they do?
 b. Compare the sites of action of Singulair and Xolair.

5. a. Although we call persons with type O blood universal donors and those with type AB blood universal recipients, what problems might accompany transfusions involving O donors or AB recipients?
 b. What is the truly universal donor and recipient blood type? (Include Rh type.)

6. Explain how cleaner living and medical advances could contribute to allergies.

Visual Challenge

1. From **chapter 15, figure 15.9.** Explain how this phenomenon pertains to allergies.

(a) Hapten — No antibody

(b) Hapten bound to carrier molecule — Antibody formed in response to hapten

2. Observe the blood typing results shown here and label the blood types, including Rh. Which types are the most and least common?

Anti-A Anti-B Anti-Rh

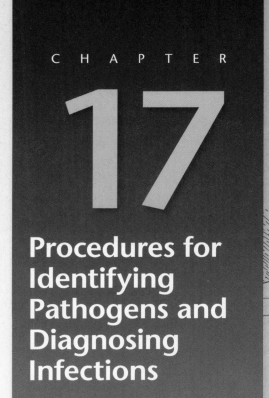

17

Procedures for Identifying Pathogens and Diagnosing Infections

Like tiny sausages, these vibrios (13,000×) cause the majority of food infections that originate in seafood.

Pathogenic vibrios often form biofilms on the exoskeletons of shellfish.

"It appeared that shrimp or other seafood used to prepare meals may not have been adequately refrigerated."

CASE FILE 17 A Nosocomial Outbreak of Seafood Sickness

Personnel working in a Taiwanese hospital during a spring heat wave were alerted to a possible outbreak of nosocomial infections. The first evidence came in a report from the nursing staff about nine patients in two adjacent wards with severe diarrhea, abdominal pain, fever, nausea, and vomiting. An infection control team was immediately brought in to monitor and record all hospital-wide cases of gastroenteritis. Stool samples were sent to their clinical laboratory and the local public health department for testing, and food served prior to and during the outbreak was refrigerated for testing. Preliminary observations over the next 48 hours revealed 16 additional patients spread among 5 wards and two separate buildings.

Patient and food specimens were screened for common enteric pathogens, including *Salmonella* spp, *Shigella* spp, *Staphylococcus aureus*, *Vibrio* spp, *Campylobacter jejeuni*, *Bacillus cereus*, and *Clostridium* spp. Plating for initial isolation was performed on blood agar, MacConkey agar, and other selective and differential media. Gram stains of the stool samples were performed early on, but they were inconclusive, yielding mostly gram-negative rods mixed with some gram-positive rods and cocci. After two days, colonies grew in the presence of air on the blood agar and on selective media containing bile. These were used to make pure cultures for testing. A second Gram stain indicated that the isolates were gram negative rods and a hanging drop slide viewed with darkfield microscopy revealed curved or comma-shaped rods with darting motility.

Morphological and physiological tests used to narrow the preliminary identification are summarized in the table below for the remaining potential pathogens, compared to the results for patient isolates in the far right column.

▶ *Explain the reason for using bile as a selective agent and the basis for the preliminary biochemical tests.*

▶ *Given the data so far, identify the possible infectious agents that best fit the pathogen profile.*

To continue the case, go to page 523.

CASE FILE TABLE 1

TEST	Salmonella	Shigella	Vibrio	Campylobacter	Patient and Food Isolate Results
Motility	+	−	+	+	+
Shape	straight rods	straight rods	curved rods	curved, spiral rods	curved rods
Oxidase	−	−	+	+	+
Growth on 6% salt medium	−	−	+	+	+
Oxygen requirements	facultative	facultative	aerobic	microaerophilic	aerobic
Fermentation of glucose	+	+	+	−	+

17.1 An Overview of Clinical Microbiology

Expected Learning Outcomes

1. Describe what is involved in the main categories of laboratory identification of infectious agents.

2. Explain several techniques in specimen collection and the concerns in acquiring good cultures.

3. Summarize the main procedures in isolation, cultivation, and identification of pathogens.

Chapters 18 through 25 survey the most clinically significant bacterial, fungal, parasitic, and viral diseases. These chapters cover prevalent infectious conditions, the organisms that cause them, and some elements of diagnosis. In preparation for these chapters, we introduce basic background information on identification and diagnostic testing. This being an overview, most of the detailed tests pertinent to a specific group are covered in later chapters with the discussion of the individual pathogenic microbes.

For many microbiology students and professionals, the most pressing topic is *how to identify unknown bacteria* in specimens or in cultures. Methods microbiologists use to identify bacteria to the level of genus and species fall into three main categories: **phenotypic,** which includes a consideration of morphology (microscopic and macroscopic) as well as bacterial physiology or biochemistry; **immunologic,** which entails analysis of the blood, or serology; and **genotypic** (or genetic) techniques.

Data from a cross section of such tests can produce a unique profile of each bacterium. Increasingly, genetic means of identification are being used as a sole resource for identifying bacteria. As universally used databases become more complete because of submissions from scientists and medical personnel worldwide, genetic analyses provide a more accurate and speedy way of identifying microbes than was possible even a decade ago. There are still many organisms, however, that must be identified in the traditional way—via biochemical, serological, and morphological means. Serology is a highly reliable procedure for some diseases and is still favored as a screening method for large numbers of people. The primary methods emphasized here are phenotypic, serological, and genetic analysis. See appendix D for a summary of the basic methods used to identify bacteria.

Phenotypic Methods

Microscopic Morphology

Traits that can be valuable aids to identification are combinations of cell shape and size, Gram stain reaction, acid-fast reaction, and special structures, including endospores, granules, and capsules. Electron microscope studies can pinpoint additional structural features (such as the cell wall, flagella, pili, and fimbriae).

Microscopic analysis is also used to evaluate specimen quality prior to culturing. This can often narrow the causative agent as bacterial, fungal, or protozoan.

Macroscopic Morphology

Traits that can be assessed with the naked eye are also useful in diagnosis. These include the appearance of growth, including texture, size, shape, pigment of colonies, speed of growth, and reactions to special types of selective and differential media.

Physiological/Biochemical Characteristics

These have been the traditional mainstay of bacterial identification. Enzymes and other biochemical properties of bacteria are fairly reliable and stable expressions of the chemical identity of each species. Dozens of diagnostic tests exist for determining the presence of specific enzymes and for assessing nutritional and metabolic activities. Examples include tests for fermentation of sugars; capacity to digest or metabolize complex polymers such as proteins and polysaccharides; production of gas; presence of enzymes such as catalase, oxidase, and decarboxylases; and sensitivity to antimicrobic drugs. Special rapid identification test systems that record the major biochemical reactions of a culture have streamlined data collection.

Chemical Analysis

This involves analyzing the types of specific structural substances that the microorganism contains, such as the chemical composition of peptides in the cell wall and lipids in membranes.

Genotypic Methods

Examining the genetic material itself has revolutionized the identification and classification of bacteria. There are many advantages of genotypic methods over phenotypic methods, when they are available. The primary advantage is that actually culturing the microorganisms is not always necessary. In recent decades, scientists have come to realize that there are many more microorganisms that we can't grow in the lab compared with those that we can (see Insight 3.3). Another advantage is that genotypic methods are increasingly automated, and results are obtained very quickly, often with more precision than with phenotypic methods.

Immunologic Methods

Bacteria and other microbes have surface and other molecules called antigens that are recognized by the immune system. One immune response to antigens is the production of antibodies that are designed to bind tightly to the antigens. The nature of the antibody response can be determined from a blood (or other tissue) sample. The presence of specific antibodies to a suspected pathogen is highly suggestive of infection. This is often easier than testing for the microbe itself, especially in the case of viral infections. Most HIV testing entails examination of a person's blood for presence of antibody to the virus. Laboratory kits based on this technique are available for immediate identification of a number of pathogens. These methods are covered in greater detail in Section 17.4.

On the Track of the Infectious Agent: Specimen Collection

Regardless of the method of diagnosis, specimen collection is the common point that guides the health care decisions of every member of a clinical team. Indeed, the success of identification and treatment depends on how specimens are collected, handled, and stored. Specimens can be taken by a clinical laboratory scientist or medical technologist, nurse, physician, or even by the patient himself. However, it is imperative that general aseptic procedures be used, including sterile sample containers and other tools to prevent contamination from the environment or the patient. **Figure 17.1a** delineates the most common sampling sites and procedures.

In sites that normally contain resident microbiota, care should be taken to sample only the infected site and not surrounding areas. One common type of collection system is the sterile swab transport tube **(figure 17.1b).** This allows for sampling, transporting, and maintaining the specimen in its original condition. When using a swab to take a sample from throat and nasopharyngeal regions, the swab should not touch the tongue, cheeks, or saliva. Saliva is an especially undesirable contaminant because it contains millions of bacteria per milliliter, most of which are normal microbiota. Sputum, the mucus secretion that coats the lower respiratory surfaces, especially the lungs, is discharged by coughing or taken by catheterization to avoid contamination with saliva. Depending on the nature of the lesion, skin can be swabbed or scraped with a scalpel

to expose deeper layers. The mucous lining of the vagina, cervix, or urethra can be sampled with a swab or applicator stick.

Urine is taken aseptically from the bladder with a thin tube called a catheter. Another method, called a "clean catch," is taken by washing the external urethra and collecting the urine midstream. The latter method inevitably incorporates a few normal microbiota into the sample, but these can usually be differentiated from pathogens in an actual infection. Sometimes diagnostic techniques require first-voided "dirty catch" urine.

Sterile materials such as blood, cerebrospinal fluid, and tissue fluids must be taken by sterile needle aspiration. Antisepsis of the puncture site is extremely important in these cases. Additional sources of specimens are the vagina, eye, ear canal, nasal cavity (all by swab), and diseased tissue that has been surgically removed (biopsied).

After proper collection, the specimen is promptly transported to a lab and stored appropriately (usually refrigerated) if it must be held for a time. Nonsterile samples in particular, such as urine, feces, and sputum, are especially prone to deterioration at room temperature. The overgrowth of normal microbiota in these samples could prevent the isolation of the pathogens. It could also alter the numbers and proportions of cells, making analysis more difficult. Many transport devices contain nonnutritive maintenance media (so the microbes do not grow), a buffering system, and an anaerobic environment to prevent possible destruction of oxygen-sensitive bacteria.

A summary of specimen collection guidelines can be found in **table 17.1.**

Figure 17.1 Specimen Collection (a) Sampling sites and methods of collection for clinical laboratories. **(b)** Sterile, transport swab with carrier—sometimes called a culturette.

TABLE 17.1 General Guidelines for Specimen Collection*

Type of Specimen	Collection Method	Comments
Abcess of skin or membrane or decubitus ulcer	Debride surface with sterile wipe; suction fluid with sterile needle and syringe.	Aspiration of tissue samples preferred to swab; transfer to anaerobic transport system.
Anaerobic cultures	Collect from deeper tissue below the surface, using sterile syringe.	Sample must not be exposed to air; transported immediately in anaerobic system to lab
Blood	Prep area of skin with iodine; use vacutainer blood collection tubes.	Inoculate blood culture bottle immediately.
Bone marrow	Prepare site for surgical incision; extract sample with special needle.	Most samples taken from sternum or ilium; placed in blood bottle for culture
Cerebrospinal fluid (CSF)	Remove CSF aseptically by lumbar puncture.	Place CSF into sterile tubes; do not refrigerate; transport immediately to lab for processing.
Feces	Take a small specimen into sterile container, cover, and transport.	Refrigerate samples held over 1 hour; culture is mainly to rule out enteric pathogens; special kits available to detect cysts and trophozoites of protozoans and ova and larvae of intestinal worms
Genital/urinary tract	Sterile swab of cervical mucus; swab urethral membrane or insert into lumen and twist.	Plate directly onto selective culture media with high CO_2 atmosphere or collect with anaerobic transport swab system; process immediately.
Respiratory tract, lower	Have patient cough to loosen phlegm and expectorate sputum into sterile cup for transport.	Patient should brush teeth and rinse mouth with water prior to any sampling. If coughing does not work, induce sputum by having patient inhale sterile saline from a nebulizer.
Respiratory tract, upper	Cleanse and debride oral cavity membranes; vigorously swab area of lesion; throat cultures are taken by swabbing the posterior pharynx, tonsils, and inflamed sites.	Tissue aspirates of infected tissue preferred; throat samples are processed for strep throat; culturing for other throat pathogens (except viruses) is done when specifically requested; care must be taken to avoid normal tissue and saliva.
Urine (two methods)	Sampling by clean midstream catch into a sterile container	Clean external genitalia, separate labia (female) or pull back foreskin (male) and urinate for a short time, followed by collection of 100–200 cc sample (midstream); samples not immediately processed must be refrigerated.
	Sampling by sterile catheter inserted into bladder; urine is drained into sterile container.	Clean and rinse the urethral opening; only a specimen catheter should be used to sample for bladder infection; patients with indwelling catheters will always have microbes in their bladders.

*For a detailed account of collecting and managing specimens, log on to http://www.medicine.uiowa.edu/path_handbook/Appendix/Micro/mi and http://www.dshs.state.tx.us/LAB/bac_guidelines.shtm#guidelines.

TAKE NOTE: OPTIONS IN LABORATORY ANALYSIS

The routes taken in specimen analysis come under the heading of (1) direct tests using microscopic, immunologic, or genetic methods that provide immediate clues as to the identity of the microbe or microbes in the sample; and (2) cultivation, isolation, and identification of pathogens using a wide variety of general and specific tests that may require several days (figure 17.2).

Most test results fall into two categories: presumptive data, which place the isolated microbe (isolate) in a preliminary category such as a genus; and more specific, confirmatory data, which provide more definitive evidence for a species. Some tests are more important for some groups of bacteria than for others. The total time required for analysis ranges from a few minutes in a streptococcal sore throat to several weeks in tuberculosis.

Some diseases are diagnosed without the need to identify microbes from specimens. Serological tests on a patient's serum can detect signs of an antibody response. One method that clarifies whether a positive test indicates current or prior infection is to take two samples several days apart to see if the antibody titer is rising. Skin testing can pinpoint a delayed allergic reaction to a microorganism. Immunological tests are also important in screening the general population for exposure to an infectious agent such as rubella or tuberculosis.

Because diagnosis is both a science and an art, the ability of the practitioner to interpret signs and symptoms of disease can be very important. AIDS, for example, is usually diagnosed by serological tests and a complex of signs and symptoms without ever isolating the virus. Some diseases (athlete's foot, for example) are diagnosed purely by the typical presenting symptoms and may require only microscopic examination of a specimen.

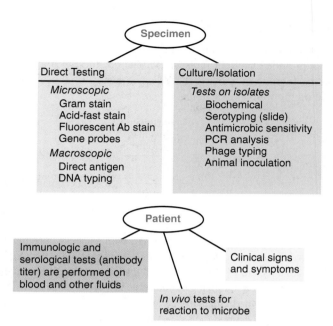

Figure 17.2 **A scheme of specimen isolation and identification.**

17.2 Phenotypic Methods

Expected Learning Outcomes

4. Describe some direct methods of testing a specimen.
5. Summarize the aims in selection of culture techniques and media.
6. Explain the basis of biochemical and other methods of testing.
7. Describe how flowcharts and comparison tables are used in identification of bacterial pathogens.

Immediate Direct Examination of Specimen

Direct microscopic observation of a fresh or stained specimen is one of the most rapid methods of determining presumptive and sometimes confirmatory characteristics. Stains most often employed for bacteria are the Gram stain (see Insight 3.2) and the acid-fast stain (see figure 19.15). For many species, these ordinary stains are useful, but they do not work with certain organisms. Direct fluorescent antibody (DFA) tests can highlight the presence of the microbe in patient specimens by means of labeled antibodies **(figure 17.3).** DFA tests are particularly useful for bacteria, such as the syphilis spirochete, that are not readily cultivated in the laboratory or if rapid diagnosis is essential for the survival of the patient.

Another way that specimens can be analyzed is through *direct antigen testing.* This technique is similar to direct fluorescent antibody testing in that known antibodies are used to identify antigens on the

Figure 17.3 **Direct fluorescent antibody test. (a)** Results for *Treponema pallidum,* the syphilis spirochete, and an unrelated spirochete. **(b)** Photomicrograph of this technique used on a chancre from a syphilitic patient reveals numerous fluorescing spirochetes.

surface of bacterial isolates. But in direct antigen testing, the reactions can be seen with the naked eye (see figure 18.14*b*). Quick test kits that greatly speed clinical diagnosis are available for *Staphylococcus aureus, Streptococcus pyogenes, Neisseria gonorrhoeae, Haemophilus influenzae, Neisseria meningitidis,* and *Salmonella.* However, when the microbe is very sparse in the specimen, direct testing is less effective, and identification will require more sensitive methods.

Cultivation of Specimen

Isolation Media

Such a wide variety of media exist for microbial isolation that a certain amount of preselection must occur, based on the nature of the specimen. When the sample contains such small numbers of the pathogen that it could be easily lost or overgrown, the specimen is often cultured in enrichment media that can amplify the pathogen. In this chapter's Case File, for instance, 1% to 2% salt is added to favor the growth of any *Vibrio* that may be present. In specimens such as feces, mucus, or urine that may have high bacterial counts and a diversity of species, selective media are chosen to make the target microbes more prominent by inhibiting microbes that are part of the normal microbiota. These media can be moderately inhibitory, such as those that grow only gram-positive or gram-negative bacteria, or highly inhibitory, being specific to a single group or species. One example is a medium used to screen clinical specimens that will grow only *Listeria monocytogenes.* It contains six selective agents, three of which are antibiotics. These media have the benefit of rapidly pulling the pathogen out of the sample in a single step.

In most cases, specimens are also inoculated into differential media that define such characteristics as reactions in blood (blood agar) and fermentation patterns (mannitol salt and MacConkey agar). A patient's blood is usually cultured in a special bottle of broth that can be periodically sampled for growth. Numerous other examples of isolation, differential, and biochemical media were presented in chapter 3. So that subsequent steps in identification will be as accurate as possible, all work must be done from isolated colonies or pure cultures, because working with a mixed or contaminated culture gives misleading and inaccurate results. From such isolates, clinical microbiologists obtain information about a pathogen's microscopic morphology and staining reactions, cultural appearance, motility, oxygen requirements, and biochemical characteristics.

Biochemical Testing

The physiological reactions of bacteria to nutrients and other substrates provide excellent indirect evidence of the types of enzyme systems present in a particular species. Many of these tests are based on the following scheme:

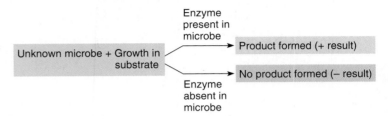

The microbe is cultured in a medium with a special substrate and then tested for a particular end product. The presence of the end product indicates that the enzyme is expressed in that species; its absence means it lacks the enzyme for utilizing the substrate in that particular way. These types of reactions are particularly meaningful in bacteria, which are haploid and generally express their genes for utilizing a given nutrient.

Among the prominent biochemical tests are carbohydrate fermentation (acid and/or gas); hydrolysis of gelatin, starch, and other polymers; enzyme actions such as catalase, oxidase, and coagulase; and various by-products of metabolism. Many are presently performed with rapid, miniaturized systems that can simultaneously determine up to 23 characteristics in small individual cups or spaces (**figure 17.4**). An important plus, given the complexity of biochemical profiles, is that such systems are readily adapted to computerized analysis.

Common schemes for identifying bacteria are somewhat artificial but convenient. They are based on easily recognizable characteristics such as motility, oxygen requirements, Gram stain reactions, shape, spore formation, and various biochemical reactions. Schemes can be set up as flowcharts (**figure 17.5**) or keys that trace a route of identification by offering pairs of opposing characteristics (positive versus negative, for example) from which to select. Eventually, an endpoint is reached, and the name of a genus or species that fits that particular combination of characteristics appears. Diagnostic tables structured like this chapter's Case File that provide more complete information are preferred by many laboratories because variations from the general characteristics used on the flowchart can be misleading. Both systems are used in this text.

Figure 17.4 Rapid Identification tests. The API 20E manual biochemical system for microbial identification. **(a)** Positive results for the 20 tests and **(b)** negative results for the same tests.

CONTINUING CASE FILE 17

The microbiologists at the medical center were able to identify the pathogen as belonging to the genus *Vibrio* based on the fact that *Campylobacter*, the only other genus that would have fit, is known to exhibit spiral-shaped cells and is microaerophilic and nonfermentative in its physiology. The next step was to perform additional tests that could differentiate among the three major species in genus *Vibrio*, all of which are associated with intestinal infections. The panel of standard tests performed on the patient and food isolates is provided in table 2, with expected results from the three pathogens for comparison.

CASE FILE TABLE 2

TEST	*Vibrio cholerae*	*Vibrio parahaemolyticus*	*Vibrio vulnificus*	Patient Isolate	Food Isolate
Sucrose*	+	−	−	−	−
Lactose*	−	−	+	−	−
Arabinose*	−	+	−	+	+
ONPG**	+	−	+	−	−
Growth in 0 salt	+	−	−	−	−
Growth in 6% salt	−	+	+	+	+

* These are sugars for testing fermentation.
** ONPG tests for slow lactose fermentation.

■ Which species is responsible for the outbreak of gastroenteritis, and what additional confirmatory tests can verify this identification?

■ Why is it so important to identify the infectious agent to the level of species?

For a wrap-up, see the Case File Perspective on page 536.

(a)

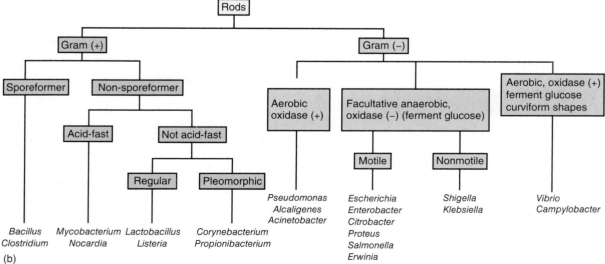

(b)

Figure 17.5 **Flowchart to separate genera of gram-positive and gram-negative bacteria often isolated from specimens. (a)** Cocci and **(b)** rods commonly involved in human diseases. Most of these genera are covered in greater detail in chapters 18, 19, 20, and 21.

Miscellaneous Tests

When morphological and biochemical tests are insufficient to complete identification, other tests come into play.

Bacteria host viruses called bacteriophages that are very species- and strain-specific. Such selection by a virus for its host is useful in typing some bacteria, primarily *Staphylococcus* and *Salmonella*. The technique of phage typing involves inoculating a lawn of cells onto a Petri dish, mapping it off into blocks, and applying a different phage to each block. Cleared areas corresponding to lysed cells indicate sensitivity to that phage. Phage typing is chiefly used for tracing strains of bacteria in epidemics.

Animals may be needed to cultivate bacteria such as *Mycobacterium leprae* and *Treponema pallidum,* whereas avian embryos and cell cultures are used to grow rickettsias, chlamydias, and viruses. Animal inoculation is also occasionally used to test bacterial or fungal virulence.

Antimicrobial sensitivity tests are not only important in determining the drugs to be used in treatment (see figure 12.18), but the patterns of sensitivity can also be used in presumptive identification of some species of *Streptococcus, Pseudomonas,* and *Clostridium.* Antimicrobials are also used as selective agents in many media.

Determining Clinical Significance of Cultures

Questions that can be difficult but necessary to answer in this era of debilitated patients and opportunists are: Is an isolate clinically important, and how can one know whether it is a contaminant or just part of the normal microbiota? The number of microbes in a sample is one useful criterion. For example, a few colonies of *Escherichia coli* in a urine sample can simply indicate normal microbiota, whereas several hundred can mean active infection. In contrast, the presence of a single colony of a true pathogen such as *Mycobacterium tuberculosis* in sputum or an opportunist in sterile sites such as cerebrospinal fluid or blood is highly suggestive of its role in disease. Furthermore, the repeated isolation of a relatively pure culture of any microorganism can mean it is an agent of disease, though care must be taken in this diagnosis. Another challenge facing the clinical laboratory is that of differentiating a pathogen from species in the normal microbiota that are similar in morphology from their more virulent relatives. Fortunately, medical technologists have the required expertise to tackle the decisions that arise, even with the most challenging diagnoses.

17.3 Genotypic Methods

Expected Learning Outcome

8. Explain the different variations on genetic testing and how they can be used in identification and diagnosis.

DNA Analysis Using Genetic Probes

The exact order and arrangement of the DNA code is unique to each organism. With a technique called *hybridization,* it is possible to identify a bacterial species by analyzing segments of its DNA. This requires small fragments of single-stranded DNA (or RNA) called **probes** that are known to be complementary to the specific sequences of DNA from a particular microbe. The test is conducted by extracting unknown test DNA from cells in specimens or cultures and binding it to special blotter paper. After several different probes have been added to the blotter, it is observed for visible signs that the probes have become fixed (hybridized) to the test DNA. The binding of probes onto several areas of the test DNA indicates close correspondence and makes positive identification possible (see figures 10.5 and 10.6).

Another genetic-based technique often brought in to analyze outbreaks or epidemics is pulse-field gel electrophoresis (PFGE). This rapid method of DNA typing makes a restriction fragment fingerprint of the suspected pathogens on a single gel that identifies and compares the isolates alongside each other. It is especially useful to trace the causative agents to a particular source or reservoir (see Insight 27.1).

Roles of the Polymerase Chain Reaction and Ribosomal RNA in Identification

The polymerase chain reaction (PCR) is one of the most valuable high technology tools for DNA and RNA analysis. A review of chapter 10 (figure 10.7) will remind you how it can rapidly synthesize hundreds of thousands and even millions of copies of a particular segment of DNA. It is both highly sensitive and specific, amplifying minute quantities from unknown samples that would otherwise be lost. It is one of the most powerful techniques in forensics and molecular biology, but it can also be applied to the identification of microbes. With this method, laboratories can often completely dispense with traditional culturing techniques and go directly to diagnosis in a few hours.

PCR-based tests are used routinely for detecting HIV, Lyme disease, the human papilloma virus, tuberculosis, hepatitis, and a number of other bacterial and viral infections (**figure 17.6**). In many laboratories, the PCR or a similar rapid test is the method of choice for identifying gonorrhea and chlamydiosis. One version of this technique called realtime PCR is able to pull DNA from a sample register the amount that is formed, and identify it.

One particularly versatile aspect to PCR is that it can be combined with other technologies for additional applications. For example, if the sample being analyzed contains mainly RNA (from an RNA virus or ribosome), the initial introduction of reverse transcriptase will convert the RNA in a sample to DNA, and this DNA can then be amplified by PCR. This technique—RT-PCR—is valuable for verifying RNA-based viruses such as rabies virus (see figure 17.17e). DNA of PCR origin can be analyzed with the other tools of technology: it can be probed, hybridized, fingerprinted, and sequenced.

One of the most viable indicators of evolutionary relatedness and affiliation is comparison of the sequence of nitrogen bases in ribosomal RNA, a major component of ribosomes. Ribosomes have the same function (protein synthesis) in all cells, and they tend to remain more or less stable in their nucleic acid content over long periods. Thus, any major differences in the sequence, or "signature," of the rRNA is likely to indicate some distance in ancestry. This technique is powerful at two levels: It is effective for differentiating general group differences (it was used to separate the three superkingdoms of life discussed in chapter 1), and it can be fine-tuned to identify at the species level (for example, in *Mycobacterium* and *Legionella*).

Figure 17.6 PCR testing for identification of pathogens.
A PCR gel fingerprint indicates 18 samples with patterns typical for *Vibrio parahaemolyticus*. These strains were isolated from both clinical specimens and seawater. Lanes 6 and 20 represent molecular size standards.

Check & Assess Sections 17.2 and 17.3

✔ The main phenotypic methods include the direct examination of specimens, observing the growth of specimen cultures on enriched, selective, and differential media, and biochemical testing of cultures.

✔ The use of genotypic methods has been increasing rapidly. These include hybridization, nucleic acid sequencing, rRNA analysis, and PCR-based methods.

5. Describe what is involved in direct specimen testing, presumptive and confirmatory tests, cultivating and isolating the pathogen, biochemical testing, and gene probes.
6. Summarize the important aspects for determining if a clinical isolate is involved in infection.
7. Describe the applications of PCR in identification techniques.

17.4 Immunologic Methods

Expected Learning Outcomes

9. Describe the background aims of immunological testing.
10. Identify how antigen-antibody reactions are detected and quantified.

11. Differentiate between agglutination and precipitation, and describe how they are used in diagnosis.
12. Explain how the Western blot and complement fixation tests are performed.
13. Interpret the outcome of direct and indirect immunofluorescent antibody testing.

The antibodies formed during an immune reaction are important in combating infection, but they hold additional practical value. Characteristics of antibodies such as their quantity or specificity can reveal the history of a patient's contact with microorganisms or other antigens. This is the underlying basis of serological testing. **Serology** is the branch of immunology that traditionally deals with in vitro diagnostic testing of serum. Serological testing is based on the familiar concept that antibodies have extreme specificity for antigens, so when a particular antigen is exposed to its specific antibody, it will fit like a hand in a glove. The ability to visualize this interaction by some means provides a powerful tool for detecting, identifying, and quantifying antibodies—or for that matter, antigens. The scheme works both ways, depending on the situation. One can detect or identify an unknown antibody using a known antigen, or one can use an antibody of known specificity to help detect or identify an unknown antigen (**figure 17.7**). Modern serological testing has grown into a field that tests more than just serum. Urine, cerebrospinal fluid, whole tissues, and saliva can also be used to determine the immunologic status of patients. These and other immune tests are helpful in confirming a suspected diagnosis or in screening a certain population for disease.

General Features of Immune Testing

The strategies of immunologic tests are diverse, and they underline some of the brilliant and imaginative ways that antibodies and antigens can be used as tools. We summarize them under the headings of agglutination, precipitation, complement fixation, fluorescent antibody tests, and immunoassay tests. First, we overview the general characteristics of immune testing, and we will then look at each type separately.

The most effective serological tests have a high degree of specificity and sensitivity (**figure 17.8**). Specificity is the property of a test to focus upon only a certain antibody or antigen and not to react with unrelated or distantly related ones. Sensitivity means that the test can detect even very small amounts of antibodies or antigens that are the targets of the test. New systems using monoclonal antibodies have greatly improved specificity, and those using radioactivity, enzymes, and electronics have improved sensitivity.

Visualizing Antigen-Antibody Interactions

The primary basis of most tests is the binding of an antibody (Ab) to a specific molecular site on an antigen (Ag). Because this reaction cannot be readily seen without an electron microscope, tests involve some type of endpoint reaction visible to the naked eye or with regular magnification that tells whether the result is positive or negative. In the case of large antigens such as cells, Ab binds to Ag and creates large clumps or aggregates that are visible macroscopically or microscopically (**figure 17.9a**). Smaller Ab-Ag complexes that do not

(a) In serological diagnosis of disease, a blood sample is scanned for the presence of antibody using an antigen of known specificity. A positive reaction is usually evident as some visible sign, such as color change or clumping, that indicates a specific interaction between antibody and antigen. (The reaction at the molecular level is rarely observed.)

Figure 17.7 Basic principles of serological testing using antibodies and antigens.

(b) An unknown microbe is mixed with serum containing antibodies of known specificity, a procedure known as serotyping. Microscopically or macroscopically observable reactions indicate a correct match between antibody and antigen and permit identification of the microbe.

Figure 17.8 Specificity and sensitivity in immune testing. (a) This test shows specificity in which an antibody (Ab) attaches with great exactness with only one type of antigen (Ag). **(b)** Sensitivity is demonstrated by the fact that Ab can pick up antigens even when the antigen is greatly diluted.

result in readily observable changes will require special indicators in order to be visualized. Endpoints are often revealed by dyes or fluorescent reagents that can tag molecules of interest. Similarly, radioactive isotopes incorporated into antigens or antibodies constitute sensitive tracers that are detectable with photographic film.

An antigen-antibody reaction can be used to read a **titer,** or the quantity of antibodies in the serum. Titer is determined by serially diluting a sample in tubes or in a multiple-welled microtiter plate and mixing it with antigen **(figure 17.9b).** It is determined by observing the highest dilution of serum that produces a visible reaction with an

Agglutination*

Epitope

Whole cell

Antigen + Antibody

Microscopic appearance of clumps

Precipitation*

Cell-free molecule in solution

Epitope

Antigen + Antibody

Microscopic appearance of precipitate

(a) Agglutination involves clumping of whole cells; precipitation is the formation of antigen-antibody complexes in cell free solution. Both reactions can be observed by noticeable clumps or precipitates in test tubes (see (b) and figure 17.10a).

*Although IgG is shown as the Ab, IgM is also involved in these reactions.

The Tube Agglutination Test

Reaction + + + + + + + − −

Unagglutinated cells

Agglutinated cells

Dilution 1/20 1/40 1/80 1/160 1/320 1/640 Control (no serum)

(b) The tube agglutination test. A sample of patient's serum is serially diluted with saline. The dilution is made in a way that halves the number of antibodies in each subsequent tube. An equal amount of the antigen (here, blue bacterial cells) is added to each tube. The control tube has antigen, but no serum. After incubation and centrifugation, each tube is examined for agglutination clumps as compared with the control, which will be cloudy and clump-free. The titer is equivalent to the denominator of the dilution of the last tube in the series that shows agglutination.

Figure 17.9 **Cellular/molecular view of agglutination and precipitation reactions that produce visible antigen-antibody complexes.**

antigen. The titer is the denominator of this dilution. The more a sample can be diluted and yet still react with antigen, the greater is the concentration of antibodies in that sample and the higher is its titer. Possible interpretations of test results are discussed in **Insight 17.1.**

Agglutination and Precipitation Reactions

The essential differences between agglutination and precipitation are in size, solubility, and location of the antigen. In agglutination, the antigens are whole cells such as red blood cells or bacteria with determinant groups on the surface. In precipitation, the antigen is a soluble molecule. In both instances, when Ag and Ab are optimally combined so that neither is in excess, one antigen is interlinked by several antibodies to form an insoluble, three-dimensional aggregate so large that it cannot remain suspended and it settles out.

Agglutination Tests

Agglutination is discernible because the antibodies cross-link the antigens to form visible clumps. **Agglutination** tests are performed routinely by blood banks to determine ABO and Rh (Rhesus) blood types in preparation for transfusions. In this type of test, antisera containing antibodies against the blood group antigens on red blood cells are mixed with a small sample of blood and read for the presence or absence of clumping. The **Widal test** is an example of a tube agglutination test for diagnosing salmonelloses and undulant fever. In addition to detecting specific antibody, it also gives the serum titer.

Numerous variations of agglutination testing exist. The rapid plasma reagin (RPR) test is one of several tests commonly used to test for antibodies to syphilis. The cold agglutinin test, named for antibodies that react only at lower temperatures (4°C to 20°C), was developed to diagnose *Mycoplasma* pneumonia. The *Weil-Felix reaction* is an agglutination test sometimes used in diagnosing rickettsial infections.

In some tests, special agglutinogens have been prepared by affixing antigen to the surface of an inert particle. In **latex agglutination** tests, the inert particles are tiny latex beads. Kits using latex beads are available for assaying pregnancy hormone in the urine, identifying *Candida* yeasts and bacteria (staphylococci, streptococci, and gonococci), and diagnosing rheumatoid arthritis.

Precipitation Tests

In precipitation reactions, the soluble antigen is precipitated (made insoluble) by an antibody. This reaction is observable in a test tube

(a)

I.

Side view

I. In one method of setting up a double-diffusion test, wells are punctured in soft agar, and antibodies (Ab) and antigens (Ag) are added in a pattern. As the contents of the wells diffuse toward each other, a number of reactions can result, depending on whether antibodies meet and precipitate antigens.

II.

Control Ab

Test serum 1

Ag

Precipitation bands

Test serum 2

II. Example of test pattern and results. Antigen (Ag) is placed in the center well and antibody (Ab) samples are placed in outer wells. The control contains known Abs to the test Ag. Note bands that form where Ab/Ag meet. The other wells (1, 2) contain unknown test sera. One is positive and the other is negative. Double bands indicate more than one antigen and antibody that can react.

III.

III. Actual test results for detecting infection with the fungal pathogen *Histoplasma*. Numbers 1 and 4 are controls and 2, 3, 5, and 6 are patient test sera. Can you determine which patients have the infection and which do not?

Figure 17.10 Precipitation reactions.
(a) A tube precipitation test for streptococcal group antigens. Specific antiserum has been placed in the bottom of tubes and antigen solution carefully overlaid to form a zone of contact. The left-hand tube has developed a band of precipitate; the right-hand tube is negative. **(b)** Double diffusion (Ouchterlony) tests in a semisolid matrix.

(b)

in which an antigen sample has been carefully laid over an antiserum solution. At the point of contact, a cloudy or opaque zone forms.

One example of this technique is the **VDRL** (Venereal Disease Research Laboratory) test that also detects antibodies to syphilis. Although it is a good screening test, it contains a heterophilic antigen (cardiolipin) that may give rise to false positive results. Although precipitation is a useful detection tool, the precipitates are so easily disrupted in liquid media that most precipitation reactions are carried out in agar gels. These substrates are sufficiently soft to allow the reactants (Ab and Ag) to freely diffuse, yet firm enough to hold the Ab-Ag precipitate in place. One technique with applications in microbial identification and diagnosis of disease is the double diffusion (Ouchterlony) method **(figure 17.10).** It is called double diffusion because it involves diffusion of both antigens and antibodies. The test is performed by punching a pattern of small wells into an agar medium and filling them with test antigens and antibodies. A band forming between two wells indicates that antibodies from one well have met and reacted with

antigens from the other well. Variations on this technique provide a means of identifying unknown antibodies or antigens.

Immunoelectrophoresis constitutes yet another refinement of diffusion and precipitation in agar. With this method, a serum sample is first electrophoresed to separate the serum proteins, as described in chapter 15. Antibodies that react with specific serum proteins are placed in a trough parallel to the direction of migration, forming reaction arcs specific for each protein **(figure 17.11).** This test is widely used to detect disorders in the production of antibodies.

The Western Blot for Detecting Proteins

The **Western blot test** is somewhat similar to the previous tests because it involves the electrophoretic separation of proteins, followed by an immunoassay to detect these proteins. This test is a counterpart of the Southern blot test for identifying DNA, described in chapter 10. It is a highly specific and sensitive way to

INSIGHT 17.1

When Positive Is Negative: How to Interpret Serological Test Results

What if a patient's serum gives a positive reaction—is **seropositive**—in a serological test? In most situations, it means that antibodies specific for a particular microbe have been detected in the sample. But one must be cautious in proceeding to the next level of interpretation. The mere presence of antibodies does not necessarily indicate that the patient has a disease but only that he or she has possibly had contact with a microbe or its antigens through infection or vaccination. In screening tests for determining a patient's history (rubella, for instance), knowing that a certain titer of antibodies is present can be significant, because it shows that the person has some protection against that infectious agent.

When the test is being used to diagnose ongoing infection or disease, however, it is necessary to show a rising titer of antibodies. The accompanying figure indicates how such a test can be used to diagnose patients who have nonspecific symptoms that could fit several diseases. Lyme disease, for instance, can be mistaken for arthritis or viral infections. In the first group, note that the antibody titer against *Borrelia burgdorferi*, the causative agent, increased steadily over a 6-week period. A control group that shared similar symptoms did not exhibit a rise in titer for antibodies to this microbe. Clinicians call samples collected early and late in an infection *acute* and *convalescent* sera.

Another important consideration in testing is the occasional appearance of biological false positives. These are results in which a patient's serum shows a positive reaction even though, in reality, he is not or has not been infected by the microbe. False positives, such as those in syphilis

and HIV testing, arise when antibodies or other substances present in the serum cross-react with the test reagents, producing a positive result. Such false results may require retesting by a method that greatly minimizes cross-reactions.

What would be the case in a false negative reaction, and what might cause such a reaction? *Answer available at* http://www.mhhe.com/talaro8

Weeks from onset of symptoms

Step 1 Serum sample separated next to trough

(+) Direction of electrophoresis ←——————— (−)

Serum sample

Step 2 Antiserum added to trough

Albumin Alpha globulin IgM Gamma globulin
 (α globulin) (γ globulin)(IgG)

Figure 17.11 Immunoelectrophoresis of normal human serum. *Step 1.* Proteins are separated by electrophoresis on a gel. *Step 2.* To identify the bands and increase visibility, antiserum containing antibodies specific for serum proteins is placed in a trough and allowed to diffuse toward the bands. This diffusion produces a pattern of numerous arcs representing major serum components.

identify or verify a particular protein (antibody or antigen) in a sample **(figure 17.12).** First, the test material is electrophoresed in a gel to separate out particular bands. The gel is then transferred to a special blotter that binds the reactants in place. The blot is developed by incubating it with a solution of antigen or antibody that has been labeled with radioactive, fluorescent, or luminescent labels. Sites of specific binding will appear as a pattern of bands that can be compared with known positive and negative samples. This is usually the second test used to verify the status of people who are antibody-positive for HIV in the ELISA test (described in section 17.5), because it tests more types of antibodies and is less subject to misinterpretation than are other antibody tests. The technique has significant applications for detecting microbes and their antigens in specimens.

Complement Fixation

An antibody that requires complement to complete the lysis of its antigenic target cell is termed a **lysin** or cytolysin. When lysins act in conjunction with the intrinsic complement system on red blood cells, the cells hemolyze (lyse and release their hemoglobin). This lysin-mediated hemolysis is the basis of a group of tests called complement fixation, or CF **(figure 17.13**, page 531**).**

Complement fixation testing uses four components—antibody, antigen, complement, and sensitized sheep red blood cells—and it is conducted in two stages.

In the first stage, the test antigen is allowed to react with the test antibody (at least one must be of known identity) in the absence of complement. If the antibody and antigen are specific for each other, they form complexes. To this mixture, purified complement proteins from guinea pig blood are added. If the antibody and antigen have complexed during the previous step, they attach, or fix, the complement to them, thus preventing it from participating in further reactions.

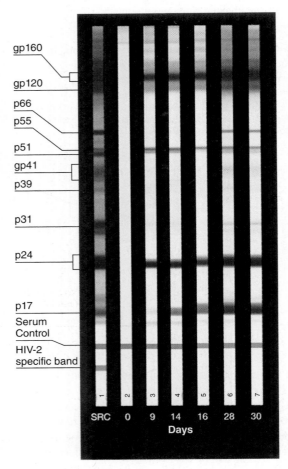

Successive tests on an HIV+ patient over 30 days reveals an increase in band intensity over time. This is due to continued formation of anti-HIV antibodies.

Figure 17.12 **The Western blot procedure.** Examples shown here test for antibodies to several HIV antigens. The test strips are prepared by electrophoresing the major HIV surface and core antigens and then blotting them onto special filters. The test strips are incubated with a serum sample and developed with a radioactive or colorimetric label. Sites where HIV antigens have bound antibodies show up as bands. A positive control strip (SRC) containing antibodies for all HIV antigens serves as a comparison.

Interpretation of bands to report a positive test
Labels correspond with glycoproteins (gp) or proteins (p) that are part of HIV-1 antigen structure.

• The test is considered positive if bands occur at two locations: gp160 or gp120 and p31 or p24.
• The test is considered negative if no bands are present for any HIV antigen.
• The test is considered indeterminate if bands are present but not at the primary locations. This result may require retesting at a later date.

The extent of any complement fixation is determined in the second stage by means of sheep red blood cells (RBCs) with surface lysin molecules. The sheep RBCs serve as an indicator complex that can also fix complement. Contents of the stage 1 tube are mixed with the stage 2 tube and observed with the naked eye for hemolysis.

• If hemolysis *does not* occur, it means that the complement was used up by the first stage Ab-Ag complex and that the unknown antigen or antibody was indeed present. This result is considered positive.
• If hemolysis *does* occur, it means that unfixed complement from tube 1 reacted with the RBC complex instead, thereby causing lysis of the sheep RBCs. This result is negative for the antigen or antibody that was the target of the test. Complement fixation tests are invaluable in diagnosing influenza, polio, and various fungal infections.

The antistreptolysin O (ASO) titer test measures the levels of antibody against the streptolysin toxin, an important hemolysin of group A streptococci. It employs a technique related to complement fixation. A serum sample is exposed to known suspensions of streptolysin and then allowed to incubate with RBCs. Lack of hemolysis indicates antistreptolysin antibodies in the patient's serum that have neutralized the streptolysin and prevented hemolysis. This is an important verification procedure for scarlet fever, rheumatic fever, and other related streptococcal syndromes (see chapter 18).

Miscellaneous Serological Tests

A test that relies on changes in cellular activity as seen microscopically is the ***Treponema pallidum* immobilization (TPI)** test for syphilis. The impairment or loss of motility of the *Treponema* spirochete in the presence of test serum and complement indicates that the serum contains anti-*Treponema pallidum* antibodies. In **toxin neutralization** tests, a test serum is incubated with the microbe that produces the toxin. If the serum inhibits the growth of the microbe, one can conclude that antitoxins are present.

Serotyping is an antigen-antibody technique for identifying, classifying, and subgrouping certain bacteria into categories called serotypes, using antisera for cell antigens such as the capsule, flagellum, and cell wall. It is widely used in typing *Salmonella* species and strains and is the basis for identifying the numerous serotypes of streptococci. The Quellung test, which identifies serotypes of the pneumococcus, involves a precipitation reaction in which antibodies react with the capsular polysaccharide. Although the reaction makes the capsule seem to swell, it is actually creating a zone of Ab-Ag complex on the cell's surface.

Fluorescent Antibody and Immunofluorescent Testing

The property of dyes such as fluorescein and rhodamine to emit visible light in response to ultraviolet radiation was discussed in chapter 3. This property of fluorescence has found numerous applications in diagnostic immunology. The fundamental tool in immunofluorescent testing is a fluorescent antibody—a monoclonal antibody labeled by a fluorescent dye (fluorochrome).

Stage 1		Reaction System	Stage 2

Figure 17.13 Complement fixation test. In this example, two serum samples are being tested for antibodies to a certain infectious agent. In reading this test, one observes the cloudiness of the tube. If it is cloudy, the RBCs are not hemolyzed and the test is positive. If it is clear and pink, the RBCs are hemolyzed and the test is negative.

The two ways that fluorescent antibodies (FABs) can be used are shown in **figure 17.14.** In **direct testing,** an unknown test specimen or antigen is fixed to a slide and exposed to a fluorescent antibody solution of known composition. If the antibodies are complementary to antigens in the material, they will bind to it. After the slide is rinsed to remove unattached antibodies, it is observed with the fluorescent microscope. Fluorescing cells or specks indicate the presence of Ab-Ag complexes and a positive result. These tests are valuable for identifying and locating antigens on the surfaces of cells or in tissues and in identifying the disease agents of syphilis, gonorrhea, chlamydiosis, whooping cough, Legionnaires' disease, plague, trichomoniasis, meningitis, and listeriosis.

In **indirect testing** methods, the fluorescent antibodies are antibodies made to react with the Fc region of another antibody (remember that antibodies can be antigenic). In this scheme, an antigen of known character (a bacterial cell, for example) is combined with a test serum of unknown antibody content. The fluorescent antibody solution that can react with the unknown antibody is applied and rinsed off to visualize whether the serum contains antibodies that have affixed to the antigen. A positive test shows fluorescing aggregates or cells, indicating that the fluorescent antibodies have combined with the unlabeled antibodies. In a negative test, no fluorescent complexes will appear. This technique is frequently used to diagnose syphilis and various viral infections.

Check & Assess Section 17.4

✔ Serological tests can test for either antigens or antibodies. Most are in vitro assessments of antigen-antibody reactivity from a variety of body fluids. The basis of these tests is an antigen-antibody reaction made visible through the processes of agglutination, precipitation, immunodiffusion, complement fixation, fluorescent antibody, and immunoassay techniques.

✔ One measurement is the *titer*, described as the concentration of antibody in serum. It is the highest dilution of serum that gives a visible antigen-antibody reaction. The higher the titer, the greater the level of antibody present.

✔ Agglutination reactions occur between antibody and antigens bound to cells. This results in visible clumps caused by large antibody-antigen complexes. In viral hemagglutination testing, the antibody reacts with the antigen and inhibits it from agglutinating red blood cells.

✔ In precipitation reactions, soluble antigen and antibody react to form insoluble, visible precipitates. Precipitation reactions can also be visualized by adding radioactive or enzyme markers to the antigen-antibody complex.

✔ In immunoelectrophoresis techniques such as the Western blot, proteins that have been separated by electrical current are identified by labeled antibodies. HIV infections are verified with this method.

Visible fluorescence

Fluorescent microscopy

Unknown antigen (usually cell or tissue)

Antibody labeled with fluorescent dye

(a) Direct Testing

Ab2 fluorescently labeled; specific for Ab1

Ab1 in serum

Known Ag

No Ab in serum

Ab2 attaches to Ab1—visible fluorescence

Positive

Ab2 cannot attach—no fluorescence

(b) Indirect Testing

Negative

(c) Indirect Immunofluorescence of specimen

Figure 17.14 **Immunofluorescent testing.** **(a)** Direct: Unidentified antigen (Ag) is directly tagged with fluorescent Ab. **(b)** Indirect: Ag of known identity is used to assay unknown Ab; a positive reaction occurs when the second Ab (with fluorescent dye) affixes to the first Ab. **(c)** An indirect immunofluorescent stain of cells infected with two different viruses. Cells with green fluorescing nuclei contain cytomegalovirus; cells with yellow fluorescing nuclei contain adenovirus.

✔ Complement fixation involves a two-part procedure in which complement fixes to a specific antibody, if present, or to red blood cell antigens, if antibody is absent. Lack of RBC hemolysis is indicative of a positive test.

✔ Serological tests can measure the degree to which host antibody binds directly to disease agents or toxins. This is the principle behind tests for syphilis and rheumatic fever.

✔ Direct fluorescent antibody tests indicate presence of an antigen and are useful in identifying infectious agents. Indirect fluorescent tests indicate the presence of a particular antibody and can diagnose infection.

8. What is the basis of serology and serological testing?
9. Differentiate between specificity and sensitivity.
10. Describe several general ways that Ab-Ag reactions are detected.
11. What do *seropositivity* and *seronegativity* mean?
12. What is a false positive test result and what are some possible causes?
13. Explain how agglutination and precipitation reactions are alike and how they are different.
14. Make a drawing of the manner in which antibodies cross-link the antigens in agglutination and precipitation reactions.
15. Give examples of several tests that employ agglutination and precipitation reactions.
16. What is meant by complement fixation? What are cytolysins?
17. Describe the principles behind direct and indirect fluorescent antibody tests.

17.5 Immunoassays: Tests of Great Sensitivity

Expected Learning Outcomes

14. Describe the concepts behind the main types of immunoassays, and discuss their uses in diagnosis.

15. Describe some immunological tests that differentiate T and B cells.

The elegant tools of the microbiologist and immunologist are being used increasingly in athletics, criminology, government, and business to test for trace amounts of substances such as hormones, metabolites, and drugs. But traditional techniques in serology are not refined enough to detect a few molecules of these chemicals. Extremely sensitive alternative methods that permit rapid and accurate measurement of trace antigen or antibody are called **immunoassays.** Examples of the technology for detecting an antigen or antibody in minute quantities include radioactive isotope labels, enzyme labels, and sensitive electronic sensors. Many of these tests are based on specifically formulated monoclonal antibodies.

Radioimmunoassay (RIA)

Antibodies or antigens labeled with a radioactive isotope can be used to pinpoint minute amounts of a corresponding antigen or antibody. Very complex in practice, these assays compare the amount of radioactivity present in a sample before and after incubation with a known,

labeled antigen or antibody. The labeled substance competes with its natural, nonlabeled partner for a reaction site. Large amounts of a bound radioactive component indicate that the unknown test substance was not present. The amount of radioactivity is measured with an isotope counter or a photographic emulsion (autoradiograph). Radioimmunoassay has been employed to measure the levels of insulin and other hormones and to diagnose allergies, chiefly by the radioimmunosorbent test (RIST) for measurement of IgE in allergic patients and the radioallergosorbent test (RAST) to standardize allergenic extracts.

Enzyme-Linked Immunosorbent Assay (ELISA)

The **ELISA,** also known as enzyme immunoassay (EIA), contains an enzyme-antibody complex that can be used as a color tracer for antigen-antibody reactions. The enzymes used most often are horseradish peroxidase and alkaline phosphatase, both of which release a dye (chromogen) when exposed to their substrate. This technique also relies on a solid support such as a plastic microtiter plate that can *adsorb* (attract on its surface) the reactants **(figure 17.15).**

(a) **Indirect ELISA,** comparing a positive vs. negative reaction. This is the basis for HIV screening tests.

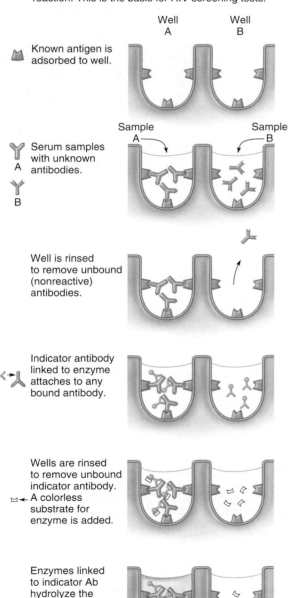

(b) **Microtiter ELISA Plate with 96 Tests for HIV Antibodies**
Colored wells indicate a positive reaction.

(c) **Capture or Antibody Sandwich ELISA Method.**
Note that an antigen is trapped between two antibodies. This test is used to detect the measles virus.

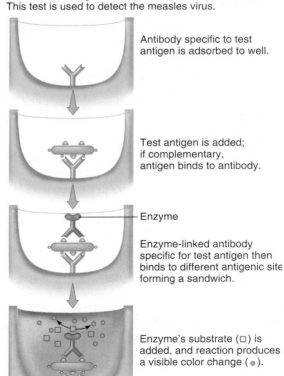

Figure 17.15 Methods of ELISA testing.

The **indirect ELISA** can capture antibodies in a serum sample. As with other indirect tests, the final positive reaction is achieved by means of an antibody-antibody reaction. The indicator antibody is complexed to an enzyme that produces a color change with positive serum samples (**figure 17.15a, b**). The starting reactant is a known antigen that is adsorbed to the surface of a well. To this, the unknown test serum is added. After rinsing, an enzyme-Ab reagent that can react with the unknown test antibody is placed in the well. The substrate for the enzyme is then added, and the wells are scanned for color changes. Color development indicates that all the components reacted and that the antibody was present in the patient's serum. This is the common screening test for the antibodies to HIV, various rickettsial species, hepatitis A and C, the cholera vibrio, and *Helicobacter,* a cause of gastric ulcers. Because false positives can occur, a verification test may be necessary (such as Western blot for HIV).

In **capture ELISA** (or sandwich) tests, a known antibody is adsorbed to the bottom of a well and incubated with a solution containing unknown antigen (**figure 17.15c**). After excess unbound components have been rinsed off, an enzyme-antibody indicator that can react with the antigen is added. If antigen is present, it will attract the indicator-antibody and hold it in place. Next, the substrate for the enzyme is placed in the wells and incubated. Enzymes affixed to the antigen will hydrolyze the substrate and release a colored dye. Thus, any color developing in the wells is a positive result. Lack of color means that the antigen was not present and that the subsequent rinsing removed the enzyme-antibody complex. The capture technique is used to detect antibodies to hantavirus, rubella virus, and *Toxoplasma.*

Tests That Differentiate T Cells and B Cells

So far, we have concentrated on tests that identify antigens and antibodies in samples, but techniques also exist that differentiate between B cells and T cells and can quantify subsets of each. Obtaining information on the types and numbers of lymphocytes in blood and other samples is a common way to evaluate immune dysfunctions such as those in AIDS, immunodeficiencies, and cancer. A simple method for identifying T cells is to mix them with untreated sheep red blood cells. Receptors on the T cells bind the RBCs into a flowerlike cluster called a **rosette formation (figure 17.16a)**. Rosetting can also occur in B cells if one uses Ig-coated bovine RBCs or mouse erythrocytes.

For routine identification of lymphocytes, rosette formation has been replaced by fluorescent techniques. These techniques can differentiate between T cells and B cells, as well as subgroup them (**figure 17.16b**). These subgroup tests utilize monoclonal antibodies produced in response to specific cell markers. B-cell tests categorize different stages in B-cell development and are very useful in characterizing B-cell cancers. Tests that can help differentiate the CD4, CD8, and other T-cell subsets are important in monitoring AIDS and other immunodeficiency diseases.

In Vivo Testing

Probably the first immunologic tests were performed not in a test tube but on the body itself. A classic example of one such technique

(a) (b)

Figure 17.16 Tests for characterizing T cells and B cells. (a) Photomicrograph of rosette formation that identifies T cells. **(b)** Plasma cells in the tonsils highlighted by fluorescent antibodies (green).

is the **tuberculin test,** which uses a small amount of purified protein derivative (PPD) from *Mycobacterium tuberculosis* injected into the skin. The appearance of a red, raised, thickened lesion in 48 to 72 hours can indicate previous exposure to tuberculosis (shown in figure 16.14). In practice, in vivo tests employ principles similar to serological tests, except in this case, an antigen or an antibody is introduced into a patient to elicit some sort of visible reaction. Like the tuberculin test, some of these diagnostic skin tests are useful for evaluating infections due to fungi (coccidioidin and histoplasmin tests, for example) or allergens. Allergic reactions and other immune system disorders were covered in chapter 16.

17.6 Viruses as a Special Diagnostic Case

Expected Learning Outcome

16. Summarize the protocols in identifying the causative agent in a viral infection.

All of the methods discussed so far—phenotypic, genotypic, and immunologic—are applicable to many different types of microorganisms. Viruses can present special difficulties because they are not cells and they are more labor intensive to culture in the laboratory. **Figure 17.17** presents an overview of various techniques available to diagnose viral infections. These range from observing symptoms and direct microscopic examination to cultivation, and serological and genetic analysis.

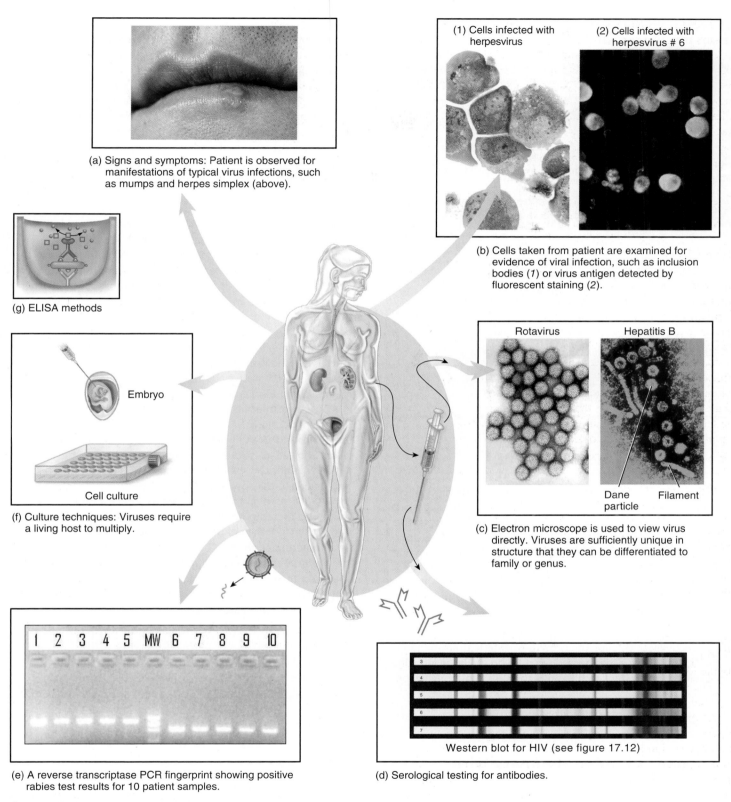

(a) Signs and symptoms: Patient is observed for manifestations of typical virus infections, such as mumps and herpes simplex (above).

(1) Cells infected with herpesvirus (2) Cells infected with herpesvirus # 6

(b) Cells taken from patient are examined for evidence of viral infection, such as inclusion bodies (1) or virus antigen detected by fluorescent staining (2).

(g) ELISA methods

Embryo

Cell culture

(f) Culture techniques: Viruses require a living host to multiply.

Rotavirus Hepatitis B

Dane Filament
particle

(c) Electron microscope is used to view virus directly. Viruses are sufficiently unique in structure that they can be differentiated to family or genus.

1 2 3 4 5 MW 6 7 8 9 10

(e) A reverse transcriptase PCR fingerprint showing positive rabies test results for 10 patient samples.

Western blot for HIV (see figure 17.12)

(d) Serological testing for antibodies.

Figure 17.17 Summary of methods used to diagnose viral infections.

Check & Assess Sections 17.5 and 17.6

✔ Immunoassays can detect very small quantities of antigen, antibody, or other substances. Radioimmunoassay uses radioisotopes to detect trace amounts of biological substances.

✔ The ELISA test uses enzymes and dyes to detect antigen-antibody complexes. It is widely used to detect viruses, bacteria, and antibodies in HIV infection.

✔ In vivo serological testing, such as the tuberculin test, involves subcutaneous injection of antigen to elicit a visible immune response in the host.

✔ Technicians use precise assays to differentiate between B and T cells and to identify subgroups of these cells for disease diagnosis.

18. Explain the differences between direct and indirect procedures in immunoassay tests, giving examples.

19. How is the reaction in a radioimmunoassay detected?

20. How does a positive reaction in an ELISA test appear? How many wells are positive in figure 17.15*b*?

21. Explain how a clinical technologist would have determined that the causative agent in this chapter's Case File was a virus instead of *Vibrio* bacteria.

22. Observing figure 17.17, indicate whether each method (a) through (g) is phenotypic, genotypic, or immunologic.

CASE FILE 17 PERSPECTIVE

This case provides some insight into the protocol behind an actual identification process. Preliminary laboratory investigations are designed to narrow the scope of possible infectious agents using as few tests as possible. The selective media and the second Gram stains informed the technologists relatively early on that the gram-positive *Bacillus cereus, Staphylococcus aureus,* and *Clostridium* were not involved in the outbreak, even though two of them are common hospital pathogens. Being able to eliminate them saved the lab from doing unnecessary specialized testing.

Bile is an agent added to media such as MacConkey agar that effectively selects for the growth of enteric gram-negative rods. The rationale is that bacteria that live in the large intestine are likely to be resistant to bile since they are constantly in contact with it in this environment. Another substance used to block the growth of gram-positive bacteria are dyes, which can be used in differentiation as well.

The oxidase test detects the enzyme cytochrome oxidase, which is part of the electron transport chain that adds oxygen to hydrogen ions and forms water. Because many bacteria lack this enzyme, it can be helpful in narrowing the choices. Fermentation tests detect the breakdown of a carbohydrate in the absence of oxygen with the formation of acid, gas, and other products such as alcohols. Taking into account the preliminary results, the most likely genera responsible for the outbreak were *Campylobacter* and *Vibrio,* which share several traits. But its lack of spiral cells and having fermentative and aerobic metabolism all point to *Vibrio* as the causative genus.

The last series of biochemical tests confirmed that the pathogen isolated from patient specimens and food was *Vibrio parahaemolyticus.* The growth requirement for salt, the fermentation of arabinose, and being ONPG negative were key tests in this determination. Although all three *Vibrio* species live in the ocean and are associated with raw or insufficiently cooked seafood, *V. parahaemolyticus* is the leading cause of enteritis associated with shellfish consumption worldwide. Its relative *V. cholerae* is the causative agent of cholera, and *V. vulnificus* causes enteritis, wound infection, and septicemia.

The most specific evidence that verified the final identification were serological tests using slide agglutination (see figure 17.7*b*) and DNA fingerprinting of restriction fragments. The results showed all isolates were serotype K6, which is the capsular antigen specific to *V. parahaemolyticus.* A PCR test produced a pattern unique to this species (see figure 17.6).

It was essential to identify to species because the types of drugs and therapy can vary with the type of *Vibrio.* It is also important to track the epidemiology of the infection. Investigators never determined the exact source of the infection, since *V. parahaemolyticus* was isolated from more than one type of seafood. It appeared that shrimp or other seafood used to prepare meals may not have been adequately refrigerated, allowing the pathogen to grow to infectious levels.

For more information on *Vibrio* detection, go to http://www.fda.gov/Food/ScienceResearch/LaboratoryMethods/BacteriologicalAnalytical ManualBAM/UCM070830

Additional coverage of *Vibrio* diseases appears in chapter 21.

Chapter Summary with Key Terms

17.1 An Overview of Clinical Microbiology

A. **Phenotypic** methods are those that assess microscopic morphology, macroscopic morphology, physiological and biochemical characteristics, and chemical composition.

B. **Genotypic** methods examine the genetic composition of a microorganism.

C. **Immunologic** methods exploit the host's antibody reaction to microbial antigens for purposes of diagnosis.

D. Diagnosis begins with accurate **specimen collection,** which means the sampling of body sites or fluids that are suspected to contain the infective agent.

E. Laboratory methods yield either presumptive data or confirmatory data. Sometimes pathogens are diagnosed based solely on signs and symptoms in the patient.

17.2 Phenotypic Methods

A. The most obvious phenotypic characteristic of a microbe is what it looks like. Many infections can be diagnosed by microscopic examination of fresh or stained microorganisms from specimens. Direct antigen or antibody testing can also be performed on fresh specimens.

B. Cultivation of specimens allows the examination of colony morphology, growth characteristics, or both, which are useful for identifying microbes.

C. Biochemical tests determine whether microbes possess particular enzymes or biochemical pathways that can serve to identify them.

D. A variety of other tests are available to test specific phenotypic characteristics of particular suspected microorganisms.

E. It is of utmost importance to consider whether an identified microbe is actually causing the disease or is simply a bystander, or member of the normal microbiota.

17.3 Genotypic Methods

Genetic methods of identification are being used increasingly in diagnostic microbiology.

A. Probes can detect specific genetic patterns, and advanced technology can characterize unique DNA sequences.

B. An increasingly useful procedure is the use of *polymerase chain reaction* to amplify DNA present in a sample in order to detect small quantities of microbial sequences.

C. RNA in ribosomes and other isolated cultures can be converted to DNA and analyzed.

17.4 Immunologic Methods

A. **Serology** is a science that attempts to detect signs of infection in a patient's serum such as antibodies specific for a microbe.

B. The basis of serological tests is that Abs specifically bind to Ag in vitro. An Ag of known identity will react with antibodies in an unknown serum sample. The reverse is also true; known antibodies can be used to detect and type antigens.

C. These Ab-Ag reactions are visible in the form of obvious clumps and precipitates, color changes, or the release of radioactivity. Test results are read as positive or negative.

D. Desirable properties of tests are high specificity and sensitivity.

E. Types of Tests

1. In **agglutination** tests, antibody cross-links whole-cell antigens, forming complexes that settle out and form visible clumps in the test chamber.
 a. Examples are tests for blood type, some bacterial diseases, and viral diseases.
 b. These tests can determine the amount of antibody in serum, or **titer.**
2. Double diffusion precipitation tests involve the diffusion of Ags and Abs in a soft agar gel, forming zones of precipitation where they meet.
3. In immunoelectrophoresis, migration of serum proteins in gel is combined with precipitation by antibodies.
4. The **Western blot** test separates antigen into bands. After the gel is affixed to a blotter, it is reacted with a test specimen and developed by radioactivity or with dyes.
5. A **complement fixation** test detects **lysins**—antibodies that fix complement and can lyse target cells. It involves first mixing test Ag and Ab with complement and then with sensitized sheep RBCs. If the complement is fixed by the Ab-Ag, the RBCs remain intact, and the test is positive. If RBCs are hemolyzed, specific antibodies are lacking.

6. In direct assays, known marked Ab is used to detect unknown Ag (microbe).
 a. In indirect testing, known Ag reacts with unknown Ab, and the reaction is made visible by a second Ab that can affix to and identify the unknown Ab.
 b. Immunofluorescent testing uses fluorescent antibodies (FABs tagged with fluorescent dye) either directly or indirectly to visualize cells or cell aggregates that have reacted with the FABs.

17.5 Immunoassays: Tests of Great Sensitivity

Immunoassays are highly sensitive tests for Ag and Ab.

A. In *radioimmunoassay*, Ags or Abs are labeled with radioactive isotopes and traced.

B. The **enzyme-linked immunosorbent assay (ELISA)** can detect unknown Ag or Ab by direct or indirect means. A positive result is visualized when a colored product is released by an enzyme-substrate reaction.

C. Tests are also available to differentiate B cells from T cells and their subtypes.

D. With in vivo testing, Ags are introduced into the body directly to determine the patient's immunologic history.

7.6 Viruses as a Special Diagnostic Case

A. Viral diseases are diagnosed through patient observation supported by microscopic, cultural, serological, and genetic tests.

Multiple-Choice Questions

Select the correct answer from the answers provided. For questions with blanks, choose the combination of answers that most accurately completes the statement.

1. **Multiple Matching.** Evaluate and match each of the following culture results to the most likely interpretation.
 a. probable infection
 b. normal microbiota
 c. environmental contaminant

 _____ 1. Isolation of two colonies of *E. coli* on a plate streaked from the urine sample
 _____ 2. Fifty colonies of *Streptococcus pneumoniae* on culture of sputum
 _____ 3. A mixture of 80 colonies of various streptococci on a culture from a throat swab
 _____ 4. Colonies of black bread mold on selective media used to isolate bacteria from stool
 _____ 5. Blood culture bottle with heavy growth

2. Which of the following methods is most sensitive for identifying different strains of a microbe?
 a. microscopic examination
 b. radioimmunoassay
 c. DNA typing
 d. agglutination test

3. In agglutination reactions, the antigen is a _____; in precipitation reactions, it is a _____.
 a. soluble molecule, whole cell
 b. whole cell, soluble molecule
 c. bacterium, virus
 d. protein, carbohydrate

4. Which reaction requires complement?
 a. hemagglutination
 b. precipitation
 c. hemolysis
 d. toxin neutralization

5. A patient with a _____ titer of antibodies to an infectious agent generally has greater protection than a patient with a _____ titer.
 a. high, low
 c. negative, positive
 b. low, high
 d. old, new

6. Direct immunofluorescent tests use a labeled antibody to identify.
 a. an unknown microbe
 c. fixed complement
 b. an unknown antibody
 d. agglutinated antigens

7. The Western blot test can be used to identify
 a. unknown antibodies
 c. specific DNA
 b. unknown antigens
 d. both a and b

8. An example of an in vivo serological test is
 a. indirect immunofluorescence
 b. radioimmunoassay
 c. tuberculin test
 d. complement fixation

9. Which of the following specimens must be removed by sterile techniques?
 a. feces
 c. upper respiratory tract
 b. urine
 d. blood

Case File Questions

1. On what basis were both *Shigella* and *Salmonella* eliminated as possible causes of the nosocomial pathogen?
 a. Gram reaction
 c. oxidase results
 b. fermentation of glucose
 d. motility

2. Which *Vibrio* test results does *V. parahaemolyticus* most resemble?
 a. *V. cholerae*, salt tolerance
 c. *V. cholerae*, sucrose fermentation
 b. *V. vulnificus*, salt tolerance
 d. *V. vulnificus*, ONPG

Writing to Learn

These questions are suggested as a *writing-to-learn* experience. For each question, compose a one- or two-paragraph answer that includes the factual information needed to completely address the question. All Check & Assess questions can be used for writing-to-learn exercises.

1. Why do we interpret positive hemolysis in the complement fixation test to mean negative for the test substance?

2. What does the antibody titer tell us about the immune status of a person?

3. Briefly describe the principles and give an example of the use of a specific test using immunoelectrophoresis, Western blot, complement fixation, fluorescent testing (direct and indirect), and immunoassays (direct and indirect ELISA).

4. Differentiate between the serological tests used to identify isolated cultures of pathogens and those used to diagnose disease from patients' serum.

5. How would the staining and cultural tests have differed if *Staphylococcus* or *Clostridium* had been the cause of the outbreak of gastroenteritis described in Case File 17?

6. What is another form of genetic testing from this chapter that could differentiate the *Vibrio* pathogens?

7. What would be needed to perform a serological test that identifies *Vibrio*?

Concept Mapping

Appendix E provides guidance for working with concept maps.

1. Supply your own linking words or phrases in this concept map, and provide the missing concepts in the empty boxes.

Critical Thinking Questions

1. How would you explain to a biology class that in the next decade, some diseases currently thought to be noninfectious will probably be found to be caused by microbes?

2. In what way could the extreme sensitivity of the PCR method be a problem when working with clinical specimens?

3. Why do some tests for antibody in serum (such as for HIV and syphilis) require backup verification with additional tests at a later date?

4. a. Look at figure 17.9*b*. What is the titer as shown?
 b. If the titer had been 40, would a different interpretation be made as to the immune status of the patient?
 c. Which patients are positive in figure 17.10*b*?
 d. What would it mean if a test 2 weeks later revealed a titer of 280?
 e. What would it mean if no agglutination had occurred in any tube?

5. Observe figure 17.15 and make note of the several steps in the indirect ELISA test. What four essential events are necessary to develop a positive reaction (besides having antibody A)? Hint: What would happen without rinses?

6. Explain how an immunoassay method could use monoclonal antibodies to differentiate between B and T cells and between different subsets of T cells.

Visual Challenge

1. From **chapter 20, figure 20.11*a*.** What biochemical characteristic does this figure illustrate? How could this characteristic be used to begin the identification of the isolated bacteria in this chapter's Case File?

Lactose (+) Lactose (−)

2. Starting with figure 17.5 and using data from the Case File tables 1 and 2, complete a flowchart that outlines the steps in identifying the species in Case File 17.

Exponents

Dealing with concepts such as microbial growth often requires working with numbers in the billions, trillions, and even greater. A mathematical shorthand for expressing such numbers is with exponents. The exponent of a number indicates how many times (designated by a superscript) that number is multiplied by itself. These exponents are also called common *logarithms,* or logs. The following chart, based on multiples of 10, summarizes this system.

Converting Numbers to Exponent Form

As **Table A.1** shows, using exponents to express numbers can be very economical. When simple multiples of 10 are used, the exponent is always equal to the number of zeros that follow the 1, but this rule will not work with numbers that are more varied. Other large whole numbers can be converted to exponent form by the following operation: First, move the decimal (which we assume to be at the end of the number) to the left until it sits just behind the first number in the series (example: $3568. = 3.568$). Then count the number of spaces (digits) the decimal has moved; that number will be the exponent. (The decimal has moved from 8. to 3., or 3 spaces.) In final notation, the converted number is multiplied by 10 with its appropriate exponent: 3568 is now 3.568×10^3. **Table A.2** represents an alternate logarithm system, based on multiples of 2. It is useful in calculating population growth and size.

Rounding Off Numbers

The notation in the previous example has not actually been shortened, but it can be reduced further by rounding off the decimal fraction to the nearest thousandth (three digits), hundredth (two digits), or tenth (one digit). To round off a number, drop its last digit and either increase the one next to it or leave it as it is. If the number dropped is 5, 6, 7, 8, or 9, the subsequent digit is increased by one (rounded up); if it is 0, 1, 2, 3, or 4, the sub-sequent digit remains as is. Using the example of 3.528, removing the 8 rounds off the 2 to a 3 and produces 3.53. If further rounding is desired, the same rule of thumb applies, and the number becomes 3.5. Other examples of exponential conversions are shown below.

Number	Is the Same As	Rounded Off, Placed in Exponent Form
16,825.	$1.6825 \times 10 \times 10 \times 10 \times 10$	1.7×10^4
957,654.	$9.57654 \times 10 \times 10 \times 10 \times 10 \times 10$	9.58×10^5
2,855,000.	$2.855000 \times 10 \times 10 \times 10 \times 10 \times 10 \times 10$	2.86×10^6

TABLE A.1	Exponential Notation for Base 10				
Number	**Quantity**	**Exponential Notation***	**Number Arrived at By:**	**One Followed By:**	
1	One	10^0	Numbers raised to zero power are equal to one	No zeros	
10	Ten	10^{1}**	10×1	One zero	
100	Hundred	10^2	10×10	Two zeros	
1,000	Thousand	10^3	$10 \times 10 \times 10$	Three zeros	
10,000	Ten thousand	10^4	$10 \times 10 \times 10 \times 10$	Four zeros	
100,000	Hundred thousand	10^5	$10 \times 10 \times 10 \times 10 \times 10$	Five zeros	
1,000,000	Million	10^6	10 times itself 6 times	Six zeros	
1,000,000,000	Billion	10^9	10 times itself 9 times	Nine zeros	
1,000,000,000,000	Trillion	10^{12}	10 times itself 12 times	Twelve zeros	
1,000,000,000,000,000	Quadrillion	10^{15}	10 times itself 15 times	Fifteen zeros	
1,000,000,000,000,000,000	Quintillion	10^{18}	10 times itself 18 times	Eighteen zeros	

Other large numbers are sextillion (10^{21}), septillion (10^{24}), and octillion (10^{27}).

*The proper way to say the numbers in this column is 10 raised to the *nth* power, where *n* is the exponent. The numbers in this column can also be represented as 1×10^n, but for brevity, the $1 \times$ can be omitted.

**The exponent 1 is usually omitted in numbers at this level.

Negative Exponents

The numbers we have been using so far are greater than 1 and are represented by positive exponents. But the correct notation for numbers less than 1 involves negative exponents (10 raised to a negative power, or 10^{-n}). A negative exponent says that the number has been divided by a certain power of 10 (10, 100, 1,000). This usage is handy when working with concepts such as pH that are based on very small numbers otherwise needing to be represented by large decimal fractions—for example, 0.003528. Converting this and other such numbers to exponential notation is basically similar to converting positive numbers, except that you work from left to right and the exponent is negative. Using the example of 0.003528, first convert the number to a whole integer followed by a decimal fraction and keep track of the number of spaces the decimal point moves (example: 0.003528 = 3.528). The decimal has moved three spaces from its original position, so the finished product is 3.528×10^{-3}. Other examples are:

Number	Is the Same As	Rounded Off, Expressed with Exponents
0.0005923	$\dfrac{5.923}{10 \times 10 \times 10 \times 10}$	5.92×10^{-4}
0.00007295	$\dfrac{7.295}{10 \times 10 \times 10 \times 10 \times 10}$	7.3×10^{-5}

TABLE A.2	Logarithmic Table for Base 2

Use this to calculate population growth as described on page •••.

Powers of 2 from 0 through 20	
Log 2^n	Number
2^0 =	1
2^1 =	2
2^2 =	4
2^3 =	8
2^4 =	16
2^5 =	32
2^6 =	64
2^7 =	128
2^8 =	256
2^9 =	512
2^{10} =	1024
2^{11} =	2048
2^{12} =	4096
2^{13} =	8192
2^{14} =	16384
2^{15} =	32768
2^{16} =	65536
2^{17} =	131072
2^{18} =	262144
2^{19} =	524288
2^{20} =	1048576

TABLE A.3	Significant Events in Microbiology		
Date	**Discovery/People Involved**	**Date**	**Discovery/People Involved**
1546	Italian physician Girolamo Fracastoro suggests that invisible organisms may be involved in disease.	1884	Koch outlines his postulates. Elie Metchnikoff,* a Russian zoologist, lays groundwork for the science of immunology by discovering phagocytic cells. The Danish physician Hans Christian Gram devises the Gram stain technique for differentiating bacteria.
1590	Zaccharias Janssen, a Dutch spectacle maker, invents the first compound microscope.		
1660	Englishman Robert Hooke explores various living and nonliving matter with a compound microscope that uses reflected light.	1885	Pasteur develops a special vaccine for rabies.
		1887	Julius Petri, a German bacteriologist, adapts two plates to form a container for holding media and culturing microbes.
1668	Francesco Redi, an Italian naturalist, conducts experiments that demonstrate the fallacies in the spontaneous generation theory.	1890	A German, Emil von Behring,* and a Japanese, Shibasaburo Kitasato, demonstrate the presence of antibodies in serum that neutralize the toxins of diphtheria and tetanus.
1676	Antonie van Leeuwenhoek, a Dutch linen merchant, uses a simple microscope of his own design to observe bacteria and protozoa.		
1776	An Italian anatomist, Lazzaro Spallanzani, conducts further convincing experiments that dispute spontaneous generation.	1892	A Russian, D. Ivanovski, is the first to isolate a virus (the tobacco mosaic virus) and show that it could be transmitted in a cell-free filtrate.
1796	English surgeon Edward Jenner introduces a vaccination for smallpox.	1895	Jules Bordet,* a Belgian bacteriologist, discovers the antimicrobial powers of complement.
1838	Phillipe Ricord, a French physician, inoculates 2,500 human subjects to demonstrate that syphilis and gonorrhea are two separate diseases.	1898	R. Ross* and G. Grassi demonstrate that malaria is transmitted by the bite of female mosquitoes. Germans Friedrich Loeffler and P. Frosch discover that "filterable viruses" cause foot-and-mouth disease in animals.
1839	Theodor Schwann, a German zoologist, and Matthias Schleiden, a botanist, formalize the theory that all living things are composed of cells.		
1847–1850	The Hungarian physician Ignaz Semmelweis substantiates his theory that childbed fever is a contagious disease transmitted to women by their physicians during childbirth.	1899	Dutch microbiologist Martinus Beijerinck further elucidates the viral agent of tobacco mosaic disease and postulates that viruses have many of the properties of living cells and that they reproduce within cells.
1853–1854	John Snow, a London physician, demonstrates the epidemic spread of cholera through a water supply contaminated with human sewage.	1900	The American physician Walter Reed and his colleagues clarify the role of mosquitoes in transmitting yellow fever. An Austrian pathologist, Karl Landsteiner,* discovers the ABO blood groups.
1857	French bacteriologist Louis Pasteur shows that fermentations are due to microorganisms and originates the process now known as pasteurization.		
1858	Rudolf Virchow, a German pathologist, introduces the concept that all cells originate from preexisting cells.	1903	American pathologist James Wright and others demonstrate the presence of antibodies in the blood of immunized animals.
1861	Louis Pasteur completes the definitive experiments that finally lay to rest the theory of spontaneous generation.	1905	Syphilis is shown to be caused by *Treponema pallidum,* through the work of German bacteriologists Fritz Schaudinn and E. Hoffman.
1867	The English surgeon Joseph Lister publishes the first work on antiseptic surgery, beginning the trend toward modern aseptic techniques in medicine.	1906	August Wasserman, a German bacteriologist, develops the first serologic test for syphilis. Howard Ricketts, an American pathologist, links the transmission of Rocky Mountain spotted fever to ticks.
1869	Johann Miescher, a Swiss pathologist, discovers in the cell nucleus the presence of complex acids, which he terms nuclein (DNA, RNA).		
1876–1877	German bacteriologist Robert Koch* studies anthrax in cattle and implicates the bacterium *Bacillus anthracis* as its causative agent.	1908	The German Paul Ehrlich* becomes the pioneer of modern chemotherapy by developing salvarsan to treat syphilis.
1881	Pasteur develops a vaccine for anthrax in animals. Koch introduces the use of pure culture techniques for handling bacteria in the laboratory. Walther and Fanny Hesse introduce agar as a solidifying gel for culture media.	1910	An American pathologist, Francis Rous,* discovers viruses that can induce cancer.
		1915–1917	British scientist F. Twort and French scientist F. D'Herelle independently discover bacterial viruses.
1882	Koch identifies the causative agent of tuberculosis.	1928	Frederick Griffith lays the foundation for modern molecular genetics by his discovery of transformation in bacteria.

continued

Date	Discovery/People Involved	Date	Discovery/People Involved
1929	A Scottish bacteriologist, Alexander Fleming,* discovers and describes the properties of the first antibiotic, penicillin.	1983	Isolation and characterization of human immunodeficiency virus (HIV) by Luc Montagnier of France and Robert Gallo of the United States.
1933–1938	Germans Ernst Ruska* and B. von Borries develop the first electron microscope.		The polymerase chain reaction is invented by Kerry Mullis.*
1935	Gerhard Domagk,* a German physician, discovers the first sulfa drug and paves the way for the era of antimicrobic chemotherapy.	1987	The molecular genetics of antibody genes is worked out by Susumu Tonegawa.*
	Wendell Stanley* is successful in inducing tobacco mosaic viruses to form crystals that still retain their infectiousness.		First release of recombinant strain of *Pseudomonas* to prevent frost formation on strawberry plants.
1941	Australian Howard Florey* and Englishman Ernst Chain* develop commercial methods for producing penicillin; this first antibiotic is tested and put into widespread use.	1989	Cancer-causing genes called oncogenes are characterized by J. Michael Bishop, Robert Huber, Hartmut Michel, and Harold Varmus.
1944	Oswald Avery, Colin MacLeod, and Maclyn McCarty show that DNA is the genetic material.	1990	First clinical trials in gene therapy testing.
			Vaccine for *Haemophilus influenzae,* a cause of meningitis, is introduced.
	Joshua Lederberg* and E. L. Tatum* discover conjugation in bacteria.	1991	Development of transgenic animals to synthesize human hemoglobin.
	The Russian Selman Waksman* and his colleagues discover the antibiotic streptomycin.	1994	Human breast cancer gene isolated.
1953	James Watson,* Francis Crick,* Rosalind Franklin, and Maurice Wilkins* determine the structure of DNA.	1995	First bacterial genome fully sequenced, for *Haemophilus influenzae.*
1954	Jonas Salk develops the first polio vaccine.	1996	Dr. David Ho develops a "cocktail" of drugs to treat AIDS.
1957	Alick Isaacs and Jean Lindenmann discover the natural antiviral substance interferon.	1997	Medical researchers at Case Western University construct human artificial chromosomes (HACs).
	D. Carleton Gajdusek* postulates the involvement of "slow viruses" in infections.		Scottish researchers clone first mammal (a sheep) from adult nuclei.
1959–1960	Gerald Edelman* and Rodney Porter* determine the structure of antibodies.	1999	Heat-loving bacteria discovered 2 miles beneath earth in African gold mine.
1972	Paul Berg* develops the first recombinant DNA in a test tube.	2000	A rough version of the human genome is mapped.
			250-million-year-old bacterium unearthed by Pennsylvania team.
1973	Herb Boyer and Stanley Cohen clone the first DNA using plasmids.	2001	Mailed anthrax spores cause major bioterrorism event.
1975	A technique for making monoclonal antibodies is developed by Cesar Milstein, Georges Kohler, and Niels Kai Jerne.	2002	New York virologists create polio virus in a test tube.
		2003	New roles for small nuclear RNAs discovered.
1979	Genetically engineered insulin is first synthesized by bacteria.	2004	Human papillomavirus (HPV) vaccine is found effective in controlling cervical cancer.
1982	Development of first hepatitis B vaccine using virus isolated from human blood.	2005	Deep ocean bacteria are found to undergo photosynthesis using light from chemical reactions.
	Barry Marshall* and J. Robin Warren* discover *Helicobacter pylori* and demonstrate its role in gastric disease.	2007	Construction of revolutionary superlenses that provide very high magnification of live cells and viruses using light waves.
	Stanley Prusiner* discovers prions and their involvement in transmissible spongiform encephalopathies.		Marketing of first screening tests that detect risks for genetic diseases.
		2009	Italian medical researchers successfully cured severe combined immunodeficiency disease using engineered viruses to deliver normal genes.

*These scientists were awarded the Nobel Prize for their contributions to the field.

Detailed Steps in the Glycolysis Pathway

Steps in the Glycolysis Pathway The first portion of glycolysis involves activation of glucose, which is followed by oxidation reactions of the glucose fragments, the synthesis of ATP, and the formation of pyruvic acid. *Although each step of metabolism is catalyzed by a specific enzyme, we will not mention it for most reactions.* The following outline lists the principal steps of glycolysis. To see an overview of glycolysis, refer to **figure 8.18,** page 234.

1. **Glucose** is phosphorylated by means of an **ATP** acting with the enzyme hexokinase. The product is glucose-6 phosphate. (Numbers in chemical names refer to the position of the phosphate on the carbon skeleton.) This is a way of "priming" the system and keeping the glucose inside the cell.

2. Glucose-6-phosphate is converted to its isomer, fructose-6-phosphate, by phosphoglucoisomerase.

3. Another ATP is spent in phosphorylating the first carbon of fructose-6-phosphate, which yields fructose-1,6-diphosphate.

Up to this point, no energy has been released, no oxidation-reduction has occurred, and, in fact, 2 ATPs have been used. In addition, the molecules remain in the 6-carbon state.

4. Now doubly activated, fructose-1,6-diphosphate is split into two 3-carbon fragments: glyceraldehyde-3-phosphate (G-3-P) and dihydroxyacetone phosphate (DHAP). These molecules are isomers, and DHAP is enzymatically converted to G-3-P, which is the more reactive form for subsequent reactions.

The effect of the splitting of fructose diphosphate is to double every subsequent reaction, because where there was once a single molecule, there are now two to be fed into the remaining pathways.

5. Each molecule of glyceraldehyde-3-phosphate becomes involved in the single oxidation-reduction reaction of glycolysis, a reaction that sets the scene for ATP synthesis. Two reactions occur simultaneously and are catalyzed by the same enzyme, called glyceraldehyde-3-phosphate dehydrogenase. The coenzyme **NAD**$^+$ picks up hydrogen from G-3-P, forming **NADH.** This step is accompanied by the addition of an inorganic phosphate (PO_4^{3-}) to form an unstable bond on the third carbon of the G-3-P substrate. The product of these reactions is diphosphoglyceric acid.

In aerobic organisms, the NADH formed during step 5 will undergo further reactions in the electron transport system, where the final H acceptor will be oxygen, and an additional 3 ATPs will be generated per NADH. In organisms that ferment glucose anaerobically, the NADH will be oxidized back to NAD^+, and the hydrogen acceptor will be an organic compound.

6. One of the high-energy phosphates of diphosphoglyceric acid is donated to ADP via substrate-level phosphorylation, resulting in a molecule of ATP. The product of this reaction is 3-phosphoglyceric acid.

9. In the final reaction of glycolysis, phosphoenolpyruvic acid gives up its high-energy phosphate to form a second ATP via substrate-level phosphorylation. This reaction, catalyzed by pyruvate kinase, also produces pyruvic acid (pyruvate), a compound with many roles in metabolism.

7, 8. During this phase, a substrate for the synthesis of a second ATP is made in two substeps. First, 3-phosphoglyceric acid is converted to 2-phosphoglyceric acid through the shift of a phosphate from the third to the second carbon. Then, the removal of a water molecule from 2-phosphoglyceric acid produces phosphoenolpyruvic acid, which is charged to release a high-energy phosphate.

The 2 ATPs formed during steps 6 and 9 are examples of substrate-level phosphorylation, in that the high-energy phosphate is transferred directly from a substrate to ADP. These reactions are catalyzed by kinases. Because both molecules that arose in step 4 undergo these reactions, an overall total of **4** ATPs is generated in the partial oxidation of a glucose to 2 pyruvic acids. However, 2 ATPs were expended for steps 1 and 3, so the net number of ATPs available to the cell from these reactions is **2**.

Tests and Guidelines

Methods for Testing Sterilization and Germicidal Processes

Most procedures used for microbial control in the clinical setting are either physical methods (heat, radiation) or chemical methods (disinfectants, antiseptics). These procedures are so crucial to the well-being of patients and staff that their effectiveness must be monitored in a consistent and standardized manner. Particular concerns include the time required for the process, the concentration or intensity of the antimicrobic agent being used, and the nature of the materials being treated. The effectiveness of an agent is frequently established through controlled microbiological analysis. In these tests, a biological indicator (a highly resistant microbe) is exposed to the agent, and its viability is checked. If this known test microbe is destroyed by the treatment, it is assumed that the less resistant microbes have also been destroyed. Growth of the test organism indicates that the sterilization protocol has failed. The general categories of testing include:

Heat The usual form of heat used in microbial control is steam under pressure in an autoclave. Autoclaving materials at a high temperature (121°C) for a sufficient time (15–40 minutes) destroys most bacterial endospores. To monitor the quality of any given industrial or clinical autoclave run, technicians insert a dried test strip of *Bacillus stearothermophilus,* a spore-forming bacterium with extreme heat resistance. After autoclaving, the ampule is incubated at 56°C (the optimal temperature for this species) and checked for growth.

Radiation The effectiveness of ionizing radiation in sterilization is determined by placing special strips of dried spores of *Bacillus sphaericus* (a common soil bacterium with extreme resistance to radiation) into a packet of irradiated material.

Filtration The performance of membrane filters used to sterilize liquids may be monitored by adding *Pseudomonas diminuta* to the liquid. Because this species is a tiny bacterium that can escape through the filter if the pore size is not small enough, it is a good indicator of filtrate sterility.

Gas sterilization Ethylene oxide gas, one of the few chemical sterilizing agents, is used on a variety of heat-sensitive medical and laboratory supplies. The most reliable indicator of a successful sterilizing cycle is the sporeformer *Bacillus subtilis,* variety *niger.*

Germicidal assays United States hospitals and clinics commonly employ more than 250 products to control microbes in the environment, on inanimate objects, and on patients. Controlled testing of these products' effectiveness is not usually performed in the clinic itself, but by the chemical or pharmaceutical manufacturer. Twenty-five standardized *in vitro* tests are currently available to assess the effects of germicides. Most of them use a non-sporeforming pathogen as the biological indicator.

The Phenol Coefficient (PC)

The older disinfectant phenol has been the traditional standard by which other disinfectants are measured. In the PC test, a water-soluble phenol-based (phenolic) disinfectant (amphyl, lysol) is tested for its bactericidal effectiveness as compared with phenol. These disinfectants are serially diluted in a series of *Salmonella choleraesuis, Staphylococcus aureus,* or *Pseudomonas aeruginosa* cultures. The tubes are left for 5, 10, or 15 minutes and then subcultured to test for viability. The PC is a ratio derived by comparing the following data:

$$PC = \frac{\text{Greatest dilution of the phenolic that kills test bacteria in 10 min but not in 5 min}}{\text{Greatest dilution of phenol giving same result}}$$

The general interpretation of this ratio is that chemicals with lower phenol coefficients have greater effectiveness. The primary disadvantage of the PC test is that, because it is restricted to phenolics, it is inappropriate for the vast majority of clinical disinfectants.

Use Dilution Test

An alternative test with broader applications is the use dilution test. This test is performed by drying the test culture (one of those species used in the PC test) onto the surface of small stainless steel cylindrical carriers. These carriers are then exposed to varying concentrations of disinfectant for 10 minutes, removed, rinsed, and placed in tubes of broth. After incubation, the tubes are inspected for growth. The smallest concentration of disinfectant that kills the test organism on 10 carrier pieces is the correct dilution for use.

Filter Paper Disc Method

A quick measure of the inhibitory effects of various disinfectants and antiseptics can be achieved by the filter paper disc method. First, a small (½-inch) sterile piece of filter paper is dipped into a disinfectant of known concentration. This is placed on an agar

Figure C.1 A filter paper disc containing hydrogen peroxide produces a broad clear zone of inhibition in a culture of *staphylococcus aureus.*

medium seeded with a test organism (*S. aureus* or *P. aeruginosa*), and the plate is incubated. A zone devoid of growth (zone of inhibition) around the disc indicates the capacity of the agent to inhibit growth (**figure C.1**). Like the antimicrobic sensitivity test (see figure 12.18, p. 379), this test measures the minimum inhibitory concentration of the chemical. In general, chemicals with wide zones of inhibition are effective even in high dilutions.

Tests for Water Analysis

Figure C.2 and **Table C.1** give standards used in analyzing water samples using the most probable number (MPN) method, as described in chapter 26. The figure indicates the steps in the laboratory procedure and the table provides estimated counts for coliforms, based on the number of tubes with growth.

Guidelines for Universal Precautions in the Clinical Setting

The significance of these procedures for infection control is introduced in chapter 13, page 413.

1. Barrier precautions, including masks and gloves, should be taken to prevent contact of skin and mucous membranes with patients' blood or other body fluids. Because gloves can develop small invisible tears, double gloving decreases the risk further. For protection during surgery, venipuncture, or emergency procedures, gowns, aprons, and other body coverings should be worn. Dental workers should wear eyewear and face shields to protect against splattered blood and saliva.

2. More than 10% of health care personnel are pierced each year by sharp (and usually contaminated) instruments. These accidents carry risks not only for AIDS but also for hepatitis B, hepatitis C, and other diseases. Preventing inoculation infection requires vigilant observance of proper techniques.
 - All disposable needles, scalpels, or sharp devices from invasive procedures must immediately be placed in puncture-proof containers for sterilization and final discard.
 - Under no circumstances should a worker attempt to recap a syringe, remove a needle from a syringe, or leave unprotected used syringes where they pose a risk to others.
 - Reusable needles or other sharp devices must be heat-sterilized in a puncture-proof holder before they are handled.
 - If a needlestick or other injury occurs, immediate attention to the wound, such as thorough degermation and application of strong antiseptics, can prevent infection.

3. Dental handpieces should be sterilized between patients, but if this is not possible, they should be thoroughly disinfected with a high-level disinfectant (peroxide, hypochlorite). Blood and saliva should be removed completely from all contaminated dental instruments and intraoral devices prior to sterilization.

4. Hands and other skin surfaces that have been accidently contaminated with blood or other fluids should be scrubbed immediately with a germicidal soap. Hands should likewise be washed after removing rubber gloves, masks, or other barrier devices.

5. Because saliva can be a source of some types of infections, barriers should be used in all mouth-to-mouth resuscitations.

6. Health care workers with active, draining skin or mucous membrane lesions must refrain from handling patients or equipment that will come into contact with other patients. Pregnant health care workers risk infecting their fetuses and must pay special attention to these guidelines. Personnel should be protected by vaccination whenever possible.

Isolation procedures for known or suspected infections should still be instituted on a case-by-case basis. See Table 13.12, page 414.

Inoculate 15 tubes: 5 with 10 ml of sample, 5 with 1.0 ml of sample, and 5 with 0.1 ml of sample

Presumptive

10 10 10 10 10 (ml) 1.0 1.0 1.0 1.0 1.0 (ml) 0.1 0.1 0.1 0.1 0.1 (ml)

Lactose or laurel tryptose broth

Negative presumptive result. The absence of gas in broth tubes indicates coliforms are absent. Incubate an additional 24 hours to be sure.

24 ± 2 hours 35°C

Negative result Positive result

After 24 hours of incubation, examine the tubes of lactose broth for gas production.

Confirmed

Positive test for gas production. Use positive confirmed tubes to determine MPN.

Use all positive presumptive cultures to inoculate tubes of brilliant green lactose bile broth. Incubate for 48±3 hours at 35°C.

Completed

Use coliform colonies to inoculate nutrient agar slant and a broth tube.

EMB

Endo

Streak plates of Levine's EMB or endo agar from positive tubes and incubate at 35°C for 24±2 hours.

Brilliant green lactose broth

Nutrient agar slant

After 24 hours of incubation, make a Gram-stained slide from the slant. If the bacteria are gram-negative non-spore-forming rods that produce gas from lactose, the completed test is positive.

Figure C.2 The most probable number (MPN) procedure for determining the coliform content of a water sample. In the presumptive test, each set of tubes of broth is inoculated with a water sample reduced by a factor of 10. After incubation, the sets of tubules are examined and rated for gas production (for instance, 0 means no tubes with gas, 1 means one tube with gas, 2 means two tubes with gas). Applying this result to the MPN table in appendix C will indicate the probable number of cells present in 100 ml of the water sample. Confirmation of coliforms can be achieved by confirmatory tests on additional media, and complete identification can be made through selective and different media and staining (see pages ••• and •••).

TABLE C.1	Most Probable Number Chart for Evaluating Coliform Content of Water						
The Number of Tubes in Series, with Positive Growth at Each Dilution							
10 ml	1 ml	0.1 ml	MPN* per 100 ml of Water	10 ml	1 ml	0.1 ml	MPN* per 100 ml of Water
0	1	0	0.18	5	0	1	3.1
1	0	0	0.20	5	1	0	3.3
1	0	0	0.40	5	1	1	4.6
2	0	0	0.45	5	2	0	4.9
2	0	1	0.68	5	2	1	7.0
2	2	0	0.93	5	2	2	9.5
3	0	0	0.78	5	3	0	7.9
3	0	1	1.1	5	3	1	11.0
3	1	0	1.1	5	3	2	14.0
3	2	0	1.4	5	4	0	13.0
4	0	0	1.3	5	4	1	17.0
4	0	1	1.7	5	4	2	22.0
4	1	0	1.7	5	4	3	28.0
4	1	1	2.1	5	5	0	24.0
4	2	0	2.2	5	5	1	35.0
4	2	1	2.6	5	5	2	54.0
4	3	0	2.7	5	5	3	92.0
5	0	0	2.3	5	5	4	160.0

*Most probable number of cells per sample of 100 ml.

Classification of Major Microbial Disease Agents by System Affected, Site of Infection, and Routes of Transmission

Tables D.1 through **D.7** provide a cross-reference that combines an organismic and taxonomic approach for presenting pathogens. The tables list the primary organs and systems that serve as targets for infectious agents and the sources and routes of infection. Many infectious agents affect both the system they first enter and others. (For example, polio infects the gastrointestinal tract and also inflicts major damage on the central nervous system.) It is for this reason that agents with major effects on more than one system are noted by a star (*). The chapter that contains detailed information on the listings is shown in parentheses in the far right column of each table.

Table D.8 outlines the major techniques for identifying and classifying bacteria. **Table D.9** summarizes the most current taxonomy for the prokaryotes (Domain Archaea and Domain Bacteria) according to *Bergey's Manual of Systematic Bacteriology*.

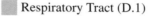

Respiratory Tract (D.1)

Gastrointestinal Tract (D.2)

Skin and Wounds (D.3)

STDs and GU Tract (D.4)

Multiple Systems (Circulatory and Lymphatic (D.5)

Nervous System (D.6)

Eye and Ear (D.7)

TABLE D.1	Infections of the Respiratory Tract (RT)			
Name of Microbe	**Name of Disease**	**Specific Targets of Infection**	**Source/Mode of Acquisition**	**(Ch)**
Bacterial pathogens				
Streptococcus pyogenes★	Streptococcal pharyngitis	Upper RT membranes	Droplets	(18)
Viridans streptococci	Sinusitis	Upper RT and saliva	Normal flora	(18)
Streptococcus pneumoniae★	Pneumococcal pneumonia; otitis media, sinusitis meningitis	Lower RT, lungs; meninges	Droplets, mucus	(18)
Bacillus anthracis★	Pulmonary anthrax	Upper and lower RT	Spores inhaled from animals/products	(19)
Corynebacterium diphtheriae★	Diphtheria	Upper RT membranes; also skin	Droplets	(19)
Haemophilus influenzae★	Pharyngitis	Upper RT membranes	Droplets	(20)
Mycobacterium tuberculosis★	Tuberculosis	Lower RT, lungs	Droplets; droplet nuclei	(19)
Nocardia brasiliensis	Nocardiosis (a form of pneumonia)	Lower RT, lungs	Spores inhaled from environment	(19)
Legionella pneumophila	Legionnaire disease, Pontiac fever	Lower RT, lungs	Aerosols from aquatic habitats	(20)
Brucella spp.★	Brucellosis (undulant fever)	Respiratory tract	Aerosols from infected animals	(20)
Bordetella pertussis	Whooping cough	Upper, lower RT	Droplets	(20)
Klebsiella pneumoniae Pseudomonas aeruginosa	Opportunistic pneumonias	Lower RT, lungs	Droplets, normal flora, aerosols, fomites	(20)
Yersinia pestis★	Pneumonic plague	Lower RT, lungs	Droplets, animal aerosols	(20)
Mycoplasma pneumoniae	Primary atypical pneumonia	Lower RT, lungs	Droplets, mucus	(21)
Chlamydiophila pneumoniae	Chlamydial pneumonitis	Lower RT, lungs	Droplets, mucus	(21)
Chlamydiophila psittaci	Ornithosis (psittacosis)	Lower RT, lungs	Aerosols, droplets from birds	(21)
Coxiella burnetii	Q (for query) fever	Lower RT, lungs	Ticks, aerosols from animals, dust	(21)

continued

TABLE D.1 *continued*

Name of Microbe	Name of Disease	Specific Targets of Infection	Source/Mode of Acquisition	(Ch)
Fungal pathogens (chapter 22)				
Histoplasma capsulatum★	Histoplasmosis	Lower RT, lungs	Spores inhaled in dust	
Blastomyces dermatitidis★	Blastomycosis	Lower RT, lungs	Spores inhaled	
Coccidioides immitis★	Coccidioidomycosis (Valley fever)	Lower RT, lungs	Spores inhaled	
Cryptococcus neoformans★	Cryptococcosis	Lower RT, lungs	Yeasts inhaled, can be normal flora	
Sporothrix schenckii★	Sporothrix pneumonia	Lower RT, lungs	Spores on vegetation	
Pneumocystis (carinii) jiroveci	Pneumocystis pneumonia (PCP)	Lower RT, lungs	Spores widespread, normal flora	
Aspergillus fumigatus★	Aspergillosis	Lower RT, lungs	Spores widespread in dust, air	
Parasitic (protozoan, helminth) pathogens (chapter 23)				
Paragonimus westermani	Adult lung fluke	Lung tissue; intestine	Larvae ingested in raw or under-cooked crustacea	
Viral pathogens				
Influenza viruses	Influenza A, B, C	Mucosal cells of trachea, bronchi	Inhalation of droplets, aerosols from active cases	(25)
Hantavirus	Hantavirus pulmonary syndrome	Lung epithelia	Inhalation of airborne rodent excrement	(25)
Respiratory syncytial pneumovirus	RSV pneumonitis, bronchiolitis	Nasopharynx; upper and lower RT	Droplets and secretions from active cases	(25)
Parainfluenza virus Paramyxovirus	Parainfluenza	Upper, lower RT	Direct hand contact with secretions; inhalation of droplets	(25)
Adenovirus	Common cold	Nasopharynx; sometimes lower RT	Close contact with secretions from person with active infection; hand to nose	(24)
Rhinovirus	Common cold	Same as above	Same as above	(25)
Coronavirus	Common cold	Same as above	Same as above	(25)
Coxsackievirus	Common cold	Same as above	Same as above	(25)
Mumps virus★	See table D.2			
Rubella virus★	See table D.3			
Measles virus★	See table D.3			
Chickenpox virus (VZV)★	See table D.3			

TABLE D.2 Diseases of the Gastrointestinal Tract (Food Poisoning and Intestinal, Gastric, Oral Infections)

Name of Microbe	Name of Disease	Specific Targets of Disease	Source/Mode of Acquisition	(Ch)
Bacterial pathogens				
Staphylococcus aureus	Staph food intoxication	Small intestine	Foods contaminated by human carriers	(18)
Streptococcus mutans	Dental caries	Tooth surface	Normal flora; saliva	(21)
Clostridium perfringens	Perfringens food poisoning	Small intestine	Rare meats; toxin produced in the body	(19)
Clostridium difficile	Antibiotic-associated colitis	Large intestine	Spores ingested; can be normal flora; caused by antibiotic therapy	(19)
Listeria monocytogenes	Listeriosis	Small intestine	Water, soil, plants, animals, milk, food infection	(19)
Actinomyces israelii★	Actinomycosis	Oral membranes, cervicofacial area	Normal flora; enter damaged tissue	(19)
Escherichia coli	Traveler's diarrhea O157:H7 enteritis	Small intestine	Food, water, rare or raw beef	(20)

continued

TABLE D.2 Diseases of the Gastrointestinal Tract (Food Poisoning and Intestinal, Gastric, Oral Infections)

Name of Microbe	Name of Disease	Specific Targets of Disease	Source/Mode of Acquisition	(Ch)
Salmonella typhi★	Typhoid fever	Small intestine	Human carriers; food, drink	(20)
Salmonella species	Food infection, enteric fevers	Small intestine	Cattle, poultry, rodents; fecally contaminated foods	(20)
Shigella spp.	Bacillary dysentery	Large intestine	Human carriers; food, flies	(20)
Yersinia enterocolitica *Y. paratuberculosis*	Yersiniosis (gastroenteritis)	Small intestine, lymphatics	A zoonosis; food and drinking water	(20)
Vibrio cholerae	Epidemic cholera	Small intestine	Natural waters; food contaminated by human carriers	(21)
Helicobacter pylori	Gastritis, gastric intestinal ulcers	Esophagus, stomach, duodenum	Humans, cats (?)	(21)
Campylobacter jejuni	Enteritis	Jejunum segment of small intestine	Water, milk, meat, food infection	(21)
Fusobacterium, *Treponema, Bacteroides*	Various forms of gingivitis, periodontitis	Gingival pockets	Normal flora; close contact	(21)

Fungal pathogens (chapter 22)

Aspergillus flavus	Aflatoxin poisoning	Liver	Toxin ingested in food	
Candida albicans★	Esophagitis	Esophagus	Normal flora	

Parasitic (protozoan, helminth) pathogens (chapter 23)

Entamoeba histolytica★	Amebic dysentery	Caecum, appendix, colon, rectum	Food, water-contaminated fecal cysts	
Giardia lamblia (intestinalis)	Giardiasis	Duodenum, jejunum	Ingested cysts from animal, human carriers	
Balantidium coli	Balantidiosis	Large intestine	Swine, contaminated food	
Cryptosporidium spp.	Cryptosporoidosis	Small and large intestine	Animals, natural waters, contaminated food	
Toxoplasma gondii★	Toxoplasmosis	Lymph nodes	Cats, rodents, domestic animals; raw meats	
Enterobius vermicularis	Pinworm	Rectum, anus	Eggs ingested from fingers, food	
Ascaris lumbricoides	Ascariasis	Worms burrow into intestinal mucosa	Fecally contaminated food and water containing eggs	
Trichuris trichiura	Trichiuriasis, whipworm	Worms invade small intestine	Food containing eggs	
Trichinella spiralis★	Trichinosis	Larvae penetrate intestine	Pork and bear meat containing encysted larvae	
Necator americanus★ *Ancylostoma duodenale*★	Hookworms	Large and small intestine	Larvae deposited by feces into soil; burrow into feet	
Fasciola hepatica	Sheep liver fluke	Liver and intestine	Watercress	
Opisthorchis sinensis	Chinese liver fluke	Liver and intestine	Raw or undercooked fish or shellfish	
Taenia saginata *T. solium*	Beef tapeworm Pork tapeworm	Large intestine, Pork larva can migrate to brain	Rare or raw beef or pork containing larval forms	

Viral pathogens

Herpes simplex virus, type 1	Cold sores, fever blisters, gingivostomatitis	Skin, mucous membranes of oral cavity	Close contact with active lesions of human	(24)
Mumps virus★ *(Paramyxovirus)*	Parotitis	Parotid salivary gland	Close contact with salivary droplets	(25)
Hepatitis A virus	Infectious hepatitis	Liver	Food, water contaminated by human feces; shellfish	(25)
Rotavirus	Rotavirus diarrhea	Small intestinal mucosa	Fecally contaminated food, water, fomites	(25)
Norwalk agent (enterovirus)	Gastroenteritis	Small intestine	Close contact; oral-fecal, water, shellfish	(25)
Poliovirus★	Poliomyelitis	Intestinal mucosa, systemic	Food, water, mechanical vectors	(25)

TABLE D.3	Infections of Skin and Skin Wounds			
Name of Microbe	**Name of Disease**	**Specific Targets of Infection**	**Source/Mode of Transmission**	**(Ch)**
Bacterial pathogens				
Staphylococcus aureus★	Folliculitis Furuncles (boils) Carbuncles Bullous impetigo Scalded skin syndrome	Glands, follicles, glabrous skin	Close contact, nasal and oral droplets; associated with poor hygiene, tissue injury	(18)
Streptococcus pyogenes★	Pyoderma (impetigo) Erysipelas Necrotizing fasciitis	Skin and underlying tissues	Contact with human carrier; invasion of tiny wounds in skin	(18)
Bacillus anthracis★	Cutaneous anthrax	Necrotic lesion in skin (eschar)	Spores from animal secretions, products, soil	(19)
Clostridium perfringens★	Gas gangrene	Skin, muscle, connective tissue	Introduction of spores into damaged tissue	(19)
Propionibacterium acnes	Acne	Sebaceous glands	Normal flora; bacterium is lipophilic	(19)
Erysipelothrix rhusiopathiae	Erysipeloid	Epidermal, dermal regions of skin	Normal flora of pigs, other animals; enters in injured tissue	(19)
Mycobacterium leprae★	Leprosy (Hansen disease)	Skin, underlying nerves	Long-term contact with infected person; skin is inoculated	(19)
Mycobacterium marinum	Swimming pool granuloma	Superficial skin	Skin scraped against contaminated surface	(19)
Pasteurella multocida	Pasteurellosis	Localized skin abscess	Animal bite or scratch	(20)
Bartonella henselae	Cat-scratch disease	Subcutaneous tissues, lymphatic drainage	Cat's claws carry bacteria, inoculate skin	(21)
Fungal pathogens (chapter 22)				
Blastomyces dermatitidis	Blastomycosis	Subcutaneous	Accidental inoculation of skin with conidia	
Trichophyton, Microsporum, Epidermophyton	Dermatomycoses: tineas, ringworm, athlete's foot	Cutaneous tissues (epidermis, nails, scalp, hair)	Contact with infected humans, animals, spores in environment	
Sporothrix schenckii★	Sporotrichosis	Epidermal, dermal, subcutaneous	Puncture of skin with sharp plant materials	
Madurella	Madura foot, mycetoma	Subcutaneous tissues, bones, muscles	Traumatic injury with spore-infested objects	
Cladosporium, Fonsecaea	Chromoblastomycosis	Deep subcutaneous tissues of legs, feet	Penetration of skin with contaminated objects	
Candida albicans★	Onychomycosis	Skin, nails	Chronic contact with moisture, warmth	
Malassezia furfur	Tinea versicolor	Superficial epidermis	Normal flora of skin	
Parasitic (protozoan, helminth) pathogens (chapter 23)				
Leishmania★ *tropica L. mexicana*	Cutaneous leishmaniasis	Local infection of skin capillaries	Sand fly salivary gland harbors infectious stage; transmitted through bite	
Dracunculus medinensis	Dracontiasis	Starts in intestine, migrates to skin	Microscopic aquatic arthropod harbors larval form of guinea worm; is ingested	
Viral pathogens				
Varicella-zoster virus★	Chickenpox Shingles	Skin of face and trunk; becomes systemic and latent in nerves	Spread by respiratory droplets; recurrences from latent infection	(24)
Herpesvirus-6	Roseola infantum	Generalized skin rash	Close contact with droplets	(24)
Parvovirus B19	Erythema infectiosum	Skin of cheeks, trunk	Close contact with droplets	(24)
Measles virus★ (*Morbillivirus*)	Measles, rubeola	Skin on head, trunk, extremities	Contact with respiratory aerosols from active case	(25)
Rubella virus★ (*Rubivirus*)	Rubella	Skin on face, trunk, limbs	Respiratory droplets, urine of active cases	(25)
Human papillomavirus (See also table D.4)	Warts, verruca	Skin of hands, feet, body	Contact with warts, fomites	(24)
Molluscum contagiosum	See table D.4			(24)
Herpes simplex, type 2	See table D.4			(24)

TABLE D.4 — Sexually Transmitted Diseases (STDs) and Infections of the Genitourinary (GU) Tract

Name of Microbe	Name of Disease	Specific Targets of Infection	Source/Mode of Acquisition	(Ch)
Bacterial pathogens				
Neisseria gonorrhoeae★	Gonorrhea	Vagina, urethra	Mucous secretions; STD	(18)
Haemophilus ducreyi	Chancroid	Genitalia, lymph nodes	Skin lesions; STD	(20)
Treponema pallidum★	Syphilis	Genitalia; mucous membranes	Chancres, skin lesions; STD	(21)
Chlamydia trachomatis★	Chlamydiosis (non-gonococcal urethritis), lymphogranuloma venereum	Vagina, urethra, lymph nodes	Mucous discharges; STD	(21)
Escherichia coli★	Urinary tract infection (UTI)	Urethra, bladder	Normal flora; intestinal tract	(20)
Leptospira interrogans★	Leptospirosis	Kidney, liver	Animal urine; contact with contaminated water, soil	(21)
Gardnerella vaginalis	Bacterial vaginosis	Vagina, occasionally penis	Mixed infection; normal flora; intimate contact	(20)
Fungal pathogens (chapter 22)				
Candida albicans	Yeast infections, candidiasis	Vagina, vulva, penis, urethra	Discharge from infected membranes; STD	
Parasitic (protozoan, helminth) pathogens (chapter 23)				
Trichomonas vaginalis	Trichomoniasis	Vagina, vulva, urethra	Genital discharges, direct contact; STD	
Viral pathogens				
Molluscum contagiosum poxvirus	Molluscum contagiosum	Skin, mucous membranes	Pox lesions; STD	(24)
Herpes simplex, type 2	Genital herpes	Genitalia, perineum	Contact with vesicles, shed skin cells; STD	(24)
Human papillomavirus	Genital warts, condylomata	Membranes of the vagina, penis	Direct contact with warts; some fomites; STD	(24)
Hepatitis B virus★	Hepatitis B	Liver	Semen, vaginal fluids, blood; STD	(24)
Human immunodeficiency virus (HIV)★	AIDS	White blood cells, brain cells	Blood, semen, vaginal fluids; STD	(25)

TABLE D.5 — Infections Affecting Multiple Systems (Circulatory and Lymphatic)

Name of Microbe	Name of Disease	Specific Targets of Disease	Source/Mode of Acquisition	(Ch)
Bacterial pathogens				
Staphylococcus aureus★	Toxic shock syndrome	Vagina, uterus, kidney, liver, blood	Normal flora; grow in environment created by tampons	(18)
	Osteomyelitis	Interior of long bones	Focal from skin infection	
Streptococcus pyogenes★	Scarlet fever	Skin, brain	Toxemia; complication of strep pharyngitis	(18)
	Rheumatic fever	Heart valves	Autoimmune reaction to strep toxins	
	Glomerulonephritis	Kidney	Immune reaction blocks filtration apparatus	
Streptococcus spp. (viridans)	Subacute endocarditis	Lining of heart and valves	Normal flora invades during dental work; colonizes heart	(18)
Corynebacterium diphtheriae★	Diphtheria	Heart muscle, nerves, occasionally skin	Droplets from carriers; local infection leads to toxemia	(19)
Mycobacterium tuberculosis★	Disseminated tuberculosis	Lymph nodes, kidneys, bones, brain, sex organs	Reactivation of latent lung infection	(19)
Yersinia pestis★	Plague	Lymph nodes, blood	Bite of infected flea	(20)

continued

Name of Microbe	Name of Disease	Specific Targets of Disease	Source/Mode of Acquisition	(Ch)
Bacterial pathogens *continued*				
Francisella tularensis	Tularemia Rabbit fever	Skin, lymph nodes, eyes, lungs	Blood-sucking arthropods (ticks, mosquitoes); direct contact with animals and meat; aerosols	(20)
Borrelia recurrentis	Borreliosis, relapsing fever	Liver, heart, spleen, kidneys, nerves	Human lice; influenced by poor hygiene, crowding	(21)
B. hermsii	Same as above	Same as above	Soft tick bite; reservoir is wild rodents	(21)
B. burgdorferi	Lyme disease	Skin, joints, nerves, heart	Larval *Ixodes* ticks; reservoir is deer and rodents	(21)
Rickettsia prowazekii	Epidemic typhus	Blood vessels in many organs	Human lice, bites, feces, crowding, lack of sanitation	(21)
R. rickettsii	Rocky Mountain spotted fever	Cardiovascular, central nervous, skin	Wood and dog ticks; picked up in mountain or rural habitats	(21)
R. typhi	Endemic typhus	Similar to epidemic typhus	Rat flea vector; rodent reservoir; occupational contact	(21)
Coxiella burnetii★	See table D.1			
Ehrlichia chaffeensis	Human monocytic ehrlichosis (HME)	Macrophages, monocytes	Lone star tick bite	(21)
Anaplasma phagocytophilum	Human granulocytic anaplasmosis	Neutrophils, granulocytes, liver, kidney	Ticks carried by deer and mice	(21)

Fungal pathogens (chapter 22)

Many pathogenic fungi (*Histoplasma, Coccidioides, Blastomyces,* and *Cryptococcus*) have a systemic component to their infection, especially in immunocompromised patients. Because their primary infection involves the lungs, information on those fungi is listed in table D.1.

Parasitic (protozoan, helminth) pathogens (chapter 23)

Leishmania spp.	Systemic leishmaniasis (kala-azar)	Spleen, liver, lymph nodes	Bite by sand (phlebotamine) fly; reservoir is dogs, rodents, wild carnivores	
Plasmodium spp.	Malaria	Liver, blood, kidney	Female *Anopheles* mosquito; human reservoir, mainly tropical	
Toxoplasma gondii★	Toxoplasmosis	Pharynx, lymph nodes, nervous system	Cysts from cats and other animals are ingested; many hosts	
Strongyloides stercoralis	Strongyloidiasis Threadworm	Intestine, trachea, lungs, pharynx, subcutaneous; migrate through systems	Larvae deposited into soil or water by feces; burrow into skin	
Schistosoma spp.	Schistosomiasis, blood fluke	Liver, lungs, blood vessels in intestine, bladder	Larval forms develop in aquatic snails; penetrate skin of bathers	
Wuchereria bancrofti	Filariasis Elephantiasis	Lymphatic circulation, blood vessels	Various mosquito species carry larval stage; human reservoir	

Viral pathogens

Cytomegalovirus (CMV)	CMV mononucleosis, congenital CMV	Lymph glands, salivary glands	Human saliva, milk, urine, semen, vaginal fluids; intimate contact	
Epstein-Barr virus	Infectious mononucleosis	Lymphoid tissue, salivary glands	Oral contact, exposure to saliva, fomites	(24)
Hepatitis B virus★	See table D.4			(24)
Flavivirus	Yellow fever	Liver, blood vessels, kidney, mucous membranes	Various mosquito species; human and monkey reservoirs	(25)
	Dengue fever (breakbone fever)	Skin, muscles, and joints	Several mosquito vectors	(25)
Filovirus	Ebola fever	Every organ and tissue except muscle and bone	Close contact with blood and secretions of victim; monkey reservoir?	(25)
Arenavirus	Lassa fever	Blood vessels, kidney, heart	Inhalation of airborne virus from rodent excreta	(25)
HIV★	See table D.4			
HTLV I	Adult T-cell leukemia Sézary-cell leukemia	T cells, lymphoid tissue	Blood, blood products, sexual intercourse, IV drugs	(25)

TABLE D.6　Diseases of the Nervous System

Name of Microbe	Name of Disease	Specific Targets of Disease	Source/Mode of Acquisition	(Ch)
Bacterial pathogens				
Neisseria meningitidis★	Meningococcal meningitis	Membranes covering rain, spinal cord	Contact with respiratory secretions or droplets of human carrier	(18)
Clostridium tetani	Tetanus	Inhibitory neurons in spinal column	Spores from soil introduced into wound; toxin formed	(19)
Clostridium botulinum★	Botulism	Neuromuscular junction	Canned vegetables, meats; an intoxication	(19)
Mycobacterium leprae★	Leprosy (Hansen disease)	Peripheral nerves of skin	Close contact with infected person; enters skin	(19)
Haemophilus influenzae★	Acute meningitis	Meninges, brain	Contact with carrier; normal flora of throat	(20)
Fungal pathogens (chapter 22)				
Cryptococcus neoformans★	Cryptococcal meningitis	Brain, meninges	Pigeon droppings; inhalation of yeasts in air and dust	
Parasitic (protozoan, helminth) pathogens (chapter 23)				
Naegleria fowleri	Primary acute meningoencephalitis	Brain, spinal cord	Acquired while swimming in fresh, brackish water; amoeba invades nasal mucosa	
Acanthamoeba	Meningoencephalitis	Eye, brain	Amoeba lives in fresh water; can invade cuts and abrasions	
Trypanosoma brucei	African trypanosomiasis, sleeping sickness	Brain, heart, many organs	Bite of tsetse fly, vector	
Trypanosoma cruzi	South American; trypanosomiasis Chagas disease	Nerve ganglia, muscle	Feces of reduviid bug rubbed into bite	
Viral pathogens (chapters 24 and 25)				
Poliovirus★	Poliomyelitis	Anterior horn cells; motor neurons	Contaminated food, water objects	(25)
Rabies virus	Rabies	Brain, spinal nerves, ganglia	Saliva of infected animal enters bite wound	(25)
Encephalitis viruses	St. Louis encephalitis Western equine encephalitis Eastern equine encephalitis West Nile fever	Brain, meninges, spinal cord	Bite of infected mosquito	(25)
Bunyavirus	California encephalitis	Brain, meninges	Mosquito bite; reservoir is wild birds	(25)
HIV★ (see table D.4)	AIDS dementia	Destroys brain cells, peripheral nerves	Blood, semen, vaginal fluids; STD	(25)
JC polyomavirus	Progressive multifocal leukoencephalopathy	Oligodendrocytes in cerebrum	Not well defined	(24)
Prions	Spongiform encephalopathies Creutzfeldt-Jakob disease	Brain, neurons	Intimate contact with infected tissues	(25)

TABLE D.7　Infections of the Eye and Ear

Name of Microbe	Name of Disease	Specific Targets of Infection	Source/Mode of Transmission	(Ch)
Bacterial pathogens				
Neisseria gonorrhoeae★	Ophthalmia neonatorum	Conjunctiva, cornea, eyelid	Vaginal mucus enters eyes during birth	(18)
Streptococcus pneumoniae★	Otitis media	Middle ear	Nasopharyngeal secretions forced into eustachian tube	(18)
Haemophilus influenzae, type aegyptius	Pinkeye	Conjunctiva, sclera	Contaminated fingers, fomites	(20)
Chlamydia trachomatis★	Inclusion conjunctivitis, trachoma	Conjunctiva, inner eyelid, cornea	Infected birth canal, contaminated fingers	(21)
Fungal pathogens (chapter 22)				
Aspergillus fumigatus	Aspergillosis	Eyelids, conjunctiva	Contact lens contaminated with spores	
Fusarium solani	Corneal ulcer, keratitis	Cornea	Eye is accidentally scratched, contact lens is contaminated	
Parasitic (protozoan, helminth) pathogens (chapter 23)				
Onchocerca volvulus	Onchocerciasis, river blindness	Eye, skin, blood	Black flies inject larvae into bite	
Loa loa	African eye worm	Subcutaneous tissues, conjunctiva, cornea	Bites by dipteran flies	
Viral pathogens				
Coxsackievirus	Hemorrhagic conjunctivitis	Conjunctiva, sclera, eyelid	Contact with fluids, secretions of carrier	(25)

General Classification Techniques and Taxonomy of Bacteria

TABLE D.8 Summary of Techniques Used in Bacterial Identification and Classification, Chapter 4, page 111

- **Microscopic Morphology** Traits that can be valuable aids to identification are combinations of cell shape and size, Gram stain reaction, acid-fast reaction, and special structures, including endospores, granules, and capsules. Electron microscope studies can pinpoint additional structural features (such as the cell wall, flagella, pili, and fimbriae).

- **Macroscopic Morphology** Appearance of colonies, including texture, size, shape, pigment, speed of growth and patterns of growth in broth and gelatin media.

- **Physiological/Biochemical Characteristics** Enzymes and other biochemical properties of bacteria are fairly reliable and stable expressions of the "chemical identity" of each species. Diagnostic tests determine the presence of specific enzymes and assess nutritional and metabolic activities. Examples include fermentation of sugars; digestion of complex polymers such as proteins and polysaccharides; presence of catalase, oxidase, and decarboxylases; sensitivity to antimicrobic drugs.

- **Chemical Analysis** Analyzing the types of specific structural substances that the bacterium contains, such as the chemical composition of peptides in the cell wall and lipids in membranes.

- **Serological Analysis** Bacteria display molecules called antigens that are recognized by the immune system. One immune response to antigens is to produce antibodies that are designed to bind tightly to the antigens. This response is so specific that antibodies can be used as a means of identifying bacteria in specimens and cultures.

- **Genetic and Molecular Analysis** Examining the genetic material itself has revolutionized the identification and classification of bacteria.

G + C base composition The overall percentage of guanine and cytosine (the G + C content as compared with A + T content) in DNA is a general indicator of relatedness because it is a trait that does not change rapidly. Bacteria with a significant difference in G + C percentage are less likely to be genetically related. See the revised classification scheme that is partly based on this percentage (table 4.4).

DNA analysis using genetic probes It is possible to identify a bacterial species by analyzing its genetic material. This uses small fragments of DNA (or RNA) called probes that are complementary to the specific sequences of DNA from a particular microbe (see figure 10.3, page 294). Unknown test DNA from specimens or cultures is bound to a special surface. After several different probes have been added, the plate is observed for visible signs that the probes have become fixed to the test DNA. The binding of probes onto areas of the test DNA indicates close correspondence and makes identification possible.

Nucleic acid sequencing and rRNA analysis One of the most valuable indicators of evolutionary relatedness is the sequence of nitrogen bases in ribosomal RNA, a major component of ribosomes. Ribosomes have the same function (protein synthesis) in all cells, and they tend to remain stable in their nucleic acid content over long periods. Thus, any major differences in the sequence, or "signature," of the rRNA is likely to indicate some distance in ancestry.

This technique is effective for differentiating general group differences (it was used to separate the three superkingdoms of life discussed in chapter 1), and it can identify species. Elements of these and other identification methods are presented in more detail in chapters 10, 16, and 17.

TABLE D.9 Organization of *Bergey's Manual of Systematic Bacteriology,* Second Edition

Taxonomic Rank	Representative Genera
Volume 1. The Archaea and the Deeply Branching and Phototrophic Bacteria	
Domain Archaea	
Phylum Crenarchaeota	*Thermoproteus, Pyrodictium, Sulfolobus*
Phylum Euryarchaeota	
Class I. Methanobacteria	*Methanobacterium*
Class II. Methanococci	*Methanococcus*
Class III. Halobacteria	*Halobacterium, Halococcus*
Class IV. Thermoplasmata	*Thermoplasma, Picrophilus, Ferroplasma*
Class V. Thermococci	*Thermococcus, Pyrococcus*
Class VI. Archaeoglobi	*Archaeoglobus*
Class VII. Methanopyri	*Methanopyrus*

continued

TABLE D.9 *continued*

Taxonomic Rank	Representative Genera
Volume 1. The Archaea and the Deeply Branching and Phototrophic Bacteria *continued*	
Domain Bacteria	
Phylum Aquificae	*Aquifex, Hydrogenobacter*
Phylum Thermotogae	*Thermotoga, Geotoga*
Phylum Thermodesulfobacteria	*Thermodesulfobacterium*
Phylum "Deinococcus-Thermus"	*Deinococcus, Thermus*
Phylum Chrysiogenetes	*Chrysogenes*
Phylum Chloroflexi	*Chloroflexus, Herpetosiphon*
Phylum Thermomicrobia	*Thermomicrobium*
Phylum Nitrospira	*Nitrospira*
Phylum Deferribacteres	*Geovibrio*
Phylum Cyanobacteria	*Prochloron, Synechococcus, Pleurocapsa, Oscillatoria, Anabaena, Nostoc, Stigonema*
Phylum Chlorobi	*Chlorobium, Pelodictyon*
Volume 2. Domain Bacteria: The Proteobacteria	
Phylum Proteobacteria	
Class I. Alphaproteobacteria	*Rhodospirillum, Rickettsia, Caulobacter, Rhizobium, Brucella, Nitrobacter, Methylobacterium, Beijerinckia, Hyphomicrobium*
Class II. Betaproteobacteria	*Neisseria, Burkholderia, Alcaligenes, Comamonas, Nitrosomonas, Methylophilus, Thiobacillus*
Class III. Gammaproteobacteria	*Chromatium, Leucothrix, Legionella, Pseudomonas, Moraxella, Acinetobacter, Azotobacter, Vibrio, Escherichia, Klebsiella, Proteus, Salmonella, Shigella, Yersinia, Haemophilus*
Class IV. Deltaproteobacteria	*Desulfovibrio, Bdellovibrio, Myxococcus, Polyangium*
Class V. Epsilonproteobacteria	*Campylobacter, Helicobacter*
Volume 3. Domain Bacteria: The Low G + C Gram-Positive Bacteria	
Phylum Firmicutes	
Class I. Clostridia	*Clostridium, Peptostreptococcus, Eubacterium, Desulfotomaculum, Heliobacterium, Veillonella*
Class II. Mollicutes	*Mycoplasma, Ureaplasma, Spiroplasma, Acholeplasma*
Class III. Bacilli	*Bacillus, Caryophanon, Paenibacillus, Thermoactinomyces, Lactobacillus, Streptococcus, Enterococcus, Listeria, Leuconostoc, Staphylococcus*
Volume 4. Domain Bacteria: The High G + C Gram-Positive Bacteria	
Phylum Actinobacteria	
Class Actinobacteria	*Actinomyces, Micrococcus, Arthrobacter, Corynebacterium, Mycobacterium, Nocardia, Actinoplanes, Propionibacterium, Streptomyces, Thermomonospora, Frankia, Actinomadura, Bifidobacterium*
Volume 5. Domain Bacteria: The Planctomycetes, Spirochaetes, Fibrobacteres, Bacteriodetes, and Fusobacteria	
Phylum Planctomycetes	*Planctomyces, Gemmata*
Phylum Chlamydiae	*Chlamydia, Chlamydiophila*
Phylum Spirochaetes	*Spirochaeta, Borrelia, Treponema, Leptospira*
Phylum Fibrobacteres	*Fibrobacter*
Phylum Acidobacteria	*Acidobacterium*
Phylum Bacteriodetes	*Bacteroides, Porphyromonas, Prevotella, Flavobacterium, Sphingobacterium, Flexibacter, Cytophaga*
Phylum Fusobacteria	*Fusobacterium, Streptobacillus*
Phylum Verrucomicrobia	*Verrucomicrobium*
Phylum Dictyoglomus	*Dictyoglomus*

An Introduction to Concept Mapping

Concept maps are visual tools for presenting and organizing what you have learned. They can take the place of an outline, though for most people they contain much more meaning and can illustrate connections and interconnections in ways that ordinary outlines cannot. They are also very flexible. If you are creating a concept map, there is a nearly infinite number of ways that they can be put together and still be "correct." Concept maps are also a way to incorporate and exploit your own creative impulses, so that you are not stuck inside a rigid framework but can express your understanding of concepts and their connections in ways that make sense to you.

This is an example of a relatively large concept map:

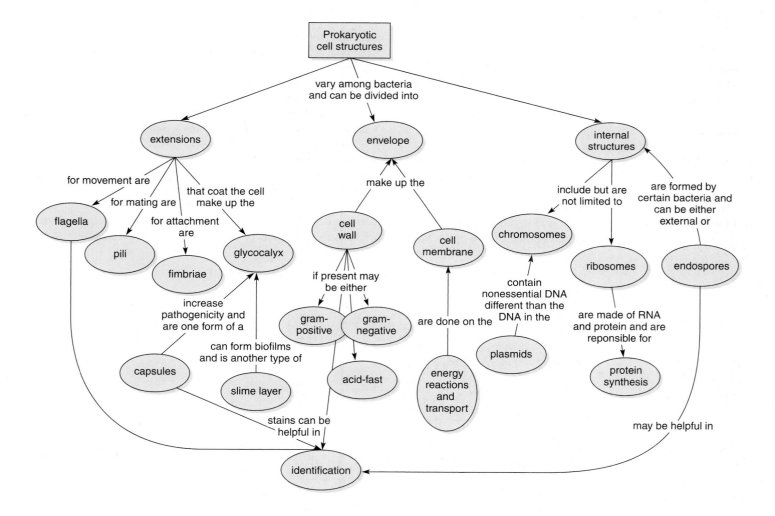

There is a wide variety of different ways to work with concept maps, such as using them as an introductory overview of material or using them as an evaluation tool. There are even software programs that enable concept mappers to create elaborate maps, complete with sound bytes and photos. Some of these will even convert an outline into a concept map for you. In the end-of-chapter materials in this book, we use only three different methods, all of them fairly simple. These three are explained and illustrated here.

All concept maps are made of two basic components:

1. Boxes or circles, each containing a single *concept,* which is most often a noun. The boxes are arranged on the page in vertical, horizontal, or diagonal rows or arrangements. They may also be arranged in a more free-form manner.
2. Connecting lines that join each concept box to at least one other box. Each connecting line has a word or a phrase associated with it—a linking word. These words/phrases are almost never nouns—but are verbs (like "requires") or adjectives or adverbs (like "underneath").

In the end, a picture is created that maps what you know about a subject. It illustrates which concepts are bigger and which are details. It illustrates that multiple concepts may be connected. Experts say that concept maps almost always lead us to conclude that all concepts in a subject can be connected in some way. This is true! And nowhere is it truer than in biology. The trick is to get used to finding the right connecting word to show how two concepts are, indeed, related. When you succeed, you will know the material in a deeper way than is possible by simply answering a single question or even a series of questions.

The first kind of concept map used in this book is the "fill-in-the-blank" version. In these concept maps, you are provided with all the boxes and most of the concepts in the boxes. Some boxes may be blank for you to fill in with the appropriate concept. You will do this by looking at the concepts close to the box and examining the connecting word. In these maps, you will also encounter blanks for linking words/phrases. Sometimes all the blanks will be filled in, but there will be no connecting lines or phrases and you will have to supply these. In a few of these maps, you may be asked to draw the linking lines themselves.

This is an exercise most like answering a simple question. In the example below, for instance, say to yourself "Enzymes are _____ by pH and temperature." You might ask yourself, "What relation do pH and temperature have with enzymes?" Either way, you would probably end up with a linking phrase like *are affected by* or *can be regulated by.* There is some variation in what is a correct answer, but not wide variation.

The second kind of concept map you will see is one in which you will be provided a list of words to be used as concepts. You will be asked to draw the boxes and fill in the words to make sense. Here, there will be a lot of variability based on your view of how the concepts might relate to each other. After you put the concepts in your own boxes, you will need to add linking words/phrases. By the time you have drawn your boxes and added the concepts, you will have many ideas about what kind of linkers you want.

The last type of concept map in this book is the "freestyle" version. You will simply be asked to choose 6 to 10 key words from the Chapter Summary and create a map—complete with linking words.

Many students report that their first experiences with concept mapping can be frustrating. But when they have invested some time in their first few concept maps, many of them find they can never "go back" to organizing information in linear ways. Maps can make the time fly when you're studying. And creating concept maps with a partner or a group is also a great way to review material in a meaningful way. Give concept maps a try. Let your creative side show!

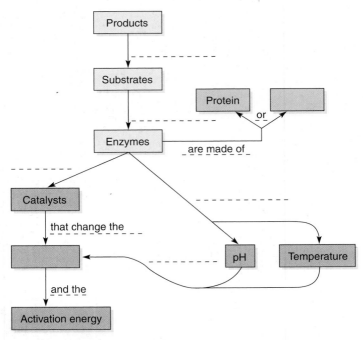

Acknowledgment: Pat Johnson, Palm Beach Community College, supplied information and a concept map for this Appendix.

Answers to Multiple-Choice, Matching, and Case File Questions

Chapter 1
1. d
2. c
3. d
4. c
5. d
6. a
7. c
8. b
9. d
10. c
11. c
12. b
13. d
14. 1st column:
 3, 7, 4, 2
 2nd column
 8, 5, 6, 1
15. c
16. c, not cellular
 and not alive

Case File Questions
1. d
2. c

Chapter 2
1. c
2. c
3. b
4. c
5. a
6. d
7. c
8. b
9. c
10. b
11. b
12. d
13. b
14. c
15. c
16. c
17. a
18. d
19. d
20. b

Case File Questions
1. d
2. b

Chapter 3
1. c
2. b
3. c
4. d
5. b
6. d
7. b
8. b
9. c
10. c
11. a
12. b
13. d
14. d
15. b
16. c
17. abf, df, abf, ef,
 def, af, bf, bf

Case File Questions
1. d
2. b

Chapter 4
1. a
2. d
3. d
4. a
5. c
6. a
7. c
8. b
9. d
10. b
11. d
12. c
13. c
14. c
15. c
16. d

Case File Questions
1. c
2. b

Chapter 5
1. b
2. d
3. d
4. a

5. b
6. c
7. c
8. b
9. c
10. a
11. d
12. b
13. c
14. d
15. a
16. b
17. e
18. c
19. h
20. g
21. j
22. i
23. d
24. a
25. f

Case File Questions
1. f
2. c

Chapter 6
1. c
2. d
3. d
4. b
5. d
6. a
7. a
8. d
9. b
10. b
11. c
12. d
13. a
14. d

Case File Questions
1. b
2. e

Chapter 7
1. c
2. a
3. a
4. c

5. c
6. b
7. a
8. a
9. b
10. b
11. c
12. d
13. c
14. c
15. b
16. c

Case File Questions
1. c
2. e

Chapter 8
1. b
2. a
3. c
4. d
5. d
6. b
7. b
8. b
9. b
10. a
11. c
12. a
13. d
14. b
15. c
16. c
17. c
18. c
19. c
20. c
21. c, a, b, c, b, a,
 c, c

Case File Questions
1. a
2. c

Chapter 9
1. b
2. e
3. b
4. b
5. c

6. b
7. c
8. a
9. b
10. a
11. b
12. d
13. b
14. d
15. b
16. c

Case File Questions
1. c
2. e

Chapter 10
1. c
2. c
3. d
4. a
5. b
6. c
7. c
8. d
9. d
10. h, c, f, a, g, b,
 e, d

Case File Questions
1. b
2. c

Chapter 11
1. d
2. c
3. b
4. a
5. c
6. b
7. d
8. b
9. d
10. c
11. b
12. d
13. c
14. d
15. a

16. b
17. c

Case File Questions
1. c
2. c

Chapter 12
1. b
2. c
3. d
4. b
5. c
6. c
7. b
8. d
9. a
10. c
11. c
12. b

Case File Questions
1. c
2. c
 d, g, b, f, h, c, e

Chapter 13
1. a
2. d
3. b
4. d
5. c
6. d
7. b
8. c
9. c
10. c
11. d
12. c
13. c
14. a
15. a
16. d

Case File Questions
1. b
2. c

Chapter 14
1. b
2. b
3. d

4. b
5. b
6. b
7. c
8. b
9. a
10. a
11. c
12. d
13. c
14. d
15. c

Case File Questions
1. f
2. b

Chapter 15
1. b
2. a
3. c
4. a
5. c
6. b
7. b
8. c
9. a
10. c
11. a
12. c
13. e
14. b
15. c
16. d
17. b
18. c
19. IgG bfgh
 IgA ac
 IgD bj
 IgE bi
 IgM deh

Case File Questions
1. b
2. f

Chapter 16
1. d
2. d
3. d
4. c

5. b
6. c
7. b
8. a
9. d
10. d
11. d
12. a
13. b
14. d
15. d
16. a

Case File Questions
1. d
2. d

Chapter 17
1. b, a, b, c, a
2. c
3. b
4. c
5. a
6. a
7. d
8. c
9. d

Case File Questions
1. c
2. b

Glossary

AB toxin A combined microbial toxin with part B that binds to the cell and part A that disrupts cell function.

abiogenesis The belief in spontaneous generation as a source of life.

abiotic Nonliving factors such as soil, water, temperature, and light that are studied when looking at an ecosystem.

ABO blood group system Developed by Karl Landsteiner in 1904; the identification of different blood groups based on differing isoantigen markers characteristic of each blood type.

abscess An inflamed, fibrous lesion enclosing a core of pus.

abyssal zone The deepest region of the ocean; a sunless, high-pressure, cold, anaerobic habitat.

acellular vaccine A vaccine preparation that contains specific antigens such as the capsule or toxin from a bacterial cell and not the whole pathogen. *Acellular* means without a cell.

acid-fast A term referring to the property of mycobacteria to retain carbol fuchsin even in the presence of acid alcohol. The staining procedure is used to diagnose tuberculosis.

acidic A solution with a pH value below 7 on the pH scale.

acidic fermentation An anaerobic degradation of pyruvic acid that results in organic acid production.

acquired immune deficiency syndrome See **AIDS**

acquired or adaptive immunity Refers to specific immunity developed by lymphocytes responding to antigens.

actin Fine protein fibers within cells that contribute to structure and support.

actinomycetes A group of filamentous, funguslike bacteria.

activated lymphocyte A T or B cell that has received an immune stimulus, such as antigens or cytokines, and is undergoing rapid synthesis and proliferation.

active immunity Immunity acquired through direct stimulation of the immune system by antigen.

active site The specific region on an apoenzyme that binds substrate. The site for reaction catalysis.

active transport Nutrient transport method that requires carrier proteins in the membranes of the living cells and the expenditure of energy.

acute Characterized by rapid onset and short duration.

acyclovir A synthetic purine analog that blocks DNA synthesis in certain viruses, particularly the herpes simplex viruses.

adapt/adaptation The act of successfully adjusting to a new environment, often made possible by random genetic changes that provide an advantage under changed environmental conditions.

adenine (A) One of the nitrogen bases found in DNA and RNA, with a purine form.

adenosine deaminase (ADA) deficiency An immunodeficiency disorder and one type of SCIDS that is caused by an inborn error in the metabolism of adenine. The accumulation of adenine destroys both B and T lymphocytes.

adenosine triphosphate (ATP) A nucleotide that is the primary source of energy to cells.

adenovirus Nonenveloped DNA virus that causes colds; transmissible via respiratory and ocular secretions.

adhesion The process by which microbes gain a more stable foothold at the portal of entry; often involves a specific interaction between the molecules on the microbial surface and the receptors on the host cell.

adjuvant In immunology, a chemical vehicle that enhances antigenicity, presumably by prolonging antigen retention at the injection site.

adsorption A process of adhering one molecule onto the surface of another molecule.

aerobe A microorganism that lives and grows in the presence of free gaseous oxygen (O_2).

aerobic respiration Respiration in which the final electron acceptor in the electron transport chain is oxygen (O_2).

aerosols Airborne suspensions of fine dust or moisture particles airborne that contain live pathogens.

aflatoxin From *Aspergillus flavus* toxin, a mycotoxin that typically poisons moldy animal feed and can cause liver cancer in humans and other animals.

agammaglobulinemia Also called hypogammaglobulinemia. The absence of or severely reduced levels of antibodies in serum.

agar A polysaccharide found in seaweed and commonly used to prepare solid culture media.

agglutination The aggregation by antibodies of suspended cells or similar-size particles (agglutinogens) into clumps that settle.

agglutinin A specific antibody that cross-links agglutinogen, causing it to aggregate.

agglutinogen An antigenic substance on cell surfaces that evokes agglutinin formation against it.

agranulocyte One form of leukocyte (white blood cells), having globular, nonlobed nuclei and lacking prominent cytoplasmic granules.

AIDS Acquired ImmunoDeficiency Syndrome. The complex of signs and symptoms characteristic of the late phase of human immunodeficiency virus (HIV) infection.

alcoholic fermentation An anaerobic degradation of pyruvic acid that results in alcohol production.

algae Photosynthetic, plantlike organisms that generally lack the complex structure of plants; they may be single-celled or multicellular and inhabit diverse habitats such as marine and freshwater environments, glaciers, and hot springs.

allele A gene that occupies the same location as other alternative (allelic) genes on paired chromosomes.

allergen A substance that provokes an allergic response.

allergy The altered, usually exaggerated, immune response to an allergen. Also called hypersensitivity.

alloantigen An antigen that is present in some but not all members of the same species.

allograft Relatively compatible tissue exchange between nonidentical members of the same species. Also called homograft.

allosteric Pertaining to the altered activity of an enzyme due to the binding of a molecule to a region other than the enzyme's active site.

alternative pathway Pathway of complement activation initiated by nonspecific components of tissue or bacterial substances such as endotoxin and polysaccharides.

amastigote The rounded or ovoid nonflagellated form of the *Leishmania* parasite.

Ames test A method for detecting mutagenic and potentially carcinogenic agents based upon the genetic alteration of nutritionally defective bacteria.

amination The addition of an amine (—NH_2) group to a molecule.

amino acids The building blocks of protein. Amino acids exist in 20 naturally occurring forms that impart different characteristics to the various proteins they compose.

aminoglycoside A complex group of drugs derived from soil actinomycetes that impairs ribosome function and has antibiotic potential. Example: streptomycin.

ammonification Phase of the nitrogen cycle in which ammonia is released from decomposing organic material.

amphibolism Pertaining to the metabolic pathways that serve multiple functions in the breakdown, synthesis, and conversion of metabolites.

amphipathic Relating to a compound that has contrasting characteristics, such as hydrophilic-hydrophobic or acid-base.

amphitrichous Having a single flagellum or a tuft of flagella at opposite poles of a microbial cell.

amplicon DNA strand that has been primed for replication during polymerase chain reaction.

anabolism The energy-consuming process of incorporating nutrients into protoplasm through biosynthesis.

anaerobe A microorganism that grows best, or exclusively, in the absence of oxygen.

anaerobic digesters Closed chambers used in a microbial process that converts organic sludge from waste treatment plants into useful fuels such as methane and hydrogen gases. Also called bioreactors.

anaerobic respiration Respiration in which the final electron acceptor in the electron transport chain is an inorganic molecule containing sulfate, nitrate, nitrite, carbonate, etc.

analog In chemistry, a compound that closely resembles another in structure.

anamnestic response In immunology, an augmented response or memory related to a prior stimulation of the immune system by antigen. It boosts the levels of antibodies.

anaphylaxis The unusual or exaggerated allergic reaction to antigen that leads to severe respiratory and cardiac complications.

anion A negatively charged ion.

anneal Characteristic of changing binding properties in response to heating and cooling.

anoxygenic Any reaction that does not produce oxygen; usually in reference to the type of photosynthesis occurring in anaerobic photosynthetic bacteria.

antagonism Relationship in which microorganisms compete for survival in a common environment by taking actions that inhibit or destroy another organism.

anthrax A zoonotic disease of herbivorous livestock. The anthrax bacillus is a facultative parasite and can infect humans in a number of ways. In its most virulent form, it can be fatal.

antibiotic A chemical substance from one microorganism that can inhibit or kill another microbe even in minute amounts.

antibody A large protein molecule evoked in response to an antigen that interacts specifically with that antigen.

anticodon The trinucleotide sequence of transfer RNA that is complementary to the trinucleotide sequence of messenger RNA (the codon).

antigen Any cell, particle, or chemical that induces a specific immune response by B cells or T cells and can stimulate resistance to an infection or a toxin. See **immunogen**.

antigen binding site Specific regions at the ends of the antibody molecule that recognize specific antigens. These sites have numerous shapes to fit a wide variety of antigens.

antigen-presenting cell (APC) A macrophage or dendritic cell that ingests and degrades an antigen and subsequently places the antigenic determinant molecules on its surface for recognition by CD4 T lymphocytes.

antigenic determinant The precise molecular group of an antigen that defines its specificity and triggers the immune response.

antigenic drift Minor antigenic changes in the influenza A virus due to mutations in the spikes' genes.

antigenic shift Major changes in the influenza A virus due to recombination of viral strains from two different host species.

antigenicity The property of a substance to stimulate a specific immune response such as antibody formation.

antihistamine A drug that counters the action of histamine and is useful in allergy treatment.

anti-idiotype An anti-antibody that reacts specifically with the idiotype (variable region or antigen binding site) of another antibody.

antimetabolite A substance such as a drug that competes with, substitutes for, or interferes with a normal metabolite.

antimicrobic A special class of compounds capable of destroying or inhibiting microorganisms.

antiseptic A growth-inhibiting agent used on tissues to prevent infection. Use of such an agent is antisepsis.

antiserum Antibody-rich serum derived from the blood of animals (deliberately immunized against infectious or toxic antigen) or from people who have recovered from specific infections.

antitoxin Globulin fraction of serum that neutralizes a specific toxin. Also refers to the specific antitoxin antibody itself.

apoenzyme The protein part of an enzyme, as opposed to the nonprotein or inorganic cofactors.

apoptosis The genetically programmed death of cells that is both a natural process of development and the body's means of destroying abnormal or infected cells.

appendages Accessory structures that sprout from the surface of bacteria. They can be divided into two major groups: those that provide motility and those that enable adhesion.

aquifer A subterranean water-bearing stratum of permeable rock, sand, or gravel.

arboviruses Arthropod-borne viruses including togaviruses, reoviruses, flaviviruses, and bunyaviruses. These viruses generally cause mild, undifferentiated fevers and occasionally cases of severe encephalitis and hemorrhagic fever.

archaea Prokaryotic single-celled organisms of primitive origin that have unusual anatomy, physiology, and genetics, and live in harsh habitats; when capitalized (**Archaea**), the term refers to one of the three domains of living organisms as proposed by Woese.

arthropod A phylum of invertebrate animals with jointed exoskeletons. Includes insects such as mosquitoes and arachnids such as ticks that may be vectors of infectious diseases.

arthrospore A fungal spore formed by the septation and fragmentation of hyphae.

Arthus reaction An immune complex phenomenon that develops after repeat injection. This localized inflammation results from aggregates of antigen and antibody that bind, complement, and attract neutrophils.

artificial chromosome A large packet of DNA from yeasts (YAC), bacteria (BAC), or humans (HAC) that can be used to carry, transfer, or analyze isolated foreign DNA.

artificial immunity Specific protection acquired through medical means.

ascospore A spore formed within a saclike cell (ascus) of Ascomycota following nuclear fusion and meiosis.

ascus Special fungal sac in which haploid spores are created.

asepsis A condition free of viable pathogenic microorganisms.

aseptic technique Methods of handling microbial cultures, patient specimens, and other sources of microbes in a way that prevents infection of the handler and others who may be exposed.

asthma An allergic respiratory disease marked by extreme sensitivity to allergens.

asymptomatic An infection that produces no noticeable symptoms even though the microbe is active in the host tissue.

asymptomatic carrier A person with an inapparent infection who shows no symptoms of being infected yet is able to pass the disease agent on to others.

atmosphere That part of the biosphere that includes the gaseous envelope up to 14 miles above the earth's surface. It contains gases such as carbon dioxide, nitrogen, and oxygen.

atom The smallest particle of an element to retain all the properties of that element.

atomic number (AN) A measurement that reflects the number of protons in an atom of a particular element.

atomic weight The average of the mass numbers of all the isotopic forms for a particular element.

atopy Allergic reaction classified as type I, with a strong familial relationship; caused by allergens such as pollen, insect venom, food, and dander; involves IgE antibody; includes symptoms of hay fever, asthma, and skin rash.

ATP synthase A unique enzyme located in the mitochondrial cristae and chloroplast grana that harnesses the flux of hydrogen ions to the synthesis of ATP.

attenuate To reduce the virulence of a pathogenic bacterium or virus by passing it through a non-native host or by long-term subculture.

AUG (start codon) The codon that signals the point at which translation of a messenger RNA molecule is to begin.

autoantibody An "antiself" antibody having an affinity for tissue antigens of the subject in which it is formed.

autoantigen Molecules that are inherently part of self but are perceived by the immune system as foreign.

autoclave A sterilization chamber that allows the use of steam under pressure to sterilize materials. The most common temperature/pressure combination for an autoclave is 121°C and 15 psi.

autograft Tissue or organ surgically transplanted to another site on the same subject.

autoimmune disease The pathologic condition arising from the production of antibodies against autoantigens. Example: rheumatoid arthritis. Also called autoimmunity.

autosome A chromosome of somatic cells as opposed to a sex chromosome of gametes.

autotroph A microorganism that requires only inorganic nutrients and whose sole source of carbon is carbon dioxide.

axenic A sterile state such as a pure culture. An axenic animal is born and raised in a germ-free environment. See **gnotobiotic**.

axial filament A type of flagellum (called an endoflagellum) that lies in the periplasmic space of spirochetes and is responsible for locomotion. Also called periplasmic flagellum.

azidothymidine (AZT) A thymine analog used in the treatment of HIV infection. AZT inhibits the action of the reverse transcriptase enzyme, making it specific for HIV.

azole Five-membered heterocyclic compounds typical of histidine, which are used in antifungal therapy.

bacillus Bacterial cell shape that is cylindrical (longer than it is wide).

back-mutation A mutation that counteracts an earlier mutation, resulting in the restoration of the original DNA sequence.

bacteremia The presence of viable bacteria in circulating blood.

bacteria (plural of *bacterium*) Category of prokaryotes with peptidoglycan in their cell walls and a single, circular chromosome. This group of small cells is widely distributed in the earth's habitats.

Bacteria When capitalized refers to one of the three domains of living organisms proposed by Woese, containing all non-archaea prokaryotes.

bacterial chromosome A circular body in bacteria that contains the primary genetic material. Also called **nucleoid**.

bacterial meningitis Inflammation of the meninges as a result of bacterial infection.

bactericide An agent that kills bacteria.

bacteriocin Proteins produced by certain bacteria that are lethal against closely related bacteria and are narrow spectrum compared with antibiotics; these proteins are coded and transferred in plasmids.

bacteriophage A virus that specifically infects bacteria.

bacteriostatic Any process or agent that inhibits bacterial growth.

bacterium A tiny unicellular prokaryotic organism that usually reproduces by binary fission and usually has a peptidoglycan cell wall, has various shapes, and can be found in virtually any environment.

barophile A microorganism that thrives under high (usually hydrostatic) pressure.

basement membrane A thin layer (1–6 μm) of protein and polysaccharide found at the base of epithelial tissues.

basic A solution with a pH value above 7 on the pH scale.

basidiospore A sexual spore that arises from a basidium. Found in basidiomycota fungi.

basidium A reproductive cell created when the swollen terminal cell of a hypha develops filaments (sterigmata) that form spores.

basophil A motile polymorphonuclear leukocyte that binds IgE. The basophilic cytoplasmic granules contain mediators of anaphylaxis and atopy.

Bdellovibrio A bacterium that preys on certain other bacteria. It bores a hole into a specific host and inserts itself between the protoplast and the cell wall. There it elongates before subdividing into several cells and devouring the host cell.

benign tumor A self-contained growth within an organ that does not spread into adjacent tissue.

benthic zone The sedimentary bottom region of a pond, lake, or ocean.

beta oxidation The degradation of long-chain fatty acids. Two-carbon fragments are formed as a result of enzymatic attack directed against the second or beta carbon of the hydrocarbon chain. Aided by coenzyme A, the fragments enter the tricarboxylic acid (TCA) cycle and are processed for ATP synthesis.

beta-lactamase An enzyme secreted by certain bacteria that cleaves the beta-lactam ring of penicillin and cephalosporin and thus provides for resistance against the antibiotic. See penicillinase.

binary fission The formation of two new cells of approximately equal size as the result of parent cell division.

binomial system Scientific method of assigning names to organisms that employs two names to identify every organism—genus name plus species name.

bioamplification The concentration or accumulation of a pollutant in living tissue through the natural flow of an ecosystem.

biochemistry The study of organic compounds produced by (or components of) living things. The four main categories of biochemicals are carbohydrates, lipids, proteins, and nucleic acid.

bioenergetics The study of the production and use of energy by cells.

bioethics The study of biological issues and how they relate to human conduct and moral judgment.

biofilm A complex association that arises from microorganisms growing together and interacting on the surface of a habitat.

biogenesis Belief that living things can only arise from others of the same kind.

biogeochemical cycle A process by which matter is converted from organic to inorganic form and returned to various nonliving reservoirs on earth (air, rocks, and water) where it becomes available for reuse by living things. Elements such as carbon, nitrogen, and phosphorus are constantly cycled in this manner.

bioinformatics A field that uses computer technology to manage and analyze large sets of biological data, including genome and protein sequences.

biological vector An animal that not only transports an infectious agent but plays a role in the life cycle of the pathogen, serving as a site in which it can multiply or complete its life cycle. It is usually an alternate host to the pathogen.

biomes Particular climate regions in a terrestrial realm.

bioremediation The use of microbes to reduce or degrade pollutants, industrial wastes, and household garbage.

biosphere Habitable regions comprising the aquatic (hydrospheric), soil-rock (lithospheric), and air (atmospheric) environments.

biotechnology The use of microbes or their products in the commercial or industrial realm.

biotic Living factors such as parasites, food substrates, or other living or once-living organisms that are studied when looking at an ecosystem.

blocking antibody The IgG class of immunoglobulins that competes with IgE antibody for allergens, thus blocking the degranulation of basophils and mast cells.

blood cells Cellular components of the blood consisting of red blood cells, primarily responsible for the transport of oxygen and carbon dioxide, and white blood cells, primarily responsible for host defense and immune reactions.

B lymphocyte (B cell) A white blood cell that gives rise to plasma cells and antibodies.

booster The additional doses of vaccine antigen administered to increase an immune response and extend protection.

botulin *Clostridium botulinum* toxin. Ingestion of this potent exotoxin leads to flaccid paralysis.

bradykinin An active polypeptide that is a potent vasodilator released from IgE-coated mast cells during anaphylaxis.

broad spectrum A term to denote drugs that affect many different types of bacteria, both gram-positive and gram-negative.

Brownian movement The passive, erratic, nondirectional motion exhibited by microscopic particles. The jostling comes from being randomly bumped by submicroscopic particles, usually water molecules, in which the visible particles are suspended.

brucellosis A zoonosis transmitted to humans from infected animals or animal products; causes a fluctuating pattern of severe fever in humans as well as muscle pain, weakness, headache, weight loss, and profuse sweating. Also called undulant fever.

bubo The swelling of one or more lymph nodes due to inflammation.

bubonic plague The form of plague in which bacterial growth is primarily restricted to the lymph and is characterized by the appearance of a swollen lymph node referred to as a bubo.

budding See exocytosis.

bulbar poliomyelitis Complication of polio infection in which the brain stem, medulla, or cranial nerves are affected. Leads to loss of respiratory control and paralysis of the trunk and limbs.

bulla A large, bubblelike vesicle in a region of separation between the epidermis and the subepidermal layer. The space is usually filled with serum and sometimes with blood.

calculus Dental deposit formed when plaque becomes mineralized with calcium and phosphate crystals. Also called **tartar.**

Calvin cycle The recurrent photosynthetic pathway characterized by CO_2 fixation and glucose synthesis. Also called the dark reactions.

cancer Any malignant neoplasm that invades surrounding tissue and can metastasize to other locations. A carcinoma is derived from epithelial tissue, and a sarcoma arises from proliferating mesodermal cells of connective tissue.

capsid The protein covering of a virus' nucleic acid core. Capsids exhibit symmetry due to the regular arrangement of subunits called capsomers. See icosahedron.

capsomer A subunit of the virus capsid shaped as a triangle or disc.

capsule In bacteria, the loose, gellike covering or slime made chiefly of simple polysaccharides. This layer is protective and can be associated with virulence.

capsule antigen (K or fimbrial antigen) The capsular cell surface antigen that can be used to type or identify gram-negative enteric bacteria.

carbohydrate A compound containing primarily carbon, hydrogen, and oxygen in a 1:2:1 ratio.

carbon cycle That pathway taken by carbon from its abiotic source to its use by producers to form organic compounds (biotic), followed by the breakdown of biotic compounds and their release to a nonliving reservoir in the environment (mostly carbon dioxide in the atmosphere).

carbon fixation Reactions in photosynthesis that incorporate inorganic carbon dioxide into organic compounds such as sugars. This occurs during the Calvin cycle and uses energy generated by the light reactions. This process is the source of all production on earth.

carbuncle A deep staphylococcal abscess joining several neighboring hair follicles.

carcinoma Cancers originating in epithelial tissue.

carotenoid Yellow, orange, or red photosynthetic pigments.

carrier A person who harbors infections and inconspicuously spreads them to others. Also, a chemical agent that can accept an atom, chemical radical, or subatomic particle from one compound and pass it on to another.

caseous lesion Necrotic area of lung tubercle superficially resembling cheese. Typical of tuberculosis.

catabolism The chemical breakdown of complex compounds into simpler units to be used in cell metabolism.

catalyst A substance that alters the rate of a reaction without being consumed or permanently changed by it. In cells, enzymes are catalysts.

catalytic site The niche in an enzyme where the substrate is converted to the product (also active site).

cation A positively charged ion.

CD4 cluster of differentiation, type 4 An antigen associated with helper T-lymphocytes.

CD8 cluster of differentiation, type 8 An antigen associated with cytotoxic T lymphocytes.

cecum The intestinal pocket that forms the first segment of the large intestine. Also called the appendix.

cell An individual membrane-bound living entity; the smallest unit capable of an independent existence.

cell-mediated immunity The type of immune responses brought about by T cells, such as cytotoxic, suppressor, and helper effects.

cellulitis The spread of bacteria within necrotic tissue.

cellulose A long, fibrous polymer composed of β-glucose; one of the most common substances on earth.

cephalosporins A group of broad-spectrum antibiotics isolated from the fungus *Cephalosporium.*

cercaria The free-swimming larva of the schistosome trematode that emerges from the snail host and can penetrate human skin, causing schistosomiasis.

cestode The common name for tapeworms that parasitize humans and domestic animals.

chancre The primary sore of syphilis that forms at the site of penetration by *Treponema pallidum.* It begins as a hard, dull red, painless papule that erodes from the center.

chancroid A lesion that resembles a chancre but is soft and is caused by *Haemophilus ducreyi.*

chemical bond A link formed between molecules when two or more atoms share, donate, or accept electrons.

chemical mediators Small molecules that are released during inflammation and specific immune reactions that allow communication between the cells of the immune system and facilitate surveillance, recognition and attack.

chemiosmotic hypothesis An explanation for ATP formation that is based on the formation of a proton (H^+) gradient across a membrane during electron transport. Movement of the protons back across an ATP synthase causes the formation of ATP.

chemoautotroph An organism that relies upon inorganic chemicals for its energy and carbon dioxide for its carbon. Also called a chemolithotroph.

chemoheterotroph Microorganisms that derive their nutritional needs from organic compounds.

chemokines Chemical mediators (cytokines) that stimulate the movement and migration of white blood cells.

chemostat A growth chamber with an outflow that is equal to the continuous inflow of nutrient media. This steady-state growth device is used to study such events as cell division, mutation rates, and enzyme regulation.

chemotactic factors Chemical mediators that stimulate the movement of white blood cells. See chemokines.

chemotaxis The tendency of organisms to move in response to a chemical gradient (toward an attractant or to avoid adverse stimuli).

chemotherapy The use of chemical substances or drugs to treat or prevent disease.

chemotroph An organism that oxidizes compounds to feed on nutrients.

chickenpox A papulo-pustular rash caused by infection by varicella-zoster virus.

chimera A product formed by the fusion of two different organisms.

chitin A polysaccharide similar to cellulose in chemical structure. This polymer makes up the horny substance of the exoskeletons of arthropods and certain fungi.

chloramphenicol A potent broad-spectrum antibiotic that is no longer widely used because of its potential for causing lethal damage to the blood-forming tissues.

chlorophyll A group of mostly green pigments that are used by photosynthetic eukaryotic organisms and cyanobacteria to trap light energy to use in making chemical bonds.

chloroplast An organelle containing chlorophyll that is found in photosynthetic eukaryotes.

cholesterol Best-known member of a group of lipids called steroids. Cholesterol is commonly found in cell membranes and animal hormones.

chromatin The genetic material of the nucleus. Chromatin is made up of nucleic acid and stains readily with certain dyes.

chromophore The chemical radical of a dye that is responsible for its color and reactivity.

chromosome The tightly coiled bodies in cells that are the primary sites of genes.

chronic Any process or disease that persists over a long duration.

chronic fatigue syndrome A collection of persistent, nonspecific symptoms possibly linked to immune deficiency and/or viral infection.

cilium (Plural: *cilia*) Eukaryotic structure similar to flagella that propels a protozoan through the environment.

class In the levels of classification, the division of organisms that follows phylum.

classical pathway Pathway of complement activation initiated by a specific antigen-antibody interaction.

clonal selection theory A conceptual explanation for the development of lymphocyte specificity and variety during immune maturation.

clone A colony of cells (or group of organisms) derived from a single cell (or single organism) by asexual reproduction. All units share identical characteristics. Also used as a verb to refer to the process of producing a genetically identical population of cells or genes.

cloning host An organism such as a bacterium or a yeast that receives and replicates a foreign piece of DNA inserted during a genetic engineering experiment.

cluster of differentiation (CD) Receptors on mature T cells that serve in cell communication; react with MHC receptors for immune recognition by antigen processing cells.

coagulase A plasma-clotting enzyme secreted by *Staphylococcus aureus*. It contributes to virulence and is involved in forming a fibrin wall that surrounds staphylococcal lesions.

coagulase-negative staphylococci (CNS) Opportunistic species that are usually part of the normal flora of the skin and mucous membranes. They cause infection when host defenses are low and lack many virulence factors such as coagulase.

coccobacillus An elongated coccus; a short, thick, oval-shaped bacterial rod.

coccus A spherical-shaped bacterial cell.

codon A specific sequence of three nucleotides in mRNA (or the sense strand of DNA) that constitutes the genetic code for a particular amino acid.

coenzyme A complex organic molecule, several of which are derived from vitamins (e.g., nicotinamide, riboflavin). A coenzyme operates in conjunction with an enzyme. Coenzymes serve as transient carriers of specific atoms or functional groups during metabolic reactions.

cofactor An enzyme accessory. It can be organic, such as coenzymes, or inorganic, such as Fe^{2+}, Mn^{2+}, or Zn^{2+} ions.

cold sterilization The use of nonheating methods such as radiation or filtration to sterilize materials.

coliform A collective term that includes normal enteric bacteria that are gram-negative and lactose-fermenting.

colinear Having corresponding parts of a molecule arranged in the same linear order as another molecule, as in DNA and mRNA.

colitis Inflammation and necrosis of the colon that may be caused by infection. Mild cases lead to diarrhea, while more severe cases can result in perforation of the caecum.

colony A macroscopic cluster of cells appearing on a solid medium, each arising from the multiplication of a single cell.

colostrum The clear yellow early product of breast milk that is very high in secretory antibodies. Provides passive intestinal protection.

commensalism An unequal relationship in which one species derives benefit without harming the other.

common (seed) warts Painless, elevated rough growths on the fingers or occasionally other parts of the body.

communicable infection Capable of being transmitted from one individual to another.

community The interacting mixture of populations in a given habitat.

competitive inhibition Control process that relies on the ability of metabolic analogs to control microbial growth by successfully competing with a necessary enzyme to halt the growth of bacterial cells.

complement In immunology, serum protein components that act in a definite sequence when set in motion either by an antigen-antibody complex or by factors of the alternative (properdin) pathway.

complementary DNA (cDNA) DNA created by using reverse transcriptase to synthesize DNA from RNA templates.

complex (nonsynthetic) medium Culture medium containing at least one component that is chemically undefined.

compounds Molecules that are a combination of two or more different elements.

concentration The expression of the amount of a solute dissolved in a certain amount of solvent. It may be defined by weight, volume, or percentage.

condylomata acuminata Extensive, branched masses of genital warts caused by infection with human papillomavirus.

congenital Transmission of an infection from mother to a fetus in utero.

congenital rubella Transmission of the rubella virus to a fetus in utero. Injury to the fetus is generally much more serious than it is to the mother.

conidia Asexual fungal spores shed as free units from the tips of fertile hyphae.

conjugation In bacteria, the contact between donor and recipient cells associated with the transfer of genetic material such as plasmids. Can involve special (sex) pili. Also a form of sexual recombination in ciliated protozoans.

conjunctivitis Sometimes called "pinkeye." A *Haemophilus* infection of the subconjunctiva that is common among children, is easily transmitted, and is treated with antibiotic eyedrops.

constitutive enzyme An enzyme present in bacterial cells in constant amounts, regardless of the presence of substrate. Enzymes of the central catabolic pathways are typical examples.

consumer An organism that feeds on producers or other consumers. It gets all nutrients and energy from other organisms (also called **heterotroph**). May exist at several levels, such as primary (feeds on producers), and secondary (feeds on primary consumers).

contagious Communicable; transmissible by direct contact with infected people and their fresh secretions or excretions.

contaminant An impurity; any undesirable material or organism. A culture into which unknown microbes have been introduced is contaminated.

control locus The region of an operon that regulates the transcription of the structural elements of the operon.

convalescence Recovery; the period between the end of a disease and the complete restoration of health in a patient.

corepressor A molecule that combines with inactive repressor to form active repressor, which attaches to the operator gene site and inhibits the activity of structural genes subordinate to the operator.

covalent bond A chemical bond formed by the sharing of electrons between two atoms.

Creutzfeld-Jakob disease (CJD) A spongiform encephalopathy caused by infection with a prion. The disease is marked by dementia, impaired senses, and uncontrollable muscle contractions.

crista The infolded inner membrane of a mitochondrion that is the site of the respiratory chain and oxidative phosphorylation.

culture The visible accumulation of microorganisms in or on a nutrient medium. Also, the propagation of microorganisms with various media.

curd The coagulated milk protein used in cheese making.

cutaneous Second level of skin, including the stratum corneum and occasionally the upper dermis.

cutaneous anthrax The mildest form of anthrax, caused by the entrance of bacterial spores into small nicks or openings in the skin, their germination, and the formation of a dark necrotic lesion called an eschar.

cutaneous candidiasis A mycosis occurring in the skin and membranes of compromised patients and neonates as a result of infection by *Candida albicans*.

cyst The resistant, dormant, but infectious form of protozoans. Can be important in spread of infectious agents such as *Entamoeba histolytica* and *Giardia lamblia*.

cysteine Sulfur-containing amino acid that is essential to the structure and specificity of proteins.

cysticercus The larval form of certain *Taenia* species, which typically infest muscles of mammalian intermediate hosts. Also called bladderworm.

cystine An oxidation product of two cysteine molecules in which the —SH (sulfhydryl) groups form a disulfide union. Also called dicysteine. HOOC—$CH(NH_2)$—CH_2—S—S—CH_2—CH (NH_2)COOH.

cytochrome A group of heme protein compounds whose chief role is in electron and/or hydrogen transport occurring in the last phase of aerobic respiration.

cytokine A chemical substance produced by white blood cells and tissue cells that regulates development, inflammation, and immunity.

cytopathic effect The degenerative changes in cells associated with virus infection. Examples: the formation of multinucleate giant cells (Negri bodies), the prominent cytoplasmic inclusions of nerve cells infected by rabies virus.

cytoplasm Dense fluid encased by the cell membrane; the site of many of the cell's biochemical and synthetic activities.

cytosine (C) One of the nitrogen bases found in DNA and RNA, with a pyrimidine form.

cytoskeleton A fine intracellular network of fibers, filaments, and other strands that function in support and shape of cells.

cytotoxic Having the capacity to destroy specific cells. One class of T cells attacks cancer cells, virus-infected cells, and eukaryotic pathogens. See **killer T cells.**

death phase End of the cell growth due to lack of nutrition, depletion of environment, and accumulation of wastes. Population of cells begins to die.

debridement Trimming away devitalized tissue and foreign matter from a wound.

decomposer A consumer that feeds on organic matter from the bodies of dead organisms. These microorganisms feed from all levels of the food pyramid and are responsible for recycling elements (also called saprobes).

decomposition The breakdown of dead matter and wastes into simple compounds that can be directed back into the natural cycle of living things.

decontamination The removal or neutralization of an infectious, poisonous, or injurious agent from a site.

deduction Problem-solving process in which an individual constructs a hypothesis, tests its validity by outlining particular events that are predicted by the hypothesis, and then performs experiments to test for those events.

defined medium Culture medium containing components whose chemical makeup is completely known. Same as synthetic medium.

definitive host The organism in which a parasite develops into its adult or sexually mature stage. Also called the **final host.**

degeneracy The property of the genetic code that allows an amino acid to be specified by several different codons.

degerm To physically remove surface oils, debris, and soil from skin to reduce the microbial load.

degranulation The release of cytoplasmic granules, as when cytokines are secreted from mast cell granules.

dehydration synthesis During the formation of a carbohydrate bond, the step in which one carbon molecule gives up its OH group and the other loses the H from its OH group, thereby producing a water molecule. This process is common to all polymerization reactions.

denature/denaturation The loss of normal characteristics resulting from some molecular alteration. Usually in reference to the action of heat or chemicals on proteins whose function depends upon an unaltered tertiary structure.

dendritic cell A large, antigen-processing cell characterized by long, branchlike extensions of the cell membrane.

denitrification The end of the nitrogen cycle when nitrogen compounds are returned to the reservoir in the air.

dental caries A mixed infection of the tooth surface that gradually destroys the enamel and may lead to destruction of the deeper tissue.

deoxyribonucleic acid (DNA) The nucleic acid often referred to as the "double helix." DNA carries the master plan for an organism's heredity.

deoxyribose A 5-carbon sugar that is an important component of DNA.

dermatophytoses Superficial mycoses such as athlete's foot and ringworm associated with certain fungi with an affinity for the skin, hair, and nails. These fungi are called **dermatophytes.**

desensitization See **hyposensitization.**

desiccation To dry thoroughly. To preserve by drying.

desquamate To shed the cuticle in scales; to peel off the outer layer of a surface.

diabetes mellitus A disease involving compromise in insulin function. In one form, the pancreatic cells that produce insulin are destroyed by autoantibodies, and in another, the pancreas does not produce sufficient insulin.

diapedesis The migration of intact blood cells between endothelial cells of a blood vessel such as a venule.

differential medium A single substrate that discriminates between groups of microorganisms on the basis of differences in their appearance due to different chemical reactions.

differential stain A technique that utilizes two dyes to distinguish between different microbial groups or cell parts by color reaction.

diffusion The dispersal of molecules, ions, or microscopic particles propelled down a concentration gradient by spontaneous random motion to achieve a uniform distribution.

DiGeorge syndrome A genetic defect that causes the loss of thymus function; results in abnormally low or absent T cells and other developmental abnormalities.

dimorphic In mycology, the tendency of some pathogens to alter their growth form from mold to yeast in response to rising temperature.

dinoflagellate A marine algae whose toxin can cause food poisoning. Overgrowth of this algae is responsible for the phenomenon known as "red tide."

diphtheria Infection by *Corynebacterium diphtheriae.* It is transmitted by human carriers or contaminated milk, and the primary infection is in the upper respiratory tract. Several forms of this infection may be fatal if untreated. Vaccination on the recommended schedule can prevent infection.

diphtherotoxin Exotoxin responsible for the effects seen in diphtheria. The toxin affects the upper respiratory system, peripheral nervous system and the heart.

diplococcus Spherical or oval-shaped bacteria, typically found in pairs.

diploid Somatic cells having twice the basic chromosome number. One set in the pair is derived from the father and the other from the mother.

direct, or total, cell count 1. Counting total numbers of individual cells being viewed with magnification. 2. Counting isolated colonies of organisms growing on a plate of media as a way to determine population size.

disaccharide A sugar containing two monosaccharides. Example: sucrose (fructose + glucose).

disease Any deviation from health, as when the effects of microbial infection damage or disrupt tissues and organs.

disinfection The destruction of pathogenic nonsporulating microbes or their toxins, usually on inanimate surfaces.

disseminated cytomegalovirus Systemic disease brought on by infection with cytomegalovirus. Often seen in AIDS patients and others susceptible to opportunistic infection.

division In the levels of classification, an alternate term for *phylum.*

DNA See **deoxyribonucleic acid.**

DNA fingerprint A pattern of restriction enzyme fragments that is unique for an individual organism.

DNA polymerase Enzyme responsible for the replication of DNA. Several versions of the enzyme exist, each completing a unique portion of the replication process.

DNA sequencing Determining the exact order of nucleotides in a fragment of DNA. Most commonly done using the Sanger dideoxy sequencing method.

DNA vaccine A newer vaccine preparation based on inserting DNA from pathogens into host cells to encourage them to express the foreign protein and stimulate immunity.

domain In the levels of classification, the broadest general category to which an organism is assigned. Members of a domain share only one or a few general characteristics.

droplet nuclei The dried residue of fine droplets produced by mucus and saliva sprayed while sneezing and coughing. Droplet nuclei are less than 5 μm in diameter (large enough to bear a single bacterium and small enough to remain airborne for a long time) and can be carried by air currents. Droplet nuclei are drawn deep into the air passages.

drug resistance An adaptive response in which microorganisms begin to tolerate an amount of drug that would ordinarily be inhibitory.

dry heat Air with a low moisture content that has been heated from 160 to several thousand degrees Celsius.

dysentery Diarrheal illness caused by exotoxins.

dyspnea Difficulty in breathing.

ecchymoses Large purple bruises caused by hemorrhage into the skin.

ecosystem A collection of organisms together with its surrounding physical and chemical factors.

ectoplasm The outer, more viscous region of the cytoplasm of a phagocytic cell such as an amoeba. It contains microtubules but not granules or organelles.

eczema An acute or chronic allergy of the skin associated with itching and burning sensations. Typically, red, edematous, vesicular lesions erupt, leaving the skin scaly and sometimes hyperpigmented.

edema The accumulation of excess fluid in cells, tissues, or serous cavities. Also called swelling.

electrolyte Any compound that ionizes in solution and conducts current in an electrical field.

electromagnetic radiation A form of energy that is emitted as waves and is propagated through space and matter. The spectrum extends from short gamma rays to long radio waves.

electron A negatively charged subatomic particle that is distributed around the nucleus in an atom.

electrophoresis The separation of molecules by size and charge through exposure to an electrical current.

electrostatic Relating to the attraction of opposite charges and the repulsion of like charges. Electrical charge remains stationary as opposed to electrical flow or current.

element A substance comprising only one kind of atom that cannot be degraded into two or more substances without losing its chemical characteristics.

ELISA Abbreviation for enzyme-linked immunosorbent assay, a very sensitive serological test used to detect antibodies in diseases such as AIDS. Two types are **indirect** and **capture ELISA.**

emerging disease Newly identified diseases that are becoming more prominent.

encystment The process of becoming encapsulated by a membranous sac.

endemic disease A native disease that prevails continuously in a geographic region.

endergonic reaction A chemical reaction that occurs with the absorption and storage of surrounding energy. Antonym: **exergonic.**

endocarditis An inflammation of the lining and valves of the heart. Often caused by infection with pyogenic cocci.

endocytosis The process whereby solid and liquid materials are taken into the cell through membrane invagination and engulfment into a vesicle.

endoenzyme An intracellular enzyme, as opposed to enzymes that are secreted.

endogenous Originating or produced within an organism or one of its parts.

endoplasmic reticulum (ER) An intracellular network of flattened sacs or tubules with or without ribosomes on their surfaces.

endospore A small, dormant, resistant derivative of a bacterial cell that germinates under favorable growth conditions into a vegetative cell. The bacterial genera *Bacillus* and *Clostridium* are typical sporeformers.

endosymbiosis Relationship in which a microorganism resides within a host cell and provides a benefit to the host cell.

endotoxin A bacterial intracellular toxin that is not ordinarily released (as is exotoxin). Endotoxin is composed of a phospholipid-polysaccharide complex that is an integral part of gram-negative bacterial cell walls. Endotoxins can cause severe shock and fever.

energy of activation The minimum energy input necessary for reactants to form products in a chemical reaction.

energy pyramid An ecological model that shows the energy flow among the organisms in a community. It is structured like the food pyramid, but shows how energy is reduced from one trophic level to another.

enriched medium A nutrient medium supplemented with blood, serum, or some growth factor to promote the multiplication of fastidious microorganisms.

enteric Pertaining to the intestine.

enterohemorrhagic *E. coli* (EHEC) A strain of *E. coli,* type 0157: H7, that secretes toxins that provoke secretory diarrhea and intestinal hemorrhage.

enteroinvasive Predisposed to invade the intestinal tissues.

enteropathogenic Pathogenic to the alimentary canal.

enterotoxigenic *E. coli* (ETEC) Strains of *E. coli* that produce toxins that attack the GI tract; effects are similar to cholera.

enterotoxin A bacterial toxin that specifically targets intestinal mucous membrane cells. Enterotoxigenic strains of *Escherichia coli* and *Staphylococcus aureus* are typical sources.

enveloped virus A virus whose nucleocapsid is enclosed by a membrane derived in part from the host cell. It usually contains exposed glycoprotein spikes specific for the virus.

enzyme A protein biocatalyst that facilitates metabolic reactions.

enzyme induction One of the controls on enzyme synthesis. This occurs when enzymes appear only when suitable substrates are present.

enzyme repression The inhibition of enzyme synthesis by the end product of a catabolic pathway.

eosinophil A leukocyte whose cytoplasmic granules readily stain with red eosin dye.

epidemic A sudden and simultaneous outbreak or increase in the number of cases of disease in a community.

epidemic parotitis (mumps) Disease caused by infection with *Paramyxovirus* and marked by swelling in and around the parotid salivary glands.

epidemiology The study of the factors affecting the prevalence and spread of disease within a community.

epimastigote The trypanosomal form found in the tsetse fly or reduviid bug vector. Its flagellum originates near the nucleus, extends along an undulating membrane, and emerges from the anterior end.

episome A small fragment of DNA such as a plasmid that may exist as an extra-chromosomal entity.

epitope The portion of a foreign cell or virus that is the precise stimulus for an immune response. Also called the antigenic determinant.

Epstein-Barr virus (EBV) Herpesvirus linked to infectious mononucleosis, Burkitt lymphoma, and nasopharyngeal carcinoma.

erysipelas An acute, sharply defined inflammatory disease specifically caused by hemolytic *Streptococcus*. The eruption is limited to the skin but can be complicated by serious systemic symptoms.

erysipeloid An inflammation resembling erysipelas but caused by *Erysipelothrix,* a gram-positive rod. The self-limited cellulitis that appears at the site of an infected wound, usually the hand, comes from handling contaminated fish or meat.

erythema An inflammatory redness of the skin.

erythroblastosis fetalis Hemolytic anemia of the newborn. The anemia comes from hemolysis of Rh-positive fetal erythrocytes by anti-Rh maternal antibodies. Erythroblasts are immature red blood cells prematurely released from the bone marrow.

erythrocytes (red blood cells) Blood cells involved in the transport of oxygen and carbon dioxide.

erythrogenic toxin An exotoxin produced by lysogenized group A strains of β-hemolytic streptococci that is responsible for the severe fever and rash of scarlet fever in the nonimmune individual. Also called a pyrogenic toxin.

eschar A dark, sloughing scab that is the lesion of anthrax and certain rickettsioses.

essential nutrient Any ingredient such as a certain amino acid, fatty acid, vitamin, or mineral that cannot be formed by an organism and must be supplied in the diet. A growth factor.

ester bond A covalent bond formed by reacting carboxylic acid with an OH group:

$$R-\overset{\overset{\textstyle O}{\|}}{C}-O-R'$$

Olive and corn oils, lard, and butter are examples of triacylglycerols—esters formed between glycerol and three fatty acids.

estuary The intertidal zone where a river empties into the sea.

ethylene oxide A potent, highly water-soluble gas invaluable for gaseous sterilization of heat-sensitive objects such as plastics, surgical and diagnostic appliances, and spices. Potential hazards are related to its carcinogenic, metagenic, residual, and explosive nature. Ethylene oxide is rendered nonexplosive by mixing with 90% CO_2 or fluorocarbon.

etiologic agent The microbial cause of disease; the pathogen.

eubacteria Term used for nonarchaea prokaryotes; stands for "true bacteria."

Eukarya One of the three domains (sometimes called superkingdoms) of living organisms, as proposed by Woese; contains all eukaryotic organisms.

eukaryotic cell A cell that differs from a prokaryotic cell chiefly by having a nuclear membrane (a well-defined nucleus), membrane-bounded subcellular organelles, and mitotic cell division.

eutrophication The process whereby dissolved nutrients resulting from natural seasonal enrichment or industrial pollution of water cause overgrowth of algae and cyanobacteria to the detriment of fish and other large aquatic inhabitants.

evolution Scientific principle that states that living things change gradually through hundreds of millions of years, and these changes are expressed in structural and functional adaptations in each organism. Evolution presumes that those traits that favor survival are preserved and passed on to following generations, and those traits that do not favor survival are lost.

exanthem An eruption or rash of the skin.

exergonic A chemical reaction associated with the release of energy to the surroundings. Antonym: endergonic.

exfoliative toxin A poisonous substance that causes superficial cells of an epithelium to detach and be shed. Example: staphylococcal exfoliatin. Also called an epidermolytic toxin.

exocytosis The process that releases enveloped viruses from the membrane of the host's cytoplasm.

exoenzyme An extracellular enzyme chiefly for hydrolysis of nutrient macromolecules that are otherwise impervious to the cell membrane. It functions in saprobic decomposition of organic debris and can be a factor in invasiveness of pathogens.

exogenous Originating outside the body.

exon A stretch of eukaryotic DNA coding for a corresponding portion of mRNA that is translated into peptides. Intervening stretches of DNA that are not expressed are called introns. During transcription, exons are separated from introns and are spliced together into a continuous mRNA transcript.

exotoxin A toxin (usually protein) that is secreted and acts upon a specific cellular target. Examples: botulin, tetanospasmin, diphtheria toxin, and erythrogenic toxin.

exponential Pertaining to the use of exponents, numbers that are typically written as superscripts to indicate how many times a factor is to be multiplied. Exponents are used in scientific notation to render large, cumbersome numbers into small, workable quantities.

extrapulmonary tuberculosis A condition in which tuberculosis bacilli have spread to organs other than the lungs.

extremophiles Organisms capable of living in harsh environments, such as extreme heat or cold.

facilitated diffusion The passive movement of a substance across a plasma membrane from an area of higher concentration to an area of lower concentration utilizing specialized carrier proteins.

facultative Pertaining to the capacity of microbes to adapt or adjust to variations; not obligate. Example: The presence of oxygen is not obligatory for a facultative anaerobe to grow. See **obligate.**

family In the levels of classification, a mid-level division of organisms that groups more closely related organisms than previous levels. An order is divided into families.

fastidious Requiring special nutritional or environmental conditions for growth. Said of bacteria.

fecal coliforms Any species of gram-negative lactose-positive bacteria (primarily *Escherichia coli*) that live primarily in the intestinal tract and not the environment. Finding evidence of these bacteria in a water or food sample is substantial evidence of fecal contamination and potential for infection (See **coliform**).

feedback inhibition Temporary end to enzyme action caused by an end product molecule binding to the regulatory site and preventing the enzyme's active site from binding to its substrate.

fermentation The extraction of energy through anaerobic degradation of substrates into simpler, reduced metabolites. In large industrial processes, fermentation can mean any use of microbial metabolism to manufacture organic chemicals or other products.

fermentor A large tank used in industrial microbiology to grow mass quantities of microbes that can synthesize desired products. These devices are equipped with means to stir, monitor, and harvest products such as drugs, enzymes, and proteins in very large quantities.

fertility (F′) factor Donor plasmid that allows synthesis of a pilus in bacterial conjugation. Presence of the factor is indicated by F^+, and lack of the factor is indicated by F^-.

filament A helical structure composed of proteins that is part of bacterial flagella.

filariasis Illnesses caused by infection by nematodes. These illnesses include river blindness, elephantiasis, and eye worm.

fimbria A short, numerous surface appendage on some bacteria that provides adhesion but not locomotion.

first line of defense Inborn nonspecific immunologic barriers that block the invasion of pathogens at the portal of entry, thereby keeping them out of sterile body compartments.

fixation In microscopic slide preparation of tissue sections or bacterial smears, fixation pertains to rapid killing, hardening, and adhesion to the slide, while retaining as many natural characteristics as possible. Also refers to the assimilation of inorganic molecules into organic ones, as in carbon or nitrogen fixation.

flagellum A structure that is used to propel the organism through a fluid environment.

flora Beneficial or harmless resident bacteria commonly found on and/or in the human body.

fluid mosaic model A conceptualization of the molecular architecture of cellular membranes as a bilipid layer containing proteins. Membrane proteins are embedded to some degree in this bilayer, where they float freely about.

fluorescence The property possessed by certain minerals and dyes to emit visible light when excited by ultraviolet radiation. A fluorescent dye combined with specific antibody provides a sensitive test for the presence of antigen.

focal infection Occurs when an infectious agent breaks loose from a localized infection and is carried by the circulation to other tissues.

folliculitis An inflammatory reaction involving the formation of papules or pustules in clusters of hair follicles.

fomite Virtually any inanimate object an infected individual has contact with that can serve as a vehicle for the spread of disease.

food chain A simple straight-line feeding sequence among organisms in a community.

food fermentations Addition to and growth of known cultures of microorganisms in foods to produce desirable flavors, smells, or textures. Includes cheeses, breads, alcoholic beverages, and pickles.

food infection A form of food-borne illness associated with ingesting living pathogenic microbes that invade the intestine. The damage caused by microbes growing in the body cause the main symptoms, which are usually gastroenteritis (example is salmonellosis).

food intoxication A form of food-borne illness that is the result of ingesting microbial toxins given off by bacteria growing in the food. The toxins cause the main symptoms (example is botulism).

food poisoning Any form of illness acquired from ingesting foods; sources include microbes, chemicals, plants, and animals (also food-borne disease).

food pyramid A triangular summation of the consumers, producers, and decomposers in a community and how they are related by trophic, number, and energy parameters.

food web A complex network that traces all feeding interactions among organisms in a community (see **food chain).** This is considered to be a more accurate picture of food relationships in a community than a food chain.

formalin A 37% aqueous solution of formaldehyde gas; a potent chemical fixative and microbicide.

frameshift mutation An insertion or deletion mutation that changes the codon reading frame from the point of the mutation to the final codon. Almost always leads to a nonfunctional protein.

fructose One of the carbohydrates commonly referred to as sugars. Fructose is commonly fruit sugars.

functional group In chemistry, a particular molecular combination that reacts in predictable ways and confers particular properties on a compound. Examples: —COOH, —OH, —CHO.

fungicide A chemical that can kill fungal spores, hyphae and yeast.

fungus Heterotrophic unicellular or multicellular eukaryotic organism that may take the form of a larger macroscopic organism, as in the case of mushrooms, or a smaller microscopic organism, as in the case of yeasts and molds.

furuncle A boil; a localized pyogenic infection arising from a hair follicle.

Gaia theory The concept that biotic and abiotic factors sustain suitable conditions for one another simply by their interactions. Named after the mythical Greek goddess of earth.

GALT Abbreviation for **gut-associated** lymphoid tissue. Includes **Peyer's patches.**

gamma globulin The fraction of plasma proteins high in immunoglobulins (antibodies). Preparations from pooled human plasma containing normal antibodies make useful passive immunizing agents against pertussis, polio, measles, and several other diseases.

gamma interferon A protein produced by a virally infected cell that induces production of antiviral substances in neighboring cells. This defense prevents the production and maturation of viruses and thus terminates the viral infection.

gas gangrene Disease caused by a clostridial infection of soft tissue or wound. The name refers to the gas produced by the bacteria growing in the tissue. Unless treated early, it is fatal. Also called myonecrosis.

gastroenteritis Inflammation of the lining of the stomach and intestine. May be caused by infection, intestinal disorders, or food poisoning.

gel electrophoresis A laboratory technique for separating DNA fragments according to length by employing electricity to force the DNA through a gellike matrix typically made of agarose. Smaller DNA fragments move more quickly through the gel, thereby moving farther than larger fragments during the same period of time.

gene A site on a chromosome that provides information for a certain cell function. A specific segment of DNA that contains the necessary code to make a protein or RNA molecule.

gene probe Short strands of single-stranded nucleic acid that hybridize specifically with complementary stretches of nucleotides on test samples and thereby serve as a tagging and identification device.

gene therapy The introduction of normal functional genes into people with genetic diseases such as sickle-cell anemia and cystic fibrosis. This is usually accomplished by a virus vector.

generation time Time required for a complete fission cycle—from parent cell to two new daughter cells. Also called **doubling time.**

genetic engineering A field involving deliberate alterations (recombinations) of the genomes of microbes, plants, and animals through special technological processes.

genetics The science of heredity.

genital warts A prevalent STD linked to some forms of cancer of the reproductive organs. Caused by infection with human papillomavirus.

genome The complete set of chromosomes and genes in an organism.

genotype The genetic makeup of an organism. The genotype is ultimately responsible for an organism's phenotype, or expressed characteristics.

genus In the levels of classification, the second most specific level. A family is divided into several genera.

germ free See **axenic.**

germicide An agent lethal to non-endospore-forming pathogens.

germ theory of disease A theory first originating in the1800s that proposed that microorganisms can be the cause of diseases. The concept is actually so well established in the present time that it is considered a fact.

giardiasis Infection by the *Giardia* flagellate. Most common mode of transmission is contaminated food and water. Symptoms include diarrhea, abdominal pain, and flatulence.

gingivitis Inflammation of the gum tissue in contact with the roots of the teeth.

glomerulonephritis An inflammatory autoimmune condition in response to streptococcal toxins.

gluconeogenesis The formation of glucose (or glycogen) from noncarbohydrate sources such as protein or fat. Also called glyconeogenesis.

glucose One of the carbohydrates commonly referred to as sugars. Glucose is characterized by its 6-carbon structure.

glutaraldehyde A yellow acidic liquid used in antimicrobial control. Because glutaraldehyde kills spores, it is considered a sterilant.

glycan A polysaccharide.

glycerol A 3-carbon alcohol, with three OH groups that serve as binding sites.

glycocalyx A filamentous network of carbohydrate-rich molecules that coats cells.

glycogen A glucose polymer stored by cells.

glycolysis The energy-yielding breakdown (fermentation) of glucose to pyruvic or lactic acid. It is often called anaerobic glycolysis because no molecular oxygen is consumed in the degradation.

glycosidic bond A bond that joins monosaccharides to form disaccharides and polymers.

gnotobiotic Referring to experiments performed on germ-free animals.

Golgi apparatus An organelle of eukaryotes that participates in packaging and secretion of molecules. Also known as **Golgi complex.**

gonococcus Common name for *Neisseria gonorrhoeae,* the agent of gonorrhea.

graft Live tissue taken from a donor and transplanted into a recipient to replace damaged or missing tissues such as skin, bone, blood vessels.

graft versus host disease (GVHD) A condition associated with a bone marrow transplant in which T cells in the transplanted tissue mount an immune response against the recipient's (host) normal tissues.

Gram stain A differential stain for bacteria useful in identification and taxonomy. Gram-positive organisms appear purple from crystal violet-mordant retention, whereas gram-negative organisms appear red after loss of crystal violet and absorbance of the safranin counterstain.

grana Discrete stacks of chlorophyll-containing thylakoids within chloroplasts.

granulocyte A mature leukocyte that contains noticeable granules in a Wright stain. Examples: neutrophils, eosinophils, and basophils.

granuloma A solid mass or nodule of inflammatory tissue containing modified macrophages and lymphocytes. Usually a chronic pathologic process of diseases such as tuberculosis or syphilis.

granzymes Cytotoxic enzymes produced by cytotoxic T cells and natural killer cells that participate in cell death.

Grave's disease A malfunction of the thyroid gland in which autoantibodies directed at thyroid cells stimulate an overproduction of thyroid hormone (hyperthyroidism).

Gray The basic measurement of ionizing radiation (replaces the rad).

greenhouse effect The capacity to retain solar energy by a blanket of atmospheric gases that redirects heat waves back toward the earth.

group translocation A form of active transport in which the substance being transported is altered during transfer across a plasma membrane.

growth curve A graphical representation of the change in population size over time. This graph has four periods known as lag phase, exponential or log phase, stationary phase, and death phase.

growth factor An organic compound such as a vitamin or amino acid that must be provided in the diet to facilitate growth. An essential nutrient.

guanine (G) One of the nitrogen bases found in DNA and RNA in the purine form.

Guillain-Barré syndrome A neurological complication of influenza vaccination. Approximately 1 in 100,000 recipients of the vaccine will develop this autoimmune disorder, which is marked by varying degrees of weakness and sensory loss.

gumma A nodular, infectious granuloma characteristic of tertiary syphilis.

gut-associated lymphoid tissue (GALT) A collection of lymphoid tissue in the gastrointestinal tract, which includes the appendix, the lacteals, and **Peyer's patches.**

gyrase An enzyme that functions in the packaging of DNA in the chromosomes of bacteria.

HAART Short for highly active antiretroviral therapy, the drug cocktail that is used as a treatment for HIV infection and AIDS.

habitat The environment to which an organism is adapted.

halogens A group of related chemicals with antimicrobial applications. The halogens most often used in disinfectants and antiseptics are chlorine and iodine.

halophile A microbe whose growth is either stimulated by salt or requires a high concentration of salt for growth.

Hansen's disease A chronic, progressive disease of the skin and nerves caused by infection by a mycobacterium that is a slow-growing, strict parasite. Hansen disease is the preferred name for leprosy.

H antigen The flagellar antigen of motile bacteria. *H* comes from the German word *hauch,* which denotes the appearance of spreading growth on solid medium.

haploid Having a single set of unpaired chromosomes, such as occurs in gametes and certain microbes.

hapten An incomplete or partial antigen. Although it constitutes the determinative group and can bind antigen, hapten cannot stimulate a full immune response without being carried by a larger protein molecule.

Hashimoto's thyroiditis An autoimmune disease of the thyroid gland that damages the thyroid follicle cells and results in decreased production of thyroid hormone (hypothyroidism).

hay fever A form of atopic allergy marked by seasonal acute inflammation of the conjunctiva and mucous membranes of the respiratory passages. Symptoms are irritative itching and rhinitis.

helical Having a spiral or coiled shape. Said of certain virus capsids and bacteria.

helminth A term that designates all parasitic worms.

helper T cell A class of thymus-stimulated lymphocytes that facilitate various immune activities such as assisting B cells and macrophages. Also called a **T helper cell.**

hemagglutinin (H) A molecule that causes red blood cells to clump or agglutinate. A major surface receptor on the influenza virus needed for entry.

hemolysin Any biological agent that is capable of destroying red blood cells and causing the release of hemoglobin. Many bacterial pathogens produce exotoxins that act as hemolysins.

hemolytic disease of the newborn Incompatible Rh factor between mother and fetus causes maternal antibodies to attack the fetus and trigger complement-mediated lysis in the fetus.

hemolytic uremic syndrome (HUS) A kidney disease caused by infection with *E.coli* 0157:H7.

hemolyze When red blood cells burst and release hemoglobin pigment.

hemopoiesis The process by which the various types of blood cells are formed, such as in the bone marrow. Also called hematopoesis.

hepadnavirus Enveloped DNA viruses with a predisposition to affect the liver. Hepatitis B is the most serious form.

hepatitis Inflammation and necrosis of the liver, often the result of viral infection.

hepatitis A virus (HAV) Enterovirus spread by contaminated food responsible for short-term (infectious) hepatitis.

hepatitis B virus (HBV) Hepadnavirus that is the causative agent of serum hepatitis.

hepatitis C virus (HCV) A blood-borne RNA virus that is the third most common cause of hepatitis.

hepatocellular carcinoma A liver cancer associated with infection with hepatitis B virus.

hepatocyte A liver cell.

herd immunity The status of collective acquired immunity in a population that reduces the likelihood that nonimmune individuals will contract and spread infection. One aim of vaccination is to induce herd immunity.

heredity Genetic inheritance.

hermaphroditic Containing the sex organs for both male and female in one individual.

herpes zoster A recurrent infection caused by latent chickenpox virus. Its manifestation on the skin tends to correspond to dermatomes and to occur in patches that "girdle" the trunk. Also called shingles.

herpetic keratitis Corneal or conjunctival inflammation due to herpesvirus type 1.

heterophile antigen An antigen present in a variety of phylogenetically unrelated species. Example: red blood cell antigens and the glycocalyx of bacteria.

heterotroph An organism that relies upon organic compounds for its carbon and energy needs.

hexose A 6-carbon sugar such as glucose and fructose.

hierarchies Levels of power. Arrangement in order of rank.

histamine A cytokine released when mast cells and basophils release their granules. An important mediator of allergy, its effects include smooth muscle contraction, increased vascular permeability, and increased mucus secretion.

histiocyte Another term for macrophage.

histone Proteins associated with eukaryotic DNA. These simple proteins serve as winding spools to compact and condense the chromosomes.

histoplasmin The antigenic extract of *Histoplasma capsulatum,* the causative agent of histoplasmosis. The preparation is used in skin tests for diagnosis and for conducting surveys to determine the geographic distribution of the fungus.

HLA An abbreviation for human leukocyte antigens. This closely linked cluster of genes programs for cell surface glycoproteins that control immune interactions between cells and is involved in rejection of allografts. Also called the major histocompatibility complex (MHC).

holoenzyme An enzyme complete with its apoenzyme and cofactors.

hops The ripe, dried fruits of the hop vine (*Humulus lupulus*) that is added to beer wort for flavoring.

host Organism in which smaller organisms or viruses live, feed, and reproduce.

host range The limitation imposed by the characteristics of the host cell on the type of virus that can successfully invade it.

human diploid cell vaccine (HDCV) A vaccine made using cell culture that is currently the vaccine of choice for preventing infection by rabies virus.

human herpesvirus-6 (HHV-6) The herpesvirus that causes roseola and may possibly be linked to chronic neurological diseases.

human immunodeficiency virus (HIV) A retrovirus that causes acquired immunodeficiency syndrome (AIDS).

human papillomavirus (HPV) A group of DNA viruses whose members are responsible for common, plantar, and genital warts.

humoral immunity Protective molecules (mainly antibodies) carried in the fluids of the body.

hybridization A process that matches complementary strands of nucleic acid (DNA-DNA, RNA-DNA, RNA-RNA). Used for locating specific sites or identifying exact sequences of nucleic acids.

hybridoma An artificial cell line that produces monoclonal antibodies. It is formed by fusing (hybridizing) a normal antibody-producing cell with a cancer cell, and it can produce pure antibody indefinitely.

hydatid cyst A sac, usually in liver tissue, containing fluid and larval stages of the echinococcus tapeworm.

hydration The addition of water as in the coating of ions with water molecules as ions enter into aqueous solution.

hydrogen bond A weak chemical bond formed by the attraction of forces between molecules or atoms—in this case, hydrogen and either oxygen or nitrogen. In this type of bond, electrons are not shared, lost, or gained.

hydrologic cycle The continual circulation of water between hydrosphere, atmosphere, and lithosphere.

hydrolase An enzyme that catalyzes the cleavage of a bond with the additions of —H and —OH (parts of a water molecule) at the separation site.

hydrolysis A process in which water is used to break bonds in molecules. Usually occurs in conjunction with an enzyme.

hydrophilic The property of attracting water. Molecules that attract water to their surface are called hydrophilic.

hydrophobic The property of repelling water. Molecules that repel water are called hydrophobic.

hydrosphere That part of the biosphere that encompasses water-containing environments such as oceans, lakes, rivers.

hypertonic Having a greater osmotic pressure than a reference solution.

hyphae The tubular threads that make up filamentous fungi (molds). This web of branched and intertwining fibers is called a mycelium.

hypogammaglobulinemia An inborn disease in which the gamma globulin (antibody) fraction of serum is greatly reduced. The condition is associated with a high susceptibility to pyogenic infections.

hyposensitization A therapeutic exposure to known allergens designed to build tolerance and eventually prevent allergic reaction.

hypothesis A tentative explanation of what has been observed or measured.

hypotonic Having a lower osmotic pressure than a reference solution.

icosahedron A regular geometric figure having 20 surfaces that meet to form 12 corners. Some virions have capsids that resemble icosahedral crystals.

immune complex reaction Type III hypersensitivity of the immune system. It is characterized by the reaction of soluble antigen with antibody, and the deposition of the resulting complexes in basement membranes of epithelial tissue.

immune surveillance The continual function of macrophages, cytotoxic T cells, and natural killer cells in identifying and destroying cancer cells within the body.

immune tolerance The ability of the immune system to remain nonreactive to normal cell molecules.

immunity An acquired resistance to an infectious agent due to prior contact with that agent.

immunoassays Extremely sensitive tests that permit rapid and accurate measurement of trace antigen or antibody.

immunocompetence The ability of the body to recognize and react with multiple foreign substances.

immunodeficiency Immune function is incompletely developed, suppressed, or destroyed.

immunodeficiency disease A form of immunopathology in which white blood cells are unable to mount a complete, effective immune response, which results in recurrent infections. Examples would be AIDS and agammaglobulinemia.

immunogen Any substance that induces a state of sensitivity or resistance after processing by the immune system of the body.

immunoglobulin (IG) The chemical class of proteins to which antibodies belong.

immunology The study of the system of body defenses that protect against infection.

immunopathology The study of disease states associated with overreactivity or underreactivity of the immune response.

immunotherapy Preventing or treating infectious diseases by administering substances that produce artificial immunity. May be active or passive.

immunotoxin An artificially prepared toxin attached to a monoclonal antibody specific to a tumor cell. Immunotoxins are used in cancer therapy to destroy tumors.

impetigo Contagious skin infection marked by pustules and vesicles; caused by *S. aureus* and *S. pyogenes*. Also known as **pyoderma.**

IMViC Abbreviation for four identification tests: **i**ndole production, **m**ethyl red test, **V**oges-Proskauer test (**i** inserted to concoct a wordlike sound), and **c**itrate as a sole source of carbon. This test was originally developed to distinguish between *Enterobacter aerogenes* (associated with soil) and *Escherichia coli* (a fecal coliform).

incidence In epidemiology, the number of new cases of a disease occurring during a period.

incineration Destruction of microbes by subjecting them to extremes of dry heat. Microbes are reduced to ashes and gas by this process.

inclusion A relatively inert body in the cytoplasm such as storage granules, glycogen, fat, or some other aggregated metabolic product. Also called **inclusion body.**

incubate To isolate a sample culture in a temperature-controlled environment to encourage growth.

incubation period The period from the initial contact with an infectious agent to the appearance of the first symptoms.

indicator bacteria In water analysis, any easily cultured bacteria that may be found in the intestine and can be used as an index of fecal contamination. The category includes coliforms and enterococci. Discovery of these bacteria in a sample means that pathogens may also be present.

induced mutation Any alteration in DNA that occurs as a consequence of exposure to chemical or physical mutagens.

inducible enzyme An enzyme that increases in amount in direct proportion to the amount of substrate present.

inducible operon An operon that under normal circumstances is not transcribed. The presence of a specific inducer molecule can cause transcription of the operon to begin.

induction Process by which an individual accumulates data or facts and then formulates a general hypothesis that accounts for those facts.

infantile diarrhea An infection in newborns frequently caused by *E.coli* infection and often having a high mortality rate.

infection The entry, establishment, and multiplication of pathogenic organisms within a host.

infectious disease The state of damage or toxicity in the body caused by an infectious agent.

infectious dose The estimated number of microbial cells or units required to establish an infection.

inflammation A natural, nonspecific response to tissue injury that protects the host from further damage. It stimulates immune reactivity and blocks the spread of an infectious agent.

inoculation The implantation of microorganisms into or upon culture media.

inorganic chemicals Molecules that lack the basic framework of the elements of carbon and hydrogen.

interferon (IFN) Naturally occurring polypeptides produced by fibroblasts and lymphocytes that can block viral replication and regulate a variety of immune reactions.

interleukin A macrophage agent (interleukin-1, or IL-1) that stimulates lymphocyte function. Stimulated T cells release yet another interleukin (IL-2), which amplifies T-cell response by stimulating additional T cells. T helper cells stimulated by IL-2 stimulate B-cell proliferation and promote antibody production.

intermittent sterilization (tyndallization) A sterilization technique for items that cannot withstand the temperature/pressure extremes of an autoclave. Materials are subjected to free-flowing steam (which kills vegetative cells), followed by an incubation period (which allows endospores to germinate). The cycle is repeated three times with the hope that, at completion, all spores will have germinated and all vegetative cells will have died.

intoxication Poisoning that results from the introduction of a toxin into body tissues through ingestion or injection.

intracellular parasite A microbe that invades within a cell's structure to carry out its life cycle.

intron The segments on split genes of eukaryotes that do not code for polypeptide. They can have regulatory functions. See **exon.**

in utero Literally means "in the uterus"; pertains to events or developments occurring before birth.

in vitro Literally means "in glass," signifying a process or reaction occurring in an artificial environment, as in a test tube or culture medium.

in vivo Literally means "in a living being," signifying a process or reaction occurring in a living thing.

iodophor A combination of iodine and an organic carrier that is a moderate-level disinfectant and antiseptic.

ion An unattached, charged particle.

ionic bond A chemical bond in which electrons are transferred and not shared between atoms.

ionization The aqueous dissociation of an electrolyte into ions.

ionizing radiation Radiant energy consisting of short-wave electromagnetic rays (X ray) or high-speed electrons that cause dislodgment of electrons on target molecules and create ions.

irradiation The application of radiant energy for diagnosis, therapy, disinfection, or sterilization.

irregular (in shape) Refers to bacteria that stain unevenly or display cell to cell variation in size and/or shape (pleomorphic) within a single species.

irritability Capacity of cells to respond to chemical, mechanical, or light stimuli. This property helps cells adapt to the environment and obtain nutrients.

isograft Transplanted tissue from one monozygotic twin to the other; transplants between highly inbred animals that are genetically identical.

isolation The separation of microbial cells by serial dilution or mechanical dispersion on solid media to create discrete colonies.

isotonic Two solutions having the same osmotic pressure such that, when separated by a semipermeable membrane, there is no net movement of solvent in either direction.

isotope A version of an element that is virtually identical in all chemical properties to another version except that their atoms have slightly different atomic masses.

jaundice The yellowish pigmentation of skin, mucous membranes, sclera, deeper tissues, and excretions due to abnormal deposition of bile pigments. Jaundice is associated with liver infection, as with hepatitis B virus and leptospirosis.

J chain A small molecule with high sulfhydryl content that secures the heavy chains of IgM to form a pentamer and the heavy chains of IgA to form a dimer.

Kaposi's sarcoma A malignant or benign neoplasm that appears as multiple hemorrhagic sites on the skin, lymph nodes, and viscera and apparently involves the metastasis of abnormal blood vessel cells. It is a clinical feature of AIDS.

keratitis Inflammation of the cornea.

keratoconjunctivitis Inflammation of the conjunctiva and cornea.

killed or inactivated vaccine A whole cell or intact virus preparation in which the microbes are dead or preserved and cannot multiply but are still capable of conferring immunity.

killer T cells A T lymphocyte programmed to directly affix cells and kill them. See **cytotoxic.**

kingdom In the levels of classification, the second division from more general to more specific. Each domain is divided into kingdoms.

Koch's postulates A procedure to establish the specific cause of disease. In all cases of infection: (1) The agent

must be found; (2) inoculations of a pure culture must reproduce the same disease in animals; (3) the agent must again be present in the experimental animal; and (4) a pure culture must again be obtained.

Koch-Weeks bacillus (*Haemophilus aegypticus*) One of the agents of bacterial conjunctivitis (pinkeye).

Koplik's spots Tiny red blisters with central white specks on the mucosal lining of the cheeks. Symptomatic of measles.

Krebs cycle Metabolic cycle that converts acetyl groups generated from pyruvic acid to CO_2; provides $NADH_1$, $FADH_2$, and ATP; also known as TCA cycle and citric acid cycle.

kuru A spongiform encephalopathy once prevalent among New Guinea natives caused by a prion. Advanced disease is marked by profound loss of coordination and mental deterioration.

labile In chemistry, molecules or compounds that are chemically unstable in the presence of environmental changes.

lacrimation The secretion of tears, especially in profusion.

lactose One of the carbohydrates commonly referred to as sugars. Lactose is commonly found in milk.

lactose *(lac)* operon Control system that manages the regulation of lactose metabolism. It is composed of three DNA segments, including a regulator, a control locus, and a structural locus.

lager The maturation process of beer, which is allowed to take place in large vats at a reduced temperature.

lagging strand The newly forming 59′ DNA strand that is discontinuously replicated in segments (Okazaki fragments).

lag phase The early phase of population growth during which no signs of growth occur.

latency The state of being inactive and not multiplying. Example: a latent virus or latent infection.

leading strand The newly forming 39′ DNA strand that is replicated in a continuous fashion without segments.

leaven To lighten food material by entrapping gas generated within it. Example: the rising of bread from the CO_2 produced by yeast or baking powder.

Legionnaire's disease Pulmonary infection by *Legionella* bacteria. Gram-negative rods survive inside amoebas in aquatic habitats. Some conditions may be fatal.

lepromas Skin nodules seen on the face of persons suffering from lepromatous leprosy. The skin folds and thickenings are caused by the overgrowth of *Mycobacterium leprae.*

lepromatous leprosy Severe, disfiguring leprosy characterized by widespread dissemination of the leprosy bacillus in deeper lesions.

leprosy See **Hansen's disease.**

lesion A wound, injury, or some other pathologic change in tissues.

leukocidin A heat-labile substance formed by some pyogenic cocci that impairs and sometimes lyses leukocytes.

leukocytes White blood cells. The primary infection-fighting blood cells.

leukocytosis An abnormally large number of leukocytes in the blood, which can be indicative of acute infection.

leukopenia A lower-than-normal leukocyte count in the blood, which can be indicative of blood infection or disease.

leukotriene An unsaturated fatty acid derivative of arachidonic acid. Leukotriene functions in chemotactic activity, smooth muscle contractility, mucus secretion, and capillary permeability.

L form L-phase variants; wall-less forms of some bacteria that are induced by drugs or chemicals. These forms can be involved in infections.

lichen An early colonizer of rocks and other primary habitats that is an association between algae or cyanobacteria and fungi.

ligase An enzyme required to seal the sticky ends of DNA pieces after splicing.

light-dependent reactions The series of reactions in photosynthesis that are driven by the light energy (photons) absorbed by chlorophyll. They involve splitting of water into hydrogens and oxygen, transport of electrons by NADP, and ATP synthesis.

light-independent reactions The series of reactions in photosynthesis that can proceed with or without light. It is a cyclic system that uses ATP from the light reactions to incorporate or fix carbon dioxide into organic compounds, leading to the production of glucose and other carbohydrates (also called the Calvin cycle).

limnetic zone The deep-water region beyond the shoreline.

lipid A term used to describe a variety of substances that are not soluble in polar solvents such as water but will dissolve in nonpolar solvents such as benzene and chloroform. Lipids include triglycerides, phospholipids, steroids, and waxes.

lipase A fat-splitting enzyme. Example: Triacylglycerol lipase separates the fatty acid chains from the glycerol backbone of triglycerides.

lipopolysaccharide A molecular complex of lipid and carbohydrate found in the bacterial cell wall. The lipopolysaccharide (LPS) of gram-negative bacteria is an endotoxin with generalized pathologic effects such as fever.

listeriosis Infection with *Listeria monocytogenes,* usually due to eating contaminated dairy products, poultry, or meat. It is usually mild in healthy adults but can produce severe symptoms in neonates and immunocompromised adults.

lithoautotroph Bacteria that rely on inorganic minerals to supply their nutritional needs. Sometimes referred to as chemoautotrophs.

lithosphere That part of the biosphere that encompasses the earth's crust, including rocks and minerals.

lithotroph An autotrophic microbe that derives energy from reduced inorganic compounds such as N_2S.

littoral zone The shallow region along a shoreline.

lobar pneumonia Infection involving whole segments (lobes) of the lungs, which may lead to consolidation and plugging of the alveoli and extreme difficulty in breathing.

localized infection Occurs when a microbe enters a specific tissue, infects it, and remains confined there.

loci A site on a chromosome occupied by a gene.

lockjaw See **tetanus.**

logarithmic or log phase Maximum rate of cell division during which growth is geometric in its rate of increase. Also called **exponential growth phase.**

lophotrichous Describing bacteria having a tuft of flagella at one or both poles.

lumen The cavity within a tubular organ.

lymphadenitis Inflammation of one or more lymph nodes. Also called lymphadenopathy.

lymphatic or lymphoid system A system of vessels and organs that serve as sites for development of immune cells and immune reactions. It includes the spleen, thymus, lymph nodes, and GALT.

lymphocyte The second most common form of white blood cells.

lymphokine A soluble substance secreted by sensitized T lymphocytes upon contact with specific antigen.

About 50 types exist, and they stimulate inflammatory cells: macrophages, granulocytes, and lymphocytes. Examples: migration inhibitory factor, macrophage activating factor, and chemotactic factor.

lymphotoxin A chemical (lymphokine) produced by cytotoxic T lymphocytes that kills target cells directly through perforating their cell membrane and causing lysis.

lyophilization Freeze-drying; the separation of a dissolved solid from the solvent by freezing the solution and evacuating the solvent under vacuum. A means of preserving the viability of cultures.

lyse To burst.

lysin A complement-fixing antibody that destroys specific targeted cells. Examples: hemolysin and bacteriolysin.

lysis The physical rupture or deterioration of a cell.

lysogeny The indefinite persistence of bacteriophage DNA in a host without bringing about the production of virions. A lysogenic cell can revert to a lytic cycle, the process that ends in lysis.

lysosome A cytoplasmic organelle containing lysozyme and other hydrolytic enzymes.

lysozyme An enzyme that attacks the bonds on bacterial peptidoglycan. It is a natural defense found in tears and saliva.

MALT(Mucosal-Associated Lymphoid Tissue) Active lymphocytes distributed within the mucosal surfaces of respiratory and other systems.

macromolecules Large, molecular compounds assembled from smaller subunits, most notably biochemicals.

macronutrient A chemical substance required in large quantities (phosphate, for example).

macrophage A white blood cell derived from a monocyte that leaves the circulation and enters tissues. These cells are important in nonspecific phagocytosis and in regulating, stimulating, and cleaning up after immune responses.

macroscopic Visible to the naked eye.

macule A small, discolored spot or patch of skin that is neither raised above nor depressed below the surrounding surface.

major histocompatibility complex (MHC) See **HLA.**

malignant tumor Cancerous growth characterized by uncontrolled growth or abnormal cells within normal tissue.

malt The grain, usually barley, that is sprouted to obtain digestive enzymes and dried for making beer.

maltose One of the carbohydrates referred to as sugars. A fermentable sugar formed from starch.

Mantoux test An intradermal screening test for tuberculin hypersensitivity. A red, firm patch of skin at the injection site greater than 10 mm in diameter after 48 hours is a positive result that indicates current or prior exposure to the TB bacillus.

mapping Determining the location of loci and other qualities of genomic DNA.

marker Any trait or factor of a cell, virus, or molecule that makes it distinct and recognizable. Example: a genetic marker.

mash In making beer, the malt grain is steeped in warm water, ground up, and fortified with carbohydrates to form mash.

mass number (MN) Measurement that reflects the number of protons and neutrons in an atom of a particular element.

mast cell A nonmotile connective tissue cell implanted along capillaries, especially in the lungs, skin, gastrointestinal tract, and genitourinary tract. Like a basophil, its granules store mediators of allergy.

mastitis Inflammation of the breast or udder.

matrix The dense ground substance between the cristae of a mitochondrion that serves as a site for metabolic reactions.

matter All tangible materials that occupy space and have mass.

maximum temperature The highest temperature at which an organism will grow.

mechanical vector An animal that transports an infectious agent but is not infected by it, such as houseflies whose feet become contaminated with feces.

medium (plural: *media*) A nutrient used to grow organisms outside of their natural habitats.

meiosis The type of cell division necessary for producing gametes in diploid organisms. Two nuclear divisions in rapid succession produce four gametocytes, each containing a haploid number of chromosomes.

memory (immunologic memory) The capacity of the immune system to recognize and act against an antigen upon second and subsequent encounters.

memory cell The long-lived progeny of a sensitized lymphocyte that remains in circulation and is genetically programmed to react rapidly with its antigen.

meningitis An inflammation of the membranes (meninges) that surround and protect the brain. It is often caused by bacteria such as *Neisseria meningitidis* (the meningococcus) and *Haemophilus influenzae*.

merozoite The motile, infective stage of an apicomplexan parasite that comes from a liver or red blood cell undergoing multiple fission.

mesophile Microorganisms that grow at intermediate temperatures.

mesosome The irregular invagination of a bacterial cell membrane that is more prominent in gram-positive than in gram-negative bacteria. Although its function is not definitely known, it appears to participate in DNA replication and cell division; in certain cells, it appears to play a role in secretion.

messenger RNA (MRNA) A single-stranded transcript that is a copy of the DNA template that corresponds to a gene.

metabolic analog Enzyme that mimics the natural substrate of an enzyme and vies for its active site.

metabolism A general term for the totality of chemical and physical processes occurring in a cell.

metabolites Small organic molecules that are intermediates in the stepwise biosynthesis or breakdown of macromolecules.

metachromatic Exhibiting a color other than that of the dye used to stain it. Metachromatic granules are stored crystals of phosphate that stain purple or red when stained with methylene blue.

metagenomics The analysis of genomes recovered from mixed environmental samples that provides information on the diversity of microbe types without isolating or culturing them.

metastasis In cancer, the dissemination of tumor cells so that neoplastic growths appear in other sites away from the primary tumor.

methanogens Microbes that produce methane gas.

MHC Major histocompatibility complex. See **HLA.**

MIC Abbreviation for minimum inhibitory concentration. The lowest concentration of antibiotic needed to inhibit bacterial growth in a test system.

microaerophile An aerobic bacterium that requires oxygen at a concentration less than that in the atmosphere.

microbe See **microorganism.**

microbial loop A process occurring in natural aquatic habitats that maintains the cycling of nutrients from producers to consumers and back to producers. It is powered primarily by mineralizing and decomposing bacteria and protozoa.

microbiology A specialized area of biology that deals with living things ordinarily too small to be seen without magnification, including bacteria, archaea, fungi, protozoa, and viruses.

microfilaments Cellular cytoskeletal element formed by thin protein strands that attach to cell membrane and form a network though the cytoplasm. Responsible for movement of cytoplasm.

micronutrient A chemical substance required in small quantities (trace metals, for example).

microorganism A living thing ordinarily too small to be seen without magnification; an organism of microscopic size.

microscopic Invisible to the naked eye.

microscopy Science that studies structure, magnification, lenses, and techniques related to use of a microscope.

microtubules Long tubular structures made of protein that function in the structure, shape, and movement of eukaryotic cells.

miliary tuberculosis Rapidly fatal tuberculosis due to dissemination of mycobacteria in the blood and formation of tiny granules in various organs and tissues. The term *miliary* means resembling a millet seed.

mineralization The process by which decomposers (bacteria and fungi) convert organic debris into inorganic and elemental form. It is part of the recycling process.

minimum inhibitory concentration (MIC) The smallest concentration of drug needed to visibly control microbial growth.

minimum temperature The lowest temperature at which an organism will grow.

miracidium The ciliated first-stage larva of a trematode. This form is infective for a corresponding intermediate host snail.

missense mutation A mutation in which a change in the DNA sequence results in a different amino acid being incorporated into a protein, with varying results.

mitochondrion A double-membrane organelle of eukaryotes that is the main site for aerobic respiration.

mitosis Somatic cell division that preserves the somatic chromosome number.

mixed acid fermentation An anaerobic degradation of pyruvic acid that results in more than one organic acid being produced (e.g., acetic acid, lactic acid, succinic acid, etc.).

mixed culture A container growing two or more different, known species of microbes.

mixed infection Occurs when several different pathogens interact simultaneously to produce an infection. Also called a synergistic infection.

moist heat Heat in the form of hot water, boiling water or steam, generally ranging from 60°C to 135°C.

molecule A distinct chemical substance that results from the combination of two or more atoms.

molluscum contagiosum Poxvirus-caused disease that manifests itself by the appearance of small lesions on the face, trunk, and limbs. Often associated with sexual transmission.

monoclonal antibody An antibody produced by a clone of lymphocytes that respond to a particular antigenic determinant and generate identical antibodies only to that determinant. See **hybridoma.**

monocyte A large mononuclear leukocyte normally found in the lymph nodes, spleen, bone marrow, and loose connective tissue. This type of cell makes up 3% to 7% of circulating leukocytes.

monomer A simple molecule that can be linked by chemical bonds to form larger molecules.

mononuclear phagocyte system A collection of monocytes and macrophages scattered throughout the extracellular spaces that function to engulf and degrade foreign molecules.

monosaccharide A simple sugar such as glucose that is a basic building block for more complex carbohydrates.

monotrichous Describing a microorganism that bears a single flagellum.

morbidity A diseased condition

morbidity rate The relative incidence of disease in a community.

mordant A chemical that fixes a dye in or on cells by forming an insoluble compound and thereby promoting retention of that dye. Example: Gram's iodine in the Gram stain.

morphology The study of organismic structure.

mortality rate Total number of deaths in a population attributable to a particular disease.

most probable number (MNP) Coliform test used to detect the concentration of contaminants in water supplies.

motility Self-propulsion.

multidrug-resistant (MDR) pumps Cellular pumps that actively transport drugs and other chemicals out of cells, providing drug resistance to the cells that possess them.

multinucleate giant cell A cytopathic effect that arises from the fusion of the cytoplasmic membranes of several infected cells. Characteristic of infection with paramyxoviruses.

mumps See **epidemic parotitis.**

must Juices expressed from crushed fruits that are used in fermentation for wine.

mutagen Any agent that induces genetic mutation. Examples: certain chemical substances, ultraviolet light, radioactivity.

mutant strain A subspecies of microorganism that has undergone a mutation, causing expression of a trait that differs from other members of that species.

mutation A permanent inheritable alteration in the DNA sequence or content of a cell.

mutualism Organisms living in an obligatory but mutually beneficial relationship.

mycelium The filamentous mass that makes up a mold. Composed of hyphae.

mycetoma A chronic fungal infection usually afflicting the feet, typified by swelling and multiple draining lesions. Example: maduromycosis or Madura foot.

mycoplasma The smallest, self-replicating microorganisms. Mycoplasma naturally lack a cell wall. Most species are parasites of animals and plants.

mycorrhizae Various species of fungi adapted in an intimate, mutualistic relationship to plant roots.

mycosis Any disease caused by a fungus.

mycotoxicosis Illness resulting from eating poisonous fungi.

myonecrosis Another name for gas gangrene.

NAD/NADH Abbreviations for the oxidized/reduced forms of nicotinamide adenine dinucleotide, an electron carrier. Also known as the vitamin niacin.

nanobes Cell-like particles, found in sediments and other geologic deposits, that some scientists speculate are the smallest bacteria. Short for **nanobacteria.**

narrow spectrum Denotes drugs that are selective and limited in their effects. For example, they inhibit either gram-negative or gram-positive bacteria but not both.

nasopharyngeal carcinoma A malignancy of epithelial cells that occurs in older Chinese and African men and is associated with exposure to Epstein-Barr virus.

natural immunity Specific protection from disease acquired through normal, nonmedical life processes.

natural selection A process in which the environment places pressure on organisms to adapt and survive changing conditions. Only the survivors will be around

to continue the life cycle and contribute their genes to future generations. This is considered a major factor in evolution of species.

necrosis A pathologic process in which cells and tissues die and disintegrate.

necrotizing faciitis Severe invasive disease of skin, muscle, and connective tissue, usually caused by group A streptococci.

negative feedback Enzyme regulation of metabolism by the end product of a multienzyme system that blocks the action of a "pacemaker" enzyme at or near the beginning of the pathway.

negative stain A staining technique that renders the background opaque or colored and leaves the object unstained so that it is outlined as a colorless area.

nematode A common name for helminths called roundworms.

neoplasm A synonym for tumor.

neuraminidase A glycoprotein found in the envelope of influenza virus that facilitates release of new viruses from the host cell.

neurotropic Having an affinity for the nervous system. Most likely to affect the spinal cord.

neutralization The process of combining an acid and a base until they reach a balanced proportion, with a pH value close to 7.

neutron An electrically neutral particle in the nuclei of all atoms except hydrogen.

neutrophil A mature granulocyte present in peripheral circulation, exhibiting a multilobular nucleus and numerous cytoplasmic granules that retain a neutral stain. The neutrophil is an active phagocytic cell in bacterial infection.

neutrophil extracellular traps (NETS) An innate capture defense provided by phagocytes called neutrophils that is deployed to trap and kill pathogens in the extracellular spaces.

niche In ecology, an organism's biological role in or contribution to its community.

night soil An archaic euphemism for human excrement collected for fertilizing crops.

nitrification Phase of the nitrogen cycle in which ammonium is oxidized.

nitrogen base A ringed compound of which pyrimidines and purines are types.

nitrogen cycle The pathway followed by the element nitrogen as it circulates from inorganic sources in the nonliving environment to living things and back to the nonliving environment. The longtime reservoir is nitrogen gas in the atmosphere.

nitrogen fixation A process occurring in certain bacteria in which atmospheric N_2 gas is converted to a form (NH_4) usable by plants.

nitrogenous base A nitrogen-containing molecule found in DNA and RNA that provides the basis for the genetic code. Adenine, guanine, and cytosine are found in both DNA and RNA, whereas thymine is found exclusively in DNA and uracil is found exclusively in RNA.

nomenclature A set system for scientifically naming organisms, enzymes, anatomical structures, etc.

noncoliforms Lactose-negative enteric bacteria in normal flora.

noncommunicable An infectious disease that does not arrive through transmission of an infectious agent from host to host.

nonionizing radiation Method of microbial control, best exemplified by ultraviolet light, that causes the formation of abnormal bonds within the DNA of microbes, increasing the rate of mutation. The primary limitation of nonionizing radiation is its inability to penetrate beyond the surface of an object.

nonpolar A term used to describe an electrically neutral molecule formed by covalent bonds between atoms that have the same or similar electronegativity.

nonself Molecules recognized by the immune system as containing foreign markers, indicating a need for immune response.

nonsense codon A triplet of mRNA bases that does not specify an amino acid but signals the end of a polypeptide chain. Also called **stop codon.**

nonsense mutation A mutation that changes an amino acid-producing codon into a stop codon, leading to premature termination of a protein.

nonsynthetic medium See **complex medium.**

normal flora The native microbial forms that an individual harbors.

Norwalk agent One of a group of Calciviruses, Norwalk agent causes gastrointestinal distress and is commonly transmitted in schools, camps, cruise ships, and nursing homes.

nosocomial infection An infection not present upon admission to a hospital but incurred while being treated there.

nucleocapsid In viruses, the close physical combination of the nucleic acid with its protective covering.

nucleoid The basophilic nuclear region or nuclear body that contains the bacterial chromosome.

nucleolus A granular mass containing RNA that is contained within the nucleus of a eukaryotic cell.

nucleosome Structure in the packaging of DNA. Formed by the DNA strands wrapping around the histone protein to form nucleus bodies arranged like beads on a chain.

nucleotide The basic structural unit of DNA and RNA; each nucleotide consists of a phosphate, a sugar (ribose in RNA, deoxyribose in DNA), and a nitrogenous base such as adenine, guanine, cytosine, thymine (DNA only) or uracil (RNA only).

nucleus The central core of an atom, composed of protons and neutrons.

numerical aperture (NA) In microscopy, a measure of the light passing from the object and into the objective to maximize optical clarity and resolution.

nutrient Any chemical substance that must be provided to a cell for normal metabolism and growth. Macronutrients are required in large amounts, and micronutrients in small amounts.

nutrition The acquisition of chemical substances by a cell or organism for use as an energy source or as building blocks of cellular structures.

obligate Without alternative; restricted to a particular characteristic. Example: An obligate parasite survives and grows only in a host; an obligate aerobe must have oxygen to grow; an obligate anaerobe is destroyed by oxygen.

Okazaki fragment In replication of DNA, a segment formed on the lagging strand in which biosynthesis is conducted in a discontinuous manner dictated by the $5' \rightarrow 3'$ DNA polymerase orientation.

oligodynamic action A chemical having antimicrobial activity in minuscule amounts. Example: Certain heavy metals are effective in a few parts per billion.

oligonucleotides Short pieces of DNA or RNA that are easier to handle than long segments.

oligotrophic Nutrient-deficient ecosystem.

oncogene A naturally occurring type of gene that when activated can transform a normal cell into a cancer cell.

oncology The study of neoplasms, their cause, disease characteristics, and treatment.

onychomycosis A fungal infection of the nail and nailbed. Also called tinea unguium.

oocyst The encysted form of a fertilized macrogamete or zygote; typical in the life cycles of apicomplexan parasites.

operator In an operon sequence, the DNA segment where transcription of structural genes is initiated.

operon A genetic operational unit that regulates metabolism by controlling mRNA production. In sequence, the unit consists of a regulatory gene, inducer or repressor control sites, and structural genes.

opportunistic infection A situation where ordinarily nonpathogenic or weakly pathogenic microbes cause disease in an immunologically compromised host.

opportunistic pathogen See **opportunistic infection.**

opsonization The process of stimulating phagocytosis by affixing molecules (opsonins such as antibodies and complement) to the surfaces of foreign cells or particles.

optimum temperature The temperature at which a species shows the most rapid growth rate.

orbitals The pathways of electrons as they rotate around the nucleus of an atom.

order In the levels of classification, the division of organisms that follows class. Increasing similarity may be noticed among organisms assigned to the same order.

organelle A small component of eukaryotic cells that is bounded by a membrane and specialized in function.

organic chemicals Molecules that contain the basic framework of the elements carbon and hydrogen.

ornithosis Worldwide zoonosis carried in the latent state in wild and domesticated birds; caused by *Chlamydia psittacosis.*

osmophile A microorganism that thrives in a medium having high osmotic pressure.

osmosis The diffusion of water across a selectively permeable membrane in the direction of lower water concentration.

osteomyelitis A focal infection of the internal structures of long bones, leading to pain and inflammation. Often caused by *Staphylococcus aureus.*

otitis media Acute infection of the middle ear, usually by the pneumococcus, that causes severe pain and drainage.

outer membrane A structure made of lipids in the outer layer of the cell wall of gram-negative bacteria.

oxidation In chemical reactions, the loss of electrons by one reactant. An oxidizing agent is a molecule that causes electron loss or removes electrons from a substrate.

oxidation-reduction Redox reactions, in which paired sets of molecules participate in electron transfers.

oxidative phosphorylation The synthesis of ATP using energy given off during the electron transport phase of respiration.

oxygenic Any reaction that gives off oxygen; usually in reference to the result of photosynthesis in eukaryotes and cyanobacteria.

PAMP Short for **pathogen-associated molecular pattern.** Unique molecular structures on microbes that are recognized by the immune system.

palindrome A word, verse, number, or sentence that reads the same forward or backward. Palindromes of nitrogen bases in DNA have genetic significance as transposable elements, as regulatory protein targets, and in DNA splicing.

palisades The characteristic arrangement of *Corynebacterium* cells resembling a row of fence posts and created by snapping.

pandemic A disease afflicting an increased proportion of the population over a wide geographic area (often worldwide).

pannus The granular membrane occurring on the cornea in trachoma.

papilloma Benign, squamous epithelial growth commonly referred to as a wart.

papule An elevation of skin that is small, demarcated, firm, and usually conical.

parainfluenza A respiratory disease (croup) caused by infection with Paramyxovirus.

parasite An organism that lives on or within another organism (the host), from which it obtains nutrients and enjoys protection. The parasite produces some degree of harm in the host.

parenteral Administering a substance into a body compartment other than through the gastrointestinal tract, such as via intravenous, subcutaneous, intramuscular, or intramedullary injection.

paroxysmal Events characterized by sharp spasms or convulsions; sudden onset of a symptom such as fever and chills.

particle radiation Radiation in which high energy particles such as protons, neutrons, and electrons provide energy that can be used for antimicrobial purposes.

passive carrier Persons who mechanically transfer a pathogen without ever being infected by it, for example, a health care worker who doesn't wash his/her hands adequately between patients.

passive immunity Specific resistance that is acquired indirectly by donation of preformed immune substances (antibodies) produced in the body of another individual.

passive transport Nutrient transport method that follows basic physical laws and does not require direct energy input from the cell.

pasteurization Heat treatment of perishable fluids such as milk, fruit juices, or wine to destroy heat-sensitive vegetative cells, followed by rapid chilling to inhibit growth of survivors and germination of spores. It prevents infection and spoilage.

pathogen Any agent—usually a virus, bacterium, fungus, protozoan, or helminth—that causes disease.

pathogen-associated molecular pattern See **PAMP.**

pathogenicity The capacity of microbes to cause disease.

pathogen recognition receptor (PRR) Surface molecules on white blood cells that can detect and interact with surface molecules on bacteria and viruses. Toll-like receptors are one example of PRRs.

pathology The structural and physiological effects of disease on the body.

pellicle A membranous cover; a thin skin, film, or scum on a liquid surface; a thin film of salivary glycoproteins that forms over newly cleaned tooth enamel when exposed to saliva.

pelvic inflammatory disease (PID) An infection of the uterus and fallopian tubes that has ascended from the lower reproductive tract. Caused by gonococci and chlamydias.

penicillinase An enzyme that hydrolyzes penicillin; found in penicillin-resistant strains of bacteria.

penicillins A large group of naturally occurring and synthetic antibiotics produced by *Penicillium* mold and active against the cell wall of bacteria.

pentose A monosaccharide with five carbon atoms per molecule. Examples: arabinose, ribose and xylose.

peptidase An enzyme that can hydrolyze the peptide bonds of a peptide chain.

peptide Molecule composed of short chains of amino acids, such as a dipeptide (two amino acids), a tripeptide (three), and a tetrapeptide (four).

peptide bond The covalent union between two amino acids that forms between the amine group of one and the carboxyl group of the other. The basic bond of proteins.

peptidoglycan A network of polysaccharide chains cross-linked by short peptides that forms the rigid part of bacterial cell walls. Gram-negative bacteria have a smaller amount of this rigid structure than do gram-positive bacteria.

perforin A protein secreted by cytotoxic and natural killer cells that attacks the cell membrane of infectious agents and forms pores.

perinatal In childbirth, occurring before, during, or after delivery.

period of invasion The period during a clinical infection when the infectious agent multiplies at high levels, exhibits its greatest toxicity, and becomes well established in the target tissues.

periodontal Involving structures that surround and support the tooth called the **periodontium.**

periplasmic space The region between the cell wall and cell membrane of the cell envelopes of gram-negative bacteria.

peritrichous In bacterial morphology, having flagella distributed over the entire cell.

petechiae Minute hemorrhagic spots in the skin that range from pinpoint- to pinhead-size.

pertussis Infection by *Bordetella pertussis*. A highly communicable disease that causes acute respiratory syndrome. Pertussis can be life-threatening in infants, but vaccination on the recommended schedule can prevent infection. Also called **whooping cough.**

Peyer's patches Oblong lymphoid aggregates of the gut located chiefly in the wall of the terminal and small intestine. Along with the tonsils and appendix, Peyer's patches make up the **gut-associated lymphoid tissue** that responds to local invasion by infectious agents.

pH The symbol for the negative logarithm of the H ion concentration; p (power) or $[H^+]_{10}$. A system for rating acidity and alkalinity.

phage A bacteriophage; a virus that specifically parasitizes bacteria.

phagocytosis A type of endocytosis in which the cell membrane actively engulfs large particles or cells into vesicles. A **phagocyte** is a cell specialized for doing this.

phagolysosome A body formed in a phagocyte, consisting of a union between a vesicle containing the ingested particle (the phagosome) and a vacuole of hydrolytic enzymes (the lysosome).

phenetic Based on phenotype, or expression of traits.

phenotype The observable characteristics of an organism produced by the interaction between its genetic potential (genotype) and the environment.

phlebotomine Pertains to a genus of very small midges or blood-sucking (phlebotomous) sand flies and to diseases associated with those vectors such as kala-azar, Oroya fever, and cutaneous leishmaniasis.

phosphate An acidic salt containing phosphorus and oxygen that is an essential inorganic component of DNA, RNA, and ATP.

phospholipid A class of lipids that compose a major structural component of cell membranes.

phosphorylation Process in which inorganic phosphate is added to a compound.

photic zone The aquatic stratum from the surface to the limits of solar light penetration.

photoautotroph An organism that utilizes light for its energy and carbon dioxide chiefly for its carbon needs.

photolysis Literally, splitting water with light. In photosynthesis, this step frees electrons and gives off O_2.

photon A subatomic particle released by electromagnetic sources such as radiant energy (sunlight). Photons are the ultimate source of energy for photosynthesis.

photophosphorylation The process of electron transport during photosynthesis that results in the synthesis of ATP from ADP.

photosynthesis A process occurring in plants, algae, and some bacteria that traps the sun's energy and converts it to ATP in the cell. This energy is used to fix CO_2 into organic compounds.

phototrophs Microbes that use photosynthesis to feed.

phylogenetic A classification system based on evolutionary relationships. Also called phyletic.

phylum In the levels of classification, the third level of classification from general to more specific. Each kingdom is divided into numerous phyla. Sometimes referred to as division.

physiology The study of the function of an organism.

phytoplankton The collection of photosynthetic microorganisms (mainly algae and cyanobacteria) that float in the upper layers of aquatic habitats where sun penetrates. These microbes are the basis of aquatic food pyramids and together with zooplankton, make up the plankton.

pili (singular, pilus) Small, stiff filamentous appendages in gram-negative bacteria that function in DNA exchange during bacterial conjugation.

pinocytosis The engulfment, or endocytosis, of liquids by extensions of the cell membrane.

plague Zoonotic disease caused by infection with *Yersinia pestis*. The pathogen is spread by flea vectors and harbored by various rodents.

plankton Minute animals (zooplankton) or plants (phytoplankton) that float and drift in the limnetic zone of bodies of water.

plantar warts Deep, painful warts on the soles of the feet as a result of infection by human papillomavirus.

plaque In virus propagation methods, the clear zone of lysed cells in tissue culture or chick embryo membrane that corresponds to the area containing viruses. In dental application, the filamentous mass of microbes that adheres tenaciously to the tooth and predisposes to caries, calculus, or inflammation.

plasma The carrier fluid element of blood.

plasma cell A progeny of an activated B cell that actively produces and secretes antibodies.

plasmids Extrachromosomal genetic units characterized by several features. A plasmid is a double-stranded DNA that is smaller than and replicates independently of the cell chromosome; it bears genes that are not essential for cell growth; it can bear genes that code for adaptive traits; and it is transmissible to other bacteria.

platelets Formed elements in the blood that develop when megakaryocytes disintegrate. Platelets are involved in hemostasis and blood clotting.

pleomorphism Normal variability of cell shapes in a single species.

pluripotential Stem cells having the developmental plasticity to give rise to more than one type. Example: undifferentiated blood cells in the bone marrow.

pneumococcus Common name for *Streptococcus pneumoniae*, the major cause of bacterial pneumonia.

Pneumocystis (carinii) jiroveci pneumonia A severe, acute lung infection caused by a fungus that is the leading cause of morbidity and mortality of AIDS patients.

pneumonia An inflammation of the lung leading to accumulation of fluid and respiratory compromise.

pneumonic plague The acute, frequently fatal form of pneumonia caused by *Yersinia pestis*.

point mutation A change that involves the loss, substitution, or addition of one or a few nucleotides.

polar Term to describe a molecule with an asymmetrical distribution of charges. Such a molecule has a negative pole and a positive pole.

poliomyelitis An acute enteroviral infection of the spinal cord that can cause neuromuscular paralysis.

polyclonal antibodies In reference to a collection of antibodies with mixed specificities that arose from more than one clone of B cells.

polymer A macromolecule made up of a chain of repeating units. Examples: starch, protein, and DNA.

polymerase An enzyme that produces polymers through catalyzing bond formation between building blocks (polymerization).

polymerase chain reaction (PCR) A technique that amplifies segments of DNA for testing. Using denaturation, primers, and heat-resistant DNA polymerase, the number can be increased several million-fold.

polymorphonuclear leukocytes (PMNLs) White blood cells with variously shaped nuclei. Although this term commonly denotes all granulocytes, it is used especially for the neutrophils.

polymyxin A mixture of antibiotic polypeptides from *Bacillus polymyxa* that are particularly effective against gram-negative bacteria.

polypeptide A relatively large chain of amino acids linked by peptide bonds.

polyribosomal complex An assembly line for mass production of proteins composed of a chain of ribosomes involved in mRNA transcription.

polysaccharide A carbohydrate that can be hydrolyzed into a number of monosaccharides. Examples: cellulose, starch, and glycogen.

population A group of organisms of the same species living simultaneously in the same habitat. A group of different populations living together constitutes the community level.

porin Transmembrane proteins of the outer membrane of gram-negative cells that permit transport of small molecules into the periplasmic space but bar the penetration of larger molecules.

portal of entry Characteristic route of entry for an infectious agent; typically a cutaneous or membranous route.

portal of exit Characteristic route through which a pathogen departs from the host organism.

positive stain Technique in which dye affixes to a specimen and imparts color to it. It takes advantage of the ready binding of bacterial cells to dyes.

potable Describing water that is relatively clear, odor-free, and safe to drink.

pox The thick, elevated pustular eruptions of various viral infections. Also spelled *pocks*.

poxvirus A group of large complex DNA viruses that create pox skin lesions. Agents of smallpox and molluscum contagiosum.

PPNG Penicillinase-producing *Neisseria gonorrhoeae.*

prevalence The total cumulative number of cases of a disease in a certain area and time period.

primary infection An initial infection in a previously healthy individual that is later complicated by an additional (secondary) infection.

primary pulmonary infection (PPI) Disease that results from the inhalation of fungal spores and their germination in the lungs. This may serve as a focus of infection that spreads throughout the body.

primary response The first response of the immune system when exposed to an antigen.

primary structure Initial protein organization described by type, number, and order of amino acids in the chain. The primary structure varies extensively from protein to protein.

primers Synthetic oligonucleotides of known sequence that serve as landmarks to indicate where DNA amplification will begin.

prion A concocted word to denote "proteinaceous infectious agent"; a cytopathic protein associated with the slow-virus spongiform encephalopathies of humans and animals.

probes Small fragments of single-stranded DNA (RNA) that are known to be complementary to the specific sequence of DNA being studied.

probiotic A microbe or mixture of microbes that may be used to balance the normal flora and prevent infections.

prodromium A short period of mild symptoms occurring at the end of the period of incubation. It indicates the onset of an infection.

producer An organism that synthesizes complex organic compounds from simple inorganic molecules. Examples would be photosynthetic microbes and plants. These organisms are solely responsible for originating food pyramids and are the basis for life on earth (also called autotroph).

proglottid The egg-generating segment of a tapeworm that contains both male and female organs.

progressive multifocal leukoencephalopathy (PML) An uncommon, fatal complication of infection with JC virus (polyomavirus).

prokaryotic cell Small cells, lacking special structures such as a nucleus and organelles. All prokaryotes are microorganisms.

promastigote A morphological variation of the trypanosome parasite responsible for leishmaniasis.

promoter Part of an operon sequence. The DNA segment that is recognized by RNA polymerase as the starting site for transcription.

promoter region The site composed of a short signaling DNA sequence that RNA polymerase recognizes and binds to commence transcription.

properdin pathway A normal serum protein involved in the alternate complement pathway that leads to nonspecific lysis of bacterial cells and viruses.

prophage A lysogenized bacteriophage; a phage that is latently incorporated into the host chromosome instead of undergoing viral replication and lysis.

prophylactic Any device, method, or substance used to prevent disease.

prostaglandin A hormonelike substance that regulates many body functions. Prostaglandin comes from a family of organic acids containing 5-carbon rings that are essential to the human diet.

protease inhibitors Drugs that act to prevent the assembly of functioning viral particles.

protein Predominant organic molecule in cells, formed by long chains of amino acids.

proton An elementary particle that carries a positive charge. It is identical to the nucleus of the hydrogen atom.

protoplast A bacterial cell whose cell wall is completely lacking and that is vulnerable to osmotic lysis.

protozoa A group of single-celled, eukaryotic organisms.

pseudohypha A chain of easily separated, spherical to sausage-shaped yeast cells partitioned by constrictions rather than by septa.

pseudomembrane A tenacious, noncellular mucous exudate containing cellular debris that tightly blankets the mucosal surface in infections such as diphtheria and pseudomembranous enterocolitis.

pseudopodium A temporary extension of the protoplasm of an amoeboid cell. It serves both in amoeboid motion and for food gathering (phagocytosis).

pseudopods Protozoan appendage responsible for motility. Also called "false feet."

psychrophile A microorganism that thrives at low temperature (0°C–20°C), with a temperature optimum of 0°C–15°C.

pulmonary Occurring in the lungs. Examples include pulmonary anthrax and pulmonary nocardiosis.

pulmonary anthrax The more severe form of anthrax, resulting from inhalation of spores and resulting in a wide range of pathological effects, including death.

pure culture A container growing a single species of microbe whose identity is known.

purine A nitrogen base that is an important encoding component of DNA and RNA. The two most common purines are adenine and guanine.

purpura A condition characterized by bleeding into skin or mucous membrane, giving rise to patches of red that darken and turn purple.

pus The viscous, opaque, usually yellowish matter formed by an inflammatory infection. It consists of serum exudate, tissue debris, leukocytes, and microorganisms. *Purulent* refers to a puslike condition.

pyogenic Pertains to pus formers, especially the pyogenic cocci: pneumococci, streptococci, staphylococci, and neisseriae.

pyrimidine Nitrogen bases that help form the genetic code on DNA and RNA. Uracil, thymine, and cytosine are the most important pyrimidines.

pyrimidine dimer The union of two adjacent pyrimidines on the same DNA strand, brought about by exposure to ultraviolet light. It is a form of mutation.

pyrogen A substance that causes a rise in body temperature. It can come from pyrogenic microorganisms or from polymorphonuclear leukocytes (endogenous pyrogens).

Q fever A disease first described in Queensland, Australia, initially dubbed Q for "query" to denote a fever of unknown origin. Q fever is now known to be caused by a rickettsial infection.

quaternary ammonium compounds See **quats.**

quaternary structure Most complex protein structure characterized by the formation of large, multiunit proteins by more than one of the polypeptides. This structure is typical of antibodies and some enzymes that act in cell synthesis.

quats A byword that pertains to a family of surfactants called **quaternary ammonium** compounds. These detergents are only weakly microbicidal and are used as sanitizers and preservatives.

Quellung test The capsular swelling phenomenon; a serological test for the presence of pneumococci whose capsules enlarge and become opaque and visible when exposed to specific anticapsular antibodies.

quinolone A class of synthetic antimicrobic drugs with broad-spectrum effects.

quorum sensing An activity among bacteria in a biofilm in which the members signal each other and coordinate their functions.

rabies The only rhabdovirus that infects humans. Zoonotic disease characterized by fatal meningoencephalitis.

radiation Electromagnetic waves or rays, such as those of light given off from an energy source.

radioactive isotopes Unstable isotopes whose nuclei emit particles of radiation. This emission is called radioactivity or radioactive decay. Three naturally occurring emissions are alpha, beta, and gamma radiation.

radioimmunoassay (RIA) A highly sensitive laboratory procedure that employs radioisotope-labeled substances to measure the levels of antibodies or antigens in the serum.

reactants Molecules entering or starting a chemical reaction.

reactive oxygen intermediate (ROI) Any charged oxygen radical (e.g. superoxide ions, peroxide) given off during oxygen metabolism that can be toxic to microbial and other cells.

real image An image formed at the focal plane of a convex lens. In the compound light microscope, it is the image created by the objective lens.

receptor In intercellular communication, cell surface molecules involved in recognition, binding, and intracellular signaling.

recombinant DNA A technology, associated with genetic engineering that deliberately modifies the genetic structure of an organism to create novel products, microbes, animals, plants, and viruses.

recombination A type of genetic transfer in which DNA from one organism is donated to another. The resultant cell is termed *recombinant*.

recycling A process that converts unusable organic matter from dead organisms back into their essential inorganic elements and returns them to their nonliving reservoirs to make them available again for living organisms. This is a common term that means the same as *mineralization* and *decomposition*.

redox Denoting an oxidation-reduction reaction.

reduction In chemistry, the gain of electrons. A reducing agent is one that causes the addition of electrons to a substrate.

reemerging disease Previously identified disease that is increasing in occurrence.

refraction In optics, the bending of light as it passes from one medium to another with a different index of refraction.

regular (in shape) Refers to bacteria with uniform staining properties and that are consistent in shape from cell to cell within a particular species.

regulator DNA segment that codes for a protein capable of repressing an operon.

regulatory site The location on an enzyme where a certain substance can bind and block the enzyme's activity.

rennin The enzyme casein coagulase, which is used to produce curd in the processing of milk and cheese.

reovirus Respiratory enteric orphan virus. Virus with a double-stranded RNA genome and both an inner capsid and an outer capsid. Not a significant human pathogen.

replication In DNA synthesis, the semiconservative mechanisms that ensure precise duplication of the parent DNA strands.

replication fork The Y-shaped point on a replicating DNA molecule where the DNA polymerase is synthesizing new strands of DNA.

replicon A piece of DNA capable of replicating. Contains an origin of replication.

reportable disease Those diseases that must be reported to health authorities by law.

repressible operon An operon that under normal circumstances is transcribed. The buildup of the operon's amino acid product causes transcription of the operon to stop.

repressor The protein product of a repressor gene that combines with the operator and arrests the transcription and translation of structural genes.

reservoir In disease communication, the natural host or habitat of a pathogen.

resident microbiota The deeper, more stable microflora that inhabit the skin and exposed mucous membranes, as opposed to the superficial, variable, transient population.

resistance (R) factor Plasmids, typically shared among bacteria by conjugation, that provide resistance to the effects of antibiotics.

resolving power The capacity of a microscope lens system to accurately distinguish between two separate entities that lie close to each other. Also called resolution.

respiratory chain In cellular respiration, a series of electron-carrying molecules that transfers energy-rich electrons and protons to molecular oxygen. In transit, energy is extracted and conserved in the form of ATP.

respiratory syncytial virus (RSV) An RNA virus that infects the respiratory tract. RSV is the most prevalent cause of respiratory infection in newborns.

restriction endonuclease An enzyme present naturally in cells that cleaves specific locations on DNA. It is an important means of inactivating viral genomes, and it is also used to splice genes in genetic engineering.

reticuloendothelial system (RES) Also known as the mononuclear phagocyte system, it pertains to a network of fibers and phagocytic cells (macrophages) that permeates the tissues of all organs. Examples: Kupffer cells in liver sinusoids, alveolar phagocytes in the lung, and microglia in nervous tissue.

retrovirus A group of RNA viruses (including HIV) that have the mechanisms for converting their genome into a double strand of DNA that can be inserted on a host's chromosome.

reverse transcriptase (RT) The enzyme possessed by retroviruses that carries out the reversion of RNA to DNA—a form of reverse transcription.

Reye's syndrome A sudden, usually fatal neurological condition that occurs in children after a viral infection. Autopsy shows cerebral edema and marked fatty change in the liver and renal tubules.

Rh factor An isoantigen that can trigger hemolytic disease in newborns due to incompatibility between maternal and infant blood factors.

rhabdovirus Family of bullet-shaped viruses that includes rabies.

rheumatic fever Chronic group A *Streptococcus*-associated syndrome that damages the heart valves.

rhinoviruses Picornaviruses associated with the common cold. Transmission is through human-to-human contact, and symptoms typically are short-lived. Most effective control is effective hand washing and care in handling nasal secretions.

rhizobia Bacteria that live in plant roots and supply supplemental nitrogen that boosts plant growth.

rhizosphere The zone of soil, complete with microbial inhabitants, in the immediate vicinity of plant roots.

ribonucleic acid (RNA) The nucleic acid responsible for carrying out the hereditary program transmitted by an organism's DNA.

ribose A 5-carbon monosaccharide found in RNA.

ribosome A bilobed macromolecular complex of ribonucleoprotein that coordinates the codons of mRNA with tRNA anticodons and, in so doing, constitutes the peptide assembly site.

ribozyme A part of an RNA-containing enzyme in eukaryotes that removes intervening sequences of RNA called introns and splices together the true coding sequences (exons) to form a mature messenger RNA.

rickettsias Medically important family of tiny, intracellular parasitic bacteria, commonly carried by ticks, lice, and fleas. Cause of Rocky Mountain spotted fever and typhus.

Rifampin An antibiotic used primarily in the treatment of mycobacterial infection.

ringworm A superficial mycosis caused by various dermatophytic fungi. This common name is actually a misnomer.

RNA polymerase Enzyme process that translates the code of DNA to RNA.

rolling circle An intermediate stage in viral replication of circular DNA into linear DNA.

root nodules Small growths on the roots of legume plants that arise from a symbiotic association between the plant tissues and bacteria (Rhizobia). This association allows fixation of nitrogen gas from the air into a usable nitrogen source for the plant.

roseola Disease of infancy, usually self-limiting, caused by infection with human herpesvirus-6.

rosette formation A technique for distinguishing surface receptors on T cells by reacting them with sensitized indicator sheep red blood cells. The cluster of red cells around the central white blood cell resembles a little rose blossom and is indicative of the type of receptor.

Rotavirus Virus with a double-stranded RNA genome and both an inner capsid and an outer capsid. Transmitted by fecal contamination, it is common in areas with poor sanitation. *Rotavirus* typically causes diarrheal disease and can be fatal.

rough endoplasmic reticulum (RER) Microscopic series of tunnels that originates in the outer membrane of the nuclear envelope and is used in transport and storage. Large numbers of ribosomes, partly attached to the membrane, give the rough appearance.

rubella Commonly known as German measles. Rubella is caused by *Rubivirus*, a member of the togavirus family. Postnatal rubella is generally a mild condition; congenital rubella poses a risk of birth defects and results when virus passes from infected mother to fetus.

rubeola (red measles) Acute disease caused by infection with *Morbillivirus*.

SALT (Skin-Associated Lymphoid Tissue) Specialized lymphatic cells dispersed through the skin layers.

saccharide Scientific term for sugar. Refers to a simple carbohydrate with a sweet taste.

salmonelloses Illnesses caused by infection of the noncoliform *Salmonella* pathogens. The cause of typhoid fever, *Salmonella* food poisoning, and gastroenteritis.

salpingitis Inflammation of the fallopian tubes.

sanitize/sanitization To clean inanimate objects using soap and degerming agents so that they are safe and free of high levels of microorganisms.

saprobe A microbe that decomposes organic remains from dead organisms. Also known as a saprophyte or saprotroph.

sarcina A cubical packet of 8, 16, or more cells; the cellular arrangement of the genus *Sarcina* in the family Micrococcaceae.

sarcoma A fleshy neoplasm of connective tissue. Growth, usually highly malignant, is of mesodermal origin.

satellite phenomenon or satellitism A type of commensal relationship in which one microbe produces growth factors that favor the growth of its dependent partner. When plated on solid media, the dependent organism appears as surrounding colonies. Example: *Staphylococcus* surrounded by its *Haemophilus* satellite.

saturation The complete occupation of the active site of a carrier protein or enzyme by the substrate.

scarlet fever Serious group A *Streptococcus* infection that causes redness, shedding of the skin, and high fever.

schistosomiasis Infection by blood fluke, often as a result of contact with contaminated water in rivers and streams. Symptoms include fever, chills, diarrhea, and cough. Infection may be chronic.

schizogony A process of multiple fission whereby first the nucleus divides several times, and subsequently the cytoplasm is subdivided for each new nucleus during cell division.

scientific method Principles and procedures for the systematic pursuit of knowledge, involving the recognition and formulation of a problem, the collection of data through observation and experimentation, and the formulation and testing of a hypothesis.

scolex The anterior end of a tapeworm characterized by hooks and/or suckers for attachment to the host.

SCP Abbreviation for single-cell protein, an expression for microbial protein intended for human and animal consumption.

sebaceous glands The sebum- (oily, fatty) secreting glands of the skin.

second line of defense An inborn, nonspecific system of cells (phagocytes) and fluids (inflammation) that acts to detect and destroy foreign substances rapidly once they have entered the body.

secondary immune response The more rapid and heightened response to antigen in a sensitized subject due to memory lymphocytes.

secondary infection An infection that compounds a preexisting one.

secondary response The rapid rise in antibody titer following a repeat exposure to an antigen that has been recognized from a previous exposure. This response is brought about by memory cells produced as a result of the primary exposure.

secondary structure Protein structure that occurs when the functional groups on the outer surface of the molecule interact by forming hydrogen bonds. These bonds cause the amino acid chain either to twist, forming a helix, or to pleat into an accordion pattern called a β-pleated sheet.

secretion The process of actively releasing cellular substances into the extracellular environment.

secretory antibody The immunoglobulin (IgA) that is found in secretions of mucous membranes and serves as a local immediate protection against infection.

selective media Nutrient media designed to favor the growth of certain microbes and to inhibit undesirable competitors.

selectively toxic Property of an antimicrobic agent to be highly toxic against its target microbe while being far less toxic to other cells, particularly those of the host organism.

self Natural markers of the body that are recognized by the immune system.

self-limited Applies to an infection that runs its course without disease or residual effects.

semiconservative replication In DNA replication, the synthesis of paired daughter strands, each retaining a parent strand template.

semisolid media Nutrient media with a firmness midway between that of a broth (a liquid medium) and an ordinary solid medium; motility media.

semisynthetic Drugs that after being naturally produced by bacteria, fungi, or other living sources, are chemically modified in the laboratory.

sensitizing dose The initial effective exposure to an antigen or an allergen that stimulates an immune response. Often applies to allergies.

sepsis The state of putrefaction; the presence of pathogenic organisms or their toxins in tissue or blood.

septic shock Blood infection resulting in a pathological state of low blood pressure accompanied by a reduced amount of blood circulating to vital organs. Endotoxins of all gram-negative bacteria can cause shock, but most clinical cases are due to gram-negative enteric rods.

septicemia Systemic infection associated with microorganisms multiplying in circulating blood.

septum A partition or cellular cross wall, as in certain fungal hyphae.

sequela A morbid complication that follows a disease.

sequencing Determining the actual order and types of bases in a segment of DNA.

serology The branch of immunology that deals with in vitro diagnostic testing of serum.

seropositive Showing the presence of specific antibody in a serological test. Indicates ongoing infection.

serotonin A vasoconstrictor that inhibits gastric secretion and stimulates smooth muscle.

serotyping The subdivision of a species or subspecies into an immunologic type, based upon antigenic characteristics.

serum The clear fluid expressed from clotted blood that contains dissolved nutrients, antibodies, and hormones but not cells or clotting factors.

serum sickness A type of immune complex disease in which immune complexes enter circulation, are carried throughout the body, and are deposited in the blood vessels of the kidney, heart, skin, and joints. The condition may become chronic.

severe acute respiratory syndrome (SARS) A severe respiratory disease caused by infection with a newly described Coronavirus.

severe combined immunodeficiencies A collection of syndromes occurring in newborns caused by a genetic defect that knocks out both B-and T-cell types of immunity. There are several versions of this disease, termed **SCIDS** for short.

sewage Liquid runoff or wastewater from communities.

sex pilus A conjugative pilus.

sexually transmitted disease (STD) Infections resulting from pathogens that enter the body via sexual intercourse or intimate, direct contact.

shigellosis An incapacitating dysentery caused by infection with *Shigella* bacteria.

shingles (also herpes zoster) A skin condition occurring on the trunk or head that is caused by the reactivation of a latent varicella-zoster virus.

side effects Unintended, usually harmful consequences of chemotherapy including damage to organs, allergy, and disruption of normal microbiota.

sign Any abnormality uncovered upon physical diagnosis that indicates the presence of disease. A sign is an objective assessment of disease, as opposed to a symptom, which is the subjective assessment perceived by the patient.

silent mutation A mutation that, because of the degeneracy of the genetic code, results in a nucleotide change in both the DNA and mRNA but not the resultant amino acid and thus, not the protein.

simple stain Type of positive staining technique that uses a single dye to add color to cells so that they are easier to see. This technique tends to color all cells the same color.

smallpox Disease caused by infection with variola virus. Smallpox has been eradicated worldwide, and variola virus exists today only in a few government laboratories.

smooth endoplasmic reticulum (SER) A microscopic series of tunnels lacking ribosomes that functions in the nutrient processing function of a cell.

solution A mixture of one or more substances (solutes) that cannot be separated by filtration or ordinary settling.

solvent A dissolving medium.

somatic (O or cell wall) antigen One of the three major antigens commonly used to differentiate gram-negative enteric bacteria.

source The person or item from which an infection is immediately acquired. See **reservoir.**

Southern blot A technique that separates fragments of DNA using electrophoresis and identifies them by hybridization.

species In the levels of classification, the most specific level of organization.

specificity Limited to a single, precise characteristic or action.

spheroplast A gram-negative cell whose peptidoglycan, when digested by lysozyme, remains intact but is osmotically vulnerable.

spike A receptor on the surface of certain enveloped viruses that facilitates specific attachment to the host cell.

spirillum A type of bacterial cell with a rigid spiral shape and external flagella.

spirochete A coiled, spiral-shaped bacterium that has endoflagella and flexes as it moves.

split gene Any gene in which the coding sequence is interrupted by intervening sequences that will not be translated (introns).

spontaneous generation Early belief that living things arose from vital forces present in nonliving, or decomposing, matter.

spontaneous mutation A mutation in DNA caused by random mistakes in replication and not known to be influenced by any mutagenic agent. These mutations give rise to an organism's natural, or background, rate of mutation.

sporadic Description of a disease that exhibits new cases at irregular intervals in unpredictable geographic locales.

sporangium A fungal cell in that asexual spores are formed by multiple cell cleavage.

spore A differentiated, specialized cell form that can be used for dissemination, for survival in times of adverse conditions, and/or for reproduction. Spores are usually unicellular and may develop into gametes or vegetative organisms.

sporicide A chemical agent capable of destroying bacterial endospores.

sporotrichosis A subcutaneous mycosis caused by *Sporothrix schenckii;* a common cause is the prick of a rose thorn.

sporozoite One of many minute elongated bodies generated by multiple division of the oocyst. It is the infectious form of the malarial parasite that is harbored in the salivary gland of the mosquito and inoculated into the victim during feeding.

sporulation The process of spore formation.

start codon The nucleotide triplet AUG that codes for the first amino acid in protein sequences.

starter culture Pure cultures of bacteria, molds, or yeasts inoculated into substrates for bulk processing, as in the preparation of fermented foods, beverages, and pharmaceuticals.

stasis A state of rest or inactivity; applied to nongrowing microbial cultures. Also called microbistasis.

stationary growth phase Survival mode in which cells either stop growing or grow very slowly.

stem cells Pluripotent, undifferentiated cells.

sterile Completely free of all life forms, including spores and viruses.

sterilization Any process that completely removes or destroys all viable microorganisms, including viruses, from an object or habitat. Material so treated is sterile.

STORCH Acronym for common infections of the fetus and neonate. STORCH stands for Syphilis, Toxoplasmosis, Other diseases (hepatitis B, AIDS and chlamydiosis), Rubella, Cytomegalovirus and Herpes simplex virus.

strain In microbiology, a set of descendants cloned from a common ancestor that retain the original characteristics. Any deviation from the original is a different strain.

streptolysin A hemolysin produced by streptococci.

strict, or obligate anaerobe An organism that does not use oxygen gas in metabolism and cannot survive in oxygen's presence.

stroma The matrix of the chloroplast that is the site of the dark reactions.

structural gene A gene that codes for the amino acid sequence (peptide structure) of a protein.

subacute Indicates an intermediate status between acute and chronic disease.

subacute endocarditis An infection of the heart lining due to invasion of the blood by oral streptococci.

subacute sclerosing panencephalitis (SSPE) A complication of measles infection in which progressive neurological degeneration of the cerebral cortex invariably leads to coma and death.

subcellular vaccine A vaccine against isolated microbial antigens rather than against the entire organism.

subclinical A period of inapparent manifestations that occurs before symptoms and signs of disease appear.

subculture To make a second-generation culture from a well-established colony of organisms.

subcutaneous The deepest level of the skin structure.

substrate The specific molecule upon which an enzyme acts.

subunit vaccine A vaccine preparation that contains only antigenic fragments such as surface receptors from the microbe. Usually in reference to virus vaccines.

sucrose One of the carbohydrates commonly referred to as sugars. Common table or cane sugar.

sulfonamide Antimicrobial drugs that interfere with the essential metabolic process of bacteria and some fungi. Also known as **sulfa drugs.**

superantigens Bacterial toxins that are potent stimuli for T cells and can be a factor in diseases such as toxic shock.

superficial mycosis A fungal infection located in hair, nails, and the epidermis of the skin.

superinfection An infection occurring during antimicrobic therapy that is caused by an overgrowth of drug-resistant microorganisms.

superoxide A toxic derivative of oxygen ($O_2{}^-$).

suppressor T cell A class of T cells that inhibits the actions of B cells and other T cells.

surfactant A surface-active agent that forms a water-soluble interface. Examples: detergents, wetting agents, dispersing agents, and surface tension depressants.

sylvatic Denotes the natural presence of disease among wild animal populations. Examples: sylvatic (sylvan) plague, and rabies.

symbiosis An intimate association between individuals from two species; used as a synonym for mutualism.

symptom The subjective evidence of infection and disease as perceived by the patient.

syncytium A multinucleated protoplasmic mass formed by consolidation of individual cells.

syndrome The collection of signs and symptoms that, taken together, paint a portrait of the disease.

synergism The coordinated or correlated action by two or more drugs or microbes that results in a heightened response or greater activity.

syngamy Conjugation of the gametes in fertilization.

syphilis A sexually transmitted bacterial disease caused by the spirochete *Treponema pallidum.*

systemic Occurring throughout the body; said of infections that invade many compartments and organs via the circulation.

tartar See **calculus.**

taxa Taxonomic categories.

taxonomy The formal system for organizing, classifying, and naming living things.

temperate phage A bacteriophage that enters into a less virulent state by becoming incorporated into the host genome as a prophage instead of in the vegetative or lytic form that eventually destroys the cell.

template The strand in a double-stranded DNA molecule that is used as a model to synthesize a complementary strand of DNA or RNA during replication or transcription.

teratogenic Causing abnormal fetal development.

tertiary structure Protein structure that results from additional bonds forming between functional groups in a secondary structure, creating a three-dimensional mass.

tetanospasmin The neurotoxin of *Clostridium tetani,* the agent of tetanus. Its chief action is directed upon the inhibitory synapses of the anterior horn motor neurons.

tetanus A neuromuscular disease caused by infection with *Clostridium tetani.* Usual portals of entry include puncture wounds, burns, umbilical stumps, frostbite sites, and crushed body parts. Vaccination repeated at the recommended times can prevent infection. Also called lockjaw.

tetracyclines A group of broad-spectrum antibiotics with a complex 4-ring structure.

tetrads Groups of four.

T helper cell See **helper T cell.**

theory A collection of statements, propositions, or concepts that explains or accounts for a natural event.

therapeutic index (IT) The ratio of the toxic dose to the effective therapeutic dose that is used to assess the safety and reliability of the drug.

thermal Related to temperature.

thermal death point The lowest temperature that achieves sterilization in a given quantity of broth culture upon a 10-minute exposure. Examples: 55°C for *Escherichia coli,* 60°C for *Mycobacterium tuberculosis,* and 120°C for spores.

thermal death time The least time required to kill all cells of a culture at a specified temperature.

thermal dimorphism A property of certain fungal pathogens of converting from hyphal to yeast form in response to increased temperature.

thermocline A temperature buffer zone in a large body of water that separates the warmer water (the epilimnion) from the colder water (the hypolimnion).

thermoduric Resistant to the harmful effects of high temperature.

thermophile A microorganism that thrives at a temperature of 50°C or higher.

third line of defense An acquired, very focused immunologic defense that depends on the recognition of specific foreign materials and microbes by B and T lymphocytes. It is the primary source of immunities.

thrush *Candida albicans* infection of the oral cavity.

thylakoid Vesicles of a chloroplast formed by elaborate folding of the inner membrane to form "discs." Solar energy trapped in the thylakoids is used in photosynthesis.

thymine (T) One of the nitrogen bases found in DNA but not in RNA. Thymine is in a pyrimidine form.

thymus Butterfly-shaped organ near the tip of the sternum that is the site of T-cell maturation.

tincture A medicinal substance dissolved in an alcoholic solvent.

tinea Ringworm; a fungal infection of the hair, skin, or nails.

tinea versicolor A condition of the skin appearing as mottled and discolored pigmentation as a result of infection by the yeast *Malassezia furfur.*

titer In immunochemistry, a measure of antibody level in a patient, determined by agglutination methods.

T lymphocyte (T cell) A white blood cell that is processed in the thymus gland and is involved in cell-mediated immunity.

toll-like receptors (TLRs) Small surface receptors on phagocytes that are specialized to bind and react to pathogens.

tonsils A ring of lymphoid tissue in the pharynx that acts a repository for lymphocytes.

topoisomerases Enzymes that can add or remove DNA twists and thus regulate the degree of supercoiling.

toxemia An abnormality associated with certain infectious diseases. Toxemia is caused by toxins or other noxious substances released by microorganisms circulating in the blood.

toxic shock syndrome Toxemia caused by *Staphylococcus aureus* and associated with use of tampons.

toxigenicity The tendency for a pathogen to produce toxins. It is an important factor in bacterial virulence.

toxin A specific chemical product of microbes, plants, and some animals that is poisonous to other organisms.

toxinosis Disease whose adverse effects are primarily due to the production and release of toxins.

toxoid A toxin that has been rendered nontoxic but is still capable of eliciting the formation of protective antitoxin antibodies; used in vaccines.

trace elements Micronutrients (zinc, nickel, and manganese) that occur in small amounts and are involved in enzyme function and maintenance of protein structure.

tracheostomy A surgically created emergency airway opening into the trachea.

trachoma Strain of *Chlamydia trachomatis* that attacks mucous membranes of the eye, genitourinary tract, and lungs in humans.

transcript A newly transcribed RNA molecule.

transcription mRNA synthesis; the process by which a strand of RNA is produced against a DNA template.

transduction The transfer of genetic material from one bacterium to another by means of a bacteriophage vector.

transfection Introduction of genetic material into an animal or plant cell using a viral or bacterial vector.

transfer RNA (tRNA) A transcript of DNA that specializes in converting RNA language into protein language.

transformation In microbial genetics, the transfer of genetic material contained in "naked" DNA fragments from a donor cell to a competent recipient cell.

transfusion Infusion of whole blood, red blood cells, or platelets directly into a patient's circulation.

transgenic technology Introduction of foreign DNA into cells or organisms. Used in genetic engineering to create recombinant plants, animals, and microbes.

transients In normal microbiota, the assortment of superficial microbes whose numbers and types vary depending upon recent exposure. The deeper-lying residents constitute a more stable population.

translation Protein synthesis; the process of decoding the messenger RNA code into a polypeptide.

transmissible spongiform encephalopathies (TSE) Progressive brain infections caused by unusual pathogens called prions, leading to loss of brain structure and function.

transposon A DNA segment with an insertion sequence at each end, enabling it to migrate to another plasmid, to the bacterial chromosome, or to a bacteriophage.

traveler's diarrhea A type of gastroenteritis typically caused by infection with enterotoxigenic strains of *E. coli* that are ingested through contaminated food and water.

trematode A fluke or flatworm parasite of vertebrates.

tricarboxylic acid cycle (TCA) or Krebs cycle The second pathway of the three pathways that complete the process of primary catabolism. Also called the citric acid cycle.

trichinosis Infection by the *Trichinella spiralis* parasite, usually caused by eating the meat of an infected animal. Early symptoms include fever, diarrhea, nausea, and abdominal pain that progress to intense muscle and joint pain and shortness of breath. In the final stages, heart and brain function are at risk, and death is possible.

trichomoniasis Sexually transmitted disease caused by infection by the trichomonads, a group of protozoa. Symptoms include urinary pain and frequency and foul-smelling vaginal discharge in females or recurring urethritis with a thin milky discharge in males.

triglyceride A type of lipid composed of a glycerol molecule bound to three fatty acids.

trimethoprim A sulfa drug often used in the treatment of *Pneumocystis* (*carinii*) *jiroveci* pneumonia in AIDS patients. (Often combined with sulfa drugs.)

triplet See **codon.**

trophozoite A vegetative protozoan (feeding form) as opposed to a resting (cyst) form.

true pathogen A microbe capable of causing infection and disease in healthy persons with normal immune defenses.

trypomastigote The infective morphological stage transmitted by the tsetse fly or the reduviid bug in African trypanosomiasis and Chagas disease.

tubercle In tuberculosis, the granulomatous well-defined lung lesion that can serve as a focus for latent infection.

tuberculin A glycerinated broth culture of *Mycobacterium tuberculosis* that is evaporated and filtered. Formerly used to treat tuberculosis, tuberculin is now used chiefly for diagnostic tests.

tuberculin reaction A diagnostic test in which an intradermal injection of a purified protein (tuberculin) extract from *M. tuberculosis* elicits an immune response, seen as a small red bump, in those persons previously exposed to tuberculosis.

tuberculoid leprosy A superficial form of leprosy characterized by asymmetrical, shallow skin lesions containing few bacterial cells.

tularemia Infection by *Francisella tularensis*. A zoonotic disease of mammals common to the northern hemisphere. Occasionally called rabbit fever. Portal of entry and symptoms are varied.

turbid Cloudy appearance of nutrient solution in a test tube due to growth of microbe population.

tyndallization Fractional (discontinuous, intermittent) sterilization designed to destroy spores indirectly. A preparation is exposed to flowing steam for an hour, and then the mineral is allowed to incubate to permit spore germination. The resultant vegetative cells are destroyed by repeated steaming and incubation.

typhoid fever Form of salmonelloses. It is highly contagious. Primary symptoms include fever, diarrhea, and abdominal pain. Typhoid fever can be fatal if untreated.

typhus Rickettsia infection characterized by high fever, chills, frontal headache, muscular pain, and a generalized rash within 7 days of infection. In more severe cases, a personality change, low urine output, hypotension, and gangrene can cause complications. Mortality is high in older adults.

ultraviolet (UV) radiation Radiation with an effective wavelength from 240 nm to 260 nm. UV radiation induces mutations readily but has very poor penetrating power.

uncoating The process of removal of the viral coat and release of the viral genome by its newly invaded host cell.

undulant fever See **brucellosis.**

universal donor In blood grouping and transfusion, a group O individual whose erythrocytes bear neither agglutinogen A nor B.

universal precautions (UPS) Centers for Disease Control and Prevention guidelines for health care workers regarding the prevention of disease transmission when handling patients and body substances.

uracil (U) One of the nitrogen bases in RNA but not in DNA. Uracil is in a pyrimidine form.

urban plague Plague passed to humans through contact with domestic animals or other humans.

urinary tract infection (UTI) Invasion and infection of the urethra and bladder by bacterial residents, most often *E. coli.*

vaccination The process of inoculating with a vaccine.

vaccine Originally used in reference to inoculation with the cowpox or vaccinia virus to protect against smallpox. In general, the term now pertains to injection of whole microbes (killed or attenuated), toxoids, or parts of microbes as a prevention or cure for disease.

vacuoles In the cell, membrane-bounded sacs containing fluids or solid particles to be digested, excreted, or stored.

valence The combining power of an atom based upon the number of electrons it can either take on or give up.

variable (v) region The antigen-binding fragment of an immunoglobulin molecule, consisting of a combination of heavy and light chains whose molecular conformation is specific for the antigen.

varicella See **chickenpox.**

varicella-zoster virus Herpesvirus responsible for the diseases chickenpox and shingles.

variolation A hazardous, outmoded process of deliberately introducing smallpox material scraped from a victim into the nonimmune subject in the hope of inducing resistance.

VDRL A flocculation test that detects syphilis antibodies. An important screening test. The abbreviation stands for **V**enereal **D**isease **R**esearch **L**aboratory.

vector An animal that transmits infectious agents from one host to another, usually a biting or piercing arthropod such as the tick, mosquito, or fly. Infectious agents can be conveyed mechanically by simple contact or biologically whereby the parasite develops in the vector.

vector$_2$ A genetic element such as a plasmid or a bacteriophage used to introduce genetic material into a cloning host during recombinant DNA experiments.

vegetative In describing microbial developmental stages, a metabolically active feeding and dividing form, as opposed to a dormant, seemingly inert, nondividing form. Examples: a bacterial cell versus its spore; a protozoan trophozoite versus its cyst.

vehicle An inanimate material (solid object, liquid, or air) that serves as a transmission agent for pathogens.

verruca A flesh-colored wart. This self-limited tumor arises from accumulations of growing epithelial cells. Example: papilloma warts.

vesicle A blister characterized by a thin-skinned, elevated, superficial pocket inflated with serum.

vibrio A curved, rod-shaped bacterial cell.

viremia The presence of a virus in the bloodstream.

virion An elementary virus particle in its complete morphological and thus infectious form. A virion consists of the nucleic acid core surrounded by a capsid, which can be enclosed in an envelope.

viroid An infectious agent that, unlike a virion, lacks a capsid and consists of a closed circular RNA molecule. Although known viroids are all plant pathogens, it is conceivable that animal versions exist.

virtual image In optics, an image formed by diverging light rays; in the compound light microscope, the second, magnified visual impression formed by the ocular from the real image formed by the objective.

virucide A chemical agent that inactivates viruses, especially on living tissue.

virulence In infection, the relative capacity of a pathogen to invade and harm host cells.

virulence factor A product of microbes such as an enzyme or toxin that increases the invasiveness or pathogenicity.

virus Microscopic, acellular agent composed of nucleic acid surrounded by a protein coat.

vitamins A component of coenzymes critical to nutrition and the metabolic function of coenzyme complexes.

vulvovaginal candidiasis (VC) Vaginal STD caused by *Candida albicans* and often associated with disruption of the normal vaginal flora.

wart An epidermal tumor caused by papillomaviruses. Also called a verruca.

Western blot test A procedure for separating and identifying antigen or antibody mixtures by two-dimensional electrophoresis in polyacrylamide gel, followed by immune labeling.

wheal A welt; a marked, slightly red, usually itchy area of the skin that changes in size and shape as it extends to adjacent area. The reaction is triggered by cutaneous contact or intradermal injection of allergens in sensitive individuals.

whey The residual fluid from milk coagulation that separates from the solidified curd.

white piedra A fungus disease of hair, especially of the scalp, face, and genitals, caused by *Trichosporon beigelii*. The infection is associated with soft, mucilaginous, white-to-light-brown nodules that form within and on the hair shafts.

whitlow A deep inflammation of the finger or toe, especially near the tip or around the nail. Whitlow is a painful herpes simplex virus infection that can last several weeks and is most common among health care personnel who come in contact with the virus in patients.

whole blood A liquid connective tissue consisting of blood cells suspended in plasma.

whooping cough See **pertussis.**

Widal test An agglutination test for diagnosing typhoid.

wild type The natural, nonmutated form of a genetic trait.

wort The clear fluid derived from soaked mash that is fermented for beer.

xenograft The transfer of a tissue or an organ from an animal of one species to a recipient of another species.

yaws A tropical disease caused by *Treponema pertenue* that produces granulomatous ulcers on the extremities and occasionally on bone but does not produce central nervous system or cardiovascular complications.

yellow fever Best-known arboviral disease. Yellow fever is transmitted by mosquitoes. Its symptoms include fever, headache, and muscle pain that can proceed to oral hemorrhage, nosebleeds, vomiting, jaundice, and liver and kidney damage.

zoonosis An infectious disease indigenous to animals that humans can acquire through direct or indirect contact with infected animals.

zooplankton The collection of nonphotosynthetic microoganisms (protozoa, tiny animals) that float in the upper regions of aquatic habitat and together with phytoplankton comprise the plankton.

zygospore A thick-walled sexual spore produced by the zygomycete fungi. It develops from the union of two hyphae, each bearing nuclei of opposite mating types.

Credits

Photographs

Chapter 1

Opener: NASA; Opener (inset): S. J. Giovannoni, E. F. DeLong, T. M. Schmidt, and N. R. Pace, "Tangential flow filtration and preliminary phylogenetic analysis of marine picoplankton," *Appl. Environ. Microbiol.* 1990. 56:2572–2575; TB 1.1A: James Gathany/CDC; TB 1.1B: CDC; TB 1.1C: Photo courtesy of Sartorius Stedim Biotech; TB 1.1D: Jack Dykinga, USDA/ARS; 1.3A: Janice Carr/CDC; 1.3B: © Tom Volk; 1.3C: © Charles Krebs Photography; 1.3D: CDC; 1.3E: © Yuuji Tsukii, Protist Information Server; 1.3F: CDC; 1.5A(1): © Kathy Park Talaro; 1.5A(2): © www.microbiological-garden.net; 1.5B(1): © Kathy Park Talaro; 1.5B(2): © University of Adelaide, Australia - Dr. David Ellis; 1.5C(1): © Peter Doran/University of Illinois, Chicago; 1.5C(2): Image courtesy of the Priscu Research Group, Montana State University, Bozeman; 1.6A(1): © El Paso Times; 1.6A(2): © Yuuji Tsukii, Protist Information Server; 1.6B(1–2): Courtesy: Pacific Northwest National Laboratory; 1.8: © Bettmann/Corbis; 1.9A(1–2) © Kathy Park Talaro/Visuals Unlimited; 1.9B: © Science VU//Visuals Unlimited; 1.11: © AKG/Photo Researchers, Inc.; 1.12: © Bettmann/Corbis.

Chapter 2

Opener: NASA, ESA, and The Hubble Heritage Team (STScI/AURA), M. Wolff (Space Science Institute) and J. Bell (Cornell University); Opener (inset-left): NASA; Opener (inset-right): Photo by David McKay/NASA; 2.6D: © Kathy Park Talaro; 2.11A-C: © John W. Hole; 2.22: Image courtesy RCSB Protein Data Bank, www.pdp.org; p. 57(left): P. Cooke, Agriculture Research Service (ARS), USDA; p. 57(right): Courtesy of Professor Robert J. Linhardt, Rensselaer Polytechnic Institute.

Chapter 3

Opener: CDC; 3.2: © Kathy Park Talaro; 3.3: © Leica Microsystems Inc.; 3.8A: © Carolina Biological Supply/Phototake; 3.8B-C: © Abbey/Visuals Unlimited; 3.9A: CDC; 3.9B: © Dr. Klaus Hausmann; 3.10: © Invitrogen Corporation; 3.11: © Anne Fleury; 3.13A: P.R. Hazelton, and H.R. Gelderblom, "Electron microscopy for rapid diagnosis of emerging infectious agents," *Emerging Infectious Diseases.* 2003. 9:294–303; 3.13B: Courtesy Dr. Richard Allen; 3.14A: © Dr. Klaus Hausmann; 3.14B: Image copyright Dr. Markus Geisen; p. 70: Courtesy of IBM Corporation. Almaden Research Center. Unauthorized use not permitted; TB 3.4: CDC; 3.16A: © Harold J. Benson; 3.16B(1A-B): CDC; 3.16B(2): © Jack Bostrack/Visuals Unlimited; 3.16B(3): © Steven R. Spilatro, Department of Biology, Marietta College, Marietta, OH; 3.16C(1): © A.M. Siegelman/Visuals Unlimited; 3.16C(2): © David Frankhauser; p. 75: © Dr. Rita Colwell & Dr. Munirul Alam; 3.18B-3.19C: © Kathy Park Talaro; 3.20A: Courtesy Hardy Diagnostics; 3.21B–3.27B: © Kathy Park Talaro; 3.28A: © Harold J. Benson; p. 88: CDC.

Chapter 4

Opener: Image reprinted with permission from eMedicine.com, 2010. Available at: http://emedicine.medscape.com/article/1612368-overview; Opener (inset): Janice Carr/CDC; 4.3A: © Dr. Jeffrey C. Burnham; 4.3B: From J. Reichelt and P. Baumann, "Taxonomy of the marine, luminous bacteria," *Arch. Microbiol.*, 1973, 94:283–330. With kind permission of Springer + Business Media; 4.3C: From Noel R. Krieg in *Bacteriological Reviews*, March 1976, Vol. 40(1):87, fig. 7. © ASM; 4.3D: From Preer et al., *Bacteriological Reviews*, June 1974, 38(2):121, fig. 7. © ASM; 4.7A: © Eye of Science/Photo Researchers, Inc.; 4.7B: Dr. S. Knutton from D.R. Lloyd and S. Knutton, *Infection and Immunity*, January 1987, p 86–92. © ASM; 4.8: © L. Caro/SPL/Photo Researchers, Inc.; 4.10A: © Kathy Park Talaro; 4.10B: © John D. Cunningham/Visuals Unlimited; 4.11: Janice Carr/CDC; 4.12A: © S.C Holt/Biological Photo Service; 4.12B : © T. J. Beveridge/Biological Photo Service; 4.15: Courtesy of Dr. Heinrich Lünsdorf, HZI.; 4.17: Courtesy of Michael J. Daly; 4.19A: © Paul W. Johnson/Biological Photo Service; 4.19B: © D. Balkwill and D. Maratea; 4.20: © Rut CARBALLIDO-LOPEZ/I.N.R.A. Jouy-en-Josas, Laboratoire de Génétique Microbienne; 4.21: Courtesy Kit Pogliano and Mark Sharp/USCD; 4.22A: SJ Jones, CJ Paredes, B Tracy, N Cheng, R Sillers, RS Senger, ET Papoutsakis, "The transcriptional program underlying the physiology of clostridial sporulation," *Genome Biol.*, 2008. 9:R114; 4.23A(2)-B(2): Janice Carr/CDC; 4.23C(2): From Jacob S. Teppema, "In vivo adherence and colonization of Vibrio cholerae strains that differ in hemagglutinating activity and motility," *Journal of Infection and Immunity*, 55(9): 2093–2102, Sept. 1987. Reprinted by permission of American Society for Microbiology; 4.23D(2): Photo by De Wood. Digital colorization by Chris Pooley; 4.23E(2): © VEM/Photo Researchers, Inc.; 4.23F(2): © Science VU/Frederick Mertz/Visuals Unlimited; 4.24A: © A.M. Siegelman/Visuals Unlimited; TB 4.3A: Maryland Astrobiology Consortium, NASA and STScI; TB 4.3B: © Angelicque E. White; TB 4.3C: Courtesy: Pacific Northwest National Laboratory; TB 4.3D: CDC; TB 4.3E: Janice Carr/CDC; TB 4.3F: Courtesy: Pacific Northwest National Laboratory; TB 4.3G: CDC; TB 4.3H: Janice Carr/CDC; TB 4.3I: © Science VU/Frederick Mertz/Visuals Unlimited; TB 4.3J: Janice Carr/CDC; TB 4.3K: CDC; TB 4.3L: Janice Carr/CDC; p. 115(left): © Heide N. Schulz/Max Planck Institute for Marine Microbiology; p. 115(right): © Dr. Philippa Uwins, Whistler Research Pty./Photo Researchers, Inc.; 4.28A(1): © Roger Burks (University of California at Riverside), Mark Schneegurt (Wichita State University) and Cyanosite (www.cyanosite.bio.purdue.edu); 4.28A(2): © Angelicque E. White; 4.28A(3): Courtesy Jason Oyadomari; 4.28B: © Richard Behl; 4.28C: J. William Schopf, Univ. Calif. Los Angeles. Reprinted with permission from *Science*, 260, April 30, fig. g, p. 643. © 1993 by the American Association for the Advancement of Science; 4.28D: © John Waterbury, Woods Hole Oceanographic Institute; 4.29A: © Kathy Park Talaro; 4.29B: © DJ Patterson/MBL, http://microscope.mbl.edu; 4.30: GBF-German Research Center for Biotechnology, Braunschweig, Germany; 4.31: Baca and Paretsky, *Microbiological Reviews*, 47(20);133, fig. 16, June 1983 © ASM; 4.32A: NASA Johnson Space Center/ISS007E8738/ (http://eol.jsc.nasa.gov); 4.32B: © Dr. Mike Dyall-Smith, University of Melbourne; 4.33: From Kazem Kashefi and Derek R. Lovley, "Extending the upper temperature limit for life," *Science*, 2003, 301:934. Image courtesy Kazem Kashefi; p. 122: © Kathy Park Talaro.

Chapter 5

Opener: NASA Earth Observatory; Opener (insets): CDC; 5.1A-B: © Andrew Knoll; p. 125: © Image by D. J. Patterson (provided by micro*scope http://microscope.mbl.edu);
5.3A-B: Courtesy Richard Allen; 5.5A–5.11B: © Don Fawcett/Visuals Unlimited; 5.13B: Courtesy of Life Technologies, Carlsbad, CA; 5.15A: © George Barron, University of Guelph, CANADA; 5.15B: © Dr. Judy A. Murphy, San Joaquin Delta College, Department of Microscopy, Stocton, CA; 5.16B: Janice Carr/CDC; 5.17A: © Kathy Park Talaro; 5.17B: © New Zealand Dermatological Society; p. 139(left): © George Barron, University of Guelph, CANADA; p. 139(right): © Neil Carlson, Department of Environmental Health and Safety; 5.18D: © George Barron, University of Guelph, CANADA; 5.22B: © Gregory M. Filip; 5.23: © Kathy Park Talaro; 5.24A-B: © George Barron, University of Guelph, CANADA; 5.25: © Joyce E. Longcore, University of Maine; 5.26B: © Jan Hinsch/Photo Researchers, Inc.; 5.26C-D: © Kathy Park Talaro; 5.28B: Janice Carr/CDC; 5.29B: © David Patterson/MBL/Biological Discovery in Woods Hole; 5.30A(2): © Eric Russell, BioMEDIA ASSOCIATES; 5.30B: © Yuuji Tsukii, Protist Information Server; 5.31B: Michael Riggs et al, *Infection and Immunity*, Vol. 62, #5, May 1994, p. 1931 © ASM.

Chapter 6

Opener: Illustrator: Dan Higgins/CDC; p. 160: © Kathy Park Talaro; 6.2: © Didier Raoult, M.D., PhD; 6.3A: Schaffer et al, *Proceedings of the National Academy of Science*, 41:1020, 1955; 6.3B: CDC; 6.6B: © Dennis Kunkel/CNRI/Phototake; 6.6D: Dr. F. A. Murphy/CDC; 6.7D: © Dr. Linda Stannard, UCT/Photo Researchers, Inc.; 6.8A(1): Fred P. Williams, Jr./EPA; 6.8A(2): © Dr. Linda Stannard, UCT/Photo Researchers, Inc.; 6.8B: © Eye of Science/Photo Researchers, Inc.; 6.9B: © Bin Ni, Chisholm Lab, MIT; 6.14B: © Chris Bjornberg/Photo Researchers, Inc; 6.15A(1-2): CDC; 6.15B: © Massimo Battaglia, INeMM CNR, Rome Italy; 6.18: © K.G. Murti/Visuals Unlimited; 6.19A-B: CDC; 6.19C: © Visuals Unlimited; p. 178: Cynthia Goldsmith/CDC; 6.20A: © State Hygenic Laboratory at The University of Iowa; 6.21A: © James King-Holmes/Institute of Animal Health/Photo Researchers, Inc.; 6.21B: Dr. Art Davis/CDC.

Chapter 7

Opener: © Judy A Mosby; Opener (inset): © Yuuji Tsukii, Protist Information Server; 7.1B: © Kathy Park Talaro; 7.1C: Courtesy: Pacific Northwest National Laboratory; 7.2A: Courtesy of Everly Conway de Macario and Alberto J.L. Macario; 7.2B: © Electron Microscope Lab, UC Berkeley; 7.10A- B(2): Image courtesy Nozomu Takeuchi; p. 200A: © Montana State University/Susan Brumfield, Dr. Mark Young; p. 200B: © Dea Slade; 7.11A: Photo by Keith Weller, USDA/ARS; 7.11B: Courtesy and © Becton, Dickinson and Company; 7.12: © Terese M. Barta, Ph.D.; p. 204: Courtesy Arthur Hauck (Germany); 7.13: © Science VU/Fred Marsik/Visuals Unlimited; p. 205(all): © Dr. Jared R. Leadbetter; 7.19A: © Kathy Park Talaro/Visuals Unlimited.

Chapter 8

Opener (left): Sharon Franklin. Colorization by Stephen Ausmus, USDA/ARS; Opener (right): Image courtesy of the RCSB PDB (www.pdb.org); PDB ID 1G9G; Parsiegla G, Reverbel C, Tardif C, Driguez H, Haser R, "Structures of mutants of cellulase Cel48F of Clostridium cellulolyticum in complex with long hemithiocellooligosaccharides give rise to a new view of the substrate pathway during processive action," *J Mol Biol.* 2008 Jan 11; 375(2):499-510; p. 243: © Kathy Park Talaro; p. 253: © Terese M. Barta, Ph.D.

Chapter 9

Opener (left): © L. Caro/SPL/Photo Researchers, Inc.; Opener (right): © Pamela McTurk/Photo Researchers, Inc.; Opener (bottom): Janice Carr/CDC; 9.1A: © PhotoLink/Photodisc/Getty Images RF; 9.3: © K.G. Murti/Visuals Unlimited; p. 258: © A. Barrington Brown/Photo Researchers, Inc.; p. 268: © Randy L. Jirtle; 9.16C: © Steven McKnight and Oscar L Miller, Department of Biology, University of Virginia.

Chapter 10

Opener: Genome Management Information System, http://genomics.energy.gov; Opener (inset): © Robert Rathe; 10.4B: © Kathy Park Talaro; 10.6: Jaana Tammur, Hiljar Sibul, Ene Ustav, Mart Ustav, Andres Metspalu, "A bovine papillomavirus-1 based vector restores the function of the low-density lipoprotein receptor in the receptor deficient CHOldlA7 cell line," BMC Molecular Biology, 2002, 3:5; p. 298(top): Genome Management Information System, http://genomics.energy.gov; 10.9B: Linda Bartlett/National Cancer Institute; 10.9C: Jack Dykinga, USDA/ARS; p. 306: © Marilyn Humphries; TB 10.3A: © Andrew Davis and Sue Bunnewell/John Innes Centre; TB 10.3B: Jack Dykinga, USDA/ARS; 10.14: © Brigid Hogan, Dept Cell Biology Duke University; TB 10.4A: H. Kubota and R. L. Brinster, School of Veterinary Medicine, University of Pennsylvania, "Growth factors essential for self-renewal and expansion of mouse spermatogonial stem cells," Cell Biology, November 23, 2004, Vol. 101 no. 47:16489–16494. National Academy of Sciences, U.S.A.; TB 10.4B: Courtesy Holly Steinkraus/University of Wyoming; 10.17C: © Dr. Michael Baird; 10.18: © Tyson A. Clark, University of California, Santa Cruz; p. 318: Courtesy Genelex Corporation.

Chapter 11

Opener: © Richard Shock/Photodisc/Getty RF; Opener (inset): CDC; p. 321: © Bettmann/Corbis; 11.6A: © Kathy Park Talaro; TB 11.4A: © Goodnature Products; TB 11.4B-11.9B: © Kathy Park Talaro; TB 11.5A: Courtesy: Pacific Northwest National Laboratory; TB 11.5B: Courtesy Violight; 11.11: Photo courtesy of ITT Water & Wastewater; 11.12B: Janice Carr/CDC; TB 11.7A: © Gregg Eckhardt; TB 11.7B: used with permission from 3M; 11.14: Photo Courtesy of Steris Corporation © 2010 Steris Corporation, All rights reserved; TB 11.9A: © Photo by Peter Morenus, University of Connecticut; TB 11.9B: © Kathy Park Talaro; TB 11.10A: Courtesy AirClean Systems; TB 11.10B: Photo Courtesy of Steris Corporation © 2008 Steris Corporation, All rights reserved; 11.19: © Kathy Park Talaro/Visuals Unlimited; p. 345(top): © Kathy Park Talaro; TB 11.11: © Los Angeles Times Photo by Robert Durell; p. 350: SJ Jones, CJ Paredes, B Tracy, N Cheng, R Sillers, RS Senger, ET Papoutsakis, "The transcriptional program underlying the physiology of clostridial sporulation," Genome Biol., 2008. 9:R114.

Chapter 12

Opener: Larry Ostby/National Cancer Institute; Opener (inset): C. Goldsmith/CDC; 12.1: Photo courtesy of Dr. Ramón Santamaría/IMB. CSIC/USAL. Salamanca. Spain; p. 353: © Bettmann/Corbis; 12.3A(2): Janice Carr/CDC; 12.3B(2): © CNRI/Photo Researchers, Inc.; p. 361: © Kathy Park Talaro/Visuals Unlimited; p. 370: © Kathy Park Talaro; p. 375: © Crown copyright material is reproduced with the permission of the Controller of HMSO and Queen's Printer for Scotland; 12.16: © Kenneth E. Greer/Visuals Unlimited; 12.18B: © Prof. Patrice Courvalin, MD, FRCP, Unité des Agents Antibactériens, Institut Pasteur, Paris, France; 12.18C: © Kathy Park Talaro; 12.19: © Copyright AB BIODISK 2008. Re-printed with permission of AB BIODISK; 12.21: Image courtesy David Ellis; p. 385A: © Eye of Science/Photo Researchers, Inc.

Chapter 13

Opener: CDC; Opener (insets): Rocky Mountain Laboratories, NIAID, NIH; 13.3B: Janice Carr/CDC; p. 393(both): From R. Randal Bollinger, Andrew S. Barbas, Errol L. Bush, Shu S. Lin, William Parker, "Biofilms in the large bowel suggest an apparent function of the human vermiform appendix," Journal of Theoretical Biology, Volume 249, Issue 4, 21 December 2007, Pages 826–831. Copyright © 2007 Elsevier Ltd All rights reserved; p. 395: © Eurelios/Phototake; p. 400: US Army; 13.19: Marshall W. Jennison, Massachusetts Institute of Technology, 1940; 13.21A-C(2): CDC; p. 423(both left): Rocky Mountain Laboratories, NIAID, NIH; p. 423(both right): CDC.

Chapter 14

Opener: Setareh Mamishi, Kamiar Zomorodian, Farshid Saadat, Mohsen Gerami-Shoar, Bita Tarazooie, and Seyad Ahmad Siadati, "A case of invasive aspergillosis in CGD patient successfully treated with Amphotericin B and INF," Annals of Clinical Microbiology and Antimicrobials, 2005, 4:4, © BioMed Central; Opener (inset): CDC; 14.2B: © Louisa Howard/Dartmouth College; 14.7B(all): © Harold J. Benson; 14.17B: © Dr. Brinkmann/Max Planck Institute for Infection Biology; p. 451(2-4): © Harold J. Benson; p. 451B: © Cabisco/Visuals Unlimited; p. 451D: CDC.

Chapter 15

Opener: Paul Cryan/U.S. Geological Survey; Opener (inset): CDC; 15.11B: Dr. Richard Feldmann/National Cancer Institute; 15.17A-B: Susan Arnold, Dr. Raowf Guirguis/National Cancer Institute; 15.18A-D: © Photodisc/Getty RF.

Chapter 16

Opener: U.S. Department of Health & Human Services, National Institutes of Health; 16.2B: Courtesy Sandia National Laboratories; 16.2C: © Louisa Howard/Dartmouth College; 16.5A: © Kenneth E. Greer/Visuals Unlimited; 16.5B: © Kathy Park Talaro; 16.6A: © STU/Custom Medical Stock Photo; 16.10: © Claude Revy/Phototake; p. 503(left-middle): Robert H. Mohlenbrock @ USDA-NRCS PLANTS Database/USDA SCS. 1991. Southern wetland flora: Field office guide to plant species. South National Technical Center, Fort Worth; p. 503(right): Jennifer Anderson. United States, IA, Scott Co., Davenport, Nahant Marsh. 2001; 16.14: © Kathy Park Talaro; 16.15B: © Kenneth E. Greer/Visuals Unlimited; 16.17A: © Diepgen TL, Yihume G et al. Dermatology Online Atlas published online at: www.dermis.net. Reprinted with permission; 16.17B: © SIU/Visuals Unlimited; 16.20: © Reprinted from R. Kretschmer, New England Journal Of Medicine, 279:1295, 1968 Massachusetts Medical Society. All rights reserved; p. 512: © Baylor College of Medicine, Public Affairs.

Chapter 17

Opener: U.S. Fish and Wildlife Service; Opener (inset): Janice Carr/CDC; 17.3B: © Fred Marsik/Visuals Unlimited; 17.4A-B: © Analytab Products, a division of Sherwood Medical; 17.6: © Dr. Anwar Huq and Bradd J. Haley; 17.10A: Gillies and Dodd's Bacteriology illustrated, 5th edition, fig. 14, Elsevier; 17.10B: © National Institute Slide Bank/The Welcome Centre for Medical Sciences; 17.12: © Genelabs Diagnostics Pte Ltd.; 17.14C: CHEMICON® International, Inc; 17.15B: © Hank Mogan/Science Source/Photo Researchers, Inc.; 17.16A: © Phototake; 17.16B: B. Huard, T. McKee, C. Bosshard, S. Durual, T. Matthes, S. Myit, O. Donze, C. Frossard, C. Chizzolini, C. Favre, R. Zubler, J. P. Guyot, P. Schneider and E. Roosnek, "APRIL secreted by neutrophils binds to heparan sulfate proteoglycans to create plasma cell niches in human mucosa," J Clin Invest. 2008;118(8):2887–2895. Copyright © 2008, American Society for Clinical Investigation; 17.17A: CDC/Dr. Hermann; 17.17B(1–2): Zaki Salahuddin. Laboratory of Tumor Cell Biology/National Cancer Institute; 17.17C(1): Bryon Skinner/CDC; 17.17C(2): CDC; 17.17D: © Genelabs Diagnostics Pte Ltd.; 17.17E: © Rolan Davis and Mike Moore/Rabies Laboratory, Kansas State University; p. 538:© Kathy Park Talaro.

Line Art

Chapter 1

Table 1.2: Data from latest "World Health Report."

Chapter 6

Figure 6.10: Detailed Structure of Complex Viruses, from Westwood, et al., Journal of Microbiology, 34:67. Reprinted with permission of General Microbiology.

Chapter 8

Figure 8.21: The Reaction of of a Single Turn of the TCA Cycle, adapted from Life: The Science of Biology, 7/e, by Purfes, Sadava, Orions, & Heller. © Sinauer Associates, Inc. Reprinted with permission.

Chapter 10

Insight 10.1, page 298: Cartoon by John Chase. Reprinted by permission.

Chapter 15

Table 15.6: CDC National Immunization Program; **Figure 15.15;** Reprinted by permission of Joseph Bellanti.

Inside Back Cover

Table: Recommended Immunization Schedule, United States 2010. CDC: National Immunization Program.

Index

Note: In this index, page numbers followed by a t designate tables; page numbers followed by an f refer to figures; page numbers set in **boldface** refer to definitions of terms or introductory discussions: page numbers preceded by a letter refer to the appendixes.